BEHAVIOUR OF OFF-SHORE STRUCTURES

Sponsored by

Delft University of Technology
Imperial College of Science and Technology
Massachusetts Institute of Technology
Norwegian Institute of Technology

BEHAVIOUR OF OFF-SHORE STRUCTURES

Proceedings of the
Third International Conference

Volume 2

Edited by
Chryssostomos Chryssostomidis
and
Jerome J. Connor

both of
Massachusetts Institute of Technology

HEMISPHERE PUBLISHING CORPORATION
Washington   New York   London

DISTRIBUTION OUTSIDE THE UNITED STATES
McGRAW-HILL INTERNATIONAL BOOK COMPANY

Auckland	Bogotá	Guatemala	Hamburg	Johannesburg	Lisbon
London	Madrid	Mexico	Montreal	New Delhi	Panama	Paris
San Juan	São Paulo	Singapore	Sydney	Tokyo	Toronto

Proceedings of the Third International Conference on the Behaviour of Off-Shore Structures, August 2-5, 1982, held at the Massachusetts Institute of Technology, Cambridge, Massachusetts

Secretariat:
Seminar Office
Center for Advanced Engineering Study
Massachusetts Institute of Technology
Cambridge, Massachusetts 02139

BEHAVIOUR OF OFF-SHORE STRUCTURES

Copyright © 1983 by Hemisphere Publishing Corporation. All rights reserved. Printed in the United States of America. Except as permitted under the United States Copyright Act of 1976, no part of this publication may be reproduced or distributed in any form or by any means, or stored in a data base or retrieval system, without the prior written permission of the publisher.

1 2 3 4 5 6 7 8 9 0   B R B R   8 9 8 7 6 5 4 3 2

Library of Congress Cataloging in Publication Data

International Conference on the Behaviour of Off-shore
  Structures (3rd : 1982 : Massachusetts Institute of
  Technology)
  Behaviour of off-shore structures.

  Sponsored by Delft University of Technology and others.
  1. Offshore structures—Design and construction—Congresses. I. Chryssostomidis, Chryssostomos. II. Connor, J. J. (Jerome J.) III. Technische Hogeschool Delft.
IV. Title
TC1665.157 1982   627'.98   82-11749
ISBN 0-89116-343-3

Contents

Preface xiii

International Organizing Committee xv

Session Chairmen xvii

VOLUME 1

GENERAL SESSION

THE STATE OF THE ART

G2 Design and Construction of Deep Water Jacket Platforms 3
    G. C. Lee

G5 Alternate Deep Water Concepts for Northern North Sea Extreme Conditions 18
    J. M. Huslid, O. T. Gudmestad, and A. Alm-Paulsen

G6 Oil and Gas Production Facilities for Very Deep Water 50
    G. M. Chateau

G7 Hydrocarbon Extraction in Arctic Frontiers 71
    B. J. Watt

TECHNICAL LECTURE SESSIONS

FOUNDATIONS

Evaluation of Pile Foundation Response

F1 Analysis of the Pile Foundation System for a North Sea Drilling Platform 95
    C. J. F. Clausen, P. M. Aas, and I. B. Almeland

F2 A Laboratory Study of Axially Loaded Piles and Pile Groups including Pore Pressure Measurements 105
    H. Matlock, D. Bogard, and L. Cheang

Experimental Results Related to Pile Foundations

F3 Pile Foundation Design Considerations for Deepwater Fixed Structures 125
    R. G. Bea, A. R. Dover, and J. M. E. Audibert

F4 Basic Data for the Design of Tension Piles in Silty Soils 141
    A. A. Puech

Evaluation of Design Parameters for Soil

F5 Cone Penetration and Engineering Properties of the Soft Orinoco Clay 161
    A. A. Azzouz, M. M. Baligh, and C. C. Ladd

F6 The Role of Soil Fabric Studies in the Evaluation of the Engineering Parameters of Offshore Deposits 181
    A. Marsland, A. Prince, and M. A. Love

Analysis of Pile Foundations

F7 Prediction of Offshore Pile Behaviour 203
    S. Nordal, L. Grande, and N. Janbu

F8 Skin Friction on Piles 221
    P. Lagoni

Foundation Issues for Gravity Structures

F9 Review of Foundation Design Principles for Offshore Gravity Platforms 243
    K. H. Andersen, S. Lacasse, P. M. Aas, and E. Andenaes

F10 Synthesis of Results from Geotechnical Instrumentation of Two Different Frigg Field Gravity Structures (North Sea) as Recorded from 1978 to 1981 262
    J. P. Mizikos and P. Y. Hicher

Numerical Modeling of Foundation Behaviour

F11 A Numerical Study of Skin Friction around Driven Piles 283
    J. P. Martins and D. M. Potts

Numerical Methods for Foundation Analysis

F12 A Numerical Model to Predict Permanent Displacement from Cyclic Loading of Foundations 297
    W. A. Marr, A. Urzua, and G. Bouckovalas

F13 Offshore Platform Foundation Shakedown Analysis 313
    A. K. Haldar, D. V. Reddy, and M. Arockiasamy

Pore Pressure in the Seabed

F14 Wave Induced Pressures Underneath a Caisson: A Comparison Between Theory and Large Scale Tests 337
    J. Lindenberg, J. H. Swart, C. J. Kenter, and K. den Boer

F15 Analytical Theories for the Interaction of Offshore Structures with a Poro-elastic Sea Bed 358
    C. C. Mei

TECHNICAL LECTURE SESSIONS
HYDRODYNAMICS

Wave Forces on Cylinders and Jacket Platforms

H1 Force Coefficient Evaluation for Offshore Structure Inclined Members 373
    P. Kaplan, C.-W. Jiang, and F. J. Dello Stritto

H2	Heavily Roughened Horizontal Cylinders in Waves 387
	J. H. Nath

Hydrodynamic Issues Related to Tension Leg Platforms

H3	Theoretical and Experimental Investigations of Tension Leg Platform Behaviour 411
	O. I. Faltinsen, I. J. Fylling, R. van Hooff, and P. S. Teigen
H4	Hydrodynamic Loading on Multi-component Bodies 424
	R. Eatock Taylor and J. Zietsman

Hydrodynamic Considerations in the Design of Semi-submersibles

H5	The Low Frequency Motions of a Semi-submersible in Waves 447
	J. A. Pinkster and R. H. M. Huijsmans
H6	Drift Forces on One- and Two-Body Structures in Regular Waves 467
	K. Kokkinowrachos, L. Bardis, and S. Mavrakos

Nonlinear Wave Kinematics and Prediction of Extreme Responses

H7	The Non-linear Properties of Random Wave Kinematics 493
	K. Anastasiou, R. G. Tickell, and J. R. Chaplin
H8	On the Prediction of Extreme Responses by the Envelope Method 516
	A. Naess

Design Considerations for Guyed Towers

H9	Dynamic Behaviour of Guyed Towers 529
	A. K. Basu and A. Dutta
H10	The Statics and Dynamics of the Mooring Lines of a Guyed Tower for Design Applications 546
	M. S. Triantafyllou, G. Kardomateas, and A. Bliek

Exploration and Production of Marine Risers

H11	An Experimental Procedure for the Prediction of the Dynamic Behaviour of Riser Type Systems 565
	C. Chryssostomidis and N. M. Patrikalakis
H12	On the Dynamics of Production Risers 599
	M. H. Patel and S. Sarohia
H13	Current-induced Motions of Multiple Risers 618
	G. Moe and T. Overvik

Environmental Design Considerations

H14	An Integrated Approach to Setting Environmental Design Criteria for Floating Production Facilities 643
	S. J. Leverette, M. S. Bradley, and A. Bliault
H15	Plumes from Subsea Well Blowouts 659
	J. H. Milgram and R. J. Van Houten
	Author Index 685

VOLUME 2

TECHNICAL LECTURE SESSIONS

STRUCTURES

Fatigue of Steel Jacket Structures

S1   Allowable Stresses for Fatigue Design   3
       P. W. Marshall and W. H. Luyties

S2   The Fatigue Strength of Welded Connections Subjected to North Sea Environmental and Random Loading Conditions   26
       R. Holmes and J. Kerr

Structural Stability

S3   Ultimate Strength of Steel Offshore Structures   39
       V. A. Zayas, S. A. Mahin, and E. P. Popov

S4   Research in Great Britain on the Stability of Circular Tubes   59
       P. J. Dowling and J. E. Harding

Response of Tension Leg Platforms

S5   Interwire Slippage and Fatigue Prediction in Stranded Cables for TLP Tethers   77
       R. E. Hobbs and M. Raoof

S6   Dynamic Response of Tension Leg Platform   100
       D. C. Angelides, C.-Y. Chen, and S. A. Will

Design Issues of Alternate Systems

S7   Design Concepts and Strategies for Guyed Tower Platforms   123
       D. Nair and P. S. Duval

S8   Probabilistic Reliability Analysis for the Fatigue Limit State of Gravity and Jacket-Type Structures   147
       H. Karadeniz, S. van Manen, and A. Vrouwenvelder

Pipeline Systems

S9   A Method of Analysis for Collapse of Submarine Pipelines   169
       P. E. de Winter

S10   On the Initiation of a Propagating Buckle in Offshore Pipelines   187
       S. Kyriakides and C. D. Babcock

Behavioural Issues for Concrete Structures

S11   Implications of Fatigue in the Design of Some Concrete Offshore Structures   203
       A. H. Tricklebank, W. I. J. Price, and E. C. Hambly

S12   Uniaxial Tensile Strength of Concrete as Influenced by Impact and Fatigue   223
       H. W. Reinhardt

S13   Behavior of Concrete in Biaxial Cyclic Compression   235
       O. Buyukozturk and J. G. Zisman

Long Term and Extreme Loads

S14 Analysis of Ship/Platform Impacts   257
    T. H. Søreide, T. Moan, J. Amdahl, and J. Taby

S15 The Sensitivity of Fatigue Life Estimates to Variations in Structural Natural Periods, Modal Damping Ratios, and Directional Spreading of the Seas   279
    J. K. Vandiver

TECHNICAL POSTER SESSIONS

FOUNDATIONS

Pile Foundations under Static Loads

PF1.1 Pore Water Pressure Development during Pile Driving and Its Influence on Driving Resistance   295
    M. Datta

PF1.2 Rod Shear Interface Friction Tests in Sands, Silts and Clays   305
    A. R. Dover, S. R. Bamford, and L. F. Suarez

PF1.3 The Design Concept and Driving Capacity of a 40 Ton Above/Under Water Hydraulic Hammer   315
    G. D. Ellery (Part A), R. M. Elliott (Part A), and E. L. James (Part B)

PF1.4 Parameter Measurements Required to Calculate the Axial Capacity of Piles   325
    J. Fournier, P. Y. Hicher, J. P. Mizikos, and F. Ropers

PF1.5 Non Linear Consolidation Analyses around Pile Shafts   338
    M. Kavvadas and M. M. Baligh

PF1.6 Lateral Stiffness of a Pile Subject to Static Monotonic Loading   348
    S. N. Pollalis, J. J. Connor, and M. J. Kavvadas

Performance of Foundations

PF2.1 Lateral Stress Measurements during Static and Cyclic Direct Simple Shear Testing   363
    R. Dyvik and T. F. Zimmie

PF2.2 Reduction of Pore Water Pressure beneath Concrete Gravity Platforms   373
    O. Eide, A. Andresen, R. Jonsrud, and E. Andenaes

PF2.3 Analysis of Pile Groups Subjected to Lateral Loads   383
    M. Hariharan and K. Kumarasamy

PF2.4 Cyclic Behaviour of Soils and Application to Offshore Foundation Calculation   391
    F. Lassoudiere, Y. Meimon, J. C. Hujeux, and D. Aubrey

PF2.5 The Collapse of a Flare Jacket Subjected to Ice Loads   401
    N. Janbu, X. Jizu, and L. Grande

Geotechnical Issues in the Seabed

PF3.1 Time and Cost Planning for Offshore Soil Investigations   417
    T. Amundsen and R. Lauritzsen

PF3.2 Soil Flows Generated by Submarine Slides—Case Studies and Consequences   425
    L. Edgers and K. Karlsrud

PF3.3　Factors Influencing the Interpretation of In Situ Strength Tests in Insensitive Low Plasticity Clays　438
　　　　D. W. Hight, A. Gens, T. M. P. De Campos, and M. Takahashi

PF3.4　Uplift Capacity of Embedded Anchors in Sand　451
　　　　H. B. Sutherland, T. W. Finlay, and M. O. Fadl

PF3.5　A New Method to Determine the Undrained Shear Strength of Muds Based on Plasticity Theory　464
　　　　L. E. Vallejo and P. A. Tarvin

TECHNICAL POSTER SESSIONS
HYDRODYNAMICS

Hydrodynamic Issues Related to Deep Water Structures

PH1.1　Effect of Wave Spectral Shape and Directional Variabilities on the Design and Analysis of Marine Structures　477
　　　　H. Chen

PH1.2　Estimates of Cross-spectral Densities of Wave Forces on Inclined Cylinders　486
　　　　M. C. Deo and S. Narasimhan

PH1.3　Response Analysis of Tension Leg Platform with Mechanical Damping System in Waves　497
　　　　M. Katayama, K. Unoki, and E. Miwa

PH1.4　The Dynamic Response of an Underwater Column Hinged at the Sea Bottom under the Wave Action　523
　　　　D. Qui and Q. Zuo

PH1.5　Studies on Pressures and Forces Due to Irregular Wind Generated Waves on Circular Cylinder—A Spectral Analysis Approach　532
　　　　H. Raman and P. S. V. Rao

PH1.6　Modelling of Wind-Wave Spectra in Laboratory Waves　541
　　　　G. Smith, G. Baron, and I. Grant

Floating Vessels

PH2.1　Dynamic Response of Offshore Platform Cranes Using Physical and Mathematical Models　551
　　　　J. A. D. Balfour and A. O. Bowcock

PH2.2　Seaquakes: A Potential Threat to Offshore Structures　561
　　　　K. Hove, P. B. Selnes, and H. Bungum

PH2.3　On Hydrodynamic Reaction of Water to Motion to Offshore Structures in Calm Sea　572
　　　　A. V. Ivanov and A. N. Kulikova

PH2.4　Study of Wave Action on Floating Structures Using Finite Elements Method. Comparison Between Numerical and Experimental Results　580
　　　　A. Lejeune, J. Marchal, Th. Hoffait, S. Grilli, and P. Lejeune

PH2.5　Drift Forces and Damping in Natural Sea States: A Critical Review of the Hydrodynamics of Floating Structures　592
　　　　H. Lundgren, S. E. Sand, and J. Kirkegaard

PH2.6　On the Large Amplitude and Extreme Motions of Floating Vessels　608
　　　　S. N. Smith and M. Atlar

Risers and Mooring Systems

PH3.1　Non-linear Analysis of Mooring Systems　621
　　　　C. Ganapathy Chettiar and S. C. Nair

PH3.2 The Dynamic Behaviour of Taut-Moored Single-Point Mooring Buoys   631
A. R. Halliwell and I. J. A. Jardine

PH3.3 Vibrations of Pipe Arrays in Waves   641
N.-E. Ottesen Hansen

PH3.4 Dynamic Behaviour of Anchor Lines   651
C. M. Larsen and I. J. Fylling

PH3.5 The Response and the Lift-Force Analysis of an Elastically-Mounted Cylinder Oscillating in Still Water   671
K. G. McConnell and Y. S. Park

PH3.6 Measurement of Fluctuating Forces on an Oscillating Cylinder in a Cross Flow   681
M. J. Moeller and P. Leehey

PH3.7 Oscillations of Cylinders in Waves and Currents   690
R. L. P. Verley and D. J. Johns

TECHNICAL POSTER SESSIONS

STRUCTURES

Behaviour of Tubular Members

PS1.1 Global Buckling Design Criteria for Risers   705
M. M. Bernitsas and T. Kokkinis

PS1.2 Non Linear Behaviour of Steel Structural Connections   716
A. Colson

PS1.3 The Effect of Variable Chord Wall Thickness on the Stresses in a Light, Cast $90°-45°$ K Tubular Joint   726
C. D. Edwards and H. Fessler

PS1.4 An Experimental and Analytical Investigation of the Fatigue Behaviour of Monopod Tubular Joints   736
S. S. Gowda, D. V. Reddy, M. Arockiasamy, D. B. Muggeridge, and P. S. Cheema

PS1.5 On the Significance of Non-linear Random Wave Loading in the Fatigue Analysis of Jacket Type of Structures   747
H. B. Kanegaonakar, R. M. Belkune, and C. K. Ramesh

PS1.6 Automation in Underwater Welding   756
K. Masubuchi and V. J. Papazoglou

Structural Response under Extreme Loads

PS2.1 Reliability of Stiffened Steel Cylinders to Resist Extreme Loads   769
P. K. Das, P. A. Frieze, and D. Faulkner

PS2.2 Extreme Wave Dynamics of Deepwater Platforms   784
R. D. Larrabee

PS2.3 The Generalized Modular Analysis in Offshore Engineering Structures   792
S. Lin, C. Qiu, and Z. Ji

PS2.4 Numerical Simulation for the Collapse Analysis of Large Stiffened Steel Assemblages   799
R. S. Puthi, F. S. K. Bijlaard, and H. G. A. Stol

PS2.5 The Hydroelastic Response of the Submerged Structures with Arbitrary Located Multi-cylindrical Piles of Different Diameters, Subjected to Earthquake and Vibration   810
N. Qu

PS2.6   Long-Term Random Sea-State Modelling   817
        S. Shyam Sunder and J. J. Connor

Design and Reliability Issues for Alternate Structural Concepts

PS3.1   Nondestructive Testing of Off-Shore Structures by Shearography   831
        Y. Y. Hung and R. M. Grant

PS3.2   Probabilistic Strength of Steel Fiber Reinforced Concrete   837
        M. Matsuishi and S. Iwata

PS3.3   Development of the Tripod Tower Platform Design   847
        F. C. Michelsen and J. Meek

PS3.4   Nonlinear Analyses and Experiments of Reinforced Concrete Plates Subjected to Concentrated Moments   864
        D. G. Morrison

PS3.5   Resistance of Prestressed Concrete Slabs to Extreme Loads   874
        S. H. Perry and I. C. Brown

PS3.6   Design Aspects of Artificial Sand-Fill Islands   884
        W. M. K. Tilmans, K. den Boer, and J. Lindenberg

        Author Index   899
        Subject Index   901

Preface

The BOSS Conference is an international endeavor sponsored every three years by the Delft University of Technology (The Netherlands), the Massachusetts Institute of Technology (U.S.A.), the Norwegian Institute of Technology (Norway), and the University of London (United Kingdom). BOSS '76 was held at Trondheim, Norway, in August 1976, and Imperial College of Science and Technology in London hosted BOSS '79 in August 1979. After two very successful conferences in Europe, the Organizing Committee elected to hold BOSS '82 on the campus of the Massachusetts Institute of Technology in August 1982.

Planning BOSS '82 has been a joint effort involving faculty and staff within the sponsoring universities, practicing professionals in the U.S. offshore industry, and the staff of the Seminar Office in the Center for Advanced Engineering Study at M.I.T. Their effort is most appreciated. We are indebted to the authors for their outstanding contributions to the proceedings. We especially appreciate the efforts of Ms. M. Chryssostomidis, Mr. Griff Lee, and Dr. Stephen Montgomery for their enthusiasm and encouragement during the initial planning stage of this conference.

BOSS '82 addresses the critical research and development issues for the new generation of offshore structures to be built during the 1980s. General state-of-the-art papers, combined with specialized technical papers, provide both a broad coverage and in-depth treatment of critical hydrodynamic, structural, and geotechnical questions for present-day systems such as deep-water steel jackets, tension leg platforms, guyed towers, and semisubmersibles. General papers on arctic structures and very deep water production systems are intended to identify future challenges.

Professor Jerome J. Connor, Sc.D.
Professor Chryssostomos Chryssostomidis, Ph.D.

International Organizing Committee

Chryssostomos Chryssostomidis
Massachusetts Institute of Technology, U.S.A.

Jerome J. Connor
Massachusetts Institute of Technology, U.S.A.

E. W. Bijker
Delft University of Technology, The Netherlands

Torgeir Moan
Norwegian Institute of Technology, Norway

Peter L. Pratt
Imperial College of Science and Technology, England

Session Chairmen

GENERAL SESSION

THE STATE OF THE ART

Design and Construction of Deep Water Jacket Platforms   *J. J. Connor*

Geotechnical Issues in Off-Shore Engineering   *J. J. Connor*

Evolution of Tension Leg Platform Technology   *K. A. Blenkarn*

Alternate Deep Water Concepts for Northern North Sea Extreme Conditions   *A. L. Guy*

Oil and Gas Production Systems for Very Deep Water   *R. I. Walker*

Hydrocarbon Extraction in Arctic Frontiers   *R. I. Walker*

TECHNICAL LECTURE SESSIONS

FOUNDATIONS

Evaluation of Pile Foundation Response   *J. A. Focht, Jr.*

Experimental Results Related to Pile Foundations   *H. Matlock*

Evaluation of Design Parameters for Soil   *D. Sangrey*

Analysis of Pile Foundations   *H. G. Poulous*

Foundation Issues for Gravity Structures   *N. F. Braathen*

Numerical Modeling of Foundation Behaviour   *S. Tsien*

Numerical Methods for Foundation Analysis   *J. Roesset*

Pore Pressures in the Seabed   *T. W. Lambe*

TECHNICAL LECTURE SESSIONS
HYDRODYNAMICS

Wave Forces on Cylinders and Jacket Platforms   R. G. Dean

Hydrodynamic Issues Related to Tension Leg Platforms   O. H. Oakley

Hydrodynamic Considerations in the Design of Semi-submersibles   T. A. Loukakis

Nonlinear Wave Kinematics and Prediction of Extreme Responses   C. Petrauskas

Design Considerations for Guyed Towers   N. A. Brown

Exploration and Production Marine Risers   R. W. van Hooff

Environmental Design Considerations   J. B. Gregory

TECHNICAL LECTURE SESSIONS
STRUCTURES

Fatigue of Steel Jacket Structures   J. H. Sybert

Structural Stability   D. Faulkner

Response of Tension Leg Platforms   D. I. Karsan

Design Issues of Alternate Systems   F. P. Dunn

Pipeline Systems   R. G. Bea

Behavioural Issues for Concrete Structures   O. Furnes

Long Term and Extreme Loads   D. Liu

Technical Lecture Sessions

STRUCTURES

Fatigue of Steel Jacket Structures

ALLOWABLE STRESSES FOR FATIGUE DESIGN

P. W. Marshall
Shell Oil Company, USA

W. H. Luyties
Shell Oil Company, USA

SUMMARY

Limits on peak environmental stresses - 20 ksi nominal, 60 ksi hot spot, and related punching shear values - are provided in API RP 2A as a means of avoiding fatigue failures in tubular members and joints of conventional Gulf of Mexico template type platforms. Similar, but lower, allowable stresses have also been found to be useful in the initial fatigue design of platforms in rougher areas like the North Sea, and in deep water where dynamic amplification must be considered. These allowables are based on preliminary, generic dynamic and fatigue analysis for the type of platform and wave climate. Typically, the structure also receives a more detailed fatigue analysis in the final stages of design.

This paper will review the development and basis for these fatigue allowable stresses, considering:

Fatigue S-N Curves

Stress Concentration Factors (SCF)

Wave Climate Data

Method of Analysis

Long Term Stress

Sensitivity to Approximations

Reliability Considerations

In addition, the impact of recent research results on the foregoing will be examined, as will such topical issues as the shoaling paradox and multiplanar loading. As appropriate, changes to existing criteria will be proposed. The emphasis will be on practical design solutions.

INTRODUCTION

The end objective of structural design criteria is to dimension the structure so that it performs satisfactorily in service, achieving a reasonable balance between economy and risk of failure.

With regard to failure by overload, the traditional working stress approach has been to dimension the structure so as to keep the acting stresses within allowable limits, using design loads which are also specified. These stresses and loads are complementary parts of the design algorithm or code, which reflects theoretical research, testing, practical experience, and value judgements.

The recent Limit State Design (LSD) approach seeks to use more realistic measures of both ultimate resistance and applied load, as well as partial safety factors applied to each which reflect their bias and uncertainty (Moses, 1980). Using the First-Order, Second-Moment theory of structural reliability, these partial safety factors can be related to the Safety Index, β_s, which is in turn a measure of the "notional" probability of failure. The word "notional" denotes that these calculations only include the bias and uncertainty which we know about, and that the actual probability of failure may be quite different from what we calculate. Thus, in practice, the new LSD criteria must be calibrated against the old working stress codes to properly reflect the experience and value judgements incorporated therein.

Generally, engineers need not look beyond the target safety index established in this calibration, nor concern themselves with the arcane art of prediction for unmentionable events like failure, in order to achieve a satisfactory design. It is only when we undertake an extrapolation from familiar applications to untried new ones that the predictive aspects of the design algorithm really get tested. The extrapolation is more likely to be successful if the design criteria are comprehensive and rational--i.e., relatively free of hidden loopholes and biases which only hold for the familiar applications.

Similarly, in designing fixed offshore platforms against failure by fatigue, the primary objective is to make the joint-cans and brace-ends thick enough, not to predict how many years the structure will last. This is fortunate, because the overall uncertainty in fatigue life prediction is so large as to render the actual numbers almost meaningless.

Target lifetime safety index is often taken to be in the range of 2 to 3. For design against overload by infrequent natural events, this corresponds to a 30-50% range in member size (Marshall and Bea, 1976); while for fatigue design, it corresponds to a 20-40% range in joint-can thickness (Marshall, 1976).

However, the corresponding range of uncertainty in fatigue life is 600% for short lives, and over 3000% for long lives (where most of the stress cycles are below the knee of the S-N curve). Furthermore, modest changes in fatigue computation procedure can produce enormous amounts of additional bias and uncertainty in the results (Wirsching, 1981). In view of these uncertainties in fatigue life, it is most appropriate that the initial design of a structure be considered in the context of an overall fracture control strategy which includes Verification, Monitoring, Inspection, and Repair (Marshall, 1979).

In this paper we shall focus on defining practical criteria for fatigue design in familiar applications, rather than the more dubious task of fatigue life prediction, or the challenging but uncertain area of extrapolation. Starting with existing criteria, we shall examine a variety of structural, environmental, and computational variables. We shall present a recent probabilistic design format to examine results of parameter studies on existing platforms and present simplified criteria for use in the design of tubular joints.

PRESENT FATIGUE DESIGN CRITERIA

Fatigue has long been recognized as an important consideration in the design of offshore structures. The present criteria can be traced back to research and development carried out in the late 1960's.

An early S-N curve (Carter et al, 1969) is shown in Figure 1. The hot spot strain amplitude is used to enter the curve; this corresponds to strain gage readings taken in areas of geometrical stress concentration, and places many different connection geometries on a common basis. Since practical as-welded structures inevitably contain flaws, notches, altered microstructure, and residual stresses in these same critical areas, fatigue test data which reflect these conditions are used to establish the design curve.

The first use of such an S-N curve has been generic fatigue analysis, performed early in the design process. Given the general structural characteristics and wave climate data, and assuming that the stress in primary lateral bracing is proportional to base shear, the long term stress distribution can be developed from a simplified analysis, as shown in Figure 2.

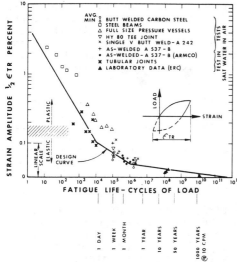

FATIGUE LIFE VS STRAIN AMPLITUDE
FIGURE 1

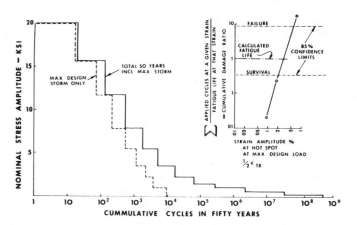

GULF OF MEXICO FATIGUE SPECTRUM
FIGURE 2

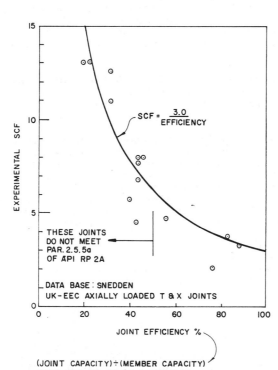

SCF VS EFFICIENCY
FIGURE 3

Several trial designs are then assumed, each with different specified hot spot stress in the design wave (60 ft. for Gulf of Mexico, according to criteria prevailing at the time). The inset in Figure 2 shows the results of calculations in terms of cumulative damage ratio vs. design hot spot strain. A design hot spot strain of .002 (or stress of 60 ksi) was found to correspond to the lower bound of the fatigue data scatter band (85% survival).[1]

Well-designed connections which at ultimate develop 100% of the branch member yield strength generally have a stress concentration factor (SCF) of about 3, or less. The corresponding design nominal stress for members would be 20 ksi. This limitation on wave-induced nominal stress was adopted as a fatigue design guideline in the 3rd edition of API RP 2A (1971). However, connections which develop less than 100% efficiency in terms of static strength also have SCF higher than 3, as shown in Figure 3. Thus, fatigue criteria more closely related to the relevant connection parameters were needed.

Ideally, the hot spot stress should be used as the basis of fatigue design for welded tubular connections. However, to perform a finite element or experimental stress analysis for each connection, in order to obtain the hot spot SCF, would require a disproportionate amount of effort in practical design situations. Fortunately, parametric equations for SCF are available (e.g., Kuang, 1977). The method of computing hot spot stress and the design S-N curve should be compatible with each other. Thus, comparisons in the form of Figure 4 are particularly relevant to practical design. Here the hot spot strain range is computed from specimen dimensions, applied loads, and the Kuang equations, and plotted against cycles to terminal failure.

Two data bases are shown (Rodabaugh, 1980 and Snedden, 1981). Rodabaugh's WRC data base represents American data from cooperative research of the late 1960's. Snedden's represents data from the UK-EEC Offshore Steels Research Program, conducted in recent years at a much higher level of sophistication (and more than ten times the cost). The ability of hot spot stress to bring a wide variety of connection geometries to a common basis (here K and T joints) is reaffirmed by the grouping of both sets of data in a common scatter band. However, the more recent data extend to lower stress levels and larger specimen thickness, and encroach upon the API-X design curve. Whether this size effect is due to the larger notches in the full size weld profiles tested, or is simply a function of plate thickness, is still being debated. Both profile and thickness appear to be important, and effects to understand their interaction are underway (Radenkovic, 1981). Meanwhile, either API curve X-prime or the proposed DOE fatigue design curve fall on the safe side of the data (possible unsafe application of the DOE curve in the very low cycle range could be prevented by using Neuber's rule[2] and the plastic strain ε_{TR} when stresses are beyond yield, or by also checking a criterion based on fraction of ultimate strength, e.g., the dashed line in Figure 1).

In practice, the initial fatigue design is done in terms of an allowable hot spot stress to be used with the peak design wave load, e.g., 60 ksi for a 60 ft. wave, with stresses due to static gravity loads excluded. For important structures, the design is subsequently checked by a detailed fatigue analysis.

To illustrate this procedure, the joints of an 8-pile platform for approximately 200 feet of water in the Gulf of Mexico were initially sized using an allowable hot spot stress of 75 ksi for the 70-foot design wave. Then 24 joint cans, including 336 hot spot locations, were subjected to a detailed fatigue analysis. This included a spectral fatigue analysis (with dynamics) for sea states in the annual directional wave scatter diagram, as described by Kinra and Marshall, together with a deterministic (static) fatigue analysis for larger hurricane waves having significant drag force nonlinearities. Cumulative damage from the two analyses were added, with the hurricane waves typically contributing 25 to 50% of the total damage.

Results for the three S-N curves of Figure 4 are presented in Figure 5. These histograms represent the worst hot spot in each of the joint cans. Using API curve X, all of the joint cans except one had a calculated fatigue life in excess of 40 years (acceptable for an intended service life of 20 years), with lives ranging up to 10,000 years. While reasons for this tremendous scatter will be examined later, the allowable hot spot stress remains a viable, if imperfect design rule.

---

[1] Recent authors (Marshall, 1976; Wirsching, 1981) have pointed out that other parts of the fatigue analysis contribute additional bias and uncertainty. The early analysis referred to here introduced considerable bias on the safe side by assuming that peak stresses correspond to half the total range (i.e., R = -1.0).

[2] $\varepsilon_{TR} = SNCF \cdot (f_r)_{nom}/E$ with $SNCF = (SCF)_{theor} \dfrac{(S_r)_{theor}}{2F_y}$

where $(f_r)_{nom}$ is the cyclic range of nominal member stress, E is Young's modulus, $(S_r)_{theor}$ is the theoretical elastic hot spot strain range, and $2F_y$ is the range from plus to minus yield, compatible with the cyclic stress-strain curve at ε_{TR}.

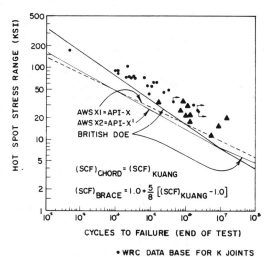

HOT SPOT S N CURVES
FIGURE 4

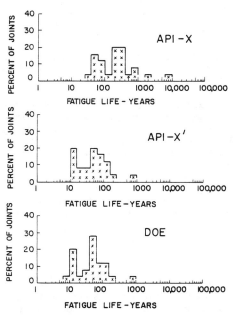

HISTOGRAMS OF FATIGUE LIFE FOR
JOINT CANS IN GULF OF MEXICO PLATFORM
FIGURE 5

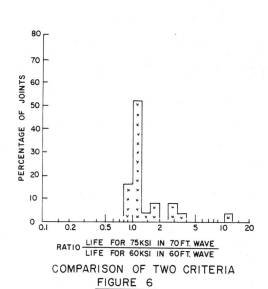

COMPARISON OF TWO CRITERIA
FIGURE 6

The two other S-N curves gave shorter calculated fatigue lives--so short that one is tempted to speculate that they are inconsistent with up to three decades of satisfactory service experience with over a thousand Gulf of Mexico structures. However, since the median time to collapse of a redundant structure has been estimated as 2 to 3 times the calculated life of the worst joint (Marshall, 1976), we would not expect many total failures yet. Also, most of the older structures avoided the use of heavy sections. Thus, such speculation cannot be used to justify extrapolating use of the old curve X to massive frontier area structures.

Since the 60 ksi rule was first derived, design waves in the Gulf of Mexico have generally been increased, while the long term wave statistics governing fatigue have remained essentially the same. For example, in 200 ft. of water, Shell's design wave height is 70 ft., slightly more than the API reference level wave height. Using a design hot spot stress allowable of 75 ksi for these higher waves is consistent with the old 60 ksi rule. As shown in Figure 6, the resulting fatigue lives come out the same, except for a few joints in which the differences are on the safe side.

In a number of cases, use of the parametric SCF equations becomes impractical. For example, some joints may fall outside their range of validity, so that unreliable results would be obtained; or the equations, which are unwieldy for repetitive hand calculations, may not have been implemented in a joint design post-processor to the structural analysis program. Since punching shear stresses are generally calculated as part of the static strength design, these can also serve as a convenient basis for fatigue checking.

Figures 7 and 8 show design S-N curves for T and K joints, respectively, along with test data plotted on the basis of cyclic punching shear. Recent data from the UK-EEC research effort have been added to the original data base; again we see results from large scale, low stress, high cycle tests encroaching on the design curve. Note that punching shear does not bring the two types of joints onto a common basis; the design curves are different by a factor of two. Also, the cyclic punching shear criteria do not reflect the important influence of chord diameter-to-thickness ratio (D/T), and should be adjusted downward for D/T ratios exceeding 30 or 40 (the range represented by test data).

The punching shear curve for K joints is lower than hot spot design curve X by a factor of 7, implying that hot spot stresses are typically 7 times the punching shear. Conversely, if we seek to limit the hot spot stress to 60 ksi, we should limit the punching shear to roughly 1/7 as much, or 9 ksi for axially loaded K-joints. If we consider that the typical design case involves axial load in combination with in-plane bending (which has a lower SCF and contributes less to hot spot stress), then an allowable peak punching shear of 10 ksi becomes appropriate.

For T and Y joints, the tolerance for punching shear is half that of K joints for axial load, and about the same for bending. Again, considering that combinations of axial load and bending will be present, an allowable peak punching shear stress of 7 ksi is appropriate.

For X joints, the SCF are even higher (and the tolerance for punching shear lower), as shown in Figure 9. Also, X joints are often subject to large out-of-plane bending moments, which approach axial load in their severity in terms of SCF. Thus, the allowable peak punching shear should be reduced again by a factor of 1.4, to 5 ksi. However, as β (the ratio of branch diameter to chord diameter) approaches unity, the SCF become less severe, and the 7 ksi punching allowable remains appropriate.

Although the acting punching shear computed according to the rules for strength design correctly reflects the overall statics and geometry of the connection, a better reflection of the localized cyclic stresses causing fatigue damage is given by the AWS expression for cyclic punching shear (Marshall & Graff).

$$\text{cyclic } V_p = \tau \sin\theta \left[\alpha f_a + \sqrt{(\tfrac{2}{3} f_{by})^2 + (\tfrac{3}{2} f_{bz})^2} \right] \qquad (\text{eq. 1})$$

where f_a is axial nominal stress range, f_{by} represents in-plane bending, and f_{bz} represents out-of-plane bending. The term α is used to merge former design S-N curves K and T into the single design curve K. In the AWS code, α is assigned a value of 1.0 for K joints and 2.0 for T, Y, and cross joints.

Figure 10 presents an empirical expression for computed α, which appears to be a straightforward and convenient way of expressing the influence of load patterns on the severity of joint can ovalizing, appropriate for the degree of approximation which punching shear represents. Results for the classic joint types, consistent with punching shear allowables in the range of 10:7:5, are as follows:

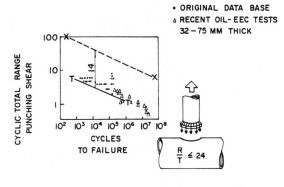

FIGURE 7

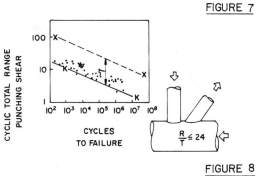

FIGURE 8

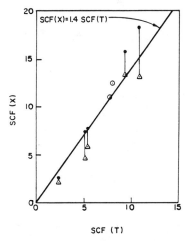

COMPARISON OF STRESS CONCENTRATION FACTORS
IN CROSS JOINTS VERSUS TEE JOINTS
HAVING OTHERWISE SIMILAR GEOMETRY
FIGURE 9

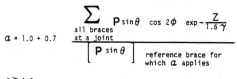

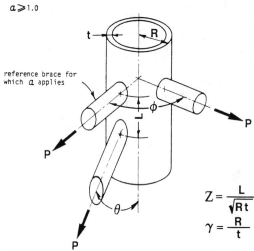

EQUATION AND DEFINITION OF TERMS FOR COMPUTED ALPHA
FIGURE 10

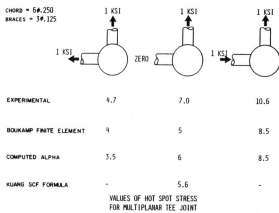

VALUES OF HOT SPOT STRESS
FOR MULTIPLANAR TEE JOINT
FIGURE 11

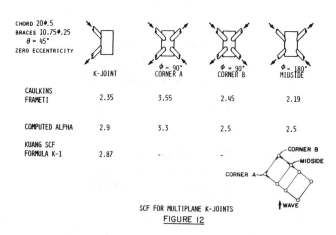

SCF FOR MULTIPLANE K-JOINTS
FIGURE 12

TYPE OF JOINT	VALUE OF COMPUTED ALPHA
balanced K joint	1.0 to 1.2 (depends on spacing)
T & Y joints	1.7
cross joints	2.4

Intermediate loading patterns produce intermediate values of computed α. In addition, computed α can easily accommodate more complex multiplanar load patterns, for which existing parametric SCF equations are inapplicable, as shown in Figures 11 and 12. In these comparisons, the "computed alpha" hot spot stress range $(S_r)_{max}$ is given by the modified Kellogg formula,

$$(S_r)_{max} = 1.8 \sqrt{\gamma} \cdot \text{cyclic } V_p \qquad (eq. 2)$$

where $\gamma = \dfrac{0.5D}{T}$.

The allowable fatigue stresses for a typical Gulf of Mexico structure are summarized in Table I, along with similar allowables derived for deeper water (Cognac), a more severe wave climate (the North Sea), and for a milder area where the design storm comes from the same population as the normal weather (Southern California). In each case, these allowable stresses were derived from preliminary generic fatigue analysis and verified by follow-up detailed analyses. In all these other cases, lower allowables and/or lower sensitivities (ksi hot spot stress per foot of design wave height) are required than for the Gulf of Mexico. This is because dynamic amplification and/or different wave statistics change the shape of the long term stress distribution so that it does not sag as much as the one shown in Figure 2, and the millions of cycles are no longer at less than 5% of the design value.

NEW DESIGN FORMAT

We shall now examine a probability based fatigue design format being studied under the API PRAC-15 project (Wirsching 1981). In this approach, the maximum hot spot stress is again limited to a design allowable. The allowable hot spot stress, however, must be determined by considering the actual long term distribution of stress ranges.

Let us begin by getting some definitions out of the way. We define an S-N curve (which extends indefinitely), using empirical constants m and K, as

$$NS^m = K \qquad (eq. 3)$$

and D as the cumulative damage ratio in the sense of Miner's Law.

For a deterministic fatigue analysis, in which the long term stress history is broken down into blocks, e.g., n_i cycles of stress level S_i, we can write

$$D = \frac{1}{K} \sum_{all_i} n_i S_i^m \qquad (eq. 4)$$

For a spectral fatigue analysis, the corresponding expression is

$$D = \frac{1}{K} T_e \Omega \qquad (eq. 5)$$

where T_e is elapsed time in consistent units, and Ω is obtained by summing over the various sea-states:

$$\Omega = \lambda(m) \frac{\Gamma(\frac{m}{2} + 1)}{(2)^{m/2}} \sum_{all_i} \gamma_j f_j (S_{rj})_{sig}^m \qquad (eq. 6)$$

In the foregoing:

$\lambda(m)$ is a rainflow correction factor, 1.0 for a narrow band random processes following the Rayleigh distribution, and ranging from 0.75 to 0.90 for typical wide band stress histories in marine structures responding linearly. If we use λ to correct for the distribution of stress cycles which results from a nonlinear response (e.g., drag dominated), λ can range up to 1.5 for m = 3, or 3.0 for m = 4.4.

Γ denotes the gamma function, where for positive integer x, $\Gamma(x+1) = x!$

γ_j is the fraction of time spent in seastate j.

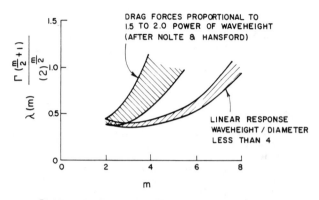

PLOT OF LEADING TERM IN EQUATION (6)
FIGURE 13

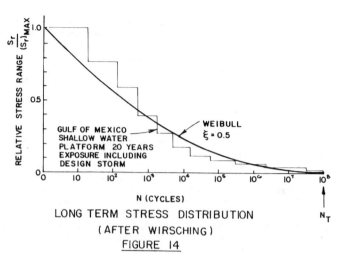

LONG TERM STRESS DISTRIBUTION
(AFTER WIRSCHING)
FIGURE 14

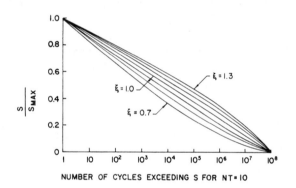

Type of Ship	Name of Ship	Weibull Load Shape, ξ	S_{max} (ksi) Stress Change at Probability of Exceedance 10^{-8}
Dry Cargo	Wolverine State	1.2	16.5
	California State	1.0	18.0
	Mormacscan, N. Atlantic service	1.3	12.0
	Mormacscan, S. America service	1.0	10.0
Large Tankers	Idemitsu Maru	1.0	12.3
	R. G. Follis	0.8	30.0
	Esso Malaysia	0.8	21.8
	Universe Ireland	0.7	18.7
Bulk Carrier	Fotini L.	0.9	29.5
SL-7 Containerships	Fleet of 8	1.2	34.1

SHIP LOADING HISTORIES
MODELED BY WEIBULL DISTRIBUTIONS
(AFTER MUNSE)
FIGURE 15

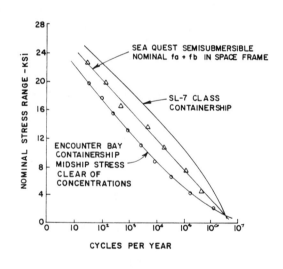

MEASURED STRESSES IN MARINE STRUCTURES
WHICH EXPERIENCED FATIGUE CRACKS AFTER
SHORT PERIODS OF SERVICE
FIGURE 16

f_j is the zero crossing frequency of stresses resulting from seastate j.

$(S_{rj})_{sig}$ is the significant stress range resulting from seastate j; that is, $(S_{rj})_{sig} = 4\sqrt{m_o}$ where m_o is the area under the response spectrum. This is analogous to significant wave height as used by oceanographers.

The term outside the summation is plotted as a function of m in Figure 13. It is a conservative and not unreasonable approximation to simply take this whole mess as unity, when the significant stress range is used for each seastate.

Finally, if we can represent the entire long term stress distribution by the Weibull distribution,[3] with maximum stress range $(S_r)_{max}$, shape parameter ξ (squiggly), and N_T total cycles, we can write (taking λ as unity):

$$D = \frac{1}{K} N_T (S_r)_{max}^m [\ln N_T]^{-\frac{m}{\xi}} \Gamma(\frac{m}{\xi} + 1) \qquad \text{(eq. 7)}$$

Examples of long term stress distributions for which the Weibull seems appropriate can be seen in Figures 14-16.

Where fatigue damage calculations have been done by deterministic or spectral methods, and we wish to relate the results to the Weibull format, it is useful to define:

$$G(\xi) = \log_{10}\left[\frac{N_T (S_r)_{max}^m}{KD}\right] \qquad \text{(eq. 8a)}$$

$G(\xi)$ has the following properties: If a structure is actually subjected to N_T cycles of $(S_r)_{max}$, the bracketed term is unity and $G(\xi)$ is zero. However, for structures subjected to a distribution of stresses, which are for the most part much less than $(S_r)_{max}$, then $G(\xi)$ expresses the number of log cycles by which fatigue damage would be overestimated by assuming N_T cycles of $(S_r)_{max}$.

Referring back to equation 7, we find that we can also write:

$$G(\xi) = \log_{10}\left[\frac{(\ln N_T)^{m/\xi}}{\Gamma(\frac{m}{\xi} + 1)}\right] \qquad \text{(eq. 8b)}$$

Thus, $G(\xi)$ can be calculated using equation 8a; using numbers available from the fatigue calculations; then m/ξ can be read from Figure 17, which represents equation 8b. Given m (which must be the same as used in the fatigue calculations), ξ is easily determined. To be consistent, $(S_r)_{max}$ should occur once in N_T cycles, and both should refer to the same time interval.

For calibration, let us consider a Gulf of Mexico structure which will just meet the API RP 2A requirements for detailed fatigue analysis, having the following particulars:

- Design wave with 100-year recurrence interval is to be used.

- Service life of 20 years, for which a damage ratio of 0.5 is allowed (this corresponds to D = 2.5 in 100 years).

- $\xi = 0.57$ and $N_T = 7.5 \times 10^8$. These values were chosen as representative for the Gulf of Mexico by the API PRAC-15 committee.

- The API-X design S-N curve is to be used having $K = 2.44 \times 10^{11}$ and $m = 4.38$.

Applying equation 7, we get $(S_r)_{max} = 100$ ksi, which is remarkably consistent with our design peak hot spot stress, S_{peak}, of 75 ksi, since

---

[3] As used herein, the Weibull cumulative distribution function for stress is given by:

$$\text{pr}\{S \leq s\} = 1 - \exp[-(s/\delta)^\xi] \quad \text{with} \quad \delta = \frac{(S_r)_{max}}{(\ln N_T)^{1/\xi}}$$

The Rayleigh distribution corresponds to $\xi = 2$.

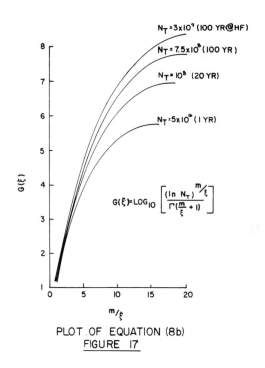

PLOT OF EQUATION (8b)
FIGURE 17

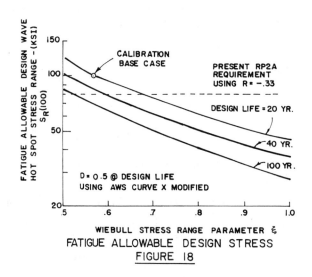

FATIGUE ALLOWABLE DESIGN STRESS
FIGURE 18

$$S_{peak} = \frac{(S_r)_{max}}{1 - R} \qquad \text{(eq. 9)}$$

and the cycle ratio (R = trough force/peak force) is -0.33 for the design wave.

Using this calibration (D = 2.5 in 100 years) as a point of departure, we can use equation 7 to derive the allowable $(S_r)_{max}$ to be used with the 100-year design wave, as a function of the Weibull parameter ξ and the intended service life, as shown in Figure 18.

Now all we need is ξ, which will depend on such factors as

(a) wave climate (e.g., North Sea vs. Gulf of Mexico)

(b) water depth (shoaling changes the relative size of design waves vs. everyday waves)

(c) type of structure (drag vs. inertial dominated)

(d) dynamic amplification (high cycle bulge)

(e) position in the structure (e.g., waterline braces responding to local wave pressure vs. those deeper in the structure responding to base shear)

The subject of Weibull parameter estimation (or, How's your squiggly?) is explored in the following section.

Estimates of the degree of reliability which is attached to the Gulf of Mexico calibration case, and thus to the derived criteria, vary. Wirsching, in transmitting his results to the API PRAC-22 committee, stated his "best estimate" for the safety index β_s as 2.36. However, the work leading to his 1981 report produced a variety of estimates of β_s, ranging from 1.4 to 5.9.

PARAMETER STUDIES

In order to get a handle on the Weibull parameter ξ, a number of parameter studies were conducted using a 200-foot, 8-pile, Gulf of Mexico platform as a base case. This is the same platform as described earlier in connection with Figure 5.

However, rather than re-running the full detailed fatigue analysis for each parameter variation, a simpler model of the structure was developed, which only represents members where they cross the waterline, as shown in Figure 19. This "candelabra" model is appropriate for the many small fatigue waves whose influence does not extend much beyond the waterline, and permits essentially closed form spectral calculations of applied wave force and dynamic response, e.g., base shear and overturning moment.

As a first step, dynamic base shear transfer functions (TF) for the simplified candelabra model were compared with those from the detailed space frame, as shown in Figure 20. Obviously, only the gross characteristics have been retained. Fortunately, a spectral fatigue analysis involves integration over a range of frequencies for each sea state, and then a summation over many sea states, so some of the differences in detail tend to average out. Nevertheless, results from such a simplified analysis should be viewed in a comparative, rather than absolute, sense.

Two computer programs were used with the candelabra model. DYNA3S is essentially the same linear program as used for preliminary fatigue analysis of the 1000-foot Cognac platform (Marshall, 1977), modified to accommodate finite water depth Airy wave theory. A second program was written for execution on the TRS-80 microcomputer, and includes the nonlinear drag force. The formulation of Borgman (1965) was modified to yield best estimates of the significant total range response (rather than RMS level) using $C_D = 0.8$ and $C_M = 1.7$. The spectral density for force on a unit length member at the water line, $S_{\phi\phi}$, becomes:

$$S_{\phi\phi}(\omega) = 8[C_D \frac{\rho}{2} D]^2 \text{var}\{v\} S_{vv}(\omega) + [\frac{\pi}{4}D^2 \rho C_M]^2 S_{AA}(\omega) \qquad \text{(eq. 10)}$$

where $S_{vv}(\omega)$ is the spectrum for water particle velocity

$S_{AA}(\omega)$ is the spectrum for water particle acceleration

$\text{var}\{v\} = \int S_{vv}(\omega) d\omega$

ρ = mass density of water

D = member diameter

ω = radian frequency

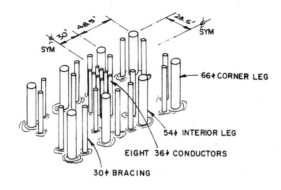

CANDELABRA MODEL OF 8-PILE
GULF OF MEXICO STRUCTURE
FIGURE 19

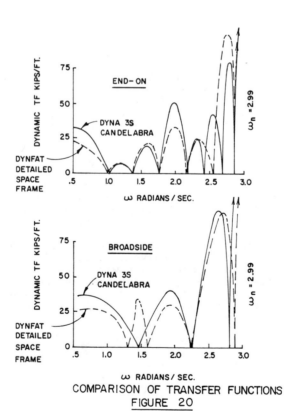

COMPARISON OF TRANSFER FUNCTIONS
FIGURE 20

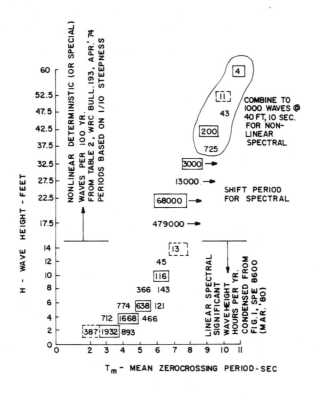

WAVE CLIMATE - GULF OF MEXICO
FIGURE 21

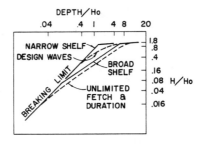

H = SIG. WAVE HEIGHT @ GIVEN DEPTH
Ho = DEEPWATER WAVE HEIGHT

SHOALING RELATIONSHIPS
FIGURE 22

Integration through the water column, combining the various cylinders in the array, and dynamic amplification follow Borgman, resulting in base shear response. Hot spot stress was assumed to be proportional to base shear, using a sensitivity factor (ksi per foot of wave height) derived from a Stokes wave static analysis of the same candelabra model, with extra forces added to represent the design wave hitting horizontal bracing, walkways, etc. at the top of the jacket. This latter step is consistent with use of allowable fatigue stresses in connection with the design wave analysis.

A condensed, unidirectional, scatter diagram, Figure 21, was used to represent Gulf of Mexico wave climate data, to reduce the burden of nonlinear spectral calculations for the TRS-80 fatigue analyses. Comparative studies using the linear DYNA3S analysis indicated that ξ for the condensed scatter diagram ranged from .02 lower to .04 higher than ξ obtained using the full directional scatter diagram with 16° RMS spreading--remarkably little difference.

For studies in various water depths, deepwater wave heights from the scatter diagram were modified using the shoaling relationships of Figure 22. Two limiting cases were assumed. "Narrow Shelf" corresponds to the currently fashionable assumption that no significant reduction in wave height occurs until the breaking limit is reached (wave height 78% of water depth). The "Broad Shelf" shoaling curve lies between that of Shell's standard design wave (very similar to API Reference Level), and Bretschneider's relationship for unlimited fetch and duration. It closely follows results obtained when an 80-foot deepwater hurricane wave is brought across the Eugene Island area, offshore Louisiana (Bretschneider, 1957).

TRS-80 fatigue analyses were performed for the candelabra model in water depths of 20, 50, 100, 200, and 400 feet (6, 15, 30, 60, and 120m). Results, in terms of the long term stress distribution for each case, are shown in Figure 23.

Although, for each structure, the sensitivity had been designed to yield a hot spot stress range of 100 ksi in the "standard" design wave, the spectral analysis did not always match this target at the once-per-100-years recurrence. The 200-foot base case is right on the money. The 400-foot structure, with a 3.0 second period, exhibits significant dynamic amplification, which one is permitted to neglect in the design analysis. The 100-foot and shallower structures, for "Broad Shelf" locations, never actually see the "standard" design wave because of shoaling. For the "Narrow Shelf" assumption, truncation of the wave distribution at the design breaking wave produces a large number of these waves per century. The "shoaling paradox" is that structures in very shallow water, which are designed for a once-in-a-lifetime wave which is greatly reduced by shoaling, are then more sensitive to the everyday waves which are not so much reduced.

Values of ξ derived from these results are shown in Figure 24. To complete the design algorithm (Figure 18 and equation 9), the cycle ratio R for the design wind and wave are also shown. Since the long term wave climate is unchanged by the somewhat unrelated choice of hurricane design wave, choosing the slightly lower API reference level design wave would result in slightly higher ξ. The full range of the grey area of API guideline wave heights would be about ± 5 times the variation shown.

Figure 24 also shows the range of ξ for the 200-foot base case structure. A fairly small range results from DYNA3S analyses (directional scatter diagram plus deterministic storm seas) in which hot spot stress is assumed to be proportional to the various gross responses which can be obtained from the simplified candelabra model--e.g., broadside shear, end-on shear, resultant shear, the foregoing three overturning moments, and corner pile load. A much larger range is obtained for individual members using the detailed space frame model, which should serve to remind us not to place too much faith in a simplified analysis.

The variation in ξ between individual members is examined in greater detail in Figure 25. Results are shown for the governing member at each of the joint cans studied. The actual stress range in the design wave, rather than the design allowable, was used as the basis for calculating ξ. This filters out one cause of the tremendous scatter when member fatigue lives are plotted (as in Figure 5), namely that many members are not right at the design limit to fatigue, either because some other condition governs, or because of the practicalities of construction (e.g., finite jumps in member size).

Member types are identified, but do not reveal any clear trend in terms of ξ. However, the effect of position in the structure is strong, with members near the waterline having high values of ξ, reflecting adverse fatigue loading. It is perhaps fortuitous that these same members tend to have little or no stress reversal, introducing conservatism into a design based on peak stress.

It should be noted that some of the very high values of ξ are caused by details of the initial design. Some members, in particular horizontals and conductor framing members near the waterline, were designed for the wave positioned for maximum base shear rather than the worst localized loading. Therefore, the actual stress ranges under the design wave are higher than the intended allowables. A second reason for some of the high ξ values is the effect of dynamics.

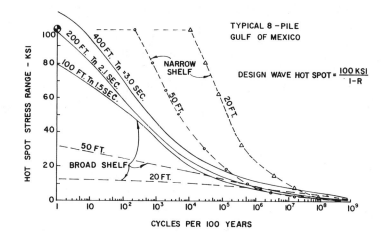

LONG TERM STRESS DISTRIBUTION
FIGURE 23

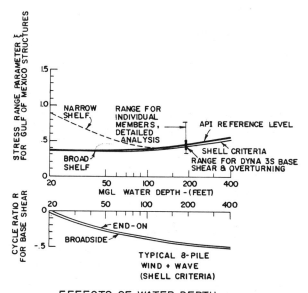

EFFECTS OF WATER DEPTH
FIGURE 24

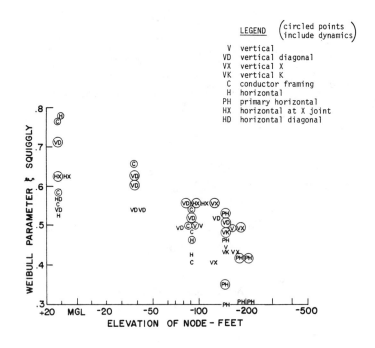

RESULTS FROM DETAILED
FATIGUE ANALYSIS
FIGURE 25

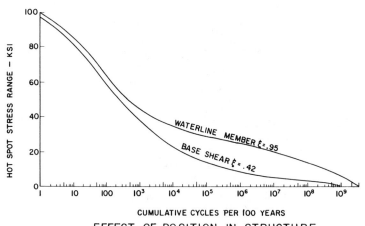

EFFECT OF POSITION IN STRUCTURE
ON LONG TERM STRESS DISTRIBUTION
FIGURE 26

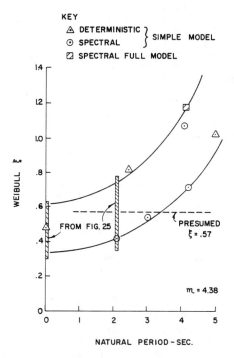

EFFECT OF NATURAL PERIOD
ON WEIBULL ξ
FOR GULF OF MEXICO STRUCTURES
FIGURE 27

NS = NORTH SEA
SC = SOUTHERN CALIFORNIA
B = BORNEO
BS = BASS STRAIT

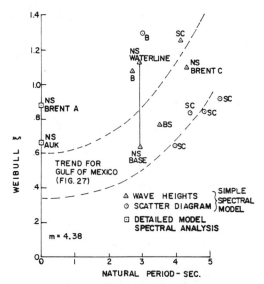

EFFECT OF NATURAL PERIOD ON
WEIBULL ξ
FOR HOMOGENEOUS WAVE CLIMATES
FIGURE 28

21

Although the structure natural period is only 2.1 seconds, there is enough energy in the seas at that period to excite the structure and cause acceleration of the heavy deck mass. Since dynamics was not considered in the initial design, members responding dynamically will result in higher ξ values. This can be seen in Figure 25 by comparing the results with those from a static analysis. If the high ξ values due to local forces near the waterline and dynamics are ignored, the maximum ξ values are closer to .6 than .8, which is very close to the .57 value that was previously described as representative for the Gulf of Mexico.

Figure 26 compares the long term stress distribution for a waterline member (stresses proportional to unit force as given by equation 10), versus that obtained earlier for base shear. The extremely adverse ξ of .95 results from the high cycle bulge--up to 1000 times as many cycles at 20% of the design stress range. Members near the top of the structure, which respond to both localized forces and gross structural action, have ξ between these two results; while members near the base more closely match the ξ values for base shear.

Increasing the natural period of the structure can have a profoundly adverse effect on fatigue life and ξ, as shown in Figure 27. These Gulf of Mexico results come from fatigue analyses in our files, dating back to 1967, some of which provided the basis for present API guidelines. For natural periods greater than 3 seconds, the presumed ξ of 0.57 would always be unconservative. For shorter periods, the influence of dynamics can also be significant.

The effect of different types of wave climate can be seen in Figure 28. A homogeneous wave climate is one in which the design storm is simply the extreme of frequently occurring weather, so that the long term wave distribution is a straight line on a semi-log plot ($\xi_{wave} = 1.0$), as opposed to the sagging distribution typical of the Gulf of Mexico. Structures in these areas tend to have higher ξ, and require lower fatigue design allowables, than their Gulf of Mexico counterparts. Paradoxically, this applies to mild wave climates like Southern California, Borneo, and the Persian Gulf, as well as to rough areas like the North Sea, North Atlantic, and Gulf of Alaska. However, the adverse affect of dynamic amplification is deferred to longer natural periods, where longer period waves and swell dominate the wave climate.

In the foregoing parameter studies, the ξ values were derived using m of 4.38, corresponding to the AWS curve X1 (modified). If all the long term stress distributions were perfect Weibulls, then these ξ values would be equally valid for other S-N curves with different slopes. Some of the fatigue calculations were in a form which permitted the computation of ξ based on equivalent damage, for a variety of S-N curves, where m = 3.74 corresponds to AWS X2 modified, and m = 3.0 corresponds to the upper slope of the DOE fatigue curve. Figure 29 shows the results. Despite some fairly wild departures from true Weibull shapes (e.g., Figures 23 and 26), the derived ξ seem reasonably consistent with each other.

SUMMARY AND CONCLUSIONS

Given that the end objective of design is a properly sized structure, rather than prediction, the Weibull parameter ξ conveys more useful information than does the calculated fatigue life.

Recent work of the API PRAC-15 project has identified a potentially useful improvement to the old blanket fatigue allowable stress, by incorporating ξ into the design process.

Parameter studies have begun to give some insight as to how ξ varies with water depth, location in the structure, wave climate, and natural period. This variation is more comprehensible than the often wildly random variation of calculated fatigue lives. Very tentatively, ξ could be selected as the larger of the values given by A, B, or C in Figure 31.

The new design procedure would then be as follows:

1) Based upon wave climate, shelf profile, structure natural period, and member location, determine the design value of ξ from Figure 31.

2) Using either Figure 18 where $(S_r)_{max}$ is a function of ξ and design life, or Figure 30 where $(S_r)_{max}$ is a function of ξ and S-N curve, determine $(S_r)_{max}$.

3) Compute design hot spot stress from equation 9:

$$\text{allowable } S_{peak} = \frac{(S_r)_{max}}{1 - R}$$

4) Values of ξ, $(S_r)_{max}$, R, and S_{peak} should be determined on a member-by-member basis.

5) S_{peak} is then used in place of the traditional blanket allowable hot spot stress (e.g., 60 ksi for 60 ft. waves in the Gulf of Mexico).

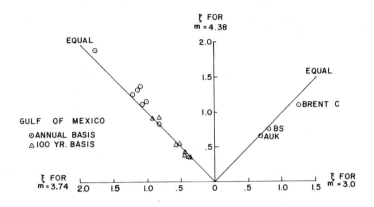

COMPARISON OF WEIBULL ξ
DERIVED FOR DIFFERENT
S-N CURVES
FIGURE 29

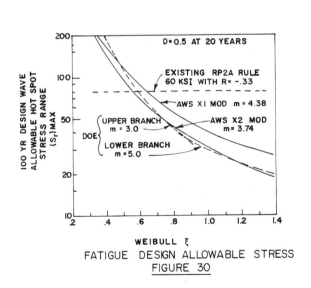

FATIGUE DESIGN ALLOWABLE STRESS
FIGURE 30

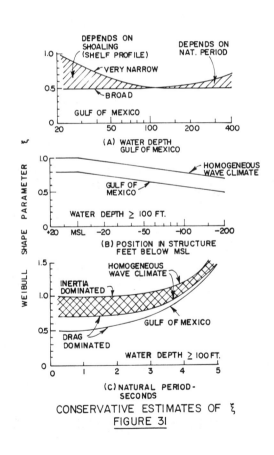

CONSERVATIVE ESTIMATES OF ξ
FIGURE 31

Recent work of the UK-EEC Offshore Steels Research Project has indicated that uncontrolled weld profile, heavy sections, or both, result in lower high-cycle fatigue performance, as indicated by new AWS X2, API 'X', and DOE fatigue curves. Extrapolation of shallow Gulf of Mexico experience to more hostile environments or deeper water, both of which involve higher values of ξ, exacerbates the difference. Thus, a deepwater structure designed for 100-year fatigue life (D=1) using the old AWS-X-modified wave may turn out to be just adequate for a 20-year service life (D=0.5) with the new curves.

These criteria are conservatively drawn based on a limited study. They are intended for initial sizing tubular joints as part of a design wave analysis. Where a more detailed fatigue analysis is to be conducted within a time frame which permits easy changes to the design, a less conservative bias would be appropriate.

More work is needed before these guidelines can be considered for inclusion in API RP 2A. Values of ξ from the work of other designers are needed to broaden the data base. The effect of a cutoff stress, below which little or no fatigue damage occurs, needs to be fully considered; this has the potential of raising the design allowables by 20 to 40 percent. Finally, the format, and the specific numbers, need to be tested more thoroughly in practical applications.

REFERENCES

API, 1981, Recommended Practice for Planning, Designing, and Constructing Fixed Offshore Platforms, API RP 2A, Twelfth Edition, American Petroleum Institute, Dallas, Texas.

BORGMAN, Leon, 1965, "The Spectral Density for Ocean Wave Forces," Proceedings, Santa Barbara Coastal Engineering Conference, Santa Barbara, CA, October 1965, pp. 147-182. New York: American Society of Civil Engineers.

BRETSCHNEIDER, C. L., 1957, "Hurricane Design Wave Practices," Journal of the Waterways and Harbors Division, ASCE, May 1957. New York: American Society of Civil Engineers.

CARTER, R. M., P. W. Marshall, T. M. Swanson, and P. D. Thomas, 1969, "Materials Problems in Offshore Platforms," Proceedings of the 1st Offshore Technology Conference, Houston, Texas, 1969, paper OTC 1043.

CAULKINS, D. W., 1968, Parameter Study for FRAMETI Elastic Stress in Tubular Joints, Shell Oil Company, CDG Report 15.

KINRA, R. K. and P. W. MARSHALL, "Fatigue Analysis of the Cognac Platform," Journal of Petroleum Technology, March 1980, p. 374. Dallas, Texas: Society of Petroleum Engineers.

KUANG, J., et al, "Stress Concentration in Tubular Joints," SPE Journal, August 1977. Dallas, Texas: Soceity of Petroleum Engineers.

MARSHALL, P. W., 1976, "Failure Modes of Offshore Platforms -- Fatigue," Proceedings, 1st International Conference on the Behaviour of Off-shore Platforms, (BOSS '76), Volume 2, paper 6.4, pp. 234-248. Trondheim, Norway: NTH.

MARSHALL, P. W., 1977, "Preliminary Dynamic and Fatigue Analysis Using Directional Spectra," Journal of Petroleum Technology, Volume 29, p. 715-722, June 1977. Dallas, Texas: Society of Petroleum Engineers.

MARSHALL, P. W., 1979, "Strategy for Monitoring, Inspection, and Repair for Fixed Offshore Structures, Proceedings, 2nd International Conference on the Behaviour of Off-Shore Structures (BOSS '79), London, England, 1979, Volume 2, paper 74, pp. 369-390. Cranfield, Bedford, England: BHRA Fluid Engineering.

MARSHALL, P. W., and R. G. BEA, 1976, "Failure Modes of Offshore Platforms," Proceedings, 1st International Conference on the Behaviour of Off-shore Platforms, (BOSS '76), Trondheim, Norway, 1976, Volume 2, contribution 21.3, pp. 579-635. Trondheim, Norway: The Norwegian Institute of Technology (NTH).

MARSHALL, P. W., and W. J. GRAFF, 1976, "Limit State Design of Tubular Connections," i.b.i.d. (BOSS '76), Volume 1, paper 4.6, pp. 634-651.

MOSES, Fred and Larry RUSSELL, 1980, Applicability of Reliability Analysis in Offshore Design Practice, Final Report, American Petroleum Institute, API-PRAC Project 79-22.

MUNSE, W. H., 1981, "Fatigue Criteria for Ship Details," Proceedings, Extreme Loads Response Symposium, Arlington, VA, 1981, paper R, pp. 231-248. New York, NY: Society of Naval Architects and Marine Engineers.

NOLTE, K. G., and J. E. HANSFORD, 1976, "Closed Form Expressions for Determining the Fatigue Damage of Structures Due to Ocean Waves," Proceedings, Offshore Technology Conference, Houston, Texas, May 1976, Paper 2606.

RADENKOVIC, V., 1981, "Stress Analysis in Tubular Joints," Proceedings, International Conference on Steel in Marine Structures, Paris, France, Plenary Session 2, pp. 53-96. Paris France: Institut de Recherches de la Siderurgie Francaise.

RODABAUGH, E. C., 1980, Review of Data Relevant to the Design of Tubular Joints for Use in Fixed Offshore Platforms, WRC Bulletin 256, January, 1980. New York: The Welding Research Council.

SNEDDEN, N. W., 1981, Background to Proposed New Fatigue Design Rules for Steel Welded Joints in Offshore Structures, Marine Technology Support Unit, AERE, Harwell, Oxfordshire, England.

WIRSCHING, Paul H., 1981, Probability-Based Fatigue Design Criteria for Offshore Structures, Final Report, Second-Year Effort, American Petroleum Institute, API-PRAC Project 80-15.

ACKNOWLEDGEMENTS

We are indebted to participants in the API PRAC-15 projects, particularly Paul Wirsching and Nick Zettlemoyer, whose ideas have been freely mixed with those of the authors. Fred Fisher performed the initial, detailed fatigue analysis of the base case 8-pile platform. MHP Systems Engineering, a division of Moonshire Hill Proprietary, furnished 30 hours of TRS-80 computer time. Shell Oil Company furnished a like amount of time on a much bigger facility.

TABLE I

EXAMPLES OF ALLOWABLE FATIGUE STRESSES (ksi) TO BE USED WITH PEAK CYCLIC DESIGN WAVE LOADS

Location	Design Wave	Nominal $f_a + f_b$	Hot-Spot API-X	Punching Shear			Hot Spot Sensitivity (ksi/ft)
				K	T and Y	X	
Gulf of Mexico under 400 ft.	60 ft.	20	60	10	7	5	1.0
Gulf of Mexico Cognac 1000 ft.	75 ft.	18	46	7	4	3	0.6
North Sea	100 ft.						
Waterline		15	34				
Elsewhere		25*	58	8			0.6
Southern California	49 ft.						
Waterline**		15	34	5			0.7
Elsewhere		19	47	7.5			1.0

1 ft. = 0.3m
1 ksi = 7MPa
  f_a = axial stress
  f_b = bending stress

*Using high strength steel and thickened stub ends.
**Including effect of 6 in. (150mm) marine growth.

THE FATIGUE STRENGTH OF WELDED CONNECTIONS
SUBJECTED TO NORTH SEA ENVIRONMENTAL AND
RANDOM LOADING CONDITIONS

R Holmes and J Kerr
National Engineering Laboratory
East Kilbride, Scotland, United Kingdom

SUMMARY

Offshore structures currently operational in the North Sea have been designed for fatigue using existing standards and codes. The basis for the derivation of these codes bears little relationship to the service conditions of a jacket platform. In order to bridge this gap in design methods, fatigue and corrosion fatigue tests have been carried out on cruciform and tubular welded joints under loading and environmental conditions closely simulating the hostile conditions prevailing in the North Sea.

Fatigue tests in air and in a simulated seawater environment have been used to investigate the effects of different types of narrow band random loading spectra with and without high tensile mean stress. Additionally, the influence of free-corrosion, cathodic protection and intermittent immersion on fatigue strength has been evaluated.

The results give a valuable insight into the effects of load spectra on the performance of welded joints and the accuracy of the techniques used in design codes for the prediction of fatigue life under random loading conditions.

INTRODUCTION

One of the many problems affecting the structural integrity of offshore installations and associated equipment is that of corrosion fatigue. This complex phenomenon has long been observed but recently an upsurge of interest in the subject has resulted in the reasonable understanding of the mechanisms of corrosion. However, the combination of the corrosion process with static and random variations in loading, as occurs on offshore structures in service, results in a corrosion fatigue situation involving numerous variables. Hence, estimates of the safe working life of installations are necessarily approximate unless validated by relevant data. Prior to the interim technical papers[1-4] produced from the United Kingdom Offshore Steels Research Project[5] these data have been extremely scarce.

Offshore structures currently operational in the North Sea have been designed using methods presently specified in design standards and codes[6] for the assessment of fatigue. The experimental data from which these codes have been derived bear little relationship to the service conditions experienced by an offshore platform throughout its working life. It was the realization of the difficulties of predicting the service life of offshore installations with confidence, which led to the inclusion in UKOSRP of the work described here. The work is a preliminary study to provide welded joint stress endurance data for environmental conditions similar to those of offshore structures in the North Sea. Research has already been carried out using basic constant amplitude sinusoidal load inputs to provide fatigue and corrosion fatigue data for simple plate joints[7] and fatigue data for a range of tubular joints[8] with a chord diameter of up to 2 m. The investigation described here extends this work to variable amplitude load inputs and provides an insight to the validity of the Palmgren-Miner linear cumulative damage rule as used in the current design codes for steel welded joints in offshore structures.

The complete spectra of loads experienced by welded connections on offshore platforms are wide ranging in nature and magnitude and it is inappropriate to attempt to include all of these in a laboratory test history. However, the major source of loading on such connections is derived from the action of seawaves and this wave action is considered here.

When the programme of work was defined, little information was available regarding the in-service stress histories of offshore structures. In the absence of reliable power spectral density information to enable the definition of bandwidths and other random process parameters the loading was assumed statistically stationary in the short term and narrow band in nature. These assumptions were made in the full understanding that only the force loading effect of the sea action was considered and not the structural response of the platform.

The sea conditions, mechanical/geometric properties of the particular connection under investigation, in addition to the dynamic characteristics of the complete jacket, greatly effect the load spectra experienced by a particular welded connection on a structure. In-service stress measurements suggest that at present there is no appropriate standardized load history for offshore applications. However, the narrow band random load histories used in this study do provide a useful initial step in the understanding and development of a data base of variable amplitude information applicable to offshore structures.

STRESS HISTORY

Theoretical Development

In early 1975 it was necessary to select variable amplitude load histories for use in investigating the validity of current procedures recommended for fatigue life assessment of welded connections. These load histories were to be broadly representative of those experienced by offshore structures during service in the North Sea. The derivations and discussions leading to the development of standardized load histories used in this investigation are summarized below.

The service load spectra of offshore installations derive mainly from the action of sea waves, whose behaviour is essentially a narrow band random process in which individual cycles of waves can be identified. In the short term the process is statistically stationary and the peak heights approach the Rayleigh distribution. In the long term the process is not stationary because of calms, storms and seasonal variations. As the relationship between wave height and load on a tubular structure is in general non-linear, wave height data cannot be used to obtain stress histories without extensive detailed analysis. Such an analysis was carried out at Liverpool University on behalf of NEL[9] and no account was taken of structural resonances.

Detailed examination of the results revealed that they could be divided into three broad groups in terms of immersion depth. At the surface the peak load distribution approximated to the Rayleigh distribution. Deeply immersed, 18.3 m, peak loads approached the Laplace distribution and at intermediate depth 6.1 m, the distributions were between these two extremes. On this basis, a set of results for a 3.66 m diameter member immersed to 6.1 m depth was selected as representative of the intermediate depth data and designated S1. The Rayleigh and Laplace distributions were proposed as representative of surface and deeply immersed data and designated S2 and S3 respectively.

Practical Development

The long-term distribution of peak loads does not contain all the information required to define a practical load history suitable for the fatigue tests. Assumptions have to be made on the short term variations of the controlling stress parameters and on the statistical correlation between a peak and its immediate predecessor.

For reproducibility the load history used in this programme was pseudo-random, repeating after a certain number of cycles defined as block length. Accurate definition of the specimen life would suggest a short block length but the inclusion of the high loads requires a long block length. A block length of 100 000 cycles was taken as a suitable compromise. For North Sea wave frequencies this represents approximately one week of testing in real time which is small compared with the design life of offshore structures. As a result of this block length selection, peak stresses with a probability of less than 10^{-5} were not included. Calculations using linear summation techniques[10] have shown that truncation at this level for the spectra under consideration, should have relatively little effect on the fatigue life.

Long term peak distributions can conveniently be built up from several Rayleigh distributions having varying levels. Breakdowns of the S1 distribution and the S3 distribution are given in Fig. 1 and the Rayleigh distribution, S2, requires no breakdown. These six levels, four at different root mean square amplitudes, can conveniently be arranged in ascending and descending order as a rough representation of calms and storms. As each level is itself a narrow band random process, some of the cycles in the low level will be higher than some in the higher level.

EXPERIMENTAL

Material

All specimens were manufactured from steel typical of that used offshore and conforming to BS 4360 Grade 50 specification. The plate specimens were Grade 50D node quality and the tubular joints Grade 50C and Grade 50E. Specimen configurations are summarized in Fig. 2. Details of methods of manufacture, material composition and mechanical properties have been described elsewhere[11][12].

Strain Gauging

To ensure correct loading conditions and strain levels every specimen was strain gauged prior to testing. With the T-joints a comprehensive pattern of gauges, including T-gauges and rosettes, was applied to one specimen from each of the geometries tested. The positions of these gauges were as agreed by the European Coal and Steel Communities Marine Technology Executive Working Group III after extensive analysis of the earlier strain distribution data. Values of hot spot strains were obtained by the currently adopted method of extrapolation. Further details of this strain gauge procedure and extrapolation are given in Reference 13.

Test Systems

All fatigue tests in air were carried out using a variety of purpose-built test rigs employing servo-hydraulic systems in the load control mode. An example of the rigs for tubular joint testing is shown in Fig. 3. Frequencies of loading were in the range 1.5 to 10 Hz, depending on the dynamic stress levels required. Environmental tests were carried out in the cantilever bending mode as shown in Fig. 4, at a test frequency of 1/6 Hz, corresponding to the mean sea wave frequency obtained for the geographical area under consideration. The tests were carried out in an environmental chamber through which artificial seawater, manufactured to the ASTM specification D1141-52, was circulated at a temperature of 5-8°C. The seawater circulation was arranged to give the three environmental conditions required for the investigation. These comprised of free-corrosion, immersion in seawater for 6 hours then exposure to air for 6 hours and impressed current cathodic protection at a potential of -0.85 V with respect to a silver/silver chloride reference electrode.

The random control signals specified earlier were produced for the air tests using an analogue digital hybrid system consisting of a pseudo-random binary sequence generator, analogue bypass filters and logic circuitry, whereas, the signals for the environmental tests were generated continuously by a micro-processor using fast Fourier transform digital filtering techniques.

Two levels of mean stress were investigated with the cruciform joints: firstly at a stress ratio of $R = -1$, and secondly at a high stress ratio in an attempt to reproduce the high mean stresses experienced by tubular joints owing to the presence of positive residual stresses introduced during the manufacturing process.

CONTROLLING STRESS PARAMETER

As a result of fatigue data obtained on welded joints during the past decade, current design codes[6] specify applied stress range irrespective of mean stress as the controlling stress parameter in the design of welded joints. For the simple planar joint this applied range is a nominal stress range whereas in a tubular joint it is the hot spot stress range.

When the welded joint is subjected to a variable amplitude load, stress range is no longer a viable criterion. In analysis of the fatigue data presented in this paper a second order averaging, that is the root mean square value of stress, is used in the absence of a universal parameter. Selection of this parameter to characterize the random load spectra investigated, was as a result of the random drive signals derived earlier being defined by the r.m.s. level and, earlier research[14], which showed r.m.s. stress to be suitable for the type of load histories considered in this investigation. A higher order averaging could have been employed. For the material and joints considered here the exponent of the Paris power law is normally taken as 3, leading to the effective stress being defined as the root mean cube of the spectrum. Extensive consideration will be given to this and other averaging techniques when a full analysis is carried out on all the UKOSRP data. However, the major advantage of employing r.m.s. stress to characterize the fatigue results at this stage is that it is not biased by fatigue characteristics, ie the slope of the S/N line, and it is an important statistical parameter.

RESULTS

Relevant experimental fatigue results obtained from previous research[13] using constant amplitude loading have been summarized in terms of r.m.s. stress by reducing the applied stress range by a factor of $2\sqrt{2}$ to give the equivalent r.m.s. stress. The results obtained from the planar joints in air are presented in Fig. 5 and the corresponding environmental data are shown in Figs 6 and 7. The definition of failure varies between rigs because of actuator stroke limitations. With the planar joints, all zero mean stress tests were taken to complete severance. In the high mean stress tests, failure was considered to have occurred when the actuator stroke limitation was reached; this corresponded to a crack depth of half the plate thickness. This difference in cycles to failure is negligible as crack growth rate was very rapid by this stage.

With tubular joints the definition of failure was even more complicated because of the vast variation in crack orientations in a T-joint, depending on the mode of loading and geometric characteristics of the joint. For this analysis and comparisons with existing constant amplitude data, failure of the tubular joint has been taken as the first through-thickness crack.

DISCUSSION

Fatigue-Planar Joints

Fatigue results obtained from both constant and variable amplitude loadings are shown in Fig. 5 and demonstrate the minimal scatter obtainable in simple welded joint fatigue testing by the use of modern servo-hydraulic control systems combined with strain gauges installed at critical positions. The results demonstrate on an r.m.s. basis the increased damage caused by the narrow band random loading spectrum S1 compared to constant amplitude loading and confirm the accepted view of zero influence of mean stress level on fatigue strength. Two additional tests were carried out at a lower truncation level (defined as maximum peak stress divided by long term r.m.s. stress) than that normally required for S1. However both tests, at a level of 4 as opposed to the normal level of 8, showed no significant effect of truncation level. This result may only apply to the type of narrow band random load signals considered in this investigation where one peak is not significantly different from the following peak.

Corrosion Fatigue - Planar Joints

Corrosion fatigue tests were carried out at a mean sea wave frequency. As a result, test times were lengthy and large numbers of tests could not be undertaken. The variable amplitude corrosion fatigue programme consisted therefore of only three tests for each combination of mean stress, environmental condition, and load spectrum. The results discussed below and subsequent conclusions drawn must therefore be tentative until further data are available.

Environment

Cruciform welded joints exposed to free and intermittent corrosion exhibit no effect of mean stress, confirming the results obtained in air. A relatively small but detectable influence of the environment is observed and this minimal influence, about 40 per cent in life, is substantially lower than that expected from previous crack growth studies. However, examination of fracture surfaces suggests that the initial stress intensity for this type of specimen may be at an extermely low level, where other research[15] has shown seawater to have negligible influence.

Additionally, with the intermittent immersion condition, there may be a trend at low stresses and long lives for the environment to have a larger influence.

Cathodic Protection

Impressed current cathodic protection at a recommended potential of -0.85 V produced a very small improvement at high mean stress and low dynamic stress. With a zero mean stress the joints exhibited a significant and substantial improvement in life as a result of the protection technique, particularly at low stresses where the final life to failure was in excess of the corresponding results in air. Fractographic examination[16], using the striations observed on the fracture face due to the non-stationary random load history, has confirmed crack growth rates in these tests to be lower than that in air. This beneficial effect experienced by the joints, due possibly to the plugging effect of the calcareous deposit which forms at the crack tip, may be mean stress dependent causing a shift in the level of stress at which the beneficial influence becomes fully effective.

Fatigue - Tubular Joints

The range of geometric variations included in the small number of T-joints tested to date makes it difficult to identify significant trends. For convenience the 168 mm joints tested under axial loading have been divided into two groups depending on the beta ratio. Mean lines for the corresponding constant amplitude experiments are shown to provide a relevant base data curve but it should be noted that some of these base lines consist of only three data points for each combination of beta ratio, tau ratio and mode of loading. As can be seen from Fig. 8, when compared on an r.m.s. basis, the increased damage effect of the narrow band random load input is demonstrated for joints with a beta ratio of 1.0. Joints with beta of less than 1.0 tend to exhibit significantly more scatter, due possibly to the unrealistic geometric characteristics which may have led to larger variations in residual stress.

With the small amount of data available for the 457 mm diameter joints, it is premature, at this stage, to extend the analysis without a more comprehensive data base. The results (Fig. 9) are therefore included for completeness. However, it is to be noted that all results to date can be described adequately by the current design procedures[6].

One final observation to be considered is the possibility that under certain conditions of low variable amplitude stresses and a particular strain distribution, it may be possible for the crack to develop to the through-thickness stage as normal, then gradually slow down and stop as the crack front moves away from the original hot spot. This phenomenon has been observed in two experiments to date in the low stress, long life region, but further verification will be required before any conclusions can be drawn.

Additionally, as the large diameter joints tested under constant amplitude loading produce results marginally above the existing design line, it will be interesting to see if the same geometries under variable amplitude loading produce data points which are also described adequately by the design procedures. A second stage of UKOSRP will include such testing and will also include comparative work using a broad band load sequence which is being developed employing recent data obtained from installations in service.

CURRENT DESIGN PROCEDURE

The transverse load-carrying cruciform joint is specified by the Class F classification and the tubular T-joint by the Class Q classification, in the current fatigue design rules applicable to offshore installations[6]. A linear cumulative fatigue damage summation was carried out on both types of joint to give a calculated design curve for the spectra considered. As recommended in DD55, the mean minus two standard deviation Class F design line with a cut off at 2×10^8 cycles was used for the planar joints and the Class Q design curve for the tubular joints. A damage summation value of $\Sigma(n/N) = 1.0$ was employed throughout the analysis.

The predicted lines for the planar joints shown in Figs 5 to 7, were not significantly affected by artificial clipping the excessive rare peak stresses, caused by the introduction of the high mean stresses, as most of the calculated fatigue damage was accumulated at low and intermediate r.m.s. stress. All cruciform joint failures to date have occurred at an endurance in excess of twice the calculated design life and it would appear therefore, for the narrow band random spectra investigated, that the current linear cumulative damage summation procedures recommended are adequate for the simple plate joints.

With the tubular joint the situation is much more complicated due to the introduction of significant size/geometric effects and mode of loading effects resulting in crack orientation variation, and the absence of a universal failure criterion. On the basis of life to first through-crack, which is a reasonable assumption from the designer's point of view as at this stage the crack could be repaired and no load shedding to other nodes would have taken place, it would appear that the current recommended procedures are also adequate for the joint sizes considered here.

DAMAGE SUMMATION VALUE

When the damage summation value of 1.0 is employed to calculate the fatigue design life on the variable amplitude loading as described in the design procedures the results appear to be adequately conservative. This type of comparison however gives limited validity to the assumption that the factor of 1.0 is correct. The accuracy of this value has already been considered in an earlier paper[17] describing analytical work not included in the programme and is currently being extended to include the latest UKOSRP data.

Results obtained from constant amplitude tests in air, see Fig. 5, were used as the base data to predict the fatigue life of similar cruciform specimens tested in the same rig under spectrum S1, and using a damage summation value of 1.0. As demonstrated, this value may not be adequate, especially in the case of low truncation levels, to maintain the degree of conservatism which exists in the design of joints for a constant amplitude service loading condition. Individual summation values of 0.5 to 1.1 were obtained for the random loading tests on cruciform joints in air using actual feedback truncation levels. The values obtained for the tubular joints were much more wide-ranging which reflects the complexity of the tubular joints when combined with very limited constant amplitude data and the extremely limited variable amplitude data points. Individual summation values are summarized in the histogram Fig. 10.

The Palmgren-Miner Linear Damage Summation Rule, employing a summation value of unity, may require modification in the future to maintain for random loading conditions the degree of safety currently available in welded joints designed for constant amplitude service load conditions. Where large peak loads do not occur, as a result of low truncation levels, the beneficial interaction effects, suggested by some results, may not occur resulting in poorer fatigue life predictions. With tubular joints the individual summation values were dependent on joint geometry, confirming the constant amplitude conclusion of a considerable size effect.

CONCLUSIONS

Fatigue and corrosion fatigue tests have been carried out on tubular and non-tubular welded joints under random loading and environmental conditions intended to represent broadly the service conditions experienced by offshore platforms in the North Sea. As the tubular joint variable amplitude data and the corrosion data are limited, the following conclusions may be considered as tentative.

Non-tubular Joints

No effect of the mean stress level was detectable with the narrow band random variable amplitude loading under environmental conditions of air and free corrosion. The influence of the free-corrosion environment reduced the fatigue life of the welded joints by a maximum of 40 per cent on life. This influence was not increased significantly by alternate immersion/exposure conditions.

The impressed current cathodic protection at a potential of -0.85 V significantly improved the corrosion fatigue performance of the joints particularly at low mean and dynamic stresses.

All three types of load spectra considered gave similar fatigue lives when plotted in terms of r.m.s. stress.

Linear cumulative damage summation with $\Sigma(n/N) = 1.0$ and using the Class F design curve adequately described the results obtained. However, further examination of the Miner's summation factor resulted in individual values of 0.5 to 1.1 for air tests and 0.9 to 1.5 for seawater tests.

Tubular Joints

Tubular welded joint fatigue data from T-joints with chord diameters of 168 mm and 457 mm subjected to the narrow band random load spectrum characterized by S1 remain adequately described by the current design procedures. As the design curves which will be used in future are based on two standard deviations below the experimental mean level, further tests will be essential to form a statistically viable data base to verify if the scatter of the variable amplitude test data is significantly different from that obtained from corresponding constant amplitude data.

ACKNOWLEDGEMENTS

The research in this paper forms part of the United Kingdom Offshore Steels Research Project commissioned by the Department of Energy through the Offshore Energy Technology Board. The research was carried out with financial aid from the European Coal and Steel Community. This paper is presented by permission of the Director, National Engineering Laboratory, UK Department of Industry. It is British Crown copyright reserved.

REFERENCES

1 HOLMES, R., 1980, "The Fatigue Behaviour of Welded Joints under North Sea Environmental and Random Loading Conditions", Twelfth Annual Offshore Conference, Houston, Texas, 1980, OTC-3700.

2 BOOTH, G. S., and HOLMES, R., 1981, "Corrosion Fatigue of Welded Joints under Narrow Band Random Loading", Proc. Conf. Fatigue in Offshore Structural Steels, London, 1981, pp 17-22. London: Institution of Civil Engineers.

3 HOLMES, R. and BOOTH, G. S., 1981, "Fatigue and Corrosion Fatigue of Welded Joints under Narrow Band Random Loading", Conf. Steel in Marine Structures, Paris, 1981, Vol. 2, Paper 7.2. Luxemburg: Commission of Eur. Communities.

4 MacDONALD, A., BROWN, G. and KERR, J., 1981, "The Influence of Geometric and Loading Parameters on the Fatigue Life of Tubular Joints", Conf. Steel in Marine Structures, Paris, 1981, Vol. 2, Paper 10.2. Luxemburg: Commission of Eur. Communities.

5 CRISP, H. G., 1974, Description of UK Offshore Steels Research Project: Objectives and Current Progress, Department of Energy, London, MAP 01SG(1)/20-2.

6 BSI, 1978, Fixed Offshore Structures, British Standards Institution, London, DD55.

7 BOOTH, G. S., 1979, "The Influence of Simulated North Sea Environmental Conditions on the Constant Amplitude Fatigue Strength of Welded Joints", Eleventh Annual Offshore Conference, Houston, Texas, 1979, OTC-3420.

8 WYLDE, J. G. and MacDONALD, A., 1979, "The Influence of Joint Dimensions on the Fatigue of Welded Joints", 2nd Int. Conf. Behaviour of Offshore Structures, London, England, 1979, Paper No 42. Cranfield, Beds: BHRA Fluid Engineering.

9 POOK, L. P., 1978, "An Approach to Practical Load Histories for Fatigue Testing Relevant to Offshore Structures", J. Society of Environmental Engineers, Vol. 7, No. 7, pp 22-35.

10 POOK, L. P., 1974, Fracture Mechanics Analysis of the Fatigue Behaviour of Welded Joints, National Engineering Laboratory, East Kilbride, Glasgow. NEL Report No. 561.

11 HOLMES, R., 1978, "Fatigue and Corrosion Fatigue of Welded Joints under Random Load Conditions", Proc. Eur. Offshore Steels Select Seminar, Cambridge, Vol. 1., pp 288-335. Cambridge: Welding Institute.

12 MARTIN, T., 1978, "The Fatigue Strength of Welded Tubular T-Joints with a Large Diameter Ratio", Proc. Eur. Offshore Steels Select Seminar, Cambridge, Vol. 1, pp 741-781. Cambridge: Welding Institute.

13 IRVINE, N. M., BROWN, G., and MacDONALD, A., 1981, Tubular Joint Fatigue Data obtained at the National Engineering Laboratory, Department of Energy, London, Interim Test Report UKOSRP 4/02.

14 OVERBEEKE, J. L., 1977, "The Fatigue Behaviour of Heavy Duty Spot Welded Lap Joints under Random Loading Conditions", Welding Research International, Vol. 7, No. 3, pp 254-275.

15 BRISTOLL, P. and ROELEVELD, J., 1978, "Fatigue of Offshore Structures: Effect of Seawater on Crack Propagation in Structural Steel", Proc. Eur. Offshore Steels Select Seminar, Cambridge. Vol. 2, pp 439-458. Cambridge,: Welding Institute.

16 POOK, L. P., 1981, "Fatigue Crack Growth in Cruciform Welded Joints under Non-stationary Narrow Band Random Loading", Symp. Residual Effects in Fatigue, Phoenix, Arizona, 1981. Philadelphia, Pa: American Society for Testing and Materials.

17 POOK, L. P. and HOLMES, R., 1981, "Forecasting the Fatigue Life of Welded Joints under Narrow Band Random Loading", Advances in Fracture Research, (ICF5), Vol. 5, pp 2079-2091. Oxford: Pergamon.

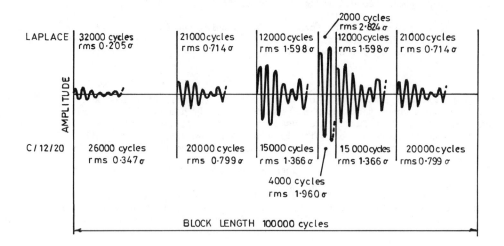

FIG 1 BREAKDOWN OF SIX LEVEL SEQUENCE

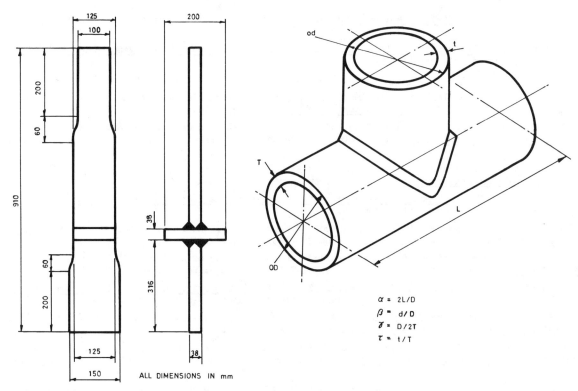

FIG 2a CRUCIFORM SPECIMEN CONFIGURATION

FIG 2b TUBULAR SPECIMEN CONFIGURATION

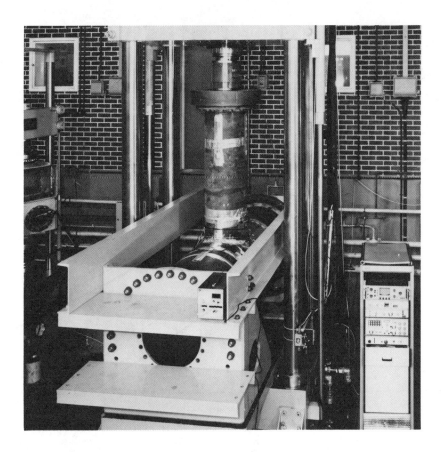

FIG. 3   TUBULAR SPECIMEN TEST FACILITY

FIG. 4   CRUCIFORM SPECIMEN ENVIRONMENTAL TEST FACILITY

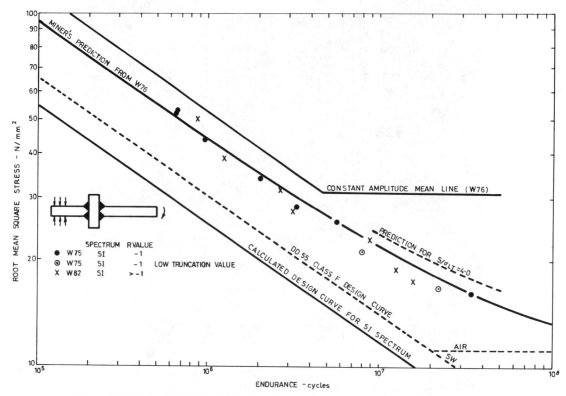

FIG 5 VARIABLE AMPLITUDE RESULTS OF CRUCIFORM SPECIMENS TESTED IN AIR

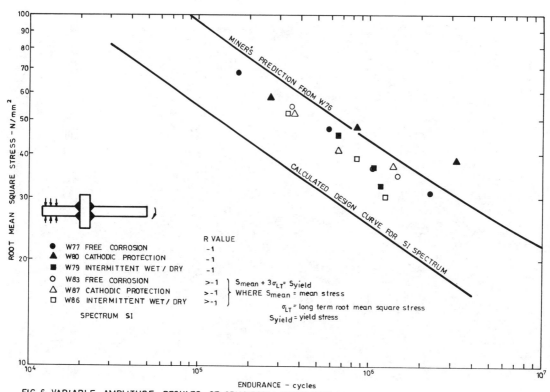

FIG 6 VARIABLE AMPLITUDE RESULTS OF CRUCIFORM SPECIMENS UNDER VARIOUS ENVIRONMENTAL CONDITIONS

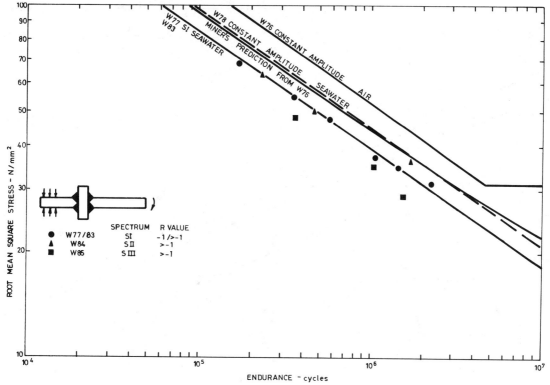

FIG 7 ENVIRONMENTAL RESULTS OF CRUCIFORM SPECIMENS UNDER VARIOUS LOAD SPECTRA

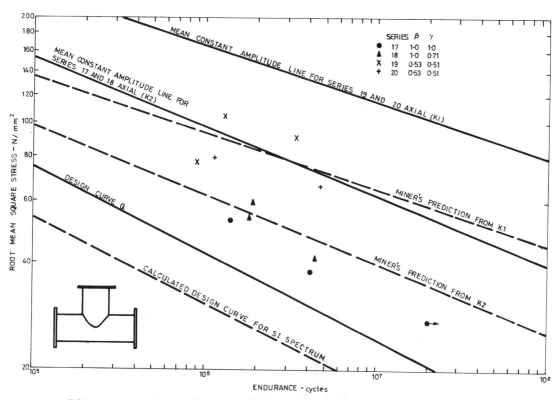

FIG 8 VARIABLE AMPLITUDE RESULTS OF 168 mm DIA CHORD TUBULAR SPECIMENS

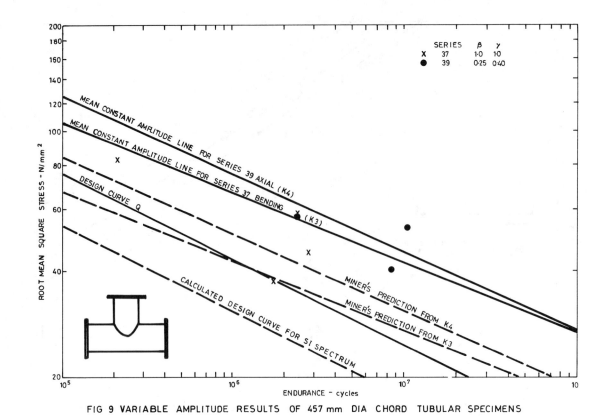

FIG 9 VARIABLE AMPLITUDE RESULTS OF 457 mm DIA CHORD TUBULAR SPECIMENS

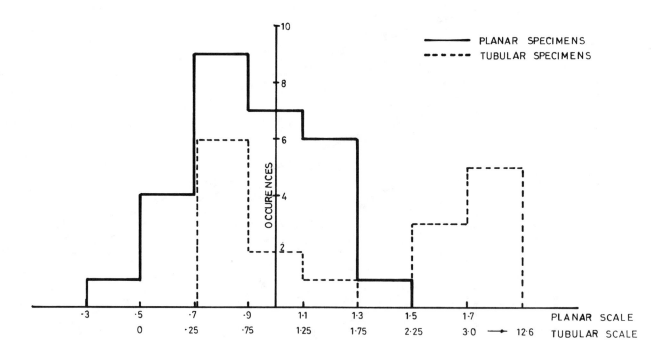

FIG 10 HISTOGRAM OF CUMULATIVE DAMAGE SUMMATION VALUES

Technical Lecture Sessions
—
STRUCTURES

Structural Stability

ULTIMATE STRENGTH OF STEEL OFFSHORE STRUCTURES

Victor A. Zayas
Earl and Wright Consulting Engineers
San Francisco, CA

Stephen A. Mahin
University of California
Berkeley, CA

Egor P. Popov
University of California
Berkeley, CA

SUMMARY

The response of braced offshore structures subjected to severe overloads depends on their nonlinear inelastic behaviour. Results are summarized from a series of experimental and analytical investigations on the cyclic inelastic behaviour of tubular brace members and tubular X-braced frames.

The experimental braces and frames are one-sixth scale models of those of a Southern California offshore platform designed according to American Petroleum Institute wave and earthquake criteria. The braces and frame specimens are subjected to imposed cyclic displacements causing buckling, post-buckling strength deterioration, tensile re-straightening, and tensile stretching of the braces. Data from the lateral force-displacement response of the frames, and the axial force-displacement response of the braces are examined. Observations are presented on: when the primary buckling of the braces in the frames occurred; which members were affected; the extent of damage as it progressed through the structure; and the occurrence of local buckling in members.

Analytical methods used for nonlinear inelastic analysis of offshore structures are evaluated and assessed by comparing experimental and theoretical results. The principal idealizations and parameters for analytical modeling are identified. A particular modeling is selected and implemented in a manner representing the best of the available idealizations. From the assessment of available analytical models, recommendations are made for nonlinear analysis methods, and improvements to the existing analytical models.

The analytical and experimental results on the inelastic behaviour of tubular steel braced frames indicate that ultimate strength concepts can be effectively used in the design and analysis of offshore structures. The damage and associated cost related to the inelastic response can be quantitatively assessed and compared with the cost of stronger initial design.

INTRODUCTION

An offshore facility may be subjected to severe accidental or environmental loads that are substantially greater in magnitude than those expected under normal service conditions. Examples of such loading conditions include the effects of collision with ships and barges, and unusually severe storm, earthquake, or sea ice loadings. Designing fixed offshore platforms for such unusual and rare events on an elastic basis may prove uneconomical and in many instances may not be necessary. In a properly designed and detailed structure, localized yielding and buckling may not adversely affect structural integrity, since the strength and deformation of such a structure can substantially increase in the post-elastic range.

It is recognized, for example, that the full plastic moment capacity of a tubular steel section is considerably larger than that at which yielding first occurs. For such a section to develop this capacity and deform inelastically, it must be fabricated from ductile materials and detailed to delay the onset of local buckling. In addition, such ductile yielding at a section in a statically indeterminate member or structure will redistribute internal forces so that larger loads can be applied. Loads can be increased until yielding occurs at sufficient locations to create a collapse mechanism. Once a collapse mechanism occurs, deformations increase disproportionately and the load capacity can increase, decrease, or remain stable depending on the member force-deformation characteristics and the structure's configuration. While inelastic deformations result in permanent structural damage that may not be acceptable under normal loading conditions, they may also result in substantial energy absorption, which is useful in resisting transient collision, wave, and earthquake loads.

Current API recommendations [R1] permit designers to take advantage of the ability of fixed tubular steel platforms to absorb energy through inelastic deformations during severe earthquake excitations. While such design and analysis methods accounting for inelastic behaviour have been successfully applied to many onshore structures, offshore construction incorporates many features not found in building structures and relatively little experimental and analytical verification for offshore structures has been available until recently. Thus, care must be exercised in evaluating the ultimate strength and behaviour of offshore structures.

In this paper some recent experimental and analytical research on the inelastic behaviour of tubular steel braced frames will be reviewed with the intent of identifying factors that should be considered in assessing their ultimate strength and behaviour. In the first part of the paper, experimental data related to the inelastic behaviour of individual braces, braced panels, and complete braced frames are reviewed with emphasis on recent results obtained at Berkeley. Implications for design are presented. In the second part of the paper, analytical models for predicting the ultimate behaviour of tubular braced frames are examined, and analytical and experimental results are compared. The use of inelastic methods of analysis for offshore structures is assessed.

PREVIOUS EXPERIMENTAL STUDIES

When ductile plastic response is desired, braced frames are usually designed so that the primary members that suffer inelastic deformations under lateral loads are the braces. Thus, experimental research has been performed to assess the inelastic behaviour of individual tubular braces. A number of other tests have been performed on braced panels and braced frames to assess the influence of boundary conditions and the interaction that occurs between members. In this section, various tests on individual braces and braced frames made from steel tubes of the type used in offshore structures are briefly summarized. Subsequent sections are devoted to discussions of specific results obtained in quasistatic tests at Berkeley.

Until recently most tests have concentrated on the behaviour of braces monotonically loaded in compression. For example, References R2 and R3 considered realistic large-scale fabricated tubular columns. More recently, tests have been carried out at the Admiralty Marine Technology Establishment to assess the effects of damage in the form of local dents and lack-of-straightness on brace buckling load and post-buckling behaviour [R4, R5].

At the University of California, Berkeley, one-sixth scale tubular steel structures were subjected to repeated cycles of reversing axial displacement, including inelastic buckling followed by tensile yielding [R6, R7]. The effects of different slenderness ratios (end fixities), tube diameter-to-thickness (D/t) ratios, and material properties on inelastic response were investigated.

A number of tests directed towards assessing the buckling and local buckling behaviour of tubular members have also been performed at the University of Wisconsin, Milwaukee [R8, R9, R10, R11]. Thirty tubular braces were subjected to monotonically and cyclically applied axial displacements through the post-buckling range. Various magnitudes of constant transverse loading were considered in these tests. Also, twenty-nine cantilever tubes with different magnitudes of constant axial load were subjected to monotonically and cyclically applied bending moments.

The ultimate strength of X-braced systems, typical of those in offshore platforms, have been investigated through tests on single braced panels and on frames with multiple panels. At the Southwest Research Institute, Austin, Texas, single panel X-braced sub-assemblages were tested at one-sixth scale [R12]. One X-braced sub-assemblage was tested to failure under monotonic loading conditions, and another two were subjected to repeated inelastic cycles.

At the University of California, Berkeley, two multi-panel, X-braced frames were subjected to cyclic inelastic loadings of reversing lateral displacements [R13, R14, R15]. These one-sixth scale bents were modeled from an example four-leg production facilities platform (Figure 1) designed for 100 ft water depths, considering wave and seismic criteria applicable to Southern California [R16]. The quasi-statically applied cyclic loadings were representative of the types of displacements that may occur during an extreme earthquake. A similiar multi-panel frame, at a 5/48 scale of the same Southern California example structure, was dynamically tested on the Berkeley shaking table [R17]. The model was subjected to strength level, ductility level, and maximum credible simulated earthquake ground motions.

INELASTIC BEHAVIOR OF TUBULAR BRACES

The results from the member tests in Reference R6 are examined in this section to highlight consideration related to ultimate strength and post-buckling behaviour of tubular braces. The tubular braces considered are one-sixth scale models of braces of the example Southern California example structure [R16]. Six braces (two pinned-end, two fixed-end heat treated braces, and two pinned-end braces with as-received steel properties) were subjected to quasi-statically applied cycles of reversed axial displacements. The two end conditions treated--both ends pinned, and both ends fixed--represent bounds on the possible end restraint within the frame. It was not possible in these specimens to realistically simulate the residual stresses present in full-scale fabricated tubes.

A typical experimentally obtained force-deformation relationship of a brace subjected to buckling followed by tensile stretching is shown in Figure 2. Load point B is the ultimate strength of the member loaded in compression, undergoing lateral buckling. Loading B-C is the post-buckling phase during which there may be a significant reduction in load carrying capacity. Loading D-E is the tensile re-straightening phase.

The cyclic force-deformation response of pinned-end and fixed-end members made of mild ductile steel are illustrated in Figure 3. The pinned and fixed end conditions result in slenderness ratios (kL/r) of 54 and 25 for the test specimens, respectively. The characteristically different appearance of these hysteretic loops is attributed to the difference in slenderness ratios. The lower the slenderness ratio, in general, the less will be the post-buckling reduction in compressive load capacity. The slenderness ratios of 54 and 25 both fall into a catagory of low slenderness ratios (kL/r <60), where the buckling load will be close to the compressive yield load.

The AISC [R18] buckling formula with safety factors removed generally gave a good prediction of the initial buckling loads. However, it must be recognized that the formula is for columns that have not been previously yielded, and with initial chambers within the code allowances.

An important observation during the cyclic buckling tests was that the buckling load reduces with each successive load cycle. Tensile yielding during each cycle (Figure 3) was sufficient to re-straighten the members such that the residual midspan deflections were equivalent to those existing at the beginning of the first cycle. The reduced buckling loads observed in successive cycles can be attributed to changes in the material properties (Bauschinger effects) associated with inelastic strain reversals. These damages in material properties are illustrated in the cyclic load-strain curves from a test on a short length of tube (Figure 4). The load-strain curve for first yielding in compression is noted to be basically elasto-plastic. Upon loading in tension, and subsequent compressive loading, the load-strain curve becomes rounded in nature. The severity of this rounding depends on the magnitude of the previous inelastic strain excursions. The buckling load of a column depends on the tangent modulus of the stress-strain curve. The more rounded the stress-strain curve, in general, the lower the buckling load will be.

Cyclic material stress-strain curves were used in a generalized tangent modulus procedure to predict the buckling loads of the brace specimens [R6]. The predicted reductions in buckling loads were found to agree with those experimentally observed in successive cycles (Figure 5). It should be noted that because of Bauschinger effects the buckling load of straight braces yielded in tension and of damaged braces that have been re-straightened will be lower than predicted by AISC formula.

The inelastic buckling of a brace results in the development of plastic hinges; that is, regions where inelastic curvatures occur. The number of plastic hinges that will form in a brace depends on the end restraint conditions. One plastic hinge forms in an idealized pinned-end brace and three hinges form in an idealized fixed-end brace. As expected inelastic curvatures were observed in the tests to concentrate in such plastic hinges. It is also significant to note that inelastic centroidal axial strains also concentrate in these plastic hinges as illustrated in Figure 6 and are not uniformly distributed along the brace. This is true for both the tensile and compressive yielding phases. This behaviour can be explained considering the axial force-moment interaction curve (Figure 7). The axial load capacity of the cross sections is reduced because of the presence of a moment. The cross sections with higher moments become plastic at lower axial loads, and both inelastic curved and axial deformations concentrate there. Once yielded, Bauschinger effects also tend to concentrate deformations in the plastic hinges.

The brace members were cycled until their axial strength was exhausted. Local buckling of the tube walls was generally observed to occur after a few cycles (Figure 8a). The onset of local buckling is followed by a rapid deterioration in the overall response of a brace. The test braces had D/t ratios 33 and 48. Local buckling was delayed in braces with D/t ratios of 33 compared to those with 48. In addition, the braces with the lower D/t ratios retained a greater percentage of their original strength with repeated cycles.

Once formed, local buckles tended to straighten out under tensile loads, but to reform when compressive loads were re-applied. The large local strains associated with this behaviour led to the development of tears in the steel (Figure 8b).

The API "Recommended Practice for Planning, Designing, and Constructing Fixed Offshore Platforms" [R1] indicates that tubular members may be considered compact and suitable for plastic design with redistribution of moments when $D/t \leq 1300/F_y$. For tubular members with $1300/F_y < D/t \leq 1900/F_y$ (semi-compact), API indicates that the full plastic load and moment capacities may be developed but that limited plastic rotation capacity may be presumed. Where $D/t > 3300/F_y$, a reduced allowable stress considering local buckling is recommended.

From the brace experiments [R6] it was observed that the use of compact sections ($D/t = 33 = 990/F_y$) did not prevent formation of local buckles under repeated inelastic cyclic displacement excursions. However, for compact members, substantial plastic rotation capacities were observed prior to local buckling. For the semi-compact members ($D/t = 48 = 1440/F_y$), full plastic load capacities and reduced plastic rotation capacities were observed, as anticipated.

INELASTIC BEHAVIOR OF BRACED PANELS

Two hypothetical braced panels are illustrated in Figure 9. The panels are designed such that under transverse (horizontal) shear, inelastic action will occur primarily in the diagonal bracing. The jacket legs at their intersection with the bracing and the brace cross-joints are stiffened, as necessary, so that the joints remain elastic. Similarly, the horizontals are sized such that they will avoid buckling through formation of the collapse mechanism and subsequent required deformations.

Inherent in this design of the panel, the collapse mechanism will initiate with compressive buckling or tensile yielding of the diagonal(s). At considerably increased displacements, the jacket legs or horizontals would form flexural plastic hinges to complete the mechanism. The transverse shear force versus lateral deformation relationship for the panel can be determined from knowledge of brace post-buckling response and the deformation characteristics of the legs and horizontals. However, the response of the braces is sensitive to the boundary conditions developed in the panel; this is uncertain especially in the case of the X-braced panel. Some results [R13] from quasi-static tests on one-sixth scale models of the Southern California Example Structure are examined in this section to assess the nonlinear behaviour of braces within a panel.

PANEL BRACING

In both panels in each frame tested, the tension braces were found to provide adequate lateral support at the cross-joint for the compression braces. In no instance did buckling of the full diagonal, by out-of-plane motion of the cross-joint, control the maximum compressive load capacity of a diagonal. The cross-joint tended to move out-of-plane with cycling but this did not significantly affect the buckling mode. Thus, each full diagonal can be idealized as two colinear braces connected at the cross-joint.

An important observation during tests [R12, R13] of the X-braced configuration was that buckling tended to concentrate in one brace (one-half) of the compression diagonal. Lateral buckling would be expected to initiate and be greater in one brace due to differences in initial imperfections or secondary moment effects. This concentration of inelastic deformations in one-half of the diagonal has two consequences. First, for the same lateral drift in the panel, the compressive load resistance will have deteriorated more than if the deformations were equally divided between the two compressive braces. Second, under successive cycles of deformation, the previously buckled brace will have a lower buckling load than the brace with lesser damage due to

Bauschinger effects. Thus, damage will continue to concentrate in the previously damaged brace. Although the straighter brace of the pair retains a higher capacity, the lowered buckling load of the damaged brace prevents it from contributing effectively to the panel's shear resistance.

This concentration of inelastic deformations in one-half of a full diagonal can have adverse effects on the response of the panel after the ultimate strength has been reached. It also exacerbates the tendency of the braces to local buckle and tear under reversed loading.

The direction of brace buckling within the panel, i.e., in-plane versus out-of-plane, is dependent on many factor including the type of end restraint, the initial member camber the direction of rotation of the jacket leg, the orientation of buckling in the adjoining diagonal brace, and the loading history. In the test frames, the elastic end restraint conditions were such that the braces would be expected to buckle out-of-plane in an S-shaped pattern because of the lesser rotational restraint at the cross-joint to out-of-plane buckling compared to in-plane. On the other hand, because of the in-plane end rotations induced by jacket legs, the braces would tend to buckle in-plane. All braces that buckled had some lateral displacement component both in-plane and out-of-plane. However, the buckling orientations observed were generally larger in-plane and corresponded to the direction induced by rotation of the jacket leg, and not necessarily in the direction of initial camber.

Deformations at the jacket leg-brace joint due to flexibility of the jacket leg pipe wall were measured since they could have an effect on the brace and frame responses. The jacket leg joints were designed to remain elastic but were not stiffened. The indentation of the jacket wall into the pipe was measured at the intersection of selected braces. The measured jacket leg indentations were very small compared to the brace axial displacements in the inelastic range.

The component of jacket leg indentation along the axis of the braces was subtracted from the brace axial displacements, and the hysteretic loops obtained were found to be essentially equivalent to the original hysteretic loops. Thus, elastic joint indentations had little effect on the inelastic response of brace members. Rotations at the brace ends relative to the jacket legs were also measured. These elastic rotations of the jacket leg wall were an order of magnitude smaller than the inelastic rotations occurring in the plastic hinges at the brace ends. The inelastic rotations were required to accommodate the midspan's lateral deflections occurring in the members. Consequently, three plastic hinges generally formed in braces undergoing post-buckling deformations, consistent with the mode of buckling observed in X-braced panels.

INELASTIC BEHAVIOUR OF BRACED FRAMES

The two one-sixth scale tubular frames tested under quasi-static loading are illustrated in Figure 10. These X-braced frames were subjected to cycles of reversing lateral displacements imposed at the deck level. The specimens were generally designed and detailed in accordance with API design practices. Members and connections were detailed so that inelastic deformations would primarily occur in the diagonal bracing members. In Frame I, the nominal member D/t ratios were 48 and in Frame II they were 33.

Lateral deck load versus deck displacement measured for both test frames are illustrated in Figure 11. For both frames, member buckling initiated in the upper X-braced panel, then progressed to those of the lower panel.

Significant loss of the capacity of these frames to resist lateral loads is mainly associated with the deterioration of the strength of the braces. Development of local buckling in a brace is particularly deleterious. Test results indicated that Frame I was prone to earlier and more severe local failures than Frame II. This is attributed to the higher D/t ratios of the bracing members in Frame I. Local buckles did not develop in any member of either frame prior to attaining the maximum frame loads.

Following damaging environmental overloads, it would be desirable for an offshore structure to remain stable under the effects of service level lateral loading resulting from current, waves, and earthquake aftershocks that can be reasonably expected to occur before repairs can be effected. To illustrate this concept, retention of lateral capacity above the initial linear elastic limit, H_e, is chosen as an indication that the frame models can maintain this lateral stability. For the frames, H_e was taken as the lateral load predicted from analysis to cause first yield in a brace based on combined axial and flexural stresses. (This was predicted to occur prior to member buckling.) The frame lateral strength capacities in both push and pull of the test cycles are plotted in Figure 12. Frame I retains capacities above H_e through cycle 14, which includes frame displacement ductilities up to 3.1. Frame II retains a capacity above H_e through cycle 17, which includes frame displacement ductilities up to 5.0.

The low slenderness ratios of the brace members and the ductile material properties were important factors contributing to the good cyclic inelastic behaviour of the frame models. Also, the use of braces with relatively low D/t ratios for both frames resulted in delayed local buckling and retained load capacity during cycles of high lateral displacement ductilities.

Low amplitude displacement cycles (Nos. 2, 6, 10, and 15) were included at key points in the loading history to check the degradation of overall lateral stiffness of the frames (Figure 13). The deterioration of stiffness of the structure after being subjected to various numbers of inelastic cycles is an index of permanent damage. This is particularly important from the point of view of dynamic loading, since the period of vibration of the structure depends directly on the stiffness. In order to properly predict displacement response, an analytical model should take into account any deterioration of structural stiffness that occurs. The stiffnesses relative to cycle 1 (a linear elastic cycle) and to analytically predicted linear elastic stiffness are noted in Figure 13. Inelastic response of the frames initiated in cycle 2 due to localized yielding at the braces from combined axial and flexural stresses. Reduced stiffness compared to that analytically predicted was already notable during this cycle.

AVOIDANCE OF WEAK LINKS

One of the primary considerations for a designer wishing to take advantage of the inelastic behaviour of offshore structures is to avoid weak links, i.e., local regions in the structure where inelastic deformations will concentrate and cause premature failure. Examples of such weak links are given below.

In the tubular brace tests [R6] two members with high strength as-received steel properties were included. The members were attached to the loading apparatus with full penetration welds. Both members failed in tension prior to reaching their nominal tensile yield strength. The failures occurred in the heat-affected zone adjacent to the welds. The material yield stress in this short length of pipe (about twice the wall thickness) was reduced due to the annealing effect of the welding process. The ultimate stress of the material in the heat-affected zone was less than the yield stress for the main length of tubing. The stress-strain curves of as-received and annealed material are shown in Figure 14.

The deformations that would normally be distributed along the plastic hinge region consequently concentrated in this short heat-affected zone, resulting in very high local strains. As a result, the brace fractured in tension at displacements lower than those corresponding to first yielding of the full length of the brace; a ductility of less than one. Thus, the effects of welding on member material properties must be considered when designing for full strength and ductility of the members.

These members, even without the welding-associated problems, would be expected to demonstrate poor ductility since the elongation capacities of the work hardened steel was low (less than 3%). The other individual braces tested [R6], and the braces in the frame models [R13], had elongation capacities of about 30%. The use of high elongation capacity steel is recommended when plastic deformations of a member may be required.

Joints with high stress concentrations may also result in weak links in a structure. As loads are increased, high local inelastic strains can occur at the stress concentration points, resulting in a joint failure. Such a failure, limiting the strength and ductility of the frame, was observed at the cross-joint of one of the X-braced subassemblages tested [R12]. Avoidance of weak links is essential to good ductile performance of a structure.

FRAME DESIGN IMPLICATIONS

The collapse mechanism of a frame is highly dependent on the nonlinear interaction among the members. To determine the collapse mechanism, and to thereby determine the ultimate strength of a frame system, the redistributions of internal forces and internal deformations must be considered.

The Berkeley X-braced frame models provide examples of these redistributions (Figure 10). In the elastic range, the horizontal members would have nearly zero axial load under horizontal loading on the frame since the horizontal component of the forces in the braces in the upper and lower panels would tend to cancel each other. Consequently, these members might be designed primarily for bending forces. However, once buckling of the main compression diagonals occurs, the diagonal braces undergo a displacement-dependent reduction in their compressive force capacity. At the joints of the center horizontal, the horizontal components of the tension diagonals are no longer balanced by the compression diagonals. Thus, the larger horizontal force components of the tension diagonals result in a compressive force in the center horizontal, which could cause buckling of this member even before the tension diagonals yield.

If the center horizontal buckles, a collapse mechanism forms for which the resulting plastic deformations in members are mostly flexural. Consequently, a large loss of ultimate frame lateral load and deformation capacity results. The center horizontal should be sized to account for the forces that would result from the difference between the forces in the tension and compression diagonals at the maximum required frame lateral displacement. Buckling is thus avoided and the inelastic deformations are forced to occur in the tension diagonal braces. Such a collapse mechanism is efficient because the main diagonals are utilized in an effective ductile manner.

In the tests, axial displacements were noted to concentrate along one brace of a compression diagonal. Similarly, at the structure level concentrations of shear distortions can occur in one panel of a structure. This was noted to occur in the upper panel of Frame I when the upper panel diagonal braces lost strength due to local buckling. Because of the strength and stiffness deterioration in this damaged panel, subsequent shear deformations also concentrated there. In fact, tension tearing had initiated, and the upper panel diagonal braces were close to tensile fracture before the lower panel diagonals buckled. Had the upper diagonals fractured, the lower diagonals might never have buckled. Such concentrations of damage can place unusually large ductility demands on a single panel and in this way this panel could lose its strength much more rapidly than expected under cyclic loading. This behaviour contributed to the deterioration of the hysteretic loops for Frame I in later cycles (Figure 11). To achieve a more efficient collapse mechanism, care must be taken to ensure that plastic deformations are distributed throughout the structure. The moment-resisting action of the jacket legs and horizontals is the principal vehicle for distributing shear deformations among the panels following yielding and buckling of the braces.

Changes in the equilibrium due to deformations of a structure may also be a necessary consideration. Frame displacements during inelastic response are considerably larger than those associated with elastic behaviour. A consequence of increased lateral frame displacements is that geometric nonlinearities (i.e., P-Δ effects) can result in instability and collapse of the structure under gravity loads. Consequently, it may become advantageous or necessary to control lateral displacements during overload conditions.

The lateral load-resisting capacity of the brace members is generally developed at considerably smaller displacements than the capacity of the moment-resisting frame. This effect is illustrated in Figure 15 where it is observed that the lateral resistance provided by the braces has peaked and begun to decrease well before the moment-resisting action develops its maximum capacity. However, the moment frame action helps maintain stability since the decrease in brace strength is partially or wholly offset by the increasing resistance of the frame.

EFFECTIVE LENGTH DETERMINATION

The buckling load and the post-buckling force-deformation response of a brace are dependent on the brace slenderness. A slenderness factor λ may be defined as:

$$\lambda = \frac{kL/r}{C_c} \tag{1}$$

where

$$C_c = \sqrt{\frac{2\pi^2 E}{F_y}}$$

$$r = \sqrt{\frac{I}{A}}$$

k = effective length factor

Three slenderness ranges are useful to consider: low ($\lambda < .4$), intermediate ($.4 < \lambda < 1$), and high ($\lambda > 1$). According to the AISC buckling formula [R18], in the low range the buckling load will be within 8% of the compressive yield load. The high range corresponds to Euler elastic buckling. In the low and intermediate ranges inelastic material responses affect the buckling load. The majority of offshore structure braces fall in the low and intermediate range. In general, the post-buckling deterioration of compressive load capacity is less severe for members with lower slenderness.

Within the test frames [R13] two bounds on the brace end conditions may be assumed: both ends pinned and both ends fixed. Assuming these two ideal conditions the upper panel diagonal braces have slenderness factors of 0.39 and 0.18, respectively. For these low slenderness factors assumptions regarding the brace end conditions are relatively unimportant for computing the initial buckling loads. However, there can be a substantial difference in the post-buckling brace response depending on the end conditions.

The response of a brace from the frames is compared to the responses of similiar braces with idealized pinned and fixed end conditions in Figure 16.

Typically, the force-deformation response for the frame brace closely follows that of the fixed end brace. Such a post-buckling response could be expected since it was observed that three plastic hinges formed in the frame members that buckled. The post-buckling force-deformation response depends on the number of hinges that form. From the experimental brace behaviour, it can be concluded that the unsupported lengths of braces within the frame models may be taken as the distance from the jacket legs to the center of the cross-joints, and a post-buckling effective length factor, k, of 0.5 could be assumed.

Computation of the appropriate effective length factor depends on the relative flexural stiffness of a brace compared to that of the joints and other adjoining members. For a brace within a frame it is useful to define three effective lengths: one for elastic, ultimate compressive strength, and post-buckling behaviour. The elastic effective length factor is computed based on assumptions of linear elasticity throughout the frame. For braces in the high slenderness range ($\lambda>1$) the elastic buckling load is the ultimate compressive load, and those two effective lengths are the same. The elastic effective length factor is not strictly applicable for predicting the ultimate compressive strength of frame braces in the low and intermediate slenderness ranges. As load on the brace is increased, local yielding within the member changes the member's flexural stiffness and therefore changes its stiffness relative to the frame. If the joints and adjoining members remain elastic the brace effective length is reduced by this inelastic behaviour. The effective length, in general, is continuously changing as yielding occurs. The effective length existing at the attainment of the brace ultimate compressive load is referenced to herein as the ultimate compressive strength effective length. Computation of this effective length is an iterative procedure which accounts for the reduction in the brace and adjoining members' stiffnesses with increasing displacements.

After the ultimate compressive load has been reached a brace buckling mechanism generally forms. This mechanism effective length is referred to as the post-buckling effective length. It should be noted that when the mechanism has formed, the incremental flexural stiffness in the brace plastic hinges tends to zero. Thus, the response tends toward that of a fixed-end beam, and the effective length factor approaches the corresponding value of 0.5.

ANALYSIS AND MODELING OF BRACED STRUCTURES

Analytical models for predicting the behaviour of braces and braced frames in the inelastic range have been proposed and applied. The reliability of predictions of structure behaviour depends in large part on the accuracy of the brace model used. An ideal brace model is one that has the capability to properly describe axial force-deformation hysteretic loops, and that accounts for damage and loss of load capacity under repeated cyclic loading. Most useful are brace models that are practical for computer analysis of complete structures.

Although models based on finite element or theoretical solutions have been used to predict brace behaviour, phenomenological brace models are presently the most commonly used type for nonlinear computer analysis of complete braced structures. Most of the available models are similar in concept and application, however, they offer differences in features, refinements, and in the number and range of cases for which input parameters have been defined. The basis of these models is to pre-define the shape of the axial force-displacement response of a truss element representing the brace by employing either mathematical or empirical results.

Models of this type have been developed by Higginbotham [R19], Nilforoushan [R20] Singh [R21], Marshall [R22], Roeder [R23], Jain [R24], and Maison [R25]. Marshall's model [R22] is the post-buckling brace element used in the offshore industry nonlinear analysis program INTRA [R26]. This model employs a seven segment piece-wise linear representation of a brace axial force-axial deformation hysteretic loop (Figure 17) and has an algorithm which defines failure of the brace based on estimating the onset of local buckling. Maison's model [R24] employs a nine-segment hysteretic loop representation with gradual buckling load deterioration capabilites, and a parameter to account for growth in brace length during buckling and restraightening (Figure 18).

Accurate representations of member responses are possible with phenomenological models when the shapes of the brace hysteretic loops are known.

The brace hysteretic behavior may be represented experimentally or analytically using the effective length concept. Using this concept a brace can be represented by a pinned-end member with equal slenderness ratio, and the axial force-displacement hysteretic loops are normalized with respect to yield force and yield displacement. This greatly simplifies analysis because braces can be modeled by single degree-of-freedom phenomenological elements.

Advantages of these phenomenological-type models are: they are practical for analysis of complete structures; large displacement analysis is not required; there are a small number of degrees of freedom; and small storage is required for element property parameters. The inavailability of appropriate experimental hysteretic loops or accurate analytical hysteretic loops for selecting input parameters is a limitation of these models.

The principal member parameters to be considered when analytically modeling the cyclic inelastic response of a brace are: effective length; yield load; yield displacement; susceptibility to local buckling (D/t ratio); and material property characteristics. When input parameters for a brace model based on experimental results are being chosen, the material property characteristics influence the response of a brace in a variety of ways. Bauschinger effects, as discussed before, cause a reduction in buckling load during repeated inelastic cycles. The initial shape of the stress-strain curve, the fracture strain, and the ultimate stress also influence the occurrence of local buckling, tearing, and the ultimate loss of capacity of the brace. Realistic analytical models must account for these effects.

EVALUATION OF ANALYTICAL MODELS

To assess the ability of nonlinear analysis methods to predict the behaviour of tubular steel frames, the two one-sixth scale models of the Southern California Example Structure tested at Berkeley were analyzed [R27] using the ANSR-I [R28] program incorporating the Maison [R25] brace model.

The model used to analyse these frames is shown in Figure 19. Modeling of these frames is greatly aided by the availability of results for the fixed-end structures [R6] which have similar properties to those of the frame braces. Since the major limitation of phenomenological models is the availability of appropriate brace hysteretic loops for defining the input parameters, the brace modeling described herein represents the best that is obtainable using such elements.

One brace of each full diagonal was modeled as a post-buckling element, and the other as an elastic truss element. This modeling agrees with the observed physical response that buckling and inelastic deformations tend to concentrate in the weaker brace. This is expected to have a significant effect on the accuracy of the results since the brace deterioration is displacement dependent. Had buckling been permitted in both halves of a diagonal, inelastic deformations would be distributed between them and deterioration would be much less severe. Moreover, serious numerical problems were encountered during analysis when both braces were modeled as co-linear post-buckling elements.

Cyclic displacements were imposed at the deck level of the analytical model to match those of the experiments, so that direct comparisons could be made. A step-by-step procedure without iterations, and with path dependent state determination, was used to analyze the frames. Unbalanced loads occurred at the nodal points whenever there was a load reversal or an element changed state. These unbalanced loads were corrected during the next step by adding balancing loads to the joints. The magnitude of the imbalances was controlled by using a small step size.

For path-dependent iteration strategies like the Newton-Raphson method, the loading and unloading of the element that may occur during iterations can result in the elements following improper paths. Since substantial changes in stiffness occur between some regions of this multilinear model, constant stiffness iteration schemes have convergence problems also. The most suitable scheme was the step-by-step method without iterations.

The overall behaviour of the analytical and experimental frame models may be conveniently presented as lateral load-lateral displacement hysteretic loops (Figures 20 and 21). The shapes and the areas inside the hysteretic loops are similar for the analytical and experimental results.

The analytical model tends to overestimate the energy dissipation and maximum frame loads in the early cycles (5 and 7), and underestimate them in the later cycles (12 and 14). The maximum analytical and experimental loads in each cycle for Frame II are compared in Figure 22. The occurrences of brace buckling in the analysis of Frame II were similar to the observed experimental events.

The principal limitations of the analytical results were concerned with the incorrect internal distribution of inelastic displacements. The numerical problems associated with using two co-linear buckling elements interfere with obtaining more realistic results. In addition, the brace model neglects the observed softening and inelastic deformations that occur prior to attaining maximum compressive or tensile forces. Moreover, loss of tensile capacity is neglected.

Although internal inelastic deformation in the frame analyses is not correct, the shapes and areas inside the hysteretic loops are similar for the analytical and experimental results. The frame model as implemented appears sufficiently accurate for general inelastic analysis, provided the brace behaviour is properly represented.

CONCLUDING REMARKS

The considerable analytical and experimental results of the inelastic behaviour of tubular steel braced frames indicate that ultimate strength concepts can be effectively used in the design and analysis of offshore structures. Properly designed and detailed tubular steel frames were formed in tests to be able to develop significant inelastic deformations and dissipate substantial amounts of energy. In addition to material properties, the inelastic cyclic behaviour of tubular steel braces depends on their effective slenderness ratio for the stage of loading under consideration and their D/t ratio. Braces and frames with lower D/t ratios exhibited significantly less deterioration in strength, stiffness, and energy dissipation than those with higher values. An X-braced frame with a nominal D/t ratio of $1440/F_y$ showed no significant deterioration in strength up to a frame lateral displacement ductility of 3. A similar frame with a D/t of $990/F_y$ behaved well through a frame ductility of 5. For the cyclic loading history considered, lateral and axial displacements tended to concentrate in one-half of a diagonal brace and the hysteretic behaviour of these buckled braces resembled that of individual braces with fixed ends.

Several generations of phenomenological models for braces have been developed incorporating several refinements and features. These have made it possible to predict the overall inelastic behaviour of offshore towers. However, these models must rely on judgment and limited experimental data to specify appropriate input parameters. Methods for improving models of braces have been proposed [R27].

Based on the ductile behaviour that can be exhibited by tubular steel frames, it appears reasonable to apply nonlinear analysis concepts to situations involving environmental and accidental overloads. For example, when structural members have been damaged (by supply boats, corrosion, or fracture), plastic analysis can be used to assess the residual strength of the damaged members and that of the tower itself. In addition, new data or experience may indicate that environmental loads can be larger on an existing structure than those used in its original design. Ultimate strength techniques may be effectively used to evaluate the actual capacity of such structures. In the design of new facilities, extreme but rare storm, earthquake, or collision loadings, which have very low probability of occurring during the life of the structure, may be designed considering inelastic response. The damage and associated costs related to the inelastic response can be quantitatively assessed and compared with the cost of a stronger initial design. New harsh environments, deeper waters, ice loadings, and new structural systems that are as yet unproven provide additional opportunities for incorporating ultimate strength methods and design.

REFERENCES

1. 1979, Recommended Practice for Planning, Designing and Constructing Fixed offshore Platforms, 10th Ed., Dallas, Texas: American Petroleum Institute.

2. BOUWKAMP, J.G., 1975, "Buckling and Post-Buckling Strength of Circular Tubular Sections," Proceedings, Offshore Technology Conference, Houston, Texas, 1975.

3. CHEN, W.F., ROSS, D.A., 1977, "Tests of Fabricated Tubular Columns," Proceedings, ASCE (Structural Division), Volume 103, March 1977.

4. SMITH, C.S., SOMERVILLE, W.L., SWAN, J.W., 1981, "Residual Strength and Stiffness of Damaged Steel Bracing Members," Proceedings, OTC 3981, Offshore Technology Conference, Houston, Texas, May 1981.

5. SMITH, C.S., KIRKWOOD, W., AND SWAN, J.W., 1979, "Buckling Strength and Post-Collapse Behavior of Tubular Bracing Members including Damage Effects," Proceedings, 2nd International Conference on Behaviour of Off-Shore Structures, London, England, 1979, Cranfield, Bedford, England: BHRA Fluid Engineering.

6. ZAYAS, V.A., POPOV, E.P., AND MAHIN, S.A., 1980, Cyclic Inelastic Buckling of Tubular Steel Braces, Earthquake Engineering Research Center, University of California, Berkeley, Report No. UCB/EERC-80116.

7. POPOV, E.P., ZAYAS, V.A., AND MAHIN, S.A., 1979, "Cyclic Inelastic Buckling of Thin Tubular Columns," Journal of the Structural Division, ASCE, Volume 105, No. ST11, pp. 2261-2277.

8. SHERMAN, D.R., 1978, "Cyclic Inelastic Behavior of Beam Columns and Struts," Preprint 3302, ASCE Convention and Exposition, Chicago, 1978.

9. SHERMAN, D.R., 1979, Experimental Study of Post Local Buckling Behavior in Tubular Portal Type Beam-Columns, Proprietary Report, Shell Oil Company, Houston, Texas.

10. SHERMAN, D.R., 1980, Post Local Buckling Behavior of Tubular Strut Type Beam-Columns: An Experimental Study, Proprietary Report, Shell Oil Company, Houston, Texas.

11. SHERMAN, D.R., 1980, *Interpretive Discussion of Tubular Beam-Column Test Data*, Proprietary Report, Shell Oil Company, Houston, Texas.

12. BRIGGS, M.J. AND MAISON, J.R., 1978, "Test of X-Braced Subassemblage," *Preprint 3302, ASCE Convention and Exposition*, Chicago, 1978.

13. ZAYAS, V.A., MAHIN, S.A., POPOV, E.P., 1980, *Cyclic Inelastic Behavior of Steel Offshore Structures*, Earthquake Engineering Research Center, University of California, Berkeley, Report No. UCB/EERC-80/27.

14. POPOV, E.P., MAHIN, S.A. AND ZAYAS, V.A., 1980, "Inelastic Cyclic Behavior of Tubular Braced Frames," *Preprint, ASCE Convention and Exhibit*, Portland, Oregon, 1980.

15. MAHIN, S.A., POPOV, E.P., AND ZAYAS, V.A., 1980, "Seismic Behavior of Tubular Steel Offshore Platforms," *Proceedings, Offshore Technology Conference*, Houston, Texas, 1980, Volume 3, pp. 247-258.

16. GATES, W.E., MARSHALL, P.W., AND MAHIN, S.A., 1977, "Analytical Methods for Determining the Ultimate Earthquake Resistance of Fixed Offshore Structures," *Proceedings, OTC 2751, Offshore Technology Conference*, Houston, Texas, 1977.

17. GHANAAT, Y., CLOUGH, R.W., 1982, *Shaking Table Tests of a Tubular Steel Frame Model*, Earthquake Engineering Research Center, University of California, Berkeley, Report No. UCB/EERC-82/02.

18. 1970, *Manual of Steel Construction*, 7th Edition, New York: American Institute of Steel Construction.

19. HIGGINBOTHAM, A.B., 1973, *The Inelastic Cyclic Behavior of Axially-Loaded Steel Members*, Dissertation, University of Michigan, Ann Arbor.

20. NILFOROUSHAN, R., 1973, *Seismic Behavior of Multistory K-Braced Frame Structures*, University of Michigan, Research Report UMEE 73R9.

21. SINGH, P., 1977, *Seismic Behavior of Braces and Braced Steel Frames*, Dissertation, University of Michigan, Ann Arbor.

22. MARSHALL, P.W., 1978, "Design Considerations for Offshore Structures Having Nonlinear Response to Earthquakes," *Preprint, ASCE Annual Convention and Exposition*, Chicago, 1978.

23. ROEDER, C.W. AND POPOV, E.P., 1972, *Inelastic Behavior of Eccentrically Braced Frames Under Cyclic Loading*, Earthquake Engineering Research Center, University of California, Berkeley, Report No. 77-18.

24. JAIN, A.K. AND GOEL, S.C., 1978, *Hysteresis Models for Steel Members Subjected to Cyclic Buckling or Cyclic End Moments and Buckling*, University of Michigan, Report UMEE 78R6.

25. MAISON, B. AND POPOV, E.P., 1980, "Cyclic Response Prediction for Braced Steel Frames," *Journal of the Structural Division*, ASCE.

26. LITTON, R.W., PAUSEY, S.F., STOCK, D.J., WILSON, B.M., 1978, "Efficient Numerical Procedures for Nonlinear Seismic Response of Braced Tubular Structures," *Preprint, ASCE Annual Convention and Exposition, Chicago, 1978*.

27. ZAYAS, V.A., SHING, P.S.B., MAHIN, S.A., POPOV, E.P., 1981, *Inelastic Structural Modeling of Braced Offshore Platforms for Seismic Loading*, Earthquake Engineering Research Center, University of California, Berkeley, Report No. UCB/EERL-81/04.

28. MONDKAR, D.P. AND POWELL, G.H., 1975, *ANSR-I General Purpose Program for Analysis of Nonlinear Structural Response*, Earthquake Engineering Research Center, University of California, Berkeley, Report No. 75/37.

ACKNOWLEDGMENTS

The financial support of the American Petroleum Institute in performing the frame tests and the subsequent analyses is greatfully acknowledged. The API Production Research Advisory Committee, chaired by P.W. Marshall, provided valuable advice and greatly aided in directing these projects toward their objectives. The individual brace tests reported were performed as part of an NSF-sponsored investigation of the cyclic inelastic behaviour of columns. The presentation of this paper at the BOSS '82 conference has been funded by Earl and Wright Consulting Engineers.

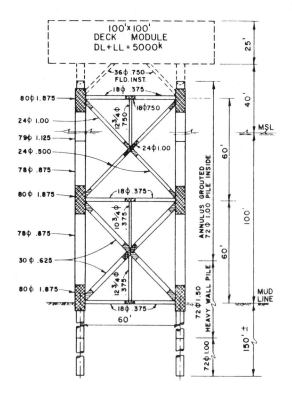
a) Southern California Example Structure

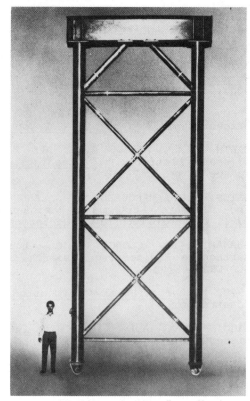
b) Photograph of Test Frame II

Figure 1
Example Structure and Test Model

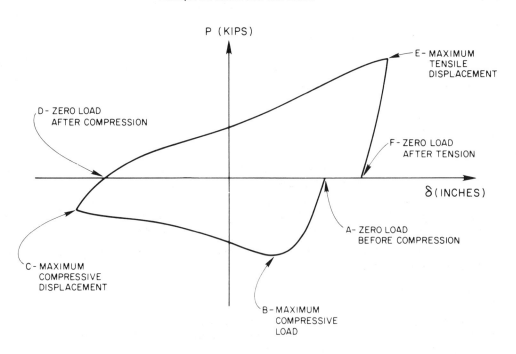

Figure 2
Typical Load-Deformation Relationship of a Brace Member

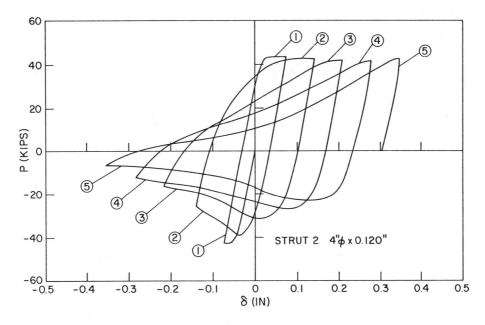

a) Both Ends Pinned (kL/r = 54)

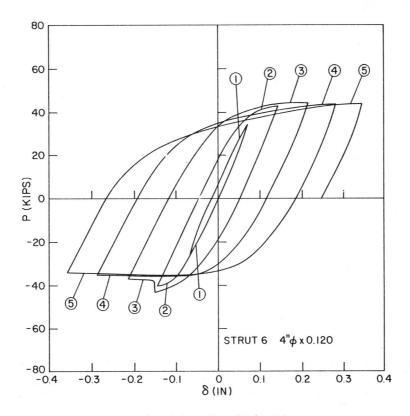

b) Both Ends Fixed (kL/r = 25)

Figure 3
Axial Load vs Axial Displacement Hysteretic Loops
(1 kip = 4.45 kN; 1 in. = 25.4 mm)

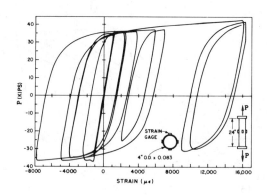

Figure 4
Cyclic Material Test, Axial Load vs Average Strain at Mid-Length;
4 in. x .083 in. Pipe Section

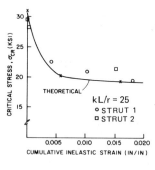

a) Struts 1 and 2

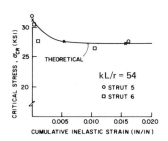

b) Struts 5 and 6

Figure 5
σ_{cr} vs CIS (1 ksi = 6.9 MPa)

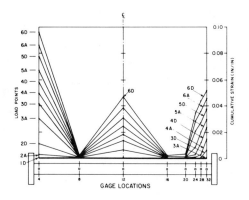

Figure 6
Cumulative Inelastic Axial Strain — Fixed End Strut 6

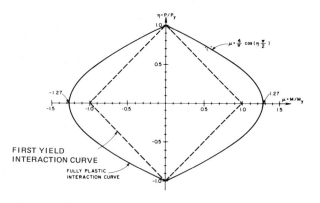

Figure 7
Axial Load-Bending Moment Interaction Curves for
Elastic-Perfectly Plastic Thin Tubular Pipe Section

a) Severe Local Buckle

b) Tearing in Tension

Figure 8
Local Buckling at Brace Center

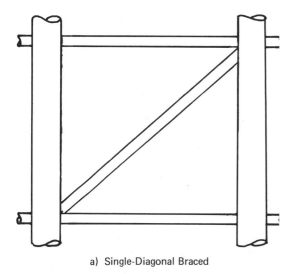
a) Single-Diagonal Braced

b) X-Braced

Figure 9
Panel Bracing

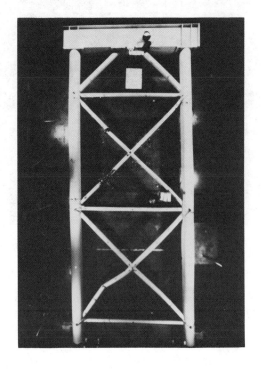
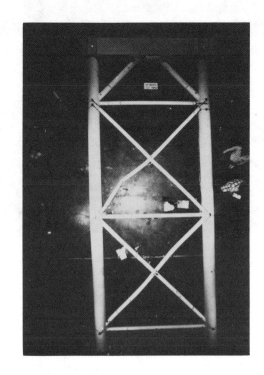

a) Frame I

b) Frame II

Figure 10
Overhead Photographs of the Frames Taken Near the Conclusion of Testing

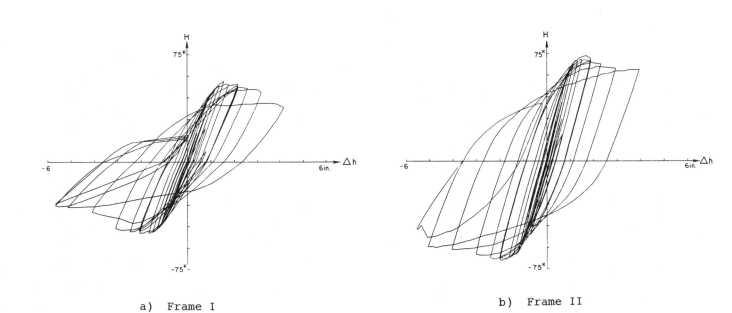

a) Frame I

b) Frame II

Figure 11
Lateral Load vs Lateral Displacement Hysteretic Loops

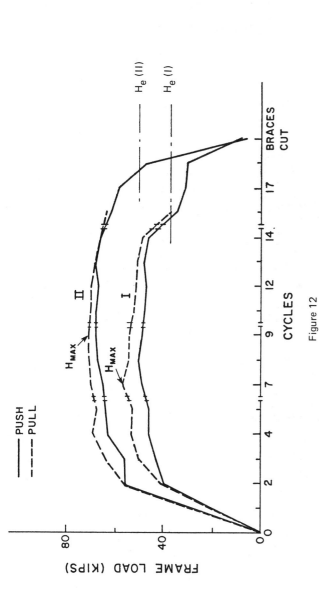

Figure 12
Frame Loads at Maximum Cycle Displacement

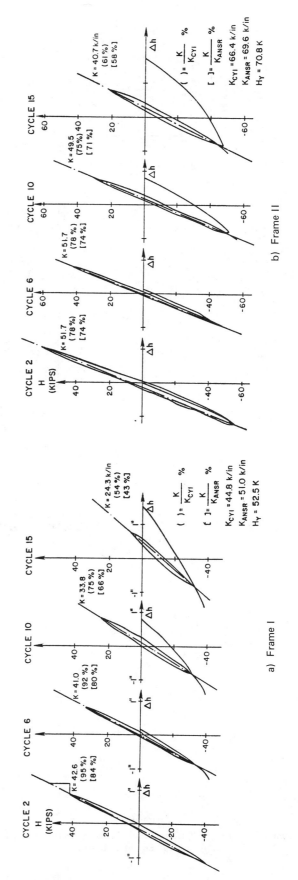

Figure 13
Frame Stiffnesses During Low Displacement Cycles

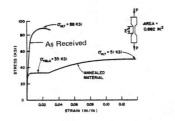
a) 0.083 in. Wall Thickness

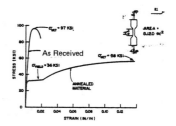
b) 0.120 in. Wall Thickness

Figure 14
Coupon Test Results — Annealed and Unannealed at Large Strains
(1 ksi = 6.9 MPa)

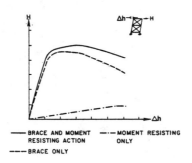

Figure 15
Typical Resistance to Lateral Loads from
Combined Brace and Moment-Resisting Action

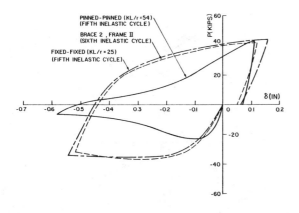

Figure 16
Comparison of Brace 2, Cycle 9, Frame II with Idealized Pinned
and Fixed End Condition Braces

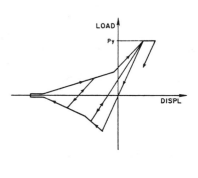

Figure 17
Intra Element [R21]

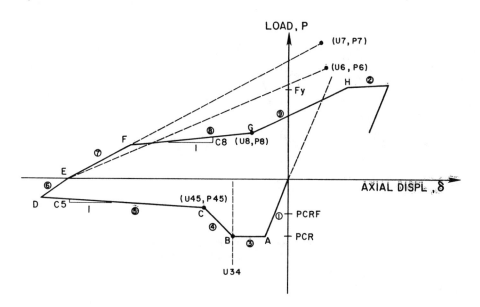

Figure 18
Maison Model

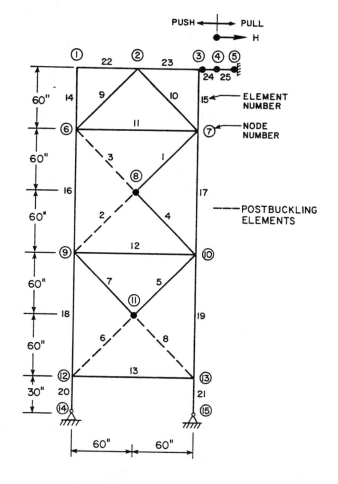

Figure 19
Frame Analytical Model

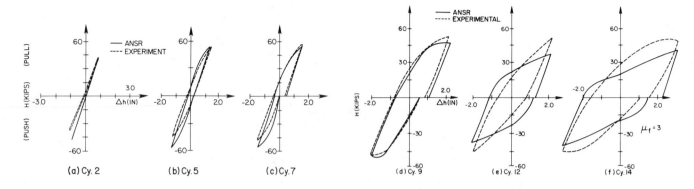

Figure 20
Comparison on Frame Hysteresis for Frame I
(1 kip = 4.45 kN; 1 in. = 25.4 mm)

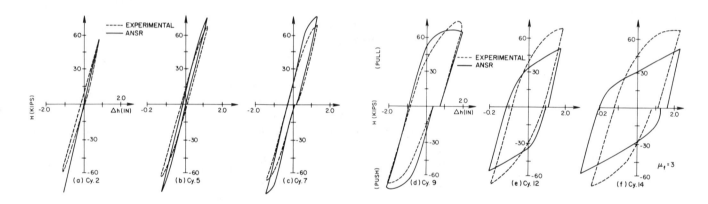

Figure 21
Comparison of Frame Hysteresis for Frame II
(1 kip = 4.45 kN; 1 in. = 25.4 mm)

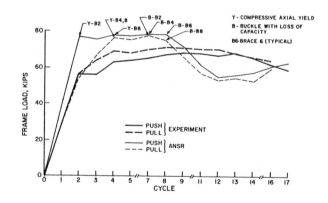

Figure 22
Load at Maximum Displacements for Frame II
(1 kip = 4.45 kN)

RESEARCH IN GREAT BRITAIN ON THE STABILITY OF

CIRCULAR TUBES

P. J. Dowling
Imperial College of Science and Technology
U.K.

J. E. Harding
Imperial College of Science and Technology
U.K.

SUMMARY

This paper reviews recent research in the U.K. on the buckling performance of cylindrical members as used in offshore installations. Unstiffened tubulars as well as ring stiffened, longitudinally stiffened and orthogonally stiffened cylinders are covered within the scope of the research. A nationally coordinated program of laboratory tests sponsored by the Department of Energy together with theoretical studies sponsored mainly by the Science and Engineering Research Council are discussed. The loss of strength in damaged unstiffened tubes has been studied along with the effect of initial imperfections caused by manufacturing processes on load carrying capacity of stiffened and unstiffened cylinders. The research has made significant progress in establishing techniques for the manufacture of small scale models which give results representative of large scale ones. It has gone a long way towards establishing the validity of various analytical techniques ranging from simple methods to comprehensive computer based numerical methods incorporating material and geometric non-linearity, and has provided reliable and well-documented data against which existing and new design rules can be calibrated. Finally, several areas needing further research have been identified.

1. INTRODUCTION

The topic of fatigue is perhaps the one which has received most attention from researchers in the context of the behaviour of offshore steel structures. Indeed it has been (and still is to a large extent) extremely difficult to persuade rig owners, operators and designers that there are requirements for buckling research on offshore rigs. A careful study of design guidance on buckling that existed when the first rigs were being installed in the North Sea would reveal, for those who were sufficiently interested, that there were deficiencies, inconsistencies and large gaps in the available information.

One such study (Ho, 1976) showed quite clearly that there was a need for a better understanding of buckling behaviour of cylindrical members of the types commonly used in offshore jacket construction. Since then the need has become even more acute as there is a marked tendency towards the use of more slender structures to minimise construction costs and more particularly to reduce dead weight for floating platforms and the new generation of Tension Leg Platforms (TLPs). At the time of writing the first of the new TLPs is under construction for the North Sea Hutton Field.

Although it has been customary in the United States to design rigs on an allowable stress basis, the custom in Europe for fixed rigs has been to use limit state design philosophy. Such an approach is increasingly being adopted by regulatory authorities worldwide and requires designers to consider explicitly the ultimate limit state. An understanding of buckling failure and its sensitivity to imperfections is central to a rational application of limit state design as on it depends the choice of strength formulae and appropriate partial safety factors.

This situation is recognised within the UK by government departments such as the Department of Energy and the main engineering research sponsoring body, the Science and Engineering Research Council, and the majority of the research outlined below has been sponsored by these bodies. Many of the results obtained so far have been reported in Reference 2 (Harding et al, 1982a) which contains the papers presented at an international conference held at Imperial College in 1981. The research is discussed in a sequence which reflects the increasing complexity of stiffening to cylindrical components as used offshore.

2. UNSTIFFENED TUBULARS

Unstiffened tubulars are used not only in pipelines but are extensively used as primary and secondary framing members within steel platforms. They are generally of fairly robust construction (D/t < 60) and the behaviour up to collapse of such tubes has been fairly well researched, particularly under the action of compressive axial and eccentric loading. More complex loading conditions such as combined axial and lateral pressurization have received less attention until relatively recently. However, the main problem receiving attention in the UK relates to the residual strength of damaged tubes. Damage is likely to be caused during launching or installation, by collisions involving supply boats or by dropping of heavy objects from deck level. The need to assess the implications on strength of damage incurred is clear.

Smith et al (Admiralty Marine Technology Establishment) have examined this problem and reported their findings in several papers (Smith et al, 1979, 1981, 1982). Both full-scale and small scale tests were carried out. The full-scale tests were carried out on four bracing members removed from the BP West Sole platform. These tubes were some 8m long and had D/t's ranging from 29 to 40. Sixteen smaller scale ($\frac{1}{4}$-scale) tests were carried out on tubes with D/ts ranging from 29 to 88. In addition four small scale tests roughly matching those of the full-scale ones were performed to assess the scaling effect on results.

Two types of damage were simulated. One consisted of a dent, typically induced in the experimental work by applying a load through a knife edge. The second consisted of a combined dent and overall bend. The dents ranged up to four times the wall thickness, whilst the bends typically were a bow with a maximum amplitude of 0.005 times the length.

The sixteen small scale tests were on a series of cold-drawn steel tubes which were divided into four sets of nominally identical geometries (although a complete survey of actual geometries was made before testing). In each group of nominally identical tubulars one with no intentional damage was tested under axial compressive loading and another under eccentric compressive loading. The remaining two were tested under axial compression with various types and magnitudes of damage.

The results of these tests showed that whilst slight bending could cause significant strength loss (15% or more), severe denting (4 times the thickness) can cause large strength losses of the order of 50% in the medium range of D/t's (D/t = 45) used in practice. It was also concluded that by providing small scale models with representative geometries and material properties (obtained by heat treating the cold-drawn steel) small scale testing provided a satisfactory way of assessing the effects of damage on tubular bracing members or other unstiffened cylinders with similar ranges of slenderness. (See Tables 1 and 2).

Analytical methods have been developed to allow the rapid evaluation of damage effects. Elasto-plastic beam-column analysis provides an accurate estimate of residual strength and stiffness for members which have been bent (Smith, 1982). It has been shown that the loss of strength depends primarily on the maximum amplitude of the bow but is relatively insensitive to its location along the length of the tubular. An approximate representation of dent effects using a similar analysis has also shown satisfactory correlation with test data, although further work is needed on dents of different configuration to establish confidence in the technique.

Figure 1 illustrates the effect of damage on the stress-strain responses of bracing members. The pre-collapse loss of stiffness and in some cases the unstable post-collapse behaviour have relevance to the behaviour of the platform as a whole.

Figure 2 shows the effect of dent location and indicates that loss of strength is not very sensitive to dent position γ is a factor relating the position of the dent to the span of the member. A value of 0.5 for example represents a dent at the centre of the span. The dimensions considered in this figure are $L/r = 58$, $D/t = 41$ and dent depth $= 0.13D$.

More extensive analytical work on unstiffened cylinder stability has been done recently (Harding, 1978; Batista and Croll, 1979). In Reference 12 (Harding, 1978) the finite-difference form of the large deflection equations for an initially distorted cylinder are solved by dynamic relaxation and account is taken of initial residual stressing and material non-linearity. Although the results in the paper related to cylinders with idealised end boundary conditions appropriate to rigid ring stiffeners, the results are also of direct interest in relation to unstiffened thin walled tubulars where the buckling mode is a multi-lobe one rather than an axi-symmetric bulge type failure. It was concluded that the mode of initial imperfection particularly in the axial rather than circumferential direction, played a very important role in determining the collapse strength and certainly more than has been generally recognised to date. Such slender unstiffened tubulars are uncommon in offshore construction at the moment and indeed designers should be wary of using such tubes to minimise weight as they are highly imperfection sensitive and show catastrophic post-collapse behaviour.

The authors of Reference 8 (Batista and Croll, 1979) have developed a reduced stiffness analysis for axially loaded tubulars which provides lower bounds to this imperfection sensitive problem. Extensive comparisons made between theory and test have shown that these are valid lower bounds.

Although much work has been done in connection with pipeline buckling and will not be covered here, it is worth noting the simple solutions provided in Reference 19 (Hobbs, 1981) for submarine pipelines in which axial compressive loads are induced by frictional restraint of axial extensions due to temperature changes or internal pressure. Both the upward (from sea-bed) mode and the lateral snaking mode are considered within the analysis.

3. RING STIFFENED CYLINDERS

A substantial amount of research has been concentrated on the behaviour of ring stiffened cylinders. This might seem surprising in view of the extensive research relating to submarines which has been going on for many decades. Indeed it is true to say that the case of externally pressurised capped ring stiffened cylinders is well understood, and covered by existing codes.

However, the behaviour of ring stiffened cylinders under axial compression and bending has received relatively little attention until recently. It is only because these offer potential constructional advantages (even for predominantly axial loading) that they are now the topic of intense investigation. Clearly the possibility of reducing costs and time by more extensive use of automated welding procedures because of the regular stiffening arrangement and the absences of costly welded intersections encountered with orthogonal stiffening is attractive to clients and fabricators alike. This form of construction is used for the columns of floating structures such as semi-submersibles and has been used in the new Hutton Field TLP. It also has potential advantages for the more conventional type of jacket structure. There is considerable uncertainty about the behaviour of such cylinders under the combined loading for which they must be designed. Load combinations include axial compression, bending, shear, torsion and external pressure.

Their behaviour has been investigated (Harding and Dowling, 1981a; Harding 1981b; Dowling and Harding 1982; and Harding and Dowling 1982b), both experimentally and theoretically, although studies reported to date have concentrated on shell collapse between frames. Work is at present being carried out on failure involving the ring stiffeners.

Three large scale cylinders with shell plating approximately 3mm thick have been tested at Imperial College (Dowling and Harding 1982). Two cylinders were nominally identical and had D/t ratios of 426. They were three bay cylinders in which the end bays were made of high yield steel in order to induce collapse to occur in the mid-bay and thus away from the end effects. (In the first test the end bay collapsed as the steel supplied was slightly thinner than specified and imperfections were larger - but it produced an extra experimental point!) The third cylinder had a

D/t of 256. All cylinders were extensively surveyed for imperfections prior to testing and strains and deflexions were measured throughout the test. Imperfection coefficients were produced by Fourier analysis for use in correlating test and analytical results. The collapse loads measured varied between 0.54 and 0.86 of the cylinder squash load. The slender cylinders failed near the ring stiffeners very suddenly. In each case outward bulging (not visible to the eye) of the cylinder shell was occurring adjacent to the rings prior to collapse. This bulging, produced by a Poisson expansion restrained by the presence of the rings, can be seen in Figure 3. The maximum deflection increased rapidly as failure approached but was barely visible by eye. At the instant of failure these deflections snapped through to a lower energy multi-lobe internal mode. The sections shown in the Figure are equally spaced around the circumference of the cylinder. The residual strength of each of the two cylinders was only about forty per cent of the peak loads - which were surprisingly close for the centre bay failures of these nominally identical geometries. The failure mode of the stocky cylinder was a progressive outward plastic bulging between rings which were at relatively much closer centres than those in the other two cylinders.

Walker et al (1982a) tested ten small scale thin steel stiffened cylinders. Some of the geometries were duplicated and the number of bays in each model varied between 1 and 5. The thickness of the plate used was only 0.81mm and the D/t's were either 300 or 500. These smaller scale models were not as extensively monitored for out-of-plane deflexions or strains as were the larger scale ones. As in the case of the large scale models only concentric axial loading was used. The two collapse modes observed in the large scale specimens also were picked up in this test series. In the majority of shells with D/t = 300 an axisymmetric localised bulge mechanism formed at or shortly before collapse while with the two of the three shells with D/t = 500 the final buckled mode was a cyclic symmetric or multilobed form. In the third the rings were so closely spaced as to break up the mechanism associated with unstiffened cylinder failure and forced it to fail by the formation of a short wave length outward bulge around the cylinder and adjacent to a ring.

When the results of the large and small scale tests were taken together they showed a high degree of imperfection sensitivity for axially loaded ring stiffened cylinders for the range of slendernesses studied. This is illustrated in Table 3.

Analyses using geometrical and material non-linear finite difference programs have been used to study the response of ring stiffened shells for combined as well as axial loading (Harding, 1981b). Discrete ring stiffeners have not been modelled exactly in these studies, but boundary conditions appropriate to their presence but assuming inter-frame buckling to be the critical failure mode have been used. Information on the effect of ring spacing on collapse has been produced by these studies although the analytical results for cylinders under combined compression and external pressure, and eccentric axial loading are probably the most interesting data produced apart from the data on axially loaded cylinders.

Figure 4 shows an interesting phenomenon produced by the analyses. For the mode of imperfection considered cylinder strengths can fall as ring spacing decreases. This effect is related to the elastic critical buckling wavelength of the shell in the longitudinal direction.

Figure 5 shows a sample set of results for cylinders under external pressure and axial compression. The large drop in compressive strength with increase in pressure, even for relatively stocky cylinders, can readily be seen. Sridharan et al (1981) have produced a mechanism approach and an elasto-plastic finite-element axi-symmetric program (Walker, et al, 1982b) to study the collapse of ring stiffened cylinders. The latter theory when used with DnV tolerances (rather than measured imperfections) gives reasonable correlation with the small scale tests. Figure 6 shows the small scale cylinder results produced by University College together with their analytical curve based on DnV tolerances. However as both the analyses of References 12 and 34 show the problem to be very imperfection sensitive further validation using actual measured imperfections would seem to be highly desirable. Attempts to do this have produced encouraging but limited results in Reference 13 (Harding and Dowling, 1981a). Figure 7 taken from the work of Harding and Dowling (1982b) shows the extreme variation of cylinder strengths that can be obtained by varying the imperfection mode.

Correlation of test data with the results of two established programs, STAGS-C1 and BOSOR 5 has been undertaken on behalf of Lloyd's Register of Shipping (Richards, 1982). Severe restrictions were imposed on the correlation exercise by the available budget and only some general conclusions can be drawn with confidence. In principle, the sophisticated STAGS-C1 program which is a finite element program allowing for shell structures of general shape and for geometrical and material non-linearities should be capable of predicting accurately the strengths of stiffened cylinders. The main limitation is an economic one. Ring stiffened shells were modelled with 90-120° segments and imperfections were included using 10 biaxial Fourier terms. No attempt was made to model residual stresses. Good agreement was obtained nonetheless for the three large scale cylinders of Reference 7 with the theoretical and actual ultimate loads differing by less than ten per cent. BOSOR 5 which accepts only axisymmetric imperfections was reported to give poor agreement with the same three tests. For the reader to draw his own conclusions he should, of course, study the way in which both of these programs were used to model the tests.

A problem which has been studied theoretically in a systematic fashion is that of a ring-stiffened shell under combined axial compression and external pressure. Reference 14 (Harding, 1981b) presents a parametric study in which allowance was made for initial imperfections, plasticity and residual stresses. The results have been presented in the form of interaction curves. Currently a test program is underway at Imperial College and Surrey University to provide some experimental data against which the validity of the theoretically generated curves can be judged. A recent paper (Galletley, 1982) has just reviewed the problem with particular attention being paid to the way it is treated within existing or proposed design rules.

Meanwhile Ellinas and Croll have extended the reduced stiffness analysis developed earlier for isotropic shells to stiffened shells including ring stiffeners (Ellinas and Croll, 1981). They have attempted to obtain a more fundamental understanding of overall shell behaviour by studying the role the membrane and flexural stiffnesses in both the axial and circumferential directions play in influencing the stability of cylindrical shells.

4. LONGITUDINALLY STIFFENED CYLINDERS

Walker et al (References 30, 31 and 35) have studied the problem of the buckling of longitudinally stiffened cylinders both experimentally and theoretically. Because of the high costs involved in fabricating large specimens using normal materials and processes they used very thin steel sheeting (less than 1mm) which was specially rolled for small scale ultimate load testing of steel bridges at Imperial College (Owens et al, 1982). They developed special welding techniques which have been described in detail in Reference 22 (Walker and Davies, 1977a). Using the same sheeting Dowling et al (1981, 1982) developed alternative fabrication techniques to check the usefulness of such small models as a source of reliable data for the elasto-plastic buckling of stiffened cylinders of the type used in offshore rigs. All of the small scale models tested were one bay long and represented the longitudinally stiffened portion between ring stiffeners in an orthogonally stiffened cylinder. The models tested by the two teams attempted to reproduce boundary conditions at each end of the model which were fixed against rotation.

Figure 8 shows a short wide panelled model from the Imperial College test series after failure. The position of the stringer stiffeners can clearly be seen due to the welding heat effect on the plating. The steel bands were attached to the model by resin to provide a clamped condition.

Some of the models tested at Imperial College duplicated those tested by Walker. Those tested by the latter, however, usually contained residual welding stresses, which were measured in some cases, and the plating was not heat treated either before or after fabrication with the result that the stress-strain curve of the material exhibited a reduced but appreciable stiffness after yield had occurred. In the Imperial College tests the models were stress relieved after fabrication and the material had a stress-strain curve representative of normal structural steels with a distinct yield plateau. Both methods of manufacture produced models which had tolerances of the order of ones normally specified within rules such as the DnV Rules.

In general models had either twenty or forty longitudinal stiffeners. The narrow-panelled cylinders generally failed in overall buckling while the broad-panelled cylinders failed by sudden local panel buckling. Three of the tests in Reference 7 (Dowling and Harding, 1982) were loaded eccentrically and exhibited similar behaviour to their concentrically loaded counterparts on the side where maximum combined bending and axial stressing occurred. There was a little reserve of strength above that obtained by using the maximum stressing from simple elastic bending calculations and applying it as concentric loading, but because the failure was of such a localised nature the redistribution capacity was less than fifteen per cent.

Extensive data on initial imperfections for many of these small scale tests are contained in References 6 and 28 as are the complete details of the small scale test procedures, results and conclusions. The differences between the results obtained from nominally identical models manufactured in two quite different ways were not very great in terms of collapse load achieved but the collapse modes of comparable models were different. In the models of Reference 28 (Sridharan and Walker, 1980) local inter-stringer panel buckling occurred at low loads, a phenomenon not observed in the tests of Reference 6 (Dowling et al, 1981). This is thought to be due to the large residual stresses and relatively higher local imperfections in the models described in the former reference. The end support detail is also believed to have affected the experimental results, particularly those of the longer models.

Two large scale models of single bay longitudinally stiffened cylinders were tested at the University of Glasgow (Green and Nelson, 1982; Nelson and Green, 1980). These were larger scale replicas of two small scale models tested at University College, London. One model had eight stringers while the other had twenty. The latter failed by local panel buckling but the former failed by a plastic mechanism involving both panels and stiffeners. It should be noted however that the eight stiffener model had relatively thicker walls than the twenty stiffener one. Although there was a difference in response during loading between the large scale and equivalent small scale models the ultimate loads expressed as fractions of the squash loads were in reasonable agreement (See Table 4). Figure 10 shows the load-deformation response of the twenty stiffener Glasgow model.

Recently an experimental program on the problem of combined axial and external pressure loading of stringer stiffened shells has been completed in a joint program carried out by Imperial College and the University of Surrey for DnV (Mc Call and Walker, 1981; Agelidis, Harding and Dowling, 1981).

Analytical work relating to stringer stiffened models has been done by both the University College and Imperial College teams. The work at University College has been concerned with the development of simplified analytical procedures and is described in References 32 and 36. On the other hand the work at Imperial College has concentrated on the development of more detailed numerical procedures which can be used to validate the tests and as a basis for computer studies which can supply the data needed for the formulation of simple design rules. Dowling and Ho (1979c) examined the effect of local curved panel buckling representative of that which occurs between stringers and later extended it to include stiffeners, (Harding and Dowling 1982b).

Currently work is nearing completion on a specially designed finite-element inelastic buckling program, NL-ASAS, developed jointly by Imperial College, W.S. Atkins and AMTE. As well as generally curved shell elements a powerful new stiffener element which can account for stiffener tripping type failures of open stiffeners of general form has been developed. This program will be used to study theoretically some of the problems of stiffened shells still needing attention and which are listed later.

5. ORTHOGONALLY STIFFENED CYLINDERS

In the recent UK research on shells only two tests, both at large scale, have been done on orthogonally stiffened cylinders. One model was tested as Glasgow (Nelson et al, 1980) and the other at Imperial College (Dowling et al, 1981). They were both three bay models and were designed to fail in the centre bay in a mode not involving the ring stiffeners. The information available from the series is, therefore, very limited.

The Imperial College cylinder had twenty longitudinal stiffeners and failed at eighty seven per cent of its squash load. Initial panel buckles appeared shortly before collapse so that only a small load margin existed between the first buckle and final collapse. This behaviour is in sharp contrast to the slender large scale ring stiffened cylinders which were tested where no such distinction could be made. In addition, the model had a considerable load capacity (eighty per cent of peak load) after collapse. (See Table 4).

The Glasgow cylinder failed suddenly at sixty nine per cent of the squash load due, it is thought, to a premature welding failure. (See Table 4).

Measurements of geometrical imperfections and residual stresses in actual orthogonally stiffened cylinders used in offshore construction have been reported by Dwight (1982). His team at Cambridge have also developed methods for predicting the level of residual stresses in stiffened cylinders. This work is essential input to any research program on the stability of offshore tubulars.

Little analytical work involving overall buckling of orthogonally stiffened shells has been carried out recently in the UK.

6. NEED FOR FURTHER RESEARCH

Some items needing further research are listed below in no particular order of priority:

(i) Data on imperfections existing in actual cylindrical components of offshore rigs are urgently needed both to aid the specification of rational tolerances and the generation of useful design curves and formulae.

(ii) More research is needed on the collapse of orthogonally stiffened shells, particularly in relation to buckling involving the orthogonal stiffening system.

(iii) Local buckling of stiffeners, both ring and longitudinal, in the tripping mode needs attention.

(iv) The influence of ring frame spacing and sizing on imperfection sensitivity under axial and bending loads needs attention.

(v) The strength of ring stiffened and orthogonally stiffened cylinders subjected to combined axial loading and external pressure needs to be assessed experimentally as well as theoretically.

(vi) The residual strength of damaged unstiffened cylinders needs to be researched further while little information exists in relation to damaged stiffened cylinders.

(vii) Some true optimisation studies involving cost as well as structural considerations need to be carried out on the form of stiffening which is most appropriate for new and future generation rigs. For example, orthogonal stiffening of the legs of TLP's may be preferable to ring stiffening in view of the load combinations to which they are subjected.

(viii) Design rules which take advantage of the recent research work in the UK and elsewhere are needed for stringer-stiffened and orthogonally stiffened cylinders under various loadings and combinations of loading. A thorough review of existing new rules would be valuable and would reveal several inconsistencies and gaps which still exist.

(ix) Interaction between local and overall buckling in thin unstiffened tubulars has not received adequate attention.

(x) A need exists to continuously update and expand a comprehensive data bank of the type started by the SSRC task group so that maximum benefit can be obtained from the research being conducted internationally in the most efficient way possible. Many existing tests are reported in insufficient detail with respect to imperfections, residual stress levels, boundary conditions and loading conditions as to make them of little use to researchers, code drafters and designers alike.

7. CONCLUSIONS

1. Experimental and theoretical methods which quantify the effect of certain types of damage on unstiffened tubulars have been developed successfully.

2. New small modelling techniques have been developed for unstiffened and stiffened steel cylinders of the type used in offshore construction which are relatively cheap. The models can be used to provide experimental data to verify design rules. Comparisons between these small scale tests and more realistic larger scale ones suggest that the latter are more suitable for verifying the sophisticated numerically based computer programs which have been produced to predict inelastic stiffened shell buckling.

3. The data which have been provided by the experimental and theoretical work carried out in the UK should help calibrate existing design rules of stringer stiffened and axially stiffened shells. They can also be used to produce a more rational set of rules when taken together with recent research carried out in other countries.

ACKNOWLEDGEMENTS

The authors wish to thank the UK Department of Energy and The Science and Engineering Research Council for the funding provided for most of the work described. They particularly wish to acknowledge the cooperation of their friends and colleagues Professor A.C. Walker, Mr. H.M. Nelson, Dr. D.M. Richards, Dr C.S. Smith, Dr. P.A. Frieze, Mr. J.B. Dwight and Mr. N.W. Snedden.

REFERENCES

1. AGELIDIS, N., HARDING, J.E. and DOWLING, P.J. Buckling Tests on Stringer Stiffened Cylinder Models Subject to Load Combinations, Vol. 1, Report to DnV, Imperial College, London.

2. BATISTA, R.C. and CROLL, J.G., 1979 "A Design Approach for Axially Compressed Unstiffened Cylinders", Stability Problems in Engineering Structures and Components, (Edited by T.H. Richards and P. Stanley, Applied Science Publishers, London.

3. DOWLING, P.J. and HARDING, J.E., 1979a, "Strength of Steel Jacket Leg Components". Proceedings of Offshore Technology Conference, Houston, Texas.

4. DOWLING, P.J. and HARDING, J.E. 1979b, "Current Research into the Strength of Cylindrical Shells Used in Steel Jacket Construction". Proceedings of 2nd International Conference on Behaviour of Offshore Structures (BOSS '79), Imperial College, London.

5. DOWLING, P.J. and HO, T.K. 1979c, "Effect of Initial Deformations on the Strength of Axially Compressed Cylindrical Panels". Proceedings of Conference on Significance of Deviations from Design Shapes, Applied Mechanics Group, Institution of Mechanical Engineers, London.

6. DOWLING, P.J., HARDING, J.E., FAHY, W. and AGELIDIS, N., 1981, Report on the Testing of Large and Small Scale Stiffened Shells under Axial Compression, CESLIC Report SS1, Imperial College, London.

7. DOWLING, P.J. and HARDING, J.E. 1982, "Experimental Behaviour of Ring and Stringer Stiffened Shells". pp 73-107 of Reference (15) below.

8. DWIGHT, J.B. 1982, "Imperfection Levels in Large Stiffened Tubulars". pp.393-412 of Reference (15) below.

9. ELLINAS, C.P. and CROLL, J.G.A., 1981, "Overall Buckling of Ring Stiffened Cylinders", Proceedings Institution of Civil Engineers, Part 2, Vol. 71, pp.637-661.

10. GALLETLEY, G.D., 1982, "Buckling of Fabricated Cylinders Subjected to Compressive Axial Loads and/or External Pressure; a Comparison of Several Codes", to be presented at ASME Pressure Vessel Design Symposium, PVP Conference, Orlando, Florida.

11. GREEN, D.R., and NELSON, H.M., 1982, "Compression Tests on Large-scale, Stringer-stiffened Tubes", pp 25-43 of Reference (15) below.

12. HARDING, J.E., 1978, "The Elastic-plastic Analysis of Imperfect Cylinders." Proceedings Institution of Civil Engineers, Part 2, Vol. 65.

13. HARDING, J.E. and DOWLING, P.J., 1981a, "Correlation between Experimental and Analytical Results for the Buckling of Ring Stiffened Shells", Proceedings of 2nd Conference on Integrity of Offshore Structures, Applied Science, London and New Jersey.

14. HARDING, J.E., 1981b, "Ring-stiffened Cylinders under Axial and External Pressure Loading". Proceedings Institution of Civil Engineers, Part 2, Vol. 71, pp.863-878.

15. HARDING, J.E., DOWLING, P.J. and AGELIDIS, N. (Editors), 1982a, Buckling of Shells in Offshore Structures, Granada, London; also published as Offshore Structures Engineering 111: Buckling of Shells in Offshore Structures, Gulf Publishing, Houston, Texas.

16. HARDING, J.E. and DOWLING, P.J., 1982b, "Analytical Results for the Behaviour of Ring and Stringer Stiffened Shells". pp.231-256 of Reference (15) above.

17. HO, T.K., 1976, Collapse of Singly-Curved Shell Structures. MSc Thesis, Imperial College, University of London.

18. HO, T.K., 1982, Collapse of Stiffened Cylindrical Shells, PhD Thesis, submitted to University of London.

19. HOBBS, R.E., 1981, "Pipeline Buckling Caused by Axial Loads". Journal of Constructional Steel Research, Vol. 1 No.2, Granada, London.

20. McCALL, S., and WALKER, A.C., 1981, Buckling Tests on StringerStiffened Cylinder Models Subject to Load Combinations, Vol. 2, Report to DnV, University of Surrey.

21. NELSON, H.M., GREEN, D.R. and PHILLIPS, D.V., 1980, Buckling Tests of Offshore Structures: Preliminary Tests of Stringer-stiffened Cylinders, Report, University of Glasgow.

22. OWENS, G.W., DOWLING, P.J. and HARGREAVES, A.C., 1982, "Experimental Behaviour of a Composite, Bifurcated Box Girder Bridge", to be presented at International Conference on Short and Medium Span Bridges, Toronto, Canada.

23. RICHARDS, D.M., 1982, "Shell Buckling Research and Design Appraisal Procedures", pp.109-136 of Reference (15) above.

24. SMITH, C.S., KIRKWOOD, W., and SWAN, J.W., 1979, "Buckling Strength and Post-collapse Behaviours of Tubular Members including Damage Effects", Proceedings of 2nd International Conference on the Behaviour of Offshore Structures (BOSS '79), Imperial College, London.

25. SMITH, C.S., SOMERVILLE, W.L. and SWAN, J.W., 1981a, "Residual Strength and Stiffness of Damaged Steel Bracing Members", Proceedings of Offshore Technology Conference, Houston.

26. SMITH, C.S. and DOW, R.S., 1981b, "Residual Strength of Damaged Steel Ships and Offshore Structures", Journal of Constructional Steel Research, Volume 1, Number 4, Granada, London.

27. SMITH, C.S., 1982, "Strength and Stiffness of Damaged Tubular Beam Columns", pp.1-23, Reference (15) above.

28. SRIDHARAN, S. and WALKER, A.C., 1980, Experimental Investigation of the Buckling Behaviour of Stiffened Cylindrical Shells, Report, University College, London.

29. SRIDHARAN, S., WALKER, A.C. and ANDRANICOU, A., 1981, "Local Plastic Collapse of Ring-Stiffened Cylinders". *Proceedings of the Institution of Civil Engineers*. Part 2, Volume 70.

30. WALKER, A.C. and DAVIES, P., 1977a, "The Collapse of Stiffened Cylinders". *Steel Plated Structures* (Edited by Dowling, P.J., Harding, J.E. and Frieze, P.A.), Crosby Lockwood, London.

31. WALKER, A.C. and ELSHARKOWI, K, 1977b, "Small Scale Model Studies of the Buckling of Welded Thin Plate Structures". *Proceedings of International Conference on the Behaviour of Slender Structures*, The City University, London.

32. WALKER, A.C., and SRIDHARAN, S., 1980, "Analysis of the Behaviour of Axially Compressed Stringer-stiffened Cylindrical Shells". Proceedings of the Institution of Civil Engineers, Part 2, Vol. 69.

33. WALKER, A.C., ANDRONICOU, A., and SRIDHARAN, S., 1982a, "Experimental Investigation of the Buckling of Stiffened Shells using Small Scale Models". pp.45-71 of Reference (15) above.

34. WALKER, A.C., ANDRONICOU, A., and SRIDHARAN, S., 1982b, "Theoretical Analysis of Stringer and Ring Stiffened Shells". pp. 183-208 of Reference (15) above.

35. WALKER, A.C., ANDRONICOU, A., and SRIDHARAN, S., 1982c, "Experimental Investigation of the Buckling of Stiffened Shells using Small Scale Models". pp.45-72 of Reference (15) above.

36. WALKER, A.C., ANDRONICOU, A. and SRIDHARAN, S., 1982d, "Theoretical Analysis of Stringer and Ring Stiffened Shells", pp.183-208 of Reference (15) above.

Test Reference	t (mm)	R/t	L/R	Eccentricity of axial loading	Intentional damage before testing	Collapse load / Squash load
AMTE ($\frac{1}{4}$-scale)						
(1) 1	2.11	14.6	69.9	0	None	0.84
2	2.12	14.5	69.9	0.32R	None	0.49
3	2.11	14.6	69.9	0	Bend and dent	0.48
4	2.12	14.5	69.9	0	Bend only	0.50
(2) 5	1.74	22.3	55.3	0	None	1.00
6	1.71	22.7	55.3	0.26R	None	0.60
7	1.72	22.6	55.3	0	Bend and severe dent	0.52
8	1.70	22.9	55.3	0	Bend and slight dent	0.61
(3) 9	1.66	30.1	43.0	0	None	1.10
10	1.73	28.9	43.0	0.20R	None	0.58
11	1.72	29.0	43.0	0	Dent	0.76
12	1.73	28.9	43.0	0	Dent	0.84
(4) 13	1.02	43.6	48.3	0	None	0.75
14	1.01	44.0	48.3	0.34R	None	0.50
15	1.03	43.1	48.3	0	Severe dent	0.53
16	1.05	42.4	48.3	0	Dent	0.64

Notes:

(1) R/t at the lower end of the range found offshore. Local damage would not normally be expected to influence performance significantly.

(2) Medium R/t in practice.

(3) R/t at the top end of the practical range. Local buckling would be expected to influence post-collapse behaviour.

(4) High strength, thin-walled tubes - of interest where weight-saving is important in structural design.

TABLE 1 Geometry and Results of Unstiffened Tubulars.

Test reference	t (mm)	R/t	L/R	Intentional damage before testing	Collapse load / Squash load
AMTE (full-size)					
17	10.6	14.7	49.8	None	0.85
18	10.0	15.7	49.3	Bend	0.73
19	9.9	20.0	39.1	None	0.86
20	9.9	20.0	39.2	Slight bend & severe dent	0.57
AMTE ($\frac{1}{4}$-scale and 1/6-scale)					
21	2.63	15.2	49.2	None	0.82
22	2.66	15.0	49.2	Bend	0.63
23	1.60	20.3	41.3	None	0.89
24	1.59	20.4	40.7	Bend & severe dent	0.47

TABLE 2   Geometry and Results of Full-Scale Tubulars and Corresponding Models.

Test reference	t (mm) (1)	R/t (1)	L/R (1)(2)	No. of bays	Steel used (1)	Collapse load / Squash load
Imperial College ($\frac{1}{4}$-scale)(3)						
25A(4)	2.87	261	1.0	3	Grade 50	0.54(4)
25B	3.52	213	1.0	3	Mild (Grade 43)	0.61
26	3.52	213	1.0	3	Mild (Grade 43)	0.86
27	3.52	128	0.2	3	Mild (Grade 43)	0.86
University College (1/20-scale)(3)						
28	0.81	150	1.0	3	Mild	0.68
29	0.81	150	1.0	3	Mild	0.69
30	0.81	150	1.0	3	Mild	0.82
31	0.81	150	1.0	5	High tensile	0.65
32	0.81	150	1.0	5	High tensile	0.69
33	0.81	150	0.58	1	Mild	0.81
34	0.81	150	0.58	1	Mild	0.74
35	0.81	250	0.33	3	High tensile	0.57
36	0.81	250	1.0	3	High tensile	0.43
37	0.81	250	1.0	3	High tensile	0.56

Notes:

(1) Dimensions, ratios and specifications refer to the bay where failure took place

(2) L is the length between ring stiffeners

(3) Some cylinder geometries are duplicated

(4) The extra test result came from an initial, unplanned, failure in one of the end bays of cylinder 25. It was retested after the end bays had been stiffened up.

TABLE 3   Geometry and Results of Ring Stiffened Cylinders.

Test reference	t (mm)	R/t	L/R	No. of stringers	No. of bays	Eccentricity of axial loading	Collapse load / Squash load
University College (1/20-scale)							
38	0.81(1)	94	1.33	8	1	0	0.93
39	0.81	200	0.4	20	1	0	0.82
40	0.81	200	1.11	20	1	0	0.76
41	0.81(1)	280	0.78	20	1	0	0.61
42	0.81(1)	280	1.11	20	1	0	0.60
43	0.81(1)	280	1.56	20	1	0	0.54
44	0.81	360	1.11	20	1	0	0.51
45	0.81	200	1.11	30	1	0	0.96
46	0.81	200	0.4	40	1	0.5R	0.93
47	0.81(2)	200	1.11	40	1	0	1.03
48	0.81(2)	200	1.11	40	1	0	1.04
49	0.81(3)	280	1.11	40	1	0	0.65
50	0.81(3)	280	1.11	40	1	0	0.86
51	0.81(1)(4)	280	1.56	40	1	0	0.82
52	0.81(1)(4)	280	1.56	40	1	0	0.82
53	0.81	360	1.11	40	1	0	0.66
Imperial College (2/4 scale)							
54	3.53	170	1.11	20	3	0	0.87
(1/20 scale)							
55	0.84	190	0.41	40	1	0	0.96
56	0.84	190	0.41	20	1	0	0.96
57	0.84	190	1.13	40	1	0	0.95
58	0.84	190	0.41	40	1	0.25R	0.78
59	0.84	190	0.41	20	1	0.4R	0.54
60	0.84	190	1.13	40	1	0.4R	0.51
Glasgow University (1/8-scale)							
61	6.0(5)	94	1.33	8	1	0	0.89
62	3.0(5)	200	1.10	30	3	0	0.69
63	2.0(5)	280	1.56	20	1	0	0.57

Notes:

(1) Cylinders tested in earlier SRC-funded programme.

(2) Similar overall dimensions but with different stringer proportions.

(3) Similar overall dimensions but with different stringer proportions.

(4) Similar overall dimensions but with different stringer proportions.

(5) Nominal steel thickness.

TABLE 4 Geometry and Results of Stringer Stiffened Cylinders.

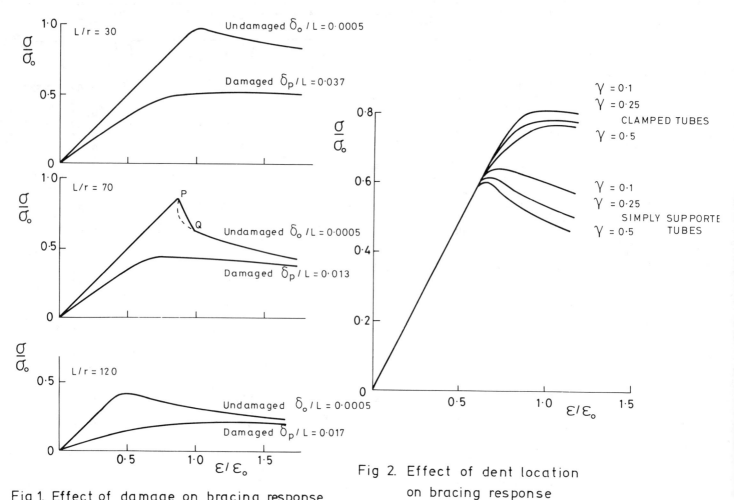

Fig 1. Effect of damage on bracing response

Fig 2. Effect of dent location on bracing response

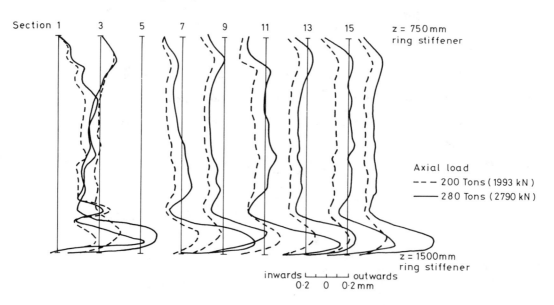

Fig 3. Deflection of the cylinder showing poisson expansion adjacent to the rings

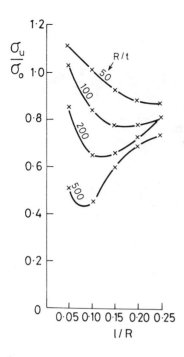
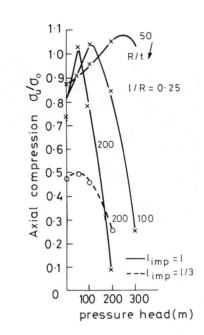

Fig 4. Effect of ring spacing on cylinder strength

Fig 5. Effect of external pressure on cylinder compressive strength

Specimen	l/t	σ_o (N/mm²)
□ UC 12 – 3 bay	120	245
● UC 12 – 1 bay	70	254
× UC 12 – 5 bay	120	315
△ UC 19 – 3 bay	60	325
○ UC 20 – 3	250	325

1^d = damaged specimens

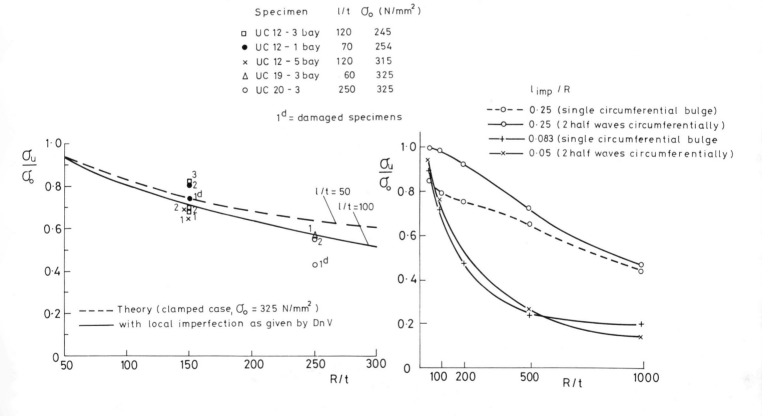

Fig 6. Comparison of University College tests on analytical results

Fig 7. Effect of imperfection mode on cylinder strength

Fig 8. One of the small scale Imperial College models after failure

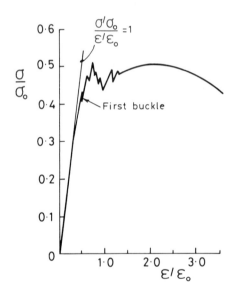

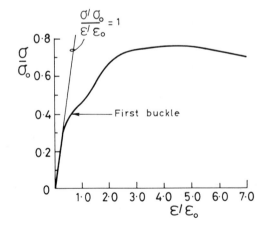

Fig 9 Stress-strain responses of two
of the University College models

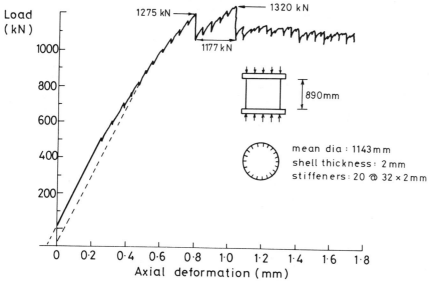

Fig 10 Load-deflection response of one of the Glasgow University tests

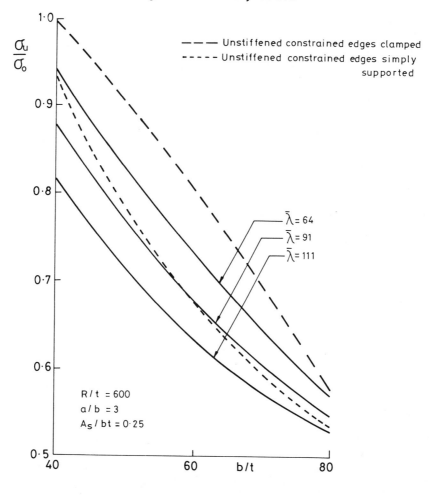

Fig 11. Effect of panel and column slenderness on strength

Technical Lecture Sessions
STRUCTURES

Response of Tension Leg Platforms

INTERWIRE SLIPPAGE AND FATIGUE PREDICTION IN
STRANDED CABLES FOR TLP TETHERS

R.E. Hobbs   M. Raoof
Imperial College, U.K.   Imperial College, U.K.

SUMMARY

Analytical work backed by experimental studies on a 6 m long, 38 mm diameter, 91 wire strand with a nominal breaking load of 1.3 MN is reported.

The theoretical work extends the classical twisted rod theories for the behaviour of helically laid wires to include a stronger set of kinematic compatibility conditions. This theory is compared with the results of a novel treatment of the layers of wires in a strand as a series of orthotropic sheets. The stiffnesses of each sheet in its principal directions are defined by reference to contact stress theory including, most importantly, shear deformation. Transformed into the direction corresponding to the strand axis, results are obtained for the tangent axial stiffness of the strand as a function of load level, and, via contact stress theory, for contact forces and slippage.

Full slip histories are predictable, from the micro-slips in the periphery of contact patches at low loads, to gross slip at higher loads and even beyond. In addition, the hysteresis in the strand under cyclic loading regimes is predicted.

The experimental work concentrates on measurements of wire stress, wire-to-wire (centre line) slip, and overall hysteresis and is in substantial agreement with the theory.

It is concluded that theoretical predictions of the interwire forces and slippage in large strands such as those envisaged as tension leg members in buoyant platforms are feasible, and this information is of obvious value as an input to a fracture mechanics analysis of the fatigue behaviour of the strand away from its terminations.

1. INTRODUCTION

The analysis of a group of helically-laid wires with a common axis - a spiral strand - appears straightforward, and the derivation of the properties of a group of such strands laid up as a wire rope only slightly more onerous. However, appearances are deceptive, and the rational analysis and design of this important group of components for engineering structures is far from simple. Indeed, many unresolved problems remain in the field.

There are remarkably few theoretical references on the static and dynamic behaviour of spiral strand and ropes. In the main, strands consisting of only six or seven wires and ropes consisting of six such strands have been considered, and even for these simplified configurations the analyses have had to include further sweeping assumptions.

Hruska (1951) assumed that each wire in a strand was subjected to tension alone and demonstrated that (for small deformations) the tensile stresses vary in proportion to the square of the cosine of the lay angle. He also derived a relationship between the radical clench forces and the axial force in the strand (Hruska, 1952) and later discussed the origin and consequences of the torque induced in a helical strand by an axial load (Hruska, 1953). However, he did not consider the actual stresses at the contact points between the wires. Chi (1971a, 1971b) extended the work of Hruska, and compared elongations and strains predicted theoretically with some experimental results for steel strands. By analysis of experimental data, Drucker and Tachau (1945) demonstrated the importance of stresses at the contact points. Their work was followed by Leissa (1959) and Starkey and Cress (1959) who attempted a theoretical study of interwire contact stresses. The work of Gibson et al. (1969) also followed essentially the same approach as Hruska.

Studies by Durelli et al. (1972, 1973) which included an experimental study of an oversized epoxy resin model of a six wire strand, emphasised the importance of bending and twisting moments in the six individual wires. However, it is doubtful whether these effects would be significant in a practical strand (which might have 90 wires or more) where the wire diameter to strand diameter ratio is much smaller.

Phillips and Costello (1973, 1974, 1976), Costello and Sinha (1977a, 1977b) and Costello and Miller (1979) have examined the non-linearity of strands when very large load-induced changes in the helix angle are considered. They treated the equilibrium of individual wires and contact forces on them in extenso, while neglecting the substantial frictional forces which must certainly be present if the helix angle changes significantly. Nowak (1974) and Knapp (1979) addressed the problem of the frictionless but geometrically non-linear properties of strands with compressible cores. Their work is an example of the related interest in predicting the properties of electrical cables and offshore umbilicals which has enjoyed considerable attention in recent years. In this field, Kasper (1973a, 1973b) used finite element techniques to analyse a multi-conductor armoured electrical cable. He determined the interactions between the cable components, and demonstrated a satisfactory correlation between experiment and theory.

Jones and Christodoulides (1979) used the equations developed by Phillips and Costello (1973) to calculate the static plastic collapse of a highly idealised single layered strand. The group at the Catholic University of Washington (Durelli et al, 1973) mention another important influence on strand and rope properties, namely the manufacturing process and its inherent variability, and the initial imperfections that result. The initial plastic strains, first in wire-drawing, and then in spinning a strand are also important. Yoshida (1952) analysed the plastic deformations produced in individual wires during strand laying on a theoretical basis. Winkelmann (1972), in an important theoretical study, examined the strains generated in practical strands and ropes by the manufacturing process. In spite of these contributions, it must be recognised that rope-making is still as much an art as a science.

There are few references (Kawashima and Kimura, 1951, Bechtloff , 1963, Wilson, 1967, Ridley, 1973) which treat the dynamic properties of strand. Nonetheless, the axial dynamic stiffness and the related hysteretic behaviour are important in structures as diverse as guyed towers or tension leg platforms and cable stayed bridges, while the torsional and bending dynamic properties are pertinent to a number of aero and hydrodynamic problems. The related problem of the fatigue lives of strands under various loadings is reviewed elsewhere (Hobbs and Smith, 1982). Indeed, these fatigue studies, which provided one of the present authors with an exceptional opportunity to study the dynamic properties of a long specimen of a full size strand, provided the stimulus which led to the work reported in this paper.

2. STRAND THEORY

2.1 Introduction

In both torsional and extensional regimes, two limiting cases can be identified: for small perturbations no interwire slip occurs while for large enough disturbances where slip occurs interwire friction forces become negligible compared to force changes in the wires themselves. In what follows it is postulated that each layer of wires in a strand has enough wires for the properties to be averaged so that the layer can be treated as an orthotropic sheet. The elastic properties of the sheet are then derived (as a function of the perturbation) with reference to principal axes parallel and perpendicular to the wires. It is then simple (Hearmon, 1961) to transform them to values parallel and perpendicular to the strand axis. Along the axis of the wires, the Young's modulus for the material applies, subject to an allowance for the net area of the circular wire being $\pi/4$ times the elemental area. Perpendicular to the wire axis, it is necessary to allow for the overall stiffness between wires in line contact. This problem was solved by Hertz (Roark and Young, 1975) as a function of the mean load on the contact line. This mean load can be estimated from the mean axial load in the wire. The shear stiffness is obtained from other results in contact stress theory (Mindlin, 1949, Mindlin and Deresiewicz, 1953, Deresiewicz, 1957).

Unfortunately there are uncertainties in calculating the line contact stress even on a nominal basis, and additional variability due to irregularities in the fit of the wires in the strand must be expected. A discussion of the contact stresses thus precedes detailed analysis of the anisotropic sheet, the assembly of several shells of wires into a strand, and predictions of strand axial and torsional stiffness and associated damping characteristics.

2.2 Kinematics of a Helix

The parametric representation of a helix is as follows:

$$x = r \cos \phi$$
$$y = r \sin \phi$$
$$z = (r \cot \alpha)\phi \quad (1)$$

where α is the lay angle, ϕ the polar angle and r is the helix radius. It is assumed that the centre-line of a wire in a strand forms a helix both before and after axial or torsional deformation. Opening out the helix, a right-angled triangle is formed (Figure 1). The deformation of a wire in a strand with a free end will be composed of an axial extension, a radial contraction and a rotation (Knapp, 1975). Treating a strand length δ, these components are $\varepsilon_c \delta$, $\phi \Delta r$ and $\phi(r - \Delta r)$ where ε_c is the strand axial strain, $\Delta \phi$ the end rotation and Δr the change in the radius. ε_h denotes the wire axial strain.

Before and after deformation, equation (1) gives

$$\phi = \delta / r \cot \alpha$$
$$\phi' = \delta' / r' \cot \alpha' \quad (2)$$

Assuming that all the wires are the same length and that plane sections remain plane which at least is reasonable for sections remote from the ends of the strand

$$\delta' = \delta(1 + \varepsilon_c) \quad (3)$$

Combining equations (2) and (3),

$$\frac{\Delta \phi}{\delta} = \frac{1 + \varepsilon_c}{r' \cot \alpha'} - \frac{1}{r \cot \alpha} \quad (4)$$

an expression for the rotation per unit length of strand.

In the case of a strand subjected to pure tension whose ends are not permitted to rotate, $\Delta \phi / \delta$ is zero and equation (4) yields

$$\frac{\tan \alpha'}{\tan \alpha} = \frac{r'}{r(1 + \varepsilon_c)} \quad (5)$$

From Figure (1)

$$\cos \alpha = \frac{\delta}{\ell} \quad \text{and} \quad \cos \alpha' = \frac{\delta(1 + \varepsilon_c)}{\ell(1 + \varepsilon_h)}$$

where ℓ is the undeformed wire length and hence

$$\frac{\cos \alpha'}{\cos \alpha} = \frac{1 + \varepsilon_c}{1 + \varepsilon_h} \quad (6)$$

For a strand with non-rotating ends (Chi, 1971b)

$$\varepsilon_h = \frac{\ell' - \ell}{\ell} = \cos \alpha \left((1 + \varepsilon_c)^2 + \left(\frac{r'}{r} \tan \alpha\right)^2 \right)^{\frac{1}{2}} - 1 \quad (7)$$

Assuming $\frac{r'}{r} \simeq 1$ and using a binomial expansion (7) is simplified to:

$$\varepsilon_h = \varepsilon_c \cos^2 \alpha \quad (8)$$

which is the expression first derived by Hruska (1951). Substituting (8) into (6) we have:

$$\cos(\alpha - d\alpha) = \cos \alpha (1 + \varepsilon_c)(1 + \varepsilon_c \cos^2 \alpha)^{-1} \quad (9)$$

From (9), ignoring second order terms, it is possible to obtain the simple relation:

$$d\alpha \simeq \varepsilon_c \sin\alpha \cos\alpha \quad (10)$$

where $d\alpha$ is the change in the lay angle due to axial cable strain ε_c.

A more exact method for determining the change in the lay angle including the effects of changes in radius is presented later.

2.3 Equilibrium Equations for a Helical Rod

The force and moment equilibrium equations for a curved rod were derived by Love (1927). For the final state of equilibrium they are:

$$\frac{dN}{dS} - N'\tau_1 + T\kappa_1' + X = 0$$

$$\frac{dN'}{dS} - T\kappa_1 + N\tau_1 + Y = 0$$

$$\frac{dT}{dS} - N\kappa_1' + N'\kappa_1 + Z = 0$$

$$\frac{dG}{dS} - G'\tau_1 + H\kappa_1' - N' + K = 0$$

$$\frac{dG'}{dS} - H\kappa_1 + G\tau_1 + N + K' = 0$$

$$\frac{dH}{dS} - G\kappa_1' + G'\kappa_1 + \Theta = 0 \quad (11)$$

in which T, N, N', G, G' etc are the external and internal force and moment resultants in the normal, binormal and tangential directions (Figure 2). For a helical wire, the initial curvatures and twist are

$$\kappa_o = 0, \quad \kappa_o' = \frac{\sin^2\alpha}{r}, \quad \tau_o = \frac{\sin\alpha \cos\alpha}{r} \quad (12)$$

Once the strand is loaded, the new wire geometry is given by

$$\kappa_1 = 0, \quad \kappa_1' = \frac{\sin^2\alpha'}{r'}, \quad \tau_1 = \frac{\sin\alpha' \cos\alpha'}{r'} \quad (13)$$

For wire cross sections remote from the clamping points, it is assumed that simple bending and torsion theory holds and

$$G = A(\kappa_1 - \kappa_o), \quad G' = A(\kappa_1' - \kappa_o'), \quad H = C(\tau_1 - \tau_o) \quad (14)$$

where A and C are bending and torsion moduli for the wire.

Durelli et al. (1972) were the first to emphasise the importance in a six wire strand of bending and twisting moments in the individual wires, as defined by equations (14). Phillips and Costello (1973) ignored friction between the wires, and assuming κ, κ' and dT/dS to be zero, used equations (11) - (14) to show that

$$X = \left[C(\tau_1 - \tau_o)\kappa_1' - A(\kappa_1' - \kappa_o')\tau_1\right]\tau_1 - T\kappa_1' \quad (15)$$

However, for large diameter steel strands under working loads the changes in the lay angle and radius are extremely small and equations (15) may be simplified to

$$X = -T\sin^2\alpha/r \quad (16)$$

This is the relation obtained by Hruska (1952) using hoop tension formulae.

2.4 Wire Accommodation in Each Layer

This problem has been addressed by several authors. Nowak (1974) considered n wires of identical circular cross sections wound helically on a cylindrical core. He showed that for large n and small lay angles, it is reasonable to assume that, in a transverse section of the strand itself, wire cross-sections are elliptical. However, for small n and/or large lay angles, the wire section is kidney shaped, as Figure 3, an enlarged section of a five wire strand, shows.

Nowak also took into account the gaps between the wires in each layer and derived analytical expressions for their size. However, he assumed a uniform distribution of gaps, and inspection of well bedded-in strand samples suggests that this may be an unrealistic assumption. Indeed the gaps seem to accumulate as one or two large gaps with the other wires in a layer in contact.

For an idealised strand it is now assumed that the wires in each layer are just touching in the unstressed configuration, and as the theory is developed for strands with a large number of wires it is further assumed that wire cross sections in a normal section of the strand may be treated as ellipses (Figure 4).

Following Phillips and Costello (1973), for a single layer of wires

$$\frac{r}{R} = \left[\frac{1 + \tan^2(\pi/2 - \pi/n)}{\cos^2\alpha}\right]^{1/2} \quad (17)$$

where R is the wire radius, and there are n wires in the layer. For the loaded configuration

$$\frac{r'}{R} = \left[\frac{1 + \tan^2(\pi/2 - \pi/n)}{\cos^2\alpha'}\right]^{1/2} \quad (18)$$

or

$$r' = r(1 + S'_{2R}) \quad (19)$$

where S'_{2R} is the radial strain in the strand cross section due to charge of lay angle, with compression regarded as negative. From equations (17) to (19)

$$1 + S'_{2R} = \left[\frac{\cos^2\alpha(\cos^2\alpha' + \tan^2(\pi/2 - \pi/n))}{\cos^2\alpha'(\cos^2\alpha + \tan^2(\pi/2 - \pi/n))}\right]^{1/2} \quad (20)$$

For n large,

$$\tan^2(\pi/2 - \pi/n) \gg \cos^2\alpha', \cos^2\alpha \quad (21)$$

and using equations (6) and (21), equation (20) reduces to

$$1 + S'_{2R} = \frac{1 + \varepsilon_h}{1 + \varepsilon_c} \quad (22)$$

and ignoring second order terms

$$S'_{2R} = -(\varepsilon_c - \varepsilon_h) \quad (23)$$

For a single layer strand with a hemp core, or no core wire, the change in diameter under load is composed of three separate mechanisms:

(i) the Poisson's ratio effects of the wire material;
(ii) deformation due to contact stresses between individual wires in line contact, and
(iii) reduction in radius due to changes in lay angle as defined in equation (20).

In a multi-layered strand, effects (i) and (ii) are operative, but (iii) is assumed to be negligible. Moreover, the diametral contraction of individual layers due to (ii) is believed to be fairly uniform over the cable cross-section and hence it is postulated that (i) does not affect the contact stresses between individual wires in different layers.

In what follows the large diameter strand is modelled as an assembly of a number of concentric partly self-prestressed orthotropic cylinders. The prestress in each layer is partly due to the body forces in the wires composing the cylinder and partly due to the clenching forces acting on the cylinder from the neighbouring cylinders. While it is recognised that the problem is essentially three-dimensional, the simplifying assumptions made are supported by the favourable agreement found later between experimental and theoretical results.

The main assumption is that twisting and bending movements in individual wires are negligible, i.e., that the wire carries a pure tension, or G, G' and H as defined in equations (14) are all zero. Plane sections are assumed to remain plane during strand deformation; in consequence the analysis is applicable only to sections remote from the clamping points. Nearer the clamps, shear deformations between cable layers may be significant.

2.5 Contact Forces in a Strand

Using the notation of Figure 5, consider a wire element of length dL, mean radius in the strand r, subtending an angle $d\phi$ on a perpendicular cross-section of the strand. For a mean axial stress F_o, and effective radial and circumferential contact stresses F_R and F_N based on gross elemental areas (eg. $r\Delta r\Delta\phi$), the radial inward equilibrium equation becomes

$$\frac{F_o}{\cos\alpha} \frac{\sin^2\alpha}{r} \frac{dL}{\cos\alpha} r\,dr\,d\phi + F_N \cos\alpha \frac{dL}{\cos\alpha} dr\,d\phi$$

$$- (F_R + r\frac{dF_R}{dr}) dL\,dr\,d\phi = 0 \quad (24)$$

$$\text{or} \quad F_o \tan^2\alpha + F_N - (F_R + r\frac{dF_R}{dr}) = 0 \quad (25)$$

F_o is positive while F_N and F_R are expected to be negative (compressive).

In normal cable constructions, the lay angles in different layers are very similar and the assumption of a uniform F_o over the whole cross-section is a reasonable one. Moreover, changes in α and r are assumed to be very small.

The first term represents the clench force generated by tension in the layer, the other two the circumferential and radial reactions to this force. The strand make-up includes a subtle variation of r and α in the various layers based on commercial experience, and the proportion of the clenching effect resisted by circumferential contact forces to radial contact forces has not previously been obtained either by theory or experiment.

In contact problems stresses in general are not linearly proportional to the load and consequently the ratio F_N/F_R for a given layer is not a constant and varies as a function of the axial load on the cable. For a given axial load various layers inside the strand are expected to have different F_N/F_R ratios.

Although it is reasonable to assume plastic flow at the contact points between individual wires for the first few cycles of loading, it is unrealistic to assume this for cases when the elements make a large number of contacts during their life: plastic flow may occur at first, but this is followed by a steady state in which the load is supported elastically (Archard, 1957).

The ideal elastic stiffness in the radial direction across the trellis crossing of cylinders (wires) is clearly smaller than in the hoop direction across line-contact between cylinders. It is further assumed that F_N and F_R are of the same order of magnitude. Consequently, the radial movements of different layers due to contact forces will be governed by the interaction of wires in line-contact in individual layers. The interaction between different layers will then take place in such a way as to be compatible with those in the hoop direction.

2.6 Properties of the Orthotropic Sheet

Using Hearmon's (1961) notation stresses T_i and engineering strains S_i referred to the axes of orthotropy are related by compliances S_{ij} where

$$S_1 = S_{11}T_1 + S_{12}T_2$$
$$S_2 = S_{12}T_1 + S_{22}T_2$$
$$S_6 = S_{66}T_6 \quad (26)$$

Taking axis 1 parallel to the wire axes,

$$S_{11} = 4/\pi E \quad (27)$$

and

$$S_{12} = -\nu S_{11} \quad (28)$$

where ν is Poissons ratio. The other two compliances may be obtained from contact stress theory.

For the line contact of two cylinders, Roark and Young (1975) gives the total width of the contact area 2b as

$$2b = 1.6 \left(\frac{pD(1 - \nu^2)}{E}\right)^{1/2} \quad (29)$$

where p is the load per unit length, and D the wire diameter. Roark also gives the diametral deflection, δ, as

$$\delta_n = \frac{4p(1 - \nu^2)}{\pi E} \left(1/3 + \ln \frac{D}{b}\right) \quad (30)$$

Substituting for p and b, using $\nu = 0.28$ again, and re-arranging

$$\delta_n = p/E \left[0.390 + 1.17 \ln \left(\frac{2}{1.54} \sqrt{\frac{ED}{p}}\right)\right] \quad (31)$$

Then the compliance in the diametral direction of the equivalent orthotropic sheet, S_{22}, is given by

$$S_{22} = \frac{1}{D} \frac{d\delta_n}{d(p/D)} = 1/E \left(0.11 + 0.59 \ln \frac{ED}{p}\right) \quad (32)$$

For tangential compliance the initial situation considered is that of two metallic bodies pressed together by a force normal to their surfaces of contact and acted upon subsequently by a tangential force tending to cause one to slide upon the other. The three dimensional case has been solved independently by Cattaneo (1938) and Mindlin (1949).

Cattaneo considered the case of a single force acting in the plane of contact and directed parallel to one of the principal axes of the contact ellipse. For a monotonically increasing force T, superposed on the constant normal force P he found that slip in the direction of T, takes place between the two bodies. Under the action of $T < \mu P$, where μ is the coefficient of limiting friction, the elliptical area of contact was shown to be divided into a central elliptical region, homothetic with the contact ellipse, where there was no relative slip between the surfaces and an annular region of slip. The distribution of tangential traction in the slip region was (reasonably) assumed to be equal to the normal pressure, as given by Hertz, multiplied by a constant coefficient of friction μ. Deresiewicz (1957) evaluated the constant displacement of the adhered region for Cattaneo's problem in the form

$$\Delta_\ell = \frac{3\mu P(2 - \nu)}{16 G a} \left[1 - \left(1 - \frac{T}{\mu P}\right)^{2/3}\right] \phi$$

$$\phi = \begin{cases} \left[4b/\pi b(2 - \nu)\right]\left[(1 - \nu/k^2)K + \nu E/k^2\right], & a < b \\ 1, & a = b \\ \left[4/\pi(2 - \nu)\right]\left[(1 - \nu + \nu/k_1^2)K_1 - \nu E_1/k_1^2\right], & a > b \end{cases} \quad (33)$$

K and E are respectively complete elliptic integrals of the first and second kind of argument $k = (1 - a^2/b^2)^{1/2}$; K_1 and E_1 are similar integrals of argument

$$k = (1 - b^2/a^2)^{1/2}$$

a is the half-width of the principal axis parallel to T and $\Delta \ell$ is the displacement per body in the direction of the applied tangential load of any point in the adhered region. The constant ϕ depends solely on the parameters of the normal Hertz problem.

Differentiating (33) with respect to T and taking the case of two bodies in contact

$$\frac{d\delta_\ell}{dT} = \frac{2 - \nu}{4Ga} \left(1 - \frac{T}{\mu P}\right)^{-1/3} \phi \quad (34)$$

where $\delta_\ell = 2\Delta_\ell$

Gross slip occurs when $T = \mu P$. Thus from (33) at the onset of gross slip

$$\Delta_{\ell max} = \frac{3\mu P(2 - \nu)}{16 Ga} \phi \quad (35)$$

For two cylinders in line contact with the tangential force directed along the line of contact the value of ϕ in (33) may conveniently be expressed in a form due to Mindlin who gives results for the ratio of initial tangential compliance to the normal compliance for bodies having the same elastic properties. For the limiting case of two parallel cylinders the ratio takes the particularly simple form of:

$$\frac{S_{66}}{S_{22}} = \frac{1}{1-\nu} \quad (36)$$

where S_{66} and S_{22} are the initial tangential and normal compliances respectively of two bodies in contact.

Putting $T = 0$ into (34) gives

$$S_{66} = \frac{1}{D} \frac{d\varepsilon_\ell}{d(T/aD)}$$

$$= \frac{d\varepsilon_\ell}{d(T/a)} = \frac{2-\nu}{4G} \Phi \quad (37)$$

where $(\frac{T}{a})$ is the tangential force per unit length of the contact line. Then from (36) and (37) finally

$$\Phi = \frac{S_{22}(4G)}{(1-\nu)(2-\nu)} \quad (38)$$

Using (35) and (33)

$$1 - \frac{T}{\mu P} = (1 - \frac{\Delta\ell}{\Delta_{\ell max}})^{3/2} \quad (39)$$

Substituting (39) and (38) into (34):

$$S_{66} = \frac{d\varepsilon_\ell}{d(T/a)} = \frac{S_{22}}{1-\nu} (1 - \frac{\Delta\ell}{\Delta_{\ell max}})^{-\frac{1}{2}} \quad (40)$$

When $\frac{\Delta\ell}{\Delta_{\ell max}} = 1$ the adhered portion of the contact surface has shrunk to zero and any further increase in T will then result in a rigid body sliding over the whole contact surface.

Mindlin and Deresiewicz (1953) considered the unloading problem of two like spheres in which the tangential force T is reduced after having reached a value T' where $0 < |T'| < \mu P$. Their work was later extended by Deresiewicz (1957) to the contact of non-spherical bodies.

For the case when T oscillates between $\pm$ T', $|T'| < \mu P$, they showed that the plot of T against Δ after the first quarter of a cycle forms a closed loop, (Figure 6) whose area represents the frictional work done during each cycle. The permanent set OR in Figure 6 is due to the presence of an initial state of stress and relative displacement at the outset of the unloading phase.

For the unloading phase:

$$\frac{d\varepsilon\mu}{dT} = \frac{2-\nu}{4Ga} (1 - \frac{T'-T}{2\mu P})^{1/3} \Phi \quad (41)$$

A subsequent increase of T from $-T'$ to T' was shown to be accompanied by the same events as occurred during the decrease from T' to $-T'$, except for reversal of sign - i.e., the loop is skew symmetric.

It can be shown that for the case of two bodies under pure tangential loading the tangential compliance may in general be reasonably expressed as:

$$\frac{d\varepsilon}{dT} = \frac{(2-\nu)}{4Ga} \Phi (1 - \delta/V)^{-\frac{1}{2}}$$

where $V = \frac{3\lambda\mu P(2-\nu)}{8Ga} \Phi$

$$= 2\lambda\Delta\ell_{max} \quad (42)$$

λ is a constant whose magnitude is determined by the initial conditions of stress and displacement at the outset of loading. That is equation (40) may be modified to:

$$S_{66} = \frac{d\varepsilon}{d(T/a)} = \frac{S_{22}}{1-\nu} (1 - \frac{\Delta\ell}{\lambda\Delta\ell_{max}})^{-\frac{1}{2}} \quad (43)$$

Equation (43) takes into account the initial conditions at the outset of loading. Note that the initial tangential compliance is always a constant given by (43) with $\dot{\Delta} = 0$, irrespective of the initial state of stress and relative displacement.

2.7 Transformation of Properties

Using Hearmon's notation again, the stresses T_i' and strains S_i' referred to arbitrary axes aligned at an angle α to the axes of orthotropy are related by

$$S_1' = S_{11}'T_1' + S_{12}'T_2' + S_{16}'T_6'$$

$$S_2' = S_{12}'T_1' + S_{22}'T_2' + S_{26}'T_6'$$

$$S_6' = S_{16}'T_1' + S_{26}'T_2' + S_{66}'T_6' \quad (44)$$

where, denoting $\cos \alpha$ by m and $\sin \alpha$ by n,

$$S_{11}' = m^4 S_{11} + 2m^2 n^2 S_{12} + n^4 S_{22} + m^2 n^2 S_{66}$$

$$S_{12}' = m^2 n^2 S_{11} + (m^4 + n^4) S_{12} + m^2 n^2 S_{22} - m^2 n^2 S_{66}$$

$$S_{16}' = -2m^3 n S_{11} + 2mn(m^2 - n^2) S_{12} + 2mn^3 S_{22} + mn(m^2 - n^2) S_{66}$$

$$S_{22}' = n^4 S_{11} + 2m^2 n^2 S_{12} + m^4 S_{22} + m^2 n^2 S_{66}$$

$$S_{26}' = -2mn^3 S_{11} - 2mn(m^2 - n^2) S_{12} + 2m^3 n S_{22} - mn(m^2 - n^2) S_{66}$$

$$S_{66}' = 4m^2 n^2 S_{11} - 8m^2 n^2 S_{12} + 4m^2 n^2 S_{22} + (m^2 - n^2) S_{66} \quad (45)$$

2.8 Kinematics of the wire layer "continuum"

If each layer of wires is treated as a cylinder with a rigid core, it is possible to solve the resulting statically indeterminate problem. Working with the unwrapped centre lines of the layer of wires, consider a rectangular element ABCD in the unstressed condition (Figure 7). A'B'C'D' corresponds to the deformed state of ABCD under the axial cable strain S_1'. Note that due to changes in the lay angle in the absence of the rigid core the wires would experience a rigid body movement resulting in a radial strain S_{2R}' in the cable cross section given by equation (20). Due to this radial strain, wires in the corresponding layer experience a slight change in their longitudinal strain $d\varepsilon_h$ which may be found by setting $\varepsilon_c = 0$ and $r' = r(1 + S_{2R}')$ in equation (7).

This leads to:

$$d\varepsilon_h = \cos \alpha \sqrt{1 + (1 + S_{2R}')^2 \tan^2 \alpha} - 1 \quad (46)$$

From Figure 7 the slip between two wires, β, in the presence of the rigid core is:

$$\beta = D \left[\tan \alpha (1 + \varepsilon_h - d\varepsilon_h) - \tan \alpha' \right] \quad (47)$$

For the orthotropic membrane $S_1 = \varepsilon_h$ and from (6):

$$1 + S_1 = \frac{\cos \alpha (1 + S_1')}{\cos \alpha'} \quad (48)$$

Note that inclusion of the rigid core will result in a slight increase in the axial wire strain.

The tensorial shear strain S_{6T} will then be:

$$S_{6T} = \tfrac{1}{2}\gamma$$

$$= \beta/2D'$$

$$= \frac{1}{2(1 + S_2)} \left[\tan \alpha (1 + S_1 - d\varepsilon_h) - \tan \alpha' \right] \quad (49)$$

where γ is the engineering shear strain and $S_2 = \dfrac{D' - D}{D}$ corresponds to the approaching strain between distant points in two wires in line contact.

From (5)

$$S'_{2C} = (1 + S'_1) \frac{\tan \alpha'}{\tan \alpha} - 1 \quad (50)$$

where S'_{2C} is the magnitude of the radial strain in the layer with the core removed. With the core included, the compatibility of movements in the radial direction gives

$$S'_2 = S'_{2C} - S'_{2R} \quad (51)$$

where S'_2 is the radial strain in cable cross section used for the orthotropic sheet.

The two dimensional element in its final deformed state may always be rotated through an angle α' with the strains on it considered as a second order tensor.

Thus

$$S_2 = -\frac{S'_1 - S'_2}{2} \cos(2\alpha') + \frac{S'_1 + S'_2}{2} \quad (52)$$

Moreover, for a cable with rotationally fixed ends, $S_6 = 0$. In other words the cable axis is coincident with the principal axis of the element, hence:

$$S_{6T} = \left(\frac{S'_1 - S'_2}{2}\right) \sin(2\alpha') \quad (53)$$

2.9 Method of Solution

For a given cable axial strain S'_1, equations (20), (46) and (48) - (53) may be solved as a set of non-linear simultaneous equations to yield values of S_1, α', S'_{2C}, S'_{2R}, S'_2, S_2, S_6 and $d\varepsilon_h$ — providing a set of compatible strains in the anisotropic cylinder with the rigid core. The method adopted by the writers involves treating S_1 as the primary unknown and using a Newton iteration (with the derivative approximated by a central finite difference form).

Once convergence has occurred, using equation (16) with the wire force $T = ES_1$, the clench force may be found. Additionally, the diametral deflection δ_n in equation (30) is very nearly

$$\delta_n = S_2 D \quad (54)$$

where D is the wire diameter.

Equations (29), (30) and (54) may be solved by a Newton iteration to give values of the contact force P_{RC} per unit length between wires in the layer as a function of S'_1, for this rigid core configuration.

2.10 Determination of the Radial Force Exerted on the Rigid Core

The angle, β, (Figure 4) which locates the lines of action of the line contact loads, P, is (Phillips and Costello, 1974)

$$\cos \beta = \frac{1}{\sin^2 \alpha}\left(\sqrt{1 + \frac{\tan^2(\pi/2 - \pi/n)}{\cos^2 \alpha}} - \sqrt{\tan^2(\pi/2 - \pi/n)}\left[1 + \frac{1}{\cot^2 \alpha \cos^2(\pi/2 - \pi/n)\left[\cos^2\alpha + \tan^2(\pi/2 - \pi/n)\right]}\right] + \cos^4 \alpha \right) \quad (55)$$

The net normal force per unit length of the wire carried by the core is

$$X_R = X_{RC} - 2P_{RC} \cos \beta$$

$$= \frac{ES_1 \sin^2 \alpha}{r} - 2P_{RC} \cos \beta \quad (56)$$

Note that in (55) and (56) changes in the lay angle α due to axial cable strains, S'_1, are assumed small enough to be ignored — see equation (10).

2.11 Radial Load Transfer in a Multi-layered Strand

Until now, a single layered cylinder with a rigid core has been considered. If α, n and R are specified, it is then possible to determine all the deformations and forces in the membrane for any given cable axial strain S_1' using the above analysis. In particular it is possible to obtain line contact forces between the individual wires in the layer, P_{RC}, and radial body forces X_{RC}, for different strains S_1'.

In a multi-layered strand, part of the radial force exerted on any layer is due to the radial body forces in that layer and part due to the clenching effects of the outer layers (see section 2.5). These effects grow radially inwards and may be simply described starting from the outermost layer, layer 1. Noting that the relationship between line contact force p_{MS} and radial force X_{MS} in the outer layer is identical to that between P_{RC} and X_{RC}, it is possible to calculate the line contact forces in the other layers as follows: for a given cable strain S_1' the clenching force provided by each wire in layer i acting on layer i + 1 is given by

$$X_{Ri} = E \frac{S_{1i} \sin^2 \alpha_i}{r_i^2} - 2 p_{MSi} \cos \beta_i \quad (57)$$

where the subscript i refers to the i th layer.

Each wire in layer j = i + 1 thus experiences a total radial force

$$X_{MSj} = E \frac{S_{1j} \sin^2 \alpha_j}{r_j^2} + X_{Ri} \frac{n_i}{n_j} \quad (58)$$

where n_i and n_j are the numbers of wires in the two layers.

Using the previously calculated P_{RC}/X_{RC} data for layer i + 1, it is then possible to find corresponding values for p_{MS} and X_{MS} for layer i + 1. The process is then repeated, moving in another layer each time until the whole strand has been analysed.

2.12 Application of Strand Theory

Table 1 gives details of the 92 wire 39 mm diameter strand used for experimental and numerical studies. Column (5) of the Table gives results from equation (18) for the theoretical radius at which N wires just touch and should be compared with the fully bedded-in measurements in column (6). Column (7) gives the net wire area for layer i in the cable transverse section based on

$$A_{ni} = n_i \sec \alpha_i \frac{\pi D_i^2}{4} \quad (59)$$

The gross area used for the orthotropic sheet model, A_{gi}, is simply

$$A_{gi} = 4 A_{ni}/\pi \quad (60)$$

The individual high tensile steel wire are hard drawn and galvanised, and after allowing for the thickness of the galvanising, the effective Young's Modulus based on the gross area of the wire is 200 kN/mm². The galvanising will clearly increase the compliances derived from contact stress theory (Section 2.6), but it is felt that this effect will be small.

Using the methods described in Section 2, results have been obtained for the wire axial and shear strains S_1 and S_{6T} for strand axial strains S_1' up to 5×10^{-3} for the various layers of the strand. The relations are essentially linear up to this strain level and on this basis slip between the wires may be obtained from

$$\delta_S = 2DS_{6T} = 2DKS_1' \quad (61)$$

where the constant K can be taken from the graphs. Results for interwire line contact forces P_{RC} and radial body force X_{RC} for a layer of wires on a rigid core are presented in Figure 8, while (from Section (2.11)) the line contact forces P_{MS} in the multi-layer strand are shown in Figure 9 as a function of cable axial strain. Finally, Figure 10 relates wire stress to the mean normal stress on the wire.

3. TORSIONAL STIFFNESS OF STRAND

A direct experimental check of the interwire contact forces is probably impossible, and indirect comparisons must be used. One relatively straight-forward comparison is via the torsional stiffness in the presence of a steady axial load on the strand. The torsional stiffness is also of interest in certain aero and hydrodynamic problems. Intuitively, the tangent torsional stiffness of a strand would be expected to fall as the applied torque increases, because at small torques interwire friction would suffice to prevent interwire movements while at higher torques the friction would be overcome leading to greater flexibility (Wyatt, 1977). Indeed this degradation of stiffness is progressive as slipping starts in the outer layer and spreads towards the core at higher torques.

It is postulated that in each layer of wires the perturbation axial and hoop strains S_1' and S_2' on applying the torque to the axially preloaded strand are zero. Using the first and second of equations (44), T_1' and T_2' can be expressed in terms of T_6' and hence the shear flexibility, S_6'/T_6', can be obtained:

$$\frac{S_6'}{T_6'} = S_{66}' + \frac{2S_{12}' S_{16}' S_{26}' - S_{22}' S_{16}'^2 - S_{11}' S_{26}'^2}{S_{11}' S_{22}' - S_{12}'^2} \quad (62)$$

As $S_1' = S_2' = 0$,

$$S_6 = S_6' \cos 2\alpha = \frac{r d\phi}{d\ell} \cos 2\alpha \quad (63)$$

where $d\phi/d\ell$ is the twist per unit length in the strand which is assumed to be the same for all layers in the strand. Slip over one wire diameter (per body) is then given by

$$\Delta = D S_6' \cos 2\alpha \quad (64)$$

In the fully bedded-in condition for a given mean axial load T, the methods of Section (2) give the normal contact forces and the normal and shear compliances in the individual layers. Equations (62) - (64) may then be used to determine T_6' as a function of $d\phi/d\ell$. Integration of this function leads to results for shear stress τ in the cable transverse section, again in terms of $d\phi/d\ell$. Then for any given twist per unit length the total torque on the cable may be obtained from

$$M = \sum_i (\tau A_g r)_i \quad (65)$$

where A_g is given by equation (60) and all layers are included.

Figure 11 compares experimental torque and twist measurements and theoretical predictions for the strand of Table 1 at two different mean load levels. Good agreement is found using a psudo-coefficient of friction $\mu_s = \lambda\mu = 0.25$ (where λ is defined in equation (42)) although the results are relatively insensitive to the value taken

It follows that the high non-linearity observed in the torque twist relation for this practical strand is not associated with changes in lay angle.

4. AXIAL STIFFNESS OF STRAND

4.1 Introduction

Another indirect test of the theory of Section (2) is to use it to predict the perturbation axial stiffness of strand. This, too, has obvious practical utility. For example, the legs of a TLP, or the hangers of a suspension bridge are both subject to relatively small variations of axial load superimposed on the mean axial tension and a knowledge of this tangent stiffness as a function of the perturbation size is a useful input to the dynamic response analysis of the platform or bridge. Again, the stiffness is much larger for small perturbations which do not initiate interwire slip than for larger perturbations which are associated with slippage on the contact lines.

4.2 Analysis

It is postulated that a change in the axial stress T_1' is associated with a perturbation of the hoop stress T_2' where

$$T_2' = T_1' \frac{T_2}{T_1} \quad (65)$$

If it is further assumed that the shear strain S_6' is zero, then the first and third of equations (44) lead to the desired axial flexibility of the membrane

$$\frac{S_1'}{T_1'} = S_{11}' + K S_{12}' - \frac{S_{16}'}{S_{66}'}(S_{16}' + K S_{26}') \quad (66)$$

where $K = T_2/T_1$. K is a function of the cable axial strain S_1' and for a given S_1' is equal to the slope of the curves in Figure 10.

Equation (43) only gives the tangential compliance between two cylinders in line contact with the oscillating force acting along the interface. With an axial load perturbation on the strand, the oscillating force is actually at an angle to the common normal to the wires. The writers have not traced a full analysis of this situation, although some progress has been made. Here, the effects of the normal component of the inclined force are ignored and equation (43) is used: Raoof (1982) presents a fuller discussion including the derivation of an alternative compliance for comparison purposes.

The limiting case of full interwire slip is addressed first. In this situation, the shear compliance S_{66} for each sheet of wires tends to infinity. The relationship between cable tension and axial strain is obtained by the following procedure:

(i) For a known cable strain S_1', calculate the interwire forces in the multi-layer strand, p_{ms}, by the methods of Section (2). The values of K for the various layers are found from Figure 10.

(ii) Calculate S_{22} by equation (32).

(iii) Using equations (45) and then (66), values of the changes in S_1'/T_1' can be obtained for various strains S_1'.

(iv) Integration of T_1' with respect to S_1' leads to values of cable axial stress for each layer as a function of the changes in S_1', and summation over the layers gives cable axial force change as a function of S_1'.

For the experimental strand, over the working load range, this last relationship was very nearly linear. Consequently, rather than using different K and S_{22} values for different cable axial strains, it was possible and easier to choose a constant K and corresponding S_{22} value and compute the full slip stiffness of the cable associated with them.

Then, for perturbations of the mean axial load it is still reasonable to use constant K and S_{22} values corresponding to the mean cable axial strain.

Turning now to the calculation of the degradation of the perturbation axial stiffness as the size of the disturbance increases, i.e. following the development of slip in the various layers, a slightly different procedure is necessary:

(i) For a given mean axial cable strain find the ratio K and hence S_{22}.

(ii) For a range of perturbations of cable axial strain, calculate values of S_{66} using equations (43) and (61)

(iii) and (iv). As above for the full slip calculation.

4.3 Results

Figure 12 presents and correlates theoretical and experimental values for the perturbation axial stiffness of the 92 wire, 39 mm strand, presented in terms of the effective Young's Modulus of the strand based on net wire area. Results for two mean loads are given. The agreement between the results is encouraging particularly for the most interesting small perturbation ratios. In the full slip regime, the effective modulus appears to be insensitive to mean load.

5. HYSTERESIS UNDER CYCLIC LOADING

5.1 Introduction

It has long been appreciated that spiral strand exhibits significant hysteresis in its longitudinal load/displacement behaviour. It also (Hobbs, et al, 1978) exhibits significant damping of torsional oscillations in the presence of an axial load. This dissipation of energy is potentially a major contribution to the total damping of the structure supported by the strands (Wyatt, 1977).

In the following, the rubbing movement between the wires in each layer is assumed to absorb much more energy than the rotational movement on the "trellis" contact points between (counter-laid) wires in different layers. As a first approximation, this trellis movement is equal to the sum of the changes in the lay angles, and these changes have been shown to be extremely small. It is further assumed that the friction between wires within a given layer is controlled by the clenching action of the mean load. No distinction is drawn between static and dynamic coefficients of friction, and the hysteresis within the metal of the wires themselves is neglected. The analysis is developed for a fully bedded-in condition, i.e. regular repeated oscillations, although it is known from torsional experiments that the damping might be as much as twice as large for the first application of a given disturbance.

5.2 Analysis

For pure tangential loading of two non-spherical bodies in contact, Deresiewicz (1957) gives the energy dissipation per cycle for one body, ΔE, in the partial slip regime to be:

$$\Delta E = \frac{9(2-\nu)\mu^2 P^2}{10 \, Ga} \left[1 - (1 - \frac{T'}{\mu P})^{5/3} - \frac{5}{6} \frac{T'}{\mu P} \left[1 + (1 - \frac{T'}{\mu P})^{2/3} \right] \right] \phi \quad (67)$$

where $T' \leq \mu p$, G is the modulus of rigidity and a is the half-width of the contact parallel to the applied tangential force.

For two parallel cylinders ϕ is given by equation (38), and inserting this value in (67) gives

$$\Delta E = \frac{18}{5} \frac{\mu^2 p^2}{(1-\nu)} S_{22} \left[1 - (1 - \frac{T'}{\mu p})^{5/3} - \frac{5}{6} \frac{T'}{\mu p} \left[1 + (1 - \frac{T'}{\mu p})^{2/3} \right] \right] \quad (68)$$

where p is the normal contact force per unit length of the contact line and hence ΔE corresponds to the dissipation per unit length of contact. From equation (39)

$$\frac{T'}{\mu p} = \left[1 - (1 - \frac{\Delta \ell}{\Delta \ell_{max}})^{3/2} \right] \quad (69)$$

For axial damping, $\Delta \ell$ is given by equation (61) with S' representing the axial cable strain change for half the perturbation load range. $\Delta \ell_{max}$ is given by equation (35) with ϕ from equation (38):

$$\Delta \ell_{max} = \frac{3}{4} \frac{\mu p}{(1-\nu)} S_{22} \quad (70)$$

Once $\frac{T'}{\mu p}$ reaches unity, gross sliding takes place between the wires and the force displacement loop is as shown in Figure 13. During partial slip, the force must change from that needed to cause sliding in one direction to that causing sliding the other way, a total change of $2\mu p$. Once gross sliding starts the force remains at μp. The area of the hysteresis loop is then

$$\Delta E = \frac{3}{5} \frac{\mu^2 p^2}{(1-\nu)} S_{22} + 8D\mu p K(S'_1 - S'_{1max}) \quad (71)$$

where S'_{1max} is the cable axial strain corresponding to the onset of gross sliding, and

$$S'_{1max} = \Delta \ell_{max} / 2KD \quad (72)$$

The first term in equation (71) represents the loop area at the onset of gross sliding, which is given also by equation (68) with $\frac{T'}{\mu p} = 1$. The second term represents the work done in the gross sliding regime.

The energy dissipation in the strand is then, per unit length of strand,

$$\Delta U = \sum_i (\Delta E n / \cos \alpha)_i \quad (73)$$

summing over all layers, with n wires in layer i. The energy input per cycle per unit length is

$$U = \tfrac{1}{2} (\frac{\text{load range}}{2}) S'_1 \quad (74)$$

Thus, the energy dissipation may be found by the following procedure:

(i) For the given mean axial load, use the procedure of Section (4) to find the change in S'_1 corresponding to any assumed axial load perturbation.

(ii) Use equations (61) and (70) to check whether $\Delta \ell / \Delta \ell_{max}$ is greater or less than one, corresponding to the full or partial slip cases respectively.

(iii) In the partial slip regime, use equations (68) and (69), or otherwise use equations (71) and (72) to determine the energy dissipated per millimetre of a wire in layer i.

(iv) Use equations (73) and (74) to obtain the energy loss ratio as a function of the load range to mean ratio.

5.3 Results and Discussion

Figure 14 gives theoretical values for the energy dissipation in each layer of the experimental strand, while Figure 15 compares experimental data due to Hobbs et al (1978) with theoretical hysteresis values for two mean load values. Overall, the agreement between experiment and theory is encouraging.

Various measurements of energy dissipation on single contacts between spheres (e.g. Johnson, 1955) have indicated that although the assumption of Coulomb friction is a reasonable approximation over that portion of the annulus of slip where relative displacement is large, it is a poor approximation for small relative displacements. For very small tangential force changes the latter regime predominates and it is not surprising that the present theory fails to predict the energy dissipation at low perturbations at all well. If the damping forces were viscous rather than frictional in nature, the energy loss expressed as $\Delta E/T_1'^2$ would be constant. Johnson found this was approximately true for small T_1'.

Hobbs et al (1978) used the decay of free torsional oscillations in the bedded-in condition to find the logarithmic decrement, η, of the cable defined as

$$\eta = \frac{1}{m} \ln (a_n/a_{n+m}) \quad (75)$$

where a_n and a_{n+m} are response peaks m cycles apart. The log. dec. was found to fall as the amplitude fell, reaching a constant for very small amplitudes. Constant log. dec. is a well known property of viscously damped systems for which

$$\eta = \Delta U/2U \quad (76)$$

For the present strand, $\Delta U/U$ for very small oscillations was about 0.05 and constant - see Figure 15. Raoof (1982) describes a procedure for calculating the torsional hysteresis rather similar to that outlined above for the axial damping.

6. DISCUSSION AND CONCLUSIONS

The theoretical analysis of a strand using orthotropic sheet theory with appropriate compliances derived from results in contact stress theory has been developed for axial and torsional loadings. In each case, predictions of wire strains, interwire forces and slippage, strand tangent stiffness and hysteresis have been made. Very encouraging comparisons have been made between experimental measurements on a 92 wire 39 mm diameter strand and the theoretical predictions.

The main novelty in the theoretical work has been to treat the layers of wires in a strand as orthotropic sheets - and if the method is to be applicable there must be enough wires in a layer to avoid serious effects from the magnitude of the polar angle between wires, or other inaccuracies. However, in the kind of large strands used for bridge hangers (and even more so for the tethers in a TLP or TBP) it appears that the orthotropic sheet approach is a valid and useful one.

For a given mean axial load, the torsional and axial stiffnesses were found to be functions of the applied torque and axial load perturbations respectively. The stiffness for small load changes was found to be larger than for large perturbations, because small disturbances do not induce interwire slippage. The theory predicts the bounds to the stiffnesses and describes the variation between the limits.

The large value of the no-slip modulus compared to the full-slip modulus may have significant design implications, raising the natural frequency of the structure. The ability to predict values of contact forces and interwire movements in large strands is of obvious value as an input to the fatigue design of, for example, the tethers of tension leg/tethered buoyant platforms.

Torque-twist predictions are pertinent to a number of aero- and hydodynamic problems such as ice accretion and galloping: they may also prove useful in the analysis of ropes as an assembly of strands and in predicting kink formation in cables.

Prediction of the energy dissipation quotient, $\Delta U/U$, under continued uniform cyclic loading is now possible and is in satisfactory agreement with experimental data: with a peak around 0.15 it is much lower than previously believed. However, in a more realistic random loading situation it is reasonable to expect rather higher values.

7. ACKNOWLEDGEMENTS

The writers are glad to acknowledge the large part played by Dr T.A. Wyatt (Imperial College, London) in the early stages of the work described here. They are also grateful for the support of the (UK) Science and Engineering Research Council.

REFERENCES

ARCHARD, J.F., 1957, "Elastic Deformation and the Laws of Friction," Proceedings of the Royal Society of London, Volume A243, pp. 190-205.

BECHTLOFF, G., 1963, "Longitudinal Elongation and Transverse Contraction of a Six-Stranded Wire Rope Under Tensile Load," Wire World International (English Translation from DRAHT-WELT), Volume 5, Number 6, pp. 247-252.

CATTANEO, C., 1938, "Sul Contatto di Due Corpi Elastici," Accademia dei Lincei, Rendiconti, Series 6, folio 27 Part I : pp. 342-348, Part II: pp. 434-436, Part III: pp. 474-478.

CHI, M., 1971a, Analysis of Multi-Wire Strands in Tension and Combined Tension and Torsion, Catholic University of America, Tech. Report No. 71-9.

CHI, M., 1971b, Analysis of Operating Characteristics of Strands in Tension Allowing End-Rotation, Catholic University of America, Tech. Report No. 71-10.

COSTELLO, G.A. and J.W. PHILLIPS, 1974, "A More Exact Theory for Twisted Wire Cables," Journal of the Engineering Mechanics Division, ASCE, Volume 100, Oct., pp. 1096-1099.

COSTELLO, G.A. and J.W. PHILLIPS, 1976, "Effective Modulus of Twisted Wire Cables," Journal of the Engineering Mechanics Division, ASCE, Volume 102, Feb., pp. 171-181.

COSTELLO, G.A. and S.K. SINHA, 1977a, "Static Behaviour of Wire Rope," Advances in Civil Engineering Through Engineering Mechanics, Proceedings Second Annual Engineering Mechanics Division Specialty Conference, North Carolina State University, Raleigh, N.C., May, 1977, pp. 475-478.

COSTELLO, G.A. and S.K. SINHA, 1977b, "Torsional Stiffness of Twisted Wire Cables," Journal of the Engineering Mechanics Division, ASCE, Volume 103, pp. 766-770.

COSTELLO, G.A. and R.E. MILLER, 1979, "Lay Effect of Wire Rope," Journal of the Engineering Mechanics Division, ASCE, Volume 105, Aug., pp. 597-608.

DERESIEWICZ, H., 1957, "Oblique Contact of Nonspherical Bodies," Journal of Applied Mechanics, Volume 24, pp. 623-624.

DRUCKER, D.C. and H. TACHAU, 1945, "A New Design Criterion for Wire Rope," Journal of Applied Mechanics, Transactions, American Society of Mechanical Engineers, Volume 12, Number 1, pp. A-33-A-38.

DURELLI, A.J., S. MACHIDA and V.J. PARKS, 1972, "Strains and Displacements on a Steel Wire Strand," Naval Engineering Journal, Volume 84, Number 6, pp. 85-93.

DURELLI, A.J. and S. MACHIDA, 1973, "Response of Epoxy Oversized Models of Strands to Axial and Torsional Loads," Experimental Mechanics, Vol. 13, July, pp. 313-321.

GIBSON, P.T., CRESS, H.A., KAUFMAN, W.J. and W.E. GALLANT, 1969, Torsional Properties of Wire Rope, ASME Paper No. 69-DE-34, May 1969.

HEARMON, R.F.S., 1961, Applied Anisotropic Elasticity. OXFORD: Oxford Univesity Press.

HOBBS, R.E. and K. GHAVAMI, with T.A. WYATT, 1978, "Fatigue of Socketed Cables: In-line and transverse tests on 38 mm specimens; Internal damping in 38 mm specimens. Imperial College, Civil Engineering Department, CESLIC Report SC2.

HOBBS, R.E. and B.W. SMITH, 1982, "The Fatigue Performance of Socketed Structural Strands," Awaiting Publication.

HRUSKA, F.H., 1951, "Calculation of Stresses in Wire Ropes," Wire and Wire Products, Volume 26, pp. 766-767, 799-801.

HRUSKA, F.H., 1952, "Radial Forces in Wire Ropes," Wire and Wire Products, Volume 27, pp. 459-463.

HRUSKA, F.H., 1953, "Tangential Forces in Wire Rope," Wire and Wire Products, Volume 28, pp. 455-460.

JOHNSON, K.L., 1955, "Surface Interaction Between Elastically Loaded Bodies Under Tangential Forces," Proceedings of the Royal Society of London, Volume A230, pp. 531-548.

JONES, N. and J.C. CHRISTODOULIDES, 1980, "Static Plastic Behaviour of a Strand," International Journal of Mechanical Sciences, Volume 22, Number 3-D, pp. 185-195.

KASPER, R.G., 1973a, "A Structural Analysis of Multi-Conductor Cable," Naval Underwater Systems Center, Distributed by National Technical Information Service, Report No. AD-767 963.

KASPER, R.G., 1973b, "Cable Design Guidelines Based on A Bending, Tension and Torsion Study of An Electro-Mechanical Cable," Naval Underwater Systems Centre, Distributed by National Technical Information Service, Report No. AD-769 212.

KAWASHIMA, S. and H. KIMURA, 1952, "Measurements of the Internal Friction of Metal Wire Ropes Through the Longitudinal Vibration, Mem. Fac. Engng. Kyushu Univ. 13, 1, pp. 119-130.

KNAPP, R.H., 1979, "Derivation of a New Stiffness Matrix for Helically Armoured Cables Considering Tension and Torsion," International Journal for Numerical Methods in Engineering, Volume 14, Number 4, pp. 515-529.

LEISSA, A.W., 1959, "Contact Stresses in Wire Ropes," Wire and Wire Products, Volume 34, Number 3, pp. 372-373.

LOVE, A.E.H., 1927, Treatise on the Mathematical Theory of Elasticity. 4th Ed. Cambridge, (and Dover Publications, New York, 1944).

MINDLIN, R.D., 1949, "Compliance of Elastic Bodies in Contact," Journal of Applied Mechanics, Volume 16, Transactions ASME, Volume 71, pp. 259-268.

MINDLIN, R.D. and H. DERESIEWICZ.,1953, "Elastic Spheres in Contact Under Varying Oblique Forces," Journal of Applied Mechanics, Volume 20, Transactions ASME, Volume 75, pp. 327-344.

NOWAK, G., 1974, "Computer Design of Electro-Mechanical Cables for Ocean Applications", Proceedings, 10th Ann. MTS Conf., Washington D.C., pp. 293-305.

PHILLIPS, J.W. and G.A. COSTELLO.,1973, "Contact Stresses in Twisted Wire Cables," Journal of the Engineering Mechanics Division, ASCE, Volume 99, April, pp. 331-341.

RAOOF, M., 1982, Official title to be approved, Ph.D Thesis, London University.

RIDLEY, S.A., 1973, "The Damping Capacities of Helically Wound Multi-Strand Steel Cables," M.Sc. Thesis, Imperial College, Department of Civil Engineering.

ROARK, R.J. and C. YOUNG., 1975, Formulas for Stress and Strain. International Student Edition: McGraw-Hill.

STARKEY, W.L. and H.A. CRESS.,1959, "An Analysis of Critical Stresses and Mode of Failure of a Wire Rope," Journal of Engineering for Industry, Transactions, American Society of Mechanical Engineers, Volume 81, Series B, Number 4, pp. 307-316.

WILSON, B.W., 1967, "Elastic Characteristics of Moorings," Journal of the Waterways and Harbors Division, ASCE, Volume 93, November, pp. 27-55.

WINKELMANN, W., 1972, "Beitrag Zur Berechnung der Verseilspannungen," Wissenschaftliche Zeitschrift der Technischen Hochschule Otto von Guericke Magdeburg 16(1972) Heft 4.

WYATT, T.A., 1977, "Mechanics of Damping," Symposium on Dynamic Behaviour of Bridges. Crowthorne, Berks, UK., 1977, pp. 10-21, Crowthorne, Berks; Transport and Road Research Laboratory Supplementary Report 275.

YOSHIDA, S., 1952, "On the Moments Applied to Wires During the Stranding Process," Proceedings, 2nd Japan National Congress for Applied Mechanics, pp. 109-112.

92 Wire Strand, 39.0 mm O.D

Layer (1)	Number of wires N (2)	Wire diameter D(mm) (3)	Lay length (mm) (4)	Pitch circle radius (theo) r(mm) (5)	Pitch circle radius (prac) r(mm) (6)	Net wire area A_n(mm^2) (7)
King	1	5.05	-	-	-	20.00
5	7	3.54	95	4.19	4.17	71.08
4	12	3.54	178	7.04	7.54	121.6
3	18	3.54	241	10.57	10.95	183.2
2	24	3.54	305	14.10	14.35	245.0
1	30	3.54	348.7	17.73	17.75	308.4

Note: Column 6, measured outside radius = 19.5 mm. Direct addition of wire diameters gives 20.2 mm. Thus bedding at each interface = (20.2 - 19.5)/5 = 0.14 . Then data in Column 6 built-up from centre using 0.15 mm bedding at each interface.

STRAND MAKE-UP

TABLE 1

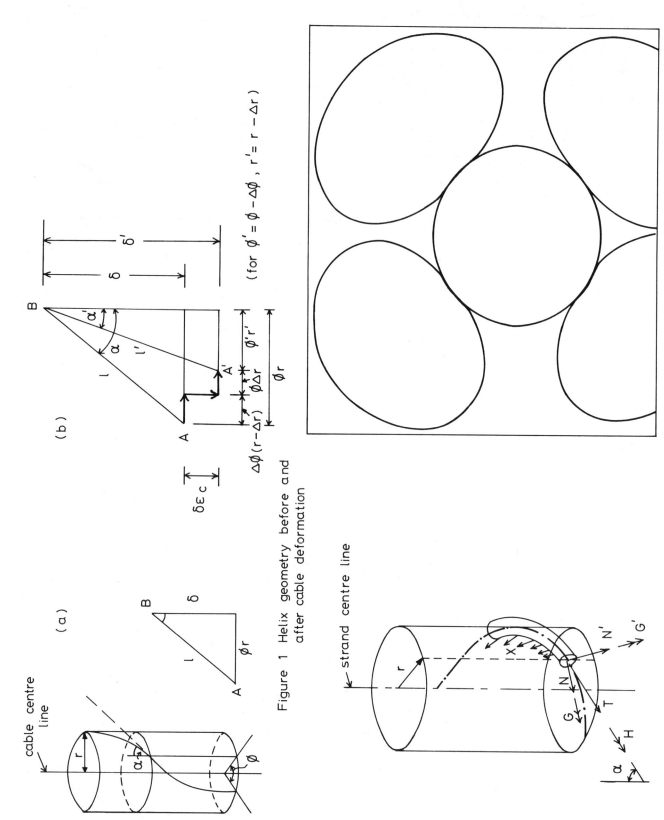

Figure 1 Helix geometry before and after cable deformation

Figure 2 Force resultants acting on a single wire

Figure 3 Enlarged wire cross sections in a five wire strand

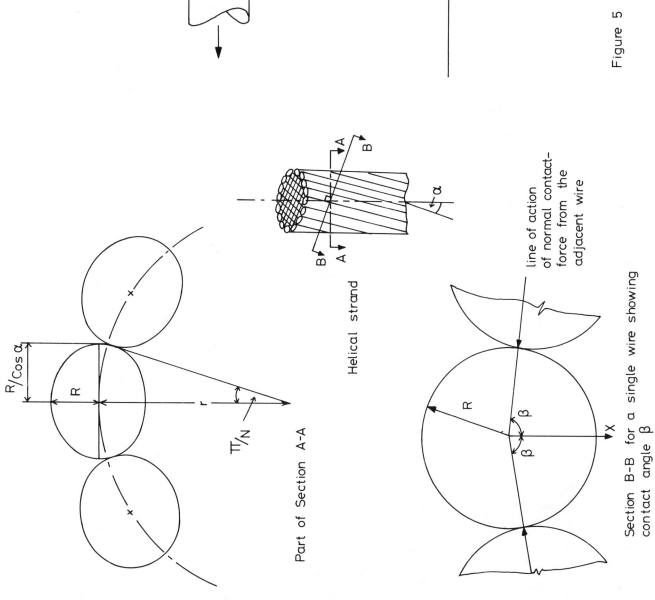

Figure 5 Strand under pure tension treated as an axially symmetrical plane strain problem

Figure 4 Single layered strand with N wires

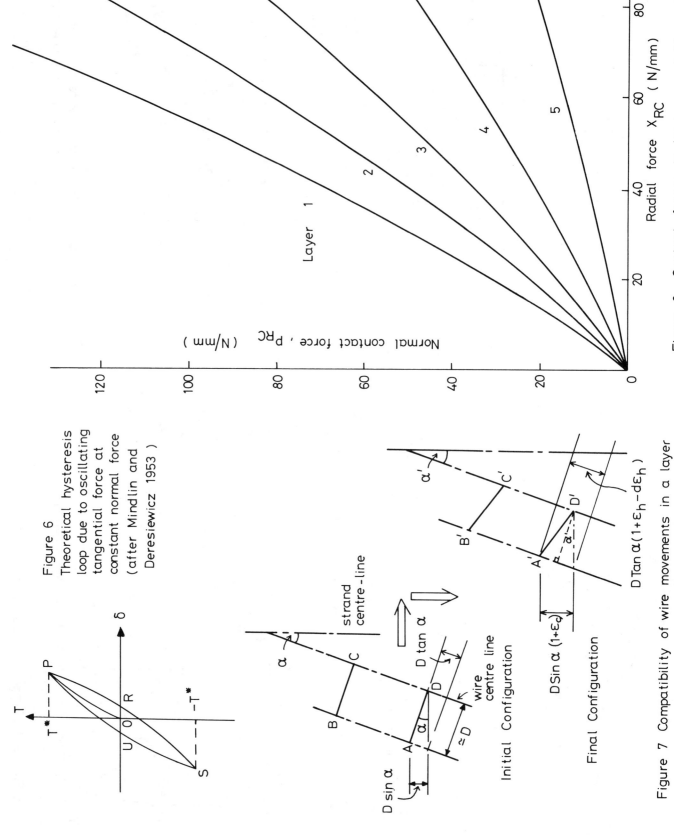

Figure 6 Theoretical hysteresis loop due to oscillating tangential force at constant normal force (after Mindlin and Deresiewicz 1953)

Figure 7 Compatibility of wire movements in a layer

Figure 8 Contact forces in layers on rigid cores

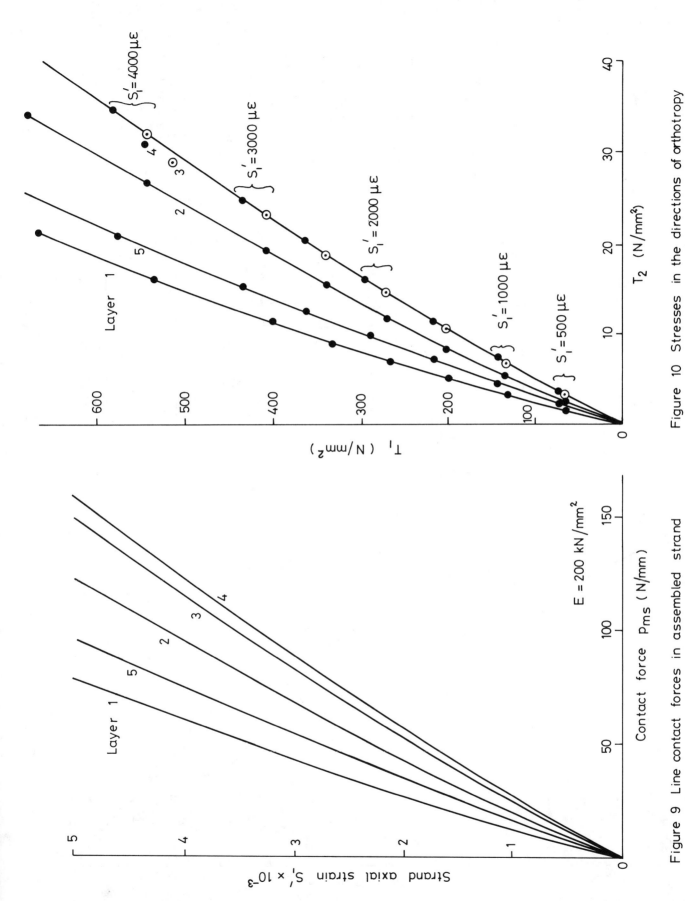

Figure 10 Stresses in the directions of orthotropy

Figure 9 Line contact forces in assembled strand

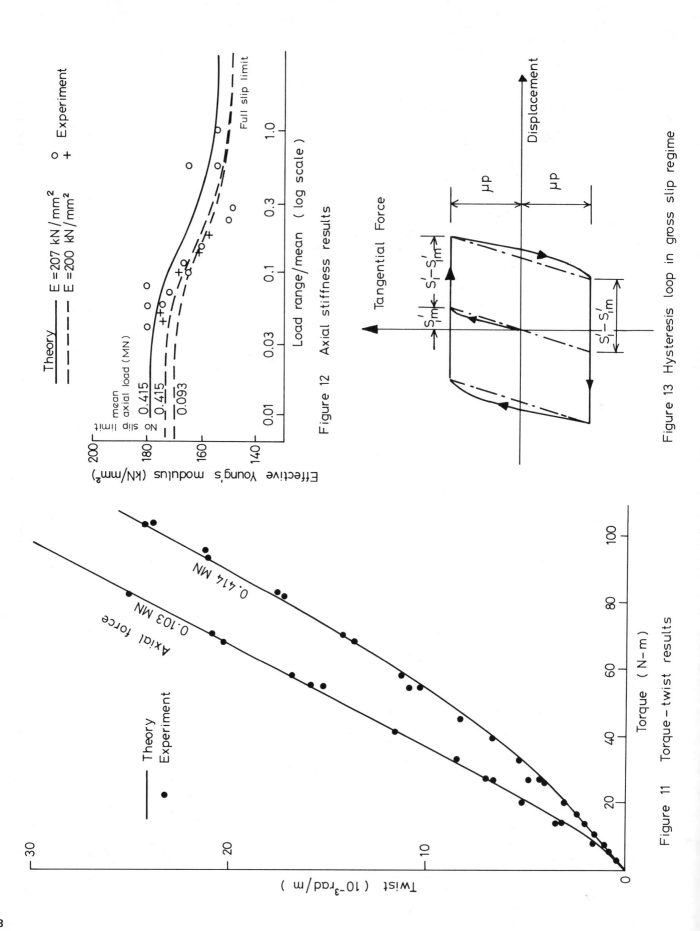

Figure 11 Torque-twist results

Figure 12 Axial stiffness results

Figure 13 Hysteresis loop in gross slip regime

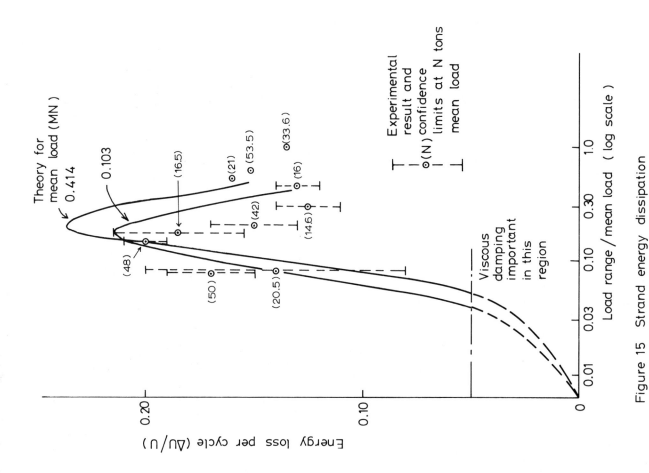

Figure 15 Strand energy dissipation

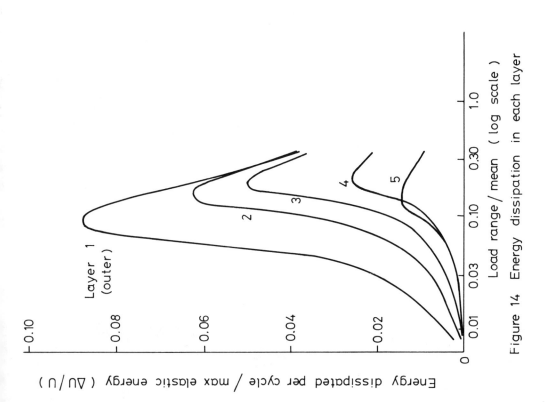

Figure 14 Energy dissipation in each layer

DYNAMIC RESPONSE OF TENSION LEG PLATFORM

Demosthenes C. Angelides	Cheng-Yo Chen	Stephen A. Will
McDermott Incorporated	McDermott Incorporated	McDermott Incorporated
New Orleans, Louisiana	New Orleans, Louisiana	New Orleans, Louisiana
U.S.A.	U.S.A.	U.S.A.

SUMMARY

The effect of some design characteristics on the dynamic response of the tension leg platform (TLP) are considered in this paper. The objectives are the understanding of the system behavior and the establishment of trends useful for an optimum design.

In particular, the influence of hull geometry, values for force coefficients, water depth, pretension, and tether axial stiffness are studied for TLPs with representative weight, displaced volume, and overall dimensions.

Extreme sea states only are considered in this investigation. Therefore, conclusions drawn here are applicable only to such a wave environment and should not be extrapolated to small wave periods.

The tool used for the analysis of the response of the floating body is a six degree of freedom time domain model. Large translations and small rotations are assumed in the formulation. The tethers are modeled as axial springs and for the evaluation of the wave loading the modified Morison's equation, which takes into account relative motion, is used.

INTRODUCTION

The move into deeper waters by the offshore industry is continuing with striking advances, as witnessed by the fact that wells are now being drilled in water depths reaching 5,000 ft. Larger water depths mean larger dynamic effects, larger total base moment, and consequently more steel. Taller structures also impose more serious installation problems. Installation can be either in sections or in one piece. The former can be quite an expensive operation; the latter is constrained to the size of the available barge.

The cost increase of fixed offshore structures with increased depth encouraged the development of compliant offshore structures as an alternative concept. The idea behind these types of structures is the minimization of the resistance of the structure to environmental loads (waves, currents, etc.), by making the structure flexible. There are several types of compliant structures: articulated single column structures (SPM, SALM, etc.), guyed tower structures, floating production platforms (catenary moored) and tension leg platforms. The tension leg platform family encompasses several different designs, e.g., vertically moored platform (VMP), tethered production platform (TPP), and tethered buoyant platform (TBP).

In all TLP structures (the term TLP is used in this paper in a general sense and without reference to any specific design) the principle is the same: a floating structure of semi-submersible type is moored by vertical tethers under pretension imposed by excess buoyancy. The excess buoyancy is created by ballasting the structure at the site, connecting the vertical mooring system, then deballasting the structure. Excess buoyancy, and therefore, tether pretension acts as restoring force the same way as gravity acts on a pendulum. The horizontal component of the tether pretension also counteracts the steady horizontal loads on the structure.

The TLP is, in principle, an extension of the traditional catenary moored semi-submersible structures; however, it offers the advantage of smaller heave, roll and pitch motions. Proper choice of volume distribution of TLP floating body in the vertical direction is important in reducing heaving dynamic response but mostly in reducing tension variation of the tethering system.

Studies on response of the TLP have been published by several investigators, e.g., Paulling (1977), Horton, et al (1972), Albrecht, et al (1978), Denise and Heaf (1979), Paulling (1981), Salvensen, et al (1982). Rainey (1977), Yoneya and Yoshida (1981) investigated, in particular, the effect of Mathieu-type instability in the response of the TLP. Angelides, et al (1981) demonstrated with a three degree of freedom model that, for realistic mass of the TLP floating body, this phenomenon can occur only for an unrealistically shallow water depth and that the response is very sensitive to the value of the C_D coefficient used in Morison's equation. Further studies, not published here, with the six degree of freedom model of this paper reconfirmed the conclusions derived with the three degree of freedom model.

In this paper analysis approaches are discussed along with several design considerations for the TLP. Parametric studies are conducted with the objective of defining grounds for an optimum TLP dynamic response. Geometry variations are considered for structures with a realistic weight and overall dimensions. Critical parameters affecting the dynamic response of the TLP are investigated through parametric studies.

DESIGN CONSIDERATIONS FOR TLP

Several TLP designs have been developed on the basis of a vertical mooring system under pretension due to excess buoyancy. In Figs. 1 and 2, some of the most common designs are presented. As can be seen, four of the five designs have two axes of symmetry, while one is axisymmetric. The buoyancy in two of them is provided mainly by the vertical bottle shaped members, while in the other three designs, a significant amount of the total buoyancy is provided by the horizontal members as well.

The hull of the tension leg platform is composed of stiffened shells. It is divided into compartments for controlling flooding due to possible damage of certain parts, and for controlling ballasting during installation. Weight of deck and hull directly affect the pretension of the tethers and is a controlling factor of the TLP economics.

The tethering system can be of cables or pipes. From a strength reliability point of view, an advantage of using cables is that cables are continuous members with multiload paths. Cables are vulnerable to corrosion in the ocean environment. This can be overcome by encapsulation of the cables with blocking materials and sheathing. Special care must be taken for the protection of the cables against fatigue due to the extreme bending at the end connections. Pipes present the advantage that they occupy less deck space than the cables, and their installation can be carried out with conventional drilling riser procedures; however, the installation time increases when compared to cables, especially with increasing water depth. Pipes present the disadvantage of having many connections. The use of pipes reduces the problems of corrosion and excess motions (especially heave). Pipes can be more easily inspected by accoustic devices to be lowered down the hole; also their maintenance is less involved. A combination of pipe and cable for a tethering system is questionable, due to the difference in elongation of pipes as opposed to cables.

The tethers of the TLP are arranged in groups (bundles), each under a leg of the floating structure. These tethers can satisfy the purpose of mooring action only, or can combine mooring and production purposes. The VMP design (see Fig. 2) integrates the function of marine risers and tethering system in the same pipe-type components. Alternately, tethers used only for mooring may be combined in the same group with other members which are exclusively production risers (see axisymmetric structure of Fig. 2). Finally, some designs keep the function of the tethering system distinctly separated from the function of the marine risers by grouping the marine risers in a separate moonpool arrangement (see Fig. 1).

There are two alternative designs for tethers and/or risers of the same group. One type utilizes horizontal spacers along the length of the tether/riser group (see Fig. 2). The purpose of the spacers is to reduce the relative motion of the members, and to avoid the impact of each member against the other; the members of a group move, more or less, as a unit. On the other hand, the distance between the successive spacers has a direct effect on the natural period of the individual members. Controlling the natural period of these members within a certain range can prevent vortex induced vibrations and locking-on phenomena. The other design leaves the tethers/risers free from each other. The absence of horizontal spacers eliminates the problem associated with fatigue at the connections of tethers/risers and spacers, as well as it reduces installation problems. The members are spaced further apart from each other to prevent the impact of each member against the other, since individual member motion increases in this type of configuration. For larger water depths, where large flexibility is present, the horizontal spacers may be a requirement.

In the TLP designs of Fig. 1 the risers are longer than the tethers. Therefore, the risers tend to become slack while the tethers remain under tension. This problem and any relative displacement of the top of the risers and the floating body can be overcome with riser tensioners. In defining the capacity of such riser tensioners, the horizontal deflection due to steady forces (current, steady drift) on the risers should be taken into account. Additional considerations should be temperature variations, buoyancy, weights, etc.

The anchoring of the tethering system of the TLP varies among the several designs. Some of the designs have one base template, while others have one template for each leg of the floating structure. The use of one single template eliminates the problems of relative positioning and orientation associated with the individual templates. However, one single large template is more cumbersome from the transportation point of view (Mercier, et al, 1982). The anchoring of the tethering system is usually achieved by tension piles. One approach is to transfer the load from each tether directly to the corresponding pile. A second approach is to connect the pile heads corresponding to the tethers under a leg of the floating structure by a rigid cap at the level of the template. These piles then behave as a fixed head pile group, such that the bending moment from the piles and tethers is reacted by axial load couples in the piles. In both cases, the template serves only as guidance for drilling and installation of the piles. If a single template is used for the entire tethering system, a third approach is to connect all pile and tether groups directly to the template. Tension piles can be mainly of two types: (1) driven, (2) drilled and grouted. There are certain advantages, disadvantages and behavioral unknowns with each of these two types. An alternative to the exclusively piled foundation is a gravity-type foundation transferring the load from the tethers to the ocean floor. The holding capacity of this gravity-type foundation can be improved in case of weak soil conditions by pile driving (see Berman and Blenkarn 1978; Beynet et al 1978).

METHODOLOGY FRAMEWORK FOR TLP ANALYSIS

The three major TLP components which require analysis and design are: the floating part of the structure (hull, deck), the tethers/risers, and the foundation system.

The external loading imposed on the floating part of the structure is due to the effect of the waves (first and second order effects), currents, wind. The floating part of the TLP is composed of large diameter and small diameter members relative to the incident wave length. For the evaluation of wave loading on the configuration of large diameter members, a 3-D diffraction-radiation problem must be solved. In such an analysis added mass, radiation damping, first order wave forces, and second order wave drift forces are evaluated. For the evaluation of wave loading on small diameter members, the modified Morison's equation (including relative motion) is applied.

The loading on the tethers/risers is attributed to the effect of the waves, currents, internal waves and response of the floating part of the structure. The effect of the waves is primarily concentrated at the top of the tethers/risers. For the evaluation of forces in-line with the flow (steady or oscillatory) Morison's equation is applied. The validity of the equation and the magnitudes of the associated coefficients are questionable due to the large amplitude motion (Moe and Verley, 1980) and clustering. The response of the tethers/risers is, in general, three dimensional. Wave, current of arbitrary directions, lift forces acting perpendicular to the current and wave direction give rise to a three dimensional response. The lift forces, attributed to vortex formation and shedding, lead to flow induced vibrations and fatigue failures. These types of vibrations can be associated more with the presence of steady flow than with the wave action. This problem is better understood for the steady current case than for the oscillatory flow case. For the case of steady flow, clustering may, under certain conditions, destroy the formation and shedding of vortices. However, it may lead to instabilities of the response identified as "whirling" and "wake-induced vibrations" (see Blevins, 1977). The geometric and dynamic characteristics of each tether/riser should be chosen adequately, relative to the flow velocity magnitude in order to (1) avoid the onset of these instabilities and (2) reduce vortex induced vibrations. Experiments are required to calibrate any computer development for modeling forcing and response of the tethers/risers.

In applying Morison's equation for monochromatic wave the C_D, C_M coefficients can be determined from data as functions of Reynolds number, Keulegan-Carpenter number and relative roughness (see Sarpkaya and Isaacson 1981). The value of C_M for rather large diameter members can be further calibrated by diffraction analysis. When the sea state is random, the C_D, C_M coefficients can be determined from periodic flow data by defining Reynolds number and Keulegan-Carpenter number based on the r.m.s. value of velocity (see Angelides and Connor, 1980). For steady current the drag coefficient can be determined as a function of Reynolds number. The choice of C_D values is an enigma when current and waves occur simultaneously. Application of Morison's equation with a single C_D value is probably not sufficient. Modification of Morison's equation and of the basis for defining the values for C_D have been suggested by Pijfers and Brink (1977) and Salvensen, et al (1982). Experimental support and calibration of such approaches is important.

For certain geographic locations, a source of forcing on the tethers/risers can be attributed to internal waves associated with density differences in the water. A field study on this kind of waves has been carried out by Exxon Production Research (Osborne, et al, 1978, 1980). Internal waves are characterized by massive movement of water at a certain depth, and can have a significant effect on the tethers/risers (Ocean Oil Weekly Report 1981).

The analysis of the foundation, i.e., tension piles or gravity foundation, is an issue that requires significant investigation. Present technology does not adequately address the case of cyclically loaded tension piles. Field and laboratory experiments and analysis are currently in progress by several groups, with the objective to assess holding capacity and values for equivalent springs required for the analysis of the TLP.

Nonlinearities associated with the TLP response are attributed to: (1) free surface effects, (2) coupling of degrees of freedom, (3) the tethering system, and (4) nonlinear drag force and associated hydrodynamic damping on slender members. Requirements for handling nonlinearities, realistic representation of the sea state, and restrictions imposed by certain methodologies lead to two procedures of analysis for the TLP: the frequency domain analysis and the time domain analysis.

Frequency Domain Analysis
This is the proper approach for a fatigue analysis of the TLP. Frequency domain analysis requires linearization of the system and the forcing function. The wave environment is described either by a simple harmonic or by a wave spectrum; linear wave theory is used. Wave forces on the floating part of the structure are evaluated by the approaches discussed previously where distinction of large and small diameter numbers is made. For tethers/risers, Morison's equation is applied. Derivation of the statistics of the TLP response with frequency domain analysis is straightforward.

Time Domain Analysis
This approach is suitable for the analysis of the response of the TLP to an extreme wave, the "design wave". It is a step-by-step integration in time. Time domain analysis is able to treat all nonlinearities in a full manner. Periodic waves are used for the description of the sea state. Wave forces are evaluated with application of Morison's equation in all members (this is reasonable due to the range of the ratio of member diameter to the wave length of extreme waves). For the floating body members, the inertia coefficient in this equation can be further calibrated, if necessary, versus the value of the added mass calculated through a 3-D diffraction-radiation analysis in frequency domain. Long-term reliability of the TLP can be assessed in terms of probability of exceedance (design wave approach). A more rigorous analysis, using a number of response time records generated from time records of the sea state is computationally very expensive.

MODEL FOR TLP FLOATING BODY RESPONSE ANALYSIS

The floating part of the TLP is modeled as a rigid body with six degrees of freedom: three translations x, y, z (surge, heave, sway) and three rotations α, β, γ (roll, yaw, pitch). Large translations and small rotations are assumed in the solution. The rotations are defined with respect to the axes of a system X, Y, Z, with origin at the center of mass of the body and axes parallel to the global system x, y, z. The definition of the coordinate systems used in this study is provided in Fig. 3. The tethers of the TLP are represented by linear axial springs. The wave is propagating at a given arbitrary direction.

Wave forces are evaluated by application of a modified expression of Morison's equation (relative motion is taken into account) on the three dimensional structure. The hydrodynamic force coefficients C_M, C_D in Morison's equation are specified as constants for all the members, or are made dependent upon the shape of the member according to Det Norske Veritas (1981). Applicability of Morison's equation is limited for D/L approximately less than 0.2 (D is representative dimension of member cross-section, e.g., diameter of cylindrical member; L wave length). For values of this ratio larger than 0.2, diffraction effects become important. Wave forces on the tethers are neglected.

The equations of motion are:

$$m \ddot{x}_1 + K_x = F'_x \tag{1}$$

$$m \ddot{y}_1 + K_y = F'_y + s + F_B \tag{2}$$

$$m \ddot{z}_1 + K_z = F'_z \tag{3}$$

$$I_{\bar{x}\bar{x}} \ddot{\alpha} + I_{\bar{x}\bar{y}} \ddot{\beta} + I_{\bar{x}\bar{z}} \ddot{\gamma} + (KM)_X = M'_{\bar{x}} + (MW)_X \tag{4}$$

$$I_{\bar{y}\bar{x}} \ddot{\alpha} + I_{\bar{y}\bar{y}} \ddot{\beta} + I_{\bar{y}\bar{z}} \ddot{\gamma} + (KM)_Y = M'_{\bar{y}} \tag{5}$$

$$I_{\bar{z}\bar{x}} \ddot{\alpha} + I_{\bar{z}\bar{y}} \ddot{\beta} + I_{\bar{z}\bar{z}} \ddot{\gamma} + (KM)_Z = M'_{\bar{z}} + (MW)_Z \tag{6}$$

m is the real mass of the structure. x_1, y_1, z_1, define the instantaneous position of the center of gravity of the TLP with respect to x, y, z, coordinate system. I_{ij} with i, j, representing $\bar{x}$, $\bar{y}$, $\bar{z}$ are the mass moments and products of inertia with respect to the system fixed with the floating body. K_x, K_y, K_z are the tether forces. $(KM)_X$, $(KM)_Y$, $(KM)_Z$ are the tethering system moments with respect to system X, Y, Z. s is the static buoyancy minus the weight of the structure. F_B is the variation of buoyancy with time. F'_x, F'_y, F'_z are the hydrodynamic forces described by Eq. 7. $M'_{\bar{x}}$, $M'_{\bar{y}}$, $M'_{\bar{z}}$ are the moments of F'_x, F'_y, F'_z with respect to system $\bar{x}$, $\bar{y}$, $\bar{z}$. $(MW)_X$, $(MW)_Z$ are the moments of $s+F_B$ with respect to system X, Z.

The expression for the forces F'_x, F'_y, F'_z is given as follows:

$$F'_\lambda = \sum_j [\frac{1}{2} \rho C_{D_j} D_j \ell_j |\dot{r}_{jn}| \dot{r}_{jn} + \rho \frac{\pi}{4}(C_{M_j} - 1) D_j^2 \ell_j \ddot{r}_{jn}]_\lambda +$$

$$\sum_j \rho \frac{\pi}{4} D_j^2 \ell_j \ddot{u}_{j\lambda} \qquad \text{with } \lambda = x, y, z \tag{7}$$

In the above equation, D_j, ℓ_j are diameter and length of member j; $\dot{r}_{jn}$, $\ddot{r}_{jn}$ are the resultant relative (orbital fluid minus structural) velocity and acceleration respectively, projected on a plane perpendicular to member j; $\ddot{u}_{j\lambda}$ is the orbital fluid acceleration projected in the λ direction. The quantity inside the brackets represents a force perpendicular to the member j and the subscript λ symbolizes the component of this force in direction λ. Σ is the summation over all members j of the floating structure. The last term in Equation 7 represents the Krylov force in the λ direction.

The structural acceleration terms incorporated in $\ddot{r}_{jn}$ are moved to the left hand side of Equations 1, 2, 3, 4, 5, 6. The coefficients of these terms define the added mass, and added mass moment of inertia.

It should be noted that the moments of F'_x, F'_y, F'_z are taken with respect to $\bar{x}$, $\bar{y}$, $\bar{z}$, system in order to retain symmetry in the added mass matrix. Moments of these forces with respect to X, Y, Z system would result in non-symmetric added mass matrix due to the structure of the transformation imposed by the small rotational angle assumption.

A fourth order Runge-Kutta time integration scheme is used for the solution of these equations. The displaced position of the structure at each time step and the effect of the free surface variation are taken into account in the evaluation of wave forces, buoyancy and added mass. Airy theory is applied and its expressions for the evaluation of fluid kinematics are assumed valid above the still water level. For purposes of reducing the effect of free vibration and accelerating the convergence to steady state, the wave height is assumed to increase gradually (ramp shape) from zero to its final value with the following expressions:

$$H_r = H[\frac{1}{2}(1 - \cos(\frac{\pi t}{TN}))] \qquad \text{for } t \leq TN$$

$$H_r = H \qquad \text{for } t > TN \tag{8}$$

where N is the number of wave cycles in the ramp, H is the final wave height, T is the wave period, t is the absolute time, and H_r is the wave height within the ramp.

This model of the response of the TLP is characterized by two main simplifications: (1) application of Morison's equation for the evaluation of the wave forces, and (2) neglect of dynamic effects of the tethers. The first assumption applies mostly for extreme waves, i.e., when the wave length is large in comparison to the diameter of the members. With respect to the second assumption, Jefferys and Patel (1981) have demonstrated analytically that neglecting the dynamics of the tethers of the TLP has an insignificant effect on the motions of the floating body.

PARAMETRIC STUDIES ON TLP DYNAMIC RESPONSE

Results of parametric studies on TLP response to extreme sea states are presented in this section. These studies were conducted by using the floating body dynamics model outlined in the previous section. The exact geometry of the structures used is given in Fig. 4 along with Table 1. The force coefficients, C_M and C_D, are according to Det Norske Veritas (1981) specifications or are taken as constants with values 2.0 and 1.0 respectively. The choice of C_M, C_D values is indicated in Table 1. In all cases, except the one of Fig. 13, a wave of height 75 ft. and period 12 sec is propagating in the positive x direction. In the case of Fig. 13 a wave of height 98.4 ft. and period 17 sec is propagating along the diagonal direction of the TLP.

First, the effect of volume distribution with draft of the TLP is analyzed. The water depth considered is 1500 ft. All the models investigated have the same characteristics, i.e., displaced volume, weight, draft, center of gravity, moment of inertia.

In Fig. 5 the variation of buoyancy, vertical hydrodynamic force (as given by Eq. 7) and the summation of these two components are presented for models 1, 2, 3, and 4. For models 1, 2, 3 the vertical hydrodynamic force is composed of the Krylov force only, due to the presence of only vertical members. For the design wave (long wave) buoyancy and a vertical hydrodynamic force are the main cause of tether tension variations. Structural inertia in the vertical direction and pitching moments have a less significant effect. Model 1, having a large volume close to the wave zone, exhibits large buoyancy variation. Model 4 experiences large hydrodynamic force due to the pontoon contribution. Model 2 shows a smaller summation of the two main components of vertical force.

The surge motion for models 1, 2, 3, and 4 is shown in Fig. 6. Model 3 presents the smallest surge motion. The large difference between model 1 and 3, 4 is attributed to larger horizontal force associated with model 1. A larger volume and diameter close to the wave zone is the reason for the larger horizontal force on model 1.

The heave and pitch response for the same models is presented in Fig. 7. Heave response for model 4 shows a phase difference compared to models 1, 2, and 3. This is attributed to the drag force on the horizontal pontoon of model 4 which results in damping in the vertical motion. For heave motion, model 3 is optimum as in the case of surge motion. This model picks up larger vertical force than model 2 (see Fig. 5). However, it gives less heave motion than model 2. The reason behind this phenomenal inconsistency is that heave motion is controlled by surge motion due to the large axial stiffness of the tethers (pendulum type effect). Therefore, for large axial stiffness of tethers, optimization in heave motion is matched with optimization in surge motion. The pitch rotation is very small. Models 1 and 2 give the smaller pitch rotation. This is consistent with the pitch moments (not presented here).

The two tether bundle dynamic axial forces are plotted in Fig. 8. Model 2 is the optimum configuration. This is consistent with the optimization of vertical force of Fig. 5.

In Fig. 9 the effect of C_M and C_D values is investigated for models 1, 2, 3, and 4. The DnV values are presented in Fig. 5. For all models, constant values ($C_M = 2$, $C_D = 1$) give a larger mean value in predicting surge. This larger steady drift is due to the larger drag force. For each model, the double amplitude of surge remains practically the same. The trends are different for the case of tether bundle dynamic axial forces. For both tether bundles of models 2 and 3 and for tether bundle 1 of model 4, the DnV values give slightly more conservative force estimates.

The effect of water depth is shown for model 6 in Fig. 10. For surge and heave the response double amplitude remains independent of the water depth. This is obvious for surge, since the surge natural period is far away from the wave period for the range of water depth considered. For heave response, only the portion of heave attributed to the axial vibration of the tethers should vary with water depth. However, this portion is masked by the portion of heave that results from coupling with surge (pendulum type effect). Therefore, only a slight increase of the double amplitude of heave response occurs. The pitch double amplitude increases with water depth. This occurs because the natural period of pitch increases with water depth and comes closer to the wave period. The dyanmic variation of tether axial forces increases with water depth. The effect of pitch on the tether force double amplitude can be noticed. In the same figure the total pretension on the platform is indicated. The mean value of surge increases with water depth, as does the coupled mean value of heave. This is expected because the mean values of horizontal and vertical forces, and, therefore, their ratio remains practically independent of the water depth. Then, the static equilibrium requires the slope of the tethers, with respect to the vertical, to be practically independent of water depth. Therefore, larger mean surge and mean heave occur with increasing water depth and tether length. The negative mean value of pitch rotation is related to the bias of the pitch moment, which again is related to the bias of the horizontal force.

The effect of total still water pretension on TLP is analyzed in Fig. 11 by using model 5 and a value for EA/L_O = 10,000 kips/ft. per tether bundle (L_O = tether length, A = cross section area of tether bundle). Different values of pretension (static buoyancy minus weight) are applied by varying the water depth and, therefore, the draft. With increasing pretension the mean value of the surge motion decreases because the system becomes stiffer in the horizontal direction. At the mean time a slight increase of the surge double amplitude is observed as a larger horizontal force is exerted on the floating body due to the imposed larger draft. The natural period of surge is away from the wave period so dynamic amplification effects are not significant. The heave motion is strongly coupled with the surge motion due to the high axial stiffness of tethers (pendulum-type effect). As a result, the mean value of the heave motion decreases with increasing pretension as does the double amplitude of the dynamic heave motion, because with increasing pretension the mean surge is smaller and the body oscillates around a mean position closer to the vertical. The dynamic axial force of the tether bundles is larger with decreasing pretension because the system exhibits larger vertical motion; therefore, it picks up a larger variation in the vertical force. Furthermore, some trends of the variation of bundle total tension with EA/L_O for three values of total still water pretension are demonstrated in Fig. 12. It can be seen that the double amplitude of the dynamic force (MAX., MIN. curves) is practically independent of EA/L_O except for smaller values of pretension. In the latter case the "MIN" curve tends towards zero bundle tension (for bundle 1 this is more obvious). However, the MAX curves are limited to the left hand side by the axial capacity of the bundle, therefore, for a given total still water pretension, failure occurs before the bundle becomes slack. From the same figure it can be concluded that the value of total still water pretension has more significant effect than the value of EA/L_O of the bundle in reducing the variation of tether tensions.

Typical sample of some response quantities of model 6 are given in Fig. 13. A wave of 98.4 ft. height and 17 sec period in water depth of 485 ft. is propagating along the diagonal direction of the TLP.

CONCLUSIONS

The limited number of parametric studies in this paper highlight some very important characteristics of the dynamic response of the tension leg platform subjected to extreme sea states. These characteristics can be summarized as follows:

o The volume distribution of the floating body plays an important role in optimizing the motions and tether axial forces of the TLP. Motion minimization does not necessarily follow the same trends with the minimization of the dyanmic tether axial forces.

o The dynamic tether axial forces are controlled primarily by the vertical time-varying force on the floating body.

o Reduction of tether tension variations can be achieved by proper distribution of the volume of the floating body with draft or by increasing the still water pretension. The second alternative leads to the requirement of higher buoyancy and, therefore, more steel.

o The choice of C_M and C_D coefficients can affect the response. Larger values for these coefficients do not mean more conservative predictions in the response of the TLP.

o For the structural characteristics of the floating body, the tether axial stiffness, and the extreme wave conditions used in this study, a strong coupling between heave and surge motions is observed; heave is controlled by the surge motion (pendulum-type effect). Axial vibration of tethers is masked by this pendulum-type effect. Tension variation in the tethers increases with increasing water depth, primarily due to the pitch contribution. For the range of EA/L_O of the tethers studied here, it is demonstrated that the dynamic axial force of the tethers remains practically invariant and the smallest value of EA/L_O for a given static pretension is dictated by the axial capacity requirement of the tethers and not by the requirement to avoid slacking of the tethers. Still water pretension is more efficient than larger values of EA/L_O of the tethers in reducing variations of tether tensions.

For all the cases analyzed in this paper an extreme wave environment is considered. This means that wave period is away from all the natural periods of the tethered floating body. For small wave periods, which are closer to the heave, roll and pitch periods, the characteristics of the response are anticipated to be different. Furthermore, for small wave periods the pitch moment will probably have a more important role in the tether tension variations. Therefore, additional similar studies are suggested by using small wave periods and heights which are of significance in fatigue considerations.

ACKNOWLEDGEMENTS

The authors are very much appreciative to McDermott Marine Construction, McDermott Inc. for permission to publish this paper. Comments by Mr. Charles Young have been very valuable. Special thanks are extended to Dr. Carl Bitzer and Mr. Henry Mire for their excellent job in coding the floating body time domain analysis model. Stimulating discussions with Mr. Edward Gaines, Mr. Robert Figgers and Dr. Carlos Llorente of McDermott are very much appreciated.

REFERENCES

1. Albrecht, H. G., Koenig, D., and Kokkinowrachos, K., 1978, "Non-linear Dynamic Analysis of Tension-Leg Platforms for Medium and Greater Depths", *Proceedings, Offshore Technology Conference*, Houston, 1978, Volume I, OTC 3044, pp. 7-15.

2. Angelides, D.C., and Connor, J.J., 1980, "Response of Fixed Offshore Structures in Random Sea" *Earthquake Engineering and Structural Dynamics*, Volume 8, number 6, pp. 503-526.

3. Angelides, D.C., Will, S.A., and Figgers, R.F., 1981, "Design and Analysis Framework of Tension Leg Platforms", *Offshore Structures Engineering, Volume 4, Proceedings, International Symposium on Offshore Engineering*, Rio de Janeiro, Brazil, 1981. Houston, Texas: Gulf Publishing Company.

4. Berman, M.Y., and Blenkarn, K.A., 1978, "The Vertically Moored Platform for Deepwater Drilling and Production", *Proceedings Offshore Technology Conference*, Houston, 1978, Volume I, OTC 3049, pp. 55-64.

5. Beynet, P.A., Berman, M.Y., and Aschwege, von J.T., 1978, "Motion, Fatigue, and the Reliability Characteristics of a Vertically Moored Platform", *Proceedings, Offshore Technology Conference*, Houston, 1978, Volume IV, OTC 3304, pp. 2203-2212.

6. Blevins, R.D., 1977, *Flow-induced Vibration*. New York, N.Y.: Van Nostrand Reinhold Co.

7. Denise, J-P.F., and Heaf, N.J., 1979, "A Comparison between Linear and Non-linear Response of a Proposed Tension Leg Production Platform", *Proceedings, Offshore Technology Conference*, Houston, 1979, Volume III, OTC 355, pp. 1743-1754.

8. Det Norske Veritas, 1981, *Rules for Classification of Mobil Offshore Units*.

9. Horton, E.E., McCammon, L.B., Murtha, J.P., and Paulling, J.R., 1972, "Optimization of Stable Platform Characteristics", *Proceedings, Offshore Technology Conference*, Houston, 1972, Volume I, OTC 1553, pp. 417-428.

10. Jefferys, E.R., and Patel, M.H., 1981, "Dynamic Analysis Models of the Tension Leg Platform", *Proceedings, Offshore Technology Conference*, Houston, 1981, Volume III, OTC 4075, pp. 99-107.

11. Mercier, J.A., Goldsmith, R.G., and Curtis, L.B., 1982, "The Hutton TLP: A preliminary Design", *Journal of Petroleum Technology*, January, pp. 208-216.

12. Moe, G., and Verley, R.L.P., 1980, "Hydrodynamic Damping of Offshore Structures in Waves and Currents", *Proceedings, Offshore Technology Conference*, Houston, 1980, Volume III, OTC 3798, pp. 37-44.

13. Natvig, B.J., and Pendered, J.W., 1979, "Motion Response of Floating Structures to Regular Waves", *Offshore Structures Engineering, Volume 2, Proceedings, International Symposium on Offshore Engineering*, Rio de Janeiro, Brazil, 1981. Houston, Texas: Gulf Publishing Company.

14. *Ocean Oil Weekly Report*, 1981, Volume 15, Number 16.

15. Osborne, A.R., and Burch, T.L. - 1980, "Internal Solitons in the Andoman Sea, "*Science*, Volume 208, May, pp. 451-460.

16. Osborne, A.R., Burch, T.L. and Scarlet, R.I., 1978, "The Influence of Internal Waves in Deep Water Drilling", *Journal of Petroleum Technology*, October, pp. 1497-1504.

17. Paulling, J.R., 1981, "The Sensitivity of Predicted Loads and Responses of Floating Platforms to Computational Methods", *Proceedings, Second International Symposium on Integrity of Offshore Structures*, Glasgow, Scotland, 1981, Paper 4, pp. 51-69. Englewood, N.J.: Applied Science Publishers Inc.

18. Paulling, J.R., 1977, "Time Domain Simulation of Semisubmersible Platform Motion with Application to the Tension-Leg Platform, *Proceedings, Spring Meeting/STAR Symposium*, San Francisco, 1977, pp. 303-314. New York, N.Y.: SNAME

19. Perrett, G.R., Webb, R.M. 1980, "Tethered Buoyant Platform Production System, "*Proceedings, Offshore Technology Conference*, Houston, 1980, Volume IV, OTC 3881, pp. 261-273.

20. Pijfers, J.G.L., and Brink, A.W., 1977, "Calculated Drift Forces of two Semisubmersible Platform Types in Regular and Irregular Waves", *Proceedings, Offshore Technology Conference*, Houston, 1977, Volume IV, OTC 2877, pp. 155 164.

21. Rainey, R.C.T., 1977 "The Dynamics of Tethered Platforms", *Proceedings, Meeting of the Royal Institution of Naval Architects*, London, England, 1977, pp. 59-80. London, England: The Royal Institution of Naval Architects.

22. Salvensen, N., Kerczek, von C.H., Yue, D.K., and Stern, F. 1982, "Computations of Nonlinear Surge Motions of Tension Leg Platforms", *Proceedings, Offshore Technology Conference*, Houston, 1982, OTC 4393.

23. Sarpkaya, T. and Isaacson, M. 1981, *Mechanics of Wave Forces on Offshore Structures*. New York, N.Y.: Van Nostrand Reinhold Co.

24. Unknown Author, 1980, "Gulf TLP to feature in-leg wells", *Offshore*, April, pp. 136-138.

25. Yoneya, T., and Yoshida, K., 1981, "The Dynamics of Tension Leg Platforms in Waves", Proceedings, *Energy-Sources Technology Conference and Exhibition*, Houston, 1981, New York, N.Y.: American Society of Mechanical Engineers.

TYPICAL TLP DESIGNS

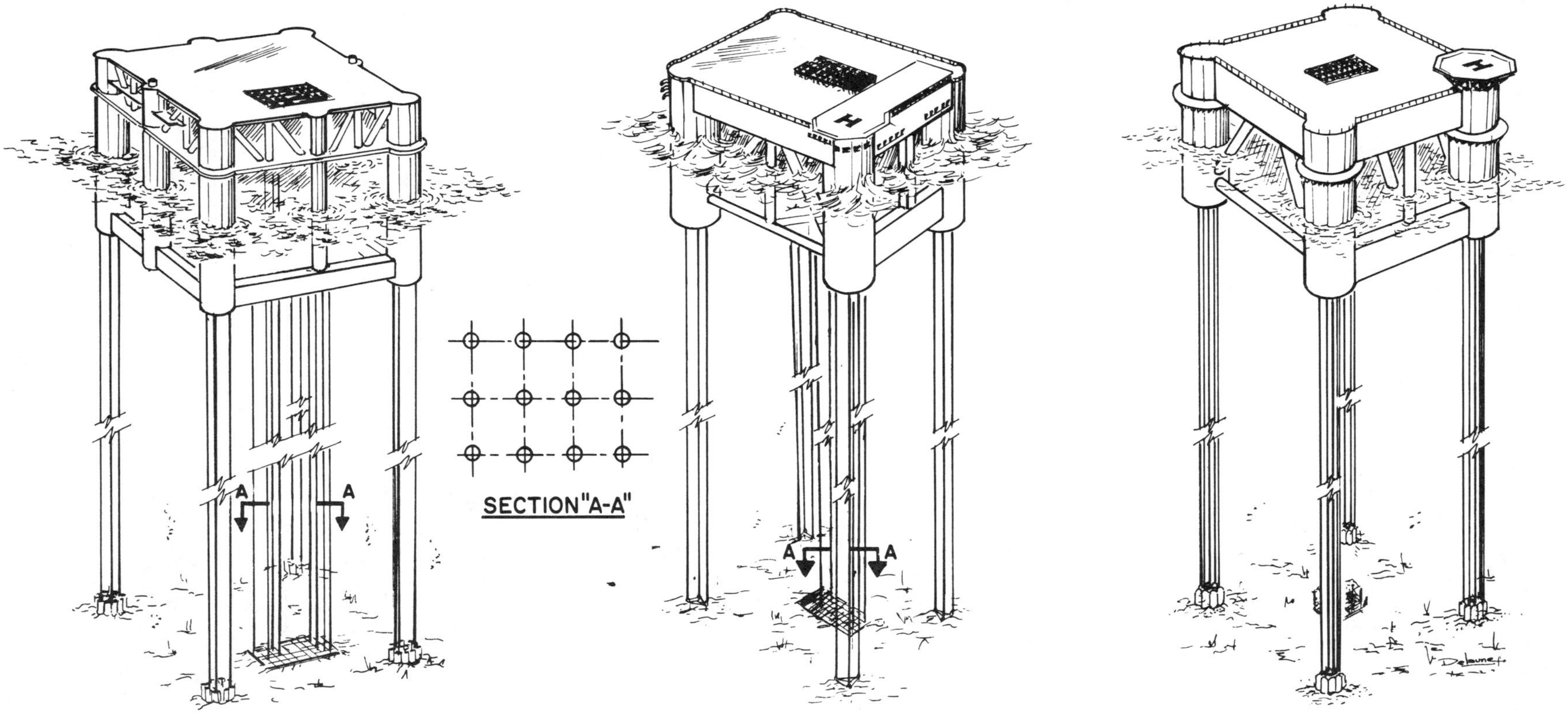

(Ref. 11) HUTTON TLP

(Ref. 13) TPP

(Ref. 19) TBP

SECTION "A-A"

FIGURE 1: TYPICAL TLP DESIGNS

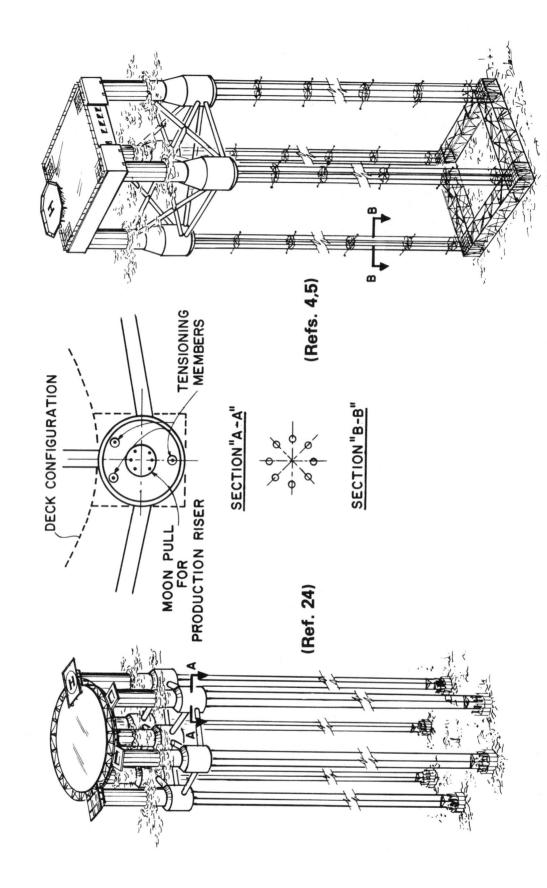

FIGURE 2: TYPICAL TLP DESIGNS

MODEL	DIMENSION (FT)								MEMBER SIZE (O.D. IN FT / NUMBER)				10^{-3} * MASS (SLUGS)	TETHER LENGTH (FT)	10^{-3} * EA (KIPS)	C_M C_D
	A	B	C	D	E	F	G	H	1	2	3	4				
1	300	300	193	131.98	-	193	-	-	-	46/6	-	46/2	2757.5	1368.02	6976	CONST OR DNV
2	300	300	193	131.98	60	133	-	-	60/8	30/6	-	30/2	2757.5	1368.02	6976	CONST OR DNV
3	300	300	193	131.98	71	122	-	-	60/8	21/6	-	21/2	2757.5	1368.02	6976	CONST OR DNV
4	300	300	193	131.98	-	153	40	21	-	21/6	40/4	21/2	2757.5	1368.02	6976	CONST OR DNV
5	260	260	180	100.85	-	180	-	-	-	53/6	-	53/2	2757.5	1399.15	6976	CONST
6	242.8	255.92	176.35	99.77	-	141.7	34.65	-	-	47.9/6	34.65/4	26.25/2	2765.9	-	6976	CONST

E: YOUNG'S MODULUS
A: TOTAL CROSS SECTION AREA OF EACH TETHER BUNDLE

TABLE I: MODELS FOR PARAMETRIC STUDIES

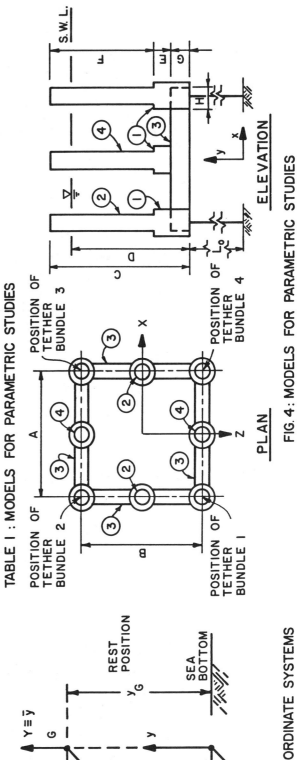

FIG. 3: COORDINATE SYSTEMS

FIG. 4: MODELS FOR PARAMETRIC STUDIES

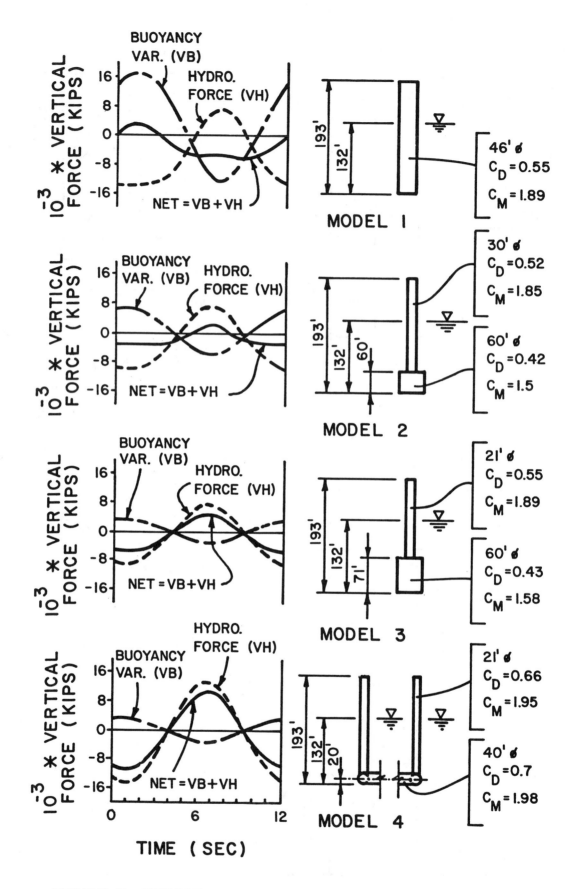

FIGURE 5: EFFECT OF LEG SHAPE ON VERTICAL FORCE

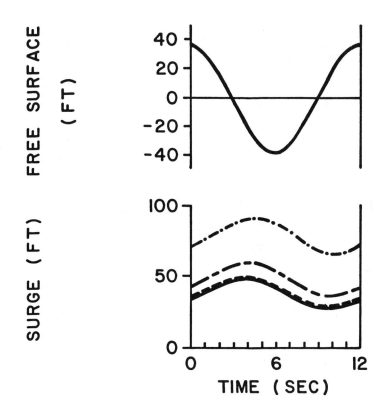

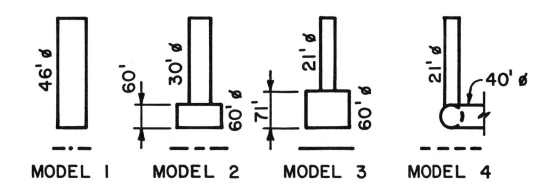

FIGURE 6: EFFECT OF LEG SHAPE ON SURGE MOTION

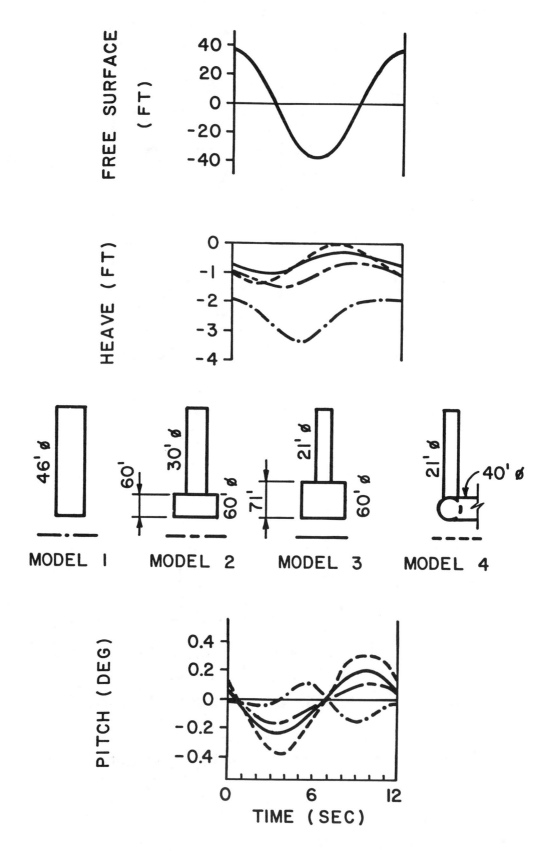

FIGURE 7: EFFECT OF LEG SHAPE ON HEAVE AND PITCH MOTIONS

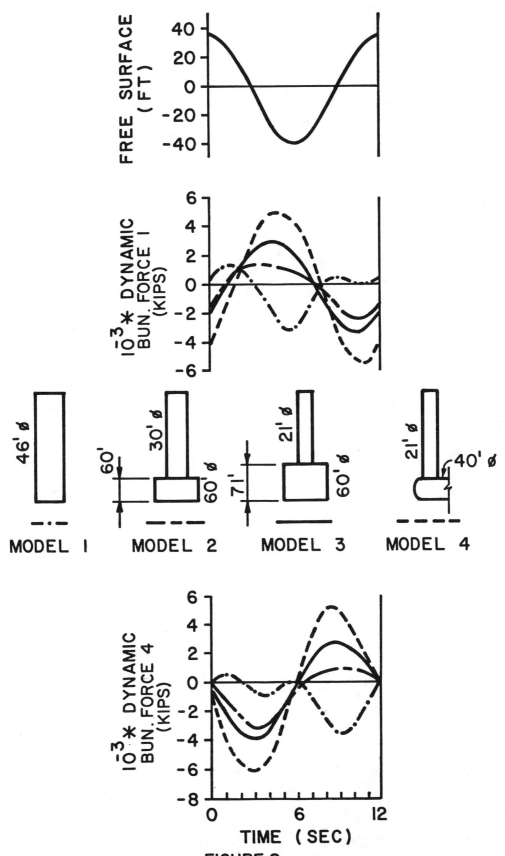

FIGURE 8:
EFFECT OF LEG SHAPE ON DYNAMIC TENSION OF BUNDLES 1 & 4

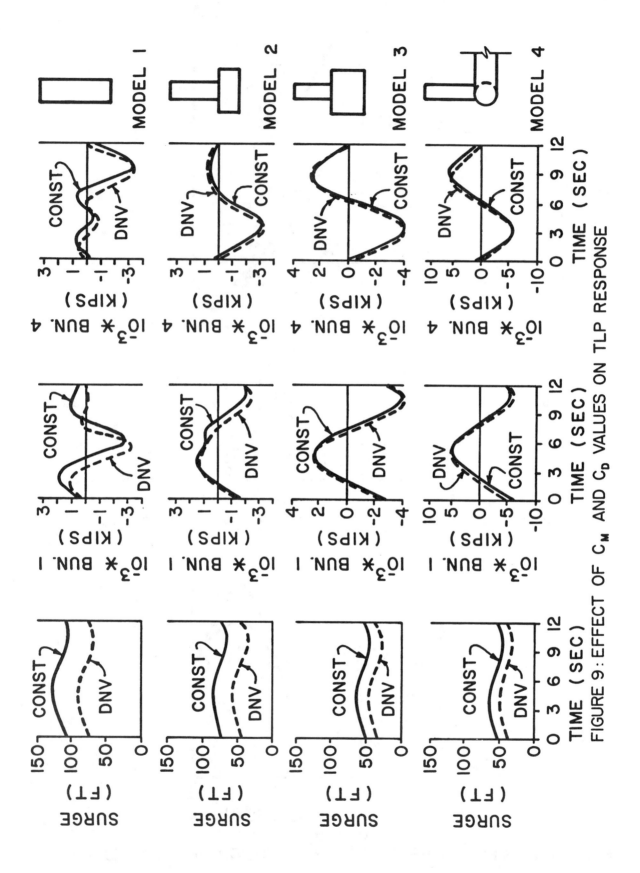

FIGURE 9: EFFECT OF C_M AND C_D VALUES ON TLP RESPONSE

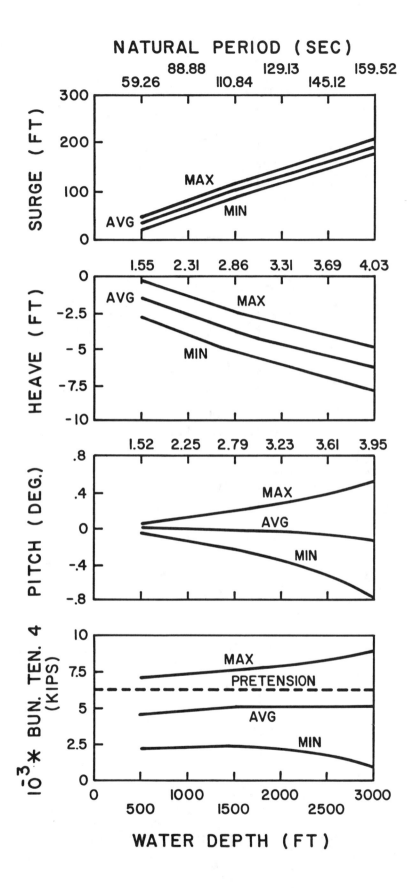

FIGURE 10: EFFECT OF WATER DEPTH ON TLP RESPONSE

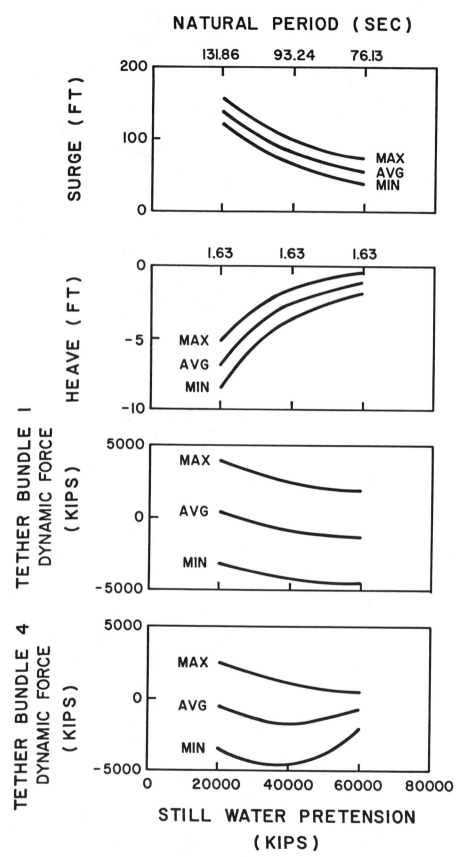

FIGURE II: EFFECT OF TOTAL STILL WATER PRETENSION ON TLP RESPONSE

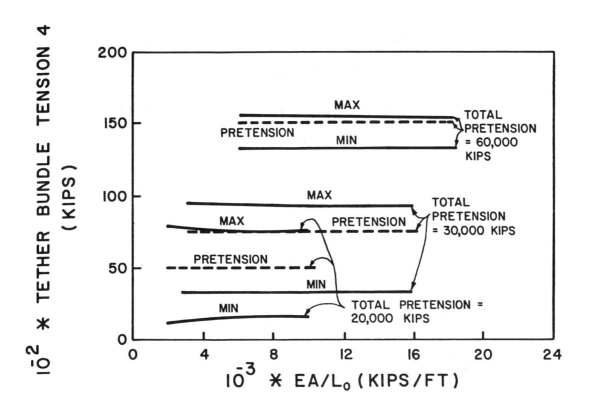

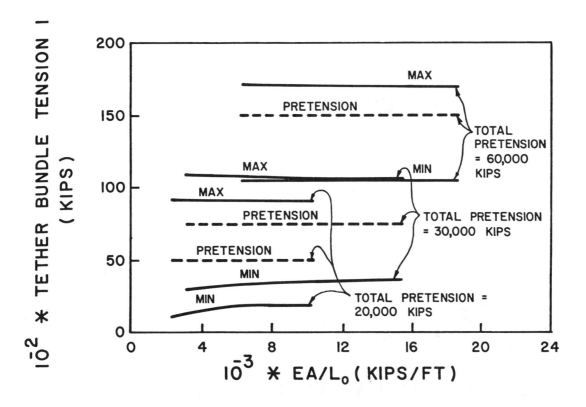

FIGURE 12:
VARIATIONS OF TOTAL TETHER BUNDLE TENSIONS WITH EA/L_0

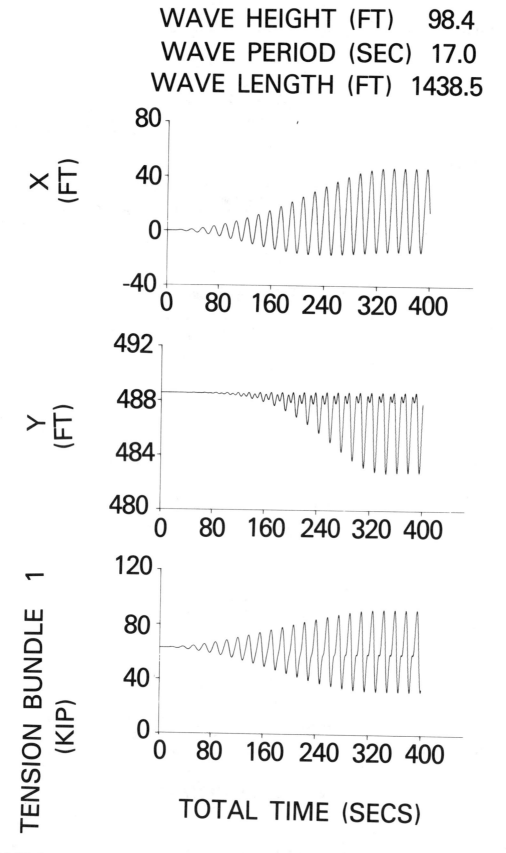

FIGURE 13: TYPICAL RESPONSE VARIATION WITH TIME

Technical Lecture Sessions
STRUCTURES

Design Issues of Alternate Systems

DESIGN CONCEPTS AND STRATEGIES FOR GUYED TOWER PLATFORMS

Damodaran Nair
Brown & Root, Inc., U. S. A.

Pierre S. Duval
Brown & Root, Inc., U. S. A.

SUMMARY

The fundamental concepts related to the compliant behavior of guyed tower platforms are reviewed, and design strategies for the platform components including the tower, foundation and mooring systems are discussed in this paper.

In the first part of the paper the structural action of a pile supported guyed tower is discussed, and the key factors affecting the behavior of a guyed tower are identified. Contrast is made to the fixed platform to illustrate the basic ideas. Secondly, the dynamic properties and behavior of the guyed tower are examined. Vibrational modes which govern the platform design are identified, namely sway, flexural, torsional and vertical modes. The platform parameters which govern the dynamic properties are identified and their importance in design emphasized. Simple methods and concepts which can be easily used to compute the dynamic properties in preliminary design stages are presented.

The design strategies for a guyed tower are discussed next with emphasis on loadings, critical loading conditions and design considerations for the tower, foundation and mooring system. Mooring systems can be designed with or without clumpweights. The mooring system with clumpweights will exhibit a nonlinear softening load-deflection relationship while the system without clumpweights will show a hardening behavior. Design guidelines for each of the systems are discussed, and the influence of key parameters such as support geometry, pretension, and the intensity and magnitude of the clumpwieght is examined.

INTRODUCTION

Economical development of oil and gas fields in the deeper waters of the ocean is a challenge facing the offshore engineering profession. A number of new deepwater concepts are actively being developed (guyed tower, tension leg platform, buoyant tower and subsea production system). The guyed tower platform is described in this paper (Figure 1).

The guyed tower platform has been under development for a number of years. Earlier work on this concept is described in References 1-3. These papers deal with the use of a spudcan as the foundation system. Recently a pile foundation has been proposed and is described in Reference 4.

The pile founded guyed tower can be thought of as a derivative of the conventional template type jacket. The guyed tower platform consists of four components: deck, tower, foundation and mooring system. The deck supports the drilling and production facilities. The tower supports the deck, protects the conductors and risers and serves as a template for driving the piles. However, there are two significant differences between a fixed platform and the guyed tower. These are: (1) the guyed tower is dynamically compliant in the lateral direction, and (2) the guyed tower is supported laterally by a mooring system.

DYNAMIC COMPLIANCY

It is accepted practice to design shallow water platforms using static methods of analysis. The static approach is adequate since the fundamental natural period of such platforms is much smaller than the predominant periods of the waves at which significant energies are contained. However, as the ratio of the natural period of the platform to the period associated with the significant energy in the design sea state increases, inertial forces become important. These concepts are further illustrated with reference to specific examples.

The action of lateral wave forces due to the design storm on a shallow water fixed platform is known to be static. The distribution of wave forces and associated inertial loads on a deepwater fixed platform is shown in Figure 2(a). Finally both the wave forces and inertial loads on a guyed tower are shown in Figure 2(b). The following important observations can be made. The inertial forces acting on the deepwater platform are quite significant. Furthermore, the inertial forces act in the same direction as the wave forces and the total force for which the platform must be designed is increased. The guyed tower exhibits an interesting phenomenon in which the inertial forces act in a sense opposite to the wave forces and hence decrease the magnitude of the lateral forces for which the platform must be designed. The first example is typical of most structures but the behavior of the guyed tower is quite different from that usually encountered in engineering practice. While in most instances dynamic response leads to amplified design forces, in the case of guyed tower dynamic action is utilized to reduce the design forces. The behavior of the guyed tower system is commonly referred to as compliancy. An alternate designation would be dynamic deamplification.

The above concepts are graphically illustrated in Figure 3. The ordinate in this figure represents the ratio of the dynamic lateral wave force to the force computed assuming the platform is rigid. The abscissa represents the ratio of the fundamental natural period of the platform to the period of the exciting forces assuming the latter to be periodic. For purposes of discussion this amplification diagram is divided into four regions. In region I, the amplification of the wave forces is negligible. Shallow water platforms fall into this category. The period of the platform must be less than about twenty percent of the design wave period so that the wave forces can be assumed to act in a static manner. Region II is characterized by dynamic amplification. Deepwater fixed platforms fall under this region. The upper limit of this region is governed by a number of factors including fatigue, practical design and construction considerations and above all platform cost. At the present state-of-the-art it is believed that the platform period can be as high as forty to fifty percent of the period of the design wave. Region III is characterized by high dynamic amplifications. Economic considerations discourage design and construction of such structures. Compliant structures such as a guyed tower, buoyant tower or tension leg platform belong to region IV. Note in particular that the design forces for structures in this region are only a fraction of the forces computed assuming static behavior of the structure.

MOORING SYSTEM BEHAVIOR

The mooring system is a distinguishing feature of the guyed tower and is the primary contributor of the lateral stiffness to the platform. Besides providing the required lateral stiffness, the mooring system with clumpweights also acts as a relief valve under extreme loads. The load-deflection relationship for a typical mooring system for lateral response is shown in Figure 4. It may be noted that the load-deflection behavior is essentially linear up to a certain level (Point A) of tower deflection which covers a wide range of operating environmental conditions. However, during the design storm the deflections of the tower are such that the tower responds in the nonlinear softening part of the load-deflection curve. The implications of the mooring system behavior on the tower response can be interpreted as follows. The decrease in the mooring system stiffness leads to an apparent increase in the instantaneous period of the platform and this places the tower response farther to the right in region IV of the response spectrum shown in Figure 3. This further reduces the dynamic amplification of environmental loads. Stated differently, unlike in a linear structure an increase in environmental loads does not produce a corresponding increase in the tower response, especially in the resultant cable forces which support the tower. Thus the mooring system can be thought of as a relief valve which provides protection against overloads. The static behavior of the mooring system will be examined in more detail in a later section.

Analysis of guyed towers reported to date model the mooring system as a massless spring having nonlinear force displacement behavior derived from a static analysis of the mooring system. Such an idealization which uncouples the dynamic behavior of the guyed tower and the mooring system is adequate for predicting overall response of the guyed tower and has been verified through the at sea model test reported in Reference 2. Analytical investigations currently in progress at Rice University also confirm this conclusion. However, the dynamic behavior of the individual guylines should be explicitly considered to predict the variation of tensions accurately (2).

PILE FOUNDED GUYED TOWER

The basic idea of the guyed tower is that it acts as a rigid tower pinned at the base in its fundamental sway mode. In the original development this compliancy was achieved by the so-called spudcan foundation (1, 3) which provided the necessary vertical and lateral support and also acted as a pivot permitting rotations of the tower. The major disadvantages of this foundation system are that the installation of the spudcan is difficult and that the long term vertical settlement of the foundation system is of concern. More recently, it has been demonstrated that the same structural behavior can be achieved with the use of a pile foundation. In the design of conventional deepwater fixed platforms the increased stiffness and foundation capacity are obtained by increasing the size of the jacket and by spacing the piles as far apart as possible. In the case of the guyed tower an opposite approach is used to decrease the stiffness and thereby increase the natural period. Specifically, the tower is designed to be slender, and the piles are closely spaced and ungrouted. This design strategy virtually uncouples the lateral stiffness of the tower and the bending stiffness of the piles. The tower can be thought of as being suspended from the piles since the pile terminations where the load transfer is effected are located above the mean water level. The axial flexibility of the piles in combination with their close spacing leads to the required pivotal action.

Figure 5(a) shows a two dimensional representation of the guyed tower. Figure 5(b) shows a physical model which simulates the action of the piles. The piles are replaced with springs having the same stiffness as the axial stiffness of the piles. Finally Figure 5(c) shows the analytical model in which the tower is represented as a rigid bar with masses lumped at various locations. In this model both the axial stiffness of the piles and the rotational stiffness provided by the foundation are represented by a single rotational spring. The rotation stiffness provided by the foundation is only a small fraction of that provided by the axial flexibility of the ungrouted piles. The above analytical model is designed to illustrate how the compliant behavior is achieved in a pile founded guyed tower.

DYNAMICS OF GUYED TOWER

The fundamental properties which provide insight into the dynamic behavior of a structure responding linearly are its natural periods and associated mode shapes. Strictly speaking, guyed towers are not linear systems; however, their behavior can be linearized over a wide range of response parameters. The principal sources of nonlinearity in a guyed tower are the nonlinear load-deflection relationship of the mooring system and the nonlinear hysteretic behavior of the soil-pile system. The behavior of the array of cables is linear for small to moderate values of tower deflection (Figure 4). Soil behavior is nonlinear over a wide range of response conditions. However, linearization of the foundation stiffness is routinely performed to predict the overall response of fixed platforms. Such linearization when performed at the level of response expected under the particular loading conditions is adequate for design applications.

The dynamic behavior of a guyed tower platform subjected to a wave excitation is governed by three types of modes of vibration. These are the sway, flexural and torsional modes. The sway mode is the fundamental mode in a particular lateral direction and is basically a rigid body mode with little or no bending. The natural period of this mode is governed by the height of the tower, the magnitude and distribution of the mass and above all by the lateral stiffness of the mooring system. Under wave excitations the tower movements are controlled by the sway mode. In typical guyed tower designs the period of the sway made is about twice the predominant period of the design sea state.

The second mode in the same lateral direction is a bending (flexural) mode. The period and the associated mode shape of the second mode are primarily governed by the magnitude and distribution of the mass and the stiffness of the tower as well as the lateral stiffness of the foundation. The stiffness of the mooring system has practically no effect on the flexural period of the guyed tower. A good approximation to the flexural period of the guyed tower can be obtained by idealizing it as a beam pinned at the base and free at the other end. However, the mode shape is strongly influenced by the lateral restraint provided by the foundation.

The properties of the flexural mode are important for the prediction of stresses and deformations in the tower. The tower should be proportioned such that its flexural period is much shorter than the predominant period of the design wave so that excessive stresses in the tower can be avoided. Another consideration is the fatigue behavior of the tower. Since fatigue is controlled by the smaller waves having shorter periods, a prudent approach is to design the tower to behave as a stiff structure in its flexural mode.

The third category of vibrational mode which governs the guyed tower design is torsion. The primary source of torsional stiffness is the foundation. The design should minimize the torsional period so that dynamic amplifications in torsional excitation can be avoided, especially under the frequently occuring smaller waves.

A fourth type of platform made which is quite important for earthquake type excitations is the vertical mode. The vertical period is much less than the flexural and torsional periods. The first three modes of a guyed tower are shown in Figure 6.

Approximate Computation of Sway Period

Even though computerized procedures and softwares are available to compute the periods and mode shapes of complex structures, simplified procedures are valuable especially in early design stages. The natural period of the sway mode can be easily computed using the Rayleigh method. The basis of this procedure is to assume a mode shape and then compute the maximum kinetic and potential energies under free vibration. By equating the two, the period of vibration can be computed. The mode shape can be assumed to be a straight line, with the base of the tower being pinned. Referring to Figure 5(c)

Let M_i = mass at the ith level
H_i = height of mass i from the base
K_c = cable stiffness
H_c = height of location of cable from the base

The period of the sway mode is computed from the relation

$$T = 2\pi \sqrt{\frac{\sum M_i H_i^2}{K_c H_c^2}}$$

The accuracy of the above expression is quite satisfactory for most preliminary applications. The natural period computed using the above relation will be smaller than the period computed by more accurate procedures, since the platform is constrained to vibrate in a particular shape in this approximation.

It is the usual practice in engineering computations to neglect the effect of the geometric stiffness when computing platform periods and mode shapes. Consideration of the geometric stiffness will result in a decrease in stiffness and hence an increase in period. This increase in period due to geometric stiffness is negligibly small for shallow water platforms. It is of some importance for the tall slender guyed towers. The sway period was found to increase by about 10 percent due to geometric stiffness for a guyed tower in a 1600 ft. water depth. The following simplified relation can be used to evaluate the effect of geometric stiffness on sway mode.

$$\frac{T}{T_o} = \frac{1}{\sqrt{1-r}}$$

in which T_o = Period computed without the effect of geometric stiffness
 T = Period considering geometric stiffness

$$r = \frac{W H}{K_c H_c^2}$$

W = Net vertical load on the platform

Method of Analysis

Methods of analysis for linear and nonlinear structures are well established. Linear methods can be used to predict the behavior of guyed tower for the linear ranges of mooring system behavior. Nonlinear methods are required to predict the response under design conditions. Direct integration of the equations of motion is discussed in Reference 1. A modified form of modal analysis can also be used to account for the mooring system nonlinearities (5, 6, 7).

Fundamental to the time domain approach is the use of realistic wave force histories. Irregular representation of the sea state which contains proper distribution of energy associated with the various wave components that make up the sea must be used in such dynamic analysis. Since the dynamic response of structures is sensitive to the specific features of the exciting wave forces, a number of wave force histories of representative characteristics should be used in design.

A technique that combines the design wave approach used in conventional static analysis and the irregular representation of the sea state for dynamic analysis is often used in practice. In this approach the inertial loads due to platform motions are computed from the dynamic analysis using an irregular representation of the sea state. These loads are next combined with static wave forces based on nonlinear wave theories such as Stokes V or Stream Function.

Since the guyed tower is dynamically sensitive to a number of environmental effects the interaction of various parameters should be investigated. The effect of wave-current interaction is discussed in Reference 1. Dynamic sensitivity of guyed tower platforms to wind excitations is discussed in Reference 8.

Behavior of Guyed Tower under Earthquake

Studies of long period structures idealized as single degree of freedom systems subjected to earthquake excitations conclude that such structures are subjected to very little lateral force. Long period structures experience deflections which are of the same magnitude as the free field displacement of the ground during an earthquake. However, behavior of real structures such as guyed towers is more complex. In particular, the higher bending modes which have much shorter periods fall in the amplified regions of the response spectrum and can produce substantial amounts of force in the platform elements.

Another important consideration is the behavior of the guyed tower in the vertical direction. The period of the vertical mode is typically one to two seconds, and substantial vertical forces on the platform may be generated. The large vertical forces can influence the design of the tower, mooring system and foundation system. The vertical stiffness of the mooring system must be considered in the computation of the vertical period.

DESIGN GUIDELINES

Design concepts for guyed tower platforms are reviewed in the following sections. Various types of loadings to be considered as well as critical design conditions are first discussed. Important aspects of tower design, foundation design and mooring system design are presented in subsequent sections.

Loadings and Critical Design Conditions

The design of the guyed tower platform follows procedures for a fixed platform (9) and should consider dead and live loads, environmental loads and construction loads. The dead and live loads will include the deck payload, deadweight of the platform and its appurtenances, buoyancy and hydrostatic forces. Environmental forces are due to wind, wave, current and earthquake induced ground motions, if applicable. Construction loads result from fabrication, load out, transportation, launch and upending of the tower. The vertical component of pretension in the cables should be considered in all inplace design conditions. Depending upon the particular design approach, all or some of the following loading conditions should be investigated: (1) Operating condition, (2) Design condition, (3) Damaged condition and (4) Extreme condition.

The operating condition usually includes some nominal environmental loadings. This condition may govern the design of a number of platform elements because of the lower allowable stresses. Furthermore, the smaller wave periods associated with the operating waves lead to higher amplifications in the flexural and torsional modes which in turn produce most of the stresses in the tower. The required axial capacity of the piles may also be controlled by the operating condition because of the higher factor of safety specified for this condition and due to the fact that the contribution of the design environmental loads to foundation forces is relatively small. The motions of the platform under operating conditions must be analyzed to ensure comfort of personnel.

Design condition refers to the behavior of the platform subjected to the 100 year storm. Dynamic analysis of the platform must be performed to determine member forces, deflections and foundation reactions. The overall design of the guyed tower is dictated by this condition. Satisfactory performance of the conductors is an important design consideration (1, 3).

Damaged condition analysis refers to the investigation of the safety of the guyed tower assuming failure of one or more critical elements of the system. Since the mooring system is vital to the integrity of the guyed tower, the performance of the guyed tower must be investigated assuming failure of one or more of the mooring lines. Deflections of the platform should not be excessive and the forces in the platform elements such as the cables, structural members and piles should be within allowable limits. The overall integrity of the platform with particular reference to its overturning tendency must also be examined.

Finally, performance of the guyed tower should be assessed for overload conditions. In the design of conventional platforms, the safety of the structure under extreme environmental conditions is ensured by the use of appropriate factors of safety in the strength and capacity of platform elements. Examples are the use of allowable stresses less than the failure stress of the material and the use of appropriate factors of safety in determining foundation capacity. Besides the use of factors of safety, the integrity of the guyed tower must be examined for specific overload conditions since the behavior of the guyed tower is nonlinear and explicit analyses are needed to identify the failure modes of a guyed tower. The overall failure could be due either to overstressing of the piles or to the P-Delta forces which are the overturning moments produced by gravity loads acting through the lateral deflections. Hence the preferred approach should be to limit the lateral deflections of the platform, limit the stresses in the piles and ensure that the mooring lines can accommodate the excursions of the tower without reaching their breaking strength. Stated differently, while the behavior of conventional fixed platforms is controlled by loads, the guyed tower behavior is governed by deflections.

Tower

The overall design of the tower should follow procedures used in the design of jackets (9). The cross sectional dimensions of the tower are governed by the following considerations: (1) the tower should be large enough to support the deck; (2) the tower should provide sufficient space for the conductors and the foundation system consisting of a number of piles; and (3) the tower should be sufficiently stiff so that its flexural period is less than about six seconds to avoid amplification of wave loads in that mode and hence minimize member forces.

The cross section of the guyed tower can be kept uniform since an increased base is not required. The bracing patterns and member sizes are selected using fixed platform design practice. The member sizes in the upper part of the tower are governed by gravity loads and the pretension in the mooring system. In the bottom part of the tower the member sizes are governed by hydrostatic loadings. In the middle portion of the tower, the flexural mode of the tower produces most of the stresses. Fabrication and installation considerations will govern the design of members locally.

Foundation

The foundation system for the guyed tower consists of a number of deep penetrating vertical piles. The number and size of piles is primarily dicated by the total foundation load to be carried, pile capacity and installation considerations. Group efficiency of piles must also be considered in selecting the foundation configuration. The axial loads on the piles are due to three components: (1) the net effect of structural dead weight plus the deck payload less the buoyancy; (2) the vertical component of the forces in the mooring system; and (3) the axial forces due to environmental loadings. The first two items produce about sixty to eighty percent of the axial load on the most heavily loaded pile. The vertical load due to pretension in the mooring system is quite significant.

The gravity loads and the contribution of the pretension produce constant axial loads on the piles. Environmental loads produce oscillatory axial forces. Since the oscillatory axial loads are less than the constant compressive loads, the piles in a guyed tower are always under compression. This is in contrast to fixed platforms in which the piles are also subjected to tension under design environmental conditions.

Since most of the environmental loads are resisted by the mooring system, the shear and moment at pile head are not significant. Hence an optimum design of the foundation system should reduce the axial load on the piles. A good design strategy is to reduce the static component of vertical load on the foundation by use of additional buoyancy tanks in the tower. Such buoyancy tanks must be located sufficiently below the mean water level so that the wave forces can be reduced.

The design of the pile foundation can be based on conventional design procedures, except that the dynamic behavior of the soil-pile system must be considered. In particular the cyclic degrading behavior of the soil under lateral loads should be considered. Besides the stress criteria, the fatigue of the pile should also be investigated.

Torsional rigidity is an important consideration in the design of the foundation system. Since the torsional rigidity of the guyed tower is provided by the foundation system, additional rigidity can be provided by using shear piles. Conductors will also contribute to the shear and torsion capacity and hence should be included in the analytical model.

MOORING SYSTEM

The selection of a mooring system is of primary importance in a guyed tower design since the lateral resistance to the environmental loadings is almost totally provided by the mooring lines. In designs of guyed towers to date, a starting point for the mooring system design has been to obtain an estimate of the required initial lateral stiffness of the mooring system which will guarantee satisfactory tower response under operational sea states. The magnitude of this stiffness, referred to as the initial stiffness, is governed by the following factors: (1) the sway period of the tower should be approximately twice the period of the design wave; (2) the stiffness should be such that the tower deflections are acceptable under static environmental loads, namely wind and current loads; and (3) the mooring system should be sufficiently redundant such that the platform behavior is acceptable even with a specified number of mooring lines out of service when subjected to the design storm.

In addition to the required initial lateral stiffness an estimate of the required strength level of the mooring system is useful in selecting a preliminary mooring configuration. The term, strength level, is used to refer to the total lateral resistance that the mooring system must provide under extreme loading conditions before any significant reduction in the mooring array stiffness occurs. Factors influencing the required strength level are: (1) the maximum wind load occurring during the extreme storm condition; (2) the nonzero mean of the combined current and wave loadings; and (3) the P-Delta effects associated with the weight of the deck and tower structure.

Mooring systems for guyed towers may consist of sixteen to twenty-four mooring lines. Two systems which utilize twenty-four mooring lines are shown in Figures 7 and 8. The system shown in Figure 7 is arranged in twenty-four evenly spaced radial directions. In Figure 8 the mooring lines are paired in twelve radial directions. The choice of the number of mooring lines in the system depends on many practical considerations which, in addition to the redundancy requirement mentioned previously, include items such as the limitations and availability of installation equipment, the cost of installation and materials, the material availability, and the long term operation considerations. The choice between single and paired lines is governed by such factors as the topography in the area of the platform, the time required for the installation of the mooring lines, the lifting capacities of the installation equipment, and the clear area requirements surrounding the platform.

Figure 9 shows the arrangement of a single mooring line. It consists of an anchoring system, anchor line, clumpweight, and guyline. As will be discussed later a distributed clumpweight offers many advantages. If a system of single lines is utilized, a single anchor pile per line may be used (1, 3). The anchor line termination is located below the mudline near the midpoint of the pile. Locating the termination below the mudline provides a very efficient means of lateral load transfer by minimizing the bending moments in the pile. In addition the pull-out resistance is significantly increased over an arrangement with the termination at the mudline. If the mooring lines are paired so that a single clumpweight is used for two lines then the installation tolerance required to ensure that both lines are evenly loaded requires that an anchor template (4) be positioned on the ocean floor. The anchor piles are then driven through this template. In this arrangement the anchor line termination is at the mudline which would create a greater demand on a single pile; however, multiple piles may be used to anchor each line since the anchor template will serve to distribute each line load to more than one pile. Because of the large mean tension level in the anchor lines which will be sustained throughout the design life of the structure the long term effects of lateral creep in the soil must be considered in the design of the anchor piles.

The elevation at which the guyline enters the tower is chosen based on installation and operational considerations and is kept sufficiently below the water surface to avoid interference with service vessels. In Reference 1 the optimum location is recommended to be the level of the centroid of the design wave loadings on the structure. The guyline as shown in Figure 9 actually consists of two sections. The first section extends from the clumpweight to an outboard connection located several hundred feet away from the tower. The second section sometimes referred to as the pendant section extends from the outboard connection to the entry elevation. From the entry elevation, the pendant is directed through the tower using fairleads or hawse pipes to an elevation where the tensioning and holding devices are located. Either chain or wire rope may be used for the pendant section. In both cases the abrasion, corrosion and bending fatigue are important design considerations for the pendant section.

Behavior of a Single Mooring Line

In order to design an efficient mooring system a thorough knowledge and understanding of the design parameters which influence the mooring behavior is required. Fundamental to the array behavior is the behavior of a single mooring line. Traditionally in catenary spread mooring systems the design approach has been to increase the unit weight and length of the mooring lines until an acceptable combination is found. As will be shown below the guyed tower type mooring is a more complicated system than the conventional catenary mooring, and additional design parameters other than the unit weight and the length of the mooring line are available for use in its design. The following parameters govern the load deflection behavior of a single mooring line.

1. The pretension applied to the line.
2. The clumpweight intensity or weight per unit length.
3. The total weight of the clumpweight.
4. The angle of inclination of the guyline or in other words the distance the clumpweight is placed away from the tower.
5. The length of the anchor line.

The length of the anchor line contributes to the mooring line behavior in two ways. First, under extreme loading conditions, excessive mooring line tensions are likely to occur unless the line behavior becomes soft. The length of the anchor line governs the range of the deflections over which the soft behavior of the mooring line will be seen after the clumpweight has lifted. Once all of the anchor line is lifted off the sea floor, the mooring line tensions will begin to increase rapidly. The possibility of overloading a mooring line is reduced by making the anchor line longer. However under moderate tensions when the clumpweight is still partially resting on the sea bottom, the anchor line is fully supported and behaves as a lateral spring support for the mooring line. It therefore reduces the overall stiffness of the mooring line. Since the spring value of the anchor line is reduced as its length increases, the length of the anchor line should be no longer than necessary. A minimum length of one water depth plus several hundred feet is required for installation purposes so that the anchor system may be lowered separate from the clumpweight. In the designs to date this minimum length has provided a sufficient soft behavior range.

The lateral stiffness of a single mooring line consisting of an anchor line, distributed clumpweight and guyline is shown in Figure 10 for various values of clumpweight intensity and angles of inclination measured from the sea floor. The cable chosen for the guyline and anchor line in this example is 5 inch diameter wire rope, and the point of attachment to the tower is 1500 feet from the sea bottom. The length of the anchor line is 1800 feet. The stiffness values shown in this figure assume that the guyline is very taut and thus are the maximum values which may be expected for a 5 inch cable system.

In Figure 10 it is seen that the stiffness of the mooring line increases as the intensity of the clumpweight increases and as the angle of inclination decreases. However, note that the length of the guyline increases rapidly as the angle of inclination decreases below the 25 to 30 degree range. Practical limits for the minimum angle of inclination may be obtained from the maximum available length of a particular cable and the relative cost of handling a higher intensity clumpweight versus purchasing additional lengths of wire rope. The intensity of the clumpweight is limited by the allowable soil bearing pressures and the maximum desired width of the clumpweight. A first guess for possible ranges of the clumpweight intensity and the angle of inclination which provide a given initial stiffness for each individual mooring line may be obtained using Figure 10.

A third parameter which governs the load-deflection behavior of the mooring line is the magnitude of the pretension applied to the mooring line. The lateral stiffness of the guyline segment alone is shown as a function of the guyline tension in Figure 11. Curves for a wide range of angles of inclination are given. In Figure 11 the stiffness of the guyline is normalized with respect to the taut line stiffness value which was assumed when computing the curves for anchor line - clumpweight - guyline stiffness given in Figure 10. The taut line lateral stiffness may be found from the simple expression

$$K_T = \frac{AE}{L} \cos^2 \theta_b = \frac{AE}{Z} \cos^2 \theta_b \sin \theta_b$$

where AE is the equivalent rope modulus, L is the approximate length of the guyline and θ_b is the angle of inclination with respect to the sea bottom.

In Figure 11 one may note that as the angle of inclination decreases larger percentages of the breaking strength of the rope are required to achieve the same ratio of actual guyline stiffness to the taut line stiffness. Furthermore, for the 5 inch wire rope and water depth considered in this example, to actually achieve the taut line stiffness value requires tensions around the breaking strength of the rope. Therefore the effect of the guyline tension must be included when attempting to predict the stiffness of a guyed tower mooring line particularly for the smaller values of the angle of inclination.

The initial stiffness of a single mooring line with an angle of inclination of 25 degrees is given in Figure 12. In this figure the single line stiffness is plotted as a function of the clumpweight intensity for various ratios of pretension to the rope breaking strength. For a given value of pretension, the stiffness of the anchor line - clumpweight - guyline system increases as the clumpweight intensity increases up to a limiting value determined primarily by the stiffness of the guyline section.

In the following figure, Figure 13, the value of the pretension has been set equal to 30 percent of the breaking strength of the guyline. The initial stiffness of the mooring line is plotted as a function of clumpweight intensity and angles of inclination. Similar curves could be drawn for other values of pretension with higher pretensions generally producing larger values of initial stiffness. One point to note in comparing possible combinations of pretension and angle of inclination is that larger pretensions and angles of inclination will produce higher amounts of vertical load on the platform. This aspect will be discussed further in a later section.

Behavior of the Mooring System

Thus far the discussion has primarily centered around the estimation of the initial stiffness of an individual mooring line. However in the design requirements the known desired stiffness quantity is the overall stiffness of the entire mooring system. In order to determine the requirement for a single mooring line, a qualitative understanding of the behavior of the mooring array in terms of the individual mooring lines is required. Without going into elaborate details consider the simple example of two opposing mooring lines shown in Figure 14.

The instantaneous lateral stiffness of the two line system at a particular value of deflection is equal to the sum of the individual stiffness of each line and is a function of the instantaneous tension of each line. For example, consider that both lines are tensioned very highly then the initial stiffness of the array would be twice the limiting stiffness of an individual line. As the tower deflects to the right increased tension in the left side line will not produce any larger stiffness, since it is already at the limit, and the decreasing tension on the right side will reduce the stiffness of that line eventually to a very low value. Therefore, the array exhibits a softening behavior with the limiting array stiffness for the deflected tower being equal to the limiting stiffness for one line which is one-half of its initial value.

Now consider that both lines have a very low value of pretension such that the initial stiffness of each line is zero for practical purposes. As the tower deflects the line on the left side will tighten, and its stiffness will increase eventually to its limit if the deflection is great enough. The line on the right side will remain slack and will not contribute to the array stiffness. Therefore the limit of the array stiffness for the deflected tower is again equal to that of a single line, and the array shows a hardening behavior. It stands to reason that for some moderate tension, between the very taut and the very slack cases, the combination of the two lines will produce a combined stiffness which varies only slightly about the limit for a single line as shown in Figure 14. It should be noted that the softening or hardening behavior discussed above is due to the mooring line pretension and is seen in the initial portion of the mooring array response curves.

If the angle between each mooring line and the direction of deflection is denoted by α_i and the instantaneous stiffness of each line in its radial direction is k_i then the instantaneous array stiffness is given by

$$K_I = \sum_{i=1}^{N} k_i \cos^2 \alpha_i$$

where N is the number of mooring lines.

Assuming that all the lines are pretensioned identically the stiffness of the array at zero tower offset for the 24 cable array shown in Figures 7 and 8 is found to be twelve times the stiffness of a single line. As an example consider that the array stiffness sought is 275 kip/ft then the required initial single mooring line stiffness is 23 kip/ft, assuming a quasi-linear response for the opposing line pairs. From the data given in Figures 12 and 13, one choice for a 5 inch mooring line configuration would be an angle of inclination of 25 degrees, a clumpweight intensity of 3 kip/ft and a pretension of 30 percent of the breaking strength of the cable.

As a first check of the behavior of the above mooring line as the tension changes recall the data for the guyline section given in Figure 11. For an angle of inclination of 25 degrees and pretension of 30 percent of the breaking strength (T_u), the initial stiffness of the guyline section is approximately 0.5 of the limiting value (k_{max}). If the tension increases to 50 percent of T_u the stiffness increases to approximately 0.85 k_{max} while if the tension decreases to around 20 percent T_u the stiffness decreases to about 0.2 k_{max}. Therefore as the tower deflects the opposing lines should interact to maintain the stiffness within a reasonable amount of the initial value.

As a further check of the choice of mooring line parameters before proceeding with an actual analysis, the data given in Figure 12 may be replotted as shown in Figure 15. In this figure the horizontal stiffness of the 5 inch mooring system has been plotted versus the maximum line tension for a single value of clumpweight intensity, the chosen 3 kip/ft value. As seen in this curve the starting tension of 30 percent T_u produces the desired single line stiffness of 23 kip/ft. At a maximum line tension of 50 percent of the breaking strength the stiffness has increased to 30 kip/ft. The tower offset which will produce a change in tension from 30 to 50 percent is estimated to be 20 ft. For a 20 ft. deflection the tension in the slacking line is estimated to be 19 percent T_u with a corresponding stiffness of 12 kip/ft. If only two opposing lines are considered the combined initial stiffness is 46 kip/ft which will result in an initial array stiffness of 275 kip/ft. As the tower deflects until the line with increasing tension reaches 50 percent T_u, the instantaneous stiffness of the two lines decreases to 42 kip/ft. Therefore the array will be slightly softening; however, the choice of mooring line still appears feasible.

Thus far we have defined the size of the cables (5 inch), the angle of inclination (25°), the length of the anchor line (1800 ft.), and the intensity of the clumpweight (3 kip/ft). The remaining quantity to be defined is the total weight or length of the clumpweight. Two factors influencing this choice are (1) the maximum allowable tension and (2) the required strength level of the mooring system.

The total weight of the clumpweight to limit the tension to a value of T_m may be estimated using the formula given below,

$$W = \sqrt{(T_m - q_G Z)^2 - (T_m - \frac{q_G Z}{2}) \cos^2 \theta_b}$$

where q_G is the submerged unit weight of the guyline and Z is the vertical projection of the guyline. In most cases Z may be taken to be equal to the distance from the tower fairlead to the ocean floor since the height of the upper end of the clumpweight above the sea floor is usually small compared to the depth.

For a maximum line tension of 1450 Kips, which is one-half the breaking strength of the 5 in. rope, the weight of the clumpweight would be on the order of 510 Kips resulting in a length of 170 ft. for the clumpweight. Lifting of the clumpweight will occur at a tower offset of 19 ft. for a 510 kip clumpweight. Due to the strength requirements of the tower for which this mooring was designed it was found that the size of the clumpweight could be reduced somewhat. The length of the clumpweight was reduced to 157 ft. with a total weight of 471 Kips. Total lift will then occur at a tower of offset of 16 ft. and a maximum line tension of 47 percent of the breaking strength.

The design of the mooring line thus far has been accomplished by using approximate relationships for the line stiffness and a few special solutions such as the solution for the tension and offset at the point of total lift of the clumpweight (10). It now remains to perform the mooring analysis which will determine the actual static mooring response and verify the design. For purposes of this discussion, only dead weight forces are considered to act along the cables. The variation of cable tension as a function of the tower movement is shown in Figure 16.

The horizontal force required to move the system for various values of horizontal displacement is shown plotted in Figure 17. The single line mooring exhibits a hardening behavior for deflections up to about 16 ft. and then the stiffness of the line begins to reduce due to the lifting of the clumpweight. The relative position of the clumpweight for various values of tower movement is shown in Figure 18. Initially, for no lateral deflection of the tower the front portion of the clumpweight is off the sea bottom. About 80 percent of the clumpweight is off the sea bottom when the tower moves 10 ft. laterally, and finally full lift of the clumpweight occurs for tower movements greater than 16 ft.

The discussion of the mooring system's behavior thus far has assumed that no forces are acting which would restrain the lifting of the clumpweight. These forces, called suction forces, will develop if the clumpweight becomes embedded in the sea floor due to settmelent or other factors and should be considered in the design of the clumpweight. If the clumpweight were a single concentrated weight, soil suction could result in a significant increase in the predicted maximum line tension since the vertical force needed to lift the clumpweight off the bottom would increase by an amount equal to the suction forces. In contrast very little difference is found in the predicted tensions for mooring lines using distributed clumpweights when suction is included since the suction forces act at discrete points along the distributed clumpweight. Due to the lifting action of the distributed clumpweight only one of the discrete suction forces must be overcome at any one time. The value of the discrete suction force depends on the clumpweight geometry along its length and the rate of pull; however, these forces are generally small when compared to the total weight of the clumpweight.

The force-deflection behavior of a mooring system consisting of 24 guylines is shown in Figure 19. The ordinate in this figure shows the horizontal force necessary to produce the corresponding lateral deflections shown on the abscissa. Note that the force-deformation behavior of the mooring system is essentially linear up to a deflection of about 16 ft. For larger deflections, the clumpweights begin to lift off from sea bottom resulting in a softening of the force-deflection behavior. There is a smooth transition between the stiff and soft regions of the curve; however, the general nature of the curve is such that a bilinear representation is adequate for preliminary design studies of the tower.

An undesirable aspect of the mooring system is the vertical component of the cable reactions. These forces are transferred to the tower at the point of attachment of the cables, and the foundation system must be designed to carry these loads. The vertical force versus the horizontal deflection of the tower is shown in Figure 20. It may be noted that the vertical component of cable reaction is a significant design parameter, but the change in vertical force due to tower movement is not significant. This means that the vertical cable reaction is primarily governed by the pretension with higher pretensions resulting in larger vertical forces. Thus the optimum mooring system is the one which minimizes the vertical component of cable reaction, maximizes initial stiffness and has the desired strength level providing an adequate factor of safety against failure under predicted values of tower deflections produced by the design storm.

Another important design consideration is the behavior of the mooring system under damaged conditions. Assuming that two adjacent mooring lines are damaged, the force-deflection behavior of the system is shown in Figures 19 and 20. Since the lines diametrically opposite to the damaged lines are intact, nonzero initial deflection is noted, and both the initial and final stiffnesses of the mooring system are reduced. The platform behavior should be investigated for the damaged condition of the mooring system to ensure acceptable platform response.

ALTERNATE MOORING SYSTEMS

There are primarily two types of mooring systems. The first is commonly referred to as a catenary mooring system. In a catenary mooring, the mooring line exits the structure or vessel at a fairlead, hangs in the shape of a catenary until at some distance out from the fairlead it contacts the sea floor and is <u>tangent</u> to the sea floor at this point. The line then runs along the sea floor to an anchor which fixes the end of the mooring line. This type system is characterized by the fact that increases in line tension caused by increased forces at the fairlead are resisted by picking up additional line from the sea floor, and thus its behavior is governed by their unit weight.

The distributed clumpweight mooring can be viewed as an optimized conventional catenary mooring. The stiffness of a conventional mooring is determined by the sag of the suspended section which is always at a maximum due to the lifting action. The most effective way to increase the stiffness is to reduce the sag of the suspended section by reducing the unit weight of that section. In the moorings considered thus far this was accomplished by using wire rope, which is very light when compared to the weight per unit length of the clumpweight, which was pretensioned to 30 percent of its breaking strength to reduce the sag to an acceptable amount.

The second class of mooring systems may be referred to as a tether system. In the tether type mooring the line is suspended between the tower and anchoring system as shown in Figure 21. The anchoring system supports the tether both horizontally and vertically. The tether type system has an obvious economic attractiveness in that it eliminates the material and installation costs associated with the clumpweights and anchor lines. The design of a tether system using wire rope is discussed below. Designs using synthetic ropes are also possible but will not be considered in this report.

One approach to begin the design of a tether system without clumpweights and anchor lines is to anchor the guyline section to the sea floor at the location of the clumpweight as shown in Figure 21. But such a system will result in excessive tensions in the cable under design level deflections of the tower.

The stiffness of a 5 in. wire rope with a pretension of 30 percent T_u is 38 kip/ft which is greater than the required 23 kip/ft for a 24 cable array. Therefore it is possible to increase the flexibility of the tether line and obtain the required stiffness. Three possible actions which will increase the tether flexibility are (1) lower the pretension which will introduce additional sag, (2) place the anchor position further from platform which will increase the cable length, and (3) select a different rope construction which has a lower rope modulus.

For the particular cable under consideration it was found that changing only the pretension was required. The force deflection behavior of a pair of opposing lines is shown in Figure 22 for various values of pretension. The force-deflection behavior for a mooring system consisting of 24 lines pretensioned to 20 percent of their breaking strength is shown in Figure 23. On the same figure is shown the force-deflection behavior of the mooring system with clumpweights. The behavior of both mooring systems are essentially the same for small values of tower deflections. However the behavior of the two systems is very different for the larger values of the deflections. The system with clumpweights and anchor lines exhibits a softening behavior. As the tower deflection increases the stiffness of the mooring system decreases. The implications of this softening behavior are that the apparent natural period of the tower increases and that the tension in the individual line does not increase with tower deflections. The clumpweight acts as a safety valve which limits the tension in the cable and the resultant forces on the tower. On the other hand, the mooring system without clumpweights exhibits a hardening behavior since as the tower deflection increases the stiffness also increases. The instantaneous value of the tower period shortens and both the tension in the cable, and the resultant tower forces increase at a faster rate than for small deflections.

REFERENCES

1. Finn, L. D., "A New Deepwater Offshore Platform - The Guyed Tower," OTC 2688, Offshore Technology Conference, May 1976.

2. Finn, L. D. and Young, K. E., "Field Test of a Guyed Tower," OTC 3131, Offshore Technology Conference, May 1978.

3. Finn, L. D., Wardell, J. B., and Loftin, T. D., "The Guyed Tower as a Platform for Integrated Drilling and Production Operations," Proceedings European Offshore Petroleum Conference and Exhibition, London, Oct. 1978.

4. Mangiavacchi, A., Abbott, P. A., Hanna, S. Y., and Suhendra, R., "Design Criteria of a Pile Founded Guyed Tower," OTC 3882, Offshore Technology Conference, May 1980.

5. Molvar, A. J. Vashi, K. M., and Gay, G. W., "Applications of Normal Mode Theory and Pseudo Force Methods to Solve Porblems with Nonlinearities", Jl. of Pressure Vessel Technology, Transactions of the ASME, May 1976.

6. Shah, V. N., Bohan, G. J., and Nahavandi, A. N., "Modal Superpositon Method for Computationally Economical Nonlinear Structure Analysis," American Society of Mechanical Engineers, Paper No. 78-PVP-70.

7. Hanna, S. Y., Mangiavacchi, A., and Suhendra, R., "Nonlinear Dynamic Analysis of Guyed Tower Platforms," ASME Winter Annual Meeting, Paper No. 81-WA/OCE-9, Washington, D.C., 1981.

8. Pike, P. J., and Vickery, B. J., "Wind Tunnel Test of a Guyed Tower Deck," OTC 4288, Offshore Technology Conference, May, 1982.

9. American Petroleum Institute, "Recommended Practice for Planning, Designing and Constructing Fixed Offshore Platforms," Twelfth Edition, January, 1981.

10. Duval, P.S., "Studies of Cable Systems", Thesis to be submitted in partial fulfillment of the requirements for the Ph.D degree, Rice University, Houston, Texas.

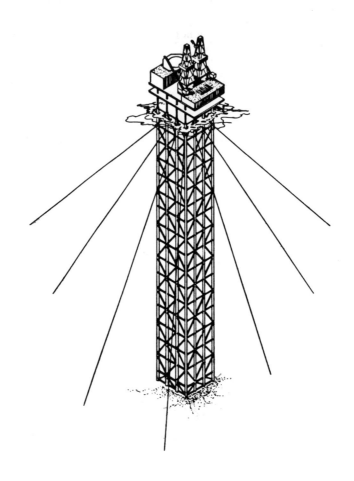

GUYED TOWER

FIGURE 1

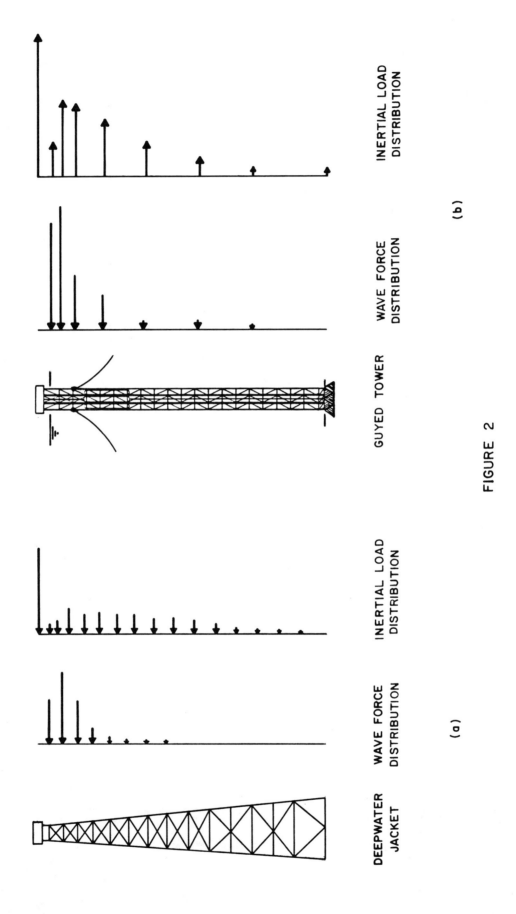

FIGURE 2

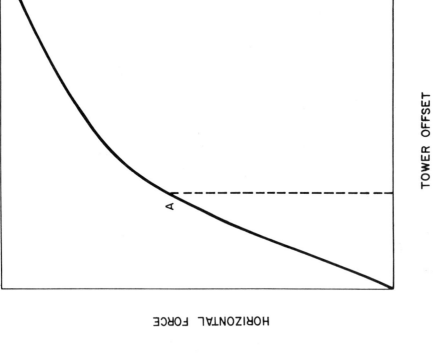

MOORING SYSTEM BEHAVIOR

FIGURE 4

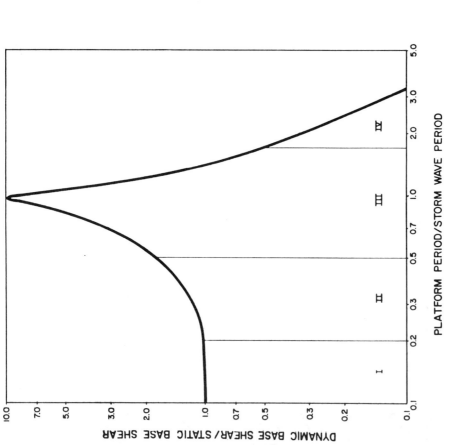

DYNAMIC BEHAVIOR OF PLATFORMS

FIGURE 3

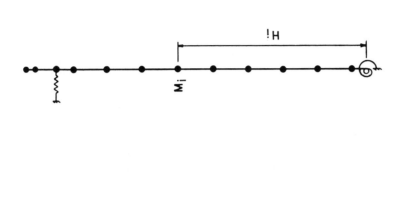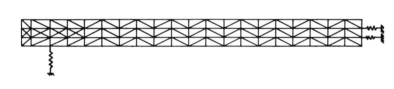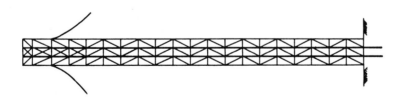

FIGURE 5

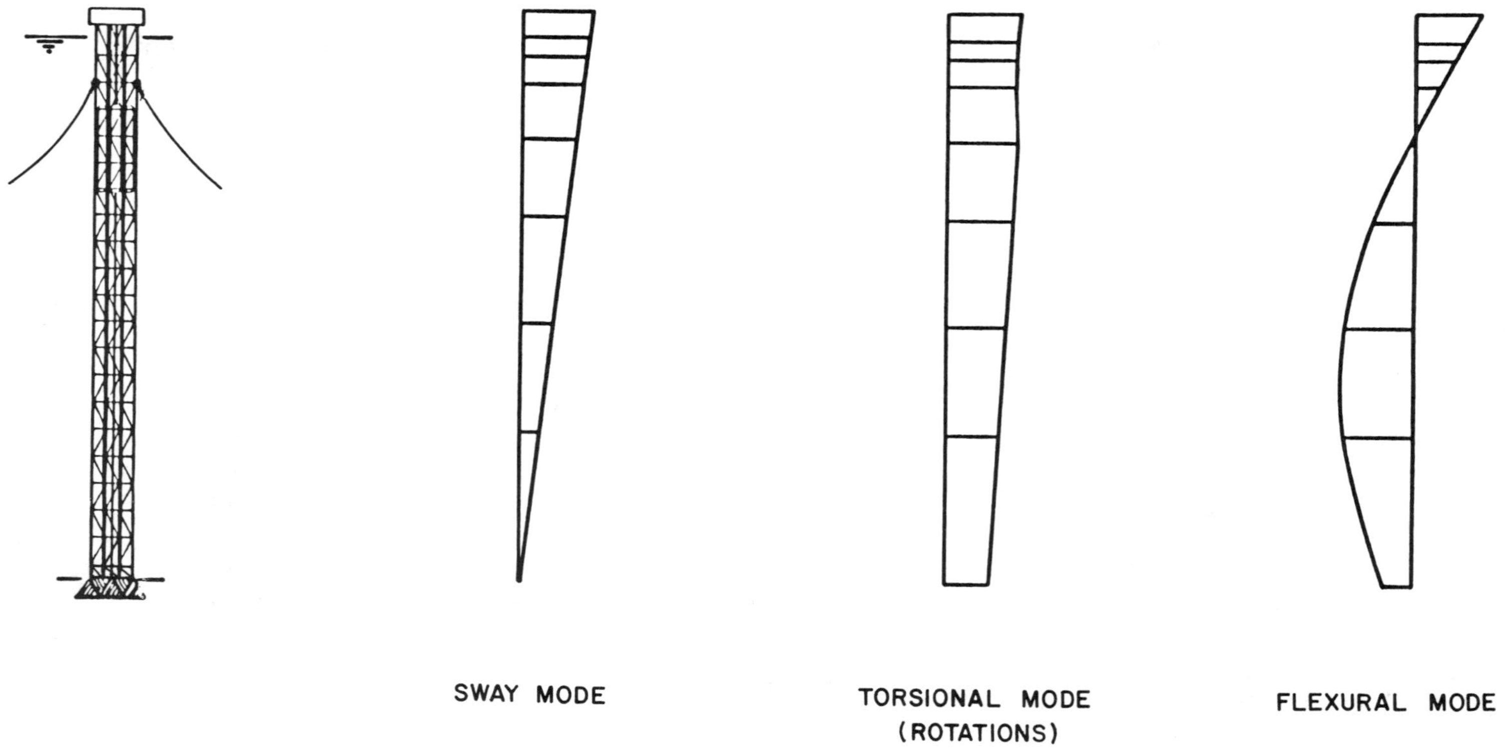

SWAY MODE

TORSIONAL MODE
(ROTATIONS)

FLEXURAL MODE

FIGURE 6

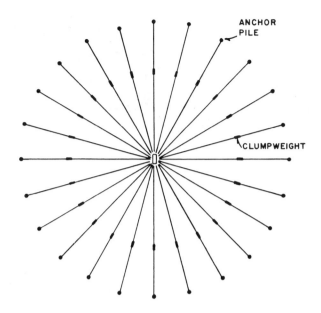

**TWENTY FOUR MOORING LINE ARRAY
SINGLE LINE ARRANGEMENT
FIGURE 7**

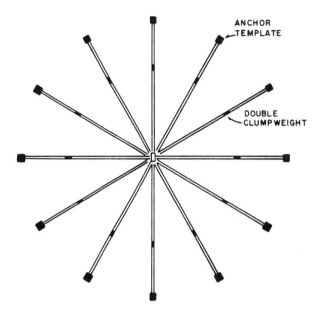

**TWENTY FOUR MOORING LINE ARRAY
PAIRED LINE ARRANGEMENT
FIGURE 8**

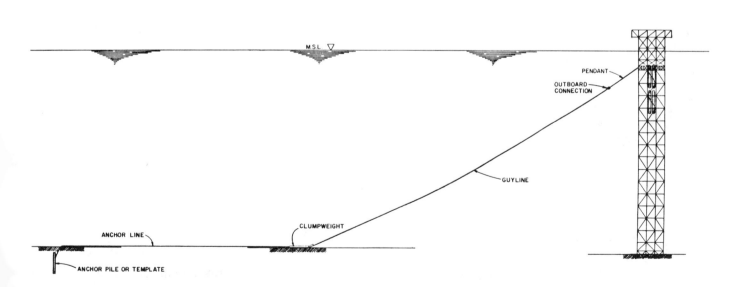

**ELEVATION VIEW OF A TYPICAL GUYED TOWER MOORING
FIGURE 9**

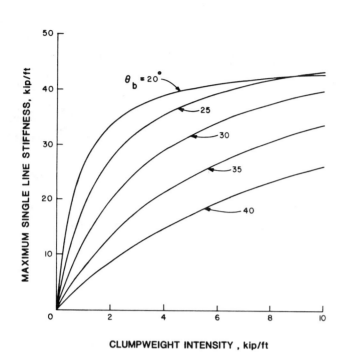

MAXIMUM STIFFNESS OF AN ANCHOR LINE-CLUMPWEIGHT-GUYLINE SYSTEM
FIGURE 10

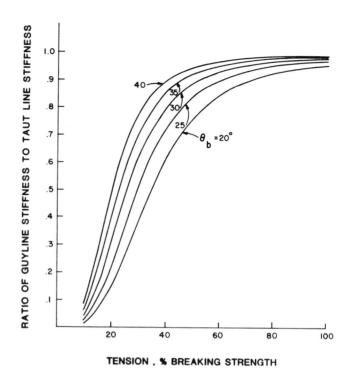

EFFECT OF TENSION ON GUYLINE STIFFNESS
FIGURE 11

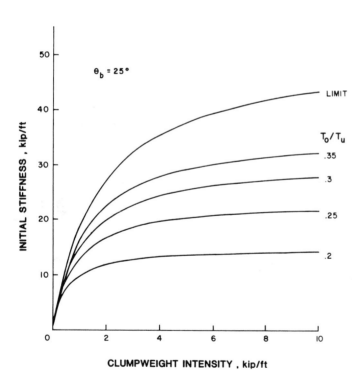

EFFECT OF PRETENSION ON THE INITIAL SINGLE LINE STIFFNESS
FIGURE 12

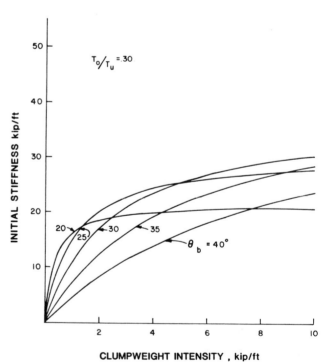

INITIAL SINGLE LINE STIFFNESS 30% PRETENSION
FIGURE 13

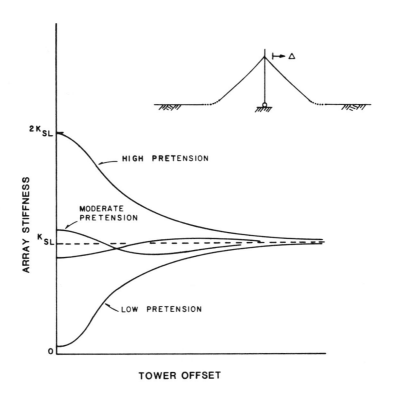

EFFECT OF PRETENSION ON ARRAY BEHAVIOR
FIGURE 14

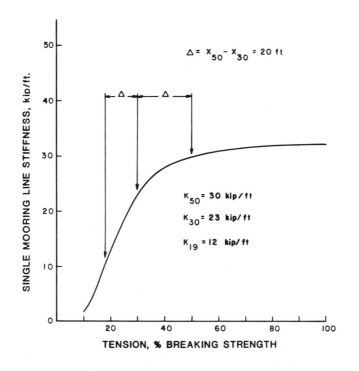

SINGLE MOORING LINE STIFFNESS AS A FUNCTION OF THE GUYLINE TENSION
FIGURE 15

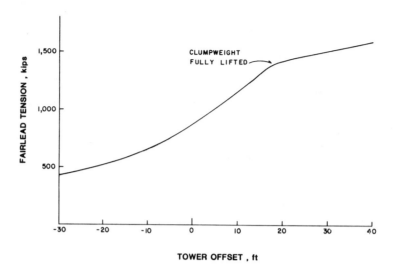

FIGURE 16 MAXIMUM SINGLE LINE TENSION

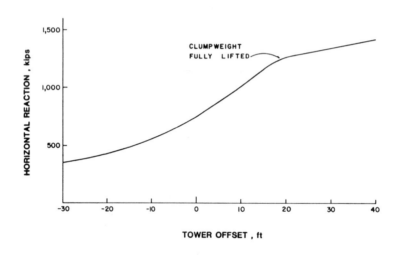

FIGURE 17 SINGLE LINE HORIZONTAL REACTION

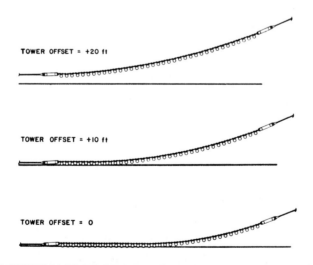

FIGURE 18 CLUMPWEIGHT PROFILE AT VARIOUS TOWER OFFSETS

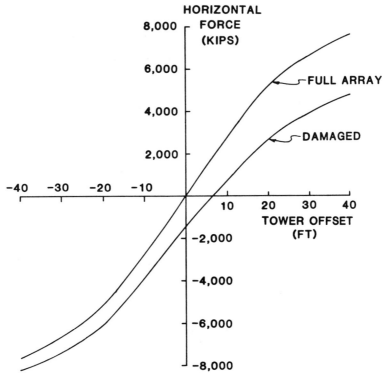

HORIZONTAL ARRAY REACTION
FIGURE 19

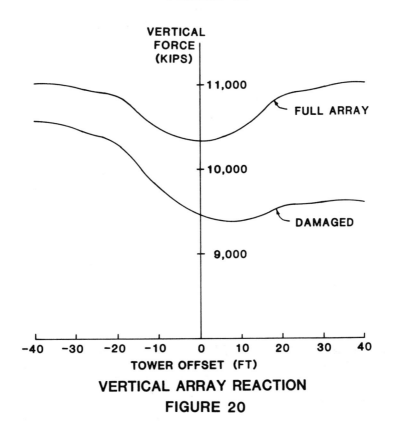

VERTICAL ARRAY REACTION
FIGURE 20

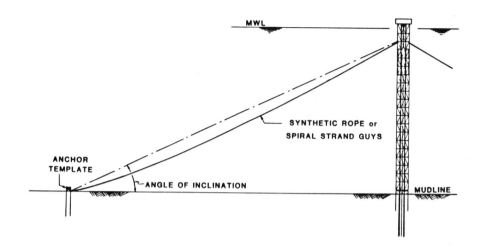

ELEVATION VIEW OF A TETHER MOORING LINE

FIGURE 21

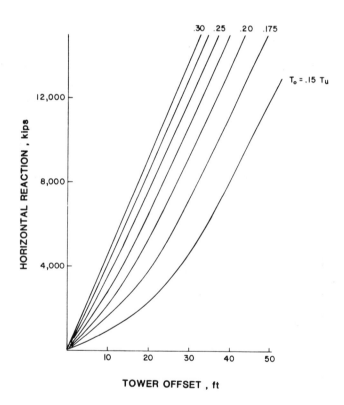

EFFECT OF PRETENSION ON TETHER ARRAY RESPONSE

FIGURE 22

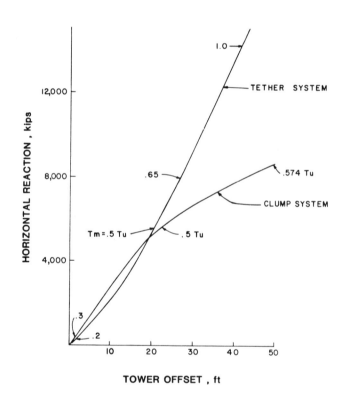

COMPARISON OF TETHER AND CLUMPWEIGHT SYSTEMS

FIGURE 23

PROBABILISTIC RELIABILITY ANALYSIS FOR
THE FATIGUE LIMIT STATE OF GRAVITY AND JACKET-TYPE STRUCTURES

H. Karadeniz,
Delft University of Technology

The Netherlands

S. van Manen,
TNO Institute for Building Materials
and Building Structures
The Netherlands

A. Vrouwenvelder,
TNO Institute for Building Materials
and Building Structures
The Netherlands

SUMMARY

A probabilistic reliability analysis is presented for a four-leg concrete gravity structure and a four-leg steel jacket-type platform. The sea actions as well as the structural response are considered as being of a random nature.

The sea is modelled as a piecewise stationary Gaussian process having a multi-directional Pierson-Moskowitz spectrum. As for the long-term statistics, the significant wave height is assumed to be Weibull distributed, while the principal wave direction is simply taken as uniform. In the paper it is argued that, because of insufficient data, the parameters in the PM spectrum as well as those in the Weibull distribution cannot be accurately known in any practical case. Therefore these parameters are also introduced as stochastic variables in the analysis.

The wave force analysis is based on Morison's equation. For the drag force part a general linearization procedure is used, corresponding approximately to Borgmann's linear term for the vertical shaft in a uni-directional sea state. Special care is taken in the transformation from continuous wave load on the members to loads concentrated at the nodes. The uncertainty with respect to the wave force model as a whole is taken into account by assuming standard deviations both for the inertia and drag force coefficients.

The structural response is calculated on the basis of a modal analysis. In determining the fatigue damage, Miner's rule is applied, for both steel and concrete. The main stochastic variables involved are the natural frequencies, the damping ratios and the material fatigue properties.

In order to calculate the probability of failure for the planned lifetime, spectral analysis techniques, reliability level II approximate methods and level III full integration procedures are combined. As a result, insight into the importance of the various stochastic variables with regard to failure is obtained. This insight can be utilized in practical level I design procedures, but serves also as guidance for future research activities.

1. INTRODUCTION

A fatigue analysis for offshore structures can be described in general terms as a calculation procedure in which, starting from the waves, the fatigue damage occuring in the material or in the joints (structural connections) is determined. The links between the wave data and the damage in the structure are formed by mathematical models for the wave forces, the structural behaviour and the material behaviour. In view of the stochastic and dynamic character of the waves it is an obvious choice to apply spectral analysis methods to the fatigue problem. A spectral fatigue analysis of this kind, though it makes use of probabilistic methods, is not a reliability analysis, however. For that purpose it is necessary to assume that the structural and material properties are also stochastic and to take account of the uncertainties in the (mechanical as well as statistical) models. In recent years, practicable computation techniques for this reliability analysis have been developed, particularly the so-called level II first-order second-moment approximations. Those approximations are mostly applied in cases where the loads are assumed to be static. For carrying out a reliability analysis for the fatigue of an offshore structure the problem therefore consists in combining the two probabilistic techniques - "spectral" and "level II" - into one analysis (Fig. 1). In the case of the gravity even a more complicated problem arises resulting from the fact that the reliability analysis partly is carried out on level III (full integration), due to strongly non-linear relationships.

In this paper, using an analysis as outlined above, the probability of the occurence of serious fatigue damage somewhere in a jacket-type or a gravity structure will be determined. This project is a continuation of earlier research in which a fatigue analysis was carried out for a simple structure consisting of a single pile in the sea (Bouma, 1979; Stupoc V, 1980). The difference between the analysis for a simple structure of that kind and a more complex one lies mainly in the computing time problem. For a jacket-type or a gravity structure a spectral fatigue analysis is in itself an elaborate procedure. In a reliability analysis that procedure has to be repeated, wholly or partly, a number of times in order to ascertain the sensitivity of the structure to the various stochastic variables. In order to limit the total computing time, it is therefore necessary to devote attention to efficient set-up of the program. This is discussed in detail in Section 5. Prior to that, however, the two basic components are dealt with in Section 3 and 4, namely, the spectral fatigue analysis and the level II reliability analysis. These two components as such contain few new features, but have been included in this paper for the sake of completeness. The structures investigated are very briefly described in Section 2. Finally, the results of the analysis are given in Section 6.

2. THE STRUCTURES INVESTIGATED

The structures in question are shown in Figs. 2 and 3, together with their computer schematizations.

The <u>gravity structure</u> (Fig. 2) comprises a caisson, four prestressed concrete columns and a deck. The depth of water is 100 m. In the schematization for the finite element method the deck and the caisson have been introduced as single elements. This is not unreasonable because the stress distributions within these elements are not of interest in the context of the present investigation. The subsoil has been schematized to a number of springs at the corners of the caisson. The columns are each subdivided into three elements, with one intermediate node at mean sea level. The analysis for the limit state of fatigue is performed at the fixed cross-section of the column at its junction with the caisson. The principal data (mean values in so far as stochastic variables are concerned) of the gravity structure are given in a table which is included in Fig. 2.

The steel <u>jacket-type structure</u> (Fig. 3) has an overall height of 65 m and is installed in 50 m depth of water. It comprises a four-leg platform with three storeys. The two lower storeys have X-bracing, while the top storey has K-bracing. The structure is supported on a piled foundation. The finite element analysis is carried out with a schematization substantially corresponding to the physical nodes and elements (joints and structural members), except that some additional nodes have been introduced at mean sea level. In all, there are 43 nodes. As in the case of the gravity structure, the deck has been introduced as a single element and the foundation has been schematized to a set of springs. For further data see the table in Fig. 3.

3. MATHEMATICAL MODEL FOR SPECTRAL FATIGUE ANALYSIS

The starting point for a spectral fatigue analysis is the assumption that the motion of the sea surface η can, for short periods (sea states), permissibly be conceived as a stationary Gaussian process. Such a process can be completely characterized by its spectrum. If this wave spectrum is known, it is possible to carry out a spectral analysis that finally yields the fatigue damage D for the critical points in the structure.

The mathematical model for this purpose, as employed in this research, is shown schematically in Fig. 4. The three main parts of the overall analysis -i.e., the wave force analysis, structural analysis and fatigue analysis- will now be considered in greater detail.

3.1 Wave force analysis

A multi-directional Pierson-Moskowitz spectrum is adopted as the spectrum characterizing a single sea state:

$$S_{\eta\eta}(\omega,\Phi) = [\tfrac{2}{\pi} \cos^2(\Phi-\Phi_s)][\alpha\ g^2\ \omega^{-5}\ \exp\{4\alpha\ g^2\ H_s^{-2}\ \omega^{-4}\}] \quad \ldots(1)$$

In this equation Φ_s is the principal direction of the sea state in which individual waves with direction

Φ simultaneously occur ($|\Phi - \Phi_s| \leq \pi/2$); furthermore ω is the frequency of the wave motion, H_s is the significant wave height, and $\alpha = 0.008$ and $g = 10$ m/s^2 (gravitational acceleration) are constants. With regard to the long-term behaviour it is assumed that the principal direction Φ_s has a uniform distribution and that H_s has a Weibull distribution. The density function for Φ_s is therefore:

$$f_{\Phi_s}(\phi_s) = 1/2\pi \qquad \ldots(2)$$

The Weibull-distribution for H_s can most simply be presented by means of the distribution function:

$$F_{H_s}(h) = 1 - \exp\{((h - A)/B)^C\} \qquad \ldots(3)$$

Equation (3) holds for $h > A$; the following numerical values have been adopted, corresponding to measurements at the Sevenstones location off the south coast of England (Battjes, 1972): $A = 0.60$ m, $B = 1.67$ m and $C = 1.2$.

Starting from the spectrum for the waves per sea state, it is possible to determine the spectra (auto-spectra and cross-spectra) for the wave forces with the aid of:

$$S_{F_i F_j} = \int H^*_{F_i \eta} H_{F_j \eta} S_{\eta\eta} \, d\Phi \qquad \ldots(4)$$

where $H_{F_i \eta}$ is the transfer function from water elevation to wave force, while the asterisk (*) denotes the complex conjugate. In establishing these transfer functions the linear wave theory and Morison's formula, applied to the undisturbed wave, were used. For the higher wave frequencies a correction based on diffraction theory was applied. The non-linear drag force term was linearized, the criterion chosen being that the area under the spectrum of the linearized wave force was equal to that of the exact wave force. This procedure, which will be dealt with in a separate paper, proved to be well practicable, more particularly for the inclined structural members in a multi-directional sea state.

By way of illustration, Fig. 5 gives some results for the gravity structure (only the inertia effects, as the drag force is negligible for gravity structures). For a multi-directional sea state with its principal direction in the x-direction and $H_s = 5.0$ m the following are indicated: the auto-spectrum of the force in the x-direction on one of the columns, the auto-spectrum of the force in the y-direction on the same column, and the cross-spectrum of the forces in the x-direction on two different columns.

3.2 Structural analysis

The structural analysis is based on the differential equation:

$$M \underline{\ddot{u}} + C \underline{\dot{u}} + S \underline{u} = \underline{F} \qquad \ldots(5)$$

where $\underline{u}$ is the displacement vector, $\underline{F}$ is the load vector and M, C and S are the mass, damping and stiffness matrix respectively. A solution with the aid of a modal analysis was chosen, in which equation (5) is reduced by means of the transformation $\underline{u} = V \underline{q}$ to a set of decoupled differential equations:

$$m_i \ddot{q}_i + 2 \zeta_i m_i \omega_i \dot{q}_i + m_i \omega_i^2 q_i = f_i \qquad \ldots(6)$$

where q_i is the modal coordinate, m_i de modal mass, ζ_i the damping ratio and ω_i the ith natural frequency. The modal loadings f_i are obtained from:

$$\underline{f} = V^T \underline{F} \qquad \ldots(7)$$

The transformationmatrix V is equal to the mode shape matrix $V = \{\underline{v}_1, \underline{v}_2, \ldots \underline{v}_n\}$ where $\underline{v}_i$ denotes the mode shape vector corresponding to the i-th natural frequency. Finally, a relation must be established between the modal coordinates $\underline{q}$ and the stresses $\underline{s}$ at the significant points in the structure:

$$\underline{s} = T \underline{q} \qquad \ldots(8)$$

Component T_{ij} of the matrix T indicates the stress at point i for a normalized oscillation in oscillation mode j.

Summarizing: in the mathematical model outlined above, the analysis from force to stress is divided into three steps:

(i)   from overall force F to modal loading f;
(ii) from modal loading f to modal coordinate q;
(iii) from modal coordinate q to stress s.

Accordingly, the spectra for f, q and s are successively obtained from:

(i) $$S_{f_i f_j} = \Sigma \Sigma V_{ik} V_{j\ell} S_{F_k F_\ell} \qquad \ldots(9)$$

(ii) $S_{q_i q_j} = H^*_{q_i f_i} H_{q_j f_j} S_{f_i f_j}$ ...(10)

(iii) $S_{s_i s_j} = \Sigma \Sigma\ T_{ik} T_{j\ell} S_{q_k q_\ell}$ ...(11)

The transfer function $H_{q_i f_i}$ is expressed by:

$$H_{q_i f_i} = \frac{1}{m_i \omega_i^2} \left\{ \frac{\omega_i^2}{\omega_i^2 - \omega^2 + 2i\zeta_i \omega \omega_i} \right\}$$ ...(12)

It is to be noted that the transfer function $H_{q_i f_j}$ is equal to 0 for $i \neq j$, this being associated with the decoupled character of the differential equations (6).

Equations (9), (10) and (11) yield the spectra and cross-spectra for the stresses in the structure. For the further fatigue analysis, however, we are not interested in the spectrum as a whole, but only in two characteristics thereof, namely, the area and the second moment:

$$m_o(s_i) = \sigma^2(s_i) = \int_0^\infty S_{s_i s_i}\, d\omega$$ ...(13)

$$m_2(s_i) = \int_0^\infty \omega^2 S_{s_i s_i}\, d\omega$$ ...(14)

As appears from (13), the area of the spectrum is equal to the variance $\sigma^2(s_i)$ of the stress. The second moment serves to establish the mean frequency and thus the number of stress cycles. Since we are interested only in these quantitites, it is possible to introduce a time-saving modification into the computational procedure. What this modification comes down to is that the integration over ω is done already after step (ii):

$$\text{cov}(q_i q_j) = \int_0^\infty S_{q_i q_j}\, d\omega$$ ...(15)

where $\text{cov}(q_i q_j)$ is the covariance of q_i and q_j. The variance of the stress is finally obtained from:

$$\sigma^2(s_i) = \Sigma_i\ T_{ik} T_{i\ell}\ \text{cov}(q_k q_\ell)$$ ...(16)

Corresponding formulations can be set up for the second moment. The advantage of this modified procedure is that step (iii) does not need to be calculated for every value of ω because integration has already been performed over ω.

To illustrate the formulas that have been given here, the spectrum $S_{q_1 q_1}$ for the uni-directional and for the multi-directional sea state is shown in Fig. 6. It is not possible to show the spectrum for the stresses, since this has -as explained above- not been calculated. In the main, however, the stress spectrum will be similar in shape to that of q_i.

Finally, at the end of this structural analysis step, the following comments are offered.

- For an exact solution the number of mode shapes n to be taken into account must be equal to the number of degrees of freedom of the structure. In general, however, a much smaller number will suffice. In the present study, for reasons of computing time, n = 3 or = 5 has been chosen.
  In so far as the dynamic effects are concerned, this number is certainly sufficient, because the higher natural frequencies are situated in the tail of the spectrum. The question is, however, whether the quasi-static part of the response is calculated sufficiently accurately with this small number of oscillation forms. For the time being, this question has been ignored as non-essential. Improvement of the accuracy of the static portion can be achieved in various ways without necessitating a disproportionate increase in computing time (Thomson, 1972; Vugts, et al, 1979).

- In determining the natural frequencies and the associated natural oscillation forms the undamped system (C = [0]) is taken as the starting point. The damping matrix C must therefore satisfy certain conditions in order that the decoupled set (6) is obtained.

- Besides receiving a contribution from the structure, the mass and the damping also receive a contribution from the surrounding water: the hydrodynamic mass and the hydrodynamic damping. The hydrodynamic mass of the water around the tubes or columns has been taken into account as a quantity of water with a volume equal to that of the water in the tube, multiplied by a coefficient γ ; according to the potential theory: $\gamma = 1$; by choosing a value $\gamma \neq 1$ any deviations from this theory can be allowed for.

3.3 Fatigue model

The fatigue model serves to establish the relationship between the variance of the stresses and the damage of the material. Both for steel and for concrete the basis is provided by Miner's rule, which states in essence that every stress cycle i results in a degree of damage D_i equal to:

$$D_i = 1/N_i \qquad \ldots(17)$$

where N_i is the number of cycles at which failure occurs if the same stress cycle is repeated over and over again. This number of cycles is expressed by the so-called S-N (stress-number) curves:

$$N_i = \left[\frac{2 \hat{s}_i}{s_F}\right]^{-k} \qquad \ldots(18)$$

where $2\hat{s}_i$ is the stress range (equal to twice the amplitude) and s_F and k are constants which can be determined with what are known as constant-amplitude tests. Accordingly, under constant-amplitude testing, failure occurs by definition when the total damage $D_{tot} = \Sigma\, D_i$ attains the value 1. However, with variable-amplitude random loading, influences associated with the "loading history" may cause failure to occur at a value not equal to 1. The value of D_{tot} at which failure occurs will be further designated as D_F.

For determining $D_{tot} = \Sigma\, D_i$ we shall now first determine the mean damage $\bar{D}_i$ associated with a single cycle in one sea state with a given spectrum. The amplitudes of the stress cycles are assumed to conform to a Rayleigh distribution:

$$f_{\hat{s}}(s) = \frac{s}{\sigma^2(s)} \exp\left\{-\frac{s^2}{2\sigma^2(s)}\right\} \qquad \ldots(19)$$

where $\sigma(s)$ is the standard deviation of the stress at the point under consideration in the structure; it is obtained from equation (16) and may, depending on the fatigue model adopted, have to be multiplied by a stress concentration factor. It is to be noted that the subscript i in (16) has a different meaning (denoting location in the structure) than the subscript i in (17) and (18) (number of the stress cycle).

Applying a Rayleigh distribution (19) does, however, raise some questions. This distribution provides an excellent approximation if the stochastic process has a narrow-band spectrum. In the present case, on the other hand, the spectrum distinctly displays multiple peaks, with one peak at the mean wave frequency of the structure (see Fig. 6). Unfortunately, no fatigue model that can take account of this in a simple manner is at present available.

Adopting equation (19) for the distribution of the amplitudes, the mean fatigue damage per cycle within one sea state can be determined as:

$$\bar{D}_i = \int_0^\infty D_i(s)\, f_{\hat{s}}(s)\, ds \qquad \ldots(20)$$

$$\bar{D}_i = \int_0^\infty \left(\frac{2s}{s_F}\right)^k \cdot \frac{s}{\sigma^2(s)} \exp\left\{-\frac{s^2}{2\sigma^2(s)}\right\} ds \qquad \ldots(21)$$

On working out the integral, we obtain:

$$\bar{D}_i = (2\sqrt{2})^k\, \Gamma\left(1 + \frac{k}{2}\right) \left(\frac{\sigma(s)}{s_F}\right)^k \qquad \ldots(22)$$

where $\Gamma(..)$ is the gamma function, which for even values of k is $\Gamma(1 + \frac{k}{2}) = (\frac{k}{2})\,!$

The total damage $D_{tot} = \Sigma\, D_i$ is obtained by summing the mean damage over the service life (planned lifetime) of the structure, taking account of the long-term distribution for the sea states:

$$D_{tot} = \int_0^\infty \int_{-\pi}^{+\pi} \frac{T_L}{T_m}\, \bar{D}_i\, f_{H_s}(h)\, f_{\phi_s}(\phi_s)\, dh\, d\phi_s \qquad \ldots(23)$$

In this expression T_L is the total lifetime and T_m the mean period of a stress cycle within the sea state, determinable as follows:

$$T_m = \frac{2\pi}{\omega_0} = 2\pi\, \sqrt{\frac{m_0(s)}{m_2(s)}} \qquad \ldots(24)$$

where $m_0(s)$ and $m_2(s)$ are obtained from (13) and (14). Furthermore, in equation (23), $f_{H_s}(\ldots)$ and $f_{\phi_s}(\ldots)$

represent the probability density functions for the significant wave height and the predominant wave direction corresponding to (2) and (3).

Finally, the following comment must be made. The program for the jacket-type structure deviates in one respect from the above description: for this structure $S_{F_iF_j}$ (equation (4)) is not determined explicitly; the calculation is carried through directly to $S_{f_if_j}$ (see Fig. 4). The reason for this lies in the way in which the integration is performed from distributed wave loading to concentrated nodal forces. In the case of the gravity structure this is a simple analytically practicable integration. On the other hand, in the case of the jacket-type structure, for which it is necessary to integrate not only in the vertical but also in the horizontal direction, this integration is performed numerically. It has been found advantageous to perform this integration per oscillation mode (Karadeniz, 1980). The drawback is that the wave force analysis and the structural analysis are not entirely seperated, but this is more than offset by the advantages offered by a quick and accurate integration procedure.

4. RELIABILITY ANALYSIS

The procedure described in Section 3 is a spectral fatigue analysis comprising deterministic structural properties and deterministic mathematical models. The analysis results (see Fig. 4) in a particular degree of damage D_{tot} that is to be expected at the end of an intended lifetime for the structure. In this section it will be indicated how, with the aid of chiefly "level II" techniques (Rackwitz, 1976; Ciria, 1977) the deterministic fatigue analysis can be extended into a probabilistic reliability analysis for fatigue.

To begin with, a reliability function Z must be so chosen that $Z < 0$ corresponds to failure of the structure or structural member (in this case: fatigue failure) and $Z > 0$ to non-failure. In this analysis it would be an obvious choice to define the reliability function as $Z = D_F - D_{tot}$. However, for technical reasons of computation, associated with the nature of the problem, the following is preferred:

$$Z = - \ln \{D_F/D_{tot}\} \qquad \ldots(25)$$

The reliability function Z is a function of the problem variables, including a number -say n- of stochastic variables (basic variables), here provisionally given the general designation $X_1, X_2 \ldots X_n$. It is assumed that the variables X_i are mutually independent and that their mean values $\mu(X_i)$ and standard deviations $\sigma(X_i)$ are known. The mean value and the standard deviation of Z can then be approximated by the following expressions:

$$\mu(Z) \sim Z\{\mu(X_1), \mu(X_2) \ldots \mu(X_n)\} \qquad \ldots(26)$$

$$\sigma^2(Z) \sim \sum_{i=1}^{n} \left\{\frac{\partial Z}{\partial X_i} \sigma(X_i)\right\}^2 \qquad \ldots(27)$$

The derivatives $\partial Z/\partial X_i$ are evaluated at the point $\underline{X} = \mu(\underline{X})$. Starting from a normal distribution for Z, we can then determine the failure probability from:

$$P\{Z < 0\} = \Phi_N(-\beta) \qquad \ldots(28)$$

$$\beta = \mu(Z)/\sigma(Z) \qquad \ldots(29)$$

where β is the reliability index and $\Phi_N(..)$ the distribution function for a normally distributed variable with mean value zero and standard deviation equal to unity.

The level II reliability analysis as embodied in equations (26) to (29) is what is known as the level II mean value variant. In the actual calculations, however, the more accurate AFDA (Approximate Full Distribution Approach) has been employed, for a description of which the literature (Rackwitz, 1976; Ciria, 1977) should be consulted.

Level II approximations, both Mean Value and AFDA, are based on the assumption that the reliability function Z can be linearised. However, as a result of the oscillating character of the force spectra (for example see S_{FF} in fig. 4) the Z-function is also an oscillating function of the natural frequencies ω_i. For the jacket structure this effect is not very pronounced because of the large number of members and because of the high value of the lowest natural frequency. In the case of the gravity, however, the effect cannot be neglected and therefore it has been decided to treat the ω_i as level III variables, assuming that all ω_i are fully correlated with the lowest natural frequency ω_1. The basic equation for the analysis is then given by:

$$P\{Z < 0\} = \int_0^\infty P\{Z < 0 \mid \omega_1 = \omega\} f_{\omega_1}(\omega) d\omega \qquad \ldots(30)$$

Equation (30) is a direct application of the Total Probability Theorem. In the program of course the integration is carried out numerically. The failure probabilaties $P\{Z < 0 \mid \omega_1 = \omega\}$ at fixed ω_i are evaluated by level II procedures as before; furthermore the probability density function of ω_1, not being a basic variable itself, is also found by first-order second-moment approximation. A problem concerning equation (30) is that, associated with a variation in ω_i, there also is a variation in the mode shape vector $\underline{v}_i$. In the next section, however, it will be explained that this variation can be disregarded.

5. COMPUTATIONAL PROCEDURE FOR RELIABILITY ANALYSIS

For the determination of $\sigma^2(Z)$ in accordance with (27) we must have the partial derivatives $\partial Z/\partial X_i$ at our disposal. If, as in the present study, these partial derivatives cannot be determined analytically, this will have to be done numerically. In consequence, a total of (n + 1) values for Z have to be determined, namely, once for the determination of $\mu(Z)$ and n times for the determination of the derivatives $\Delta Z/\Delta X_i$. The AFDA variant is even more extensive and, in the case of the gravity, the whole procedure must be repeated quite a number of times because of equation (30). Now the once-only determination of Z in accordance with the scheme presented in Fig. 4 is in itself a very time-consuming process, so that the total computing time required for a reliability analysis threatens to become very long. Fortunately, on closer examination, it turns out not to be necessary to repeat the whole computational procedure for each evaluation of Z. Quite often it is sufficient merely to repeat particular parts, because other parts are either invariant or linear in particular parameters. A good example is provided by the parameters of the fatigue model. If variation occurs in the fatique parameters, we can still base ourselves on the same stress data, so that only the step from stress to damage is changed. Similar considerations are found to be valid for many other parameters. In Fig. 7 it is indicated, separately for gravity and jacket, what part of the total computational program has to be repeated. In all, 15 stochastic variables are involved, the meanings of which are stated in the tabel in the centre of Fig. 7. Further comments on the choice of these variables and their statistical modelling will be given in Section 6.

For the fatigue parameters D_F and s_F only a very short portion of the computational process has to be restarted, namely, the portion after the determination of D_{tot}. For D_F this will be immediately evident, since D_F first appears only in the formula for the reliability function Z (25). For s_F the situation is somewhat more complex. The parameter s_F appears already in equation (22) for the determination of $\bar{D}_i$; yet it is not necessary to go back all the way to that formula. This is so because a variation in s_F equally affects all the stress cycles and can therefore be excluded from the integration (23). For the third fatigue parameter k, however, this not so, and for this parameter it is therefore necessary to repeat the determination of $\bar{D}_i$ and the integration over all the cycles.

The parameters A, B and C of the Weibull distribution for the significant wave height H_s have likewise been introduced as stochastic variables in this study. Clearly, these parameters exercise an influence only on the integration (23) for determining D_{tot}.

For the gravity structure the portion assignable to drag force is negligible, and therefore all the forces and stresses are proportional to C_M. This variable can, for this reason, be treated in the same manner as s_F. In the case of the jacket-type structure the drag force does have to be taken into account; hence strictly speaking for the jacket a complete loop has to be executed both for C_D and for C_M. However, in the program this has been solved very efficiently. In the determination of the transfer function $H_{f\eta}$ the drag portion and the inertia portion are kept carefully separated, so that actually the wave force analysis need not be repeated.

For all other variables the program has to be repeated from $S_{f_i f_j}$. This is perhaps not so obvious for the shape parameter α because this parameter directly affects $S\eta\eta$. The results up to $S_{f_i f_j}$ are, however, entirely proportional to $S\eta\eta$. Therefore it is not necessary to repeat the entire wave force analysis; instead, $S_{f_i f_j}$ can be fairly easily adapted.

Finally it is of interest to consider how the variation in the structural variables E, t, m, k and γ is ascertained. Because these variables also affect the natural frequencies, modal masses, and mode shapes, the eigenvalue analysis would in principle have to be restarted for each evaluation of Z. But the eigenvalue analysis is an extensive procedure, and for this reason a different solution was chosen. The basic idea is that (as easily can be shown by calculation) the quasi static part of the stress spectrum hardly is affected by a variation of the parameters mentioned, except of course by the direct effect of the wall thickness t which must be treated separately. This means that a good approximation can be obtained by assuming that only the frequency dependent part of $H_{x_i f_i}$ (factor between brackets in eq. (12) will be changed while the static value of $H_{x_i f_i}$ (= $1/m_i \omega^2_i$) and the eigenvectors $\underline{v}_i$ remain unaltered. The frequency dependent part of $H_{x_i f_i}$ is fully governed by the value of the natural frequency ω_i, which easily can be estimated by Rayleigh's method:

$$\omega_i^2 = \frac{\underline{v}_i^T S \underline{v}_i}{\underline{v}_i^T M \underline{v}_i} = \frac{\underline{v}_i^T S_o \underline{v}_i + (\Delta X) \underline{v}_i^T S' \underline{v}_i}{\underline{v}_i^T M_o \underline{v}_i + (\Delta X) \underline{v}_i^T M' \underline{v}_i} \qquad \ldots(31)$$

S' and M' are the derivatives of the stiffness matrix and the mass matrix with respect to the relevant stochastic variable X. Determining these derivatives involves some extra work, but this is nevertheless inconsiderable in comparison with the effort that redoing an eigenvalue analysis requires. An associated advantage of the assumption that $\underline{v}_i$ remains unchanged is that the spectra for the modal loadings $S_{f_i f_j}$ (see eq. (9)) need not be recalculated.

6. NUMERICAL DATA AND RESULTS

The fatigue reliability analysis as outlined above has been applied to the example structures of Fig. 2 and 3, discussed in section 2. For the gravity structure the probability of fatigue failure has been calculated at a cross section located at the lower end of one of the columns; the fatigue failure probability for the jacket structure is related to one of the bottom joints, based on the nominal leg stress multiplied by a stress concentration factor equal to 2.

The principal input data for the reliability analysis with the associated results have been collected in Table 1. For each stochastic variable are stated the type of distribution applied, the mean value, the coefficient of variation (standard deviaton divided by mean value) and the relative contribution α_i^2 of those variables to the variance of Z (see (27)):

$$\alpha_i^2 = \frac{\Delta\,\sigma_i^2(Z)}{\sigma^2(Z)} = \frac{\{\frac{\partial Z}{\partial X_i}\sigma(X_i)\}^2}{\sum_j \{\frac{\partial Z}{\partial X_j}\sigma(X_j)\}^2} \qquad \ldots(32)$$

For the gravity structure (32) is not directly applicable, resulting from the fact that the analysis has been carried out partly on level III (see equation (30)). In appendix A is explained how in this case the α_i^2-values have been obtained. Furthermore it should be noted that at the time of preparing this paper the results were still subject to verification. Consequently conclusions have to be considered as preliminary.

In analogy with the tripartite set-up of the analysis the stochastic variables can likewise be subdivided into three groups, namely, "wave parameters", "structural parameters" and "fatigue parameters". The respective variables will now be briefly explained in this order. However, it should be noted first that the reliability indices β calculated for the gravity structure and for the jacket-type structure are 4.8 and 5.2 respectively. Such values, corresponding to failure probabilities of 10 to the power -6.1 and -7.0, indicate entirely acceptable designs.

The stochastic parameters of the wave force model are the parameters A, B and C of the Weibull distribution (3) for H_s, the parameter α of the Pierson-Moskowitz spectrum (1), and the C_M and C_D coefficients of the Morison wave formula. The reason for introducing the parameters A, B and C as stochastic variables is that in general not enough data at the appropriate location are available to enable these parameters to be ascertained with "deterministic" certainty: in other words, there is uncertainty with regard to the statistical model. Similar considerations apply to the shape parameter α of the Pierson-Moskowitz spectrum. The values for the coefficient of variation σ/μ have been adopted on a purely intuitive basis. In view of the minor contributions that these parameters make to the variance of Z (see the α^2 column in Table 1), a more detailed investigation for them appears to be of little interest.

The C_M and C_D values for the jacket-type structure have been assigned a coefficient of variation of 20% for each frequency; 15% has been adopted for the gravity structure. These values are assumed to allow for all uncertainty in the wave force model applied. The C_D parameter turns out to be rather unimportant, whereas the effect of C_M is considerable, both for the gravity (α^2 = 25%) and for the jacket-type structure (α^2 = 34%).

The stochastic variables relating to the structure are the wall thicknesses, the modulus of elasticity (for concrete only), the rigidity of the foundation, the mass of the deck, the allowance for water mass, and finally the damping. The coefficients of variation in all these cases represent in principle the intuitively assessed uncertainty in these factors in the design stage. For a number of variables it would, after completion of the structure, be possible to carry out observations and thus to adjust mean values and reduce coefficients of variation. Via an updated analysis it would then be possible to make a fresh estimate of the reliability, which could, if necessary, lead to the application of supplementary measures of precautions.

With regard to the modulus of elasticity it has been assumed to be completely correlated over the whole structure. A similar assumption has been made for the wall thickness t. In principle, of course, this is not correct; partial correlation would have to be adopted instead. However, such an approach would require an increase in the number of stochastic variables and thus add considerably to the computing time involved. The consequence of the assumed complete correlation is mainly that the scatter of the natural frequencies is somewhat over-estimated because no statistical averaging takes place. In the case of the modulus of elasticity this can be offset by the choice of a somewhat lower value for the coefficient of variation. For the wall thickness t this is not possible because it also plays a major part in the step proceeding from normal force and bending moment to local stress, and this is a step in which no averaging effect occurs.

The most striking result according to tabel 1 is the high contribution of the damping ratio ζ to the variance of Z in the case of the gravity. The reason is that the respons of the gravity is almost fully dynamic, which must be combined with the high value of k (slope S-N curve) in the concrete fatigue model. Indeed, a check run having k = 4 for concrete resulted in a damping contribution equal to only 9%. Another interesting result is the contribution of the foundation stiffness which is substantial for the jacket but negligible for the gravity. This difference can be explained by the large mass of the gravity caisson preventing the foundation to play a dominant role in the determination of the natural frequency.

There remains the third group of stochastic variables to consider, namely, the parameters of the fatigue model. Both for the gravity structure and for the jacket-type structure this group of parameters is found to be important, together accounting for 23% and 29% of the variance of Z respectively. The statistical modelling, which has been assumed for these parameters will be outlined below.

The scatter in constant amplitude fatigue tests on plain concrete can be described reasonably well by deterministic k and lognormal s_F with a coefficient of variation roughly the same as for the cylinder compressive strength (Van Leeuwen and Siemes, 1979). The concrete in offshore structures usually has a relatively high quality and a value $V(s_F) = 0.10$ has been adopted. In order to check if Miner's rule could be applied to concrete a number of tests have been carried out based on program loading and variable amplitude loading (Siemes, 1980). Analysing the results using Miner's rule values for D_F in the range of 0.38 to 13.87 have been observed, depending on the type of loading. However, it proved very difficult to deduce exactly what loading characteristics should be held responsable for an increase or decrease in the fatigue resistance. For that reason D_F has been modelled in this study as a random variable, having a lognormal distribution, a mean value of 3.0 and a coefficient of variation equal to 1.2

In the steel fatigue model the uncertainty resulting from applicaton of Miner's rule has been considered as relatively low and a deterministic value $D_F = 1$ has been adopted. The uncertainty in k and s_F furthermore has been assumed to originate from two distinct sources. The first source again is the scatter as observed in the test results of indentical test specimen, which can be described by

$V_1(S_F) = 0.14$ and $V_1(k) = 0.00$

In Fig. 8a an average S-N curve and the $\pm 2\sigma$ bands are shown.

The second source of uncertainty is related to the choice of an S-N curve in a real design situation. In principle every joint has an (average) S-N curve of its own depending on the geometry, the dimensions, loading type, etc. In Fig. 8b several S-N curves (based on hot spot stress) have been drawn (de Back, et al, 1981). Broadly, the dispersion in these curves can be characterized by $V(S_F) = 0.40$ and $V(k) = 0.15$, while S_F and k are correlated very close to unity. Now of course, this dispersion does not result from a random mechanism, but can be fully explained in terms of the joint characteristics. However, when dealing with some specific joint, which does not corresponds exactly in all aspects to one of the test joints, the selection of the S-N curve cannot be performed without introducing uncertainty. For the experienced designer it will be assumed that this uncertainty equals 10 to 20% of the total dispersion, resulting in:

$V_2(S_F) = 0.07$ and $V_2(k) = 0.025$

The correlation $\rho_2(S_F, k)$ can be put equal to 1. Fig. 8c shows the average and the $\mu \pm 2\sigma$ curves. For stresses of 100 MPa the test scatter turns out to be dominant; for lower stresses, however, the curve selection uncertainty becomes more and more important.

Finally the two sources of uncertainty have to be combined. For the coefficient of variation this leads to:

$V^2(S_F) = V_1^2(S_F) + V_2^2(S_F) = (0.14)^2 + (0.07)^2 = (0.16)^2$

$V^2(k) = V_1^2(k) + V_2^2(k) = (0.00)^2 + (0.025)^2 = 0.025)^2$

For the combined uncertainties the coefficient of correlation amounts to $\rho(S_F, k) = V_2(S_F)/V(S_F) = 0.44$. In the program k and S_F have been transformed into two uncorrelated variables. As a result it is impossible to separate the contributions of the two parameters to $\sigma^2(Z)$ and in table 1 only their combined contribution is presented.

7. SUMMARY AND CONCLUSIONS

This paper describes a probabilistic fatigue analysis for a concrete gravity structure and a steel jacket-type structure. The analysis is based on a spectral model with a modal analysis for the structure. The reliability analysis is mainly based on the Approximate Full Distribution Approach, this being the most accurate level II first-order second moment technique. For the gravity structure a mixed level II - level III analysis had to be carried out, due to strong nonlinear behaviour of the reliability function.

In general, reliability analyses are expensive because the calculations have to be performed many times. Thanks to the good modular construction of the program, however, it proved possible to reduce the extra calculations to a minimum by not repeating more of the calculations per variation than strictly necessary.

In all, 15 stochastic variables were introduced into the analysis, subdivided into three groups, namely: the variables in the wave force model, the variables in the macro response model for the structure, and the variables in the fatigue model. The parameters of the fatigue model, the Morison inertia coefficient, and in the case of the gravity also the damping ratio were found to be mainly responsible for the uncertainty with regard to fatigue failure.

The object of the research was primarily to investigate whether an analysis as performed here would be possible for more realistic structures. It can be stated that the feasibility of the analysis has indeed been proved and that the results can claim to be valuable as an aid in establishing practical design rules.

NOTATION

cov $(X_1 X_2)$	covariance of X_1 and X_2
g	gravitational acceleration
$f_X(\ldots)$	probability density function for X
f_i	modal loading for i^{th} oscillation mode
k	fatigue constant
m	mass of the deck
m_i	modal mass of ith oscillation mode
$m_0(s), m_2(s)$	moments of the stress spectrum
q_i	modal coördinate of ith oscillation mode
s_i	stress at point i
$\hat{s}_i$	amplitude of s_i; amplitude at i^{th} cycle
s_F	fatigue parameter
t	wall thickness
u_i	displacement of node i
$\underline{v}_i$	mode chape vector for i^{th} natural frequency
x,y,z	coördinate system
A,B,C	parameters for distribution of significant wave height
C	damping matrix
C_D, C_M	drag coefficient and inertia coefficient
D	diameter
D_i	damage at i^{th} stress cycle
$\bar{D}_i$	mean damage per wave in one sea state
D_{tot}	total damage in lifetime
D_F	damage at which failure occurs
$F_X(\ldots)$	probability distribution function for X
F_i	loading on node i
H_s	significant wave height
$H_{XY}(\ldots)$	transfer function from Y to X
M	mass matrix
N_i	number of cycles to failure associated with $\hat{s}_i$
$P(\ldots)$	probability of ...
S_{XX}	autospectrum of X
S_{XY}	cross-spectrum of X and Y
S	stiffness matrix
T	relation matrix from modal coordinate to stress
T_L	lifetime (service life) of the structure
T_m	mean period between two zero crossings
V	transformation from overall to modal coordinate
X, X_i	stochastic basic variable
Z	reliability function
α	parameter in the PM spectrum
α_i	influence coefficient of X_i

β	reliability index
γ	coefficient for associated water mass
ζ_i	damping ratio for oscillation mode i
$\mu(X)$	mean value of X
$\sigma(X)$	standard deviation of X
ω	circular frequency
ω_i	natural frequency of ith oscillation mode
$\Gamma(\ldots)$	gamma function
Φ	wave direction
Φ_s	principal direction of a sea state
$\Phi_N(\ldots)$	distribution for normal distribution

APPENDIX - CONSTITUTION OF THE VARIANCE FOR MIXED LEVEL II/III APPROACH

The reliability function Z in the mixed level II/III gravity analysis depends on a number of level II type variables, which will be called here $X_1 \ldots X_n$, and the level III type variable ω_1:

$$Z = Z(X_1 \ldots X_n, \omega_1) \qquad \ldots (A-1)$$

The probability of failure then is given by:

$$P(Z < 0) = \int_0^\infty P(Z < 0 | \omega_1 = \omega) f_{\omega_1}(\omega) d\omega \qquad \ldots (A-2)$$

For the mean value $\mu(Z)$ of Z a similar formula holds:

$$\mu(Z) = \int_0^\infty \mu(Z | \omega_1 = \omega) f_{\omega_1}(\omega) d\omega \qquad \ldots (A-3)$$

Equation (A-2) follows directly from the total probability theorem. Equation (A-3) on the other hand has to be proved in a different way:

$$\mu(Z) = \int \int Z(\xi_1 \ldots \xi_n, \omega) f_{X_1 \ldots X_n}(\xi_1 \ldots \xi_n) f_{\omega_1}(\omega) d\xi_1 \ldots d\xi_n d\omega$$

$$= \int \{ \int \int Z(\xi_1 \ldots \xi_n, \omega) f_{X_1 \ldots X_n}(\xi_1 \ldots \xi_n) d\xi_1 \ldots d\xi_n \} f_{\omega_1}(\omega) d\omega$$

$$= \int \mu(Z | \omega_1 = \omega) f_{\omega_1}(\omega) d\omega$$

In the same way it can be proved that:

$$m_2(Z) = \int m_2(Z | \omega_1 = \omega) f_{\omega_1}(\omega) d\omega$$

From $m_2 = \mu^2 + \sigma^2$ it follows that:

$$\mu^2(Z) + \sigma^2(Z) = \int \{ \mu^2(Z | \omega_1 = \omega) + \sigma^2(Z | \omega_1 = \omega) \} f_{\omega_1}(\omega) d\omega$$

Now write $\sigma^2(Z | \omega_1 = \omega) = \Sigma \alpha_i^2(\omega_1) \sigma^2(Z | \omega_1 = \omega)$, then finally:

$$\sigma^2(Z) = \{ \int \mu^2(Z | \omega_1 = \omega) f_{\omega_1}(\omega) d\omega - \mu^2(Z) \} + \sum_{i=1}^{n} \int \alpha_i^2(\omega_1) \sigma^2(Z | \omega_1 = \omega) f_{\omega_1}(\omega) d\omega \qquad \ldots (A-4)$$

The first term is the contribution of ω_1 to the variance of Z; the terms under the summation sign represent the contributions of $X_1 \ldots X_n$.

X	designation	type	GRAVITY ($\beta = 4.8$)			JACKET ($\beta = 5.2$)		
			$\mu(X)$	σ/μ	α^2	$\mu(X)$	σ/μ	α^2
A	parameter of H_s distribution	N	0.60 m	0.10	0 %	0.60 m	0.10	0 %
B	parameter of H_s distribution	N	1.67 m	0.10	0	1.67 m	0.10	2
C	parameter of H_s distribution	N	1.2	0.10	0	1.2	0.10	2
α	parameter PM-spectrum	LN	0.008	0.15	5	0.008	0.15	1
C_M	inertia coefficient (function of ω)	LN	0-2	0.15	25	0-2	0.20	34
C_D	drag force coefficient	LN	-	-	-	1	0.20	1
t	wall thickness	N	0.40 m	0.05	3	divers	0.05	7
E	modulus of elasticity	N	30 GPa	0.10	7	-	-	-
$k_{v,h}$	rigidity of foundation	LN	divers	0.20	1	divers	0.25	11
m	mass of deck	N	10800 t	0.05	3	4800 t	0.10	9
γ	addes mass factor	N	0.9	0.10	0	0.9	0.10	0
ζ	damping ratio	LN	0.02	0.40	33	0.01	0.40	4
s_F	fatigue parameter (intersection S-N curve)	LN	45 MPa	0.10	11	6400 MPa	0.16*	29
k	fatigue parameter (slope S-N curve)	N	9.5	-	-	3.8	0.025*	
D_F	failure value of Miner sum	LN	3.0	1.20	12	1	-	-
					100%			100%

Table 1 Summary of the stochastic variables for the analysis of the gravity structure and the jacket-type structure (At the time of preparing this paper the figures were still subject to verification).

Note: N_2= normal; LN = log-normal; * correlated with $\rho = 0.44$;
α^2= relative contribution to uncertainty with regard to failure (see eq. (32)).

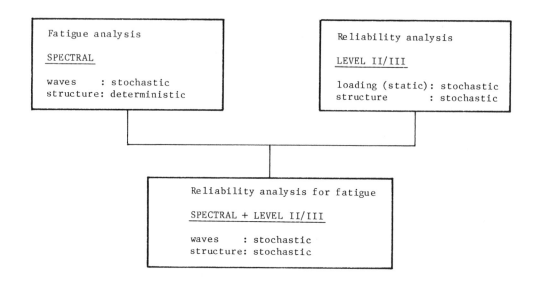

Fig. 1 In carrying out a reliability analysis for the fatigue of offshore structures spectral and level II/III reliability techniques have to be combined

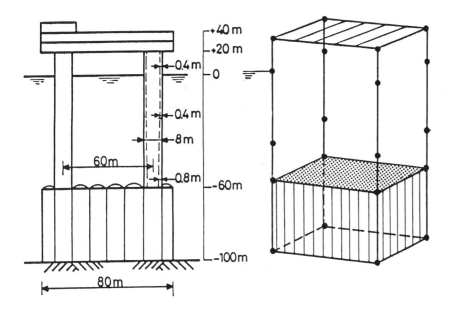

Fig. 2 Gravity structure with computer schematisation and structural data

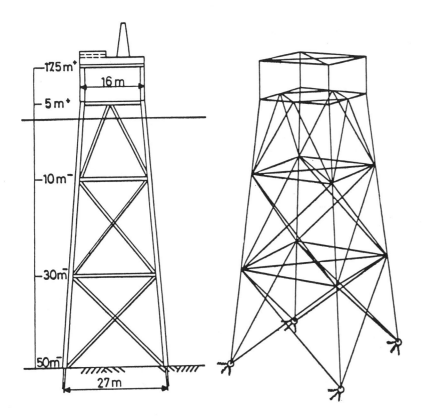

Fig. 3 Jacket type structure with computer schematisation and structural data

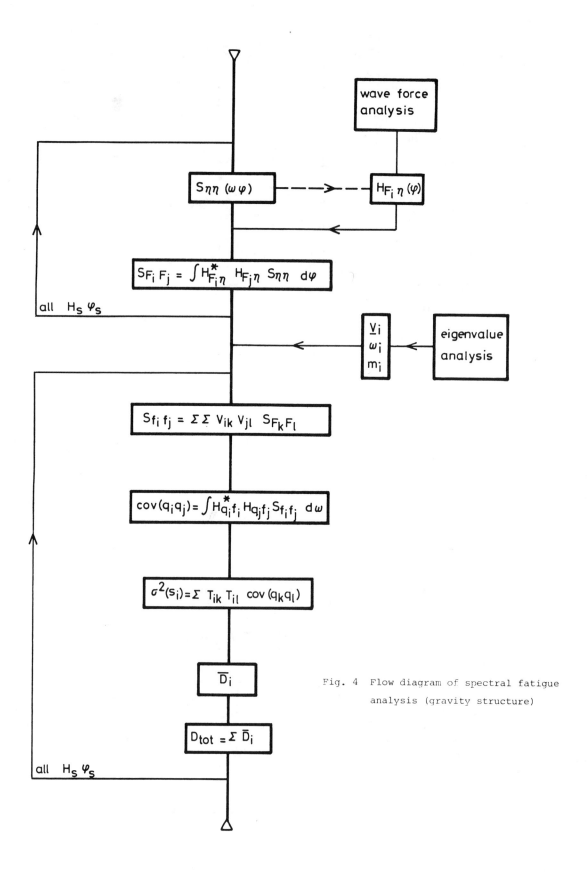

Fig. 4 Flow diagram of spectral fatigue analysis (gravity structure)

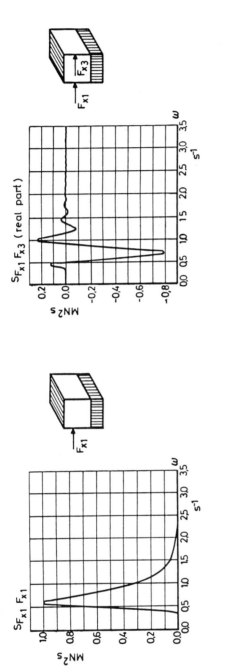

Fig. 5 Spectra and cross-spectra for the wave forces F_{x1}, F_{y1} and F_{x3} for sea state with $H_s = 5.0$ m and $\phi_s = 0$ (x-axis is principal direction)

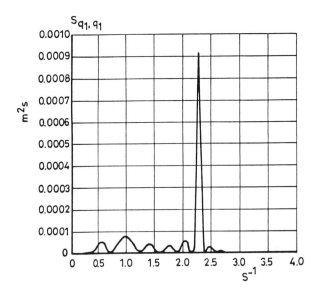

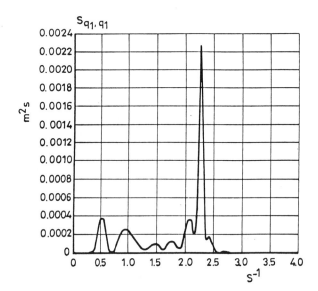

Fig. 6 Spectrum $S_{q_1 q_1}$ for the modal coordinate of the first vibration mode in a uni-directional (upper) and a multi-directional (lower) sea state (H_s = 5.00 m, ϕ_s = 0.0 rad)

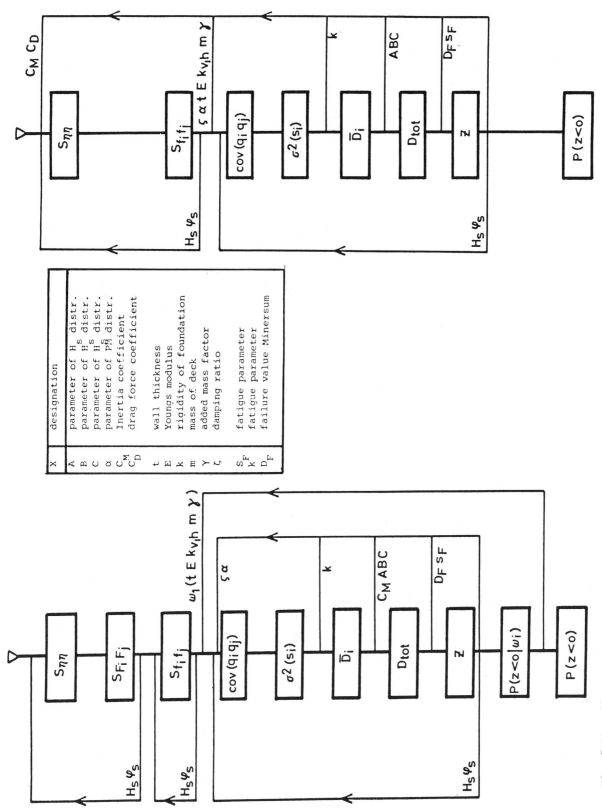

Fig. 7 Flow diagrams for gravity (left hand side) and jacket (right hand side) reliability analysis

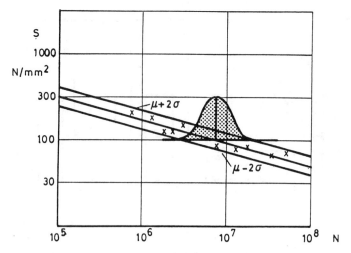

Fig. 8a Hot spot S-N diagram for specific detail (average and scatter band)

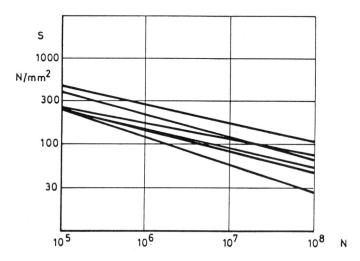

Fig. 8b Average Hot spot S-N curves for various details

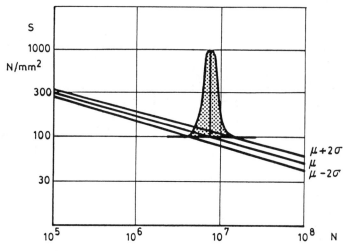

Fig. 8c Mean and uncertainty of the estimated average S-N curve for arbitrary joint.

REFERENCES

BACK, J. DE, VAESSEN, B.H.B., 1981, <u>Fatigue and Corrosion fatigue behaviour of offshore steel structures</u>, ECSC Convention 7210-KB/6/602 (J.7.1 f/76), final report.

BATTJES, J.A., 1972, "Long-term wave height distribution at seven stations around the British Isles", <u>Deutsche Hydrografische Zeitschrift 25</u>, No. 4.

BOUMA, A.L., MONNIER, Th., VROUWENVELDER, A., 1979, "Probalistic reliability analysis", <u>Proceedings Second International Conference on the Behaviour of Offshore Structures</u>, London, England, 1979, Volume 2, Paper 85, pp 521-542. Cranfield, Bedford, England: BHRA Fluid Engineering.

CIRIA, 1979, <u>Rationalisation of safety and reliability factors in structural codes</u>, Ciria Report 63, 6 Storey's Gate, London, <u>SW1P 3AU</u>.

LEEUWEN, J. VAN, SIEMES, A.J.M., 1979, "Miner's rule with respect to plain conrete", <u>Proceedings Second International Conference on the Behaviour of Offshore Structures</u>, London, England, 1979, Volume 1, paper 45, pp 591-610, Cranfield, Bedford, England: BHRA Fluid Engineering.

KARADENIZ, H., 1980, <u>An integration procedure to calculate response spectral moments of offshore structures</u>, Delft University of Technology, Department of Civil Engineering, Applied Mechanics Division.

RACKWITZ, R., 1979, <u>Principles and methods for a practical probabilistic approach to structural safety</u>, JCSS, Subcommittee for First-Order Reliability Concepts for Design Codes. Published by CEB.

SIEMES, A.J.M., 1982, Fatigue of plain concrete in uniaxial compression - random loading tests, <u>Proceedings of the IABSE Colloquium on Fatigue of Steel and Concrete Structures</u>, Lausanne, 1982.

STUPOC V, 1980, <u>Probalistic reliability analysis for offshore structures</u>, IRO publication, Delft.

THOMSON, W.T., 1972, <u>Theory of vibration with applications</u>, Engelwood Cliffs, Prentice Hall.

VUGTS, J.H., HINES, I.M., NATARAJA, R., SCHUMM, W., 1979, "Modal superposition v. direct solution techniques in dynamic analysis of offshore structures", <u>Proceedings Second International Conference on the Behaviour of Offshore Structures</u>, London, England, 1979, Volume 2, paper 49, pp. 23-42, Cranfield, Bedford, England: BHRA Fluid Engineering

ACKNOWLEDGEMENTS

This paper contains some of the results of a research project on the safety of offshore structures which is being carried out within the framework of Marine Technological Research (MaTS). The MaTS projects in the Netherlands are being conducted under the auspices of the Ministry for Economic Affairs and with financial contributions from the public authorities and from industry.

The safety project is being carried out under the responsibility of a steering group and is more directly serviced by a project group. These committees were constituted as follows:

<u>Steering Group</u>: Prof.Ir. A.L. Bouma, Chairman (Delft University of Technology); Prof.Dr.Ir. E.W. Bijker (Delft University of Technology); Ir. R.W. de Sitter (Hollandse Beton Groep, Rijswijk); Ir. J. Wolters (Protech International, Schiedam).

<u>Project Group</u>: Ir. Th. Monnier (TNO Institute for Building Materials and Building Structures, Rijswijk), Chairman; Ir. E. Calle (Delft Soil Mechanics Laboratory); Dr. H. Karadeniz (Delft University of Technology); Ir. S.E. van Manen (TNO Institute for Building Materials and Building Structures, Rijswijk); Ir. W. Meyer (TNO Institute for Building Materials and Building Structures, Rijswijk); Dr.Ir. G. van Oortmerssen (Netherlands Ship Model Basin, Wageningen); Ir. A. Paape (Delft Hydraulics Laboratory); Ir. F.P. Smits (Delft Soil Mechanics Laboratory); Dr.Ir. J. Strating (Protech International, Schiedam); Ir. A. Vrouwenvelder (TNO Institute for Building Materials and Building Structures, Rijswijk); Ir. C. v.d. Zwaag (Fugro, Leidschendam).

Technical Lecture Sessions
STRUCTURES

Pipeline Systems

A METHOD OF ANALYSIS FOR COLLAPSE OF SUBMARINE PIPELINES

Peter E. de Winter
Institute for Building Materials and Building Structures (IBBC)
of the Netherlands Organization for Applied Scientific Research (TNO), The Netherlands

SUMMARY

In this paper an approximate method of analysis is discussed on the circumferential stability of pipes, which are loaded by a combination of forces. In a relative simple manner a good insight is gained into the relation between pipe parameters and ultimate load combinations and deformations. The method of analysis is based on an analytical (kinematic) model, with which calculations both in the elastic and the plastic range can be carried out. With a relatively small amount of arithmatical work limit state interaction formulae can be derived for a number of loading cases.

INTRODUCTION

In deep water the loads and the deformations that a pipeline can safely withstand are reduced by external water pressure. The magnitude of such reduction depends to a large extent on the pipe parameters and the combination of loads acting upon the pipe. If the load carrying capacity or the deformation capacity is exeeded, the pipe will collapse.
This paper presents an approximate method of analysis enabling the relation between pipe parameters, load combinations and the deformation capacity to be determined (see table).

pipe parameters	load combinations	deformations
diameter wallthickness initial-out-of-roundness yield stress	bending external liquid pressure axial force	curvature ovalization axial strain

Parameters in the mathematical model

In 1975, Protech International first commenced experimental work on this subject. Later, the Netherlands Organization for Applied Scientific Research (IBBC-TNO), Protech International and the Universities of Delft and Eindhoven carried out further investigations to determine the mechanics of pipeline collapse, constantly aiming for easily applicable, limit state interaction formulas. Subject investigations have led to the developement of an approximate method describing collapse that may occur in submarine pipelines.

The method can be explained with reference to a physical interpretation: The pipe cross section is replaced by two elements:
- a bending element for modelling the circumferential stiffness.
- a rocker element to simplify the geometric non-linear behaviour of a tube.

The collapse load of such a model can be calculated in a relatively simple manner. In frame analysis a similar method is being used for the calculation of the buckling load of framed structures. To facilitate explanations, a simplified yield function has been adopted for the description of the physical non-linear behaviour. For practical application within the pipeline industry preference might be given to more advanced yield functions.

The method is demonstrated for three different loading cases:
- collapse under combined external pressure and axial force, including initial out-of-roundness
- buckle propagating pressure, including initial out-of-roundness and axial force
- collapse curvature under combined bending and external pressure.

It is the aim of this paper to attract attention for the method of analysis because it is considered very well possible that this method is also applicable to other stability problems apart from tubes and frames.

DISCRIPTION OF THE MODEL

To establish computational rules that describe the behaviour of a pipe is by no means a simple matter, because it involves simplifying the physical reality to a model. To do this, all sorts of assumptions have to be made and limitations imposed. Before a choice can be made, it must first be ascertained what requirements are to be applied to the model.

The method of analysis to be used must give insight into the load carrying capacity and deformation capacity of thick-walled pipelines. This capacity is limited by the occurrence of large deformations (instability of the pipe wall) and furthermore by cracking of the material of the pipe wall or the welds.
Hence it follows that, with the model to be formulated, it should be possible to perform stability calculations in the plastic range of behaviour. In addition, the shape deviations in the pipe material, which are liable to play an important part, will also have to be incorporated in the model.
Besides a number of 'technical' requirements, some requirements of a 'didactic' nature are applicable to the model. Thus, it must be simple to understand, and the amount of arithmetical work should be limited.

From experimental simulation it was observed that almost all deformation in the collapsed cross section of a pipe was concentrated in four "plastic hinges" (see fig. 1 and 2). Therefore the following model for ovalization has been adopted:
- four rigid quarters of a circle
- mutually connected by pin joints

This part of the model is called the rocker element. The geometrical non-linear deformation in circumferential direction can now simply be determined with only one parameter; the ovalization angle β. By virtue of the four hinges the rocker element can be deformed without any resistance. Therefore also a flexural element, representing the circumferential stiffness is needed.
The stiffness of the pipe cross section can be modelled by either rational springs or by a bending element (see fig. 3 and 4). Rotational springs have been used to calculate the buckling load of a column (Shanley, 1946). This same method can be used for pipes, using quarters of a circle instead of a column. Another possibility was published by Wilhelm where he used a bending element for the stability analysis of portal frames (Wilhelm, 1964). This model can also be applied to pipes.
In principle these kinematic models produce upperbound solutions. The model proposed by Wilhelm however allows, in principle, for lower bound solutions (Vrouwenvelder and Witteveen, 1975). In view of future research, preference is given to this last model (see fig. 4).

For a good understanding it is essentially to know that only at the hinges the rocker element is connected to the bending element. When ovalization occurs, the rocker element transmits point loads to the bending element, so that resistance to the ovalizing deformation is developed. For clarity of representation the two rings are shown in fig. 4 as having different diameters. For the purpose of analysis they coincide (see figure 5).
Primarily, the load is applied to, and supported by, the rocker element. The bending element only serves to resist ovalization. However, its stiffness is affected by the loading that the rocker element supports. It is endeavoured to manage with 'manual' calculations, i.e. not having recourse to a computer. This means that derived functions must already be linearized at

an early stage, which in turn means that the model is valid only for small ovalizations. To describe the equilibrium of the model the principle of virtual work will be used, which remains valid in the plastic as well as in the elastic range of behaviour.

The fundamentals for the whole analysis are: plane sections remain plain, and the cross section remains perpendicular to the axis of the pipe. In the appendix, the static, kinematic and constitutive (elastic and rigid plastic) equations for the model are presented for the combined loading case of external pressure, axial force and bending. In the next sections some loading cases are worked through in order to demonstrate the calculation method.

COLLAPSE UNDER ACTION OF HYDROSTATIC PRESSURE AND AXIAL FORCE

The model will be elaborated for the interaction of normal force (tension or compression) and external liquid pressure. The effect of initial out-of-roundness will also be considered. Pressure ovalization curves will be determined for both elastic and for plastic behaviour. The greatest pressure that the pipe can resist - the pressure at which implosion occurs - is attained at the intersection of the elastic and the plastic curve (see fig. 9). In reality the critical pressure will be somewhat less because of elasto-plastic material behaviour.

The deformation behaviour is described with the aid of the equations derived in the appendix (43, 44, 45, 46). For a straight pipe ($\kappa = 0$) these equations reduce to:

$$M = 0 \tag{1}$$

$$F = \int \sigma_{aa} \, dA \tag{2}$$

$$8PR^2\beta = 4fR \tag{3}$$

$$f = \frac{2m_r}{R} \tag{4}$$

In order to simplify the calculations, the stresses in the pipe wall are, at an early stage, summed to plate forces and moments*).

The normal force n_r is assumed to be constant around the circumference. For a circular section this is exactly the case, but in an ovalized section it involves a small error in the order of magnitude of the ovalization angle β.

On the assumption that the magnitude of the normal force in axial direction is also the same at all cross-sections this expression gives:

$$F = \int \sigma_{aa} dA = \int n_a ds = 2\pi R n_a \tag{5}$$

The <u>elastic</u> pressure-ovalization curve can be determined by assuming the bending element to behave elastically. Combining the above equations (3, 4) with the elastic equation for the bending element (47) the following relation between ovalization and pressure is obtained:

---

*) Note: By plate forces is understood the stress resultant per unit length.

$$\beta = \frac{\alpha P_E}{\alpha P_E - P} \beta_o \qquad (6)$$

where: α constant*) $\qquad \alpha = \dfrac{2}{3(\frac{\pi}{2} - 1)} \simeq 1.17$

P_E the elastic collapse pressure

β_o the initial out-of-roundness

From the above formula can be concluded that, in the elastic range, the ovalization is unaffected by axial force. This means that the elastic collapse pressure ($P_{cr} \rightarrow \alpha P_E$, $\beta \rightarrow \infty$) is also unaffected by axial force. This result is in agreement with the exact solution (Flügge, 1960). Recent test results on high strength oil well casing (Kyogoku et.al., 1981) support these results.

The above expression gives an over-estimation ($P_{cr} = \alpha P_E$) with respect to the exact solution for the collapse pressure ($P_{cr} = P_E$). A more accurate solution is found by increasing the number of hinges in the rocker element. Incidentally it is of importance to realize that for practical cases the over-estimation will be less, for in deep water the plastic collapse load (P_p) for thick-walled pipes is smaller than the elastic collapse load (P_E) (see fig. 9).

The <u>plastic</u> pressure-ovalisation curve can be determined by assuming rigid-plastic material behaviour for the bending element. For explanatory reasons a simplified yield function (49) is used for the relation between biaxial normal force and bending. On working out the equations for rocker element and bending element (3, 4, 49) the relation between ovalization and pressure for plastic material behaviour, is obtained:

$$\beta = \frac{m_p}{PR^2} \left(1 - \left(\frac{n_a}{n_p}\right)^2 - \left(\frac{n_r}{n_p}\right)^2 + \frac{n_r n_a}{n_p}\right) \qquad (7)$$

where n_a = stress resultant in axial direction per unit length

n_r = stress resultant in circumferential direction

n_p = fully plastic stress resultant

$$m_p = \tfrac{1}{4} t^2 \sigma_e, \quad n_r = -PR, \quad n_p = t\sigma_e$$

Now two equations have been derived which represent the relation between ovalization and pressure. The first of these (6) describes an elastic stable state of equilibrium with the buckling load αP_E as its boundary. The second equation (7) describes the plastic unstable state of equilibrium. The critical pressure is attained at the intersection of the elastic and the plastic curve. After substitution of the stress resultants by the external forces the following equation holds:

$$\boxed{\left(\frac{P}{P_p}\right)^2 + \left(\frac{P}{P_p}\right)\left\{\left(\frac{R}{t}\right)\frac{4\alpha P_E}{\alpha P_E - P}\beta_o + \left(\frac{F}{F_p}\right)\right\} + \left(\frac{F}{F_p}\right)^2 - 1 = 0} \qquad (8)$$

Collapse under the action of external pressure and axial tensile (or compressive) force (pipes without endcaps)

*)Note: The constant α has been introduced as an abbreviated notation in order to simplify the formula. In addition, α is a measure of the over-estimation with respect to the exact solution (see below)

A reduced solution (F = 0, α = 1) has been presented before (Haagsma and Schaap, 1981).

The axial force F is defined as the sum of all the stresses in the axial direction. For a pipe with end caps this means that part of this force F is caused by pressure of the liquid on the end caps. The remaining external force N, the 'effective' force, will then be smaller. This can be formulated as follows:

$$\frac{F}{F_p} = \frac{N}{N_p} - \frac{1}{2}\left(\frac{P}{P_p}\right) \tag{9}$$

where $F_p = N_p = 2\pi R t \sigma_e$

Substitution into the above interaction formula gives:

$$\boxed{3/4 \left(\frac{P}{P_p}\right)^2 + \left(\frac{P}{P_p}\right)\left(\frac{4\alpha P_E}{\alpha P_E - P}\right)\left(\frac{R}{t}\right)\beta_0 + \left(\frac{N}{N_p}\right)^2 - 1 = 0} \tag{10}$$

Collapse under the action of hydrostatic pressure and axial
tensile (or compressive) force including end cap effect

For practical application in pipeline industry preference might be given to this latter formula. A graphical example is presented in fig. 10.

BUCKLE PROPAGATION

If the external water pressure is small a once initiated buckle will be limited to a local section of the pipeline. On the other hand, if the external water pressure is high enough, the buckle may be driven along a great length of the pipeline. If no structural precautions have been taken, propagation of the buckle may flatten the pipeline until shallower waters are reached.
Extensive investigations have been made (Palmer, 1975; Johns et.al., 1976; Mesloh et.al., 1976) into determining the minimum pressure required just to sustain propgation of the buckle. However little has been published about the influence of initial-out-of-roundness or axial force on the magnitude of the propagating pressure. A fundamental approach seems to be too complex for a convenient solution to this problem. With the calculation method presented here a simplified approach might be possible. Considering the propagationfront to consist of a series of deformed rings an interaction formula including the effect of initial out-of-roundness and axial force can be derived.

For calculation purposes the pipe is considered to consist of three different parts (fig. 11). The undamaged part, the flattened part and the propagating part in between. Considering the propagating part to consist of a series rings, the plastic unstable state of equilibrium will be formulated with the equations presented before. For the propagating part with length ℓ the equilibrium equation of the rocker element becomes:

$$\int_0^\ell 8PR^2\beta \, dx = \int_0^\ell 4fR \, dx \tag{11}$$

Only cross-sectional deformation is considered. Axial bending effects in the pipe wall are neglected. This means that in the virtual work equations a lower value is obtained for the internal work. Therefore this will give a lower value for the propagating pressure.

For convience the deformation (β) of the rings in the propagating part is assumed to be linear between the undamaged section and the flattened section.

$$\beta(x) = \beta_1 + \frac{(\beta_2 - \beta_1)}{\ell} x \tag{12}$$

For the undamaged section the relation between external pressure and ovalization, which has been derived, earlier on holds (6). For the flattened section maximum ovality is assumed.

$$\beta_1 = \frac{\alpha P_E}{\alpha P_E - P} \beta_o, \qquad \beta_2 = \frac{\pi}{4} \tag{13}$$

The same simplified yield function (49) will be used as has been done for the previous loading case. On working out the equations (4, 11, 12, 13) the following interaction formula holds:

$$\boxed{(P/P_p)^2 + 2 \left(\frac{R}{t}\right)\left(\frac{P}{P_p}\right) \left[\frac{\alpha P_E}{\alpha P_E - P} \beta_o + \frac{\pi}{4}\right] + \left(\frac{F}{F_p}\right)\left(\frac{P}{P_p}\right) + \left(\frac{F}{F_p}\right)^2 - 1 = 0} \tag{14}$$

Propagation pressure for pipes without end caps

Accounting for the effect of end caps the interaction formula becomes:

$$\boxed{3/4\,(P/P_p)^2 + 2\left(\frac{R}{t}\right)\left(\frac{P}{P_p}\right)\left[\frac{\alpha P_E}{\alpha P_E - P} \beta_o + \frac{\pi}{4}\right] + \left(\frac{N}{N_p}\right)^2 - 1 = 0} \tag{15}$$

Propagation pressure for pipes with end caps, including
the effects of initial-out-of-roundness and axial tensile
(or compressive) force

In case of zero out-of-roundness the propagating pressure becomes:

$$P/P_p = \frac{2}{3}\left\{-\frac{\pi}{2}\left(\frac{R}{t}\right) + \sqrt{\frac{\pi^2}{4}\left(\frac{R}{t}\right)^2 + 3 - 3\left(\frac{N}{N_p}\right)^2}\right\} \tag{16}$$

Approximation of the square root with a Taylor series simplifies this formula to:

$$\boxed{P/P_p = \frac{2}{\pi}\left(\frac{t}{R}\right)\left\{1 - \left(\frac{N}{N_p}\right)^2\right\}} \tag{17}$$

Propagation pressure for a circular pipe

Solutions for $N = 0$ are to be found in the literature (Richardson and Chang, 1980). These differ from one another only in so far as the constant is concerned (here: $2/\pi$). Despite the very rough approximations adopted, it may be concluded that the solution presented here is in reasonable good agreement with those in literature

From the interaction formulas presented above it can be concluded that the influence of initial-out-of-roundness is small and probably may be neglected. However there is no experimental evidence available to support that conclusion.

The axial force reduces the propagating pressure. The interaction formula suggests a parabolic relation between propagation pressure and axial tension. So limited axial tensile force has only a minor effect on the propagating pressure. High ratios of axial tension however substantially reduce the propagating pressure.

Because of the simplifications made, the above formulae have their limitations. Presently no test results are available for the investigation of the reliability and the limits to be set to the range of application.

A graphical example of the above formula is presented in fig. 12.

BENDING UNDER EXTERNAL PRESSURE

One of the loading cases that occur in connection with the installation of undersea pipelines is the load due to external pressure and bending. The pipe emerges from the pipelaying vessel at a slope and has to curve to the horizontal position in order to settle down on the sea bed. The deformation associated with this change of direction is imposed upon the pipe. Because of this imposed character it is of considerable importance to know what maximum curvature the pipeline can safely absorb.

In the preceding sections cases of ultimate loading have been considered. In this section a case of ultimate deformation will be considered. For completeness, formulae for both the elastic and the plastic curvature at collapse, will be derived.

Starting point is the equations which have been derived for the rocker element and the bending element (43, 44, 45, 46).

For elastic bending the following geometric non-linear moment - curvature relationship can be derived by equating (43, 45, 46, 47).

$$M = \left(\frac{c - EI\kappa^2}{c + EI\kappa^2}\right) EI \cdot \kappa \qquad (18)$$

where: $c = 8R^2(\alpha P_E - P)$, $EI = \pi ER^3 t$

A linear and a non-linear part are to be distinguished in this equation. The factor $EI \cdot \kappa$ constitutes the known physical linear part of the equation. The non-linear part is caused by ovalization of the pipe.

A stable state of equilibrium exists until the point is reached where the tangent to the load-deformation diagram is horizontal. For the moment-curvature diagram this point is found for $dM/d\kappa = 0$. The curvature at that point is expressed by:

$$\boxed{\kappa_{cr} = \sqrt{\frac{c(\sqrt{5}-2)}{EI}} = \sqrt{\frac{0,24c}{EI}}} \qquad (19)$$

curvature at collapse (elastic bending)

For pure bending (p=0) this relation reduces to:

$$\boxed{\kappa_{cr} = 0{,}42 \frac{t}{R^2}} \qquad (20)$$

Similar solutions for elastic bending are to be found in the literature (Brazier, 1927; Chwalla, 1933; Reissner and Weinitschke, 1963). These differ from one another only in so far as the constant is concerned (here: 0,42).

The range of application is limited to elastic bending. In case of zero pressure (p = 0) this means limitation to the curvature at which yielding starts in the extreme fibers ($\kappa_{cr} < \kappa_e$). This can be formulated as follows:

$$R/t > 0{,}42 \frac{E}{\sigma_e} \qquad (21)$$

For $E = 210000$ N/mm^2 and $\sigma_e = 320$ N/mm^2 this becomes $R/t > 276$ or $D/t > 550$. This value is larger by a factor of more than 10, with respect to the value commonly associated with undersea pipelines - which means that in order to determine the critical curvature it is not sufficient to base oneself on an elastic analysis.

The <u>plastic</u> analysis will be limited to a so-called "rigid plastic" analysis. In the case of rigid-plastic material behaviour deformations occur only when the yield point of the material has been reached. The main advantage of this schematization of the behaviour is that it substantially reduces the amount of arithmetical effort. A disadvantage is that there is then no way of directly relating stress to strain. Calculation of the deformations will have to be based on applying the normality principle as conceived in plastic theory.
The following fundamental points have been adopted for the purpose of the rigid-plastic analysis:
(1) The external pressure (P) is kept constant, while the curvature is increased from zero to the critical value.
(2) A three-dimensional yield surface for bending moment (M), external pressure (P) and rocker forces (f) is adopted.
(3) The normality principle is used for calculating the critical curvature
(4) The analysis is performed for pipes with end caps (hydrostatic pressure).

Starting from the circumferential stress distribution the (fully plastic) stress distribution in the axial direction is determined. For convenience the stresses in the pipe wall are summed up into plate forces and moments ($\int \sigma_{aa} \, dA = \int n_a ds$) The distribution of forces in the circumferential direction is described with the aid of two parameters, namely, the normal force (n_r) due to hydrostatic pressure (P) and the bending moment (m_r) due to ovalisation.
The circumferential bending moment varies from maximum at the point where the plastic hinges are formed to zero at the points of zero bending moment (fig. 13). For these 'extremes' (zero and maximum) the magnitude attainable by the axial force is determined.

The same simplified yield function which has been used in the preceding loading cases will be used for the fully plastic relation between normal force and bending acting upon two plane surfaces.

$$\frac{n_a}{n_p} = \frac{1}{2} \left\{ \frac{n_r}{n_p} \pm \sqrt{4 - 3\left(\frac{n_r}{n_p}\right)^2 - 4\frac{m_r}{m_p}} \right\} \qquad (22)$$

Since the bending moment (m_r) is a function of the angle ϕ (fig. 13), the axial force ($n_a = n_a(\phi)$) is also a function of that angle. It is of course possible to determine the magnitude of the forces in the axial direction as a function of the bending moment (m_r). However, integration of that function (this being necessary for determining the magnitude of the external bending moment M and external force F) can then no longer be done analytically. For this reason a sine interpolation - which is integrable - will be applied to the distribution of the plate forces in the axial direction. Points of reference are: $\phi = 0$ and $\phi = \frac{\pi}{2}$ where the moment m_r attains its maximum and the point $\phi = \frac{\pi}{4}$ where the moment m_r equals zero.

$$\frac{n_a(\phi)}{n_p} = \frac{1}{2} \left\{ \frac{n_r}{n_p} \pm W_1 + (W_2 - W_1) \sin 2\phi \right\} \tag{23}$$

where: $W_1 = \sqrt{4 - 3 \left(\frac{n_r}{n_p}\right)^2 - 4 \frac{m_r}{m_p}}$

$W_2 = \sqrt{4 - 3 \left(\frac{n_r}{n_p}\right)^2}$

On substitution the above expression in the equilibrium equation of the rocker element (44) the force in axial direction is obtained:

$$\boxed{F = \pi n_r R = - \pi P R^2} \tag{24}$$

Force in axial direction (= pressure on end cap)

This expression is readily recognizable as representing the force caused by the pressure acting on the end cap. The bending moment can be found by equating (23, 43):

$$\boxed{M = (1-\beta) M_p \left\{ \frac{1}{6} W_1 + \frac{1}{3} W_2 \right\}} \tag{25}$$

Yield surface for the bending moment

In this equation the factor $(1-\beta)$ constitutes the geometric non-linear part that is caused by ovalization of the pipe. The factor $(\frac{1}{6} W_1 + \frac{1}{3} W_2)$ describes the reduction of the bending moment by ovalization forces and by hydrostatic pressure.
With the aid of the above function the three-dimensional yield surface (M, f, P) is fairly simple to determine. It is shown in normalized geometric-linear form in Fig. 14.

The analysis has been established for rigid-plastic material behaviour. There is then no direct relation between stress and strain. Because ovalization remains limited ($\beta \leq 0.1$) it is assumed that the normality principle can permissibly be applied here. The increase in the deformations is determined with the aid of this principle. This is conditional upon a geometric-linear analysis, which is the reason why the three-dimensional yield surface consists only of the geometric-linear terms of the functions that describe the bending moment M.

In the analysis given here the pressure P is kept constant and the curvature is increased from zero until collapse of the pipe occurs. This means that it will be necessary only to consider a straight section through the three-dimensional yield surface (see fig. 14, and 15). The deformation parameter κ is associated with the bending moment M; the displacement parameter $R\Delta\beta$ is associated with the ovalization force f. Making use of the normality principle, the relation between increase in deformation and force is:

$$\frac{4R d\beta}{d\kappa} = - \frac{dM}{df} \tag{26}$$

On solving this differential equation and evaluating for point C the quotient $d\beta/d\kappa$ tends to ∞. This means that ovalization greatly increases (implosion) while the curvature remains constant.

The analysis can be substantially simplified if the yield contour (the line BC in fig. 15) is straight, for then the ratio dM/df is a constant, so that the above differential equation is simple to solve. For a linearized surface the following equation holds:

$$\frac{4R d\beta}{d\kappa} = \frac{\frac{1}{6} M_p \sqrt{4 - 3(P/P_p)^2}}{\frac{1}{4} f_p (4-3(P/P_p)^2)} \tag{27}$$

On working out the above differential equation and taking the integration constants κ_o and β_o as zero (the pipes under consideration are assumed initially to be truly circular and straight) the relationship between curvature and ovalization is obtained:

$$\beta = \frac{4}{3} \frac{R^2}{t} \frac{1}{\sqrt{4-3(P/P_p)^2}} \kappa \tag{28}$$

For point C, where the critical curvature is attained, the following equations are applicable (22, 25, 28, 45).
On working out these equations the function for the critical curvature is obtained:

$$\boxed{\kappa_{cr} = \frac{3}{4} \frac{t}{R^2} \frac{(1 - \frac{3}{4}(P/P_p)^2)^{1.5}}{\{1 + 2(\frac{P}{P_p})(\frac{R}{t}) - \frac{3}{4}(P/P_p)^2\}}} \tag{29}$$

Plastic collapse curvature

The above relation is graphically represented in fig. 16

Because of the assumptions made the above formula has a limited range of application. From the condition that the ovalization angel β is assumed to be small ($\beta < 0.1$) it follows for zero pressure that:

$$\frac{\kappa}{\kappa_e} < 0.075 \frac{tE}{R\sigma_e} \tag{30}$$

For $E = 210000$ N/mm^2 and $D/t = 25$ the validity of the above formula is limited to $\kappa_{cr} < 9 \kappa_e$

CONCLUDING REMARKS

For reasons of consistency, throughout this paper the same yield function (49) has been used. This yield function applies to a ring ($m_a = 0$). For practical application the results can be improved by using more advanced yield functions and accounting for axial plate bending in the pipewall ($m_a \neq 0$) (De Winter, 1981b).
It has been demonstrated that the proposed method of analysis yields results in a relative simple manner. The elementary loading cases have been checked, succesfully, against known solutions from literature.

ACKNOWLEDGEMENT

This contribution forms part of the Marine Technological Research Program sponsored by the Dutch government. The work is carried out jointly by the Institute for Building Materials and Building Structures of the Netherlands Organization for Applied Scientific Research (IBBC-TNO), Protech International and the Universities of Delft and Eindhoven. The author wishes to thank the members forming the offshore pipelines working group for their support and valuable comments.

REFERENCES

BRAZIER, L.G., 1927, "On the flexure of thin cylindrical shells and other thin sections," Proceedings Royal Society, London, 1927, volume 116, pp 104-114.

CHWALLA, E., 1933, "Reine Biegung slanker, dünwandiger Röhre mit gerader Achse", Zeitschrift für angewandte Mathematik und Mechanik, vol 13, number 1, pp 48-53.

DE WINTER, P.E., 1981a, "Deformation Capacity of Steel Tubes in Deep Water", Proceedings 13th Annual Offshore Technology Conference, Houston, 1981, Paper 4035.

DE WINTER, P.E., 1981b, Strength and Deformation Properties of Pipelines in Deep Water, Second stage MaTS project, Industriele Raad voor de Oceanologie, Delft, The Netherlands.

FLÜGGE, W., 1960, Stresses in shells, Berlin, Springer-Verlag.

HAAGSMA, S.C. and SCHAAP, D., 1981, Collapse resistance of submarine lines studied", Oil and Gas Journal, number 2, pp 86-95.

JOHNS, T.G. et al., 1976, "Propagating Buckle Arrestors for Offshore Pipelines", Proceedings, 8th Annual Offshore Technology Conference, Houston, 1976, Paper 2680.

KYOGOKU, T., et al., 1981, "Experimental study of the effect of axial tension load on the collapse strength of oil well casing", Proceedings 13th Annual Offshore Technology Conference, Houston, 1981, paper 4108.

MESLOH, R. et al., 1976, "The propagating buckle", Proceedings, 1e International Conference on the Behaviour of Off-Shore Structures, 1976

PALMER, A.C., 1975, "Buckle propagation in submarine pipelines", Nature, volume 254, number 5495, pp 46-48.

REISSNER, E. and WEINITSCHKE, H.J., 1963, "Finite pure bending of circular cylindrical tubes", Quarterly of Applied Mathematics, pp 305-319.

RICHARDSON, T.W.G. and CHANG, H.H., 1980, "Buckle arrestors for offshore pipe lines", Pipe line industry, volume 51, number 7, pp 47-51.

SHANLEY, F.R., 1946, "The column paradox", Journal of Aeronautical Science, volume 13, number 12, pp 578.

TIMOSHENKO, S.P. and GERE, J.M., 1961, Theory of elastic stability, Tokyo: McGraw-Hill Kogakusha.

VROUWENVELDER, A. and WITTEVEEN, J., 1975, "Lower bound approximation for elastic buckling loads", HERON Volume 20, number 4, Delft

WILHELM, P., 1964, "Stability analysis of portal frame structures in the elastic and the plastic range (in Dutch)", Constructies (De Vries-Robbé), Gorinchem, The Netherlands.

APPENDIX

For the kinematic model, three degrees of freedom will be considered (see figure 6)
- axial deformation (ε_{ao})
- curvature (κ)
- ovalisation (β)

The external loads considered are:
- axial force (F)
- bending moment (M)
- external pressure (P)

Basic assumptions are: Bernoullis hypothesis is valid and the cross section remains perpendicular to the axis of the tube. For the sake of simplicity, formulas are linearised at an early stage and the ovalisation angle β is considered to be small.

Rocker element
Geometry   (see figure 8)

undeflected: z	$= R \sin \theta$	(31)
deflected : $z+u_z$	$= R(\sin(\theta+\beta)-\sin\beta)$	(32)
linearised : $z+u_z$	$= z(1-\beta)$	(33)

Displacements along the axes:

u_z	$= R(\cos\beta-\sin\beta-1)$	(34)
u_x	$= R(\cos\beta+\sin\beta-1)$	(35)

Linearised:

$u_z = -R\beta$ $1°$ variation $\delta u_z = -R\delta\beta$ (36)

$u_x = R\beta$ $1°$ variation $\delta u_x = R\delta\beta$ (37)

Area change of the cross section

$$\Delta A = 4R^2 \sin^2\beta \tag{38}$$

Linearised:

$\Delta A = 4R^2\beta^2$ $1°$ variation $\delta\Delta A = 8R^2\beta\delta\beta$ (39)

Kinematic equations

Strain in axial direction:

$$\varepsilon_{aa} = \varepsilon_{ao} + \kappa \cdot (z+u_z) = \varepsilon_{ao} + \kappa \cdot z(1-\beta) \tag{40}$$

$$1° \text{ variation } \delta\varepsilon_{aa} = z(1-\beta)\delta\kappa - \kappa z \delta\beta + \delta\varepsilon_{ao} \tag{41}$$

Static equations

Equilibrium of the cross section is expressed using virtual work theory (see figure 6)

$$\int \sigma_{aa} \cdot \delta\varepsilon_{aa} dA = M\delta\kappa + F\delta\varepsilon_{ao} + P\delta\Delta A - 4fR\delta\beta \tag{42}$$

After substitution of the above derived variations, an equation is found which is valid for random variation of the three degrees of freedom. Therefore, three equilibrium equations may be written:

$$M = \int \sigma_{aa} \, z(1-\beta) \, dA \tag{43}$$

$$F = \int \sigma_{aa} \, dA \tag{44}$$

$$8PR^2\beta = -\int \sigma_{aa} \cdot \kappa \cdot z \, dA + 4fR \tag{45}$$

The first two equations can be easily recognised. The external bending moment (M) equals the resulting bending moment caused by axial stresses in the deformed model. The same applies for the axial force (F). The third equation shows the connection between both external pressure and bending with the ovalisation of the tube.

Bending element

The bending element is loaded only by those (opposite) contact forces (f) which act on the rocker element. Stiffness of the bending element is nevertheless influenced by the load acting on the rocker element.
The starting point is a linear analysis in which the displacements on the horizontal and vertical axis are of equal magnitude.

Static equation

Because of symmetry it is sufficient to consider the equilibrium of a quarter of the cross section (see figure 7)

$$f = \frac{2m_r}{R} \tag{46}$$

Deformation of the bending element

For elastic deformation only, the following relationship between diameter change and bending in the circumferential direction may be derived (Timoshenko and Gere, 1961):

$$R(\beta - \beta_o) = u_x = -u_z = \frac{R^2 (\frac{\pi}{2} - 1) m_r}{2EI} \tag{47}$$

For rigid plastic material behaviour, a simplified yield function is used. The yield function describes the interaction between bi-axial bending and extension.
On the assumption of three zones in an elementary block of material, each satisfying the Von Mises yield condition, the following equation can be derived:

$$\left(\frac{m_r}{m_p}\right)^2 + \left(\frac{m_a}{m_p}\right)^2 - \frac{m_r m_a}{m_p^2} - \left\{1 - \left(\frac{n_r}{n_p}\right)^2 - \left(\frac{n_a}{n_p}\right)^2 + \frac{n_r n_a}{n_p^2}\right\}^2 = 0 \tag{48}$$

where: $m_p = \frac{1}{4} t^2 \sigma_e$, $n_p = t\sigma_e$, $n_r = -PR$

For a ring the plate bending moment in axial direction equals zero: $m_a = 0$. The bending moment in circumferential direcion m_r then becomes:

$$m_r = m_p \left\{1 - \left(\frac{n_a}{n_p}\right)^2 - \left(\frac{n_r}{n_p}\right)^2 + \left(\frac{n_a}{n_p}\right)\left(\frac{n_r}{n_p}\right)\right\} \tag{49}$$

The normality rule is applied for deriving the deformation in the plastic area (see interaction between bending and external pressure).
Based on the relationships presented above, the interaction between bending, external pressure and axial force can be derived.

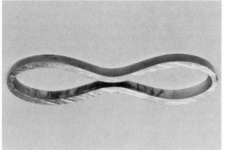
fully deformed section after collapse

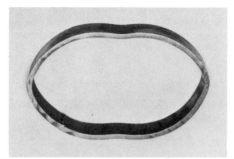
deformed section after the onset of collapse

theoretical idealization

figure 1

figure 2

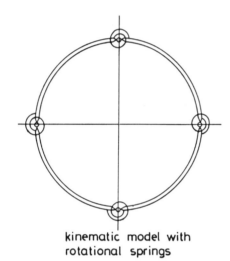
kinematic model with rotational springs

figure 3

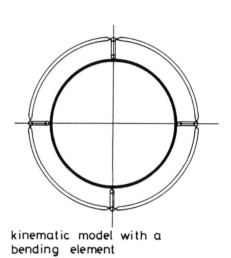
kinematic model with a bending element

figure 4

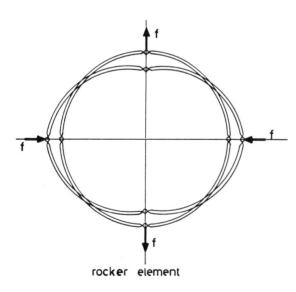
rocker element

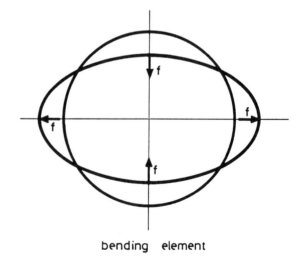
bending element

figure 5

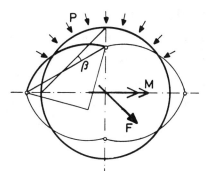

ovalisation due to external pressure and bending

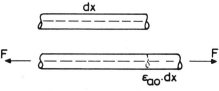

axial deformation due to axial force

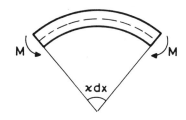

curvature due to bending moments

figure 6

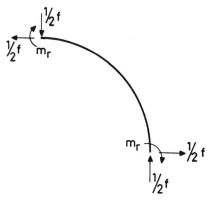

equilibrium of the bending element

figure 7

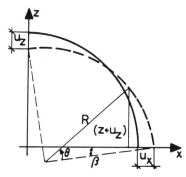

deformation of the rocker element

figure 8

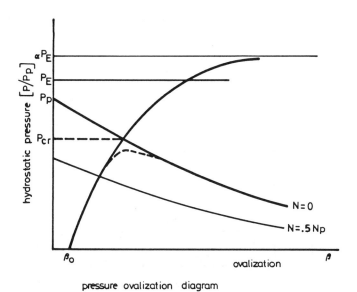

pressure ovalization diagram

figure 9

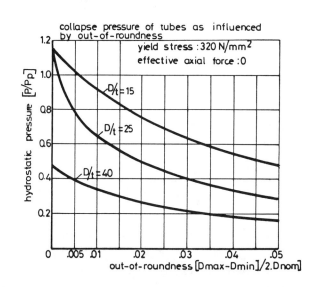

figure 10

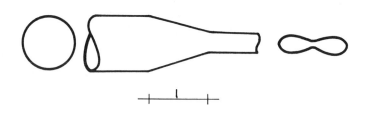

assumed deformation in the propagating section

figure 11

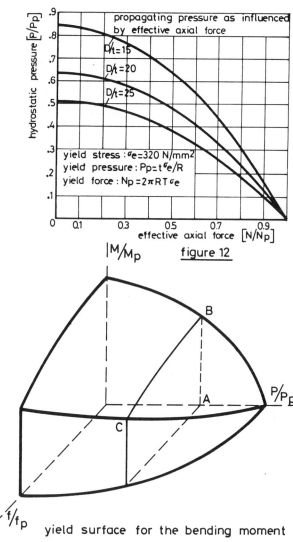

figure 12

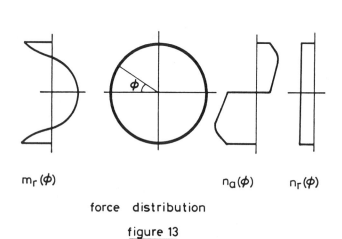

$m_r(\phi)$ $n_a(\phi)$ $n_r(\phi)$

force distribution

figure 13

yield surface for the bending moment

figure 14

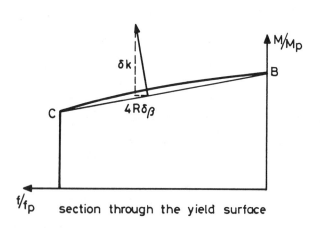

section through the yield surface

figure 15

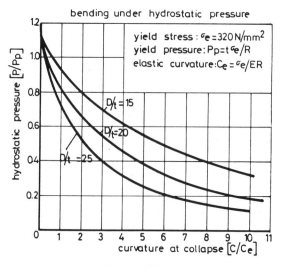

figure 16

ON THE INITIATION OF A PROPAGATING BUCKLE IN OFFSHORE PIPELINES

S. Kyriakides
University of Texas at Austin

C.D. Babcock
California Institute of Technology

SUMMARY

Initiation of a propagating buckle in an undersea pipeline is studied in this work. This type of buckling can occur at any pressure above the propagation pressure which is defined as the minimum pressure that can sustain this mode of buckling after initiation. The pressure at which this type of failure is initiated depends upon the loading and the condition of the pipe. In this work, the initiation is studied for the cases of (1) combined bending and external pressure acting upon a nominally perfect pipe, (2) uniform pressure acting on a damaged section of a pipeline and (3) uniform pressure acting on a pipe that has been buckled by a bending moment. The damage considered is that which is likely to occur during the installation or operational life of a pipeline. This is typically caused by accidents occuring during installation or by objects striking the pipeline.

For local damage caused by external objects, it was found that the data for the initiation pressure could be coalesced if the initiation pressure was normalized by the propagation pressure and the damage was characterized by the dimensions of the most damaged section. Both point-type damage and line-type damage (normal to the axis of the pipe) were considered in the experimental program. For the case of bending buckle damage, it was found that the pressure at which a propagating buckle initiated was dependent upon the post buckled bend angle of the pipe but that this dependence was not too strong. The combined bending and external pressure results showed the transition from a local buckle to a propagating buckle as had been previously reported. It was interesting, however, that this transition pressure was not much higher than the propagation pressure.

I. INTRODUCTION

As the exploration and exploitation of the mineral and oil deposits of the continental shelf and the deeper waters beyond expands, pipelines laid in these offshore areas have become the main means of transportation of these important raw materials. Worldwide, over 18,000 miles of offshore pipeline is in place. The demand for large diameter pipelines in ever increasing depths applies a continuous pressure on the existing technology. High external pressures, more severe currents, rougher seas during the laying process, unstable and unexplored sea floors, are all factors that have to be considered in the design of the pipeline. The laying process, the trenching process, the testing and pressurization process and the pipeline operation over a long period of time must be taken into consideration in the design.

One of the problems that gains more prominence in deeper waters is that of the "Propagating Buckle" [1-3]. Propagating buckle is the name given to a pipe collapse failure that can occur if a pipe is locally dented. Under the influence of the external pressure, the geometric imperfection induced causes local buckling. If the external pressure is high enough this buckle propagates, flattening the pipeline. Such a propagating buckle stops only if the external pressure is reduced below a critical value or if it is physically arrested. Arresting devices used in practice are described in ref. 4-6. The most commonly used devices are ring stiffeners that locally increase the wall bending rigidity of the pipe, thus prohibiting a buckle from passing beyond that point.

The lowest pressure which will sustain a buckle in propagation is given the name <u>Propagation Pressure</u> (P_p). It is a characteristic pressure of the pipe and was shown in ref. 1 to be dependent on the diameter to thickness ratio of the pipe, the yield stress and the post yield slope of the pipe material stress-strain behavior. An empirical relationship is given by

$$P_p \simeq \sigma_o [10.7 + 0.54 (\frac{E'}{\sigma_o})] (\frac{t}{D})^{2.25}$$

For the range of material and geometric parameters used in offshore pipeline applications, the propagation pressure ranges between one-half to one-tenth of the pipe Buckling Pressure (P_C). The buckling pressure is determined by considering the buckling of an infinitely long pipe under the action of uniform external pressure [7]. If the pipe has a large diameter-to-thickness ratio (D/t) and is relatively free of geometrical initial imperfections, this pressure is accurately given by

$$P_C = \frac{2}{1-\nu^2} E (\frac{t}{D})^3$$

Corrections to this formula are available to take into account thick wall effects, initial out of roundness, plasticity, etc. [8]. However, these corrections do not change the basic nature of this buckling or collapse pressure. Thus a propagating buckle, once initiated, can propagate at any pressure bounded by the buckling pressure (upper limit) and the propagation pressure (lower limit). However, the pipe must first be locally damaged to initiate the propagation.

The causes of local damage, in an adverse environment such as the ocean floor, are numerous [9,10,11]. They can occur during all stages of the pipeline's life, from the laying operation to the time the pipeline completes its operational life span. During the laying operation, sudden loss of tension or other off-design conditions can cause the bending moment to exceed the maximum allowable value under the combined effect of bending, tension and external pressure [12]. Such an event causes the pipe to locally buckle, thus generating a possible initiation point for a propagating buckle. References 9, 13 and 14 give examples of such occurrences. Similar problems have been recorded during the trenching operation [7].

Anchors, fishing gear and direct impact by foreign objects have caused catastrophic failures [10,11,15] in both buried and unburied pipelines. Many such accidents have been recorded in the North Sea as well as the Gulf of Mexico. Recent pipelaying activities in arctic waters have exposed pipelines to the additional hazard of moving blocks of ice [16]. Pipelines in the area of the Mississippi River delta and other similar offshore areas with unstable sea floor conditions have been heavily damaged by landslides [17,18]. Local damage (dents) produced by such causes as the ones listed above are potential initiation points for a propagating buckle. This paper deals with the practical problem of assessing the probability that a given local damage will initiate a propagating buckle.

The pressure at which local geometric imperfections transform themselves into the correct profile and develop into a propagating buckle is given the name <u>Initiation Pressure</u> (P_I). This value is bounded by the propagation and buckling pressures of the pipe ($P_p < P_I < P_C$). Whereas P_C and P_p are characteristic pressures of the pipe, P_I is a pressure that is dependent upon the particular damage considered. Clearly, the shape and magnitude of the geometric imperfection, as well as the material parameters of the pipe, influence the value of P_I. Obviously the geometry of imperfections inflicted upon the pipe can vary in an infinite number of ways. This paper examines three classes of geometric imperfections which are thought to apply to a high percentage of the cases encountered in practice. These are geometric imperfections caused by point indentors, line indentors and local buckles produced by pure bending. The work presented is experimental in nature. Empirical relationships for determining the initiation pressure are derived. the purpose is to give the field engineer a way to measure whether a given damage on a pipeline has the potential of being an initiation point for a propagating buckle.

2. PROPAGATING BUCKLE DUE TO LOCALIZED DAMAGE

This section is concerned with determining the initiation pressure for pipes that are locally damaged. This damage is assumed to occur from some object contacting the pipe. A typical example would be a piece of equipment which has been dropped overboard and then strikes an unburied pipeline. The variety of this type of damage is great and it seemed necessary to look at two extreme types. For this purpose, the damage inflicted on the model pipes was produced by either a point type indentor or a knife edge indentor.

After the damage was inflicted on the pipe, careful measurements were taken in order to quantify the imperfection. Complete imperfection scans were carried out and the data recorded automatically. Next the pipes were buckled and the initiation pressure determined. The details of the experimental program and the data correlation are given in the next sections.

2.1 Experimental Procedure

The pipes used for these experiments were made of steel. The bulk of the data was taken on model-size pipes with diameters ranging from 1.250 in. to 1.500 in. Two tests were performed on larger diameter pipe to determine if any scaling problems existed. The steel used for the pipes was 1018 or 1010. The material properties of the steel were not directly measured. Instead, the propagation pressure of each pipe size was found experimentally and this quantity was used to correlate the data. This seemed preferable to the use of a defined yield stress and an empirical equation for the propagation pressure. The dimensions of the test pipes and the propagation pressure are given in Table 1.

The damage was inflicted on the pipes using a knife edge type indentor or a point indentor. The point indentors are actually rods of different diameters varying from 1/4 in. to 5/8 in., the ends of which have been radiused. The different point indentors are shown in figure 1. The knife edge indentors likewise have finite radii which vary from 3/32 in. to 5/4 in. The dimensions of the ten different indentors are given in Table II.

The pipes were damaged using a conventional testing machine. The point indentors were pushed normal to the pipe surface. Indention depth was varied by controlling the displacement of the testing machine. For the knife edge damage, the edge of the indentor was aligned normal to the pipe axis. Indentor depth was controlled in the same manner as for the point indentor. A range of pipe knife indentor damage is shown in fig. 2.

The character of the knife edge and point damage is dissimilar in the neighborhood of the indentor but has much of the same character otherwise. Use will be made of this similarity in the data correlation. A comparison of these two types of damage is shown in two views in fig. 3.

Each pipe that was damaged and pressure tested to determine its initiation pressure was measured to quantify its initial imperfection. The measurements were carried out using a scanning device constructed for this purpose. The device, shown in fig. 4, clamps the pipe in a vertical position. A rotating table revolves around the pipe clamping. The measuring transducer, an LVDT, is attached to the rotating table on a vertical slide. The slide and rotating table and displacement transducer comprise a cylindrical coordinate system. The pipe shape is then measured referenced to this coordinate system.

The LVDT was calibrated prior to measurements using a micrometer. The vertical position is adjusted by hand. Measurements were taken every 0.5 inches. The circumferential motion is hand provided and the circumferential position is read by a photo diode device indicating off a black and white tape. The tape can be seen in figure 4. The indicating device is also used to trigger the data acquisition system to record the data during each circumferential scan. Data points at 102 locations are taken during each circumferential scan.

The data acquisition system used in this experiment, (fig. 5), can convert and store data at a rapid rate. The data acquisition calculator was also used to convert the voltage readings to displacements and plot the data in the form of contour plots. Typical contours are shown in figures 6 and 7. Figure 6 shows the damage resulting from the point indentor. Each contour is displaced 0.5 inches in the axial direction from the previous contour. Knife edge damage is shown in fig. 7. The difference between the point and knife damage is evident in the vicinity of the contact area. Outside of the immediate region the damage is similar.

After the imperfections were measured, the pipes were buckled to determine their initiation pressures. The ends of the pipes were closed by end caps. The pipes were then placed in a pressure tank and pressurized with water. A constant volume test was achieved by using a piston type water pump. The buckling could be detected by the drop off in pressure when the buckle began to propagate.

The propagating buckle which initiates from the local damage is the typical dog bone type buckle reported earlier (ref. 4). A few diameters from the local damage it is impossible to detect the nature of the initial cause of the buckle. Two examples of propagation from a point and knife damage are shown in fig. 8.

2.2 Data Correlation

The initiation pressure tests provided 43 usable pipe tests for correlation. The information obtained consists of the initial pressure, the propagation pressure and the imperfection survey for each test pipe. The correlation problem is to somehow relate the initiation pressure to the damage. Unfortunately there does not exist an analysis to help in this correlation. It is necessary to use judgement and trial and error in carrying out the correlation.

The goal of the correlation is to provide a simple and reasonably accurate relation between the initiation pressure and the parameters of the imperfection. The parameters of the imperfection are quite diverse. In addition, in a situation in practice, the amount of information about the damage would most likely be limited. It is, therefore, desirable to characterize the damage by the fewest number of parameters possible.

It is in this spirit that the characterization was undertaken. To begin with it was decided to find a relationship between the initiation pressure and the characteristics of the damage rather than the characteristics of the indentor. It seemed that the details of the damage should not be important. To illustrate this, several pipes of the same dimensions were damaged with approximately the same amount of damage by indentors of different diameters. The results of these tests are shown in figure 9 where the initiation pressure is shown as a function of the indentor diameter. The data are fairly independent of the indentor diameter with a slight decrease as the diameter increases. Some of this decrease can be explained by a closer examination of the damage, which shows that the pipes damaged with the larger indentors have somewhat larger damage. In any event, the data substantiates the hypotheses that the initiation pressure should not be a strong function of the indentor parameters, at least to the accuracy achievable in this correlation.

Next it was decided to attempt a data correlation based upon the parameters of the "most damaged" cross section of the pipe. This cross section was taken to be the cross section under the indentor. In most cases this cross section was visibly the most damaged.

Once this decision was made, the next problem was to somehow characterize this cross section. A typical cross section is given in fig. 10. Several parameters are obvious candidates for consideration. Among these are the maximum diameter, D_{max}, and the minimum diameter, D_{min}. Since the cross section is not necessarily convex, these dimensions are defined as measured between parallel lines. In addition, the section may not have an axis of symmetry so that the D_{max} and D_{min} may not be at right angles to each other.

Other parameters that were tried in the correlation were the diameters of the circumscribed and inscribed circles that have a common center and the area inside the contour. These parameters were not as successful in coalescing the data and will not be further discussed.

The correlation was achieved by plotting on log paper the initiation pressure as a function of the various parameters. This was done for the test data points for each pipe size in order to establish an adequate power law relation. If the correlation showed sufficient promise the remaining test points were similarly plotted and then a master curve of all data constructed.

The correlation parameter that appears most promising is the ratio of D_{min} to D_{max}. The data for pipe #2 is shown in fig. 11 for both the knife edge and point indentors. It is seen from this figure that there appears to be very little difference in the data for the knife edge and point indentor damage when plotted against this parameter. The overall correlation shows some scatter but the coalescence is reasonable.

The data is next plotted for all the families of pipes. In order to bring the various families into agreement, the initiation pressure is normalized by the propagation pressure. It was also found that a diameter-to-thickness ratio parameter used to shift the horizontal axis was helpful in correlation of the data. The improvement in this was slight and can not be substantiated by the narrow range of diameter-to-thickness tested. The data for all tests is shown in fig. 12. There is some scatter in the data but in appears reasonable for the type of problem considered.

In the spirit of seeking the fewest number of parameters for the correlation, the initiation pressure data are presented in fig. 13 as a function of the minimum diameter as divided by the original diameter. This correlation is not quite as good as the previous one but may be useful for some purposes. The data for the large diameter pipe (#4) are also given in this figure. The agreement with the small diameter pipe is reasonable although the large diameter results lie on the lower bound of the data. The propagating pressure for this pipe was calculated since it could not be measured in the initiation tests because the pipe fractured at the initiation pressure. A photograph of the propagated buckle and the fracture is shown in fig. 14. The fracture can be seen along the edge of the collapsed section of the pipe.

2.3 Conclusion

The pressure at which a local damage initiates a propagating buckle has been studied experimentally. The local damaged area was produced by point and knife edge indentors of different diameters. It was shown that the diameter of the indentor did not have much influence on the initiation pressure. Tests with pipes of different dimensions showed that the data could be successfully correlated using the propagating pressure to normalize the initiation pressure and characterizing the damaged area by the geometric properties of the most damaged section. The parameter that gave the best correlation was the minimum diameter divided by the maximum diameter which is a common measurement of out of roundness. For the application at hand the data are fairly well correlated for large out of roundness but considerable scatter exists for small out of roundness. Other simple correlations were attempted and one, minimum diameter divided by nominal diameter, was fairly successful.

Two tests of large diameter pipe were carried out to assess scale effects. No significant differences were noted in these tests as far as the initiation pressure is concerned. It should, however, be emphasized that the present tests are confined to local damage. Tests of pipes with damaged areas that are elongated in the axial direction will produce lower initiation pressures.

3. BUCKLING DUE TO BENDING AND EXTERNAL PRESSURE

The implications of local dents on pipelines under external pressure were discussed in section 2. One of the other most common types of pipeline damage is bending buckles. These can be due to earthquakes, sediment instabilities, severe sea floor currents that tend to move the pipeline, anchors pulling on the pipeline, as well as buckles induced during the laying process. Since the pipe is under external pressure, such local damage can initiate a propagating buckle. The pressure at which a local damage transforms itself into the correct profile to become a propagating buckle is called the initiation pressure, P_I. Its value depends on the geometric characteristics of the dent. These include both the geometry as well as the amplitude of the damage. To establish these, it is necessary to find the conditions that lead to buckling and then to examine the post-buckling configuration that results.

Although a great deal of work has been done on pure bending as well as combined bending and external pressure of cylindrical shells [19-25], only relatively thin elastic structures have been considered (D/t>100). For pipelines, the diameter to thickness range is 15<D/t<80. For this range of D/t plastic effects are unavoidable. In what follows, the experimental equipment and methodology developed for studying the response, stability and post buckling behavior of pipes having D/t in the above range given are described. The response under combined bending and pressure is recorded, the point of instability established and the stability envelope as a function of both pressure and moment is obtained. In addition, the conditions under which a local buckle initiates a propagating one are studied for the two classes of damages identified in these experiments.

3.1 Combined Loading Test Facility

The experimental facility for the combined loading experiment was designed and built in house. It consists of a pure bending device which slides into a pressure tank and can be operated remotely while under external pressure (fig. 15). Two pairs of sprockets, 9 in. in diameter, symmetrically placed 36 in. apart on a beam frame provide the bending moment. The distance that separates the two sets of sprockets can be varied so that different lengths of pipe can be tested. The sprockets support two rollers which apply point loads in the form of a couple at each end of the test specimen. Chains running over the sprockets are connected to a load cell and a hydraulic cylinder as shown in the figure.

Rotation of the sprockets can be achieved by contracting or extending the hydraulic cylinder. A hydraulic control system, which allows reversal of the flow, was designed and built in order to carefully control the movement of the cylinder. A load cell in series with the cylinder continuously monitors the tension in the chains. The applied moment can be found by multiplying the tension load by the sprocket radius.

The rotation of each sprocket is measured by an LVDT as follows. A thin steel cable runs over the hub of the sprocket and is in turn connected to the LVDT core. The LVDT is fixed at a position such that its axis is aligned with the horizontal tangent of the sprocket hub. Angular movement of the sprockets produces linear movement of the LVDT core. The cable is kept in tension by inserting a soft spring in series. Each LVDT independently measures the rotation angle of each sprocket. The two signals are electronically added and the resultant voltage is directly proportional to the curvature. (Note that this is a pure bending device so the curvature is constant across the length of the test specimen.)

The load is transferred to the test specimen through two rollers that produce a couple on each end of the pipe. The rollers allow free horizontal movement of the test specimen. This is necessary in order to avoid axial loads in the pipe. The test specimen is closed at both ends by solid steel plugs as shown in fig. 16. These plugs extend about 12 in. on either side of the pipe specimen so that the rollers are always in touch with the solid bars rather than the thin pipe. This was necessary in order to avoid loading the thin pipe directly through point loads which can prematurely cause local dimpling of the pipe. The end plugs were designed in such a way so as to reduce the stress concentration due to the discontinuity in the pipe cross section. The plug and pipe assembly is positioned in the device as shown in fig. 15. Pipes having diameters from 0.75-1.5 in. and lengths from 25-50 in. can be tested.

The testing machine is capable of applying moments up to 10,000 in.-lb. The whole apparatus was designed in a manner to ensure a "rigid" machine situation as far as the test specimen is cocnerned. This type of design is necessary in order to guarantee minimum effect from the testing machine on the pipe postbuckling configuration. The energy absorbed by the testing machine is less than 5% of the energy in the deformed pipe.

The whole device with the test specimen slides into a pressure tank. The tank has a diameter of 18 in., length of 8 ft. and working pressure of 800 psi. It can be pressurized by water or air or filled with water and pressurized with air. Each of the three pressurization methods establishes different experimental conditions. The bending device can be remotely controlled from outside the tank. The LVDT's and load cell are hermetically sealed and can operate under high pressure in water.

3.2 Experimental Procedure

Two series of experiments were carried out. In both cases aluminum 6061-T6 tubes were used due to their availability in good quality. In the first series, the tubes had a nominal diameter, D, of 1.215 in. and thickness, t, of 0.035 in. and in the second series D= 0.980 in. and t= 0.020 in. These two sizes were chosen because of the differences exhibited in their post-buckling behavior. All tubes tested had an effective length of 29 in.

The tubes were sealed on both sides using the long plugs described earlier (see fig. 16). The ends of the plugs were epoxied to the tubes to ensure atmospheric internal pressure at all stages of the experiment. After positioning the pipe in the testing machine, the instruments were zeroed and the bending device was placed in the tank (see fig. 15). As explained earlier, the tank can be pressurized in any of three different ways, each of which establishes different experimental conditions. For most experiments the tank was filled with water leaving an air pocket that was subsequently pressurized by air. This arrangement simulates a "nearly" constant pressure condition similar to the physical situation that the pipe encounters in the ocean. On buckling, the volume displaced by the pipe suddenly decreases and the air pocket expands to fill the extra volume generated. If the ratio of the volume of the air pocket to the volume generated by buckling is large enough, the pressure in the tank remains practically constant. On buckling under combined loading, a propagating buckle may be initiated from the resultant local damage. This must be determined without opening the tank so the following procedure was devised. As soon as buckling occurred (noise and sudden drop in the moment), loading was stopped and the pressure transducer output checked. The air pocket in the tank was arranged to be of such a size that if the local buckle propagated, collapsing the total length of the pipe, then the pressure in the tank would drop by 2-3 psi. Such a discontinuity in pressure can easily be observed on the pressure-time record kept during the experiment. If such a step change was observed the experiment was considered completed. If not, the pressure was increased until a propagating buckle was finally initiated from the damaged section of the pipe.

For experiments where only the moment-curvature response was sought, the tank was completely filled and pressurized with water using a single piston pressure pump. The pressure tank and pressurization system are arranged in such a way so that the whole system can be vented and practically all the air bled out of the system. For pressures below 500 psi the volume expansion of the tank is negligible. The pipe specimen tested under such conditions sees a relatively rigid system. This was taken to represent a constant volume experimental condition. Any sudden change in the geometry of the pipe causes a drop in the pressure due to the relative stiffness of the loading system.

For all the experiments presented the following loading procedure was followed irrespective of the way the tank was pressurized. After positioning the testing machine and specimen in the tank, the system was pressurized to the required level and the pipe was subsequently deformed by rotating the sprockets. The sprocket rotation was manually controlled and was very slow. The process can be considered to be quasistatic. The pipe was continuously deformed until buckling occurred. The moment, curvature and pressure were recorded during the experiment on strip chart recorders. On buckling, the moment dropped drastically and the LVDT output increased suddenly. The LVDT's reading after buckling was used to approximately calculate the post-buckling configuration of the pipe.

3.3 Experimental Results

A number of investigators, [24-27], have shown that there are two types of instability in the case of pure bending of a circular pipe. The first is a limit-point type of geometric instability (i.e., $dM/d\kappa \to 0$) and the second is a bifurcation type of instability typical of many thin structures under compressive stresses. Figure 17 shows a number of typical responses obtained from combined loading experiments on pipes with D/t= 34.7. For the case of zero pressure the response follows a linear path for low curvatures and as expected it becomes nonlinear for higher values of curvature. The maximum value of moment achieved appears to be a limit point. For external pressures higher than zero, the path is similar but buckling occurs much earlier. In these cases it is not possible to establish from the experimental results what type of instability occurred.

One way of physically looking at the interaction failure is as follows. By bending the pipe in a circular arc, the cross sections of the pipe tend to ovalize, which causes a reduction in the bending rigidity of the pipe. The application of an external pressure to such an ovalized pipe causes the already non-circular cross section to become more oval, which further reduces the bending rigidity of the pipe and eventually causes buckling under the applied moment.

The second mechanism involved can be described as follows. It has been shown in ref. 28 that initial geometric imperfections in the cross section of an elastoplastic pipe will cause a reduction in its buckling pressure. The applied bending causes the cross sections to ovalize and when the ovalization reaches a critical value the pipe buckles under the external pressure. Clearly, the two mechanisms interact. However, one can say that the first mechanism dominates for low pressures and the second one for pressures closer to the buckling pressure of a circular pipe.

The pressures at buckling, normalized by the buckling pressure of the pipe, P_c, are plotted vs. the moment at buckling normalized by the elastic limit load, M_m^e for the two different D/t in figures 18 and 19. The corresponding pressure-curvature plots are shown in figures 20 and 21. A limit-type load analysis carried out recently in ref.29 found the experimental results for D/t= 34.7 to be in good agreement with the predicted limit loads. On the other hand for D/t= 49.0, considerable difference was observed particularly in the case of the pressure-curvature interaction plot. This reinforces the earlier expressed opinion that the thinner pipes exhibited bifurcation failure before the limit load was reached. Due to the features of the M-κ curves for higher values of curvature, bifurcation does not cause a pronounced reduction of the values of moment predicted by the limit load analysis. The limit and bifurcation values of curvature however, can be substantially different. The buckles that initiated a propagating buckle are indicated on all figures. The initiation pressure for dents caused by local buckling under combined loading is clearly very close to the pipe propagation pressure in both cases. This contradicts efforts by previous investigators [2] to define the initiation pressure as another characteristic pressure of the pipe.

Figures 22 and 23 show two typical buckles obtained by pure bending of the pipes for the two D/t ratios considered. The difference in the mode of buckling is obvious. For D/t= 34.7 a single corsswise dent occurs. The damage is restricted to about 5 diameters on either side of the dent. The geometry of the damage is doubly symmetric. For D/t= 49.0 the pipe buckles in a diamond mode typical of buckling of elastic shells under axial load. The buckle affects a smaller length of pipe and has only one plane of symmetry. This difference in behavior has consistently been observed in pipes of other sizes as well (e.g. D/t= 32.1 and 68.7). Further experiments are necessary in order to establish the exact D/t at which the mode changes.

The difference in the mode of failure seems to also affect the initiation pressure of the buckled pipes. As mentioned in the introduction, it was felt that the initiation pressure should be a function of the type of damage inflicted to the pipe as well as its severith. The buckle usually propagates in a dog bone type of collapse so that any damage that tends to easily progress to that mode will have a low initiation pressure.

In an effort to establish a parameter that measures the amplitude of the damage for different pipes, the angle θ_c was chosen (this angle is defined in figure 27). Pipes bent to different values of θ_c are shown in figure 24. A series of experiments was carried out for pipes having D/t= 49.0 and 34.7 where the initiation pressure was found as a function of the post-buckling angle θ_c. Each pipe was first buckled under pure bending. By further rotating the sprockets the required angle θ_c was established. The bending device was subsequently locked in that position and placed into the pressure tank. The experiments were carried out under constant pressure conditions. The pressure was increased until a propagating buckle was initiated. Fig. 25 shows a buckle initiated from such damage. For the bigger values of θ_c the damaged pipe locally buckled under pressure without propagating (see fig. 26). In such cases pressurization was continued until the buckle finally propagated.

The results of these experiments are shown in fig. 27 and 28. For D/t= 34.7 the initiation pressure increases with θ_c. For the smallest buckling angle obtained ($\theta_c= 4.5°$), $P_I= 1.17\ P_p$. The results can be fitted with a smooth curve. For D/t= 49.0, the results show more scatter. If any conclusion can be drawn in this case, it is that the buckling angle θ_c is not one of the important parameters that affect the initiation pressure. This can be expected because the geometry of the damage varies much more from pipe to pipe than in the case of D/t= 34.7. However, the lowest initiation pressure obtained in this case was $P_I= 1.49\ P_p$. This indicates that for more irregular and complex postbuckling configurations, the propagating buckle is more difficult to initiate.

3.4 Conclusion

Combined pressure-moment buckling tests have been carried out for pipes with two different diameter-to-thickness ratios. The buckling behavior varies from a limit-type buckling to a bifurcation-type buckling. The failure mode changes along the interaction envelope. For low pressur and high moment, the failure mode is a local dent. For pressures somewhat higher than the propagati pressure, the failure mode includes a complete collapse of the pipe. This transition point is highe than the propagation pressure but the difference is not as great as that reported by previous investigators. The influence of post buckled configuration on this transition pressure was further studied. It was shown that the post buckling bend angle was not highly influential in the determina tion of the initiation pressure.

REFERENCES

1. KYRIAKIDES, S., and BABCOCK, C.D., Nov. 1981, "Experimental Determination of the Propagating Pressure of Circular Pipes," J. of Pressure Vessel Techn., Trans. of the ASME, Vol. 103.
2. MESLOH, R., JOHNS, T.G., and SORENSON, J.E., 1976, "The Propagating Buckle," Boss 76, Proc. Battelle-Columbis Laboratories, Vol. 1, pp. 787-797.
3. PALMER, A.C., March 6, 1975, "Buckle Propagation in Submarine Pipelines," Nature, Vol. 254, No. 5495, pp. 46-48.
4. KYRIAKIDES, S., and BABCOCK, C.D., May 1980, "On the 'Slip-On' Buckle Arrestor for Offshore Pipelines," ASME Journal of Pressure Vessel Techn., Vol. 102, pp. 188-193.
5. KYRIAKIDES, S., and BABCOCK, C.D., March 1982, "The Spiral Arrester - A New Buckle Arrestor Design for Offshore Pipelines," J. of Energy Resources Techn., Trans. of the ASME, Vol. 104.
6. JOHNS, T.G., MESLOH, R.E., SORENSON, J.E., May 1978, "Propagating Buckle Arrestors for Offshore Pipelines," J. of Pressure Vessel Techn., Trans. of the ASME, Vol. 100, pp. 206-214.
7. TIMOSHENKO, S.P., and GERE, J.M., 1961, Theory of Elastic Stability, McGraw-Hill.
8. API Bulletin on Performance Properties of Casing, Tubing and Drill Pipe, April 1970, API Bulletin C52, 12th Edition.
9. STRATING, J., May 1981, "A Survey of Pipelines in the North Sea Incidents During Installation, Testing and Operations," Proc. Offshore Techn. Conf., OTC 4069, pp. 25-32.
10. BROWN, R.J., May 1972, "Pipelines can be Designed to Resist Impact from Dragging Anchors and Fishing Boards," Proc. Offshore Techn. Confl., OTC 1570, pp. 579-580.
11. HEMPHILL, D.P., MILZ, E.A., and LUKE, R.R., "Repair of Offshore Pipelines in Water Depths to 3,000 ft.," Proc. Offshore Techn. Conf., OTC 1939, pp. 55-64.
12. KYRIAKIDES, S., and SHAW, P.K., "Response and Stability of Elastoplastic Circular Pipes Under Combined Bending and External Pressure," Int. Journ. of Solids & Structures, in print.
13. WALKER, D.B.L., May 1976, "A Technical Review of the Fortie Field Submarine Pipeline," Proc. Offshore Techn. Conf., OTC 2603, pp. 819-829.
14. LALLIER, L., and JEGON, A., May 1977, "A Technical Review of the Frigg Pipelines Construction," Proc. Offshore Techn. Conf., OTC 2915, pp. 303-314.
15. GJORSVIK, O., and KJELDSEN, S.P., "Influence of Bottom Trawl Gear on Submarine Pipeline," Proc. Offshore Techn. Conf., OTC 2280, pp. 337-346.
16. HARTIG, E.P., NOTTINGHAM, D., SWANSON, J.E., and TISDALE, B.C., "Reburial Considerations for an Exposed Pipeline," ASCE Report.
17. COLEMAN, J.M., and PRIOR, D.B., May 1978, "Submarine Landslides in the Mississippi River Delta, Proc. Offshore Techn. Conf., OTC 3170, pp. 1067-1074.
18. REIFEL, M.D., "Storm Related Damage to Pipelines, Gulf of Mexico," ASCE Report.
19. BRAZIER, L.G., 1927, "On the Flexture of Thin Cylindrical Shells and Other "Thin" Sections," Proc. Royal Society Series A, Vol. 106, pp. 104-114.
20. REISSNER, E., 1961, "On Finite Pure Bending on Cylindrical Tubes," Osterr. Ing. Arch., Vol. 15.
21. SEIDE, P. and WEINGARTEN, V.I., 1961, "On the Buckling of Circular Cylindrical Shells Under Pure Bending," J. of Appl. Mech., Vol. 28, pp. 112-116.
22. STEPHENS, V.B., STARNES, J.H., and ALMROTH, B.O., 1975, "Collapse of Long Cylindrical Shells Under Combined Bending and Pressure Loads," AIAA Journal, Vol. 13, pp. 20-25.
23. AXELRAD, E.L., 1965, "Refinement of Critical Load Analysis for the Tube Flexure by Way of Considering Precritical Deformation," lzv. An SSSR, ONT, Mech.:Mash. n. pp.123-129, in Russian.
24. AXELRAD, E.L., "Flexible Shells," Proc. 15th IUTAM, Aug. 17-23, 1980, Toronto, Canada, pp.45-56
25. FABIAN, O., 1977, "Collapse of Cylindrical, Elastic Tubes Under Combined Bending: Pressure and Axial Loads," Int. J. of Solids & Structures, Vol. 13, pp. 1257-1270.
26. TUGGU, P. and SCHROEDER, J., 1979, "Plastic Deformation and Stability of Pipes Exposed to External Couples," Int. J. of Solids & Structures, Vol 15, pp. 643-658.
27. GELLIN, S., 1979, "The Plastic Buckling of Long Cylindrical Shells Under Pure Bending," Int. J. of Solids & Structures, Vol. 16, pp. 397-407.
28. KYRIAKIDES, S., and BABCOCK, C.D., 1981, "Large Deflection Collapse Analysis of an Inelastic Inextensional Ring Under External Pressure," Int. J. of Solids & Structures, Vol. 17, No.10.
29. KYRIAKIDES, S., and SHAW, P.K., "Response and Stability of Elastoplastic Circular Pipes Under Combined Bending and External Pressure," Proc. of the 5th Annual Energy-Sources Technology Conf., ASME, March 7-10, 1982.

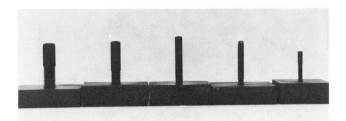

FIG. 1 POINT INDENTORS

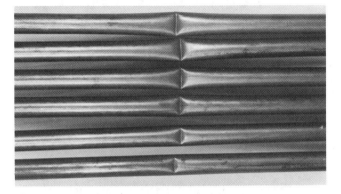

FIG.2 RANGE OF KNIFE EDGE INDENTOR DAMAGE

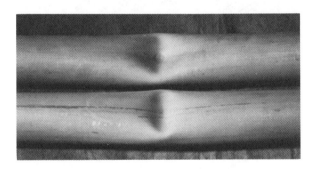

FIG.3 COMPARISON OF KNIFE EDGE AND POINT DAMAGE

FIG.4 IMPERFECTION SCANNING SYSTEM

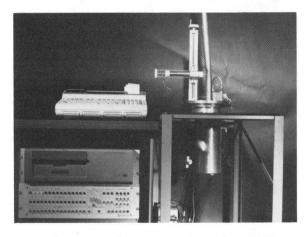

FIG.5 IMPERFECTION MEASURING DEVICE WITH DATA ACQUISITION SYSTEM

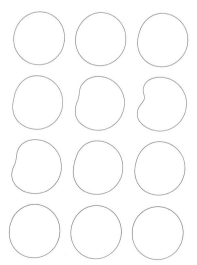

PIPE CROSS SECTIONS — MODERATE POINT DAMAGE

FIG. 6

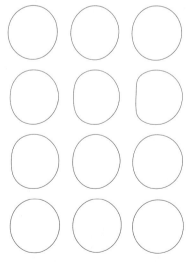

PIPE CROSS SECTIONS — MODERATE KNIFE EDGE DAMAGE

FIG. 7

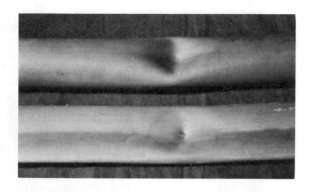
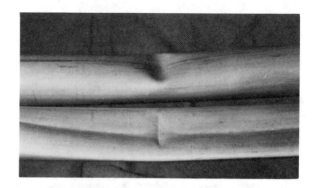

FIG. 8 PROPAGATING BUCKLE AS INITIATED FROM POINT AND KNIFE EDGE DAMAGE

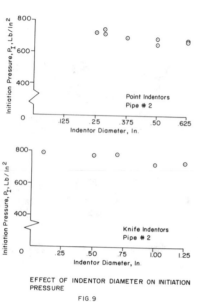

EFFECT OF INDENTOR DIAMETER ON INITIATION PRESSURE

FIG. 9

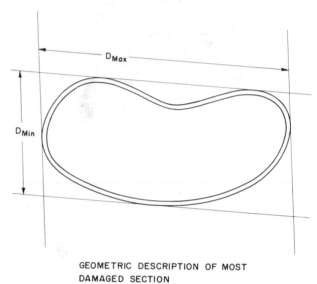

GEOMETRIC DESCRIPTION OF MOST DAMAGED SECTION

FIG. 10

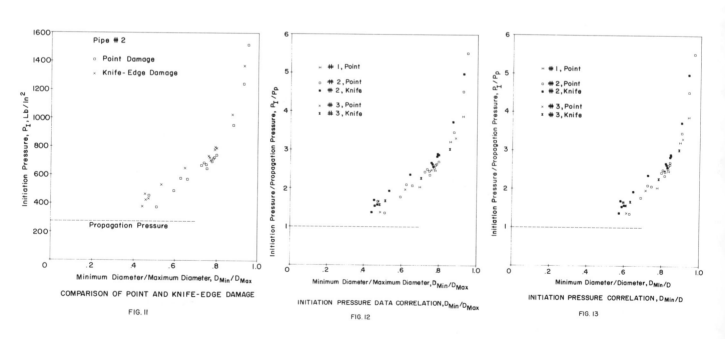

COMPARISON OF POINT AND KNIFE-EDGE DAMAGE

FIG. 11

INITIATION PRESSURE DATA CORRELATION, D_{Min}/D_{Max}

FIG. 12

INITIATION PRESSURE CORRELATION, D_{Min}/D

FIG. 13

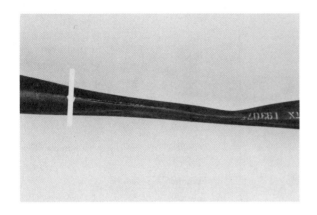

FIG. 14 PROPAGATING BUCKLE THAT LED TO A FRACTURE

FIG. 15 COMBINED BENDING - PRESSURE TEST FACILITY

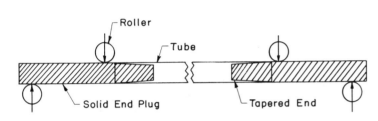

TEST SPECIMEN AND END-PLUG ASSEMBLY

FIG. 16

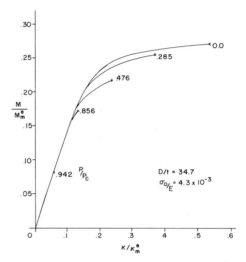

MOMENT-CURVATURE RESPONSE AS A FUNCTION OF PRESSURE

FIG. 17

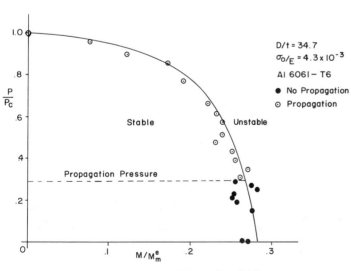

MOMENT-PRESSURE INTERACTION, D/t = 34.7

FIG. 18

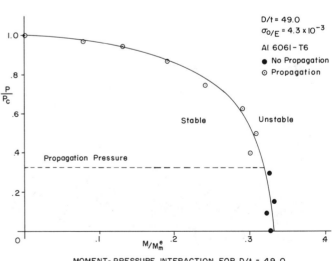

MOMENT-PRESSURE INTERACTION FOR D/t = 49.0

FIG. 19

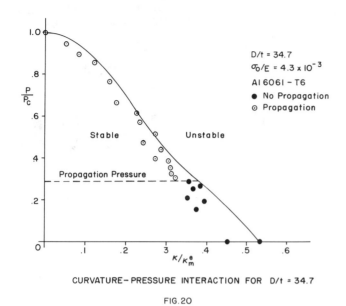

CURVATURE–PRESSURE INTERACTION FOR D/t = 34.7

FIG. 20

PRESSURE–CURVATURE INTERACTION FOR D/t = 49.0

FIG. 21

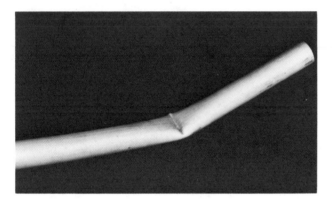

FIG. 22   BENDING BUCKLE (D/t = 34.7)

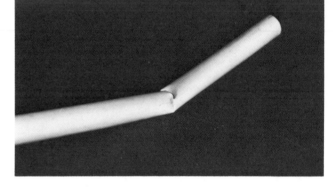

FIG. 23   BENDING BUCKLE (D/t = 49.0)

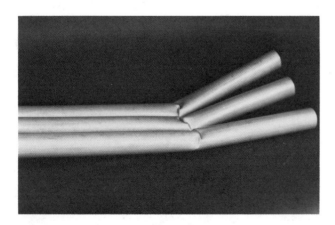

FIG. 24   PIPES WITH DIFFERENT BEND ANGLE (D/t = 49.0)

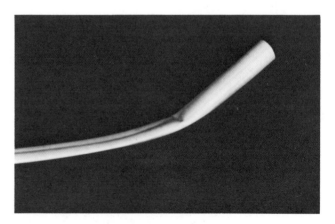

FIG. 25   PROPAGATING BUCKLE INITIATED FROM A BENDING BUCKLE (D/t = 34.7)

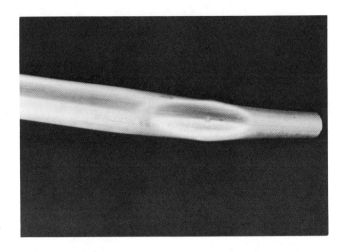

FIG. 26 LOCAL COLLAPSE UNDER EXTERNAL PRESSURE (D/t = 34.7)

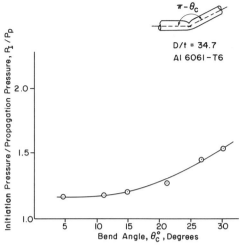

VARIATION OF INITIATION PRESSURE WITH BEND ANGLE θ_c

FIG. 27

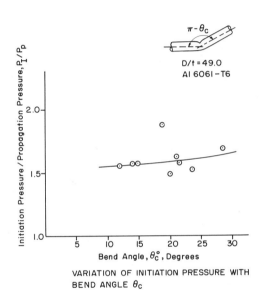

VARIATION OF INITIATION PRESSURE WITH BEND ANGLE θ_c

FIG. 28

NOMENCLATURE

D	pipe mean diameter
E	Young's modulus
E'	strain hardening modulus
M	bending moment
M^e_m	elastic limit load, $0.493 \dfrac{E}{\sqrt{1-\nu^2}} Dt^2$
P	pressure
P_c	buckling pressure, $\dfrac{2E}{1-\nu^2}\left(\dfrac{t}{D}\right)^3$
P_I	initiation pressure
P_p	propagation pressure, $\sigma_0\left[10.7+.54\dfrac{E'}{\sigma_0}\right]\left(\dfrac{t}{D}\right)^{2.25}$
t	pipe thickness
θ_c	bend angle
κ	curvature
κ^e_m	elastic limit curvature, $\dfrac{.943}{\sqrt{1-\nu^2}}\dfrac{t}{D}$
ν	Poisson's ratio
σ_0	yield stress

TABLE I. INITIATION PRESSURE TEST PIPES

Pipe	Material (steel)	Diameter (in.)	Thickness (in.)	D/t	P_p (psi)
1	1018	1.250	.038	32.9	360
2	1018	1.375	.038	36.2	275
3	1010	1.500	.035	42.9	198
4	1018	5.252	.121	43.4	207*

* Calculated

TABLE II. INDENTOR GEOMETRY

Point Indentor #	Diameter (in.)	Knife Indentor #	Diameter (in.)
1	0.250	1	0.094
2	0.281	2	0.500
3	0.375	3	0.688
4	0.500	4	1.000
5	0.625	5	1.250

ACKNOWLEDGEMENT

This work was supported by the Department of Transportation under Contract DTSR 5680-C-00010. This aid is gratefully acknowledged. The authors would also like to acknowledge the assistance of Erdal Arikan, Sam Chang, Dov Elyada, Marta Nyiri and Betty Wood in carrying out and documenting this work.

Technical Lecture Sessions
—
STRUCTURES

Behavioural Issues for Concrete Structures

IMPLICATIONS OF FATIGUE IN THE DESIGN OF SOME CONCRETE OFFSHORE STRUCTURES

A H Tricklebank
Gifford and Partners U.K.

W I J Price
Gifford and Partners U.K.

E C Hambly
Consulting Engineer U.K.

SUMMARY

The paper describes the results of a design study into the significance of fatigue in concrete offshore structures. These include oil production platforms, floating petrochemical barges and wave energy converters of both raft and spine form.

For the oil industry structures a design wave height spectrum for the North Sea was used to evaluate loadings and forces. For wave energy converters, governing moments and shears were determined by means of tests on small scale models in a wide tank which accurately reproduced fully developed random multi-directional seas representative of North Atlantic conditions. The methods used to determine and apply the fatigue stress spectra are described.

The data for the fatigue response of concrete, reinforcement and prestressing tendons were obtained either from current design standards or from recent tests undertaken for the "Concrete in the Oceans" programme on behalf of the Department of Energy and the UK Offshore Industry. This design study also formed part of the same programme and was initiated to determine the needs for future experimental research on fatigue.

The results confirm that fatigue is not likely to be a dominating criterion in the design of current types of gravity platforms, except for sensitive locations where stresses are raised by the formation of large holes or, locally, by the attachment of supporting brackets for pipework and services. Fatigue is a more important consideration in the design of wave energy converters because of the higher ratio of cyclic environmental loading to total design loading.

It is likely that the main structural elements of wave energy converters will be prestressed longitudinally by post-tensioning and these may be liable to fatigue problems if the full economic benefits of partial prestressing are to be exploited. Transversely, the structures are likely to be of reinforced construction and significant fatigue damage to reinforcement may occur under local wave pressure loading, with corner details being particularly vulnerable. The sensitivity of the fatigue assessment to characteristic strengths, S-N curves, stress ranges and mean stress levels is examined.

INTRODUCTION

The important characteristic difference between offshore structures and other structures lies in the severity of the environment in which they are situated. Several features of this environment and the particular duties of these structures create problems which, because of their scale, cannot be resolved by previous experience without fundamental reappraisal and research.

One particular aspect concerns the nature of the loading. Offshore structures are placed for prolonged periods, often their entire service life, in an energetic wave climate which subjects them to loads of randomly varying magnitude and direction. Extreme storm conditions exert loadings which are the dominant consideration in determining the required strength of the structure.

For North Sea and North Atlantic conditions where completely calm conditions are exceptional, the major proportion of waves have periods in the 6 to 10 second range and simple calculation reveals that 10^8 cycles of loading will be applied during a 25 year service life. Under these circumstances fatigue damage must also be regarded as a serious concern. Indeed, it is well known that fatigue problems have been experienced in a number of steel offshore structures. There is however, little if any recorded experience of fatigue problems with any type of concrete structure. Nevertheless, it may be misleading to conclude from this that fatigue is not a problem, particularly as research in recent years has highlighted the reduced fatigue performance of component materials and reinforced concrete elements when immersed in sea water. A very considerable research effort has been made and is continuing in this area, but understanding of the phenomena is far from complete. For designers and researchers alike, it is difficult to achieve a broad perspective on the problem. As part of the 'Concrete in the Oceans' programme funded by the Department of Energy and U.K. Offshore Industry, a design review study (Price et al, 1981) has been carried out to determine how serious a problem fatigue is in the design of Offshore structures, and to highlight research topics that would be of immediate interest to the designer.

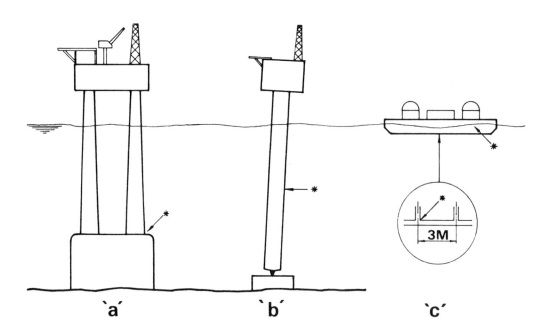

Fig. 1. Offshore Oil industry structures
(Critical locations are starred)

This paper describes some of the work carried out in this study and also in engineering studies undertaken as part of the Department of Energy's programme for Extraction of Energy from Sea Waves. It is published by permission of the Management Committees for both of these programmes. The review considers two broad categories of structures, namely those for the Offshore Oil Industry and those for Wave Energy Converters.

Typical designs for the following structures were examined and are indicated in Figures 1 and 2. They were:-

    a.   Prestressed and Reinforced Concrete Tower Platforms.

    b.   Prestressed Concrete Articulated Column Platform.

    c.   Prestressed Concrete Pontoon Hull for Petrochemical Plant.

    d.   Wave Energy Converter - Spine Hull-Reinforced and Prestressed Concrete.

    e.   Wave Energy Converter - Raft Hull-Reinforced and Prestressed Concrete.

The loadings, material properties and evaluation of fatigue damage are briefly discussed, followed by consideration of some of the results of the assessments. Some general conclusions are drawn on the implications of fatigue in offshore structures and on areas where further work is important.

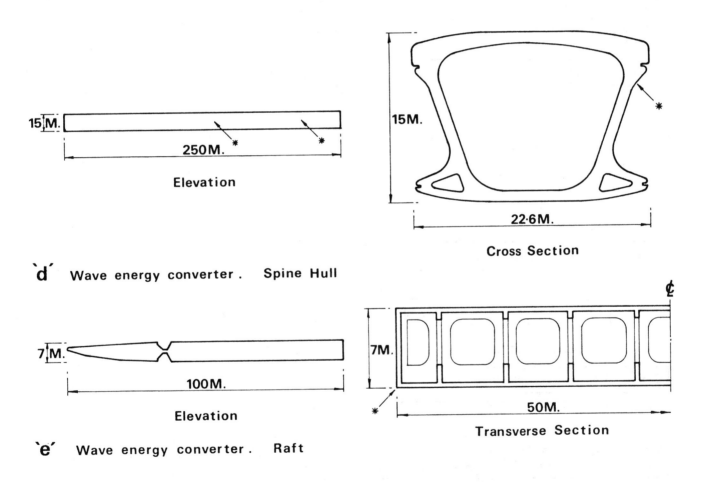

Fig. 2. Wave energy structures (Critical locations are starred)

FATIGUE ANALYSIS PROCEDURE

The calculation of fatigue damage to a structure involves:-

        a.   Assessment of the loading history
        b.   Calculation of the Structural Response
        c.   Knowledge of Material Fatigue Characteristics
        d.   Selection of Damage Criteria

The procedure within the design process is represented by the simple flow diagram in Figure 3.

a. Assessment of Load History

As offshore structures are subjected to fluctuating loads from a variety of sources, in a rigorous analysis it may be necessary to consider various combinations of loads from:-

    waves, current, variable differential hydrostatic pressures, mooring loads,
    wind, temperature effects, cranes, machinery, deck live loads.

In general, these other load effects will be secondary contributors to fatigue damage of the primary structure by virtue of their relatively lower magnitude (1 order of magnitude lower) and much lower frequency (3-4 orders of magnitude lower). However, it is common practice to consider only wave loading as fluctuating in the fatigue analysis, and this approach has been applied in the study. Although other loads may be secondary as far as direct contributions to fatigue damage is concerned, they may be of primary importance in their influence on the design strength of the structure, thus indirectly affecting the cyclic stress range and possibly the mean stress about which cyclic stressing takes place.

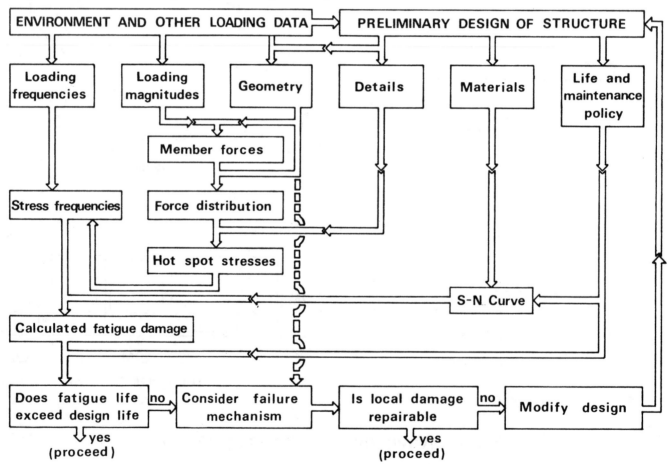

Fig. 3. Fatigue analysis procedure

A primary concern therefore is the estimation of wave loading and its frequency of occurrence. The turbulent climates of the North Sea and North Atlantic were considered for the assessment of the Oil Industry Structures and Wave Energy Structures respectively. Typical wave statistics are readily available for these locations. However, in the case of the Wave Energy devices, the envisaged location within 15kms of land and the crucial importance of relevant statistics for power generation assessments has required a special programme of wave measurement. The statistical data gathered as part of this programme has been used in the model testing work, part of which was specifically aimed at providing information on structural response for ultimate strength and fatigue strength evaluations. The procedures adopted are referred to later in this paper.

b. Calculation of Structural Response

The choice of appropriate method to be used will depend upon particular characteristics of the structure being examined. When the structural response is not sensitive to the excitation frequency, as for instance when the natural periods of the structure are much shorter than the period of the exciting forces, then a quasi-static analysis may be employed with reasonable confidence. This choice can only be made with some knowledge of the dynamic behaviour of the structure. If the fluctuating load excites a resonant response from the structure, implying magnification of stresses, or if the structures natural period is much longer than the period of the exciting forces, implying reduced response and lower stresses, then it will be necessary to use a dynamic analysis.

As the number of degrees of freedom increases as one moves from fixed to floating structures taking account of response to random waves, the problems of computation increase considerably and model testing coupled with spectral analysis becomes the most suitable, if not the only way, of obtaining the desired information.

Thus, a variety of approaches were used in the study for estimating structural response, and are mentioned again later.

c. Material Fatigue Characteristics

Results of fatigue tests display a large scatter which can be represented by a probability distribution (usually a log- normal distribution). For design purposes, S-N curves are generally used with the design curve drawn to represent the 97.5% survival level.

Despite the very substantial amount of testing work on various aspects of material behaviour, much of the information is not available in a form that can safely or conveniently be used by the designer and indeed there are few published sources of such information. The most comprehensive source is the DNV Rules for the Design, Construction and Inspection of Offshore Structures 1977, and these have been used as the basic reference. Some North Sea platforms have been designed to T.N.O procedures (Monnier 1975) and so comparative calculations were done to establish whether there was any significant difference in practice.

Additionally, recent tests carried out by John Laing Ltd (Paterson, et al 1981) as part of the 'Concrete in the Oceans' programme have confirmed the absence of an endurance limit for cold-worked reinforcement in concrete beams in sea water and demonstrate the effects of crack blocking and corrosion. Suitable design curves were derived from these test results by regression analysis. They differ significantly from the fatigue provisions in the DNV rules and the study provided an opportunity to assess the likely significance of this difference. A further interesting difference in conclusions between the Laing tests and the DNV results, (Waagard 1977) is that load cycling under simulated 30 metre depth immersion had no significant influence on the extent of surface spalling of the cracks in the concrete, perhaps reflecting the different conditions under which the tests were carried out. The scope for discussion of materials behaviour and the ramifications of the various characteristic curves is virtually limitless and a number of references (Hawkins 1976), (Waagard 1981) provide a very full background to these questions. This paper will confine itself to highlighting a few points.

A general point worth making is that reinforced and prestressed concrete basically have good fatigue characteristics. Concrete is not a notch-sensitive material, redistributing high stresses by micro-cracking and creep with steel reinforcement providing numerous alternative load paths. Failure of one element will rarely lead to catastrophic collapse and overall failure is preceeded by the appearance of major cracking and loss of stiffness with significant additional cycling required to bring about final collapse. It therefore has a ductile failure mode.

(i) Concrete

Under cyclic loading, plain concrete fails by progressive internal microcracking initiating at the aggregate/paste boundaries and fracture is associated with the development of ultimate strains which are more than twice those at failure under static loading. Variables such as mix proportions, aggregate type, curing conditions and so on affect fatigue strength and static strength similarly thus allowing universal S-N curves to be developed. This reduces the number of variables that have to be considered in addition to stress range, minimum stress and stress gradient. The beneficial effect of a flexural stress gradient on the fatigue strength of concrete in compression is allowed for by a factor on the characteristic strength ($1 \leq \alpha \leq 1.26$) in the DNV rules. Figure 4 shows design curves for concrete used in the study. It is generally thought that concrete does not show an endurance limit in sea water and therefore the TNO curve probably underestimates the fatigue damage.

(ii) Steel Reinforcement

The fatigue strength of reinforcing steel is relatively unaffected by minimum stress level and for purposes of fatigue calculations stress range is the important parameter.

Four curves shown in Figure 5 were compared. Curve R1 is based upon the DNV rules and incorporates an endurance limit. Curve R1A is a suggested bilinear curve with the upper portion based upon the DNV curve and the lower portion fixed as a compromise between the T.N.O. curve R2 and the Laing data curve R3. Curve R3 has been derived from tests carried out by Laings (Paterson et al) referred to earlier.

(iii) Prestressing Steel

There is less information on the fatigue performance of prestressing steel than for reinforcing steel particularly in marine conditions. All the fatigue tests reviewed have been performed on tendons in a non-corrosive environment. Although in general prestressing steel will be more deeply embedded than reinforcement and cracking will be absent or limited, a safe approach would be to assume that any tendency towards an endurance limit in laboratory results is likely to be absent. A log S- log N curve is likely to represent this situation rather better than the more usual S- log N form. Figure 6 shows a convenient plot for design purposes of results from a number of tests on strand.

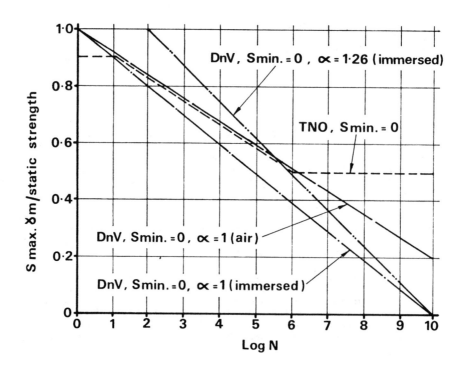

Fig 4. Design curves for concrete

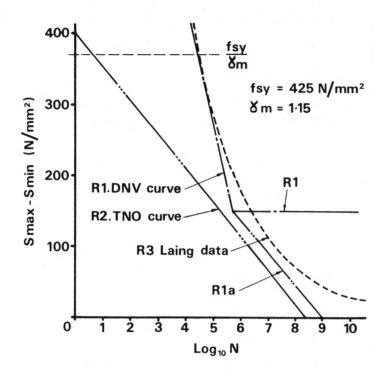

Fig 5. S-N Curves for steel reinforcement

(iv) Damage Criteria

Cumulative Damage under a variable amplitude loading is usually assessed using Miners Sum

$$\sum n/N \leq K \qquad - (1)$$

The ordinate of stress range (or related parameter) is divided into a number of blocks. n is the number of cycles during the service life for each block whilst N is the limiting number of cycles for the mean stress range in each particular block as obtained from the design S-N curves.

Factor K is normally taken as 1 for steel structures, whereas (Waagard 1981) suggests that, for concrete, K lies in the range 0.2 to 0.5. Work by (Van Leeuwen et al 1979) shows that the value can be described by a log-normal distribution with a mean value less than 1. A fixed value for K implies that all of the load cycles for any particular stress range, whether they be the first or the last few before failure, cause identical damage and similarly that the order of application of cycles of different stress ranges has no effect on the rate of damage. It is known that these assumptions are not entirely valid for steel, with fracture mechanics providing some explanation for the varying rate of crack growth with increasing cycling and with stress interactions in the region of the crack tip due to programme and variable amplitude loadings. Similarly for concrete it is known that loading sequence can have a considerable effect, with high loadings applied early giving greater damage than with high loadings applied later.

Despite the apparent shortcomings of the method, Miner's Sum remains attractive for its simplicity, and in the study K = 0.2, as stipulated in the DNV rules, was used as the basis for assessment. In the majority of situations for current offshore concrete structures the question of precision in damage assessment is not of great practical interest as Miner's Sum values are of one or two orders of magnitude below the critical range.

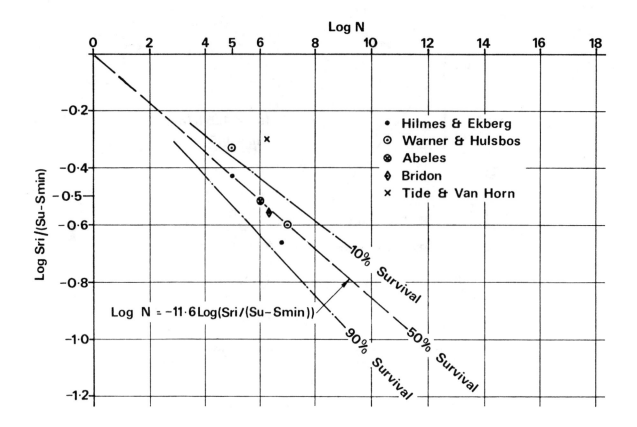

Fig. 6 Summary of results of fatigue tests on prestressing strand

OFFSHORE OIL INDUSTRY STRUCTURES

The results of the damage calculations for these structures are summarised in Table 1.

For the North Sea Climate, the relationship between wave height H (metres) and incidence n was obtained from increased wave exceedance curve given by

$$\log n_e = 8.3 (1 - H/30) \qquad - (2)$$

where n_e indicates the number of waves occurring in all directions which exceed a given height H during a 50 year period at a given location. Stress range at a point on the structure may only be sensitive to waves in particular directions and in the case of the gravity platform for example, it was assumed that 50% of the waves contributed to fatigue damage.

Quasi-static analysis was applied to the gravity platform. Monitoring of the behaviour of the Brent B platform has shown this to be a reasonable approach where the natural period of the structure (approximately 2 secs in this case) lies below the wave periods containing significant energy. (Skaar and Hansvold 1979).

Similarly Quasi-static wave balancing was considered to give a satisfactory assessment of response for the pontoon. The articulated tower response was obtained from model test information which showed enhanced response in certain lower wave periods.

For prestressing tendons it is easily established that calculated damage is minimal for fully bonded cables in a structure which is sufficiently prestressed to remain uncracked at extreme loads or which is likely to be cycled beyond cracking stresses on only a few occassions during its life. For instance, assuming a modular ratio for Youngs Moduli of steel and concrete of 7, the maximum stress range in the prestressing steel of the tower platform is about 125 N/mm^2. This is only about 8% of the characteristic strength of the steel and the majority of cycles are over a very much smaller range.

TABLE 1.

Structure	Material	Critical Location	Fatigue Damage $\sum n/N$ in 100 years.					
			Concrete			Reinforcement		
			f_k N/mm^2	C_1	C_2	R_1	R_2	R_3
Tower Platform	Prestressed Concrete	1. Base of Tower	45	0.00	0.01			
		2. Base of Tower Hot Spot	45	5.3	6.2			
	Reinforced Concrete	3. Base of Tower (Flooded)	45	0.00	0.01			
		4. Base of Tower (Flooded)				0.00	0.56	0.00
Column Platform (Articulated)	Prestressed Concrete	5. Mid-Height of Column	45	0.00	2.7			
Pontoon Hull	Prestressed Concrete (Longitudinal)	6. Base Slab Longitudinal Bending	40	0.00	6.8			
	Reinforced Concrete (Transverse)	7. Base Slab Longitudinal Bending	40	0.00	0.03	0.00	3.2	0.16
		8. Side Wall Shear	40	V1 0.00	V2 15.0			

The Miner's Sum at critical location '2' suggests a potential problem. This calculation is set out in Figure 7, and represents the situation at the base of the tower where a pipe hole has been included at the section where wave bending is a maximum. The assumption has been made that locally concrete compressive stresses will be increased by a stress concentration factor of 2. In practice the prestress would not be enhanced by an SCF=2 as due to creep it could be expected to approach the average value on the section. This would tend to introduce tensile stresses locally under extreme loads and reduce the extreme compressive stresses. Even so, it is interesting to note that most of the damage is due to the block of waves 'a'.

With wave heights greater than 27.5m, cycling the concrete close to its static strength, a 20% increase in concrete strength would reduce the Miners Sum value to 0.04 and 0.17 for curves C1 and C2 respectively, illustrating the extreme sensitivity of this low cycle fatigue calculation to assumed values of parameters. Concrete normally behaves as a notch-insensitive material capable of stress redistribution and given that it would be associated with reasonable quantities of steel reinforcement, it is unlikely that the stresses assumed would actually occur in practice.

Nevertheless, there remains a possibility that at least local damage will occur at the low cycle end of the spectrum. Whilst prudent design avoids siting holes or other weaknesses at critical sections, it does sometimes occur and it would be useful to have reliable data on performance in these circumstances. For a reinforced concrete tower - critical location 4 - calculations using TNO curve R2 suggested significant damage to the reinforcement over a 100 year period. Over a 25 year service life the sum would be less than 0.2 and so may be regarded as satisfactory. However curve R2 is very severe in the high cycle/low stress range region, as curve R3 suggests.

The dynamic response function derived for the articulated column tower shows stresses for certain lower, shorter waves approaching those for the higher waves. These stresses are below the endurance limit for Curve C1. Curve C2 is the more appropriate and gives significant damage in the high cycle region. The calculations indicate that fatigue could be a significant problem for this type of structure, requiring particularly careful analysis of the response and careful choice of operating stresses.

The maximum damage of the pontoon hull in longitudinal bending was similarly shifted to the higher cycle end of the spectrum as maximum response occurs in the medium wavelength waves around the 100m length corresponding to the hull length. The high values of Miner's Sum are less significant here as such a barge would normally be moored in a much calmer climate than the North Sea, and a service life of only 0.1 years, representing an ocean towage, would correspond to the assumed loading.

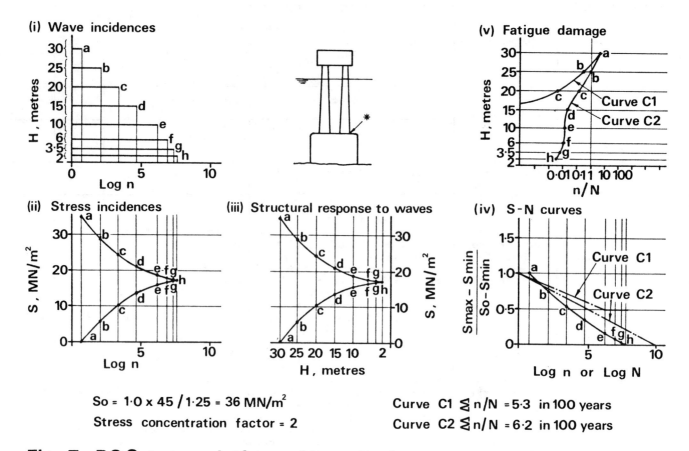

Fig. 7. P.S.C. tower platform. Hot spot in concrete at base of tower

WAVE ENERGY STRUCTURES

(a) General Description

Offshore Oil industry structures are now well known, with many examples in existence, whereas wave energy structures are not and some general introduction is appropriate.

Devices for tapping the enormous energy resources of ocean waves are currently under development in a number of countries. In the United Kingdom, the Department of Energy has been funding a research programme for some years and, since 1978, extensive testing and engineering studies, looking at all aspects of system design and construction, have been undertaken on a number of promising devices.

The particular wave energy converter (WEC) structures considered in this paper are two types of floating hull.

    i. The Cockerell Raft
    ii. The Lancaster Flexible Bag (Spine)

These devices have been sized for permanent mooring 12-15km offshore from the Outer Hebrides in the fierce climate of the North East Atlantic. A general impression of the structures can be gained from Figures 8 and 9.

The raft consists of front flaps hinged to a back section, all of concrete prestressed longitudinally and transversely with overall dimensions of 100m x 50m x 7m and a displacement of 13,500 tonnes. Power is taken off by pumps linked across the hinge line.

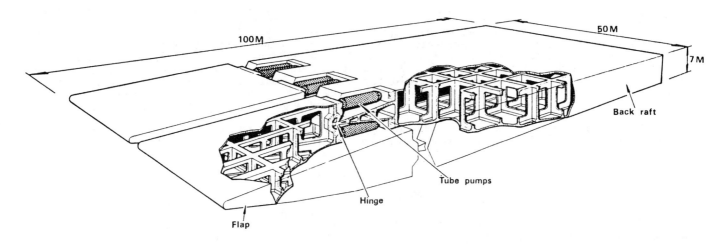

Fig. 8. Prestressed concrete hull. Cockerell raft

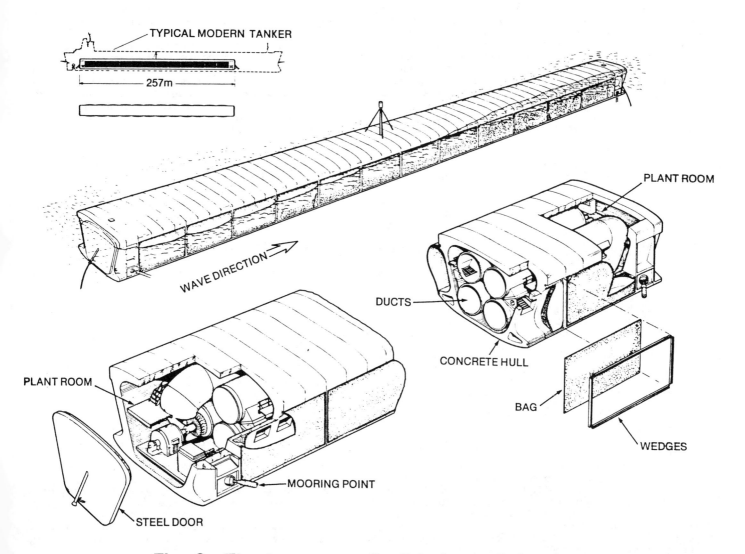

Fig. 9. The Lancaster flexible bag. Spine hull

The LFB is a long concrete spine with a number of large rubber bags or diaphragms arranged along each side. The bags are pressurised and linked through valves to a closed air circuit such that a wave crest travelling along the spine progressively collapses the bags which are then reinflated by air from the low pressure side of the turbo generator as the wave trough passes. For optimum performance the bags are set so as to be just fully immersed in still water conditions. This device has a considerably higher output than the raft and is more massive with an overall length of 250 metres, sufficiently long to remain stable in the energy rich, long wavelength, ocean swells. It is 22 metres wide by 15 metres deep with a displacement of 64,000 tonnes. The concrete spine is prestressed longitudinally and reinforced transversely.

These devices will be needed in great numbers with 250 to 300 LFB hulls envisaged for the Hebridean Chain. A modular segmental design has been developed to facilitate mass production on this scale, with concrete segments glued with epoxy resin and stressed together.

(b) Loading

Whereas most hulls have to be designed for substantial still water bending moments to allow for a variable distribution of cargo or imposed load, wave energy hulls have no cargo carrying function and it is possible to design for much smaller still water forces relative to wave loading, still water bending being less than 10% of extreme wave loading. As both the raft and spine have small freeboard - approximately 3 metres - wind loading is negligible. Wave loading dominates the ultimate strength design as well as the fatigue design. Wave energy hulls therefore provide an extremely severe example of fatigue loading.

Because of the importance of determining device behaviour and performance in realistic sea conditions, a mixed sea basin for model testing was built at Cadnam near Southampton to serve the wave energy programme. The tank was commissioned during 1980. The large volume of development testing has placed limits on the time available for exploring the structural response of the hull, essentially limiting thorough testing to the overall bending moments. Some measurement of local panel pressures at a number of points on the hull cross-section has been possible, but extrapolation using the bending moment results as guidance has been necessary to derive the load history for these pressures. These assumptions are discussed briefly later. The Cadnam mixed sea basin can generate deterministic (or predictable) seas which are in other respects random like real seas, allowing tests to be repeated but maintaining realistic modelling. It also allows singular events such as extreme or breaking waves to be deliberately generated within an irregular sea.

The seas are generated by a system of 60 absorbing wavemakers driven by computer which compiles up to 100 component sine waves to obtain desired sea parameters such as spectral density function, spreading function (typically Cos N or Mitsuyasu), height and period. Real time processing of up to 64 channels of measured data produces information on any specified statistical parameters such as mean, maximum, minimum, standard deviation, turning point or bivariate histograms for any combination of the data channels.

(c) Prediction of Extreme Loads and Fatigue Loadings from Model Tests

The values of extreme moments and the prediction of load history have been derived from an extensive series of tests (Hancock 1982) on 1/125 and 1/150 scale model spines strain gauged at mid-length allowing the measurement of vertical, horizontal and torsional moments. At 1/150th scale, the tank can reproduce extreme North Atlantic conditions, and the models were tested over a wide range of sea states for a range of headings to the principal sea direction, in seas with a Pierson-Moskowitz spectrum and Cos 5 spreading function.

The procedure for deriving design information from test data is illustrated in Figure 10. For a given heading angle, and a range of sea states, the RMS bending moment is measured for each sea state. Values beyond the tested sea states (and the peak response) are suitably extrapolated according to the asymptotic behaviour under quasi-static conditions.

For each sea state in the long term scatter diagram, the frequency distribution of the moment is computed from the appropriate RMS value using a Longuet-Higgins distribution or a Raleigh distribution (which were shown to give a good fit to measured histograms). The long term scatter diagram was generated from the recorded 2 year scatter diagram by first normalising it into probability form and then extending and smoothing it using a log normal distribution over most of the range with a Weibull distribution for higher sea states. A wave steepness cut off $Hs/\lambda z < 1/18$ was applied (λz = wave length for zero cross period). Multiplying the frequency distribution of moment with the probability from the long term scatter diagram for each sea state and totalling gives the overall distribution. This can be conveniently plotted as a log exceedance curve for a given length of time. Figure 11 shows such a curve for the LFB for a 25 year period (the minimum design service life). These curves can be used to predict extreme design moments and the load history.

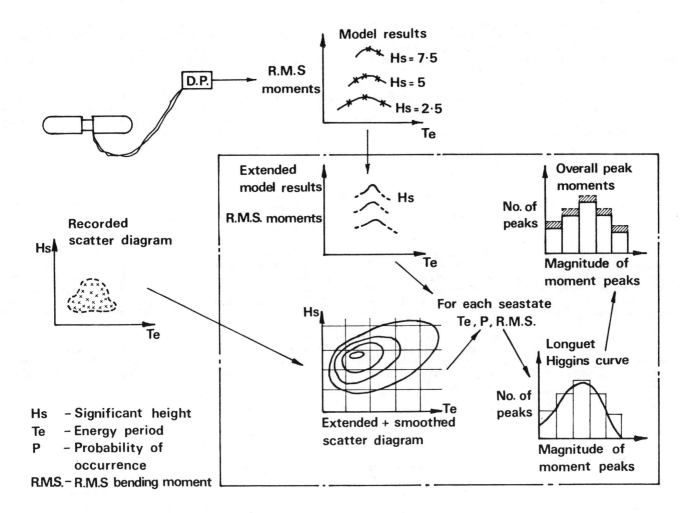

Fig. 10. Calculation of peak bending moment distribution

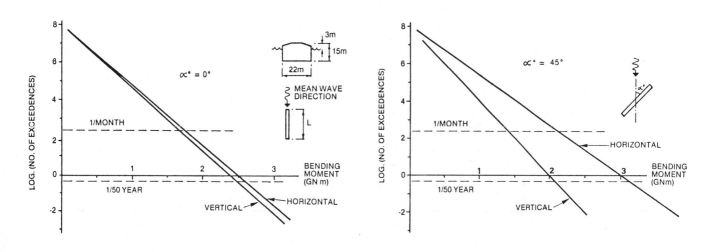

Fig. 11. Twenty five year peak bending moments. Spine `L´ 250m.

For instance, the zero ordinate (log n=0) corresponds to the moment which is on average exceeded once every 25 years (or is the most probable maximum moment in 25 years). The one in 50 year moment which is the value used for ultimate strength design (with a $\gamma f = 1.3$) corresponds to the ordinate (n = ½) log n = -0.30 which is approximately 4% greater than the 25 year maximum moment.

It is interesting to reflect that this technique of load prediction is likely to provide a more reliable prediction of load history for environmentally loaded structures than is possible for most land based structures, as on land patterns of loading are statistically less stationary due to changing patterns of usage.

(d) <u>Fatigue Calculations</u>

Table 2 sets out the principal cases considered and summarises the Miner's Sum values calculated. These were calculated for designs which have been determined by normal ultimate and serviceability limit state criteria, but in which no particular regard has been paid to the fatigue limit state. Thus where Miner's Sum exceeds 0.2, there is an indication that fatigue is a critical design criterion. The calculations have been based upon a service life of 25 years.

The table illustrates that fatigue damage is a significant concern in the design of wave energy hulls, and particular care is needed in the critical locations to ensure that premature failure is avoided. Overall bending strength of the spine hull is provided by fully bonded prestressing with a level of prestress sufficient to ensure that the cracking strength of the hull is not exceeded under the most probable once in 50 year loading. Under these circumstances fatigue damage from bending stresses is minimal. The precise location of the most critical point for fatigue is difficult to determine, as strictly the stress history at a number of points around the cross-section must be calculated taking into account the maximum membrane compressive stresses due to the combined horizontal bending and vertical bending moments whose phase correlation is not sufficiently well known from the series of tests carried out so far. However, with the assumption that they are always adversely in phase, a conservative result is produced.

TABLE 2.

Structure	Material	Location	Critical Effect	Fatigue Damage $\sum n/N$ in 25 years.				
				Concrete		Reinforcement		
				f_k N/mm^2	C_1	R_1	R_2	R_3
Spine Hull LFB	PSC	Mid Length	Overall Bending (Longitudinal)	60	0.02			
	PSC/RC	Quarter Point	Longitudinal Shear	60	0.38			
	RC	Fore/Aft Section, Upper corner on cross-section.	Transverse (Panel) Bending	60	0.04	0.07	0.87	0.08
	RC	Fore/Aft Section	Transverse Panel Shear	60	V2 295			
Raft Hull	RC	Aft Section	Panel Bending	60	0.10	0.40	3.36	0.37
	RC	Aft Section	Panel Shear	60	V2 45.30			

It is generally considered in the design of prestressed concrete hulls that the hull should be designed as uncracked even under the extreme wave moment to maintain serviceability and watertightness. This requires high levels of prestress to guard against infrequent occurrences. There could be important cost savings if this level of prestress could be lowered with the structure being allowed occassionally to cycle into a cracked condition. This question is discussed again later in the paper.

Maximum shear forces associated with the overall bending occur in the quarter length regions of the hull. As with the bending moments, the shears are fully reversing. The shear strength may be considered as the sum of three component strengths due to concrete, prestress and reinforcement.

$$V_r = V_{cr} + V_{pr} + V_{sr} \qquad - (3)$$

Under reversing loading the DNV rules recommend that the concrete component V_{cr} is neglected. If the shear strength of the glued segmental hull is considered as being solely provided by the longitudinal prestressing, and a uniform level of prestressing force is maintained along the length of the hull, as dictated by midlength bending moment, then the damage calculation, according to DNV rules, with a Wöhler curve representation of shear fatigue strength, indicates that the fatigue strengh of the hull may be lower than is desirable. In the (NPD 1977) rules the components of shear strength are dealt with separately and the component strength due to axial force or prestress is not considered to be affected by fatigue. This is an important difference which requires clarification. The segmental units making up the hull are substantially reinforced although only prestressing tendons cross the glue line. The shear calculation relates to the performance of the hull under principal diagonal tensile and compressive stresses rather than pure shear across the glue line, and it seems reasonable to consider the reinforcement as providing a useful additional strength component. However the authors have no knowledge of fatigue tests relating to this particular question.

The wave pressures producing overall bending and shear also produce significant local structural actions which are similarly cyclic in nature. Whilst it is perhaps desirable to prestress panels against local bending and shear actions, it is frequently difficult to achieve in practice and reliance has to be placed upon reinforcement to provide strength. In the cases considered for both the spine hull and raft hull the critical sections are reinforced and are at the fore and aft ends where the apparent wave height is enhanced by pitching motions. The pressures were calculated using a semi-empirical relationship between apparent wave height h and the actual wave height H with upper limits as given by the DNV rules. The curves used for the raft are shown in Figure 12. The assumption is made that pressures are due simply to the hydrostatic head of the waves and that radiation and diffraction pressures can be neglected. The limited amount of data on pressures from the tank tests is in reasonable agreement with these assumptions.

A sample calculation for the reinforcement at the base of a side panel of the raft is shown in Figure 13. The panel is assumed to be spanning vertically. This calculation also illustrates the effect of increasing the percentage of reinforcement over what is required for purely "static" strength.

The critical section considered shown in Figure 14b was immediately adjacent to the corner detail. Unless large radius bends (> 25 x bar diameter) are provided to the tensile reinforcement at the corner, the fatigue strength is likely to be severly reduced - DNV rules indicating that life is reduced by a factor of 10. Equally the accommodation of anchorages in this region for the transverse prestressing (Figure 14a) of the bottom slab will induce complex stress states and raises questions of whether fatigue strength of the joint will be adversely affected or whether cyclic loading will lead to increased cracking and reduced corrosion resistance.

The results set out in Table 2 also indicate that shear strength will be adversely affected and that substantial shear reinforcement across the panel thickness will be required despite the concrete strength component being adequate for ultimate strength. In the case of the spine hull, the critical location for panel bending is situated close to the upper corner and the particular cross-sectional geometry has the effect of reducing stress range due to wave pressure, making fatigue slightly less critical for the tensile reinforcement. Shear strength however is reduced requiring the employment of substantial shear reinforcement.

As previously mentioned, it is well established that fatigue damage to prestressed concrete is negligible provided the structure is not cycled beyond the cracking moment. Prestressed structures subject to wave loading are normally designed so that the extreme design moment does not exceed the cracking moment, that is, tensile stresses are limited to be not greater than the flexural tensile strength of the concrete.

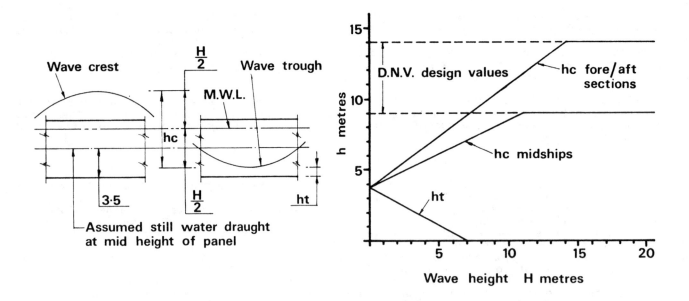

Fig. 12. Apparent wave height 'h' against actual wave height 'H' Assumed relationship for determination of hydrostatic panel loading for wave energy converter

This extreme design moment is based upon the most probable highest value within the return period considered (50 years). There remains a reasonable probability that this moment may be exceeded by a small amount during the life of the structure. If it is assumed that this exceedance cracks the structure, and that autogenous healing does not restore the tensile strength at the cracked section, the question arises as to whether cycling beyond the now reduced cracking moment with enhanced stresses in the tendons significantly reduces the fatigue life of the structure. Analysis indicates that cracking could increase the maximum stress range in the prestressing strand typically for a well proportioned section from 5% for an uncracked section up to about 15% of the ultimate tensile strength of the strand.

With a minimum stress in the strand of 55% UTS:-

$$\text{Log} \frac{S_{ri}}{S_u - S_{min}} = \log \frac{15\%}{100\% - 55\%} = -0.48$$

From Figure 6 log N = 4.4 therefore N = 25,000.

This relatively small number of cycles when compared with the performance of individual wires is a reflection of the reduced performance of strand due to fretting action in the twisted wires. Nevertheless the performance is well above the low cycle fatigue range (N=100). Considering the data for the spine hull, the ratio of the 1 month moment to the 50 year moment is in the region of:-

$$\frac{M \ 1 \ \text{Month}}{M \ 50 \ \text{Year}} = 0.7$$

There are 300 probable exceedances of the 1 month moment in the 25 year service life. With a uniform prestress of 9 N/mm² and an allowed tensile stress in the concrete of 2 N/mm², the ratio of the moment producing zero tension Mto = to the cracking moment Mcr is:-

$$\frac{M_{to}}{M_{cr}} = \frac{9}{11} = 0.8$$

Thus the number of cycles producing enhanced stresses in the tendons is likely to be less than 600 and the fatigue damage to the tendons from cycling the cracked structure is very small.

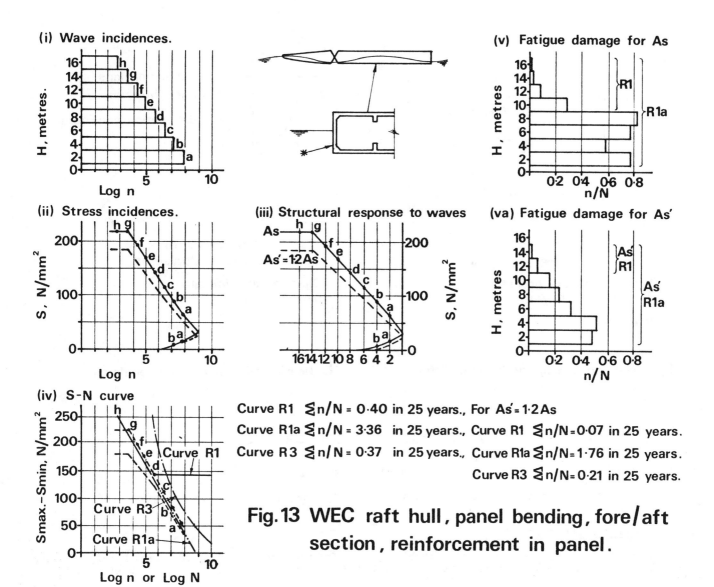

Fig. 13 WEC raft hull, panel bending, fore/aft section, reinforcement in panel.

For wave energy devices, the cost of prestressing forms a very considerable proportion of the structural cost. As the devices are essentially unmanned, questions of human safety do not arise as they do for most offshore structures, and putting aside the possible problems of watertightness and durability, there may be useful economic benefits from saving prestressing strand if it could be demonstrated that partial prestressing - allowing M 50 > Mcr - does not produce fatigue problems. For most current offshore structures cost savings achieved in this way are likely to be marginal because of the relatively lower proportion of the structural cost to the total cost and the lower proportion of prestressing cost within that structural cost.

(e) <u>Sensitivity of fatigue calculations to uncertainty in parameters.</u>

The calculations were examined for sensitivity by noting the effect on Miner's sum for simultaneous advantageous or disadvantageous variations in the main parameters of ± 20%. This showed clearly how the log/arithmic nature of the S-N curve makes the Miner's Sum value particularly sensitive to the calculated values of the stress range and to the chosen S-N curve. Differences of several orders of magnitude in the Miner's Sum can be produced in this manner.

The examples considered suggested that generally the greatest proportion of calculated damage is produced by the high cycle end of the response curve, and this therefore highlights the care needed when choosing the log N intercept on the S-N curve.

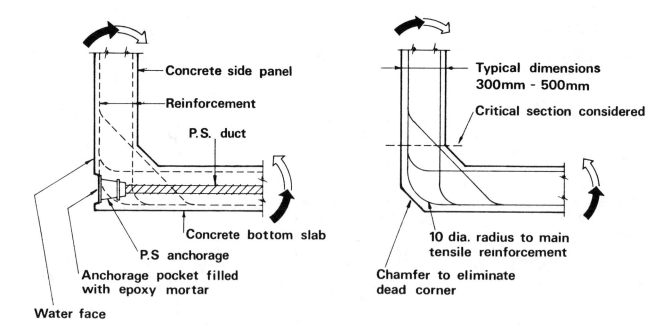

a Anchorage at edge of bottom slab   **b** Reinforced concrete corner detail

Fig 14. Raft corner details

CONCLUSIONS

It is convenient to draw conclusions separately for the two broad categories of structure considered in the study.

(a) <u>Oil Industry Structures</u>

The study indicated that the three types of structure considered were likely to have adequate fatigue strength for their operational life on the North Sea climate except under the following conditions:-

i. Where stresses at a stress concentration (such as around a hole or sudden change of cross section) approach the static strength under serviceability limit state conditions. Such a stress level is unlikely to be acceptable if it is identified but it is not general practice in concrete design to identify hot spot stresses. Some doubt exists whether such stresses occur in concrete structures and research to clarify this issue would be valuable.

ii. Where stresses in a structure are as sensitive to the action of small shorter waves as to the action of larger waves. This was shown to be a possibility for the Articulated Column type of structure, indicating that particularly careful dynamic analysis is appropriate in this situation. This is also a possibility for the pontoon hull, but normally such structures will have very limited exposure to open sea conditions.

(b) <u>Wave Energy Structures</u>

Examination of the performance of two hull configurations - the raft and spine hulls - moored for their 25 year operational life in the North Atlantic climate indicates that fatigue is an important consideration in the design of certain elements and could have an important effect on the shape and economy of particular hull designs. In future similar structures may be needed for other purposes, and the conclusions may have broader interest beyond wave energy applications.

In particular, cyclic local bending and shear forces induced by wave pressures can cause significant fatigue damage. As high local bending may occur at corners of hulls and as such areas may be difficult to prestress effectively, strength may be provided by unstressed reinforcement. The bends in this reinforcement may reduce fatigue strength further. Such locations are critical areas requiring very careful attention to ensure satisfactory fatigue strength.

The overall (membrane) shear strength of the prestressed spine hull girder is also subject to appreciable fatigue damage. The different approaches in the calculation of the effect of cyclic reversing shear forces on the component of shear strength, due to the longitudinal prestress, alters the results of this calculation dramatically. Further clarification would be useful of the extent to which load cycling reduces the strength of flexurally uncracked prestressed sections in marine conditions.

Examination of the post-cracking behaviour of a prestressed spine supports the view that cracking under exceptional overload is unlikely to increase the fatigue damage to the tendons sufficiently to reduce the life of the structure, as fully prestressed. Significant economies could be achieved in the level of prestress, if the spine were designed as partially prestressed. There is reason to believe that the design could be pushed some way in this direction before fatigue becomes critical. However further examination of this question is required before accepting partial prestressing as a safe design. A particular concern would be the performance of glued joints in segmental construction where no additional reinforcement is present to assist in crack control. Other issues such as the effect of durability and watertightness would also need careful attention.

In general, where fatigue of steel reinforcement is a strong possibility, the sensitivity of the fatigue calculation to the existence of an endurance limit is an important issue. As recent tests on reinforced concrete beams immersed in sea water provide evidence that no such limit exists, further information on the performance of reinforcement in the very high cycle range is desirable.

REFERENCES

PRICE W. I. J., HAMBLY E. C., & TRICKLEBANK A. H., 1981.
"Review of fatigue in Concrete Marine Structures"
Concrete in the Oceans Phase II C.I.R.I.A./U.E.G. London Report P5A.

D.N.V.
Rules for the Design, Construction and Inspection of Offshore Structures.
1977 - Appendix D
Det Norske Veritas, Oslo.

MONNIER T.,
Fatigue Strength Procedure for Concrete Offshore Structures.
1975. T.N.O. Report. Delft.

PATERSON W. S., DILL M. J., & NEWBY R. 1981.
"Fatigue Strength of Reinforced Concrete in Sea Water"
Concrete in the Oceans, C.I.R.I.A./U.E.G. Report P6.

WAAGARD K., 1977.
"Fatigue of Offshore Concrete Structures - Design and Experimental Investigations"
Proceedings, 9th Annual Offshore Technology Concrete. Houston, Texas 1977.

HAWKINS N. M., 1975
"Fatigue Considerations for Concrete Ships and Offshore Structures"
Proceedings, Concference on Ships and Concrete Floating Structures
University of California, 1975 pp. 136-148.

WAAGARD K., 1981
"Fatigue Strength Evaluation of Offshore Concrete Structures"
Proceedings ACI Convention, Dallas 1981

LEEUWEN J. VAN & SIEMES A. J. M., 1979,
"Miner's Rule with Respect to Plain Concrete",
Proceedings, 2nd International Conference on Behaviour of Offshore Structures
Boss '79, London.

SKAAR J. T., & HANSVOLD C. H., 1979
"The Brent B Instrumentation Project"
SUT Seminar/Symposium 1979 London.

HANCOCK M. 1982
Wave Induced Bending Moments - A Prediction Method using an Advanced Mixed Sea Basin
Wavepower Ltd., Southampton, England. 1982. Report No. WP 141.

N.P.D. 1977
Regulations for the Structural Design of Fixed Structures on the Norwegian Continental Shelf.
Norwegian Petroleum Directorate 1977.

UNIAXIAL TENSILE STRENGTH OF CONCRETE AS INFLUENCED BY IMPACT AND FATIGUE

Prof.Dr.-Ing. H.W. Reinhardt, Stevin Laboratory,
Delft University of Technology, Delft, The Netherlands

SUMMARY

Although the tensile strength of concrete is neglected in normal design, it is the property which largely determines the cracking behaviour, crack width and crack spacing, the bond between steel and concrete and the shear capacity of structural elements without stirrups. Furthermore, it is one of the parameters of two- and three-dimensional failure envelopes of concrete which are essential for advanced analysis of reinforced concrete.

Besides static loading, offshore structures are exposed to cyclic loading due to current, waves and wind forces and sometimes also to impact loading due to ship collision, explosions and impacting objects. The tensile behaviour of concrete under this circumstances has hardly been studied so far.

The following will deal with the experimental methods, the results and the analysis of a research project containing single and repeated impact tests on concrete and cyclic loading of concrete under tensile and alternating tensile-compressive loading.

Impact loading with stress rates between 2000 and 60 000 $N/mm^2 s$ has been achieved by use of a modified Split-Hopkinson-bar-technique consisting of two aluminium bars between which the cylindrical concrete specimen has been sandwiched. The loading pulse has been generated by a drop weight. A variety of concrete mixes, cement types and climatic conditions have been investigated under single and repeated impact loading. The results show a significant influence of the stress rate on the tensile strength of concrete: the strength increases with increasing stress rate. This increase is higher for higher water-cement-ratio, for higher cement content, for less coarce aggregate, and the loading direction perpendicular to the casting direction.

Repeated impact loading leads to a decrease of strength in such a way that the strength gain due to high stress rate is already lost after 10 to 20 cycles. This strength decrease is attributed to crack propagation which occurs the more easy the more brittle the material is. This means that high strength concrete is more affected by repeated impact loading than concrete with lower strength.

Cyclic sinusoidal tests have been carried out in a stress controlled hydraulic loading equipment with constant frequency. The remarkable feature of this equipment are prestressed swivel joints which enable uniaxial tension-compression loading without lack of contact. The results as achieved with various upper and lower fatigue stresses show a clear influence of the stress level on the tensile fatigue strength of concrete. As usual, the results follow a logarithmic relation between stress level and number of cycles to failure. It seems that alternating stresses are considerably more detrimental for concrete than pure tensile loading.

Design values of the tensile strength of concrete as influenced by impact and fatigue are proposed.

1. INTRODUCTION

The tensile strength of concrete is regarded as a minor and unreliable quantity and is therefore neglected in the usual design of reinforced concrete structures. Actually, the tensile strength is only a fraction of the compressive strength, and non-uniform shrinkage and temperature often consume a great part of it, so that the effective tensile strength is even less. Therefore, only in unreinforced concrete structures, such as concrete roads, does the tensile strength form a design criterion, whereas in reinforced structures the tensile strength is neglected as far as bending is concerned. As soon as shear due to bending or torsion is considered, the load bearing capacity of the structure relies on the tensile strength if no shear reinforcement is used. The same is true of punching shear in slabs and shells.
Even though tensile strength is neglected in normal design, it plays a dominant part in the deformation and rigidity of reinforced structures. The occurrence of cracks depends upon the tensile strength; the number of cracks, crack spacing and crack width are governed by it. Because cracks influence the deformation more than any other factor does, it can be stated that deformation and rigidity are a function of tensile strength.
Advanced methods of analysis need constitutive laws for two and three dimensional states of stress of concrete. One of the decisive parameters is the tensile strength, which cannot be neglected without a major loss of accuracy of the analysis.

It is well known that the type of loading has a significant influence on the mechanical properties of concrete. The relation between compressive strength, fatigue and impact is fairly well established, whereas the tensile strength has received little attention. But offshore structures in particular, with their complicated shapes and complex states of stress, are analysed by advanced methods, and therefore it is of importance to know the tensile strength of concrete. Abnormal loading cases may occur, e.g., impact of falling objects (Jensen, 1980), collision of ships or aircraft crashes, which cause high stress rates in the material. On the other hand, wind and wave loads occur many times and can lead to fatigue phenomena such as strength decrease and cracking. If the safety of the structure or of a part of it is governed by shear, tensile strength plays an important part.
Vibration and deformation of a structure depend upon its rigidity, which is - at least in the cracked state - a function of the tensile strength.

Thus, there are reasons enough why the tensile strength of concrete as a function of impact and fatigue should be known. As the knowledge of it was very scarce, an experimental investigation was carried out, which yielded results that can be used in practical engineering.

2. SCOPE OF THE INVESTIGATION

The impact experiments had to cover a wide variety of parameters because very little information was available so far. Therefore the following variables were considered:
- cement type: Portland A, B, C blastfurnace slag HoA, HoB
- cement content: 300, 325, 375 kg/m^3
- water-cement ratio: 0.40; 0.45; 0.50
- maximum particle size: 8; 16; 24 mm
- specimen humidity: dry, wet
- loading-casting direction: parallel, perpendicular
- stress rate: 2 to 60 N/mm^{2}ms
- number of cycles to failure: 1 to 1000

Of course the main parameter was the stress rate, which was varied between wide limits and was very high compared with normal "static" loading at about 0.1 N/mm^{2}s. The composition of the concrete was such that a workable mix was achieved and that the maximum grain size was less than a third of the smallest dimension of the specimen. The cube compressive strength at 28 days was between 37 and 61 N/mm^2.
Because the combination of all the variables would have involved a vast program, the variables have been split up into a preliminary and a final program. This allowed us to omit some combinations which appeared not to be significant. It is thought that the number of about 300 single impact tests and 90 fatigue tests were sufficient for a reliable study of the impact tensile behaviour of concrete.

The fatigue tests with normal stress rate (sine wave, 6 Hz) were performed on one concrete mix with a water-cement ratio of 0.50, 325 kg of Portland cement B/m³, river gravel with maximum particle size of 16 mm. The cube compressive strength was 48 N/mm² at 28 days. 250 constant amplitude tests were executed with various combinations of lower and upper stress-strength levels. The lower levels were 40, 30 and 20 per cent of the tensile strength and 10, 20 and 30 per cent of the compressive strength, while the upper level varied from 40 to 90 per cent of the tensile strength.

3. SINGLE IMPACT TESTS

3.1. Experimental Procedure

The uniaxial tensile test was chosen because of its uniform stress distribution and the well defined boundary conditions. The specimens were 74 mm diameter cylinders, 100 mm long, drilled from a larger concrete block. They were sandwiched between two 74 mm diameter aluminium bars with a total length of 3,50 and 5,50 m respectively. The whole composite bar was mounted vertically. By means of a drop weight a pulse was generated in the lower bar which loaded the specimen within a given period of time in the range of 100 μs to 2 ms to failure. The load was measured on the elastic aluminium bars, while the deformation of the concrete was measured on the concrete by strain gauges. Details of the experimental procedure, which is based on the Split Hopkinson bar method, are given in (Reinhardt and de Vries, 1978).

3.2. Results and Discussion

From each concrete mix a set of static results (splitting tests with stress rate of 0.1 N/mm²s) and some results of impact test with stress rates between 2 and 60 N/mm²ms were available. For each parameter the results were statistically evaluated, and statistical relations were derived also for a set of parameters. The best relation appeared to be a double logarithmic function (Zielinski et al., 1981).

$$\ln f = A + B \ln \dot{\sigma} \tag{1}$$

where f is the tensile strength, $\dot{\sigma}$ the stress rate, A and B coefficients depending upon the concrete mix and the environmental conditions.
A similar relation was used by Mihashi and Wittmann (Mihashi and Wittmann, 1980) which was arrived at on the basis of probabilistic considerations.

Formula (1) expresses the fact that the tensile strength increases with increasing stress rate. This result was found for each mix. The coefficients A and B determine the magnitude of the rate influence. They depend upon the concrete mix. The larger the value of B, the greater is the influence of the stress rate on the tensile strength. Two examples are given in Fig. 1a and b illustrating the influence of stress rate, cement type, water-cement ratio and humidity on the strength.

Although the specific influence of the mix parameters is relatively small compared with the overall influence of the stress rate, some conclusions can be drawn from the results: there is a tendency that higher cement content and higher water-cement ratio lead to a greater influence of the stress rate, whereas larger aggregate size reduces that influence. The type of cement and the state of humidity caused no significant difference. The strength perpendicular to the casting direction is higher than parallel to that direction.

When the structure of concrete is concerned as a composite material, it emerges that a brittle matrix (hardened cement paste and fine material) is less affected by the stress rate and that a more ductile matrix is more sensitive to rate influences. This behaviour must be attributed to crack arresting effects of pores and aggregate, while the increase of tensile strength with higher loading rate as such seems to be caused by multiple cracking and forced failure of aggregate (Zielinski and Reinhardt, 1982).

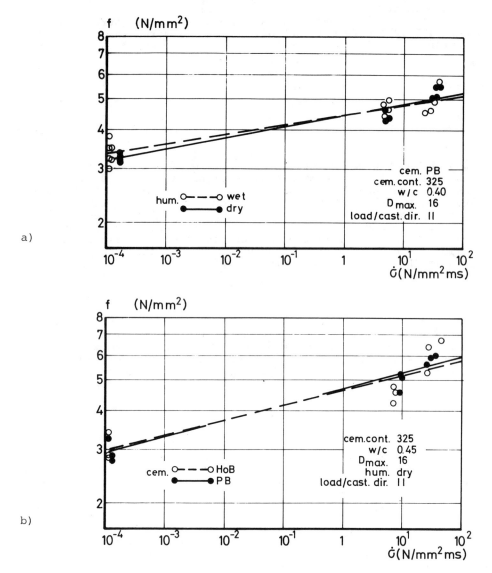

Fig. 1 Influence of several parameters on the tensile strength of concrete

For engineering purposes there is not much point in differentiating between the various concretes. With some loss fo accuracy all experiments can be treated as one population. Then one diagram shows the influence of the loading rate more clearly, Fig. 2, together with the 90% confidence band. The horizontal axis gives the stress rate, where 10^{-4} corresponds to the static test.

A more general interpretation of the results is possible if the ratio f/f_o (f_o static strength) is plotted against the ratio $\dot{\sigma}/\dot{\sigma}_o$ ($\dot{\sigma}_o$ stress rate of static test). Hence the static case is denoted by unity. From these results it can be concluded that the average increase of tensile strength is 80% if the stress rate increases by a factor of 10^6. 5 out of 100 cases increased by 40% and 5 out of 1000 reached at least the average static strength. In general it can be stated that a higher concrete strength can be expected if the stress rate increases. This phenomenon has also been found to occur in impact testing of structural elements (Brandes, 1980).

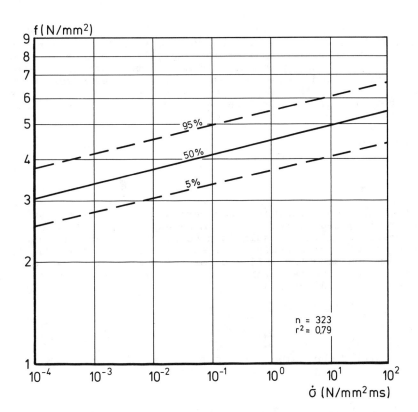

Fig. 2 Tensile strength vs. stress rate

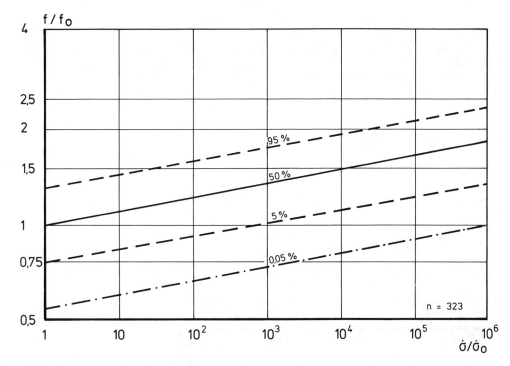

Fig. 3 Relative tensile strength vs. stress rate

4. IMPACT FATIGUE TESTS

4.1. Experimental Procedure

The loading equipment was the same as for single impact tests, with the only difference that the drop height was smaller because of the smaller loading pulse. Whereas in single impact loading a surplus of energy was put into the equipment in order to fracture the specimen, in repeated loading the energy had to be limited to such a level that the upper stress was just reached. A pneumatic lifting device lifted the drop weight and released it 16 times per minute until failure occurred. The number of cycles to failure was recorded by a counter. Because it was not possible to keep the loading rate constant during all the cycles, the stress rate varied between 2 and 6 N/mm²ms. In view of the logarithmic relation of strength and stress rate this variation was felt to be acceptable.

4.2. Results and Discussion

The lower stress limit in all the experiments was zero and the upper stress limit was so determined that different number of cycles to failure could be expected. As the tests were performed as constant-amplitude impact tests, the results can be plotted as an S-N curve. It must be pointed out that the maximum number of cycles was restricted to about 3000 for practical reasons. The results obtained for various mixes show the influence of the number of cycles on the strength in a rather similar way: the strength decreases with increasing number of cycles to failure, Fig. 4. The relation can be described with the equation

$$\sigma_{max} = A_1 + B_1 \ln N$$

where σ_{max} is the upper stress limit, N the number of cycles to failure, and A, B coefficients depending on the concrete mix and the humidity.

The interpretation of the results shows some significant effects of the various parameters on the endurance curve: higher water-cement ratio results in higher impact fatigue strength, concrete with higher cement content behaves better than lean concrete, the humidity shows no effect, and the loading-to-casting direction does not affect the slope of the endurance curve. These respective effects were not as significant as the main effect, namely, the number of cycles to failure. These results are in agreement with the expectation based on the structure of concrete conceived as a composite material. The more ductile the matrix of the concrete, the higher is the endurance impact strength. Whereas in brittle material cracks lead to high stress concentrations, ductile material equalizes these stress peaks and consequently has a higher endurance strength. In terms of concrete mix the results mean that a concrete with low compressive strength, which is normally less brittle, will show a relatively higher endurance limit than a high-strength concrete. In the absolute sense, the high-strength concrete will still have a higher fatigue strength.

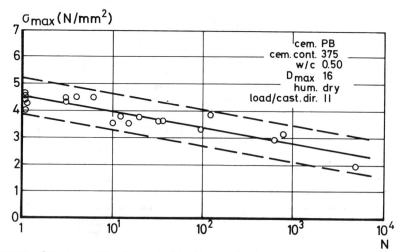

Fig. 4 S-N curve for tensile impact loading of one mix

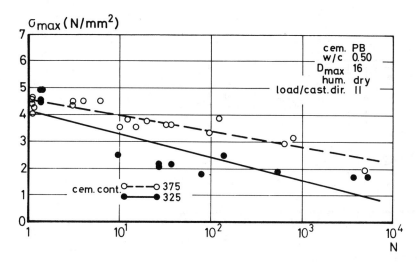

Fig. 5 S-N curve showing the influence of cement content

For practical purposes all the results can be worked out in one relation showing the average values and the 90% band, Fig. 6. The decrease in strength is very obvious, the average strength after one thousand cycles being only 2 N/mm², with a single impact strength of 4.3 N/mm². The relative strength, Fig 7, i.e., the ratio between impact fatigue strength and static strength, decreases from circa 1.4 to 0.7 or by 50%. Repeated impact loading has a detrimental effect on the tensile strength of concrete which is much severer than in "normal" fatigue as will be shown below.

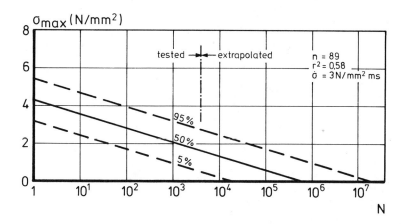

Fig. 6 Impact endurance curve for all mixes

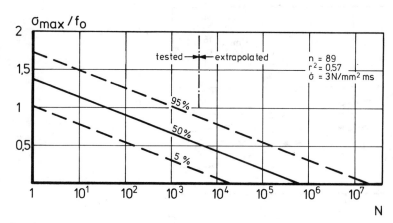

Fig. 7 Relative impact endurance curve for all mixes

5. CYCLIC FATIGUE TESTS

5.1. Experimental Procedure

This chapter deals with cyclic fatigue tests with "normal" stress rates, i.e., comparable static loading. The load is applied according to a sine wave with 6 Hz. Tests with tensile upper and lower levels were performed, as well as tests with tensile upper and compressive lower stress, in order to obtain repeated and alternating loading conditions. Uniaxial loading was applied just as in the impact tests, the difference being the size of the specimen. Here this was a tapered cylinder 120 mm in diameter and 300 mm long. The specimen was fixed to steel plates which were connected by aid of prestressed swivel heads to the hydraulic actuator. Fig. 8 shows the loading and controlling systems with the adjacent unit for data acquisition (Cornelissen and Reinhardt, 1982). The strain was measured with LDVTs along the specimen. The force and elongation were measured eight times during one cycle and stored in a computer memory.

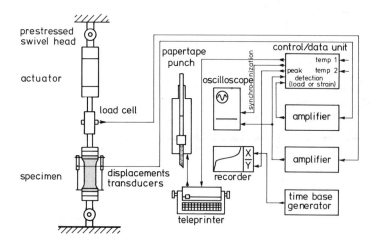

Fig. 8 Loading and controlling system (schematic)

5.2. Results and Discussion

The results have been plotted in Wöhler diagrams (S-N diagrams) showing the influence of the number of cycles to failure on the maximum stress at a certain minimum stress. Two examples are given in Figs. 9 and 10, the first with tensile upper and lower stress and the second with tensile upper and compressive lower stress; these diagrams show the test results and the 90% confidence limits. The scatter is rather large, but not unusual in fatigue tests. The S-N curves were calculated by linear regression analysis and gave the following results for the average relation if $\sigma_{min} \geq 0$

$$^{10}\log N = 15.02 - 14.90 \frac{\sigma_{max}}{f_{tm}} + 3.13 \frac{\sigma_{min}}{f_{cm}} \tag{3}$$

and if $\sigma_{min} < 0$

$$^{10}\log N = 9.46 - 7.71 \frac{\sigma_{max}}{f_{tm}} - 3.78 \frac{\sigma_{min}}{f_{cm}} \tag{4}$$

where N is the number of cycles to failure, f_{tm} the mean static tensile strength, and f_{cm} the mean cube compressive strength. The 90% confidence regions are $^{10}\log N \pm 1.84$ and $^{10}\log N \pm 1.45$ respectively.

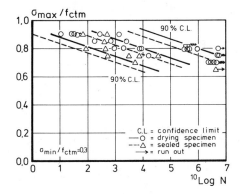

Fig. 9   S-N diagram for $\sigma_{min}/f_c = 0.3$

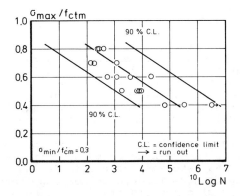

Fig. 10   S-N diagram for $\sigma_{min}/f'_c = 0.3$

Figs. 9 and 10 and equation (3) and (4) indicate the marked influence of the number of cycles on the strength; besides, they indicate that according as the minimum stress is lower, the number of cycles to failure at a given maximum stress is also lower.

6. USE OF RESULTS IN DESIGN PRACTICE

In order to make use of the information obtained in the experimental investigation, the results should be worked out in such a way that the main conclusion is evident. The formulation should be as simple and as illustrative as possible. Furthermore, the presentation should follow known patterns which already have proved useful in practice. After this general statement it will be attempted to evaluate the results of this research.

Because it is not usual to measure the tensile strength of concrete, it would be better to refer to the compressive strength which is tested in standard tests. There are relations between tensile and compressive strength by means of which the tensile strength can be deduced from the compressive strength. These relations are valid for static loading and usual environment. If impact is to be considered, the relations must be adjusted to high stress rates.

A familiar relation given by CEB (CEB, 1978), is

$$f_{tm} = 0.30 \, f_{ck}^{2/3} \tag{5}$$

where f_{tm} denotes the tensile strength and f_{ck} the characteristic cube compressive strength, both in N/mm^2. Since $f_{tk} = 0.75 \, f_{tm}$ (CEB, 1978) eq. (5) can be written as

$$f_{tk} = 0.225 \, f_{ck}^{2/3} \tag{6}$$

Thus, two corresponding quantities are related in a simple manner.
Assuming the same relation to be valid for the mean value, it follows that

$$f_{tm} = 0.225 \, f_{cm}^{2/3} \tag{7}$$

which is valid for static loading. If <u>single impact loading</u> is considered, it would be attractive to retain the type of relation and to <u>adjust only the coefficient</u>. This would mean that static concrete compressive strength and impact tensile strength are closely related and that no additional influence from loading rate would occur. A critical scruting of the results shows this not to be really true, but for practical purposes it is permissible to adopt a lower limit which is conservative.

By processing the data in this way, the following expressions were derived

- static loading, $\dot{\sigma} = 0.1$ $N/mm^2 s$ $f_{tm} = 0.225 \, f_{cm}^{2/3}$ (8)
- impact, $\dot{\sigma} = 0.1 \cdot 10^3$ $N/mm^2 s$ $f_{tm} = 1.10 \, f_{cm}^{1/3}$ (9)
- impact, $\dot{\sigma} = 0.1 \cdot 10^6$ $N/mm^2 s$ $f_{tm} = 3.10 \, f_{cm}^{1/10}$ (10)

Fig. 11 gives these relations, together with the impact results. It can be seen that the static tensile strength is predicted rather well, whereas for higher stress rates the relation is rather conservative. This can be justified by the fact that the results show rather a lot of scatter if the static compressive strength is taken as the reference value. Nevertheless, for the sake of simplicity these relations are proposed for use in practice. The two main features are clearly depicted: first, the increase of strength with increasing stress rate and, secondly, the fact that this increase is smaller for higher concrete quality. This is a consequence of the brittleness of concrete, as has been pointed out.

<u>Fatigue behaviour</u> can be depicted in a Goodman diagram which contains all the information from S-N curves with various values of the lower stress. As we are here concerned with impact fatigue, there are only results available with zero lower stress, whereas for cyclic fatigue six levels have been examined. Fig. 12 shows a modified Goodman diagram for cyclic loading (round dots) and the few results of impact fatigue. The pure tension cyclic results are in good agreement with relations found for compressive loading by other researchers. As we are here concerned with reversed cyclic loading and impact, the author is not aware of corresponding results that could be compared. Two points are notable: reversed load is more detrimental than cyclic tensile loading, and impact loading appears to be much more detrimental than cyclic loading. In judging abnormal

loading conditions these two results should be borne in mind.

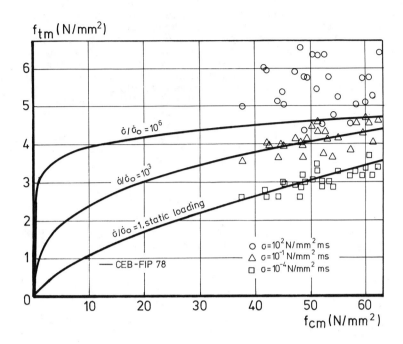

Fig. 11 Relation between static cube compressive strength and tensile strength at three rates of loading

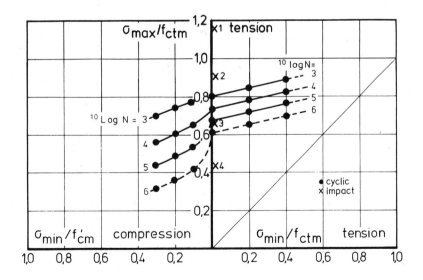

Fig. 12 Modified Goodman diagram for tensile upper stresses

7. CONCLUSION

Tensile strength is a property of concrete which is decisive in bending shear, punching shear, bond, cracking, deformation and stiffness of reinforced concrete structures. Furthermore, it is an important parameter in constitutive relations for concrete.
On the other hand, offshore structures are subject to fatigue and impact loading. Therefore it was considered necessary to investigate the influence of impact and cyclic loading on the tensile strength in order to be able to judge the behaviour and safety of offshore structures under those conditions. It has been shown that the tensile strength increases with increasing rates of loading. On the basis of the CEB relation between tensile and compressive strength a similar stress rate dependent formulation has been given which can easily be applied in engineering practice.
Cyclic tensile tests and alternating stress tests gave results which showed the detrimental influence of alternating loading on the tensile strength. Tensile impact fatigue tests revealed an even more detrimental effect on the tensile strength. The values found are lower than in the CEB code. These facts should be kept in mind when the behaviour of an offshore structure has to be estimated under these loading conditions.

REFERENCES

JENSEN, J.J., 1980, "Impact of falling loads on submerged concrete structures," *Proceedings, International Symposium Offshore Structures*, Rio de Janeiro, pp. 1.215-1.231. London, Pentech Press.

REINHARDT, H.W. and A.W. de Vries, 1978, "Bestimmung der Zugfestigkeit von Beton unter schlagartiger Beanspruchung," *Materialprüfung*, Volume 20, Number 11, pp. 427-430.

ZIELINSKI, A.J., H.W. REINHARDT and H.A. KÖRMELING, 1981, "Experiments on concrete under uniaxial impact tensile loading," *Matériaux et Constructions*, Volume 14, Number 80, pp. 103-112.

MIHASHI, H. and F.H. WITTMANN, 1980, "Stochastic approach to study the influence of rate of loading on strength of concrete," *Heron*, Volume 25, Number 3.

ZIELINSKI, A.J. and H.W. REINHARDT, 1982, "Stress-strain behaviour of concrete and mortar at high rates of tensile loading," to be published in *Cement and Concrete Research*.

CORNELISSEN, H.A.W. and H.W. REINHARDT, 1982, "Fatigue of plain concrete in uniaxial tension and in alternating tension-compression loading," *Proceedings, IABSE Colloquium on Fatigue of Steel and Concrete Structures*, Lausanne.

CEB-FIP, 1978, *Model Code for Concrete Structures*. London.

TEPFERS, R. and Th. KUTTI, 1979, "Fatigue strength of plain, ordinary and lightweight concrete," *ACI Journal*, Volume 76, Number 5, pp. 635-652.

HSU, Th.T.C., 1981, "Fatigue of plain concrete," *ACI Journal*, Volume 78, Number 4, pp. 292-305.

BRANDES, K., 1980, "Versuche zum mechanischen Verhalten von Stahlbetonbauteilen unter Stosseinwirkung," *Proceedings Forschungskolloquium Stossartige Belastung von Stahlbetonbauteilen*, Dortmund.

8. ACKNOWLEDGEMENT

The author wishes to thank H.A.W. Cornelissen, H.A. Körmeling, G. Timmers and A.J. Zielinski for their valuable contribution to this investigation. Furthermore, the support and encouragement of the Netherlands Committee for Research, Codes and Specifications for Concrete is gratefully acknowledged.

BEHAVIOR OF CONCRETE IN BIAXIAL CYCLIC COMPRESSION

O. Buyukozturk
Massachusetts Institute of Technology
USA

J. G. Zisman
Massachusetts Institute of Technology
USA

SUMMARY

An experimental program was conducted to study the behavior of concrete under biaxial cyclic compression. Biaxial loading was achieved by subjecting square concrete plates to in-plane loading where compressive stress was applied in one direction while confining the specimen in the orthogonal direction.

Four main types of tests were performed: monotonic loading to failure, cyclic loading to a limiting envelope curve, cyclic loading to prescribed values of stress, and fatigue cycling (repeated cycles of stress between zero and a pre-determined value). Each type of test was performed on both unconfined specimens and specimens under different levels of strain confinement. Complete stress-strain histories were recorded, analyzed and compared in order to assess the effects of confinement on concrete behavior under different loading conditions. The cyclic stress-strain histories are the basis for describing biaxial cyclic behavior and can be used for the development of constitutive relations.

INTRODUCTION

In recent years, considerable interest has developed in the strength of concrete members under repeated loadings. There are several reasons for this interest. First, the widespread adoption of ultimate strength design procedures and the use of higher strength materials require that structural concrete members perform satisfactorily under high stress levels. Hence, there is concern about the effects of repeated loads on safety critical concrete structures. Second, new or different uses are being made of concrete members and systems. Reliable information on strength, failure mode, ductility and energy absorbtion capacity is required for the design of complex systems such as offshore gravity and other reinforced concrete structures, particularly for seismic loading conditions. Third, there is new recognition of the effects of repeated loading on a member, even when repeated loading does not cause a fatigue failure. Repeated loading may lead to diagonal tension cracking in prestressed concrete members at lower than expected loads, or repeated loading may cause cracking in component materials of a member that alters the static load carrying characteristics.

Offshore concrete gravity structures are subjected to cyclic loadings such as those generated by wind, waves and earthquakes. Stress cycling at relatively low levels results in an approximately linear response. These cyclic forces may cause failure to the structures at load levels below the ultimate static strength of reinforced concrete members. Repeated loads causing cycling at high stress levels which result in nonlinear concrete response may also occur. There are many examples of severe damage and failure of reinforced and prestressed concrete structures under a relatively small number of cycles of high amplitude which is characterized as low-cycle high amplitude failure.

In all offshore and many other structural applications, complex behavior of concrete under biaxial or triaxial cyclically varying stresses occur. At present, only extremely limited experimental and prototype experience on such cyclic behavior is available; somewhat more but again significantly limited information is available on monotonically loaded concrete under biaxial and triaxial conditions. Material and structural testing is the cornerstone of any investigation to remedy this situation.

The main objective of the investigation reported herein was to experimentally study the behavior of concrete in biaxial cyclic compression. Cyclic compressive stresses were generated by the application of repeated loadings on the specimen in one direction while the specimen was subjected to predetermined levels of strain confinement in the second direction. For this purpose, square, flat concrete specimens were subjected to two loading phases: (1) initial confining phase where the specimen was loaded horizontally to a predetermined strain value, and (2) application of cyclic compressive stress in vertical direction during which the horizontal strain imposed in phase one was kept constant. To establish a comparative basis for the biaxial cyclic data, separate tests were performed to investigate the behavior of the concrete specimens under uniaxial cyclic, and monotonically increasing biaxial stresses.

In this paper, first, a brief review of the behavior of concrete under uniaxial cyclic loading is presented. The present experimental program on biaxial cyclic loading of concrete is then described; test results and their evaluation are presented, and finally, findings of the investigation are summarized.

BEHAVIOR OF PLAIN CONCRETE UNDER CYCLIC LOADING

Most of the early research on plain concrete subjected to variable load histories was aimed toward obtaining a fatigue limit for the material. Fatigue tests of plain concrete showed that fatigue limits were generally from 40% to 60% of the static cylinder strength. A decrease in the tangent modulus and the Poisson's ratio with increased number of cycles of loading was observed. This early work has been reviewed in detail elsewhere [Nordby, 1958].

Experiments on plain concrete subjected to uniaxial cyclic loadings have been reported in [Sinha, et al, 1964, Karsan and Jirsa, 1969, Shah and Chandra, 1970, Lam, 1980]. The reported test results indicate that the stress-strain relationships of concrete under compressive load histories possess an envelope curve, which may be considered unique and identical with the stress-strain curve obtained from constantly increasing strain (there are indications, however, that the envelope may be different for different strain rates). This behavior under cyclic compressive loading is shown in Fig. 1 where the stress-strain curve for monotonic loading forms an envelope curve for cyclic loading. For small cycles of compressive stress (not shown) the hysteresis curves are closed loops. For large cycles of compressive stress, the hysteresis loops intersect the envelope curve, and additional inelastic strain is accumulated during each cycle, eventually leading to failure. With increasing numbers of cycles the slopes of the hysteresis curve decreases (stiffness degradation) and the energy dissipated, interpreted as the area within the hysteresis loop, decreases.

During uniaxial cyclic loading of a specimen there exists a point of intersection where the reloading portion of any cycle will cross the unloading curve of that cycle. This point has been defined as a common point [Sinha, et al, 1964]. Stresses above the common point will introduce additional strains, but stresses below the common point cause no further strain and cause the stress-strain history to go into a loop repeating the previous cycle. Considerable amount of scatter exist in the experimental common point data. It would appear that a common point would exist at a certain high stress level (common point limit) and with subsequent cycles evantually converge to a lower stress limit (stability limit), thereafter the stress-strain relationship would remain in an identical hysteresis loop with additional strain past the stability limit is implemented by increased loading.

Available experimental data indicates that the behavior of concrete in cyclic loading is influenced greatly by microcracking [Shah and Winter, 1966a, 1966b]. Furthermore, the maximum stress of the common point curve corresponds to the load at which the concrete volume no longer decreases and microcracking sharply increases. It appears that cyclic loading greatly increases crack propagation at stress levels in excess of approximately 70% of ultimate strength [Shah and Chandra, 1970]. The cracking occurs in both the interfaces and the concrete matrix itself. It would seem that if concrete is subjected to cyclic stress levels below the maximum stability limit, failure will not occur [Karsan and Jirsa, 1969]. A recent investigation indicated that the behavior of concrete under uniaxial cyclic loading is highly controlled by the nonlinear behavior of mortar constituent [Maher and Darwin, 1980].

Previous experiments and analytical studies have produced information on the uniaxial cyclic loading condition; information on biaxial and triaxial cyclic load conditions is lacking. A recent experimental investigation on pressure-confined plain concrete under cyclic compression provides very limited information about the behavior of concrete in cyclic multi-axial stress state [Costello, 1980].

EXPERIMENTAL PROGRAM

In this section the scope of the experimental program is defined. Test specimens, materials, and test procedures are briefly described.

Test Specimens

The specimens tested were 5"x5" flat concrete plates with 1" thickness. In all tests only one type of concrete was used having a water-cement ratio of 0.67 by weight, sand-cement ratio of 2.5, and gravel-cement ratio of 3.5. Walbro pea gravel was used with maximum size of 0.263 in., and Portland Type III (high early strength) cement was used in the mix. Specimens were cured in water until about 2 days prior to testing at 7 days. Average uniaxial strength of the specimens was approximately 4200 psi.

Loading Arrangement

The specimens were loaded in two orthogonal directions. A 50 metric ton capacity Automatic Materials Testing System (MTS) with accompanying PDP-11 computer was used for applying vertical loads. An independent loading frame was constructed for applying load in the horizontal direction. Horizontal loads were applied by a manually operated 15 ton capacity screw jack. The specimens were loaded using brush-like bearing platens developed for a previous investigation [Buyukozturk, et al, 1971]. In this arrangement the thin steel plates were flexible enough that they offered little restraint against flexural deformation in the direction normal to the applied load, but were such as to transmit the load to the concrete without buckling. (No reduction in restraint was provided in the third direction). Swivel mechanism, by means of steel ball bearing, were implemented in the loading arrangement to provide proper allignment of the applied loads.

Strain Measurements

Resistance strain gages, mounted on both faces of the specimen were used to measure vertical and horizontal strains. Monitoring and plotting equipment included digital strain indicators, two X-Y recorders and the MTS X-Y recorder.

Test Operations

Each biaxial test consisted of two loading phases: (1) initial confining phase where, by manual operation of the jack, the specimen was loaded horizontally to a predetermined strain value, and (2) application of cyclic (or monotonic) stress in the vertical direction through the MTS loading ram, during which the initially applied horizontal strain remained constant. In all tests, the MTS was operated under stroke (displacement) control. The stroke rate imposed in the monotonic uniaxial and biaxial tests was such that failure of the specimens occurred in 15 to 20 minutes. In the cyclic tests, duration of a typical load-unload cycle was about 2 1/2 to 5 1/2 minutes depending on the test series.

Test Series

A total of thirty specimens were tested in four different series. The test series are classified according to the type of loading history imposed on the specimen. (See Table I).

In all tests the horizontal confinement levels were characterized by the values of predetermined horizontal strain ε_{hc} which remained constant during each test, and the initial value of horizontal stress σ_{hi} introduced by the application of ε_{hc} during the first phase of loading. Horizontal stresses fluctuated as a function of the applied vertical load during the second phase of loading. In order to establish a basis for comparison tests on unconfined specimens were also performed. For each test series, concrete batch properties were determined by performing uniaxial, monotonic load tests. (f'_c = peak stress, ε_o = strain corresponding to peak stress).

Complete stress-strain histories were recorded. For unconfined specimens vertical stress (σ_v) vs. vertical strain (ε_v), and vertical strain vs. horizontal strain (ε_h) relationships were established. For the confined specimens vertical stress vs. vertical strain and horizontal stress (σ_h) vs. vertical strain relationships were recorded.

Test Series 1: Monotonic Loading to Failure. Unconfined and confined specimens were tested to failure by monotonically increasing the vertical strain. Under these boundary conditions, vertical and horizontal stresses increased monotonically but in a non-proportional fashion. Note that in biaxial tests of concrete previously reported [Kupfer, et al, 1969, Liu, et al, 1972, Tasuji, et al, 1978] proportional loading schemes were adopted.

Test Series 2: Cycles to the Envelope Curve. Confined specimens were loaded vertically up to a given value of vertical strain, unloaded to zero vertical stress, and then reloaded until the stress-strain curve followed the trend of the previous loading portion of the curve. The specimen was, again, unloaded to zero vertical stress and the procedure was repeated in the same fashion until failure occurred.

Test Series 3: Cycles to Prescribed Values of Vertical Stress. Unconfined and confined specimens were subjected to a given vertical cyclic stress history. Thereafter, the specimens were loaded monotonically to failure. These tests allowed the evaluation of the effects of different confinements on the stress-strain behavior.

Test Series 4: Cycles Between Constant Maximum and Minimum Stresses. Confined specimens were subjected to load cycles between zero and constant maximum stress levels a number of times equal to that required for failure of an unconfined specimen cycled between zero and a stress of $0.82 f'_c$. After the required number of cycles were performed, the confined specimens were loaded monotonically to failure. These tests provided information regarding the effects of confinement on stress-strain behavior of concrete subjected to repeated stress cycles of constant amplitude.

TEST RESULTS AND EVALUATION

In this section, test results obtained from this investigation will be summarized. Some selected, representative data will be given and basic behavioral trends highlighted. Complete data analysis and details may be found in [Zisman, 1982]

Test Series 1: Monotonic Loading to Failure

Stress Strain Behavior. The confined specimens initially exhibited a linear behavior with an effective modulus (slope of the σ_v vs. ε_v curve) equal to or slightly higher than that corresponding to the unconfined specimen. At a value of vertical stress approximately equal to the uniaxial elastic limit, the confined concrete started exhibiting an inelastic behavior with a stiffer response compared to that of the uniaxial case. Vertical stress and strain at failure were higher than those for the uniaxial compression test. Both stiffness and strength increased with the level of confinement.

For confined specimens the $\sigma_h - \varepsilon_v$ curves were fairly linear (Fig. 2). The initial slope of the curve was nearly the same for specimens confined by stress levels below $0.50 f'_c$ and decreased at higher values of σ_{hi}. The initial ratio of horizontal to vertical stress increments was found to be considerably lower than that predicted by linear elastic isotropic theory, according to which $\Delta\sigma_h/\sigma_v$ is equal to the initial Poisson's ratio ν of the uniaxial compression test, which is approximately equal to 0.20. For example in test 1.5, the initial ratio $\Delta\sigma_h/\sigma_v$ was equal to 0.08.

The observed behavior which is related to microcracking in concrete suggests that the concrete under confinement does not behave like an isotropic material but rather may be assumed as a stress-induced orthotropic material with properties in the vertical and horizontal directions dependent on the initial confining conditions.

Stresses and Strains at Failure. In the present tests, where stresses varied non-proportionally, strains at failure were up to 23% higher than the uniaxial peak strain ε_o, while stresses at failure were up to 43% higher than f_c'. Stress and strain states at failure, did not coincide with those observed in previous biaxial tests, [See for example Kupfer, et.al, 1969], where stresses varied proportionally. In such tests the maximum increase in strength due to confinement was less than 30% of the uniaxial strength. These differences can be attributed to the different boundary conditions and loading paths adopted in the reported investigations.

Test Series 2: Cycles to the Envelope Curve

Uniaxial Cyclic Test. Figs. 3 and 4 show the σ_v vs. ε_v and ε_v vs. ε_h curves, respectively, of the uniaxially cycled specimen 2.4. General behavior exhibited in this test resembles that reported in other uniaxial cyclic tests carried out by different investigators. The concrete exhibited a typical hysteretic behavior, [Karsan, 1969, and Sinha, et.al., 1964] where the area within the hysteresis loops representing the energy dissipated during a cycle became larger as vertical strain increased. It was observed (Fig. 3) that the reloading curves were nearly linear along their length up to the common points (intersection with the unloading curve of the cycle) after which they decreased in slope. The unloading curves were slightly nonlinear along most of their length, and exhibited a marked increase in curvature as they approached zero vertical stress. This curvature increased with the number of cycles. A continuous degradation of the concrete was observed as evidenced by the decrease of the slopes of the reloading curves. The envelope to the cyclic curve was seen to coincide with the uniaxial monotonic curve.

For uniaxial cyclic tests, ε_v vs. ε_h relations are shown in Fig. 4. The points where the reloading curves intersect the unloading curves in a cycle may be termed common strain points. These points do not coincide with the common points of the vs. curves.

An upper envelope to the ε_v vs. ε_h curve may be formed by joining the upper peaks of the curve while a lower envelope may be formed by joining the lower peaks. The lower peaks correspond to non-recoverable (or plastic) strain states attained upon completion of unloading to zero vertical stress. The lower envelope, thus defined, may be attributed as a lower bound to uniaxial cyclic strain histories. The upper envelope may possibly coincide with the monotonic $\varepsilon_v - \varepsilon_h$ curve. More tests are required however to examine the uniqueness of these upper and lower envelopes.

Biaxial Cyclic Tests. Figs. 5,6, and 7 show the σ_v vs. ε_v curves for confined cyclic tests and Figs. 8,9 and 10 show the corresponding $\sigma_h - \varepsilon_v$ curves. The general $\sigma_v - \varepsilon_v$ behavior of confined specimens subjected to cyclic stress was found to be similar to that of specimen under uniaxial cyclic loading. The initial linear portion of the envelope curves had at all confinement levels, nearly the same slope which was slightly higher than the initial modulus in uniaxial loading. In the inelastic ranges, however, lateral confinement induced much stiffer behavior than that obtained under uniaxial conditions, as evidenced by the rise of the biaxial cyclic envelopes above the uniaxial cyclic envelope curve (Fig. 11).

In all cases reloading was practically linear (except at very low stress levels) up to the common points (Figs. 3,4,5). In all confined tests the slope of the first reloading curve was slightly higher than that of the first loading branch, indicating a hardening effect on the concrete produced by the first load-unload cycle. This behavior is the same as that observed in the uniaxial cyclic tests. The reloading slopes progressively decreased as the number of cycles increased. A comparison of reloading curves corresponding to different tests and starting at the same values of vertical strain shows that they coincide in slope up to the common point, regardless of the level of confinement (including zero confinement). Thus, apparently the slopes of reloading curves in their linear portion are not dependent on the level of confinement and can be defined on the basis of the plastic strain level at which they originate.

Unloading curves corresponding to different tests (including the uniaxial test) and starting at the same vertical strain levels σ_{vun} were found to overlap when superimposed at their lowest point ($\sigma_v = 0$); the only difference being in their lengths which were greater for tests at higher confinement levels, where unloading occurred at higher values of stress. Therefore the unloading curves too, appear to be basically non-dependent on the level of confinement, and a major criterion for determining their shape is the value of vertical strain at which unloading occurs. An important factor influencing the slopes of both reloading and unloading curves may be the initial uniaxial modulus of the concrete. However this effect was not studied in the present investigation.

The common point location was found to rise with increasing confinement. The decrease in slope of the reloading curve at stress levels higher than the common point, was less pronounced at higher confinement levels, which was responsible for the fact that the envelope curves were higher at higher confinement levels.

Nonrecoverable (or plastic) strains are those corresponding to a zero stress level on the unloading stress strain curve. Fig. 12 shows the relationship between the plastic strain and the envelope strains at unloading ε_{vun}, for the uniaxial and biaxial cyclic tests. Quantities are normalized with respect to ε_o. From the figure the plastic strains appear to be dependent mainly on the vertical strain at unloading and do not seem to be significantly affected by the stresses and strains in the horizontal direction. Only at values of $\varepsilon_{vun}/\varepsilon_o$ higher than 0.60 do the plastic strains in the confined tests appear to be slightly reduced with respect to those in the unconfined tests. However, the difference is not significant and for practical purposes unique values may be considered. For comparison, the curve obtained by [Karsan and Jirsa, 1969] based on their uniaxial cyclic tests is shown on Fig. 12. It coincides initially with the data obtained in this investigation for low values of $\varepsilon_{vun}/\varepsilon_o$. At higher values there is some deviation which may be attributed to the differences in experimental techniques, boundary conditions, form and size of the specimens.

Generally, the confined specimens exhibited a degradation process similar to that of the uniaxially cycled specimens. Such a process can be characterized by the progressive decline of the initial slopes of the reloading curves, and has been generally attributed in the uniaxial cyclic tests to a steady increase in microcracking as cycling proceeds. However, in the biaxial tests, microcracking was inhibited due to confinement and, thus, it appears that the degradation process may not be solely attributed to the microcracking at the mortar-aggregate interface. The nonlinear behavior of mortar itself and its effect on the degradation process should also be considered. A recent investigation [Maher and Darwin, 1980] has reported that mortar specimens subjected to uniaxial monotonic and cyclic loading exhibited a very similar behavior to that of concrete specimens under the same loading conditions.

In Figs. 8,9, and 10 σ_h vs. ε_v relations are shown. A peculiar characteristic of these relations is that reloading curves initially coincide with the unloading curves in the unload-reload cycles. This might be related to the fact that the horizontal stress does not drop to zero at the end of each cycle and thus reloading occurs at a relatively high value of stress. From uniaxial cyclic tests it was observed that if reloading in a given cycle occurs at a relatively high value of stress, the reloading curve would then tend to follow more closely the unloading curve thus exhibiting a reduced hysteretic behavior. Thus, in the biaxial tests the observed deformation behavior in the horizontal direction may be considered analogous to that of the uniaxial cyclic behavior at high stresses.

The upper and lower peaks of the $\sigma_h - \varepsilon_v$ curves correspond to the peaks of the $\sigma_v - \varepsilon_v$ curves. The envelopes of these peak points are shown in Fig. 13. The initial linear portion of the upper envelope coincides with the corresponding biaxial monotonic curve. At higher values of strain the upper envelope falls below the monotonic curve. This deviation is greater at higher confinement levels. The lower envelope falls below the initial horizontal stress level. Its deviation from this level is greater at higher confinement levels. The initial confining conditions therefore influence the variations of horizontal stress during confined cyclic compression.

Stresses and strains at the peak of the $\sigma_v - \varepsilon_v$ curve (failure), in the biaxial cyclic tests are shown in Figs. 14 and 15. Strains at failure in the biaxial cyclic test practically lie on the strain failure envelope corresponding to the monotonic tests of test Series 1. On the other hand, stress at failure in the biaxial cyclic tests show in general good agreement with the monotonic stress failure envelope. The vertical stress at failure was up to 48% higher than f_c'. It thus appears that concrete failure occurs upon attainment of given stress-strain states represented by pairs of points on the envelope, independently of the stress-strain path (cyclic or monotonic) leading to such states.

Test Series 3: Cycles to Prescribed Values of Vertical Stress

In this test Series an unconfined specimen, 3.3, was cycled to the envelope curve until failure (Fig. 16). Then specimens under different confinement levels were cycled to the same vertical stress levels as the unconfined specimens, after which they were loaded monotonically to failure. Figs. 17 and 18 show the $\sigma_v - \varepsilon_v$ curves of confined specimens 3.4 and 3.6. Fig. 19 shows the $\sigma_h - \varepsilon_v$ curve of specimen 3.6. Vertical stress-strain behavior was significantly affected by the confinement. Total and plastic strain accumulation in the confined tests at the end of the cycles were considerably lower than that in the uniaxial test. Such total and plastic strain accumulations were about 70% and 46% of those in the uniaxial tests, respectively, also, the ratio of total to plastic strain was considerably lower in the confined tests, thus indicating a more elastic behavior.

The curves of tests 3.4 and 3.6 when superimposed show an excellent agreement. Reloading takes place up to about the same values of vertical strain, and, therefore, the envelopes to both curves practically coincide throughout the cyclic portion of the test. The difference between the curves exists in the subsequent monotonic branches where the curve of the test at lower confinement falls below that of the test at higher confinement; this may be expected according to the results of the test Series 2. Behavior of the reloading and unloading curves follow a pattern similar to that in test Series 2. the confinement does not seem to affect the shapes and slopes of reloading and unloading curves originating at given values of vertical strain. The reloading slopes in the confined tests do not vary much from one cycle to another because cycling takes place in a small strain range. the same trend is exhibited by the unloading curves.

The $\sigma_h - \varepsilon_v$ behavior during the cyclic portion of the tests was found to be similar to that observed in Test Series 2. Upon monotonic loading the last reloading curve gradually decreased in slope, adopted the initial loading slope and continued its course until failure, thus exhibiting a behavior paralell to that of the corresponding biaxial monotonic curve (Fig. 18). However failure occurs at a value of horizontal stress lower than that corresponding to the monotonic biaxial test, especially a high confinement levels. Values of vertical and horizontal stresses and strain at the peak of the $\sigma_v - \varepsilon_v$ curve for tests 3.3, 3.4 and 3.6 are plotted in Figs. 13 and 14, for comparison with the data obtained in the previous test Series. It would seem that both stress and strain states at failure fall in the same regions as found in previous tests.

Test Series 4: Cycles between Fixed Maximum and Minimum Stresses

Very limited number of tests were performed in this series. The tests in this series indicated extended fatigue life for confined specimens. It would seem that confined specimens would not fail when cycled the same number of times and between the same stress levels as an unconfined specimen cycled to failure. For example, an unconfined specimen, 4.3, was continually cycled between zero and a vertical stress of $0.82 f'_c$. The number of cycles to failure was 35. Confined specimens 4.5 and 4.6, were then cycled 35 times between zero and vertical stress levels of $0.96 f'_c$ and $1.21 f'_c$, respectively. The confined specimens were far from failing after 35 cycles, at the end of which they were loaded monotonically to failure.

The confinement seemed to considerably reduce total and plastic strain accumulation. In confined tests 4.5 and 4.6, for example, total and plastic strain accumulation at the end of the cycles were reduced to an average of 36% and 43% respectively, of the corresponding values in uniaxial tests. In the confined tests it was found that after a certain number of cycles, no significant additional straining occurred and the stress strain curves went into a closed loop. Evidently, failure would have never occurred if cycling proceeded normally to the given stress levels. This is especially significant in test 4.5 where the maximum stress level was 84% of the confined strength (stress at peak of the curve). Previous investigators [Karsan and Jirsa, 1969, Shah and Chandra, 1970] reported that failure of the specimens under uniaxial cyclic loading occurred when they were cycled continuously to a maximum stress level higher than about 70% of its strength. This lower bound does not seem valid and needs to be established for confined cyclic loading conditions.

Vertical and horizontal stress and strains at failure by increased loads followed a trend similar to that observed in the other test series. The maximum vertical stress at failure was 46% higher than f'_c, and occurred in test 4.5.

Failure occurred suddenly once peak stress was attained. In uniaxial cyclic and monotonic tests failure occurred by splitting in a plane paralell to the load and perpendicular to the free faces of the specimens. In the biaxial monotonic and cyclic tests failure occurred with the occurrence of a major crack running perpendicular to the major loading direction at about specimens mid-height and in a plane forming an angle of about 25 degrees to the free surface of the specimen.

SUMMARY AND CONCLUSION

A total of 30 square, flat, concrete specimens were tested in four different test series: (1) Monotonic loading to failure, (2) Cycles to envelope curves, (3) Cycles to prescribed values of vertical stress, and (4) cycles between fixed maximum and minimum stress levels. Each test Series was performed on unconfined specimens, and on specimens under different confinement levels. In the confined tests the specimens were first loaded horizontally up to a predetermined strain value and then vertical load was applied, either cyclically or monotonically, while the initially imposed horizontal strain remained constant. Complete stress-strain histories were recorded.

The results obtained from this investigation can be summarized as follows:

1. Concrete specimens confined in one direction behaved orthotropically when loaded in a direction normal to the confinement. Concrete stress and strain characteristics in the vertical and horizontal directions were dependent upon the initial confining conditions. Under the implemented non-proportional monotonic stress conditions initial stiffness of the specimen in the linear range of loading was not significantly affected by the level of confinement; however in the nonlinear range of loading, confinement resulted in stiffness increases of the specimen greater than those found in previous investigations where proportional biaxial load conditions were employed.

2. Envelopes to the confined and unconfined cyclic stress strain curves corresponding to the direction of loading coincided during the initial stage of loadings. At higher values of strain the envelope curves were higher at greater confinement levels.

3. In both uniaxial and biaxial cyclic tests to the envelope the slope of the first reloading curve was greater than that of the loading branch indicating a hardening effect on the concrete produced by the first load-unload cycle. As cycling proceeded there was a continuous degradation of elastic moduli indicated by the progressive decrease of the slopes of the reloading curves. Reloading curves up to the common point may be defined on the basis of plastic strain at which they originate, regardless of the confinement level. The decrease in slope of the reloading curve at stress levels higher than the common point was less pronounced at higher confinement levels.

4. Unloading curves corresponding to different tests including the uniaxial test and starting at the same values of vertical strain overlaped when superimposed at their final zero-stress point.

5. Plastic strain magnitudes due to unloading from the envelope curve were not significantly affected by the level of confinement, and may be defined as a function of the envelope strain at unloading.

6. The degradation process of concrete under cyclic loading before peak stress, appeared to be affected by the inelastic behavior of mortar.

7. Stress strain states at failure under biaxial loading, whether cyclic or monotonic, appeared to be unique, i.e, independent of the stress strain paths followed.

8. Confined concrete specimens subjected to the same loading history as unconfined ones accumulated considerably lower total and plastic strains.

The investigation reported in this paper represents an initial effort to study the behavior of concrete under biaxial cyclic loading. The number of tests performed was limited. Further research to verify the obtained results and to include the effect of confinement levels other than those reported in this study is needed. Furthermore, additional tests are required to study the effects of other variables such as concrete strength, rate of loading, different levels of maximum and minimum stresses during cycling, and different stress paths.

ACKNOWLEDGEMENT

The authors would like to thank T-M. Tseng and J. Calvo for their assistance in the test program. They also acknowledge the cooperation of the Department of Structural Engineering, Cornell University, in providing the brush bearing platens used in the tests reported in this paper.

REFERENCES

BUYUKOZTURK, O., A.H. NILSON, and F.O. SLATE, 1971, "Stress-Strain Response and Fracture of a Concrete Model in Biaxial Loading," Journal of the American Concrete Institute Volume 68, pp. 590-599.

COSTELLO, S.D., 1980, "Behavior of Confined Concrete Under Cyclic Loading," M.S. Thesis, Massachusetts Institute of Technology.

KARSAN, I.D. and J.O. JIRSA, 1969, "Behavior of Concrete Under Compressive Loadings," *Journal of the Structural Division,* Volume 95, Number 5112, pp. 2453-2563.

KUPFER, H., H.K. HILSDORF, and H RUSCH, 1969, "Behavior of Concrete Under Biaxial Stresses," *Journal of the American Concrete Institute,* Volume 66, Number 8, pp. 656-666.

LAM, Y., 1980, "Behavior of Plain Concrete Under Cyclic Compressive Loading, *M.S. Thesis,* Massachusetts Institute of Technology.

LIU, T.C.Y., A.H. NILSON, and F.O. SLATE, 1972, "Stress-Strain Response and Fracture of Concrete in Uniaxial and Biaxial Compression," *Journal of the American Concrete Institute,* Volume 69, Number 5, pp. 291-295.

MAHER, A. and D. DARWIN, 1980, "Mortar Constituent of Concrete Under Cyclic Compression," *SM Report,* The University of Kansas Center for Research, Inc., Lawrence, Kansas, Number 5.

NORDBY, G.M., 1958, "Fatigue of Concrete-A Review of Research," *Journal of the American Concrete Institute,* Volume 55, Number 2.

SHAH, S.P. and G.M. STURMAN, 1965, "Microcracking and Inelastic Behavior of Concrete," *The International Symposium,* Miami, Florida.

SHAH, S.P. and G. WINTER, 1966a, "Inelastic Behavior and Fracture of Concrete," *Journal of the American Concrete Institute,* Volume 63, Number 9.

SHAH, S.P. and G. WINTER, 1966b, "Response of Concrete to Repeated Loadings," *RILEM, International Symposium on the Effects of Repeated Loading on Materials and Structural Elements,* Mexico City, 1966.

SHAH, S.P. and S. CHANDRA, 1970, "Fracture of Concrete Subjected to Cyclic and Sustained Loading," *Journal of the American Concrete Institute,* Volume 67, Number 9, pp. 816-825.

SINHA, B.P., K.H. GERSTLE, and L.G. TULIN, 1964, "Stress-Strain Relations for Concrete Under Cyclic Loading," *Journal of the American Concrete Institute* Volume 61, Number 2, pp. 195-211.

TASUJI, M.E., F.O. SLATE and A.H. NILSON, 1978, "Stress-Strain Response and Fracture of concrete in Biaxial Loading," *Journal of the American Concrete Institute,* Volume 75, Number 7, pp. 306-312.

ZISMAN, J.G., 1982, "Behavior of Concrete Under Biaxial Cyclic Compression", *M.S. Thesis,* Massachusetts Institute of Technology.

TABLE I - Test Program

Test Series 1 Monotonic Loading to Failure		Test Series 2 Cycles to Envelope		Key
Test	Specimen	Test	Specimen	
U/M	1.1	U/M	2.1	U=unconfined
C/M $\sigma_{hi} = 0.07f'_c$	1.2	U/M	2.2	C=confined
U/M	1.3	U/M	2.3	M=monotonic
U/M	1.4	U/C$_y$	2.4	C$_y$=cyclic
C/M $\sigma_{hi} = 0.40f'_c$	1.5	U/C$_y$	2.5	
C/M $\sigma_{hi} = 0.60f'_c$	1.6	C/C$_y$ $\sigma_{hi}= 0.20f'_c$	2.6	
		C/C$_y$ $\sigma_{hi}= 0.40f'_c$	2.7	
		C/C$_y$ $\sigma_{hi}= 0.60f'_c$	2.8	
		U/C$_y$	2.9	
		C/C$_y$ $\sigma_{hi}= 0.18f'_c$	2.10	
		C/C$_y$ $\sigma_{hi}= 0.37f'_c$	2.11	
		C/C$_y$ $\sigma_{hi}= 0.55f'_c$	2.12	
Test Series 3 Cycles to Prescribed Stresses		Test Series 4 Cycles between fixed stresses		
Test	Specimen	Test	Specimen	
U/M	3.1	U/M	4.1	
U/M	3.2	U/M	4.2	
U/M	3.3	U/C$_y$	4.3	
C/C$_y$ $\sigma_{hi} = 0.18f'_c$	3.4	C/C$_y$ $\sigma_{hi} = 0.24f'_c$	4.4	
C/C$_y$ $\sigma_{hi} = 0.37f'_c$	3.5	C/C$_y$ $\sigma_{hi} = 0.48f'_c$	4.5	
C/C$_y$ $\sigma_{hi} = 0.55f'_c$	3.6	C/C$_y$ $\sigma_{hi} = 0.72f'_c$	4.6	

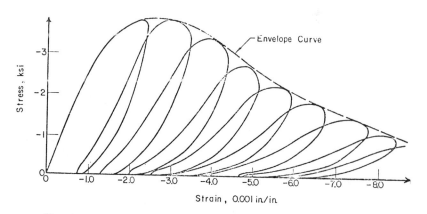

Fig. 1 - Stress-Strain Curve for Concrete in Uniaxial Cyclic Compression (Sinha et al., 1964)

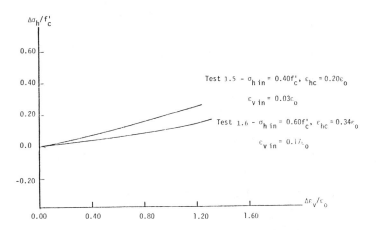

Fig. 2 - Horizontal Stress Variation Under Confined Monotonic Loading

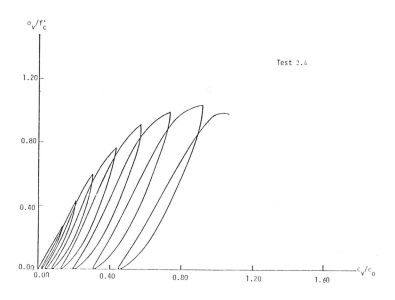

Fig. 3 - Uniaxial Cyclic Loading to the Envelope Curve. Test Series 2

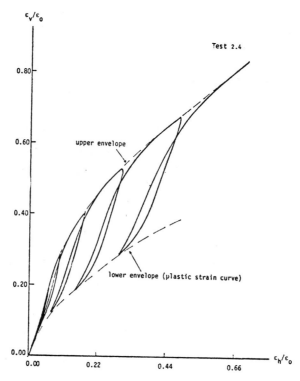

Fig. 4 - Relationship Between Vertical and Horizontal Strain Under Uniaxial Cyclic Loading

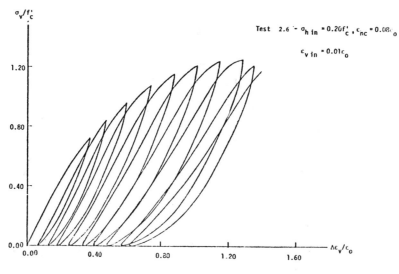

Fig. 5 - Confined Cyclic Loading to the Envelope Curve Test Series 2

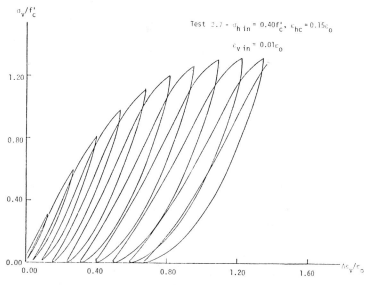

Fig. 6 - Confined Cyclic Loading to the Envelope Curve.
Test Series 2

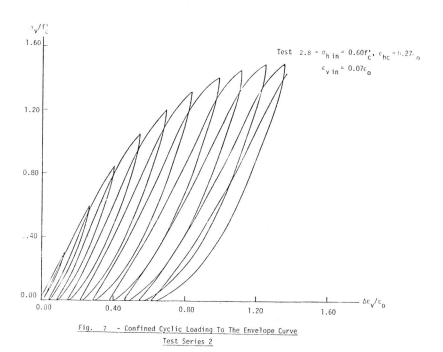

Fig. 7 - Confined Cyclic Loading To The Envelope Curve
Test Series 2

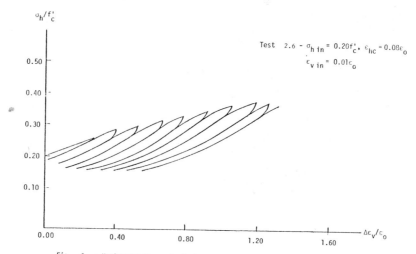

Fig. 8 - Horizontal Stress Variation Under Confined Cyclic Loading To The Envelope Curve

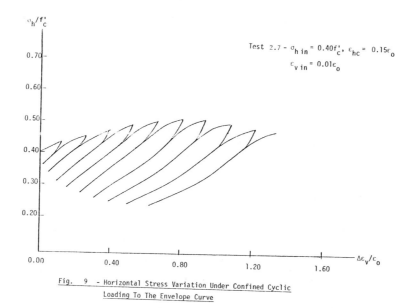

Fig. 9 - Horizontal Stress Variation Under Confined Cyclic Loading To The Envelope Curve

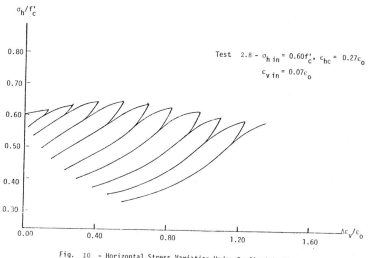

Fig. 10 - Horizontal Stress Variation Under Confined Cyclic Loading To The Envelope Curve

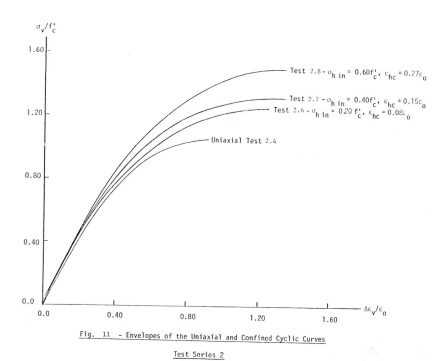

Fig. 11 - Envelopes of the Uniaxial and Confined Cyclic Curves

Test Series 2

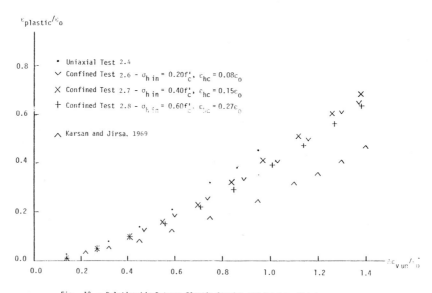

Fig. 12 - Relationship Between Plastic Strains and Envelope Strains at Unloading   Test Series 2

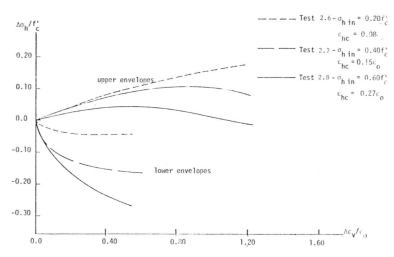

Fig. 13 - Upper and Lower Envelopes of σ_h-ε_v Curves Corresponding to Confined Cyclic Tests to the Envelope

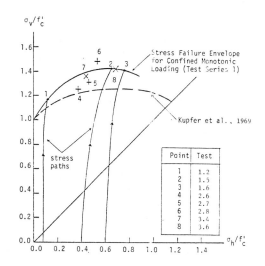

Fig. 14 - Biaxial Strength of Concrete

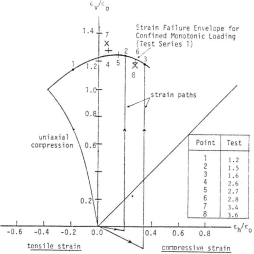

Fig. 15 - Strains at Failure Under Biaxial Loading

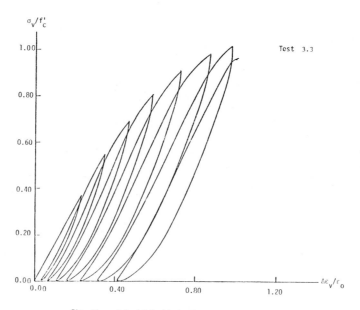

Fig. 16 — Uniaxial Cyclic Loading to the Envelope Curve
Test Series 3

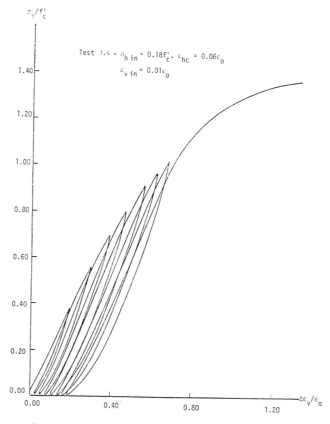

Fig. 17 — Confined Cyclic Loading to Same Vertical Stress Levels of Uniaxial Cyclic Test 3.4.
Test Series 3

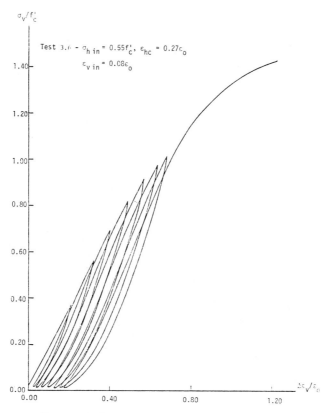

Fig. 18 - Confined Cyclic Loading to Same Vertical Stress Levels of Uniaxial Cyclic Test 3.6.
Test Series 3

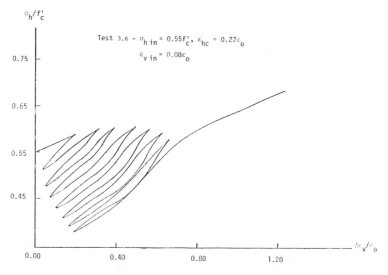

Fig. 19 - Horizontal Stress Variation Under Confined Cyclic Loading to Same Vertical Stress Levels of Uniaxial Cyclic Test 3.6

Technical Lecture Sessions
STRUCTURES

Long Term and Extreme Loads

ANALYSIS OF SHIP/PLATFORM IMPACTS

T.H. Søreide   T. Moan   J. Amdahl   J. Taby
The Norwegian Institute of Technology, Trondheim, Norway

SUMMARY

The paper deals with ship collision against platform. A general description of the collision mechanisms is given and various methods available for analyzing the collision problem are discussed. Models of energy absorption of steel platforms are identified and compared.

Simple computer programs based on plastic yield line theory has been developed for the analysis of local energy absorption, i.e. energy associated with deformation work in the vicinity of the point of impact. For studying energy absorption by beam bending of the impact element between adjacent joints yield hinge models accounting for axial restrictions as well as F.E.M. programs are applied. Constraints on the energy absorbing capability caused by local buckling, ovalization and tubular joint capacity are discussed.

Theoretical load-deformation predictions derived from the methods above are compared with experimental results from local denting and global bending of small scale tube models, showing reasonably well agreement.

The results of an introductory study with axial crushing of radially stiffened cylinders are presented with special reference to ship deformation characteristics in bow and stern collisions.

Comparison is given between analytical models and experiments on axial capacity of dented tubes. The main theory behind an efficient computer program for predicting post-damage strength of tubular members is described.

1. INTRODUCTION

Many types of steel structures which are employed in the offshore petroleum activity are composed of tubular members. These structures are subjected to various types of loads. Besides the normal functional loads and environmental loads, loads due to accidental events such as collisions, falling objects might occur. Statistics for world wide operation platforms during 1/1-70 to 31/12-80 indicate 9 cases with total or severe damage of the platform among 114 incidents with similar result /1/. At the same time many impact incidents with minor consequences have been reported. As the number and size of vessels used in offshore operations (especially in the North Sea) increase, the collision risk should be seriously considered in the design of platforms. It is therefore of interest to assess the damage caused by such loads.

Furthermore, both in connection with design and operation it is of interest to know the residual strength of a structure damaged due to accidental or other types of loads. During operation such knowledge is necessary to make decision regarding repair in order to balance the costs and safety.

This paper presents recent findings regarding the structural behaviour during impact between ships and unstiffened tubular members of an offshore platform. In Chapter 2 the fundamental impact mechanics is briefly summarized. Chapter 3 is devoted to the load-deflection characteristics of tubular members and ship-bows necessary to determine the maximum impact force and/or damage. In Chapter 4 the problem of residual strength of a damaged tubular member is addressed.

2. IMPACT MECHANICS

The derivation of a mathematical model of the ship/platform impact is based on two criteria

- the energy conservation law
- force equilibrium

The condition of energy conservation is expressed by the following equation for collision against a fixed platform

$$E_k = E_s + E_p + E_f + E_r \tag{1}$$

where

E_k = kinetic energy of the striking vessel immediately before impact

E_s = energy absorbed by ship

E_p = energy absorbed by platform

E_f = energy absorbed by fenders

E_r = rotational kinetic energy of the ship after impact

In case of eccentric impact some of the kinetic energy may remain as rotational energy of the ship after impact. For design purposes the worst case is a central impact which means that the line through the centre of gravity of the ship and the contact point coincides with the direction of ship motion.

For an unfendered structure the two last terms of Eq. (1) disappear and the energy conservation condition comes out as:

$$\tfrac{1}{2}mv_0^2 = E_s + E_p \tag{2}$$

where

m = mass + added mass of ship

v_0 = impact velocity

Force equilibrium is fulfilled simply by assuming that the same force P is acting in each medium. Given the load-deformation characteristics for ship and platform it is possible to calculate the energy absorption at different load levels satisfying force equilibrium.

The choice of design impact situations must be done under the consideration of probability of occurrence /3/. The size of the design vessel is to be determined on the basis of the vessels intended to operate in the area such as service vessels, tankers for offshore loading and by-passing ships. The usual criterion is to consider a supply vessel drifting sideways and causing a central impact on the platform.

3. LOAD-DEFORMATION CHARACTERISTICS OF PLATFORM BRACING AND SHIP BOW

3.1 Energy Absorption in Platform Bracing

The deformation modes of the platform at impact may conveniently be categorized as

- local deformation of bracing/leg at point of impact
- global beam deformation of bracing and leg element
- overall deformation of platform

The transition between local and global deformation of a bracing element is difficult to set. As a dent is created in the tube wall the neutral axis changes, a phenomenon that is more associated with global beam deformation. While the two first modes involve considerable plastic energy absorption, the global response is mainly elastic, possibly involving some dynamic effects.

3.1.1 Local Deformation of Tube Wall

The extent and form of local damage in the wall of a bracing element depends on the nature of impact. A head-on collision gives a more concentrated force than a sideways impact and results in a larger amount of local energy absorption for given mass and velocity of the vessel. Due to this complexity it is impossible to present one single analytical model for establishing local energy absorption. Several types of models have to be considered related to different collision cases.

For sideways impact a simple yield line model is presented by Furnes and Amdahl /3/, see also Fig. 1. The deformed surface is bounded by a series of yield lines and the following plastic effects are included

- rotation of surface at yield lines
- flattening of surface between yield lines
- tension work due to elongation of generators

The theoretical model gives fairly good agreement with experimental results for small and medium indentations. The model can form the basis for possible design curves when further verification against experiments has been performed.

3.1.2 Analytical Techniques for Beam Deformation of Bracing Element

The rigid-plastic methods of analysis /4, 5/ provide simple analytical results, often with acceptable accuracy and are appropriate for design situations. The simplest approach to the beam type of deformation is the three hinge mechanism, Fig. 2. In case of axially restrained ends the load carrying capacity of the beam increases considerably as the beam undergoes finite deflections due to the development of membrane tension forces. For a centrally loaded tubular beam the load-deflection relation is given by /4/

$$\frac{P}{P_0} = \sqrt{1 - \left(\frac{w}{D}\right)^2} + \frac{w}{D} \arc \sin \frac{w}{D}; \quad \frac{w}{D} \leq 1 \qquad (3)$$

$$\frac{P}{P_0} = \frac{\pi}{2} \frac{w}{D} \quad ; \quad \frac{w}{D} > 1 \qquad (4)$$

where w is the central deflection at the point of impact and D is the tube diameter. P_0 is the plastic collapse load according to linear yield theory.

The above expressions are based on the assumption that the ends have full axial restraint. In a real frame system like a jacket the bracing element sustains a certain degree of elastic support from the adjacent elements. Besides, local deformations occur in the tubular joints. Such elastic restrictions can be included by extending Hodge's method /6/ for the case of tubular members.

It is a major requirement for the validity of the present simple theory that no buckling of the tube wall takes place so that the full plastic capacity of the cross section is retained during deformation. Thus, restrictions must be set on maximum D/t-ratio for which the rigid-plastic theory can be used. Sherman /7, 8/ on the basis of tests on steel tubes in bending concluded that for members with D/t of 35 or less the full plastic moment is activated and sustained during deformation.

The API rules /9/ prescribe $D/t < 9000/\sigma_y$ (σ_y in N/mm^2) to maintain full capacity through plastic deformation. In the range $9000/\sigma_y < D/t < 15200/\sigma_y$ only a limited plastic rotation capacity can be presumed.

For the clamped ideally plastic element the absorbed plastic energy at any level of deflection w is found by integration of the load-displacement expressions in Eqs. (3, 4). The following energy expression

$$E = P_0 D \left\{ \frac{3}{4} \frac{w}{D} \sqrt{1 - \frac{w^2}{D^2}} + \frac{(1+2\frac{w^2}{D^2})}{4} \arcsin\frac{w}{D} \right\}; \quad \frac{w}{D} \leq 1 \tag{5}$$

$$E = \frac{\pi}{8} P_0 D (1 + 2\frac{w^2}{D^2}) \; ; \quad \frac{w}{D} > 1 \tag{6}$$

3.1.3 Computational Techniques for Bracing Elements

In an effort to come up with an efficient tool for collision analysis of bracing elements a simple finite element beam program IMPACT has been modified to take care of local effects such as moment capacity reductions due to indentation and buckling of tube wall.

The deformation model at the point of impact is shown in Fig. 3. It is assumed that the indented area is flat and that the remaining part of the cross section has constant radius of curvature. The shape of the cross section can now be determined for any indentation by requiring the area to be constant and the reduced plastic section modulus comes out from simple integration over the deformed cross section. The beam elements at point of impact are modified continuously during deformation so as to satisfy this reduced plastic moment capacity.

In order to calculate the amount of indentation at different load levels, the technique described in Sect. 3.1.1 is applied. The program system now consists of three integrated parts

- a finite element beam program with elasto-plastic material behaviour and large deflection effects incorporated

- a yield line program for predicting local indentation

- a program for calculating reduced moment capacity due to local deformation at point of impact

The load is applied in increments with equilibrium interaction within each load step. A further documentation of the numerical technique is found in Ref. /10/.

In order to simulate the real deformation pattern in the tubular member a full shell analysis must be performed in which material as well as geometric nonlinearities are incorporated. Fig. 4 shows the finite element model of a bracing element analysed by the computer program TUBBUC /11/. In the central region close to the point of impact a fine mesh of triangular thin shell elements is used while the tube wall outside the local indentation area is represented by a coarser mesh of rectangular elements. The total number of elements is 370.

Strain hardening is accounted for by combining the socalled sublayer technique with the flow theory of plasticity. Geometric nonlinaritiy is modelled by the updated Lagrangian formulation /10/ in which element strains are referred to local element axes.

The displacement pattern along the bracing and deformation of midsection are shown in Fig. 5 as calculated by TUBBUC and the corresponding load/displacement curve is illustrated in Fig. 6. The stiffening effect of membrane forces is clearly demonstrated.

It is clear that the above technique is time consuming and costly. For design purposes a simpler numerical tool should be available. However, modification of the computational method also reduces the accuracy of the predictions. The shell program can be used to study special effects in the member performance eg. stress distribution close to the point of impact and local instability phenomena.

3.1.4 Behaviour of Tubular Joints

A major condition for the validity of the above models is that the restraints from the joints on the bracing elements can be incorporated. In the elastic regime the linear stiffness of the joint can be calculated by shell programs for different load cases such as axial tension, axial compression, in-plane bending and out-of-plane bending. The finite element beam program of Sect. 3.1.1 is modified so as to take in translational and rotational springs at the ends of each element.

However, in order to simulate the energy absorption under extreme loading it is also essential to model the nonlinear load-deformation characteristics of the joints. Further, failure criteria for tubular joints should be given as deformation limits for the springs. Typically, a joint subjected to bending may fail by buckling of the chord wall on the compression side resulting in a reduction in joint stiffness and strength. On the tension side fracture through the chord wall is the most probable mode of failure. It is clear that the combination of membrane forces and moments at the ends of the tubular elements should be checked against empirical data on tubular joint capacity. The simplest alternative is to use springs for which the load-deformation characteristic is given as input, either in the form of an ideal elasto-plastic type of relation or a general nonlinear curve including unloading in the plastic regime.

Valuable information on the nonlinear behaviour of unstiffened tubular joints has been presented by Yura et al. /12/. Capacity formulas based on experimental data are also given in design codes /9, 13/.

3.1.5 Laboratory Tests of Bracing Elements

The experiments presented are part of a series of tests performed to study the energy absorption capability and post-damage strength of tubular members. Characteristic cross-sectional dimensions of bracing elements in the water plane are

$1.0 < D < 2.0$ m
$20 < D/t < 100$
$10 < L/d < 30$

In the study of impact capability of bracing elements two series of tests were performed, namely with no axial restraint at the ends and with horizontally fixed ends. In both cases the specimens were clamped against rotation. Strain gauges were placed at point of impact and at the ends in order to get information about the very complicated deformation patterns in these areas. Extensiometers recorded horizontal movements at the ends and lateral deflection of tube wall in top and bottom of the central cross section.

Geometric and material data for the models are given in Table 1. The range of variation is

$63 < D < 125$
$22 < D/t < 61$
$10 < L/D < 20$
$204 < \sigma_Y < 328$ N/mm^2

Displacement control of the hydraulic jack was applied with a displacment rate of 0.15 mm/sec. for the quasi-static tests and 54 mm/sec. for the dynamic tests.

For a more detailed description of the experiments, see Ref. /14/. A short presentation of the most interesting results will be given here.

First, attention is given to the effect of membrane forces by considering the two specimens IAI and IAIII with equal geometric and material properties. Specimen IAI has horizontally free end conditions and IAIII is clamped against axial movement. The corresponding load-displacement curves are given in Fig. 7, where comparison is made with simple mechanism theory and with computer predictions by the modified beam program of Sect. 3.1.3. The effect of membrane forces is clear in the post-collapse domain of deformation. For specimen IAI the experiment gave a drop in lateral load in the plastic region. This phenomenon is due to a reduction in moment capacity at the ends caused by local crippling of the compression side of the tube wall, Fig. 8. For case IAIII the membrane tension field prevents this type of local damage. Instead, the ends of IAIII fail by fracture close to the weld, Fig. 9.

The stretching effect of membrane forces is also illustrated by the pictures in Fig. 10 and 11. The ovalization of the tube is of a much more local character for IAI than the overall deformation of IAIII.

The conclusion to be drawn is that the modified beam computer program predicts well the load-displacement characteristics. The simple mechanism theory gives too high capacity at small deflections due to the neglection of local deformation at point of impact. For large deflections the mechanism model is conservative neglecting the effect of strain hardening.

3.2 Energy Absorption in Ship Bow

The configuration of a ship structure is very complex, being comprised of an outer shell stiffened by a grid of stringers, frames and profiles. In a collision, the interaction between the ship and the platform may cause a complicated deformation pattern with cutting, puncturing and rupture of both structures. At the Division of Marine Structures, NTH, a series of tests on small-scale models of ship bows is presently conducted, in close cooperation with Det norske Veritas.

In this section some introductory experiments with ring-stiffened cylindrical shells are reported. Bulbous bows bear some resemblance to simple cylindrical shell. In a collision with a concrete tower of a platform or a bridge pier it is likely that the bulb will suffer the greater part of the deformation. Thus axial compression tests with cylindrical shells provide valuable information with regard to the properties of bulbs.

3.2.1 Shell Buckling

The initial buckling phase of deformation contributes to a minor extent to the energy absorption of the cylinder. However, the buckling mode may predict the type of plastic mechanism. In most cases the buckling mode develops further into the plastic regime and forms the energy absorbing mechanism. Thus, it is of great importance in the study of energy absorbing capability of cylindrical shells also to predict the initial buckling mode.

The elastic buckling behaviour of a ring-stiffened cylindrical shell may conveniently be related to the Batdorf-parameter /15/

$$Z = \frac{\ell^2}{Rt}\sqrt{1 - \nu^2} \tag{7}$$

where ℓ is the stiffener spacing, R is radius of cylinder and t is wall thickness, see also Fig. 12.

For cylindrical shells under axial loading and bending shape imperfections may significantly reduce the elastic buckling stress. Thus the critical stress according to classical shell buckling theory should be reduced by a knockdown factor as given by design rules /13/.

3.2.2 Plastic Collapse Models

After initial buckling the shell continuous to deform into the plastic regime. The post-buckling behaviour may take on two forms, one with the shell in axisymmetric convolutions, the other in which asymmetric folds are developed.

The analysis is carried out by assuming the mode of collapse of the tube and calculating the initial work required to achieve this mode. Equating this work to the external work the mean collapse load is derived. To simplify the analysis elastic strains are neglected and initially a rigid-plastic material without strain hardening is assumed. The axisymmetric collapse model is shown in Fig. 13.

There are two contributions to the internal energy absorption; namely the work associated with bending at the hinge lines and stretching in the circumferential direction, respectively.

According to Alexander /16/ the average collapse load is given by the expression

$$P_{av} = \frac{\pi \sigma_y t^2}{\sqrt{3}}(3.3\sqrt{\frac{D}{t}} + 1) \tag{8}$$

where D and t is the shell diameter and thickness respectively.

It is interesting to note that the plastic buckling half length given by

$$h = 0.953\sqrt{Dt} \tag{9}$$

is quite close to the elastic buckling length /15/.

$$h' = 1.22\sqrt{Dt} \tag{10}$$

Fig. 14 shows that the theoretical plastic collapse load approaches infinity with the formation of a new fold. However, in practice the load level is limited by the initial buckling strength of the cylinder wall.

The mechanism mode of the asymmetric collapse mode shown in Fig. 15, is comprised of a grid of hinge lines in which angular rotation is concentrated. The tube surface is transformed into plane triangles. In this mode there is no stretching of the material around the circumference in the axial direction. The following terms contribute to the internal energy absorption

- removal of shell curvature
- bending about horizontal yield lines
- bending about oblique yield lines
- travelling horizontal yield lines

The last term derives from the assumption that a new buckle is created by a horizontal hinge travelling down from the last hinge to the position of the new one. During this cycle the material is bent and rebent in the hinge line.

The average collapse load according to Johnson et al /17/ reads

$$\bar{P}_{av} = \frac{2\pi M_p}{1-2\frac{r}{h}}(1 + \frac{n}{tg(\frac{\pi}{2n})} + \frac{n}{sin(\frac{\pi}{2n})} + \frac{D}{r}) \tag{11}$$

where n is the number of triangles around the circumference, D is the shell diameter, r is the rolling radius of travelling yield lines and M_p is the plastic moment of the shell pr. unit width.

The unknown rolling radius r can be found formally by minimizing the load with respect to r. The load decreases with decreasing n. However, the number of circumferential waves is influenced by the initial buckling pattern.

In real collisions the strain rate sensitivity of the yield stress may be of great importance for the energy absorption in steel members. Static considerations of impacts may highly underestimate the energy absorption capability.

A simple method of including the dynamic material behaviour is to adjust the yield stress value according to the average strain rate $\dot{\varepsilon}$. The following power type formula has been suggested by Cowper and Symonds /18/

$$\frac{\sigma_y'}{\sigma_y} = 1 + (\frac{\dot{\varepsilon}}{\dot{\varepsilon}_0})^{\frac{1}{n}} \tag{12}$$

where $\dot{\varepsilon}_0$ and n are material constants determined to get the best fit of experimental data. Recommended values for mild steel are $\dot{\varepsilon}_0 = 40$ sec^{-1} and n = 5.

The average strain rate in the shell wall may be approximated by /19/

$$\dot{\varepsilon} = \frac{\dot{u}t}{4hr} \tag{13}$$

where $\dot{u}$ is the loading velocity, t is the wall thickness, h is the half-length of the buckling wave $(=\ell/2)$ and r is the radius of rolling.

The present experiments indicate t/r around 1.0. Thus, Eq. (13) simplifies to

$$\dot{\varepsilon} = \frac{\dot{u}}{4h} \tag{14}$$

which is used in subsequent calculations of test models.

3.2.3 Experiments

The series of collision tests with axially loaded cylindrical shells contains ten specimens, four of which are machined and six fabricated cylinders. The four machined shells have been tested to initial buckling /20/ and in the present study they are plastically compressed to total collapse. The specimens are machined from 12 mm thick seamless tubes that are stress relieved before being machined. These cylinders are almost ideal in the sense that residual stresses and shape imperfections are small. In the subsequent text these machined specimens are denoted MA1-MA4.

The six fabricated cylinders are rolled from 2 mm plate, and then closed by a longitudinal butt weld. The ring stiffeners are machined from 25 mm seamless tubes and attached to the shell by intermittent fillet welds. After welding, the cylinders are stress released by heating to 550-600°. In the following, these fabricated models are denoted FA1-FA6.

The diameter is 400 mm for all cylinders. Data for thickness, stiffener spacing and yield stress are given in Table 2. The initial bukcling stresses are presented in Table 3.

The initial buckling stresses are presented in Table 3.

Comparison is made with theoretical elasto-plastic buckling stresses. It is seen that the analysis underestimates the capacity of the machined models. One major reason for this result is that they are almost ideal specimens as far as shape imperfections and residual stresses are concerned. The two specimens FA1 and FA5 are seen to have test results considerably below the analytical prediction. The low experimental stresses are caused by difficulties arising during the loading of FA1 and FA5. Strain measurements around the cylinders indicated a skew load distribution coming from inaccuracy in the end geometries of the two specimens. However, in the subsequent plastic deformation the specimens FA1 and FA5 behave normally.

The low R/t ratio of 100 for the fabricated models implies low sensitivity for shape imperfections. Thus, the four successful models FA2, FA3, FA4, and FA6 have buckling stresses considerably higher than predicted.

The conclusion from the present study of initial buckling is that the theory of sec. 3.3.1, which forms the basis of the DnV rules of 1977 /11/ seems to include a reasonable degree of safety. However, it should be emphasized that more and larger-scale tests have to be performed before any clear conclusion can be drawn. The buckling tests of Odland /20/ give a more reliable basis for such considerations.

An example of a measured load axial displacement curve in the plastic collapse phase is shown in Fig. 16. The shape of the curve is explained by the development of plastic mechanisms. The initial buckling implies axisymmetric deformation of the cylinder wall between two stiffeners. A sudden drop in the axial load is observed after the initial buckling. During continuing axial compression this region deforms. The load drops to a minimum and rises again until two stiffeners overlap and a new region buckles. Due to the shape disturbances from the collapsed region, the successive buckling loads are much lower than the initial one.

The successive development of axisymmetric plastic mechanisms is also illustrated by the pictures in Fig. 17. Fig. 18 shows the plastic deformation of specimen FA6 for which four asymmetric folds are developed around the circumference.

In order to check the validity of the proposed mechanism theory it is convenient to compare the average collapse load P_{av} with the experimental values. The latter are found by numerical integration of the load/displacement curves. Because of hardening effects during plastic deformation an average value of the uniaxial tensile stress is introduced as characteristic yield stress.

Table 4 indicates some discrepancy between analytic predictions and test results. In all cases the experimental values lie above the analytical predictions. Several factors can explain this discrepancy.

First, the deformation patterns in the tests differ somewhat from the configurations assumed in the mechanism models. For the machined models the ring stiffeners are rather weak so that they undergo twisting. The deformed stiffeners prevent the development of an ideal mechanism in the shell wall.

The second effect of importance is that in the theoretical models the plastic hinges are supposed to be stationary. However, during the tests it is easily observed that the location of plastic zones changes during deformation. Such a motion of plastic hinges gives an extra contribution to plastic energy absorption and thereby raises the average load P_{av}. The theoretical models do not account for this effect.

The choice of realistic yield stress to be used in the capacity formulae is another factor of uncertainty. The yield stress σ_y is modified for plane strain conditions and material hardening. The question is how good these approximations are, and above all, how representative are the uniaxial tensile tests for describing the material behaviour in the shell wall.

3.2.4 Analysis of Bulbous Bow

Among the variety of shapes, bulbs may also take on a cylindrical cross-section, thus showing great similarity to the tested cylindrical shells. However, there is one important difference. In addition to transverse frames, the bulbs normally also contain a longitudinal stiffening system, consisting of stringers and a centerline bulkhead. This will, indeed, influence the deformation pattern as well as the energy absorption of the bulb. For the time being, the effects of the stringers and bulkheads can be included in a simplified and inaccurate way. This fact must be taken into account when evaluating the results.

Unfortunately, there are very little data available on the crushing strength of bulbs. The only information is provided by the tests conducted by Woisin /21/. Some of these tests involved the bow model of the 195,000 dwt. tanker "Esso Malaysia" to scale 1:12 and the side barrier of a nuclear ship. The barrier was of a resistance type, confining almost all damage to the bow structure. In one of the tests only the bulb part of the bow suffered the deformation.

The force-deformation relationship of this test, derived from the acceleration recordings, is reproduced in Fig. 19 /22/. The absorbed energy, found by integrating the load curve, comes out to be 28% less than the initial ramming energy applied in the test, indicating that the acceleration measurements were inaccurate. In the same diagram the load prediction by means of Gerard's method is depicted /23/. In this procedure the cross-section is cut into simple elements for which the crippling strength is determined on the basis of experimental data from the aerospace industry.

From Fig. 19 it is seen that Gerard's method agrees reasonably well with tests measurements while the yield line theory underestimates the load level significantly.

The discrepancy between yield line theory and test results is not surprising. First of all, the assumed deformation mode deviates from the actual one. Probably a considerable amount of energy is absorbed by membrane straining of bulkheads and stringers.

The other factor is the dynamic nature of the experiments. Inertia forces may introduce a deformation pattern differing from the static one. The strain rate sensitivity is also important. A rough estimate suggests that the dynamic increase in yield strength might well have exceeded 50%.

4. RESIDUAL STRENGTH OF DAMAGED BRACING ELEMENTS

4.1 General

The present section deals with axial strength of tubular members with damage in the form of local denting. The theoretical background of an efficient computer program for solving such problems is presented and comparison is made between analysis and test results on pin-ended columns.

A tubular member that has been dented without bending has local eccentricity in the dented region as well as additional residual stresses. The influence on the buckling load of these additional local imperfections is expected to be small, since buckling is an overall phenomenon of the whole member. However, the plastic collapse load of the member may be reduced to a value much lower than the buckling load, depending on the geometry of the dent and the properties of the member.

The behaviour of the dented member stressed until its ultimate strength may be divided into two phases.

Phase 1: Dent plastification.

When a small axial compressive load is applied to a dented tubular member, axial compressive stresses develop in the tube shell. At the dent these stresses are supported by compression and bending of the dented shell.

Considering the dented region, as shown in Fig. 20, axial compressive strain and bending deformation occur. Bending deformation which takes place mainly in the short flattened part at the middle of the dented shell is restrained by the rest of the tube and may be neglected. Only axial compressive strain is considered. As a result of this simplification, uniform shortening of the tube takes place and axial compressive stresses develop uniformly in the tube shell. In the dent, bending stresses are superimposed. As the axial load increases, yielding starts at the middle of the dent. Yielding continous until a full plastic hinge line is formed as shown in Fig. 20. Integrating the full plastic stress distribution at the middle of the dent and considering the equilibrium of the dented portion, the following expression is obtained

$$\sigma_{dp} = \sigma_y[\sqrt{4(\eta/t)^2 + 1} - 2(\eta/t)] \tag{15}$$

where

σ_{dp} = The uniform compressive stress in the tube shell when a full plastic hinge line is formed at the middle of the dent.

σ_y = the yield stress of the material,

η = as defined in Fig. 20 and

t = the thickness of the tube shell.

The axial compressive load acting on the tube at this time, P_{dp}, may then be given by

$$P_{dp} = P_p[\sqrt{4(\eta/t)^2 + 1} - 2(\eta/t)] \tag{16}$$

where

P_p is the full plastic compressive strength of the tubular member under consideration.

Phase 2: Ultimate Strength.

After plastification at the middle of the dent, the stiffness of the dented region is reduced to a large extent. Increment ΔP of the axial load P above the value P_{dp} of Eq. (16) is supported in the dented part of the tube primarily by the arc πD-S in Fig. 20. In the parts of the tube from its ends to close up to the dent, the axial force is supported by the total cross-section. Adjacent to the dent, the axial stresses acting in the fibers which pass through the dent are transferred gradually to the arc πD-S by the action of longitudinal shear in the tube shell. Thus, the eccentricity of the load increment starts to grow gradually a short distance before the dent to a maximum value at the middle of the dent and decrease again until it vanishes a short distance after the dent. This distance is dependent on the material of the tube and the proportions of the tube and the dent. The bending moment M_e caused by this eccentricity may then have a distribution as that shown in Fig. 21. A further simplification is obtained when a uniform distribution of M_e along the dent is assumed as shown in Fig. 21. M_e is given by

$$M_e = \Delta P a \tag{17}$$

where

a is the distance between the line of ΔP (center line of the tube) and the center of the area of the arc πD-S.

The bending moment M_e causes an overall bending of the member which is magnified by the axial load.

The deflection curve resulting from the moment M_e without the effect of the axial load is shown in Fig. 21. The part along the dent is an arc of a circle with radius ρ, since the moment M_e is assumed constant along the dent. The rest is straight lines since there is no bending moment acting in these regions.

The bending equation may be written as

$$\frac{E}{\rho} = \frac{M_e}{I^*} \tag{18}$$

where

E is the modulus of elasticity.

I^* the 2nd moment of area of the arc πD-S about its neutral axis.

The deflection, δ_b, at the middle of the dent may be found approximately from the geometry as

$$\delta_b \simeq \ell_1 \theta_1 \simeq \ell_2 \theta_2 \tag{19}$$

where

θ_1, θ_2, ℓ_1 and ℓ_2 are defined in Fig. 21

The relation between the bending radius ρ and the angles of slope at the ends, θ_1 and θ_2 may be found from geometry in terms of the length of the dent, ℓ_d, as

$$\ell_d = \rho(\theta_1 + \theta_2) \tag{20}$$

Eqs. (17, 18, 19 and 20) yield expression for δ_b

$$\delta_b = \ell_1 \ell_d a \Delta P / [EI^*(1+\ell_1/\ell_2)] \tag{21}$$

Taking the magnifying effect of the axial force into account, the deflection at the middle of the dent may be expressed as

$$\delta = \delta_b \cdot mf \tag{22}$$

where

mf is the magnification factor,

$$= 1 - \gamma(1-c)/(1-\gamma) \tag{23}$$

$$\gamma = (P_{dp} + \Delta P)/P_E \tag{24}$$

P_E is the Euler buckling load,

c is a geometrical factor

$$= [\ell/2\ell_2 + \ell_d)] \cdot \sin[\pi(\ell_1 + \ell_d/2)/\ell] \tag{25}$$

The total bending moment acting on the arc $\pi D-S$ at the middle of the dent is then given by

$$M = \Delta P(a+\delta) + P_{dp}\delta \tag{26}$$

The stress distribution across the tube at the middle of the dent may be assumed linear across the arc $\pi D-S$ as shown in Fig. 22.

The stress σ at the fibers adjacent to the dent is the sum of the uniform stress σ_{dp}, given by Eq. (15), the compressive stress σ_c caused by the force ΔP, and the bending stress σ_b caused by the bending moment M of Eq. (26), as shown in Fig. 22.

$$\sigma = \sigma_{dp} + \sigma_c + \sigma_b \tag{27}$$

When the fibers adjacent to the dent start to yield, the stiffness of the dented region decreases rapidly. As the lateral deflection develops, not much increase of the axial load is expected and the ultimate strength may be considered to be reached at this time. Denoting the load increment ΔP at this time by ΔP_U, Eq. (27) yields the following expression

$$\Delta P_U \left(\frac{1}{A^*} + \frac{a+\delta_U}{Z^*}\right) + P_{dp}\frac{\delta_U}{Z^*} = \sigma_y - \sigma_{dp} \tag{28}$$

where

A^* is the cross-sectional area of the arc $\pi D-S$,

Z^* is the modulus of section of the arc $\pi D-S$, and

δ_U is the lateral deflection at the dent when the ultimate strength is reached, as given by Eq. (22).

δ_U is a function of ΔP_U and Eq. (28) may be solved for ΔP_U and δ_U by iteration.

The ultimate compressive strength P_U is then given by

$$P_U = P_{dp} + \Delta P_U \tag{29}$$

4.2 Post-Ultimate Strength Behaviour

After the member has reached its ultimate strength, a more complicated behaviour of the dented part of the tube takes place. However, it may be assumed that the axial load remains almost constant for some time. The increase of the bending moment at the location of the dent due to the increasing lateral deflection is more or less overcome by the increasing capacity of the arc $\pi D-S$ to form a full plastic hinge. As lateral deflection increases, the axial load obviously starts to decrease before a full plastic hinge is formed. However, it is assumed that the plastic hinge is formed at the ultimate value of the axial load. Also as the lateral deflection increases, the dent starts to grow deeper in the tube shell. This growth is rather slow in the deformation range of interest for this study /24/ and the effect on the strength of the cross-section may be neglected. The ultimate strength interaction relationship of the axial load to the bending moment at the dent may then be simply constructed according to Fig. 23. This interaction relationship may be considered as plastic potential and a plastic hinge may be inserted in the location of the dent /25/.

The incremental load-deformation relationship of the member, with plastic hinge inserted, may now be constructed in the form of a stiffness matrix. Load increments (negative) may be applied in steps and the corresponding deformation obtained.

In order to obtain the stiffness matrix in its incremental form after a plastic hinge is formed at the dent, the member is divided at the middle of the dent into two beam-column elements. Three degrees of freedom is allowed at each nodal point as shown in Fig. 24. The elastic non-linear stiffness matrix of each element is established. The form of this stiffness matrix may be found in standard texts /26/. In this stiffness matrix the effect of bending deflection on axial deformation is not considered. Having these two matrices the tangential forms are derived. Then a plastic hinge is inserted at the position of the dent in one element and the tangential elastic-plastic stiffness matrix of this element is derived according to the plastic flow theory following the formulation presented by Ueda et al. /27/. The two elements are then assembled to obtain the global stiffness matrix. As deformation increases some nonlinear terms which are neglected in the elastic stiffness matrix become larger. They are included by establishing a global system of axes and a local one for each element. The local systems are updated after each load step.

4.3 Computer Program

A computer program, DENTA, has been developed to perform the analysis. Examples of results are presented in Fig. 8. Ultimate load and accompanying deformations are calculated analytically while the post-ultimate strength behaviour is calculated numerically as explained in preceeding sections. The load and deformation in each case have been traced in the post-ultimate strength range from the ultimate load down to 0.2 of its value in about 20 steps. A UNIVAC 1108 computer was used for the analysis. Computer time has been recorded less than 0.8 second (CPU) per tube.

4.4 Experimental and Numerical Investigation

Axial compression tests were carried out on a series of 21 tubes after dented. The aim of the tests was to gain a better understanding of the behaviour of dented tubular members. The analytical model is based on this understanding and checked agains the results of the tests.

The tests were carried out on DIN 2391 ST 35 BK high precision tubes. Dimensions, summarized in Table 1, were chosen to represent, to scale, the range of tubes usually met in offshore practice. The tubes were stress relived in order to eliminate unknown residual stresses by heating to 550° for one hour followed by slow cooling.

Following heat treatment, test tubes were cut and machined at the ends. Two or three tensile test specimens were cut from each tube. Thickness, circularity and material properties of tubes were then surveyed. Thickness was measured at 20 points on each tube using an ultrasonic transducer. The measurements were checked against micrometer readings at the ends of test tubes. Outer diameters are measured at 5 stations along each tube, 4 diameters spaced at 45 degrees are measured at each station.

Tensile tests was carried out under displacement control at a strainrate of about 3×10^{-7}/sec. Average values of the results are summarized in Table 5 together with geometric properties.

Test tubes are then dented at 3/8th of their length in order to differentiate between buckling collapse, and plastic collapse at the dent. Denting was carried out in a hydraulic press through a knife edge with 5 mm noze radius. The back of each tube was supported in a cylindrical wooden bed having the same diameter as the tube in order to prevent lateral deflection of the tube as far as possible. After denting, the straightness of each tube was measured optically in the x-z plane, see Fig. 25. Results are summarized in Table 5. Each tube was fitted at its ends with sliding fit steel plugs to prevent premature local buckling at the ends and spherical heads of hardened high strength steel to simulate simple supports. Axial compressive load was applied under displacment control at an average strain rate of about 3×10^{-7} using an electronically controlled hydraulic jack of 2500 kN capacity through a calibrated load cell. End shortening, lateral deflection in x-direction at the dent and the growth of the dent were continously recorded at about 10 seconds intervals by displacement transducers. Displacement signals were automatically translated and printed out.

Strains at stations 2D from each end and at midlength were measured. Three electric strain gauges at 120° were fitted at each section. Axial load and bending moment were automatically calculated from strain measurements and printed out together with those obtained from the load cell and deflection measurements. The load, measured by the load cell, is plotted against end shortening on an X-Y plotter.

Results of these are summarized in Table 5 and Fig. 26 together with theoretical results calculated by the present theoretical model. Typical load shortening curves are presented in Fig. 27. Satisfactory agreement of results of analysis by the present model with those of the tests exists for the ultimate load as well as the post-ultimate strength behaviour.

5. CONCLUSIONS

Simple methods to estimate the energy absorption characteristics of tubular steel members of platforms and ship bows have been assessed by comparison with experiments.

In tubular members energy is dissipated due to the local denting at the loaded part of the member and overall hinge mechanism of the actual member. Up to now the methods developed have been based on plastic mechanisms, neglecting local effects. This initial work has also been limited to the members themselves, without considering explicitly the behaviour of the tubular joints where crippling or brittle fracture might significantly affect the energy absorption properties. Simple methods for collapse analysis of tubular beams have been modified so as to reproduce the behaviour predicted by experiments and refined numerical techniques.

Analytical expressions for the energy absorption in cylindrical shells for various collapse modes have been outlined and related to the collision behaviour of bulbous bows. A series of tests on ring-stiffened cylinders under static and dynamic loading have been described and compared with analytical solutions. Some discrepancy is found between analysis and test results, and in all cases the analytical methods come out to be conservative in the sense that they give capacities below the experimental values.

A simple numerical model for predicting post-damage strength of dented tubular members has been outlined. Good agreement is obtained with experimental studies and the computer time requirement for the analysis is extremely short. The technique has proven to be an efficient tool for checks of individual dented tubular members.

The present work should be understood as an introductory study of ship/platform impacts. It is clear from the above considerations that future improvements in theoretical models must be accompanied by experimental verifications. In the study of deformation characteristics of ships all energy absorbing components must be taken into account including decks and bulkheads. For bracing elements special attention should be given to the performance of tubular joints under ultimate loading.

REFERENCES

/1/ Moan, T. and Holand, I., 1981, "Risk Assessment of Offshore Structures. Experience and Principles", Structural Safety and Reliability (ed. Moan and Shinozuka), Proceedings of ICOSSAR'81, Elsevier Scientific Publishing Company.

/2/ Costa, F. Vasco, 1964, "The Berthing Ship", The Dock and Harbour Authority, Vol. XLV.

/3/ Furnes, O. and Amdahl, J., 1980, "Ship Collisions with Offshore Platforms", Intermaritec, Hamburg.

/4/ Søreide, T.H., 1981, Ultimate Load Analysis of Marine Structures, Tapir Publishing Company, Trondheim, Norway.

/5/ Jones, N., 1976, "Plastic Behaviour of Ship Structures", Transactions Society of Naval Architects and Marine Engineers, Vol. 84, pp. 115-145.

/6/ Hodge, Ph.G., 1974, "Post-Yield Behaviour of a Beam with Partial End Fixity", *International Journal of Mechanical Sciences*, Vol. 16, pp. 385-388.

/7/ Sherman, D.R. and Glass, A.M., 1974, "Ultimate Bending Capacity of Circular Tubes", *OTC 2119*, pp. 901-910.

/8/ Sherman, D.R., 1976, "Tests of Circular Steel Tubes in Bending", *ASCE J. Struct. Div.*, Vol. 102, No. ST11, pp. 2181-2195.

/9/ American Petroleum Institute, 1979, *Recommended Practice for Planning, Designing and Constructing Fixed Offshore Platforms*, API RP 2A.

/10/ Remseth, S.N., Holthe, K., Bergan, P.G. and Holand, I., 1978, "Tube Buckling Analysis by the Finite Element Method", *Finite Elements in Nonlinear Mechanics*, Tapir Publishing Company, Trondheim, Norway.

/11/ Søreide, T.H., 1977, *Collapse Behaviour of Stiffened Plates using Alternative Finite Element Formulations*, Dr.ing. Thesis, Division of Structural Mechanics, The Norwegian Institute of Technology, Trondheim, Norway.

/12/ Yura, J.A., Zettlemoyer, N. and Edwards, I.F., 1980, "Ultimate Capacity Equations for Tubular Joints", *OTC 3690*.

/13/ Det norske Veritas, 1977, *Rules for the Design, Construction and Inspection of Offshore Structures*.

/14/ Søreide, T.H. and Amdahl, J., 1982, "Energy Absorption in Bracings", *Norwegian Maritime Research*.

/15/ Odland, J., 1978, "Buckling Resistance of Unstiffened and Stiffened Circular Cylindrical Shell Structures", *Norwegian Maritime Research*, Vol. 6, No. 3, pp. 2-22.

/16/ Alexander, H.M., 1960, "An Approximate Analysis of the Collapse of Thin Cylindrical Shells Under Axial Loading", *Quart. Journal of Mech. Appl. Math.*, Vol. XIII, No. 4.

/17/ Johnson, W., Soden, P.D. and Al-Hassani, S.T.S., 1977, "Inextensional Collapse of Thin-Walled Tubes Under Axial Compression", *J. Strain Analysis*, Vol. 12, No. 4.

/18/ Cowper, G.R. and Symonds, P.S., 1957, *Strain Hardening and Strain Rate Effects in the Impact Loading of Cantilever Beams*, Brown Univ. Techn., Report No. 28.

/19/ Wierzbicki, T. and Abramowicz, W., 1979, "Crushing of Thin-Walled Strain Rate Sensitive Structures", paper presented at *Euromech Colloquium No. 121*.

/20/ Odland, J., 1981, *On the Strength of Welded Ring Stiffened Cylindrical Shells Primarily Subjected to Axial Compression*, Dr.Ing. Report, Division of Marine Structures, The Norwegian Institute of Technology, Trondheim.

/21/ Woisin, G., 1976, "Die Kollisionsversuche der GKSS", *Jahrbuch der Schiffbautechnischen Gesellschaft*, 70.8, Hamburg.

/22/ Minorsky, V.U., 1977, *Bow Loading Values (Bulbous Bow)*, U.S. Mar. Ad., Dept. of Commerce, Washington, Rep. MA-RD-920-78035.

/23/ Gerard, G., 1957, "The Crippling Strength of Compression Elements", *Am. Inst. of Aeronautics and Astronautics*, May 1957.

/24/ Taby, J. and Rashed, S.M.H., 1980, *Experimental Investigation of the Behaviour of Damaged Tubular Members*, Division of Marine Structures, Norwegian Institute of Technology, Trondheim.

/25/ Taby, J., Moan, T. and S.M.H. Rashed, 1981, "Theoretical and Experimental Study of the Behaviour of Damaged Tubular Members in Offshore Structures", *Norwegian Maritime Research*, Vol. 9, No. 2.

/26/ Livesley, R.K., 1975, *Matrix Methods of Structural Analysis*, Pergamon Press.

/27/ Ueda, Y. et al., 1967-1968, *Elastic-Plastic Analysis of Framed Structures Using the Matrix Method*, 1st and 2nd reports. Journal of the Society of Naval Architects of Japan. Vols. 124 and 126. (In Japanese).

SPECI-MEN	OUTER DIAMETER (mm)	WALL THICKNESS (mm)	D/t	LENGTH (mm)	L/d	YIELD STRESS (N/mm^2)
IAI	125.15	2.04	61.33	1244.3	9.94	204
IAII	125.13	2.04	61.34	1245.9	9.96	211
IAIII	125.19	2.04	61.37	1244.8	9.94	207
IBI	125.14	2.50	50.06	1244.9	9.95	251
IBII	125.19	2.51	49.88	1245.8	9.95	230
IBIII	125.11	2.50	50.04	1245.5	9.96	268
ICI	125.11	3.07	40.75	1245.9	9.96	260
ICII	125.14	3.10	40.37	1240.5	9.91	328
ICIII	125.09	3.06	40.88	1246.6	9.97	256
IDI	114.5	3.25	35.23	1240.5	10.83	235
IDII	114.5	3.25	35.23	1240.5	10.83	235
IEI	88.5	3.0	29.50	1240.5	14.02	235
IEII	88.5	3.0	29.50	1240.5	14.02	235
IFI	63.4	2.9	21.86	1240.5	19.57	235
IFII	63.4	2.9	21.86	1240.5	19.57	235

Table 1. Data for bracing models.

Test specimen	MA1	MA2	MA3	MA4	FA1	FA2	FA3	FA4	FA5	FA6
Wall thickness t mm	0.97	1.22	0.99	0.98	2.03	2.06	2.06	2.04	2.05	2.05
Stiffener spacing ℓ mm	23.0	37.0	33.5	33.5	47.0	57.0	67.0	97.0	117.0	
Yield stress σ_y N/mm^2	267	267	267	267	236	236	236	228	228	228

Table 2. Wall thickness and stiffener spacing for ringstiffened cylinders.

Test specimen	MA1	MA2	MA3	MA4	FA1	FA2	FA3	FA4	FA5	FA6
Theoretical solution N/mm^2	224	218	202	204	222	220	200	205	206	206
Experimental results N/mm^2	259	293	260	267	142	230	247	261	135	209

Table 3. Initial buckling stresses for ringstiffened cylinders.

Test specimen	MA1	MA2	MA3	MA4	FA1	FA2	FA3	FA4	FA5	FA6
Experiments MN	68.1	103.2	80.5	62.5	158.2	147.8	143.6	139.9	147.4	101.4
Static theory	55.0	65.7	48.8	48.8	111.8	109.3	107.3	112.5	124.8	71.7
Static theory/Exper.	0.81	0.64	0.61	0.78	0.71	0.74	0.75	0.80	0.85	0.71
Dynamic theory/Exper.	0.93	0.73	0.69	0.89	0.80	0.83	0.84	0.90	0.96	0.81

Table 4. Analytical and experimental values for average load P_{av} for ringstiffened cylinders.

Specimen No.	Outer-Diameter D mm		Thickness t mm		Length ℓ mm	Young's Modulus E N/mm² x10⁵	Yield stress σ_y N/mm²	$\frac{D}{t}$	$\frac{L}{r}$	λ	Max. Deviation from straightness δ_0/ℓ	Depth of Dent DD mm	$\frac{DD}{D}$	Buckling[3] Load (Undamaged) σ_u/σ_y	Theoretical max. load σ_u/σ_y	Experimental max. load σ_u/σ_y
	mean	cov.	mean	cov.												
I A I	125.145	.00111	2.04	.02875	3500	1.88	204	61.3	80.40	0.839	.00074	6.25	.050	.834	.635	.667
I A II	125.127	.00082	2.04	.03071	"	1.96	211	61.3	80.42	0.839	.00183	12.6	.101	.834	.476	.529
I B I	125.138	.00071	2.50	.03030	"	2.11	250	50.1	80.70	0.885	.00054	6.3	.050	.811	.641	.638
I B II	125.191	.00110	2.51	.03183	"	1.98	230	49.9	80.68	0.874	.00151	12.5	.100	.816	.484	.528
I C I	125.107	.00045	3.07	.01503	"	2.01	290	40.8	81.09	0.981	.00057	6.25	.050	.756	.647	.656
I C II (1)	125.136	.00059	3.095	.03470	"	1.98	328	40.4	81.09	1.050	.00206	12.25	.098	.713	.484	.505
II A I	160.202	.00181	2.525	.01881	"	2.00	351	63.4	62.78	0.838	.00023	8.00	.050	.835	.631	.683
II A II	160.161	.00136	2.52	.02220	"	1.97	392	63.6	62.79	0.874	.00166	16.08	.100	.816	.471	.437
II A III	160.171	.00132	2.525	.01720	"	1.94	314	63.4	62.79	0.773	.00106	3.2	.020	.865	.801	.731
II B I	160.162	.00138	3.06	.03135	"	1.94	330	52.3	63.00	0.795	.00120	7.9	.049	.855	.643	.544
II B II	160.141	.00054	3.075	.03452	"	1.94	233	52.1	63.02	0.668	.00194	16.0	.100	.908	.488	.548
II B III	160.089	.00059	3.065	.02765	"	1.94	258	52.2	63.03	0.721	.00051	3.2	.020	.907	.816	.796
II C I	160.110	.00050	4.065	.01443	"	1.96	470	39.4	63.42	0.979	.00091	8.6	.054	.758	.629	.607
II C II (2)	160.230	.00092	4.10	.04293	"	2.01	457	39.1	63.38	0.985	.00217	16.1	.100	.754	.483	.478
II C III	160.074	.00084	4.07	.01457	"	1.92	384	39.3	63.44	0.885	.00077	3.15	.020	.811	.833	.791
III A I	250.259	.00045	4.23	.01936	"	2.02	500	59.2	40.23	0.639	.00060	12.6	.050	.917	.638	.595
III A II	250.225	.00055	4.27	.02191	"	2.02	502	58.6	40.24	0.638	.00100	25.5	.102	.918	.477	.463
III B I	250.245	.00096	5.20	.02003	"	2.00	470	48.1	40.39	0.625	.00010	13.6	.054	.923	.632	.714
III B II	250.257	.00132	5.21	.01960	"	1.97	433	48.0	40.39	0.603	.00200	26.0	.104	.930	.481	.533
III C I	250.373	.00112	6.02	.02271	"	1.98	472	41.6	40.67	0.629	.00087	12.6	.050	.922	.657	.734
III C II	250.457	.00201	6.00	.01659	"	2.02	465	41.7	40.65	0.624	.00183	25.0	.100	.923	.495	.556

(1) Collapse started at the dent but continued at a different location.
(2) The theoretical max. load is higher than the buckling load. The tube buckled outside the dent as expected.
(3) The buckling load is calculated by table 14.2 and Fig. 14.10 (a), Chen, W.F. and Atsuta, T., Theory of Beam Columns, Vol. 1, McGraw-Hill Inc.

Table 5. Summary of tests on residual strength of tubular members.

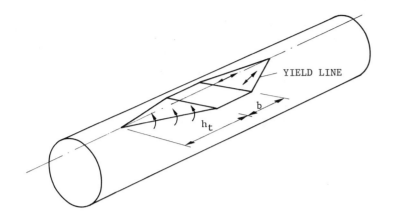

Fig. 1 Plastic mechanism for sideways impact by supply vessel on jacket leg.

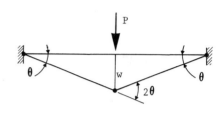

Fig. 2 Collapse mechanism for bracing element.

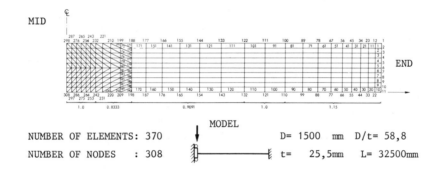

Fig. 3 Finite element shell model of bracing element.

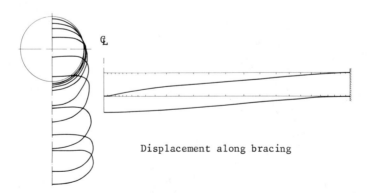

Deformation of midsection

Displacement along bracing

Fig. 4 Deformation patterns from shell analysis.

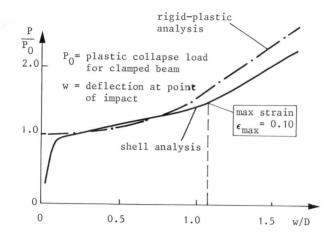

Fig. 5 Load-displacement relation for bracing element. Finite element shell analysis.

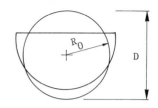

Fig. 6 Deformation model of cross-section at point of impact for calculating reduced section modulus.

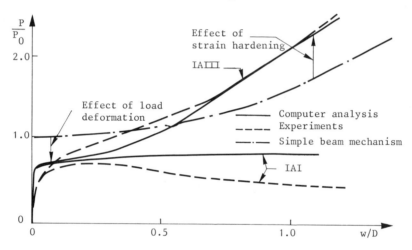

Fig. 7 Load-deflection characteristics for specimens IAI and IAIII.

Fig. 8 Local crippling of tube wall at end of specimen IAI.

Fig. 9 Fracture at end of specimen IAIII.

$P/P_0 = 0.67$

$P/P_0 = 0.63$

$P/P_0 = 0.53$

Fig. 10 Deformation of horizontally free specimen IAI.

$P/P_0 = 3.41$

Fig. 11 Deformation of horizontally fixed specimen IAIII.

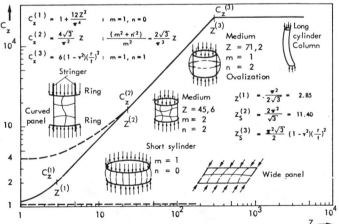

Fig. 12 Buckling coefficients for ring-stiffened cylinders.

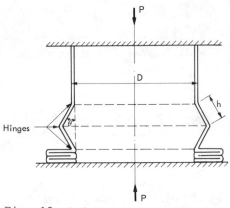

Fig. 13 Axisymmetric collapse model for ring-stiffened cylinder.

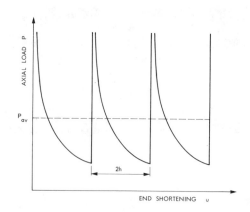

Fig. 14 Theoretical load-displacement curve for ring-stiffened cylinder.

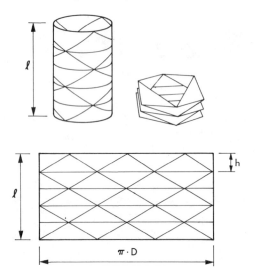

Fig. 15 Asymmetric collapse model for cylindrical shell.

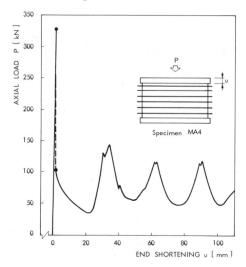

Fig. 16 Load-displacement curve for specimen MA4.

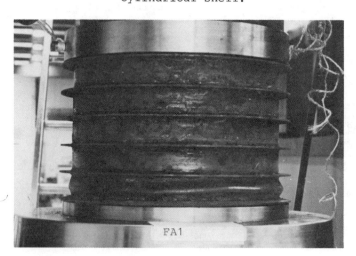

u = 10 mm

u = 120 mm

Fig. 17 Axisymmetric collapse of specimen FA1.

u = 80 mm

u = 115 mm

Fig. 18   Asymmetric collapse of specimen FA6.

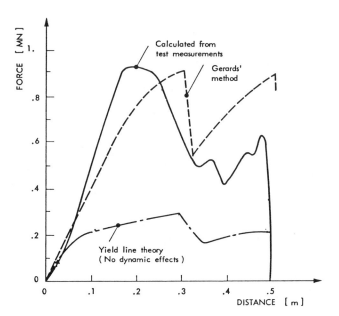

Fig. 19   Comparison between experiments and theory on tanker bow model.

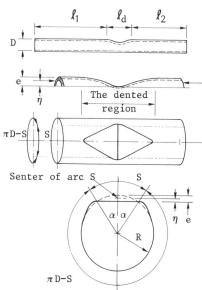

Section through the middle of the dent

Fig. 20   The dented tubular member.

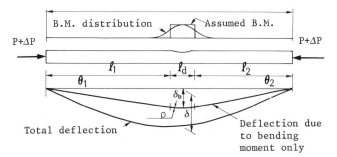

Fig. 21   Bending moment and deflection curves.

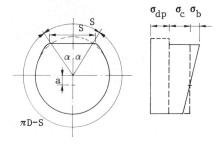

Fig. 22   Stress distribution at dented region during phase II.

277

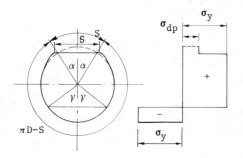

Fig. 23 Assumed stress distribution at ultimate strength of dented region.

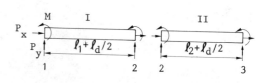

Fig. 24 Beam-column elements with parameters.

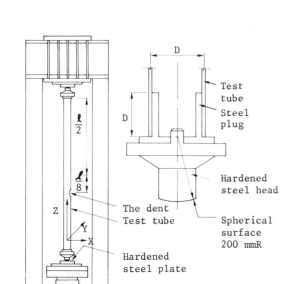

Fig. 25 General arrangement of test rig and tube head arrangement.

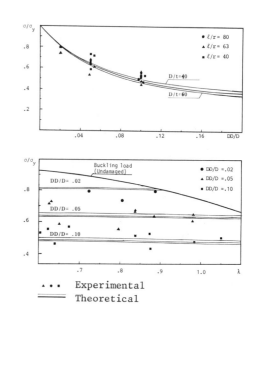

Fig. 26 Theoretical and experimental values on ultimate loads.

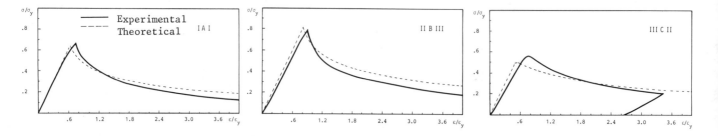

Fig. 27 Typical load-shortening curves.

THE SENSITIVITY OF FATIGUE LIFE ESTIMATES
TO VARIATIONS IN STRUCTURAL NATURAL PERIODS,
MODAL DAMPING RATIOS, AND DIRECTIONAL SPREADING
OF THE SEAS

by J. Kim Vandiver
Associate Professor of Ocean Engineering
Massachusetts Institute of Technology
Cambridge, Massachusetts

SUMMARY

Structures with natural periods in the range of four to ten seconds will be susceptible to hi cycle-low stress fatigue damage due to resonant structural response in commonly occurring sea conditions. It is shown that the computed fatigue life of a structure is extremely sensitive to the designer's estimate of the natural period - varying by as much as the natural period raised to the minus eighteenth power. A 10% error in the estimated natural period may result in a factor of six error in computed fatigue life. Damping ratio estimates are very prone to error. Fatigue life is shown to vary as approximately the square of the estimated damping ratio.

It is known that directional spreading of wave energy has a mitigating effect on fatigue damage. This is quantified in a parameter variation study. A new wave spreading model is proposed that as a result of informal communication is already being adopted by oceanographers for the description of observed sea states.

NOMENCLATURE

A	constant of proportionality
b, c	constants of the SN fatigue life curve
C_x	factor which accounts for spreading of waves
$D(\theta)$	spreading function
E	Young's modulus
e	eccentricity
$F(\ )$	rate of fatigue damage accumulation
g	acceleration of gravity
$H_{\eta s}(\ )$	stress transfer function
K_x	modal stiffness
m_x, M_x	modal mass
N	number of cycles
$R_r(\omega)$	radiation damping
$R_T(\omega)$	total damping
$S_{max}(\omega)$	Krogstad upper bound wave spectrum
$S_\eta(\omega)$	point wave amplitude spectrum
$S_\eta(\omega, \theta)$	directional wave amplitude spectrum
$S_s(\omega)$	stress spectrum
γ	Wirsching correction factor
$\Gamma(\ )$	Gamma function
$\delta(\ )$	Delta function
ν_o^+	average zero upcrossing frequency in Hz
ω	frequency in radians per second
ω_p	frequency of the peak of the wave spectrum
ω_x	natural frequency of mode x
ξ	modal damping ratio
ρ_w	density of water
σ_d^2	mean square dynamic response
σ_q^2	mean square static response'
σ_s^2	mean square stress
σ_x^2	mean square deflection for mode x
θ	angle of wave incidence

INTRODUCTION

The purpose of this analysis is to investigate the sensitivity of fatigue life calculations to variations in natural frequencies, modal damping ratios, and directional spreading of the wave spectrum. The results of such an analysis may be used to reveal the extent to which uncertainties in the estimates of such parameters will affect the estimated fatigue life of offshore structures excited by waves.

This analysis does not consider the uncertainties in material properties or the fatigue damage accumulation models themselves. This area is left to the materials specialists. This study also leaves to others the analysis of the uncertainties associated with the description of the sea states to be encountered by the structure.

The influence of wave spreading is considered for a given wave spectrum, and a new single parameter spreading function is introduced. A structural model and its idealization are selected and one method of wave force estimation is used. The wave force model assumes that drag exciting forces are negligible and that finite wave amplitude effects are not significant. In any specific application these two assumptions can and should be checked. However, for the computation of high cycle-low stress fatigue damage on large deepwater structures these assumptions are usually valid.

For the case that drag excitation cannot be neglected, the results of some recent research at MIT are mentioned. With these results the second order statistics of response may be estimated including non-linear drag exciting forces.

The exclusion of finite wave amplitude effects is probably valid for large deepwater structures in low to moderate seas, which contribute the most to high cycle-low stress fatigue damage. The governing non-dimensional parameter is likely the ratio of wave amplitude to water depth for slender bottom mounted structures. However, this is an area in which some additional research is justified.

THE FATIGUE ACCUMULATION MODEL

For the purpose of this study the assumed form of the fatigue damage accumulation model is that used by Crandall and Mark, 1973, when the stress history is assumed to be described by a narrow band random process. This formulation implicitly assumes a Palgren-Miner rule for damage accumulation. Equation (1) describes the mean rate of accumulation of the fatigue damage index for a location β in the structure due to a directionally spread random sea with mean direction θ_o.

$$F(\beta,\theta_o) = \frac{\nu_o^+}{c} (2^3 \sigma_s^2)^{b/2} \Gamma(1+b/2) \tag{1}$$

$F(\beta,\theta_o)$ = the mean rate of accumulation of the fatigue damage index at position β, due to a wave field with nominal direction of propagation θ_o.

σ_s^2 = the mean square stress at position β.

ν_o^+ = the average zero upcrossing rate of the stress process in Hz.

$\Gamma(\ )$ = the Gamma function

b, c = constants of the S-N fatigue curve of the material as defined by Equation (2), where N is the number of cycles to failure with a stress range S.

$$NS^b = c \tag{2}$$

This model, and the material constants b and c are assumed fixed. This leaves ν_o^+ and σ_s^2 as variables to be considered.

ν_o^+ depends on the frequency content of the wave spectrum as well as the wave amplitude to stress transfer function for the structure. If the structure has no natural frequencies in the region of significant wave force, then the response is generally quasi-static in nature and ν_o^+ is governed primarily by the frequency content of the wave spectrum. When the stress is primarily due to the response at a natural frequency, then ν_o^+ is strongly dependent on the natural frequency.

In both of the cases the response is approximately narrow band and the use of Equation (1) is appropriate. In the case that the response spectrum is composed of significant quasi-static and dynamic response peaks then it may be necessary to modify the above equation. One such modification is the use of a final correction factor, such as proposed by Wirsching, 1979, in which rain flow cycle counting procedures are used to obtain a correction factor to account for broad band stress spectra. The use of such a correction factor is assumed to be valid here.

The task is then to investigate the sensitivity of the mean square stress σ_s^2 and the average zero upcrossing frequency, ν_0^+, to variations in structural natural frequency and modal damping.

QUASI-STATIC AND DYNAMIC CONTRIBUTIONS TO MEAN SQUARE STRESS

In this study, it is assumed that mean square stress at a point in a structure may be approximated by the sum of a quasi-static component due to low frequency waves and a dynamic component due to the damping controlled response of natural modes of the structure excited by the higher frequency components of the wave spectrum. This is comparable to the procedure of supplementing a full static finite element solution with the dynamic contributions of the significantly responding natural modes.

In this analysis the response is assumed to be quasi-static up to within one half power bandwidth of the lowest natural frequency of the structure. Furthermore, the lowest natural frequency is not allowed to be less than the peak frequency of the wave spectrum. The computation of the mean square stress is then accomplished by summing the mean square static component with the dynamic contributions.

The quasi-static component of stress at a specific location is assumed uncorrelated with the dynamic components. However, for closely spaced natural frequencies, correlation between the stress components of two or more natural modes may have to be considered. The partitioning of static and dynamic contributions to the total stress is illustrated in Figure 1, a stress spectrum with a quasi-static stiffness controlled peak and one damping controlled resonant peak.

The quasi-static mean square stress σ_q^2, is obtained by integrating the stress spectrum up to $\omega_c = \omega_1(1-2\xi)$ where ω_1 is the lowest natural frequency and ξ is the modal damping ratio of that mode.

$$\sigma_q^2 = \int_0^{\omega_c} S_s(\omega) d\omega \tag{3}$$

where $S_s(\omega)$ is the stress spectrum.

For a complex structure σ_q^2 could be computed from a static finite element model. The calculation of the static mean square stress may include the influence of drag forces, in which an equivalent linearization procedure has been used or a more accurate non-linear wave force spectrum has been computed using the results of Dunwoody, 1981. Drag forces are neglected in the examples of this report.

This static approximation does neglect any dynamic amplification at frequencies below the cut off.

The average zero upcrossing frequency of the static component of stress is computed from the zero and second order moments of the truncated spectrum.

$$\omega_q^2 = \frac{\int_0^{\omega_c} \omega^2 S_s(\omega) d\omega}{\int_0^{\omega_c} S_s(\omega) d\omega} = \frac{1}{\sigma_q^2} \int_0^{\omega_c} \omega^2 S_s(\omega) d\omega \tag{4}$$

σ_q^2 and ω_q^2 for the example calculations are assumed to be provided for the purposes of the remaining discussions.

The dynamic or damping controlled contributions to the mean square stress are computed separately. The area under the stress spectrum as shown in Figure 1 for $\omega_1(1-2\xi) \leq \omega \leq \omega_1(1+2\xi)$ is defined as the mean square dynamic response for mode 1.

There may be more than one mode which has significant dynamic response. The dynamic contribution of each must be separately evaluated. In this report the mean square dynamic response of all significant modes will be computed using techniques described by Vandiver, 1980. In this reference it is shown that the mean square dynamic response of an individual mode x is given by:

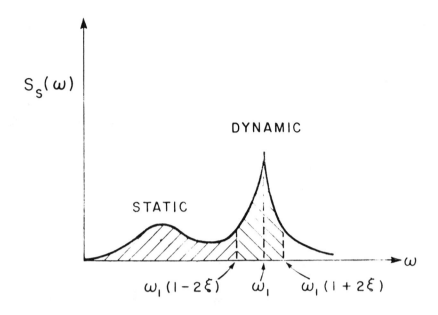

Figure 1. The partitioning of stress into static and dynamic components.

$$\sigma_x^2 = \frac{2.5 C_x \rho_w g^3}{m_x \omega_x^5} S_\eta(\omega_x) \frac{R_r(\omega_x)}{R_T(\omega_x)} \tag{5}$$

where:

- σ_x^2: mean square dynamic deflection of the xth normal mode
- m_x: modal mass
- ω_x: natural frequency
- $S_\eta(\omega_x)$: wave amplitude spectrum evaluated at ω_x
- ρ_w: density of water
- g: acceleration of gravity
- $\frac{R_r(\omega_x)}{R_T(\omega_x)}$: ratio of the radiation (wave making) to total modal damping evaluated at ω_x

This result is valid for lightly damped modes excited by linear wave forces. The constant C_x depends on structural geometry and wave spreading and is assumed to have been evaluated as as described in Vandiver, 1980. Through knowledge of the mode shape and structural details, the mean square stress at a specific location can be related to σ_x^2.

If there is more than one mode contributing in a significant way to the dynamic response then the stress at any specific location in the structure will depend upon the superposition of stresses from each mode. If the natural frequencies of each responding mode are different, (at least so that their damping controlled peaks as defined in Figure 1 do not overlap), then the stresses contributed by each may be assumed to be uncorrelated and the total mean square stress will be the sum of the mean square stresses due to each individual mode. This is a consequence of the fact that waves and hence wave forces of different frequencies are uncorrelated. If two peaks overlap then the correlation between stress components must be included.

The mean zero upcrossing frequency for mode x is simply $\omega_x/2\pi$. The mean upcrossing frequency for the combined static and dynamic stress history may be computed as a weighted sum of the individual contributions as shown below for a system with a single dynamic component.

$$\nu_o^+ (H_z) = \frac{1}{2\pi} \left[\frac{\omega_q^2 \sigma_q^2 + \omega_1^2 \sigma_{d1}^2}{\sigma_q^2 + \sigma_{d1}^2} \right]^{1/2} \qquad (6)$$

where ω_q^2 and σ_q^2 reflect the static response and ω_1^2 and σ_{d1}^2 are the natural frequency and mean square dynamic stress contributed by mode 1.

THE EFFECT OF NATURAL FREQUENCY ON FATIGUE

If the fundamental flexural natural period of a steel jacket structure was taken to be 3.5 seconds for the purpose of fatigue life computation, and the as-installed natural period turned out to be 4.0 seconds, how much would the estimated fatigue life be reduced? Recalling equation (1) and adding γ, a Wirsching type correction factor to account for broadbanded spectral effects, yields

$$F = \gamma \frac{\nu_o^+}{c} (2^3 \sigma_s^2)^{b/2} \Gamma(1 + b/2) \qquad (7)$$

Assuming that wave spreading effects have been taken into consideration, then a variation in the estimated natural period of a mode will influence three parameters in the above equation: γ, ν_o^+, and σ_s^2. σ_s^2 will change because its dynamic component will change. This is because the wave spectrum is a rapidly changing function of frequency, and as can be seen in Equation 4, the mean square dynamic response is proportional to the wave spectrum divided by the natural frequency raised to the fifth power. ν_o^+ will change as can be seen in Equation 6 because it depends on the natural frequency as well as on the mean square dynamic stress; γ may change because the broadbandedness of the stress spectrum may change. If an asterisk is used to denote the result with a shifted natural frequency, then the ratio of fatigue damage between two cases may be expressed as:

$$\frac{F^*}{F} = \frac{\gamma^*}{\gamma} \left(\frac{\nu_o^{+*}}{\nu_o^+} \right) \left(\frac{\sigma_s^{2*}}{\sigma_s^2} \right)^{b/2} \qquad (8)$$

The two extreme cases are simple to evaluate. The first is when the estimated and actual natural periods are so short that the dynamic component of σ_s^2 is negligible. This is true for most structures when the lowest natural frequency corresponds to a period of 2.4 seconds or less. In this case $F^*/F = 1.0$.

The more interesting extreme is when σ_{d1}^2, the dynamic component of stress of a single natural mode is assumed to be much larger than the static component. This may not always be the case, but provides a useful upper bound on the variation of fatigue with natural frequency. One way to estimate this case is through the ratio of fatigue damage at two different natural frequencies.

$$\frac{F^*}{F} = \frac{\nu_o^{+*}}{\nu_o^+} \left(\frac{\sigma_{d1}^*}{\sigma_{d1}} \right)^b = \frac{\omega_1^*}{\omega_1} \left(\frac{\sigma_{d1}^*}{\sigma_{d1}} \right)^b \qquad (9)$$

Because the process is narrow banded, the Wirsching correction factor reduces to 1.0 for both cases, and the upcrossing frequency reduces to the natural frequency divided by 2π.

$$\nu_o^+ = \frac{\omega_1}{2\pi} \qquad (10)$$

The only remaining step is to evaluate the frequency dependence of σ_{d1}^2, the mean square stress from dynamic response of the mode. This is quite easy and may be estimated directly from Equation (5), with one minor modification. In normal mode formulations, the product of the modal mass and the natural frequency squared is simply the modal stiffness.

$$M_1 \omega_1^2 = K_1 \qquad (11)$$

If the natural frequency varies because the modal stiffness is different than expected then the effect on mean square stress should be evaluated using equation (5). However, if the modal mass varies, then the effect on mean square stress should be evaluated after substituting Equation (11) into Equation (5), as follows.

$$\sigma_1^2 = \frac{2.5 C_1 \rho_w g^3}{K_1 \omega_1^3} S_\eta(\omega_1) \frac{R_r(\omega_1)}{R_T(\omega_1)} \tag{12}$$

If it is assumed for small variations in natural frequency that the ratio between mean square modal deflection and mean square stress at a location of concern remains constant, then the frequency dependence of the mean square stress is the same as that for mean square deflection as given in Equations (5) or (12). This is essentially an assumption that the mode shape does not change, which is not true, but is adequate here for the purpose of a simple check on sensitivity to changes in natural frequency. Therefore stress and deflection may be related as shown.

$$\sigma_{d1}^2 = A^2 \sigma_1^2 \tag{13}$$

If there is any substantial wave spreading, such as cosine squared, then C_1 is only weakly dependent on frequency and is assumed not to vary. Similarly the ratio of radiation to total damping is assumed constant in comparison to other sources of variation. Lumping all constant quantities into A^2 in Equation (13), two expressions for σ_{d1}^2 result, depending on whether the source of change was mass or stiffness.

$$\sigma_{d1}^2 = \frac{A^2}{M_1} \frac{S_\eta(\omega_1)}{\omega_1^5} \quad \binom{\text{stiffness}}{\text{changes}} \tag{14}$$

$$\sigma_{d1}^2 = \frac{A^2}{K_1} \frac{S_\eta(\omega_1)}{\omega_1^3} \quad \binom{\text{mass}}{\text{changes}} \tag{15}$$

It remains only to evaluate the frequency dependence of the wave spectrum.

Krogstad, 1979, has presented evidence that wind driven wave spectra may be modeled at frequencies higher than the frequency of the peak in the wave spectrum as given below:

$$S_{max}(f) = 1.62 \times 10^{-3} f^{-4.6} \quad m^2 - sec \tag{16}$$

This is the upper bound curve for spectral values, but possesses the frequency dependence characteristic of the high frequency side of wind driven wave spectra.

Expressed as a function of ω, Equation (16) takes the form

$$S_{max}(\omega) = \frac{1}{2\pi} \times 1.62 \times 10^{-3} \left(\frac{\omega}{2\pi}\right)^{-4.6} \tag{17}$$

Assuming all of the constants in this spectrum are absorbed into the constant A^2 in Equations (14) or (15) yields

$$\sigma_{d1}^2 = \frac{A^2}{M_1} \frac{1}{\omega_1^{9.6}} \quad \binom{\text{stiffness}}{\text{changes}} \tag{18}$$

$$\sigma_{d1}^2 = \frac{A^2}{K_1} \frac{1}{\omega_1^{7.6}} \quad \binom{\text{mass}}{\text{changes}} \tag{19}$$

Substituting each of these expressions into Equation (9) and setting the slope, b, of the S-N curve equal to 4.1 for welded tubular joints yields

$$\frac{F^*}{F} = \left(\frac{\omega_1^*}{\omega_1}\right)^{-14.6} \quad \binom{\text{mass}}{\text{changes}} \tag{20}$$

$$\frac{F^*}{F} = \left(\frac{\omega_1^*}{\omega_1}\right)^{-18.7} \qquad \text{(stiffness changes)} \qquad (21)$$

Therefore, if the natural frequency is 10% greater than predicted, then the fatigue life will be increased by a factor of 4.02 or 5.94 depending on the source of the error.

These examples were upper bound situations in which the quasi-static contributions to mean square stress were assumed small. In most cases of practical interest both contributions will be of importance and the sensitivity to natural frequency variation will not be so extreme.

THE EFFECT OF DAMPING ON FATIGUE

A variation in the estimated damping of a normal mode influences the mean square dynamic contribution to the total stress directly, and the average upcrossing frequency indirectly, because of its dependence on the mean square dynamic stress.

To place an upper bound on the significance of an error in the prediction of modal damping an analysis similar to the previous section may be performed. If only the dynamic component of a single mode is presumed to contribute to the total mean square stress, then proceeding as before leads immediately to the following conclusion:

$$\frac{F^*}{F} = \left\{ \left(\frac{R_r(\omega_1)}{R_T(\omega_1)}\right)^* \Big/ \left(\frac{R_r(\omega_1)}{R_T(\omega_1)}\right) \right\}^{b/2} \qquad (22)$$

All terms involving frequency directly cancel out because the natural frequency does not change in the example.

The method of computing mean square dynamic stress used in this analysis is somewhat unconventional and not widely used in the industry. Therefore, to reflect conventional practice the same upper bound on the sensitivity of fatigue damage calculations to variations in estimated total damping may be expressed as follows:

$$\frac{F^*}{F} = \left(\frac{\xi_T}{\xi_T^*}\right)^{b/2} \qquad (23)$$

when ξ_T and ξ_T^* are the estimated and actual total modal damping ratios, which are commonly estimated in the range from 1% to 5%.

It is the position of the author that the uncertainty in estimating the ratio of the radiation to total damping is much less than the uncertainty in estimating the total modal damping itself. Furthermore, the use of Equation (12) leads to estimates of mean square dynamic stress which are bounded because the ratio of radiation to total damping is at most 1.0. No such upper bound exists when conventional methods of computing dynamic response are used.

Furthermore, conventional methods of estimating response require independent estimates of the modal wave force spectrum and the total modal damping. This ignores the fact that the modal radiation damping and the linear modal wave force spectrum are proportional to one another. Thus two sources of uncertainty enter the calculations where only one exists.

For the sake of example, suppose in either formulation the damping is underestimated by a factor of 2.0. This will lead to an overestimate of the fatigue life by a factor of

$$(2)^{b/2} = 4.14 \text{ for } b = 4.1 \qquad (24)$$

for the extreme case of no static contribution to the stress.

THE EFFECT OF WAVE SPREADING ON FATIGUE

For a given sea state the stress time history at any particular point on the structure will depend on the directional distribution of wave energy. When the stress is linearly dependent on the wave amplitudes, the stress spectrum at a point designated by the character β may be expressed in terms of a transfer function.

$$S_s(\beta,\omega,\theta) = |H_{\eta s}(\beta,\omega,\theta)|^2 S_\eta(\omega,\theta) \qquad (25)$$

where

$H_{\eta s}(\beta,\omega,\theta)$ - wave amplitude to stress transfer function

$S_{\eta s}(\omega,\theta)$ - directional wave amplitude spectrum

θ - angle of incidence of various wave components

Spreading Functions

The directional wave amplitude spectrum has the property that integration over all possible angles of incidence must yield the point wave amplitude spectrum.

$$S_\eta(\omega) = \int_0^{2\pi} S_\eta(\omega,\theta)d\theta \tag{26}$$

In general, the amount of spreading for a given sea state will depend on wave frequency. However, most commonly used models assume, for mathematical conveniences, that the wave spreading for each sea state is independent of frequency. The use of such simplified models is acceptable because at this time the ability to predict more complex descriptions of the sea is not available. In this paper spreading models will be of the frequency independent form as shown in the following equation.

$$S_\eta(\omega,\theta) = S_\eta(\omega)D(\theta) \tag{27}$$

There are two simple limiting forms of the spreading function, $D(\theta)$. The first is the uni-directional spectrum in which waves come from a single direction θ_o, and the second is the totally diffuse or omni-directional spectrum in which waves come from all directions with equal probability. These cases are given below.

Uni-directional

$$S_\eta(\omega,\theta) = S_\eta(\omega)\delta(\theta-\theta_o) \tag{28}$$

Omni-directional

$$S_\eta(\omega,\theta) = S_\eta(\omega)/2\pi \tag{29}$$

The most common non-trivial spreading function is known as the 'cosine squared'. It is given below.

$$S_\eta(\omega,\theta) = S_\eta(\omega)\frac{2}{\pi}\cos^2(\theta-\theta_o) \tag{30}$$

$$\text{for } \pi/2 \leq \theta-\theta_o \leq \pi/2$$
$$= 0 \text{ otherwise}$$

The cosine squared model is awkward to use in a sensitivity analysis because the extent of the spreading cannot be continuously varied from uni-directional to omni-directional by simple variations of a single parameter. An equally valid and much more flexible spreading model is introduced in the next section.

The Elliptical Spreading Model

The elliptical spreading function was initially suggested by Dunwoody and is described here for the first time. The function is given below (Dunwoody, 1979).

$$D(\theta-\theta_o) = \frac{\sqrt{1-e^2}}{2\pi(1-e\cos(\theta-\theta_o))} \tag{31}$$

In polar coordinates, $D(\theta-\theta_o)$ describes a family of ellipses based on the eccentricity parameter e. One of the focii of the ellipse lies on the origin of the coordinate system and the other focus lies along the direction θ_o. The eccentricity parameter can take on any value between zero and one. Zero corresponds to a completely diffuse sea with equal amplitudes of waves propagating in all directions. The spreading function, $D(\theta-\theta_o)$, is suitably normalized so that the point wave amplitude spectrum, computed by integrating the directional spectrum over all angles, equals the original point spectrum. This angular spreading function has been chosen over other possibilities because the amount of spreading is a smooth function of a single parameter. The parameter, e, can be used as the measure of spreading in the computation of fatigue resistance. The parameter e may also be easily fitted to experimental wave spreading data.

Relative Rates of Fatigue Damage

Variation in the extent of wave spreading may change the rate of fatigue damage, F, as expressed in Equation 7, because of resulting changes in the mean square stress, the zero upcrossing frequency or the Wirsching correction factor. Two different spreading models may be compared by taking the ratio of the appropriate expressions for the rates of fatigue damage. The result will in general have the form of Equation 8. No simple generalizations can be made as to the effect of spreading on fatigue, with the exception that fatigue damage rates based on a worst case direction in a uni-directional sea will be reduced by spreading. Not much more can be concluded without evaluating a particular structure. To add some insight to this discussion, the particular example of a single vertical cylinder is presented in detail.

A simple caisson structure is shown in Figure 2. The structural properties are assumed to

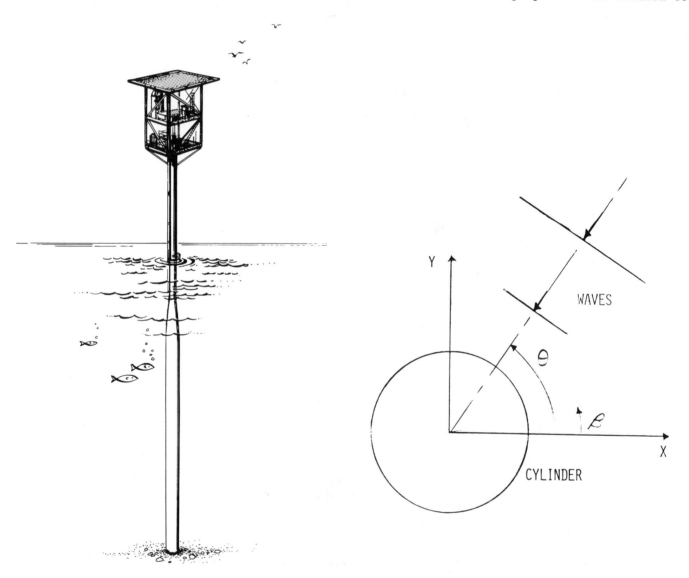

Figure 2 Caisson Production Platform   Figure 3 Coordinate System Definition

be symmetric with respect to the longitudinal axis. At any particular level on the structure, such as the mud line, the stress transfer function at a point on the perimeter defined by the angle β as defined in Figure 3, is given by the following equation.

$$H_{\eta s}(\beta,\omega,\theta) = H_{\eta s}(\omega) \cos(\theta-\beta) \tag{32}$$

where θ is the incidence angle of a regular wave component at the frequency ω.

To demonstrate the influence of spreading in this example the rate of fatigue damage F, corresponding to a spreading function $D(\theta)$ will be compared to the fatigue damage rate F_o, corresponding to a uni-directional spectrum incident on the structure from the angle θ_o. $D(\theta)$ is assumed to be a frequency independent spreading function which is symmetric about the mean incidence angle, θ_o.

For these conditions, each point on the caisson will have stress spectra whose frequency dependence will be independent of the amount of spreading. Put another way, if only the spreading function is varied, all the resulting stress spectra at a point will be proportional to one another. As a consequence the mean zero upcrossing frequency and the Wirsching correction factor will not change with spreading. The ratio of the fatigue damage rate, F, with spreading to the uni-directional case, F_o, will simplify to

$$\frac{F}{F_o} = \left(\frac{\sigma_s^2}{\sigma_{so}^2}\right)^{b/2} \tag{33}$$

The two relevant expressions for mean square stress are given below. Due to the axial symmetry of the caisson the mean incidence angle, θ_o, can be set to zero with no loss of generality. This has been done in all subsequent calculations.

$$\sigma_s^2 = \int_0^\infty |H_{\eta s}(\omega)|^2 S_\eta(\omega) d\omega \int_0^{2\pi} \cos^2(\theta-\beta) D(\theta) d\theta \tag{34}$$

$$\sigma_{so}^2 = \int_0^\infty |H_{\eta s}(\omega)|^2 S_\eta(\omega) d\omega \int_0^{2\pi} \cos^2(\theta-\beta) \delta(\theta) d\theta \tag{35}$$

Substitution into Equation 33 leads to

$$\frac{F}{F_o} = \frac{\int_0^{2\pi} \cos^2(\theta-\beta) D(\theta) d\theta}{\cos^2(\beta)} \tag{36}$$

The above ratio can be evaluated for various spreading functions $D(\theta)$. This is shown below for the cosine squared and elliptical spreading functions.

Cosine squared:

$$\frac{F}{F_o} = [\tfrac{3}{4} + \tfrac{1}{4} \tan^2\beta]^{b/2} \tag{37}$$

Hence, for the worst case direction $\beta = \theta_o = 0$, the fatigue damage rate is reduced to

$$\frac{F}{F_o} = [\tfrac{3}{4}]^{b/2} \tag{37a}$$

$$\text{For } b/2 = 2.05 \tag{37b}$$

$$\frac{F}{F_o} = .554 \tag{37c}$$

and the fatigue life

$$\frac{1}{F} = 1.8 \times \frac{1}{F_o} \tag{37d}$$

The fatigue life is increased by a factor of 1.8.

Elliptical spreading:

$$\frac{F}{F_o} = [G + \tan^2\beta(1-G)]^{b/2} \qquad (38)$$

where
$$G = \int_0^{2\pi} \frac{\cos^2\theta \ \sqrt{1-e^2}}{2\pi(1-e\cos\theta)} d\theta \qquad (39)$$

Again for the case $\beta = \theta_o = 0$ a simplified expression is obtained.

$$\frac{F}{F_o} = G^{b/2} \qquad (40)$$

The ratio of the corresponding fatigue lives is simply

$$\frac{L}{L_o} = G^{-b/2} \qquad (41)$$

These results are shown in the following table for various values of the spreading parameter e, and $b/2 = 2.05$.

TABLE 1

Spreading parameter e versus G, F/F_o and L/L_o for $b/2 = 2.05$

e	G	$\frac{F}{F_o}$	$\frac{L}{L_o}$	Description of Wave Spreading
0	0.5	.24	4.2	omni-directional
.5	0.53	.27	3.7	
.7	0.58	.33	3.0	
.8	0.62	.38	2.6	
.85	0.65	.41	2.4	
.9	0.69	.47	2.1	
.95	0.76	.57	1.8	approximate cosine squared
.99	0.87	.75	1.3	
1.0	1.0	1.0	1.0	uni-directional

These results show that cosine squared spreading extends the fatigue life by a factor of 1.8 while omni-directional spreading would increase the life to 4.2 times the uni-directional result. The possibility that omni-directional spreading might happen in nature may seem remote. However for any linear stress transfer function for a structure of arbitrary shape

$$|H_{\eta s}(\beta,\omega,\theta)| = |H_{\eta s}(\beta,\omega,\theta+\pi)| \qquad (42)$$

It is therefore only necessary that the waves be uniformly distributed over π radians to achieve the maximum extension of fatigue life over the uni-directional spectrum coming from the worst case direction. Structural symmetry may also reduce the total angle over which the waves must be evenly spread to achieve the maximum benefits. In some cases, a realistic amount of spreading, such as cosine squared is sufficient to derive the maximum benefit. An example is the fatigue caused by the heave response of a square planform tension leg platform as discussed in Vandiver, 1980.

CONCLUSIONS

By means of general formulations and a specific example, the dependence of fatigue on the uncertainties related to natural frequencies and damping ratios have been demonstrated.

Uncertainties related to the prediction of structural natural frequencies are primarily related to the structural idealizations or models used in the design process. The greatest weakness is probably in the area of foundation modelling. The behavior of soil under cyclic loading conditions remains a rather uncertain field. Assumptions regarding soils stiffness have dramatic impact on the estimation of structural natural frequencies.

The uncertainties related to damping estimates have several sources. One of the greatest is a general lack of accurate estimates of damping on existing structures. This issue and a method for obtaining improved measurements of damping on existing sturctures are addressed by Campbell, 1980. The second reason for uncertainty is that direct estimation of individual components of damping are rarely made, and the knowledge required for making such estimates is not widely available in the industry. To understand the complete damping problem one must understand the fluid mechanics, the soil mechanics, the structural mechanics, and their interaction. A final source of misuse of damping is that the relationships between exciting forces and damping mechanisms are too frequently ignored. A versatile single parameter wave spreading function has been introduced and used to demonstrate for a particular example, the importance of wave spreading in fatigue calculations.

The purpose in this study was to highlight the significance that estimation of natural frequencies, damping ratios, and wave spreading has in the calculation of the fatigue life of a structure. The results are in a subjective sense quite general, even though a specific fatigue damage accumulation rule was assumed. Of the various high cycle damage accumulation rules proposed to date, none are so different that the qualitative insights contained in this paper would be invalidated. These insights should be of help to the designer in judging the relative importance of the various factors which must be considered in the performance of a fatigue life calculation.

These results might be extended by means of a sensitivity analysis on an actual numerical model of an offshore platform intended for use in, for example, the North Sea or the Gulf of Mexico. In a very recent paper (Vugts, 1981), the sensitivity of fatigue damage rate to variations in water depth, damping ratios and several other structural parameters has been investigated and is recommended to the reader.

REFERENCES

CAMPBELL, R.B., and Vandiver, J.K., 1980: The Estimation of Natural Frequencies and Damping Ratios of Offshore Structures", <u>Proceedings of the 1980 Offshore Technology Conference</u>, Paper No. OTC 3861, pp. 53-61, Houston.

CRANDALL, S.H., and Mark, W.D., 1973: <u>Random Vibration in Mechanical Systems</u>, Academic Press.

DUNWOODY, A.B., and Vandiver, J.K., August 1981: "The Influence of Separated Flow Drag on the Dynamic Response of Offshore Structures to Random Waves", Presented at the International Symposium on Hydrodynamics in Ocean Engineering, Trondheim, Norway.

DUNWOODY, A.B., 1979: Personal communication with the author, MIT.

KROGSTAD, H.E., et al., 1979: "Analysis of Wave Spectra from the Norwegian Continental Shelf", <u>Proceedings of the POAC</u>, Trondheim, Norway.

VANDIVER, J.K., February 1980: "Prediction of the Damping Controlled Response of Offshore Structures to Random Wave Excitation", <u>Society of Petroleum Engineers Journal</u>.

VUGTS, J.H., November 1981: "Dynamic Response and Fatigue Damage with Increasing Water Depths", Presented at the conference on "Safety of Deepwater Oil and Gas Production", Det Norske Veritas.

WIRSCHING, P.H. and Light, M.C., November 1979: "Probability-Based Fatigue Design Criteria for Offshore Structures", Report to the American Petroleum Institute, API-PRAC Project H15.

Technical Poster Sessions

FOUNDATIONS

Pile Foundations under Static Loads

PORE WATER PRESSURE DEVELOPMENT DURING PILE DRIVING AND ITS INFLUENCE ON DRIVING RESISTANCE

Manoj Datta
Indian Institute of Technology, Delhi
New Delhi, India.

SUMMARY

Pile foundations for jacket type fixed offshore structures are usually installed by driving. It is observed that resistance to redriving of piles after stoppages is usually many times higher than the resistance encountered prior to suspension of driving operations. This behaviour is thought to occur on account of development of pore water pressures in the foundation soil during driving and their subsequent dissipation during stoppages. This paper analyses data available in literature on pore water pressure response during and after driving and presents a comprehensive summary on the influence of driving pore pressures on soil strength and bearing capacity. It further suggests a method for incorporating reduction in bearing capacity of soil due to induced pore water pressures in the wave equation analysis for estimating driving resistance.

Pore water pressures induced during pile driving in fine grained soils increase linearly with depth and exceed the effective overburden stress. These pore water pressures decrease rapidly with distance from pile surface and become negligible beyond a distance of fifteen pile diameters. The size of pile does not have a significant influence on the magnitude of pore water pressure developed. Reductions in the shear strength and bearing capacity of fine grained soils on account of the driving pore pressures are observed to be of the order of 25 and 75 percent respectively. Skin friction resistance offered by the foundation soil reduces markedly but the tip resistance is only marginally affected.

In pure coarse grained soils the magnitude of pore water pressures induced by pile driving is usually less than 20 percent of the effective overburden stress. However, if coarse grained soil strata are interbedded with fine grained soils, or if they contain significant finer fraction, their behaviour during pile driving appears to be akin to fine grained soils.

The final part of the paper shows that by appropriately reducing the static soil resistance values used in wave equation analysis one can get good correlation between theoretical estimates of driving resistance and actual values observed during pile installation.

INTRODUCTION

Pile foundations for jacket type fixed offshore structures are usually installed by driving. Driving of piles is not a continuous process since it is periodically interrupted for welding new sections. Bad weather at site, changing of hammers, minor breakdowns etc. also sometimes cause stoppages in pile driving operations. Resistance to redriving of piles after stoppages has been observed to be many times higher than the resistance encountered during continuous driving. This behaviour is thought to occur on account of development of pore water pressures in the foundation strata during driving and their subsequent dissipation during stoppages. Information on pore water pressure development during pile driving is scarce and a satisfactory method for identifying the influence of induced pore water pressures on the driving resistance is yet to be evolved.

This paper analyses data available in literature on pore water pressure response during and after pile driving. It presents a comprehensive summary about the variations in the magnitude of pore water pressure developed as a function of depth, distance from pile surface, pile size, and time after driving stops. It further identifies the influence of the pore water pressures developed during driving on the shear strength of the foundation soil and on the static resistance offered by the soil to the pile penetration. On the basis of this study, a method has been suggested for estimating the driving resistance using the wave equation analysis by suitably modifying the values of the static soil resistances used.

PORE WATER PRESSURES INDUCED BY PILE DRIVING

The phenomenon of development of pore water pressure when piles are driven into soil has been recognised since long. However, serious attempts to analyse the problem theoretically and to obtain actual data from the field began only in the early sixties (Bjerrum et.al.(1958), Bjerrum and Johannessen (1961), Lo and Stermac (1965)), and bulk of the information on this subject has been reported only in the recent past. A study of the data reported by various investigators reveals that, since pore water pressure development during pile driving is a dynamic phenomenon, the observations recorded are influenced by factors such as type of piezometer used, response time of the piezometer and method of installation of piezometer into the soil. On the other hand, records of pore water pressures induced during slow jacking of cone penetrometers fitted with pore pressure probes are very reliable and correlate well with those observed during pile driving as is discussed hereafter. Hence, in the following text, data available from such cone penetration tests has been used to supplement data available from instrumented pile driving tests.

In Fine Grained Soils

Magnitude and distribution:

Lo and Stermac (1965) observed that during driving of a pile, pore water pressure at any level starts to increase rapidly when the tip of the pile is a little above that level and it attains its maximum value when the pile tip just passes that level. Blanchet et.al. (1980) showed that the induced pore pressure reduces when the pile tip penetrates below the level of observation and gradually attains a stable value. This is diagrammatically shown in Fig.1. These observations correlate well with results reported by Baligh et.al.(1978) who observed that during static cone penetration tests, pore water pressure at the cone tip is always greater than the pore water pressure on the shaft behind unenlarged cones.

At any fixed distance from the pile wall, the pore water pressures induced during pile driving increase linearly with depth along the length of the pile,(Bjerrum and Johannessen(1961), Lo and Stermac (1965)). This is true both for the maximum and the stable residual values of pore water pressure recorded. This is shown in Fig.2 based on data reported by Roy et.al.(1981). Because of the linear increase of induced pore water pressure with depth, the ratio of the induced pore water pressure to effective overburden pressure is a constant at a fixed radial distance from the pile wall, as is shown in Table 1. Results of pore water pressures observed by Baligh et.al.(1978) during static cone penetration tests are also shown in the same table and they agree well with the results of Roy et.al.(1981).

The variation in magnitude of induced pore water pressure with distance from the pile wall is shown in Fig.3 which presents bulk of the data reported in literature for single piles. A glance at the figure reveals the following:
(a) At the pile surface, the pore water pressures induced have a magnitude of 1.5 to 4 times the effective overburden stress. It appears that the higher values recorded correspond to the pore water pressures at the pile tip whereas the lower values apply to the pile shaft.
(b) The induced pore water pressures become negligible beyond a distance of about fifteen pile diameters from the pile wall.

One cannot readily compare results of different investigators to identify the influence of pile size, i.e., the volume of soil displaced on the magnitude of induced pore water pressure because of differences in instrumentation used as already discussed earlier. Nevertheless, it appears appropriate to tentatively conclude that the magnitude of pore water pressure is not significantly affected by the pile size on the basis of the following two observations:

(a) In Table 1, one notes that the magnitude of induced pore water pressures reported by Roy et.al.(1981) and Baligh et.al.(1978) are strikingly similar even though in the former case the pile has a tip area of 314 cm^2 as against a tip area of 10 cm^2 of the cone in the latter case.

(b) Table 2 lists the maximum pore water pressures recorded by various investigators at the centre of pile groups. One notes from the table that these pore water pressures lie within the range observed for single piles in Fig.3. Quite clearly, though pile groups displace more soil than single piles they do not induce higher pore water pressure. The influence of larger pile diameter appears to be restricted to the fact that pore water pressures are induced in a larger zone surrounding the pile.

Dissipation :

At the end of driving, excess pore water pressures induced during driving begin to dissipate. Flow occurs from points of high pore water pressure to points of lower pressure i.e. radially away from the pile. Some flow may also occur in the vertical direction and if the pile is porous it may soak up water from the soil immediately adjacent to it (Taylor (1948)). Observations by various investigators on the process and rate of dissipation of excess pore water pressure reveal the following:

(a) At a particular level the pore water pressure near the pile surface begins to dissipate rapidly at the end of driving and there is a redistribution of pore water pressure. As a consequence of this redistribution, pore water pressure away from the pile surface continues to rise for some time even after driving has stopped before dissipation begins. This is evident from Table 3 which presents data reported by Roy et.al.(1981).

(b) For single piles, the time taken for fifty percent dissipation of excess pore water pressure to occur (t_{50}) in different types of soils, (summarised in Table 4), depends predominantly on the permeability of the soil. Piles in clays and silty clays, have much larger t_{50} values in comparison to those in silts and sands.

(c) The size of zone of soil affected by pile driving also significantly influences the rate of dissipation. This is evident from a comparison of t_{50} for single piles and pile groups in Table 4, which shows that dissipation in pile groups is much slower in comparison to single piles. Orrje and Broms (1967) also observed that dissipation of pore water pressure in smaller pile groups is faster in comparison to larger pile groups.

Theoretical models :

The first attempt to theoretically estimate pore water pressures induced during driving was made by Lo and Stermac (1965). Though the values arrived at using their theory correlate well with the maximum pore water pressures observed in the field, their method does not give the distribution of pore water pressure away from the pile surface and it also does not apply to overconsolidated soils. More recently theories of expansion of cavities in an infinite soil mass developed by Ladanyi(1963) and Vesic(1972) have been used to predict the induced pore water pressures around piles. Results presented by Appendino et.al.(1980) and Roy et.al.(1981) (Table 5) show that around the pile wall, pore water pressure distribution can be estimated using the theory of expansion of a cylindrical cavity but around the pile tip, theory of expansion of a spherical cavity gives better results.

The process of dissipation of pore water pressures has not been treated theoretically by many investigators. Soderberg (1962) was the first to analyse this problem using Terzaghi's consolidation theory. He showed that the time required for dissipation of pore pressure is proportional to the square of the pile size. Currently, finite element techniques have been developed which estimate changes in pore water pressure (i) during driving by using the cavity expansion approach and (ii) after the end of driving by using the consolidation theory (Carter et.al.(1978), Desai (1978)). Steenfelt et.al.(1981) have conducted model pile load tests in the laboratory and shown that such techniques give good agreement between measured and predicted changes of pore water pressures.

In Coarse Grained Soils

Though many investigators, e.g. Swiger (1948), Chellis (1951), have suggested the possibility of development of pore water pressures during driving of piles in sand, no serious attempt has been made to measure them in the field. The data reported by Plantema (1948) is perhaps the only of its kind available in literature. It shows that, similar to the behaviour observed in fine grained soils, pile driving induces pore water pressures as the pile tip penetrates into sand. However, at any level, the induced excess pore water pressures reduce as the pile tip penetrates below the level and finally become zero as depicted in Fig.4. The time taken for pore water pressures to decrease to their original value was observed to be 5 minutes for coarse sand and 45 minutes for silty fine sand.

Recently, Moller and Bergdahl (1981) have measured positive and negative pore water pressures during driving of model piles in dense to very dense fine sand in laboratory tests. Their study also reveals that during driving, dynamic pore water pressures are set up beneath the pile tip and along the pile wall upto seven times the pile diameter above the pile tip. These pore pressures dissipate completely after each blow and there is no overall rise in the static pore water pressure.

Table 6 shows the variation in induced pore water pressure observed by Plantema(1948) and Moller and Bergdahl (1981). The important observation from this table is that the induced pore water pressures do not cause a change of more than 20 percent in the effective overburden stress. This is in contrast to the behaviour observed in fine grained soils which exhibit $\Delta u/\bar{\sigma}_{vo}$ values of greater than unity. Such high values are possible in sand strata interspersed with fine grained soil strata as indicated by Lacy (1981).

INFLUENCE OF INDUCED PORE WATER PRESSURES ON SHEAR STRENGTH AND BEARING CAPACITY

In Fine Grained Soils

Shear strength:
Variation in the undrained shear strength of fine grained soils immediately after pile driving has been studied by Orrje and Broms (1967), Flaate (1972), Fellenius and Samson (1976), Bozuzuk et.al.(1978) and Roy et.al.(1981). From their observations one can highlight the following points.
(a)   The undrained strength measured immediately after pile driving is about 60 to 85 percent of the strength measured before piles are installed.
(b)   This reduction in strength is confined to a small zone of about 1.5 to 4 pile diameters from the pile surface. Beyond this zone the undrained strength is not influenced by pile driving and remains equal to the original value.
(c)   There is lack of agreement regarding regain of strength after pile driving has ended. Orrje and Broms (1967) observed that shear strength increases slightly with time after driving but does not attain its original value. However, Roy et.al.(1981) observed that after complete dissipation of excess pore water pressures induced by pile driving, the undrained strength achieves a value equal to the original strength of the soil. In contrast, Flaate (1972) showed that after reconsolidation of the soil around a pile, the strength of the stratum becomes larger than the strength observed before piling is undertaken.

It is interesting to note that even though very high pore water pressures are set up in a zone of about 8 pile diameters from the pile surface as shown in Fig.3, the reduction in strength is only of the order of 25 percent within a small zone of 2.5 pile diameters around the pile surface. This difference is apparently on account of the fact that as the pile penetrates into soil it displaces the soil outwards; this movement results in a large increase in the total radial stresses as well as in the pore water pressures. The effective stresses, which govern the shear strength, are only marginally changed. Stress distributions around cylindrical and spherical cavities derived by Ladanyi (1963) confirm that total radial stresses and pore water pressures increase very markedly due to cavity expansion but the effective stresses undergo a small change only. Hence one does not observe a drastic reduction in the soil strength even though very high pore water pressures are developed.

Bearing capacity:
The phenomenon of increase in axial capacity of piles with time after driving is now well recognised. Bulk of the data available in literature on this phenomenon is presented in Fig.5 which shows the increase in bearing capacity, expressed as a percentage of the long term or ultimate bearing capacity, with time. A study of the figure and the relevant references indicated on it reveal the following.
(a)   The bearing capacity of soil immediately after the end of pile driving is about 10 to 30 percent of the ultimate bearing capacity.
(b)   Pile groups (Bjerrum et.al.(1958)) and long, large diameter single piles (McClelland (1969)) show slower gain in bearing capacity in comparison to other piles.
(c)   Significant gain in bearing capacity may continue to occur even after one month after pile installation.
(d)   There appears to be a direct correlation between decrease in pore water pressure and increase in bearing capacity (Bjerrum et.al.(1958), Vesic (1977)).

In fine grained soils, since the contribution of point resistance to the total bearing capacity is usually very small, the increase in bearing capacity observed in Fig.5 reflects the increase in skin friction with time. Studies performed by Bartolomey et.al.(1979) and Roy et.al. (1981) show that point resistance does not change by more than 15 percent with time after driving whereas there is a very marked increase, almost to 200 to 500 percent, in the skin friction offered by the soil. Apparently, the tip resistance, which reflects the work done in creating a new cavity each time the pile penetrates the soil, is not influenced by the pore water pressures developed during driving.

In Coarse Grained Soils

Chellis (1951) and others have pointed out that driving and redriving of piles in sands often indicate time effects - an increase or decrease in driving resistance may be observed upon redriving loose or dense deposits. However, field records which actually show variations in strength and bearing capacity of sands with time after driving are not available in literature. Pore water pressure measurements reviewed in the previous section indicate that driving pore pressures in clean sands alter the effective overburden stresses by less than 20 percent and hence the bearing capacity should be influenced only marginally. However, if sand strata are interbedded with fine grained soils, data presented by Lacy (1981) and Moe et. al. (1981) show that bearing capacity increases with time after driving in a manner akin to that observed in fine grained soils. More field data is required to arrive at a quantitative estimate of the influence of driving pore water pressures on bearing capacity of coarse grained soils.

INFLUENCE OF INDUCED PORE WATER PRESSURES ON DRIVING RESISTANCE

A comparison of the driving resistance observed during continuous driving with the resistance observed after stoppages reflects the influence of pore water pressures developed during pile driving on the soil resistance to driving. McClelland et.al. (1969), Heerema (1979) and others have shown that soil resistance to driving increases with time during stoppages in normally consolidated and overconsolidated fine grained soils. Aggarwal et. al. (1977) have reported similar behaviour for piles driven in sands interspersed with clays.

The most widely used method for estimating driving resistance is the wave equation analysis proposed by Smith in 1960. In this method the soil resistance to driving is represented by two components - the static resistance and the dynamic resistance. The former is estimated from soil properties using soil mechanics theories whereas the latter is a function of the static soil resistance, the pile penetration velocity, and the damping coefficient of the soil. The static soil resistance estimated using soil mechanics theories corresponds to the ultimate bearing capacity of the foundation strata. It has already been highlighted in Fig. 5 that the static soil resistance during driving, which is equal to the bearing capacity measured immediately after driving, is a small fraction of the ultimate bearing capacity because of the pore water pressures induced. Hence, the static soil resistance values used in the wave equation analysis should be lower than those estimated from soil mechanics theories as depicted in Fig. 6.

Pile driving resistances at two sites - one onshore (site A) and one offshore (site B) - were estimated using the wave equation analysis by appropriatly reducing skin friction values as indicated in Table 7. The theoretically estimated values are compared with actual field records in Figs. 7 and 8. A study of these figures reveals that during continuous pile driving the static skin friction resistance is of the order of 30 to 60 percent of the ultimate static skin friction resistance. This ratio increases to more than 75 percent during stoppages. These results clearly show that static soil resistance values used in the wave equation analysis must be lower than the ultimate static resistance values estimated from soil mechanics theories for arriving at satisfactory estimates of pile driving resistance.

CONCLUSIONS

The static resistance offerred by soil during pile driving is markedly different from the ultimate static resistance of the soil because of pore water pressures induced during driving. Pile driving resistance can be satisfactorily estimated by suitably modifying the static soil resistance values used in the wave equation analysis.

REFERENCES

AGGARWAL, S.L., MALHOTRA, A.K., and BANERJEE, R., 1977, "Engineering Properties of Calcareous Soils Affecting the Design of Deep Penetration Piles". *Proceedings, Offshore Technology Conference*, Houston, Vol. 3, pp 503-512.

APPENDINO, M., JAMIOLKOWSKI, M., and LANCELLOTTA, R., 1980, "Pore Pressure of NC Soft Clay around Driven Displacement Piles" *Proceedings, Conference on Recent Developments in Design and Construction of Piles*, ICE London, pp 169-175.

BALIGH, M.M., VIVATRAT, V., and LADD, C.C., 1978, *Exploration and Evaluation of Engineering Properties for Foundation Design of Offshore Structures*, Massachusetts Institute of Technology, Department of Civil Engineering, Report No. R 78-40.

BARTOLOMEY, A.A., YUSHKOV, B.S., ROUKAVISHNIKOVA, N.E., 1979, "Stress-Strain Conditions in Active Zone of Pile Foundations," *Proceedings 6th Asian Regional Conference on Soil Mechanics and Foundation Engineering*, Singapore, Vol. I, pp 269-272.

BLANCHET, R., TAVENAS, F., and GARNEAU, R., 1980, "Behaviour of Friction Piles in Soft Sensitive Clays," Canadian Geotechnical Journal, Vol. 17, No. 2, pp 203-224.

BOZOZUK, M., FELLENIUS, B.H., and SAMSON, L., 1978, "Soil Disturbance from Pile Driving in Sensitive Clay," Canadian Geotechnical Journal, Vol. 15, No. 3, pp 346-361.

BJERRUM, L., HANSEN, J.B., and SEVALDSON, R., 1958, Geotechnical Investigations for a Quary Structure in Horten, Norwegian Geotechnical Institute, Report No. 28.

BJERRUM, L., and JOHANNESSEN, I.J., 1961, "Pore Pressures Resulting from Driving Piles in Soft Clay", Proceedings, Conference on Pore Pressure and Suction in Soils, London, pp 108-111.

BRENNER, R.P., BALASUBRAMANIAM, A.S. CHOTIVITTAYATHANIN, R., and PANANOOKOOLN, 1979, "Pore Pressures from Pile Driving in Bangkok Clay" Proceedings, 6th Asian Regional Conference on Soil Mechanics and Foundation Engineering, Singapore, Vol. 1, pp 133-136.

CARTER, J.P., RANDOLPH, M.F., and WROTH, C.P., 1978, Stress and Pore Pressure Changes in Clay During and After Expansion of a Cylindrical Cavity, University of Cambridge, Department of Engineering, Report No. 51.

CHELLIS, R.D., 1951, Pile Foundations, McGraw-Hill, New York.

CROOKS, J.H.A., MATYAS, E.L., and McKAY, H.M., 1980, Excavation Slope Stability Related to Pore Water Pressure Variation during Piling," Canadian Geotechnical Journal, Vol. 17, pp 225-235.

DESAI, C.S., 1978, "Effects of Driving and Subsequent Consolidation of Behaviour of Driven Piles", International Journal for Numerical and Analytical Methods in Geomechanics, Vol.2, pp 283-301.

FELLENIUS, B.H., and SAMSON, L. 1976, "Testing of Drivability of Concrete Piles and Disturbance to Sensitive Clay," Canadian Geotechnical Journal, Vol. 13, No.2, pp 139-160.

FLAATE, K., 1972, "Effects of Pile Driving in Soft Clay", Canadian Geotechnical Journal, Vol. 9, No. 1, pp 81-88.

HEEREMA, E.P., 1979, "Pile Driving and Static Load Tests on Piles in Stiff Clay", Proceedings, Offshore Technology Conference, Houston, OTC 3490, pp 1135-1145.

ISMAEL, N.F. and KLYM, T.W., 1979, "Pore-Water Pressure Induced by Pile Driving", ASCE Journal of Geotechnical Engineering Division, Vol. 105, No. GT 11, pp 1349-1354.

LACY, H.S., 1981, "Pile Integrity and Capacity Determined by Redriving", Proceedings, Tenth International Conference on Soil Mechanics and Foundation Engineering, Stockholm, Vol.2, pp 781-786.

LADANYI, B. 1963, "Expansion of a Cavity in a Saturated Clay Medium," ASCE Journal of Soil Mechanics and Foundation Division, Vol. 89, No. SM 4, pp 127-161.

LO, K.Y., and STERMAC, A.G., 1965, "Induced Pore Pressures During Pile Driving Operations", Proceedings, Sixth International Conference on Soil Mechanics and Foundation Engineering, Montreal, Vol. 2, pp 285-289.

McCLELLAND ENGINEERS, 1969, Additional Soil Investigations, Pile Research Program, Eugene Island Area, Report to Shell Oil Co.

McCLELLAND. B., FOCHT, J.A., and EMRICH, W.J., 1969, "Problems in Design and Installation of Offshore Piles", ASCE Journal of the Soil Mechanics and Foundation Division, Vol. 95, SM 6, pp 1491-1513.

MOE, D., ARVESON, H., and HOLM, O.S., 1981, "Friction Bearing Pipe Piles at Calabar Port", Proceedings, Tenth International Conference on Soil Mechanics and Foundation Engineering, Stockholm, Vol. 2, pp 781-786.

MOLLER, B,, and BERGDAHL, U., 1981, "Dynamic Pore Pressure During Pile Driving in Fine Sand", Proceedings, International Conference on Soil Mechanics and Foundation Engineering, Vol.2, pp 791-794.

ORRJE, O., and BROMS, B.B, 1967, "Effects of Pile Driving on Soil Properties", ASCE Journal of Soil Mechanics and Foundation Division, Vol. 93, No. SM 5, pp. 59-73.

PLATEMA, G., 1948, "The Occurrence of Hydrodynamic Stresses in the Pore Water of Sand Layer during Driving of Piles", Proceedings, Second International Conference on Soil Mechanics and Foundation Engineering, Rotterdam, Vol. 4, pp 127-128.

ROY, M., BLANCHET, R., TAVENAS, F., and LaROCHELLE. P., 1981, "Behaviour of Sensitive Clay During Pile Driving, Canadian Geotechnical Journal, Vol. 18, No. 1, pp 67-85.

SODERBERG, L.O., 1962, "Consolidation Theory Applied to Foundation Piles", Geotechnique, pp 217-225.

TAYLOR, D.W., 1948, "Fundamentals of Soil Mechanics, Asia Publishing House, Bombay.

TOMLINSON, M.J., 1977, "Pile Design and Construction Practice, Viewpoint Publication, London.

VESIC, A.S., 1972, "Expansion of Cavities in Infinite Soil Mass", ASCE Journal of Soil Mechanics and Foundation Division, Vol. 98, No. SM 3, pp 265-290.

TABLE 1: Ratio of Induced Pore Water Pressure to Effective Overburden Stress ($\Delta u / \bar{\sigma}_{vo}$) as a Function of Depth.

	Roy et.al.(1981)*			Baligh et.al.(1978)**	
Depth (m)	$\Delta u / \bar{\sigma}_{vo}$		Depth (m)	$\Delta u / \bar{\sigma}_{vo}$	
	Pile tip	Pile wall		Cone tip	Shaft behind cone tip
2	3.73	2.07	10	3.68	1.84
4	3.78	2.07	20	3.47	1.80
6	4.08	2.14	30	3.60	2.00

*22cm dia. pile driven into silty clay
**18° unenlarged cone jacked into Boston Blue clay

TABLE 2: Ratio of Induced Pore Water Pressure to Effective Overburden Stress ($\Delta u / \bar{\sigma}_{vo}$) for Pile Groups.

Reference	Maximum value of $\Delta u / \bar{\sigma}_{vo}$ observed
Bjerrum et.al.(1958)	~1.0
Bjerrum and Johannessen (1961)	~1.0
Lo and Stermac (1965)	1.0 to 1.3
Orrje and Broms (1967)	1.1 to 1.2
Appendino et.al.(1980)	1.2 to 2.3
Brenner et.al.(1979)	1.0 to 1.5
Crooks et.al. (1980)	~1.0

TABLE 3: Dissipation of Induced Pore Water Pressure After the End of Pile Driving (as per Roy et.al.(1981)).

Time elapsed (hrs)	Excess pore water pressure (kg/cm^2) at a radial distance of		
	0.5D	1.5D	4.5D
0.25	0.60	0.35	0.15
1	0.59	0.34	0.17
10	0.47	0.28	0.18
100	0.14	0.12	0.10

TABLE 4: Time for Fifty Percent Dissipation of Excess Pore Water Pressure

Reference	Soil type	LL	PL	PI	Single pile or pile group	t_{50}
Plentema (1948)	Coarse sand	-	-	-	single	5 min
Ismael and Klym (1979)	Clayey sandy silt	20.5	10.4	10.1	single	7 hrs
Blanchet et.al (1980)	Silty clay	65 to 75	20 to 25	45 to 50	single	1.5 days
Roy et.al.(1981)	Silty clay to clayey silt	35 to 50	20 to 25	15 to 25	single	3 days
Bjerrum and Johannessen (1961)	Clay	45 to 55	20 to 25	20 to 30	group	3 weeks
Brenner et.al. (1979)	Silty clay	80 to 110	25 to 40	40 to 70	group	3 to 4 weeks
Crooks et.al. (1980)	Silt	16	11.5	4.5	group	1 day

TABLE 5 Comparison of Predicted and Observed Pore Water Pressures Induced by Pile Driving

Reference	$\Delta u/\bar{\sigma}_{vo}$ at zero distance from pile			
	Observed at		Estimated from theory of expansion of cavity	
	Pile tip	Pile wall	Sperical cavity	Cylindrical cavity
Appendino et.al. (1980)	1.9 to 2.3	1.2 to 1.7	1.8 to 2.84	1.4 to 1.8
Roy et.al. (1981)	2.7 to 3.6	1.3 to 1.9	3.95	1.8

TABLE 6 Variation in Induced Pore Water Pressure during Pile Driving in Sands

Reference	Soil type	Variation in $\Delta u/\bar{\sigma}_{vo}$ observed
Plantema (1948)	Coarse sand	0.0 to 0.1
Plantema (1948)	Silty fine sand	0.0 to 0.2
Moller and Bergdahl (1981)	Fine sand	-0.2 to 0.15

TABLE 7 Soil Parameters Used in Wave Equation Analysis

Parameters	Site A (onshore)		Site B (offshore)	
Water depth (m)	–		70	
Soil type	Silty CLAY		Strata of silty CLAY alternating with calcareous SAND	
Pile type	Cylindrical steel pile		Cylindrical steel pile	
Pile penetration (m)	24		70	
Ultimate static resistance estimated by	API method		API method	
Static soil resistance during driving (as a percentage of ultimate static resistance)	Skin Friction	Tip resistance	Skin friction	Tip resistance
Case I	25	100	25	100
Case II	50	100	50	100
Case III	75	100	75	100
Case IV	100	100	100	100
Quake (cm)	0.254		0.254	
Damping (sec/cm)				
Side	0.0049		0.0049	
Tip	0.0033		0.0033	

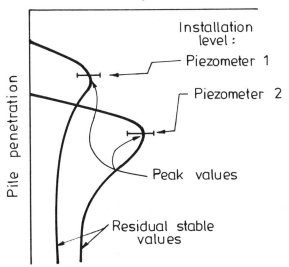

FIG.1. PORE WATER PRESSURES INDUCED BY PILE DRIVING IN FINE GRAINED SOILS

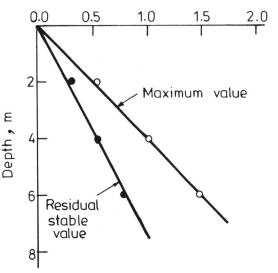

FIG.2. VARIATION OF INDUCED PORE WATER PRESSURE WITH DEPTH (As per Roy et.al., 1981)

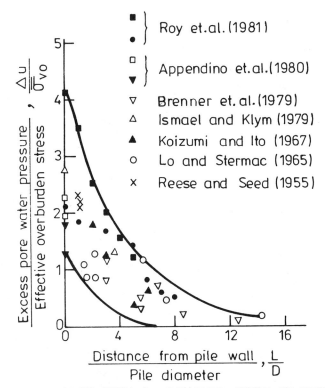

FIG.3. VARIATION OF INDUCED PORE WATER PRESSURE WITH DISTANCE FROM PILE WALL

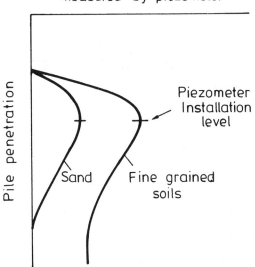

FIG.4. PORE WATER PRESSURE INDUCED BY PILE DRIVING IN SAND

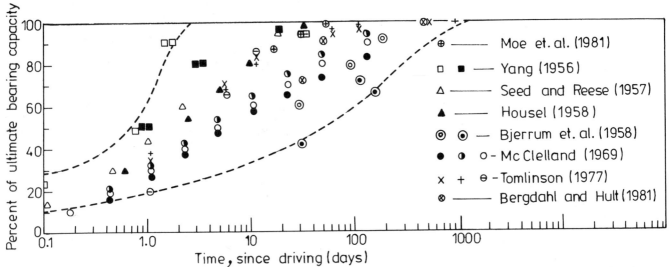

FIG.5. FIELD DATA ON INCREASE OF BEARING CAPACITY WITH TIME AFTER DRIVING

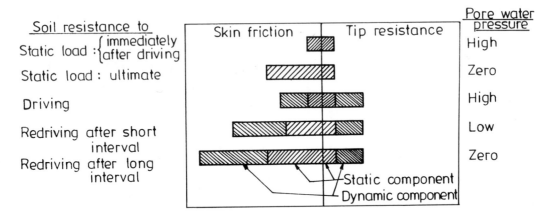

FIG.6. RELATIVE RESISTANCES DURING VARIOUS LOADINGS ON PILE

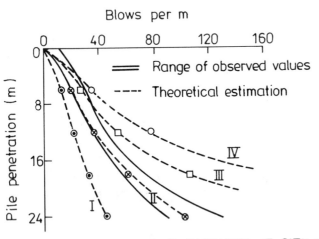

FIG.7. PILE DRIVING RESISTANCE AT SITE A

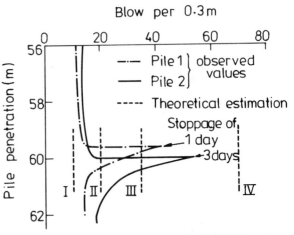

FIG.8. PILE DRIVING RESISTANCE AT SITE B

ROD SHEAR INTERFACE FRICTION TESTS IN SANDS, SILTS AND CLAYS

A. R. Dover
Woodward-Clyde Consultants
Houston, Texas

S. R. Bamford
Woodward-Clyde Consultants
Houston, Texas

L. F. Suarez
Woodward-Clyde Consultants
Houston, Texas

SUMMARY

The paper presents the results of rod shear interface friction tests in sands, silts and clays, and discusses use of the data in axial pile capacity analysis and design. The rod shear test consists of a steel rod within a cylindrical soil specimen which is confined in a modified triaxial chamber. The test series presented covers a wide range of effective confining stresses, soil types and loading rates. In addition to sands and silts, both normally-consolidated and overconsolidated clay soils were tested.

The results of these tests show that local skin friction values for all soils that were tested exhibit no limit with increasing effective confining stress level. Rod roughness was found to be a significant factor for most soils. These results are discussed in terms of their potential importance in axial pile design, especially for longer, higher capacity offshore piles.

INTRODUCTION

Estimation of the frictional resistance along the soil-pile interface is a very important requirement in design of pile foundations. Large structural loads and limited number of piles for offshore structures has created the need for very long and large-diameter piles. For these piles, an "accurate" estimate of the soil-steel and/or soil-soil interface friction becomes very important if a safe and economical design is to be achieved.

The rod shear test is a laboratory test that was developed to model and estimate the interface friction at the soil-pile interface. The test consists of a steel rod within a cylindrical soil specimen which is confined in a modified triaxial chamber. By increasing the total and effective confining pressure and pushing the rod through the sample, it is possible to model a discrete segment of a pile-soil system. This is one method of modeling the behavior where pile load test data for large and very long piles (> 200 ft) is limited or nonexistent.

Results of rod shear tests in sands, silts and clay soils are presented herein and are discussed relative to soil type, effective confining pressure, degree of overconsolidation (OCR), rod roughness, soil plasticity index, rod displacement at peak interface friction, rate of loading and sample cure time effects. The relationship between interface friction, f_s, and effective confining pressure, $\bar{\sigma}_c$, is a key issue of interest.

Implications on pile foundation design are also investigated. Differences between predicted pile capacity/length of pile resulting from adopting current design guidelines versus using rod shear interface friction results are illustrated. A simple cost analysis shows the economic implications.

TEST EQUIPMENT AND PROCEDURES

The test equipment, sample preparation technique and test procedures involved in performing rod shear tests have been described previously by Bea and Doyle (1975). The equipment and procedures are similar to those used to perform conventional triaxial shear tests on soils.

The rod shear test (Fig. 1) is performed on a cylindrical specimen of soil approximately 2.0 in. in diameter and 4 to 5 in. high and enclosed in a rubber membrane. A 0.495-in. hole is cored along the sample's longitudinal axis, through which a steel rod of 0.5-in. diameter is passed and set in place. The sample is isotropically consolidated in a triaxial compression cell to the desired effective pressure level. During shear, the rod is moved axially through the soil specimen, at which time the interface friction and rod displacement responses are obtained.

Two types of rod sections were generally used for interface friction tests presented in this paper:

- mild, smooth, cold-rolled steel (smooth rod)
- highly grooved steel with a knurled pattern (rough rod)

Attempts were made to use sandpaper-coated steel rods. This technique failed, as the sandpaper deteriorated during testing. One test was conducted with a spiral screw-threaded rod with deep recesses.

Following extrusion from the Shelby tube or liner, the sample was inserted in a 2 in. ID stainless steel mold. A 2 in. phenolic disk with a center hole of 0.51 in. ID was centered on top of the sample. A small stainless steel cutting tube (similar to a miniature Shelby tube) with a 0.495 in. OD and 0.04 in. maximum wall thickness was inserted through the phenolic guide and pushed into the sample approximately 1/4 to 1/2 in. The tube was removed and the small soil plug in the cutting end removed. This process was repeated until a cylindrical hole of approximately 0.5 in. diameter was constructed through the length of the sample. In some cases, sand samples were reconstructed in a mold around the rod.

The rod (smooth or knurled surface) was carefully inserted into the cylindrical hole to the desired position. After removal of the mold, the rod and sample were then placed in the cell with the bottom end of the rod pushed through the lower cell bushing to a predetermined distance. The remainder of the sample and cell preparation was conducted using the same procedures necessary for preparation of a triaxial compression test.

After consolidation at the first effective stress level, the load was applied to the rod, forcing it to shear within the sample. The load was measured with a proving ring. Rod displacement was measured with a standard dial gage. Loading rates of 0.003 in./min were typically used, although this test parameter did vary. Following measurement of the maximum interface load or stress, the rod load was removed and the consolidation stress increased to the next effective stress level, after which the test was repeated (multi-stage test).

It should be recognized that the overall stress state within the sample during shearing is not clearly defined, i.e., the effect of principal stress rotation and load shedding/transfer to the base are not taken into account. Potential sample preparation technique effects on results are also not taken into account. Finite element modeling might be used to better understand the stress state. The authors wish to explicitly state that this test is only one test that may be used to estimate localized skin friction effects of soil on steel. The results may be more applicable in certain soils than in others. However, the authors also believe that no one test is appropriate for all soil conditions, and the idea that there exists only one "unique" interface friction value is not correct - the interface value may depend upon the loading condition as much as the soil type or strength.

SCOPE OF TESTING PROGRAM

A total of 27 rod shear tests were performed during the course of five separate investigations to supplement knowledge of the frictional behavior at the pile-soil interface. Table 1 presents test data for each of the rod shear tests performed. This limited number of tests was not intended to be, and is not sufficient for, a rigorous research program, and the test results must be interpreted within this context.

A wide range of soils were tested. Rod shear tests were performed on specimens from the following locations: a) U.S. East Coast; b) Offshore Alaska; c) Gulf of Mexico; d) Offshore California; and e) Houston, Texas.

These locations have provided soil samples ranging from fine sands to silty clays.

TEST RESULTS AND INTERPRETATION

Test results are presented and discussed in this section. The results are discussed in terms of soil type, effective confining pressure, degree of overconsolidation, rod roughness, plasticity index, rod displacement at peak interface friction, rate of loading and sample cure time effects. Comparisons with previously reported investigations are also made.

Effective Confining Pressure Effect

Figure 2 shows the peak measured interface friction, f_s, versus initial effective confining pressure, $\bar{\sigma}_{ci}$, for all soils tested. There is the expected trend of higher f_s values for increasing effective confining pressures. The scatter in the data is obviously due to the wide range of soils tested. Measured values of ϕ_{ss}, defined as the slope angle of the f_s vs. $\bar{\sigma}_{ci}$ line range from 12 to 26 degrees, while the f_s intercept ranges from 1.5 to 2.5 psi. Sand samples lie close to the upper bound, while clay samples are close to the lower bound. Data for both drained and undrained tests are shown.

The data generated in this study do not indicate the existence of a limiting value on f_s at high effective stress levels. The increase in interface friction, f_s, at high confining stresses is approximately linear for each particular test.

The average trend of the test results reported by Coyle and Reese (1966) is plotted in Fig. 2 also. The samples used by Coyle and Reese were highly plastic clays with PI of approximately 47 and water contents ranging from 52 to 55 percent. The large PI and water contents appear to be responsible for the lower interface frictions reported by Coyle and Reese.

Bea and Doyle (1975) made an extensive series of rod shear tests in marine clays from the Gulf of Mexico. The confining pressures used in their tests ranged from 50 to 415 psi. The average trend of interface friction versus effective confining pressure is plotted in Fig. 2. Bea and Doyle indicated that there was no apparent limiting value of interface friction at high confining pressures.

Overconsolidation Effect

Figure 3 shows the relationship between induced laboratory overconsolidation expressed in terms of OCR and the normalized interface friction. The limited data indicate that for cohesive soils the $f_s/\bar{\sigma}_{ci}$ ratio increases with OCR. More data is needed in sands before a trend can be defined.

Rod Roughness

Influence of the roughness of the contact surface between soil and pile was studied by performing rod shear tests on rods of variable roughness. Four types of rod surfaces were used: smooth milled steel, sandpaper-coated rods, knurled rods, and a screw-threaded rod (one test). Characteristics of each rod type were given in the section on test procedures and equipment. The sandpaper-coated rod shear tests were not successful and will not be further discussed.

Figure 4 shows the stress-strain behavior for tests GOM-1 and GOM-3, run with smooth and knurled rods respectively. Both tests were run on marine silty clays from a depth of approximately 384 ft. Tests GOM-2 and GOM-4 were conducted on the same marine silty clay from a depth of approximately 408 ft and had a very similar stress-strain behavior. Comparison of peak interface friction versus effective initial confining pressure for smooth and knurled rods is presented in Fig. 5.

From these limited data on silty clays, the following observations can be made:

a. The normalized interface friction, $f_s/\bar{\sigma}_{ci}$ and the axial displacement at failure along rough surfaces are higher than for smooth surfaces. Similar findings were reported by Coyle and Reese (1966) and Bea and Doyle (1975).

b. For the marine silty clays tested, average normalized interface friction $f_s/\bar{\sigma}_{ci}$ is 0.22 and 0.31 for smooth and rough rods, respectively. The equivalent ϕ_{ss} values are 12.5 and 17.5 degrees, respectively.

The importance of the roughness effect is that the interface or near-interface shearing process is primarily a frictional phenomenon. If the pile surface were rough enough to cause the interface failure to be governed by soil-to-soil friction, the interface friction would be given by the undrained shear strength of the soil at the time of testing and for the particular state of stress existing at that time. Bea and Doyle reported much reduced values of interface friction when the rod was purposely lubricated.

Plasticity Index Effects

A correlation between normalized interface friction, $f_s/\bar{\sigma}_{ci}$, and plasticity index, (PI), was suggested by Bea and Doyle (1975). Their results indicated that the normalized interface friction, $f_s/\bar{\sigma}_{ci}$, decreases with increasing PI. The results of tests with cohesive soils and smooth steel rods are shown in Fig. 6. More data need to be obtained in order to explain some of the scatter and to suggest a more definitive trend.

Rod Displacement at Peak Interface Friction

The rod displacement, δ, at peak interface friction ranged from approximately 0.005 in. to 0.080 in. These values are smaller than those usually associated with full-scale pier or pile load tests. There is some indication that in rough rods the rod displacement at failure, δ, is larger than that in smooth rods. The same observation was noted by Coyle and Reese (1966). No correlation between peak displacement, δ, and normalized peak friction, $f_s/\bar{\sigma}_{ci}$, was found. Samples with laboratory-induced OCR showed a decrease in δ with increasing OCR, indicating a more brittle response.

Rate of Loading Effect

Clay samples from the Gulf of Mexico were tested at various loading rates. Loading rates spanned three orders of magnitude. Tests at fast loading rates showed values of high interface friction on smooth rods. Bea and Doyle (1975) observed a similar trend. The increase of interface friction at high loading rates is compatible with studies of rate of loading effects on piles (Bea, et al, 1980).

Rate of loading effects were not investigated using rough rods. It is believed that for rough rod conditions, the rate of loading effect may be very close to the rate of loading effect observed in triaxial compression tests on soil samples (e.g., Casagrande and Wilson, 1951), since soil-to-soil failure is more likely to occur in rough rod tests. There may, however, exist a complex relationship between rate of loading and failure mode. Does faster loading cause soil-soil failure regardless of rod roughness?

Sample Cure Time

Figure 7 presents the results of multi-stage tests by Bea and Doyle (1975) to study the effect of "set-up" or cure time (period of time over which the final consolidation stress is applied) on the interface friction behavior for cohesive samples. It is observed that while the peak f_s values are about the same, the initial load-displacement response is stiffer for increasing days of cure time. Also, the residual strength levels (10 to 20 percent lower than peak) are observed to be higher for the longer cure times. This effect may be in part due to the multi-stage nature of the test. Longer curing times may also allow total thixotropic soil strength regain and/or chemical bonding between steel and soil, thus contributing to the observed behavior. It is also possible that longer curing time may also affect the rheological behavior of the bushings and mask the real response to some degree.

IMPLICATIONS ON PILE FOUNDATION DESIGN

The conventional method of API RP 2A (API, 1981) used for predicting pile ultimate load carrying capacity typically relies upon results of soil-soil shear strength determinations for selecting the values of ultimate skin friction, f. For longer piles, this value of f may not accurately represent the strength at or near the pile-soil shearing interface, especially at greater penetrations. The value of f determined from rod shear tests represents another manner in which to define the interface friction for use in pile capacity analyses. It should be recognized that the rod shear test is a measure of localized skin friction and does not explicitly take into account the complex stress state that likely exists around a real pile (for example, arching effects or residual stress effects that may cause a limitation of field-measured skin friction at deep penetrations are not taken into account).

For the purposes of illustration in this paper, it is assumed that the interface friction between a pile and the surrounding soil is a function only of the lateral effective stress and the soil-steel normalized soil properties such that:

$$f_s = K \bar{\sigma}_{3i}$$

where f_s = soil-steel interface friction
K = the normalized interface friction coefficient = $f_s/\bar{\sigma}_{ci}$ = $\tan \bar{\phi}_{ss}$
$\bar{\sigma}_{3i}$ = initial lateral effective confining stress = $k_o \bar{\sigma}_{1i}$
ϕ_{ss} = soil-steel effective angle of friction

The end bearing capacity developed in a cohesive or a non-cohesive soil is not dependent upon soil-steel friction parameters and, consequently, will not be presented here.

Examples of the effect of using soil-steel interface friction values described above versus soil-soil shear strength determinations for pile capacity analysis are presented in the following section. Two pile-soil systems have been selected for comparison purposes as follows:

Case I: Cohesionless soil stratigraphy

Case II: Cohesive soil stratigraphy

A 72 in. OD pipe pile was used for both analyses. The API RP 2A method and rod shear test results were used for estimating ultimate skin friction capacity for both cases.

The results of Case I are presented in Figs. 8 and 9 for a soil stratigraphy of predominantly cohesionless material. Figure 8 shows the unit skin friction values, using conventional soil strengths and the API RP 2A code are normally limited to a certain value at depth. Although API does not specify a limiting value or depth at which the limitation begins, conventional practice among offshore geotechnical engineers has resulted in a reasonably small range of limiting values based on soil type. The rod shear test results (smooth rod) indicate no limiting values of friction vs. pressure (depth); consequently the predicted unit skin friction tends to increase approximately linearly with depth. The resulting pile capacity analysis (Fig. 9) reveals a 40% increase in predicted capacity (compression) due to the use of rod shear test results without limiting values.

The results of Case II are presented in Figs. 10 and 11. These analyses were performed using smooth and rough rod test results. The example stratigraphy and test results used consist of underconsolidated offshore sediments which present additional problems which must be viewed in the analysis.

Figure 10 presents unit skin friction versus depth. The effect of using interface friction results is dependent upon the rod roughness. The rough rod results approximate the values of conventionally-derived soil shear strengths, while the smooth rod results show reduced strengths. This trend may partially be a result of the underconsolidated state of the sediments used in this example. The in-situ stress used in this analysis was equal to the maximum past pressure, $\bar{\sigma}_{vm}$, which is less than the present overburden, ($\bar{\sigma}_{vo}$), thus the underconsolidated state. This example is a special case (underconsolidated soils) and reveals that for normally or overconsolidated soil conditions ($\bar{\sigma}_{vm} \geq \bar{\sigma}_{vo}$), f_S from rough rod shear tests would be greater than conventionally-derived interface friction values.

The effect on pile ultimate capacity is shown in Fig. 11. As would be expected from above, the calculated resistance profiles using conventional values of S_u and those from rough rod shear tests are nearly equivalent. The pile capacity profile using smooth rod shear test data indicates lower capacities at equivalent depths. The data indicate that rough rod shear tests for this soil type agree with soil-soil shear strength predicted pile capacities. The smooth rod shear tests, which may simulate the soil-steel failure mode better, predict lower capacities.

As a hypothetical example of cost implications relating to the methods of predicting pile capacity, consider a difference in pile length of 40 ft per pile (as for Case I, Fig. 9, at an ultimate capacity of 15,000 kips) depending upon the method chosen to estimate ultimate capacity. The following assumptions are reasonable, but do not reflect site-specific cases and should be considered as illustrative. Assume that 16 piles are used to support the structure, steel piling costs are $1000 per ton, the equipment installation spread costs are $85,000 per day, and pile driving time is a maximum of 1 ft per minute. Each additional pile section must be stabbed, welded and inspected in an average time of 4 hours per section.

The cost difference can be stated as follows:

Additional driving time	$ 38,000
Additional steel	240,000
Additional installation time	225,000
TOTAL	$503,000

For conventional non-deepwater platforms, this sum is not an insignificant amount. For deepwater structures, the foundation cost percentage of the total platform cost will increase; consequently, the absolute cost difference will be larger. A key element is the greatly increased cost of failure of a deepwater structure.

COMMENTS

For deeper penetrations, or in the case of high-capacity tension piles, the difference in predicted ultimate capacities may increase. The geotechnical engineer, faced with providing an estimate of ultimate pile capacity, could logically and reasonably select the rod shear test results (or direct shear test or borehole shear tests, etc.) and determine a capacity. For deep penetrations in sands, the amount of potential capacity that is ignored by using limiting values can be considerable.

The questions that arise from examining these results from imperfect, idealized but reasonable tests and their cost implications include:

a) Do we really understand the stress state around piles, especially at deeper penetrations?

b) How rough are actual steel pipe piles?

c) What does cyclic loading and/or constant compression or tension bias do to the stress state and ultimate capacity?

d) Do we know whether the failure mode is soil-soil or soil-steel? Is the rate of loading controlling the failure mode?

e) What is the risk versus economic situation when two widely different predicted capacities are considered?

f) What factor of safety is appropriate for designing piles by methods other than API RP 2A?

As a final comment, the first author's experience with pile driving dynamic measurements has indicated that "measured" pile ultimate capacities (via the dynamic measurements) may be considerably more or considerably less than those predicted by any single method.

The purpose of this paper has been to present some limited test data, illustrate how it might be interpreted and used in piling design, investigate the cost implications of using it, and ponder some of the unknowns and uncertainties that exist in piling design. The authors recognize the limited nature of the test results and some of the potential problems involved in the performance of the tests and interpretation of the results. The authors do not recommend the use of these rod shear test results over other methods of pile design. This information has been presented herein for the purposes of thought and reflection, rather than definition and quantification.

ACKNOWLEDGEMENTS

The authors are grateful to Shell Development Company for the use of the rod shear equipment and to the Professional Development Committee of Woodward-Clyde Consultants, which sponsored a portion of the testing work.

REFERENCES

Bea, R. G., Audibert, J. M. E. and Dover, A. R. (1980). "Dynamic Response of Laterally and Axially Loaded Piles," _Proceedings, Twelfth Offshore Technology Conference_, Houston, 1980, Vol. 2, OTC 2749, May.

Bea, R. G. and Doyle, E. H. (1975). "Parameters Affecting Axial Capacity of Piles in Clay," _Proceedings, 7th Offshore Technology Conference_, Houston, 1975, Vol. 2, OTC 2307.

Casagrande, A. and Wilson, S. D. (1951), "Effect of Rate of Loading on the Strength of Clays and Shales at Constant Water Content," _Geotechnique_, Vol. 2, No. 3, pp. 251-263.

Coyle, H. M. and Reese, L. C. (1966). "Load Transfer for Axially Loaded Piles in Clay," _Journal of the Soil Mechanics and Foundations Division_, ASCE, Vol. 92, SM2, March, pp. 1-26.

Coyle, H. M. and Sulaiman, I. H. (1967),"Skin Friction for Steel Piles in Sand," _Journal of the Soil Mechanics and Foundations Division_, ASCE, Vol. 93, No. SM6, November, pp. 261-278.

TABLE 1. SUMMARY OF TEST RESULTS

Test Series	Test No.	Soil	LL/PL	Rod	Stage	$\bar{\sigma}_c$ (psi)	Drainage	Peak $f_s/\bar{\sigma}_{c,i}$
U.S. East Coast	1	f.sand, silty clay, clayey silt, w/shell fragments	N/A	smooth	1 2 3	5 15 30	drained " "	0.20 0.12 0.23
	2	fine to med. sand w/ shell fragments	N/A	smooth	1 2 3	15 75 150	drained " "	0.32 0.34 0.34
	3	fine to med. silty sand w/shell fragments	N/A	smooth	1 2 3	50 100 270	drained " "	0.34 0.39 0.43
Offshore Alaska	1	silty sand to sandy silt w/shell fragments	N/A	smooth	1	20	drained	0.46
	2	fine silty sand to sandy silt w/shell fragments	N/A	rough screw thread	1 2 3	20 40 80	drained " "	0.36 0.48 0.53
	3	v. sandy silt to silty sand w/shell fragments	N/A	smooth	1 2 3 4 5	20 40 80 160 160	drained " " undrained drained	0.46 0.39 0.45 0.38 0.46
	4	f. silty sand w/shell fragments	N/A	smooth	1 2 3 4	10 20 40 20	drained " " "	0.70 0.56 0.51 0.50

TABLE 1. SUMMARY OF TEST RESULTS (CONT'D.)

Test Series	Test No.	Soil	LL/PL	Rod	Stage	$\bar{\sigma}_c$ (psi)	Drainage	Peak $f_s/\bar{\sigma}_{c,i}$
Offshore California	1	silty clay	54/28	smooth	1	83	drained	0.27
	2	silty clay	39/25	"	1	278	"	0.25
	3	clayey silt	50/23	"	1	315	"	0.24
	4	silty clay	47/26	"	1	265	"	0.23
	5	silty clay	49/26	"	1	222	"	0.29
	6	silty clay	49/26	"	1	56	"	0.43
	7	silty clay	41/16	"	1	77.5	"	0.33
	8	clayey silt	38/25	"	1	16.5	"	0.35
	9	silty clay	45/26	"	1	194	"	0.38
	10	silty clay	45/26	"	1	97	"	0.45
Houston, Texas	1	silty clay	64/24	smooth	1	14	undrained	0.49
					2	28	"	0.28
					3	28	"	0.27
					4	28	"	0.28
	2	silty clay	53/18	smooth	1	14	undrained	0.65
					2	28	"	0.43
					3	28	"	0.37
					4	56	"	0.32
	3	silty clay	53/25	smooth	1	7	undrained	0.59
					2	14	"	0.33
					3	28	"	0.26
					4	28	"	0.37
	4	silty clay	59/29	sand-paper coated	1	14	undrained	0.48
					2	28	"	0.37
	5	silty clay	37/21	sand-paper coated	1	14	undrained	0.44
					2	28	"	0.40
					3	56	"	0.35
					4	112	"	0.28
					5	28	"	0.47
	6	silty clay	53/21	rough	1	14	undrained	0.38
					2	28	"	0.32
					3	56	"	0.27
					4	112	"	0.24
					5	28	"	0.41
Gulf of Mexico	1	silty clay	51/21	smooth	1	42	undrained	0.24
					2	84	"	0.24
					3	168	"	0.22
	2	silty clay	43/19	smooth	1	42	undrained	0.26
					2	84	"	0.25
					3	184	"	0.23
	3	silty clay	51/21	knurled	1	42	undrained	0.34
					2	84	"	0.34
					3	168	"	0.32
	4	silty clay	43/19	knurled	1	42	undrained	--
					2	84	"	0.30
					3	184	"	0.30

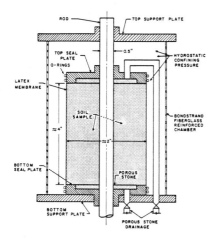

FIG. 1. ROD SHEAR DEVICE

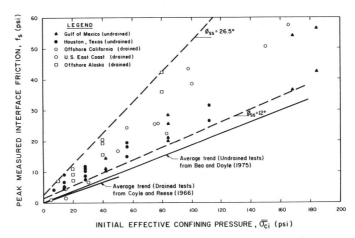

FIG. 2. PEAK MEASURED INTERFACE FRICTION VERSUS EFFECTIVE CONFINING PRESSURE

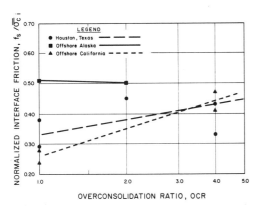

FIG. 3. NORMALIZED INTERFACE FRICTION VERSUS OVERCONSOLIDATION RATIO

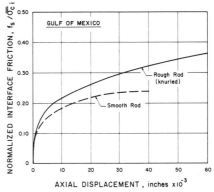

FIG. 4. EFFECT OF ROD ROUGHNESS ON STRESS-STRAIN CHARACTERISTICS

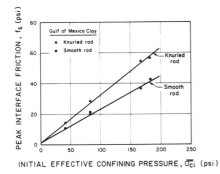

FIG. 5. PEAK INTERFACE FRICTION VERSUS INITIAL EFFECTIVE CONFINING PRESSURE

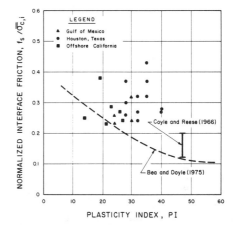

FIG. 6. NORMALIZED INTERFACE FRICTION VERSUS PLASTICITY INDEX

313

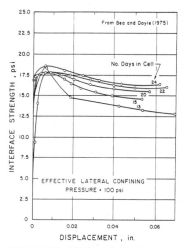

FIG. 7. INTERFACE STRENGTH VERSUS ROD DISPLACEMENT

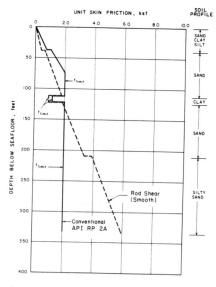

FIG. 8. CASE I UNIT SKIN FRICTION PROFILE (72-IN. OD PIPE PILE)

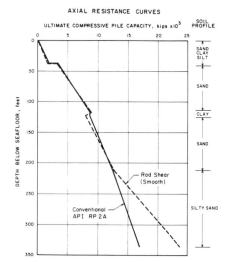

FIG. 9. CASE I AXIAL RESISTANCE CURVES (72-IN. OD PIPE PILE)

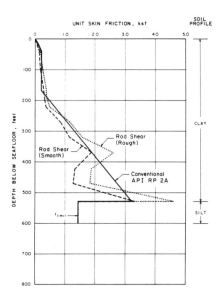

FIG. 10. CASE II UNIT SKIN FRICTION PROFILE (72-IN. OD PIPE PILE)

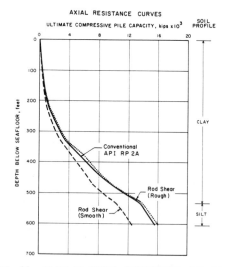

FIG. 11. CASE II AXIAL RESISTANCE CURVES (72-IN. OD PIPE PILES)

THE DESIGN CONCEPT AND DRIVING CAPACITY OF A 40 TON ABOVE/UNDER WATER HYDRAULIC HAMMER

G.D. ELLERY
BSP International Foundations Ltd.
United Kingdom

R.M. ELLIOTT
BSP International Foundations Ltd.
United Kingdom

E.L. JAMES
Fugro Limited
United Kingdom

SUMMARY

As the exploration and exploitation of the world's offshore oil and gas reserves continues in both shallow and now deeper water; we believe that there will be an increase in demand for the driving of piles underwater for applications such as the driving of conductor tubes, pinning of templates, installation of single point moorings, floating breakwaters, buoyant bridges and tension leg platforms, etc.

This has led to the development of a hammer system suitable for all the above applications along with many other piling applications and that can be used for both above and below water operation. It is envisaged that the hammer will operate in both shallow and deep water environments up to depths of 300 metres.

The design concept and modular construction gives a completely flexible system capable of being tailored to suit a particular piling requirement.

This paper is in two parts: Part A discusses the development of the Hydraulic Hammer system and the need to prove its capability, reliability and durability. A rigorous programme of tests and trials were undertaken culminating in the testing of the 40 ton (H.A) piling hammer. This was carried out by driving a 1.22m diameter steel tube pile of 25mm wall thickness driven both above and eventually below water. The pile is closed ended and will be driven to refusal. This is taking place within a circular cofferdam 6.2m diameter, approximately 4.0m deep, through soil conditions consisting mainly of chalk with occasional flint layers.

The pile has been extensively monitored and analyses of the results are taking place.

Part B of the paper describes independent measurements made on the hammer and the upper portion of the pile, together with the results of associated analyses.

The instrumentation includes stress and acceleration measurements on the pile monitored, during above water driving. The results have been examined with particular reference to:-

1. The performance of the hammer under various modes of operation.

2. Supplementing the scant information currently available on the driving of Large Diameter Piles into chalk.

The hammer performance has been assessed by examination of stress time spectra and evaluation of total energy imparted to the pile. The results of wave equation and impedence type analyses have been used in conjunction with the measurements to establish appropriate values for such drivability parameters as quake and damping for chalk.

Introduction   PART A

From the early 1970's it was evident that the exploration and exploitation of oil and gas reserves was going to enter deeper and deeper water. It was, therefore, clear that an underwater piling hammer would help with the installation of driven piles into the sea bed, etc. This concept led to the design of a range of hydraulic piling hammers for use both above and below water to considerable depths. It was decided to develop a hammer on an interchangeable modular system. The major component of the total system being the hydraulic actuator which provides the lifting mechanism for the ram weight. The actuators are of various capacities, along with varying sizes of ram weights and guide cages for numerous piling applications.

The Hydraulic Actuator

The hydraulic actuator is shown in Fig. 1 and basically provides a cylinder with an extendable piston rod which lifts the various ram weights. Either side of the cylinder is the low pressure and high pressure accumulators, thus the piston is extended in the cylinder when hydraulic fluid flows under high pressure from the power source and the high pressure accumulator through ports into the cylinder. Towards the end of the stroke the fluid from the high pressure port is shut off and the velocity of the piston thus slows to an equilibrium position and then begins the downward stroke. When this occurs, oil from the cylinder is transferred from the full bore side of the cylinder to the other side of the piston head and also into the low pressure accumulator and return line to the power source and fluid pressure is again increased in the high pressure accumulator ready for the next stroke.

Tests were first carried out on a 5 ton capacity actuator to determine its efficiency by varying flow rates, accumulator precharge and back pressures. The actuator was used to lift a drop weight of approximately 2.5 tons and this was placed on a set of conventional piling leaders to drive a 0.457m diameter steel tube pile. In all, eighty tests were recorded and the actuator and hammer efficiency under free fall calculated at 89%. From these tests it was clearly evident that there would be no major problems involved in scaling up the 5 ton actuator to the 10 and 20 ton capacities envisaged.

Fig. 1

Hydraulic   Actuator

Efficiency Tests

A series of tests were undertaken to compare the efficiency of the hammer in water with that in air, and determine the best shape possible. Firstly, theoretical calculations were undertaken. An assessment of a piling hammer operating in air is very simple, the energy capacity available from the falling ram weight at impact being generally the criteria used and can be calculated by simple laws of motion. Experience has shown that frictional loss is minimal. However, in water other factors are involved and the initial series of tests were to primarily establish these factors, relevant to assessing the kinetic energy of a drop weight falling through water, which will allow the actual energy capacity of a hammer system to be predicted and also enable the design of hammer to be chosen that will minimise the effect of these factors.

The theoretical approach has been concentrated on an idealised cylinder which is easier to evaluate. The variables under consideration were the drag coefficient, the physical proportions of the falling weight, the mass of water entrained with the falling weight together with external friction effects.

All these various factors were considered and finally by using a simple numerical technique whereby the motion of the body is advanced in increments, and considering a particular free-falling, unguided weight in water, the formula for the equation of equilibrium and motion becomes

$$(0.872 \times \text{MASS} - \text{FT}) g - 3.11 \times C_d \times \left(\frac{dy}{d\theta}\right)^2 = 1.128 \times \text{MASS} \times \left(\frac{d^2y}{d\theta^2}\right)$$

Where FT = External Friction Forces
C_d = Drag Coefficient
g = Gravity

Substitution of the values into the above equation then enables us to obtain a theoretical guide to the efficiencies of a falling weight through air or water.

To try and prove the above, model tests were undertaken with various shaped unguided weights which were allowed to fall from rest within an 800 gallon rectangular tank. The displacement of the falling weights and duration of fall from rest were monitored for three different shapes.

    a) Cylindrical
    b) Semi Elipsoid
    c) Stepped

From the tests the comparison of the performance of differing shaped weights of the same mass and sectional area falling through water are summarised in Fig. 2. and indicate the difference in performance to be quite insignificant. This infers that the drag coefficient (Cd) itself does not play a major part in the performance of the falling weight. A further assumption was made that the falling body carries with it a volume of water no greater than the volume of the body.

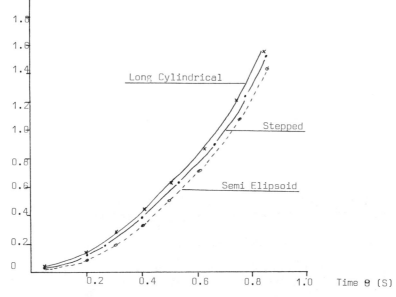

Fig. 2.

Comparison of Different Shaped Weights Falling Unguided in Water.

From all the results the following can be ascertained:-

a) The value of the external friction force (Ft) is evidently clearly contributory to the major deviance in the results expected, the solution worsening as the mass of the falling weight was reduced. This problem of the frictional forces resulted from the test arrangement used for these particular experiments. In the case of a full size piling hammer, neither the frictional forces acting on the falling weight, nor the forces retarding the weight, caused by back pressure in the hydraulic cylinders, should be ignored. The effect of the latter can be minimised by the choice of suitable hydraulic return hose, whereby fluid friction loss in the hose can be kept to a minimum. As the size and weights increase the performance should accurately reflect that which could be anticipated.

b) The value of drag coefficient (Cd) was found to have minimal effects on the shapes chosen.

c) It is noted that for the value of the effective mass (Me) no improvement resulted in the predicted performance if one considers that less water was entrained with the falling weight.

d) The relationship of length to diameter is critical and altering proportions caused performance variation.

In conclusion, whilst the actual results of the series of tests have been somewhat distorted by the scale of the models used, we have been able to recognise the major factors likely to effect performance of a weight falling in water. The next series of tests involve a full scale trial and should be able to confirm the conclusions without inherent distortions.

Testing of 9 Ton Drop Weight

As before in the model tests the procedure is to monitor the fall of the drop weight from rest with respect to a time base. The measurements of the slope of the curve of displacement against time provide us with the velocity of the falling weight of any particular point.

The tests were conducted with a 9 ton hammer operating in air and in water and subsequently repeated in order to check on the consistency of the results.

The tests were carried out within a water filled cofferdam. The cofferdam was formed from 50 sections of Larssen 2 sheet piles, each 10 metres long, pitched inside a template to form a circle. At the centre a 762mm diameter, 50 metre long, tubular pile, had already been driven, the base of a cofferdam was excavated to a depth of 1.5 metres below ground level.

As anticipated from early experiments, the results obtained in the case of this full size test were quite encouraging when compared with those anticipated by the theoretical analysis. This is mainly accounted for by the frictional losses being much smaller in proportion to the size of the falling mass than in the earlier model tests. At the typical rated stroke of 1.37 metres the following performance data is extracted.

	VELOCITY M/S	
	AIR	WATER
Theoretical	5.18	4.42
Experimental	5.16	4.30
% Error	1%	2.8%

The reduced impact energy is given by:-

$\dfrac{4.3^2}{5.16^2}$ which equals 69% for the same mass of falling weight.

If the mass of falling weight is increased to utilise the same available energy from the hydraulic cylinders then increased mass equals $\dfrac{7.83}{6.89}$ which equals 1.146 x Mass.

The reduced impact energy becomes 69% x 1.146 and is, therefore, 79% efficient below water.

The results obtained indicate that in general the shorter cylindrical shaped drop weights give suitable efficiencies both above and below water.

Investigation Of The Use of Water As a Drive Cushion

From the previous experiments, we can be reasonably certain of the velocity of a body falling through water. This enables us to assess the kinetic energy of that body at a point where its motion is arrested by impacting on a face. At impact, particularly in water, we are unsure as to the efficiency with which the kinetic energy is transmitted through to the pile in the form of impact energy. It is, however, desirable to afford some protection to both the falling drop weight and the top of the pile from excessive impact forces.

Conventionally in air this is done by transmitting the impact forces from the hammer to the pile by a resilient material, such as timber or coiled wire rope. Alternatively in sophisticated hammers the drop weight impacts onto the elements supported by hydraulic oil under gas pressure.

When a hammer operates submerged we suspect the efficiency by which the kinetic energy of the hammer is dissipated into the pile is impared by the rate in which the water contained between the closing impact faces can be displaced radially to the surrounding water.

Our next set of experiments were set to assess the degree to which this water film influences the build up of the impact forces in the pile and to set out to optimise this, by varying the geometry of the impacting faces, to render the need for a resilient material to protect the hammer and pile from excessive impact forces unnecessary.

As with the earlier test to ascertain the best shape of the ram weight, we firstly examined the behaviour of small scale models and extended the experience gained to full scale tests involving the 9 ton hammer, impacting on the pile already described. The model tests were aimed at observing the effect of three variables on the impact stresses by a weight falling onto a surface whilst underwater:-

a) The effect of the impact velocity by varying the height from which the weight has fallen.

b) Variations in the geometry of the falling weight.

c) Variations in the geometry of the impacting surfaces.

Some of the results obtained can be seen in Fig. 3.

Conclusions

The results of the tests show clearly that the standard drive cap would be unsuitable for use underwater where water would be present between the impacting faces, the degree of attenuation of the peak forces being unacceptable. If one considers the results of the tests conducted with the special form drive cap, then in terms of maximising the peak stress, then design tested, provides us with the best results we could expect, but at the same time the rate of increase of stress on the pile has been effectively reduced whereby the risk of damage to the pile head and the chassis of the hammer is minimal.

Generally, it is suggested that the hammer will operate satisfactorily underwater without the need for an air bell providing the marginal reduction in impact performance of 20/25% is acceptable. If this is not the case then the stroke of the hammer could be increased to increase the impact velocity and/or mass of the drop weight increased to increase the kinetic energy at impact, and also utilising more fully the available horsepower.

The benefit of using the water film as a cushion, however, must not be forgotten, particularly when compared with the complexity of the construction of other hydraulic underwater hammers, or the tediousness of replacing timber cushions, used in conventional above water steam hammers driving through the follower system.

All the tests proved that with the special drive cap most piles could be effectively driven underwater. As all the previous tests and experiments proved successful and effective it was decided to continue with the development programme.

Underwater Trials

After completion of the land based trials it was decided to take the 10 ton ram weight hammer offshore to drive a pile on to the sea bed. The trials took place at Loch Linnhe, Scotland, in approximately 120 metres of water.

The pile used in this instance was for consideration of pontoon stability, only 10 metre overall length x 762mm diameter, but the duration of driving was intended to be extended by a flange 2 metres from the top of the pile, being arrested by the pile guide, which has a much greater area in contact with the bed of the lake.

The hammer size used was the 10 ton version which was fully instrumented. The depth of water in which the pile and hammer were placed was measured by using the pneumofathometer method.

The pile frame, winches, power pack and hammer were all stored upon a specially designed barge, as can be seen in Fig. 4. The pile and pile template were lowered down guide wires to the sea bed and then the hammer was guided down the same wires and interlocked with the top of the pile.

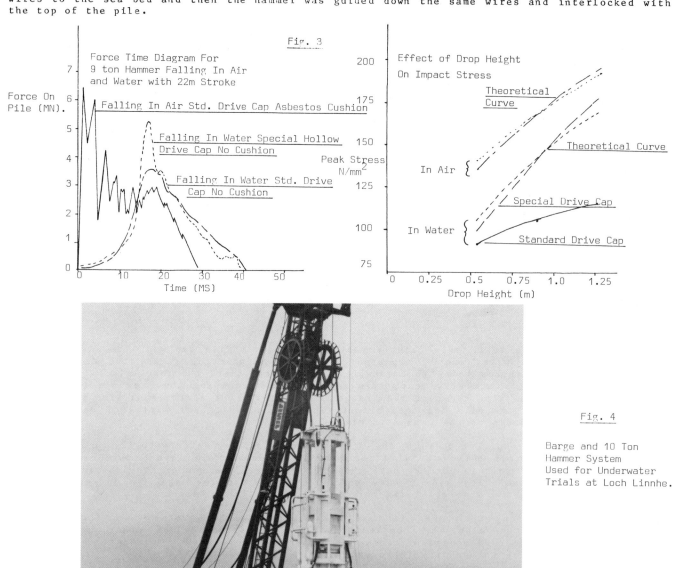

Fig. 4

Barge and 10 Ton Hammer System Used for Underwater Trials at Loch Linnhe.

A camera observed the approach of the hammer to the pile and the picture was simultaneously relayed to a monitor on the barge so that the operator could fully negotiate the mating of the pile in the cone.

The drive cap resting on the pile was indicated by the first proximity switch on the control panel lighting up.

The whole system is controlled by a remote electric hydraulic system which allows for both manual and single blow operation and for an automatic sequence. This enables the stroke to be varied from 0.3 metres to 1.5 metres during operation and allows for the blow count to be varied to suit individual circumstances.

The height of stroke and blow count will be indicated at the surface at all times and the velocity at impact was recorded on the oscillascope and provision is made to indicate the penetration per number of blows with the manometer system.

Although the verticality of the pile at Loch Linnhe was checked by a signal from an inclinometer which was situated at the top of the hammer, the extra weight on the pile when the hammer interlocked caused the pile to be pushed 4.6 metres into the bed of the lake. The pile was then given several blows with the hammer and was quickly driven to the remaining pile test length, until the flange met the pile guide.

Hard driving conditions could now be attained. However, the operations had to be curtailed as the pile and pile guide tilted to such an angle that driving was no longer possible. The hammer system spent a total of 14 hours in sea water and no ingress of water was noted within the hydraulic actuators at the end of this period.

Although the tests were unable to continue, enough information had been gained to indicate that the total package hammer and in particular the hydraulic cylinders would operate sufficiently well in a reasonable depth of water, and it was envisaged that no problems would occur at even greater depths.

From the successful underwater trials, it was decided to link up 20 ton actuators with a 40 ton ram weight and guides and also manufacture a drive cap and cushion, etc.

40 Ton Ram Weight Hammer

At our original test site within the cofferdam it was decided to ascertain the overall capability, reliability and durability of the 40 ton hammer system by driving a 1.22m closed ended steel tube pile into material mainly consisting of soft to hard chalk. The sections of pile driven are between 4.0m and 5.0m long, and are extended by welded joints.

The hammer is shown in Fig. 5 and the driving record Fig. 6. At present the pile has reached a depth of 45.36m and as yet, refusal conditions have not been reached, but this is expected shortly. The hammer has performed approximately 14000 blows on the pile at various stroke heights. Generally at the commencement of each driving period we have attempted to use a longer stroke, hence reaching high energy at the start of the drive. This has caused a "set down" effect to appear on the driving record. Monitoring of the pile has continued throughout the driving period, and some of these results are discussed in Part B of this paper. At present over 8 hours driving has taken place and the 2 No. 20 ton actuators being used have presented no problems. They were removed and examined after approximately 8000 blows and showed no sign of wear. After further driving of the pile, using the full stroke of 1.37m at refusal conditions, we will then be aware of the hammer capabilities above water. It is envisaged that after completion of the above water trials, the cofferdam will be flooded to try to attain underwater driving conditions. Further monitoring of the pile and hammer will then take place.

PART B

The installation, by driving of a 1.22m diameter pile to a depth of about 50m completely in chalk presents a unique opportunity to examine the response of chalk to driving action. Piles of such size are rarely driven onshore although they are common place on offshore structures.

In order to examine independently the performance of the BSP 40 ton hammer whilst at the same time gaining valuable data on piling in chalk, Fugro Limited offered to make measurements on the test pile. The measurements were secondary to BSP International Foundations main aim of installing a pile which would provide a realistic test bed for their hammers. Measurements were only made at one depth of penetration but further measurements are planned.

Fig. 5

40 Ton Hydraulic Hammer Driving 1.22m Tube Pile Within Cofferdam

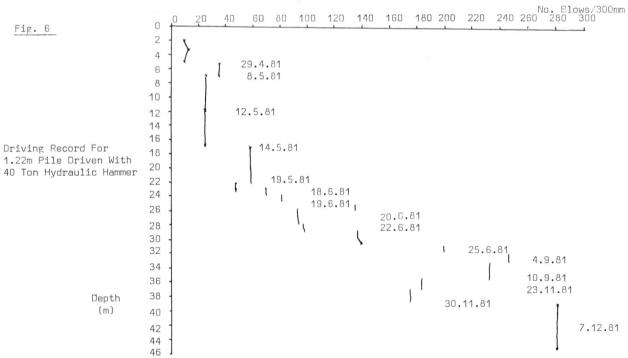

Fig. 6

Driving Record For 1.22m Pile Driven With 40 Ton Hydraulic Hammer

Measuring Techniques

Strain and acceleration measurements were made on the outer surface of the pile at a level three pile diameters below the driven head. Use was made of bolt-on transducers obtained from Pile Dynamics Inc. These sensors were mounted in pairs diagonally opposite each other. In view of the slight rake of the pile the appropriate diagonal to minimize the effects of bending was selected.

The sensors were connected to a Fugro designed signal conditions which, in addition to providing the excitation voltage, permits signal amplication averaging of signals and integration. The amplified signals were fed directly into a seven track FM recorder. Two channels of recorded signal were simultaneously examined on a digital oscilloscope.

Test Sequence

At the time of making the measurements the pile had penetrated 31m below the ground level within the caisson. At this stage the sensors were about 2.5m above ground level, the total stick up being about 6m.

The 40 ton hammer was fitted with a 200mm elm cushion on end grain. Driving was carried out at various set stroke lengths during which time the sensors were continuously monitored. Due to the nature of the test programme no measurements were possible immediately before and after a significant break in driving.

Analyses

Typical records for hammer strokes of 0.68m, 0.84m and 1.02m have been examined in detail. The analogue records were digitized and stored as computer data files. Using an approach akin to that described by Dolwin and Poskitt (Ref.1) an attempt was made to establish an acceptable mathematical model of the driving. In their paper Dolwin and Poskitt describe similar independent measurements made on the same pile and the same hammer, but using a Bongossi cushion.

The elm cushion fitted during the tests described here proved considerably softer than Bongossi which influenced both the magnitude of the initial peak force in the pile and the rise time to this peak. As a consequence the measured peak acceleration and stress were influenced by the arrival of reflections from the upper levels of the chalk. Any model set up to describe this part of the blow had therefore to include this soil influence.

Preliminary analyses indicated that the soil damping and quake parameters normally used by Fugro were inappropriate for this site. Use was made of the side and end quake and damping values established by Dolwin and Poskitt. The period of interest for assessing hammer performance (the first 6 milliseconds) is relatively insensitive to these values.

Wave equation anaylses were made for a wide range of cushion stiffness, cushion restitution hammer efficiency and soil resistance distributions. A manual optimization process was used to obtain the best fit between predicted and measured stress and acceleration profiles for the 1.02m stroke blow. The values established for these parameters for the fit shown on Figs. 7 and 8 were:

    Cushion Stiffness   1250 kN/mm
    Cushion Restitution   0.3
    Hammer Efficiency   65%

Soil resistance - triangular distribution of side friction totalling 3000 kN with additional 4000 kN applied to the pile tip.

It should be noted that the efficiency quoted represents the fraction of the total energy available at ram impact from a free fall through the full stroke of 1.37m. Since the stroke for this blow was constrained to be 1.02m the "effective" hammer efficiency amounts to 86.7%.

The parameters established above were then used for a range of hammer efficiency values to obtain best fit solutions for the measurements made for the 0.68m and 0.84m stroke blows. The 'best' solutions were obtained at overall hammer efficiency values of 54% and 44% respectively which represent 'effective' efficiency values of 87.3% and 87.0%.

The measurements thus indicate both consistant and highly efficient hammer performance over the range of stroke lengths monitored.

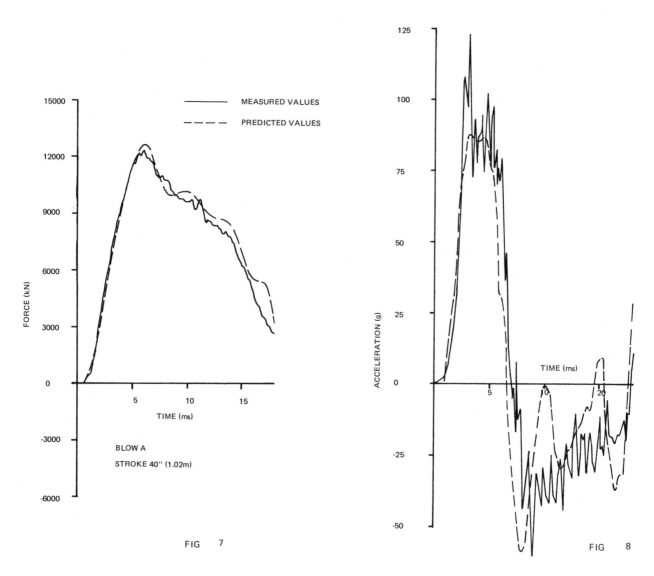

FIG 7

FIG 8

References

1. DOLWIN J. and POSKITT T.J. 1982. <u>An Optimisation Method for Pile Driving Anaylsis.</u>

   Paper to be presented at the 2nd International Conference on Numerical Methods in Offshore Piling. Austin, Texas. April 1982.

Acknowledgements

The authors wish to thank the Directors of BSP International Foundations Limited for permission to publish the information gained whilst carrying out the development of the hydraulic hammer system and to thank Fugro Limited for making funds available for performing the tests described in Part B. They also wish to acknowledge the advice given by Professor T.J. Poskitt and J. Dolwin of Queen Mary College, London, and the use of parameters established by them.

PARAMETER MEASUREMENTS REQUIRED TO

CALCULATE THE AXIAL CAPACITY OF PILES

J. FOURNIER	P.Y. HICHER	J.P. MIZIKOS	F. ROPERS
Ecole Centrale de Paris, France	Ecole Centrale de Paris, France	Elf Aquitaine Oil Company, France	Ecole Centrale de Paris, France

SUMMARY

The calculation of the axial capacity of driven piles in clays can be considerably improved based on the two following hypotheses :

- detailing of the effective stresses during the history of the pile.

- detailing of the relationship between the static axial capacity and the dynamic axial capacity.

The major difficulty encountered for devising calculation is essentially the <u>lack of reliable measurements</u> of the required parameters (mainly axial capacity, lateral stress, pore pressure and damping factor) as these parameters are not easily accessible in situ. We therefore measured this set of parameters by a model driving test in the laboratory and by a triaxial test with shocks. The evolution of effective stresses was measured at each step of the pile history and described based on the critical state concept.

The results obtained enable us to :

- understand and predict the evolution of effective stresses during the history of a pile driven into clay.

- to show that the static skin friction is reduced by roughly half during driving. This reduction is reversible in the case of clays with OCR of less than 3, and non-reversible in the case of highly overconsolidated clays.

- to show that the variations in skin friction during driving are solely a function of the variation of effective stress.

The calculation method devised in the laboratory was subsequently used successfully on actual cases in situ. The work presented here is a step forward in the reduction of costs of piles driven into the seabed by reducing the penetration depth of piles.

1 - INTRODUCTION

A general tendency is the increasing use of very long piles which are obviously stressed for the greater part under skin friction when installed in clay. As suggested by MEYERHOFF in his 1975 Terzaghi lecture [1], in order to obtain a better determination of the bearing capacity, it is necessary to develop calculations of effective stresses rather than total stresses.

The difficulties in obtaining in situ measurements which are essential for drawing up effective stress calculation formulas led to the collaboration between the Soil Mechanics Laboratory of the Ecole Centrale de Paris and the Elf Aquitaine Oil Company in a research program with the aim of improving our knowledge concerning skin friction along piles driven into clays.

The parameters for effective stress analysis were measured by simulating pile driving in the laboratory and completed by a triaxial test with shocks.

2 - PRESENTATION OF MODEL DRIVING TEST IN THE LABORATORY [2]

The clay sample was placed in a remoulded state in a metallic cylinder 20 cm in diameter and 1 m high (fig.1). The sample was then consolidated around the model tube by applying given vertical and lateral stresses. Once consolidated, the dimensions of the sample were approximately 0.7 m high and 0.17 m in diameter. The tubes used were equipped with pore pressure sensors or strain gauges. The displacements of the tube were measured at the base of the tube. Static loading tests were performed by loading the base of the tube. Driving was carried out using a free falling hollow ram, hitting the top of the tube. The soil used was mainly a kaolinitic clay (WL = 70%, WP = 40%).

On each model test, the following measurements were made :

- the displacement of the tube at each blow of the hammer, measured in $\frac{1}{10}$ mm.

- the static friction (fs), when required.

- either the evolution of the pore pressure in contact with the tube.

- or the dynamic response caused by percussion of the ram, at two levels, at the top of the tube and at the base of the tube in order to test the dynamic reaction of the soil.

- the water content of different parts of the sample at the end of the test. These measurements enable verification of the initial stress of the soil, show the final state of the soil along the length of the tube and enable measurements of the distance along which driving disturbs the soil.

The results obtained are shown by various curves :

- the characteristic curve (fig 2 a with 2 b)

- the evolution of pore pressure (fig 3 a)

- the evolution of skin friction (fig 3 b)

- the diagrams of water content (fig 4)

3 - ANALYSIS OF RESULTS OBTAINED

3.1 - Measurements of energy consumption during pile driving displacement in the model test.

For five model driving tests in clays, we plotted the relationship linking the maximum stress variation and the maximum velocity of dynamic response caused by the percussion of the ram on the top of the tube.

This relationship is $\Delta\sigma_{max} = \sigma_o + Z_c u_T$ (fig 2)

Z_c : characteristic impedance of system

$$u_T = \frac{\sigma_{max}}{\rho C_o}$$

σ_{max} : maximum stress measured at the top of the tube
ρ : density of tube material
C_o : propagation velocity in tube
σ_o : ordinate at origin of the relationship established for u_T varying between 0.5 m/s and 3 m/s

For the five tests, σ_o is equal to $q_\ell = \frac{Q_s}{S_p}$.

Q_s : total skin resistance measured by a static loading test carried out at the end of the driving sequence.
S_p : annular section

The above-mentioned relationship raises three typical problems encountered by the engineer in driving interpretation :

- how can we calculate q_ℓ, either from soil surveys or from measurements made during driving ?

- what is the significance of the slope Z_c ? Is it solely related to the mechanical properties of soils loaded by shocks ?

- what is the relationship between the adhesion coefficient α and the empirical formulas proposed in literature linking dynamic resistance to static resistance (Smith and Gobble, in particular) ?

In this article, we only treat the analysis of the static problem. The analysis of the dynamic interaction between pile and soil is being studied and the preliminary results have been published independently with comparison to real drivings of open-ended off-shore piles [3, 4].

3.2 - Validity of standard calculation formulas of skin friction in model test.

The knowledge of the state of the soil and its mechanical characteristics, the systematic measurements of skin friction by static loading tests and of pore pressures, enable the verification, during a model test, of the validity of static formulas currently used to calculate skin friction during and after driving :

$f_s = \alpha C_u$ (fig 5, [5])

$f_s = \sigma'_H \, tg\, \phi'$ (COULOMB)

$f_s = \frac{M}{2} P_f \cos\phi'_{ss}$ (ESRIG and KIRBY [6])

The angle ϕ' used in the formulas is the angle measured by undrained triaxial tests.

We present the results obtained on four model tests carried out on a normally consolidated clay (Table 1).

When skin friction occurs only on the outside wall of the pile, which is the case with our model tests, the value measured fs is never equal to the normal stress applied (C_u or $\sigma_H' \, tg\, \phi'$ or $\frac{M}{2} P_f \cos\phi'_{ss}$). These three values being similar, the ratio of proportionality measured $\alpha = \frac{f_s}{C_u} \simeq \frac{f_s}{\sigma_H' \, tg\, \phi'} \simeq \frac{f_s}{M/2 \, P_f \cos\phi'_{ss}}$ is close to that established by CAQUOT and KERISEL [5] and presented in figure 5.

This ratio is much lower than those determined by other authors such as MENARD, TOMLINSON and A.P.I. (fig 5).

In the range of consolidation stresses studied, α decreases with the normal initial stress.

This result raises two problems not solved by our tests :

a) when the material is not disturbed (state 1, fig 3 b, Table 1), what is the physical significance of the coefficient α ?

b) in the case of open-ended piles, what is the value of skin friction on the inside wall of the pile ?

During the first driving sequence, the skin friction (measured by a static loading test immediately at the end of each driving sequence) decreases by roughly half compared to its initial static value (its approximate value is therefore $\alpha C_u/2$). The only explanation for this variation, like those which follow, appears to be variations in the effective stresses (Table 1).

Taking the mean stress for calculations makes the ESRIG and KIRBY formula dependent on the Ko, which is not confirmed by our tests (Table 1).

It is more complex to perform this study on overconsolidated clays as the pore pressure measurements are not very accurate because of the desaturation of the sensors on swelling of the soil.

3.3 - <u>Reminder of main results obtained using model test.</u>

a) <u>Normally consolidated soil.</u>

The evolution of skin friction measured during a model test is summarized in figure 3 b. We can see (figs. 3, states 1.2.3.) that driving temporarily deteriorates skin friction (measured by a static test performed immediately after each driving sequence). The value of skin friction is then determined by the variations in effective stresses. After several driving and drainage sequences, skin friction returns to a value close to the initial static value (figs. 3, state 4). In this state, the soil has a void ratio much smaller than that of the initial state (figs. 3, state 1), and a new driving sequence will no longer have any effect on the pore pressure.

When the soil is left at rest after driving, in state 4, the skin friction (always measured by static loading tests) increases slowly with time to reach two to three times its initial value (state 5, fig 6) after three months at rest. This evolution with time takes place without any variation in the effective stress (thixotropy).

b) <u>Overconsolidated soils (fig 7)</u>

If kaolinite is used, when the O.C.R. is greater than 3, driving reduces the skin friction definitively to approximately 50% of its initial value.

For overconsolidated soils, the model driving test does not permit the performance of an effective stress analysis, due to inaccuracy of pore pressure measurements.

4 - DEVELOPMENT OF CALCULATION OF EFFECTIVE STRESSES

Our work consisted of continuing and specifying in greater detail, by mean of measurements, the works of ESRIG and KIRBY [6].

For this purpose, we devised a triaxial test where shocks are given on the top of the sample by means of a piston. In the same way as for driving, a series of shocks is given until variation in pore pressure is no longer recorded after a new shock (from state 1 to state 2, fig 8). The drainage circuit is then opened in state 2, and it is thus possible to measure the variation in the state of the soil during reconsolidation. When reconsolidation is finished (state 3), the drainage circuit is closed, and a new series of shocks is begun. The test is continued until a further driving sequence no longer causes variation in the pore pressure (state 4).

4.1 - Evolution of effective stresses during shocks.

a) Normally consolidated soils

The triaxial test with shocks and the model driving test give comparable increases in the pore pressure : whatever the initial state of the soil, the locus of maximum reduction in mean effective stresses obtained during the first driving sequence is, in the plane (W, log p'), a straight line with a slope Cc, thus parallel to the virgin consolidation line (fig 9).

W : water content
p' : mean effective stress
Cc : compression index

The reduction in the mean effective stress recorded in model and shock tests is greater than that measured when the soil is brought to its critical state by an undrained static triaxial test. This result is similar to those presented by ESRIG and KIRBY, but we cannot attribute this phenomenon to an increase in the total stress during driving, as in our tests boundary conditions are fixed. It is therefore necessary to look for the explanation of this phenomenon in a poor determination of the critical state based on standard triaxial tests.

Figure 10 a shows the evolution of the state of the soil during successive shocks-drainage sequences. We can see that the path followed during reconsolidations really is parallel to the paths of the swelling of the soil (Cg), whereas each driving sequence causes a decrease in the mean effective stress, leading to the straight line defined in the preceding paragraph (fig 9).

Under these conditions, the minimum final state reached by the soil around the pile due to driving (state 4, fig 10 a) can be determined graphically in the plane (W, log p'), for a given mean consolidation stress p'o, as being the water content at the limit state after driving.

b) Overconsolidated soils

The measurements performed during triaxial tests with shocks showed two different types of behaviour :

. as long as the state of the overconsolidated soil is situated on the right side of the limit state path (wet side), the pore pressures generated during driving are positive (fig 11 a), but are smaller than for a normally consolidated soil with the same water content.

. as soon as the state of the soil is situated on the left side of the limit state path (dry side), the pore pressures generated during driving are negative and tend to bring back the soil to its limit state (fig 12 a).

4.2 - Evolution of the shear strength during the triaxial test with shocks.

Similarly to the model test, we followed the evolution of the static shear strength of the clay, before the first shock sequence, and at the end of the shock sequence, by static shearing on samples in states 1 and 4 (fig 10 a). The results of these tests are presented in the plane (p',q).

p' : mean effective stress $\quad p' = \dfrac{\sigma'_1 + \sigma'_2 + \sigma'_3}{3}$

q : stress deviator $\quad q = \sigma'_1 - \sigma'_3$

a) **Normally consolidated and slightly overconsolidated clays.**

The clays presenting a decrease in the mean effective stress during driving sequences, i.e. situated on the wet side of the limit state path, show an increase in shear strength which can reach 70% in the case of normally consolidated clay (fig 10 b and 11 b).

b) **Highly overconsolidated materials.**

The clays presenting an increase in the mean effective stress during shocks, i.e. situated on the dry side of the limit state path, show a small loss of shear strength (fig 12b).

4.3 - Comments

- The variations in the state of the clay during driving are correctly predicted by both the model test and the triaxial test with shocks.

- The evolution of skin friction in the model test is in agreement with the variations of effective stresses, but not with the variations in water content.

- In the triaxial test, the clays subjected to shocks all present the behaviour of overconsolidated clays. We do not have enough tests to make a correct analysis of the evolution of shear strength.

5 - CONCLUSIONS

The numerous measurements carried out during the model driving test in the laboratory have enabled us :

- 1) to measure the evolution of skin friction during driving of piles in normally consolidated and overconsolidated clays.

- 2) to link the variations of skin friction during driving with the variations of effective stresses.

- 3) to predict variations in effective stresses by triaxial tests with shocks.

- 4) to show that, during driving, the skin friction decreases by roughly half for all clays tested. This evolution concords with the Cambridge theory (Camclay), which predicts a ratio of p'/p'o constant and equal to about 0.5 [8, 9], as it depends only on the ratio $\dfrac{C_g}{C_c}$ which is roughly constant.

  . For normally consolidated clays, this decrease is reversible : skin friction rapidly returns to a value close to the initial value at the end of driving. After several months of rest, it can reach much higher values (two to three times its initial value in our tests).
  . For overconsolidated clays, in our tests, with an O.C.R. higher than 3, deterioration of skin friction is stable : we have never observed any strength regain during the course of time.

- 5) to affirm that, after driving, on putting the pile into service, the variations in effective stresses will be smaller than usually predicted.

The results presented above were found using a remoulded kaolinitic clay, reconsolidated in the laboratory. It is therefore not possible to generalize without a more detailed study. However, triaxial tests with shocks have been performed on both normally consolidated and overconsolidated natural clayey soils. In figures 13 and 14, we can see that these materials obey the laws previously established in chapter 4.

These results have already been used for interpretation of in situ driving curves. [3, 4]. The synthesis of these operations (laboratory measurements, in situ verifications) leads to proposing the following modifications of skin friction in the case of large diameter open-ended piles, i.e. bigger than 0.5 m.

- For normally consolidated or overconsolidated clays with OCR of less than 3, we suggest :

$$fs = 2\,(\alpha C_u)$$

$$fs = 2\,(\alpha \sigma'_H \, tg\phi')$$

$$fs = 2\,(\alpha \frac{3K_o}{1+2K_o} \frac{M_1}{2} \, p'_f \cos\phi')$$

where
- α is the adhesion coefficient proposed by CAQUOT and KERISEL
- ϕ' is the internal angle of friction determined by undrained triaxial tests. It is possible to use $\delta' = \phi' - 5°$ to take into account the soil-pile friction.
- the coefficient 2 represents the friction on the inside wall of the pile. It needs to be confirmed by measurements but we consider it to be reasonable [10, 11].

- For overconsolidated soils, and for the lateral resistances during driving to be used in a pile driving calculation program (Batpil or Capwap type), we recommend taking : fs/2

6 - ACKNOWLEDGEMENTS

We wish to thank the SOCIETE NATIONALE ELF AQUITAINE (Production) for proposing and financing this study, and Professor J. BIAREZ who was kind enough to undertake its scientific direction.

REFERENCES

[1] MEYERHOFF, C.G., 1976, Bearing Capacity and Settlement of Pile Foundations, the eleventh Terzaghi Lecture, Journal of the Geotechnical Engineering Division, Vol 102, n° GT3, March 1976, pp. 197-228.

[2] FOURNIER, J., 1980, Le Frottement Latéral le long des Pieux battus dans les Argiles, Thesis of Docteur-Ingénieur, Ecole Centrale de Paris, Laboratoire de Mécanique des Sols.

[3] MIZIKOS, J.P., and FOURNIER, J., 1980, From Driving to Pile Foundation : a First Step for Clays, Seminar on the Application of Stress Wave Theory on Piles, Stockholm, 1980.

[4] MIZIKOS, J.P., and FOURNIER, J., 1982, Rheological and Characteristic Laws for Friction Resistance of Off-Shore Driven Piles, 2nd International Conference on Numerical Methods in Off-Shore Piling, Austin, 1982.

[5] LE TIRANT, P., 1976, Reconnaissance des Sols en Mer pour l'Implantation des Ouvrages Pétroliers, Paris, Edition Technip.

[6] ESRIG, M.E., and KIRBY, R.C., 1979, Advances in General Effective Stress Method for the Prediction of Axial Capacity for Driven Piles in Clay, O.T.C., 3406.

[7] TOMLINSON, M.J., 1977, Pile Design and Construction Practice, London, View Point Publications.

[8] SCHOFFIELD, A., and WROTH, P., 1968, Critical State Soil Mechanics, London, Mc Graw-Hill.

[9] ZERVOYANNIS, C., 1982, Synthèse Comparative des Propriétés Mécaniques des Sables et des Argiles, Thèse de Docteur-Ingénieur, Ecole Centrale de Paris, Laboratoire de Mécanique des Sols.

[10] HEEREMA, E.P., 1979, Pile Driving and Static Load Test on Piles in Stiff Clay, O.T.C., 3490.

[11] BURT, N.J., and TRENTER, N.A., 1981, Steel Pipe Piles in Silty Clay Soils at Belawan, Indonesia, Proceedings, X. ICSMFE, Stockholm, 1981, Volume 2., pp. 873-880.

TABLE 1 : SUMMARY OF DATA OF 4 SIMULATION TESTS

	Initial state of consolidation (kPa)	Measured f_s (kPa)	Calculated f_s (kPa)			Measured f_s / Calculated f_s			
			Esrig	Coulomb	αC_u	Esrig	Coulomb	α	
A	$\sigma_H = 50, \sigma_v = 80$	14	22.7	18	18	0.62	0.78	0.78	State 1 : the static load test is performed before the first driving sequence.
B	$\sigma_H = 120, \sigma_v = 100$	18.2	43.2	44.5	34	0.43	0.41	0.53	
C	$\sigma_H = 120, \sigma_v = 100$	20	48.9	53.7	34	0.40	0.39	0.59	
D	$\sigma_H = 120, \sigma_v = 200$	20.6	57.8	45	42	0.36	0.46	0.49	
A		8.2	16	11.1		0.51	0.74		State 2 : the static load test is performed immediately after the end of the first driving sequence.
B		11.9	30.1	32.5		0.40	0.37		
C		13.5	28.8	31.2		0.47	0.43		
D		11.6	40.8	28		0.28	0.41		
A		9.3	27	22.8	18	0.34	0.42	0.81	State 3 : the static load test is performed after the first dissipation of excess pore pressure.
B		20.3	50.2	52.1	34	0.40	0.39	0.60	
C		20.5	51.6	53.4	34	0.39	0.38	0.60	
D		18.1	66.4	53.4	42	0.27	0.34	0.43	
A		11.1	27	22.3	18	0.41	0.50	0.61	State 4 : the static load test is performed after the last driving sequence. There is no excess pore pressure.
B		18.4	51.2	52.9	34	0.31	0.35	0.54	
C		21.7	51.6	53.4	34	0.42	0.41	0.64	
D		21.6	66.4	53.4	42	0.33	0.40	0.51	

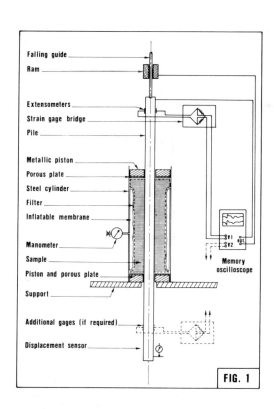

Figure 1 : Schema of model test equipment.

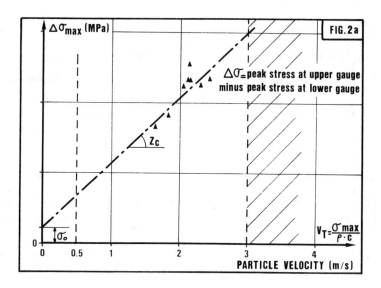

Figure 2 : Characteristic line (a) and example of recorded signals (b) ; notations are defined in text. Note the existence of when particle velocity is nul, representing half the static unit friction of clay at rest.

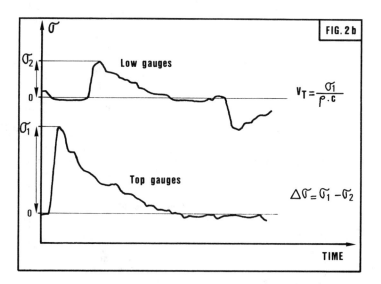

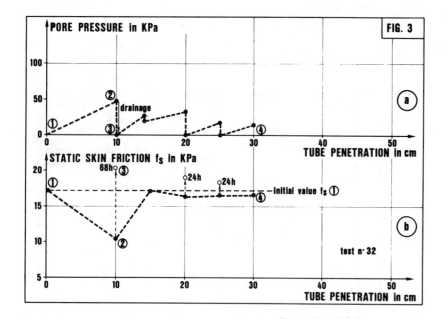

Figure 3 : Evolution of shaft adhesion and pore pressure during the pile history for a normally consolidated clay

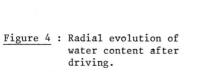

Figure 4 : Radial evolution of water content after driving.

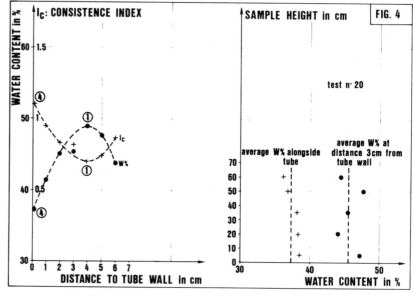

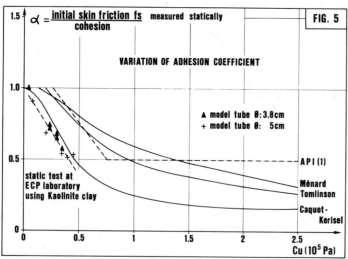

Figure 5 : Evolution of adhesion coefficient

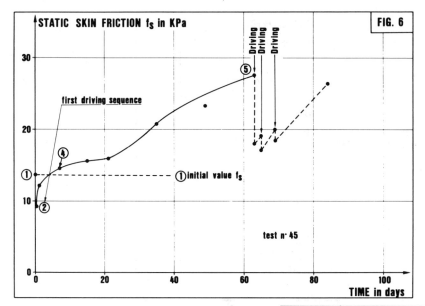

Figure 6 : Evolution of skin friction with time after driving for a normally consolidated clay.

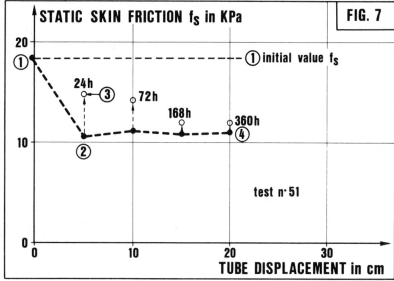

Figure 7 : Evolution of skin friction during the pile history for an overconsolidated clay.

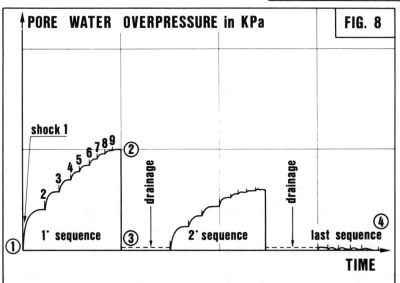

Figure 8 : Graph of pore pressure evolution versus time during a triaxial test with shocks.

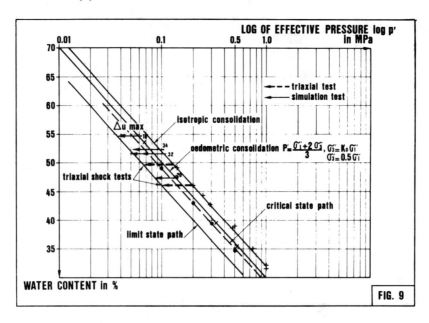

Figure 9 : Line of maximum excess pore pressures after the first driving sequence in model tests and triaxial tests

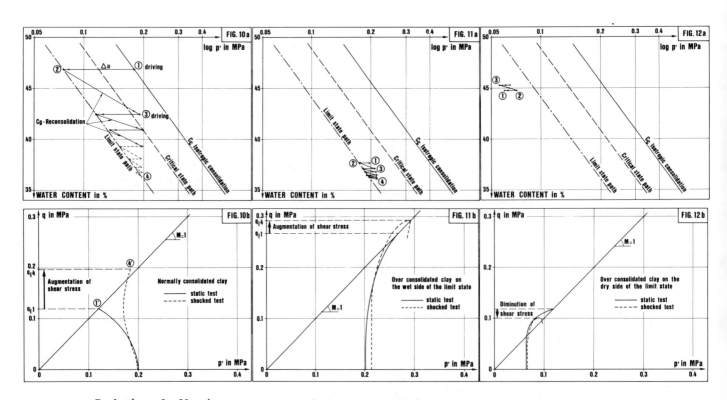

Evolution of effective mean stress and shear stress during a triaxial test with shocks.

Figure 10
For a normally consolidated clay.

Figure 11
For a slightly overconsolidated clay.

Figure 12
For a highly overconsolidated clay.

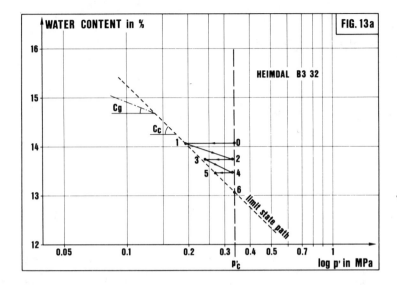

Figure 13 : Evolution of effective mean stress and shear stress during a triaxial test with shocks for an overconsolidated natural clay from the North Sea.

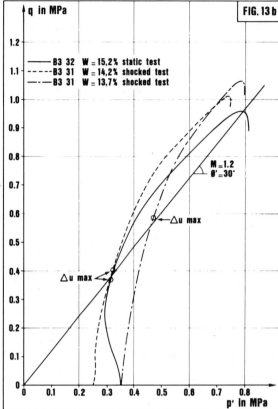

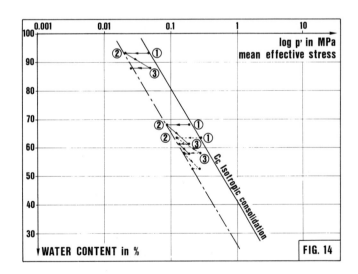

Figure 14 : Evolution of effective mean stress during a triaxial test with shocks for a normally consolidated natural clay from GULF of GUINEA.

NON LINEAR CONSOLIDATION ANALYSES
AROUND PILE SHAFTS

M. Kavvadas
Massachusetts Institute of Technology,
U.S.A.

M. M. Baligh
Massachusetts Institute of Technology,
U.S.A.

SUMMARY

This paper is concerned with the prediction of the effective stress changes during consolidation around piles, because of their importance in estimating the axial capacity of driven friction piles in clays. Linear analyses predict that, at the pile wall, the radial effective stress after consolidation is equal to the radial total stress immediately after installation. This value of the radial effective stress grossly overpredicts the axial pile capacity and makes non-linear consolidation analyses necessary.

The first part of the paper proposes a new effective stress soil model based on the Theory of Anisotropic Elasto-Plasticity and evaluates this model by comparing its predictions to the results of various laboratory tests on Boston Blue Clay. The final part of the paper uses the proposed model to estimate the pore pressure and effective stress changes during installation and subsequent consolidation around a long cylindrical pile. Pile installation is modelled via the Undrained Cavity Expansion and the Strain Path methods. Consolidation is analyzed using an incremental non-linear Finite Element scheme based on a Weighted Residuals formulation of Biot's theory of consolidation. It is shown that at the pile wall, the ratio of the radial effective stress after consolidation to the radial total stress immediately after installation is less than unity and increases with increasing overconsolidation ratio of the clay.

1. INTRODUCTION

Piles represent an important foundation type, especially in difficult soil conditions (e.g. soft clays) and hostile environments (e.g. offshore). The prediction of the ultimate axial capacity of a single pile is the first step in deep foundation design. Soil resistance is provided at the tip and along the shaft of the pile. For long piles in clays, the major portion of the soil resistance is derived from skin friction along the shaft, especially when no competent end bearing layer exists. Therefore, the limiting skin friction that can be provided by the soil is of primary importance in the design of piles. Currently, design procedures rely basically on empirical methods because of the complicated and very little understood mechanisms of soil deformation and stressing around piles. However, attempts to study these mechanisms are reported in the literature (e.g. Esrig et al, 1977, 1979; Randolph and Wroth, 1981). Such studies realize that the limiting skin friction depends on the magnitudes of the effective stresses at (and close to) the pile-soil interface during pile loading to failure and estimate these stresses by analysing the sequence of events preceding pile loading to failure, i.e., pile installation and subsequent soil consolidation. Pile installation (driving) causes intense disturbance in the soil, changes the in-situ effective stress state and induces significant excess pore pressure gradients in the vicinity of the pile. During subsequent consolidation, the excess pore pressures dissipate with time and the effective stresses around the pile increase, thus increasing its ultimate axial capacity. This effect can result in a five to ten-fold increase of the axial capacity if the pile is loaded a long time after installation (see e.g. Stermac et al, 1969). Therefore, studies of soil consolidation around piles are very important in estimating the ultimate axial capacity of driven piles.

Soil consolidation studies require the previous analysis of pile installation because the pore pressure and effective stress changes caused by pile installation represent the initial conditions for consolidation. Predictions of the effective stress changes during pile installation and soil consolidation are very sensitive to the characteristics of the soil stress-strain model used in the analysis. Hence, an effective stress constitutive model capable to describe the stress-strain-strength behavior of clays under general conditions should be used. Therefore, this paper is organized as follows: Section 2 presents a new stress-strain model for clays based on the Theory of Anisotropic Elasto-Plasticity and evaluates this model by comparing its predictions to data obtained from laboratory tests on Boston Blue Clay. Section 3 uses the proposed model to estimate the excess pore pressures and effective stress changes caused by pile installation. Finally, Section 4 uses these excess pore pressures and effective stresses as initial conditions in non-linear consolidation analyses around a long cylindrical pile, in conjunction with the proposed soil model.

2. THE PROPOSED SOIL MODEL

Most of the analyses of soil consolidation around piles reported in the literature (e.g. Miller et al, 1978; Randolph et al, 1979) used the Modified Cam-Clay (MCC) model (Roscoe and Burland, 1968) to describe the stress-strain-strength behavior of the clay. However, this model has serious limitations when used to predict the behavior of natural clays:
(1) For normally consolidated and slightly overconsolidated clays, the model does not predict strain softening in any mode of deformation.
(2) The yield function depends only on the (first and second) invariants of the effective stress tensor. Therefore, the model does not predict strength anisotropy (e.g. predicts the same value of the undrained shear strength in triaxial compression and extension) and its stress-strain anisotropy is very limited.
(3) The yield surface of the MCC model can be represented by an ellipsoid oriented along the hydrostatic axis in the effective stress space. Recent data on the lean sensitive Champlain Clay reported by Leroueil et al (1979) show that for K_o-consolidated clays, the yield surface appears to be oriented along a line close to the K_o-line.

These findings indicate that the MCC model has been developed for isotropic materials. In natural clays, deposition and preconsolidation develop under K_o-stress conditions. As a consequence, the clay structure is organized anisotropically and thus, a yield surface oriented along the K_o-line is a more realistic representation for natural clays.

Kavvadas (1982) proposes an effective stress-strain model for clays in an attempt to remedy the limitations of the MCC model. The proposed model adopts a yield surface represented by a distorted ellipsoid in an effective stress space comprised of the mean effective stress (hydrostatic) axis $\bar{\sigma}$ and the deviatoric hyperplane $\underline{s}$ (see Figure 1). The yield function is given by

$$f(\bar{\sigma}, \underline{s}; \bar{\alpha}, \underline{b}) \equiv (\underline{s} - \bar{\sigma}\underline{b}) : (\underline{s} - \bar{\sigma}\underline{b}) - c^2 \bar{\sigma}(2\bar{\alpha} - \bar{\sigma}) \qquad (1)$$

where, $\bar{\alpha}$ is a measure of the current size of the yield surface, c is a material parameter representing the ratio of the axes of the yield surface and ":" denotes the double contraction of the indices of the associated tensors. The deviatoric tensor $\underset{\sim}{b}$ specifies the orientation of the yield surface with respect to the hydrostatic axis and therefore is a measure of clay anisotropy. In the case of an isotropic clay, b=0, the yield surface is oriented along the hydrostatic axis and is identical to the yield surface of the MCC model. In the case of a K_o-consolidated clay, the yield surface of the proposed model is oriented along the K_o-line, i.e.,

$$\underset{\sim}{b} = \underset{\sim}{s} / \bar{\sigma} \qquad (2)$$

and the stress state is represented by a point located at the tip of the yield surface (point A in Figure 1).

During deformation, clay anisotropy is generally altered. The proposed model takes into consideration such changes via a kinematic hardening rule that allows continuous rotation of the yield surface by changing the magnitude and principal directions of the anisotropy tensor $\underset{\sim}{b}$. Its infinitesimal change, caused by the application of an infinitesimal increment of the plastic volumetric strain $\dot{\epsilon}^p$, is

$$\underset{\sim}{\dot{b}} = \psi \{ \frac{1}{\bar{\alpha}} (\underset{\sim}{s} - \bar{\sigma}\underset{\sim}{b}) \} \dot{\epsilon}^p \qquad (3)$$

where, ψ is a material parameter specifying the rate of change of the anisotropy. The hardening characteristics of the proposed model are supplemented by an isotropic hardening rule specifying the evolution of the size $\bar{\alpha}$ of the yield surface during plastic deformation

$$\dot{\bar{\alpha}} = \phi \bar{\alpha} \dot{\epsilon}^p \qquad (4)$$

where, ϕ is a function of the state variables given by Kavvadas (1982). This isotropic hardening rule allows to predict increasing strength by compacting the clay.

During shearing of natural clays the applied shear stress initially increases (strain hardening region) until it reaches a maximum (peak strength state), then decreases (strain softening region) and eventually the clay deforms in shear without further changes of the stresses (critical state). In soil modelling, failure criteria are used to specify stress combinations corresponding to peak strength and critical states (often called peak strength and critical state failure criteria, respectively). The proposed model adopts a non-associated flow rule and thus, peak strength and critical state failure criteria can be specified independently of the direction normal to the yield surface (i.e., the normality rule does not apply). According to this model, the peak strength (denoted with a subscript p) and the critical state (denoted with a subscript c) failure criteria are geometrically represented by two nested conical surfaces in the effective stress space given by

$$h_p(\bar{\sigma}, \underset{\sim}{s}) \equiv (\underset{\sim}{s} - \bar{\sigma}\underset{\sim}{\xi}_p) : (\underset{\sim}{s} - \bar{\sigma}\underset{\sim}{\xi}_p) - k_p^2 \bar{\sigma}^2 = 0 \qquad (5)$$

$$h_c(\bar{\sigma}, \underset{\sim}{s}) \equiv (\underset{\sim}{s} - \bar{\sigma}\underset{\sim}{\xi}_c) : (\underset{\sim}{s} - \bar{\sigma}\underset{\sim}{\xi}_c) - k_c^2 \bar{\sigma}^2 = 0 \qquad (6)$$

where, the deviatoric tensors $\underset{\sim}{\xi}_p$, $\underset{\sim}{\xi}_c$ and the scalars k_p, k_c are material parameters evaluated from the peak and critical state friction angles measured in various modes of deformation (e.g. triaxial compression and extension).

According to Plasticity Theory, the volumetric and deviatoric components of the (infinitesimal) plastic strain increment are

$$\dot{\epsilon}^p \equiv \dot{\Lambda} P \quad ; \quad \underset{\sim}{\dot{e}}^p \equiv \dot{\Lambda} \underset{\sim}{P}' \qquad (7)$$

where, $\dot{\Lambda}$ is the magnitude of the plastic strain increment and P, $\underset{\sim}{P}'$ are the volumetric and deviatoric components of the direction of the plastic strain increment. This direction is commonly determined by the flow rule. In the case of an associated flow rule, P and $\underset{\sim}{P}'$ are respectively equal to the volumetric and deviatoric components of the direction normal to the yield surface f:

$$Q = \partial f / \partial \bar{\sigma} \quad ; \quad \underset{\sim}{Q}' = \partial f / \partial \underset{\sim}{s} \qquad (8)$$

The proposed model adopts the following non-associated flow rule:

$$P \equiv 2c^2 \bar{\alpha} r_c \quad ; \quad \underset{\sim}{P'} \equiv c^2 x (\underset{\sim}{Q'} + r_c \underset{\sim}{s}) \tag{9}$$

where, r_c is the distance of the stress point from the critical state failure cone and x is a material parameter evaluated from the ratio of the horizontal to the vertical effective stress (equal to K_o) for virgin one-dimensional consolidation.

Using the previously defined quantities, the incremental stress-strain relationships are:

$$\dot{\bar{\sigma}} = K(\dot{\varepsilon} - \Lambda P) \quad ; \quad \Lambda = \frac{2G(\underset{\sim}{Q'} : \underset{\sim}{\dot{e}}) + KQ\dot{\varepsilon}}{H + 2G(\underset{\sim}{Q'} : \underset{\sim}{P'}) + KQP} \tag{10}$$

$$\underset{\sim}{\dot{s}} = 2G(\underset{\sim}{\dot{e}} - \Lambda \underset{\sim}{P'})$$

where, K and G are the elastic bulk and shear moduli and H is an elastoplastic modulus given by Kayvadas (1982). Equations (10) can be used to compute the hydrostatic and deviatoric components ($\dot{\bar{\sigma}}$ and $\dot{\underset{\sim}{s}}$) of the effective stress increment caused by a (given) strain increment having volumetric and deviatoric components $\dot{\varepsilon}$ and $\underset{\sim}{\dot{e}}$, respectively.

The proposed model is evaluated by comparing its predictions to data obtained from undrained laboratory tests on resedimented K_o-normally consolidated Boston Blue Clay. Figure 2 presents measured and predicted stress-strain curves (left-hand-side part) and the associated effective stress paths (right-hand-side part) during (a) Triaxial Compression and Extension, (b) Plane Strain Compression and Extension and (c) Direct Simple Shear (DSS) tests. For comparison, predictions of the Modified Cam-Clay (MCC) and the total stress PLB model (presented by Prevost, 1977, and modified by Levadoux and Baligh, 1980) are also shown. The proposed model predictions are in good agreement with the measurements in all modes of deformation.

3. PILE INSTALLATION

The complexity of the mechanism of pile installation requires the use of assumptions in obtaining estimates for the stress and pore pressure changes in the soil around the pile. Commonly, pile installation is modelled as the undrained expansion of a long cylindrical cavity (e.g. Randolph et al, 1979), i.e., soil is assumed to deform under plane strain and axisymmetric conditions. In this case, incompressibility yields the strains at any location around the pile and then, the effective stresses and excess pore pressures can be estimated by means of the stress-strain law and the equilibrium equation, respectively. The Cavity Expansion method is a one-dimensional model of pile installation and therefore does not take into consideration the effect of the pile tip on soil deformation. The Strain Path method (Baligh et al, 1978) attempts to remedy this limitation by introducing a two-dimensional model of pile installation. According to this model, the strains in the soil are estimated by means of the (known) velocities of an ideal fluid moving around an object having the shape of the pile (see e.g. Levadoux and Baligh, 1980) and then, stresses and pore pressures are obtained following the same procedure as in the Cavity Expansion method.

In the region far behind the pile tip (which is of main interest in estimating the skin friction along the shaft of long piles), the Cavity Expansion and the Strain Path methods predict similar strain states after pile installation. However, the corresponding strain paths leading to these strain states are very different: the Cavity Expansion method predicts a monotonic strain path between the initial (undeformed) state and the strain state after pile installation, whereas the Strain Path method predicts significant strain reversals due to the effect of the pile tip. Therefore, the two methods predict different stresses and pore pressures when used in conjunction with soil stress-strain models that take into consideration the "path dependency" of natural clays. Figures 3 and 4 show these differences for the predicted radial effective stress and excess pore pressure (i.e., in excess to the ambient pore pressure p_o) in the radial direction, respectively. The radial coordinate is normalized with the pile radius r_o, whereas the stresses and pore pressures are normalized with the value of the in-situ vertical effective stress before pile installation (at the location under consideration). This normalization is possible because the soil models used in the analyses follow the "principle of normalized behavior", i.e., the predicted stress changes are proportional to the initial value of the mean effective stress. Close to the pile wall, the Strain Path method predicts significantly smaller values of the radial effective stress because the strain reversals cause unloading and, therefore, reduced effective stresses. Furthermore, the Cavity Expansion method predicts an almost logarithmic increase of the excess pore pressure, whereas the Strain Path method gives approximately constant values close to the pile.

Both the Strain Path and the Cavity Expansion methods are approximate because of the assumptions involved. However, the Strain Path method is expected to give more realistic estimates of the stresses and pore pressures in the soil after pile installation. For the sake of comparison, consolidation analyses presented in Section 4 use initial conditions predicted by both methods.

4. CONSOLIDATION AROUND PILES

Baligh and Kavvadas (1980) prove that linear radial consolidation analyses around a cylindrical pile predict that the radial total stress at the pile wall remains constant during consolidation. This means that, after consolidation is completed, the radial effective stress at the pile wall is equal to the radial total stress immediately after pile installation (minus the ambient pore pressure). Estimates of the limiting skin friction based on this value of the radial effective stress grossly overpredict the bearing capacity of the pile (by a factor of three to five). Therefore, non-linear consolidation analyses are necessary to obtain realistic estimates of the stress changes in the soil around the pile.

Kavvadas (1982) proposes an incremental non-linear Finite Element scheme for soil consolidation based on a Weighted Residuals formulation of Biot's Theory of Consolidation. The solution is marched in time using an unconditionally stable implicit time integration scheme with increasing time step length as consolidation proceeds. An "equivalent elastic" symmetric stiffness matrix is used to improve the efficiency of the computational scheme by avoiding the costly triangularization of the non-symmetric tangent stiffness matrix in every time step. Accurate solutions are then obtained by means of equilibrium iterations in each time step, until convergence is achieved. Thus, the stiffness matrix is assembled and triangularized only when the length of the time step is changed. The proposed scheme showed fast convergence and proved very effective, especially since part of the soil unloads elastically and the boundary between the elastic and the plastic domains changes location during consolidation. This Finite Element scheme is applied to the problem of radial soil consolidation around a pile using three node bar elements with quadratic displacement and linear pore pressure interpolations (in order to achieve the same degree of accuracy for stresses and pore pressures). Figures 5 and 6 show the predicted pore pressure and radial stress changes at the pile wall during consolidation of normally consolidated Boston Blue Clay, using the Proposed and the Modified Cam-Clay models, respectively. The time variable is described by the "equivalent time factor"

$$T \equiv c t / r_o^2 \quad ; \quad c \equiv \left(\frac{1 + 2 K_o}{3}\right) \frac{k \bar{\sigma}_{vo}}{\gamma} \qquad (11)$$

where, k is the soil permeability (assumed constant) and γ is the unit weight of the pore water. These non-linear solutions (with the exception of Figure 6b) show that the radial total stress decreases during consolidation and thus the predicted radial effective stress after consolidation is smaller than the prediction of the corresponding linear solution. The method used to study pile installation has a significant effect on the predicted radial effective stress after consolidation: the Strain Path method predicts initial values smaller than the Cavity Expansion and the same trends appear in the corresponding values after consolidation. Furthermore, consolidation analyses using the proposed model predict a smaller increase of the radial effective stress compared to analyses using the Modified Cam-Clay model; this difference is due to the shape of the yield surface adopted by each model. Figure 7 shows the predicted pore pressure and radial stress changes at the pile wall during consolidation of slightly overconsolidated Boston Blue Clay. The predicted radial effective stress after consolidation increases with the overconsolidation ratio (OCR) of the clay. These analyses show that in the early stages of consolidation the soil unloads elastically and only in the later stages plastic deformations take place in a zone close to the pile. Furthermore, by increasing the OCR of the clay, the value of the time factor when plastic deformations are introduced increases (because of the larger size of the yield surface) and thus the initial elastic portion of the effective stress path is longer. Therefore, the characteristics of linear consolidation are more pronounced for higher OCRs, which explains the increasing values of the radial effective stress at the pile wall after consolidation, with increasing OCR. Figure 8a shows the inadequacy of linear consolidation analyses to realistically predict the radial stress changes at the pile wall, by comparing the prediction of linear consolidation (dashed line) with the prediction of a non-linear analysis using the proposed model (full line). However, the linear model predicts reasonably well the pore pressure dissipation rates at all times. This is because, according to the results of the non-linear analyses, two compensating mechanisms occur simultaneously during consolidation (see Figure 8b):
(a) A plastic (soft) zone develops close to the pile wall, tending to retard pore pressure dissipation rates and,
(b) The soil stiffness increases because the effective stresses increase during consolidation, which tends to accelerate pore pressure dissipation rates.

These two mechanisms are approximately compensated and thus the "effective" soil stiffness is approximately constant during consolidation. Therefore, a linear model with constant stiffness can predict reasonably well the pore pressure dissipation rates.

5. REFERENCES

BALIGH, M. M., V. VIVATRAT and C. C. LADD, 1978, *Exploration and Evaluation of Engineering Properties for Foundation Design of Offshore Structures*, Massachusetts Institute of Technology, Department of Civil Engineering, Report No. 78-40.

BALIGH, M. M. and M. KAVVADAS, 1980, *Axial Static Capacity of Offshore Friction Piles in Clays*, Massachusetts Institute of Technology, Department of Civil Engineering, Report No. 80-32.

ESRIG, M. I., R. C. KIRBY, R. G. BEA and B. S. MURPHY, 1977, "Initial Development of a General Effective Stress Method for the Prediction of Axial Capacity for Driven Piles in Clay", *Proceedings, 9th Offshore Technology Conference*, Houston, Texas, 1977, Paper 2943, Vol. 3, pp. 495-506.

ESRIG, M. I., R. C. KIRBY and B. S. MURPHY, 1979, "Advances in General Effective Stress Method for the Prediction of Axial Capacity for Driven Piles in Clay", *Proceedings, 11th Offshore Technology Conference*, Houston, Texas, 1979, Paper 3406, Vol. 1, pp. 437-448.

KAVVADAS, M., 1982, *Non Linear Consolidation around Driven Piles in Clays*, ScD Thesis, Massachusetts Institute of Technology, Department of Civil Engineering.

LADD, C. C. and J. VARALLYAY, 1965, *The Influence of Stress System on the Behavior of Saturated Clays during Undrained Shear*, Massachusetts Institute of Technology, Department of Civil Engineering, Report No. 65-11.

LADD, C. C., R. B. BOVEE, L. EDGERS and J. J. RIXNER, 1971, *Consolidated-Undrained Plane Strain Shear Tests on Boston Blue Clay*, Massachusetts Institute of Technology, Department of Civil Engineering, Report No. 71-13.

LADD, C. C. and L. EDGERS, 1972, *Consolidated-Undrained Direct Simple Shear Tests on Saturated Clays*, Massachusetts Institute of Technology, Department of Civil Engineering, Report No. 72-82.

LEROUEIL, S., et al, 1979, "Behavior of Destructured Natural Clays", *Journal of the Geotechnical Engineering Division*, ASCE, Vol. 105, No. GT6, pp. 759-778.

LEVADOUX, J.-N. and M. M. BALIGH, 1980, *Pore Pressures during Cone Penetration in Clays*, Massachusetts Institute of Technology, Department of Civil Engineering, Report No. 80-15.

MILLER, T. W., J. D. MURFF and L. M. KRAFT, 1978, "Critical State Soil Mechanics Model of Soil Consolidation Stresses around a Driven Pile", *Proceedings, 10th Offshore Technology Conference*, Houston, Texas, Paper 3307, Vol. 4, pp. 2237-2242.

PREVOST, J.-H., 1977, "Mathematical Modelling of Monotonic and Cyclic Undrained Clay Behavior", *International Journal for Numerical and Analytical Methods in Geomechanics*, Vol. 1, pp. 195-216.

RANDOLPH, M. F., J. P. CARTER and C. P. WROTH, 1979, "Driven Piles in Clay - The Effects of Installation and Subsequent Consolidation", *Geotechnique*, Vol. 29, No. 4, pp. 361-393.

RANDOLPH, M. F. and C. P. WROTH, 1981, "Application of the Failure State in Undrained Simple Shear to the Shaft Capacity of Driven Piles", *Geotechnique*, Vol. 31, No. 1, pp. 143-157.

ROSCOE, K. H. and J. B. BURLAND, 1968, "On the Generalized Behavior of Wet Clay", in *Engineering Plasticity*, pp. 535-609, Cambridge, England: University Press.

STERMAC, A. G., K. G. SELBY and M. DEVATA, 1969, "Behavior of Various Types of Piles in Stiff Clay", *Proceedings, 7th International Conference on Soil Mechanics and Foundation Engineering*, Mexico City, Mexico, 1969, Vol. 2, pp. 239-245.

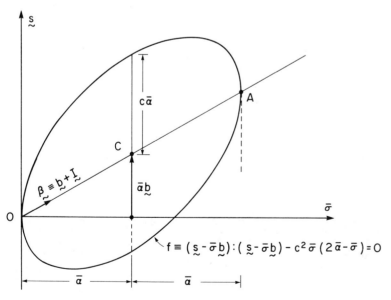

Figure 1:

The yield surface of the proposed soil model.

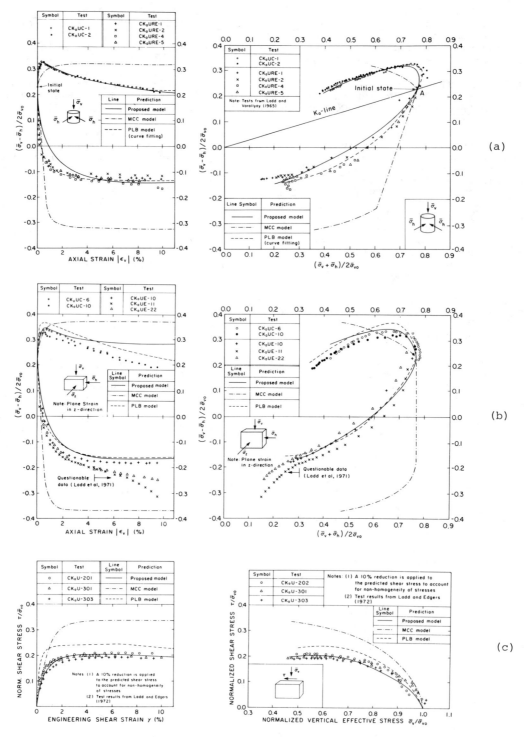

Figure 2: Measured and predicted stress-strain curves (left-hand-side) and effective stress paths (right-hand-side) during: (a) Triaxial Compression and Extension, (b) Plane Strain Compression and Extension and, (c) Direct Simple Shear tests on K_o-normally consolidated Boston Blue Clay.

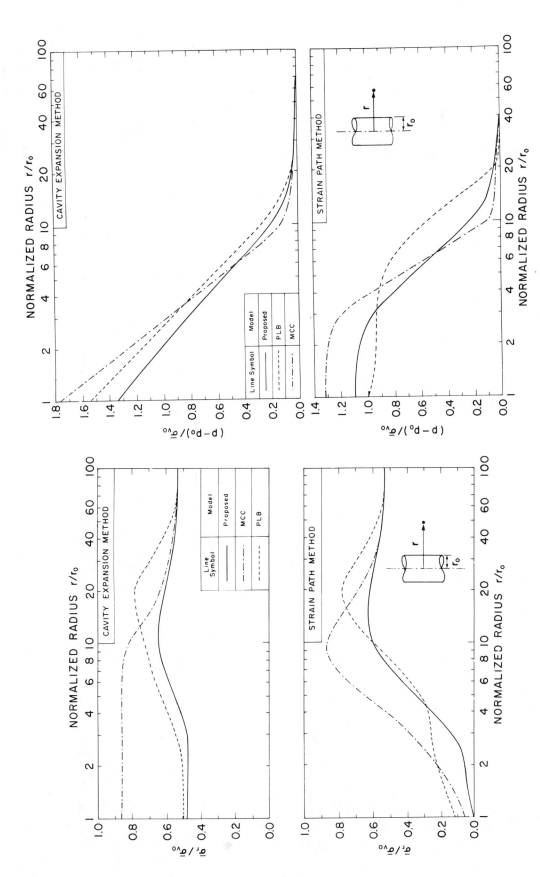

Figure 3: Radial Effective Stress far behind the pile tip, predicted according to the Cavity Expansion and the Strain Path methods.

Figure 4: Excess Pore Pressure far behind the pile tip, predicted according to the Cavity Expansion and the Strain Path methods.

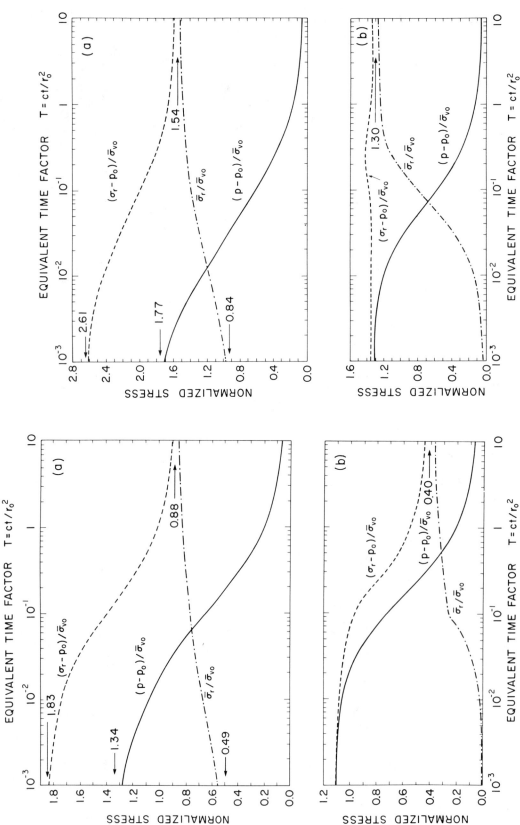

Figure 5: Proposed model predictions of the pore pressure and radial stress changes at the pile wall during consolidation of a soil with the properties of normally consolidated Boston Blue Clay.
Initial Conditions:
(a) Cavity Expansion, (b) Strain Path methods.

Figure 6: Modified Cam-Clay model predictions of the pore pressure and radial stress changes at the pile wall during consolidation of a soil with the properties of normally consolidated Boston Blue Clay.
Initial Conditions:
(a) Cavity Expansion, (b) Strain Path methods.

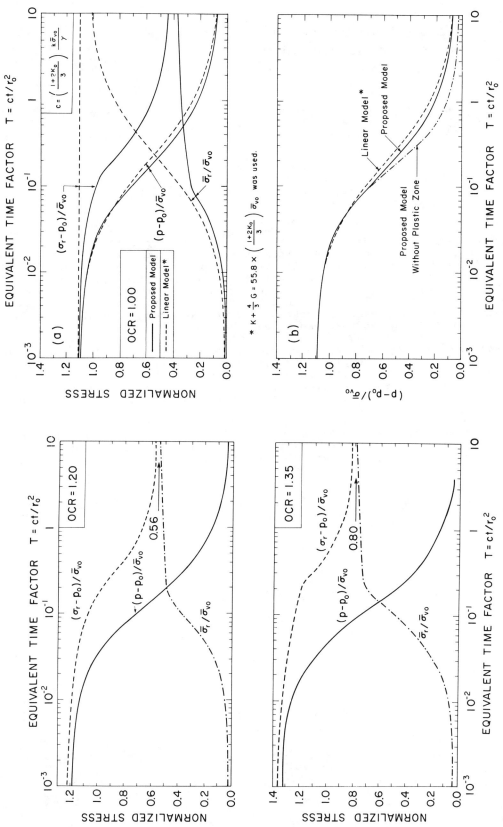

Figure 7: Proposed model predictions of the pore pressure and radial stress changes at the pile wall during consolidation of a soil with the properties of slightly overconsolidated Boston Blue Clay. Initial conditions via the Strain Path method.

Figure 8: Radial stress and pore pressure changes at the pile wall predicted by the proposed model are compared to the predictions of an 'equivalent linear' elastic soil model for a clay with the properties of normally consolidated Boston Blue Clay.

347

LATERAL STIFFNESS OF A PILE SUBJECT TO STATIC MONOTONIC LOADING

S.N. Pollalis
Massachusetts Institute of
Technology, U.S.A.

J.J. Connor
Massachusetts Institute of
Technology, U.S.A.

M. Kavvadas
Massachusetts Institute
of Technology, U.S.A.

SUMMARY

The present study concerns the prediction of the behavior of laterally loaded piles based on analytical and numerical techniques. A three-dimensional formulation is developed in a non-dimensional form which is applicable for non-linear elastoplastic representation of the soil continuum. The analytical part provides insight on behavioral issues, defines the non-dimensional parameters, and simulates the far field effects. The numerical part uses a mixed finite element formulation (displacements and pore pressures) and assumes three-dimensional non-linear elastoplastic anisotropic undrained behavior of the soil continuum. It also allows for slippage and gapping around the pile.

The analytical and numerical components are coupled in a computational model and a full scale pile experiment is simulated. The pile head displacement is applied incrementally and the load required to produce such a displacement is calculated together with the shear forces and bending moments along the pile and the stresses, strains, and pore pressures in the soil continuum. The predictions of the model apply for a specific set of non-dimensional parameters covering a wide range of pile designs and soil conditions.

Comparison with the predictions of the p-y curve methods shows the same general trend in the bending moment distribution but the proposed model values for peak moments are less. However, the deflections predicted by the two methods, and therefore the pile head stiffness, are quite different.

INTRODUCTION

Existing methods for the design of laterally loaded piles do not provide sufficient insight information (Poulos and Davis, 1980). Considering that future pile loading conditions will be more severe, and full scale experiments are expensive and time consuming, the lack of a realistic computer based model to predict pile behavior is particularly critical. The objectives of this study are to develop such a model and carry out numerical simulations which will improve our understanding of the behavior of laterally loaded piles. The model can also be used for design, although this is not our immediate objective. Although numerical models based on p-y curves (Matlock and Reese, 1960) represent the current state of the art, they do not have the potential for further analytical work. The alternative is an analytical formulation. However, the soil has to be treated as a non-linear elastoplastic material. Development of a numerical model for the continuum seems to be the most reasonable way to evaluate the behavior of laterally loaded piles. A combined approach, using the linear theory of elasticity and numerical techniques, is used. The linear elastic phase provides an understanding of the behavior of the system, defines the non-dimensional parameters, and generates a solution which is applicable for the far field. In the second phase, the non-linear elastoplastic material model is applied for the region adjacent to the pile. The non-linearity resulting from slippage and gapping is also incorporated in the numerical formulation. Both formulations, the linear elastic and the non-linear elastoplastic, are expressed in non-dimensional form, normalized with respect to characteristic length and stress measures.

ANALYTICAL SOLUTION

The three-dimensional analytical model is based on the Mindlin (1936) solution which defines the displacements in a linear elastic homogeneous half-space due to a concentrated force acting at an arbitrary point within the medium. An empirical modification proposed by Banerjee and Davies (1978) can be incorporated in the model to treat the case of a non-homogeneous continuum in an approximate sense. The interaction of the pile with the elastic half-space is defined numerically in terms of non-dimensional flexibility coefficients. Solution of the flexibility equations provides the displacements and stresses for both the pile and the elastic half-space. The Mindlin solution applies for a concentrated force acting at a point. In the case of a distributed force, one needs to integrate the Mindlin equations over the loaded surface. The pile is represented as a strip of finite width and integration of the Mindlin equation over a rectangular area is required. This integration has been performed by Douglas and Davis (1964) for points located in the plane of the loaded strip and by Pollalis (1982) for an arbitrary point in the three dimensional continuum.

The pile-elastic half-space is represented by a discrete physical model consisting of a finite beam element discretization for the pile and the analytical closed form coefficients of the soil flexibility matrix generated by integrating the Mindlin solution. In addition to the pile-soil interaction relations, the equilibrium equations for the pile need to be considered. The resulting fully populated symmetrical system of equations yields the forces between the pile and the soil. Thus the load distribution along the pile and the displacements and stresses in the whole domain are obtained. The complicated closed form solution of the integrated Mindlin equations, together with the discretized solution of the pile-soil interaction, require a numerical implementation. The numerical code NDASM-3 (Non-Dimensional Analytical Solution using the Mindlin equations in 3 dimensions as modified by Banerjee and Davies) was developed in order to meet this requirement. NDASM-3 offers several computational options. It allows for a modulus which varies with depth, solves the pile-soil interaction problem, and generates the three-dimensional distribution of stresses and displacements in the neighborhood of the pile. Numerical simulations are carried out for the elastic case and a representative range of the non-dimensional parameters. Typical results for displacement are shown in Figure 1. The pile is modelled as a strip with flexural rigidity EI. Reducing the circular cross section to a flat strip is expected to have a negligible effect on the behavior far away from the pile. However, the strip formulation results in a more flexible pile since by applying the force on the cylinder axis, there is additional displacement due to the deformation of the soil which replaces the cylinder.

The linear elastic solution is suitable for the soil region that experiences low strains. Therefore, the discrete numerical solution has to be applied only to a finite region of the half-space adjacent to the pile where the strains are high and consequently, the soil behavior is non-linear. The region is bounded by the free surface and an outer boundary, beyond which the soil is assumed to behave linearly. The linear elastic solution is used to establish the boundary conditions for the outer boundary which simulate the effect of the rest of the half-space.

NUMERICAL SOLUTION

A three-dimensional isoparametric finite element discretization is employed for the area close to the pile. The availability of realistic constitutive laws for soil that have recently been developed justify the use of such a detailed numerical model for a soil medium.

The initial stiffness matrix approach was chosen as the non-linear solution algorithm for the following reasons:

(i) Only one triangularization of the stiffness matrix is required. Iteration involves only a sequence of back-substitutions and updates on the load vector. This is particularly important for complex models with many unknowns.
(ii) It is stable and guarantees convergence under certain conditions.

The non-linear incremental expressions for displacements and stresses are:

$$\begin{aligned} \underline{U}_{t+\Delta t} &= \underline{U}_t + \Delta \underline{U}^{(i-1)} + d\underline{U}^{(i)} \\ \bar{\underline{\sigma}}_{t+\Delta t} &= \bar{\underline{\sigma}}_t + \Delta \bar{\underline{\sigma}}^{(i-1)} + d\bar{\underline{\sigma}}^{(i)} \end{aligned} \quad (1)$$

where: $\underline{U}_t$: displacement vector at time t

$\Delta \underline{U}^{(i-1)}$: total change in the displacement vector within the time interval Δt for the i-1 interation cycles

$d\underline{U}^{(i)}$: change in the displacement vector within the time interval Δt for the i'th iteration cycle

The same notation is adopted for the effective stresses, $\bar{\underline{\sigma}}$. Convergence is achieved when $d\underline{U}^{(i)} \cong \underline{0}$ and $d\bar{\underline{\sigma}}^{(i)} \cong \underline{0}$. The incompressibility condition is imposed by a "penalty" formulation (Kavvadas and Baligh, 1980) coupled with reduced integration (Bathe and Wilson, 1976; Nayak and Zienkiewicz, 1972) for the element compliance matrix. Iteration is carried out on the following incremental equations:

$$\begin{bmatrix} \underline{K}_e & \underline{H} \\ \underline{H}^T & -\frac{1}{\kappa}\underline{G} \end{bmatrix} \begin{Bmatrix} d\underline{U}^{(i)} \\ \Delta\underline{P} \end{Bmatrix} = \begin{Bmatrix} \Delta\underline{R} - \int_V \underline{B}^T \Delta\bar{\underline{\sigma}}^{(i-1)} dV \\ -\underline{H}^T \Delta\underline{U}^{(i-1)} \end{Bmatrix} \quad (2)$$

where $\underline{K}_e$: initial elastic deviatoric stiffness matrix

$\underline{H}$: couples the displacements and pore pressures

$-\underline{G}$: pore pressure compliance matrix

κ : penalization factor

$\Delta\underline{P}$: pore pressure vector increment from time t to time $t+\Delta t$

$\Delta\underline{R}$: external load vector increment from time t to time $t+\Delta t$

$\underline{B}$: displacement - strain transformation matrix

A double finite element mesh is required in order to discretize the two-phase continuum. Comparative studies of different element expansions showed that the 21-8 cubic element (21 nodes for displacement, 8 nodes for pore pressure) combined with the 17-8 (17-nodes for displacement) for elements in contact with the pile, is the most efficient combination for this problem (Pollalis, 1982).

Slippage and gapping effects are significant and dramatically reduce the lateral stiffness of piles. The modelling strategy for the horizontal directions is shown in Figure 2. Double nodes are introduced on the pile wall. Every set of nodes has initially the same location, and the nodes are connected with two very stiff springs in the radial and tangential directions. The stiffness K of these springs is set much higher than any other stiffness measure, and the nodal displacements are essentially identical as long as these springs are active. The algorithms for deactivation and activation which simulate real slippage and gapping, are defined in Figure 2. This modelling of gapping and slippage introduced a complication at the finite element level. When quadratic elements were used for the soil continuum, the mid-side node of the element was subjected to a higher force than the corner nodes. This behavior is perfectly consistent for the element but it creates problems with the gapping and slippage criteria. A comparative study with fine mesh linear expansion elements (which do not have internal nodes) showed that every "averaging" or "double pricing" system failed. The solution adopted here was to eliminate the mid-side nodes for the face in contact with the pile, i.e., to use 17-node superparametric elements at the pile-soil boundary and 21-node elements in the interior.

In the three-dimensional model, the pile is discretized with linear beam elements. The translational degrees of freedom of the beam are actually representing a set of kinematically condensed nodes, and the rotational degree of freedom is uncoupled with the soil. The pile modelling takes into account the cavity around the pile. In addition, by choosing appropriate initial conditions for the soil, installation of the pile and subsequent consolidation can be simulated.

Four different models are used to describe the soil continuum: linear elastic isotropic; elastic-perfectly plastic (Pollalis, 1982); modified Cam-Clay (Roscoe and Burland, 1968); and the model proposed by Kavvadas (1982). All models work with the effective stresses and describe the constitutive behavior of the soil skeleton. The pore pressure induced during loading is taken into consideration through the incompressibility constraint imposed on the displacements. The linear elastic isotropic model is primarily used to provide a link with the analytical solution, and to give first estimates of the effects of phenomena such as slippage, gapping, and mesh density. The elastic-perfectly plastic model is used as a first step in the elastoplastic models. The features of the model proposed by Kavvadas make it more realistic than Cam-Clay while slower in computer time. All the above models are used in a transformed stress space introduced by Kavvadas (1982), which requires a reduced number of operations to compute the stresses corresponding to specified strains.

THE MODEL "DEMETRA"

The numerical solution scheme, together with the analytical solution for the far field, are implemented in the computer code DEMETRA, named for the ancient Greek Godess of the earth. The underlying philosophy of DEMETRA is the use of numerical simulation to identify key behavioral issues and to design in-situ test programs which will provide verification for the numerical model. The data required by the model are exactly the same as for a real experiment: pile geometry, soil data, and loading conditions. The length, diameter, and flexural rigidity are sufficient to define the pile. The soil can be assumed to be layered or have a random profile with variable initial conditions. Initial conditions enter in the elasto-plastic analysis, and the soil constitutive law and element mesh for initial stress evaluation should also be used for the lateral load analysis. A tubular open-ended pile does not alter significantly the state of stress close to the surface when it is driven in a normally consolidated clay deposit provided that it is unplugged. Since the lateral pile stiffness is dependent mainly on the soil close to the surface, it is reasonable to assume that the state of stress after pile driving is the same as existed before driving. However, this would not apply for axially loaded piles.

A horizontal displacement and/or a rotation are applied to the pile head and the resulting lateral force and moment are calculated with the equilibrium equations. Listed below are the basic characteristics of the model, implementation techniques, and various options.

(i) The model takes advantage of symmetry and antisymmetry and modifies the boundary conditions accordingly.
(ii) It has the capability of imposing either zero displacements on the outer boundary or the displacements corresponding to the linear elastic solution.
(iii) One can specify either slippage or gapping or both.
(iv) The results are generated in non-dimensional form.
(v) The finite element mesh and coordinate generators are incorporated in the model. These generators allow one to vary the mesh size in the most highly stressed area, i.e., close to the pile and near the surface.
(vi) The number of integration points per element can be varied.
(vii) The imposed pile head displacement is applied incrementally in order to follow more closely the non-linear response. The driving routine of the constitutive law imposes a limit on the maximum strain increment at the soil element level. If the strain increment at any integration point exceeds a specified limiting value, the strain increment is subdivided into smaller equal increments and the constitutive law routine is applied repeatedly. In this way, the accuracy of the non-linear algorithm is guaranteed for arbitrary displacement increment.

DEMETRA may be used to predict the lateral behavior of a group of piles. Such an analysis will determine the interaction between the piles and will establish the influence zone around each pile. DEMETRA may accept more than one pile by modifying the finite element mesh and coordinate generators. A cyclically loaded pile may also be analysed with this computational model. The additional features required include a soil model that will correctly predict cyclic soil behavior and a modification of the gapping and slippage formulation in order to accomodate reverse loading. The cyclically loaded pile will require increased computer time since it has to follow the detailed stress-strain path over the loading time history. So, practical applications may be limited to only a certain number of cycles.

At this time, there is considerable interest in the behavior of axially loaded piles, especially tension piles. The framework for such an analysis is already in DEMETRA. Since the behavior of an axially loaded pile is axisymmetric, the size of the problem will be reduced significantly. However, the axial resistance of the soil is generated primarily in the region of the pile tip and information on pile installation and consolidation around the pile shaft must be available in order to carry out a realistic analysis.

Pile - Soil System

The results of DEMETRA are compared with the predictions of the p-y curve methods (Matlock, 1970; Stevens and Audibert, 1979) which are recommended for designing offshore piles by the American Petroleum Institute (API, 1980). The pile geometry for the numerical simulation is the same as used by Matlock (1970). The pile is considered driven in a normally consolidated Boston Blue Clay profile. The pile head can be either moment free or subjected to a prescribed rotation. Displacement, bending moment and the shear force are calculated at the nodes of the finite element mesh. These nodes can be interpreted as corresponding to strain gauge stations on an instrumented pile. The only difference is that fewer stations are used in the present analysis. However, the stations of DEMETRA are sufficient to give accurate bending moments and shear forces on the pile. The discretization of the pile is governed by the discretization of the soil which includes the option of generating smaller elements close to the surface. Element lengths follow a geometric progression with specified step in this study. Displacements and pore pressures are evaluated at the nodes for each increment of the pile head applied displacement. The corresponding effective and total stresses of the soil are calculated at the element integration points.

The Non-dimensional Parameters α, $\bar{\alpha}$, β, ω, $\bar{\omega}$, Q

Both the analytical and numerical solutions are developed in non-dimensional form. The objective is to create a solution scheme which is independent of units, and more importantly, to treat a set of problems having the same non-dimensional parameters with just one solution. When the soil is modelled as a linear elastic material, the shear modulus at the pile tip can be taken as the stress measure of the soil, and the following non-dimensional parameters govern the problem

$$\alpha = \frac{EI}{G_L d^4} \qquad \beta = \frac{d}{L} \qquad \omega = \frac{\bar{\sigma}_h}{G_L} = \frac{K_o \bar{\gamma} L}{G_L} \qquad (3)$$

where EI : pile rigidity

G_L : shear modulus of the soil at the pile tip

d : pile diameter

L : pile length

$\bar{\gamma}$: average effective unit weight from ground surface to pile tip

$\bar{\sigma}_h$: initial horizontal effective stress at the pile tip

K_o : ratio of horizontal to vertical initial stress

The parameter β is not expected to be of particular importance below a certain value since the rest of the pile experiences negligible deformation and practically does not contribute to the lateral stiffness. The critical length depends on the pile diameter, the parameter α and the magnitude of the applied load. The particular importance of the above observation is that for $\beta \leq \beta_{cr}$ the problem has only one length normalizer and the non-dimensional solution has a wider field of applications. Furthermore, the discretization of slender piles requires less elements. A different stress measure has to be introduced when a non-linear elastoplastic model is used. The shear modulus is inappropriate in such a case since its physical characteristic is remarkably limited. The effective mean stress at the pile tip is used instead. This is permissable because of the nature of the elastoplastic models employed in DEMETRA. The parameters adopted for the non-linear elastoplastic case are $\bar{\alpha}$ and $\bar{\omega}$.

$$\bar{\alpha} = \frac{EI}{\bar{\sigma}_L d^4} \qquad \bar{\omega} = \frac{\bar{\sigma}_h}{\bar{\sigma}_L} = \frac{3K_o}{1+2K_o} \qquad (4)$$

The model also requires a set of non-dimensional parameters $\underset{\sim}{Q}$, which define the properties of the soil in the non-linear elastoplastic constitutive laws.

Comparison of the Proposed Model to the Existing Design Methods

Boston Blue Clay is a soft clay and the Matlock p-y curve method, together with the Stevens-Audibert reformulation, will be used to compare the results of DEMETRA. The evaluation of the present model with the data of Matlock's 1970 experiment is not possible because the in situ and laboratory results have not been released. A different approach is proposed instead. The comparison of the finite element model with the p-y curve results will be carried out on the same pile geometry but a normally consolidated Boston Blue Clay profile will be used to simulate the soil. The numerical codes employed for the p-y curve method are:

(i) PYSTATIC by Chandrashekar (1982) which generates the p-y curves.
(ii) COM622 by Reese (1977) which generates the deflections and bending moments along a pile subjected to a horizontal loading. COM622 utilizes the p-y curves derived by PYSTATIC.

Before proceeding with the comparison of the results for the two methods, it is of interest to examine the p-y curve predictions for the Sabine Clay experiment. Matlock states that his method applies for normally consolidated soil profiles. In a normally consolidated clay, the strength is proportional to the vertical effective stress which generally increases with depth. If the detailed strength profile used by Matlock is replaced with a linear strength profile (Figure 3) the maximum bending moment increases by 53%. Furthermore, pile sections that experience low bending moments with the detailed soil strength profile now have a much higher moment and the pile head deflection increases by more than 100%. After further study (Pollalis, 1982), it is concluded that:

(i) Use of the detailed strength profile for the region close to the surface leads to an accurate prediction of the bending moments along the pile.
(ii) The method is very sensitive to the stiffness of the p-y curves for points close to the surface.
(iii) The Stevens-Audibert modification does not predict very well the bending moment along the pile when the detailed strength profile is used. Agreement with the measured values improves when the linear strength profile is employed.

The comparison of the finite element model with the p-y curve method is carried out on the basis that both methods represent an effort to predict the actual pile behavior. In this context, the sensitivity on the input parameters for the p-y curves is also studied. Furthermore, the input soil properties for the p-y curves are engineered quantities of a correlation design method, which the p-y curve method is intended to be. The p-y curve method for a soft clay, such as the Boston Blue Clay, requires three soil properties: the undrained shear strength, the strain at half the maximum shear stress, and the unit weight. The undrained shear strength for a normally consolidated Boston Blue Clay is $S_u = 0.32\,\bar{\sigma}_{vo}$ for the triaxial test and $S_u = 0.20\,\bar{\sigma}_{vo}$ for the direct simple shear test. The strain at half the maximum shear stress in an undrained triaxial test is $\varepsilon_c = 0.005$ with a lower value of ε_c corresponding to a soil providing greater lateral stiffness in accordance to the p-y curve method (Chandrashekar, 1982). The submerged unit soil weight, $\bar{\gamma}$, is taken constant over the depth and equal to 60 pcf. However, the effect of the unit soil weight on the lateral pile behavior using the p-y curve method is very small (Chandrashekar, 1982). Assuming that the Boston Blue Clay deposit is normally consolidated, the shear strength is, in general, linearly dependent on the vertical effective stress, $\bar{\sigma}_{vo}$:

$$S_u(z) = a\bar{\gamma}z = a\bar{\sigma}_{vo} \qquad (5)$$

The effect of the parameter 'a' for a normally consolidated soil profile (linearly varying shear strength with depth) as well as the equivalent Sabine Clay shear strength profile for the Boston Blue Clay (Figure 3, profile No. 2) are investigated in this study.

The first numerical simulation considered no rotational restraint for the pile head, a lateral load of 14.3 kips, and the shear strength given by either equation 5 with variable 'a' or the equivalent Sabine Clay shear strength profile. Comparison of the predictions based on p-y curves with the finite element model results showed the following:

(i) For a = 0.20, the maximum bending moment is higher by 45% and the deflection by more than 400% with respect to the values obtained with DEMETRA.
(ii) For a = 0.32 (Figure 9) the maximum bending moment is 25% higher for Matlock's method and 7% for the Stevens-Audibert method. Pile head deflection is greater by 280%

(Matlock) and 190% (Stevens and Audibert).
- (iii) For a = 0.64 which is unrealistically high, Matlock's method gives similar bending moments as DEMETRA, while the Stevens-Audibert method yields close agreement for the displacements with respect to DEMETRA.
- (iv) Results generated with using the equivalent Sabine Clay strength profile are shown in Figure 10. DEMETRA's deflections fall between Matlock's and Stevens-Audibert's. The bending moments are lower by 16% (Matlock) and 42% (Stevens and Audibert).

Non-Dimensional Results for Long Piles

Sophisticated numerical models such as DEMETRA can be utilized to produce p-y curves for a specific site but this effort is difficult to justify since the p-y curves would be valid only for the specific pile geometry considered. Also, a large number of nodes would be required in order to obtain p-y curves for general use. Increasing the number of nodes provides a more detailed tracing of the p-y curves but does not improve the numerical accuracy (Pollalis, 1982). The proposed method is designed to produce a set of non-dimensional solutions for a specific site by varying the non-dimensional input. The parameter β can be set equal to β_{cr} since offshore piles are generally longer than the critical length, and $\bar{\alpha}$ can be varied to cover a wide range of pile geometries. The resulting set of non-dimensional solutions provides sufficient information for an understanding of the behavior and allows one to identify the optimal pile geometry. This approach has been carried out for a normally consolidated Boston Blue Clay deposit. Figure 4 shows the non-dimensional pile head stiffness for various values of $\bar{\alpha}$ and $\beta = \beta_{cr}$. Figures 5 and 6 show the displacements and bending moments along the pile, respectively, for the same range of $\bar{\alpha}$ and β. The bending moment distributions corresponding to various values of pile head displacements applied to a specific pile ($\bar{\alpha}$ = 22560.) with no rotational restraint are shown in Figure 7. Figure 8 contains similar curves for a restrained pile. Information concerning the influence of pile diameter, pile stiffness and the rotational restraint at the pile head can be obtained by interpreting these results in terms of $\bar{\alpha}$. Detailed comments are listed below.

Pile Diameter

The parameter, $\bar{\alpha}$, is a measure of the relative stiffness of the pile vs. the soil medium. Specializing (4) for a thin walled circular cross section leads to

$$\bar{\alpha} = \frac{EI}{\bar{\sigma}_L d^4} \approx \frac{\pi}{8} \frac{E}{\bar{\sigma}_L} \frac{t}{d} \qquad (6)$$

where $\bar{\sigma}_L$ is the mean effective stress at the pile tip. Pile length is accounted for through the parameter $\beta = d/L$ and the dependence of $\bar{\sigma}_L$ on L. Our interest here is the effect of varying d. Assuming L is fixed (and therefore $\bar{\sigma}_L$ remains constant) and t varies linearly with d such that t/d remains constant, then $\bar{\alpha}$ is constant. From Figure 4, one observes that the pile head secant stiffness is proportional to the pile diameter, for a specified $\bar{\alpha}$. According to Figure 6, the bending moment varies with the diameter cubed.

A comparison study was carried out for two piles having a diameter five times greater than the Sabine test pile diameter. The ratio, t/d, for the first pile is the same as the Sabine value; the thickness is doubled for the second pile. The parameter β is assumed constant. In the first case, Matlock's p-y model overpredicts the pile head deflection by 320% and the maximum bending moment by 61% with respect to the finite element model. Stevens-Audibert's p-y model overpredicts displacement by 160% and moment by 36%. For the second pile, the overpredictions are 360% (Matlock) and 180% (Stevens-Audibert) for pile head displacement, and 57% (Matlock) and 29% (Stevens-Audibert) for maximum bending moment. A 57% overprediction corresponds to approximately 10,000 tons-ft additional moment for a typical large scale pile.

Pile Flexural Rigidity

In what follows, it is assumed that the pile rigidity is varied by changing the wall thickness and holding the outer pile diameter constant, so that $\bar{\alpha}$ varies but β remains constant. The effect of pile stiffness on the behavior can be derived from Figures 4, 5 and 6 which contain plots of pile head stiffness, bending moment, and displacement along the pile in a non-dimensional form, as a function of $\bar{\alpha}$. Analysis of a pile similar to the Sabine experimental pile, but with double the moment of inertia predicted higher values, with respect to DEMETRA, for both moment and deflection. The discrepancy for a = 0.32 is significantly higher than observed for the original Sabine pile. When a is increased to 0.64, the discrepancy decreases, and the Stevens-Audibert results are in reasonable agreement with DEMETRA. Further increase to a = 1.44 results in close agreement between DEMETRA and Matlock's results. However, a = 1.44 is not realistic. An interesting result is observed when the modified Sabine clay profile which consists of a linear strength profile with a = 0.32 for depths below 10 ft. and the actual strength for the first 10 ft. below the surface (Figure 3) is used. Good agreement between all three methods is obtained.

Pile Head Fixity

Matlock (1970) reports that the p-y curves derived from free head and fixed head experiments were very close to each other. A similar trend is observed for the numerical solutions generated with the finite element program. The variation in the bending moment distribution along a fixed head pile ($\bar{a}=22560.0$ and $\beta=0.04$) with increasing head displacement is shown in Figure 8. The Sabine pile geometry was analyzed with DEMETRA and it was found that the pile head displacement and head moment were less than the values predicted by the p-y models. Taking DEMETRA's result as the reference value, the displacements are greater by 460% (Matlock) and 265% (Stevens-Audibert). A similar trend is observed for the pile head moment: 72% (Matlock) and 47% (Stevens-Audibert). A second pile, having the same diameter but double the thickness, exhibited a similar discrepency. The above results apply for a fixed head pile (no rotation of the pile head).

CONCLUSIONS

Predictions of the behavior of laterally loaded piles based on p-y curve idealizations for lateral soil resistance are generally more conservative than the results generated with a three dimensional finite element model. The discrepency in head displacement is quite large, and the pile head stiffness obtained with a p-y discretization is lower by a factor of 3 to 4. Agreement for bending moment is closer, with only a 50% difference. However, this difference is important for large diameter piles since the actual moment is high. It is also observed that the difference between results predicted by the Matlock and the Stevens-Audibert p-y formulations is sometimes greater than the difference between the finite element result and Matlock's prediction.

The non-dimensional solutions generated with the finite element model apply for a specific soil profile and provide information for a wide range of pile designs. Thus one can obtain a global picture of the lateral load capacity for different pile geometries which is useful for optimization and preliminary study of the pile foundation.

REFERENCES

American Petroleum Institute, 1980, API Recommended Practice for Planning, Designing and Constructing Fixed Offshore Platforms, 11th edition.

BANERJEE, P.K., and DAVIES, T.G., 1978, "The Behavior of Axially and Laterally Loaded Single Piles Embedded in Non-Homogeneous Soils", Geotechnique, Volume 28, Number 3.

BATHE, K.J., and WILSON, E.L., 1976, Numerical Methods in Finite Element Analysis, Englewood Cliffs, New Jersey: Prentice Hall Inc.

CHANDRASHEKAR, K.M., 1982, Response Sensitivity of Laterally Loaded Offshore Pile Foundations Using P-Y Curves, SM Thesis, Massachusetts Institute of Technology, Department of Civil Engineering.

DOUGLAS, D.J., and DAVIS, E.H., 1964, "The Movement of Buried Footings Due to Moment and Horizontal Load and the Movement of Anchor Plates", Geotechnique, Volume 14, pp. 115-132.

KAVVADAS, M., and BALIGH, M.M., 1980, Linear and Non-Linear Finite Element Techniques for Soil Consolidation Around Pile Shafts, Massachusetts Institute of Technology, Department of Civil Engineering, Research Report R80-41.

KAVVADAS, M., 1982, Non-Linear Consolidation around Driven Piles in Clays, Sc.D. Thesis, Massachusetts Institute of Technology, Department of Civil Engineering.

MATLOCK, H., and REESE, L.C., 1960, "Generalized Solutions for Laterally Loaded Piles", ASCE, Volume 86, SM5, pp. 63-91.

MATLOCK, H., 1970, "Correlations for Design of Laterally Loaded Piles in Soft Clay", Proceedings of the 2nd Offshore Technology Conference, Volume 1, pp. 577-594 Houston, Texas.

MINDLIN, R.D., 1936, "Force at a point in the Interior of a Semi-Infinite Solid", Physics 7, 195.

NAYAK, G.C., and ZIENKIEWICZ, O.C., 1972, "Elasto-Plastic Stress Analysis. A Generalization for Various Constitutive Relations Including Strain Softening", International Journal for Numerical Methods in Engineeering, Volume 5, pp. 113-135.

POLLALIS, S.N., 1982, *Analytical and Numerical Techniques for predicting the Lateral Stiffness of Piles*, Ph.D. Thesis, Massachusetts Institute of Technology, Department of Civil Engineering.

POULOS, H.G., and DAVIS, E.H., 1980, *Pile Foundation Analysis and Design*, New York: Wiley.

REESE, L.C., et al, 1975, "Field Testing and Analysis of Laterally Loaded Piles in Stiff Clay", *Proceedings of the 7th Offshore Technology Conference*, Houston, Texas, OTC Paper 2312, pp. 671-690.

REESE, L.C., 1977, "Laterally Loaded Piles: Program Documentation", *ASCE, Journal Geotechnical Engineering Division*, GT4, pp. 287-305.

ROSCOE, K.H., and BURLAND, J.B., 1968, "On the Generalized Behavior of Wet Clays", *Engineering Plasticity*, Cambridge, University Press, pp. 535-609.

STEVENS, J.B., and AUDIBERT, J.M.E., 1979, "Re-examination of p-y Curve Formulations", *Proceedings, 11th Offshore Technology Conference*, OTC Paper 3402, pp. 397-403.

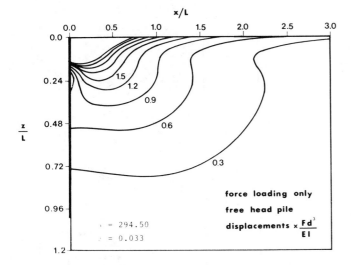

Fig. 1 Non-dimensional Contours of the Horizontal Displacement on the y=0 plane; Linear Elastic Half-Space.

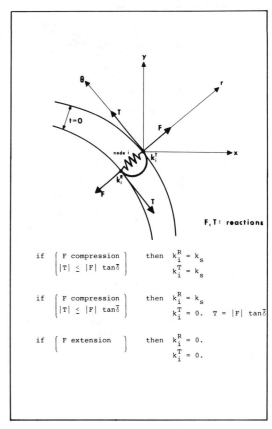

Fig. 2 Slippage and Gapping Modelling.

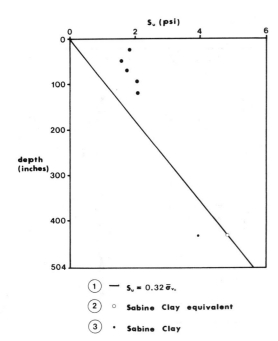

Fig. 3 The Different Shear Strength Profiles Used in the p-y Curves Method.

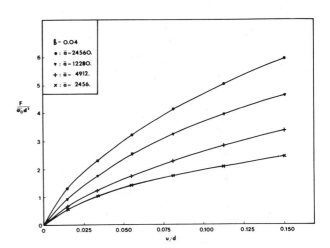

Fig. 4 Non-Dimensional Pile Head Stiffness for Different Values of $\bar{\alpha}$.

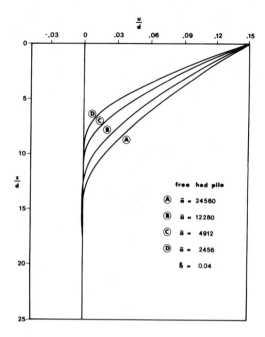

Fig. 5 Horizontal Displacements Along the Pile; Pile Head Applied Displacement 0.15d.

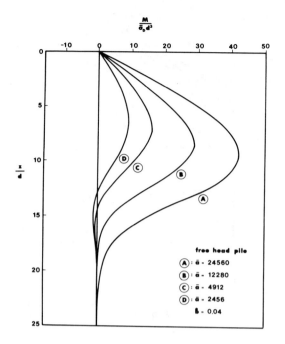

Fig. 6 Bending Moments Along the Pile; Pile Head Displacement 0.15d.

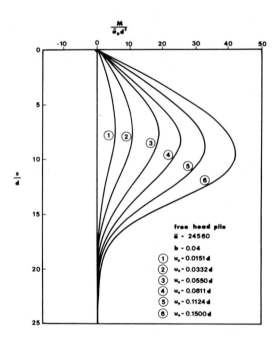

Fig. 7 Bending Moments Along the Pile for Incremental Pile Head Applied Displacement u_o.

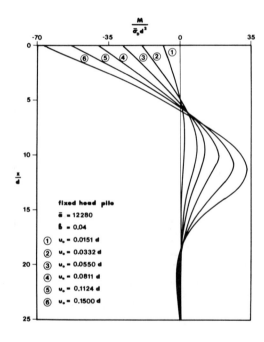

Fig. 8 Bending Moments Along the Pile for Incremental Pile Head Applied Displacement u_o.

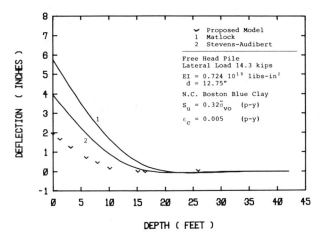

Fig. 9a Lateral Loaded Pile: Predictions of Horizontal Displacements.

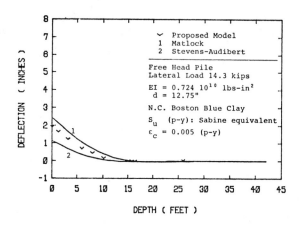

Fig. 10a Lateral Loaded Pile: Predictions of Horizontal Displacements.

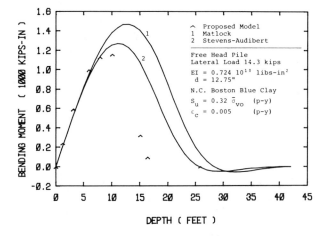

Fig. 9b Lateral Loaded Pile: Predictions of Bending Moments.

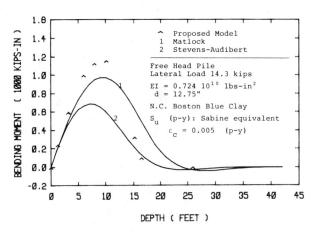

Fig. 10b Lateral Loaded Pile: Predictions of Bending Moments.

Technical Poster Sessions

FOUNDATIONS

Performance of Foundations

LATERAL STRESS MEASUREMENTS DURING STATIC AND CYCLIC DIRECT SIMPLE SHEAR TESTING

Rune Dyvik
Rensselaer Polytechnic Institute
U.S.A.

Thomas F. Zimmie
Rensselaer Polytechnic Institute
U.S.A.

SUMMARY

Consolidated, constant volume (CCV) static and cyclic laboratory shear tests were performed on marine clay soils. The Norwegian Geotechnical Institute (NGI) direct simple shear device, modified for cyclic loading (square wave) capabilities, was used for these tests. Three clays were investigated; two undisturbed samples from the Gulf of Mexico and the Gulf of Alaska, and a reconstituted Pacific Illite. All specimens were consolidated to a vertical confining stress of 0.518 kg/cm^2 and then sheared monotonically, or cyclically with complete stress reversal.

All tests included lateral stress measurements taken during consolidation and shear and these results are presented herein. These measurements were accomplished with the use of calibrated, wire-reinforced rubber membranes acting on a strain gage principle.

Lateral stress measurements greatly increase the knowledge of the state of stress within the soil specimen. During consolidation, the coefficient of lateral stress at rest, K_o, was determined experimentally (directly). The Mohr's circle state of stress was also established for a soil element at the center of the specimen during the entire shear test. The change in lateral stress ratio, the orientation of the failure planes, stress paths (q-$\bar{p}$ plots) and similar, were determined for the shear phase of the tests (similar to what is done for triaxial tests). These results are compared with the more conventional methods of analyzing shear data.

INTRODUCTION

The cyclic simple shear behavior of fine-grained soils subjected to repeated loadings such as those produced by earthquakes, wind, waves, and machine vibrations has recently been studied by many researchers (Anderson et al. 1978, Dyvik and Zimmie, 1981, Floess and Zimmie, 1979 and others).

However, lateral stresses in direct simple shear soil specimens are typically not measured. Without lateral stress measurements, the complete state of stress of the specimen is not known. Roscoe, et al. (1967), using the Cambridge Simple Shear Device have measured lateral stresses in sands during static shear. Youd (1975), Finn et al. (1978), and Tatsuoka et al (1980), have measured lateral stresses in sands during cyclic shear loading. This paper will present lateral stress results for three fine-grained soils (clays) during the consolidation stage and the static or cyclic portion of a direct simple shear test. A detailed procedure for the techniques and considerations involved in measuring and interpreting lateral stress results has been presented by Dyvik et al. (1981a), therefore this paper will primarily be a presentation of shear test results including lateral stresses.

TESTING

All tests were conducted utilizing a Norwegian Geotechnical Institute (NGI) direct simple shear apparatus (model number 4). The device was developed by NGI and is manufactured by Geonor. It was modified so that cyclic stress-controlled tests with square wave loading could be performed (test frequency was 0.25 Hz). The same NGI device is presently integrated with an MTS closed-loop servo hydraulic system and is capable of stress or strain-controlled cyclic testing in a variety of waveforms and frequencies.

During shear, undrained conditions were simulated by keeping the volume of the specimen constant. A wire reinforced membrane maintained a constant cross sectional area of the specimen. The height of the specimen was kept constant by changing the vertical (normal) stress. This change in vertical stress is assumed to be equal to the change in pore water pressure that would have occurred during an undrained test (Prevost and Høeg, 1976).

The results of tests on two undisturbed and one remolded clay samples which were part of the ongoing investigations performed at Rensselaer Polytechnic Institute will be presented herein. The undisturbed samples were a Gulf of Mexico sample obtained 225 miles (360 km) east of Corpus Christi, Texas and a Gulf of Alaska sample from the Copper River Prodelta. Both samples were 10.2 cm diameter undisturbed cores obtained from the United States Geological Survey (USGS). The reconstituted sample was a Pacific Illite dredged about 600 miles (960 km) north of Hawaii at about 5000 feet (1500 m) of depth by the University of Rhode Island and reconstituted at their Geomechanics Laboratory. Pertinent geotechnical data for the clays is listed below in Table 1. All three samples were essentially uniform and non-stratified.

TABLE 1 GEOTECHNICAL DATA

	GULF OF MEXICO CLAY	GULF OF ALASKA CLAY	PACIFIC ILLITE
WATER CONTENT	70-105	57-65	86-94
LIQUID LIMIT (AVERAGE,%)	105	50	86
PLASTIC LIMIT (AVERAGE,%)	30	27	30
SPECIFIC GRAVITY	2.71	2.84	2.71
SENSITIVITY (FALL CONE)	2.7	3.8	2.3
PERCENT SAND	2	1	0.05
PERCENT SILT	28	34	33
PERCENT CLAY	70	65	67
CONSOLIDATION HISTORY	N.C.	N.C.	N.C.

Calibrated, wire reinforced rubber membranes manufactured by Geonor for the NGI device were used in this study. These were the standard sized membranes with a specimen area of 50 cm^2.

The membranes must provide adequate lateral resistance to prevent cross sectional area change during consolidation and shear. The space between the wire reinforcement winding allows the membrane and specimen to strain vertically during consolidation.

The calibrated membranes operate on a strain gauge principle; the average lateral stress within the soil specimen is calculated from changes in the electrical resistance of the reinforcing wire due to its very small elongation. The total height of the wire windings is 3 cm. The middle third acts as the strain gauge. The soil specimen height is approximately 2 cm and lateral stresses are therefore measured at the central centimeter. The maximum allowable lateral stress in these membranes is about 1.4 kg/cm^2.

After the membranes had been calibrated (Dyvik et al. 1981a), mounted over the trimmed specimens and installed in the direct simple shear device, all specimens were consolidated to a confining stress of 0.518 kg/cm^2. This brought all the specimens to a normally consolidated state (OCR = 1.0). Although this low confining stress was used, these soft specimens consolidated vertically on the order of 10 to 20 percent.

After consolidation, the specimens were sheared either statically or cyclically (stress-controlled square wave with complete stress reversal). Lateral stresses were measured at all times during these tests and the results are presented herein.

CONSOLIDATION RESULTS

Since lateral strains in the specimen are not permitted by the wire reinforced membranes, the specimens are consolidated to at rest conditions. The measurement of lateral stresses enables one to easily, accurately and directly determine the coefficient of lateral earth pressure at rest (K_o). Figures 1a, 1b and 1c show the lateral stress results for three consolidation increments on each of the Gulf of Mexico, Gulf of Alaska clays and the Pacific Illite, respectively. The data suggests a straight line, the slope of which is the value of K_o for each sample. These K_o values compare very well with various indirect determinations (Dyvik et al. 1981a). In addition, during the consolidation phase one obtains the same compressibility and coefficient of consolidation data as obtained from the standard one-dimensional incremental consolidation test (Ladd, 1981, Schimelfenyg and Zimmie, 1981).

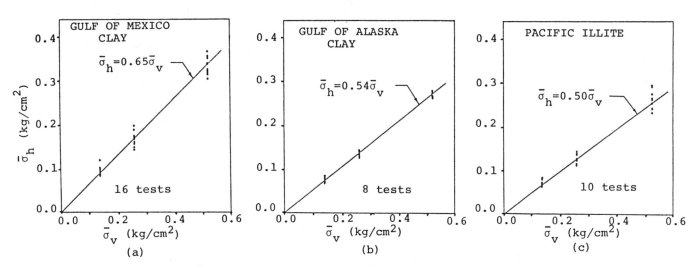

FIGURE 1   EFFECTIVE HORIZONTAL (LATERAL) STRESS VERSUS EFFECTIVE VERTICAL STRESS FOR THREE CONSOLIDATION LOAD INCREMENTS

SHEAR TEST RESULTS

The following sections will show lateral stress results obtained during the shear portions of the tests. Since the effective vertical normal stress, $\bar{\sigma}_v$, the effective horizontal normal stress, $\bar{\sigma}_h$ and the horizontal shear stress, τ_h are known at all times, a Mohr's circle state of stress can be defined for the specimen. Figure 2 shows this state of stress with equations to determine the following: $\bar{\sigma}_1$ and $\bar{\sigma}_3$, the major and minor principal stressses, respectively; θ_p, θ_q and θ_f, the angles between the horizontal plane of the specimen and the planes of the major principal stress, maximum shear stress and maximum obliquity, respectively; and q and $\bar{p}$, the coordinates of the top of the circle.

Static Shear Results

Several static shear tests were performed on each sample. The test results were very consistent and therefore only one representative static test for each of the three samples will be presented. Figure 3 is of the stress-strain-behavior with test numbers 1, 2 and 3 corresponding to the Gulf of Mexico clay, Gulf of Alaska clay and the Pacific Illite samples, respectively. The shear stresses in this Figure have been normalized to the initial vertical confining stress, $\bar{\sigma}_{vo}$ of 0.518 kg/cm^2. The ultimate static shear strengths, s_u, for tests 1, 2 and 3 are 0.129, 0.133 and 0.145 kg/cm^2, respectively.

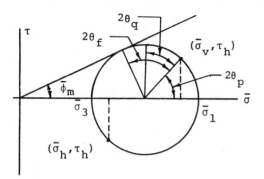

$$\bar{p} = \frac{\bar{\sigma}_v + \bar{\sigma}_h}{2} = \frac{\bar{\sigma}_1 + \bar{\sigma}_3}{2} \qquad \theta_q = 45 - \theta_p$$

$$q = \left[\frac{(\bar{\sigma}_v - \bar{\sigma}_h)^2}{4} + \tau_h^2\right]^{1/2} = \frac{\bar{\sigma}_1 - \bar{\sigma}_3}{2} \qquad \theta_f = 45 + \frac{\bar{\phi}_m}{2} - \theta_p$$

$$\theta_p = \tan^{-1}\left[\frac{\tau_h}{\frac{(\bar{\sigma}_v - \bar{\sigma}_h)}{2} + q}\right] \qquad \bar{\sigma}_1 = \bar{p} + q$$

$$\bar{\sigma}_3 = \bar{p} - q$$

FIGURE 2   MOHR'S CIRCLE STATE OF STRESS FOR AN INFINITESIMAL ELEMENT OF SOIL AT THE CENTER OF THE NGI DIRECT SIMPLE SHEAR SPECIMEN

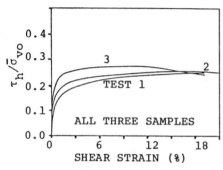

FIGURE 3   NORMALIZED STRESS-STRAIN CURVES FOR STATIC TESTS

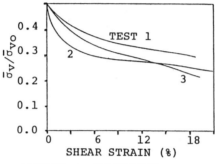

FIGURE 4   NORMALIZED VERTICAL STRESS VERSUS SHEAR STRAIN

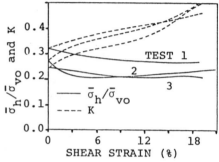

FIGURE 5   NORMALIZED LATERAL STRESS AND $K = \bar{\sigma}_h / \bar{\sigma}_v$ VERSUS SHEAR STRAIN

Figure 4 shows the change in normalized effective vertical normal confining stress, $\bar{\sigma}_v / \bar{\sigma}_{vo}$, with shear strain. This plot also reflects the normalized excess pore pressure, $\Delta u / \bar{\sigma}_{vo}$, with shear strain since the sum of $\bar{\sigma}_v / \bar{\sigma}_{vo}$ and $\Delta u / \bar{\sigma}_{vo}$ always equals a constant total vertical confining stress or 1.0 (the curves can be flipped and started at the origin). As expected for normally consolidated clays, $\bar{\sigma}_v$ monotonically decreases (or Δu monotonically increases) with increasing shear strain.

Figure 5 shows the relationship between the normalized effective horizontal normal stress, $\bar{\sigma}_h / \bar{\sigma}_{vo}$ (solid lines), as well as the coefficient of lateral stress, K (dashed lines), versus shear strain for the three samples. The coefficient K, is defined as the ratio of the effective horizontal normal stress, $\bar{\sigma}_h$, to the effective vertical normal stress, $\bar{\sigma}_v$, in the specimen. The lateral stresses monotonically decrease to a value close to the vertical confining stress at high shear strains which can be seen as the value of K approaching unity with increasing shear strains. Note that the values of $\bar{\sigma}_h / \bar{\sigma}_{vo}$ and K start at K_o at zero shear strain.

With known vertical and lateral stresses within the specimen, a complete Mohr's circle state of stress can be defined at any time during shear. Figure 6 shows the stress paths (q versus $\bar{p}$) for the three samples with q and $\bar{p}$ being defined in Figure 2. Each curve starts at the top of the circle defining K_o conditions at the end of consolidation (where $\bar{\sigma}_v$ and $\bar{\sigma}_h$ are principal stresses) and then move to the left with increasing shear. Plots of the principal stress ratio, orientation and magnitude of the maximum shear stress and principal stresses, and similar, which follow from the Mohr's circle state of stresses can easily be constructed.

A mobilized effective angle of internal friction, $\bar{\phi}_m$, for each sample can be determined from the following relation:

$$\bar{\phi}_m = \sin^{-1} q/\bar{p} \qquad (1)$$

This friction angle is computed in the same way as is done for triaxial shear tests.

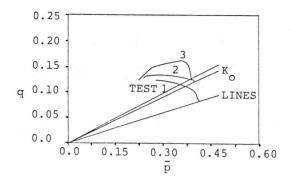

FIGURE 6 STRESS PATHS (q-p̄) FOR STATIC TESTS

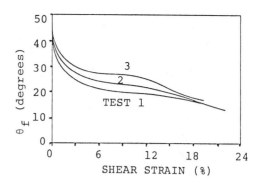

FIGURE 7 THE ANGLE BETWEEN THE HORIZONTAL PLANE OF THE SPECIMEN AND THE PLANE OF MAXIMUM OBLIQUITY VERSUS SHEAR STRAIN

The angle between the horizontal plane of the specimen and the plane of maximum obliquity will be defined as θ_f. For direct shear or direct simple shear tests without lateral stress measurements, the horizontal plane of the specimen is assumed to be the plane of maximum obliquity (i.e., $\theta_f = 0$). The angle of internal friction, $\bar{\phi}$, of the soil is then calculated by:

$$\bar{\phi} = \tan^{-1} \tau_h/\bar{\sigma}_v \qquad (2)$$

There is an important difference between the friction angles $\bar{\phi}_m$ and $\bar{\phi}$: the former includes lateral stress information and the latter does not. The values of $\bar{\phi}_m$ and $\bar{\phi}$ are, of course, only defined at a failure criterion, but if one follows the progression of each from the initiation of shear to failure, the values computed from Equation 2 would start at zero (τ_h is zero) but the values from Equation 1 would start at a non zero value (a tangent to the Mohr's circle state of stress corresponding to K_o-conditions after consolidation and before shear). Also, if one constructs a Mohr's circle tangent to the $\bar{\phi}$ line at the intersection of τ and $\bar{\sigma}_v$ (at failure), $\bar{\sigma}_v$ would be on the lower side of the center of the circle ($\bar{p}$). This would imply that the lateral stress ($\bar{\sigma}_h$) would be greater than $\bar{\sigma}_v$ and that K could always be greater than unity. This, as can be seen from Figure 5, is clearly not the case. Lambe and Whitman (1969) present a similar discussion in a qualtative manner.

Figures 8a, 8b and 8c show a comparison of values of $\bar{\phi}_m$ and $\bar{\phi}$ with progressing shear strain for each sample. As can be seen in these Figures, the values of $\bar{\phi}_m$ are always greater than $\bar{\phi}$ but are closer to each other at large shear strains. The reason for this can, perhaps, best be shown by a plot of θ_f with progressing shear strain as is shown in Figure 7 for the three samples. The assumption that θ_f equals zero, as is made in the computation of $\bar{\phi}$, is never true, but θ_f attains lower values at large shear strains and therefore $\bar{\phi}$ should converge towards $\bar{\phi}_m$.

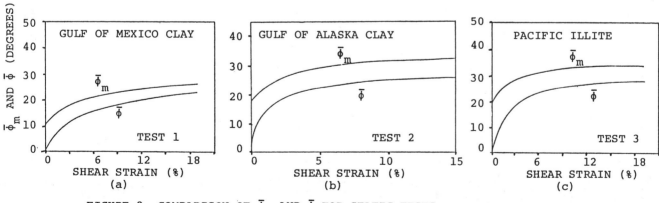

FIGURE 8 COMPARISON OF $\bar{\phi}_m$ AND $\bar{\phi}$ FOR STATIC TESTS

Cyclic Shear Results

A selected number of cyclic tests performed on each sample will be presented. All tests were performed on normally consolidated specimens with a confining pressure of 0.518 kg/cm^2 (as for the static tests). The specimens were then subjected to symmetric, complete stress reversal stress-controlled square wave cycles at a frequency of 0.25 Hz. Figures 9a, 9b and 9c show the progression of cyclic shear strain, γ_c, with number of loading cycles in the tests on the Gulf of Mexico clay (6 tests) the Gulf of Alaska clay (3 tests), and the Pacific Illite (5 tests), respectively. The reported cyclic shear strain is one half peak to peak (left to right) strain. The number next to each curve is the cyclic shear stress level expressed as a percent of the ultimate static shear strengths, s_u. Note that of the six cyclic tests presented in Figure 9a, only three cyclic stress levels were used. The two tests at 46% τ_c/s_u and the three tests at 54% τ_c/s_u indicate excellent repeatability of duplicate cyclic tests.

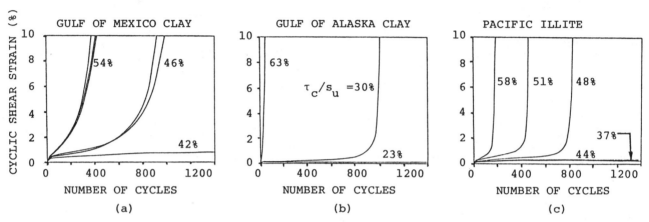

FIGURE 9 CYCLIC SHEAR STRAIN VERSUS NUMBER OF LOADING CYCLES

The relationship between the normalized effective vertical normal stress, $\bar{\sigma}_v/\bar{\sigma}_{vo}$, and number of loading cycles is shown in Figures 10a, 10b and 10c. As for the static tests, these plots also reflect the pore pressure generated in the cyclic tests. Figures 11a, 11b and 11c are of the normalized effective horizontal normal stress $\bar{\sigma}_h/\bar{\sigma}_{vo}$ (solid lines), and the lateral stress ratio, K (dashed lines) versus number of loading cycles. The effective vertical and lateral stresses decrease with increasing number of loading cycles, but for the cases where large cyclic shear strains develop (high cyclic stress level tests), $\bar{\sigma}_v$ drops down to and slightly below the level of $\bar{\sigma}_h$, making the value of K just greater than 1.0.

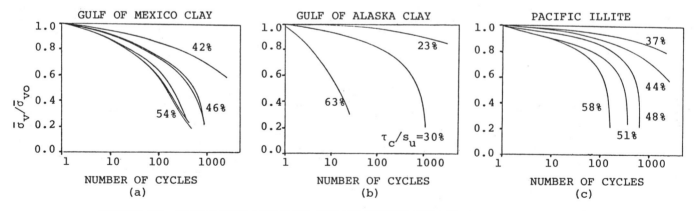

FIGURE 10 NORMALIZED VERTICAL STRESS VERSUS NUMBER OF LOADING CYCLES

Figures 12a, 12b and 12c are the stress paths (q versus $\bar{p}$) for each test on the three samples. A dashed line corresponding to K_o conditions for each sample is also shown. Each stress path starts on the K_o line and jumps almost vertically upwards in the first cycle. This is due to the fact that shear stresses are suddenly applied to the planes of $\bar{\sigma}_v$ and $\bar{\sigma}_h$ which were once principal stress and now are not. Therefore the Mohr's circle suddenly increases in diameter and then gradually decreases in diameter and moves towards the origin (to the left). The stress paths are

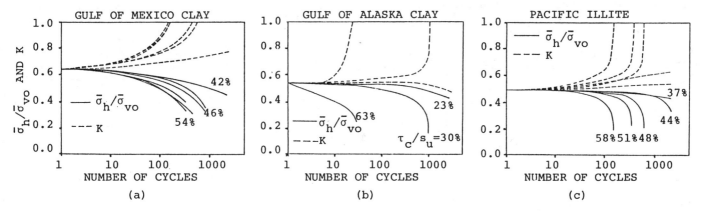

FIGURE 11 NORMALIZED HORIZONTAL (LATERAL) STRESS AND $K = \bar{\sigma}_h/\bar{\sigma}_v$ VERSUS NUMBER OF LOADING CYCLES

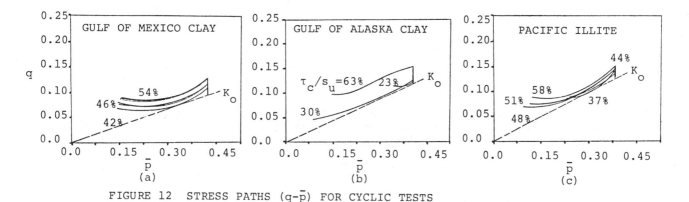

FIGURE 12 STRESS PATHS (q-$\bar{p}$) FOR CYCLIC TESTS

fairly parallel with the higher paths corresponding to higher cyclic shear stress levels.

Figures 13a, 13b and 13c show the progression of the angle between the horizontal plane of the specimen and the plane of maximum obliquity for these tests. For the same tests in which the cyclic shear strains became very large and the K values exceed unity, θ_f decreases to values close to zero.

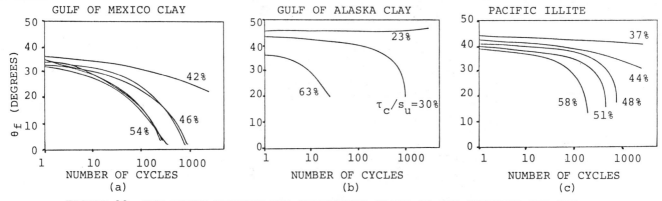

FIGURE 13 THE ANGLE BETWEEN THE HORIZONTAL PLANE OF THE SPECIMEN AND THE PLANE OF MAXIMUM OBLIQUITY VERSUS NUMBER OF LOADING CYCLES

Figures 14a, 14b and 14c show the comparison of the effective friction angles $(\bar{\phi}_m)_c$ and $(\bar{\phi})_c$ with increasing cyclic shear strain, where the subscript c represents cyclic test results. These curves are the average for all the cyclic tests on each sample (the friction angles are very similar at the same cyclic shear strains in the different cyclic tests). The slight drop in $(\bar{\phi}_m)_c$ at small strains is reflected in the q-$\bar{p}$ plots as they converge towards the K_o-line and then deviate from it. The curves for $(\bar{\phi}_m)_c$ and $(\bar{\phi})_c$ are closest at high cyclic shear strains with $(\bar{\phi}_m)_c$

the higher of the two (as in the static case). It is interesting to note that the values of $(\bar{\phi}_m)_c$ and $(\bar{\phi})_c$ are very similar to the values of $\bar{\phi}_m$ and $\bar{\phi}$ (from the static tests), respectively, up to 3 or 4 percent shear strain.

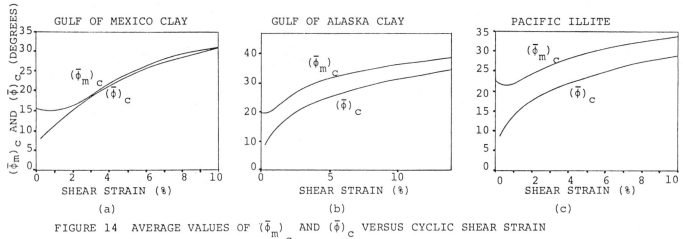

FIGURE 14   AVERAGE VALUES OF $(\bar{\phi}_m)_c$ AND $(\bar{\phi})_c$ VERSUS CYCLIC SHEAR STRAIN

SUMMARY OF LATERAL STRESSES THROUGHOUT ALL THE TESTS

In the consolidation section, the relationship between lateral stresses and vertical stresses (the K_o line) was established for each clay sample. It may be of interest to investigate this same relationship during the shear portions of the static and cyclic stress-controlled tests.

Figures 15a, 15b and 15c are plots of normalized effective lateral stress versus normalized effective vertical stress for all the static tests and cyclic tests performed on each clay sample (more tests are included in this figure than were previously introduced). As positive pore pressures are generated during shear, the effective vertical and lateral stresses decrease (i.e., the soil specimens are unloaded). The bulk of the shear tests follow a fairly unique curve which starts at the end of the K_o loading line, is above this line and concave downward. The only exceptions seem to be static tests at very high shear strains (13% and up) at which point they deviate upwards from the curve, and low cyclic stress level tests that stabilize at a small cyclic shear strain (cyclic equilibrium) which fall slightly below this curve.

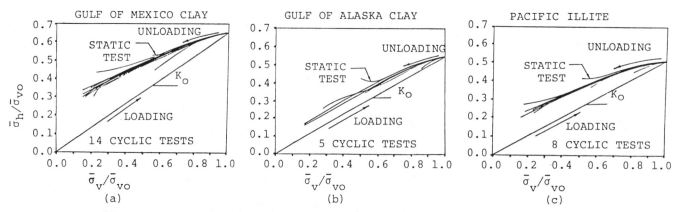

FIGURE 15   NORMALIZED EFFECTIVE HORIZONTAL (LATERAL) STRESS VERSUS NORMALIZED EFFECTIVE VERTICAL STRESS DURING LOADING (CONSOLIDATION) AND UNLOADING (STATIC AND CYCLIC SHEAR)

The general shape and location of these unloading curves is exactly what would be expected during unloading of a soil specimen in a consolidation test (Hendron, 1963). Singh et al., (1978) indicates that the build up of excess pore pressures during cyclic loading is very similar to changing (increasing) a soil's overconsolidation ratio (OCR) in the unloading portion of a consolidation test. One difference is that these specimens were under constant volume conditions during shear and consolidation test specimens are not, but this may be of little consequence, especially with soils having a high rebound modulus. This concept of "induced OCR" may be of importance in determining the static behavior of soils previously subjected to cyclic loading.

CONCLUSIONS

Calibrated wire reinforced membranes have been successfully used in the NGI direct simple shear device to measure lateral stresses within the soil specimens during consolidation and static and/or cyclic shear. Consistent test results for undisturbed Gulf of Mexico clay and Gulf of Alaska clay samples and a reconstituted Pacific Illite were obtained.

The calibrated membranes provide a convenient method for experimentally determining the coefficient of lateral stress at rest, K_O, of a soil specimen. This direct determination has been found to agree well with various indirect and other experimental determinations of K_O. If satisfactory lateral stress measurements are obtained prior to shear, the measurement of lateral stress during shear is a relatively simple matter for static, cyclic or combined shear tests.

Lateral stress measurements during shear allow one to define the complete Mohr's circle state of stress and the associated magnitudes and orientations of the principal stresses, maximum shear stress, failure plane, and the effective mobilized angle of internal friction as well as stress paths, lateral stress ratios and similar. The analysis of the shear tests becomes similar to that of triaxial test results. One important observation is that the horizontal plane of the specimen is never the plane of maximum obliquity as is often assumed in direct shear tests, but that the angle between them becomes smaller at large shear strains (as K approaches unity).

Static simple shear tests with lateral stress measurements can also be used to predict monotonic triaxial compression and extension stress-strain behavior with the use of Plasticity Theory as has been demonstrated by Prevost (1978). The comparison must, however, be made with anisotropically consolidated triaxial tests (consolidated to the same K_O conditions as determined from the first part of the simple shear test).

ACKNOWLEDGEMENT

The authors wish to express their appreciation to Dr. Carsten Floess and Mr. Paul Schimelfenyg for their research work on this project. Also, to Dr. Dwight Sangrey, Carnegie-Mellon University, for providing soil samples and many helpful comments; to Dr. Armand Silva and David Calnan, University of Rhode Island, for providing the samples of Pacific Illite; and to the United States Geological Survey for providing offshore marine samples.

Funds for this project were provided by grants from the National Science Foundation; Grant No. PFR78-18743, sponsored by the Earthquake Hazards Mitigation Program. The NSF Project Officer was Dr. William Hakala.

REFERENCES

1. Anderson, K.H., Hansteen, O.E., H\phi eg, K., and Prevost, J.H., 1978, Soil Deformations due to Cyclic Loads on Offshore Structures, NGI Publication No. 120, Oslo, Norway, pp. 1-40.

2. Dyvik, R. and Zimmie, T.F., 1981, Strain and Pore Pressure Behavior of Fine Grained Soils Subjected to Cyclic Shear Loading, NSF Final Report, Rensselaer Polytechnic Institute, Troy, N.Y. (also available through NTIS).

3. Dyvik, R., Zimmie, T.F. and Floess, C.H.L., 1981a, Lateral Stress Measurements in Direct Simple Shear Device, Special Technical Publication 740, Shear Strength of Soils, ASTM, Philadelphia, Pa., pp. 191-206.

4. Dyvik, R., Zimmie, T.F. and Schimelfenyg, P., 1981b, "Cyclic Simple Shear Behavior of Fine Grained Soils", Proceedings International Conference on Recent Advances in Geotechnical Earthquake Engineering and Soil Dynamics, St. Louis, pp. 101-106.

5. Finn, W.D. Liam, Vaid, Y.P. and Bhatia, S.K., 1978, "Constant Volume Cyclic Simple Shear Testing", 2nd International Conference on Microzonation, San Francisco, Calif.

6. Floess, C.H.L. and Zimmie, T.F., 1979, Direct Simple Shear Behavior of Fine Grained Soils Subjected to Repeated Loads, NSF Report, Rensselaer Polytechnic Institute, Troy, N.Y., (also avaliable through NTIS).

7. Hendron, A.J., Jr., 1963, The Behavior of Sand in One-Dimensional Compression, Doctoral Thesis, University of Illinois, Urbana.

8. Ladd, C.C., 1981, Discussion on Laboratory Shear Devices, Special Technical Publication 740, ASTM, Philadelphia, Pa., pp. 643-652.

9. Lambe, T.W. and Whitman, R.V., 1969, <u>Soil Mechanics</u>, John Wiley and Sons, Inc., New York, NY.

10. Prevost, J.H., 1978, "Plasticity Theory for Soil Stress-Strain Behavior", <u>Journal Engineering Mechanics Division</u>, ASCE, Vol. 104, No. EM5, pp. 1177-1194.

11. Prevost, J.H. and Høeg, K., 1976, "Reanalysis of Simple Shear Soil Testing", <u>Canadian Geotechnical Journal</u>, Vol. 13, No. 4, pp. 418-429.

12. Roscoe, K.H., Bassett, R.H. and Cole, E.R.L., 1967, "Principle Axes Observed During Simple Shear of a Sand", <u>Proceedings, Geotechnical Conference</u>, Oslo, Vol. 1, pp. 231-237.

13. Schimelfenyg, P. and Zimmie, T.F., 1981, <u>Consolidation and Stress Reversal Using the NGI Direct Shear Device</u>", NSF Interim Report, <u>Rensselaer Polytechnic Institute</u>, Troy, N.Y. (also available through NTIS).

14. Singh, R.D., Gardner, W.S. and Dobry, R., 1978, "Post Cyclic Loading Behavior of Soft Clays", <u>Proceedings, International Conference on Microzonation for Safer Construction Research and Application</u>, San Francisco, pp. 945-956.

15. Tatsuoka, F., Silver, M.L., Phukunhaphan, A., and Anestis, A., 1980, "Cyclic Undrained Strength of Sand by Simple Shear Test and Triaxial Test I (Test Procedures)", <u>Journal of the Institute of Industrial Science</u>, University of Tokyo, 32, 1, pp. 35-38.

16. Youd, T.L. and Craven, T.N., 1975, "Lateral Stress in Sands During Cyclic Loading", <u>Journal Geotechnical Engineering Division</u>, ASCE, Vol. 101, No. GT2, pp. 217-221.

REDUCTION OF PORE WATER PRESSURE
BENEATH CONCRETE GRAVITY PLATFORMS

O. Eide   A. Andresen   R. Jonsrud   E. ANDENÆS
Norwegian Geotechnical Institute, Oslo, Norway

SUMMARY

The six Condeep concrete gravity platforms installed in the North Sea during the period 1975 to 1981 are all equipped with an under base drainage system. This will also be the case with the next Condeep now under construction and scheduled for installation 1984.

The paper outlines the objectives of the drainage system, the construction details and experiences gained during operation.

1. INTRODUCTION

When the first Condeep platforms, Beryl A for Mobil and Brent B for Shell, were designed back in 1973, no one had any experience with such structures. Especially for the foundation design this was a serious matter. Both the size of the foundation area and the loading conditions were unprecedented. Compared to structures on land, loading conditions are characterized by the enormous horizontal loads from the waves acting two ways cyclicly. We did not at that time know very much about pore pressure build up and degradation of clays due to cyclic loading. In addition the soil investigation techniques were less advanced than they are today, especially with regard to the possibility of obtaining undisturbed samples.

On this background it was natural to look for every possible way of improving the foundation by methods not being too costly. Such an improvement would be to increase the effective stresses below the base by drainage, and this was proposed by NGI.

All six Condeep platforms installed in the North Sea so far, have been equipped with steel skirts penetrating 3.5 - 5 m into the sea bed. These skirts are dividing the platform base into compartments which can be drained separately. The drainage system has also been designated the antiliquefaction system. The applied underpressure varies between 10 and 20 metres.

The drainage system may also be used for injection of water into the skirt compartments during removal of the platform after the production has been terminated.

It is however, not intended to operate the drainage system during the whole lifetime of the platform. It is especially during the first storm periods before full consolidation has taken place, that the effect of the system is of greatest importance.

2. FACTORS AFFECTED BY DRAINAGE

The drainage system may be used to influence the platform behaviour in several ways.

2.1 RATE OF SETTLEMENT

The drainage system will speed up consolidation settlement. This is an advantage as all settlements taking place before conductor installations and riser connections, will have no consequences. Long term settlements will cause negative skin friction on conductors and lead to relative movements.

2.2 TILT

As the base is divided into skirt compartments, usually 6 - 10 compartments, it is possible to adjust the water pressure within each compartment separately. In this way corrections for tilt can be made. This has been done in the final stages of skirt penetration. No correction for tilt has been necessary after grouting.

2.3 STABILITY

The main purpose of the drainage system is to improve the stability. The most critical period in the lifetime of a gravity platform with respect to foundation behaviour is before consolidation has taken place. Full consolidation has been obtained after 1 to 2 years for the present platforms installed on very stiff clays.

3. FOUNDATION DESIGN

According to the regulations of the Norwegian Petroleum Directorate, NPD, a gravity platform placed on the Norwegian Continental Shelf shall exhibit sufficient foundation capacity to withstand a 100-year storm. This storm shall in the design be expected to occur during the first winter after platform installation.

Usually there will be a significant increase in the soil strength during the first months after platform installation due to the consolidation process. The amount that can be included in the design is dependent on the rate of strength increase and time available before occurrence of the storm. Thus use of underpressure to speed up the consolidation might be of great favour in the foundation design.

To illustrate this its effect on the Statfjord B platform is presented in the following.

In brief the soils at the site consist of a top sand layer with an average thickness of 1 metre. Below there is a silty clay material extending to great depths interbedded with a few silty fine sand layers. The clay has an overconsolidation ratio, OCR, decreasing from above 25 at the top to 5 at 10 metres depth, becoming normally consolidated below 50 metres.

The NPD regulations require a certain safety agains failure, normally calculated from bearing capacity considerations, effect of cyclic wave loading included. In a paper to this conference K.H. Andersen et al. recommend to include a control against large displacements under cyclic loading. This is included herein.

Figures 1 and 2 show the results from the Statfjord B design. The curves show relative increase in safety against bearing capacity failure and large displacements with and without an underpressure of 200 kN/m² in the drainage system. It is shown that during the first 6 months the time to achieve a certain level of safety is reduced to approximately the half by means of the underpressure.

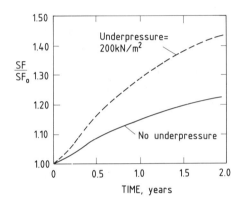

Fig. 1 Safety against bearing capacity failure versus time after platform installation

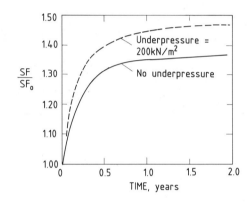

Fig. 2 Safety against large displacements under cyclic loading versus time after platform installation

4. DESIGN OF DRAINAGE SYSTEM

4.1 GENERAL

The system can either consist of drainage wells which have been drilled and installed after the platform has been put down, or it can be prefabricated filters placed below the base of the structure. Filters can be placed on the inside of the skirts, on dowels or on penetrating domes or concrete blocks. Where to place the filter is dependent on the soil conditions, whether a top sand layer is present or not.

Tubes from all filters in one skirt compartment are connected to a common standpipe in the utility shaft. The drainage level can be regulated by opening valves at 5 - 10 - 15 or 20 m below LAT. See Fig. 3. The amount of drainage water can be measured for each compartment separately, and samples may betaken to check the content of solid particles and make chemical analysis.

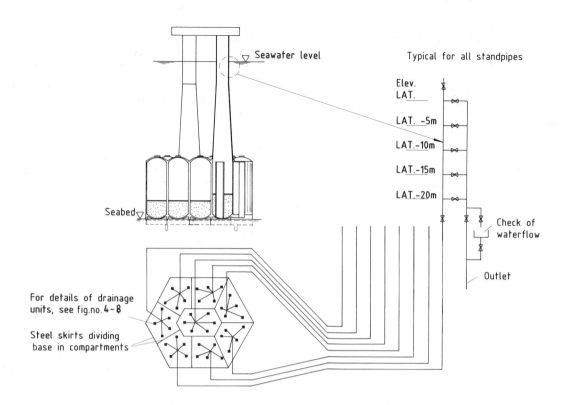

Fig. 3   Drainage system, principle drawing

4.2   DRAINAGE WELLS

On the two first Condeep platforms two drainage wells were installed to 25 m depth below sea floor to drain deeper sand and silt layers. The wells are located below the drill shafts, and they are connected to a stand pipe in the utility shaft where the drainage level can be regulated, see Figs. 3 and 4.

6" wire wrapped Johnson well screens of stainless steel AISI 316, with 0.33 mm slots were installed in prebored 14 3/4" holes stabilized by CMC mud. The holes were then filled with gravel up to the top of the filter screens. There have been some problems in establishing stable bore holes of relatively uniform diameter. Special types of drilling muds and slow penetration made it possible to overcome these difficulties.

Boring and installation of drainage wells were found time consuming and interferred with other ongoing activities on the platform. Partly for this reason, drainage wells have not been utilized on later platforms.

4.3   DRAINAGE FILTERS

Figures 4, 5, 6, 7 and 8 show in principle the different types of drains which have been utilized.

For vertical drains mounted on steel skirts and dowels, mineral wool has been used as filter material inside perforated steel casings, see Fig. 5.

The specifications of the mineral wool is as follows:

    Make: Rockwool
    Density: 150 kg/m^3
    Permeability: 1 x 10^{-1} cm/s

The filter material retains silt.

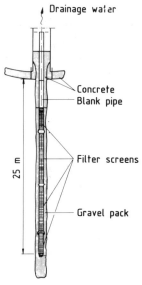

Fig. 4   Deep well

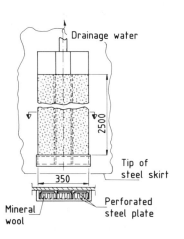

Fig. 5   Drain on steel skirt

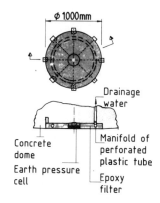

Fig. 6   Drain on concrete dome with earth pressure cell

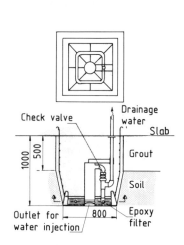

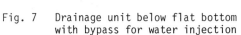

Fig. 7   Drainage unit below flat bottom with bypass for water injection

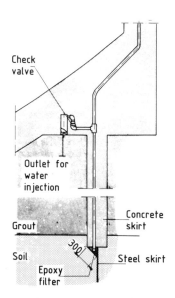

Fig. 8   Drain on skirt with bypass for water injection

For horizontal drains which are intended to penetrate only a short distance into a top sand layer, epoxy bonded sand has been used as filter material. Grain size of the sand has been 0.6 - 1.2 mm. The filter can either be made of one sand fraction or in two layers of different fractions, the less permeable layer at the surface. The filter material has a minimum compression strength of 1 MPa. As a safety precaution if the filter should crack, a filter cloth is wrapped around the manifold inside the filter.

The epoxy filters have mainly been used on concrete elements and bottom domes, see Figs. 6, 7 and 8.

During construction the drains are submerged in sea water for 1 - 2 years, and in order to prevent marine growth during this period, an antifouling liquid has been applied to the filters.

4.4 PIPE SYSTEM

The pipe systems have been dimensioned in accordance with expected flow of water and required degree of automation. Systems with level gauges, flow meters and automatic pumps have been installed. However, the simple system shown in Fig. 3 has proved to be preferable. The drainage head for each skirt group may by this system be adjusted to required level by opening or closing valves. Water flows have been measured by means of a calibrated bucket.

5. EXPERIENCE

5.1 HEAD OF OPERATION

The drainage systems on most of the platforms have been operated at a level 10 m below LAT, corresponding to 11 - 12 m below mean sea level.

5.2 RATE OF WATER FLOW

The amount of drainage water varies from one platform to another depending on the soil conditions. Also, on one platform the amount of drainage water from the different skirt compartments may vary considerably for the same drainage level. In general the water flow decreases with time to a small fraction of the initial flow, as illustrated in Figs. 9 and 10.

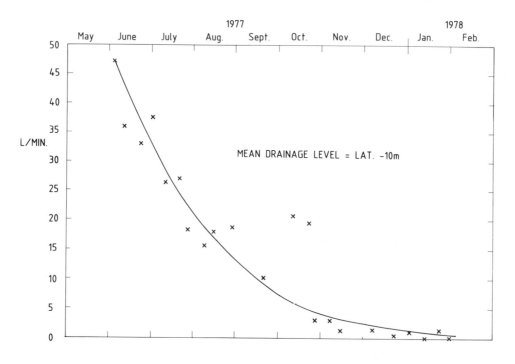

Fig. 9 Flow of water, Statfjord A

The flow of water has also been compared to the rate of settlement, see Fig. 10, where the amount of water drained is 4 - 8 times greater than that corresponding to the rate of settlement. For the last installed platform the skirts penetrated the top sand layer and into the clay. No piping took place during skirt penetration, and the sealing was perfect. At this platform very small quantities of drainage water have been measured, even for an underpressure of 20 m below LAT. The quantity of drainage water corresponds to the rate of settlement. See Figs. 11, 12 and 13. Two months after installation the rate of settlement was 0.3 mm per day. With a base area of 18,200 m^2, this corresponds to 5.5 m^3 per day. The amount of drainage water measured at that time was 4.6 m^3 per day.

A reduction in rate of settlement has been observed when drainage is stopped and hydrostatic water pressure in the skirt compartments reestablished, see Figs. 14 and 15.

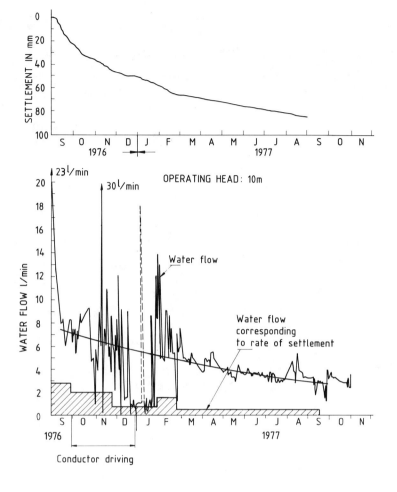

Fig. 10   Settlement and flow of water
         Brent B

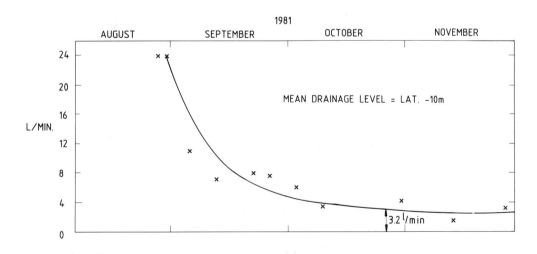

Fig. 11   Flow of water, Statfjord B

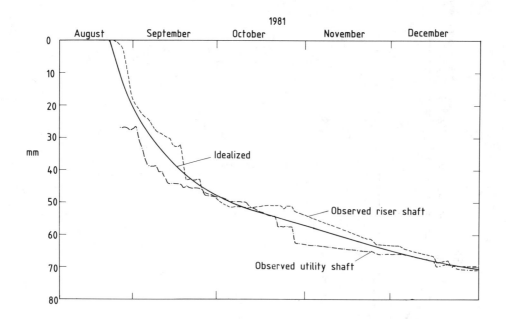

Fig. 12   Settlement, Statfjord B

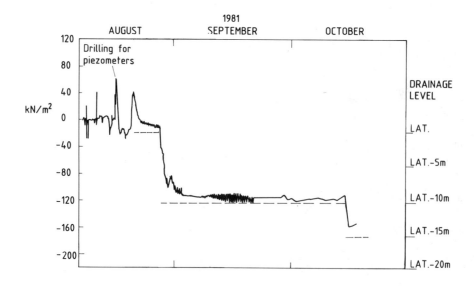

Fig. 13   Skirt pressure, Statfjord B

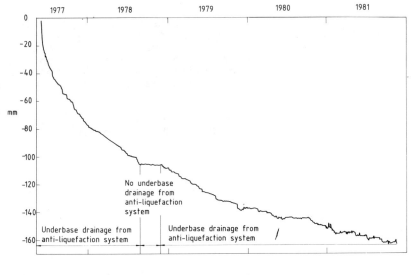

Fig. 14 Settlement, Statfjord A

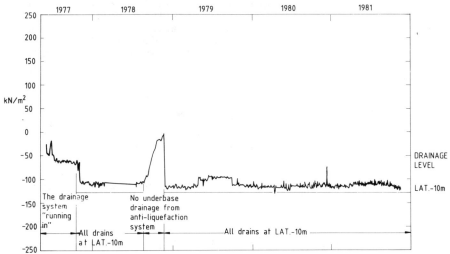

Fig. 15 Skirt pressure, Statfjord A

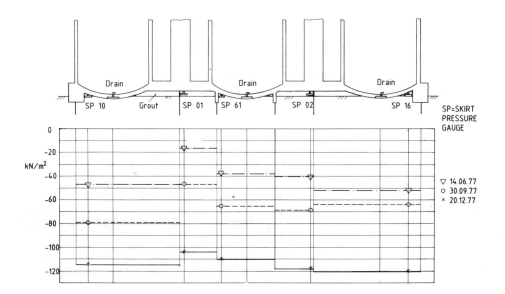

Fig. 16 Distribution of skirt water pressure across section, Statfjord A

381

5.3 CONTAMINATION OF DRAINED WATER

Contamination of the drained water varies with time. The contents of solids in suspension is usually very small. Chemicals from the underbase grouting are dominant in the beginning. The drained water has a bad smell, and very often the water has a black colour immediately after sampling. The colour disappears within a few minutes in open air. It is assumed that the black colour is due to the presence of anaerobic sulphate reducing bacterial. The smelly components have not been identified, however, there are strong indications that amines and mercaptones are responsible for the smell of the water.

In a few cases the amount of solids in the drainage water from one or two compartments have been on the borderline to what is acceptable, and the valves have been throttled for some time. A complete shut down of drainage from one compartment has also been necessary, possibly due to a damaged filter.

On two of the first platforms installed the drainage systems have now been closed down. The filter may also be clogged after 3 - 5 years in operation. At that time full consolidation has been obtained, and the foundation stability has improved considerably.

6. REMOVAL

It is usually a requirement in the job specifications that the platform should be removed after production is terminated. It is not believed that the drainage filters will be intact after 30 years. For this reason, injection of water into the skirt compartments will be done through bypass connections. The only safe way of pulling the skirts and dowels out of the ground will be by increasing the water pressure below the platform after deballasting to a neutral buoyancy.

7. CONCLUSIONS

The drainage systems, also termed antiliquefaction system, on the installed Condeep platforms have functioned satisfactorily and in accordance with the intentions.

Only a small amount of water has to be drained and pumped in order to maintain an underpressure of the order of 10 to 20 m below LAT under the base slab, see Figs. 9 and 16. The drainage has speeded up the rate of consolidation settlements and stability improvement. This has been of significance to the foundation as the time just after installation is the most critical period in the lifetime of a gravity platform.

On NGI proposal the system was applied patented by Høyer-Ellefsen A/S in 1973.

It remains to be seen how the system can be utilized during removel of the platforms after production is terminated, 25 to 30 years after installation.

ACKNOWLEDGEMENTS

The authors want to express their thanks to the client the Norwegian Contractors, who has responded positively to NGI proposals for underbase drainage on the Condeep platforms, and for the permission to publish information on the systems utilized.

REFERENCES

ANDERSEN, K.H., S. LACASSE, P.M. AAS and E. ANDENÆS, 1982, "Review of Foundation Design Principles for Offshore Gravity Platforms," To be publ. in: Proceedings, 3rd International Conference on the Behaviour of Off-Shore Structures, BOSS'82, Boston 1982.

DIBIAGIO, E., F.MYRVOLL and S.B. HANSEN, 1976, "Instrumentation of gravity platforms for performance observations," Proceedings, 1st International Conference on the Behaviour of Off-Shore Structures, BOSS'76, Trondheim 1976, Volume 1, pp. 516-527. Also publ. in: Norwegian Geotechnical Institute, Publication No.114.

EIDE, O., K.H. ANDERSEN and T. LUNNE, 1979, "Observed Foundation Behaviour of Concrete Gravity Platforms Installed in the North Sea 1973-1978", Proceedings, 2nd International Conference on the Behaviour of Off-Shore Structures, BOSS'79, London 1979, Volume 2, pp. 435-456. Also publ. in: Norwegian Geotechnical Institute, Publication No. 127.

ANALYSIS OF PILE GROUPS SUBJECTED TO LATERAL LOADS

M. Hariharan
Engineers India Limited
New Delhi, India

K. Kumarasamy
Engineers India Limited
New Delhi, India

SUMMARY

The paper is aimed at the development of a rational and simple procedure for the analysis of pile groups subjected to static lateral loads, consistent with the common p-y curve approach used in the analysis of offshore piles. A pair of multipliers, one for the load and the other for the displacement, is generated for any given pile in a group based on the distribution of stresses and deformations in a horizontal layer due to movement of the piles. The multipliers are basically functions of the configuration of piles in the group and the direction of loading. The p-y curves for any pile in the group are obtained by multiplying the p-values of the curves for a single pile by the load multiplier and the y-values by the displacement multiplier. It is proposed that an 'average' pair of multipliers may be used to analyse the group in its totality.

The procedure is verified with published experimental data on the behaviour of pile groups in clay. It is concluded that the procedure can predict the behaviour of pile groups over the entire range of loading provided the p-y curves for a single pile are established accurately. Further studies are recommended on the formulation of p-y curves for single piles, and the applicability of the proposed procedure for sand and cyclic loading effects.

INTRODUCTION

It is well known that when a group of piles is subjected to lateral loads, the group deformation is considerably larger than the deformation of a single pile under the same average load per pile. Many offshore pile foundation systems have been designed with piles in groups at a fairly close spacing. A rational method of analysis and design of the piles in a group has not yet been evolved. Several approximate methods have been developed and used, but these are not adequately supported by rigorous theory, nor confirmed by experimental evidence. This paper aims to evolve a more rational method of analysis of pile groups. The procedure developed is consistent with the conventional method of analysis of piles in the offshore industry, namely, by use of p-y curves. The procedure is verified with available experimental results. The results show that the procedure can be used to predict the load-deflection behaviour of the pile group reasonably over the entire range of loading.

CURRENT APPROACH

The current approach to the analysis of pile groups is essentially based on the work of Poulos (1971b). Poulos analysed a single pile as a beam in an infinitely elastic half space. The soil response was computed using Mindlin's equations. The pile was divided into several segments, and using the finite difference scheme the pile head deformation and the pile bending moments were obtained for various pile head loads, both for fixed and free pile head conditions. For interaction effects, two piles were considered at a given spacing. The same procedure was performed as for a single pile, this time including the influence of the adjacent pile. The increase in pile head deformation was computed, and parametric charts were developed to estimate the additional pile head deformation. While the theoretical basis is consistent in this approach, it is not quite realistic because of the assumptions of uniform elastic modulus and linear elastic soil behaviour. Also, the entire method of analysis is based on elastic theory, and its results are significantly different from those based on subgrade reaction theory (Poulos, 1971a). It should be noted that the popular p-y curve approach is essentially a subgrade reaction approach.

Focht and Koch (1973) developed a procedure based on Poulos' approach, and incorporating the p-y curve approach for the analysis of single piles. First a single pile was analysed for a given design lateral load using the p-y curves to obtain the pile head displacement. The additional displacement due to the presence of adjacent piles was then computed using the interaction curves of Poulos, and added to the single pile displacement to obtain the group displacement. A series of single pile analyses were performed with the y-values in the p-y curves multiplied by a factor Y=2, 3, 4, 5 etc., and corresponding pile head displacements were obtained. By interpolation, a value of Y corresponding to the pile head displacement equal to the computed group displacement was obtained. It was concluded that multiplying the y-values of the p-y curves for a single pile by the factor Y would take care of the group effect. This procedure is now commonly used for the design of offshore pile groups.

This approach, however, suffers from certain incompatibilities of basic assumptions. Firstly, the results of group interaction analysis based on linear elastic theory are superposed on the results of single pile analysis based on nonlinear subgrade reaction approach. Secondly, in the resulting p-y curves, the y-values are increased significantly, while the p-values remain unaltered. This is contrary to the finite element results of group behaviour, wherein the ultimate p-values are also seen to reduce (Yegian and Wright, 1973). In some examples the authors have considered a 'shadow' effect for the p-values. The basis of this is not very evident. There are some other difficulties in applying this procedure to actual design. Some of these are:

a) The interaction curves developed by Poulos are for fixed or free headed piles, while the actual condition is a partially fixed pile head.

b) The value of 'equivalent' modulus of soil to be used is rather vague and very approximate.

c) The y-factor is not constant over the range of loading of interest; it depends on the value of the design load, and

d) The procedure involves several trial analyses before a Y-factor is determined. If during the preliminary design phase, the design of the piles or the configuration of the group is modified, the entire procedure needs to be repeated.

It should be noted that these limitations were recognised when the procedure was developed.

A generalisation of Poulos' approach to three dimensions and incorporating into the p-y analysis has been proposed by O'Neill et al (1977). Again this procedure combines the nonlinear p-y approach and the linear Mindlin's equation, and thus suffers from the shortcomings mentioned earlier.

It is therefore evident that a rational approach to the analysis of group behaviour of offshore piles should be based on the commonly used p-y approach and subgrade reaction theory and not be combined with elastic theory. The procedure developed in this paper is based on this objective.

THE PRESENT APPROACH

The present approach is based on generating two multipliers to the p-y curves for the single pile, one for the load and the other for the displacement. The theoretical development of the multipliers is based on the distribution of stresses and displacements in a horizontal layer of soil due to the movement of the piles in the group at that layer. The procedure is thus consistent with the subgrade reaction approach and can therefore be directly applied to the p-y approach. Only ratios of displacements and stresses are used, eliminating numerical addition of results of linear and nonlinear analyses. The multipliers obtained are only functions of the spacing to diameter ratios of the various piles in the group and the direction of loading. They are independent of the properties of the pile or the elastic modulus of the soil.

Stresses in a Horizontal Layer

Consider a horizontal layer of soil of uniform thickness. The pile is idealised as a rigid circular inclusion of radius 'a'. The modulus of elasticity of the soil is assumed as constant, E and poisson's ratio is ν. Consider the pile to move in the X-direction by a distance d. The displacement at any point P (See Fig. 1) in the layer can be obtained from elastic analysis in the polar coordinates as

$$u = \frac{C}{E} \left[\frac{(1+\nu)^2}{8} \left(\frac{a^2}{r^2} - 1 \right) + \left(1 - \frac{(1-\nu)^2}{4} \right) \ln \frac{r}{a} \right] \cos \theta + d \cos \theta \tag{1}$$

$$v = \frac{C}{E} \left[\frac{(1+\nu)^2}{8} \left(\frac{a^2}{r^2} - 1 \right) - \left(1 - \frac{(1-\nu)^2}{4} \right) \ln \frac{r}{a} \right] \sin \theta - d \sin \theta \tag{2}$$

where u and v are the radial and tangential displacements respectively, r is the distance to point P and θ is the angle subtended by the radius vector with the positive X-axis. The constant C is to be evaluated by prescribing a suitable boundary condition. The boundary condition in this work is taken that u = 0 at a radius of R = 30a. This is similar to the one considered by Baguelin et al (1977). The stresses at P are given in the polar coordinate system by

$$\sigma_r = \frac{C}{4} \left[\frac{(3+\nu)}{r} - \frac{a^2(1+\nu)}{r^3} \right] \cos \theta \tag{3}$$

$$\sigma_\theta = \frac{C}{4} \left[\frac{a^2(1+\nu)}{r^3} - \frac{(1-\nu)}{r} \right] \cos \theta \tag{4}$$

$$\tau_{r\theta} = -\frac{C}{4} \left[\frac{a^2(1+\nu)}{r^3} + \frac{(1-\nu)}{r} \right] \sin \theta \tag{5}$$

The above results may be converted to cartesian coordinate system by the following transformations. The displacement in the X-direction is given by

$$x = u \cos \theta - v \sin \theta \tag{6}$$

and the normal stress σ_{xp} is given by

$$\sigma_{xp} = \sigma_r \cos \theta - \tau_{r\theta} \sin \theta \tag{7}$$

The stress at the periphery of the pile in the X-direction σ_{xo} is obtained by substituting r = a and θ = 0 in Eq. 3. These results for a linear elastic material may be applied to a nonlinear elastic material by treating the displacement d as an incremental displacement and the elastic modulus E as the tangent modulus at the given initial state of stress. The tangent modulus is assumed to be uniform over the entire range of influence. The above results are also applicable for the stresses and displacements at the origin for a pile movement at the point P.

A study of the final expressions reveals that the incremental displacement x is only a function of a, r, θ and ν and independent of the elastic modulus. The stress, however, is a function of the elastic modulus as well, but the ratio of the incremental stress at the point P to the incremental stress at the boundary of the pile is again independent of the elastic modulus.

Interaction of Two Piles.

Consider a group of two piles, one located at the origin and the other at the point P in Fig. 1. Consider that at any state of stress, an incremental displacement of d is applied to the two piles in the X-direction. The interaction effect results in an additional incremental displacement of x in each pile. The ratio $(d+x)/d$ is the interaction effect for displacements. Similarly, the incremental stress at the pile boundary, $\bar{\sigma}xo$, is increased to $(\bar{\sigma}xo + \bar{\sigma}xp)$. The interaction effect for stresses is then the ratio $(\bar{\sigma}xo + \bar{\sigma}xp)/\bar{\sigma}xo$. These two ratios are only functions of the geometrical properties of the pile group and are independent of the soil modulus.

Two factors are defined for the inclusion of the interaction effects in the p-y analysis. The first is the displacement multiplier, equal to the ratio $(d+x)/d$. The second is the load multiplier, equal to the inverse of the stress interaction effect, i.e., $\bar{\sigma}xo/(\bar{\sigma}xo + \bar{\sigma}xp)$. The p and y values for a single pile are multiplied by the corresponding multipliers to obtain the p-y curves for the two piles in the group. The reason for the use of the p-multiplier is as follows: The p-values are the function of stress in the soil adjacent to the periphery of the pile. The ultimate stress in this area is reached for a group at a much earlier stage than the single pile. At the ultimate state of stress, only a part of the stress contributes to the resistance to the pile motion, and this part is given by the inverse of the stress interaction factor.

The values of the p and y multipliers are seen to be only functions of the ratio r/a and the angle, θ, between the direction of loading and the line joining the pile axes. The multipliers computed as functions of r/a for $\theta = 0°$ and $\theta = 90°$ are shown in Fig. 2.

Using these multipliers, the modified p-y curves generated for an example problem considered by Yegian and Wright (1973) are compared with the results of nonlinear finite element analysis computed by them (Fig. 3). It is seen that the results are comparable. It is noted that the estimated p-multiplier is smaller for the $\theta = 0°$ case. This may be partially due to the effect of local variation in the interaction stress $\bar{\sigma}xp$ around the periphery of the pile, which is quite pronounced at small spacings between the piles.

Interaction of Several Piles

Consider a group of N piles. For pile i, the interactive stress σ_{xij} and displacement x_{ij} due to an incremental displacement d of pile j can be computed as mentioned above. The total incremental stress then the sum of the interactive stresses due to all the other piles and $\bar{\sigma}xo$. The p-multiplier for pile i is then the ratio of $\bar{\sigma}xo$ to the total incremental stress. Similarly, the total incremental displacement is the sum of the interactive displacements due to all the other piles and d. The y-multiplier is the ratio of the total incremental displacement to d. Mathematically, the p-multiplier for pile i,

$$P_i = \bar{\sigma}xo / \left(\bar{\sigma}xo + \sum_{\substack{j=1 \\ j \neq i}}^{N} \sigma_{xij}\right) \tag{8}$$

and the y - multiplier is

$$Y_i = \left(d + \sum_{\substack{j=1 \\ j \neq i}}^{N} x_{ij}\right)/d \tag{9}$$

It has been seen from several pile groups analysed that these multipliers calculated for the different piles do not vary significantly, and therefore their behaviours are not very different. This has also been observed in the experiments of Matlock et al (1980). It is therefore reasonable to assume an average value of these multipliers to apply to all the piles within a group. These are defined as the average values of the P_i and Y_i values computed for all the piles in the group. It is seen in Fig. 5 that the use of average multipliers results in the same load-displacement relation for a group of piles as the use of different multipliers for different piles. These average multipliers can be used with sufficient accuracy for design purposes.

VERIFICATION

The proposed approach is verified with the results of a limited number of experimental investigations reported on the behaviour of pile groups in clay. The comparisons show that the procedure can reasonably estimate the load deflection behaviour of the pile group over the entire range of loading. It is however, recognised that sufficient experimental investigations have not been published and further study is required in this topic.

The general procedure followed in the verification is as follows: The properties of the pile given in the publication are used and a set of p-y curves are generated to result in the load-deflection behaviour for a single pile to match with the published results. The generated p-y curves generally follow the recommendations of API RP 2A (1981), but the p-values are adjusted to result in a matching load-deflection relationship. These p-y curves are modified using the multipliers defined above for individual piles in the group and the load deflection behaviour of individual piles is obtained. Individual pile head loads for the same deflection are added to obtain the total load for the group for that deflection, and the load-deflection relation for the group is generated. This is compared with the experimental results.

Experiments of Prakash and Saran

Prakash and Saran (1967) conducted experiments with aluminium model piles 9mm in diameter and 29cm long. Single piles as well as 4-pile and 9-pile groups were subjected to lateral loads, and the load-deflection characteristics at pile head were reported. The load-deflection behaviour computed by the procedure developed in the earlier section is compared with the reported experimental results. Fig. 4 shows the comparison of results for a four-pile group for a grid spacing of 5 times the pile diameter. It is seen that the results are in good agreement. Fig. 5 shows the comparison of results for a nine-pile group for grid spacings of 3 and 4 times the pile diameter. Once again the results are comparable.

Experiments of Matlock et al

Matlock et al (1980) reported the results of experimental investigation of the behaviour of a single pile and groups of 5 and 10 piles along the periphery of a circle. The piles were 6.625 inches in diameter and 45 feet long. A certain amount of rotational restraint was provided at the pile head by the use of a special loading device. It should be noted that the experimental set up and the results reported are not exactly amenable to theoretical investigations. For these experiments, only the region below the lower support was considered. The p-y curves were generated to match the deflection at the lower support for the various pile head shear and moment combinations reported. For the 5-pile group, the bending moment at the lower support for the different piles is not known; hence, the analysis is performed using average group multipliers, assuming the same average pile shear and moment as for a single pile. The results are compared in Fig. 6. It is seen that the experimental and theoretical results are comparable, even with the approximation involved. For the 10-pile group, the pile head shear and moment values for pile no. 1 are available. The theoretical analysis is performed for pile no. 1, using the multipliers generated for this pile. The load deflection relationship for pile no. 1 is compared in Fig. 7.

The above comparisons show that the proposed procedure is able to predict the behaviour of a group of piles reasonably well over the entire range of loading. The multipliers computed for the various pile groups reported in this section are summarised in Table 1.

Some interesting observations were made during the verification process. These are summarised below :

a. The direct use of the recommendations of API RP 2A for the generation of p-y curves did not result in satisfactory load-displacement relations for a single pile. Additional points on the curves upto y_c based on cubic interpolation were necessary. Further, the ultimate p-values had still to be adjusted to obtain satisfactory load-deflection results. This calls for a reexamination of the formulation of p-y curves for single piles in clay. The modifications suggested by Stevens and Audibert (1979) were also not satisfactory for the experiments of Matlock et al.

b. The generated p-y curves based on both the API criteria and the modifications of Stevens and Audibert with adjustment of p-values to result in matching pile head displacements overestimated the maximum positive bending moment in the single pile of Matlock et al by about 10%. However, both the sets of curves could predict the group deflection to a reasonable degree of accuracy. Further, the computed positive bending moment for pile no. 1 in the group was found to be very close to the experimental results. It may therefore be deduced that the proposed procedure can reasonably estimate the group behaviour, provided the p-y curves generated for a single pile are accurate enough.

CONCLUSIONS

A procedure for predicting the behaviour of a group of piles, based on a pair of multipliers to be applied to the p-y curves for a single pile is developed and verified with available experimental results. The procedure does not suffer from the shortcomings of the earlier known methods, wherein the elastic linear analysis results are superposed on the results of nonlinear analysis based on subgrade reaction theory. The multipliers obtained are only functions of pile spacing to diameter ratio and orientation of the piles with respect to the direction of loading. The study indicates that further research is required into the formulation of p-y curves for single piles and that more experimental investigation needs to be conducted. Applicability of the procedure to piles in sand and for cyclic loading needs to be studied.

ACKNOWLEDGEMENTS

This paper is based on an in-house R&D study performed at Engineers India Limited. The management permission to publish the paper and the encouragement given by Mr. C.K.Tandon and Dr. S. L. Agarwal are gratefully acknowledged. The various discussions with the authors' colleagues were very useful in formulation of the problem and the development of the solution procedure.

REFERENCES

AMERICAN PETROLEUM INSTITUTE, 1981, <u>Recommended Practice for Planning, Designing and constructing Fixed Offshore Platforms</u>, American Petroleum Institute, API RP 2A, Twelfth Edition.

BAGUELIN, F., et al, 1977, "Theoretical Study of Lateral Reaction Mechanism of Piles", <u>Geotechnique</u>, Volume 27, Number 3, pp. 405-434.

FOCHT, J.A., Jr. AND K. J. KOCH, 1973, "Rational Analysis of the Lateral Performance of offshore Pile Groups", <u>Proceedings, Fifth Annual Offshore Technology Conference</u>, Houston, U.S.A., 1973, Volume II, Paper OTC 1896, pp. 701-708. Dallas, U.S.A. : Offshore Technology Conference.

MATLOCK, H., et al, 1980, "Field Tests of the Lateral Load Behaviour of Pile Groups in Soft Clay", <u>Proceedings, 12th Annual Offshore Technology Conference</u>, Houston, U.S.A., 1980, Volume IV, paper OTC 3871, pp. 163-174. Dallas, U.S.A. : Offshore Technology Conference.

O'NEILL, M.W., et al, 1977, "Analysis of Three Dimensional Pile Groups with Non-Linear Soil Response and Pile-Soil Interaction", <u>Proceedings, 9th Annual Offshore Technology Conference</u>, Houston, U.S.A., 1977, Volume III, Paper OTC 2838. Dallas, U.S.A. : Offshore Technology Conference

POULOS, H.G., 1971a, "Behaviour of Laterally Loaded Piles : I-Single Piles", <u>Journal of the Soil Mechanics and Foundations Division</u>, ASCE, Volume 97, Number SM5, pp. 711-731.

POULOS, H.G., 1971b, "Behaviour of Laterally Loaded Piles : II- Pile Groups",<u>Journal of the Soil Mechanics and Foundations Division</u>, ASCE, Volume 97, Number SM5, pp. 733-751.

PRAKASH, S. AND D. SARAN , 1967, "Behaviour of Laterally Loaded Piles in Cohesive Soil", <u>Proceedings, Third Asian Regional Conference on Soil Mechanics and Foundation Engineering</u>, Haifa, Israel, Volume I, paper 5/12, pp. 235-238.

STEVENS, J.B. AND J.M.E., AUDIBERT, 1979, "Re-examination of p-y curve Formulations for Clay", <u>Proceedings, 11th Annual Offshore Technology Conference</u>, Houston, U.S.A., 1979, Volume I, Paper OTC 3402, pp. 397-403.Dallas, U.S.A. : Offshore Technology Conference.

YEGIAN, M. AND S.G. WRIGHT, 1973, "Lateral Soil Resistance - Displacement Relationships for Pile Foundations in Soft Clays", <u>Proceedings, Fifth Annual Offshore Technology Conference</u>, Houston, U.S.A. 1973, Volume II, Paper OTC 1893, pp. 663-676. Dallas, U.S.A. : Offshore Technology Conference.

Table 1 : Multipliers used for the different Pile groups

S.No.	Group Description	Grid Spacing	p-multiplier	y-multiplier
1.	4-Pile Group	5D	0.787	1.661
2.	9-Pile Group (Average)	3D	0.521	3.272
		4D	0.592	2.530
3.	5-Pile Group (Average)	3.4D	0.679	2.213
4.	10-Pile Group (Average)	1.8D	0.430	4.155
	Pile No. 1	1.8D	0.402	4.047

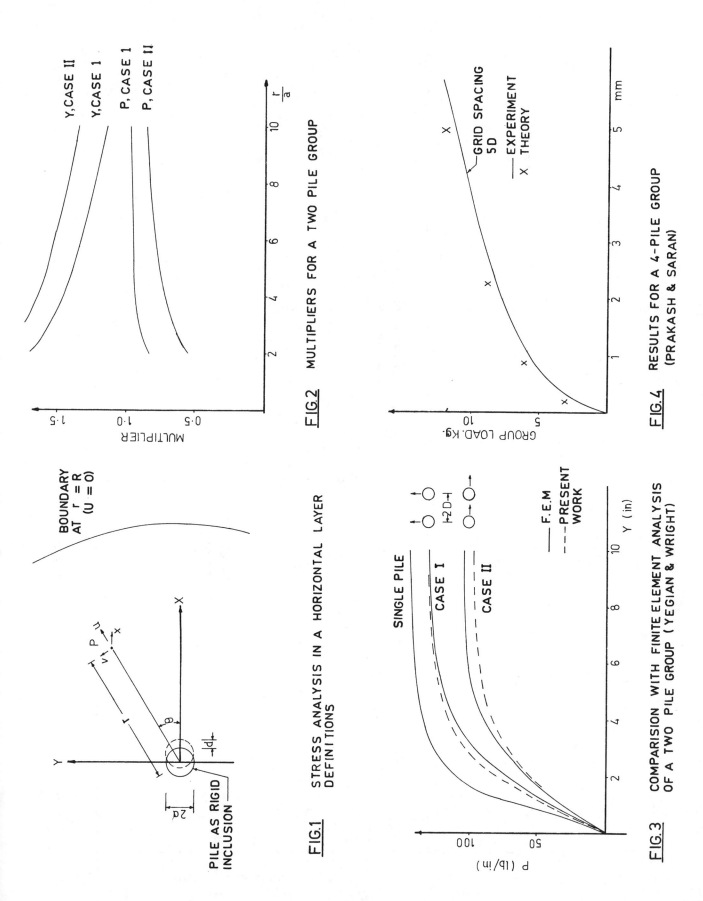

FIG.1 STRESS ANALYSIS IN A HORIZONTAL LAYER DEFINITIONS

FIG.2 MULTIPLIERS FOR A TWO PILE GROUP

FIG.3 COMPARISION WITH FINITE ELEMENT ANALYSIS OF A TWO PILE GROUP (YEGIAN & WRIGHT)

FIG.4 RESULTS FOR A 4-PILE GROUP (PRAKASH & SARAN)

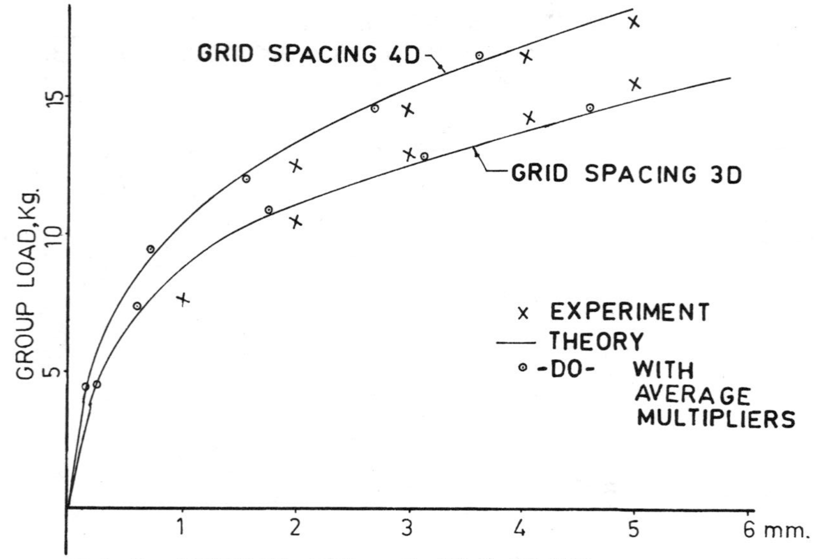

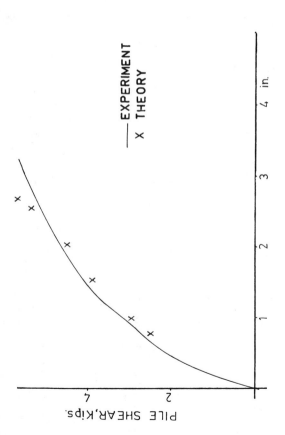

FIG.5 RESULTS FOR A 9-PILE GROUP (PRAKASH & SARAN)

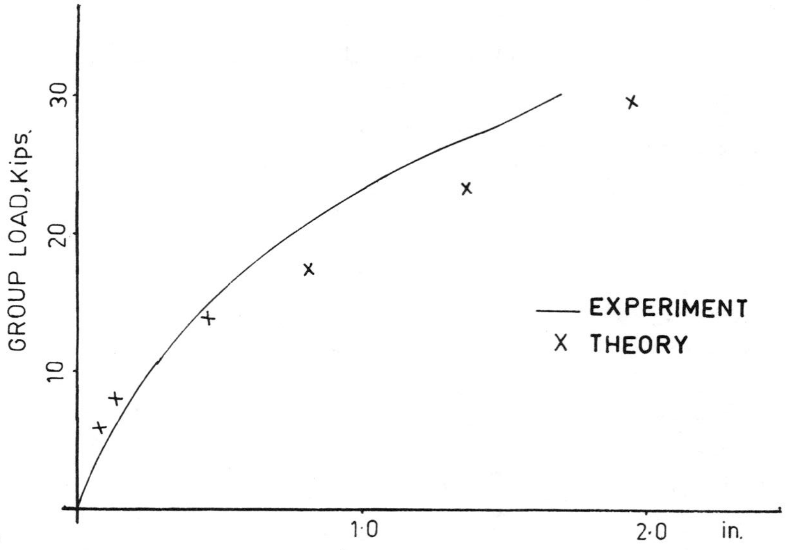

FIG.6 BEHAVIOUR OF A 5-PILE GROUP (MATLOCK ET AL)

FIG.7 BEHAVIOUR OF PILE NO.1 IN 10-PILE GROUP (MATLOCK ET AL)

CYCLIC BEHAVIOUR OF SOILS AND APPLICATION TO
OFFSHORE FOUNDATION CALCULATION

F. LASSOUDIERE
Institut Français du Pétrole
France

Y. MEIMON
Institut Français du Pétrole
France

J.C. HUJEUX
INTEVEP
Vénézuéla

D. AUBRY
Ecole Centrale de Paris
France

SUMMARY

The development of stress-strain relationships describing the behaviour of soils under cyclic loading is a way to improve the design and the safety of offshore foundations. This paper proposes a model of soil behaviour under cyclic loading. It is based on the experimental and physical approach of soil mechanics : critical state concept, non associated flow rule, separation between volumetric and deviatoric strain hardening, generalization to cyclic loading through the application of the concept of field of hardening moduli, on the deviatoric part only. A new formulation using a Mohr-Coulomb criterion has been chosen : yielding is discussed in three orthogonal planes and then, the resulting mechanisms are coupled through the volumetric strain hardening. That enables a three dimensional analysis taking into account the influence of the intermediate principal stress. The model has been tested for both static and cyclic loading. Model predictions for drained, static or cyclic triaxial tests are shown and they agree well to experimental results for sands and clays. Undrained cyclic behaviour of sand is calculated and the effect of liquefaction is well described. The stress-strain relationship has been used in a finite element program. The prediction of a tension pile behaviour under loading and unloading is compared to experimental results.

1. INTRODUCTION

The design of offshore structures by increasing water depths requires more and more accurate foundation calculations (pile anchorings of tension leg platforms, rafts of gravity platforms). Using the finite element method (F.E.M.) offshore foundations may be calculated provided that the soil behaviour under cyclic loading is well described.

A way to perform such an analysis is to use a constitutive law for the soil behaviour under monotonic loading and to summarize the cyclic soil behaviour with a relationship between the applied cyclic loads and the residual strain obtained at a given number of cycles. This so called "cyclic pseudo creep" method leads to replace in the F.E.M. calculations a great number of cyclic loads increments by some residual strain increments. However, especial laboratory test procedures, depending on the hypothesis made for the stress path evolution around the foundation during the design cyclic loading, are needed (Meimon and Hicher, 1980). On the other hand, it would be necessary to carry a great number of laboratory tests to have a sufficiently accurate behaviour modelling and this acts strongly on the F.E.M. analysis results (Puech et al., 1982).

Another way is to use a constitutive model describing the soil behaviour under three dimensional non monotonic loading. Thus, the behaviour of the foundation under cyclic loading would be calculated without any additional hypothesis. Besides the high cost of such a calculation, the main difficulty is to obtain such a model.

This paper presents firstly an elastoplastic three-dimensional constitutive law for soils, secondly one example of F.E.M. calculation of a tension pile and the comparison to experimental results.

2. PREVIOUS WORKS.

2.1. Experimental results.

More and more sophisticated models have been developed as a result of experimental evidence. For example, the derivation of the usual triaxial stress-strain curve into an elastic part and a plastic part has no rheological justification for soils both undergo plastic and elastic strains even at small strain levels. The analysis of the experimental behaviour under monotonic loading has led to define the well known critical state and the "characteristic state". At the critical state (Schofield and Wroth, 1968), reached at large strains, the irreversible volumetric strain ε_v is constant and on the failure plane the Mohr Coulomb law is satisfied. The soil characterictic state concept (Luong, 1980) may be defined by the stress state of the soil at the change of ε_v from a contraction behaviour to a dilatancy behaviour in a standard drained triaxial test. In the (p,q) plane (where p is the mean effective stress and q the deviatoric stress), the characteristic state is represented by a curve, generally a radial line on which the q/p ratio is presumed to be identical to the critical state q/p ratio. When the test exhibits a peak, this state is typical of the small strain behaviour. Finally, "true triaxial" tests (Lade and Musante, 1978) show firstly the intermediate stress effect, secondly that the irreversible strain increments tend, at failure, to be normal to the Mohr Coulomb surface in the deviatoric plane while this doesn't seem true at small strains.

Results of cyclic drained tests show the importance of the characteristic state concept and the separated influence of the deviatoric and the volumetric hardening. When cyclic loading is applied inside the characteristic domain in the (p,q) plane, strains tend to stabilize while failure occurs if cyclic loading is applied outside this domain (fig. 1) (Luong, 1980). Unloading allows to isolate the deviatoric hardening effect : this means essentially that the material keeps in memory the effect of the highest deviatoric strain undergone during the loading history.

Finally, undrained behaviour confirms the drained behaviour results. If contraction occurs under drained condition, pore pressure will increase under undrained condition - This allows to explain the liquefaction effect in soils : cyclic loading on a loose sand (which always densifies) or on a sand loaded in the domain of contraction leads to porepressure accumulation and eventually to failure by liquefaction (Ishiara, 1975)

2.2. Basic models.

Based on both elastoplastic theory hypothesis and small strains hypothesis (which allows the partition of the total strain ε into the sum of an elastic part ε^e and of a plastic part ε^p) an elastoplastic model requires to choose a yield surface, a hardening rule to describe the yield surface evolution and a flow rule to link the plastic strain increment to the stress increment and the hardening state. Based on experimental evidence, such a model would have to involve the characteristic state and critical state concepts (with a Mohr Coulomb behaviour at large strains), different hardening rules for volumetric and deviatoric hardening, a non associated flow rule at small strains that becomes associated at large strains in the deviatoric plane.

In fact, most of elastoplastic models describing the monotonic behaviour of soils are based on the Cam Clay model (Schofield and Wroth, 1968) which includes the critical state concept, an associated flow rule, a yield surface which, projected upon the deviatoric plane is a circle centered on the hydrostatic axis (Von Mises criterion), and a volumetric hardening rule. An extension to cyclic loading, using Mroz's concept of "field of hardening moduli" and Mohr Coulomb yield condition instead of Von Mises's has been established (Zienkiewicz, 1979) (fig. 2): however, flow rule is associated, and deviatoric hardening is not taken into account. A particularly interesting method has been developed (Prevost, 1978) : Volumetric and deviatoric hardening moduli are determined from experimental curves (triaxial and simple shear tests), implying the choice of the hardening parameters (to which the "field of hardening moduli" concept is applied in a "discrete" form). At last, the model described by Hujeux and Aubry (1981) may be considered as a synthesis of previous models : a critical state model with non associated flow rule and continuous field of hardening moduli in the deviatoric plane. The monotonic behaviour of a triaxial test is well predicted, and the cyclic behaviour is obtainable, from a qualitative point of view. However because of the use of Von Mises criterion in the deviatoric plane the effect of the intermediate principal stress is ignored in this model.

3. PROPOSED MODEL

3.1. Formulation principles.

The presented model is an extension of Hujeux and Aubry's model (1981) and was developed by Hujeux (1979). A detailed formulation will soon be presented. The basic idea is that yielding in soils may occur through an infinity of plastic mechanisms and that for a given stress state only some of them are activated. Here the hypothesis is made that yielding is completely described as the combination of three independent plastic mechanisms occuring in three perpendicular planes. Using Mohr Coulomb yielding condition, it is convenient to identify these planes to the three co-ordinate planes in the principal stress space. It may be easily seen that this concept generalizes the previous models where only one mechanism is involved in the deviatoric plane.

The three mechanisms are coupled for the volumetric hardening, uncoupled for the deviatoric hardening. Then, the model formulation outlines are similar to that presented formerly (Hujeux and Aubry, 1981). In each plane k, the mean stress p_k, the deviatoric stress q_k and the deviatoric plastic strain $\bar{\varepsilon}_k^p$ are defined as :

$$p_k = \frac{1}{2}(\sigma_i + \sigma_j) \qquad q_k = \frac{1}{2}(\sigma_i - \sigma_j) \qquad \bar{\varepsilon}_k^p = \left[\frac{1}{2}(\varepsilon_i - \varepsilon_j)^2 + \varepsilon_{ij}^2\right]^{1/2}$$

Where σ_i, σ_j are the i and j principal stress and ε_i, ε_j, ε_{ij} are the strains expressed in the principal stress space co-ordinate system. For the monotonic behaviour, the k-mechanism yield surface f_k and the plastic potential g_k may be represented by :

$$(I) \begin{cases} f_k(p, p_k, q_k, \bar{\varepsilon}_k^p, \varepsilon_v^p) = q_k - \sin\phi \; p_k \; (1 - b \log \frac{p}{p_c}) \; h_k(\bar{\varepsilon}_k^p) \\ \\ g_k(p, p_k, q_k, \varepsilon_v^p) = \frac{q_k}{p_k \sin\phi} + \log \frac{p}{p_c} \quad \text{(with } p_c = p_{co} \exp(\beta \cdot \varepsilon_v^p)\text{)} \end{cases}$$

where ε_v^p is the total volumetric plastic strain, ϕ, b, β, p_{co} are the model parameters, $h_k(\bar{\varepsilon}_k^p)$ which is a deviatoric hardening variable is also chosen as previously :

$$h_k(\bar{\varepsilon}_k^p) = \frac{\bar{\varepsilon}_k^p}{a + \bar{\varepsilon}_k^p} \quad \text{and} \quad \dot{h}_k = \frac{dh_k}{d\bar{\varepsilon}_k^p} = \frac{(1-h_k)^2}{a}$$

The "field of hardening moduli" concept may then be applied to each mechanism to treat the cyclic behaviour : when the stresses are reversed, the deviatoric hardening variable is reset to an initial value h_{ko}^c. The monotonic yield surface corresponding to the maximum deviatoric hardening h_k^{max} undergone by the material is kept in memory. A cyclic yield condition is calculated from the hardening state and the stress state q_{kL} reached at the reversal:

$$(II) \begin{cases} f_k^c = q_k^c - p_k \sin\phi \; (1 - b\log \frac{p}{p_c}) \; h_k^c(\bar{\varepsilon}_k^p) \\ \\ q_k^c = q_k - (q_{kL} - h_k^c \cdot p_k \cdot \sin\phi \cdot (1 - \log \frac{p}{p_c})) \end{cases}$$

Written in the initially chosen principal stress co-ordinate space, the evolution rule of h_k^c is :

$$\dot{h}_k^c = \frac{(1-h_k^c)^2}{2a}$$

When h_k^c reaches the maximum h_k^{max}, the material hardening evolution is again defined by Equations (I) representing the monotonic behaviour.

In this model, the material memory is constituted by the present stress-strain state, the volumetric plastic strain, the maximum value of the deviatoric variable obtained during the loading history and the stress state before the last loading reversal. Finally, the model requires the two usual elastic parameters and five plastic parameters to be run : parameter a describes the deviatoric hardening, b the volumetric hardening, β is the usual compressibility coefficient used in Cam Clay models, $\emptyset$ is the residual angle of friction and p_{co}, the reference critical stress, defines the material initial density state.

3.2. Comparison between laboratory test results and model prediction.

The model parameters were determined from a single drained triaxial compression test and an oedometric test, and the other tests, including cyclic tests were calculated with the same set of parameters. On figure 3 and figure 4, the model prediction is presented for drained monotonic triaxial tests on a laboratory normally consolidated clay and on a dense sand. Both extension and compression behaviours are well predicted, due to the use of the Mohr Coulomb yield condition. On figure 5 is plotted the stress strain curve predicted by the model for a simple shear test on the Hostun sand. The normal stress is kept equal to 0.4 MPa during shearing. As $\emptyset$ is 32° the failure shear stress is well predicted.

Figure 6 shows the model prediction of a drained cyclic strain-controlled triaxial test for the dense sand of figure 4. The corresponding experimental results are plotted on figure 7. Except the first loading reversal, the observed behaviour is well described by the proposed model, from both qualitative and quantitative points of view : especially, the shape of the stress-strain curve with the concavity change when going from extension to compression, the value of the volumetric strain and the tendency to stabilization of both strains and stresses.

Finally, the undrained cyclic behaviour of the same Hostun sand has been calculated with the model. Results are plotted in the (p',q) plane for two cyclic stress controlled triaxial tests. (Both tests began from p' = 0.2 MPa, q = 0.). During the first test, a 0.1 MPa cyclic shear stress is applied and shakedown with an hysteresis loop occurs at 20th cycle (figure 8) : in fact, in this test, the dilation effect obtained during a cycle when the stress point is outside the characteristic domain is sufficient to balance the pore pressure increase due to the compaction effect when the stress point is inside this domain. Figure 9 shows a liquefaction case for the same sand : in this test calculation, the cyclic shear stress applied is decreased to 0.05 MPa and the dilation effect is insufficient to stabilize the cycle behaviour. Such a behaviour is in good agreement with experimental results as Ishiara's test (Ishiara, 1975) on Fuji river sand (figure 10).

3.3. Model parameters

The presented model uses only 7 parameters, that may be determined from ordinary laboratory tests. Elastic strains are calculated by means of the usual elasticity theory coefficients (the elastic bulk modulus, K, and the shear modulus, G). The bulk modulus may be determined from the unloading part of the stress-strain curve in a drained triaxial test or an oedometric test. The shear modulus must be calculated from the bulk modulus and the Poisson's ratio, ν. Plastic strains are calculated by means of 5 parameters (cf. 3.2.). $\emptyset$ is the residual internal friction angle, that may be determined from simple shear tests, or drained triaxial shear tests. In this case, $\emptyset$ is linked to the q/p ratio when the characteristic stress state is reached (small strains) or when perfect yielding occurs (large strains) :

$$\sin \emptyset = \frac{3M}{6+M}$$

where M is the critical or characteristic q/p value. β is linked to the compressibility under hydrostatic stress : it is the slope of the (ε_v, Log p) diagram in an isotropic consolidation test. A sufficient approximation may be calculated from the C_c coefficient in an oedometric consolidation test, using the initial void ratio e_o :

$$\beta = \frac{\Delta \log P}{\Delta \varepsilon_v^P} \quad \text{or} \quad \beta = \log 10 \, \frac{1+e_o}{C_c}$$

p_{co} enables to calculate the critical hydrostatic stress p_c corresponding to a given bulk density γ, from an initial bulk density γ_o :

$$p_c = p_{c_o} \exp\left(\beta \cdot \frac{\gamma - \gamma_o}{\gamma_o}\right)$$

In a drained triaxial test at a given σ_3 value, p_{co} may be calculated from the difference between the initial bulk density γ_i and the ultimate bulk density γ_f when perfect yielding occurs :

$$p_{c_o} = \frac{\sigma_3}{1 - \frac{M}{3}} \exp\left(-\beta \frac{\gamma_f - \gamma_i}{\gamma_i}\right)$$

As perfect yielding is difficult to obtain, a sufficient approximation of p_{co} could be given by an undrained triaxial test : in the (p',q) diagram, the value of p' when the stress path crosses the q = Mp' line may be used as a p_{c_o} value. The internal hardening parameters a and b have a major influence on the volumetric strain at small strains. In a drained triaxial test on a dense sand, a governs the minimum volumetric strain abscissa, and b governs its ordinate. For a loose sand or a normally consolidated clay a and b govern the concavity of the volumetric strain curve (drained test) or the concavity of stress path in the (p',q) diagram (undrained test). Then a and b have to be chosen to fit triaxial laboratory tests. In fact, it seems that the b parameter (introduced to "lengthen" the shape of the Cam-clay yield surface along the hydrostatic axis) may be chosen around 0.3 for a large range of materials.

Figure 11 shows the influence of separate variations of β, p_{c_o}, a and b parameters in the case of a drained triaxial test on a dense sand. Each parameter has its proper influence on the shape of the curve, and is clearly linked to its own physical meaning, so that the set of parameters can easily be checked on a few laboratory tests. On the basis of a few usual laboratory tests (drained or undrained triaxial, oedometric tests), it is possible to determine and check easily a set of parameters (in a reasonable number of 7) that gives a realistic description of soil behaviour under various loading conditions.

4. F.E.M. calculation.

4.1. Formulation principles

A FEM calculation using the proposed soil behaviour model has been performed. The plane strain/axisymmetric formulation is based on the infinitesimal strain hypothesis. A Newton Raphson algorithm is used : assuming the problem is linear, unbalanced forces are calculated taking into account the material nonlinearity and are applied through an iterative process. The used convergence rule is a work criterion which is more precise than the usual displacement criterion : the work of the unbalanced forces in the incremental displacements at iteration i is compared to the work of the incremental forces at iteration 0. During the calculation, the auxiliary operator which is the elastic matrix is inversed once for all and no convergence acceleration is used. The mesh is constituted of eight nodes quadrilateral elements and strains and stresses are calculated at nine gauss points inside each element.

4.2. Tension pile calculation

The calculation presented hereafter corresponds to a cyclicly loaded tension pile experiment carried in laboratory at I.M.G. (Institut de Mécanique de Grenoble) (Puech et al, 1982). A 5.5 cm diameter, 1.62 m long rigid pile was centered in a 1.50m diameter, 2m high cell filled with dense dry Hostun sand (density of 1700 kg/m3). Applying a displacement at the pile head a cyclic test was carried and the cyclic displacement amplitude was increased till failure. The used FEM mesh model is shown on figure 12. The sand parameters used in the calculation have been derived from the set used for the triaxial behaviour prediction shown in Figure 4. The axisymmetric calculation has been carried in two stages. Firstly, initialization of hardening variables has been done by calculating an oedometric initial stress state : volumetric forces corresponding to 1700kg/m3 density have been applied from a small isotropic stress level. Such an initialization procedure describes well the experiment conditions and more generally is valid for drilled piles. For driven piles, another procedures have to be faced.

Secondly, the tension test is calculated with small load increments especially at the reversal load steps. On Figure 13, the experimental load Q obtained at the head pile is plotted versus the imposed head pile displacement yo - Figure 14 presents the same curve resulting from the FEM calculation. A good agreement is obtained between the calculation and the experiment almost till failure. The model predicts well hardening of soil after a cycle : the (Q, yo) curve has a slight tendency to stiffen. However, while the surface of each hysteresis loop is smaller in the calculation than in the experiment no effect is observed on the results after reloading. Figure 15 presents the repartition of the mobilized load Q along the pile versus depth at the maximum load of each cycle. Calculated values plotted in figure 15 are the result of shear stress integration at 3 mm from the pile. (This tends to slightly overestimate Q.) Agreement is also good with experimental values plotted in dotted line. Finally, calculated value of $K = \sigma_n/\gamma z$ (where σ_n is the radial stress at 3 mm from the pile) at z = 1.225 m show a great effect of dilatancy near the pile : K = 1.6 for 2 mm head pile displacement. This value is less than the experimental value 2.5 but it is known that dilatancy decreases very rapidly with the radius.

5 - CONCLUSIONS

Based on experimental evidences and consistent hypothesis of elastoplasticity theory, a constitutive model for soil behaviour has been developed. It involves the critical state and characteristic state concepts. The Mohr Coulomb yield condition, the effect of the intermediate principal stress, the separate effects of deviatoric hardening and volumetric hardening are taken into account. Non associated flow rule at small strains, becoming associated at large strains in the deviatoric stress plane is included in the model. A generalized formulation using three yielding mechanisms in the three principal stress planes has been developed. Cyclic behaviour is treated using the field of hardening moduli concept for each mechanism.

Model predictions agree well with experimental results for drained triaxial cyclic or monotonic tests on sand and clay. Undrained cyclic behaviour of sand is calculated and agrees well with usual results : especially, the liquefaction effect in soils may be explained.

Used in a FEM calculation of a cyclicly loaded tension pile, the model prediction agrees with experimental results for a drilled pile. Further research is planned to take in account the initial state for driven piles and to calculate offshore foundations which are submitted to a large number of repeated cyclic loads at a reasonable cost.

6. REFERENCES

DARVE. F and LABANIEH. S, 1981, "Incremental Constitutive Law for Sands and Clays, Simulations of Monotonic and Cyclic Tests", *International Journal of Numerical and Analytical Methods in Geomechanics*, Volume 5, 1981.

FLAVIGNY. E, 1978, *Trois aspects des propriétés mécaniques des sols*, Dr Ing Thesis, Institut de Mécanique de Grenoble, France.

HUJEUX. J-C, 1979, *Calcul numérique des problèmes de consolidation élastoplastique*, Dr Ing. Thesis, Ecole Centrale Paris, France.

HUJEUX. J-C and AUBRY. D, 1981, "A Critical State Type Stress-Strain Law for Monotonous and Cyclic Loading, Geotechnical and Numerical Considerations", *Implementation of Computer Procedures and Stress-Strain Laws in Geotechnical Engineering*, Chicago, USA, 1981.

ISHIARA. K, TATSUOKA. F and YASUDA. S, 1975, "Undrained Deformation and Liquefaction of Sand under Cyclic Stresses", *Soils and Foundations*, JSSMFE, Volume 15, Number 1, pp 29-44.

LADE. P-V and MUSANTE. H, 1978, "Three Dimensional Behaviour of Remolded Clay", *Journal of the Geotechnical Engineering Division*, ASCE, Volume 104, Number GT2, pp. 193-209.

LUONG. M-P, 1980, "Stress-Strain Aspects of Cohesionless Soils under Cyclic and Transient Loading", Proceeding : *International Symposium on Soils under Cyclic and Transient Loading*, Swansea, UK, 1980, Volume 1, pp. 315-324, Rotterdam, Netherlands : A-A Balkema.

MEIMON. Y and HICHER. P-Y, "Mechanical Behaviour of Clays under Cyclic Loading", Proceeding : *International Symposium on Soils under Cyclic and Transient Loading*, Swansea, UK, 1980, Volume 1, pp. 77-87, Rotterdam, Netherlands A-A. Balkema

PREVOST. J-H, 1978, "Anisotropic Undrained Stress-Strain Behaviour of Clays", *Journal of the Geotechnical Engineering Division*, ASCE, Volume 104, Number GT 8, pp. 1075-1090.

PUECH. A, BOULON. M and MEIMON. Y, 1982, "Behaviour of Tension Piles : Field Data and Numerical Modelling", *2nd International Conference on Numerical Methods in Offshore Piling*, University of Texas at Austin, 1982

SCHOFIELD. A-N and WROTH. C-P, 1968, *Critical State Soil Mechanics*, London UK : Mac Craw Hill publishing company.

THANOPOULOS. I, 1981, *Contribution à l'étude du comportement cyclique des milieux pulvérulents*, Dr Ing. Thesis, Institut de Mécanique de Grenoble, France.

ZIENKIEWICZ. O-C, 1979, "Constitutive Laws and Numerical Analysis for Soil under Static, Transient or Cyclic Loads", Proceedings : *2nd International Conference on Behaviour of Off-Shore Structures*, London UK, 1979, Volume 1, paper number 32, pp. 391-406, Cranfield, Bedford, England : BHRA Fluid Engineering.

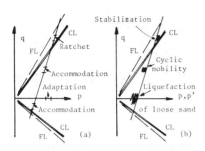

Fig. 1 - Typical behaviour of soils on drained (a) or drained (b) stress path
CL : Characteristic state line
FL : Failure line
(From Luong, 1980)

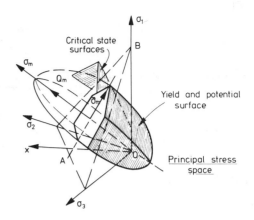

Fig. 2 - Strain hardening critical state associated plasticity with Mohr Coulomb critical state surface
(From Zienkiewicz, 1979)

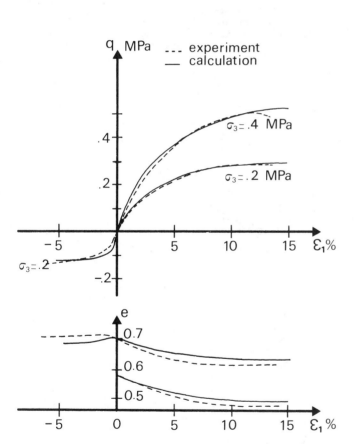

Fig. 3 - Drained triaxial tests on normally consolidated La Roche Chalais Clay.
Set of parameters used : $\phi = 25° \quad \beta = 20$
$p_{c_o} = 0.2$ MPa $\quad a = 0.004$
(Experimental data from Flavigny, 1978)

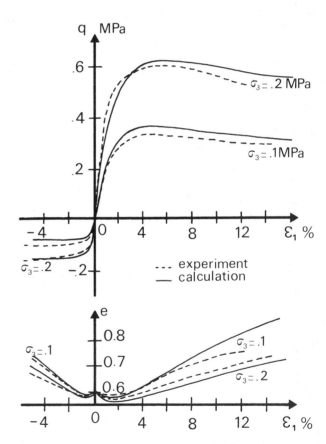

Fig. 4 - Drained triaxial tests on dense Hostun sand.
Set of parameters used : $\phi = 32° \quad \beta = 30$
$p_{c_o} = 0.8$ MPa $\quad a = 0.002$
(Experimental data from Darve and Labanieh, 1981)

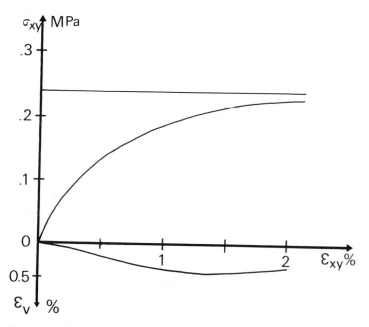

Fig. 5 - Simulation of a simple shear test on dense Hostun sand. Vertical stress : 0.4 MPa
Set of parameters used : $\phi = 32°$ $\beta = 30$ $P_{C_0} = 0.8$ MPa $a = 0.002$

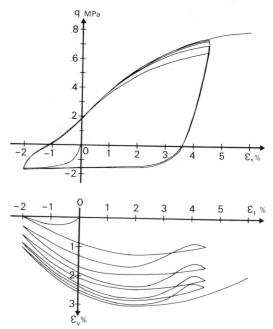

Fig. 6 - Simulation of a drained cyclic strain-controlled triaxial test on dense Hostun sand. Lateral stress $\sigma_3 = 0.2$ MPa
Set of parameters used : $\phi = 32°$ $\beta = 30$
$P_{C_0} = 0.8$ MPa $a = 0.002$

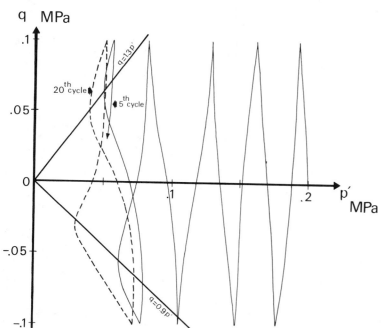

Fig. 8 - Simulation of an undrained cyclic stress-controlled triaxial test on dense Hostun sand. Deviatoric stress : ± 0.1 MPa. Lateral stress $\sigma_3 = 0.2$ MPa
Set of parameter : $\phi = 32°$ $\beta = 30$ $P_{C_0} = 0.8$ MPa $a = 0.002$

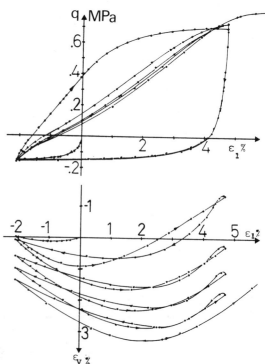

Fig. 7 - Drained cyclic triaxial test on dense Hostun sand (simulated Figure 6)
(Experimental data from Thanopoulos, 1981)

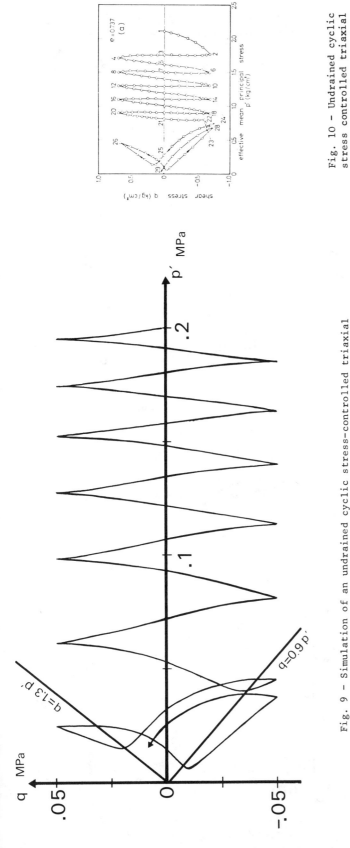

Fig. 9 - Simulation of an undrained cyclic stress-controlled triaxial test on dense Hostun sand. Deviatoric stress : ± 0.05 MPa. Lateral stress σ_3 = 0.2 MPa $\quad P_{C_0}$ = 0.8 MPa $\quad$ a = 0.002
Set of parameters : $\emptyset$ = 32° $\quad \beta$ = 30 $\quad P_{C_0}$ = 0.8 MPa $\quad$ a = 0.002

Fig. 10 - Undrained cyclic stress controlled triaxial test on Fuji River sand. (Experimental data from Ishiara, 1975).

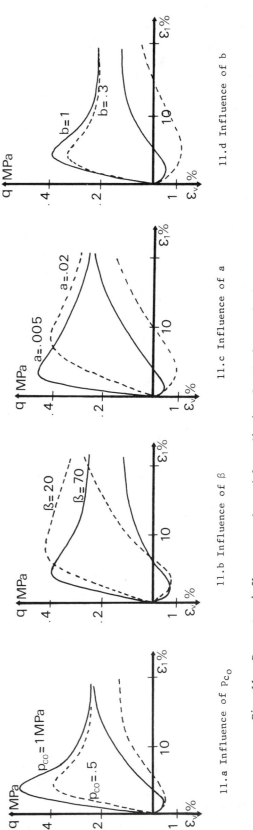

11.a Influence of p_{co} 11.b Influence of β 11.c Influence of a 11.d Influence of b

Fig. 11 - Parameter influence on the model prediction for drained triaxial test carried on a dense sand σ_3 = 0.1MPa. Basic set of parameters $\emptyset$ = 32° $\quad P_{c_0}$ = 1MPa $\quad \beta$ = 70 $\quad$ a = 0.02 $\quad$ b = .3

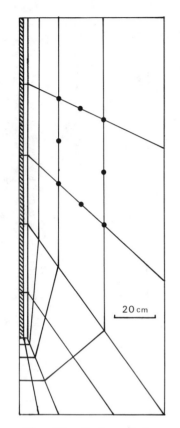

Fig. 12 - Mesh used for the FEM calculation (137 nodes-38 elements)

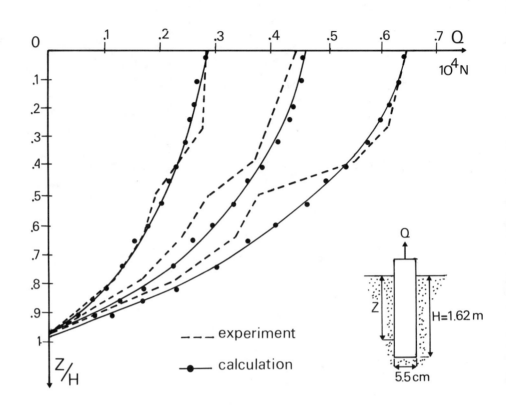

Fig. 15 - Repartition of the mobilized pulling load along the pile at the maximum of each cycle

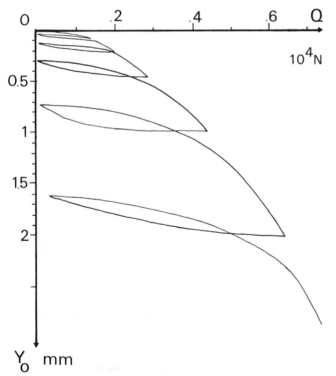

Fig. 13 - Experimental pulling curve giving the pulling load Q for an imposed head pile displacement yo (from Puech et al, 1982)

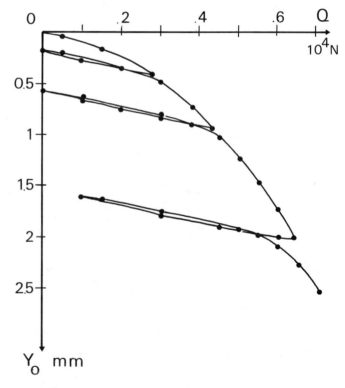

Fig. 14 - Calculated pulling curve for the experiment shown on figure 13

THE COLLAPSE OF A FLARE JACKET SUBJECTED TO ICE LOADS

N. Janbu, Professor
The Norwegian Institute
of Technology, Trondheim

XU JIZU, Associate Professor
Tianjin University, China.
The Norwegian Institute of
Technology, 1979-1981.

L. Grande, Dr.Eng.
The Norwegian Institute
of Technology, Trondheim

SUMMARY

A flare jacket, located in the Bo-hai Gulf of North China, collapsed during the ice drifting period of 1977. This case represents a unique set of information about the ultimate resistance of tubular steel piles in soft clay under lateral ice floe loads.

Herein, a comparison is made between the probable failure load for the ice, and the lateral failure load for the piles, obtained by using both the dowel theory and soil-pile interaction analyses.

The results of the analyses seem to explain fairly well the actual situation of the collapsed piles, including the ultimate lateral load, the failure mode and the position of the bending failure of the piles. On the basis of these results, analytical modelling and design of vertical piles subjected to lateral loads are discussed.

The particular soil conditions of this case record resulted in the derivation of a new dowel theory for a two-layered soil system. Because of the simplicity of this theory, the most likely failure mechanism was clearly brought out.

INTRODUCTION

The location of the flare tower is shown in Fig.1. The flare tower was a steel jacket on four steel pipe piles, one in each corner. The sole function of the tower was to burn gas, coming from an adjacent production platform. There was a simply supported bridge between the flare tower and the production platform. The orientation of the bridge axis, in horizontal plane was about 45° from the direction of ice drifting. At the time of collapse the live load on the flare tower was insignificant.

The four steel piles were driven 15 to 16m below the mudline. The outer diameter of the piles was 0.84m, while the steel thickness changed from 24mm to 12mm at about 6.5m below the mudline. Assuming the yield stress of steel to be 200-400 MPa, the yield moments of the piles would become

M_f = 2300-2750 kNm, for upper part
M_f = 1250-1500 kNm, for lower part

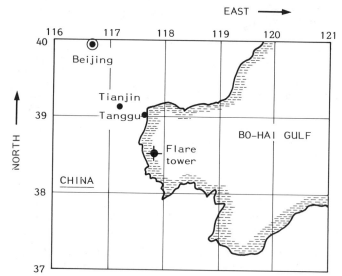

FIG. 1. Location of flare tower.

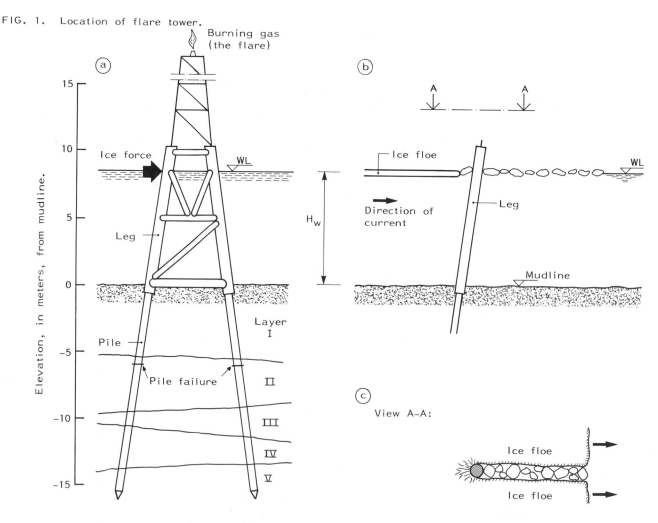

FIG. 2. The flare tower and the ice floe. Key sketches.

After the collapse of the tower the piles were found to be broken at a depth of around 6-7m under the mudline, but the lower part of the jacket was still in its integrated manner.

Fig.2a shows a cross section through the jacket and the soil layers penetrated by the piles, while Figs.2b and 2c illustrate the mechanism of the drifting ice as it shears its way through a leg. The ice drift followed the ocean currents, to and fro twice a day. At any time during the drifting, there were only 2 of the legs in contact with the ice floe.

The crushing of the ice against a vertical leg is a repeted loading process on the pile, and will thus in principle induce vibration on the structure. However, for the flare jacket, having a small mass, the deformative forces applied on the structure would be close to the maximum static ice loads (Xu and Leira, 1981) because the inertia effect of the soil mass around the piles was found to be insignificant in this case. A conventional static approach is therefore well justified.

For the conditions in the Bo-hai Gulf a maximum ice failure pressure, on a unit projected pile area, of 980kPa was established on the basis of an extensive literature survey (Xu and Leira, 1981). The ice thickness was about 0.6m in average. Hence, the maximum (ice failure) load on a single leg was about 285kN. Therefore the maximum ice thrust on the jacket was estimated to be 570kN, since no more than 2 piles were involved at any given time of the drift.

SOIL CONDITIONS

In the vicinity of the flare tower a large number of soil borings are performed. Numerous data from an extensive laboratory test program were used to obtain average values of typical soil parameters for the 6 layers identified by the soil investigations. Each of the layers varied substantially in thickness over the investigated area, as is illustrated in principle in Fig.3, at left.

The test results showed appreciable scatter. For instance, an average water content of 62% (from 71 tests) varied between 53% and 74%. An average s_u of 22kPa could easily be scattered by ± 10kPa.

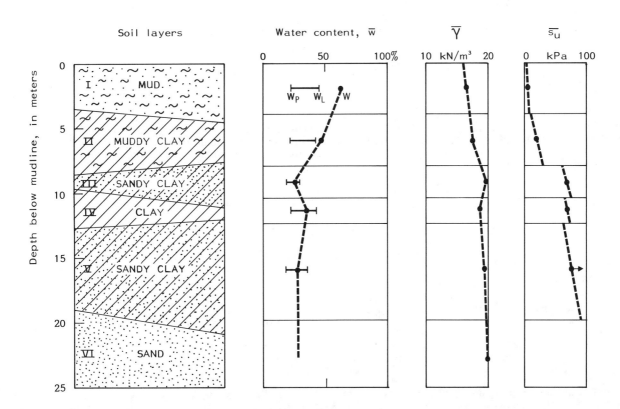

FIG. 3. Soil profile, by average parameters.

Seen as a whole, the original layers I and II, can be treated as one *upper layer*, for design purposes. Likewise, the three following layers (III, IV and V) could be considered as one *middle layer*, while layer VI represents the *lower layer* explored. These three design layers can be characterized as follows:

Upper layer: Soft, organic, sensitive clay, (mud, or muddy). Water content above liquid limit, unit weight 16-17kN/m³, and average undrained strength 3-25kPa, increasing with depth. Layer thickness = 8m ± 3m.

Middle layer: Fairly stiff sandy, silty clay with low sensitivity. Water content between plastic and liquid limit, unit weight 19-20kN/m³, and average undrained strength = 70kPa ± 30kPa, with insignificant depth variations. Layer thickness = 12m ± 3m.

Lower layer: Sand. Friction angle given at 30°-36°. Otherwise few soil data available. Below depth of interest. Thickness unknown to authors.

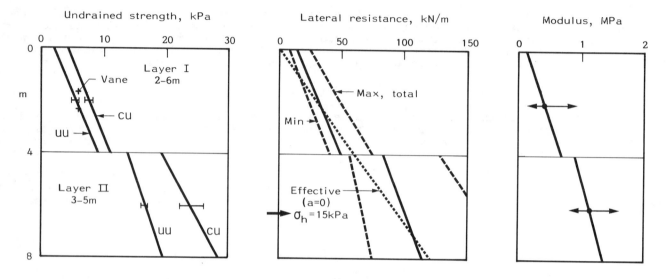

FIG. 4. Average design parameters for the two soft top layers, I+II = upper layer.

Since the pile behaviour is vastly influenced by the upper layer an average design profile is given in greater detail in Fig.4. The left diagram countains the undrained strength s_u as obtained from several UU- and CU-tests, while the right hand diagram gives the tangent moduli (M) determined by oedometer tests. The modulus of horizontal deformation (M_h) used in the succeeding analyses is evaluated from this diagram.

The middle diagram in Fig.4 shows the calculated lateral soil resistance at failure of the surrounding soil. The formuli for calculations of this resistance are presented herein.

POSSIBLE FAILURE MODES

For a complete investigation of the collapse mechanism of the flare tower several possible failure modes have been studied, Table 1. Broadly speaking, the actual case may be found somewhere between two extreme boundary cases:

- If the jacket is pushed horizontally, as a rigid block without rotation, the total lateral resistance of the jacket is determined by a two-hinge failure mode, see State 1 in Table 1. This case represents the maximum possible interaction between the jacket and the imbedded piles.
- If the jacket, and two of the upper thickwalled parts of the piles tilt semirigidly forming yield hinges at about -6.5 (State 3, Table 1), and two other piles are pulled out of the ground, the total resistance is the combined effect of pile bending and pile pullout.

The first boundary case will resist a total horizontal force at sea level of 2500kN ± 500kN, according to a simple static analysis of the two-hinge system for all four piles. Hence, each single pile in this mode of failure, can resist 600kN ± 100kN, which is more than twice the actual ice drift load, 285kN. Therefore, one can exclude mode 1 as a failure mechanism in the flare tower collapse.

TABLE 1. Load effects for different values og mudline moment M_o.

Load effects	State 1 : $M_o<0$ Integral effect dominating	State 2 : $M_o = 0$ Pile head free	State 3 : $M_o>0$ Single pile effect dominating
Moment on the pile head = M_o (Positive for clockwise rotation)	$M_o<0$	$M_o = 0$	$M_o>0$
Rotation of the pile head, θ.	$0 \leqq \theta < \theta_o$	$\theta = \theta_o > 0$	$\theta > \theta_o$
Failure mode of the pile. (Yield hinges A and B).	$\theta = 0$ at A		$H_w \geqq H > 0$
Depth of the critical section, Z_m	$Z_m > Z_{mo}$	$Z_m = Z_{mo}$	$Z_m < Z_{mo}$

The second boundary case is found to resist a total horizontal force at sea level of about 1800kN ± 300kN, while the total ice force on the jacket at any time was much smaller, 570kN. Hence, this combined case of pile bending and pile pullout can also be excluded.

The mechanism of failure must therefore be sought somewhere between states 2 and 3 in Table 1. The rest of this paper is devoted to the task of identifying more precisely the actual mode of failure. Two types of analyses will be applied; the simple dowel theory, and an advanced pile soil interaction analysis. In addition a special analysis was also performed at an early stage, to determine the interaction between the jacket and the foundation piles, the prupose of which was to obtain a reasonable estimate of the actual pile moment at the mudline, and the corresponding rotation.

THE DOWEL THEORY

The dowel theory, as applied herein, was given a first complete presentation some 10 years ago (Janbu 1973). It is based on very simple static principles, as illustrated in Fig.5.

For a single pile, acted upon by a horizontal load Q_h at an height of H above ground level, the ground level moment is $M_o = HQ_h$. The lateral soil resistance p_r may vary with depth. For a known, or calculated variation, the maximum bending moment M_{max} occurs at a depth of z_m where the shear force is zero. Hence z_m is easily determined from horizontal equilibrium, and the limit value of $M_m = M_f$ is the yield moment of the pile, or the wall. By assuming $M_{max} = fM_f$ where f = degree of mobilization, the system is statically determinate down to z_m. By idealization of p_r below z_m one can also determine the total depth D_n necessary for equilibrium of the pile below z_m. Using the tangent modulus concept M_h for horizontal deformation, it is also fairly simple to obtain a rapid forecast formula for the lateral deflection δ of the pile at ground level, δ_h. These simple principles are applied below. In a more recent paper, model (a) in Fig.5, has also been published in English (Janbu, 1980).

Lateral soil pressure.

The most important part of the dowel theory analyses is the assessment of the lateral pressure towards the pile. For more than 10 years this topic has been under continous investigations here at NTH, both by theory, by model tests, and by backcalculating available observations. The formuli in present use at our division are the following:

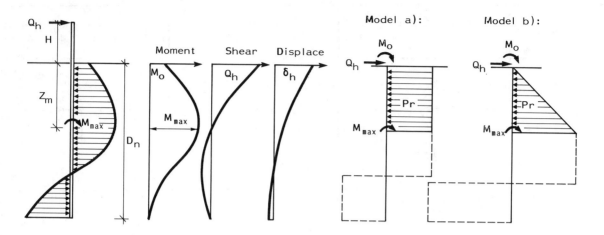

FIG. 5. Dowel theory principles, and two idealizations of lateral soil resistances; models (a) and (b).

For short-term, undrained conditions in saturated clay using a total stress approach, based on the undrained strength, s_u

$$p_r = N_{ru} \tau_a d \qquad (1)$$

wherein
- p_r = lateral pressure against the pile (kN/m)
- N_{ru} = reaction number for total, undrained conditions
- τ_a = fs_u = applied shear stress level in soil
- f = τ_a/s_u = degree of shear mobilization in soil
- d = pile diameter, or pile width

Theoreticially, the value of $N_{ru} \cong 5-9$, if the soil is always in contact with the entire pile perifery. If not $N_{ru} < 4$.

For long-term, drained conditions of the subsoil (clay, silt or sand) using an effective stress approach,

$$p_r = N_r (p'+a) d \qquad (2)$$

wherein
- N_r = reaction number for effective stress
- p' = $\gamma z - u$ = effective overburden at depth z
- a = attraction (Note: cohesion $c = a \tan\phi$)

The reaction number N_r depends primarily on mobilized friction $\tan\rho = f \cdot \tan\phi$, and to some extent on p'. For a conservative estimate one could use the approximation.

$$N_r \cong 4 \tan\rho (1+\tan\rho)^2 \qquad (3)$$

For more detailed information one is referred to the literature on the subject.

Formuli for models.

For the idealization shown in model (a) in Fig.5 the equilibrium conditions lead to a set of very simple formuli for a single vertical pile.

$$Q_h = \sqrt{2(M_m - M)\bar{p}_r} \qquad (4)$$

$$z_m = \frac{Q_h}{\bar{p}_r} \qquad D_n \geq (1+\sqrt{2}) z_m \qquad (5)$$

$$\delta_h = \frac{2\bar{p}_r}{\bar{M}_h} = \text{displacement at ground level} \qquad (6)$$

wherein $\bar{p}_r$ and $\bar{M}_h$ are average values over the depth z_m.

For model (b) the corresponding equations read; when $p_r = z\sigma_h$ for $z \leq z_m$:

$$Q_h^3 = \frac{9}{8} \sigma_h (M_m - M_0)^2 \qquad (7)$$

$$z_m = \frac{3}{2} \frac{M_m - M_0}{Q_h}, \quad D_n \geq 2z_m \qquad (8)$$

$$\delta_h \cong \frac{3\bar{p}_r}{\bar{M}_h} \qquad (9)$$

Note that for $a=0$ and $p'=\gamma' z$ one will get $\sigma_h = N_r \gamma' d$ from Eq.(2).

For estimating δ as a function of the load Q_h the average $\bar{p}_r$ is introduced as $\bar{p}_r = f\bar{p}_{rf}$ where $\bar{p}_{rf}$ = average failure resistance over z_m, which means using $\tau_a = s_u$ and $\tan\rho = \tan\phi$ in Eqs.(1) and (2). The modulus M_h corresponding to a given degree of mobilization, f can in practice be introduced as

$$M_h = (1-f^n) M_{hr} \qquad (10)$$

wherein n = exponent determined by triaxial tests, while M_{hr} = reference modulus. For rapid loading in intact soil n may vary between 2 and 4, roughly. For longterm drained conditions n may be reduced to 1.0, or even less.

The horizontal load Q_h corresponding to a given degree of mobilization is found from Eqs.(4) and (7) using $\bar{p}_r$ and $M_m = fM_f$, where M_f = yield moment. The various degrees of mobilization, f, need not be the same for soil stress and pile stress (moment), but it is beyond the scope of this article to enlarge on this question.

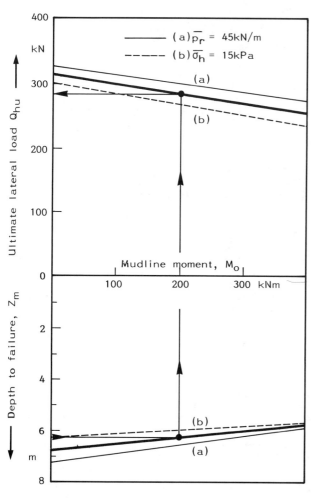

FIG. 6. Dowel theory results.

Results from the dowel theory.

The initial analyses for the interaction between the jacket and the foundation piles indicated that the maximum M_0 was 500kNm per pile. Therefore, the dowel theory was used to analyse the ultimate failure load Q_{hu} for varying M_0 from 0 to 400kNm, using the yield moment of the lower part of the pile M_f = 1250kNm, corresponding to a steel stress of about 200 MPa, in bending alone. For each value of M_0 the corresponding value of z_m was found for models (a) and (b). The results are given in Fig.6.

As an example let the actually observed position of pile bending be z_m = 6.25m. Then Fig.6 shows that $M_0 \sim$ 200kNm while $Q_{hu} \sim$ 280kN. These are very reasonable results compared to the information given for the actual failure load of the ice and the collapse of the tower.

The actual distribution of lateral resistance p_r is unknown. However, judged from the soil profiles in Fig.4 it is believed that the real distribution over z_m lies somewhere between models (a) and (b). Such a situation is represented by thick lines, between (a) and (b), in Fig.6. An overall evaluation of this figure will lead to the following conclusion for practical purposes:

Q_{hu} = 280kN ± 20kN = ultimate pile load
z_m = 6.3m ± 0.4m = depth to pile failure
M_0 = 200kNm ± 100kNm = mudline moment

The ultimate values corresponds to simulataneous failure of soil and pile, in bending. To be able to judge the possibility of a certain degree of a progressive failure mechanism, it is necessary to obtain load-deflection curves.

Using Eqs.(6), (9) and (10), and assuming (for simplicity) equal degrees of mobilization for soil, pile and M_h the $Q_h-\delta_h$ curves in Fig.7 were obtained for two different n-values, 2 and 4. The reference modulus was judged to be 500kPa, see Fig.4. For the yield stress of the steel σ_y = 200MPa was used, while $M_0 = f \cdot 200$kNm.

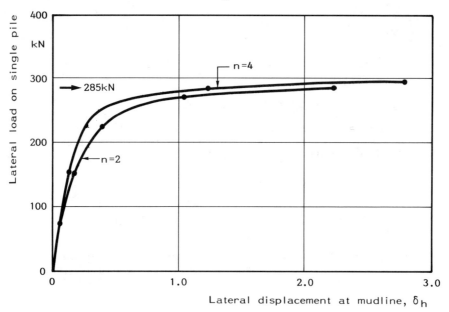

FIG. 7. Load-deflection curve, estimated by dowel theory concepts.

It is important to notice that a very rapid increas in lateral displacements takes place for Q_h ~ 250kN, which is some 15% lower than the theoretical value of the ultimate load. This observation underscores the need for establishing reasonable load deformation curves in actual design, because a mere calculation of the ultimate load, from an ideal elasto-plastic approach, may lead to an overoptimistic conclusion.

THE INTERACTION ANALYSES

Two levels of interaction analyses were performed. The initial analysis aimed at obtaining information about the actual pile moment at the mudline level. Later, supplementary analyses were carried out to study the single pile-soil interaction behaviour in greater detail.

Initial analysis.

The actual boundary conditions imposed on the pile head by the jacket structure (i.e. the rotation and its corresponding moment on the pile head) can be found by an iteration of the jacket structure behaviour and the pile-soil interaction behaviour. The program FEDA (Langen,1977) was used for the jacket analysis, while the Program SPJ (Nordal, 1977) was used for the pile-soil interaction analyses. This initial analysis was carried out shortly after the investigation of the dynamic behaviour of the tower (Xu and Leira, 1981). Therefore, the soil stiffness was judged on the basis of $G_{max} = 300\ s_u$, which is representative of a low-strain, very short-term loading.

The result of these analyses, for Q_h = 285kN on each of two piles, is shown in Fig.8. The mudline moment was found to be $M_0 \simeq 500$kNm and the corresponding rotation of the pile θ = 0.019 radians.

The distribution of the pile bending moment, and the lateral soil resistance is shown in Fig.8. It is interesting to note that over a depth of 6-10m the calculated moment exceeds the theoretical yield value M_f = 1250kNm. The maximum moment is 1650kNm and occurs around 6.5m depth where the actual failure took place. The maximum steel stress is 265MPa for bending alone.

The distribution of p_r with depth is non-uniform because of detailed soil data specification for a threelayer system. Nevertheless, in broad terms the p-diagram for a dowel theory model, midway between (a) and (b), is very close to the result of the intricate interaction analysis. As could be expected, the maximum dowel theory moment (for $M_0=200kNm$) is seen to be lower than the maximum pile moment obtained by interaction analyses (for $M_0=500kNm$. However, the depths of maximum moments are almost identical for the two types of analyses.

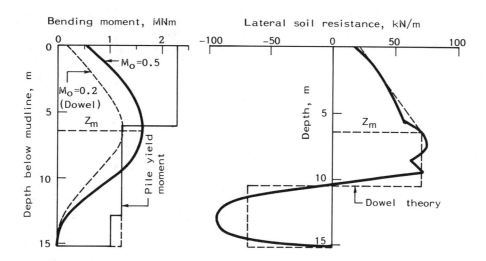

FIG. 8. Results of interaction analyses, for short-term loading.

An extension of the jacket-pile-soil interaction analysis indicate that moderate changes of input parameters may bring the mudline moment close to zero. For $M_0=0$ a pile rotation of $\theta=0.010$ radians (at the mudline) was found, using the soil resistance formulation based on G_{max}.

Single pile-soil interaction analyses

Since the M_c-value was narrowed down to 0-500kNm, it was possible to study in great detail the pile-soil interaction behaviour using the SPJ-program (Nordal, 1979). For detailed information about the basic principles and the formulations used, together with examples of application, reference is made to Grande and Nordal (1979) and Nordal et.al. (1982).

The results of these analyses are shown in Fig.9, for two different cases regarding the modulus of lateral deformation. For each case the figure shows the depth distribution of: lateral resistance, lateral deformation, horizontal shear and bending moment. Each of these diagrams have curves for the following load steps of Q_h = 60-120-180-240-300 kN. The soil was idealized by a two-layer system, where the upper layer of 8m was soft, and the underlying soil stiff. The input parameters for the stiff soil were:

M_{oed} = 4MPa, $\bar{\gamma}$ = 9.5kN/m³, K_i = 0.50(0.65) = at rest
a = 20kPa, $\tan\phi$ = 0.60, f_0 = 0.55 (at rest)

For the soft, top layer the input values were:

M_{oed} = 12$\sigma'_{vert.}$ for case (a), M_{oed} = 600kPa for case (b)
a = 5kPa, $\tan\phi$ = 0.45, f_0 = 0.50 (at rest)
K_i = 0.65

In all cases the modulus of lateral deformation was assumed to decrease linearly with increasing degree of shear mobilization in the soil.

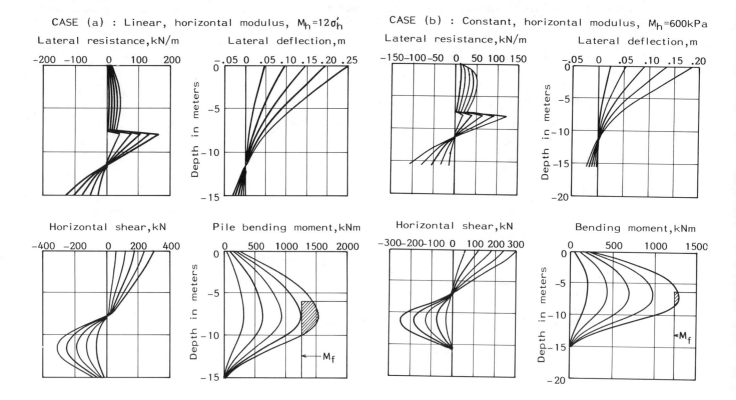

FIG. 9. Results of pile-soil interaction analysis, using the SPJ-program (Nordal, 1979).

The results in Fig.9 are almost self-explanatory, if time is allowed for scrutiny of each diagram. A few trends may be emphasized:

- The maximum moment for Q_h = 300kN is seen to be 1300-1600kNm, occuring at a depth of about 6 to 7.5m which is in close agreement with the other results reported above.
- The two-layer boundary (-8) and the change in steel thickness (-6) shows up very distinctly in the p_r-distribution, but only vaguely in the other diagrams.
- The maximum deformation at the mudline is found to be 0.10-0.15m for Q_h = 200kN, which is considerably less than the dowel theory result of 0.2-0.25m. This is due to the fact that the SPJ-analyses herein use an effective stress, drained modulus variation, while the dowel theory used an undrained, softer modulus variation, with no gain for effective stress increase during lateral load increase on the pile.

Seen as a whole the SPJ-results is in good agreement with the other analyses as far as horizontal load an moments are concerned. Moreover, the differences in average lateral soil reaction in the soft clay are moderate. However, for lateral loads above 200kN the SPJ-analysis leads to less deformation and an entirely different soil reaction in the stiff clay.

PILE-SOIL INTERACTION BY A TWO-LAYER DOWEL THEORY

The preceeding dowel theory (Figs.6 and 7) was based on a one-layer system for deformations, while the SPJ pile-soil interaction (Fig.9) used a two-layer system, closer to the real in situ conditions. The actual situation at hand is characterized by the fact that the stiffest pile section is in soft soil, while the more slender part of the pile is imbedded in stiff soil. To cope with this special case in a simple way a two-layer dowel theory has been developed.

The two-layer dowel theory; principle.

Reference is made to Fig.10, showing a pile imbedded by D_0 in a soft soil and by D_u in a stiff soil.

The total lateral deflection of the pile is partly due to a rigid body rotation δ_θ and partly elastic bending δ_e. The rotation θ in the stiff soil is counteracted by average soil reactions $\pm \bar{p}_{ru}$, above and below the point of rotation.

The resulting lateral displacement at the midpoint of the soift layer δ_{m0} will determine the average lateral soil reaction $\bar{p}_{ro}$ in the soft soil. An idealized soil reaction diagram is shown at left in Fig.10.

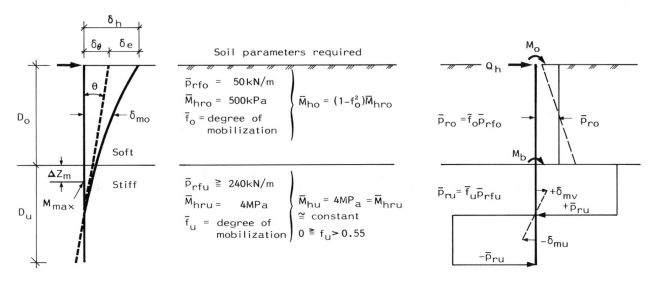

FIG. 10. Two-layer dowel theory for pile-soil interaction analysis.

The required soil parameters of the two-layer system are defined in Fig.10. The degree of shear stress mobilization in the soil layers are denoted f_0 and f_u respectively. The general relationship $\bar{p}_r = \bar{f} \bar{p}_{rf}$ applies to both layers.

Formuli for the two-layer system.

The rigid body rotation is first obtained by calculating the lateral deformations $\pm \delta_{mu}$ in the stiff soil, where $\delta_{mu} \cong \bar{p}_{ru}/\bar{M}_{hu}$, and $\theta = 4\delta_{mu}/D_u$. The bending moment at the two-layer boundary is denoted $M_b = \bar{p}_{ru} D_u^2/4$. By combining these simple relations one gets, for the stiff layer:

$$\theta = \frac{16 M_b}{D_u^3 \bar{M}_{hu}} \ ; \qquad \bar{p}_{ru} = \frac{4 M_b}{D_u^2} \ ; \qquad \bar{f}_u = \frac{\bar{p}_{ru}}{\bar{p}_{ruf}} \qquad (11)$$

The displacement $\delta_h = \delta_\theta + \delta_e$ at the mudline can be estimated by the formuli

$$\delta_\theta = \theta l, \qquad \delta_e \leq \frac{Q_h l}{3EI}, \qquad l = D_0 + \tfrac{1}{2} D_u \qquad (12)$$

By the same principles one can also obtain the lateral deformation δ_{m0} at the midpoint of the soft layer. Since

$$\delta_{m0} = \frac{\bar{p}_{r0}}{\bar{M}_{h0}} = \frac{\bar{f}_0}{1-\bar{f}_0^2} \cdot \frac{\bar{p}_{rf0}}{\bar{M}_{hr0}} = \frac{\bar{f}_0}{1-\bar{f}_0^2} \cdot 0.1 m$$

one can obtain $\bar{f}_0$ and hence $\bar{p}_{r0} = \bar{f}_0 \bar{p}_{rf}$. Horizontal equilibrium will then yield the lateral load at mudline

$$Q_h = \frac{M_b - M_0}{D_0} + \tfrac{1}{2}\bar{p}_{r0} D_0 \qquad (13)$$

The absolute maximum bending moment may not be equal to M_b. The distance to M_{max} from the layer boundary Δz_m can be calculated as follows:

$$\Delta z_m = \frac{Q_h - \bar{p}_{r0} D_0}{\bar{p}_{ru}} \qquad (14)$$

Because of the simplicity of the formuli above it has been possible to carry out a fairly comprehensive interaction analysis, for the actual case at hand.

Numerical results.

The main purpose of these numerical analyses is to incestigate which of the materials, the steel or the soil, mobilized its strength first. Besides, it is of interest to know the influence of the yield stress of the steel σ_y and the influence of variation in the soft layer thickness D_0.

In the first case, Fig.11, it is assumed that $D_0 = 6.5$m and $D_u = 9.0$m for a total imbedment of 15.5m. The ultimate lateral resistances of the two soil layers are 50kN/m and 240kN/m, respectively, see Fig.10. Moreover, it is assumed that the modulus of lateral deformation of the stiff clay layer is 4MPa, as estimated from oedometer test results.

Assuming different values of M_b one can calculate θ and $\bar{f}$ from Eq.(11) in addition to the steel stress in bending ($\sigma = M/W$). Fig.11 shows that $\theta < 0.01$ radians at the time when the steel yields (if $\sigma_y \leq 240$MPa) and still the degree of shear mobilization of the stiff clay layer is less than 35%. Hence, the steel pile yields in bending near the layer boundary long before the stiff clay.

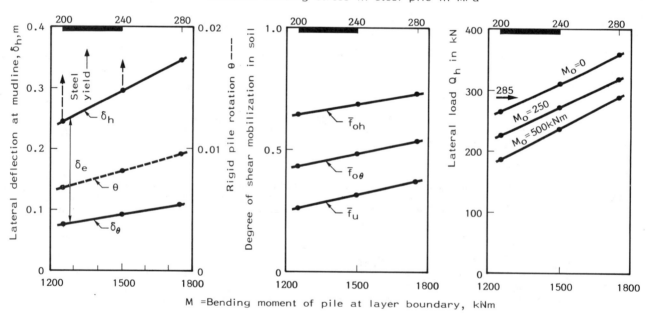

FIG. 11. Influence of pile bending moment (or yield stress) on pile-soil interaction behaviour.

The corresponding degree of mobilization in the soft soil, due to the rotation only is about 50%. Then the lateral load Q_h may be calculated according to Eq.(13) and the result is shown in Fig.11 to the right. Now, the elastic deformation due to bending is estimated. This leads to an increase of the shear mobilization in the soft clay from about 50% to about 70%. The corresponding increase in Q_h is found to be about 30kN, and the distance to the maximum moment is less than 1m below the layer boundary. Therefore M_b is very close to the actual maximum moment.

The estimated lateral load due to ice flow failure (285 kN) is seen to correspond very will with the calculated pile-soil interaction load for M_0 = 0-200kNm at steel stresses near the yield point (say 200-240MPa).

However, the main observation to be made from Fig.11 is that the yield stress of the steel due to pile bending, caused by the ice load, was reached for a very low degree of shear mobilization in the stiff clay (less than 35%) and for a moderate defree of mobilization in the soft clay (around 70%). At the time of initial pile failure by bending the lateral deflection of the pile at the mudline is estimated to be 0.3m, whereof 0.1m is due to rotation only. When this state is reached the rest is just a question of deformation due to yield under almost constant load.

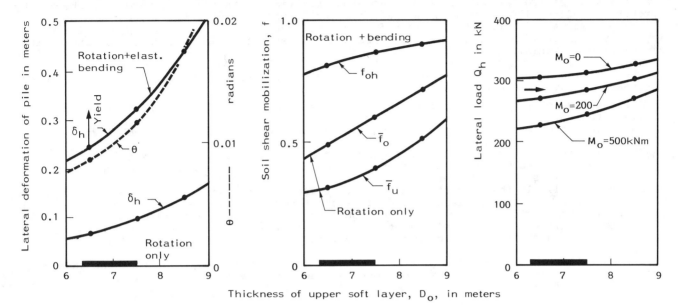

FIG. 12. Influence of soft layer thickness on pile-soil interaction.

In the next case, Fig.12, it is assumed that the yield moment of the pile $M_f = M_p$ = 1500kNm, and that the total imbedment of the pile in the two layers is 15.5m. By selecting different values of the soft layer thickness D_0 one can use Eq.(11) to estimate θ and f_u.

Fig.12 shows that even for D_0 = 8m the rotation $\theta_0 \leq 0.015$ radians $f_u \leq 0.45$, and that $Q_h \sim$ 255kN for $M_0 \sim$ 200kNm. The average degree of mobilization in the soft clay reach up towards 0.9 for a lateral deformation close to 0.4m. For D_0 = 9m it is seen that $\theta \sim 0.025$ and $f_u \sim 0.60$, $f_{oh} \sim 0.90$ for $Q_h \sim$ 310kN, and that $\delta_h \approx$ 0.5m before yield.

The main conclusion which can be drawn from Fig.12 is that the steel pile will yield in bending due to the estimated ice load for a moderate degree of shear mobilization (~50%) in the stiff clay, even if the soft soil is 8.5m thick, implying only 7.0m imbedment of pile in the stiff clay.

RESUME OF FINDINGS

In 1977 a flare tower collpsed in the Bo-hai Gulf in North China. Drift ice caused the collapse by bending of the steel piles some 6-7m below the mudline. Possible collapse mechanisms have been analysed herein.

The subsoil is layered, and each layer vary in thickness over the investigated area. Within each layer the variation of geotechnical parameters are significant. In broad terms, however, a three-layer system describe the subsoil sufficiently well for practical purposes:

- An upper layer (8m ± 3m) of soft organic clay.
- A middle layer (12m ± 3m) of fairly stiff sandy-silty clay.
- A lower layer of sand, below the depth of interest (> 20m).

The maximum total ice thrust on the jacket has been determined previously (Xu and Leira,1981). The net result was that the maximum load on a single pile did not exceed 285kN at any time. This value has now been used as a guideline for the analyses. However, the main conclusion drawn herein would not be altered even if the estimated ice load should be considerably off, say by ± 20%.

Several methods of analyses have been used. Initially, a fairly comprehensive analysis was carried out for the interaction between the steel jacket and the pile-soil system (Xu et.al.1981). This lead to the important information that at the time of yielding the pile bending moment at the mudline was 500kNm, or less. Subsequent analyses indicated a lower limit of 200kNm.

With this information at hand simple dowel theory analyses were carried out, in which the soil was first considered as one layer of homogenous material. The findings were then checked by a comprehensive pile-soil interaction analyses for the actual two-layer system, using a SPJ-computer program (Nordal, 1979). The results of these analyses inspired the development of a novel two-layer dowel theory, the results of which proved to be essential for the understanding of the actual collapse mechanism.

When all the numerical results of the various analyses are summarized, the following conclusions seem to cover the actual case:

- The steel piles yielded in bending at a time when the stiff clay was moderately stressed (less than 50 of failure).
- The calculated depth of maximum bending moment corresponds well with the observed position of bending.
- The calculated static force system required to bend a single pile agrees reasonably well with the estimated maximum ice load excerted on a pile when the drift ice sheared its way through the legs of the jacket.
- It is not unlikely that a certain degree of progressive pile failure mechanism took place, for instance so that two piles started to yield before the two others on the opposite side.
- Composite modes of failure were studied, such as a combined pile bending and pile pullout, and rotationless lateral displacement, forming two yield hinges in each pile. These modes lead to much higher lateral load resistances, and are thus considered unlikely.

The fact that the thin-walled pile was imbedded in the stiff clay, and that the change in cross section was located near the maximum moment and near the layer boundary, apperas to have been of great significance.

ACKNOWLEDGEMENT

The authors are greatly indepted to dr.ing. candidate S.Nordal for carrying out the comprehensive pile-soil interaction analyses leading to the results shown in Fig.9.

REFERENCES

GRANDE,L. and S.NORDAL, 1979, "Pile-Soil Interaction Analyses on Effective Stress Basis". Proceedings, Recent Developments in the Design and Construction of Piles. The Institution of Civil Engineers, Thomas Telford Ltd., London. Paper No.28, pp.275-285.

JANBU,N., 1973, "Statisk dimensjonering av peler i jord" Paper no.10, Kursdagene ved NTH,1973. Norwegian Institute of Technology, Trondheim, Norway.

JANBU,N., 1980, "Simple Pile Foundation Analyses". The Arne Selberg Volume; pp.177-210, in Norwegian Bridge Building, Tapir Forlag, Trondheim, Norway.

LANGEN,I., 1977, A General Dynamic Analysis Program for Linear Structures. User's Manual, Division of Structural Engineering, SINTEF, Trondheim.

NORDAL,S., 1979, SPJ-User's Manual (in Norwegian) Geotechnical Division, Norwegian Institute of Technology, Trondheim, Norway, Report F.79.05.

NORDAL,S. and L.GRANDE and N.JANBU, 1982, "Prediction of Offshore Pile Behaviour" to be published in: The 3rd International Conference on Behaviour of Offshore Structures, Massachussetts Institute of Technology, USA.

XU JIZU and B.J.LEIRA, 1981, "Dynamic Response of a Jacket Platform Subjected to Icer Flare Loads" Proceedings, 6th International Conference on Part and Ocean Engineering under Arctic Conditions, Quebec, Canada, Vol.1, pp.502-516.

XU JIZU and N.JANBU and L.GRANDE, 1981, Ultimate Lateral Resistance of an Offshore Pile. Geotechnical Division, Norwegian Institute of Technology, Trondheim, Norway. Report F.81.06.

Technical Poster Sessions

FOUNDATIONS

Geotechnical Issues in the Seabed

TIME AND COST PLANNING FOR OFFSHORE SOIL INVESTIGATIONS

T. Amundsen   R. Lauritzsen
NOTEBY - Norsk Teknisk Byggekontroll A/S [1]

SUMMARY

Soil investigations for permanent offshore structures are becoming increasingly complex and costly. This article presents field performance data based on several years of experience with soil investigations in the North Sea. The interpreted work efficiency figures, weather downtime percentages and other important parameters that influence the total cost are intended to be used as tools in scheduling and cost planning. The influence of factors like water depth, investigation program, soil conditions, type of drillship, time of the year and weather conditions are discussed. Necessary time for mobilization before the investigation starts, sailing time and demobilization are referred to. Finally the article presents average figures for consumables as well as lost and damaged equipment. The data presented here must be used with care, keeping in mind that factors like unexpected soil conditions, extreme weather, unproven equipment, etc. may have an adverse influence on field performance and total cost.

---

[1] NOTEBY is a partner of NORCONSULT A/S.

INTRODUCTION

Extensive soil investigations have been carried out in the North Sea in connection with the development of oil and gas fields in the past several years. NOTEBY's[1] position as geotechnical consultant and client's offshore representative during the planning, execution and evaluation of a majority of the investigations in the Norwegian sector (1978-82) has enabled a systematic recording of all data relevant to the evaluation of field performance.

The total cost of a soil investigation is more dependant on the time aspect than any other single factor. Therefore, a more accurate estimate of the time required to perform a given investigation programme is essential for improved cost estimates. The authors have chosen to divide the total time required to carry out a soil investigation into clearly defined activities and have attempted to quantify the effect that the most important factors have on the time required for each of the activities. These activities can be summarized as follows: mobilization/demobilization, sailing, weather downtime, working time and standby due to technical breakdown. The article concentrates on the time spent from the arrival at the site until departure, although other aspects are commented on later in the article.

The site conditions and the type of drillship are the first two major issues to be considered in the planning process when the field program has been established. For both of these, the influence on the total performance and cost must be evaluated. By dividing the total field time in two main portions, weather downtime and working time, a more detailed study of the work efficiency can be achieved. The working time (i.e. total field time minus weather downtime) is influenced by a number of factors, such as water depth, investigation program, type of drillship, seasonal variations, etc. For example, the presented material involves two basically different types of drillships: M.V. "Ferder", a conventionally anchored drillship (here called C.A. drillship) and M.V. "Pholas", a much larger dynamically positioned drillship (here called D.P. drillship). The ships' displacements are 2725 tons and 5950 tons respectively. The influence of the differences between the two types of drillships on efficiency is discussed.

The statistical material is used to quantify the influence of all these factors. The authors feel that the figures are representative for extensive soil investigations but must emphasize that application of these figures must be made with care. The parameters will be discussed below in three sections: work efficiency, weather downtime and miscellaneous. Other variables than those described in this article may influence particular investigations.

WORK EFFICIENCY

As a measure of work efficiency the following definition is used: efficiency number = hours required to perform one metre of borehole with sampling and in-situ testing. Similarly, for Cone Penetration Tests (CPT) performed with separate jacking equipment from the seafloor (example: Fugro's Seacalf, McClelland's Stingray) the efficiency is defined in terms of hours per metre of penetration. In both cases the time includes all activities related to the work, including anchoring/positioning, technical breakdown, equipment handling as well as the actual drilling, sampling or testing. The average efficiency number interpreted from the available data is 0.60 hours per metre including all variables.

To arrive at the relevant efficiency numbers two basic parameters must be established: time and production. The number of hours are found by splitting the working time in the field (total field time minus weather standby) into the different activities performed. Such an activity breakdown is shown in the following table.

Activity Breakdown 1981

% of work time	Block 34/10	Block 15/9	Block 2/4	Block 2/11	Block 31/2	Block 16/11 17/7	TOTAL AVERAGE
Type of drillship	C.A.	D.P.	D.P.	D.P.	D.P.	D.P.	
Drilling, sampling, in-situ testing	59.3	54.7	59.6	46.5	23.4	63.0	55.5
Pipe and equipment handling	20.7	23.7	23.6	19.8	31.2	19.7	21.9
Anchoring/positioning	8.7	6.5	9.9	6.2	4.1	5.1	7.4
Stand-by due to in-situ test equipment x)	3.9	5.3	4.5	22.3	33.3	7.5	8.2
" " " drillship	1.7	9.9	2.4	1.7	7.0	4.7	3.9
" " " navigation equipment	1.1	0	0	0	1.0	0	0.6
" " " outside factors	4.6	0	0	3.5	0	0	2.5
No. of hours	1494	489	199	207	236	428	3053

x) Includes soil sampling equipment.

Similar tables for the 1978, -79 and -80 seasons do not diverge significantly from the one above.

Comparing the various 1981 surveys, two of them deviate from the others. In both cases, the same malfunctioning prototype equipment was used in connection with sampling/in-situ testing. Also, in the 31/2 case, extremely soft soil conditions made the in-situ testing and sampling more time consuming than normal. The high proportion of pipe/equipment handling was directly related to the problems with the prototype equipment.

The anchoring/positioning time does not immediately appear to favour the D.P. vessel as strongly as one might expect. However, the only case where the D.P. vessel does not show favourable positioning time compared with the anchored vessel is in the Block 2/4 survey. This survey was carried out only 50-100 m from existing production platforms. Therefore, a conventionally anchored vessel would not have been able to carry out the investigation without the help of an anchor handling boat, which would most likely have resulted in an even less favourable time statistic for this activity.

The activity labelled "drilling, sampling and in-situ testing" would normally be split into two parts, separating surface CPT and drilling/sampling/wireline in-situ testing. However, due to technical difficulties with the surface CPT equipment, most of the planned surface CPT's in 1981 were carried out as down-the-hole tests. Therefore, the very few surface CPT's are not presented as a separate activity.

The activity breakdown table indicates that the navigation systems were only to a small degree the cause of delays. However, technical breakdowns did occur and only favourable circumstances prevented significant delays. Similar cases have been recorded during previous surveys. The choice of navigation system should receive increased attention at an early stage of the planning process.

The production figures are achieved by accumulating the borehole depths over a given time period, or penetration depths of surface CPT's. Redrilling aborted holes due to other reasons than weather downtime is not included, in order to avoid an artificially high efficiency.

Experience indicates that five factors influence the work efficiency significantly: time of the year, water depth, soil conditions, investigation program and type of drillship. These variables normally concern all soil investigations. Also, in certain cases, other factors such as the presence of drilling rigs, other high priority vessels (f.ex. seismic survey boats), pipelines etc. may result in lower efficiency than what may normally be expected. Below, the effect of the five main factors are discussed and quantified.

The seasonal variation of recorded efficiency for 1980 and -81 is presented in figure 1. Also the average efficiency through each of the years is shown, with 0.59 hrs/m (1980) and 0.68 (1981). The general trend is one of lower efficiency (high number) during spring and fall than during the summer period. Since the weather factor, at least theoretically, is isolated from the definition of work efficiency, this trend may seem surprising. However, the weather factor does influence the efficiency because of the continous marginal working conditions experienced during early spring and fall. Although work may continue under these conditions, efficiency is reduced due to the rough weather. Further, during the first periods of a new season, the crew are not as accustomed to the routines and operations, resulting in a necessary "warming up" period. There may also be other factors behind the recorded variation.

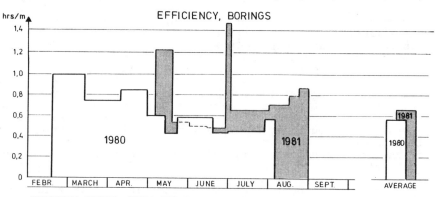

EFFICIENCY NUMBER = HOURS REQUIRED TO PERFORM ONE METRE OF BOREHOLE WITH SAMPLING AND IN-SITU TESTING.

The peak periods registered both in 1980 and 1981 are results of special occurences. The peak value for CPT's recorded in the middle of May 1980 was due to working on a steep slope exceeding the design limits of the testing equipment. For the 1981 season, the efficiency of CPT testing is of limited statistical interest due to the following reasons:

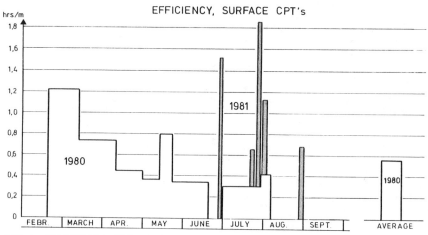

FIG.1 SEASONAL VARIATIONS IN WORK EFFICIENCY

- the first and fourth column reflect the technical problems encountered with prototype geotechnical equipment.

- the third column reflects the extremely difficult soil conditions experienced on one site in the 34/10 area.

These "exceptions" are, however, valuable information because they demonstrate how average efficiency may deteriorate seriously by factors that in advance of an investigation may not seem to be significant.

Soil conditions may also affect efficiency. Three comparable investigations (same water depth, same time of year, comparable investigation programs and same drillship) were checked. The actual soil conditions at all three sites were within what is normally found in the North Sea, none with extremely hard or soft zones. However, the proportion of sand to clay as well as relative densities varied. The efficiency numbers varied between 0.4 to 0.7 hrs. per metre, reflecting the relatively large influence of soil conditions. There are two main factors in such average soil conditions that may affect efficiency. Stiff and plastic clay can result in plugging of the drillbit. Further, loose sand may lead to partial cave-in of the hole when the pipe is lifted to take samples or perform in-situ tests. Larger variations in efficiency than those referred to above may occur with particularly difficult conditions. Extremely hard soil or very soft soil may increase the efficiency number to more than 1.0.

Water depth influences work efficiency in several ways. As an example the efficiency number versus water depth in the 1981 North Sea Investigations is shown in fig. 2. It demonstrates particularly how water depth affects time required to handle drillpipe and seafloor equipment. The figure is not intended to be used in absolute terms but rather as an illustration. The plotted results show the C.A. drillship at 140 metres water depth and the D.P. drillship for all other cases. Some reservations are attached to the application of these data. Firstly, C.A. drillships will not normally be able to anchor and perform soil investigations in water depths of more than approximately 250 meters. Exceeding this depth efficiency deteriorates significantly both due to very limited ability to move on anchors and because greater caution must be shown regarding rough weather. Also the need for a separate anchor handling vessel arises with greater water depth. However, with water depths down to 200-230 m C.A. drillships can operate with only insignificant deterioration in performance compared to a D.P. drillship, provided the investigation program does not require unusually many re-anchorings.

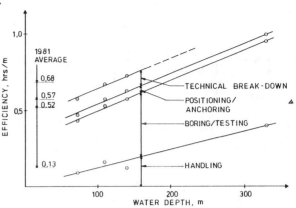

FIG. 2 FIELD HOURS REQUIRED (excl. WEATHER DOWNTIME) per METRE BOREHOLE VERSUS WATER DEPTH IN 1981 NORTH SEA SURVEY.

Investigation program. Most extensive investigation programs for gravity structures or piled structures in the North Sea involve a range of deep and shallow boreholes with sampling and/or in-situ testing as well as surface in-situ testing. The investigations referred to in this article are all rather complex and comparable in this context. However, figure 2 shows a single point at 360 m water depth with an efficiency number of 0.55 hrs. per metre, a very low number (high efficiency) at this water depth. The technical data from this investigation (America 1981) indicates that the main reason for the high efficiency is the investigation program consisting mainly of a few deep boreholes requiring relatively less time for pipe handling and positioning. Other factors such as a low percentage of technical breakdown time and favourable soil conditions have also contributed. It should therefore be kept in mind that the overall efficiency may benefit from a program consisting mainly of fewer and deeper boreholes. Conversely, complex programs involving a large proportion of shallow boreholes with frequent in-situ testing and wide spacing between boreholes will require more working time than average.

Drillship category. As mentioned in the introduction this article is based on experience with two basically different drillships. The C.A. drillship showed an overall efficiency number of 0.73 hrs. per metre in 1981 compared with 0.65 for the large D.P drillship. The difference may not seem significant. Under normal circumstances with water depths ranging from 50 to 250 m and without nearby obstructions there will be little difference between the two vessels in terms of performance. The D.P. drillship has a slightly better efficiency which can probably be credited to its significantly larger displacement and hence low response to waves and swell. However, the dynamic positioning enables the D.P. drillship to maintain its efficiency in cases involving great water depths, nearby obstructions or the requirement for frequent moving within a survey area. The C.A. drillship may also be able to carry out surveys in many of those special cases. However, performance will deteriorate with consequent cost penalty.

WEATHER DOWNTIME

A drillship's performance with respect to weather conditions may be described by limit values of wind and wave parameters. Although wind direction and wave period both affect a ship's performance, the authors have chosen to use wave height and wind speed to describe the limit working conditions. The background for this is based on two factors. Firstly, experience during the past few years with several drillships has indicated that wind speed and wave height are the dominating factors when establishing limit working conditions. There are, of course, several cases where change in wind direction and wave period have been limiting factors, but only in a minority of the weather standby cases that form the statistical background of this study. Secondly, the weather observations made on drillships normally do not include wave period and only very approximately describe wind direction.

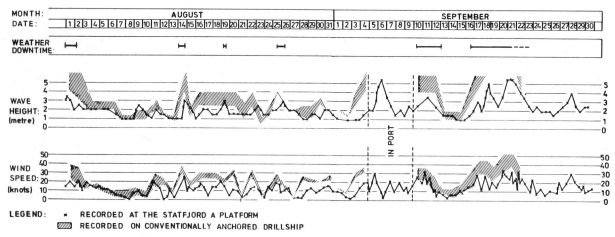

FIG. 3 WIND AND WAVE DATA. BLOCK 34/10 1981

As a check on the accuracy of the environmental observations, automatic wind and wave recordings from nearby permanent oil platforms are used as references. Figure 3 shows an example of data from Block 34/10 in 1981. A conventionally anchored drillship was used in this investigation and the operational limits were approximately 30 knots wind and 3.5 - 4.0 metre wave height. The shaded areas show the field observations from the drillship, the black line shows the automatic recordings from the platforms. The two agree well in calm weather periods. However, they diverge increasingly with increasing wind and waves. This is perhaps a consequence of the way the two types of recordings are made. The weather recordings made on the permanent oil installations are made automatically and are presented as significant wave height and wind speed. The manual recordings on a drillship tend towards maximum values particularly due to short observation time. Other sources of inaccuracy are the visual estimate of the wave height as well as the rather rare calibration of wind speed metres. Similar observations and recordings have been made for the larger D.P. drillship. Despite its displacement advantage as well as its ability to respond to changing wind direction, the limit wind and wave parameters did not exceed those of the C.A. drillship.

The environmental conditions vary considerably throughout the year. Figure 4 shows weather downtime observations through the 1978-81 seasons. The upper portion shows how the downtime percentage in a given seasonal period may vary from year to year. The lower figure shows how the weighted average for the four years varies through the year and is a parallel to the seasonal variation in efficiency number, fig. 1. Both figures are important for an estimate of likely weather downtime in a given time period. The application of these figures must be made with care, as demonstrated by the large spread in the month of May.

Finally, figure 5 shows the total annual average of weather downtime for the four years concerned. This last figure shows a general trend towards lower percentages. However, that does not necessarily mean that the general weather conditions in the North Sea have improved proportionally. The explanation is probably a combination of slightly improved performance in marginal weather conditions and favourable timing of visits to port. This last point is particularly applicable to the 1981 season.

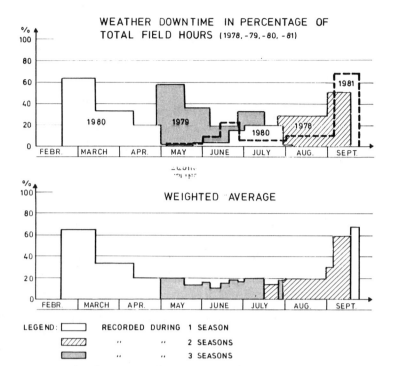
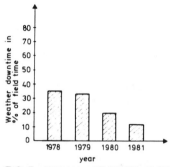

FIG. 4. SEASONAL VARIATIONS IN WEATHER DOWNTIME   FIG. 5 ANNUAL WEATHER DOWNTIME

MISCELLANEOUS

In addition to weather downtime and work efficiency, other factors also contribute to the overall cost of an offshore soil investigation. The most important are mobilization/demobilization time, sailing time, fuel and mud consumption and the cost of reimbursable items such as lost or damaged drillpipe, in-situ test equipment, etc.

Mobilization/demobilization. The time spent on these activities varies depending on such factors as the type of testing equipment to be used, what port is to be used, etc. Generally, two days are spent for mobilization when starting up a new season. When going from one survey contract to another where similar equipment is to be used, less than one day may be sufficient. Demobilization is normally completed in one day or less.

Sailing time. When transitting large distances, the ship's cruising speed may significantly influence the required transit time. However, most drillships working in the North Sea will arrive at any particular site from most ports within one to two days, even in rough weather.

Fuel comsumption. A C.A. drillship normally consumes 2-3 tonnes of fuel per day when working, slightly more when sailing. The large D.P. drillship referred to in this article consumed an average of 8.5 tons per day during approximately three months of soil investigations, 500-600 tonnes of fuel more than a C.A. drillship would consume during the same period.

Reimbursable items. The contracts normally used in the North Sea require reimbursement for lost or damaged equipment. The following table shows the average level of consumption of the most common items recorded over a two year period involving a total of 326 days of field work. The table indicates the "working life" of each item.

Item	"Working life"
CPT cones	166 m drilled borehole per cone
Conerods (1 m length)	15 m penetration per rod
Sampling tubes	9 samples per tube
Drillpipe (10 m length)	306 m drilled borehole per pipe
Anchor buoys	11 days field time per buoy
Wire rope	13 m wire rope per day

The above figures are of course average figures and should be used with care. For extensive surveys of long duration, however, the figures may serve as a guideline, not as absolute figures.

Drillmud consumption. Two different types are normally used in offshore drilling, dry mud for salt water gel and polymer mud. Consumption figures vary according to soil conditions. Average figures are 0.15 - 0.20 tonnes per metre borehole for dry mud, 0.015 tonnes per metre borehole for polymer mud. With a price difference on the order of 1 to 10 (dry mud/polymer mud) the total cost difference is insignificant.

RECOMMENDATIONS

Applying the data presented herein in the planning of offshore soil investigations that are within the limitations described in this article a basic efficiency number of 0.6 hours per metre borehole with sampling and down-the-hole testing is recommended. The influence of the investigation program may reduce this by approx. 0.1 hrs. per metre where few, deep boreholes form the dominating part of the program. Very complex investigations dominated by several shallow boreholes with unfavourable spacing and frequent down-the-hole in-situ-testing may increase the number by as much as 0.3-0.4. Having arrived at a modified basic value the effect of soil conditions may change this by at least ±0.1 hrs. per metre. Extreme soil conditions (including steep seafloor slope) may increase the number by more than 0.4 hours per metre.

The basic value is for average water depths in the region of 120-140 metre. Water depths of less than 90-100 metres may reduce the efficiency number by approx. 0.05. Similarly, deeper water increases the number by up to 0.2 for 220 metres, possibly less for D.P. drillships. Beyond approx. 240-250 metres only a D.P. drillship will work efficiently, although the available statistics do not justify a quantification of the water depth influence.

The type of drillship may influence the efficiency number in what have been described as normal conditions with respect to water depth, soil conditions, investigation program and absence of physical obstacles. A D.P. drillship of the category referred to in this article will probably be able to reduce the efficiency number by 0.05 hrs. per metre or by up to 0.1 in the spring and fall.

In the early parts of the season, particularly if the drillship to be used has been out of work for a longer period during winter, the efficiency number may increase by 0.1-0.2 hrs. per metre, more for short investigations. A similar trend (although to a lesser degree) is found in the fall period when rough weather has increased influence on performance.

Weather downtime in the North Sea averages 20% in the summer period May to August. This proportion may be applied for investigations exceeding 4-6 weeks. However, for shorter investigations one should be prepared for values as high as 40% or even more in extreme cases. Outside this "weather window" percentages may be on the order of 40 to more than 70%.

The above mentioned figures apply to borehole operations. For surface cone penetration tests similar values apply but with increased seasonal variations. Also, soil conditions in the upper layers have greater influence on performance. For extensive use of surface CPT's in the summer months our statistics indicate an average of 0.4 hrs. per metre, increasing to more than twice this figure with difficult soil conditions. The influence of other factors like investigation program, water depth etc. has not been possible with the available statistical material.

As for the other cost factors discussed in this article the presented statistic figures may be used as a guideline in connection with extensive investigations of long duration, keeping in mind that large variations will occur from one survey to another.

ACKNOWLEDGEMENTS

The authors wish to express their appreciation to the following companies for allowing publication of the stastical material in this article:

Amoco Norway Oil Company, McClelland Engineers, s.a, Phillips Petroleum Company Norway, Statoil, Den norske stats oljeselskap A/S, A/S Norske Shell.

SOIL FLOWS GENERATED BY SUBMARINE SLIDES – CASE STUDIES AND CONSEQUENCES

Lewis Edgers
Tufts University, U.S.A.

Kjell Karlsrud
Norwegian Geotechnical Institute, Norway

SUMMARY

When placing a structure on the sea bottom it is not only necessary to evaluate the slope stability situation at the specific site, but also to look at the possibility of the structure being hit by slide debris from failures up slope. Turbidity currents, soil flow, and progressive liquefaction are discussed as possible mechanisms of slide propagation.

Runout distance and velocity data are collected from a number of the best documented submarine slides. They range in size from small coastal slides which ran out a few hundreds of meters to large deepwater slides which ran out hundreds of kilometers. These data show velocities larger than 25 m/s based upon the sequence of downslope cable breaks. The data for submarine slides are compared to observations of subaerial quick clay slides which can flow as a liquid for very large distances without significant reconsolidation. An approximate relationship is established between the volume of the masses involved in these slides with the relative runout distance, L/H (L equals the maximum distance from the back scarp to the outermost reach of the slide debris, H equals the corresponding difference in elevation).

The consequences of submarine slides (runout distance, velocity, forces) for offshore structures are then discussed in view of the collected data and some simple viscous flow analyses. Approximate methods for estimating the possibility of a slide reaching a structure, and its velocity are developed.

INTRODUCTION

When placing a structure on the sea bottom it is not only necessary to evaluate the stability situation at the specific site itself, but also to look at the possibility of the structure being "hit" by slide debris from failures upslope. Such an evaluation must include answers to a number of very difficult questions, some of which are:

1. What is the extent of a possible slide, and how does the slide develop with time?

2. What will be the velocity, height, density and run out distance of the masses?

3. What forces might the moving masses exert on obstructions (structures) of different kind as they move downward?

This paper discusses the mechanisms of slide runout, presents field data on runout distance and velocity, and discusses methods for analysis of the consequences of submarine slides (runout distance, velocity, and forces) for offshore structures. Methods of engineering analysis of the stability of submarine slopes have been reviewed recently by Karlsrud and Edgers (1980) and will not be discussed herein.

SLIDE MECHANICS

The masses involved in a slide or slump might behave in a variety of ways after failure, ranging in general terms from slow creep movements, to rapid debris flows, and dilute turbidity currents. Medium to coarse sands and nonsensitive clays, will normally move as more or less rigid material with fairly low velocity, and movements will cease relatively rapidly when the slope angle is reduced below some critical level.

Soft sensitive clays and loose fine sands and silts might on the other hand run great distances and with considerable velocity on very gentle slopes. As it moves downslope, water might be mixed into the flowing mass, making it even more liquid in behaviour. Some material might also in the process be "torn loose" from the main body of the flow and turn into very dilute turbidity currents. Three possible mechanisms of run out of these unstable materials are discussed in the following paragraphs.

Turbidity Currents

In the past, turbidity currents have been emphasized as the predominant mechanism in very large and rapid submarine slides. For example, submarine cable breaks 700 km from the epicentral area of an earthquake off the Grand Banks have been attributed to an enormous deepwater turbidity current by Heezen and Ewing (1952).

An important objection to the turbidity current interpretation is that unless the flowing mass meets some obstructions, or very abrupt changes in slope geometry, creating hydraulic jumps, it is difficult to see how the main body of a flowing mass will become sufficiently dilute to turn into a low density turbidity current. This question must, however, be strongly dependent on the grain size distribution of the masses involved.

Soil Flow

Geologic evidence of submarine soil flows, in which the slide debris moves as a more or less concentrated fluid, has been found in a number of areas. Coleman and Garrison (1977) described surface mudflows in the Gulf of Mexico which often extend laterally for distances in excess of 100 km and which move at rates as much as several hundred meters per year. Based on a study of numerous subaqueous slope failures in Norwegian fjords, Bjerrum (1971) clearly expressed the view that a slide mass in loose silt and fine sand deposits could turn into a dense soil flow and run very long distances before it re-consolidated and came to rest.

Such rapid soil flows have also been observed directly in an interesting series of flume tank experiments conducted by the Delft Soil Mechanics Laboratory in their studies of retrogressive flow slides along the Dutch coast (Delft Soil Mechanics Laboratory, 1980). Failures were induced in 2.5 m high beds of loose fine sand which were then free to flow about 50 m, underwater, to the end of the flume tank. These tests clearly demonstrated the rapid flow of loose fine sands on very flat slopes.

It is quite possible that submarine flows can be very fluid and move, at least episodically, at very large velocities. Very rapid soil flow may have caused the loss of two oil drilling platforms in the Gulf of Mexico (Sterling and Strohbeck, 1973). However, there have been no direct observations of very rapidly moving submarine flows. Thus it is not known if soil flow is the predominant transport mechanism for some of the most rapidly moving (in excess of 25 m/s as discussed below) submarine slides that have been observed.

Progressive Liquefaction

In view of the very large runout distances and velocities, over very flat slopes, Terzaghi (1956) considered it inconceivable that the distant cable breaks following the Grand Banks slide could be the result of either turbidity currents or soil flow. He postulated a mechanism, termed progressive liquefaction, by which the high pore water pressures generated by an initial failure propagate along the sea bottom. These pore pressures will then induce failure in any other loose and metastable sediments which are encountered. This mechanism though cannot account for mass transport features such as graded bedded sands observed in deep waters (Heezen and Ewing, 1952).

PRESENTATION OF FIELD DATA

One approach to understanding the runout behaviour of slide masses is to look at the field data from previous submarine slides. Table 1 compiles data from the known modern and ancient submarine slides. Karlsrud and By (1981) present detailed descriptions of the cases. Most of these slides have been detected only by means of geophysical surveys of the remnants of the slide, with no direct observations. Thus, the precise age, preslide geometry, geotechnical soil properties, triggering mechanisms and sea levels at the time of the slide are not precisely known. A few of the slides (for example Grand Banks, Orkdalsfjord) have been better documented by more direct means such as by submarine cable breaks or coastal slides and, can be related to specific triggering mechanisms such as earthquake or man-made filling. However, even these have not been documented as well as would normally be the case for most slides on land.

The volume, V, has been taken as best as could be determined from the cited references, as the total volume of slide masses involved in the slide. However, it is not possible for most of these slides, to clearly distinguish between the slide pit (source area), the slide path, and the materials redeposited by the slide. Because of this, the runout distance, L, and height drop H, have been taken as the dimensions from the back-scarp of the slide pit to the toe of the slide deposits, Figure 1.

The data on Table 1 show that:

1. Submarine slides may be triggered on very flat slopes. Only one of the slopes (Orleansville) was steep as 10 degrees. Most of the slopes are flatter than 6 degrees and some of these slides have occurred on slopes with an inclination of only about 1 degree (Storegga, Kayak Trough).

2. The volumes and distances from back scarp to toe are enormous in comparison to most terrestrial slides. The Bassein and Storegga (not all slide materials have yet been mapped) slides involved about 800 cu. km of material. Cable breaks due to the Grand Banks slide were observed 700 km away from the original slide and it is not known how much farther the slide materials ran out. Many of the slides have involved volumes of material from 1 to 30 cu. km. with runout distances of from 10's to 100's of kilometers.

3. The most predominant soil types of Table 1 are silts and fine sands although some slides (for example, California) have been observed in more clayey soils. This is because of our selection process in which we have previously looked for slides where large runout distances have been indicated.

4. A major factor in the development of all of these slides is of course the availability of weak and unstable sediments. Aside from this, triggering mechanisms which have served to explain the observed features include sediment overloading, earthquakes, and seemingly minor human activities along the coast. It is interesting to note that the three largest known historical slides, Grand Banks, Messina, and Orleansville were triggered by large earthquakes, of Richter Magnitude 6.5 or greater with shallow focal depths, less than 8 km.

TABLE 1

SUMMARY OF SUBMARINE SLIDES

Slide	Year	V(m^3)	L(km)	H(m)	L/H	Soil type	Triggering mechanism	Comments	References
1. Bassein	Ancient	700-900×10^9	215	2200	98		S, EQ		Moore et al, 1976
2. Storegga	Ancient	800 × 10^9	>160	1700	94		EQ ?		Bugge et al, 1979
3. Grand Banks	1929	760 × 10^9	>750	5000	152	Sand/silt	EQ	$\bar{v}$ = 7,7 m/s based on cable breaks	Heezen and Ewing, 1952
4. Spanish Sahara	Ancient	600 × 10^9	700	3100	226	Gravelly clayey sand	S		Embley, 1976
5. Rockall	Ancient	296 × 10^9	>13,5	700	>19.3		S		Roberts, 1972
6. Walvis Bay S.W.Africa	Ancient	90 × 10^9	250	2100	119				Emery et al, 1975
7. Messina	1908	>>10^6	>220	3200	>69	Sand/silt	EQ	$\bar{v}$ = 6 m/s based on cable breaks	Heezen and Ryan, 1965
8. Orleansville	1954	>>10^6	100	2600	38		EQ	$\bar{v}$ = 5 m/s based on cable breaks	Heezen and Ewing, 1955
9. Icy Bay/ Malaspina	Ancient	32 × 10^9	12	80	150	Clayey silt	EQ		Carlson, 1978
10. Copper River	Ancient	24 × 10^9	8	85	94	Silt/sand	S		Carlson and Molnia, 1977
11. Ranger	Ancient	20 × 10^9	37	800	46	Clayey and sandy silt	S (EQ)		Normark, 1974
12. Mid. Alb. Bank	Ancient	19 × 10^9	5,3	600	9	Silty clay	EQ, S		Hampton et al, 1978
13. Wil. Canyon	Ancient	11 × 10^9	60	2800	21,4	Silty clay and silt	S		McGregor and Bennet, 1977
14. Kidnappers	Ancient	8 × 10^9	11	200	55	Sandy silt, clay	EQ (?)		Lewis, 1971
15. Kayak Trough	Ancient	5,9 × 10^9	18	150	120	Clayey silt	S, EQ		Carlson and Molnia, 1979
16. Paoanui	Ancient	1 × 10^9	7	200	35	Silt/sand	EQ (?)		Lewis, 1971
17. Mid. Atl. Cont. Slope	Ancient	4 × 10^8	3,5	300	11,7	Silty clay	S		Knebel and Carson, 1979
18. Magdalena R.	1935	3 × 10^8	24	1400	17		S	Cable breaks	Heezen, 1956, Menard, 1964
19. California	Ancient	2,5 × 10^8	3,5	150	23	Clayey and sandy silt	EQ (?)		Edwards et al, 1980
20. Suva, Fiji	1953	1,5 × 10^8	110	1800	61	Sand	EQ	Cable breaks	Houtz and Wellman, 1962
21. Valdez	1964	7,5 × 10^7	1,28	168	7,6	Gravelly silty sand	EQ	Cable breaks	
22. Orkdalsfjord	1930	2,5 × 10^7	22,5	500	45	Sand/silt	M	$\bar{v}$ = 2,6 m/s based on cable breaks	Bjerrum, 1971
23. Sokkelvik	1959	10^6 - 10^7	>5	100	>50	Quick clay + sand?		A fishing net was pulled along	Brænd, 1961
24. Sandnessjøen	1967	10^5 - 10^6	1,2	180	7		M	$\bar{v}$ = 0,8 m/s based on cable breaks	Karlsrud, 1979
25. Helsinki harbour	1936	6 × 10^3	0,4	11	13	Sand/silt	M		Bjerrum + Andersen, 1967

V = Volume of masses involved in slide
L = Maximum horizontal distance from back scarp of slide to outermost reach of soil (debris) flow
H = Corresponding maximum difference in elevation
$\bar{v}$ = Average velocity of soil (debris) flow
S = Slide induced by rapid sedimentation
M = Man-made slide
EQ = Slide induced by earth quake

x) Speculative, poorly documented

Note: See also detail description in report

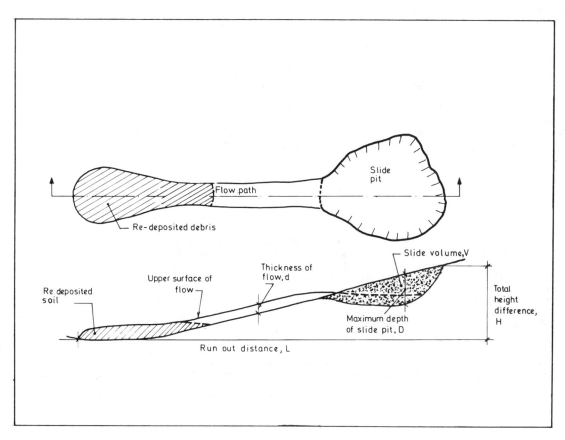

Figure 1   Definition of Slide Parameters

ANALYSIS OF SLIDE RUNOUT DISTANCES

Although the true mechanisms and nature of post failure mass movements are poorly understood, one would expect the runout distance of a slide L, to increase with the volume of the masses involved in the slide, V. This is in fact confirmed by a general review of the data on Table 1. These data also show a general tendency for increasing relative runout distance, L/H, with increasing slide volume V, plotted on Figure 2.

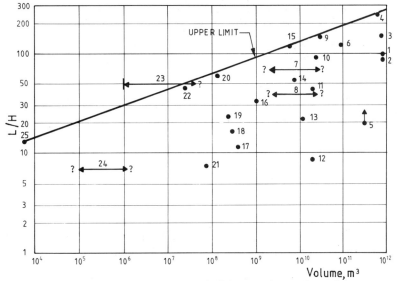

Figure 2   Relative Runout Distance (L/H) vs Slide Volume (V), Submarine Slides

Such relationships between volume and relative runout distance have been developed for landslides by for example, Scheidegger (1973) and Cruden (1980). These studies have expressed landslide runout in terms of the coefficient of friction of the slide materials, taken as H/L. Interestingly, Scheidegger (1973) observed that for large, catastropic landslides, this coefficient of friction becomes progressively smaller as the volume of the sliding mass increases. This is entirely consistent with the general observation that L/H increases with V for the subaqueous slides of Figure 2. However, as discussed below, the details of such a relationship must depend upon the excess pore pressures during sliding as well as material friction.

The plot of log L/H vs log V, Figure 2, shows a great deal of scatter. As a practical matter, this scatter means that our ability to accurately predict from these data, the runout distance of a slice, knowing the volume of the potential slide masses is extremely limited. On the other hand, a line can be drawn to envelop all of the data points. Figure 2 superimposes this "upper limit" line on the slide data. This upper limit line can be used for engineering predictions of the maximum possible runout of a slide, knowing the volume V.

The large amount of scatter below the upper limit line is due to differences in materials, the completeness with which the slides have been mapped, and topographic effects such as the presence of obstacles and the degree of channelization. These factors are discussed below.

1. Material differences - The L/H ratio must be a function of frictional resistance of slide masses as well as the pore pressures and effective stresses during sliding. Slides in free draining non collapsing coarse grained sediments will come to rest when the slide materials are more or less at their angle of repose. See for example, Terzaghi's (1956) description of a slide in sand and gravel deposits on the North Shore of the Howe Sound. On the other hand, slides in collapsing soils, such as loose fine sands, will be accompanied by significant pore pressure development and will run out much farther than the distances inferred from the frictional resistance of the material. This material effect may serve to explain, for example, the very small L/H ratio of the Valdez slide, Table 1. The Valdez slide deposits, described as gravelly silty sand, may be less prone to sudden collapse and pore pressure development than the silts and fine sands of other slides. It should be noted that this specific case is offered only as a possibility. The Spanish Sahara slide has also been described as occurring in a gravelly material but it had a very high L/H ratio. It should be kept in mind of course, that the material descriptions of Table 1 are highly approximate.

2. Mapping - The geophysical surveys or the most distant direct observations (e.g. cable breaks) fall short of the outermost reach of many of the slide deposits. For these slides the dimensions L and H correspond to the available data, and the L/H ratios would be larger if surveys extended to the toe of the slide deposits. (V would increase as well but not so rapidly as L/H). Examples of this are the Valdez, Rockall, Storrega, and Grand Banks slides. On the other hand, the Helsinki, Orkdalsfjord, Suva, Kayak Trough, Icy Bay, and Spanish Sahara slides, all of which plot on or near the upper limit line, were apparently fully mapped, from back scarp to toe.

3. Topographic obstructions - Some of these slides may have run out farther, but for the presence of topographic obstructions such as a reversal of slope. One clear example of this is the Paoanui slide.

4. Channelization - If the slide masses run out on a broad plain, then they will probably spread out and slow down over a relatively short run out distance. On the other hand, if channelized for example by submarine canyons or valleys, then the slide masses may run out very large distances. The extent to which this was a factor for most of the subaqueous slides is not very well known because of lack of survey data and the fact that many of the slides may have infilled and covered preexistent channels. Also, the extent to which slide masses may have eroded a channel in an other wise smooth sea floor is not known. One case for which channelization may have been an important factor is the Bassein slide. A mudflow from this slide poured into a distributary channel on the Bengal fan and virtually filled it for an additional 145 km. along its length.

Runout data from 8 subaerial quick clay slides have also been collected on Table 2 for comparison with the submarine slides. Quick clay slides can flow as a liquid for very large distances without significant reconsolidation. There are no other slides in the subaerial environment which give as large runout distances.

TABLE 2

SUMMARY OF SUBAERIAL QUICK CLAY SLIDES

Slide		V(m³)	L(m)	H(m)	L/H	D(m)	d(m)	Comments	References
27	Hekseberg	1.5-2 x 10⁵	700	40	18	12	3.5		Aas and Eide, 1967
28	Selnes	1.4 x 10⁵	450	25	18	10	2		Kenney, 1966
29	Borgen	1.2 x 10⁵	≥1500	29	>52	9	3-5	Large velocity in the beginning	Bjerrum, 1967
30	Baastad	1.5 x 10⁶	1200	34	35	18	7	Small velocity ($\bar{v} < c/m/s$)	Gregersen and Løken, 1979
31	Skjelstad-marken	2 x 10⁶	~2800	~45	~62	15		$\bar{v}$ ~ 0.2 m/s	Aas, 1964
32	Rissa	5-6 x 10⁶	>1100	35	>31	20	(~5 out of pit)	v = 11 m/s out of the pit	Gregersen, 1981
33	St.Jean Vianney	6.9 x 10⁶	>3880	73	>53	30	10-18	$\bar{v}$ ~ 7.2 m/s *	Tavenas and Rochelle, 1971
34	Verdal	5.5 x 10⁷	9100	70	130	30*	12*	$\bar{v}$ = 10-15 m/s out of pit * $\bar{v}$ ~ 1.7 m/s in lower part *	Verdal Turistforening, 1968

V = Volume of masses involved in slide
L = Maximum horizontal distance from back scarps of slide to outermost reach of debris flow
H = Corresponding maximum difference in height
D = Maximum depth of slide pit below original ground surface
d = Average thickness of soil flow
$\bar{v}$ = Velocity of soil flow

* Speculative, poorly documented

Note: See also detail description in report

The plot of L/H vs. slide volume, V, for the subaerial quick clay slides, Figure 3 shows considerable scatter. This occurs because of many of the same factors discussed above with reference to the submarine slides. For example, the runout of the Rissa slide was greater than the value indicated on Table 2 and Figure 3. However, the slide materials flowed into a lake downslope of the slide pit and have not yet been fully mapped.

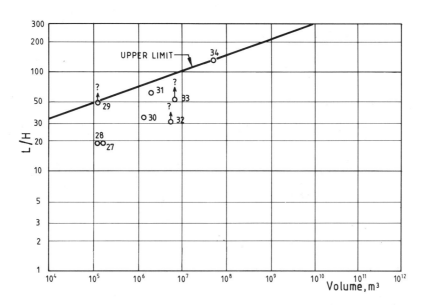

Figure 3   Relative Runout Distance (L/H) vs Slide Volume (V), Subaerial Quick Clay Slides

Comparison of the volumes of the subaerial quick clay slides, Figure 3, with those of the submarine slides, Figure 2, indicates that in general, the volume of masses involved in subaerial slides is less than that of most of the submarine slides which were considered. This is true even for the St. Jean Vianney and Verdal slides which represent two of the most catastropic and massive land slides of the last 100 years. This comparison is due to the large volumes of weak and sensitive soils which are often available in the marine environment.

The comparison of the upper limit of log L/H vs. log V for the quick clay and submarine slides is shown on Figure 4. Recognizing of course the limited data base, this comparison shows that for the same volume, the runout distance for a quick clay slide on land will be potentially larger than for a submarine slide. This comparison is entirely logical in view of the following factors:

1. The downslope gravity driving force on land is almost twice as large as in the marine environment due to buoyancy in the latter case.

2. There is no hydrodynamic drag stress at the top of the quick clay flows. This drag stress will shorten the runout distance of submarine slides, relative to subaerial flows.

3. Quick clays probably experience a more complete breakdown of structure and slower reconsolidation than do the materials involved in submarine slides of comparable size. This will result in less resistance to flow and longer runout distances for the quick clay slides. The viscous analyses of observed slide velocities described in the next section of this paper also show that the quick clay slides offer less resistance to flow than do the submarine slides.

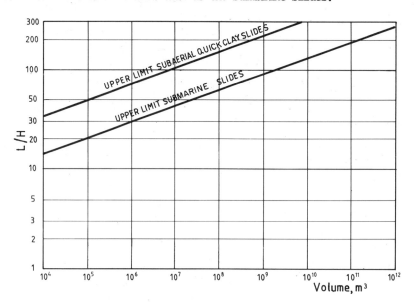

Figure 4   Comparison of Upper Limit Lines

In summary, there is a clear trend for increasing L/H with increasing volume of slide, for both the submarine and subaerial slides. The uncertainty of the observations of L/H ratio vs slide volume V for submarine slides makes accurate predictions of the runout of submarine slides from this data base difficult. On the other hand, the upper limit line of log L/H vs log V, Figure 2 is useful for making estimates of the maximum possible extent of a submarine slide and whether or not it will threaten engineering structures in its path.

ANALYSIS OF SLIDE RUNOUT VELOCITY

Five of the submarine slides of Table 1, Grand Banks, Messina, Orleansville, Orkdalsfjord, and Sandnessjøen caused submarine telephone and telegraph cable breaks sequentially downslope after the initial disturbances. The time sequence of these cable breaks can then be used to estimate runout velocities.

For example, a topographic profile and time relationships for the Grand Banks cable breaks are shown on Figure 5. Very near the earthquake epicenter, it is unclear whether cables may have broken due to initial earthquake shocks and slides or due to the subsequent mass movement. However, further downslope, the cable breaks must be associated with run out of slide debris or as hypothesized by Terzaghi (1956), with an advancing front of spontaneously liquefied material. Figure 5 shows that these disturbances have propagated at very high velocities, on the order of 6 m./s. hundreds of kilometers downslope and approaching 20 to 30 m./s. within the first 100 to 200 kilometers downslope of the initial disturbances.

Two types of theoretical models which can be used to calculate these observed velocities are turbidity current models and viscous flow models. The Grand Banks, Messina, and Orleansville slides have been attributed in the original references to turbidity currents generated by the initial slides. This interpretation has been based mostly on geologic features, such as graded sand layers and submarine canyons whose origins can be explained by turbidity currents, and some theoretical calculations by Kuenen (1952). More recent experimental and analytical studies of the hydrodynamics of turbidity currents have been presented by for instance Middleton (1966a, b, 1967), Hampton (1972), and Pantin (1979). Such models are primarily applicable to low density flows where the soil particles are carried in water by suspension (i.e. with little or no interparticle contact). There is only limited experimental evidence on the validity of these models available, and most experiments have been carried on materials with density less an 1.1 Kn/cu. m.

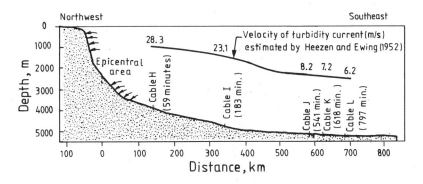

Figure 5   Profile of 1929 Grand Banks Slide

Viscous flow models have been applied to a variety of problems ranging from slow creep deformations and "mud flows" to more rapid fluid types of mass movement. Morgenstern (1967) described an early viscous flow analysis in which the soil shear resistance consisted of a velocity dependent viscous component in addition to Coulomb friction. Johnson (1970) analyzed subaerial debris flows by means of a steady state Bingham model (plastic-viscous model with soil yield resistance, k, and linear viscosity, η). Suhayda and Prior (1978) used an effective stress modification of Johnson's (1970) model to study transport chutes in the Gulf of Mexico.

The major difficulty in applying these theories to the analysis of submarine flows is the selection of the input parameters describing the viscous behaviour of soils.

Edgers and Karlsrud (1981) have developed a simple viscous flow analysis, shown in Figure 6, to estimate the runout velocities of submarine slides. They have used this analysis to backcalculate the equivalent soil viscosities for the five submarine slides described above, making reasonable estimates of flow thicknesses. As shown in Figure 7 for the Grand Banks slide, the viscous analysis provides good agreement with the available field observations. Similar agreement was found for the other four submarine slides analyzed. The range of soil viscosities backcalculated for these five submarine slides was 0.7 to 1.4 k. Pa-sec. As discussed by Edgers and Karlsrud (1981), however, the computations are sensitive to both flow thickness and viscosity. If different flow thicknesses had been estimated, then significantly different soil viscosities would have been backcalculated.

As a check, Edgers and Karlsrud (1981) also analyzed three of the quick clay slides of Table 2, including the Rissa slide which had been filmed by an amateur photographer (Gregersen, 1981). They backcalculated viscosities of 0.2 to 0.4 k. Pa-sec for the subaerial quick clay slides.

The variations in viscosity are quite small in view of the large range in size of the slides considered. The viscosities of the submarine and the subaerial quick clay flows agree remarkably, in view of material differences, and the different effects of consolidation, hydrodynamic drag on the top of the submarine flows, and channelization of flow in the two types of slide considered. This comparison increases the degree of confidence associated with the soil viscosities that have been backcalculated. This also indicates that viscous soil flow, clearly the predominant mechanism in the runout of subaerial quick clay slides, may also be an important factor in the run-out of the submarine slides as well.

The Edgers and Karlsrud (1981) viscous flow analyses modelled the observed runout of the submarine and quick clay slides very well. The study showed that viscous soil flows may theoretically, at least, attain very large runout distances and velocities, consistent with the limited available field data. It is interesting to note that for the values of soil viscosity backcalculated in these analyses, the Reynolds flow number indicates essentially laminar conditions at the soil-water interface. This would preclude therefore, the large amount of turbulent mixing necessary to maintain the flow primarily as a turbidity current.

These considerations indicate that viscous soil flows may be a very important, and perhaps underestimated, factor in the runout of sometimes very large and rapidly moving submarine slides. The viscous analysis of field data also provides some important guidance on the appropriate input soil viscosities.

SOIL FORCES

An estimate of the forces exerted by a slide against a structure in its path requires first, that an estimate be made of the thickness and velocity of the slide as it reaches the location of the structure.

However, given these quantities (thickness, velocity), there are still no general and well documented methods for estimating the corresponding forces. Two methods which have been used are based upon the ultimate capacity of laterally loaded piles and the general drag equation for flow around submerged objects

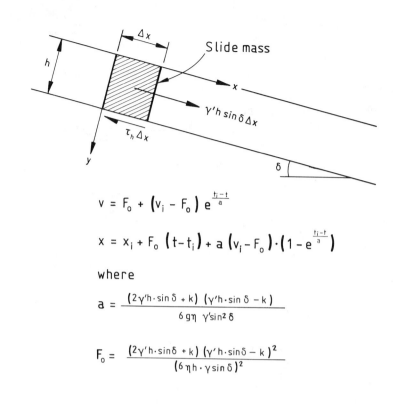

$$v = F_o + (v_i - F_o) e^{\frac{t_i - t}{a}}$$

$$x = x_i + F_o (t - t_i) + a (v_i - F_o) \cdot \left(1 - e^{\frac{t_i - t}{a}}\right)$$

where

$$a = \frac{(2\gamma' h \cdot \sin\delta + k)(\gamma' h \cdot \sin\delta - k)}{6 g \eta \ \gamma' \sin^2 \delta}$$

$$F_o = \frac{(2\gamma' h \cdot \sin\delta + k)(\gamma' h \cdot \sin\delta - k)^2}{(6 \eta h \cdot \gamma \sin\delta)^2}$$

v = average flow velocity over thickness h
γ' = effective unit weight of soil
h = thickness of the flowing mass
δ = slop angle
g = acceleration of gravity
k = soil yield resistance
η = soil viscosity
t = time
v_i, x_i, t_i = initial velocity, position, time

Figure 6 Summary of Viscous Flow Analysis

The ultimate horizontal bearing capacity of a foundation element in soft soil may be expressed as follows:

F = K . C . D (Eq. 1)

where

F = ultimate bearing force per unit length (vertical) of the foundation element.

C = soil shear strength

D = projected width of the foundation element

K = a bearing capacity coefficient, generally empirically determined

Bea and Arnold (1973) and Bea et al (1980) have described the use of this equation for estimates of the soil forces against the piles involved in the South Pass Block 70 failure in the Mississippi Delta. K has been estimated to be about 4 at the mudline to values of about 12 at depths of about 4 pile diameters based on analyses of data from field load tests on laterally loaded piles (Stevens and Audibert 1979, Bea et al 1980). Schapery and Dunlap (1978) measured rate dependent bearing capacity coefficients, termed drag factors, based upon small scale model pile experiments. Typical results are shown in Figure 8. Aside from this limited data base, there is virtually no information available on the effects of foundation shape, size, and neighbouring member interactions. It is not known, for example, whether it would be at all appropriate to apply this equation to large diameter gravity platforms.

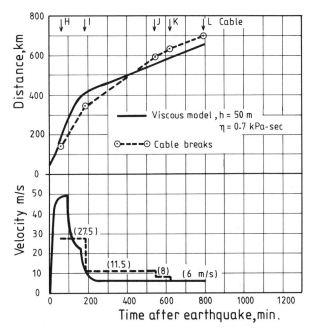

Figure 7 Computed Runout Distance and Velocity vs. Time for 1929 Grand Banks Slide

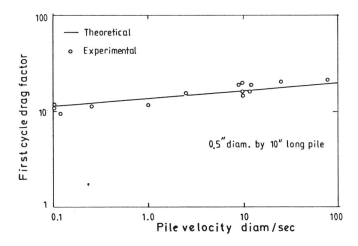

Figure 8 Drag Factor as a Function of Relative Velocity (after Schapery and Dunlap, 1978)

The forces exerted by a slide on a structure in its path may also be estimated by analogy with the hydrodynamic drag of a flow on a submerged object. The general drag equation for flow around submerged objects is:

$$F_D = C_D \cdot 1/2 \cdot \rho \cdot A \cdot v^2 \qquad \text{(Eq. 2)}$$

where

F_D = the drag force

C_D = drag coefficient depending upon the shape of the submerged object and the Reynolds number

ρ = density of the flow

A = the projected cross-sectional area of the object in the direction of flow

v = velocity of the flow

Tesaker (1980) has used this equation in an analysis of possible drag forces exerted by a turbidity current on a gravity structure in the North Sea. He also presented equations for computing the extra unbalanced force caused by the velocity head due to stagnation and the acceleration force in unsteady flow. It should be emphasized that this application of hydrodynamic theory is extremely approximate and has not been verified by field observations. Much additional work in the subject area is required.

SUMMARY AND CONCLUSIONS

The studies described in this paper suggest the following method for analysis of the consequences of submarine slides (runout distance, velocity and forces) for offshore structures.

1. The maximum possible extent of a submarine slide may be estimated from the relative runout distances observed in previous slides, Figure 2. This will require first, an estimate of the volume of the potential slide masses.

2. If it is determined that a slide may possibly reach the site of an offshore structure, then the velocity of the slide masses at the structure may be computed by means of a viscous flow model such as that described by Edgers and Karlsrud (1981). Velocities may also be estimated by turbidity current models such as those described by Middleton (1966a, b, 1967) but such models are primarily applicable to very low density flows where the soil particles are carried in water by suspension.

3. The thickness of the slide masses at the site can presently be only roughly estimated from the volume of available slide materials and some reasonable estimate of their distribution along the slide runout path.

4. The forces exerted by the slide masses against the structure may then be estimated from the equations based upon the ultimate capacity of laterally loaded piles or the general drag equation for flow around submerged objects.

There are still many questions regarding the mechanics of slides such as which soil and environmental conditions govern the transition of a slope stability failure into different types of downslope movement. Little is known about the degree to which flow geometry is influenced by progressive or retrogressive action near the slide pit, and downslope erosion, channelization, or spreading. The method of analysis described above must be used very tentatively and with great caution. Detailed field studies such as geophysical surveys and direct sampling of existing submarine slide deposits will provide considerable insight into these questions.

REFERENCES

Bea, R.G. and P. Arnold, 1973, "Movements and Forces Developed by Wave Induced Slides in Soft Clays", Offshore Technology Conf. 5, Houston 1973, Preprints, Volume 2, pp. 731-742.

Bea, R.G., S.G. Wright, P. Sircar, and A.W. Niedorada, 1980, "Wave Induced Slides in South Pass Block 70, Mississippi Delta", ASCE Convention and Exposition, October 1980, Preprint 80-506.

Bjerrum, L., 1971, "Subaqueous Slope Failures in Norwegian fjords." Norwegian Geotechnical Institute. Publication, 88. Also published in: International Conference on Port and Ocean Engineering Under Artic Conditions, 1. Trondheim 1971. Proceedings, Volume 1, pp. 24-47.

Coleman, J.M. and L.E. Garrison, 1977, "Geological Aspects of Marine Slope Stability, Northwestern Gulf of Mexico." Marine Geotechnology, Volume 2, Marine Slope Stability Volume, pp. 9-44.

Cruden, D.M., 1980, "The Anatomy of Landslides" Canadian Geotechnical Journal, Volume 17, pp. 295-300.

Delft Soil Mechanics Laboratory, 1980, "Combined Hydraulics and Soil Mechanics Research" LGM Mededelingen, Part XXI, No. 3.

Edgers, L. and Karlsrud, 1981, "Viscous Analysis of Submarine Flows - Field Case Studies", Norwegian Geotechnical Institute, Report 52207-8.

Gregersen, O., 1981, "The Quick Clay Landslide in Rissa, Norway; The Sliding Process and Discussion of Failure Modes". International Conference on Soil Mechanics and Foundation Engineering, 10. Stockholm.

Hampton, M.A., 1972, "The Role of Subaqueous Debris Flow in Generating Turbidity Currents", Journal of Sedimentary Petrology, Volume 42, Number 4, pp. 775-993.

Heezen, B.C. and M. Ewing, 1952, "Turbidity Currents and Submarine Slumps, and the 1929 Grand Banks Earthquake". American Journal of Science, Volume 250, pp. 849-873.

Johnson, A.M., 1970, Physical Processes in Geology. San Francisco, Freeman. p. 577.

Karlsrud, K. and L. Edgers, 1980, "Some Aspects of Submarine Slope Stability" NATO Workshop on Marine Slides and other Mass Movements, December.

Karlsrud, K. and T. By, 1981, "Data on Run-Out Distance and Velocity of Soil Flows Generated by Subaqueous Slides and Quick-Clay Slides", Norwegian Geotechnical Institute, Report 52207-7.

Kuenen, P.H., 1952, "Estimated Size of the Grand Banks Turbidity Current". American Journal of Science, Volume 250, pp. 874-884.

Middleton, G.V., 1966a, "Experiments on Density and Turbidity Currents, I. Motion of the Head". Canadian Journal of Earth Sciences, Volume 3, pp. 523-546.

Middleton, G.V., 1966b, "Experiments on Density and Turbidity Currents, II. Uniform Flow of Density Currents". Canadian Journal of Earth Sciences, Volume 3, pp. 627-637.

Middleton, G.V., 1967, "Experiments on Density and Turbidity Currents, III. Deposition of the Sediments". Canadian Journal of Earth Sciences, Volume 4, pp. 475-505.

Morgenstern, N.R., 1967, "Submarine Slumping and the Initiation of Turbidity Currents". Marine Geotechnique. Ed. by: A.F. Richards. University of Illinois Press, pp. 189-220.

Pantin, H.M., 1979, "Interaction Between Velocity and Effective Density in Turbidity Flow; Phase-Plane Analysis, with Criteria for Autosuspension". Marine Geology, Volume 31, Number 1/2, pp. 59-100.

Schapery, R.A. and W.A. Dunlap, 1978, "Prediction of Storm Induced Sea Bottom Movement and Platform Forces" Offshore Technology Conference 10, Houston 1978, Paper Number OTC 3259.

Scheidegger, A.A., 1973, "On the Prediction of the Reach and Velocity of Catastrophic Landslides". Rock Mechanics, Volume 5, pp. 231-236.

Sterling, G.M. and E.E. Strohbeck, 1973, "The Failure of the Southpass 70 'B' Platform in Hurricane Camille" Offshore Technology Conference 5, Houston 1973. Preprints, Volume 2, pp. 719-730.

Stevens, J.B. and J.M.E. Audibert, 1979, "Re-examination of p-y Curve Formulations" Offshore Technology Conference 11, Houston 1979, Paper Number OTC 3402, Preprints, Volume 1, pp. 397-403.

Suhayda, J.N. and D.B. Prior, 1978, "Explanation of Submarine Landslide Morphology by Stability Analysis and Rheologic Models". Offshore Technology Conference, 10. Houston 1978. Proceedings, Volume 2, pp. 1075-1082.

Terzaghi, K. 1956, "Varieties of Submarine Slope Failures". p. 41. Texas Conference on Soil Mechanics and Foundation Engineering, 8. Austin 1956. Proceedings. Also published in: Norwegian Geotechnical Institute. Publication, 25.

Tesaker, E., 1980, "Forces on Gravity Structure from Suspension Current", Norwegian Hydrodynamics Laboratories, (NHL), Report No. 2-80029.

ACKNOWLEDGEMENTS

This research was done while the senior author was a Post doctorate Research Fellow at the Norwegian Geotechnical Institute, supported by the Royal Norwegian Council for Scientific and Industrial Research. In addition, the research was supported in part by Statoil, Exxon Production and Research, Norsk/Hydro, and Elf Aquitane Norge, Marine Norsk. Trond By, of the Norwegian Geotechnical Institute assisted in the collection and interpretation of the field data.

FACTORS INFLUENCING THE INTERPRETATION OF IN SITU
STRENGTH TESTS IN INSENSITIVE LOW PLASTICITY CLAYS

D. W. Hight, A. Gens, T. M. P. De Campos, M. Takahashi
Imperial College, U. K.

SUMMARY

A detailed study has been made of the undrained behaviour of a low plasticity clay under monotonic loading at different rates. Certain features of this behaviour are described and used to illustrate the importance of consolidation history, shearing direction and disturbance in the interpretation of two in situ strength tests, the penetrometer and vane, which might be used in an offshore investigation of this type of soil.

Taking spherical cavity expansion to represent penetration at depth, it is shown that the parameter, N_K, used to relate undrained strength to penetration resistance, varies markedly with consolidation history. This is confirmed by the results of model penetrometer tests. The range of values for N_K found by calibration depends on the method used to measure shear strength.

When the vane test is interpreted in the conventional way, an undrained strength may be derived which is similar to that measured in triaxial compression tests on horizontally-cut samples. Corrections for rate effects are relatively small at all overconsolidation ratios. The vane appears to underestimate the strength of the soil in its remoulded state and is unsuitable for determining its sensitivity.

INTRODUCTION

Low plasticity soils exist at many of the sites for oil production platforms in the Northern North Sea. Their glacial origin (see, for example, Fannin, 1979) means that they will have undergone a wide variety of stress histories. For the overconsolidated soil, Kjekstad et al (1979) report overconsolidation ratios (OCRs) up to 25. In view of the large thicknesses of glacial sediments, the soils will approach a normally consolidated condition as depth increases, irrespective of their mode of deposition.

Parameters for foundation design in these soils are derived largely from:-

(i) laboratory element tests on samples which have been subjected to a sampling cycle involving a large total stress release and temperature change;

(ii) in situ tests which are of necessity carried out rapidly; and

(iii) pocket penetrometer and torvane tests performed on extruded samples (Andresen et al, 1979).

There is, up to present, no scope for in situ inspection of the soils or for taking large size intact samples, so that test data cannot be related properly to macrofabric. Furthermore, there is a lack of performance data with which laboratory and in situ test results can be correlated. In view of the different loading conditions and foundation systems used offshore, land-based correlations are unlikely to be applicable. It follows, then, that it is desirable to interpret test data in terms of fundamental soil parameters, except where direct extrapolation is possible, for example from a penetrometer test to prediction of skirt penetration resistance (Lunne and St John, 1979). Even then allowance must be made for differences in drainage and shearing rate.

In general, in situ tests do not directly measure a fundamental soil parameter. These parameters can only be derived:-

(i) by interpretation of the test, which requires knowledge of the mechanism involved and how this is affected by the in situ parameters, and how these are, in turn, affected by shearing rate, drainage and disturbance; or

(ii) by calibration, using the device in soil with known in situ properties.

In this paper the application of these two approaches is considered for the case where undrained strength parameters for a low plasticity clay are to be derived from two in situ tests, the penetrometer and vane, which might be used in an offshore investigation. For illustrative purposes, data for a particular soil, the Lower Cromer Till, is used; details of this clay, which is typical of some of the glacial sediments in the North Sea, are given by Hight et al (1979).

The behaviour of the Lower Cromer Till has been investigated using different consolidation histories and shearing rates, with and without the effects of disturbance superimposed. The essential features of the behaviour are described and their effect on the interpretation of the two in situ tests is illustrated by way of mechanisms which are generally accepted as representing the tests. The factors which need to be taken into consideration when deriving undrained strength from the in situ tests are highlighted.

Results of model penetrometer and vane tests carried out in soil beds of known properties are reported. These provide independent confirmation of the factors that need to be considered.

THE UNDRAINED BEHAVIOUR OF THE LOW PLASTICITY CLAY UNDER MONOTONIC LOADING

To determine the factors which affect the interpretation of in situ penetrometer and vane tests in a particular soil, a pre-requisite is a framework which describes the behaviour of that soil under monotonic loading. Such a framework would require, in principle, knowledge of the behaviour under generalised stress paths, which include principle stress rotation. The difficulties in achieving this mean that reliance is placed on data from triaxial and plane strain tests which can be carried out on vertical and horizontal samples.

A framework has been developed for an insensitive, reconstituted low plasticity clay, the Lower Cromer Till, using triaxial tests (Gens, 1982). The relevant stress-strain properties from this framework are now presented. Attention is given to those features which vary with stress history and to the effect on them of disturbance and shearing direction.

Fig. 1 shows the stress-strain curves from undrained triaxial compression and extension tests on anisotropically consolidated samples of the soil at OCRs[1] of 1, 2, 4 and 7; a shearing rate of 1%/day was used. The characteristics to note are:-

(i) Undrained brittleness in compression (i.e. strain-softening) reduces as OCR increases. This is emphasised in Fig. 2 where the ratio of the peak undrained strength, C_{up}, to the ultimate undrained strength reached in the triaxial test, C_{ur}, is plotted against OCR.

(ii) Brittleness is not evident in triaxial extension.

(iii) Stiffness reduces with increasing OCR, the reduction being more marked in compression then in extension. This is demonstrated in Fig.3, using the parameter E_u/C_u, where E_u is the secant modulus measured at a deviator stress midway between the initial and peak values. This large variation in E_u/C_u with OCR is a characteristic of low plasticity clays, see, for example, Duncan and Buchignani (1976).

(iv) The axial strain to failure, ε_p, in compression increases with OCR.

(v) The ultimate shear strength in triaxial compression, C_{ur}, exceeds that in triaxial extension by about 50% at all OCRs. These differences reflect both anisotropy of strength and the influence of the intermediate principal stress, σ_2.

To provide a measure of the anisotropy in undrained shear strength, comparisons have been made between triaxial compression tests on samples cut vertically and horizontally from soil blocks of different OCR. The strength measured on horizontally-cut samples, C_{uH}, is, on average, 16% lower than that measured on vertically-cut samples, independent of OCR. Slightly lower strength differences were found when comparing samples cut vertically and at 45°.

The effect of sampling, or of limited disturbance by, for example, undrained cyclic loading, is to reduce both undrained brittleness and the ratio E_u/C_u; as a consequence the variation of brittleness and E_u/C_u with OCR is also reduced. The ultimate strength, C_{ur}, is not affected provided that there is no change in water content.

There is a lack of undrained brittleness in horizontally-cut samples which may result from the effect of sampling. However, the evidence from extension tests and the marked difference in response between vertically- and horizontally-cut samples, suggests that, in any event, undrained brittleness will be small in these samples.

Complete remoulding removes all undrained brittleness so that peak and ultimate strengths coincide and are only mobilised at large strains. In triaxial tests, the strengths mobilised in remoulded soil are approximately 25% less than the ultimate strength, C_{ur}, reached in triaxial tests on the reconstituted soil. The soil can be said, therefore, to have a sensitivity of 1.33. It must be emphasised that sensitivity is not synonymous with brittleness.

This overall picture is broadly supported by the results of the plane strain tests on Lower Cromer Till.

At large displacements in the ring shear apparatus, the soil undergoes turbulent shear (Lupini, Skinner and Vaughan, 1981). Residual surfaces, marked by oriented clay, do not form within the body of the soil but may form at rigid interfaces.

The coefficient of consolidation for the normally consolidated soil ranges between 0.5 and 2.0 m^2/year. Although higher values are applicable to the soil when overconsolidated, in situ penetrometer and vane tests at typical field rates will be largely undrained.

THE INFLUENCE OF SHEARING RATE ON UNDRAINED BEHAVIOUR UNDER MONOTONIC LOADING

The influence of shearing rate on the undrained behaviour of the soil after it has undergone different consolidation histories, has been investigated by Hight et al (1982). It has been demonstrated (refer Fig. 4) that after anisotropic consolidation, samples loaded in triaxial compression show:-

---

[1] OCR is defined here as the ratio of the maximum to the current vertical effective stress.

(i) an increase in peak strength and stiffness with increase in shearing rate; the effect is more marked at low degrees of overconsolidation and at short times to failure;

(ii) that the strain at which peak strength is mobilised is unaffected by shearing rate; and

(iii) no corresponding increase in the ultimate strength with shearing rate; in fact, evidence from ring shear tests indicates that, over a certain range of increasing displacement rate, there is a reduction in ultimate undrained strength, which will, of course, be independent of OCR.

The net result is that undrained brittleness and E_u/C_u increase with rate of compression loading, more so at low OCR.

In triaxial extension, undrained brittleness does not appear to be introduced by increased rate of shearing. Both extension strength and stiffness increase, as shown in Fig.5. In a companion series of extension tests on isotropically consolidated samples at different OCR, it was found that the increase in stiffness is larger than that in undrained strength at low OCR, so that the ratio E_u/C_u increases; **at high OCR, however, there is a small reduction in E_u/C_u.** As with compression loading, the strain at failure in anisotropically consolidated **samples** loaded in extension is unaffected by rate; the increase in stiffness results from the non-linearity introduced by increased loading rate.

To check the effect of rate on the anisotropic properties, triaxial samples were cut vertically and horizontally from anisotropically consolidated soil blocks and tested in compression at different rates. The results of fast and slow compression tests are compared in Fig. 6. Two features should be noted:-

(i) the effect of loading rate on strength is far greater on vertical samples;

(ii) in contrast to the vertical samples, the horizontal samples show practically no brittleness at increased rate of loading.

It was also found that for both vertical and horizontal samples from soil blocks of different OCR, the influence of rate was similar.

The effect of sampling and other forms of disturbance, is to reduce the effect of shearing rate on both stiffness and peak strength. In this way, the effect of rate on horizontally-cut samples shown in Fig. 6 may have been suppressed somewhat, although the effect remains clearly less than on vertical samples.

When the soil is in the fully remoulded condition, evidence, including that from the ring shear tests referred to above, indicates that undrained strength reduces with increasing shearing rate up to a particular rate; above this rate it appears that there is a reversal in this pattern.

INTERPRETATION OF IN SITU PENETROMETER AND VANE TESTS IN THE LOW PLASTICITY CLAY

It can be seen from the preceding sections that in the low plasticity clay, parameters such as undrained strength, stiffness, brittleness and strain to failure are all strongly influenced by three factors, consolidation history, direction of loading and prior disturbance. The interpretation of different in situ tests is affected by these parameters by varying amounts. The two tests examined herein, the penetrometer and the in situ vane, involve different shearing directions and cause totally different failure modes; they illustrate well the varying importance of the three factors.

A. PENETROMETER

1. Determination of Theoretical N_K Values

For saturated clays the mechanism involved in the undrained penetration at depth by a cone or plate has been found to be well-represented by the expansion of a cavity (e.g. Roy et al, 1974, Schmertmann, 1975). This approach takes into account the deformability of the soil and allows for the prediction of penetration resistance, q_p, from

$$q_p = N_K C_u + p \qquad (1)$$

where p is an in situ stress which depends on the shape of the cavity assumed. For a spherical cavity p is taken as the mean normal stress; for a cylindrical cavity it is taken as the horizontal normal stress.

Baligh et al (1978) have proposed that an upper limit to the penetration resistance can be found by adding the contribution from a corrected slip line solution for penetration of a wedge to that from the expansion of a cylindrical cavity. The effect of soil stiffness is included, therefore, in the second contribution.

Since the main purpose of this paper is to assess the relative effects of different factors, the specific solution adopted is not critical. The results of Roy et al (1974) show that for sensitive clays the solution for expansion of a spherical cavity provides a reasonably good prediction of penetration resistance and will be used herein.

Two solutions are available for this form of cavity expansion:-

(i) In the first, the soil is assumed to be a linear elastic - perfectly plastic material, and to have the stress-strain curve shown in Fig. 7(a). The expression derived for N_K (e.g. Bishop et al, 1945) is

$$N_K = 1 + \frac{4}{3}\left[\ln \frac{E_u}{3C_u} + 1\right] \tag{2}$$

(ii) In the second, the soil is assumed to be linearly elastic-work softening plastic and to have the stress-strain curve shown in Fig. 7(b). The expression derived for N_K (Ladanyi, 1976) is

$$N_K = 1 + \frac{4}{3}\frac{C_{ur}}{C_{up}}\left[1 + \ln \frac{2}{3\varepsilon_r}\right] + \frac{4}{3}\left[\frac{\frac{\varepsilon_r}{\varepsilon_p} - \frac{C_{ur}}{C_{up}}}{\frac{\varepsilon_r}{\varepsilon_p} - 1}\right] \ln \frac{\varepsilon_r}{\varepsilon_p} \tag{3}$$

where the parameters C_{up}, C_{ur}, ε_p, ε_r are defined in Fig. 7(b). In equation (1), q_p is predicted from N_K and C_{up}.

Both solutions assume isotropy of strength, C_u, and of stiffness, E_u, and radially symmetric initial stresses. These conditions are rarely met in the field and are certainly not applicable to the soil whose behaviour has been described. However, the two solutions enable bounds to be established for penetration resistance by assuming isotropy of initial stresses and isotropy of the strength and stiffness corresponding to:-

(i) compression loading, and

(ii) extension loading.

The actual penetration resistance will then lie between the two bounds.

In terms of the shape of the stress-strain curves, the first solution provides a reasonable approximation to the behaviour of the soil in triaxial extension, compare Figs 5 and 7(a), and the second solution a reasonable approximation to its behaviour in triaxial compression, compare Figs 4 and 7(b). It is easy to show, following the method of Ladanyi (1963), that the two solutions can be applied when starting from unequal stress, providing that p, in equation (1), is taken as the mean normal stress.

From the form of equations (2) and (3) and from the described manner in which E_u/C_u, C_{up}/C_{ur} and $\varepsilon_p/\varepsilon_r$ vary with consolidation history, it is apparent that the bounds calculated for N_K will also vary with consolidation history. This is demonstrated in Fig. 8 where the theoretical values for N_K, calculated using the data from the slow triaxial tests presented in Figs 1, 2 and 3, has been plotted against OCR as lines A and B.

2. **Shear Rate Effects**

The strain field around a penetrometer is markedly non-uniform (see, for example, Baligh et al, 1978). As a result there will be variations in rate of strain around a penetrometer as it penetrates at a fixed rate. In view of the dependence of the soil's behaviour on rate of shear, these variations will affect the interpretation of the test; they have not, however, been taken into account in determining the theoretical N_K values.

The times to failure for elements sheared by a penetrometer depend, inter alia, on:-

(i) the position of the element relative to the path of the penetrometer,

(ii) the rate of penetration, and

(iii) the strain to failure for the soil, appropriate to the direction of shearing.

Ladanyi (1976) has suggested a method for finding an average time to failure, t_f, for elements directly in the line of a penetrometer. He gives:-

$$t_f = 1.825 \, \gamma_{af}^{-1/3} (B/\dot{s}) \qquad (4)$$

where γ_{af} is the shear strain at failure under axial symmetry, B is the diameter of the penetrometer tip and $\dot{s}$ is the penetration rate.

The expression can be rearranged in order to calculate a penetration rate to which the triaxial data shown in Figs 1, 2 and 3 is applicable. These penetration rates have been added to the theoretical N_K values shown as lines A and B in Fig. 8. The numerical values for N_K given by lines A and B are only applicable to penetrometer tests carried out at these rates; for convenience, these will be referred to as the N_K values for "slow" penetration rates.

The triaxial data shown in Figs 4 and 5 involves times to failure of 3.7, 14 and 36 seconds, making it more applicable to penetrometers pushed in at typical field rates of 2cm/sec. The modified undrained brittleness and E_u/C_u at fast rates change the theoretical N_K values calculated using equations (2) and (3). For these short times to failure, the estimated trend for the variation in theoretical N_K values with OCR is indicated by lines C and D in Fig. 8. The upper bound for N_K, i.e. line C, based on compression behaviour, is reduced at the faster shearing rates. The lower bound for N_K, i.e. line D, based on extension behaviour, is increased at low OCR and reduced at high OCR. The penetration rates to which lines C and D apply have been added to Fig. 8; N_K values given by lines C and D will be considered as N_K values for "fast" penetration rates.

Of course, the actual penetration resistance at these faster penetration rates will depend also on the accompanying changes in strength which occur; this is made evident in the following section.

3. Determination of N_K Values by Calibration

The bounds to the theoretical N_K values shown in Fig. 8 can be combined with the appropriate undrained strengths to predict the bounds to penetration resistance, q_p, of penetrometers used at "fast" and "slow" rates in soil beds of different consolidation history. These bounds to q_p are shown normalized in Fig. 9; an increase in penetration resistance with increase in rate is apparent. In a real situation, q_p would be higher because of the contribution from the term for total stress, P. The predicted penetration resistances in Fig. 9 are used to illustrate the potential range in N_K values which can result from field calibrations of the penetrometer; subsequently they are used in a comparison with model penetrometer test data.

When calibrating a penetrometer, measured penetration resistance is related to undrained strength and total stress, p. The magnitude of the empirical linking parameter depends on:-

(i) the method by which C_u was measured, if the soil is anisotropic or brittle,

(ii) the differences in time to failure for elements involved in the penetrometer test and the strength measurement, and

(iii) the value taken for p, which will be influenced by OCR.

For the overconsolidated clays in the North Sea, Kjekstad et al (1979) quote a range for N_K of 15 - 20, based on correlations between penetration resistance and triaxial compression tests on samples from the sea bed.

As shown previously, in low plasticity clay tested in triaxial compression after sampling, and without reconsolidation, the undrained brittleness is reduced so that the peak strength is not measured. The undrained strength used in correlations is then close to the ultimate, C_{ur}. The effect of this is, naturally, to make the N_K found by calibration larger than the theoretical N_K, especially at low OCR. This is demonstrated in Fig. 10 where the ranges for N_K which have been plotted have been found by dividing the bounds for predicted penetration resistance shown in Fig. 9, by the ultimate strength, C_{ur}, measured at each OCR in slow triaxial compression tests. Comparison of Figs 8 and 10 reveals the increased dependence of calibrated N_K values on OCR.

The two solutions for a spherical cavity were adopted to provide bounds to probable N_K values and penetration resistances. The anisotropy of the problem, in terms of shear strength, stiffness and initial stresses, means that the actual N_K values will lie somewhere between these bounds; the relative position will probably vary with OCR. For simplicity, an average line has been constructed for each set of bounds on Fig. 10, i.e. lines E and F. It is apparent from the figure that a wide range of N_K values determined by calibration should be anticipated in low plasticity clays. The wide separation of the bounds shown in Fig. 10 suggests strongly that an improved understanding of the test in anisotropic and brittle soils will be of immediate benefit in determining the precise values of N_K and their variation with OCR.

The magnitude of calibrated N_K values will increase further when penetrometer resistance is correlated with undrained shear strength derived from the in situ vane test; this is considered subsequently.

4. The Results of Model Penetrometer Tests in Reconstituted Soil Beds

Model penetrometer tests have been carried out in beds of the clay which were consolidated to different OCRs under controlled conditions and which had known properties (Martins, 1980). The penetrometers comprised both circular plates, 25.4mm diameter, and cones with a base diameter of 25.4mm and an apex angle of $90°$. Of necessity, the penetration rates were slow relative to typical field penetration rates.

From the measured penetration resistances, empirical N_K values could be determined by calibration. For penetrometer tests carried out at a rate of 0.025cm/sec (15mm/min), the N_K values found by correlating the measured penetration resistances with C_{ur} from slow triaxial tests are shown on Fig. 10. At the penetration rate of 0.025cm/sec, significant drainage was avoided (Martins, 1980) while reasonable comparisons could be made with the estimates for calibrated N_K values also shown in Fig. 10.

The results from the model penetrometer tests confirm the anticipated variation in calibrated N_K values with OCR. They agree well with the estimates for calibrated N_K values, falling between the average lines predicted for fast and slow penetration rates. It should be pointed out that there is some uncertainty as to the precise range for N_K determined for OCR 1 as a result of the difficulties of measuring the low strength of the soil bed in which the test was performed.

It is of interest to note that model cone and plate penetrometers gave similar calibrated values for N_K in beds of OCR 7. In a normally consolidated bed the calibrated N_K value for a plate was 10% higher than for a cone.

5. Model Penetrometer Tests in Remoulded Soil

As described, gross disturbance causing remoulding of the soil, eliminates the brittleness which occurs at low OCR in the reconstituted soil and reduces the ratio E_u/C_u and shear strength. Consequently, theoretical N_K values will be reduced in remoulded soil and penetration resistances will be lower than in intact soil at the same water content, especially when at low OCR. This was confirmed in a series of penetration tests in soil remoulded at different water contents (Sano, 1977); these showed that penetration resistances were approaching half of the value measured in corresponding beds of intact soil.

Since there is little influence of shearing rate on stiffness and strength in the remoulded soil, then, in the absence of sampling effects and anisotropy, calibrated N_K values should agree reasonably well with theoretical values. This was also confirmed in the model tests in the soil.

Clearly, if the soil is heavily disturbed before carrying out an in situ penetrometer test so that penetration resistances are reduced, strengths will be underestimated if an N_K calibrated in intact soil is used.

B. IN SITU VANE

1. Interpretation

In a vane with a height to diameter of 2:1 and for which a uniform shear distribution is assumed on the horizontal surfaces at the top and bottom, about 86% of the torque results from shear on a vertical cylindrical surface (Schmertmann, 1975). The shear strength on a vertical cylindrical surface is closely related to the shear strength measured on samples tested in plane strain with the plane strain direction being vertical. In the soil which has been investigated, it has been found that triaxial compression tests on horizontally-cut samples provide an estimate of the strength and behaviour measured in such plane strain tests.

An important feature of this behaviour is that shear is not associated with significant undrained brittleness. Progressive failure is not, then, an important consideration in interpretation. This is in marked contrast to sensitive clays in which progressive failure is very important (La Rochelle et al, 1973). The mode of failure and the lack of brittleness imply that any effect of stiffness will be small.

In the absence of brittleness and stiffness effects, and since limited disturbance does not affect the ultimate undrained strength, the measured torque should be relatable to C_{uH}. (If delays occur between installing the vane and performing the test, then water content changes can take place and these will change the strength.) The influence of rotation rate should then mirror only the effect of shearing rate on C_{uH} which has been shown to be small. The correction to be made for rate when calculating a desired undrained strength will, therefore, also be small. The only part which OCR plays in the interpretation is in terms of the varying effect of rate on C_{uH} with OCR. Since this is a relatively minor effect, a knowledge of OCR is of less importance than for interpretation of the penetrometer. Furthermore, since strength anisotropy appears to be independent of OCR in this soil, a measure of C_{ur} for compression loading as well as C_{uH} can be obtained.

2. The Results of Model Vane Tests in Reconstituted Soil Beds

Model vane tests (H = 189mm, D = 125mm) have been conducted in blocks of the soil which were consolidated to OCRs of 1, 4 and 7. Various rates of angular rotation were used. Interpreting the vane in the conventional way, i.e. assuming a vertical cylindrical shear surface of diameter equal to the vane diameter and two horizontal shear surfaces of the same diameter, shear strengths have been derived for each OCR and rate.

Shear strengths determined in this way at vane rotation rates of 630 and 30°/min are compared and plotted against OCR in Fig. 11. There is little dependence of rate effect on OCR, as anticipated in the previous discussion. The increase in measured torque with increase in rotation rate is somewhat larger than expected, based on the triaxial compression data on horizontally-cut samples. This may be, in part, related to the sampling effect on the horizontally-cut samples.

A rotation rate of 30°/min for the laboratory vane gives a tangential velocity at the perimeter of the vane equal to that in a typical field vane (D = 65mm) rotated at 6°/min. Undrained strengths back-calculated from model vane tests rotated at 30°/min were within 10% of the strengths measured in triaxial compression tests on horizontally-cut samples, C_{uH}, at all OCRs. It should be noted that the correlation found by Ladd (1975) between the strength measured in plane strain extension and the undrained strength derived from the vane in tests on Boston Blue Clay was also independent of OCR.

3. Model Vane Tests in Remoulded Soil

When model vane tests were made with the soil in an unconsolidated remoulded condition, the strength derived was an underestimate of the remoulded strength of the soil measured in triaxial tests carried out at slow shearing rates. Similarly, there was a large reduction in torque when this was measured after several rotations of the vane in consolidated beds of the soil. Since the soil has been shown to be relatively insensitive (S_t = 1.33) methods to measure sensitivity which rely on several rotations of the vane are quite inappropriate in these soils.

CORRELATIONS BETWEEN PENETROMETER AND VANE TESTS

Correlations have been made between penetration resistance and an undrained shear strength derived from an in situ vane test; for example, Lunne et al (1976) quote values between 13 and 24 for N_K calibrated this way in five marine clays. Such correlations introduce further changes to the range of calibrated N_K values which can be anticipated.

In the low plasticity clay under discussion, it has been shown that the undrained strength derived from the vane at typical field rotation rates correlates with C_{uH}. Furthermore, it has been found that C_{uH} is 16% lower than the ultimate strength measured in triaxial compression, C_{ur}. The net result is that N_K values obtained by such a calibration will be 1.2 times the N_K values shown in Fig. 10, which were found by calibrating penetration resistance with C_{ur}.

CONCLUSIONS

The undrained behaviour of the insensitive low plasticity clays in the Northern North Sea is likely to be strongly influenced by:-

(i) consolidation history,

(ii) shearing direction, and

(iii) disturbance.

These factors determine undrained strength and brittleness, stiffness, strain to failure and determine the changes made to these parameters by shearing rate. Interpretation and calibration of in situ tests in these clays will be affected by these factors by varying amounts.

The penetrometer test involves multi-directional shearing. No unique relationship between penetration resistance and an undrained shear strength should be anticipated. Theoretical N_K values calculated assuming spherical cavity expansion reduce with increasing OCR. This dependence on OCR is greater when N_K values are obtained by correlating penetration resistance with undrained strengths. The numerical values found this way for N_K depend on how and at what rate the shear strength used in the calibration was measured. Knowledge of OCR is vital for complete interpretation of the penetrometer test.

Disturbance of the soil prior to conducting the test will reduce the measured penetration resistance and could lead to an underestimate of undrained strength.

The variation in N_K with OCR may contribute to the large range in N_K values reported for a particular soil type (see, for example, the range shown by Marsland and Powell, 1979, for the Cowden Till). Similarly, it may help to explain the apparent reduction in N_K with undrained strength shown by Kjekstad et al (1979) and Eide (1974).

In the vane test, shearing in one direction dominates. When the test is interpreted in the conventional way it gives a reasonable measure of the undrained strength, C_{uH}, found in triaxial compression tests on horizontally-cut samples. In this soil, C_{uH}, can be related to triaxial compression strength, independently of OCR. Corrections for differences in rate of shear are relatively small at all OCRs. Knowledge of OCR is of less importance for interpretation of the vane than for the penetrometer.

Limited disturbance of the soil during installation of the vane will only significantly affect the derived strength if sufficient time is allowed for water content changes to take place.

The vane appears to underestimate the strength of the remoulded soil. It is quite unsuitable for measuring sensitivity in these soils.

These conclusions have been reached on the basis of data from triaxial and model penetrometer and vane tests in soil which has been reconstituted. Where important macrofabric features are present, e.g. sand or silt partings providing drainage paths or fissures with more plastic clay coating, they add to the variability of in situ calibration factors such as N_K.

ACKNOWLEDGEMENTS

The work reported has been supported by grants from the Science and Engineering Research Council to the London Centre for Marine Technology.

REFERENCES

ANDRESEN, A., BERRE, T., KLEVEN, A. and LUNNE, T., 1979, "Procedures used to obtain soil parameters for foundation engineering in the North Sea," Marine Geotechnology, Volume 3, 3, pp. 201-266.

BALIGH, M. M., VIVATRAT, V. and LADD, C.C., 1978, Exploration and evaluation of engineering properties for foundation design of offshore structures, Massachusetts Institute of Technology, Department of Civil Engineering, Res. Report R78-40.

BISHOP, R. F., HILL, R. and MOTT, N. F., 1945, "Theory of indentation and hardness tests," Proceedings of the Physical Society, Volume 57, Part III, pp. 147-159.

DUNCAN, J. M. and BUCHIGNANI, A. L., 1976, An engineering manual for settlement studies, Department of Civil Engineering, University of California, Berkeley, Geotechnical Engineering Report.

EIDE, O., 1974, "Marine soil mechanics: applications to North Sea offshore structures," Proceedings of Offshore North Sea Technology Conference, Stavanger.

FANNIN, N. G. T., 1979, " The use of regional geological surveys in the North Sea and adjacent areas in the recognition of offshore hazards," Proceedings of Conference on Offshore Site Investigation, London, pp. 5-21.

GENS, A., 1982, Stress-strain and strength characteristics of a low plasticity clay, Ph.D. Thesis, University of London (in preparation).

HIGHT, D. W., EL-GHAMRAWY, M. K. and GENS, A., 1979, "Some results from a laboratory study of a sandy clay and implications regarding its in situ behaviour," Proceedings, 2nd International Conference on the Behaviour of Off-shore Structures, London, England, Volume 1, pp. 133-150. Cranfield, Bedford, England: BHRA Fluid Engineering.

HIGHT, D. W., TAKAHASHI, M., GENS, A. and DE CAMPOS, T. M. P., 1982, "The influence of rate of undrained loading on the behaviour of a low plasticity clay." In preparation.

KJEKSTAD, O., LUNNE, T. and CLAUSEN, C. J. F., 1979, "Comparison between in situ cone resistance and laboratory strength for overconsolidated North Sea clays," Marine Geotechnology, Volume 3,1, pp. 23-26.

LADANYI, B., 1963, "Expansion of a cavity in a saturated clay medium," Journal of Soil Mechanics and Foundation Division, Proceedings ASCE, SM4, pp. 127-161.

LADANYI, B., 1967, "Deep punching of sensitive clays," Proceedings, 3rd Pan American Conference on Soil Mechanics and Foundation Engineering, Caracas, Volume 1, pp. 533-546.

LADANYI, B., 1976, "Use of the static penetration test in frozen soils," Canadian Geotechnical Journal, 13, 2, pp. 95-110.

LADD, C.C., 1975, "Measurement of in situ shear strength," Discussion, Proceedings ASCE Specialty Conference on "In Situ Measurement of Soil Properties," Raleigh, N.C., Volume II, pp. 153-160.

LA ROCHELLE, P., ROY, M. and TAVENAS, F.A., 1973, "Field measurements of cohesion in Champlain clays," Proceedings, 8th International Conference on Soil Mechanics and Foundation Engineering, Moscow, Volume 1.1, pp. 229-236.

LUNNE, T., EIDE, O. and DE RUITER, J., 1976, "Correlations between cone resistance and vane shear strength in some Scandinavian soft to medium stiff clays," Canadian Geotechnical Journal, 13, pp. 430-441.

LUNNE, T. and ST JOHN, H. D., 1979, " The use of cone penetration tests to compute penetration resistance of steel skirts underneath North Sea gravity platforms," Proceedings, 7th European Conference Soil Mechanics and Foundation Engineering, Volume 2, pp. 233-238.

LUPINI, J. F., SKINNER, A. E. and VAUGHAN, P. R., 1981, "The drained residual strength of cohesive soils," Geotechnique, 31, 2, pp. 181-213.

MARSLAND, A. and POWELL, J. M. M., 1979, "Evaluating the large scale properties of glacial clays for foundation design," Proceedings, 2nd International Conference on the Behaviour of Off-shore Structures, London, England, Volume 1, pp. 193-214. Cranfield, Bedford, England: BHRA Fluid Engineering.

MARTINS, M. C. R., 1980, Large model footing tests on a sandy clay till, Ph.D. Thesis, University of London.

ROY, M., MICHAUD, D., TAVENAS, F.A., LEROUEIL, S. and LA ROCHELLE, P., 1974, "The interpretation of static cone penetration tests in sensitive clays," Proceedings, European Symposium on Penetration Testing, Stockholm, 2-2, pp. 323-330.

SANO, Y., 1977, Influence of rate on penetration resistance of Lower Cromer Till, Internal Report, Civil Engineering Department, Imperial College, University of London.

SCHMERTMANN, J. H., 1975, "Measurement of in situ shear strength," State-of-the-Art Report, Proceedings, ASCE Specialty Conference on "In Situ Measurement of Soil Properties," Raleigh, N.C., Volume II, pp. 57-138.

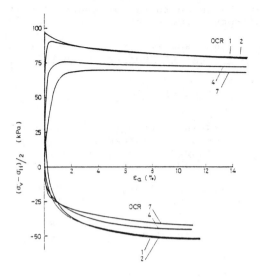

Fig. 1 Stress-strain curves from triaxial compression and extension tests on anisotropically consolidated samples of Lower Cromer Till at different OCRs.

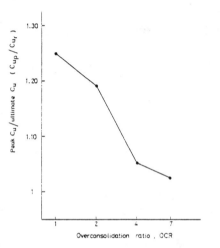

Fig. 2 The ratio of peak to ultimate undrained strength measured in triaxial compression at different OCRs.

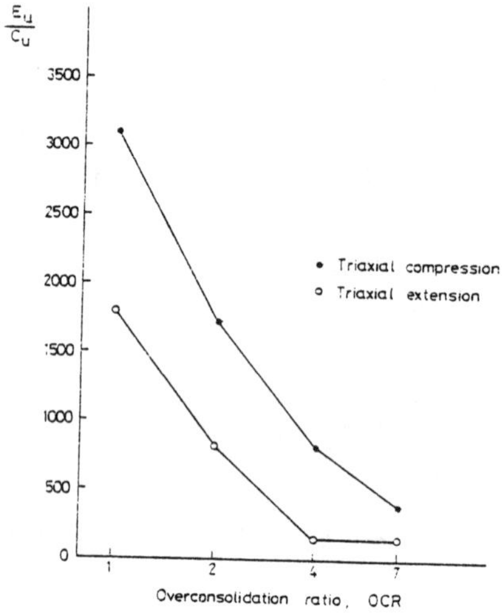

Fig. 3 The ratio of E_u/C_u measured in triaxial compression and extension at different OCRs.

Fig. 4 The effect of shearing rate on the stress-strain curve measured in triaxial compression of anisotropically consolidated soil.

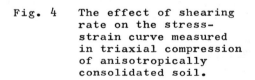
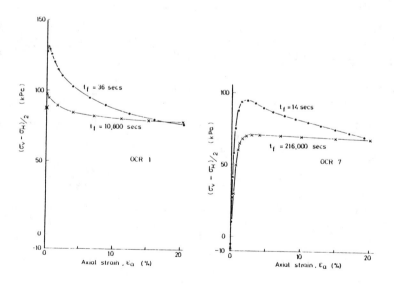

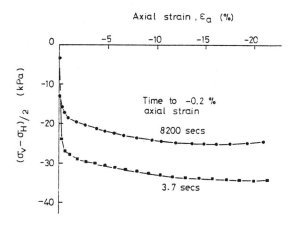

Fig. 5 The effect of shearing rate on the stress-strain curve measured in triaxial extension of anisotropically consolidated soil.

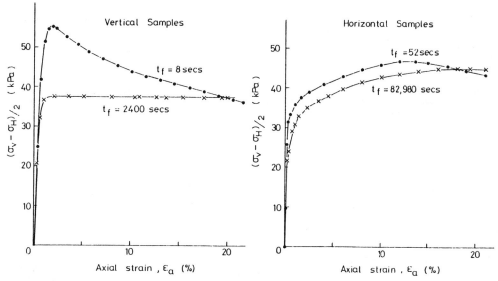

Fig. 6 The effect of shearing rate on the stress-strain curve measured in triaxial compression of vertically- and horizontally-cut samples.

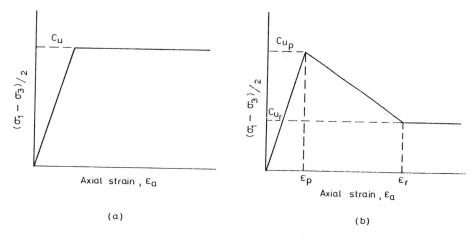

Fig. 7 Idealised stress-strain curves used in the analysis of spherical cavity expansion.

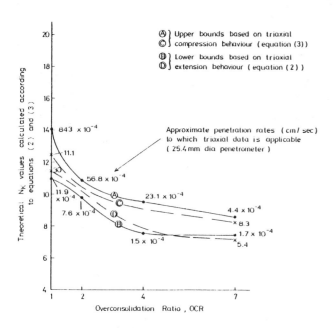

Fig. 8 Bounds to Theoretical N_K Values for "Fast" and "Slow" Penetration Rates, Based on Spherical Cavity Expansion Theory.

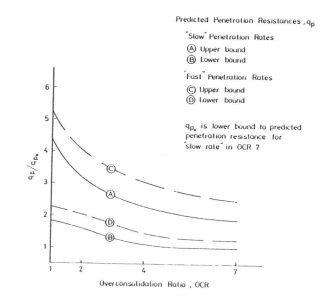

Fig. 9 Bounds to Predicted Penetration Resistances for "Fast" and "Slow" Penetration Rates in Soil Beds of Different OCR.

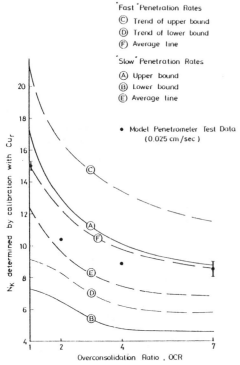

Fig. 10 Bounds to Calibrated N_K Values Found by Correlating Predicted Penetration Resistance with Ultimate Undrained Strength Measured in Triaxial Compression.

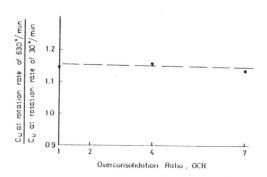

Fig. 11 Measured effect of rotation rate on undrained strength derived from vane at different OCRs.

UPLIFT CAPACITY OF EMBEDDED ANCHORS IN SAND

H.B. Sutherland,
University of Glasgow, U.K.

T.W. Finlay,
University of Glasgow, U.K.

M.O. Fadl,
University of Khartoum, Sudan.

SUMMARY

The University of Glasgow's Department of Civil Engineering has for some years been using laboratory-scale tests to study certain types of soil mechanics problems. By presenting the results of such tests in dimensionless form it has proved possible to predict full scale behaviour. This approach has been successfully used in shaft-raising from below the sea-bed at Sizewell Nuclear Power Station in the U.K.

More recently during a research project into the fundamental behaviour of embedded anchors, a large number of laboratory-scale tests have been carried out on simple disc anchors embedded in cohesionless soil, and these are reported in this paper. The tests, done in sand at relative densities of 25%, 50% and 85%, covered a range of depth/diameter ($\frac{D}{B}$) ratios up to 25, with the anchors being subjected to vertical and inclined static loads under displacement control.

The results of the tests are presented, and include anchor load v displacement, anchor load v surface displacement and ultimate anchor load v internal displacement of the sand mass, the latter being obtained using a special tracer technique involving cement powder.

A new pull-out theory, based on the test results, is presented. The theory includes parameters, particularly relative density and the critical depth ratio, which have not previously been considered by other investigators in their theories. The new theory has been applied to predict the pull-out capacity of both shallow and deep anchors used in model and full-scale tests by other investigators, and is shown to give good agreement with the experimental results used. This is in contrast to most other theories which appear often to produce results which are compatible only with the particular test conditions and materials for which they were derived, and cannot be applied satisfactorily to other situations.

Finally a design procedure, based on the new theory, is suggested.

INTRODUCTION

In offshore engineering there is an increasing use of structures which transmit not only large compressive forces but also considerable tensile forces to their foundations. Over many years, methods have been established for calculating the bearing capacity and deformations of soils subjected to compressive forces. A comparable amount of work has not been carried out on calculating the soil uplift resistance resulting from tensile foundation forces and there is no generally accepted theory. The problem occurs in both land and offshore structures. Typical of the land structures are high-mast transmission towers. On offshore structures, storm waves can produce substantial tensile forces that can be transferred to the soil through piled or similar foundations. Embedded anchor slabs could also be used to resist the pullout forces, and the problem is similar in principle to that arising in a shaft raising operation at Sizewell Nuclear Power Station, where closed-ended shafts were jacked up from the cooling water tunnels beneath the sea floor to seabed level. Sutherland (1965).

This paper considers the effects of uplift forces on shallow, deep, and inclined circular anchors embedded in sand. Existing uplift resistance theories are reviewed and compared. Recent laboratory model tests are described. A new approximate theory is developed based on the laboratory tests, a theory which embodies the influence of factors such as the relative density of the sand, a factor which had not been previously considered. The new theory is tested against the present laboratory results and also where possible, against the laboratory and field results of other investigators. Further comparisons are desirable but suggestions are made as to how the theory can be applied at this stage to design problems.

SUMMARY OF EXISTING THEORIES

Various authors have proposed theoretical solutions to the uplift capacity of footings or anchors embedded in cohesionless soils. Some of the theories have been derived for anchors at all depths, some for shallow anchors only, and others for deep anchors. Some of the theories have considered both vertical and inclined anchors.

A shallow anchor may be considered as one in which any variation in the depth of the anchor has an appreciable influence on the uplift resistance. For the shallow anchor, the failure surface in the soil mass at ultimate load extends from the anchor to the soil surface. With deep anchors, an increase in the depth of embedment has relatively little effect on uplift resistance since failure is of a local nature. The transition between shallow and deep anchor behaviour for anchors in cohesionless soils can be defined by the critical depth ratio which is a function of the depth of embedment (D) and diameter of the anchor (B).

Theories for predicting the uplift resistance of shallow and deep anchors have been developed along different lines and they can be summarised under the categories of shallow and deep anchors both vertical and inclined.

Vertical Shallow Anchors

For vertical shallow anchors, Balla (1961) was the first to adopt a systematic analysis of the problem. He assumed a circular slip failure surface extending from the anchor to ground level and calculated the vertical component of shear stress over this surface using Kotter's equation. Mariupol'skii (1965) used a curved slip failure surface and assumed that failure occurred, not by shearing, but in tension, a wedge of soil lifting away from the soil mass below at the limiting value of shear stress of the soil. Vesic (1963, 1965, 1971, 1972) based his calculations on his work with spherical cavities expanded by explosive charges. He equated uplift resistance to the pressure required to expand a spherical cavity to break through the soil surface, assuming a circular slip failure surface. Matsuo (1967, 1968) carried out an analysis assuming a combination of a logarithmic spiral slip failure surface and a plane slip failure surface. Both Vesic and Matsuo calculated the ultimate shearing resistance from the resultant of shearing and normal forces on the slip failure surface. Meyerhof and Adams (1968) derived a method of calculation based on theoretical considerations and experimental observations, assuming a cylindrical slip failure surface.

Predictions of the ultimate uplift resistance pressure (p_u) of an anchor can be made using the above theories. The predictions can be compared using a non-dimensional plot of $\frac{p_u}{\gamma D}$ against $\frac{D}{B}$ for different values of angle of internal friction (φ) of a sand. A comparison of the predictions is given in Table 1 for high and low values of φ and for the case of $\frac{D}{B} = 3$.

Theory	$\frac{p_u}{\gamma D}$ for $\frac{D}{B} = 3$	
	$\varphi = 25°$	$\varphi = 45°$
Balla	8.0	10.3
Mariupol'skii	3.3	9.2
Vesic	4.3	7.0
Matsuo	9.2	15.0
Meyerhof and Adams	3.3	14.4

Table 1. Comparison of theoretical predictions.

The variations in the predictions arise from the assumptions made regarding the shape of the failure surface. Apart from the differences obtained for one value of φ, the theories differ as to the influence of φ on the predicted uplift resistance. For example Balla predicts an increase of 29% in $\frac{Pu}{\gamma D}$ as φ increases from 25° to 45°, while Meyerhof and Adams predict an increase of 336% for the same range of φ.

Sutherland (1965) reported that no one form of failure surface could be detected which would apply to all states of density for a cohesionless soil and it was for this reason that model tests were used to obtain the uplift resistance rather than direct calculation. In the model tests dimensional similarity was established between model and prototype, and predictions from the model tests gave good results compared with measurements of uplift resistance made in the field.

Vertical Deep Anchors

Several methods have been proposed for the prediction of the ultimate uplift resistance of deep vertical anchors. Mariupol'skii (1965) presented a solution based on the assumption that the work done by the anchor during vertical displacement should equal the work needed to expand a vertical cylindrical cavity to the radius of the anchor. Vesic (1963, 1965, 1972) used the analogy of the expanding cavity in an infinite mass to obtain the limiting cavity pressure which caused a change in volume equal to the volume of the soil displaced by the anchor. Meyerhof and Adams (1968) employed Meyerhof's (1951) bearing capacity factors for deep foundations to find the ultimate uplift resistance of deep anchors.

As with the shallow anchor theories, the deep anchor theories produce predicted ultimate uplift pressures which vary widely. For high φ values, Meyerhof and Adams predict an uplift resistance twice as great as that of Vesic, while for low φ values, Vesic predicts a smaller uplift resistance than Meyerhof and Adams. The Vesic theory is relatively insensitive to the influence of φ compared with that of Meyerhof and Adams.

Inclined Anchors

Meyerhof (1973) extended his theory for vertical anchors to inclined strip anchors under axial loading, and the theoretical predictions were compared with the results of model tests and field tests. He formulated separate theories for shallow and deep anchors. Harvey and Burley (1973) considered shallow anchors. Initially they assumed the failure surface to be a circular arc perpendicular to the anchor plate and meeting the ground surface at an angle of $(\frac{\pi}{4} - \frac{\varphi}{2})$. The assumed circular arc failure surface was later taken to be a straight line failure surface and they concluded that the ultimate uplift resistance of inclined and vertical anchors embedded at the same depth in cohesionless soils were approximately the same. Their theory gave reasonable agreement with the results of laboratory experiments in shallow anchors, but overestimated the capacity of deep anchors. However their tests were carried out on only one density of sand.

PREVIOUS EXPERIMENTAL STUDIES

There have been many laboratory and field investigations of the characteristics of anchors embedded in cohesionless soils. These investigations have been directed at determining the nature of the failure surfaces developed and of the uplift resistance mobilised for vertical and inclined shallow and deep anchors. Fadl (1981) carried out an extensive review of these investigations, and commented on the different predictions from the different theories. Various authors often obtained reasonable agreement between their predicted and experimental values but these agreements were usually for particular conditions. None of the theories gave a comprehensive solution to the behaviour of anchors at failure for a wide range of soil types and conditions.

In general the failure surface and the shear behaviour of the soil along that surface was defined by the unit weight (γ) of the soil and its angle of shearing resistance (φ). However Fadl noted from previous experiments that a sand of high φ value and comparatively low relative density could give a lower uplift resistance than a sand with a moderate value of φ and a higher relative density. This indicated that the relative density of the sand, which reflects its compressibility, is an important factor which should also be included in any analysis of the behaviour of anchors embedded in cohesionless soils. Particular attention was paid to this in the present experimental work.

PRESENT EXPERIMENTAL STUDY

Tests performed

A programme of tests was carried out covering shallow and deep anchors subject to both vertical and inclined loading and conducted on a cohesionless soil placed at a range of known relative densities. It would appear that no one previous investigator had carried out such a comprehensive series of tests. Details of the tests and the equipment used are described by Fadl (1981).

The sand used was a Leighton Buzzard sand. This is a standard sand widely used in research laboratories and the results can form a basis of comparison with those from other laboratories. The particle size distribution of the sand is shown in Fig. 1.

The apparatus used for determining the uplift resistance of the anchors is shown in Fig. 2. The tank in which the sand was deposited was 762mm square by 762mm deep. It was of a size which permitted tests to be carried out on anchors with depth to diameter ratios ($\frac{D}{B}$) of up to 25 without any side effects, and could be adapted for vertical and inclined loading. Push out tests using displacement control were used. Push out tests were adopted as they eliminate shaft effects on the anchor, and permit unrestricted measurement of surface movements. El-Rayes (1965) has demonstrated that the differences between pull out and push out tests are not significant. Displacement control tests allow the observation of pre-and post-peak loads and displacements to be observed.

Substantial development work led to the adoption of an air activated sand spreader device which could produce, and reproduce, a range of relative densities of the sand. One density in each of the dense, medium and loose states was selected to represent these states, to give unit weights of 16.97 ± 0.08, 15.87 ± 0.09, and 15.18 ± 0.11 kN/m^3 respectively corresponding to relative densities of 85.2, 50.2 and 25.4%.

The anchors used consisted of brass discs 3mm thick ranging in diameter from 25.4mm to 76.4mm, and a range of depth to diameter ratios up to 25 was investigated. The anchor loads and displacements were recorded on a data logger with an electric mechanical printer. In addition to the anchor displacement, the surface displacements were continuously measured at five points during each test. A separate special series of tests was carried out to investigate the form of the rupture surface developed at failure of an anchor. Previous investigators have used semi-spatial tests to observe the behaviour of a half section of an anchor against a glass plate, the presence of which, however, can distort the deformations. X-ray techniques, Burland and Roscoe (1969) have also been used but, apart from the potential hazards, the penetration of the rays is limited to about 250mm. In the present study, three dimensional tests were carried out at different sand relative densities in a separate tank 500mm square and deep. The sand was deposited in layers each separated by a tracer layer of cement. After an anchor had been failed, the sand bed was moistened to set the cement, and then excavated to expose and measure the shape of the failure surface.

The total number of anchor tests carried out was 137, distributed as shown in Table 2.

Anchor Type	No. of tests			Range of $\frac{D}{B}$
	Dense sand	Medium sand	Loose sand	
Vertical $\psi = 0°$	17	21	33	1 to 25
Inclined $\psi = 22.5°$	11	11	11	1 to 20
Inclined $\psi = 45°$	11	11	11	1 to 20

Table 2. Summary of Anchor Tests.

Results of Tests

A major difficulty in formulating a satisfactory theory for the prediction of uplift resistance is the definition of a failure surface which will be applicable to all states of the soil and to all types of anchor, whether shallow, deep or inclined.

In the present programme of tests, measurements of anchor load versus anchor and soil surface displacements were obtained for all the tests. This data, in conjunction with the special tests using the tracer cement layers enabled an assessment to be made of the shape of the failure surface.

The special tests applied to shallow vertical anchors indicated a typical shape of failure surface for these anchors as shown in Fig. 3. The failure surface was generally curved, inclined slightly outwards as it approached the soil surface, and did not meet that surface at an angle of $(\frac{\pi}{4} - \frac{\phi}{2})°$ as assumed by Balla (1961), Vesic (1965) and Matsuo (1967). The shape shown in Fig. 3 agrees with that reported by El-Rayes (1965) and Carr (1970), and can be approximated to by a straight line inclined at an angle α to the vertical as shown. Anchor load versus surface displacement plots assisted in determining values of α. From them the radius of the failure circle at soil surface level (or ellipse in the case of inclined anchors) could be determined. Typical examples are shown in Fig. 4 for vertical anchors in dense, medium, and loose sand.

The change from the shallow to deep anchor case could also be assessed from displacement plots. For example Fig. 5 is a non-dimensional plot of $\frac{P}{\gamma D}$ versus $\frac{\partial a}{B}$ (the ratio of anchor displacement to anchor diameter) for the cases of $\frac{D}{B} = 4$ (shallow anchor) and $\frac{D}{B} = 15$ (deep anchor). In the shallow anchor case, a marked reduction in uplift capacity was measured after the peak load was reached especially for dense sand. This phenomenon, which can be explained by dilatancy effects, was not observed in the deep anchor case. The surface displacement measurements also showed (Fig. 6) how the surface heave at ultimate load decreased as $\frac{D}{B}$ increased. Where zero surface heave was reached the corresponding $\frac{D}{B}$ value was taken as the critical $\frac{H}{B}$ ratio defining the change from a shallow to a deep anchor.

Fig. 7 shows the dimensionless ratio $\frac{P_u}{\gamma D}$ plotted against $\frac{D}{B}$ for different relative densities of the sand. For shallow anchors the $\frac{Pu}{\gamma D}$ value increased at an increasing rate as $\frac{D}{B}$ increased until a change to a convex curve occurred at $\frac{D}{B}$ values of about 4.3, 7.8 and 10.5 corresponding to relative densities of 25.4, 50.2 and 85.2% respectively. These values corresponded to the critical depth ratios $\frac{H}{B}$ observed by the soil surface displacement measurements. As $\frac{H}{B}$ increased above the critical value, the rate of increase of $\frac{Pu}{\gamma D}$ decreased until $\frac{Pu}{\gamma D}$ reached an almost constant value at high $\frac{D}{B}$ values.

Analysis of the experimental work indicated that the relative density of the sand affected the assumed angle of inclination α of the failure surface and also the critical depth ratio. The relationships between the critical depth ratio $\left(\frac{H}{B}\right)$ and relative density (D_r), φ and density for the present tests in Leighton Buzzard sand are shown in Fig. 8. This relationship does not necessarily apply to other sands.

APPROXIMATE THEORY FOR THE UPLIFT RESISTANCE OF ANCHORS EMBEDDED IN SAND

Examination of the present test results and those of other investigators led to the evaluation of empirical factors incorporating relative density and other sand properties. This enabled a simplified general theory to be developed for an inclined deep anchor embedded in sand from which expressions could be derived for the separate cases of shallow and deep vertical anchors and shallow inclined anchors.

Fig. 9 shows for the general case of the deep inclined anchor the forces which were considered in deriving the total uplift resistance as a function of soil weight contained within the failure volume and the shearing resistance developed in the failure surface. In Figs. 10, 11 and 12 the failure surfaces are indicated for a shallow vertical anchor, a deep vertical anchor, and a shallow inclined anchor.

For a shallow anchor the failure surface is defined by a truncated cone with an apex angle of 2α as shown in Figs. 10 and 12. For deep anchors, the failure surface defined by α extends to a height H above the anchor as shown in Figs. 9 and 11. Above this height H, the failure surface is assumed to extend vertically to the soil surface. The height H is defined by the critical depth ratio $\frac{H}{B}$. If D is equal to or less than H, a shallow anchor condition exists.

The derivation of the general expression for the deep inclined anchor is given in the appendix as:-

R_4 = total uplift resistance in the direction of the anchor

$$= \frac{\pi\gamma}{24}\left[\cos\psi\left[Q(B + 2H\frac{\tan\alpha}{\cos\psi})^2(3D - 2H + \frac{B}{2}\frac{\cos\psi}{\tan\alpha}) - \frac{B^3}{\tan\alpha}\right] + \frac{4H\tan\alpha}{\cos^2\psi}\left[B\cos\psi(6D - 3H) + H\tan\alpha(6D - 4H)\right]\right.$$
$$\left. + 3K_0(D - H)^2(B\cos\psi + 2H\tan\alpha)(3 + 1.5Q - \sqrt{2Q})\tan\bar{c}\,\varphi\right]$$

where $Q = \cos\alpha\left[\frac{\cos(\psi - \alpha) + \cos(\psi + \alpha)}{\cos(\psi - \alpha)\cos(\psi + \alpha)}\right]$

The above formula can be modified to the other cases as follows:

1. Shallow Vertical Anchor (Fig. 10)

    R_1 = total vertical uplift resistance

    $= \frac{\pi\gamma D}{12}(8D^2\tan^2\alpha + 12\,BD\tan\alpha + 3B^2)$

2. Deep Vertical Anchor (Fig. 11)

    R_2 = total vertical uplift resistance

    $= \frac{\pi\gamma}{12}\left[8H^2(3D - 2H)\tan^2\alpha + 12\,HB(2D - H)\tan\alpha + 3DB^2 + 6K_0(D - H)^2(B + 2H\tan\alpha)\tan\bar{c}\,\varphi\right]$

3. Shallow Inclined Anchor (Fig. 12)

    R_3 = total uplift resistance in the direction of the anchor

    $= \frac{\pi\gamma}{48\tan\alpha}\left[Q(B\cos\psi + 2D\tan\alpha)^3 + 8D^2(3B\cos\psi + 2D\tan\alpha)\cos\psi\tan\alpha - 2B^3\cos\psi\right]$

The weight of the anchor G_0 is omitted in each of the above expressions as it normally has a negligible effect on the uplift resistance.

COMPARISON BETWEEN THE APPROXIMATE THEORY AND TESTS

Comparison with the present experimental work

In order to apply the theory to a particular problem, values of φ, α, D_r, $\bar{c}$, and the critical depth ratio $\frac{H}{B}$ must be known. In the present investigation all the factors could be determined.

The angle of internal friction, φ, was found from triaxial compression tests carried out at lateral cell pressures corresponding to the overburden pressure acting on the model anchors. This is an important point as the confining pressure can affect the φ value. In the model tests, where the overburden pressures are lower than those in field tests, the value of φ can be several degrees higher than those in field tests on the same sand.

The angle of inclination α of the assumed failure surface was found to be a function of D_r and φ, and Fadl (1981) established the following empirical relationship,

$$\alpha = M\varphi$$

where $M = 0.125 \left[D_r(1 + \cos^2\varphi) + (1 + \sin^2\varphi) \right] \left[1 + \cos\psi \right]$

The relative density D_r was obtained from standard test procedures, and the $\frac{H}{B}$ relationship with D_r was established as shown in Fig. 8.

The partial shear resistance developed along the upper prism of a deep anchor is assumed to be represented by a factor $\bar{c}$. Fadl's experimental results indicated that $\bar{c}$ could be represented empirically by $\bar{c} = D_r \cos\varphi$. For the case of an inclined anchor a corrected value for $\frac{H}{B}$ was derived empirically as

$$\left[\frac{H}{B}\right]_\psi = \left[\frac{H}{B}\right]_{\psi=0} \times \left[\frac{0.4 + \cos\psi}{1.4}\right]$$

Using the factors derived as above, Figs. 13, 14 and 15 give a comparison between the experimental values of uplift resistance (as represented by $\frac{p_u}{\gamma D}$) and those predicted by the approximate theory. The comparison is made for shallow and deep vertical anchors, and for anchors inclined at 22.5° and 45° to the vertical. Satisfactory agreement was obtained between the predicted and the experimental values for the Leighton Buzzard sand used in this investigation.

Comparison with other laboratory and field tests

In order to test the approximate theory against the work of others, values of φ, α, D_r, $\bar{c}$ and $\frac{H}{B}$ critical are required. There are comparatively few published test results which list all the information required to apply the theory. This particularly applies to deep foundations with respect to the $\frac{H}{B}$ value.

However it is possible to compare the predicted and experimental values for some shallow vertical anchors and reasonable agreement has been obtained as can be seen from Figs. 16 and 17 concerning laboratory and field tests by Sutherland (1965).

The only published data for deep anchors from which all the factors can be assessed are those by Esquivel-Diaz (1967) and Bemben and Kupferman (1975). Esquivel-Diaz notes the influence of cell pressure on φ in triaxial compression tests on sand, and if this is taken into account in accordance with his suggestions, then a comparison of the predicted against experimental values is as shown in Fig. 18. Predicted and experimental values for Bemben and Kupferman are shown on Fig. 19. In both cases reasonable agreement is obtained.

It may be that the other investigators have non-published information which would allow them to check the approximate theory against their experimental data.

CONCLUSIONS

1. A number of previous investigators have carried out theoretical and experimental studies on the uplift resistance of anchors embedded in sand. Substantial differences exist between the predictions of uplift resistance from the different theories.

2. Examination of previous work and the results from the present experimental work indicate that the relative density of the sand, reflecting its compressibility, has an important influence on the ultimate resistance and displacement characteristics of embedded anchors. An approximate general theory has been developed, which embodies relative density and also the critical depth ratio $\frac{H}{B}$ which signifies the transition between shallow and deep anchors.

3. The approximate theory was checked against the present experimental work, and also, where possible, against laboratory and field measurement by other investigators, and reasonable agreement was obtained.

4. Most of the information required for the application of the theory to design problems can be obtained from conventional site investigation. The critical depth ratio, $\frac{H}{B}$ is the one factor which can not be readily obtained unless laboratory experiments on models are carried out for the particular problem. At this stage, however, it is suggested that if a critical depth ratio of $\frac{H}{B} \leq 6$ is used in the theory then an acceptably conservative value of uplift resistance can be predicted. It should be emphasised that the choice of φ value used in any analysis should be based on the confining pressure corresponding to that acting in the laboratory tests or actual field problem.

APPENDIX

Approximate Theory

The derivation of a general expression for a deep inclined anchor is given in this Appendix.

Fig. 9 shows the assumed failure surface and the forces acting. The volumes bounded by the failure surface are an oblique truncated cone FGIJ starting at depth D with a diameter B equal to that of the anchor and terminating at a height H above the anchor in an elliptical horizontal surface. From this level an elliptical prism GIVW is assumed to extend to the surface. The forces considered are the weight of soil contained within those volumes, and the shearing stresses acting on the surfaces bounding the volumes; all forces act in the direction of the anchor axis which is inclined at an angle ψ to the vertical.

The analysis is as follows:

1. **Weight of soil in the truncated oblique cone FGIJ**

   The major axis of the horizontal ellipse at the top of the truncated cone has a diameter 2a where

   $$2a = \frac{B + 2L \tan \alpha}{2} \cos \alpha \left[\frac{\cos(\psi - \alpha) + \cos(\psi + \alpha)}{\cos(\psi + \alpha)\cos(\psi + \alpha)} \right]$$

   and if $Q = \cos \alpha \left[\frac{\cos(\psi - \alpha) + (\psi + \alpha)}{\cos(\psi - \alpha)\cos(\psi + \alpha)} \right]$

   then $2a = \left(\frac{B + 2L \tan \alpha}{2} \right) Q$

   The minor axis, $2b = B + 2L \tan \alpha$
   The volume of the truncated cone is then evaluated as

   $$\text{Vol} = \frac{\pi}{24} \left[(B + 2L \tan \alpha)^2 \, Q(L + \frac{B}{2 \tan \alpha}) \cos \psi - \frac{B^3}{\tan \alpha} \right]$$

   and the component of the soil weight in the direction of loading is W_5 where

   $$W_5 = \frac{\pi \gamma \cos \psi}{24} \left[(B + 2H \frac{\tan \alpha}{\cos \psi})^2 \, Q(H + \frac{B \cos \psi}{2 \tan \alpha}) - \frac{B^3}{\tan \alpha} \right]$$

2. **Weight of soil in the elliptical prism GIVW**

   The volume is π ab $(D - H)$ from which the component of the weight in the direction of loading, W_7 is obtained where

   $$W_7 = \frac{\pi \gamma (D - H) \cos \psi}{8} (B + 2H \frac{\tan \alpha}{\cos \psi})^2 Q$$

3. **Shear resistance along the surface of the oblique truncated cone FGIJ**

   The distribution of the resultant shearing stress will take the form indicated by the trapezoids FGTP and IJOU (Fig. 9). For simplicity, however, an equivalent symmetrical distribution is assumed in the form of the trapezoids FNRM and KJQS. Following Matsuo (1967) the assumed plane stress condition was taken to be applicable to the actual three-dimensional stress condition, and the following expression was derived for T_5, the shear force acting along the anchor axis

   $$T_5 = \int_0^L \tan \alpha \int_0^{2\pi} (\frac{B}{2} + x)(D - H - l \cos \psi) \gamma \tan \alpha \, d\beta \, \frac{dx}{\sin \alpha}$$

   from which, by integrating and re-arranging the terms -

$$T_5 = \frac{\pi H \gamma}{6 \cos^2 \psi} \tan \alpha \left[B \cos \psi (6D - 3H) + H \tan \alpha (6D - 4H) \right]$$

4. **Partial shear resistance along the surface of the elliptical prism GIVW**

   The elliptical prism is assumed to offer a shear resistance along its vertical perimeter as the cone below is forced upwards. The degree of shear resistance mobilised will depend on the amount of movement undergone by the prism within its height and because of this it is considered as a partial shear resistance evaluated with the angle of internal friction φ modified by an empirical factor $\bar{c}$. The horizontal pressure on the prism is assumed to be governed by K_o, the coefficient of earth pressure at rest, defined as $(1 - \sin \varphi)$. When the vertical partial shear resistance is transformed into a force acting along the anchor axis this force is evaluated as -

$$T_7 = \frac{\pi \gamma K_o \tan \bar{c} \varphi}{8} (D - H)^2 (B \cos \psi + 2 H \tan \alpha)(3 + 1.5Q - \sqrt{2Q})$$

By summing the above forces, the total ultimate uplift resistance of a deep inclined anchor is obtained as R_4 where

R_4 = total uplift resistance in the direction of the anchor

$$= \frac{\pi \gamma}{24} \left[\cos \psi \left[Q(B + 2H \frac{\tan \alpha}{\cos \psi})^2 (3D - 2H + \frac{B}{2} \frac{\cos \psi}{\tan \alpha}) - \frac{B^3}{\tan \alpha} \right] + \frac{4H \tan \alpha}{\cos^2 \psi} \left[B \cos \psi (6D - 3H) + H \tan \alpha (6D - 4H) \right] \right.$$
$$\left. + 3 K_o (D - H)^2 (B \cos \psi + 2H \tan \alpha)(3 + 1.5Q - \sqrt{2Q}) \tan \bar{c} \varphi \right]$$

where $Q = \cos \alpha \left[\frac{\cos(\psi - \alpha) + \cos(\psi + \alpha)}{\cos(\psi - \alpha) \cos(\psi + \alpha)} \right]$

Normally, $G_o \cos \psi$ would be omitted since it has a negligible effect on the uplift resistance.

The general expression R_4 reduces to the expressions R_1, R_2 and R_3 quoted earlier as follows:

(a) <u>Deep vertical anchor</u>
In this case $\psi = 0$, $\cos \psi = 1$ and $Q = 2$ and the particular expression for this case becomes R_2 as quoted earlier.

(b) <u>Shallow inclined anchor</u>
In this case $D = H$, the overlying prism does not exist, and expression R_3 results.

(c) <u>Shallow vertical anchor</u>
The expression for R_3 is used with $\psi = 0$ giving the expression R_1.

REFERENCES

Balla, A (1961). "The resistance to breaking out of mushroom foundations for pylons". <u>Proc. 5th Int. Conf. Soil Mech. and Found. Engg.</u> Paris, Vol. 1, pp 569-576.

Bemben, S.M. and Kupferman, M. (1975). "The vertical holding capacity of marine anchor flukes subjected to static and cyclic loading". <u>Proc. 7th Offshore Techn. Conf.</u>, Dallas, Vol. 1, pp 363 - 374.

Burland, J.B. and Roscoe, K.H. (1969). "Local strains and pore pressures in a normally consolidated clay layer during one-dimensional consolidation". <u>Geotechnique</u>, Vol. 19, No. 3, pp 335 - 356.

Carr, R.W. (1970). "<u>An experimental investigation of plate anchors in sand</u>". Ph.D. Thesis, University of Sheffield.

El-Rayes, M.K. (1965). "<u>Behaviour of cohesionless soils under uplift forces</u>". Ph.D. Thesis, University of Glasgow.

Esquivel-Diaz, R.F. (1967). "<u>Pullout resistance of deeply buried anchors in sand</u>". M.Sc. Thesis, Duke University, Durham, North Carolina, U.S.A.

Fadl, M.O. (1981). "<u>The behaviour of plate anchors in sand</u>". Ph.D. Thesis, University of Glasgow.

Harvey, R.C. and Burley, E. (1973). "Behaviour of shallow inclined anchorages in cohesionless sand", <u>Ground Engineering</u>, Vol. 6, No. 5, pp 48 - 55.

Mariupol'skii, L.G. (1965). "The bearing capacity of anchor foundations". <u>Soil Mech. & Found. Engg.</u> Vol. 2, 1, pp 26 - 32. Translated from Osnovaniya, Fundamentyi Mekhanika Gruntov.

Matsuo, M. (1967). "Study on the uplift resistance of footing (1)". <u>Japanese Soc. Soil Mech. & Found. Engg.</u>, Soils and Foundations, Vol. 7, No.4, pp 1 - 37.

Matsuo, M. (1968). "Study on the uplift resistance of footing". <u>Japanese Soc. Soil Mech. & Found.Engg.</u>, Soils and Foundations, Vol. 8, No. 1, pp 18 - 48.

Meyerhof, G.G. (1951). "The ultimate bearing capacity of foundations". <u>Geotechnique</u>, Vol. 2, No. 4, pp 301 - 322.

Meyerhof, G.G. (1973). "Uplift resistance of inclined anchors and piles". <u>Proc. 8th Int. Conf. Soil Mech. & Found. Engg.</u>, Moscow, Vol. 2.1, pp 167 - 172.

Meyerhof, G.G. and Adams, J.I. (1968). "The ultimate uplift capacity of foundations". <u>Canadian Geotechnical Jnl.</u>, Vol. 5, No. 4, pp 225 - 244.

Sutherland, H.B. (1965). "Model studies for shaft raising through cohesionless soils". <u>Proc. 6th Int. Conf. Soil Mech. & Found. Engg.</u>, Montreal, Vol. 2, pp 410 - 413.

Vesic, A.S. (1963). "Bearing capacity of deep foundations in sand". National Academy of Sciences, <u>Highway Research Record</u>, 39, pp 112 - 153.

Vesic, A.S. (1965). "Cratering by explosives as an earth pressure problem". <u>Proc. 6th Int. Conf. Soil Mech. & Found. Engg.</u>, Montreal, Vol. 2, pp 427 - 431.

Vesic, A.S. (1971). "Breakout resistance of objects embedded in ocean bottom". <u>Proc. A.S.C.E. Soil Mech. & Found. Div.</u>, Vol. 97, SM 9, pp 1183 - 1205.

Vesic, A.S. (1972). "Expansion of cavities in infinite soil mass". <u>Proc. A.S.C.E. Soil Mech. & Found. Div.</u>, Vol. 98, SM 3, pp 265 - 290.

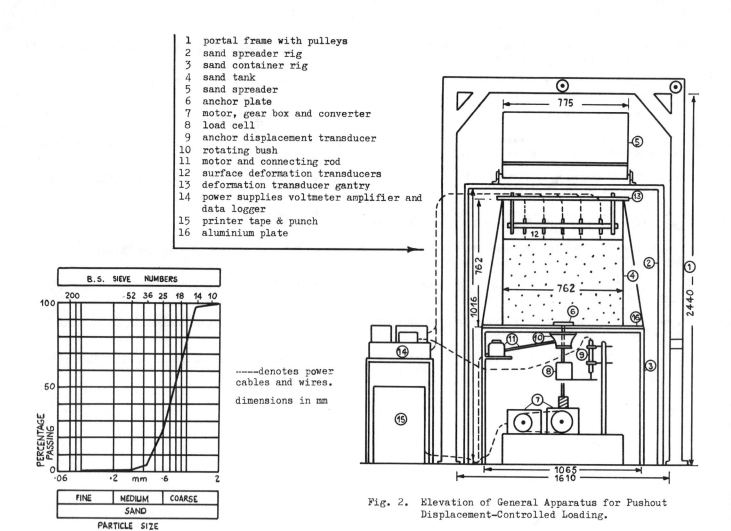

1. portal frame with pulleys
2. sand spreader rig
3. sand container rig
4. sand tank
5. sand spreader
6. anchor plate
7. motor, gear box and converter
8. load cell
9. anchor displacement transducer
10. rotating bush
11. motor and connecting rod
12. surface deformation transducers
13. deformation transducer gantry
14. power supplies voltmeter amplifier and data logger
15. printer tape & punch
16. aluminium plate

----- denotes power cables and wires.

dimensions in mm

Fig. 1. Particle Size Distribution of the Leighton Buzzard Sand

Fig. 2. Elevation of General Apparatus for Pushout Displacement-Controlled Loading.

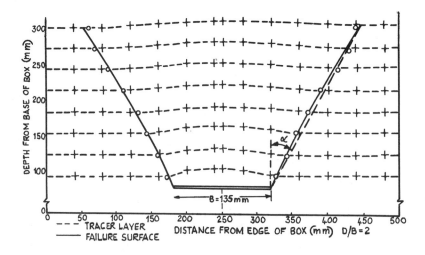

Fig. 3. Final Position of Tracer Layers.

460

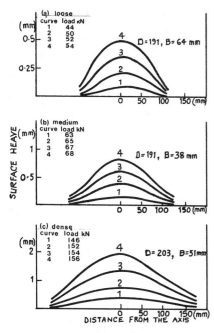

Fig. 4. Typical Surface Heave for Vertical Anchors.

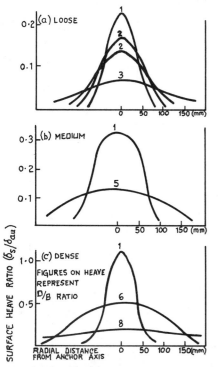

Fig. 6. Dimensionless Surface Heave for Vertical Anchors at Maximum Uplift Resistance

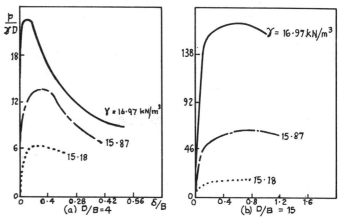

Fig. 5. Effect of Variation of γ for Constant $\psi = 0°$

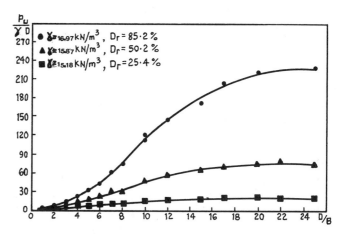

Fig. 7. $\dfrac{p_u}{\gamma D}$ v $\dfrac{D}{B}$ for Different Densities in Sand ($\psi = 0°$)

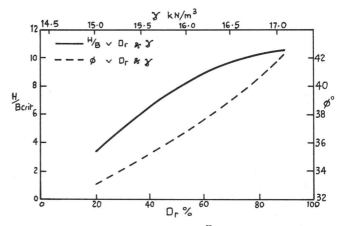

Fig. 8. Relationships between $\dfrac{H}{B}$ and φ versus D_r and γ

461

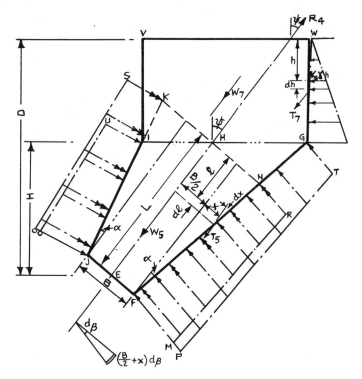

Fig. 9. Deep inclined anchor - forces contributing to uplift resistance.

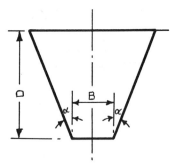

Fig.10. Shallow vertical anchor

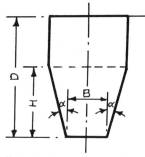

Fig.12. Deep vertical anchor

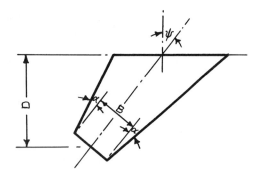

Fig.12. Shallow inclined anchor

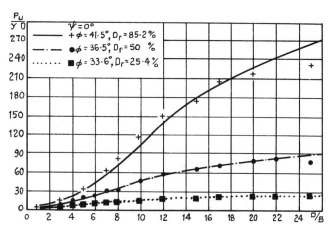

Fig. 13. Comparison of Approximate Method with Present Vertical Model Test Results.

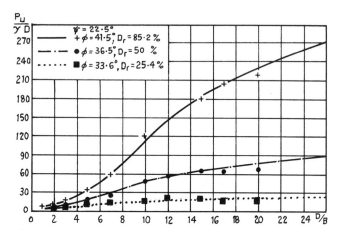

Fig. 14. Comparison of Approximate Method with Present Inclined Model Test Results.

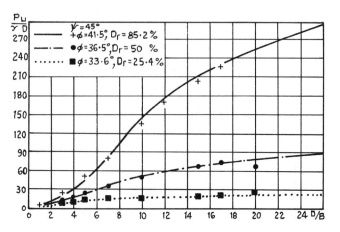

Fig. 15. Comparison of Approximate Method with Present Inclined Model Test Results.

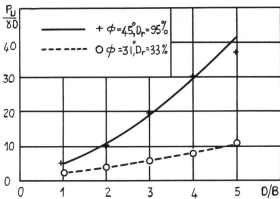

Fig.16. Sutherland (Model Tests)

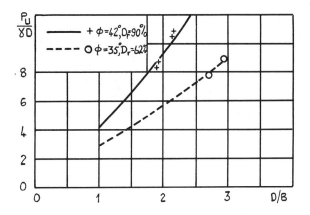

Fig. 17. Sutherland (Field Tests)

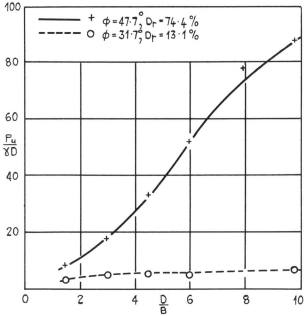

Fig.18. Esquivel-Diaz (Deep model tests)

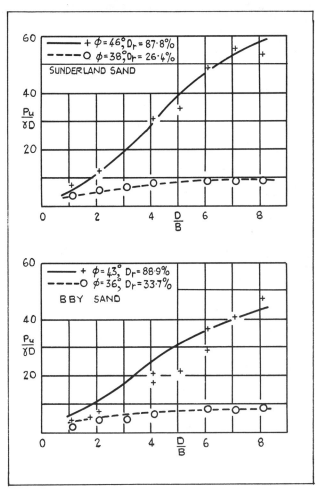

Fig. 19. Bemben & Kupferman (Deep Model Tests)

A NEW METHOD TO DETERMINE THE UNDRAINED SHEAR STRENGTH OF MUDS BASED ON PLASTICITY THEORY

L. E. Vallejo
Department of Civil Engineering
Michigan State University
East Lansing, Michigan 48824

P. A. Tarvin
Department of Civil Engineering
Michigan State University
East Lansing, Michigan 48824

SUMMARY

A new method called the "cylinder-strength meter test" designed to measure the undrained shear strength of liquid-like soil slurries (muds) is introduced. The theory behind the method and its implementation in the laboratory is presented. The new approach uses a cylinder as a measuring tool and the theory developed by Sokolovskii to calculate the indentation pressures developed in a Tresca plastic when a cylinder penetrates it.

INTRODUCTION

Submarine mudslides are powerful agents of degradation of submarine slopes and are reported to cause during their travel destruction of offshore installations such as oil platforms, pipelines and communication cables (Arnold, 1967; Bea and Audibert, 1980). Mudslides often take place at shallow depths of the ocean floor covered by very soft unconsolidated sediments or muds which have high water contents and low shear strengths (Richards and Parks, 1976; Prior and Coleman, 1978; Bea and Audibert, 1980). According to Schapery and Dunlap (1978), "the sediments are so weak that core samples sag under their own weight." Civil Engineering structures are often built on seafloor areas prone to mudslides. Therefore, a knowledge of the stability against sliding of mud-made areas where offshore installations are going to be located is of primary consideration (Tsui, 1972; Demars, et al, 1977).

The stability analysis of submarine mud slopes can be made in terms of total or effective stresses. However, an effective stress or long-term stability analysis of mud slopes is difficult to make for the following reasons. First, the materials involved with mudslides are too soft to allow samples to be taken and tested in conventional triaxial or direct shear apparatuses (Tsui, 1972; Mitchell, et al, 1972). Second, even if conventional tests can be run on mud samples, the complexity of pore water pressure response in muds to a known set of total normal and shear stresses makes the short-term analysis which uses the undrained shear strength, c_u, of the mud to be more suitable than the long-term or effective stress analysis (Kraft Jr., et al, 1976). Therefore, the stability analysis of submarine mud slopes is often made in terms of the undrained shear strength of the mud. Thus, its correct measurement either in the laboratory or in the field becomes of primary consideration when analyzing the stability of an offshore area with respect to mudslides. Let's examine conventional methods designed to measure the c_u of muds

The Measurement of the Undrained Shear Strength of Muds.

Two methods are often used to measure the undrained shear strength of very soft clays, excluding triaxial compression tests which are difficult to apply to muds. These two methods are: a) the vane shear test (Tsui, 1972; Kraft Jr., et al, 1976; Schapery and Dunlap, 1978), and b) the cone penetrometer test (Hirst, et al, 1972; Rodine, 1975). Let's examine each method in detail when applied to muds.

The Vane Shear Test.
One of the methods used by geotechnical engineers to measure the undrained shear strength of undisturbed or remoulded saturated cohesive soils under laboratory or field conditions is the vane shear test (Ladd, et al, 1977). The vane consists of four rectangular paddles at right angles to each other which are fixed at the end of a rod. The vane is forced into the soil and then rotated at a constant rate while the torque, T, is measured. This torque resistance is used in conjunction with Equation [1] to obtain the undrained shear strength, c_u, of the cohesive soils

$$c_u = \frac{2T}{\pi D^2 H \left(1 + \frac{D}{3H}\right)} \quad [1]$$

where H is the height of the vane blades and D is their diameter.

The vane shear test works on the assumption that the failure surface for mud strength determination is that of a cylinder with dimensions equal or slightly greater than those of the vane blades (Kraft Jr., et al, 1976; Ladd, et al, 1977). However, due to the fluidity of the muds, the failure surface induced by the vane can not rigorously be represented as that of a cylinder (Johnson, 1965). Also, it has been found by Krone (1963, 1976) that during the rotation of a vane-like apparatus in muds, collisions and adhesion between the particles of clay take place which causes the formation of different orders of aggregation among the particles with a resulting decrease of the mud's undrained shear strength (see Table 1).

The above two important shortcomings of the vane shear test when applied to muds, therefore, reduces its reliability when used to measure the undrained shear strength of them.

The Cone Penetrometer Test.
Another way to obtain the undrained shear strength of muds is by the use of quasistatic cone penetrometers (Hirst, et al, 1972; Rodine, 1975). The test involves the measurement of the resistance offered by a mud sample to its penetration by a cone. This resistance is then correlated to the undrained shear strength of the mud by the use of the following bearing capacity equation (Hirst, et al, 1972; Ladd, et al, 1977)

Table 1. Orders of Particle Aggregation and Changes in the Shear Strength Induced When a Vane-Like Test Device Acts on a Sample of Soft Sediment (Mud) (data from Krone, 1963).

Sediment Sample	Order of Aggregation	Unit Weight grm/cm^3	Shear Strength dyn/cm^2
San Francisco Bay	0	1.269	22
	1	1.179	3.9
	2	1.137	1.4
	3	1.113	1.4
	4	1.098	0.82
	5	1.087	0.36
	6	1.079	0.20

$$q_c = c_u N_c + \sigma_o \qquad [2]$$

where

q_c = measured cone resistance,

N_c = bearing capacity factor,

σ_o = in-situ vertical total stress at cone level (for measurements at shallow depths this vertical stress can be taken equal to zero),

c_u = undrained shear strength of the mud.

In order to obtain the value of c_u from Equation [2], the value of the bearing capacity factor N_c needs to be known. Various theoretical solutions for the calculation of N_c for cohesive soils ($\phi = 0$) have been developed on the basis of bearing capacity theories (Meyerhof, 1951; Rodine, 1975), and by the use of cavity expansion theories (Ladanyi, 1967; Vesic, 1972; Baligh, 1975). However, there is no general agreement of what values to choose for the bearing capacity factor N_c (see Table 2). The value of N_c has been found to be related to the type of soil tested and the type of cone used (Ladd, et al, 1977).

Table 2. N_c Values for $\phi = 0$ soils

Type of Soil	N_c	Reference
Young, non-fissured not highly sensitive clays	10 - 16	Schmertmann (1975)
Soft clays	5 - 70	Amar, et al (1975)
Normally consolidated marine clays	17 $\pm$ 2	Lunne, et al (1976)
Varved clay	13	Lunne, et al (1976)

Since the results of c_u obtained from the cone penetrometer test are related to the value of the bearing capacity factor N_c, and since this parameter does not have a fixed value, but has a wide range of values as Table 2 shows, Ladd, et al (1977) states that the cone penetrometer test can only be used as a "useful strength index test."

Due to the limitations of the methods described above when used to determine the undrained shear strength of muds, a new approach to measure their c_u is presented next.

A NEW APPROACH TO MEASURE THE UNDRAINED SHEAR STRENGTH OF MUDS

Theoretical Background of the New Approach.

A new approach, that will be called "the cylinder-strength meter test", to measure the undrained shear strength of muds is presented. The new approach consists in slowly lowering cylinders of known dimensions and weight into a mud sample, measuring the depth of penetration of the cylinder into the mud, and calculating the strength of the mud required to support the cylinder at that depth.

The theory underlying the new approach is based on the support mechanism of the cylinder by the mud. It considers that the weight of the cylinder tends to cause it to sink in the mud, wherear the undrained strength (bearing capacity) of the mud and its buoyancy tend to support the cylinder. Thus at equilibrium conditions, the weight of the cylinder will be resisted by a) an upward force resulting from the undrained shear strength of the mud which can be calculated from the slip-line approach of Plasticity Theory for the case of a cylinder indenting a Tresca plastic, and b) buoyancy, which is equal to the weight of the displaced mud resulting from part of the cylinder penetrating the mud sample.

The Support of the Cylinder by the Mud.

When a cylinder is slowly lowered into a mud sample, its weight will be supported by an upward force, which depends upon the undrained shear strength of the mud, and by buoyancy, this being equal to the weight of the displaced mud (Figure 1A). The weight of the cylinder is equal to

$$W = \pi R^2 L \gamma_c \qquad [3]$$

where R is the radius of the cylinder, L is its length, and γ_c the unit weight of the material forming the cylinder.

For the calculation of the upward force resulting from c_u of the mud, the solution developed by Sokolovskii (1955) for calculating the pressure developed in a Tresca plastic when a cylinder penetrates it will be used. Sokolovskii (1955) used the slip-line approach of Plasticity Theory for the obtention of his solution. According to Sokolovskii when a _smooth_ cylinder indents a Tresca plastic (in our case mud), the plastic develops an upward force which is equal to (Figure 1B)

$$P_s = 2 c_u R L \left[(\pi + 2) \sin \alpha + 2 (1 - \cos \alpha - \alpha \sin \alpha) \right] \qquad [4]$$

For the case of a _rough_ cylinder, the upward force can be obtained from (Figure 1C)

$$P_r = 2 c_u R L \left[\left(\frac{3\pi}{2} + 1 - 2\alpha\right) \sin \alpha + 2 (\sqrt{2} - \cos \alpha) - \sin \left(\frac{\pi}{2} - \alpha\right) \right] \qquad [5]$$

The parameters involved with Equations [4] and [5] are shown in Figures 1A, 1B, and 1C. The buoyancy resulting when part of the cylinder penetrates the mud can be obtained from (Fig. 1A)

$$P_b = R^2 L (\alpha - \sin \alpha \cos \alpha) \gamma_f \qquad [6]$$

where γ_f is the unit weight of the mud.

At equilibrium conditions the following equation applies

$$W = P_{s,r} + P_b \qquad [7]$$

After replacing into Equation [7] the values of the respective terms given by Equations [4], [5], and [6], the value for the undrained shear strength of the mud, c_u, can be obtained as follows. For the case of a _smooth_ cylinder

$$c_u = \frac{R \left[\pi \gamma_c - (\alpha - \sin \alpha \cos \alpha) \gamma_f \right]}{2 \left[(\pi + 2) \sin \alpha + (1 - \cos \alpha - \alpha \sin \alpha) \right]} \qquad [8]$$

For the case of a _rough_ cylinder

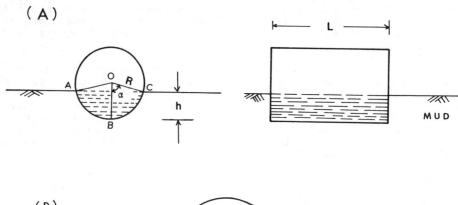

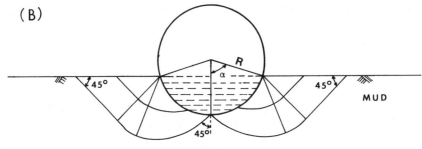

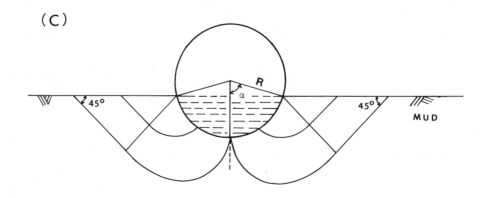

Figure 1. A) Parameters used for the cylinder-strength meter test, B) Slip-line pattern used for the case of a smooth cylinder, C) Slip-line pattern used for the case of a rough cylinder (after Sokolovskii, 1955).

$$c_u = \frac{R\left[\pi\gamma_c - (\alpha - \sin\alpha\cos\alpha)\gamma_f\right]}{2\left[(\frac{3\pi}{2} + 1 - 2\alpha)\sin\alpha + (\sqrt{2} - \cos\alpha) - \sin(\frac{\pi}{2} - \alpha)\right]} \quad [9]$$

The value of α can be obtained from (Figures 1A, 1B, 1C)

$$\alpha = \cos^{-1}\left(\frac{R-h}{R}\right) = \cos^{-1}\left(1 - \frac{h}{R}\right) \quad [10]$$

Equations [8] and [9] apply only for values of α equal or smaller than 90 degrees (Figures 1A, 1B, 1C). In Equation [10] h is the depth of penetration of the mud by the cylinder.

Implementation of the Method in the Laboratory.

For the implementation of the cylinder-strength meter test in the laboratory, mud samples having different concentrations of kaolinite clay, water, sand, glass beads and calgon (hexamethaphosphate used as a dispersive agent) were mixed in a container with dimensions of 30x20x25 cm using an electrix mixer. The type and amount of materials used are described next.

Clay-Water Mixtures. It has been found by Rodine and Johnson (1976), and Vallejo (1981a, 1981b) that good mud samples for shear strength determination in the laboratory can be prepared by mixing commercially available kaolinite clay (Hydrite 10, marketed by the Georgia Kaolin Co.), calgon and distilled water. Therefore, the above mixture was used in the preparation of mud samples that contained clay as its major component. Three samples of mud, having the same amount of water (225 grms) and calgon (36 grms) but different amounts of clay (346, 392, 426 grms) were prepared in the laboratory. The purpose of this test was to learn what effect the concentration of clay (specific gravity equal to 2.64) has on the c_u values of the mud as determined by the cylinder-strength meter test. Results of the laboratory investigation are shown in Figure 2.

Clay-Sand Mixtures. It has been reported by Hutchinson (1970), and Vallejo (1979, 1980) that mudflows in the field are usually formed of a mixture of hard clay fragments or rock pieces and a matrix of liquid-like soil slurry (mud). Therefore, in order to simulate the field conditions, granular materials in the form of sand and glass beads were added to a clay slurry made of a mixture of kaolinite clay, calgon and distilled water. The clay slurry for the clay-sand experiments was formed by mixing 697 grms of clay, 243 grms of water and 36 grms of calgon. To this clay slurry, which was used for all the experiments involving clay-sand mixtures, different amounts of a uniform, rounded, medium size (average diameter equal to 0.3 mm, specific gravity equal to 2.65) sand from Wedren, Illinois were added and the c_u of the mixtures were determined using the cylinder-strength meter test. The results are shown in Figure 3. The clay-water slurry used for the sand-clay experiments was also found to have enough cohesive strength and buoyancy to maintain the sand grains afloat and in dispersion. Therefore, no settlement of the sand grains was recorded during the testing of the sand-clay mixtures.

Clay-Glass Beads Mixtures. For the clay-glass beads experiments, a clay-water slurry with enough cohesive strength (bearing capacity) and buoyancy to mainatin afloat and in dispersion glass beads 5 mm in diameter (with specific gravity equal to 1.55) was found to be formed by mixing 730 grms of kaolinite clay, 235 grms of distilled water and 43 grms of calgon. This mixture was used for all the experiments involving glass beads-clay slurry combinations. Different amounts of beads were added to the clay-water slurry, after which the c_u of the mixture was determined using the cylinder-strength meter test and the results are shown in Figure 4.

Cylinders Used and the Measurement of the Undrained Shear Strength. For the implementation of the Sokolovskii (1955) approach to measure the undrained shear strength of muds, plexiglass cylinders (specific gravity equal to 1.2) were constructed and used in the experiments. The cylinders used had smooth surfaces and the following dimensions: Cylinder Number 1 was 3 cm in diameter and 6 cm in length; Cylinder Number 2 was 3 cm in diameter and 4.5 cm in length; and Cylinder Number 3 was 3.5 cm in diameter and 4.5 cm in length.

For measuring the undrained shear strength of the mud samples prepared in the laboratory, the cylinders were slowly lowered into the mud samples and allowed to sink into it by their own weight. Once the cylinders had indented the mud sample, the mud container was slightly jostled in order to reduce possible surface tension in the cylinder-mud contact area. After this was done, using a micrometer, the depth of penetration of the mud by the cylinder (Figure 1A) was measured. This depth of penetration (equal to h in Figure 1) was used in conjunction with Equations [8] and [10] to measure the undrained shear strength, c_u, of the muds.

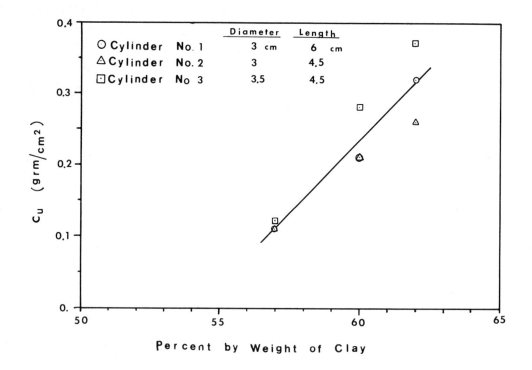

Figure 2. Undrained shear strength of the clay-water slurry versus the clay concentration in the mixture.

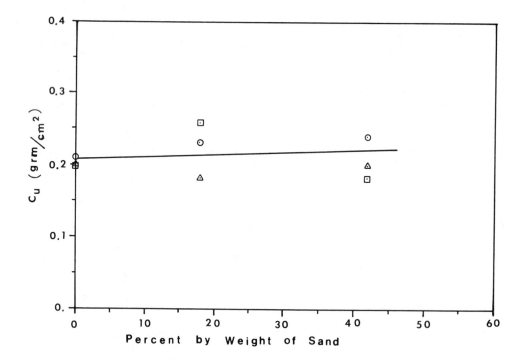

Figure 3. Undrained shear strength of the clay-sand slurry versus the sand concentration in the mixture.

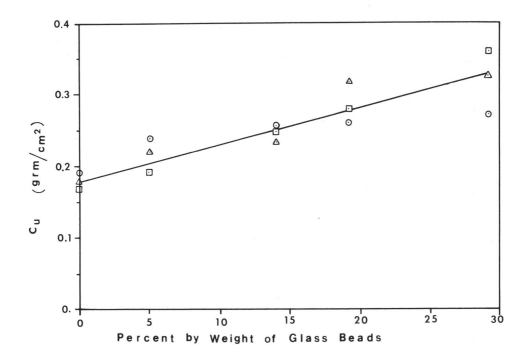

Figure 4. Undrained shear strength of the clay-glass beads slurry versus the glass-beads concentration in the mixture.

Analysis of the Results.

An analysis of the results of c_u measured using the cylinder-strength meter test depicted in Figures 2, 3 and 4 shows that some difference exists, even though small, in the measured values of c_u obtained by the three testing cylinders. An explanation for the discrepancies that become somewhat larger as the clay or granular concentration increases seems to be the result of the following factors: a) as the clay or granular content increases it becomes difficult to prepare a mud sample in the laboratory that is totally homogeneous, b) since there is time involved in testing the mud samples using the three different cylinders, some water was undoubtedly lost by evaporation from the samples. This results in somewahat different values of c_u being measured by the three different cylinders, and c) operator error is involved.

CONCLUSIONS

A new method called the "cylinder-strength meter test" designed to measure the undrained shear strength of liquid-like soil slurries (muds) is introduced. The theory behind the method and its implementation in the laboratory is presented. The new method measures the undrained shear strength of muds for which conventional geotechnical laboratory testing designed to measure shear strength is difficult to apply.

The undrained shear strength, c_u, of muds prepared in the laboratory with different concentrations of clay and granular materials was determined using cylinders of different lengths and diameters. It was found that the c_u results obtained using the cylinders were somewhat affected by the loss of water that take place during the testing, as well as the non-homogeneity of the mud samples.

REFERENCES

AMAR, S., F. BAGUELIN, J.F. JEZEQUIEL, and A. LEMEHAUTE, 1975, "In-Situ Shear Resistance of Clays," Proceedings of the American Society of Civil Engineers Special Conference on In-Situ Measurement of Soil Properties, Raleigh, North Carolina, 1975, Volume 1, pp. 22-45. New York, New York: American Society of Civil Engineers.

ARNOLD, K. E., 1967, Soil Movements and Their Effects on Pipelines in the Mississippi Delta Region, MSc Thesis, Tulane University, New Orleans, Department of Civil Engineering.

BALIGH, M. M., 1975, Theory of Deep Site Cone Penetration Resistance, Massachusetts Institute of Technology, Department of Civil Engineering, Research Report R75-56, Number 517.

BEA, R. G. and J. M. E. AUDIBERT, 1980, "Offshore Platforms and Pipelines in Mississippi River Delta," Journal of the Geotechnical Engineering Division, American Society of Civil Engineers, Volume 106, Number GT8, pp. 853-869.

DEMARS, K. R., V. A. NACCI, and W. D. WANG, 1977, "Pipeline Failure: A Need for Improved Analysis and Site Surveys," Proceedings of the Ninth Offshore Technology Conference, Houston, Texas, 1977, Paper OTC 2966, pp. 63-70. Dallas, Texas: American Institute of Mining, Metallurgical and Petroleum Engineers, Inc.

HIRST, T. J., A. F. RICHRDS, and A. L. INDERBITZEN, 1972, "A Static Cone Penetrometer for Ocean Sediments," Proceedings of the Seventy-Fourth Annual Meeting of the American Society For Testing and Materials, Atlantic City, New Jersey, 1972, Volume STP 501, pp. 69-89. Philadelphia, Pennsylvania: American Society for Testing and Materials.

HUTCHINSON, J. N., 1970, "A Coastal Mudflow in the London Clay Cliffs at Beltinge, North Kent," Geotechnique, Volume 20, Number 4, pp. 412-438.

JOHNSON, A. M., 1965, A model for Debris Flow, PhD Thesis, The Pennsylvania State University, Department of Geology.

KRAFT Jr., L. M., N. AHMAD, and J. A. FOCHT, 1976, "Application of Remote Vane Results to Offshore Geotechnical Problems," Proceedings of the Eight Offshore Technology Conference, Houston Texas, 1976, Paper OTC 2626, pp. 75-96. Dallas, Texas: American Institute of Mining, Metallurgical and Petroleum Engineers, Inc.

KRONE, R. B., 1963, A Study of Rheological Properties of Estuarial Sediments, University of California at Berkeley, Hydraulic Engineering Laboratory and Sanitary Engineering Research Laboratory, SERL Report No. 63-8.

KRONE, R. B., 1976, "Engineering Interest in the Benthic Boundary Layer," Proceedings of the NATO Science Committee Conference on the Benthic Boundary Layer, Les Arcs, France, 1976, pp. 143-156. New York, New York: Plenum Press.

LADANYI, B., 1967, "Deep Punching of Sensitive Clays," Proceedings of the Third Panamerican Conference on Soil Mechanics and Foundation Engineering, Caracas, Venezuela, Volume 1, pp. 533-546. Caracas, Venezuela: Venezuelan Society of Soil Mechanics and Foundation Engineering.

LADD, C. C., R. FOOT, K. ISHIHARA, F. SCHLOSSER, and H. G. POULOS, 1977, "Stress-Deformation and Strength Characteristics," Proceedings of the Ninth International Conference on Soil Mechanics and Foundation Engineering, Tokyo, Japan, 1977, State of the Art Reports, Volume 2, pp. 421-494. Tokyo, Japan: The Japanese Society of Soil Mechanics and Foundation Engineering.

LUNNE, T., O. EIDE, and J. DE RUITER, 1976, "Correlations Between Cone Resistance and Vane Shear Strength in Some Scandinavian Soft to Medium Stiff Clays," Canadian Geotechnical Journal, Volume 13, Number 4, pp. 430-441.

MEYERHOF, G. G., 1951, "The Ultimate Bearing Capacity of Foundations," Geotechnique, Volume 2, Number 4, pp. 301-332.

MITCHELL, R. J., K. K. TSUI, and D. A. SANGREY, 1972, "Failure of Submarine Slopes Under Wave Action," Proceedings of the Thirteenth Conference on Coastal Engineering, Vancouver, Canada, 1972, Volume 2, pp. 1515-1541. New York, New York: American Society of Civil Engineers.

PRIOR, D. B. and J. M. COLEMAN, 1978, "Submarine Landslides on the Mississippi River Delta-Front Slope," Geoscience and Man, Volume 29, Number 1, pp. 41-53.

RICHARDS, A. F. and J. M. PARKS, 1976, "Marine Geotechnology: Average Sediment Properties, Selected Literature and Review of Consolidation, Stability and Bioturbation - Geotechnical Interactions in the Benthic Layer," Proceedings of the NATO Science Committee Conference on the Benthic Boundary Layer, Les Arcs, France, 1976, pp. 157-181. New York, New York: Plenum Press.

RODINE, J. D., 1975, Analysis of the Mobilization of Debris Flows, PhD Thesis, Stanford University, Department of Geology.

RODINE, J. D. and A. M. JOHNSON, 1976, "The Ability of Debris Heavily Freighted with Coarse Clastic Material to Flow on Gentle Slopes," Sedimentology, Volume 23, Number 2, pp. 213-234.

SCHAPERY, R. A. and W. A. DUNLAP, 1978, "Prediction of Storm-Induced Sea Bottom Movement and Platform Forces," Proceedings of the Tenth Offshore Technology Conference, Houston, Texas, 1978, Paper OTC 3259, pp. 1789-1796. Dallas, Texas: American Institute of Mining, Metallurgical and Petroleum Engineers, Inc.

SCHMERTMANN, J. S., 1975, "Measurement of In-Situ Shear Strength: State of the Art Report," Proceedings of the American Society of Civil Engineers Speciality Conference on In-Situ Measurement of Soil Properties, Raleigh, North Carolina, 1975, Volume 2, pp. 57-138. New York, New York: American Society of Civil Engineers.

SOKOLOVSKII, V. V., 1955, Theorie Der Plastizitat. Berlin, Germany: Veb Verlag Technik Press.

TSUI, K. K., 1972, Stability of Submarine Slopes, PhD Thesis, Queens University, Ontario, Canada, Department of Civil Engineering.

VALLEJO, L. E., 1979, "An Explanation for Mudflows," Geotechnique, Volume 29, Number 3, pp. 351-354.

VALLEJO, L. E., 1980, "Mechanics of Mudflow Mobilization on Low Angled Clay Slopes," Engineering Geology, Volume 16, Number 1, pp. 63-70.

VALLEJO, L. E., 1981a, "Determination of the Shear Strength of Granulo-Viscous Materials Using the Vane Shear Apparatus and the Two-Phase Suspension Theory," Proceedings of the International Conference on the Mechanical Behaviour of Structured Media, Ottawa, Canada, 1981, Volume B, pp. 373-381. Amsterdam, The Netherlands: Elsevier Scientific Publishing Company.

VALLEJO, L. E., 1981b, "Stability Analysis of Mudflows on Natural Slopes," Proceedings of the Tenth International Conference on Soil Mechanics and Foundation Engineering, Stockholm, Sweden, 1981, Volume 3, pp. 541-544. Rotterdam, The Netherlands: A. A. Balkema Press.

VESIC, A.S., 1972, "Expansion of Cavities in Infinite Soil Mass," Journal of the Soil Mechanics and Foundation Engineering Division, American Society of Civil Engineers, Volume 98, Number SM3, pp. 265-290.

Technical Poster Sessions

HYDRODYNAMICS

Hydrodynamic Issues Related to Deep Water Structures

EFFECT OF WAVE SPECTRAL SHAPE AND DIRECTIONAL VARIABILITIES ON THE DESIGN AND ANALYSIS OF MARINE STRUCTURES

HENRY CHEN

BROWN & ROOT, INC.

SUMMARY

The offshore industry has traditionally used the concepts of design wave height and period or wave spectra to describe environmental extremes and operational conditions for marine structure design and analysis prediction. The design wave height and period approach is typically used in offshore jacket designs. It proved to be both attractive and satisfactory due to its simplicity and the less sensitive frequency dependent nature of typical jacket structures. The wave spectra approach, on the other hand, has been widely used in floating vessel designs (e.g., ships, barges, and semi-submersibles). This approach generally utilizes a few theoretical wave spectral formulae to describe the wave energy distribution in the frequency domain. Since the theoretical wave spectra was derived with emphasis on the frequency range of ship responses, the approach has been proven satisfactory in most design applications of floating structures.

With the recent industry emphasis of deep water production techniques, new platform concepts such as Guyed Tower and Tension Leg Platform have been under intense study. Analytical results show that the response of these types of structure is extremely sensitive to wave frequency as well as direction. The traditional approach could lead to both overestimate and underestimate of responses compared to model test data. The consequences of these uncertainties usually results in large cost increase or unwarranted risk exposure of the structure.

This paper describes an approach using the First Order Uncertainty Analysis to account for the variations of spectral shapes and directionalities which have significant influences on the wave-induced responses of marine structures. For a specific level of sea severity, the wave energy distribution in frequency as well as directions are treated as random vectors which can be characterized by its mean and covariance matrix.

The resulting characterization of the wave spectra can then be used to estimate the unbiased parameters of an assumed root-mean-square response distribution for long-term prediction. Exact results for the Rayleigh-Normal distribution have been derived using theory of random functions and the linear superposition principle of frequency domain analysis. Examples using a large number of hindcast spectra to calculate the extreme responses and fatigue life of a TLP tether confirm the assumption and demonstrate the importance of wave spectral shape and directionality in response prediction.

An approach for computing the long-term responses of marine structures is outlined in view of the availability of the large amounts of hindcast directional spectra covering more than two thousand grid points and fifteen years of continuous data for the Northern Hemisphere.

INTRODUCTION

Frequency domain analysis utilizes theoretical point spectra coupled with a spreading function to evaluate barge motion and jacket fatigue characteristics. While such theoretical spectral formulation represents the general mean wave conditions in a wind generated sea, they do not account for the random variations from its mean value in spectral shape (even though the significant wave heights are the same). This may be caused by a variety of oceanographic conditions such as duration and fetch of wind, stage of growth and decay of sea and the presence of swell, etc. Furthermore, since the spectra may vary considerably according to geographical location and season of the year, a single theoretical spectral formulation may not be realistically applied in predicting responses of a structure which may encounter an infinite variety of wave conditions during its lifetime. Hence, unless the random nature of the wave energy distributions, both in the frequency and direction, is properly reflected in the prediction technique, there exists some reservation with regard to the confidence of the predicted responses which are significantly influenced by the shape and type of spectra used.

One approach to account for such variations in wave spectral energy contents is the use of the wave spectra family concept (1, 2). Instead of using a single theoretical spectrum to represent a given sea state, several spectra of the same level of sea severity as characterized by its significant wave height and period are used. The selected spectra thus represent the variation of spectral energy contents due to long term variation of sea/swell directionality and other oceanographic phenomena.

The methods for selecting spectra may vary; some will utilize Monte Carlo techniques in generating random samples of measured/hindcast spectra within a wave height group to match the spectral properties of the group (1), while the other uses different parameters of theoretical spectral formulation based on statistical analysis of recorded data (2).

Unfortunately, while these approaches attempted to take full advantage of the available data, they may produce biased estimates unless a large number of sample spectra are used. Furthermore, the parameterization of measured spectra also eliminates the detailed information on the wave energy frequency/directional distribution.

Recent release of the 20 year Spectral Ocean Wave Model (SOWM) hindcast project tapes (3) by Fleet Numerical Oceanographic Center (FNOC) of the U.S. Navy has made this unique source of long term wave spectral data available to the public. The tapes contain millions of hindcast directional wave spectra covering close to 2000 grid points in the Northern Hemisphere.

The vast amount of data both in terms of sample size and geographical coverage thus provides a unique opportunity to investigate the probabilistic nature of wave energy distributed over the frequency as well as the direction, and consequently the effect on the short and long term prediction of wave-induced responses of marine structures.

SPECTRA FAMILY CONCEPT

The concept of spectral family was originally developed for estimating the long term extreme midship bending moment (4), and later extended to other motions (1), particularly when the responses of an offshore vessel are considered to be linear and the superposition principle is valid. The short term peak responses may be adequately described by a Rayleigh distribution for a narrow band process. By considering a family of spectra of the same seastate severity, the effect of various spectral shapes can be accounted for by the variations of mean square values of the response. Furthermore, in order to include this long term varability in the prediction technique, the probability density function of the peak responses are assumed to be a combined Rayleigh-Normal distribution (5).

In other words, the Rayleigh distribution parameter (i.e. Root-Mean-Square value of the responses) itself is normally distributed.

A combined Rayleigh-Normal distribution is given by:

$$P(R>R_o) = \int_0^\infty \exp(-R_o^2/2x^2) \, f(x) \, dx \quad \ldots \ldots \ldots \quad (1)$$

where,

$$f(x) = \frac{1}{\sqrt{2\pi} \, \sigma_x} \exp((x-\mu_x)^2 / \sigma_x^2) \quad \ldots \ldots \ldots \quad (2)$$

where,

- R_o = peak response value
- X = Root-Mean-Square Value $\sqrt{M_o}$
- μ_X = Mean of X
- σ_X = Standard deviation of X

It should be noted that the effect of parameter variation of a Rayleigh distribution could lead to significant underestimation of peak response level for the same probability level in a Rayleigh distribution as shown in Figures 1 and 2. Furthermore, the change in the probability density function would also have substantial impact on the fatigue life calculation of offshore structures.

ESTIMATION OF THE DISTRIBUTION PARAMETERS

In order to numerically compute the probability levels of the extreme response in equations (1) and (2), the parameters of the combined Rayleigh-Normal distribution must be estimated. Ideally, the mean and variance of the R.M.S. value, μ_X and σ_X^2 should be obtained statistically by superpositioning a large number of wave spectral samples of the same sea state. However, this may not be practical if the data involves hundreds and thousands of hindcast spectra in order to obtain an adequate representation of the stochastic variations of spectral energy contents over frequency as well as direction representing many years of wave conditions at a specified offshore site.

To account for the random variation of wave energy over the frequency as well as direction domains, the First Order Uncertainty Analysis technique (6) may be used. For a given sea severity characterized by a range of significant wave height, the probabilistic distribution of the spectral energy content each frequency and direction band can be approximated, up to first order, by their mean and covariance matrix. If there are N samples in the wave height group, the normalized mean spectral energy at 15 frequency and 12 direction bands are given by:

$$\overline{E}_i = \sum_{n=1}^{N} \frac{(E_i/ETOT_i)_n}{N} \quad \text{for} \quad i = 1, 2, 3 \ldots 180 \quad (3)$$

where, $ETOT_i$ = Total wave energy of the spectrum i. It is equivalent to the mean square value ($H_{1/3} = 4\sqrt{ETOT}$).

The covariance matrix of each energy pair is given by

$$\Gamma_{i,j} = \frac{1}{(N-1)} \sum_{n=1}^{N} (E_i - \overline{E}_i)(E_j - \overline{E}_j) \quad (4)$$

$$\text{for} \quad i = 1, 2 \ldots 180$$
$$j = 1, 2 \ldots 180$$

To estimate the mean and variance of the mean square value response (M_o) superposition principle is applicable for a linear, time-invariant system.

Let y denote M_o, m_y denote the estimate of the mean of y, μ_y.

Thus

$$M_y = \sum_{n=1}^{N} \frac{y_n}{N} = \sum_{n=1}^{N} \sum_{i=1}^{180} RAO_i^2 (E_i)_n / N$$

$$= \sum_{i=1}^{180} RAO_i^2 \sum_{n=1}^{N} \frac{(E_i)_n}{N} = \sum_{i=1}^{180} RAO_i^2 \overline{E}_i \quad (5)$$

where

RAO_i = Response Amplitude Operator the system at frequency/direction i.

For the variance, the estimate of σ_y, as denoted by S_y is given by:

$$S_y^2 = \sum_{n=1}^{N} (y_n - m_y)^2 / (N-1)$$

$$= \frac{1}{(N-1)} \sum_{n=1}^{N} \sum_{i=1}^{180} RAO_i^2 [(E_{in} - \overline{E}_i)]^2 \qquad (6)$$

by mathmatical induction, it can be shown that the above equation reduces to:

$$S_y^2 = \sum_{i=1}^{180} \sum_{j=1}^{180} (RAO_i^2 \times RAO_j^2 \times \Gamma_{ij}) \qquad (7)$$

where

Γ_{ij} = covariance matrix of spectral energy for various frequency/direction pairs from equation (4).

The above equations describes a method to calculate the mean and variance of the mean sequence value y. However, it is assumed that the Root-Mean-Square value, $\sqrt{y}$ follow a Normal distribution. To estimate its parameter, a new estimation procedure utilizing the moment generating functions and the unique properties of Normal distribution functions has been developed. The values of μ_x and σ_x^2 can be estimated in terms of the mean and variance of the mean square response value m_y and S_y^2 as derived above:

$$y = M_o$$

$$m_x = (m_y - S_y^2/2)^{1/4} \qquad \cdots \cdots \cdots \cdots \quad (8)$$

$$S_x^2 = m_y - (m_y^2 - S_y^2/2)^{1/2} \qquad \cdots \cdots \cdots \cdots \quad (9)$$

A detailed derivation of the above transform pair is shown in the Appendix.

The above approach thus provides an efficient mean of estimating the distribution parameters without calculating the responses for the entire sample spectra. The use of covariance matrix of spectral ordinates not only considers the correlations of wave energy over frequency and direction, but also accounts for the statistical uncertainties according to the number of samples available in each wave height group.

It should be noted, however, that since the above derivations are based on the Normal distribution of the R.M.S. value of responses, the results are exact only if the assumption holds. Before the above procedure can be fully implemented, therefore, the underlying assumption must be checked and verified.

EVALUATION OF THE ESTIMATION TECHNIQUES

With the availability of the hindcast wave spectra tape, a unique opportunity has been presented for testing the new estimation technique. Over 1400 directional spectra at grind point GP 83 Northern North sea (Longitude = 2.401 East, Latitude = 61.918 North). Were retrieved from the hindcast tapes. The data covers a period of about two years (1973 - 1974).

A set of typical tether force Response Amplitude Operator (RAO) of a Tension Leg Platform (TLP) design were used to compute the response together with the hindcast spectra. The stress RAO's were computed for 12 direction and 15 frequencies using an in-house computer program TENMOT for TLP frequency domain motion prediction.

The above derived estimation technique was implemented by grouping the wave spectra into 9 groups according to specified significant wave height range. The number of samples in each group varies from 68 in high seastates to several hundred in lower seastates. To eliminate the total spectral energy variation within the group, each directional wave spectrum was normalized by its total mean square energy. The normalized spectrum thus represents the distribution of wave energy over frequencies as well as directional domains. The mean and covariance wave spectra were then calculated based on the normalized spectra for each group.

For comparison purpose, the mean and standard deviation of the RMS responses were estimated by using the new approach as well as the straightforward superposition of all the sample spectra within the group. Table 1 shows the comparison of the results by using both methods.

TABLE 1. Comparison of Parameter Estimates for Selected Wave Height Groups

Sig. Wave Ht Range (ft)	No. of Samples	Mean RMS/H13	Std. Deviation[1] (% of Mean)	Std. Deviation[2] (% of Mean)
0 - 3	158	4.92	1.36 (27.6%)	1.13 (22.9%)
3 - 6	289	5.08	1.19 (23.4%)	1.12 (22.0%)
9 - 11	189	5.48	0.439 (8.0%)	0.447 (8.2%)
20 - 25	98	6.39	0.335 (5.25%)	0.343 (5.37%)
25 - 35	68	6.74	0.438 (6.49%)	0.443 (6.57%)

NOTE:
1. Calculated statistically by superposition of the entire sample spectra in the group.
2. Calculated by using mean and covariance energy spectra.

It is interesting to note that the new approach is quite satisfactory as far as estimating the standard deviation of the RMS response. The mean RMS response is of course exactly the same as compared to the statistical approach.

To verify the assumed Normal distribution of the RMS responses, histogram of the RMS responses were plotted for several wave height groups. Figures 3 and 4 show two typical plots.

In most cases when the Normal distribution assmption holds, the mean-covariance estimates agree well with the sample estimates. Although in certain cases assumption is less accurate. The Chi-square statistics show that they can be accepted within 95% confidence level.

The Rayleigh-Normal distribution of the peak stress responses has been implemented in a probablistic fatigue program for TLP tethers. By introducing the uncertainties of the Rayleigh distribution parameter (i.e., the RMS value of the response), the predicted fatigue life reduced by as much as 20 percent for the case tested. The reduced fatigue life is primarily due to the high tether stress responses at high wave frequencies and the increased probability level of peak responses as a consequence of the parameter uncertainty introduced in the Rayleigh distribution.

CONCLUDING REMARKS

The evaluation of a deepwater structure design require many analytical tools. Amongst the available tools, frequency domain analysis proved to be the most cost effective in preliminary design. Although the technique ideally assumes the system to be linear and time invariant, extreme response value can be readily extrapolated without the time-consuming simulations in time domain.

The paper presents a new approach to accounting for the random variation of wave energy frequency/direction distributions which play a significant role in predicting responses of marine structures in a seaway, and outlined a methodology obtaining long term response statistics for design evaluations. In applying the wave information to evaluate responses of deep water structures an estimation technique was developed in conjunction with the first order uncertainty analysis and theory of random functions. The results of numerical computations carried out using the responses of a TLP tether with a large hindcast spectral wave data base as input demonstrates the effect of such variabilities and the potential use of the data base in the design and evaluation process of marine systems. Conclusions and recommendations for continual research efforts based on the results of the present study are summarized below.

1. The first order uncertainty analysis of the wave spectral information represents one of the most efficient and comprehensive means of processing the vast number of wave spectral data. As far as the information contents are concerned, the mean and

covariance of spectral energy distribution, though is only correct to first order and incomplete, is at least consistent and meaningful in view of the superposition principle of the frequency domain approach.

2. A new approach utilizing the mean-covariance of spectral energies in each wave height group has been developed for estimating the mean and variance of the R.M.S. values to facilitate the long term prediction of wave induced responses. The approach relies on the assumption of Normal distribution R.M.S. value responses to derive the exact results of transforming the estimates from M.S. to R.M.S. domain. Figure 5 shows a schematic diagram of the approach. A comparison of the estimated mean and variance using the new approach and the straight sample estimates by superpositioning over 1400 spectra shows the effectiveness of the approach in obtaining the response statistics.

3. Since the results are correct only if the Normal distribution assumption holds, verification of the assumption has been carried out for the TLP tether response to a satisfactory degree with the availability of the large number of hindcast spectral samples. Consequently, the results have rendered, to a large extent, the rationale of the prediction method for calculating the long term wave induced response of marine systems in a seaway.

4. The advantages of the approach are obvious with its complexity order of magnitude less than the full p.d.f. description of the wave energy distribution (if obtainable) and its computational efficiency compared to the straight superpositioning of all the samples if unbiased estimates are desired. In view of the great potential use of the hindcast wave data because of its detailed spectral content and large geographical coverage, the mean-covariance representation of hindcast directional spectra at a particular location, on monthly, seasonal or yearly basis seems to be one of the most efficient ways of storing wave information and the subsequent application in design and evaluation process of marine systems.

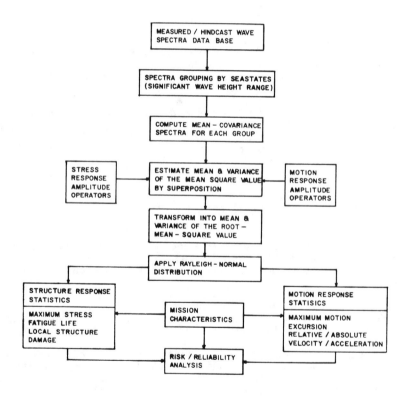

Figure 5 An approach for Long-Term response prediction and design evaluation of marine structures.

APPENDIX

TRANSFORMATION OF THE MEAN AND VARIANCE FROM M.S. TO R.M.S. DOMAIN

Since $y = g(x) = x^2$, its expected value is given by

$$E(g(x)) = \int_{-\infty}^{\infty} g(x) f(x) dx \qquad (A.1)$$

Furthermore, for a normal random variable, x, with mean, m_x, and variance S_x^2, it can be proved, using characteristic functions, that,

$$\frac{\partial^n E(g(x))}{\partial (v)^n} = \frac{1}{2^n} E\left(\frac{d^{2n} g(x)}{dx^{2n}}\right) \qquad v = S_x^2 \qquad (A.2)$$

Let the kth moment of x be defined as

$$M_k = E(x^k)$$

by using equation (A.2) recursively, it can be shown that

$$\frac{\partial M_k(m_x, v)}{\partial v} = \frac{k(k-1)}{2} M_{k-2}(m_x, v) \qquad (A.3)$$

with initial value

$$M_k(m_x, 0) = m_x^k \qquad (A.4)$$

Hence

$$M_k(m_x, v) = \frac{k(k-1)}{2} \int_0^v M_{k-2}(m_x, v) dv + m_x^k$$

Since $M_0 = 1$, the above equation yields

$$M_2 = v + m_x^2 \qquad (A.5)$$

$$M_4 = \frac{4 \times 3}{2} \int_0^v (v + m_x^2) dv + m_x^4$$
$$= 3v^2 + 6 v m_x^2 + m_x^4 \qquad (A.6)$$

By definition

$$M_2 = E(x^2) = E(y)$$
$$M_4 = E(x^4) = E(y^2)$$

Therefore, from equations A.5 and A.6 it can be seen

$$m_y = S_x^2 + m_x^2 \qquad (A.7)$$

$$S_y^2 = E(y^2) - E(y)^2 = 3S_x^4 + 6S_x^2 m_x^2 + m_x^4 - (S_x^2 + m_x^2)^2$$
$$= 2S_x^4 + 4S_x^2 m_x^2 \qquad (A.8)$$

Solving the above equations, and choosing the correct sign, the transform pair finally becomes

$$S_x^2 = m_y - (m_y^2 - \frac{S_y^2}{2})^{1/2} \qquad (A.9)$$

$$m_x = (m_y^2 - \frac{S_y^2}{2})^{1/4} \qquad (A.10)$$

REFERENCES

1. Hoffman, D. and Walden, D.A., "Environmental Wave Data for Determining Hull Structural Loadings", SSC report 268, 1977.

2. Ochi, M.K., "Wave Statistics for the Design of Ships and Ocean Structures," SNAME Annual Meeting, New York, 1978.

3. Chen, H.T., Hoffman, D. and Chen, H.H., "Implementation of the 20 Year Hindcast Wave Data in the Design and Operation of Marine Structures," OTC Paper 3644, to be presented at the Offshore Technology Conference, May 1979.

4. Band, E.G.U., "Analysis of Ship Data to Predict Long Term Trends of Hull Bending Moments," Webb report to ABS, November 1966.

5. Lewis, E.V., "Predicting Long-Term Distribution of Wave-Induced Bending Moments on Ship Hulls," SNAME Spring Meeting, 1967.

6. Benjamin, J.R., and Cornell, C.A., <u>Probability, Statistics and Decision for Civil Engineers</u>, McGraw-Hill Pub. Co., 1970.

ACKNOWLEDGEMENTS

The author wishes to thank Brown and Root management for their encouragement and support of this effort. The assistance provided by the Marine Research and Development Department is greatly appreciated.

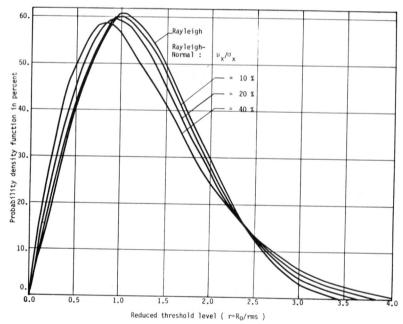

Figure 1 Effect of parameter uncertainty on the Rayleigh distribution probability density function.

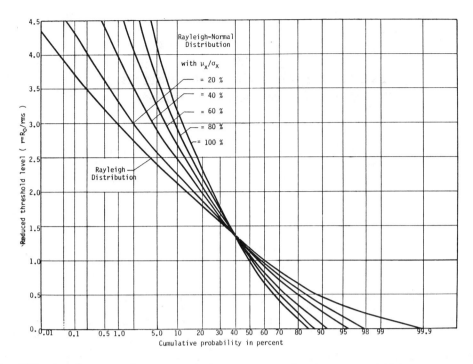

Fig. 2 Effect of parameter uncertainty on cumulative probability of Rayleigh distribution

OBSV FREQ	RELA FREQ	CUML FREQ	UPPER CELL LIMIT
0	0.0	0.0	0.4000E 01
0	0.0	0.0	0.4250E 01
0	0.0	0.0	0.4500E 01
10	0.053	0.053	0.4750E 01
16	0.085	0.138	0.5000E 01
29	0.153	0.291	0.5250E 01
44	0.233	0.524	0.5500E 01
46	0.243	0.767	0.5750E 01
23	0.122	0.889	0.6000E 01
13	0.069	0.958	0.6250E 01
3	0.016	0.974	0.6500E 01
3	0.016	0.989	0.6750E 01
2	0.011	1.000	0.7000E 01
0	0.0	1.000	0.7250E 01
0	0.0	1.000	INF

189			

TEST FOR NORMALITY

Sample mean = 5.48

Std. Dev. = 0.439

$\chi^2 = 4.59$

d.f. = 9

$\chi^2_{95} = 16.9$

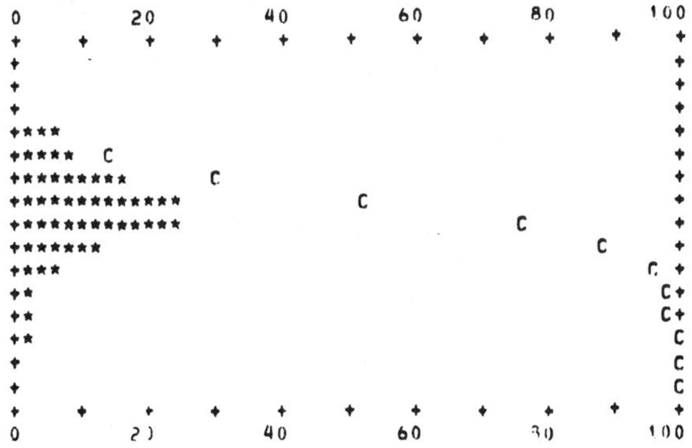

Fig. 3 Histogram of RMS/H1/3 value for wave height range 9-11 ft.

OBSV FREQ	RELA FREQ	CUML FREQ	UPPER CELL LIMIT
0	0.0	0.0	0.5250E 01
0	0.0	0.0	0.5500E 01
0	0.0	0.0	0.5750E 01
1	0.015	0.015	0.6000E 01
6	0.088	0.103	0.6250E 01
19	0.279	0.382	0.6500E 01
13	0.191	0.574	0.6750E 01
9	0.132	0.706	0.7000E 01
11	0.162	0.868	0.7250E 01
5	0.074	0.941	0.7500E 01
4	0.059	1.000	0.7750E 01
0	0.0	1.000	0.8000E 01
0	0.0	1.000	0.8250E 01
0	0.0	1.000	0.8500E 01
0	0.0	1.000	0.8750E 01
0	0.0	1.000	0.9000E 01
0	0.0	1.000	0.9250E 01
0	0.0	1.000	0.9500E 01
0	0.0	1.000	0.9750E 01
0	0.0	1.000	INF

68			

TEST FOR NORMALITY

Sample mean = 6.74

Std. Dev. =0.438

$\chi^2 = 13.8$

d.f. = 7

$\chi^2_{95} = 14.1$

Fig. 4 Histogram of RMS/H1/3 value for wave height range 25-35 ft.

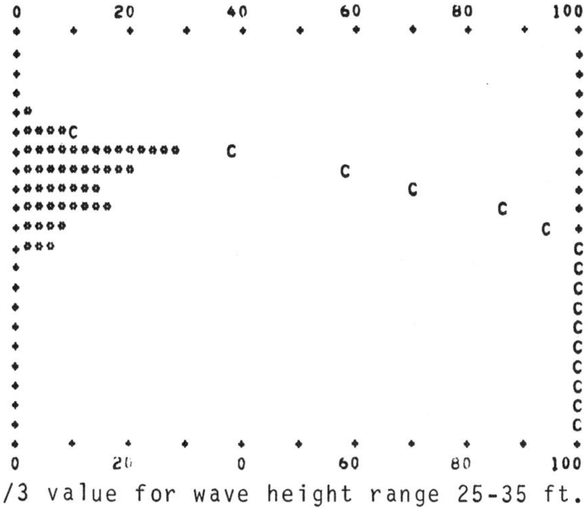

ESTIMATES OF CROSS-SPECTRAL DENSITIES OF WAVE FORCES ON INCLINED CYLINDERS

M.C. Deo
Indian Institute of Technology
Bombay

S. Narasimhan
Indian Institute of Technology
Bombay

SUMMARY

This paper gives a new mathematical procedure to estimate the cross-covariance and spectral density functions of ocean wave forces exerted at any two points on the inclined cylindrical members of the offshore structures. The cylindrical members or piles are assumed to have any random orientation with respect to the vertical. It is also assumed that they are subjected to directional seas. The necessity and scope of the present work is discussed first. The description of the formulation starts with explaining how the basic wave force equations are formed for the inclined piles along three directions. Then some expressions of cross-covariance and spectral density functions between various wave flow kinematic components in a three-dimensional space are stated. The next section describes how the equations for cross-correlation functions of wave force components can be obtained by direct multiplication of force equations. The Fourier transforms of these equations give the cross-spectra of wave forces - which is explained in the same section. The corollaries of the formulation described so far, like statistics for measurements made at a single point, unidirectional conditions, are then indicated. Based upon the preceding force spectral equations, a mathematical procedure is presented in the subsequent section to obtain estimates of the wave force coefficients, C_D and C_M, for inclined piles in directional seas for three different directions. Finally a numerical example is given to show how the theory described in this paper can be applied in practice.

1. INTRODUCTION

The three different techniques available to the designer for carrying out the structural analysis of an offshore structure are static, dynamic and stochastic. The last approach, that is variedly known as probabilistic or spectral approach is replacing the first two approaches rapidly, these days, as it is more realistic and accurate. The stochastic analysis of offshore structures requires the wave forcing function, often in the form of the spectral density relationships of forces exerted at any two points on the piles, as a major input. It is therefore obvious that such analysis will be more appropriate and meaningful if the wave forcing function is given in a more realistic and generalized form.

Borgman (1967) gave the expressions to obtain the wave force spectra from the sea surface spectra for the first time, on the assumptions that (a) the waves were unidirectional, (b) surface spectrum had only one dimension $[S(f)]$, and (c) the piles were vertical. He thus gave the expressions for cross-covariances and spectral densities along the x-direction only and for vertical cylinders. In actual practice, however, the wave energy has a directional spread. This was taken into account by Y Fantis and Borgman (1978), who obtained the cross-spectral density functions for forces on vertical piles along two directions x and y in a horizontal plane using the directional spectrum, $S(f,\theta)$, as input.

Both of these preceding formulations assume that the piles over which the forces are being exerted as purely vertical. From the structural and other considerations however the piles of offshore structures are seldom truly vertical and are often inclined. The present studies are therefore aimed at evolving a general formulation to obtain the force spectra for 'inclined' piles along three principal coordinate directions - x, y, z using a directional spectrum $S(f,\theta)$.

2. THE WAVE FORCE TIME HISTORIES

It is well known that the force exerted on a vertically placed circular cylinder that is subjected to wave action consists of a drag as well as an inertia component. Morison et al (1950) combined the two force terms in linear vector form to obtain the total wave force for unit length of the circular cylinder of diameter D as

$$\bar{f}(t) = C|\bar{v}(t)|\bar{v}(t) + K\bar{a}(t) \qquad \ldots(2.1)$$

$$\text{where } C = C_D \rho D \qquad \ldots(2.2)$$
$$K = C_M \rho \pi D^2/4 \qquad \ldots(2.3)$$

where C_D and C_M are coefficient of drag and inertia, ρ is the mass density of water and $\bar{v}(t)$ and $\bar{a}(t)$ are velocity and acceleration of water particles at time 't'. This equation assumes that the velocity, $\bar{v}(t)$ and the acceleration, $\bar{a}(t)$ of the wave particles are horizontal and exactly normal to the vertical axis of the cylinder. In case of cylinders that are not vertical, Morison's equation can be interpreted and implemented in many ways depending upon which factor mainly contributes to the effective wave force. Wade and Dwyer (1976) have shown that forces predicted by any of these methods, that are currently being used by the industry, do not differ much - leaving the choice of any one of them to the discretion of the designer.

Borgman (1958) assumed that the force normal to the axis of the inclined pile is contributed by the normal velocity and acceleration components and gave the wave force on the inclined pile as:

$$\bar{f}_n(t) = C|\bar{v}_n(t)|\bar{v}_n(t) + K\bar{a}_n(t) \qquad \ldots(2.4)$$

where the subscript 'n' refers to the normal direction. Strictly, this exact formula should be used in the derivations described subsequently. However, it was found that the nonlinear drag term involved in it makes the computations for correlations and spectral densities extremely difficult and intractable. Recourse is made, therefore, to the linearization as suggested by Borgman (1967) which, for the case of vertical cylinder is

$$\bar{f}(t) = C(8/\pi)^{\frac{1}{2}} \sigma \bar{v}(t) + K\bar{a}(t) \qquad \ldots(2.5)$$

where σ is the root-mean-square velocity. In case of inclined cylinders, it is the normal velocity and acceleration that are considered. Hence after linearization, equation (2.4) becomes

$$\bar{f}_n(t) = C(8/\pi)^{\frac{1}{2}} \sigma_n \bar{v}_n(t) + K\bar{a}_n(t) \qquad \ldots(2.6)$$

where σ_n is the root-mean-square value of the normal velocity. The linearized version is found to be sufficiently accurate for normal range of velocities (Borgman, 1967, 1976) and is therefore more appropriate in the present correlation studies that deal with such velocities only that are oriented towards their mean values. The linearized formula has been used very successfully, so far, in response studies using probabilistic techniques by several investigators (Penzien and Tseng, 1978, Y Fantis and Borgman, 1978). Moreover it is possible to get rid of the linearization error in subsequent response calculations.

The normal vectors $\bar{v}_n(t)$ and $\bar{a}_n(t)$ in the equation (2.6) could be expressed in terms of a unit vector $\bar{c}$ along the cylinder axis as follows:

$$\bar{v}_n(t) = \bar{c} \times (\bar{v}(t) \times \bar{c}) \qquad \ldots(2.7)$$

$$\bar{a}_n(t) = \bar{c} \times (\bar{a}(t) \times \bar{c}) \qquad \ldots(2.8)$$

The velocity, acceleration and unit axial vectors can be written in terms of their components $\bar{v}_x(t)$, $\bar{v}_y(t)$, $\bar{v}_z(t)$; $\bar{a}_x(t)$, $\bar{a}_y(t)$; $\bar{a}_z(t)$; $\bar{c}_x, \bar{c}_y, \bar{c}_z$ respectively along x, y and z axes of a right handed coordinate system with 'z' vertically upwards [Fig.1]. The values of $\bar{v}_n(t)$ and $\bar{a}_n(t)$ can be thereafter substituted from equations (2.7) and (2.8) into equation (2.6) to get the equations for the three wave force components as under:

$$\begin{Bmatrix} F_x(t) \\ F_y(t) \\ F_z(t) \end{Bmatrix} = C(8/\pi)^{\frac{1}{2}} \sigma_n \begin{Bmatrix} (1-c_x^2)v_x(t) + (-c_xc_y)v_y(t) + (-c_xc_z)v_z(t) \\ (-c_xc_y)v_x(t) + (1-c_y^2)v_y(t) + (-c_yc_z)v_z(t) \\ (-c_xc_z)v_x(t) + (-c_yc_z)v_y(t) + (1-c_z^2)v_z(t) \end{Bmatrix}$$

$$+ K \begin{Bmatrix} (1-c_x^2)a_x(t) + (-c_xc_y)a_y(t) + (-c_xc_z)a_z(t) \\ (-c_xc_y)a_x(t) + (1-c_y^2)a_y(t) + (-c_yc_z)a_z(t) \\ (-c_xc_z)a_x(t) + (-c_yc_z)a_y(t) + (1-c_z^2)a_z(t) \end{Bmatrix} \qquad \ldots(2.9)$$

where the numerical value of the normal velocity is given by

$$v_n(t) = \left\{ v_x^2(t) + v_y^2(t) + v_z^2(t) - [c_x v_x(t) + c_y v_y(t) + c_z v_z(t)]^2 \right\}^{\frac{1}{2}} \qquad \ldots(2.10)$$

3. COVARIANCE AND SPECTRAL DENSITY FUNCTIONS FOR WAVE KINEMATICS

The instantaneous values of velocities and acceleration components can be expressed in terms of directional spectrum $S(f,\theta)$ for frequency, f, and angle of incidence, θ; in terms of the pseudo-integrals as follows:

$$v_x(t) = 4\pi \int_0^\infty \int_0^{2\pi} [S(f,\theta) \, df \, d\theta]^{\frac{1}{2}} f \frac{\cosh k(d+z)}{\sinh kd} \cos\theta \cos(kx\cos\theta + ky\sin\theta - 2\pi ft + \phi) \qquad \ldots(3.1)$$

$$v_y(t) = 4\pi \int_0^\infty \int_0^{2\pi} [S(f,\theta) \, df \, d\theta]^{\frac{1}{2}} f \frac{\cosh k(d+z)}{\sinh kd} \sin\theta \cos(kx\cos\theta + ky\sin\theta - 2\pi ft + \phi) \qquad \ldots(3.2)$$

$$v_z(t) = 4\pi \int_0^\infty \int_0^{2\pi} [S(f,\theta) \, df \, d\theta]^{\frac{1}{2}} f \frac{\sinh k(d+z)}{\sinh kd} \sin(kx\cos\theta + ky\sin\theta - 2\pi ft + \phi) \qquad \ldots(3.3)$$

The equations for the acceleration components are obtained by differentiating the above equations with respect to time. In the above equations, 'k' is the wave number, 'd' is the water depth, and 'd+z' is the distance of the point at which above quantities are considered from the sea bottom [Fig.1], ϕ is the random phase.

Let (x_n, y_n, z_n, t) and $(x_m, y_m, z_m, t+\tau)$ be any two points in the wave field. Then the cross-covariance functions in between various combinations of velocity and acceleration components are obtained by direct multiplication of the expressions (Borgman, 1967). Though in the present theory all such combinations are used, following typical examples are presented here to indicate their nature:

$$R_{v_{xn}v_{ym}}(\tau) = 8\pi^2 (C_n C_m) \int_0^\infty \int_0^{2\pi} S(f,\theta) f^2 \cos\theta \sin\theta \cos\varphi \, d\theta \, df \qquad \ldots(3.4)$$

where $R_{v_{xn}v_{ym}}(\tau)$ = the cross-covariance function in between v_{xn} and v_{ym} for time lag 'τ',

v_{xn} = velocity in x-direction at point 'n'; v_{ym} = velocity in y-direction at point 'm'

$$C_n = \frac{\cosh k(d+z_n)}{\sinh kd} \qquad \ldots(3.5)$$

$$C_m = \frac{\cosh k(d+z_m)}{\sinh kd} \qquad \ldots(3.6)$$

$$\varphi = P - 2\pi f \tau \qquad \ldots(3.7)$$

$$P = k(x_m - x_n)\cos\theta + k(y_m - y_n)\sin\theta \qquad \ldots(3.8)$$

Similarly we have,

$$R_{v_{xn} a_{zm}}(\tau) = -16\pi^3 (C_n S_m) \int_0^\infty \int_0^{2\pi} S(f,\theta) f^3 \cos\theta \cos\varphi \, d\theta \, df \qquad \ldots(3.9)$$

where $R_{v_{xn} a_{zm}}(\tau)$ = the cross-covariance function in between v_{xn} and a_{zm} for time lag 'τ'

a_{zm} = acceleration in z-direction at point 'm'

$$S_m = \frac{\sinh k(d+z_m)}{\sinh kd} \qquad \ldots(3.10)$$

It is noted that the identities such as one mentioned below hold

$$R_{a_{xn} a_{xm}}(\tau) = (2\pi f)^2 R_{v_{xn} v_{xm}}(\tau) \qquad \ldots(3.11)$$

The expressions for the corresponding cross-spectral densities in between the various flow kinematic components can be determined from the preceding cross-covariance functions, e.g.

$$S_{v_{xn} v_{ym}}(f) = c_{v_{xn} v_{ym}}(f) - i\, q_{v_{xn} v_{ym}}(f) \qquad \ldots(3.12)$$

where the co-spectral density, $c_{v_{xn} v_{ym}}(f) = 2\pi^2 f^2 (C_n C_m) \int_0^{2\pi} S(f,\theta) \sin 2\theta \cos P \, d\theta \qquad \ldots(3.13)$

the quadrature spectral density, $q_{v_{xn} v_{ym}}(f) = 2\pi^2 f^2 (C_n C_m) \int_0^{2\pi} S(f,\theta) \sin 2\theta \sin P \, d\theta \qquad \ldots(3.14)$

Similarly we have,

$$S_{v_{xn} a_{zm}}(f) = c_{v_{xn} a_{zm}}(f) - i\, q_{v_{xn} a_{zm}}(f) \qquad \ldots(3.15)$$

where the co-spectral density, $c_{v_{xn} a_{zm}}(f) = -8\pi^3 f^3 (C_n S_m) \int_0^{2\pi} S(f,\theta) \cos\theta \cos P \, d\theta \qquad \ldots(3.16)$

the quadrature spectral density, $q_{v_{xn} a_{zm}}(f) = -8\pi^3 f^3 (C_n S_m) \int_0^{2\pi} S(f,\theta) \cos\theta \sin P \, d\theta. \qquad \ldots(3.17)$

Consistent with equation (3.11) we have

$$S_{a_{xn} a_{xm}}(f) = (2\pi f)^2 S_{v_{xn} v_{xm}}(f) \qquad \ldots(3.18)$$

The next section will indicate how the preceding covariances and spectral densities are used in getting the wave force spectra.

4. THE CROSS-COVARIANCE AND SPECTRAL DENSITY RELATIONSHIPS FOR THE WAVE FORCES

The covariance functions for the various combination of wave force components at the two points can be found in a way similar to that of velocities and accelerations mentioned in the preceding section (i.e. by direct multiplication) as follows.

Assuming that point 'n' is on the pile with direction cosines c_x, c_y, c_z and point 'm' is on another pile with direction cosines c'_x, c'_y, c'_z and with the same values of 'C' and 'K', we have,

$$E\{F_{xn}(t) \cdot F_{xm}(t+\tau)\} = R_{F_{xn}F_{xm}}(\tau) \qquad \ldots(4.1)$$

$$= E\{(8/\pi)^{\frac{1}{2}} C(\sigma_n)_n [(1-c_x^2)v_{xn} + (-c_x c_y)v_{yn} + (-c_x c_z)v_{zn}] + K[(1-c_x^2)a_{xn} + (-c_x c_y)a_{yn} + (-c_x c_z)a_{zn}]\} \times$$

$$\{(8/\pi)^{\frac{1}{2}} C(\sigma_n)_m [(1-c'^2_x)v_{xm} + (-c'_x c'_y)v_{ym} + (-c'_x c'_z)v_{zm}] + K[(1-c'^2_x)a_{xm} + (-c'_x c'_y)a_{ym} + (-c'_x c'_z)a_{zm}]\} \ldots(4.2)$$

where $E\{F_{xn}(t) F_{xm}(t+\tau)\}$ = expectation of the product of forces $F_{xn}(t)$ and $F_{xm}(t+\tau) = R_{F_{xn}F_{xm}}(\tau)$

$F_{xn}(t), F_{xm}(t+\tau)$ are forces in x-direction at point 'n' at time 't' and at point 'm' at time 't+τ' respectively

$(\sigma_n)_n$ and $(\sigma_n)_m$ are rms normal velocities at point 'n' and 'm' respectively.

The product in the equation (4.2) is expressed in a typical notation in the following equation:

$$R_{F_{xn}F_{xm}}(\tau) = \frac{8}{\pi} C^2 (\sigma_n)_n (\sigma_n)_m \begin{bmatrix} (1-c_x^2)(1-c'^2_x) & (1-c_x^2)(-c'_x c'_y) & (1-c_x^2)(-c'_x c'_z) \\ (-c_x c_y)(1-c'^2_x) & (-c_x c_y)(-c'_x c'_y) & (-c_x c_y)(-c'_x c'_z) \\ (-c_x c_z)(1-c'^2_x) & (-c_x c_z)(-c'_x c'_y) & (-c_x c_z)(-c'_x c'_z) \end{bmatrix} \begin{bmatrix} R_{v_{xn}v_{xm}} & R_{v_{yn}v_{xm}} & R_{v_{zn}v_{xm}} \\ R_{v_{xn}v_{ym}} & R_{v_{yn}v_{ym}} & R_{v_{zn}v_{ym}} \\ R_{v_{xn}v_{zm}} & R_{v_{yn}v_{zm}} & R_{v_{zn}v_{zm}} \end{bmatrix}$$

$$+ K^2 [C_{F_{xn}F_{xm}}] \begin{bmatrix} R_{a_{xn}a_{xm}} & R_{a_{yn}a_{xm}} & R_{a_{zn}a_{xm}} \\ R_{a_{xn}a_{ym}} & R_{a_{yn}a_{ym}} & R_{a_{zn}a_{ym}} \\ R_{a_{xn}a_{zm}} & R_{a_{yn}a_{zm}} & R_{a_{zn}a_{zm}} \end{bmatrix} + (8/\pi)^{\frac{1}{2}} KC(\sigma_n)_n [C_{F_{xn}F_{xm}}] \begin{bmatrix} R_{v_{xn}a_{xm}} & R_{v_{yn}a_{xm}} & R_{v_{zn}a_{xm}} \\ R_{v_{xn}a_{ym}} & R_{v_{yn}a_{ym}} & R_{v_{zn}a_{ym}} \\ R_{v_{xn}a_{zm}} & R_{v_{yn}a_{zm}} & R_{v_{zn}a_{zm}} \end{bmatrix}$$

$$+ (8/\pi)^{\frac{1}{2}} KC(\sigma_n)_m [C_{F_{xn}F_{xm}}] \begin{bmatrix} R_{a_{xn}v_{xm}} & R_{a_{yn}v_{xm}} & R_{a_{zn}v_{xm}} \\ R_{a_{xn}v_{ym}} & R_{a_{yn}v_{ym}} & R_{a_{zn}v_{ym}} \\ R_{a_{xn}v_{zm}} & R_{a_{yn}v_{zm}} & R_{a_{zn}v_{zm}} \end{bmatrix} \quad \text{for lag '}\tau\text{'}$$

$$\ldots(4.3)$$

The brackets, [...], in this equation are not the matrices in the conventional sense but used here only to denote a row-to-column multiplication. After one row-to-column multiplication is made the sum is added to the second row-to-column multiplication and so on such that the final result of multiplication of two such 'pseudo-matrices' is only a single term.

$$\text{In equation (4.3)} [C_{F_{xn}F_{xm}}] = \begin{bmatrix} (1-c'^2_x)(1-c_x^2) & (1-c_x^2)(-c'_x c'_y) & (1-c_x^2)(-c'_x c'_z) \\ (1-c'^2_x)(-c_x c_y) & (-c_x c_y)(-c'_x c'_y) & (-c_x c_y)(-c'_x c'_z) \\ (1-c'^2_x)(-c_x c_z) & (-c_x c_z)(-c'_x c'_y) & (-c_x c_z)(-c'_x c'_z) \end{bmatrix} \qquad \ldots(4.4)$$

If the 'pseudo-matrix' consisting of velocity-velocity correlations is replaced by $[R_{vv}]$, that of acceleration-acceleration correlations by $[R_{aa}]$, that of velocity-acceleration correlation by $[R_{va}]$ and that of acceleration-velocity correlations by $[R_{av}]$ we get,

$$R_{F_{xn}F_{xm}}(\tau) = (8/\pi) C^2 (\sigma_n)_n (\sigma_n)_m [C_{F_{xn}F_{xm}}][R_{vv}] + K^2 [C_{F_{xn}F_{xm}}][R_{aa}]$$

$$+ (8/\pi)^{\frac{1}{2}} KC(\sigma_n)_n [C_{F_{xn}F_{xm}}][R_{va}] + (8/\pi)^{\frac{1}{2}} KC(\sigma_n)_m [C_{F_{xn}F_{xm}}][R_{av}] \qquad \ldots(4.5)$$

It may be seen that for each term, $[R_{aa}] = (2\pi f)^2 [R_{vv}]$ ... (4.6)

$$[R_{av}] = -[R_{va}] \quad \ldots (4.7)$$

$$R_{F_{xn}F_{xm}}(\tau) = \left\{(8/\pi)C^2(\sigma_n)_n(\sigma_n)_m + K^2(2\pi f)^2\right\}\left\{[C_{F_{xn}F_{xm}}][R_{vv}]\right\} + \left\{(8/\pi)^{\frac{1}{2}}KC[(\sigma_n)_n - (\sigma_n)_m]\right\}\left\{[C_{F_{xn}F_{xm}}][R_{va}]\right\}$$
... (4.8)

The above 'pseudo-matrix' representation is specially given as it carried certain advantages in representation and programming. On similar lines we can form equations for all possible combinations of wave force components. We thus have,

$$R_{F_{xn}F_{ym}}(\tau) = \left\{(8/\pi)C^2(\sigma_n)_n(\sigma_n)_m + (2\pi f)^2 K^2\right\}\left\{[C_{F_{xm}F_{ym}}][R_{vv}]\right\} + \left\{(8/\pi)^{\frac{1}{2}}KC[(\sigma_n)_n - (\sigma_n)_m]\right\}\left\{[C_{F_{xn}F_{ym}}][R_{va}]\right\}$$
...(4.9)

where

$$[C_{F_{xn}F_{ym}}] = \begin{bmatrix} (1-c_x^2)(-c'_x c'_y) & (1-c_x^2)(1-c_y'^2) & (1-c_x^2)(-c'_y c'_z) \\ (-c_x c_y)(-c'_x c'_y) & (-c_x c_y)(1-c_y'^2) & (-c_x c_y)(-c'_y c'_z) \\ (-c_x c_z)(-c'_x c'_y) & (-c_x c_z)(1-c_y'^2) & (-c_x c_z)(-c'_y c'_z) \end{bmatrix}$$
...(4.10)

This can be generalized in the following equation (for $i = x, y, z$; $j = x, y, z$)

$$R_{F_{in}F_{jm}}(\tau) = \left\{(8/\pi)C^2(\sigma_n)_n(\sigma_n)_m + (2\pi f)^2 K^2\right\}\left\{[C_{F_{in}F_{jm}}][R_{vv}]\right\} + \left\{(8/\pi)^{\frac{1}{2}}KC[(\sigma_n)_n - (\sigma_n)_m]\right\}\left\{[C_{F_{in}F_{jm}}][R_{va}]\right\}$$
...(4.11)

The cross-spectral densities in between various force components can be found by taking the Fourier transforms of the corresponding cross-covariance functions. The cross-spectral density in between force F_{xn} and F_{xm} for frequency 'f' is thus

$$S_{F_{xn}F_{xm}}(f) = \left\{(8/\pi)C^2(\sigma_n)_n(\sigma_n)_m + (2\pi f)^2 K^2\right\}\left\{[C_{F_{xm}F_{xm}}][S_{vv}]\right\} + \left\{(8/\pi)^{\frac{1}{2}}KC[(\sigma_n)_n - (\sigma_n)_m]\right\}\left\{[C_{F_{xn}F_{xm}}][S_{va}]\right\}$$
...(4.12)

where

$$[S_{vv}] = \begin{bmatrix} S_{v_{xn}v_{xm}} & S_{v_{yn}v_{xm}} & S_{v_{zn}v_{xm}} \\ S_{v_{xn}v_{ym}} & S_{v_{yn}v_{ym}} & S_{v_{zn}v_{ym}} \\ S_{v_{xn}v_{zm}} & S_{v_{yn}v_{zm}} & S_{v_{zn}v_{zm}} \end{bmatrix} \text{ for frequency 'f'}$$
...(4.13)

$$[S_{va}] = \begin{bmatrix} S_{v_{xn}a_{xm}} & S_{v_{yn}a_{xm}} & S_{v_{zn}a_{xm}} \\ S_{v_{xn}a_{ym}} & S_{v_{yn}a_{ym}} & S_{v_{zn}a_{ym}} \\ S_{v_{xn}a_{zm}} & S_{v_{yn}a_{zm}} & S_{v_{zn}a_{zm}} \end{bmatrix} \text{ for frequency 'f'}$$
...(4.14)

The other combinations of the force spectral densities can be generalized as follows:

$$S_{F_{in}F_{jm}}(f) = \left\{\frac{8}{\pi}C^2(\sigma_n)_n(\sigma_n)_m + (2\pi f)^2 K^2\right\}\left\{[C_{F_{in}F_{jm}}][S_{vv}]\right\} + \left\{(8/\pi)^{\frac{1}{2}}KC[(\sigma_n)_n - (\sigma_n)_m]\right\}\left\{[C_{F_{in}F_{jm}}][S_{vv}]\right\}$$
...(4.15)

for $i = x, y, z$ and $j = z, y, z$.

It may be noticed from the preceding equations that all force spectra are basically obtained by multiplying the directional spectrum by a transfer function and hence if a directional spectrum (or equivalently a one-dimensional spectrum and a spreading function) is given along with rms normal velocities at two points under consideration, we can calculate all of them. One of the unique advantages of equations like (4.11) is that they are simple and uniform to express. This is unlike those presented for vertical cylinders along two directions by Y Fantis and Borgman (1978).

Special Cases

If the two points 'n' and 'm' under consideration are on the piles having different dimensions, the preceding formulae get slightly modified, e.g. the general form of the cross-wave force spectra would become (for i and j = x, y, z, and C', K' equivalent to C,K of first pile):

$$S_{F_{in}F_{jm}}(f) = \left\{\frac{8}{\pi}(\sigma_n)_n(\sigma_n)_m C C' + (2\pi f)^2 K K'\right\} \left\{[C_{F_{in}F_{jm}}][S_{vv}]\right\}$$
$$+ \left\{CK'(\sigma_n)_n - C'K(\sigma_n)_m\right\}\left\{(8/\pi)^{\frac{1}{2}}[C_{F_{in}F_{jm}}][S_{va}]\right\} \qquad \ldots(4.16)$$

If the measurements are made at a single point only, the corresponding single point statistics could be obtained by a modified - and at the same time highly simplified - version of the preceding general formulation (equation 4.15), e.g. the general cross-force spectra in such a case are (for i, j = x, y, z):

$$S_{F_iF_j}(f) = \left\{\frac{8}{\pi}C^2(\sigma_n)^2 + (2\pi f)^2 K^2\right\}\left\{[C_{F_{in}F_{jm}}][S_{vv}]_s\right\} \qquad \ldots(4.17)$$

where

$$[S_{vv}]_s = \begin{bmatrix} S_{v_x v_x} & S_{v_y v_x} & S_{v_z v_x} \\ S_{v_x v_y} & S_{v_y v_y} & S_{v_z v_y} \\ S_{v_x v_z} & S_{v_y v_z} & S_{v_z v_z} \end{bmatrix} \qquad \ldots(4.18)$$

If unidirectional wave conditions can be assumed, $S(f,\theta)$ could be replaced by $S(f)$, the one-dimensional spectrum, in all of the preceding formulae and analysis could be made in (x,z) coordinate frame. This results in a further simplification of the general formulae. Analysis for vertical cylinders can be made by means of formulae given in this section by substituting the value of direction cosine along z-direction as unity and by equating the direction cosines along other two directions to zero.

5. ESTIMATES OF WAVE FORCE COEFFICIENTS

The mathematical procedure described below to make an estimate of force coefficients, C_D and C_M, for inclined cylinders using the spectral technique has been developed consistent to the one indicated by Borgman (1972), that is meant for vertical cylinders - by assuming a different force law. This method requires measurement of surface elevations $\eta(t)$, force $F_x(t)$ [or/and $F_y(t)$, $F_z(t)$] together with rms value of normal velocity for the given sea state. These input values would then give a set of C_D, C_M values as a function of frequencies, along three coordinate directions.

We have the surface elevation time history measured at a point 'n', given by the following expression:

$$\eta_n(x,y,t) = \int_0^\infty \int_0^{2\pi} \sqrt{4\, S(f,\theta)\, d\theta\, df}\, \cos(kx\cos\theta + ky\sin\theta - 2\pi f t + \phi) \qquad \ldots(5.1)$$

The wave force time histories measured at an another point 'm' on the inclined pile subjected to directional sea are given by equation (2.9). Then, the cross-covariance function in between η_n and force components would be as follows:

$$C_{\eta_n F_{xm}}(\tau) = E[\eta_n(t)\, F_{xm}(t+\tau)] \qquad \ldots(5.2)$$

$$= (8/\pi)^{\frac{1}{2}} C(\sigma_n)_m [1-c_x^2)C_{\eta_n v_{xm}} + (-c_x c_y)C_{\eta_n v_{ym}} + (-c_x c_z)C_{\eta_n v_{zm}}]$$
$$+ K[(1-c_x^2)C_{\eta_n a_{xm}} + (-c_x c_y)C_{\eta_n a_{ym}} + (-c_x c_z)C_{\eta_n a_{zm}}] \qquad \ldots(5.3)$$

The correlation in between η and the flow kinematic components are obtained by direct multiplication as in section 3.

Consider measured time histories of η_n and F_{xm}. Equation (5.2) is equivalent to

$$C_{\eta_n F_{xm}}(\tau) = 2\int_0^\infty [c(f)]_{\eta_n F_{xm}} \cos 2\pi f\tau \, df + 2\int_0^\infty [q(f)]_{\eta_n F_{xm}} \sin 2\pi f\tau \, df \qquad \ldots(5.4)$$

From equations (5.4) and (5.3), equating co- and quadrature parts,

$$[c(f)]_{\eta_n F_{xm}} = (8/\pi)^{\frac{1}{2}} C(\sigma_n)_m [(1-c_x^2)S_{cc} + (-c_x c_y)S_{sc} + (-c_x c_z)S_s]$$
$$+ K[1-c_x^2) \, 2\pi f \, S_{cs} + (-c_x c_y) \, 2\pi f \, S_{ss} - (-c_x c_z) \, 2\pi f \, S_c] \qquad \ldots(5.5)$$

where $S_{cc} = 2\pi f \, C_m \int_0^{2\pi} S(f,\theta) \cos\theta \cos P \, d\theta \qquad \ldots(5.6)$

$S_{sc} = 2\pi f \, C_m \int_0^{2\pi} S(f,\theta) \sin\theta \cos P \, d\theta \quad \ldots(5.7) \qquad S_s = 2\pi f \, S_m \int_0^{2\pi} S(f,\theta) \sin P \, d\theta \qquad \ldots(5.8)$

$S_{cs} = 2\pi f \, C_m \int_0^{2\pi} S(f,\theta) \cos\theta \sin P \, d\theta \quad \ldots(5.9) \qquad S_{ss} = 2\pi f C_m \int_0^{2\pi} S(f,\theta) \sin\theta \sin P \, d\theta \qquad \ldots(5.10)$

$S_c = 2\pi f \, S_m \int_0^{2\pi} S(f,\theta) \cos P \, d\theta \qquad \ldots(5.11)$

$$[q(f)]_{\eta_n F_{xm}} = (8/\pi)^{\frac{1}{2}} C(\sigma_n)_m [(1-c_x^2)S_{cs} + (-c_x c_y)S_{ss} - (-c_x c_z)S_c]$$
$$- K[(1-c_x^2)] \, 2\pi f \, S_{cc} + (-c_x c_y) \, 2\pi f \, S_{sc} + (-c_x c_z) \, 2\pi f \, S_s] \qquad \ldots(5.12)$$

Equations (5.5) and (5.12) contain two unknowns 'C' and 'K' and can be easily solved.

A similar set of expressions can be obtained to measure C_D, C_M along y and z directions if we consider expressions for $C_{\eta_n F_{ym}}$, $C_{\eta_n F_{zm}}$, respectively.

6. EXAMPLE

This section indicates how the theory described in the previous sections could be applied to actual site conditions. The data utilised for this purpose pertains to one collected at Christchurch Bay, U.K., by NMI. It was proposed to obtain the predicted values of cross-spectral densities for all possible combinations of wave force components; the wave forces being assumed to be exerted at a point on a hypothetical pile subjected to directional waves at the site. The data are in the form of time histories of surface elevations and three components of velocities collected at a single point. A part of the data are shown in Fig.2. The velocity measurements were made by a perforated ball instrument that is described in detail in Bishop(1979). Basically, measurements of time histories of surface elevations and normal velocity are sufficient to predict all possible combinations of the force spectral densities at a point using equation (4.15). However, since the normal velocity measurements were not available, the same is calculated using the relationship in equation (2.10).

Two different programs are written which finally give the output in the form of wave force cross-spectra indicated by the sets of equations (4.15). One program calculates these quantities by means of predicting the cross-spectra in between the flow kinematic components by applying the transfer functions to the surface elevation values whereas other one uses the direct spectral analysis in between the kinematics. The input values of the programs were as follows: depth of water = 9.06 m, depth of point of measurement = 6.3 m. The hypothetical pile was assumed to have an inclination given by its direction cosines $C_x = 0.551$, $C_y = 0.122$ and $C_z = 0.826$. The pile diameter was assumed to be 1 m with the values of drag and inertia coefficients as 1.4 and 2.0 respectively. To perform the spectral analysis by covariance method for surface elevations and flow kinematics, the values of these quantities were sampled at an interval of 0.2362 seconds and 1200 such values are used for one separate analysis with maximum 100 number of lags. With these input quantities the cross-spectral densities for all possible combinations of wave forces were calculated using both of the programs mentioned earlier. It was found that the results given by both of them are almost identical. For discussions sake, therefore, only the second program is considered below.

Figures 3a and 3b show the intermediate results of the cross-spectra in between various flow kinematic components. Figures 4a and 4b show the predicted values of the cross-spectral densities in between various wave force components. It may be seen that the trend of the spectral curves is one which can be expected. The force density is concentrated in between 0.085 to 0.381 Hz frequency components culminating at the frequency of 0.15 Hz. Obviously the auto spectral densities along the principal wave direction $-x$, are higher than those along the vertical $-z$ and transverse $-y$ directions. This clearly indicates how the spectral densities along the vertical directions are significant in case of inclined cylinders. There appears to be a positive correlation nature in between $F_y - F_z$ components and a negative one for others. Obviously a different trend is obtainable for co-and-quadrature spectra in case of statistics for two different points under consideration. The force spectra would be very sensitive to the values of C_D, C_M, direction cosines of the pile. It may be remembered here that $S_{F_i F_j} = \overline{S}_{F_j F_i}$ where $\overline{S}_{F_j F_i}$ is the complex conjugate of $S_{F_i F_j}$ and hence Fig. 4 is sufficiently indicative for all possible wave force combinations.

The preceding example thus illustrates how the theoretical formulae indicated in this paper could be utilised to obtain the cross-spectral densities for all possible combinations of wave force components. Further investigations, covering all aspects of the theory, briefly described in earlier sections, are under progress.

7. CONCLUSIONS

1. Given a basic input, mainly in the form of time histories of wave surface elevation and normal velocity, it is possible to obtain estimates of cross-covariance and spectral density functions of wave forces exerted at one or two points on the inclined piles by means of the new theory presented. Though it incorporates the linearity approximation, it considers for the first time the three directional spectral densities together with directional wave conditions for spectrally analyzing forces on inclined piles and gives a uniform and simple formulae for all combinations of force components.

2. The force spectral formulae become increasingly simple and easy to evaluate, as indicated in section 4 if we consider the special cases of the general formulation like single point statistics, unidirectional sea conditions. Forces on vertical piles can be analysed as a special case with appropriate changes in the values of direction cosines.

3. With the help of measured time histories of surface, normal velocity and forces along desired directions, it is possible to predict values of the force coefficient C_D and C_M for inclined cylinders by means of formulae in section 5.

4. A numerical example given at the end indicates that it is possible to get meaningful and reasonable results from the theory discussed in section 4. A detailed discussion on various other aspects of the theory presented in this paper would be possible after the on-going analysis with more field data is completed.

ACKNOWLEDGEMENTS

The help of Dr J.R. Bishop and the authorities of National Maritime Institute, U.K. in the form of providing data is gratefully acknowledged.

REFERENCES

BISHOP J.R., 1979, <u>Measurements of Wave Particle Motion at the Christchurch Bay Tower Using a Perforated Ball Instrument</u>, National Maritime Institute, Feltham, Middlesex, UK, Report Number R-55.

BORGMAN L.E., 1958, "Computation of Ocean Wave Forces on Inclined Cylinders," <u>Transactions, American Geophysical Union</u>, Volume 39, Number 5, pp 885-888.

BORGMAN L.E., 1967, <u>The Spectral Density of Ocean Wave Forces</u>, Technical Report HEL 9-8, University of California, Berkeley.

BORGMAN L.E., 1972, "Statistical Models for Ocean Waves," <u>Advances in Hydroscience</u>, Volume 8 pp 139-181.

BORGMAN L.E., 1976, "Statistical Analysis of Sea Waves," <u>Coastal Wave Hydrodynamics - Theory and Engineering Applications</u>, Lecture Notes of the Summer Course, Massachusetts Institute of Technology, Department of Civil Engineering, Cambridge, Massachusetts, pp 2.1-2.38.

MORISON et al, 1950, "The Force Exerted by Surface Waves on Piles," <u>Petroleum Transactions</u>, Volume 189, Number TP 2846, pp 149-154.

PENZIEN J. and TSENG W.S., 1978, "Three Dimensional Analysis of Fixed Offshore Structures," in <u>Numerical Methods in Offshore Engineering</u> by Zienkiewicz O.C. et al, Chapter 7, John Wiley and Sons, pp 221-243.

WADE B.G. and DWYER M., 1976, "On the Application of Morison's Equation to Fixed Offshore Platforms," Proceedings, <u>Offshore Technology Conference</u>, Houston, Texas, Vol 3, Paper 2723, pp 1181-1190.

Y FANTIS E.A. and BORGMAN L.E., 1978, <u>A Statistical Theory for Directional Ocean Wave Forces on Oil Drilling Platforms</u>, Stat Lab Report No. 132, University of Wyoming, Department of Statistics, Laramie, Wyoming, USA.

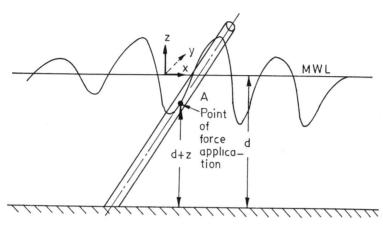

FIG.1. DEFINITION SKETCH

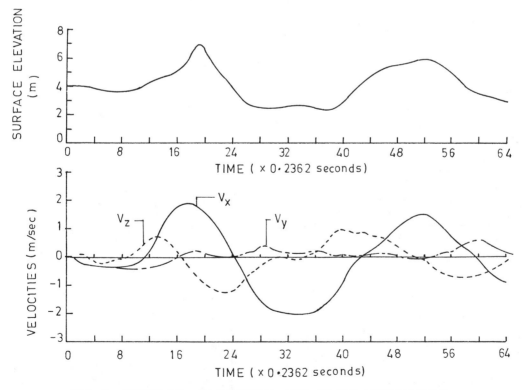

FIG.2. SURFACE ELEVATION AND FLOW KINEMATICS

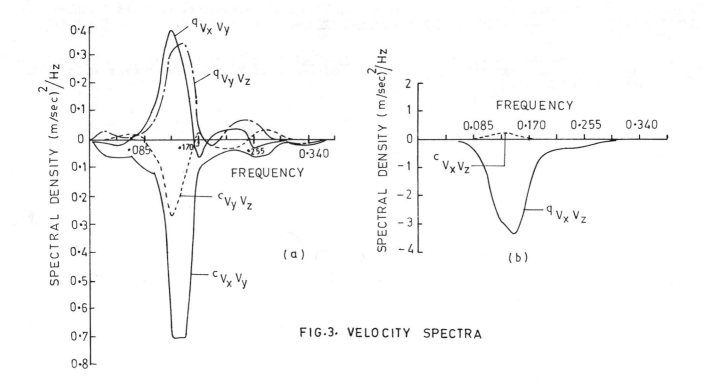

FIG. 3. VELOCITY SPECTRA

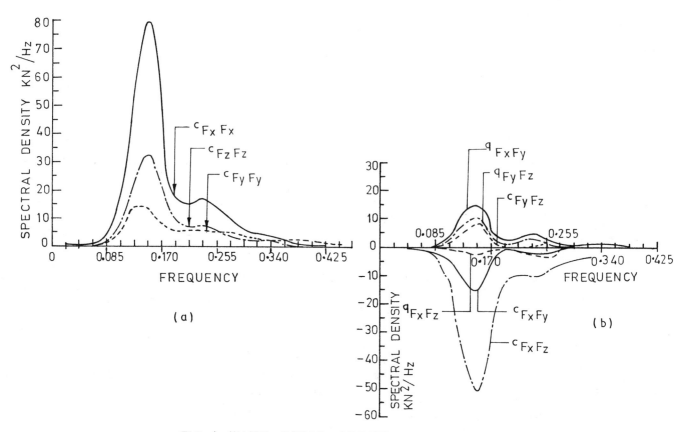

FIG. 4. WAVE FORCE SPECTRA

RESPONSE ANALYSIS OF TENSION LEG PLATFORM WITH MECHANICAL DAMPING SYSTEM IN WAVES

M. Katayama, K. Unoki and E. Miwa

Mitsubishi Heavy Industries, Ltd, Japan

Summary

Recently, the various types of offshore oil drilling and production platforms have become larger and operated in deeper areas in the ocean.

Therefore, for deep sea oil drilling and production platforms, a tension leg platform, guyed tower platform, etc. are being proposed in addition to the conventional fixed-type and gravity-type platforms.

This paper first presents on approximate response analysis method for the tension leg platform with a mechanical damping system (hydraulic damper) in waves based on the small amplitude and linearized theory. Next, the results of a series of tank tests for verifying the effect of the mechanical damping system and the accuracy in computation and the applicability to practical problems of the above response analysis method are introduced.

Finally, the response characteristics of the tension leg platform with the mechanical damping system in waves and its applicability to practical problems are discussed by pursuing the systematic numerical analysis on a prototype tension leg platform.

NOMENCLATURE

A_i	:	virtual inertia force coefficient matrix for i member
A_k	:	water plane area of k member
A_{xi}, A_{yi}, A_{zi}	:	projected area of i member in x, y and z directions
B^I	:	buoyancy force
B_i^{II}	:	wave making damping and linearized drag force coefficient matrix for i member
B_j	:	damping force coefficient matrix of external dynamical mechanism at attaching point j to upper structure
$C_{Dxi}, C_{Dyi}, C_{Dzi}$	:	drag force coefficient of i member in x, y and z directions
C_j^I	:	linear spring constant matrix of external dynamical mechanism at attaching point j to upper structure
C_k^{II}	:	restoration force coefficient matrix for k member at water plane
C_{xj}, C_{yj}, C_{zj}	:	damping force coefficient in x, y and z directions at attaching point j to upper structure (as damping force in proportion to velocity)
C_1	:	damping constant of hydraulic damper
$\bar{D}_{xi}, \bar{D}_{yi}, \bar{D}_{zi}$	:	linearized drag force coefficient of i member in x, y and z directions
F_i^B	:	buoyancy vector
F_i^C	:	current force vector
F_i^{DR}	:	wave drifting force vector
F_i^W	:	weight vector (functional load, ballast, etc.)
F_i^{WI}	:	wind load vector
F_j^T	:	initial tension vector
G_i	:	inertia force coefficient matrix of wave exciting force acting on i member
g	:	gravitational acceleration
H_i	:	damping force coefficient matrix of wave exciting force acting on i member
$H_{xc}(\omega)$	:	frequency response function of displacement
h_w	:	wave height
KB	:	height of center of buoyancy
KG	:	height of center of gravity
K_{xj}, K_{yj}, K_{zj}	:	spring constant in x, y and z directions at attaching point j to upper structure
k_1	:	spring constant of hydraulic damper
k_2	:	spring constant due to axial rigidity of leg
M_i	:	mass of i member
m_{xi}, m_{yi}, m_{zi}	:	added mass of i member in x, y and z directions
m_{ox}	:	variance of displacement
N_{xi}, N_{yi}, N_{zi}	:	wave making damping force coefficient of i member in x, y and z directions
$p(H,T)$	:	long-term probability of occurrence for sea condition
Q_R	:	long-term cumulative exceeding probability of the extreme value of response amplitude
q_R	:	exceeding probability of the extreme value of response amplitude
$R_{1/3}$	:	significant mean value of the response amplitude
$S_{xx}(\omega)$	:	power spectral density function of displacement
$S_{\zeta\zeta}(\omega)$	:	power spectral density function of incident wave
T	:	initial tension
T_i	:	displacement transfer matrix
T_i'	:	transposed matrix of T_i
T_w	:	wave period

V_i	:	volume of i member
W	:	weight
x_G, y_G, z_G	:	coordinates at the center of gravity G
x_i, y_i, z_i	:	coordinates at point i
Z_p	:	height of attaching point of initial tension
ζ_i	:	particle displacement vector of fluid at point i (based on small amplitude and linearized theory)
η	:	displacement vector at the center of gravity of overall structure (including external dynamical mechanism)
η_{static}	:	static displacement vector (corresponding to wind, current and wave drifting forces)
μ	:	wave direction
ρ	:	specific gravity of sea water
ω	:	wave circular frequency
$\cdot$	:	showing differential with respect to time $\left(\dot{x}=\dfrac{dx}{dt}\right)$

1. INTRODUCTION

A so-called compliant structure of the guyed tower or tension leg platform has been proposed for use in deep sea oil and gas production in addition to the conventional jacket-type and gravity-type platforms.[1][2]

The feature of such a compliant platform is to decrease its wave load, while admitting its motion in waves, in comparison with the conventional fixed/sit-on-bottom type platforms. Of the two types, the tension leg platform (hereinafter called "TLP") has excellent motion characteristics in waves. TLP is simimar, in principle, to the jacket-type in operability of production and ease of maintenance and repair, and yet highly economical and mobile. Since around 1970, such advantages of TLP have begun to be recognized for use in developing so-called small and medium scale marginal fields in areas of unfavorable sea and weather or geographic conditions. Research and development of TLP have been performed in various places worldwide.[3]-[15] However, no operation of TLP has yet been experienced.

In Japan, there are only a few number of taut moored structure of this type in small scale including "SOSEI II,[16]" a working platform for Honshu-Shikoku Bridge, and "Tension Leg Buoy,[17][18]" a marine observation buoy(installed in the Sea of Genkai), etc. The world's first practical large scale TLP would be the one (total displacement of 63,300 tons) that is now under construction in England to be completed in 1984 for operation at Hutton Field in the North Sea.[19]

Based on the achievements of fundamental studies[20]-[34] on various semi-submersible offshore structures including offshore oil drilling units, the authors have been conducting studies, for these several years, on the TLP's response characteristics in waves through tank tests of a small model and linearized response analyses based on the approximate evaluation of hydrodynamical force.

In research and development of TLP, it is important to study the type of structure of the leg mechanism (hereinafter called external dynamical mechanism) and the structural integrity as well as the motion characteristics in waves of TLP. The tension response characteristics in waves of the external dynamical mechanism, in particular, greatly depend on its rigidity and damping mechanism.

Noting that the damping mechanism would reduce the leg tension response of TLP in waves, the incorporation of a mechanical damping system (hydraulic damper) in the external dynamical mechanism was studied. That is, a systematic computer simulation and a series of tank tests of a small model with the damping mechanism were performed by parametrically changing the rigidity (restoring force characteristics) and the damping force characteristics of the external dynamical mechanism.

TLP thus studied by the authors is characterized by the incorporation of a mechanical damping system (hydraulic damper) in its external dynamical mechanism which will improve the structural integrity by greatly reducing the leg tension response amplitude in waves.

This paper first outlines the approximate response analysis of TLP in waves by the small amplitude and linearized theory.

Next, described are the results of a series of tank tests performed to verify the effect of the damping mechanism and the accuracy in computation and applicability of the small amplitude and linearized theory.

Finally, using a prototype model in systematic numerical analyses, the response characteristics of TLP provided with a damping mechanism in waves and its applicability to practical problems are discussed.

2. APPROXIMATE RESPONSE ANALYSIS METHOD IN WAVES

2.1 Fundamental assumptions and method of analysis

The following assumptions are made in the analysis:

(1) The overall response analysis can be carried out as a space framed structure.

(2) Fluid is assumed to be incompressible, inviscid and irrotational.

(3) Cross-sectional dimensions of the structural members are to be sufficiently smaller than the wave length and their hydrodynamical interactions can be neglected.

(4) Hydrodynamical forces acting on the external dynamical mechanism can be neglected.

(5) Such progressive waves in deep sea as small amplitude (Airy wave) are considered. Fluid force can be expressed as a linear combination of terms due to displacement, velocity and acceleration, provided non-linearized terms are considered partially with respect to displacement and velocity.

(6) Structural deformation is linear and elastic, and motion is of harmonic oscillation about the center of gravity.

(7) Wind force, current force, wave drifting force, etc., can be treated as static external forces being constant, regardless of the displacement of the overall structure.

Forces acting on various structural members are assumed to be of five types of forces due to inertia, pressure, relative displacement, relative velocity and relative acceleration. They are considered to be concentrated and/or distributed external forces at nodal points taken at the center of gravity of each member element of the space framed structure.

Of the above forces except inertia force, the terms due to displacement, velocity and acceleration of the member element are restoration force (including external dynamical mechanism), radiation force and mooring force due to external dynamical mechanism, and wave exciting forces such as Froude-Kriloff force, diffraction force, etc.

From the above, a motion equation can be formulated consisting of virtual inertia force, damping force, restoration force (including external dynamical mechanism) and wave exciting force for each wave frequency. By analyzing this motion equation, six components of motion amplitude and phase angle can be obtained. Thereby the member force to be distributed to each structural member and the leg tension can be determined.

2.2 Approximate response analysis in regular waves based on small amplitude and linearized theory

The frequency response characteristics of the overall structure (including external dynamical mechanism) can be obtained by solving the following Eqs. (1)-(3) linearized with respect to the member element i (Fig. 1).

$$\sum_i F_i^W + \sum_i F_i^B + \sum_j F_j^T = 0 \quad \cdots (1)$$

$$\sum_i F_i^{WI} + \sum_i F_i^C + \sum_i F_i^{DR} + \left\{ \sum_i (T_i^t \cdot C_i^I \cdot T) + \sum_k (T_k^t \cdot C_k^{II} \cdot T_k) \right\} \eta_{static} = 0 \quad \cdots (2)$$

$$\sum_i (T_i^t \cdot A_i \cdot T) \ddot{\eta} + \left\{ \sum_i (T_i^t \cdot B_i^I \cdot T) + \sum_j (T_j^t \cdot B_j^{II} \cdot T) \right\} \dot{\eta} + \left\{ \sum_j (T_j^t \cdot C_j^I \cdot T) \right.$$

$$\left. + \sum_k (T_k^t \cdot C_k^{II} \cdot T_k) \right\} \eta = \sum_i (T_i^t \cdot G_i) \xi_i + \sum_i (T_i^t \cdot H_i) \dot{\xi}_i \quad \cdots (3)$$

Where

F_i^W : weight vector (functional load, ballast, etc.)

F_i^B : buoyancy vector

F_j^T : initial tension vector

F_i^{WI} : wind load vector

F_i^C : Current force vector

F_i^{DR} : wave drifting force vector

η_{static} : static displacement vector (corresponding to wind, current and wave drifting forces)

A_i : virtual inertia force coefficient matrix for i member

$$= \begin{pmatrix} M_i + m_{xi} & 0 & 0 \\ 0 & M_i + m_{yi} & 0 \\ 0 & 0 & M_i + m_{zi} \end{pmatrix} \quad \cdots (4)$$

M_i : mass of i member

m_{xi}, m_{yi}, m_{zi} : added mass of i member in x, y and z directions

B_i^I : wave making damping and linearized drag force coefficient matrix for i member

$$= \begin{pmatrix} N_{xi} + \overline{D}_{xi} & 0 & 0 \\ 0 & N_{yi} + \overline{D}_{yi} & 0 \\ 0 & 0 & N_{zi} + \overline{D}_{zi} \end{pmatrix} \quad \cdots (5)$$

N_{xi}, N_{yi}, N_{zi} : wave making damping force coefficient of i member in x, y and z directions

$\overline{D}_{xi}, \overline{D}_{yi}, \overline{D}_{zi}$: linearized drag force coefficient of i member in x, y and z directions

B_j^{II} : damping force coefficient matrix of external dynamical mechanism at attaching point j to upper structure

$$= \begin{pmatrix} C_{xj} & 0 & 0 \\ 0 & C_{yj} & 0 \\ 0 & 0 & C_{zj} \end{pmatrix} \quad \cdots \cdots (6)$$

C_{xj}, C_{yj}, C_{zj} : damping force coefficient in x, y and z directions at attaching point j to upper structure (as damping force in proportion to velocity)

C_j^I : linear spring constant matrix of external dynamical mechanism at attaching point j to upper structure

$$= \begin{pmatrix} K_{xj} & 0 & 0 \\ 0 & K_{yj} & 0 \\ 0 & 0 & K_{zj} \end{pmatrix} \quad \cdots \cdots (7)$$

K_{xj}, K_{yj}, K_{zj} : spring constant in x, y and z directions at attaching point j to upper structure

C_k^{II} : restoration force coefficient matrix for k member at water plane

$$= \begin{pmatrix} 0 & 0 & 0 \\ 0 & \rho g A_k & 0 \\ 0 & 0 & 0 \end{pmatrix} \quad \cdots \cdots (8)$$

ρ : specific gravity of sea water
g : gravitational acceleration
A_k : water plane area of k member
G_i : inertia force coefficient matrix of wave exciting force acting on i member

$$= \begin{pmatrix} \rho V_i + m_{xi} & 0 & 0 \\ 0 & \rho V_i + m_{yi} & 0 \\ 0 & 0 & \rho V_i + m_{zi} \end{pmatrix} \quad \cdots \cdots (9)$$

V_i : volume of i member
H_i : damping force coefficient matrix of wave exciting force acting on i member $= B_i^I$
T_i : displacement transfer matrix

$$= \begin{pmatrix} 1 & 0 & 0 & 0 & (z_i - z_G) & -(y_i - y_G) \\ 0 & 1 & 0 & -(z_i - z_G) & 0 & (x_i - x_G) \\ 0 & 0 & 1 & (y_i - y_G) & -(x_i - x_G) & 0 \end{pmatrix} \quad \cdots \cdots (10)$$

x_i, y_i, z_i : coordinates at point i
x_G, y_G, z_G : coordinates at the center of gravity G
η : displacement vector at the center of gravity of overall structure (including external dynamical mechanism)

$$= \{x, y, z, \phi, \theta, \psi\} \quad \cdots \cdots (11)$$

$\cdot$: showing differential with respect to time $\left(\dot{x} = \dfrac{dx}{dt} \right)$

ζ_i : particle displacement vector of fluid at point i (based on small amplitude and linearized theory)

$$= \{\zeta_{xi}, \zeta_{yi}, \zeta_{zi}\} \quad \cdots \cdots (12)$$

T_i^t : transposed matrix of T_i

The linearized drag force coefficient can be obtained by

$$\overline{D}_{xi} = \frac{4}{3\pi} C_{Dxi} \cdot \rho \cdot A_{xi} | \dot{X}_i - \dot{\zeta}_{xi} |$$
$$\overline{D}_{yi} = \frac{4}{3\pi} C_{Dyi} \cdot \rho \cdot A_{yi} | \dot{Z}_i - \dot{\zeta}_{zi} | \quad \cdots (13)$$
$$\overline{D}_{zi} = \frac{4}{3\pi} C_{Dzi} \cdot \rho \cdot A_{zi} | \dot{Z}_i - \dot{\zeta}_{zi} |$$

Where

$$\{\dot{X}_i, \dot{Y}_i, \dot{Z}_i\} = T_i \cdot \dot{\eta} \quad \cdots (14)$$

Where

- $C_{Dxi}, C_{Dyi}, C_{Dzi}$: drag force coefficient of i member in x, y and z directions
- A_{xi}, A_{yi}, A_{zi} : projected area of i member in x, y and z directions

Substituting the following in Eq. (3), all elements for each coefficient matrix can be expressed as shown in Table 1.

$$A = \sum_i (T_i^t \cdot A_i \cdot T_i)$$
$$B = \sum_i (T_i^t \cdot B_i^I \cdot T_i) + \sum_j (T_j^t \cdot B_j^{II} \cdot T_j)$$
$$C = \sum_j (T_j^t \cdot C_j^I \cdot T_j) + \sum_k (T_k^t \cdot C_k^{II} \cdot T_k)$$
$$G = \sum_i (T_i^t \cdot G_i)$$
$$H = \sum_i (T_i^t \cdot H_i)$$

For TLP with an external dynamical mechanism, the initial tension of the external dynamical mechanism must be considered in the diagonal terms (C_{44}, C_{66}), corresponding to the rolling and pitching, of the restoration force coefficient matrix by the following Eq. (15):

$$B \cdot KB - W \cdot KG - T \cdot Z_p \quad \cdots (15)$$

Where

- B : buoyancy force
- W : weight
- T : initial tension
- KB : height of center of buoyancy
- KG : height of center of gravity
- Z_p : height of attaching point of initial tension

The spring constants (k_1, k_2) and the damping constant (C_1) of the external dynamical mechanism shown in Fig. 2 can be substituted by the equivalent and parallel spring constant k_e and the damping constant C_e in Eqs. (16) and (17), as:

$$k_e = k_2 \frac{k_1(k_1+k_2)+(\omega C_1)^2}{(k_1+k_2)^2+(\omega C_1)^2} \quad \cdots (16)$$

$$C_e = C_1 \frac{k_2^2}{(k_1+k_2)^2+(\omega C_1)^2} \quad \cdots (17)$$

Where

- k_2 : spring constant due to axial rigidity of leg $= EA/l$
- k_1 : spring constant of hydraulic damper
- C_1 : damping constant of hydraulic damper
- ω : wave circular frequency

In this case, $K_{xj}, K_{zj}, K_{yj}, C_{xj}, C_{yj}$ and C_{zj} can be expressed more materially, as:

$$\left.\begin{aligned} K_{xj} &= T_j/l_j \\ K_{zj} &= T_j/l_j \\ K_{yj} &= k_e \\ C_{xj} &= C_{zj} = 0 \\ C_{yj} &= C_e \end{aligned}\right\} \quad (18)$$

The motions about the center of gravity of the overall structure are expressed, as:

$$\left.\begin{aligned} \eta &= \eta_0 \cos(\omega t + \varepsilon) \\ \dot{\eta} &= -\omega \eta_0 \sin(\omega t + \varepsilon) \\ \ddot{\eta} &= -\omega^2 \eta_0 \cos(\omega t + \varepsilon) = -\omega^2 \eta \end{aligned}\right\} \quad (19)$$

Eq. (19) is substituted for in Eq. (3), which is decomposed into sine and cosine terms to form an algebraic equation of 12 x 12 order. By solving this algebraic equation, the motion responses in six degrees of freedom can be analyzed.

Also, the tension variation acting on the external dynamical mechanism can be computed by:

$$\left.\begin{aligned} F_v &= \eta_0 \sqrt{k_e^2 + \omega^2 C_e^2} \cdot \cos(\omega t + \varepsilon_v) \\ \varepsilon_v &= \tan^{-1}\left(\frac{k_e \sin \varepsilon + \omega C_e \cos \varepsilon}{k_e \cos \varepsilon - \omega C_e \sin \varepsilon}\right) \end{aligned}\right\} \quad (20)$$

In computing the hydrodynamical forces in Eq. (3), it is the normally used method that hydrodynamical forces acting on the elemental members (columns, lower hulls, braces, etc.) are obtained by the linearized potential theory or a systematic forced oscillation test for arithmetic addition.

For TLP consisting of columns, lower hulls, braces, etc., the following approximate analysis method may be also used to obtain the motion response in not-so-high regular waves:

(1) For lower hull, horizontal brace, etc., the strip method is used.

(2) The two dimensional added mass is approximated using values in the infinite fluid, neglecting the effect of free surface being small.

(3) Wave exciting force is inertia force only or inertia force and drag force (by Morison's formula).

(4) For hydrodynamical damping force, the values obtained by the free oscillation model test are used in all frequency ranges or a calculation with only nonlinearized damping force (drag force) considered is used.

(5) Hydrodynamical interactions between members are neglected.

2.3 Response analysis in irregular waves

It is assumed that the irregular sea state of waves is the superposition of many regular waves having small amplitude and random phases, that its variation with time approximates in the stationary stochastic process according to the normal distribution, and further that the frequency response of displacement of the overall structure, wave load and tension variation in the external dynamical mechanism, etc., is linear.

The frequency response function $H_{x\varsigma}(\omega)$ of displacement, member force and tension variation acting on the external dynamical mechanism, etc., can be obtained by the method as described in this Chapter 2. The power spectral density function $S_{xx}(\omega)$ and its variance m_{0x} of displacement, member force and tension variation acting on the external dynamical mechanism, etc., in the irregular sea state can be computed by:

$$\left.\begin{aligned} S_{xx}(\omega) &= |H_{x\varsigma}(\omega)|^2 S_{\varsigma\varsigma}(\omega) \\ m_{0x} &= \int_0^\infty S_{xx}(\omega) d\omega \end{aligned}\right\} \quad (21)$$

Where $S_{\zeta\zeta}(\omega)$ is the power spectral density function of incident wave.

If further assumption is made that the probability distribution of maximum and minimum of the response follows the Rayleigh distribution, the significant mean value $R_{1/3}$ of the response amplitude, for example, can be obtained by:

$$R_{1/3} = 2.002 \sqrt{m_{0x}} \tag{22}$$

Also, the exceeding probability q_R of the extreme value of response exceeding a certain value R can be computed by:

$$q_R = e^{-\frac{R^2}{2m_{0x}}} \tag{23}$$

Further, a long-term prediction can be easily computed, as necessary, based on any wave statistical data (for example, the long-term probability of occurrence for the sea condition $p(H,T)$, if available. For example, the long-term cumulative exceeding probability Q_R of the extreme value of response amplitude exceeding a certain value R can be obtained by:

$$Q_R = \int_0^\infty \int_0^\infty e^{-\frac{R^2}{2m_{0x}}} p(H,T) dH dT \tag{24}$$

3. MOTION AND LEG TENSION RESPONSE TEST IN WAVES

3.1 Method of tank test

The upper structure model of TLP used was a 1/92.1 scaled semi-submersible structure consisting of 4 columns and 4 horizontal braces (Fig. 3). For the mechanical damping system, one of the components of the external dynamical mechanism, an electro-magnetic damper (Fig. 4) was used. For leg elements, coil springs and piano wires were used (Figs. 5 and 6). The main particulars of the upper structure and the external dynamical mechanism are given in Tables 2–4.

(1) Wave exciting force test

The wave exciting force was measured by restraining the upper structure model using a 6-component force detector. Regular waves were used in the following testing ranges:

　　Wave direction　: $\mu = 0°$ and $45°$
　　Wave height　　: $h_w = 3.0$cm (2.8m), 10.0cm (9.2m) and 15.0cm (13.8m)
　　Wave period　　: $T_w = 0.6$-2.5s (5.8-24.0s)

Where the figures in parentheses show the converted values for prototype. (Same applies hereunder.)

(2) Motion and leg tension response test in regular waves

The motion in waves was measured using a 6-component motion detector set on the deck of the model and the leg tension, using a ring gauge set on the leg bottom (Fig. 6).
Regular waves were used under the following test condition:

　　Wave direction　: $\mu = 0°$ and $45°$
　　Wave height　　: $h_w = 3.3$cm (3.0m) and partially 10.0cm (9.2m)
　　Wave period　　: $T_w = 0.6$-2.5s (5.8-24.0s)

The water depth of 0.85m (78.3m) has caused a shallow water effect at the wave period $T_w > 1.2$s.

Table 6 shows the test condition including the combination of the spring and damping constants of the external dynamical mechanism.

3.2 Results of motion response test in regular waves

Typical results of the motion response test in regular waves (Case 12 without and Case 13 with mechanical damping system in oblique sea condition) are shown in Figs. 7–9. The computed values were obtained based on the small amplitude and linearized theory. The experimental values and the computed values in all motions of surging, heaving and pitching are relatively in good agreement, including those in the vicinity of the natural period of the overall structure of TLP where a greater effect of non-linearity might be predicted.

3.3 Results of leg tension response test in regular waves

Fig. 10 shows the results of the leg tension response test in regular waves of No.2 leg (Fig. 5) and the computed values based on the small amplitude and linearized theory in the same manner as in the motion response test (Cases 12 and 13 in oblique sea condition). Also, in this case, the experimental values and the computed values are relatively in good agreement, including those in the vicinity of the natural period of the overall structure of TLP where a greater effect of non-linearity might be predicted.

3.4 Effect of mechanical damping system

As clearly shown also in Figs. 7-10, the effect of the mechanical damping system is evident in the vicinity of the natural period of the overall structure of TLP. That is, in comparison between Cases 12 and 13, Case 13 with the mechanical damping system (C_1=0.02kg s/cm) shows the peak values approximately 2/3 in surging, heaving and pitching (Figs. 7–9) and approximately 2/3 in leg tension variation (Fig. 10) of those of Case 12 without the mechanical damping system.

4. STUDY ON PROTOTYPE MODEL

4.1 Prototype model

The concept of a prototype model as studied hereunder is shown in Fig. 11 with its principal particulars shown in Table 6.

The mechanical damping system for the leg mechanism to be installed on the prototype has the following functions as well as to reduce tension variation caused in the leg in waves, as described previously:

(1) As a tensioner at the time of installation/transit of TLP.

(2) For fine adjustment of initial tension of each leg element at the time of installation of TLP.

(3) For fine adjustment of inclination (trim/heel) due to environmental load variation during operation of TLP.

(4) For quick disconnection of TLP in case of emergency (for example, blow-out during oil drilling or production, etc.).

(5) As an equalizing system to secure stability of the overall structure should the leg mechanism of TLP be damged.

(6) For adjustment of the leg initial tension to buoyancy variation due to the tide, etc.

4.2 Short-term prediction of motion response in irregular waves

Of the results of short-term prediction based on the small amplitude and linearized theory by means of ISSC spectrum (1964) as irregular waves, the surging (at deck level) and heaving motions in oblique sea condition (μ=45°) are shown in Figs. 12 and 13, as examples of motion response. With respect to surging motion, almost no difference is seen between Model A (with soft spring and mechanical damping system) and Model C (with hard spring system). While, Model B shows the mean wave period T_v=3–14 s that is greater than that of the other Models. The heaving motion becomes smaller in the order of Models B, A and C. Model C with hard spring system, of course, shows very small heaving.

However, even Model A with soft spring and mechanical damping system shows the heaving motion of only 30 cm or less which may be considered sufficiently permissible for drilling pipe, riser system, etc.

4.3 Short-term prediction of leg tension response in irregular waves

Fig. 14 shows the results of short-term prediction based on the small amplitude and linearized theory of the leg tension response on weather side in oblique sea condition (μ=45°) where the greatest leg tension response has been observed. In the range of the mean wave period T_v=2-8 s, both Model A (with soft spring and mechanical damping system) and Model C (with hard spring system) show nearly the same values. In $T_v > 8$ s, Model A shows a decreasing tendency of the generated tension as the wave period increases, while Model C shows a greatly increasing tendency.

Model B (with soft spring system) shows an increasing tendency of the tension caused in the range of the relatively small mean wave period, but a decreasing tendency as the wave period increases.

From the above results, TLP with a highly rigid leg system, such as the steel pipe leg system, will cause very great tension in the range of the greater mean wave period (in storm), which will become a serious problem in strength design of the external dynamical mechanism. Therefore, it is desirable to improve the structural integrity by incorporating a suitable mechanical damping system and/or a spring system in the leg mechanism.

4.4 Long-term prediction of motion and leg tension response in irregular waves

Assuming the East China Sea for an offshore installation, the long-term cumulative exceeding probability of maximum wave height, motion and leg tension response was predicted using the statistical wave frequency data as shown in Table 7. [35] The results are shown in Figs. 15-18. From the long-term cumulative exceeding probability of maximum wave height (Fig. 15), the maximum wave height H_{max} in a 100-year storm is predicted as approximately 24.4 m.

The maximum expected value of surging motion to $Q_R = 10^{-8}$ level (20 year return period) becomes 7.5-8.8m showing no great difference among Models A, B and C (Fig. 16). In the same manner, the maximum expected value of heaving motion is smallest for Model C, 0.01m, increasing for Model A (1.06m) and Model B (2.96m) (Fig. 17). The maximum expected value of leg tension variation is smallest for Model A, 900t/leg, which is smaller than the initial tension 1125t/leg. Therefore, any slack condition or compression deformation would not be caused, nor would any snap load (Fig. 18).

Next, as to the leg tension of Model A (with soft spring and mechanical damping system), the linearized cumulative damage factor D at $Q_R = 10^{-8}$ was computed by the modified Miner's law. The S-N curve[36] by Lucht and others and the AWS-X curve[37] were used for wire rope and steel pipe, respectively (Figs. 19 and 20).

In case of wire rope, D=0.03 is obtained for the specification (for example, 4-200 mm ϕ wire rope/leg, breaking strength 2000t) to sufficiently withstand a maximum stress being caused including stationary external force (wind, current, wave drifting force, etc.) and initial tension. Thus, the fatigue strength may not be considered to cause any problem.

Also in case of steel pipe with elastic material used, in the same manner as for wire rope, D=0.05 is obtained for the specification (for example, sectional area 1500cm^2/leg, breaking strength 80 kg/mm^2) to sufficiently withstand a maximum stress, including stationary external force and initial tension, at the stress concentration factor 2.0.

The authors are confident that the incorporation of a suitable mechanical damping system and spring system in the external dynamical mechanism (leg mechanism) will greatly contribute to improvement of the overall structural integrity of TLP and to it's optimum design.

5. CONCLUSION

In this paper, the authors have outlined the response analysis method of TLP with mechanical damping system in waves applying the small amplitude and linearized theory based on the approximate evaluation of hydrodynamical force and have also described the results of a series of tank tests performed to verify the effect of the mechanical damping system, the accuracy in computation by small amplitude and linearized theory and their applicability. Further, numerical analyses have been conducted on a prototype model and the response characteristics of TLP with mechanical damping system in waves and the applicability to practical problems have been discussed.

The above results can be summarized, as follows:

(1) From the results of the tank tests and theoretical analyses of TLP with external dynamical mechanism, the applicability to practical problems of the response analysis method in waves by the small amplitude and linearized theory based on the approximate evaluation of hydrodynamical force has been described.

(2) From the results of a series of tank tests and the theoretical analyses, it has been verified that the incorporation of a mechanical damping system (hydraulic damper) in the external dynamical mechanism (leg mechanism) of TLP will greatly decrease the leg tension response amplitude in waves leading to improvement of the structural integrity.

(3) The response characteristics of TLP with mechanical damping system in waves have been clarified and a sufficient feasibility of applicability to an actual TLP has been obtained by performing numerical analyses on a prototype model.

In future, systematic tank tests and numerical analyses of TLP with mechanical damping system in very high non-linear waves will be performed for further study on applicability to practical problems of the approximate response analysis method in waves based on the finite amplitude and non-linearized theory. It is expected that the report thereon will follow on the next opportunity.

6. ACKNOWLEDGEMENT

The authors wish to express their appreciation to Dr. Y. Arita, Mr. E. Kitami and Mr. K. Ninomiya of MHI Hiroshima Shipyard & Engine Works who have generously extended their helpful instructions and advices for this study; and Mr. Y. Norimatsu who has taken the responsibility for performing the tank test and analysis.

REFERENCES

(1) Katayama M., et al., On Response Analysis of Compliant Structure for Deep Sea Oil Drilling and Production in Waves, the Society of Naval Architects of Japan, Text for the 5th Symposium on Ocean Engineering (1981-1) (in Japanese)

(2) Katayama M., et al., On Response Analysis of Guyed Tower Platform for Deep Sea Oil Drilling and Production in Waves, Mitsubishi Juko Giho, Vol. 17, No.3 (1980-5) (in Japanese)

(3) The following papers were presented at the annual Offshore Technology Conference. Concept and preliminary design of TLP; OTC3049, OTC3881 Response analysis in waves; OTC1263, OTC1553, OTC2070, OTC2494, OTC2690, OTC2796, OTC3044, OTC3304, OTC3515, OTC3555, OTC3883, OTC4070 thru 4075, Full scale field measurement; OTC2104, Material and strength of leg; OTC2924, OTC2925, OTC3850, Others; OTC3289

(4) Paulling J. R. Jr., Time Domain Simulation of Semisubmersible Platform Motion with Application to the Tension-leg Platform, SNAME Spring Meeting/STAR Symposium, San Francisco (1977-5)

(5) Kirk C. L. and Etok E. U., Dynamic Response of Tethered Production Platform in a Random Sea State, Proc. BOSS '79, Paper 57 (1979-8)

(6) Roren E. M. Q. and Steinvik B., Deep Water Resonance Problem in the Mooring System of the Tethered Platform, Offshore Structures Engineering-Proc. of the Int. Conf. on Offshore Structures-Pentech Press (1979)

(7) Yoshida K., et al., Snap Loads in Taut Moored Platforms, J. Soc. Nav. Archit. Japan, Vol. 144 (1978-12) (in Japanese)

(8) Nishihara S., et al., Parametric Excitation in Tension Leg Platform, J. Soc. Nav. Archit. Japan, Vol. 145 (1979-6) (in Japanese)

(9) Yoshida K., et al., Dynamic Response Characteristics of Taut Moored Platforms, J. Soc. Nav. Archit. Japan, Vol. 146 (1979-11) (in Japanese)

(10) Yoshida K., et al., Dynamic Response Characteristics of Taut Moored Platform (continued), J. Soc. Nav. Archit. Japan, Vol. 147 (1980-5) (in Japanese)

(11) Yoshida K., et al., Response Analysis of Tension Leg Platforms, Symp. on Computational Methods for Offshore Structures, ASME, Chicago (1980-11)

(12) Yoshida K., et al., Three Dimensional Dynamic Response Characteristics of Taut Moored Platforms, J. Soc. Nav. Archit. Japan, Vol. 150 (1981-11) (in Japenese)

(13) Kanetsuna M., et al., On Tension Leg Platforms, the Society of Naval Architects of Japan, Text for the 4th symposium on Ocean Engineering (1979-2) (in Japanese)

(14) Yoneya T., et al., Motion and Leg Strength of Taut Moored Platforms, the Society of Naval Architects of Japan, Text for the 5th Symposium on Ocean Engineering (1981-1) (in Japanese)

(15) Katayama M., et al., Response Analysis of Tension Leg Platform with Mechanical Damping System in Waves, Mitsubishi Juko Giho, Vol. 18, No. 3 (1981-5) (in Japanese)

(16) Oshima M., Sosei II, Working Vessel, Vol. 89 (1973) (in Japanese)

(17) Yamaguchi M., "Wave Observation Buoy" Tension Leg Buoy, Civil Engineering, Vol. 35, No.10 (1975-10) (in Japanese)

(18) Takahashi T., et al., Marine Observation Buoy, Mitsubishi Juko Giho, Vol. 18, No.1 (1981-1) (in Japanese)

(19) Mercier J. A. and Marshall R. W., Design of a Tension Leg Platform, Proc. RINA Offshore Eng'g Group, One-day Symposium on "Offshore Engineering" Teddington Middlesex (1981-11)

(20) Fujii H. and Takahashi T., Estimation of Hydrodynamical Force Acting on a Marine Structure, Mitsubishi Juko, Giho, Vol. 7, No.1 (1970-1) (in Japanese)

(21) Fujii H. and Takahashi T., Estimation of Hydrodynamical Forces Acting on a Maine Structure, MHI Tech. Rev., Vol. 7 No.2 (1970-5)

(22) Arita Y., et al., The Full Scale Field Measurement of the Column Stabilized Drilling Unit, HAKURYU II (MD 20 S), OCEANOLOGY INTERNATIONAL, Brighton, England (1972-3)

(23) Arita Y., et al., The Full Scale Field Measurement of the Column Stabilized Drilling Unit in Waves, Mitsubishi Juko Giho, Vol. 10, No.2 (1973-3) (in Japanese)

(24) Arita Y., et al., The Design, Construction and Operation of the Column Stabilized Drilling Unit HAKURYU II, NOR-SHIPPING SYMPOSIUM, OFFSHORE ACTIVITIES, 1973, Olso, Norway (1973-5)

(25) Hoshino M., et al., Design, Construction, Installation and Outline of Model Test on "AQUAPOLIS" at the World's First International Ocean Exposition, Mitsubishi Juko Giho, Vol. 13, No.4 (1976-7) (in Japanese)

(26) Satake M., et al., On the Design and Construction of Semi-Submersible Offshore Structures, Mitsubishi Juko Giho, Vol. 13, No.4 (1976-7) (in Japanese)

(27) Fujishima K., et al., On the Design and Construction of Semi-Submersible Offshore Structures, MHI Tech. Rev. Vol. 14 No.1 (1977-2)

(28) Satake M. and Katayama M., On Structural Response Analysis of Semi-Submersible Offshore Structures in Waves, Proc. of the Int. Res. Seminar on Safety of Structures under Dynamic Loading, Trondheim, Norway (1977-6)

(29) Katayama M., et al., On Analysis Method of Mooring System for Floating Offshore Structures, Mitsubishi Juko, Giho, Vol. 13, No.4 (1976-6) (in Japanase)

(30) Katayama M., et al., On Analysis Method of Mooring System for Floating Offshore Structures, MHI Tech. Rev., Vol. 14, No.1 (1977-2)

(31) Katayama M., et al., On Structural Response Analysis of Semi-Submersible Offshore Structures in Waves, Mitsubishi Juko, Giho, Vol. 15, No.1 (1978-1) (in Japanese)

(32) Katayama M., et al., On Structural response Analysis of Semi-Submersible Offshore Structures in Waves, MHI Tech. Rev., Vol. 15 No.2 (1978-6)

(33) Seto H., Fundamental Studies on Steady Ship Wave Problems by the Finite Element Method (The 4th Report), J. Soc. Nav. Archit. Japan, Vol. 144 (1978-12) (in Japanese)

(34) Seto H., New Hybrid Element Method for Analysis of Wave Forces acting on Offshore Structures, The 2nd Symposium on Finite Element Methods in Flow Problems, JUSE (1980-8) (in Japanese)

(35) Hogben N. and Lumb F. E., Ocean Wave Statistics, H. M. Stationary Office, London (1967)

(36) Lucht W. A. and Donecker F. W., Factors affecting wire rope life in a marine environment, OTC 2924 (1977)

(37) American Welding Society, Design of New Tubular Structures, AWS Structural Welding Code D1. 1-72, Sec. 10 (1972)

Table 1 Element of matrix for equation of motion

Row (p)	Column (q)	Element of matrix A for inertia force, a_{pq}	Element of matrix B for damping force, b_{pq}
1	1	$\sum_i (M_i + m_{xi})$	$\sum_i (N_{xi} + \bar{D}_{xi}) + \sum_j C_{xj}$
	2	0	0
	3	0	0
	4	0	0
	5	$\sum_i (M_i + m_{xi})(z_i - z_G)$	$\sum_i (N_{xi} + \bar{D}_{xi})(z_i - z_G) + \sum_j C_{xj}(z_j - z_G)$
	6	$-\sum_i (M_i + m_{xi})(y_i - y_G)$	$-\sum_i (N_{xi} + \bar{D}_{xi})(y_i - y_G) - \sum_j C_{xj}(y_j - y_G)$
2	2	$\sum_i (M_i + m_{yi})$	$\sum_i (N_{yi} + \bar{D}_{yi}) + \sum_j C_{yj}$
	3	0	0
	4	$-\sum_i (M_i + m_{yi})(z_i - z_G)$	$-\sum_i (N_{yi} + \bar{D}_{yi})(z_i - z_G) - \sum_j C_{yj}(z_j - z_G)$
	5	0	0
	6	$\sum_i (M_i + m_{yi})(x_i - x_G)$	$\sum_i (N_{yi} + \bar{D}_{yi})(x_i - x_G) + \sum_j C_{yj}(x_j - x_G)$
3	3	$\sum_i (M_i + m_{zi})$	$\sum_i (N_{zi} + \bar{D}_{zi}) + \sum_j C_{zj}$
	4	$\sum_i (M_i + m_{zi})(y_i - y_G)$	$\sum_i (N_{zi} + \bar{D}_{zi})(y_i - y_G) + \sum_j C_{zj}(y_j - y_G)$
	5	$-\sum_i (M_i + m_{zi})(x_i - x_G)$	$-\sum_i (N_{zi} + \bar{D}_{zi})(x_i - x_G) - \sum_j C_{zj}(x_j - x_G)$
	6	0	0
4	4	$\sum_i (M_i + m_{yi})(z_i - z_G)^2 + \sum_i (M_i + m_{zi})(y_i - y_G)^2$	$\sum_i (N_{yi} + \bar{D}_{yi})(z_i - z_G)^2 + \sum_i (N_{zi} + \bar{D}_{zi})(y_i - y_G)^2$ $+ \sum_j C_{yj}(z_j - z_G)^2 + \sum_j C_{zj}(y_j - y_G)^2$
	5	$-\sum_i (M_i + m_{zi})(x_i - x_G)(y_i - y_G)$	$-\sum_i (N_{zi} + \bar{D}_{zi})(x_i - x_G)(y_i - y_G) - \sum_j C_{zj}(x_j - x_G)(y_j - y_G)$
	6	$-\sum_i (M_i + m_{yi})(x_i - x_G)(z_i - z_G)$	$-\sum_i (N_{yi} + \bar{D}_{yi})(x_i - x_G)(z_i - z_G) - \sum_j C_{yj}(x_j - x_G)(z_j - z_G)$
5	5	$\sum_i (M_i + m_{xi})(z_i - z_G)^2 + \sum_i (M_i + m_{zi})(x_i - x_G)^2$	$\sum_i (N_{xi} + \bar{D}_{xi})(z_i - z_G)^2 + \sum_i (N_{zi} + \bar{D}_{zi})(x_i - x_G)^2$ $+ \sum_j C_{xj}(z_j - z_G)^2 + \sum_j C_{zj}(x_j - x_G)^2$
	6	$-\sum_i (M_i + m_{xi})(y_i - y_G)(z_i - z_G)$	$-\sum_i (N_{xi} + \bar{D}_{xi})(y_i - y_G)(z_i - z_G) - \sum_j C_{xj}(y_j - y_G)(z_j - z_G)$
6	6	$\sum_i (M_i + m_{xi})(y_i - y_G)^2 + \sum_i (M_i + m_{yi})(x_i - x_G)^2$	$\sum_i (N_{xi} + \bar{D}_{xi})(y_i - y_G)^2 + \sum_i (N_{yi} + \bar{D}_{yi})(x_i - x_G)^2$ $+ \sum_j C_{xj}(y_j - y_G)^2 + \sum_j C_{yj}(x_j - x_G)^2$

Row (p)	Column (q)	Element of matrix C for restoring force, c_{pq}
1	1	$\sum_j K_{xj}$
	2	0
	3	0
	4	0
	5	$\sum_j K_{xj}(z_j - z_G)$
	6	$-\sum_j K_{xj}(y_j - y_G)$
2	2	$\sum_j K_{yj} + \rho g \sum_k A_k$
	3	0
	4	$-\sum_j K_{yj}(z_j - z_G) - \rho g \sum_k A_k(z_k - z_G)$
	5	0
	6	$\sum_j K_{yj}(x_j - x_G) + \rho g \sum_k A_k(x_k - x_G)$
3	3	$\sum_j K_{zj}$
	4	$\sum_j K_{zj}(y_j - y_G)$
	5	$-\sum_j K_{zj}(x_j - x_G)$
	6	0
4	4	$\sum_j K_{yj}(z_j - z_G)^2 + \sum_j K_{zj}(y_j - y_G)^2 + \rho g \sum_k A_k(z_k - z_G)^2$ $+ B \cdot KB - W \cdot KG - T \cdot Z_p$
	5	$-\sum_j K_{zj}(x_j - x_G)(y_j - y_G)$
	6	$-\sum_j K_{yj}(x_j - x_G)(z_j - z_G) - \rho g \sum_k A_k(x_k - x_G)(z_k - z_G)$
5	5	$\sum_j K_{xj}(z_j - z_G)^2 + \sum_j K_{zj}(x_j - x_G)^2$
	6	$-\sum_j K_{xj}(y_j - y_G)(z_j - z_G)$
6	6	$\sum_j K_{xj}(y_j - y_G)^2 + \sum_j K_{yj}(x_j - x_G)^2 + \rho g \sum_k A_k(x_k - x_G)^2$ $+ B \cdot KB - W \cdot KG - T \cdot Z_p$

Wave exciting force

Row (p)	Column (q)	Element of matrix G for inertia force, g_{pq}	Element of matrix H for damping force, h_{pq}
1	1	$\sum_i (\rho V_i + m_{xi})$	$\sum_i (N_{xi} + \bar{D}_{xi})$
	2	0	0
	3	0	0
2	1	0	0
	2	$\sum_i (\rho V_i + m_{yi})$	$\sum_i (N_{yi} + \bar{D}_{yi})$
	3	0	0
3	1	0	0
	2	0	0
	3	$\sum_i (\rho V_i + m_{zi})$	$\sum_i (N_{zi} + \bar{D}_{zi})$
4	1	0	0
	2	$-\sum_i (\rho V_i + m_{yi})(z_i - z_G)$	$-\sum_i (N_{yi} + \bar{D}_{yi})(z_i - z_G)$
	3	$\sum_i (\rho V_i + m_{zi})(y_i - y_G)$	$\sum_i (N_{zi} + \bar{D}_{zi})(y_i - y_G)$
5	1	$\sum_i (\rho V_i + m_{xi})(z_i - z_G)$	$\sum_i (N_{xi} + \bar{D}_{xi})(z_i - z_G)$
	2	0	0
	3	$-\sum_i (\rho V_i + m_{zi})(x_i - x_G)$	$-\sum_i (N_{zi} + \bar{D}_{zi})(x_i - x_G)$
6	1	$-\sum_i (\rho V_i + m_{xi})(y_i - y_G)$	$-\sum_i (N_{xi} + \bar{D}_{xi})(y_i - y_G)$
	2	$\sum_i (\rho V_i + m_{yi})(x_i - x_G)$	$\sum_i (N_{yi} + \bar{D}_{yi})(x_i - x_G)$
	3	0	0

Remarks; A, B and C are symmertric matrix

Table 2 Principal particulars of upper structure

Item		Model	Prototype
Length (between column center)	L	0.747 m	68.8 m
Deck height	D	0.542 m	49.9 m
Column No. -O.D.		4-ϕ0.15 m	4-ϕ13.8 m
Brace No. -O.D.		4-ϕ0.09 m	4-ϕ8.29 m
Draft	d	0.326 m	30.0 m
Displacement	$\triangle$	38.8 kg	30,000 t
Weight	W	32.3 kg	25,200 t
Height of C.O.G.	KG	0.310 m	28.6 m
Initial tension	T_1	1.5 kg/leg	1,200 t/leg

Table 3 Characteristics of electro-magnetic damper

Range of damping coefficient	0.001~0.2 kg/cm/s
Characteristics of damping	Proportional to velocity v or v^2
Maximum stroke	±15 mm
Range of frequency	0~30 Hz
Control mechanism	Electro-magnetic type

Table 4 Characteristics of external dynamical mechanism

Item \ Condition		K_{51}	K_{52}	K_{53}
Spring const. per leg in column K_1		∞(piano wire)	2 lines-25.7 kg/m	2 lines-305 kg/m
lower K_2		57.1 kg/m	57.1 kg/m	57.1 kg/m
Prototype	K_1	∞	2-218 t/m	2-2,587 t/m
	K_2	484 t/m	484 t/m	484 t/m

Table 5 Test condition

Case	Spring const.	Damping coef. (kg/cm/s)	Wave direction (°)	Wave height (cm)	Wave period (s)
1	K_{51}	0	0	3.3	0.6~2.5
2	K_{51}	0	45	3.3	0.6~2.5
3	K_{53}	0	0	3.3	0.6~2.5
4	K_{53}	0.02	0	3.3	0.6~2.5
5	K_{53}	0	45	3.3	0.6~2.5
6	K_{53}	0	45	3.3	0.6~2.5
7	K_{53}	0.02	0	10.0	0.6~2.5
8	K_{53}	0.02	45	10.0	0.6~2.5
9	K_{52}	0	0	3.3	0.6~2.5
10	K_{52}	0.02	0	3.3	0.6~2.5
11	K_{52}	0.05	0	3.3	0.6~2.5
12	K_{52}	0	45	3.3	0.6~2.5
13	K_{52}	0.02	45	3.3	0.6~2.5

Table 6   Principal particulars of prototype model

Length (between column center) L	68.8 m	
Deck height D	49.9 m	
Column	4-ϕ 13.8 m	
Draft d	30.0 m	
Displacement $\triangle$	30,000 t	
Initial tension T_1	1,125 t/leg (Total 4,500 t; 15% displacement)	
Water depth h	300 m	

	Model A	Model B	Model C
External dynamical mechanism	with soft spring & damping mechanism	with soft spring	with hard spring
Spring const. of damping mechanism K_1 (t/m/leg)	500	500	∞
Damping const. of damp. mechanism C_1 (t.s/m/leg)	1,500	0	0
Spring const. of leg K_2 (t/m/leg)	1,000	1,000	50,000
Remarks	4 wire lines/leg or elastic materials + steel pipe 330t/m/leg and 1,500 cm²/leg	without damping mechanism	with steel pipe

Table 7　Wave frequency in East China Sea[35]

Wave height (m) \ Wave period (s)	~4	4~5	6~7	8~9	10~11	12~13	14~15	16~17	18~19	20~21	21~	Sum.
0.25	91	155	3	0	1	0	0	0	0	2	0	252
0.5	5	444	37	7	3	0	0	0	0	0	23	519
1.0	2	683	266	30	5	0	0	1	0	0	2	989
1.5	11	318	499	99	28	7	3	1	0	0	0	966
2.0	3	70	250	155	38	16	6	1	0	0	0	539
2.5	9	34	166	152	52	11	1	2	0	0	0	427
3.0	0	13	82	104	47	14	10	0	1	0	0	271
3.5	1	13	41	56	39	19	10	0	0	0	0	179
4.0	1	6	23	35	26	16	3	1	0	0	0	111
4.5	1	3	10	22	27	10	6	1	0	0	0	80
5.0	2	1	3	2	0	1	0	0	0	0	0	9
5.5	1	0	1	6	4	4	0	0	0	0	0	16
6.0	3	1	2	3	4	3	3	0	0	0	0	19
6.5	0	0	2	2	3	2	0	2	0	0	0	11
7.0	0	0	4	1	3	0	0	0	0	0	0	8
7.5	1	0	1	0	3	0	0	0	0	0	0	5
8.0	0	0	2	2	2	0	0	0	0	0	0	6
8.5	0	0	0	1	0	0	0	0	0	1	0	2
9.0	0	0	2	0	2	0	0	0	1	4	0	9
9.5	0	0	2	0	0	1	0	0	0	0	0	3
10.0	0	0	0	0	0	0	0	0	0	0	0	0
11.0	0	0	0	0	0	0	0	0	0	0	0	0
Sum.	131	1,741	1,396	677	287	104	42	9	2	7	25	4,421

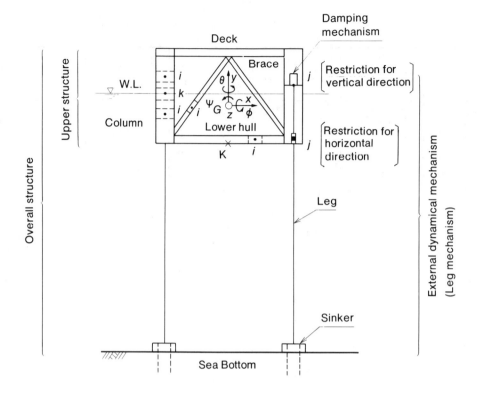

Fig. 1 Idealized model of TLP

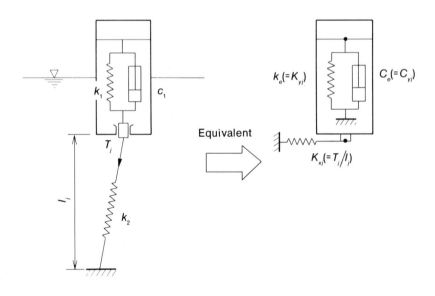

Fig. 2 Idealization of external dynamical mechanism

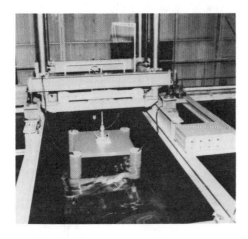

Fig. 3 1/92.1 scaled model

Fig. 4 Electro-magnetic damper

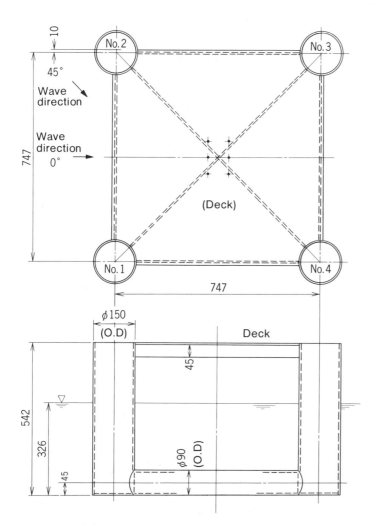

Fig. 5 Main dimension of upper structure model

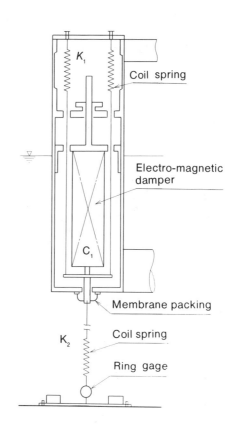

Fig. 6 Model construction of external dynamical mechanism

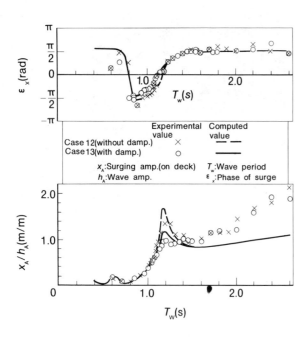

Fig. 7 Surging motion in regular waves

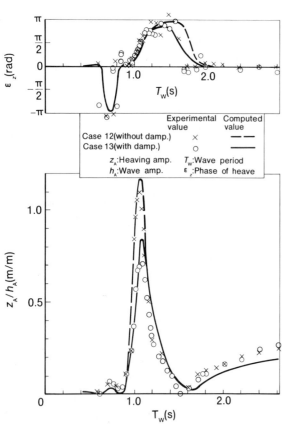

Fig. 8 Heaving motion in regular waves

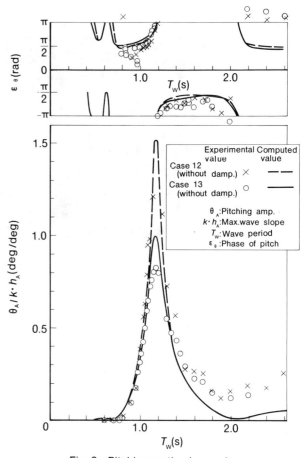

Fig. 9 Pitching motion in regular waves

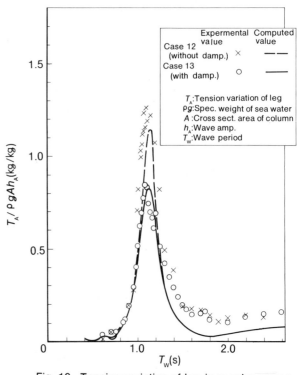

Fig. 10 Tension variation of leg in regular waves

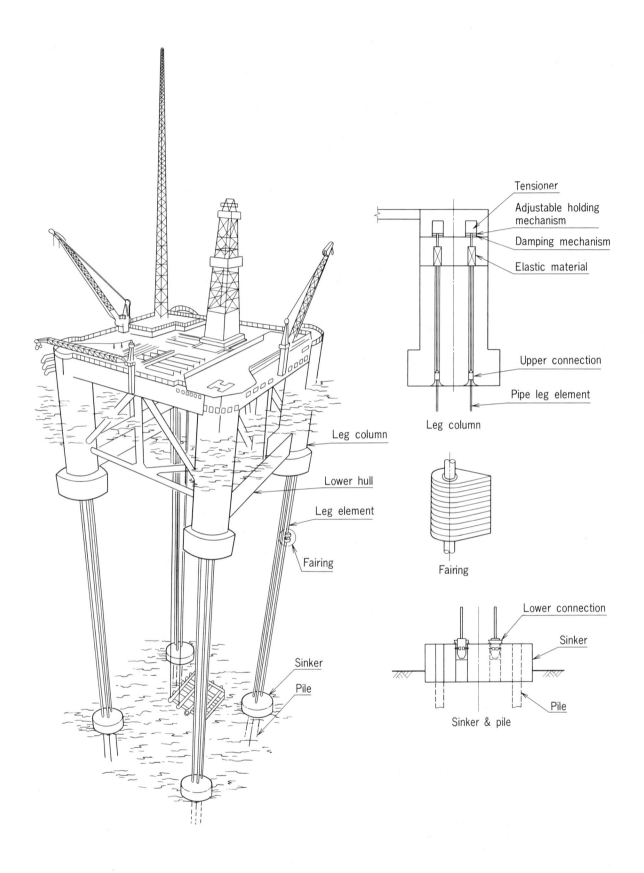

Fig. 11 Concept of prototype model

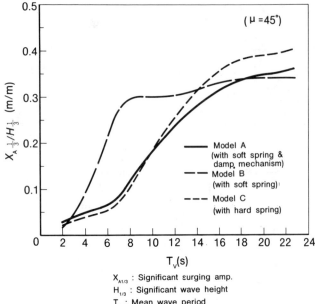

Fig. 12 Significant amplitude of surging motion in irregular waves (prototype model)

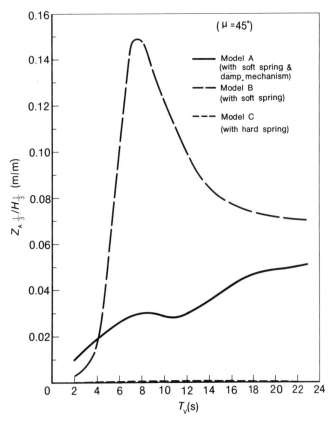

Fig. 13 Significant amplitude of heaving motion in irregular waves (prototype model)

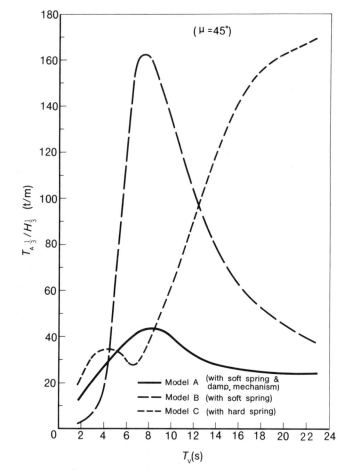

Fig. 14 Significant tension variation of leg in irregular waves (prototype model)

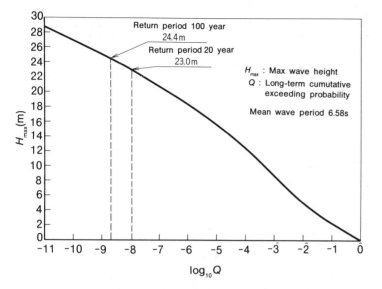

Fig. 15 Long-term cumulative exceeding probability of wave height

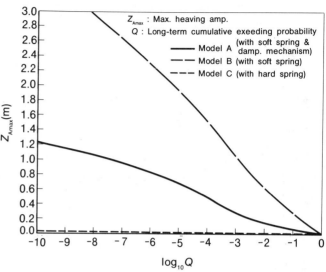

Fig. 17 Long-term cumulative exceeding probability of heaving amplitude

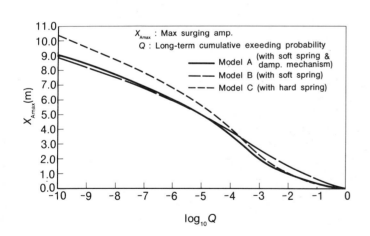

Fig. 16 Long-term cumulative exceeding probability of surging amplitude

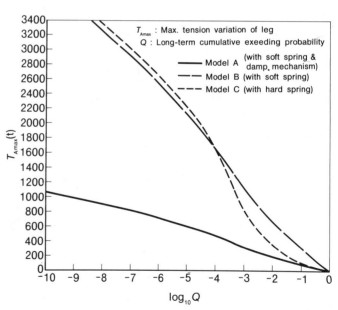

Fig. 18 Long-term cumulative exceeding probability of tension variation of leg

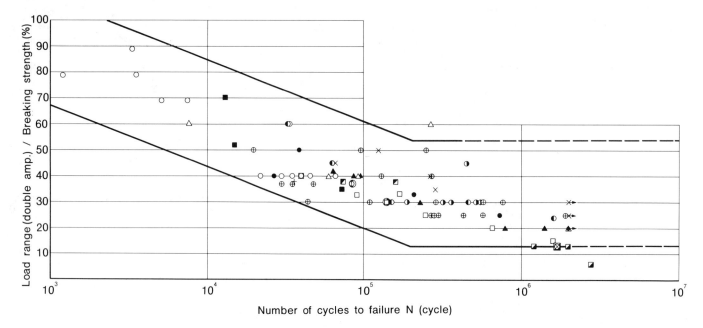

Fig. 19 S-N curve for wire rope[36]

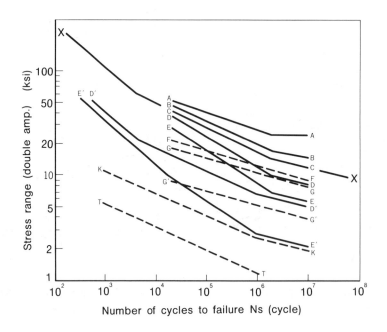

Fig. 20 S-N curve for steel pipe[37]

THE DYNAMIC RESPONSE OF AN UNDERWATER COLUMN HINGED AT THE SEA BOTTOM UNDER THE WAVE ACTION

Qiu Dahong
Dalian Institute of Technology,
The People's Republic of China

Zuo Qihua
Dalian Institute of Technology,
The People's Republic of China

SUMMARY

Under the wave action, the dynamic response of an underwater column hinged at the sea bottom is discussed in this paper. The first part deals with the theoretical consideration of the motion of such a column. Formulae are presented for calculating the inline and transverse swinging amplitude. The second part gives a brief introduction of the model tests which have been done in the wave flume and the test results of the hydrodynamic coefficients C_m, C_D, and C_l. The relations between each of them and the Keulegan-Carpenter number are given in diagrams. The final part gives the response spectrum of the swinging angle under irregular waves. Formulae for calculating spectrum density are presented, and the results calculated by these formulae are compared with those calculated by numerical simulation for several examples.

In this paper, the dynamic response of an underwater column hinged at the sea bottom is investigated. The construction of it is shown in fig(1). Under the action of the plane wave, the column rotates about the Z-axis due to the existence of the lift force. The wave force acted on the column may be resolved into 3 components F_α, F_β, F_γ along the three directions α, β, γ, see fig(2), in which the force F_α and F_β may cause the column to rotate, and F_γ only give the hinge a force, which does not make the column in motion. If we use the generalized coordinate ϕ, ψ to determine the position of the column in any moment, then from fig(2), the relation between the coordinate system (x,y,z) and (r,ϕ,ψ) can be obtained as follows.

$$x = r \sin\psi$$
$$y = r \sin\phi$$
$$z = r \sqrt{\cos^2\psi - \sin^2\phi} = r B \qquad (1)$$

And the relation between the vectors $\vec{\alpha}, \vec{\beta}, \vec{\gamma}$ and $\vec{x}, \vec{y}, \vec{z}$ can be written as

$$\begin{Bmatrix}\alpha\\ \beta\\ \gamma\end{Bmatrix} = \begin{bmatrix} B\cdot\sin\psi/\sqrt{1-B^2} & B\cdot\sin\phi/\sqrt{1-B^2} & -\sqrt{1-B^2}\\ \sin\phi/\sqrt{1-B^2} & -\sin\psi/\sqrt{1-B^2} & 0\\ \sin\psi & \sin\phi & B \end{bmatrix} \begin{Bmatrix}x\\ y\\ z\end{Bmatrix} = [A]\begin{Bmatrix}x\\ y\\ z\end{Bmatrix} \qquad (2)$$

THE BASIC EQUATION AND ITS SOLUTION

The structure can rotate about the hinge with two degrees of freedom, and the dynamic equilibrium equations of this system may be obtained by using the d'Alembert's principle as follows.

$$M_{ox} = \int_0^l (F_y \cdot z - F_z \cdot y)dr - \int_0^l (f-w)\cdot y\,dr - M_{cx} = 0$$
$$M_{oy} = \int_0^l (F_x \cdot z - F_z \cdot x)dr - \int_0^l (f-w)\cdot x\,dr - M_{cy} = 0 \qquad (3)$$

in which

F_x, F_y, F_z — the sum of the wave force and inertia force acting on unit length of the column in direction x, y, z,

w, f — the unit weight and the buoyancy force acting on unit length of the column,

M_{cx}, M_{cy} — the damping moment acting on the column in direction x, y during motion.

In order to simplify the calculation, all the forces acting on column are projected on the direction α, β, γ and by using the coordinate system r, ϕ, ψ, we obtain

$$\int_0^l (F_\alpha \cdot \sin\phi - F_\beta B \cdot \sin\psi)rdr/\sqrt{1-B^2} - M_{cx} - K^*\sin\phi = 0$$
$$\int_0^l (F_\alpha \cdot \sin\psi - F_\beta B \cdot \sin\phi)rdr/\sqrt{1-B^2} - M_{cy} - K^*\sin\psi = 0 \qquad (4)$$

in which

$$K^* = F \cdot H_b - W \cdot H_w,$$

F — the buoyancy force acting on the column,

W — the weight of the column,

H_b — the distance of the buoyancy center from the hinge,

H_w — the distance of the center of gravity from the hinge,

F_α, F_β -- the sum of wave force and inertia force acting on unit length of the column along direction α, β.

$$F_\alpha = F_{m\alpha} + F_{I\alpha} + F_{D\alpha} + F_{l\alpha}$$
$$F_\beta = F_{m\beta} + F_{I\beta} + F_{D\beta} + F_{l\beta}$$

index m denotes the inertia force, due to the acceleration $\dot{u}$ of the column,
     I denotes the wave inertia force, due to the relative acceleration $(\dot{u} - \dot{v})$ between the structure and the water particles,
     D denotes the wave drag force, due to the relative velocity $(u - v)$ between the structure and the water particles,
     l denotes the wave lift force.

$$\begin{Bmatrix} F_{m\alpha} \\ F_{m\beta} \end{Bmatrix} = m(r) \begin{Bmatrix} \dot{u}_\alpha \\ \dot{u}_\beta \end{Bmatrix} \tag{5}$$

$$\begin{Bmatrix} F_{I\alpha} \\ F_{I\beta} \end{Bmatrix} = \tfrac{1}{4}\pi D^2 \rho (C_m - 1) \begin{Bmatrix} \dot{v}_\alpha - \dot{u}_\alpha \\ \dot{v}_\beta - \dot{u}_\beta \end{Bmatrix} + \tfrac{1}{4}\pi D^2 \rho \begin{Bmatrix} \dot{v}_\alpha \\ \dot{v}_\beta \end{Bmatrix} \tag{6}$$

$$\begin{Bmatrix} F_{D\alpha} \\ F_{D\beta} \end{Bmatrix} = \tfrac{1}{2}\rho D\, C_D \begin{Bmatrix} (v_\alpha - u_\alpha)|v_\alpha - u_\alpha| \\ (v_\beta - u_\beta)|v_\beta - u_\beta| \end{Bmatrix} \tag{7}$$

$$\begin{Bmatrix} F_{l\alpha} \\ F_{l\beta} \end{Bmatrix} = \tfrac{1}{2}\rho D\, C_l \begin{Bmatrix} (v_\beta - u_\beta)|v_\beta - u_\beta| \\ (v_\alpha - u_\alpha)|v_\alpha - u_\alpha| \end{Bmatrix} \tag{8}$$

From the linear wave theory

$$\begin{Bmatrix} v_x \\ v_y \\ v_z \end{Bmatrix} = \begin{Bmatrix} 0 \\ \tfrac{1}{2}H\omega\, \mathrm{chk}(z+z_0)/\mathrm{sh}(kd) \cdot \cos(ky - \omega t) \\ \tfrac{1}{2}H\omega\, \mathrm{shk}(z+z_0)/\mathrm{sh}(kd) \cdot \sin(ky - \omega t) \end{Bmatrix}$$

$$\begin{Bmatrix} \dot{v}_x \\ \dot{v}_y \\ \dot{v}_z \end{Bmatrix} = \begin{Bmatrix} 0 \\ \tfrac{1}{2}H\omega^2\, \mathrm{chk}(z+z_0)/\mathrm{sh}(kd) \cdot \sin(ky - \omega t) \\ -\tfrac{1}{2}H\omega^2\, \mathrm{shk}(z+z_0)/\mathrm{sh}(kd) \cdot \cos(ky - \omega t) \end{Bmatrix} \tag{9}$$

in which
     z_0 -- the distance of the hinge above the sea bottom,
     H -- wave height,
     k, ω -- wave number and wave frequency.
From equation (2), get

$$\begin{Bmatrix} v_\alpha \\ v_\beta \\ v_\gamma \end{Bmatrix} = [A] \begin{Bmatrix} v_x \\ v_y \\ v_z \end{Bmatrix}, \quad \begin{Bmatrix} u_\alpha \\ u_\beta \\ u_\gamma \end{Bmatrix} = [A] \begin{Bmatrix} \dot{x} \\ \dot{y} \\ \dot{z} \end{Bmatrix}$$

$$\begin{Bmatrix} \dot{v}_\alpha \\ \dot{v}_\beta \\ \dot{v}_\gamma \end{Bmatrix} = [A] \begin{Bmatrix} \dot{v}_x \\ \dot{v}_y \\ \dot{v}_z \end{Bmatrix}, \quad \begin{Bmatrix} \dot{u}_\alpha \\ \dot{u}_\beta \\ \dot{u}_\gamma \end{Bmatrix} = [A] \begin{Bmatrix} \ddot{x} \\ \ddot{y} \\ \ddot{z} \end{Bmatrix} \tag{10}$$

In the model test, we can see that the column is swinging with the same frequency as that of the wave, if we suppose that ϕ, ψ are small quantities, then (2) can be simplified

$$\begin{Bmatrix} \alpha \\ \beta \\ \gamma \end{Bmatrix} = \begin{bmatrix} \psi/\sqrt{\psi^2 + \phi^2} & \phi/\sqrt{\psi^2 + \phi^2} & -\sqrt{\psi^2 + \phi^2} \\ \phi/\sqrt{\psi^2 + \phi^2} & -\psi/\sqrt{\psi^2 + \phi^2} & 0 \\ \psi & \phi & 1 \end{bmatrix} \begin{Bmatrix} x \\ y \\ z \end{Bmatrix} \tag{11}$$

and from (1), the following relations hold

$$\dot{x} = r\dot{\psi} \qquad , \qquad \ddot{x} = r(\ddot{\psi} - \psi\cdot\dot{\psi}^2) \doteq r\ddot{\psi}$$
$$\dot{y} = r\dot{\phi} \qquad , \qquad \ddot{y} = r(\ddot{\phi} - \phi\cdot\dot{\phi}^2) \doteq r\ddot{\phi}$$
$$\dot{z} = r(\psi\cdot\dot{\psi} + \phi\cdot\dot{\phi}) \doteq 0 \quad , \quad \ddot{z} = -r(\ddot{\psi}\cdot\psi + \ddot{\phi}\cdot\phi + \dot{\psi}^2 + \dot{\phi}^2) \doteq 0 \qquad (12)$$

After analyzing the experimental data, the motion of the column in directions ϕ, ψ may be considered as harmonic motion with the same frequency as that of the wave, and the amplitude of ψ is smaller than that of ϕ. On the basis of the previous suppositions, the equations of motion can be further simplified

$$(\int M^* r^2 dr)\cdot\ddot{\phi} + M_{cx} + K^*\phi$$
$$= C_m^* \int (\dot{v}_y - \dot{v}_z \phi) r dr + C_D^* \int r[(v_y - v_z\phi) - r\dot{\phi}]\cdot|[\cdots]|dr$$
$$(\int M^* r^2 dr)\cdot\ddot{\psi} + M_{cy} + K^*\psi$$
$$= C_m^* \int (-\dot{v}_z)\psi r dr + C_1^* \int r[(v_y - v_z\phi) - r\dot{\phi}]\cdot|[\cdots]|dr \qquad (13)$$

in which

$$M^* = m(r) + \tfrac{1}{4}\pi D^2 \rho(C_m - 1),$$
$$C_m^* = \tfrac{1}{4}\pi D^2 \rho C_m,$$
$$C_D^* = \tfrac{1}{2}\rho D C_D,$$
$$C_1^* = \tfrac{1}{2}\rho D C_1.$$

After analyzing the right-hand side of equation (13), the following considerations can be received:

(1) relative to $\dot{v}_y$, $\dot{v}_z \phi$ is a small quantity and can be neglected,
(2) relative to v_y, $v_z \phi$ is a small quantity and can be neglected,
(3) relative to $C_1^* \int (v_y - r\dot{\phi})|v_y - r\dot{\phi}| r dr$, $C_m^* \int (-\dot{v}_z)\psi r dr$ is a small quantity and can be neglected.

With the supposition $M_{cx} = C\dot{\phi}$, $M_{cy} = C\dot{\psi}$, the equation of motion now can be written as

$$J^*\ddot{\phi} + C\dot{\phi} + K^*\phi = C_m^* \int r\dot{v}_y dr + C_D^* \int r(v_y - r\dot{\phi})|v_y - r\dot{\phi}|dr$$
$$J^*\ddot{\psi} + C\dot{\psi} + K^*\psi = C_1^* \int r(v_y - r\dot{\phi})|v_y - r\dot{\phi}|dr \qquad (14)$$

in which

$$J^* = J_o + (C_m - 1)\rho\pi D^2 l^3/12$$

J_o -- the moment of inertia of the mass of the column about Ox axis.

In order to solve the problem more simply, we denote the equation of the wave profile in complex form

then
$$\eta = \tfrac{1}{2}H e^{-i\omega t}$$
$$v_y = \tfrac{1}{2}H\omega \, chk(z+z_o)/sh(kd) \, e^{-i\omega t} = H'e^{-i\omega t}$$
$$\dot{v}_y = -i\omega H' e^{-i\omega t}$$

Substitute in (14), and take $r = z$ approximately, get

$$J^*\ddot{\phi} + C\dot{\phi} + K^*\phi = C_m^* \int z(-i\omega H')e^{-i\omega t}dz + C_D^* \int z(H'e^{-i\omega t} - z\dot{\phi})|(\ldots)|dz$$
$$J^*\ddot{\psi} + C\dot{\psi} + K^*\psi = C_1^* \int z(H'e^{-i\omega t} - z\dot{\phi})|(\ldots)|dz \qquad (15)$$

The nonlinear term of the drag force in equation (15) can be linearized by using the concept of equivalent drag coefficient C_{eq}, i.e. let

$$(H'e^{-i\omega t} - r\dot{\phi})|H'e^{-i\omega t} - r\dot{\phi}| = C_{eq}(H'e^{-i\omega t} - r\dot{\phi}) \qquad (16)$$

and then equalize the work done in a wave period by the nonlinear drag force with that by the equivalent linearized drag force to determine the value of C_{eq}.

According to the experimental data, the solution of equation (15) may be supposed as

$$\phi = \phi_o e^{-i(\omega t - \varepsilon_\phi)}$$
$$\psi = \psi_o e^{-i(\omega t - \varepsilon_\psi)} \qquad (17)$$

then

$$C_{eq} = (8/3\pi)\cdot\left|H' + ir\phi_o \omega e^{i\varepsilon_\phi}\right| \qquad (18)$$

Substitute (17),(18) in equation (15), get

$$\phi_o e^{i\varepsilon_\phi} = (iP_I C_m + P_D C_D)/((K^* - \omega^2 J^*) - i(C + C_D C^*)\omega)$$

$$\psi_o e^{i\varepsilon_\psi} = (P_D C_1 + i\omega C_1 C^* \phi_o e^{i\varepsilon_\phi})/((K^* - \omega^2 J^*) - i\omega C) \qquad (19)$$

in which

$$C^* = \tfrac{1}{2}\rho D \int_o^l z^2 C_{eq} dz$$

$$P_I = -\tfrac{1}{4}\pi D^2 \rho \int_o^l zH'\omega dz$$

$$P_D = \tfrac{1}{2}\rho D \int_o^l zH' C_{eq} dz$$

THE HYDRODYNAMIC COEFFICIENT C_m, C_D, C_1 AND DAMPING COEFFICIENT C

In order to determine the hydrodynamic coefficient and the damping coefficient, experiments were made in the wave flume. The values of ϕ_o, ψ_o, ε_ϕ, ε_ψ can be measured in the experiments and from equation (19), C_m, C_D, C_1, and C can be solved.

There are two models to be tested, one is a hollow cylinder made by plexiglass with outer diameter 77mm, length 305mm, thick 3mm, weights can be added in the cylinder to change the position of the center of gravity. The other one is a hollow cylinder made by 0.5mm thick galvanized iron sheet with outer diameter 41mm and length 603mm.

Model tests were carried out in the wave flume, 21m in length and 0.75m in width. The wave profile was recorded by the oscillograph, and the motion of the column by cine-camera, the two records were connected by synchro-signal.

46 model tests made, each at least repeats 3 times, the range of the Keulegan-Carpenter number, KC= $U_m T/D$, in these tests is 1.1-19.7.

The relation between the hydrodynamic coefficient and KC number are obtained from the experiments and shown in fig(3) to fig(5).

The measured phase differences between the swinging angle and the surface elevation at the original position of the column ε_ϕ and ε_ψ give larger effect on the calculation of damping coefficient C than C_m, C_D and C_1. Since the speed of the cine-camera used in the experiments is not very high, it influences the accuracy of the measurement of ε_ϕ and ε_ψ, so there is some scatter to the results of damping coefficient C calculated from the test data. In our test range (KC = 1.1-19.7), the damping ratio $\xi = \tfrac{1}{2}C/\sqrt{J^* K^*}$ lies between 0.03-0.08, and its mean value is equal to 0.06.

In order to verify the possibility of using simplified equation (15) in the analysis of the dynamic response of the structure, comparisions between the calculated result and the measured record were made, one of them is shown in fig(6). The solid line in the figure is the result calculated by equation(15) and the dots are the test records. It is shown that the coincidence between them are quite well, therefore it is acceptable that the motion of the structure can be supposed as harmonic motion, and the other suppositions we have made above are all acceptable.

It must be emphasized that the amount of data obtained in the experiments is not so plenty and thus we can only use these data for initial estimating, further work needs to be done for such a structure.

THE RESPONSE SPECTRUM OF THE SWINGING ANGLE OF THE STRUCTURE UNDER IRREGULAR WAVE ACTION

Suppose that the wave motion be a stationary stochastic process, the elevation of the water surface may be expressed by

$$\eta = \sum_{n=1}^{N} \tfrac{1}{2} H_n e^{-i(\omega_n t - \varepsilon_n)} \qquad (20)$$

in which, H_n, ω_n, ε_n are the amplitude, frequency and random initial phase angle of the nth composite wave.

Thus, the terms of the external force in the right hand side of equation (14) may be substitute by the sum of the moments produced by the n composite waves. In order to linearize the equation, it is needed to select a new equivalent drag coefficient C'_{eq} instead of C_{eq}.

From equation (19)

$$\phi = (iM_I + M_D)\eta /((K^* - \omega^2 J^*) - i\omega C') = T_\phi \eta$$

$$\psi = (M_1 + i\omega C_1 C^* T_\phi)\eta /((K^* - \omega^2 J^*) - i\omega C) = T_\psi \eta \qquad (21)$$

in which

$$M_I = -\tfrac{1}{4}\pi D^2\omega^2\rho C_m \int_o^l z\cdot \mathrm{chk}(z+z_o)/\mathrm{sh}(kd)\cdot dz$$

$$M_D = \tfrac{1}{2}D\omega\rho C_D \int_o^l z C'_{eq}\mathrm{chk}(z+z_o)/\mathrm{sh}(kd)\cdot dz$$

$$M_1 = \tfrac{1}{2}D\omega\rho C_1 \int_o^l z C'_{eq}\mathrm{chk}(z+z_o)/\mathrm{sh}(kd)\cdot dz$$

$$C' = C + \tfrac{1}{2}D\rho C_D \int_o^l z^2 C'_{eq} dz$$

Then the spectra of the swinging angle $S_{\phi\phi}$ and $S_{\psi\psi}$ can be written as

$$S_{\phi\phi} = |T_\phi|^2 \cdot S_{\eta\eta}$$
$$S_{\psi\psi} = |T_\psi|^2 \cdot S_{\eta\eta} \tag{22}$$

In order to obtain C'_{eq}, let $\mathring{a} = v_y - z\dot{\phi}$. Since the distribution of the water surface elevation η is zero-mean gaussian, therefore v_y, $\dot{\phi}$ and thus $\mathring{a}$ also have the same distribution as η. Suppose that the mean square deviation of the distribution of $\mathring{a}$ is $\sigma_{\mathring{a}}$, we can use the least square method to determine C'_{eq}, i.e.

$$E = \int_{-\infty}^{+\infty}(\mathring{a}|\mathring{a}| - C'_{eq}\mathring{a})^2/\sqrt{2\pi}\sigma_{\mathring{a}}\cdot\exp(-\mathring{a}^2/2\sigma_{\mathring{a}}^2)d\mathring{a} = \min$$

from which, we obtain

$$C'_{eq} = \sqrt{(8/\pi)\int_o^\infty |T|^2 S_{\eta\eta} d\omega} \tag{23}$$

in which

$$T = \omega\,\mathrm{chk}(z+z_o)/\mathrm{sh}(kd) + i\omega z T_\phi \tag{24}$$

Using equations (21), (22), (23), (24), when $S_{\eta\eta}$ is known, we can calculate the spectrum of swinging angle by iteration.

In order to verify the reliability of the result calculated by equation (22), numerical simulation for a 7.5m diameter and 30m long underwater column were made.

The spectrum of the swinging angle ϕ along the wave direction are calculated from the time history of ϕ obtained by numerical simulation and also by using equation (22). The comparision between them is shown in fig(7). From the figure, it is shown that the results approach to each other, the difference between them increases as the wave height increases. From the view point of engineering practice, the application of equation (22) in estimating the swinging angle spectrum of the structure is feasible.

After determining the swinging angle spectrum, the moments M_i and the width δ of the spectrum can be calculated, thus we can get the mean swinging amplitude $\overline{\phi}_o$ along the wave direction, the swinging amplitude ϕ_{op} for different cumulative probability P, and also the expected value of the maximum swinging amplitude $\overline{\phi}_{o,max}$ and its variance $\sigma_{\overline{\phi}_{o,max}}$ for a certain duration t of the wave.

$$\overline{\phi}_o = (\pi M_o(1-\delta^2)/2)^{1/2}$$
$$\overline{\phi}_{o,max} = \alpha\sqrt{2M_o}$$
$$\sigma_{\overline{\phi}_{o,max}} = \sigma_\alpha\sqrt{2M_o}$$

in which
$$\delta^2 = 1 - M_2^2/M_o M_4$$

M_o, M_2, M_4 — the zero, 2nd, 4th moment of spectrum $S_{\phi\phi}$.

$$\alpha = (\ln(tR))^{1/2} + \tfrac{1}{2}\gamma(\ln(tR))^{-1/2}$$

$$\sigma_\alpha = \pi(\ln(tR))^{-1/2}/(2\sqrt{6})$$

$$R = \sqrt{M_2/M_o}/\pi$$

t — time duration of the wind wave
γ — Euler's constant, $\gamma = 0.57722$

For different values of tR, the values of $\alpha \pm \sigma_\alpha$ are shown in fig(8).

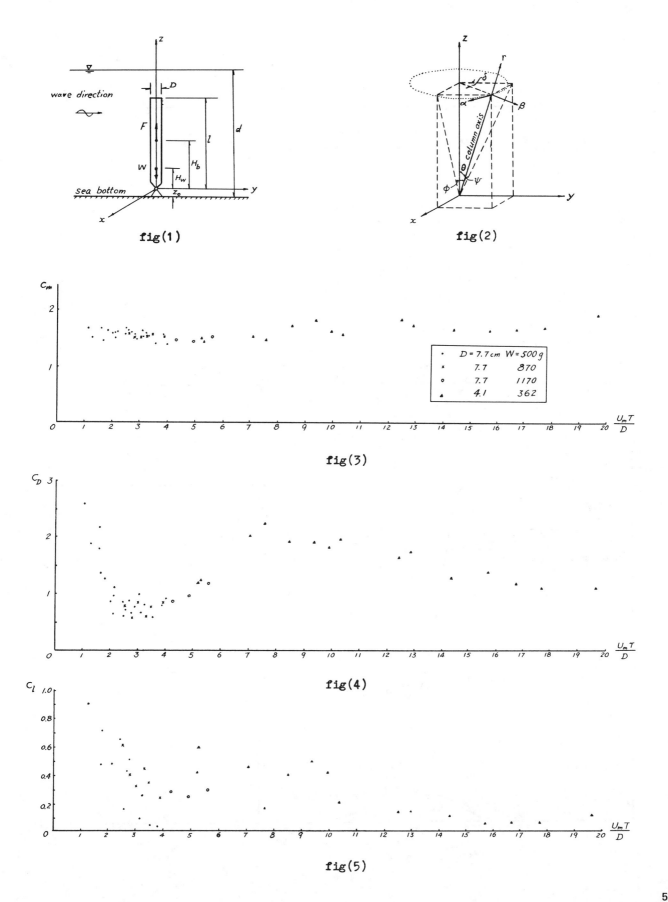

fig(1)

fig(2)

fig(3)

fig(4)

fig(5)

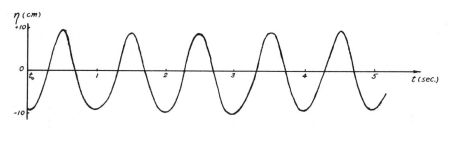
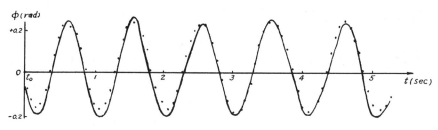
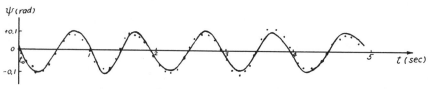

fig(6)

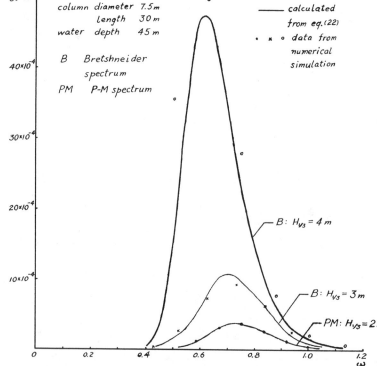

fig(7)

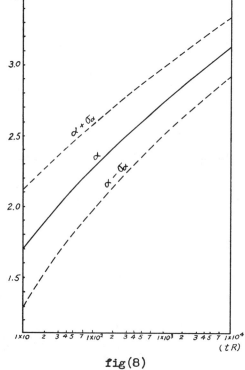

fig(8)

REFERENCES

1. CARTWRIGHT, D. E. and LONGUET-HIGGINS, M. S., 1956, "Statistical Distribution of the Maxima of a Random Function," Proceedings, Royal Society of London, 1956, Series A, Volume 237(1209), pp. 212-232.
2. CHAKRABARTI, S. K., et.al., 1978, "Analysis of a Tower-tanker System," Proceedings, 10th Annual Offshore Technology Conference, Houston, U. S. A., 1978, Paper Number OTC 3202.
3. CHAKRABARTI, S. K. and COTTER, D. C., 1980, "Transverse Motion of Articulated Tower," Journal of the Waterway, Port, coastal and Ocean Division, Proceedings, The American Society of Civil Engineering, 1980, Volume 106, Number WW1, Paper 15178, pp. 65-78.
4. DAVENPORT, A. G., 1964, "Note on the Distribution of the Largest Value of a Random Function with Application to Gust Loading," Proceedings, The Institution of Civil Engineering, 1963-64, Volume 28, Paper 6739, pp. 187-196.
5. ISAACSON, M. Q. and MAULL, D. J., 1976, "Transverse Forces on Vertical Cylinders in Waves," Journal of the Waterways, Harbors and Coastal Engineering Division, Proceedings, The American Society of Civil Engineering, Volume 102, Number WW1, Paper 11934, pp. 49-60.
6. KIRK, C. L. and JAIN, R. K., 1977, "Response of Articulated Towers to Waves and Current," Proceedings, 9th Annual Offshore Technology Conference, Houston, U.S.A., 1977, Paper OTC 2798.
7. SARPKAYA, T., 1976, "In-line and Transverse Forces on Cylinders in Oscillatory Flow at High Reynolds Numbers," Proceedings, 8th Annual Offshore Technology Conference, Houston, U.S.A., 1976, paper OTC 2531.

STUDIES ON PRESSURES AND FORCES DUE TO IRREGULAR WIND GENERATED WAVES ON CIRCULAR CYLINDER - A SPECTRAL ANALYSIS APPROACH

H. RAMAN
Professor
Civil Engg
IIT Madras

P. SAMBHU VENKATA RAO
Research Scholar
Civil Engg. IIT, Madras

SUMMARY

The pressure distribution around a circular cylinder in an irregular wave field has been studied experimentally in the laboratory. The pressures at three different elevations below still water level and various locations around the cylinder were measured with flush type pressure transducers. The spectral density of pressures have been calculated via autocorrelation functions. The total horizontal force has been derived from the pressure time history and represented in the form of spectral density.

INTRODUCTION

The irregular wave loading on offshore structures represents one of the major design considerations in off-shore engineering. In the case of regular waves, for relatively small objects (i.e. as long as diameter of the object does not affect the incident wave field), the force computation is straight forward using Morison's equation (1). The major task in using Morison's equation is to estimate suitable C_D and C_M values for the actual design. For large structures, McCamy and Funchs (2) formulated a diffraction theory for pressures and forces on circular cylinders excited by regular periodic waves. Chakraborti (3) extended the linear diffraction theory of McCamy and Fuchs (2) to Stoke's fifth order wave, treating the fifth order wave as a sum of five linear independent waves. Raman et al (4) developed a mathematical model to the non-linear diffraction problem where the incident wave is in Stoke's second order wave.

For irregular waves, using Morison equation Borgman (5) calculated the force spectrum by means of a transfer function. Information is scarce in attempting to study the pressure variation around the large structure in irregular wave field both analytically and experimentally. This is because the motion of ocean waves is a random process and no exact mathematical solution can be found as yet to compute the pressures on the surface of the cylinder excited by irregular waves. In this paper an attempt has been made to measure the pressure fluctuation on the circumference of the cylinder excited by an unidirectional irregular wind generated wave field in the laboratory.

EXPERIMENTAL TECHNIQUES AND INVESTIGATIONS

The studies have been conducted in a wind wave tank 29m long, 0.9m wide and 0.9m height at hydraulic engineering laboratory of Indian Institute of Technology, Madras, India.

A Schematic diagram showing the experimental set-up is given in Fig. 1. Air flow is created by a blower fitted at one end of the flume and a transition is provided at other end for a smooth entry of air. The top of the flume is covered with aluminium planks overlapping each other and made air tight. An absorbing sloping beach of sand with rubble to a slope of 1:8 is provided at other end to dissipate the wave energy and thus minimize the reflection of wave energy.

The test set up for measuring pressures on the cylinder is shown in Fig. 1. The model is held firmly to the floor of the flume by a cross-beam secured to the side walls of the flume by G-clamps. Pressure fluctuations are sensed by flush type inductive pressure transducers (type HBM P11 made by Hottinger Baldwin Messtechnik, West Germany with a range of pressure ± 1 kg/cm^2). The transducers are inserted into the holes of the model in a such a way that the transducer is flush with the outer surface of the cylinder. Leads from the pressure transducers are taken out of the model through the top and connected to carrier frequency amplifier. The signals thus amplified are fed to a multichannel oszilloscript (PT5108 Philips) and recorded on a graph paper moving at a predetermined speed. A wave gauge is located adjacent to the model and the side wall of the flume to obtain the incident wave elevation. This wave gauge is of resistance type and in the form of thin strip as it should not disturb the incoming wave.

The leads from wave gauge are connected to oszilloscript through bridge. The record of changes in the resistances of the gauge are used as a record of changes of water surface elevation by a calibration shown in Fig. 2.

The model consists of a PVC cylinder of 20cm diameter and 80cm length. Pressure transducers are fitted to the model at three different locations. The flume is filled with 45cm depth of water. At this position, the amplifier bridge and oszilloscript are balanced and zero pressure lines and zero elevation line are drawn on oszilloscript chart. The air blower is started. Pressures and wave profile are simultaneously recorded as analog signals. During the test the model is rotated so that the vertical line containing the pressure transducers occupy seven different positions ($\theta = 0$, $\pi/6$, $\pi/3$, $\pi/2$, $2\pi/3$ and π) with respect to the direction of wave propagation. At each position the experiment is repeated. The pressures and wave profiles are recorded.

The pressure transducers are calibrated by increasing water head and noting the corresponding shift in the recording paper. Calibration is done before and after the experiment. The calibration curves for the three transducers are shown in Fig. 2.

The analog signals of wave elevation and dynamic pressures are digitized manually at a regular interval of 0.1sec. The digitized values are processed in IBM 370/155 computer to evaluate the spectral density via autocorrelation function.

COMPUTATION OF DYNAMIC FORCE

Considering a segment between any two pressure tappings say $P_1(t)$ and $P_2(t)$ at an elevation Z, acting at angles of θ_1 and θ_2 with respect to the direction of wave motion.

$P_1(t)$ and $P_2(t)$ are intensities of pressures acting on the cylinder due to irregular wave field in time series. Then the total horizontal force per unit length on the cylinder can be

$$\sum \frac{P_2(t) + P_2(t)}{2} \cos \frac{(\theta_1 + \theta_2)}{2} \cdot r \cdot d\theta.$$

where r is the radius of the cylinder and $d\theta = \theta_1 - \theta_2$.

In the experiment pressure is calculated as $P = \gamma h$ where h is pressure head in cms of water. The total horizontal force will again be in time series form. The force spectrum is computed via autocorrelation function.

RESULTS AND DISCUSSION

The table 1 presents the statistical parameters calculated from pressure time series. The table includes the spectral parameters calculated from spectral density curves.

The variance of pressure is nearly equal to the total area under the spectral density curve (i.e. 0th spectral moment). The third central moment of pressures at elevations near still water level indicates that the occurrences are clustered to the left of central line when compared to normal distribution curve. The pressure near the bottom indicates an opposite trend. The fourth central moment reveals that all the distribution curves of the pressures are peaked and narrow.

When compared to the spectral parameters the significant pressure calculated as $4.0\sqrt{m_o}$ where m_o is 0th spectral moment, is reducing as one goes to rear side of the cylinder. The same phenomenon is noticed in mean pressure calculated as $1.77\sqrt{m_o}$. The zero crossing period of these pressures ($T_z = m_o/m_2$) is same at every angle and at all depths. The spectral width parameter is 0.7 to 0.8 for all the dynamic pressures at all depths.

The spectral density functions of pressure time series at various angles are shown in Fig. 3. The spectra of dynamic pressures at three locations below still water level around the cylinder at various angles exhibit similarities such as peak frequency where the maximum energy occurs was almost same (1.5 to 1.6 Hz) and coincides with the peak frequency of incident wave spectrum. The distribution of energy over the range of frequencies for pressure is similar to incident wave. The peak energy in the spectrum of pressure is reducing with increase in depth measured from still water level. As one proceeds to the rear side of the cylinder the peak energy in the power spectra is reduced for pressures at three elevations. The statistical and spectral parameters for horizontal forces derived from pressure series are presented in table 2. Spectral density curves for horizontal forces are shown in Fig.4. The horizontal force spectral also reduces with increase in depth measured from still water level.

TABLE - 1

θ	Variance			0th Spectral moment			3rd Central moment			4th Central moment		
	P_1	P_2	P_3	P_1	P_2	P_3	P_1	P_2	P_3	P_1	P_2	P_3
0°	2.55	0.55	0.23	2.27	0.50	0.19	-0.65	-0.53	-0.20	3.03	2.72	2.40
30°	1.98	0.52	0.20	1.95	0.54	0.17	-0.50	-0.30	0.17	2.56	2.47	2.37
60°	1.92	0.40	0.16	1.68	0.34	0.11	-0.58	-0.44	0.10	2.77	2.70	2.30
90°	1.06	0.28	0.11	0.96	0.24	0.08	-0.52	-0.32	0.37	3.28	3.00	2.38
150°	0.91	0.21	0.02	0.76	0.18	0.01	-0.11	-0.21	0.15	3.28	2.80	2.70
180°	0.88	0.24	0.16	0.78	0.20	0.10	-0.09	-0.00	0.02	2.86	2.80	2.80

θ	2nd spectral moment			4th spectral moment			Significant wave pressure			Zero Crossing period		
	P_1	P_2	P_3	P_1	P_2	P_3	P_1	P_2	P_3	P_1	P_2	P_3
0°	7.00	1.52	0.57	41.8	8.77	3.70	6.03	2.85	1.74	0.56	0.57	0.57
30°	5.50	1.50	0.52	31.3	8.70	3.67	5.30	2.70	1.66	0.56	0.55	0.57
60°	4.90	1.00	0.30	28.1	6.50	2.32	5.20	2.30	1.32	0.58	0.58	0.59
90°	2.95	0.74	0.26	19.5	5.30	2.25	3.90	1.90	1.14	0.57	0.57	0.55
150°	2.29	0.56	0.28	15.0	4.24	2.30	3.50	1.70	1.21	0.57	0.57	0.56
180°	2.37	0.58	0.30	15.4	4.10	2.41	3.50	1.79	1.32	0.57	0.58	0.59

θ	Period between crest and trough			Mean wave pressure			Spectral width parameter		
	P_1	P_2	P_3	P_1	P_2	P_3	P_1	P_2	P_3
0°	0.40	0.41	0.39	2.67	1.26	0.77	0.69	0.69	0.73
30°	0.41	0.40	0.37	2.36	1.14	0.73	0.67	0.70	0.74
60°	0.41	0.39	0.37	2.30	1.03	0.58	0.69	0.73	0.78
90°	0.38	0.37	0.34	1.73	0.87	0.50	0.73	0.75	0.79
150°	0.39	0.36	0.30	1.55	0.76	0.20	0.73	0.76	0.80
180°	0.39	0.37	0.35	1.57	0.79	0.58	0.73	0.77	0.80

TABLE - 2

Horizontal Force	Variance	0th spectral moment	3rd central moment	4th Central moment	2nd spectral moment	4th spectral moment
F1	291.38	272.50	-0.2013	3.0428	879.68	5535.88
F2	65.83	64.50	-0.0918	2.7690	213.33	1412.82
F3	21.09	20.11	-0.0100	2.6600	69.31	543.61

Horizontal Force	Significant force	Zero Crossing Period	Period between crest and trough	Mean wave force	Spectral width parameter
F1	65.90	0.55	0.39	29.16	0.69
F2	32.10	0.55	0.38	14.20	0.70
F3	17.90	0.54	0.35	7.90	0.74

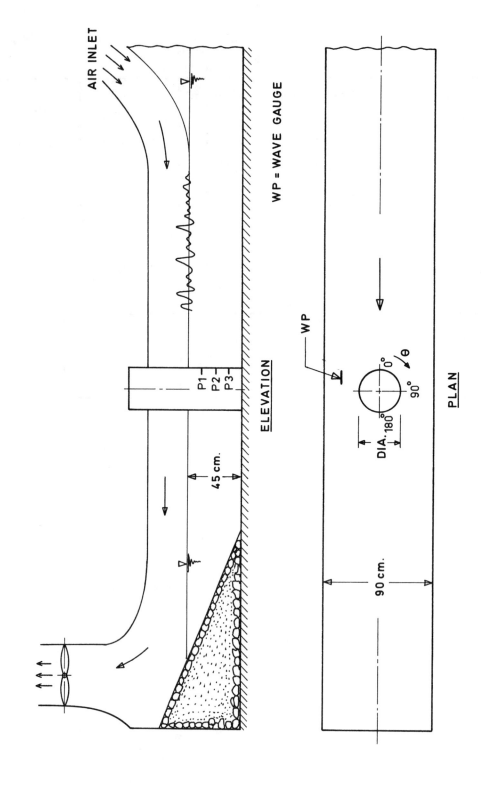

FIG. 1 — SCHEMATIC DIAGRAM OF EXPERIMENTAL SET-UP

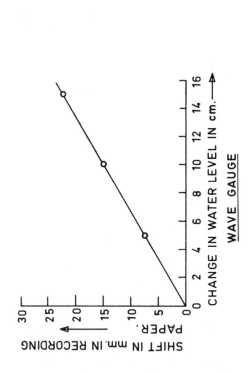
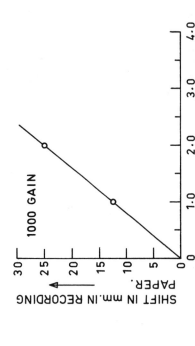
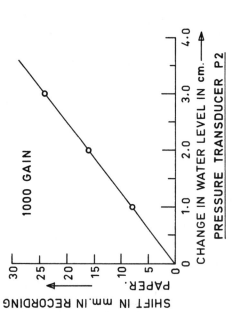
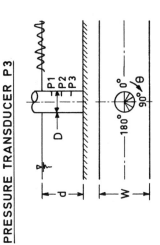

D = DIAMETER = 20 cm.
d = CONSTANT WATER DEPTH = 45 cm.
W = WIDTH OF THE FLUME = 90 cm.
P1 AT 10 cm. BELOW SWL.
P2 AT 20 cm. BELOW SWL.
P3 AT 30 cm. BELOW SWL.

FIG. 2 CALIBRATION CURVES FOR WAVE GAUGE AND PRESSURE TRANSDUCERS.

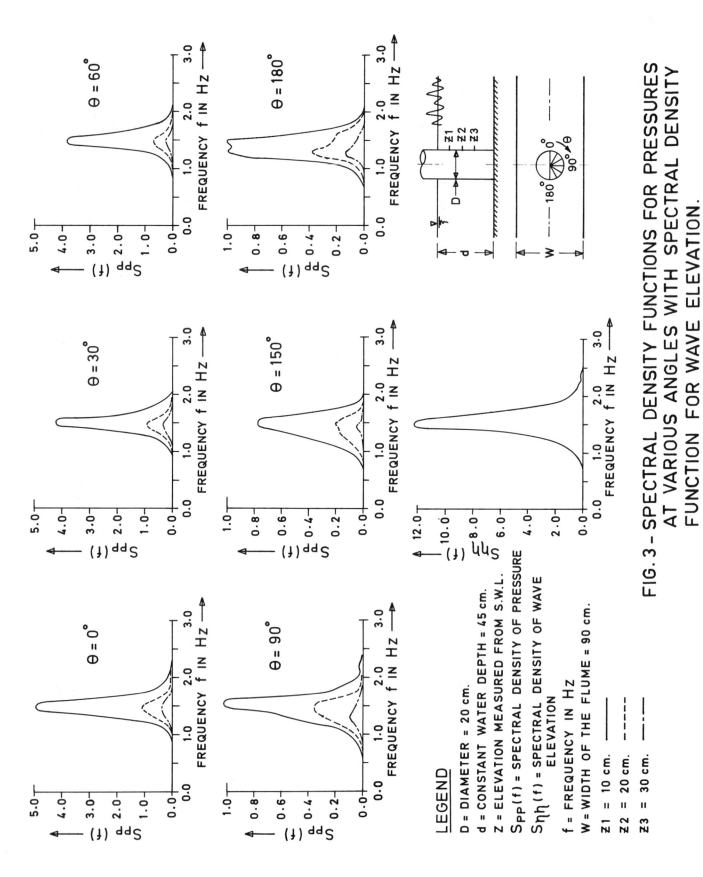

FIG. 3 – SPECTRAL DENSITY FUNCTIONS FOR PRESSURES AT VARIOUS ANGLES WITH SPECTRAL DENSITY FUNCTION FOR WAVE ELEVATION.

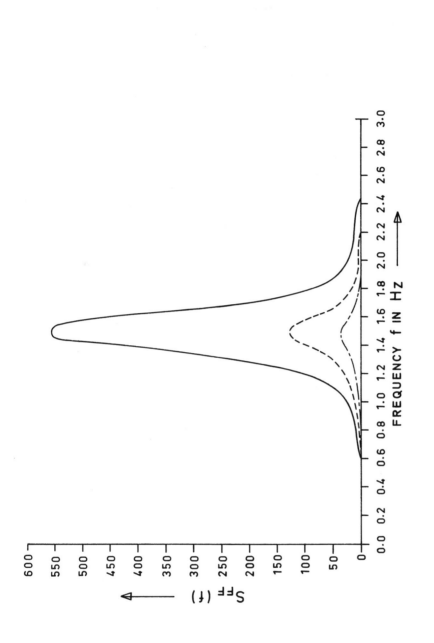
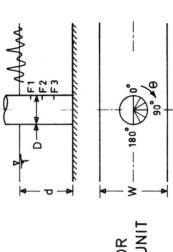

FIG. 4

SPECTRAL DENSITY FUNCTIONS FOR TOTAL HORIZONTAL FORCES PER UNIT LENGTH AT THREE ELEVATIONS.

D = DIAMETER = 20 cm.
d = CONSTANT WATER DEPTH = 45 cm
$S_{FF}(f)$ = SPECTRAL DENSITY FOR HORIZONTAL FORCE IN gm^2-sec.
f = FREQUENCY IN Hz
W = WIDTH OF THE FLUME = 90 cm.
F1 AT 10 cm. BELOW SWL. ———
F2 AT 20 cm. BELOW SWL. ----
F3 AT 30 cm. BELOW SWL. —·—

REFERENCES

1. MORISON, J.R., M.P. O'BRIEN, J.W. JOHNSON and S.A. SHAFF (1950).
   'The forces exerted by surface waves on piles' Petroleum Transactions AIME Vol.189, Technical Paper 2846, 1950, pp 149-154.

2. McCAMY, R.C. and R.A. FUCHS (1954)
   'Wave forces on piles: A Diffraction theory' Beach erosion Board Technical Memo No.69, 1954, 17 p.

3. CHAKRABORTI, S.K. (1972) 'Non-linear wave forces on vertical cylinder' Journal of Hydraulic Division, ASCE, Vol.98, No.HY 11, Proc. paper 9333, Nov. 1972 pp 1895 to 1909.

4. RAMAN, H., and P. Venkata Narasaiah, (1976)
   'Forces due to Non-linear waves on vertical cylinder'
   Journal of water ways, Harbours and Coastal Engineering Division, ASCE, Vol.102, No.WW3, August 1976, pp 301-315.

5. Borgman L.E. (1965) 'The spectral density for ocean wave forces' Proceedings of speciality conference on Coastal Engineering, Chapter 8, 1965, pp. 147-152.

MODELLING OF WIND-WAVE SPECTRA IN LABORATORY WAVES

G Smith, G Baron and I Grant
Heriot-Watt University, Edinburgh, UK

SUMMARY

Wind-wave surface displacement records taken in the laboratory wind-wave flume at Heriot-Watt University have been frequency analysed, using Fast Fourier Transform methods. The resulting energy spectra were studied to determine whether the so called JONSWAP type spectrum provided a suitable method of parameterisation. The results were taken for non-dimensional fetches in the range 10^2-10^3, fetch being scaled to the friction velocity, U_*, which ranged from $0.2 ms^{-1}$ to $0.6 ms^{-1}$. In particular, fetch relations for the JONSWAP parameters were sought and a comparison with previous results, obtained both in the laboratory and at sea, were made. The wave records were recorded using resistance wire gauges at seven different fetches ranging from 3m to 10m.

The fetch relations obtained were incorporated into a finite difference representation of the wave energy transport equation. A parametric approach of the type used by Hasselmann et al (1976) was developed to predict the change in spectral parameters. Initially only f_m, the peak frequency, and α, Phillips' "constant", were used. Comparisons between the computer model results and laboratory data have been made for simple wind conditions. Those studied were for a step change in wind speed from one equilibrium fetch limited spectrum to another. Hot wire probes were used to define the friction velocities for input into the computer model.

1. INTRODUCTION

1.1 Modelling of Sea States

In recent years the prediction of sea-states has aroused major interest. Most computer predictions are based upon some numerical solution of the energy transport equation. This equation has the form, for deep water waves

$$\frac{\partial F}{\partial t}(f,\theta) + v_i(f,\theta) \frac{\partial F(f,\theta)}{\partial x_i} = T \tag{1}$$

where F is the two-dimensional wave enrgy spectrum
V is the group velocity of the wave component, f
T is the general source term.

The numerical solution of (1) is dependent on the form which the various source terms assume. Until about 1968 the source term contained only input and output terms (Barnett, 1968). The input term consisted of two parts, a linear growth function (Phillips, 1957) and an exponential growth function (Miles, 1957; Phillips, 1966). The experimentally determined rate of growth was, however, found to be an order of magnitude greater than expected from theory. The coefficient for exponential growth was thus set empirically. The output term, which accounted for dissipative processes was also modelled to fit experimental results.

The outlook for the numerical solution of (1) changed with the work of Phillips (Phillips, 1960) and Hasselmann (Hasselmann, 1962). They have shown how important the non-linear interaction between wave components can be in determining the form of the energy spectrum and the distribution of energy within the spectrum. Indeed the difference between the theoretical and experimental value of the growth coefficient, mentioned above, was explained by the redistribution of energy, due to non-linear processes, from the region of the spectrum just above the spectral peak to that just below. The non-linear interaction has also been used to explain the shift in peak frequency and the similarity form of the spectrum, including the enhanced peak.

These phenomena, and others, have been studied in the JONSWAP experiment (Hasselmann et al, 1973). The so called JONSWAP spectrum was formulated to describe the spectral form of the energy spectra found experimentally (Figure 1).

$$E(f) = \alpha g^2 (2\pi)^{-4} f^{-5} \exp(\frac{-5}{4}(\frac{f_m}{f})^4 + \ln\gamma\exp(-\frac{(f-f_m)^2}{2\sigma^2 f_m^2})) \tag{2}$$

where α = Phillips' "constant"
f_m = frequency of spectral peak
γ = peak enhancement factor, i.e. ratio of peak spectral density to that predicted by the Pierson-Moskowitz formula
σ = peak width.

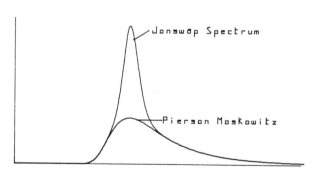

Fig 1: Jonswap and P.M. Spectra.

It was concluded that any wave prediction scheme would have to include some representation of the non-linear interaction if it were to model correctly the most influential factors affecting wave growth. Unfortunately incorporation of the theoretical equations, describing the transfer of energy in the spectrum into conventional wave prediction models is inappropriate since the time needed to perform the numerical integrations is prohibitive.

The energy balance equation was therefore projected into parameter space, in fact the five parameter JONSWAP set, and equations for the change in these parameters found.

$$\frac{\partial a_i}{\partial t} + D_{ijk} \frac{\partial a_j}{\partial x_k} = S_i \quad ; \quad i = 1, 5 \tag{3}$$

D_{ijk} = generalised propagation matrix.

Following this approach a simple parametric mode was produced (Hasselmann et al, 1976) and extended to incorporate swell (Gunther et al, 1979).

It was decided to attempt to adapt their approach to laboratory wind-waves and to this end an investigation of the suitability of the JONSWAP spectrum to laboratory waves was carried out.

2. THE MODEL

The model used in this discussion is basically the same as that due to Hasselmann (see above) which was formulated to predict sea states at large non-dimensional fetches. Some difficulty would be expected in adapting the scheme to model laboratory waves and in this case only simple changes were incorporated. The model relies on empirical data for the dependence of f_m, the non-dimensional frequency, on alpha, the Phillips' "constant". This is because the model is initially set up to model this relation for steady state conditions. All wind speeds and non-dimensional scalings were expressed in terms of U_*, the friction velocity, as this was thought to characterise the energy input into the wave field in a better manner. Also, upon consideration of the conditions for the application of the model, it can only be expected to respond correctly for slowly varying wind fields. A small step change in speed allowing the sea to relax from one equilibrium state to another is considered to be within this definition. Another simplification is that the waves in the tank are considered to be essentially one-dimensional and therefore no difficulty in proposing some form of spreading function is encountered.

Some thought must be given to whether the physical processes occurring in the two cases of sea and laboratory are the same. In this model it is the non-linear interaction which plays the predominant part and there is evidence to support the conclusion that it has a similar role to play in laboratory waves as it has in waves at sea-going fetches. In fact several of the physical phenomena occurring at sea can be found in laboratory waves, for example the decrease in peak frequency with fetch and wind speed, the existence of some type of similarity form and the so called over-shoot effect (Mitsuyasu, 1969). Mitsuyasu (1968) has compared data he acquired on energy transfer in waves with that predicted from the Hasselmann theory using a parametric form, due to Barnett (1968). It was found that near the peak of the spectrum the experimental values of energy transfer agreed well with the theoretically predicted values. At higher frequencies there was a marked disagreement in all cases which was thought to be due to the many modes of energy dissipation not included. There was also disagreement at low frequencies for high wind speeds. This is to be expected as, in this case, strong non-linear interactions due to wave breaking occur. All the present tests were thus run at low enough speeds to prevent any wave breaking to any significant degree.

2.1 The Finite Difference Scheme

The model in its present form is basically that due to Gunther et al (1979) with only two parameters being predicted, f_m and alpha. The equations to be solved were

$$\frac{\partial a_i}{\partial t} + D_{ij} \frac{\partial a_i}{\partial x_j} = S_i \quad \begin{array}{l} a_1 = f_m \\ a_2 = \alpha \end{array} \tag{4}$$

The equations are coupled through terms in the propagation matrix D_{ij} and the source functions S_i. The functional form of the D_{ij} are the same as those used by Gunther.

The finite difference scheme chosen was the McCormack Predictor-Corrector method which uses both forward and backward differences.

Stage 1: Predict approximate value for a parameter at $t + \Delta t$ and point i

$$\tilde{a}_i^{n+1} = a_i^n - \frac{\Delta t}{\Delta x} (D^n (a_i^n - a_{i-1}^n)) + t S^n$$

Stage 2: Correct to give a new value at $t + \Delta t$

$$a_i^{n+1} = (a_i^n + \tilde{a}_i^{n+1})/2 - \frac{\Delta t}{2\Delta x} (\tilde{D}^{n+1} (\tilde{a}_{i+1}^{n+1} - \tilde{a}_i^{n+1})) + \frac{1}{2} \Delta t \tilde{S}^{n+1}$$

This method was chosen for its simplicity in modelling boundary conditions. The stability of the relevant homogeneous equation is ensured if

$$\frac{\Delta x}{\Delta t} > (v_g)_{max}$$

where v_g is the maximum group velocity encountered.

At outgoing boundaries a^{n+1} was set equal to $\hat{a}^{n+1}$ as the second equation was inappropriate.

The model was run for equilibrium values for f_m and alpha and shown to come to the required fetch relation.

3. THE EXPERIMENTS

3.1 Wind Wave Facility

The experiment was conducted in an open circuit wind-wave channel recently constructed in the Department of Civil Engineering at Heriot-Watt University. The supports of the channel were constructed from metal framed modules 2.44m long attached in line to provide the basic configuration. Its sides were constructed from plywood and glass (glass was chosen in preference to perspex so that other fluid flow experiments involving laser doppler anemometry could be carried out in the channel). The design allowed the channel roof elevation to be varied. This feature meant that the pressure gradient along the channel could be varied if required.

The working section was 14m long and had a width of 0.45m. In this experiment the water depth was 0.58m, leaving an air space of 0.47m above the surface of the mean water level. The reflection of waves at the downstream end was reduced to a negligible level by a two-component beach.

The fan and motor were mounted on a frame independent of the tunnel and connected to it by a flexible rubber membrane. The fan was mounted axially and powered by a dc motor. Straighteners were incorporated in the diffuser section of the fan housing to minimise swirl of the air from the fan. Freestream turbulence was reduced by incorporating a contraction, screen and honeycomb.

In designing the tunnel for the maximum velocity reference was made to the work of Hidy and Plate (Hidy and Plate, 1966) and Grant and Barstow (Grant and Barstow, 1979). The maximum friction velocity attainable at the present was approximately $0.9 ms^{-1}$.

3.2 Experimental Arrangement (I); Derivation of Parameters and Fetch Relations

3.2.1 Measurement of Friction Velocities

Air velocities were measured using a pitot-static tube connected to a micromanometer. The air flow distribution was examined and showed the typical "law of the wall" profile for a turbulent boundary layer (Figure 2). The results obtained were consistent with those of Gad-El-Hak (1981) within experimental error and a fetch relation (5) was used to relate the friction velocity to the free stream velocity.

$$U_* = U_\infty \times 0.055 \left(\frac{gx}{U_\infty^2}\right)^{-1/10} \tag{5}$$

3.2.2 Wave Spectra

Water surface displacements were measured at 7 fetches, between 4 and 10m, using resistance type twin-wire probes. They were obtained at 3 different wind speeds. The probes were calibrated before and after the measurements had been taken. No appreciable drift was encountered.

3.2.3 Experimental Arrangement (II); Step Change in Wind Speed

Again fetch limited conditions were allowed to develop. When recording of the analogue signals were started one wave gauge was shorted for several seconds. This was to provide a method of aligning the separate records after they had been digitised. Wind speeds were measured using two hot-wires, one at each end of the measuring region. They had previously been calibrated using a pitot static tube in the normal manner. After a suitable length of record had been taken (about 10 minutes) to obtain spectral estimates a step change in wind speed was introduced. Recording continued for a further twenty minutes.

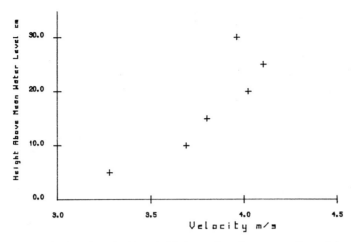

Fig 2: Plot of Wind Velocity Versus Height

3.3 Data Processing

The analogue data from the tapes was digitsed at a sampling frequency of 16Hz using a DEC PDP8 A-D converter and stored on an ICL 2972 digital computer. Spectral estimates were produced using standard Fast Fourier Transform methods. The spectra were smoothed and the parameters f_m, the peak frequency, alpha and gamma were found.

3.3.1 Parameter Fitting Technique

The parameter f_m, the frequency of the spectral peak, was found by using a least squares parabolic fit about the spectral peak, i.e. $y = a_0 + a_1 x + a_2 x^2$. The peak frequency is given by $f_m = -a_1/2a_2$.

The parameter alpha was found by one of the methods used by Mitsuyasu (1973)

$$\alpha = \int_{f_m}^{\infty} E(f) df / (g^2 (2\pi)^{-4} \int_{f_m}^{\infty} f^{-5} df) \qquad (6)$$

Further investigation into the derivation of this parameter is envisaged.

The parameter gamma was found by taking the ratio of the spectral density, at its maximum, to that found from the Pierson-Moskowitz spectrum with corresponding f_m and alpha parameters.

3.4 Results

Typical spectra are shown in Figures 3 and 4. The first depicts the variation in spectrum with fetch at constant friction velocity. The second shows the variation with friction velocity at constant fetch.

A graph of $\hat{f}_m$, the non-dimensional frequency = $f_m U_*/g$ versus $\hat{x}$, the non-dimensional fetch = xg/U_*^2 is shown in Figure 5. The fetch relation derived from this data using correlation techniques was

$$\hat{f}_m = 1.02 \hat{x}^{-0.35}$$

compared with

$$\hat{f}_m = 1 \hat{x}^{-0.33} \quad : \quad \text{Mitsuyasu (1973)}$$

$$\hat{f}_m = 1.08 \hat{x}^{-0.33} \quad : \quad \text{JONSWAP}$$

The results for alpha are more scattered. A plot of some of the alpha values obtained are shown in Figure 6. The fetch relation was found to be

$$\alpha = 0.2 \hat{x}^{-0.14}$$

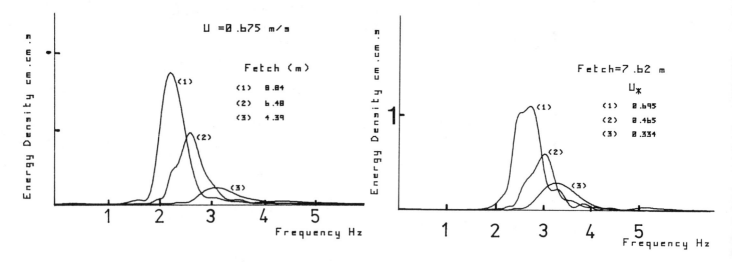

Fig 3: Variation in Spectra With Fetch.

Fig 4: Variation in Spectra With U_*.

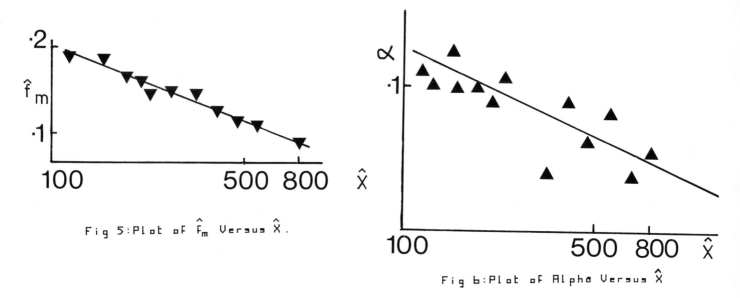

Fig 5: Plot of $\hat{f}_m$ Versus $\hat{x}$.

Fig 6: Plot of Alpha Versus $\hat{x}$

compared with

$\alpha = 0.313\, \hat{x}^{-0.22}$: JONSWAP

$\alpha = 0.589\, \hat{x}^{-0.308}$: Mitsuyasu (1973)

The values of gamma showed no discernible fetch relation and ranged from 9 to 14.

Comparsion has been made between experimentally derived spectra and JONSWAP spectra, calculated using parameters obtained from the fitting technique of Section 3.3. One such comparison is shown in Figure 7 and as can be seen the comparison is good.

Figure 8 shows the variation of a spectra due to a step change in wind speed. The second spectrum was calculated using a 10 minute record commencing 10 minutes after the step change.

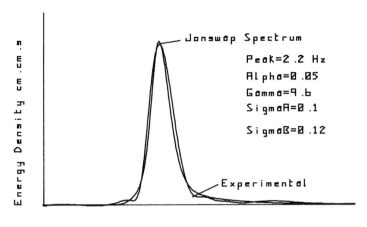

Fig 7: Comparison of Experimental and Jonswap Spectra

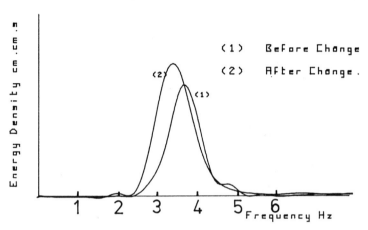

Fig 8: Effect of Step Change On Spectra

4. DISCUSSION OF MODEL RESULTS

The model has been shown to predict the correct fetch relation if allowed to relax from a non-equilibrium position. It has been run to simulate a step change in wind speed and to predict the variation of the two parameters f_m and alpha from one equilibrium fetch limited solution to another. The relations used to test the model were those derived in the JONSWAP study (scaled with respect to U_*). In the computation, the boundary parameters were changed to fit the equilibrium fetch relation for f_m and alpha, for the change in non-dimensional fetch due to the change in wind speed. The time needed for the parameters at a certain fetch to come back to equilibrium was of the same order of magnitude as the time inferred from experimental data. (This was measured by noting the change in rms values for the experimental spectra).

5. CONCLUSIONS

The spectra of wind-generated waves produced under laboratory conditions have been compared with the power spectral density curves predicted by the JONSWAP model. The non-dimensional fetches available in the experiment fell in the range 100-1000 (scaled with respect to the friction velocity, U_*). Within this range the principal spectral parameters f_m, the modal frequency, and alpha, the Phillips' "constant", agreed well with the results of other laboratory and field studies.

The dependence of f_m and alpha on fetch was inferred from a finite difference representation of the governing wave energy transport equation using JONSWAP data. Using the parametric approach developed by Hasselmann et al (1976) a scheme was devised to allow prediction of these parameters for a wide range of fetch conditions.

Current work, which is described only briefly here, involves the measurement of changes in the wave spectral form with changes in wind speed. Preliminary results are presented showing the change in spectral form induced by rapid, discontinuous shifts in windspeed.

REFERENCES

BARNETT T P, 1968, "On the Generation, Dissipation and Prediction of Ocean Wind Waves", Journal of Geophysical Research, Volume 73, pp513-530.

GAD-EL-HAK M, 1981, "Measurements of Turbulence and Wave Statistics in Wind-Waves", International Symposium on Hydrodynamics in Ocean Engineering, Trondheim, 1981.

GRANT I and BARSTOW S, 1979, Mechanics of Wave Induced Forces on Cylinders, London: Pitman Publishing Ltd, pp272-286.

GUNTHER H et al, 1979, "A Hybrid Parametrical Wave Prediction Model", Journal of Physical Oceanography, Volume 8, pp5727-5737.

HASSELMANN K, 1962, "On the Non-Linear Energy Transfer in a Gravity Wave Spectrum: 1 General Theory", Jounral of Fluid Mechanics, Volume 12, pp481-500.

HASSELMANN K et al, 1973, "Measurements of Wind-Wave Growth and Swell Decay During the Joint North Sea Wave Project (JONSWAP)", Herausgegegen vom Deutchen Hydrographischen Institut, Reihe A(8), number 12.

HASSELMANN K et al, 1976, "A Parametric Wave Prediction Model", Journal of Physical Oceanography, Volume 6, pp200-228.

HIDY G M and PLATE E J, 1966, "Wind Action on Water Standing in a Laboratory Channel", Journal of Fluid Mechanics, Volume 26, pp651-687.

MILES J W, 1957, "On the Generation of Surface Waves by Shear Flow, 1", Journal of Fluid Mechanics, Volume 1, pp185-204.

MITSUYASU H, 1968, "A Note on the Non-Linear Energy Transfer in the Spectrum of Wind-Waves", Reports of Research Institute for Applied Mechanics, Kyushu University, Volume 16, N°54, pp251-264.

MITSUYASU H, 1969, "On the Growth of the Spectrum of Wind Generated Waves (II)", Reports of Research Institute for Applied Mechanics, Kyushu University, Volume 17, N°59, pp235-248.

MITSUYASU H, 1973, "The One Dimensional Wave Spectra at Limited Fetch", Reports of Research Institute for Applied Mechanics, Kyushu University, Volume 20, N°59, pp37-53.

PHILLIPS O M, 1957, "On the Generation of Waves by Turbulent Wind", Journal of Fluid Mechanics, Volume 2, pp417-445.

PHILLIPS O M, 1960, "On the Dynamics of Unsteady Gravity Waves of Finite Amplitude, Part 1 The Elementary Interactions", Journal of Fluid Mechanics, Volume 9, pp193-217.

PHILLIPS O M, 1966, The Dynamics of the Upper Ocean, London, Cambridge University Press.

Technical Poster Sessions

HYDRODYNAMICS

Floating Vessels

DYNAMIC RESPONSE OF OFFSHORE PLATFORM CRANES
USING PHYSICAL AND MATHEMATICAL MODELS

J. A. D. BALFOUR
HERIOT-WATT UNIVERSITY

A. O. BOWCOCK
HERIOT-WATT UNIVERSITY

SUMMARY

The paper is concerned with the use of mathematical models to predict forces in offshore platform cranes while off-loading from supply vessels. Small scale physical models are used to demonstrate that mathematical models can reliably predict dynamic forces induced in the crane structure during lifting. Both physical and mathematical models are used to show that non-linear cable stiffness, acceleration of the load prior to the load lifting off the deck and non-linear retrieval of the hoist cable by the winch can significantly affect the dynamic response of the crane. A unified computational strategy to deal with these problems is outlined.

INTRODUCTION

The technology first employed in the development of oil and gas fields in the North Sea was based almost entirely upon that used to develop the relatively quiet waters off the Californian coast and in the Gulf of Mexico. The harsh environment of the North Sea soon highlighted areas where the existing technology was inadequate and where current understanding needed to be expanded. One important instance of this was in the design and operation of offshore platform cranes. Statistics show that during the early stages of development in the North Sea the incidence of failures and dangerous occurences during crane operations was unacceptably high. Prior to January 1979 there where ten fatalities directly attributable to such operations. It is now generally recognised that the root cause of many of these problems was that insufficient attention had been given to the dynamic forces induced in the crane structure during lifting. All crane operations are dynamic in nature. Even on land the fact that the load must be accelerated from rest to the lifting speed of the crane means that forces greater than the static forces are always induced in the crane structure. During offshore lifting the additional complication that the load is rarely, if ever stationary opens up the possibility of even more severe dynamic effects.

The problem of trying to lift a load from a supply vessel onto an offshore platform can be easily visualised. The crane operator, from his cab high above the water level, must observe the motion of the supply vessel and try to initiate the lift such that the load is lifted quickly and cleanly away from the vessel, thus minimising the danger to the supply boat and its crew. Generally this means that the crane operator attempts to lift the load as the supply vessel approaches a peak of its motion. Gauging the boat's vertical motion from a vantage point almost vertically above is difficult, and misjudgement of a lift such that the crane tries to lift a downwards moving load will have two undesirable consequences. Firstly, as the load has an initial momentum in a direction opposite to that intended, the shock loading of the crane will be high. Secondly, the load will take longer to clear the vicinity of the supply vessel hence increasing the danger of collision.

The regular transfer of seaborne supplies onto an offshore installation is essential for its continuous operation. This is especially true during the fabrication and drilling stages, but it also holds during the production phase. As the economic consequences of stopping or reducing the operation of an offshore platform are enormous it is understandable that the owners of these installations demand that their cranes work in all but the most severe conditions.

In recognition of the dynamic nature of offshore lifting, cranes operating in the North Sea are now subject to derating rules which set safe working load according to the prevailing sea state. The objective of any form of derating is to allow the best possible use of craneage while maintaining adequate safety levels. To establish a realistic derating criterion is a difficult task requiring account to be taken of both crane dynamics and supply boat motion and any rational derating scheme will produce a different derating factor for each combination of crane, supply vessel and sea state. Unfortunately there is little consistency between derating rules currently used in the North Sea, where the safe working load for given conditions can vary widely from one sector to another.

In order to design or derate an offshore platform crane effectively a study of the dynamic behaviour of that crane over a wide range of operating conditions must be made. Since the operation of offshore platform cranes in exposed waters was first recognised to be a problem a considerable amount of work has been done to develop techniques to improve the design and derating of such cranes. These techniques have varied from simple mathematical models capable of explicit solution (3, 4, 5) through to the use of complex finite element packages (1, 6, 7, 10) and analogue computers (8). In a previous paper (2) the authors argued strongly in favour of the use of computer programs specifically written for offshore platform cranes and results from such programs were produced for mathematical models of varying complexity. The remainder of this paper assesses the validity of these programs using physical models and investigates a number of computational problems associated with mathematical modelling. Throughout the paper a typical pedestal mounted, short tail radius, high "A" frame crane commonly used on offshore installations was considered. The results presented are for whip hoist or single part lifting only.

PHYSICAL MODELS

Description

Two 1:20 scale model of offshore platform cranes of the general form shown in Figure 1 were constructed. The principal difference between the models was that one had a jib whose mass and flexibility were scaled to represent a typical prototype jib. The other model had by comparison a very stiff and heavy jib which effectively reduced the number of degrees of freedom to two. In both models the hoist was made from nylon covered multistrand wire and the boom hoist cables were made from monofilament steel wire. The winch drum was driven by a heavy duty, 1.5 kW motor fitted with a variable speed controller. The deck of the supply boat was represented by a square steel plate attached to a vertically mounted Seasim actuator driven by a suitable analogue signal.

Instrumentation and Data Processing

The force in the hoist and boom hoist cables was measured using pairs of strain gauges attached to proving rings which were located as shown in Figure 1 and the motion of the supply vessel simulator was measured using an accelerometer. To assess bending strains in the "flexible boom" five pairs of strain gauges were placed at equal intervals along its length. The cable stiffnesses were measured using a linear voltage displacement transducer (LVDT) and are shown in Figures 2 and 3. Both plots show the stiffness to be approximately linear at high loads and highly non-linear at low loads.

Output from the instrumentation was amplified and recorded onto a multi-track tape recorder. The results were then played back through an analogue to digital converter and stored in digital form on magnetic disk. Finally the results were output on a digital plotter, this having the advantage that results from the mathematical model could be displayed on the same set of axes enabling a direct comparison to be made.

MATHEMATICAL MODELS

At any instant of time in a vibrating structure, inertia, damping and spring forces must be in balance with the applied forces. This leads to the well known matrix equation of motion. (Nomenclature is given at the end of the paper)

$$[M][\ddot{X}] + [C][\dot{X}] + [K][X] = [P] \tag{1}$$

By using Newmarks approximations to future velocity and displacement

$$\dot{x}_{s+1} = \dot{x}_s + \Delta t(\ddot{x}_s + \ddot{x}_{s+1})/2$$
$$x_{s+1} = x_s + \Delta t \dot{x}_s + (1/2 - \beta)(\Delta t^2)\ddot{x}_s + \beta(\Delta t^2)\ddot{x}_{s+1} \tag{2}$$

the following matrix relationship can be derived

$$([M] + \tfrac{1}{2}\Delta t[C] + \beta(\Delta t^2)[K_{s+1}])[X_{s+1}] =$$
$$(\Delta t^2)(\beta[P_{s+1}] + (1 - 2\beta)[P_s] + \beta[P_{s-1}]) + (2[M] - (\Delta t^2)(1 - 2\beta)[K_s])[X_s]$$
$$- ([M] - \tfrac{1}{2}\Delta t[C] + \beta(\Delta t^2)[K_{s-1}])[X_{s-1}] \tag{3}$$

or

$$[LHS][X_{s+1}] = [PN] + [R1][X_s] + [R2][X_{s-1}] \tag{4}$$

$$= [RHS] \tag{5}$$

Essentially current and past conditions are used to predict displacement one time increment into the future. Apart from the future displacements the only unknowns at any instant of time in equation (3) are the future stiffness matrix $[K_{s+1}]$ in the [LHS] matrix and the vector of future forces $[P_{s+1}]$ in the [RHS] matrix. If these can be found the solution of the resulting system of linear equations described by equation (5) yields the displacements $[X_{s+1}]$ at one time increment into the future. The problems associated with finding $[K_{s+1}]$ and $[P_{s+1}]$ depend upon whether the system is assumed to be a linear or non-linear. Algorithms have been developed for both types of system, and because the equations have to be solved at each time step, the choice of algorithm can greatly affect the program efficiency.

Linear Systems

In this model all of the structural elements within the system were assumed to behave as linear springs (ie. doubling the load doubles the deflection). Hence the structural stiffness matrix was constant for all steps and the [LHS] matrix was generated only once at the beginning of the analysis and reduced to diagonal form. Using the diagonal form of the matrix in all subsequent calculations resulted in a great saving of computer time. For a crane idealised as shown in Figure 4 there will be 2n+1 degrees of freedom prior to the load lifting off the deck and 2n+2 thereafter (Where n is the number of beam elements).

Retrieval of line by the winch was modelled by considering external forces as shown in Figure 5. These forces had a magnitude equal to the stiffness of the hoist cable times the amount line retrieved by the hoist. In a system where the line is retrieved in some predetermined manner this means that at the beginning of each time step the future external force vector $[P_{s+1}]$ can be determined and the equations need solving only once per step. If the amount of line retrieved during the future time increment depends upon future conditions (say in the case of a winch with limited power) a number of iterations may be required at each time step until the criterion for line retrieval is satisfied.

For the period of time between the hoist cable first going taut and the load lifting off the deck the system was regarded as having 2n+1 degrees of freedom. Account of the supply vessel motion was taken by algebraically adding the displacement of the supply vessel to the amount of line retrieved when calculating the external forces. At each time step prior to liftoff the magnitude of the reaction between the deck and the load was calculated (taking account of acceleration of the load) and when the reaction became zero or negative the load had parted company with the supply vessel. Thereafter the reaction was taken to be zero and the system was regarded as having 2n+2 degrees of freedom.

As cables have zero stiffness in compression they are not true linear springs. Should the stiffness matrix described above be used then the hoist and boom hoist cables would be capable of carrying compressive load. To take realistic account of cable action four [LHS] matrices were generated to cater for all possible combinations of slack and taut cables. At each time step additional computation was required to ensure that the stiffness matrix used in the [LHS] matrix was the correct one.

The algorithm used was

(i) Solve for future displacements using the [LHS] matrix from the previous time step.

(ii) From future displacements calculate cable extensions.

(iii) If assumption of cable stiffness inherent in the [LHS] matrix was correct go to the next time step, otherwise

(iv) Use [LHS] matrix appropriate to future extensions and solve again for future displacements then go to (ii).

Non-linear Systems

The most obvious non-linear structural components of an offshore platform crane are the cable stiffnesses. If the cable stiffnesses are to be regarded as non-linear then there are a number of different ways of tackling the problem. One approach is to include stiffness contributions from the cables in the structural stiffness matrix. However this means that the cable stiffness terms to be used in the future stiffness matrix $[K_{s+1}]$ are unknown when solving for future displacements $[X_{s+1}]$. To enable solution to continue the future cable stiffness must be predicted from current and past stiffnesses. The future displacements can then be found and the cable stiffnesses based upon those displacements compared with the stiffnesses used. If the difference is acceptably small then the solution moves to the next time step, otherwise the cable stiffness terms in $[K_{s+1}]$ must be adjusted and the procedure repeated. In general there are an infinite number of possible stiffness matrices which means that the solution of the simultaneous equations at each time step must start with an unreduced coefficient matrix. Consequently each analysis will make heavy demands on computer time.

A more elegant and efficient strategy is to regard the cables not as structural elements, but as external forces, as shown in Figure 6. By removing the non-linear cable terms, the structural stiffness matrix becomes constant and will contain only simple beam element terms. Hence the resulting [LHS] coefficient matrix is also constant and can be reduced to diagonal form at the beginning of the analysis with the consequent saving in computer time.

The following algorithm demonstrates how this enables the related problems of line retrieval, cable non-linearities and behaviour prior to liftoff, to be dealt with in a unified and efficient manner.

(i) Generate [K] and [M] (which will contain only constant simple beam terms), calculate the LHS and reduce to diagonal form.

(ii) Use the current and past value of displacement at freedoms 2n+1 (the jib tip) and 2n+2 (the load) to estimate future displacement.

(iii) Calculate the cable extensions by algebraically adding these displacements to the line retrieved by the hoist and boom hoist winches.

(iv) Calculate the future cable forces based upon these extensions and include in the vector of future external forces $[P_{s+1}]$.

(v) Generate the right and side vector [RHS].

(vi) Solve for the future displacements $[X_{s+1}]$ (rapid because [LHS] is in diagonal form).

(vii) If the calculated future displacement at the jib tip and load from the previous step are sufficiently close to those estimated in (ii), start next time step otherwise go to (viii).

(viii) Revise estimated future displacements at the jib tip and load (experience has shown that using displacements from step (viii) results in rapid convergence).

(ix) Subtract cable forces from the right hand side vector [RHS].

(x) Calculate new values for the cable forces and add to the 2n+1 and 2n+2 terms of [RHS] (steps (iii),(iv),(v)) then go to (vi).

Note that prior to the load lifting off the deck the system was regarded as having 2n+1 degrees of freedom (see Figure 4) as the future displacement at freedom 2n+2 is known and hence only dispacement at freedom 2n+1 need be estimated in step (ii). Also as the extension of both the boom and boom hoist cables takes account of line retrieved by their respective winches, this model can easily be used to investigate effects such as the simultaneous raising of the boom and the load.

RESULTS

Experiments were carried out lifting loads from the supply vessel simulator using both the stiff and flexible jib cranes. The lifts were made using a number of different loads, jib angles and line speeds. Although the supply vessel simulator could be driven by any signal source, a sine wave generator was used throughout the tests. When lifting from the supply vessel simulator the problems faced were very similar to those facing the offshore crane operator inasmuch that little control could be exercised over where on the wave cycle the load lifted off the deck. For a given amplitude and frequency, a large number of random tests were made which gave lift-off points on all parts of the wave cycle, thus enabling the effect of various combinations of velocity and acceleration to be studied. This procedure was carried out for a number of different wave amplitudes and frequencies.

Figure 7 shows experimental and theoretical results for a typical test using the stiff jib model. The load lifted was 0.9 kg, the line speed was 300 mm/s and the supply vessel simulator had a frequency of 1 Hz and an amplitude of 70mm. From the experimental results it can be seen that the load lifts off the deck some time after the peak of the wave. Two mathematical simulations of this lift are presented. The first shows the undamped response using the non-linear approximation to the hoist cable stiffness shown in Figure 2. This simulation can be seen to give a reasonable approximation to the first two load cycles but due to the lack of damping, subsequent cycles show less correlation. The second simulation introduces damping while still using a non-linear hoist cable stiffness and shows, as might be expected, that the introduction of realistic damping produces a better model.

Hitherto no workers in the field of offshore crane dynamics have given consideratiion to the effect of supply vessel acceleration on subsequent dynamic forces. The likely reasons for this are the lack of prototype and experimental data and the inability of most proposed simulations to model this effect. Rightly the relative velocity between the load and the hoist line has been isolated as being the variable that most significantly affects subsequent dynamic forces. However investigation of supply vessel acceleration using both physical and mathematical models showed that parameter should not be neglected. Plotting the peak dynamic factors from a series of tests in which only the position of lift-off varied, produced the graph shown in Figure 8. Inspection of the sine wave shown in Figure 8 shows that for every point between A and C there is a corresponding point between C and E which has an equal velocity. The acceleration at corresponding points is equal in magnitude, but opposite in direction and varies from zero at points A,C and E to a maximum at points B and D. Figure 8 clearly shows the greater the difference in acceleration between corresponding points the bigger is the difference in peak dynamic factor. The broken line shows the results from the undamped mathematical model. Both models show the same trend and if realistic damping were introduced to the mathematical model even better correlation would result.

Tests similar to those described for the stiff jib model were conducted on the model having a flexible jib. Figure 9 shows results from a test on the flexible jib model for initial conditions identical to those that produced the results shown in Figure 7, except that in this case the point of lift-off is at the point of maximum downward velocity. Results from multi and two-degree-of-freedom mathematical simulations are also presented in Figure 9. The multi-degree-of-freedom model, which is lightly damped, is seen to give a good prediction of the time history of the forces. The two-degree-of-freedom simulation is shown to be capable of giving a reasonable estimate of the peak forces likely to be generated in the hoist and boom hoist cables although, it cannot, of course, supply any information about bending moments.

To demonstrate how mathematical models can be used to investigate problems such as non-linear retrieval of line by the winch, the multi-degree-of-freedom model was run for the same initial conditions as those used for Figure 9, only instead of the line being retrieved at a constant rate, the line speed was accelerated from zero to its final velocity (300 mm/s) during the first one tenth of a second of the lift. Inspection of Figure 10 shows the effect of this upon the time history for the forces. Hence to realistically analyse a crane where the winch system retrieves line in a non-constant manner, the velocity profile of the line should be one of the parameters included in the mathematical model.

VALIDITY OF SMALL SCALE MODELS

Making realistic small scale models of dynamic structural systems is a difficult business.

Usually it is impractical to scale both stiffness and mass to conform with the prototype, and even should this be achieved the scaling of the damping forces is normally impossible. For these reasons the authors feel that to apply results from small scale physical models to prototype cranes cannot be justified. However physical models do have an important role to play in the validaton of mathematical models and in the promotion of the understanding of crane behaviour. While not removing the need to eventually test mathematical models against prototype results, small scale models provide a valuable testing ground during the early stages of development. Tests on small scale models are inexpensive, quick and being in a laboratory environment allow a great degree of control to be exercised over the testing conditions.

CONCLUSIONS

(i) Using Newmarks Beta Method and treating the hoist and boom hoist cables as external forces produces a mathematical model that can easily take account of retrieval of the line by the hoist and boom hoist winches, cable non-linearities and behaviour prior to the load lifting off the deck.

(ii) Acceleration of the load during the instant of time between the hoist cable first going taut and the load lifting off the deck has a significant effect upon subsequent dynamic factors and should be catered for in any mathematical simulation of crane behaviour.

(iii) Small scale physical models have an important role to play in the validation of mathematical models and in the understanding of crane behaviour.

(iv) If hoist line is retrieved by the winch in any non-constant manner it is important that this effect be taken into account in any mathematical model used.

REFERENCES

1. STRENGENAGEN,J. and GRAN,S.,1980 "Supply Boat Motion, Dynamic Response and Fatigue of Offshore Cranes," <u>Proceedings, 1980 Offshore Technology Conference</u>, Houston, Texas, Paper OTC3795, pp 519-534.
2. BALFOUR,J.A.D. and OWEN,D.G., 1980 "Dynamic Behaviour of Platform Cranes," <u>Proceedings, 1980 Offshore Technology Conference</u>, Houston, Texas, Paper OTC3794, pp 509-517.
3. CHARRET,D.E. and HYDEN,A.M., 1976 "Dynamic Factors for Offshore Cranes," Proceedings 1976 <u>Offshore Technology Conference</u>, Houston, Texas, Paper OTC2578, pp 565-576.
4. JOHNSON,K.V.,1976 "Theoretical Overload Factor Effect of Sea State on Marine Cranes," <u>Proceedings 1976 Offshore Technology Conference</u>, Houston, Texas, Paper OTC2584, pp 601-609.
5. STENHOUSE,G.A. and HILTON,R.J.,1978 "Method of Rating High Pedestal Cranes for Lifting from Supply Boats in Various Sea States," <u>Proceeding 1978 European Offshore Petroleum Conference and Exibition</u>, London, England, Paper No.71, pp 125-132.
6. WYON.,A.S. and MACKINNON,J.A.,1975, "Dynamic Response of North Sea Cranes," <u>National Engineering Laboratory Report</u>, East Kilbride, Scotland.
7. KENNEY,F.M.,1978 "Operational Dynamics of an Offshore Crane," <u>Proceedings 1980 European Offshore Conference and Exibition</u>, London, England, Paper No.69, pp 109-116.
8. WATTERS,J.A.,MOORE,D.J. and GILL,I.S.,1980, "<u>Proceedings 1980 Offshore Technology Conference</u>, Houston, Texas, Paper OTC3792, pp485-496.
9. WARBURTON,G.B.,1976, <u>The Dynamical Behaviour of Structures</u>, Exeter, England: Pergamon Press.
10. CLARKSON,J,A. and KENNY,F.M.,1980 "Offshore Crane Dynamics," <u>Proceedings 1980 Offshore Technology Conference</u>, Houston, Texas, Paper OTC3793, pp497-597.

NOMENCLATURE

β	Newmarks Beta
$[C]$	damping matrix.
$[K_{s-1}], [K_s], [K_{s+1}]$	past, current and future stiffness matrices.
$[LHS]$	left hand side matrix.
$[M]$	mass matrix.
$[P_{s-1}], [P_s], [P_{s+1}]$	vectors of past, current and future external forces.
$[PN]$	matrix combining past, current and future external forces.
$[R1]$	combination of mass and current stiffness matrices.
$[R2]$	combination of mass and past stiffness matrices.
$[RHS]$	right hand side matrix.
Δt	time increment.
x_{s-1}, x_s, x_{s+1}	past, current and future displacement.
$\dot{x}_{s-1}, \dot{x}_s, \dot{x}_{s+1}$	past, current and future velocity.
$\ddot{x}_{s-1}, \ddot{x}_s, \ddot{x}_{s+1}$	past, current and future acceleration.
$[X], [\dot{X}], [\ddot{X}]$	vectors of displacement, velocity and acceleration.
$[X_{s-1}], [X_s], [X_{s+1}]$	vectors of past, current and future displacement.

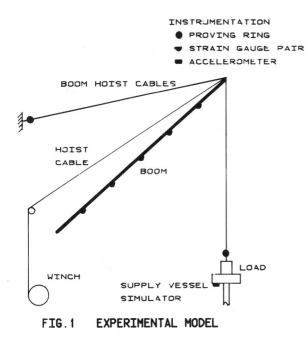

FIG.1   EXPERIMENTAL MODEL

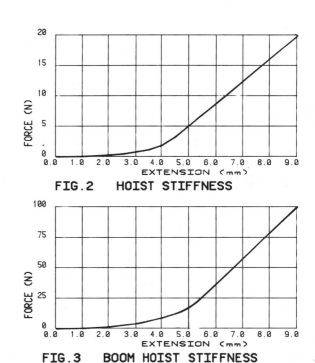

FIG.2   HOIST STIFFNESS

FIG.3   BOOM HOIST STIFFNESS

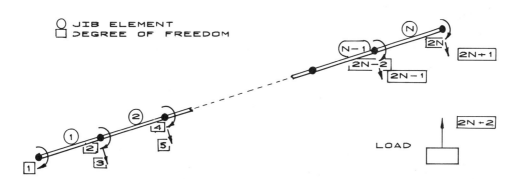

FIG.4   IDEALISATION MULTI-DEGREE OF FREEDOM MODEL

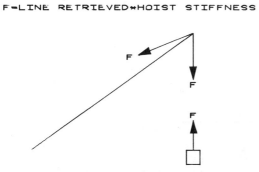

FIG.5   ADDITIONAL EXTERNAL FORCES LINEAR MODEL

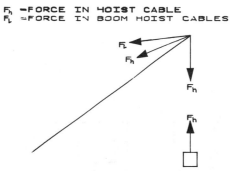

FIG.6   ADDITIONAL EXTERNAL FORCES NON-LINEAR MODEL

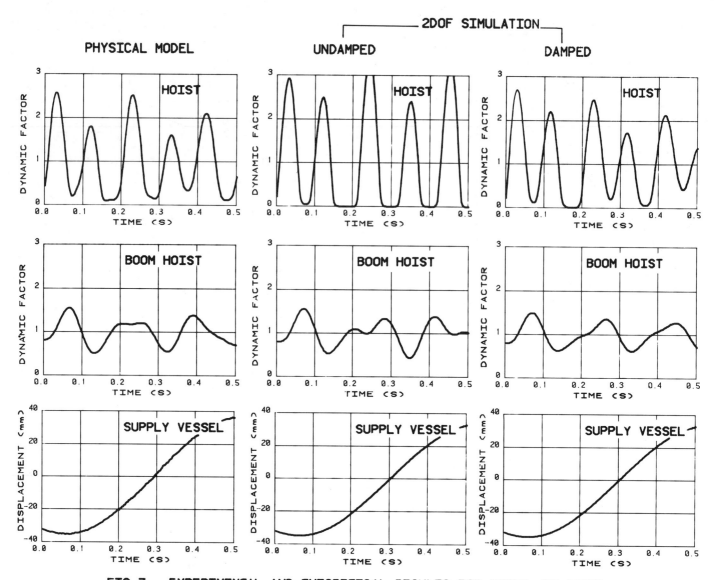

FIG. 7 EXPERIMENTAL AND THEORETICAL RESULTS FOR STIFF JIB MODEL

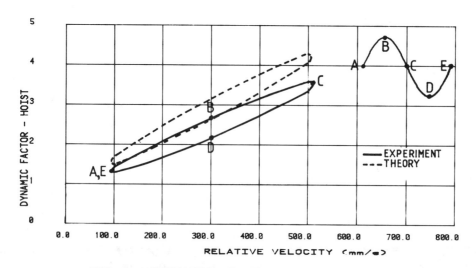

FIG. 8 VARIATION IN PEAK DYNAMIC FACTOR

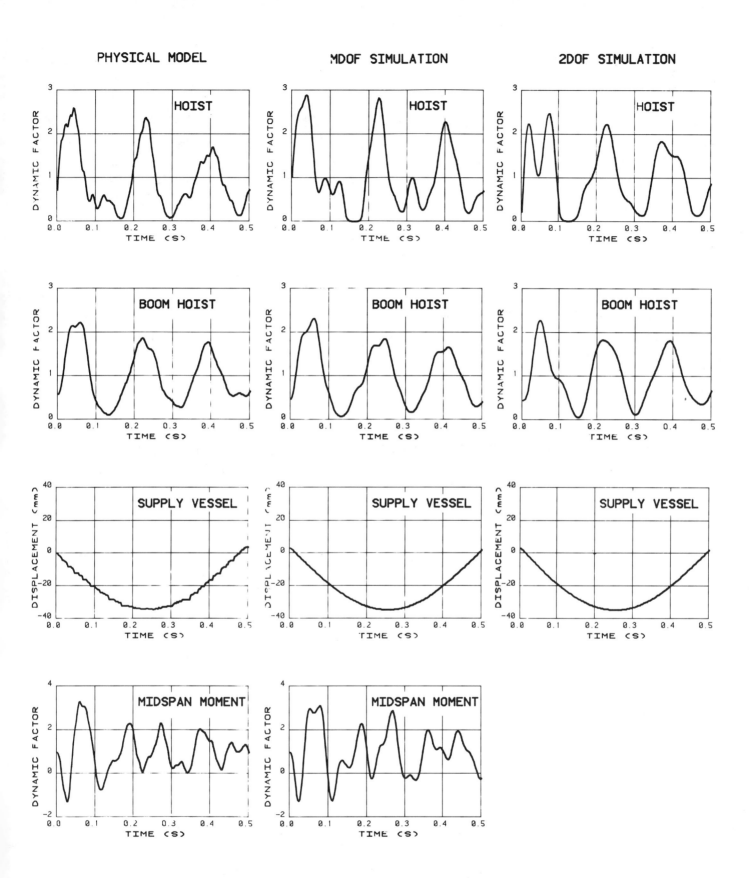

FIG.9 EXPERIMENTAL AND THEORETICAL RESULTS FOR FLEXIBLE JIB MODEL

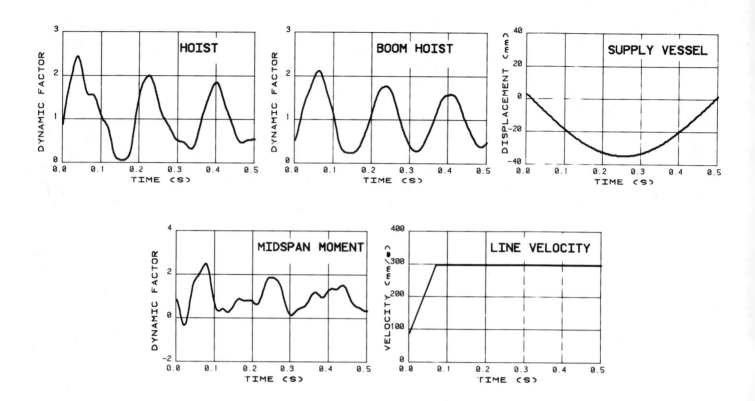

FIG. 10 RESULTS FROM MDOF MATHEMATICAL MODEL FOR VARIABLE LINE SPEED

Seaquakes: A Potential Threat to Offshore Structures

Knut Hove	Per B. Selnes	Hilmar Bungum
Det norske Veritas	Norwegian Geotechnical Institute	NTNF/NORSAR
N-1322 Hoevik, Norway	Oslo, Norway	N-2007 Kjeller, Norway

SUMMARY

A number of well documented cases showing that ships have been severely damaged or shaken by seismic waves in the open sea are presented. The potential threats of this phenomenon - called seaquakes - to offshore structures, especially floating installations in seismically active regions, are significant.

Further research should be undertaken to determine the extent to which seaquakes need to be considered in the design of offshore structures.

INTRODUCTION

The objective of this paper is, through a number of well documented cases, to call attention to the effects of propagating seismic waves in water - a phenomenon called seaquakes, Richter (1957) - and to the potential damaging effects that such waves may have on offshore structures. This especially applies to floating structures where seismic effects at present are normally not considered.

Energy from earthquakes propagates through rock and soil in the form of either compressional or shear waves. Since water has no shear resistance, only compressional waves can exist in water. The literature contains relatively little information on the propagation of such acoustic waves, or on their effects on ships or offshore structures. Rudolph (1887, 1895) was the first to study the phenomena and gives an extensive list of occurrences. Richter (1957) discusses further observations of the effects of earthquakes on surface ships, Rossi (1969) gives several more accounts of ships being shaken, Bradner and Isaacs (1972) evaluate probabilities for various levels of pressure increase on submarines due to earthquakes. More recently Hove (1981) has presented three cases of ships being severely shaken by seaquakes, the most serious being MT "Ida Knudsen", a 32,500 ton Norwegian tanker that was struck by a magnitude 8.0 event on February 28, 1969 in the Atlantic Ocean approximately 450 km west of Gibraltar. Even though the tanker was about 20 km from the epicenter in 4900 m of water, the damage was classified as total loss. All communication and navigation equipment was broken, piping was broken. Pumps and machinery were torn loose on their foundations. Frames and bottom girders were buckled and welds were cracked and torn. Some of the cross beams on the mast were shaken off, the others were bent. Typical damages are depicted in Figures 1 - 2. Furniture and loose equipment were thrown around, doors were torn off their hinges, indicating deck acceleration far above gravity.

More details on this occurrence and others are given in the following section and in Tables 1 and 2. Although the "Ida Knudsen" case is the most severe one known to the authors, there are reliable reports on many similar occurrences, although less severe. There are reasons to believe, however, that many severe cases have not been reported as earthquake generated, simply because of the fact that most sailors and other people - including engineers - are not aware that seismic waves can cause any effects on vessels at sea. It is believed that such incidents most often are explained by grounding on unknown shallows or rock, collision with a 'floating wreck', or even underwater explosions.

On vessels at sea, earthquake effects are most often described as a rumbling or boom-explosion-like sound, quickly followed by a series of shocks under which the ship may develop violent shaking, or the ship is so suddenly arrested in its course that it gives the impression of grounding upon a rock, or of collision. A number of such shock series may be received in succession, after which the ship appears to slide over the 'rock'.

When considering the chance of a ship going world-wide being hit by a seaquake it would of course be quite low compared to the chance of a seaquake hitting an offshore structure permanently located in an active seismic region. It therefore appears necessary to investigate these phenomena further with respect to

- the probability of occurrence of seaquakes,
- the characteristics of seaquakes, and
- the effects that seaquakes may have on permanently anchored, floating structures,

as this may have an impact on the design criteria and the safety of the installation.

CASE HISTORIES AND INTENSITY ESTIMATES

The principal seismic effect onboard ships is the strong vibrations induced in the hull by the arrival of the P-waves, which upon entering the water above the seafloor sediments are refracted and will propagate almost vertically to the sea surface at the velocity of sound in water. Some of the energy is transmitted into the air as sound waves and heard as rumbling - boom or explosion-like sound, depending on the frequency content and energy distribution of the incoming waves.

The duration of the vibrations may vary from a few seconds to several minutes. Reports indicate a usual duration of 10 - 60 seconds. Sometimes the amplitudes of the incoming waves are large enough to produce white-topped waves as if the water were whipped. Sometimes there may be two shocks coming right after each other. The second shock is normally the energy from the slower S-waves transmitted as P-waves up through the water, although there is reason to believe that also the surface wave train may be responsible for some of the P-waves generated at the seabottom. The epicentral distance is a critical factor in this respect. In some cases, however, the interval between shocks appears to be so long that a second earthquake may be needed for an explanation.

Nine cases, for which the information and data from ship reports can be correlated with the corresponding earthquake events, are given in the following case descriptions and in Tables 1 and 2.

Case 1

The earthquake in Nicaragua in 1926 on November 5 (M_s = 7.0), was felt by ships on the high sea and the following are the reports from two vessels, [2].

1-1, SS "Magician" (report in British Marine Observer)

The following report has been received from SS Magician, Captain P.O. Nicolas: 5th November, 1926 ... in lat. 10° 25'N, long 88° 10' W, a very severe submarine disturbance was experienced. Two distinct shocks, lasting about 10 or 15 seconds, with an interval of about 1½ minutes, were felt. The vessel shook violently, a rumbling, grating sensation was experienced. Masts, funnel, and superstructures vibrated and rattled alarmingly, giving the impression that the ship was running aground on to hard bottom and buckling fore and aft ... The chart shows 1800 to 1900 fathoms water in the vicinity. The American SS Eagle, then some 15 miles NW of our position, was later communicated with by wireless and her master replied that the shocks had been felt on board his vessel with such severity that the engines were stopped in the belief that the ship was running over something.

1-2, SS "Eagle" (report in an American newspaper)

November 5, 1926. Aboard SS Eagle in lat. 10° N long 88° W, two severe shocks were felt. They were of about one minute duration, with an interval of one minute between each. The Eagle listed four or five degrees, and the masts, booms, rigging, and stock vibrated considerably. Capt. P.O. Nicholas, of the British steamer Magician, then about 17 miles distant from the Eagle and at the same time, experienced two severe earthquake shocks with a similar effect on the vessel.

These ships were not far off the coast of Nicaragua, where this earthquake caused much damage; at Managua about half the houses were damged and there was some loss of life. The instrumentally located epicenter was on land, near Managua with an estimated depth of 160 km, thereby eliminating the possibility of surface faulting at the ocean bottom.

Case 2

An excellent example of the varied manner in which individual ships are affected at various locations around the earthquake epicenter is given in [3] with ship reports describing effects from the large Mexican earthquake of June 3, 1932 (M_s = 8.1). The epicenter was located about 40 km inland near 19.2°N, 104.2°W, and fell in the chain of volcanic mountains that traverse Mexico in an east-west trend and are, probably, a continental continuation of the long straight Clarion Fracture Zone that originates in the Central Pasific and passes through the volcanic Revilla Gigedo Islands before emerging on the Mexican coast.

The main quake caused considerable damage throughout the countryside inland of Manzanillo and inundated the immediate coastal area with a minor tsunami. Also in Hawaii and in Pago Pago, 7000 km away, tsunamis were recorded.

The seaquake effects were reported by four vessels [3] and are described in the following.

2-1, SS "SOLANA" - During the early morning of June 3, 1932 the SS SOLANA was steaming through a smooth sea with light variable winds near 18.50°N, 104.13°W. At 1037 GMT she was violently shaken for about 7 sec. The ship was then about 80 km from the epicenter in approximately 1500 m of water and did not detect any change in the state of the sea. The perpendicular distance to the fault zone was also about 80 km.

2-2, MV "SEVENOR" - A few miles to the southwest at 18.33°N, 104.53°W, the MV SEVENOR experienced, at the same time, vibrations that were less severe but of longer duration (1 min.). The SEVENOR was approximately 100 km from the epicenter. The ship reported a calm sea and slight westerly swells and detected no noticeable change in the surface of the sea.

2-3, MV "NORTHERN SUN" - Conditions aboard the MV NORTHERN SUN further to the north were entirely different. Although the vessel was 230 km from the epicenter, vibrations, commencing at 1039 GMT, continued for 3 min. and became so violent that the engines were stopped. Before the earthquake the sea had been smooth with a slight westerly swell, but by 1046 GMT the swell pattern had changed and the sea was confused.

2-4, SS "ARIZONA" - Further to the north at 20.47°N, 106.33°W, the SS ARIZONA started to vibrate at 1039 GMT and continued to do so for about 75 sec. The ship was about 260 km from the epicenter with a slight southwesterly sea and did not notice any change in the state of the sea.

After-shocks of this event continued for many days with M_s values up to 6.0. According to Rossi (1969) ship reports indicate that during the next 36 hours several strange underwater disturbances were experienced in the area. The MV SILVERWILLOW at 18.75°N, 104.57°W, began to vibrate dangerously in every part and at the same time began an uneven short pitching motion followed by heavy rolling. The disturbances commenced at 0530 GMT on June 4, and the rolling continued for 15 min. Seven hours later at 1245 GMT in 19.52°N, 105.45°W, the crew aboard the SS TALAMANGA heard a loud noise like distant gunfire, then experienced severe vibrations, and at 1337 GMT two similar reports were heard about 10 sec. apart but there were no apparent vibrations. However, 20 min. later the sea surface was littered for 8-10 km with small white oval objects, presumably dead fish. Several hours later near 19.47°N, 106.10°W, the SS HANOVER reported at 1205 ship's time violent shocks that rocked the ship as a nearby explosion might. Fifteen minutes later two more shocks were experienced with only slight vibrations. However, none of these reports can be correlated directly with any of the reported after-shocks.

Case 3

Rossi (1969) mentions in his paper a report from a ship (name unknown) near 18.05°N, 103.32°W, off the west coast of Mexico, on April 15, 1941, stating that earthquake vibrations caused "a large deckload of steel assembly, some pieces weighing 6 tons, to shift about 6 in. and jump as much as 5 to 6 in. up and down from its blocks".

This report indicates deck accelerations exceeding gravity and with a frequency content similar to earthquakes on land. The generating earthquake occurred at a reported distance of 90 km, with a depth of 100 km, and a magnitude of 7.7.

Case 4 - MV "NINGHAI"

On June 15, 1966 the MV NINGHAI (see Table 2) in the Solomon Islands reported the following damage after being shaken repeatedly at various intervals for about 2 hr. "The cathode ray tube shattered, the capillary tube in the barometer was smashed, valves were shaken out of their sockets in the wireless transmitter, the suspension wire on the gyro snapped and the azimuth mirror on the monkey island gyro repeater fell off. In addition we made some water in No. 3 double bottoms and after peak. Also the main engine fuel line was broken and the sanitary tank on the monkey island was holed. No water was made after the tremors, which suggested that as the ship was being shaken water was entering these tanks through various rivets and seams. The masts whipped about a great deal and the funnel rattled alarmingly." - Rossi (1969). The earthquake responsible occurred at a distance of 50 km and with a magnitude of 7.3.

Case 5 - USS BELMONT

A most interesting case described by Rossi (1969) is the report on the seaquake shaking of the US Navy communication research ship USS BELMONT. Rossi gives the following description:

"USS BELMONT left Callao, Peru, at 1300 GMT on October 17, 1966, proceeding at 5 kt. in good weather. Some time later, shortly before 2143 GMT, the captain ordered a course change and the speed increased to 10 kt. Suddenly, the vessel began to shake.

The BELMONT has a critical speed of 9 kt. and at first the skipper, Cdr. Scappini, thought the vibration was due to the ship's passing through this speed. But the vibration continued and became worse. The ship's mast whipped and so did her many antennas. Down on the mess deck crockery was smashed. Everything that wasn't tied down came adrift. After an estimated 12 to 18 sec. the vibration ended, just as suddenly as it had begun.

Cdr. Scappini ordered the engines stopped and general quarters sounded. "I knew we weren't aground", he later reported. "The Exec said we had 62 fathoms beneath the keel." After it was determined that there was no damage or fouling, the engines were again started and a speed of 10 kt. ordered. Only a slight shudder was felt when the critical speed of 9 kt. was reached -- as normally expected.

"I don't think there was a man aboard who wasn't scared", Cdr. Scappini commented. "I've been caught in typhoons and even the eye of a hurricane, but I've never before gone through anything like this, and I hope never do again."

Exact ship location is not reported, but estimated by us on the basis of previous location, reported speed and water depth. The earthquake data then gives an epicentral distance of approx. 170 km. No surface magnitude (M_s) has been reported for this earthquake, and we have therefore estimated it to M_s = 7.5 based on a common M_s/M_b relationship (M_b was 6.3). Tsunami was reported.

Case 6 - MT "Ida Knudsen"

MT "Ida Knudsen", a 32,500 tdw Norwegian tanker, was hit by a large earthquake (M_s = 8.0) on February 28, 1969 at 36.12°N and 10.70°W, approximately 450 km west of Gibraltar. The tanker was going in ballast in 4900 m of water at windforce 4 and seastate 4.

At 0243 GMT an extreme shaking started, giving the impression of an explosion. It was felt as heavy blows coming from beneath and going up through the ship. The mate in charge on the bridge reported that the shaking was so strong that he in fact was jumping on the floor during the shaking. Another crew member standing at the wheel noticed that the front of the ship was rising, first slowly then faster, during the shaking, then she dropped back as the shaking diminished. In the engine room the crew had difficulties standing on their feet.

In short the damages were: all communication and navigation equipment destroyed. Instruments mounted on the walls were torn off. Doors were torn off their hinges. Handrailing on stairways were shaken off. Loose equipment and furniture were thrown up into the air. Outriggers on mast were shaken off. Piping was broken. Some equipment was torn loose in way of anchorbolt failure or stretching. Leaden linings in way of machinery mountings were squeezed out. Welded connections and stiffeners had been broken. Bulkheads, hull frames and girders were buckled. Bulkheads were severely torn in one tank. All the wing tanks leaked, however the outer skin was tight except for one tank. In general the bottom plating and also the lower part of the side plating were torn away from the stringers/girders yielding gaps up to 50 mm wide. The damages in the superstructure appeared more severe at midship than at the aft peak. After 7 hours of drifting the crew got the engine started and managed to make it back to Lisboa, despite a deformed main engine frame and a misaligned propeller shaft. Some of the damages are depicted in Figures 1 and 2.

No one thought about earthquake as being the cause until a Spanish sailor the day after heard about the earthquake damage in Spain on radio.

After inspection in drydock at Lisnave the ship was condemned as total loss. The hull looked as if it had been subjected to a heavy mine explosion, the inspection report states, which indicates extreme dynamic pressure loadings. However, she was later bought and rebuilt by a Greek shipping company.

The distance to the reported epicenter was particularly small in this case, only 20 km, and the magnitude very large ($M_s = 8.0$), so peak ground accelerations in the range of gravity should be expected.

Case 7 - MS "Bergensfjord"
MS "Bergensfjord", a large Norwegian passenger liner, was hit by an earthquake off the coast of Equador in the evening of December 10, 1970. The earthquake had a magnitude of M = 7.6 with an epicenter onshore about 100 km away from the location of the vessel. The shaking was logged at 04.36 GMT. The quake occurred at 04.34 GMT.

The following is the recollection of the vessel's chief engin. O.O. Olsen as told to the first author.

Mr. Olsen was asleep and woke up because the vessel was shaking violently. Through the window he observed that the sea was completely whitetopped with waves 0.5 - 1.0 meter high as seen in the light from the vessel. His first thought was that the engine was exploding, or that they had grounded. He dressed in a hurry and ran front, from his cabin, on the upper deck up front, to the engine room at the rear of the vessel. He used the stairs from the deck down to the engine. He did not dare to take the lift as he was afraid it might get stuck due to the heavy shaking. When he came down to the engine room the shaking had stopped. Severe vibrations developed twice lasting for periods of 30-40 sec. The estimate of the time from Mr. Olsen woke up until the shaking had stopped is 3-4 minutes. During the heavy shaking Mr. Olsen also noticed that the main engine rpm increased as if the propeller was losing water.

After the shaking the engine was stopped and the whole ship was inspected:
- the sea was black as before the shaking
- no serious damage or leakage was found and there were no marks on the outside of the ship. Diving inspection was considered, but dropped upon advice from ship classification society
- furniture had been thrown around and crockery and glassware had been smashed.
- one passenger died from heart attack.

Captain Fasting described the occurrence as a most frightening experience. The cause was not understood until the next day, when the large earthquake that had struck Equador the previous evening killing about 80 people was reported on radio.

Case 8 - MS "Troyka"
On June 20, 1978 the Norwegian gas tanker "Troyka" lay at anchor in the inner harbour of Thessaloniki, Greece, approximately one mile off the breakwater. At about 11 p.m. the ship was severely shaken for a period of approx. 10 seconds. A strange sound, something between a boom and an explosion, was heard at the start of the shaking. This happened at the same time as the June 20, 1978 Thessaloniki earthquake, M = 6.4.

Captain Per Fosen, who reported the incident to the first author, first thought that an explosion had occurred onboard the ship, or that she had been hit by another vessel. Captain Fosen ran up to the bridge, then checked the fore peak, then the sides of the ship, but nothing could be seen, and no paint was damaged. The vibrations were described as frightening, but no objects had been observed jumping on the floor. Later when hearing about the earthquake damage in Thessaloniki the reason for the shaking was understood.

The earthquake caused about 50 deaths in the city and several hundred injuries, mainly resulting from the ensuing panic. Damage to buildings included one collapsed 8-storey building which contained 40 of the casualties. The intensity in Thessaloniki was IX on the Modified Mercalli scale.

a) Typical damage to navigation and communication equipment on the bridge.

b) Broken outriggers on mast.

c) Piping failure.

d) Typical fracture of plating and stiffeners.

Fig. 1 Typical damage on MT "Ida Knudsen" due to seaquake

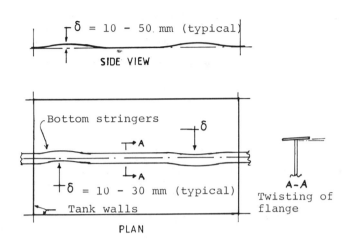

Fig. 2 Typical buckling deformations of tank bottom stringers of MT "Ida Knudsen"

Case 9

The 1904 Oslofjord earthquake of magnitude 6 - 6.5 produced seaquake effects in the Oslo harbour approximately 70 - 80 km away from the epicentral zone. The following is an excerpt from a newspaper article describing the situation in Oslo harbour [Selnes, 1981].

"The water erupted all over, as if it had started to boil, and on board ships it felt like violent heavy seas. At the same time hard blows seemed to hit against the hull of the ship. Many ships speeding on course came to a full stop such that the crew believed that they had suddenly grounded."

The MM intensity near the harbour onshore was VI, and the effects onshore and offshore are comparable for this earthquake as well:

Oslo onshore (MM VI)
Felt by everybody.
Furniture displaced.
Broken glassware, merchandise
fell off shelves, cracks in
plaster.

Oslo Harbour
Water erupted as if boiling
felt onboard like violent heavy
seas, hard blows seemed to hit
ship. Ships came to full stop,
crew believed they had
grounded.

---oOo---

ACCELERATIONS AND INTENSITIES

For the first eight cases of seaquakes presented in this paper there are reasonably accurate data, both on vessel location and damage and on the responsible earthquake. This gives us a possibility of calculating theoretically the acceleration and intensity that would be observed on the sea bottom right below the vessel. The acceleration - magnitude - distance relationship used in this estimate is of the form

$$a = b_1 \cdot e^{b_2 M} \cdot R^{-b_3}$$

where a is horizontal acceleration (in g), M is magnitude (M_s), and R is epicentral distance (in km). For the parameters the following values are used; $b_1 = 0.40$, $b_2 = 0.8$, and $b_3 = 1.7$, which are derived as follows: The b_2 parameter has been chosen in consistency with Esteva (1970) and Ahorner and Rosenhauer (1975), while b_1 and b_3 have been selected so as to be consistent with the typical inter-plate data presented by Trifunac and Brady (1976). Furthermore, an acceleration truncation for distances below 30 km is introduced.

Using the accelerations calculated in this way expected seismic intensities are calculated using the relationship

$$I = 4 \log a - 1 \quad (a \text{ in } cm/s^2)$$

which is derived by Murphy and O'Brien (1977) as a good average for world-wide data.

The results for cases 1 - 8 are shown in Table 2, where calculated accelerations are given in percentage of g, and calculated intensities together with our estimate of observed intensity on each of the vessels. The range in calculated intensities reflect the given uncertainties in epicentral distance.

It is fair to say that calculated and observed intensities show a remarkably good fit, which essentially demonstrates that the damage potential for structures in the open sea is similar to what could be expected for onshore structures at the same epicentral distances. The depth of water has no observable effect.

Case No	Earthquake Origin time (GMT)	Epicenter Lat.	Long.	Depth (km)	NST	Ref.	Magnitude Mb	Ms	Region	Comments
1	1926/11/05 07.55.33.0	12.30N	85.80W	160		ISS		7.0	Nicaragua	Casualties
2	1932/05/03 10.36.53.0	19.20N	104.20W			ISS		8.1	Mexico	Tsunami
3	1941/04/15 19.09.51.0	18.80N	103.00W	100		ISS		7.7	Mexico	
4	1966/06/15 01.32.54.0	10.20S	160.90E		178	ISC	6.1	7.3	Solomon Isl.	
5	1966/10/17 21.41.56.6	10.76S	78.63W		331	ISC	6.3		Peru	Tsunami
6	1969/02/28 02.40.31.2	35.97N	10.58W	14	402	ISC	6.5	8.0	Gibraltar	Tsunami, 7 dead
7	1970/12/10 04.34.38.0	3.97S	80.66W	15	400	ISC	6.3	7.6	Peru-Ecuador	81 dead
8	1978/06/20 20.03.21.5	40.78N	23.24E	3	451	ISC	6.1	6.4	Thessaloniki	50 dead

Table 1 Earthquake parameters for the 8 cases of 'seaquakes' discussed in this paper. The origin time is given in year, month, day, hour, minute, and second, NST is number of stations used in locating the earthquake, Ref. is reporting agency (ISS = International Seismological Summary, ISC = International Seismological Centre), and Mb and Ms is body wave and surface wave magnitude, respectively.
Event 2 was followed by a large number of aftershocks the following days (with M_s up to 6.0), and Event 7 had an aftershock with M_b = 5.3 only 15 minutes later.

Case No	Vessel name	Location Lat.	Long.	Water (m)	Distance (km)	Err. (km)	Duration (sec.)	Acc. (%g)	Intensity Cal.	Obs.
1-1	Magician	10.42N	88.17W	3400	330	100	100	< 1	1-3	5-7
1-2	Eagle	10.57N	88.32W	3400	340	100	120	< 1	1-3	5-7
2-1	Solana	18.50N	104.13W	1500	80	40	7	16	6-10	6-8
2-2	Sevenur	18.33N	104.53W	3000	100	40	60	10	6-9	5-6
2-3	Northern Sun	19.93N	106.23W	2500	230	40	180	3	4-6	6-8
2-4	Arizona	20.47N	106.33W	2500	260	40	75	2	3-5	
3	unknown	18.05N	103.32W	500	90	40		9	5-9	8-9
4	Ninghai	10.58S	161.08E	4000	50	20		20	7-10	8-9
5	Belmont	12.00S	77.75W	110	170	15	15	3	4-5	5-6
6	Ida Knudsen	36.12N	10.70W	4900	20	15	(60)	75	9-11	9-10
7	Bergensfjord	4.37S	81.43W	3000	100	10	40*	7	6-7	7-8
8	Troyka	40.60N	22.88E	20	35	10	10	15	6-9	7-8

* 2 periods of 30-40 sec.

Table 2 Vessel information referring to the 8 earthquakes listed i Table 1. The distance is between epicenter and ship, with an error estimate reflecting the combined uncertainty in epicenter and vessel location, then follows the duration of shaking and acceleration as computed from magnitude and distance, and finally two intensity values, the first one calculated from acceleration, and the second observed on board. Case 1 is from Richter (1958), cases 2 - 5 are from Rossi (1969), and cases 6 - 8 are from this paper.

Realizing that ship structures are very robust structures compared to onshore structures these few cases seem to indicate that:

- Earthquake effects are just as severe to structures in the open sea as to structures onshore.
- The damage potential has equally wide distribution in the open sea as onshore.
- The depth of water has negligible effect on the damage potential.

---oOo---

ANALYTICAL CONSIDERATIONS

The energy from the earthquake is transmitted by propagation of compressional and shear type waves through the base rock and the sediments. At the ocean floor interface the P-waves are refracted and may be assumed to propagate almost vertically up through the water to the ship or structure. A similar assumption was used in the submarine study for the evaluation of the pressure increase in water due to seismic waves Bradner and Isaacs, [1972].

Part of the energy in the S-waves is also transformed into P-waves at the ocean floor interface. However, considering the P-waves only, such a model is similar to the model used in seismic wave propagation analysis for soils, and a one-dimensional analysis of the above problem for the rock-soil-water system can be performed simply by including a water column above the soil. An example of such an analysis [Selnes, 1981] carried out with the computer program SHAKE [Schnabel et al., 1972] is shown in Fig. 3.

The P-wave velocity of the soil in the analysis was based on the expression

$$V_{ps}^2 \cdot \rho_s = K + 4/3\, G$$

where

V_{ps} = P-wave velocity of the soil

K = bulk modulus of soil-water system, kept constant independent of strain

ρ_s = mass density of soil

G = strain compatible with shear modulus of soil obtained from a separate analysis of the same deposit subjected to the horizontal component of the same earthquake.

The damping in the soil was taken as the same as the strain compatible value obtained in the analysis of the horizontal shear wave, and the damping in water was given an arbitrary, very low value.

The submarine study evaluated the effects of the acoustic waves only in terms of the increase in water pressure. However, it is believed that the damage due to the dynamic effects of the shaking may be equally or more important, as indicated by the large response amplification in the period range 0.1 - 0.5 sec. (Fig. 3) and the type of damage inflicted on MT "Ida Knudsen".

For this paper no attempt has been made to carry out more detailed analysis of the wave propagation and the ship response problem. Analysis of the two-dimensional case with the ship included should, however, be carried out. This would give a rough check on the applicability of vertical wave propagation models for such problems - whether such analyses could explain the type of damage that has been observed.

The structure frequencies of ships and floating platforms are typically in the range of 1.0 Hz and upwards, as illustrated in Fig. 4. The internal damping in these types of structures is relatively small: 0.5 - 1.0 %. Also the geometrical damping - i.e. reflection of energy back into the water - is of less importance for this type of loading. Large dynamic amplifications of the structural response and of the equipment response should therefore be expected. An illustration of this effect is given in Fig. 5, which shows typical node response spectra for steel template structures subjected to base motion. Similar amplification may be expected for floating structures in the frequency range 1 - 10 Hz.

It is also of interest to note that in some cases the ship appeared to be lifted up or out of the water by the pressure pulses, as should be expected, since the average pressure of the time history will be in the direction of propagation. Combined with the structural vibration effects this upward force could be of importance for the dimensioning of e.g. tension-leg anchoring systems.

A comprehensive analytical study of the MT "Ida Knudsen" case would be of great value for an improved understanding of the characteristics and the effects of seaquakes on floating offshore structures.

In order to broaden the data base on seaquakes a review of ship reports connected with grounding on unknown shallows in deep water should be undertaken, as there appears to be a correlation between areas with unknown shallows in deep water and regions with high seismic activity.

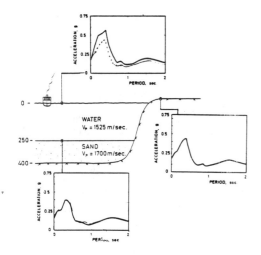

Fig. 3 One dimensional wave propagation of soil-water system [6].

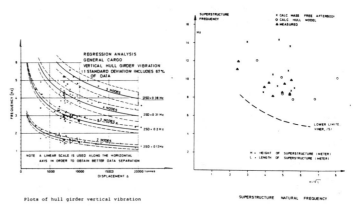

Fig. 4 Natural frequencies for ships.

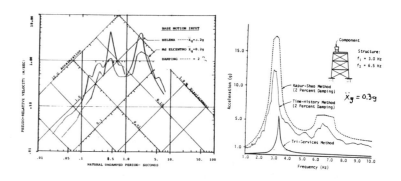

Fig. 5 Typical node response spectra for steel template structure.

SUMMARY AND CONCLUSIONS

Although the amount of data available is small, the cases and estimates presented in this paper indicate that earthquakes may impose just as severe loadings on floating structures in the open sea as on structures on land.

The damage potential appears to have equally wide distribution in the open sea as onshore.

The depth of water has negligible effect on the damage potential.

Seaquakes can cause extreme response amplifications of ship structures and will most likely cause amplifications of similar magnitude in floating offshore structures, in addition to the increase in hydrostatic water pressure.

The mechanisms behind the characteristics of seaquakes are not well established.

Further studies of seaquakes are clearly desirable with respect to energy transfer, to probable loading effects on structure and equipment, as well as hazard evaluation.

The opinions and conclusions presented in this paper are those of the authors and not necessarily those of their affiliates.

ACKNOWLEDGEMENT

This paper is based on data reported in the literature and from ship reports. Acknowledgement is given especially to Mr. K.T. Sætre of Knut Knutsen O.A.S., Haugesund, and to Mr. D. Kahrs and Mr. O.O. Olsen of Den Norske Amerikalinje A/S, Oslo, who provided the data on MT "Ida Knudsen" and MS "Bergensfjord", respectively, and to Mrs. Vibeke Wold for typing the manuscript.

REFERENCES

1. Rudolph, E.: "Über submarine Erdbeben und Eruptionen, Part I", G. Beitr. Vol. 1 (1887) pp 133-373 Part II; Vol. 2 (1895) pp 537-666; Part III Vol. 3 (1898) pp 273-336.

2. Richter, C.F., 1958: "Elementary Seismology", W.H. Freeman & Co., San Francisco.

3. Rossi, F.P., 1969: "Seaquakes: Shakers of Ships", Mariner's Weather Log, 11,(5) pp 161-164.

4. Bradner, H., Isaacs, J.D., (1972): Final Technical Report on "Overpressures Due to Earthquakes Program" of the Advanced Ocean Engineering Laboratory. SIO Reference No. 72-18. Scripps Institution of Oceanography, Univ. of Cal., San Diego.

5. Hove, K., (1981): "Earthquake Effects on Vessels in the Open Sea", Det norske Veritas Report

6. Selnes, P., (1981): "Geotechnical Problems in Offshore Earthquake Engineering", Norges Geotekniske Institutt, Report 40009-6.

7. Esteva, L. (1970): "Seismic Risk and Design Decisions", in: R.J. Hansen (ed.), "Seismic Design for Nuclear Power Plants", MIT Press, Cambridge, Mass, USA.

8. Ahorner, L. and W. Rosenhauer (1975): "Probability Distribution of Earthquake Acceleration with Application to Sites in the Northern Rhine Area, Central Europe", J. Geophys. Res., $\underline{41}$, pp 581-594.

9. Trifunac, M.D. and A.G. Brady (1976): "Correlation of Peak Acceleration, Velocity and Displacement with Earthquake Magnitude, Distance and Site Condition", Int. J. Earthquake Engin. Struct. Dynamics, 4, pp 455-471.

10. Murphy, J.R. and L.J. O'Brien (1977): "The Correlation of Peak Ground Acceleration Amplitude with Seismic Intensity and other Physical Parameters", Bull. Seism. Sve. Am., $\underline{67}$, pp 877-916.

11. Schnabel, P.B., J. Lysmer and H.B. Seed (1972): "SHAKE: A Computer Program for Earthquake Response Analysis of Horizontally Layered Sites", Univ. of Cal., Berkeley, College of Engineering, Earthquake Engineering Research Center, Report EERC-72-12.

ON HYDRODYNAMIC REACTIONS OF WATER TO MOTION TO OFFSHORE STRUCTURES IN CALM SEA

A.V.Ivanov,
Krylov Shipbuilding
Research Institute, USSR

A.N.Kulikova,
Krylov Shipbuilding
Research Institute, USSR

SUMMARY

This paper presents the results of experimental studies on hydrodynamic reactions of water acting on semi-submersible (SSDU) and jack-up (JDU) drilling units moving in still water.

Used as test subjects were "Aker" - type SSDU models and JDU models with a displacement hull of triangular shape in plan. The tests were carried out in the KSRI model basins for a number of steady towing speeds, three alternatives of loading and several drift angles. In the course of experiment studies were performed on the effect of the SSDU static angles of heel and the depth to which the JDU support columns are lowered, on hydrodynamic forces and moments.

As a result of the work performed the effect of the Reynolds numbers on the non-dimensional coefficients of hydrodynamic forces and moments was analysed. Specific considerations were made as to the possibility of scaling the model test results to full-size. Qualitative and quantitative relationships between the non-dimentional hydrodynamic coefficients and variations in the drilling unit wetted surfaces due to changing draft, submersion of the support columns and static angles of heel were established.

Comparison of the experimental results with similar data for conventional vessels revealed some features characteristic for the hydrodynamic forces acting on offshore structures.

INTRODUCTION

In dealing with seaworthiness problems of free or moored offshore structures one should be aware of the hydrodynamic forces applied to the structure moving in calm sea or in the current. The existing methods of theoretical analysis of these forces as applied to floating platforms, particularly to semi-submersible and jack-up drilling units (SSDU and JDU) which are bluff constructions of sophisticated geometry, are imperfect, since they are inefficient, e.g., for a full evaluation of the hydrodynamic interaction in flow around structural elements of the viscous and vortex effects which occur to the floating structure in motion, and of the vertical components of hydrodynamic forces originating in the process.

Therefore, the investigation of hydrodynamic reactions is based on model experiments carried out both in model tanks /3/ and aerodynamic tunnels /1/. The results obtained offer an opportunity to completely evaluate the hydrodynamic resistance of floating platforms and to assess their safety in operation.

The symbols used in the paper are as follows:

L_0, B_0, T_0 - length, breadth and depth of the SSDU pontoons, respectively;
H_0 - draft of SSDU;
D_0 - weight displacement of SSDU;
l_0 - distance between the element for measuring the hydrodynamic forces and moments and the basic plane of the SSDU model;
θ, ψ - static heeling and trimming angles of SSDU;
S, L - characteristic area and length of the drilling unit, respectively;
S_x - double area of SSDU centerline plane submerged at $\theta = \psi = 0$;
S_y - midsection area submerged at $\theta = \psi = 0$;
L_Δ - centerline length of JDU pontoon;
H_Δ, T_Δ - draft and depth of JDU pontoon, respectively;
D_Δ - weight displacement of JDU;
l_K - submersion of support columns (the distance between the basic pontoon plane and the upper plane of the supporting shoe);
l_Δ - distance between the element for measuring the hydrodynamic forces and moments and the JDU pontoon basic plane;
a - distance between the element for measuring the hydrodynamic forces and moments and an arbitrary point on the vertical axis Oz;
$C_i (i=x,y)$ - nondimensional coefficients of hydrodynamic forces;
$C_{m_i} (i=x,y)$ - nondimensional coefficients of hydrodynamic moment with respect to corresponding coordinate exes;
P_x, P_y - longitudinal and normal forces, respectively;
M_x, M_y - heeling and trimming moments, respectively;
V, β - velocity and drift angle, respectively;
g - acceleration of gravity;
ρ - mass density of fluid;
γ - kinematic viscosity coefficient of fluid.

$$Fr_{L_0} = \frac{V}{\sqrt{gL_0}}, \quad Fr_{B_0} = \frac{V}{\sqrt{gB_0}}, \quad Fr_{L_\Delta} = \frac{V}{\sqrt{gL_\Delta}} \quad \text{- Froude numbers;}$$

$$Re_{L_0} = \frac{VL_0}{\gamma}, \quad Re_{B_0} = \frac{VL_0}{\gamma}, \quad Re_{L_\Delta} = \frac{VL_\Delta}{\gamma} \quad \text{- Reynolds numbers.}$$

PURPOSE OF INVESTIGATION

The principal purpose of our studies was to experimentally determine the hydrodynamic reactions of water acting on semi-submersible and jack-up drilling units moving in calm water with different speeds and drift angles. The results obtained enabled us to establish and analyse the effect of some geometrical characteristics of floating structures on hydrodynamic reactions of water.

SUBJECTS OF INVESTIGATION

Used as test-subjects were: (1) a SSDU model consisting of two submersible pontoons of a rectangular cross-section, six stabilizing columns, cylindrical bracing connecting the pontoons to the columns, and the above-water deck (Fig.1) and (2) a JDU model with a displacement hull of triangular shape in plan and the truss-type support columns also of triangular cross-section (Fig.2).

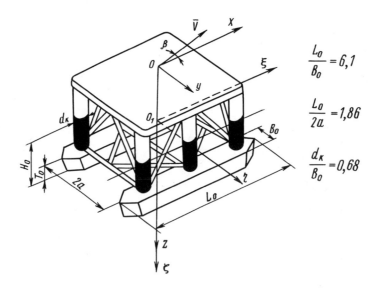

Fig. 1. Semi-submersible drilling unit. Coordinate system.

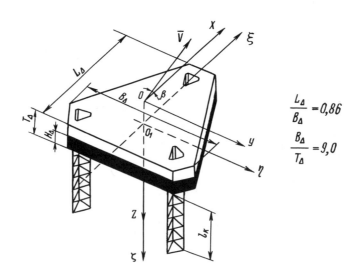

Fig. 2. Jack-up drilling unit. Coordinate system.

TEST PROGRAM

The SSDU model tests were carried out for three draft values corresponding to the following operating conditions:

towing to the site: $\dfrac{H_0}{T_0} = 0,93$

drilling: $\dfrac{H_0}{T_0} = 2,83$

during installation: $\dfrac{H_0}{T_0} = 2,0$;

and three drift angles: $0°$, $50°$, $90°$. At drift angles of $0°$ and $90°$ the effect of static heeling and trimming angles, respectively, on the hydrodynamic water reactions was investigated.

All static heeling angles (ranging from $0°$ to $15°$) of SSDU were given for incident flow.

With draft and trim being specified, the SSDU model, placed under the carriage and capable of executing vertical motions while underway was towed at speeds of $Fr_{L_0} = 0,02 \div 0,25$, $Re_{L_0} = 2 \cdot 10^5 \div 1,3 \cdot 10^6$ (with the model positioned along the basin), and $Fr_{B_0} = 0,06 \div 0,40$, $Re_{B_0} = 2 \cdot 10^4 \div 2,4 \cdot 10^5$ (with the model positioned across the basin).

In the course of tests the JDU model was towed with the carriage speed corresponding to $Fr_{L_\Delta} = 0,13 \div 0,19$, $Re_{L_\Delta} = 10^5 \div 1,4 \cdot 10^5$ while the drift angle was varied from $0°$ to $180°$ and the submersion of the columns was $\frac{L_K}{T_\Delta} = 0; 2,2; 15,0$. The drafts of the JDU displacement hull for two minimal column submersions were the same and amounted to $H_\Delta/T_\Delta = 0,5$, while the draft of the maximum submersion was 1,78 times that value.

The fluid forces acting on the models were measured by means of a standard dynamometer in the coordinate system Oxyz. The direction of Ox and Oy axes was taken to be the positive force direction. The positive directions of measured moments M_x and M_y can be seen in Fig. 1 and Fig. 2.

The drift angles are counted off from the velocity vector towards the Ox axis clockwise. It is well known that in the tests where average values of hydrodynamic forces are determined for a body in stable motion close to the free surface of uncompressed fluid the Froude and Reynolds numbers are taken as similarity parameters. Since in the course of the test it is impossible to keep the same values of these for the model and the full-size structure, the modelling was performed using the Froude number only.

TEST RESULTS

As a result of the SSDU and JDU model tests the relationships between hydrodynamic forces and moments and the drift speed were established. It is shown that the resistance force of the drifting SSDU has its maximum values while moving broadside. For instance, the maximum value of resistance force for a model drifting broadside with the speed of $Fr_{L_0} = 0,12$, drilling draft and heeling of $15°$, is about $0,07 D_0$, while for drifting with the bow first at the same values of speed, draft and trim angle, it is $0,025 D_0$. The same holds true for the moment values as well. Thus, for the above-said test conditions the heeling moment is $0,09 D_0 H_0$, and the trimming moment is $0,04 D_0 H_0$ with respect to the point with a non-dimensional coordinate $l_0/H_0 = 1,98$.

It should be noted that when a SSDU model is towed with static heel angle, its draft increases, the submersion at the initial unsteady stage of motion being much deeper than during subsequent steady-state motion.

The same may occur under real conditions, e.g. when the SSDU is exposed to squall, particularly in the presence of a static angle of heel due, say, to constant wind speed.

The following table shows the variation in the hydrodynamic force P_x related to weight displacement D and acting on the JDU while it is drifting with a speed corresponding to $Fr_{L_\Delta} = 0,135$:

β, deg \ l_K/H_Δ	0	2,2	15,0
0	-0,012	-0,023	-0,080
180	0,018	0,027	-0,090

It is clear that the hydrodynamic force P_x shows a considerable increase with the submersion of support columns, and the total resistance is mainly due to the columns resistance, since it is many times larger than that of the JDU hull.

It should be noted that the effect a $180°$ change in the direction of JDU motion has on the P_x/D_Δ value, markedly decreases with an increase in depth to which the support columns are lowered, as is seen from the table.

According to the laws of hydrodynamic similitude the hydrodynamic forces and moments can be expressed in terms of nondimensional coefficients as:

$$P_i = C_i \frac{\rho V^2}{2} S, \quad M_i = C_{m_i} \frac{\rho V^2}{2} SL, \quad (i=x,y). \tag{1}$$

In calculating the nondimensional coefficients of hydrodynamic forces the following areas were taken as the characteristic area S:

S_x - for SSDU motion at drift angles of $\beta = 50°$ and $90°$,
S_y - for the same at drift angle of $0°$.

For JDU motion the submerged centerline plane area of the JDU pontoon was taken to be the area S.

Representation of hydrodynamic forces as in equation (1) makes it possible to calculate the nondimensional coefficients and determine the qualitative and quantitative relationships between these and a number of SSDU and JDU characteristics.

With an increase in Re_{B_0} and Re_{L_0} numbers the coefficients of hydrodynamic forces acting on the SSDU asymptotically approach some constant values, which proves the fact that there is a self-similarity zone in the tests. The most characteristic curves are shown in Fig. 3.

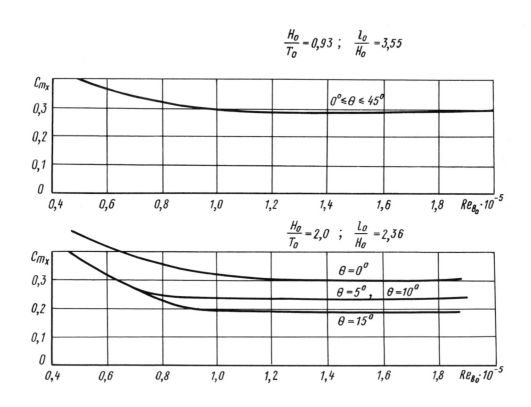

Fig. 3. Coefficient C_{m_x} versus number Re_{B_0} at $\beta = 90°$.

The experimentally obtained values of horizontal hydrodynamic force coefficients as a function of drift angles, shown in Fig. 4 for three values of Re_{L_0} numbers and characteristic of other coefficients, prove that the Reynolds numbers have but a slight effect on the coefficient value.

The above-said seems to be certain and can be physically explained.

The calm-water resistance to SSDU or JDU movement includes both wave and viscosity components. The latter makes up the principal portion of the total resistance, as the SSDU and JDU built from pontoons of rectangular cross sections, columns, and bracings, are bluff constructions with fixed lines of the boundary layer separation. The width of the vortex street behind the drifting SSDU or JDU does not depend on the drift velocity hence the viscous resistance is

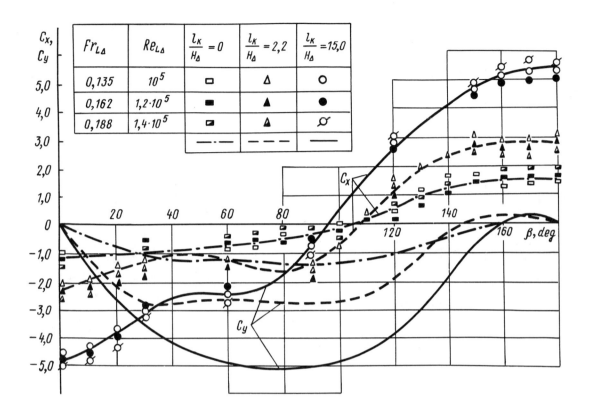

Fig. 4. Nondimensional coefficients versus drift angle for a moving JDU.

indepent of the same as well. This consideration and the insignificant part which the wave resistance plays in the total resistance /3/, indicate that the assumption of the coefficients being independent of the drift velocity can be used for practical purposes, which is experimentally confirmed.

Therefore the coefficients obtained at maximum Reynolds numbers (Figs. 4 - 7) can be recommended for the calculation of hydrodynamic forces.

It is obvious, that the above consideration requires to be experimentally confirmed, e.g., by testing a number of large-scale models with the aim of investigating the resistance crisis of the flow past the SSDU and JDU cylindrical structural elements for various roughnesses and at different submersion depths.

The tests of the SSDU model having static angles of heel made it possible to establish the relationship between the hydrodynamic forces with constant drift angle and constant draft, and the heel and trim angles (Figs. 6 and 7).

Comparing the results obtained from towing the SSDU model broadside in calm water with similar results of /2/ for ships of conventional forms, one can see substantial difference between the two. It is stated in /2/ that the nondimensional coefficient of hydrodynamic force of drifting independent of the heel angle may be taken as constant and equal to 0,8. It is not true of the SSDU nondimensional coefficient.

The test results showed that the coefficients of forces and moments, except for the heeling moment coefficient, tend to increase with an increase of the SSDU static angles of heel, or remain constant at same drafts. When the model is drifting broadside, the coefficient of the heeling moment decreases at certain drafts with an increase of the angle of heel. This is evidently associated with the vertical components of hydrodynamic force acting on the SSDU underwater elements and lead to a decrease of the heeling moments with the increasing of the static angle of heel. A similar effect is described in /1/.

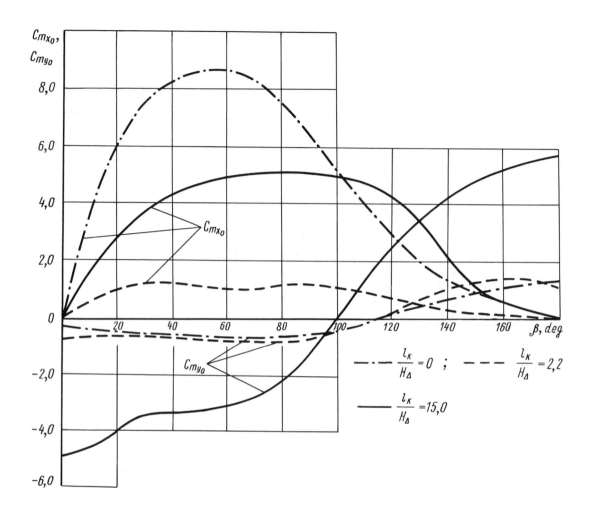

Fig. 5. Nondimensional coefficients of the trimming and heeling moments versus drift angle.

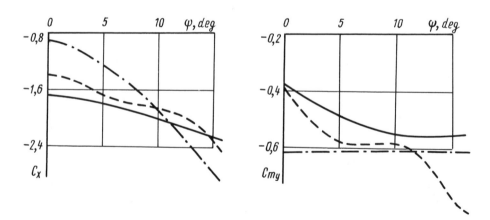

Fig. 6. Nondimensional coefficients versus trim angle for a SSDU model drifting with the bow first.

The analysis of the results obtained from measuring the hydrodynamic forces and moments which are acting on the drifting JDU has also confirmed the fact that the hydrodynamic moment is formed both by the horizontal and vertical forces. The presence of the vertical force can easily be established by comparing the signs of the corresponding horizontal force and moment components under certain test conditions. Thus, the test results obtained do not allow determining the load

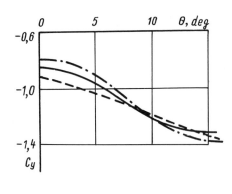
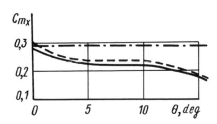

Fig. 7. Nondimensional coefficients versus angle of heel for a SSDU model drifting broadside. (For symbols see Fig. 6).

point of the hydrodynamic reaction of water while the SSDU or JDU is underway. However, they enable the heeling and trimming moments to be calculated with respect to the corresponding axis placed at an arbitrary height above the basic plane of the drilling unit, which is necessary, e.g., for the evaluation of its stability. As the value of the hydrodynamic moment induced by vertical forces parallel to the Oz axis is independent of the vertical translation of the origin of coordinates, and the moment caused by horizontal forces is varying proportionally with the value of the distance between the new origin of coordinates and that accepted for measurements, the following expressions can be written for the heeling and trimming moments with respect to the corresponding axes passing through an arbitrary point of the Oz axis:

$$C_{m_x} = C_{m_{x_0}} + C_y \frac{a}{L}, \qquad C_{m_y} = C_{m_{y_0}} - C_x \frac{a}{L},$$

where $C_{m_{x_0}}$, $C_{m_{y_0}}$ - nondimensional coefficients of hydrodynamic moments with respect to corresponding axes passing through point O (Figs. 1 and 2) and a - distance between the new origin of coordinates and the one accepted for measurements.

CONCLUSIONS

1. The experimental investigation showed that with an increase of the Reynolds number the nondimensional coefficients of hydrodynamic forces and moments tend to approach some constant value which can be recommended for use in the estimation of forces and moments.

2. The SSDU inclination towards the incident flow results in an increase of hydrodynamic forces, while the hydrodynamic component of the heeling moment which occurs as the SSDU is drifting broadside, decreases with an increase in the angle of heel. This is caused by the development of the lift forces acting on the SSDU hull.

3. When the SSDU is drifting with heel angles, the draft of the unit increases, the increase being more considerable if the motion is unsteady.

4. As the submersion of the JDU support column increases, the effect of the change in the direction of the SSDU movement (no matter bow or stern first) on the value of the longitudinal force markedly decreases.

REFERENCES

1. Bjerregaard, E., Velschou, S., Clinton, J.S.: "Wind overturning effect on a semi-submersible", Proceedings 10-th annual Offshore Technology Conference, Houston, Texas, 1978, vol.1, p. 147-154.

2. Fedyaevsky, K.K. and Firsov, G.A.: "Heeling of a ship due to wind", Sudostroyenie, 1957, vol.12, p.3-11.

3. Grekoussis, C. and Miller, N.S.: "The resistance of semi-submersibles", Transactions Institute Engineers and Shipbuilding Scotland, 1977-1978, 121, number 5, p.33-51.

STUDY OF WAVE ACTION ON FLOATING STRUCTURES USING FINITE ELEMENTS METHOD.

COMPARISON BETWEEN NUMERICAL AND EXPERIMENTAL RESULTS

LEJEUNE A.   Dr. Ir. Professeur
MARCHAL J.   Agr. Dr. Ir. - Chercheur qualifié F.N.R.S.
HOFFAIT Th. Ir. Boursier I.R.S.I.A.
GRILLI S.   Ir. Aspirant F.N.R.S.
LEJEUNE P.   Ir. Boursier I.R.S.I.A.

Institut du Génie Civil - Laboratoires d'Hydrodynamique, d'Hydraulique Appliquée et de Constructions Hydrauliques de l'Université de Liège Belgique.

ABSTRACT

In marine field, the number of floating structures is always increasing. We develop first a finite elements programm studying the diffraction-radiation problem for a body in sea waves for bi-dimensional and tridimensional cases. We obtain fluid velocities, dynamic pressures, added mass and damping coefficients and free surface amplitudes. The following approximations were made : perfect fluid (potential theory), free surface linearization, periodical movements.

Different size of elements have been tested in bi and tri-dimensional cases using Galerkin's method of integration to compare accuracy and boundary's conditions (infinite and finite fluid). We compute the coefficients in function of wave period.

To check qualitatively and quantitatively physical and mathematical models, we add an experimental study, realized in our University wave channel, on the behaviour of real parallelipipedical tank model under laboratory wave actions. We study particularly their heave and roll movements by displacements, rotations and wave height.

The combination of all the results allow to have added mass coefficients, heave and roll damping curves in function of the wave period.

1. INTRODUCTION

The sea's motions are the result of a superposition of many disturbing forces. Mathematical tools have been created to define these motions, but despite all efforts, any such model requires tremendous approximations which seldom represent the true facts with all their complexities. The problem of wind offshore can be severe on parts of a structure and is especially important regarding floating stability. Nevertheless, for most fixed structures in deeper waters the wind load represents less than 5 percent of the total environmental loading. Current loads can be severe in some locations but it is usually wave loads which dominate the designer's thinking.

2. WAVE ACTION

A. Introduction

The basic deterministic approach to analyse wave loads on a framed structure comprising linear prismatic elements is to use the well known Morison equation in which the fluid force is the vector sum of an inertial and a drag component. These are out of phase due to the phase shift between particle velocity and acceleration. Drag effects dominate for waves with a ration of characteristic structural diameter to wave length less than 0,2. This is the case for most framed towers in fully arisen seas. The calculation of the fluid loads on the entire structure is relatively straightforward, if tedious. Assumptions are required concerning the inertial and drag coefficients C_m and C_D.

The water particle kinematics are determined from the incident wave, provided its profile is assumed (e.g. Stokes Fifth) and its height and period are defined. The fluid load on the over all structure is obtained by summing the component forces on each prismatic element, making due allowance for shift in relation to wave crest and angle of attack. This is quite a straightforward mechanistic approach which has been used for many years. It requires no elaborate fluid flow analysis and the validity of the basic Morison equation has been checked many time in the laboratory in the context of frame structures.

Despite all this, there is sufficient doubt as to the accuracy of its application to complete structures for the oil industry to undertake a multimillion dollar, large scale test project in which an Ocean Test Structure is being installed and instrumented to measure fluid loads. The reasons behind some of the suspected inaccuracies concern factors such as energy spreading and directionality characteristics of real seas as opposed to those which are assumed in analysis. In summary, the primary need right now is not for an analysis method which gives a more accurate measure of fluid flow around submerged prismatic members. It is rather a matter of being able to deal with the water particle kinematics in a random three-dimensional sea.

B. Assumptions Validity Limits

Following assumptions are made for this study. Wave theory is linear Stokes theory and wave forces are divided into 2 types : inertia forces are proportional to acceleration and viscous forces are proportional to velocity square (figure 1.)

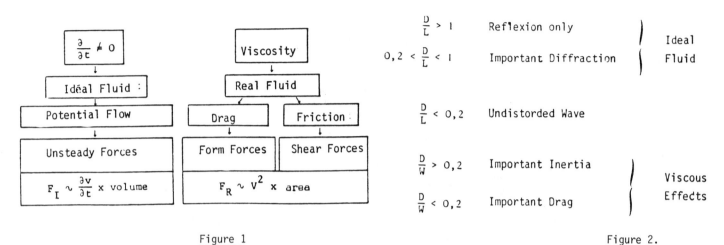

Figure 1.   Figure 2.

Waves action on a structure is summarized in figure 2 where D is structure diameter
L is wave length
W is velocity of wave particle

In our case, computing of floating structures like GNL floating tank, viscous effects are neglected, and irrotational and ideal fluid is used. It could be noticed that the ratio D/W is proportional to F_I/F_R, where F_I and F_R are respectively inertia and viscous forces due to wave action.
The complex motion of the structure on sea is analysed and studied only by separating it into components. The usual method of doing this is used here and it is possible at the end of the computing to combine all the results of the components. The three translation components are surging along the longitudinal X axis, swaying along lateral Y axis, and heaving along the vertical Z axis. The three rotational components about these axes are called rolling, pitching and Yawing respectively (figure 3).

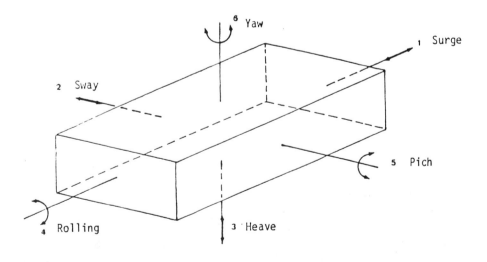

Figure 3.

If the wave is defined by linear Stokes theory we have the following relations

$$F_I \sim \frac{\partial v}{\partial t} \times \text{volume} \sim k \, \sigma \, \phi \, D^3$$

where ϕ is velocity potential
$k = \frac{2\pi}{L}$ is wave number
$\omega = \frac{2\pi}{T}$ is frequency (T is period)

We could deduce : $\phi \sim \frac{g.H}{\sigma}$

where H is wave height

$$F_I \sim \frac{2\pi}{L} \cdot \frac{2\pi}{T} \cdot \frac{g.H.T}{2\pi} \cdot D^3 = \frac{2\pi g H D^3}{L}$$

$$F_R \sim v^2 \cdot \text{surface} \sim k^2 \phi^2 L^2 = \frac{4\pi^2}{L^2} \cdot \frac{g^2 H^2 T^2}{4\pi^2} \cdot D^2 = \frac{g^2 H^2 T^2 D^2}{L^2}$$

and then

$$\frac{F_I}{F_R} \sim \frac{2\pi g H D^3}{L} \cdot \frac{L^2}{g^2 H^2 T^2 D^2} = \frac{2\pi L}{gT^2} \cdot \frac{D}{H}$$

we have to remember that

$$L = \frac{gT^2}{2\pi} \text{th} \frac{2\pi d}{L} \qquad \text{(d is water depth)}$$

$$\frac{F_I}{F_R} \sim \text{th} \frac{2\pi d}{L} \cdot \frac{D}{H} = \frac{D}{W}$$

where

$$W = \frac{H}{\text{th} \frac{2\pi d}{L}}$$

3. STUDY OF FLOATING STRUCTURE MOVEMENTS

A. Heaving

1. Movements

The equation of structure motion are (figure 4)

displacement : $\quad y = y(t)$

velocity : $\quad \dot{y} = \frac{\partial y}{\partial t}$

acceleration : $\quad \ddot{y} = \frac{\partial^2 y}{\partial t^2}$

where y is vertical movement of structure point O.
The equation of periodic wave is

$$\xi = \frac{H}{2} \cos \omega t$$

where ξ is vertical movement of the water level.

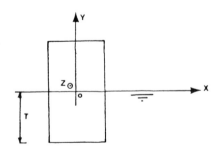

Figure 4.

2. Forces

Fluid forces are composed of four components

- Hydrostatic force is giben by

$$F_1 = -\rho \cdot g \cdot s \cdot y = -C_s \cdot y$$

- Hydrodynamic force on structure in still water is divided in 2 components :
  - The first one F_2 is in phase with the acceleration and we have no energy dissipation

$$F_2 = -\Delta m \, \ddot{y}$$

  - The second one, F_3 takes into account dissipated energy of wave generated by structure movement. In fact F_3 is in phase with the velocity and produces a damping. We call F_3 the potential damping force

$$F_3 = -C_A \cdot \dot{y}$$

where ρ is volumic mass of water
       g is gravity acceleration
       S is horizontal section of the hull
       Δm is added mass of the tank
       C_A is damping coefficient

- Wave force F_4 on structure taking into account the diffraction effects is divided too in 2 parts

$$F_4 = \frac{H}{2} C_1 \cdot \rho V \omega^2 \cos \omega t + \frac{H}{2} C_2 \cdot \rho V \omega \sin \omega t = F_0 \cos(\omega t + \sigma)$$

First part, in cosine, is Froude-Krilov force, where C_1, is added mass coefficient (see Morison O'Brien equation)
Second part, in sine, express phase displacement.
C_2 is an other damping coefficient and V is horizontal velocity of wave.

3. Dynamic equations

The equation of the movement is given by :

$$m\ddot{y} = \Sigma_i F_i = -\Delta m \ddot{y} - C_A \dot{y} - C_s y + F_0 \cos(\omega t + \sigma)$$

$$Cm \cdot m \cdot \ddot{y} + C_A \dot{y} + C_s y = F_0 \cos(\omega t + \sigma)$$

$$Cm = \frac{m + \Delta m}{m}$$

and the solution by

$$y = y_a \cos(\omega t + \varepsilon)$$

$$y_a = \frac{H \omega \rho V}{2} \left[\frac{C_1^2 \omega^2 + C_2^2}{(C_s - Cm \cdot m \cdot \omega^2)^2 + C_a^2 \omega^2} \right]^{1/2}$$

where phase angle ε is defined by

$$\varepsilon = +\arctg\left(\frac{C_2}{C_1 \omega}\right) - \arctg\left[\frac{C_A \cdot \omega}{C_s - Cm \cdot m \cdot \omega^2}\right]$$

B. Rolling

1. Movement

The rolling movement, defined by the figure 5 could be written as follow

$$I_m + \Delta I_m = \frac{\Delta}{g}(k_m + \Delta k_m)^2$$

$$C_I = \frac{I_m + \Delta I_m}{I_m}$$

where I_m is mass inertie moment
k_m is turning radius

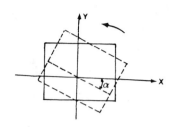

Figure 5.

2. Forces

Using stability floating structures theory, for small angles of rolling, return couple M_1, is defined by

$$M_1 = \Delta \cdot GM \cdot \sin\alpha \sim \Delta \cdot GM \cdot \alpha$$

where GM is metacentric height.

Hydrodynamic couple M_2, in still water, due to the structure movement producing waves, is expressed in the same form like in heaving

$$C_A \cdot \dot{\alpha}.$$

where C_A is a damping coefficient.

Wave couple is given by

$$M_h = C_1 \cdot \frac{H}{2} \cdot \rho \cdot I\omega^2 \cos\omega t + C_2 \cdot \frac{H}{2} \cdot \rho \cdot I\omega \cdot \sin\omega t$$

or

$$M_h = M_o \cos(\omega t + \sigma)$$

where

$$M_o = \frac{H\rho I\omega}{2}\sqrt{C_1^2 \cdot \omega^2 + C_2^2}$$

$$\sigma = \text{arctg}\left[\frac{C_2}{C_1 \omega}\right]$$

3. Dynamic Equation

Dynamic equations have the following forms

$$C_I \cdot I_m \ddot{\alpha} + C_A \dot{\alpha} + \Delta GM \alpha = M_o \cos(\omega t + \sigma)$$

and the solution will be

$$\alpha = \alpha_a \cdot \sin(\omega t + \varepsilon)$$

$$\alpha_a = M_o \left[(\Delta GM - C_I \cdot I_m \omega^2)^2 + C_A^2 \omega^2\right]^{-1/2}$$

$$\varepsilon = \text{arctg}\left(\frac{C_2}{C_1 \omega}\right) - \text{arctg}\left(\frac{C_A \omega}{\Delta \cdot GM - C_I \cdot I \cdot \omega^2}\right)$$

c. Generalisation

In generalisation, we could write the equation of movements in the following form

$$[M][\ddot{X}] + [B][\dot{X}] + [C][X] = [F]$$

and use the matrix of figure 6 in which diagonal terms represent own characteristics of each movement and other terms represent connectings between the different movements. All (M), (B) and (F) coefficients have to be computed by the method that we developpe in the chapter 4.

	1	2	3	4	5	6
1	x	o	o	o	x	o
2	o	x	o	x	o	o
3	o	o	x	o	o	o
4	o	x	o	x	o	o
5	x	o	o	o	x	o
6	o	o	o	o	o	x

Figure 6.

4. POTENTIAL THEORY

A. Introduction

To calculate forces due to wave action, we need velocity and acceleration of water particles. To do that, we use velocity potential theory i.c.

$$\vec{V} = \text{grad } \phi$$

$$\Delta \phi = 0$$

where ϕ is velocity potential with boundary conditions of

Dirichlet : $\quad \phi = \overline{\phi} \quad S_1$

Neumann : $\quad \frac{\partial \phi}{\partial n} = \overline{\frac{\partial \phi}{\partial n}} \quad S_2$

Fluid dynamic pressure P is given by generalisation of Bernoulli equation

$$\frac{\partial \phi}{\partial t} + \frac{1}{2} |\vec{V}|^2 + g y + \frac{P}{\rho} = 0$$

and with the assumption of small movements and linears equations, P will be defined finally by

$$P = - \rho \frac{\partial \phi}{\partial t}$$

For wave potential, we use linear Stokes theory (figure 7)

$$\phi(x, y, t) = + \frac{gHT}{4\pi} \frac{\text{ch } 2\pi \frac{y+d}{L}}{\text{ch } \frac{2\pi d}{L}} e^{i(\omega t + \frac{\pi}{2} - \frac{2\pi x}{L})}$$

where the equations of the free surface y_s and the wave length L are

$$y_{\text{surface libre}} = \frac{H}{2} \cos 2\pi (\frac{t}{T} - \frac{x}{L})$$

$$L = \frac{gT^2}{2\pi} \text{th } \frac{2\pi d}{L}$$

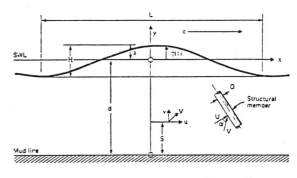

Figure 7.

B. Potential Flow

Using ideal fluid, the potential flow around a floating structure in wave is a linear combination of the following potentials.

1. ϕ_I is the potential of the incident wave, which is speeding out in the whole sea surface and then coming from the infinite. We use the linear Stokes potential defined above. The integration of dynamic pressures on the hull surface, deduced from ϕ_I, gives the Froude Krilov force.

2. ϕ_d is the potential of the diffracted wave, around the fixed structure. The integration of dynamic pressures, deduced from $\phi_I + \phi_d$, gives the total hydrodynamic force due to the wave action

3. ϕ_R is the potential due to the structure in still water. To define those potential, we have to determinate the added mass and damping coefficients of the movements. In fact, the structure has 6 degrees of freedom i.e. 6 own movements independent of wave characteristics. So the potential will be a sum of 6 potentials

$$\phi_R = \sum_{i=1}^{6} V_i \cdot \phi_{Ri}$$

where V_i is the amplitude of velocity (translation or rotation)
ϕ_{Ri} is the potential of an unit velocity following the i degree of freedom

In this condition, the total potential ϕ_t could be written as follow

$$\phi_t = \phi_I + \phi_D + \sum_{i=1}^{6} V_i \cdot \phi_{Ri}$$

C. Boundary Conditions

Boundary conditions are the following :

1. on the sea bottom : $\frac{\partial \phi}{\partial n} = 0$ 2. at the free surface : $\frac{\partial^2 \phi}{\partial t^2} + g \frac{\partial y(x,t)}{\partial t} = 0$

and then we obtain the condition at the linearized free surface : $\frac{\partial^2 \phi}{\partial t^2} + g \frac{\partial \phi}{\partial y} = 0$

and if we pose $\phi = \phi e^{i\omega t}$

$$\frac{\partial \phi}{\partial y} - \frac{\omega^2}{g} \phi = 0$$

3. on the structure : normal fluid velocity on the structure is equal to the velocity along the hull. This condition is presented in different forms

diffraction $\frac{\partial(\phi_I + \phi_D)}{\partial n} = 0$ on the hull and tank movement $\frac{\partial \phi_{Ri}}{\partial n} = \vec{u} \cdot \vec{V}$ on the hull

4. Radiation conditions : the radiation imposes that at boundary of fluid space, diffraction potential ϕ_D and radiation potential ϕ_R tend to zero

$\phi_D \to 0$ if $r \to \infty$ $\phi_R \to 0$

and using polar coordinates, it becomes : $\frac{\partial \phi}{\partial t} + c \frac{\partial \phi}{\partial r} = 0$

In fact the space fluid is limited, and we impose that at its boundaries, radiation waves are only composed by divergent cylindrical waves coming from inside.

In that case, wave celerity C is defined by $c = \frac{gT}{2\pi} \text{th} \frac{\omega \cdot d}{c}$ and if $\phi = \phi_I + \phi_D$

the above equation is changing in $\frac{\partial \phi}{\partial r} + \frac{1}{c} \frac{\partial \phi}{\partial t} = \frac{\partial \phi_I}{\partial r} + \frac{1}{c} \frac{\partial \phi_I}{\partial t}$

6. FINITE ELEMENTS METHOD

The finite elements method is used to solve the following flow differential equation $A(\phi) = 0$. The fluid space is divided in small elements in which potential ϕ is approached by the relation : $\phi_a = \underline{N} \, \underline{\phi}'$
where ϕ' is the vector of potential values at the element knots
$\underline{N}$ is the vector of ponderation functions only depending on element knots coordinates.
The studied phenomena being periodical, ϕ' must be written (using complex notation) as follow : $\underline{\phi}' = \phi e^{i\omega t}$
where ϕ is the vector of potential amplitude at the element knots
Then : $\phi_A = \underline{N}^T \, \underline{\phi} \, e^{i\omega t}$

The discretized problem formulation is obtained by the Galerkin method briefly resumed here. Suppose a system of differential equations $A(\phi) = 0$
If the field ϕ is approached by a field ϕ_a as defined here above, in general we shall have $A(\phi_a) = R \neq 0$
where R is the remainder different of zero resulting from the retained approximation.
The solution which will minimize the remainder value at any point of field V will be the best one.

An evident way consists to utilize the fact that if R is identicaly equal to zero everywhere, than

$$\int_V W.R.dV = 0$$

where W is any coordinates function whatever.
If m functions W_i linearly independant are chosen, then we can write an appropriate number of simultaneous equations in the form

$$\int_V \underline{W}^T R \, dV = 0$$

In the Galerkin method, the functions w_i are taken equal to fonctions N_i defining the approximation inside the elements. This processus gives generally the best approximation and is the most used. In the case of Laplace equation, we have thus

$$\int_V \underline{W}^T A(\underline{N}\phi) dV = 0$$

which becomes after partial integration

$$\int_V \left(\frac{\partial \underline{N}^T}{\partial x}, \frac{\partial \underline{N}^T}{\partial y}, \frac{\partial \underline{N}^T}{\partial z}\right) \begin{pmatrix}\frac{\partial \Phi}{\partial x}\\ \frac{\partial \Phi}{\partial y}\\ \frac{\partial \Phi}{\partial z}\end{pmatrix} dV = \int_S \underline{N}^T \left(\frac{\partial \Phi}{\partial n}\right) ds$$

where S is the boundary surface of fluid space. The second number of this last equation authorizes the introduction of Neumann type boundary conditions.
The Dirichlet type conditions are here implicitly verified and must effectively been imposed during the resolution of the equations system.

A. A body motions in still water : ϕ_{Ri}

a. Heaving

We shall develop here the equations for the bidimensional problem, the three dimensional case being the same but heavier for the statement. The contributions to the second member of equation (1) will be the following :

- free surface $\quad \int_{\Gamma_1} \underline{N}^T (-\frac{1}{g}) \underline{N} \, \underline{\phi} e^{i\omega t} (-\omega^2) d\Gamma_1 \qquad$ where Γ_1 is the boundary curve at the free surface

- bottom
  the condition $\frac{\partial \Phi}{\partial n} = 0 \quad$ gives no contribution. In this case we shall only stop the finite element mesh

- body surface
  The body has a periodical motion that can be written in the form $\quad y = y_m . e^{i\omega t}$

  Its speed will be : $\quad \vec{V} = \dot{y} = y_m . i . \omega . e^{i\omega t} = y_m . \omega . e^{i(\omega t + \frac{\pi}{2})}$

  The slip condition will thus have the form $\quad \frac{\partial \Phi}{\partial n} = \vec{n} . \vec{V} = \sin \theta . y_m . i\omega . e^{i\omega t}$

  where θ is the angle between OX axe and normal outside the fluid on the boundary between body and fluid.

  The contribution will thus have the following form : $\quad \int_{\Gamma_3} \underline{N}^T \sin \theta y_m i\omega e^{i\omega t} d\Gamma_3$

- radiation boundary $\quad \int_{\Gamma_4} \underline{N}^T (-\frac{1}{c}) \underline{N} \, \underline{\phi} e^{i\omega t} d\Gamma_4$

- symmetry condition

  The condition $\frac{\partial \Phi}{\partial n} = 0 \quad$ gives also no contribution to the second member, and we obtain the following system

$$\int_S \left(\frac{\partial \underline{N}^T}{\partial x}, \frac{\partial \underline{N}^T}{\partial y}\right)\begin{pmatrix}\frac{\partial \underline{N}}{\partial x} \\ \frac{\partial \underline{N}}{\partial y}\end{pmatrix}\underline{\phi}\, e^{i\omega t}\, dS = \int_{\Gamma_1} \underline{N}^T(-\frac{1}{g})\underline{N}\,\underline{\phi} e^{i\omega t}(-\omega^2) d\Gamma_1 + \int_{\Gamma_3} \underline{N}^T \sin\theta\, y_m i\omega e^{i\omega t} d\Gamma_3 + \int_{\Gamma_4} \underline{N}^T(-\frac{1}{c})\underline{N}\,\underline{\phi} e^{i\omega t} d\Gamma_4$$

In matricial form, the system is : $(-i\omega\underline{\underline{K}}_2 + \omega^2 \underline{\underline{K}}_3 - \underline{\underline{K}}_1)e^{i\omega t}\underline{\phi} + i\omega y_m \underline{P}_1 e^{i\omega t} = \underline{0}$

simplifying the time harmonic term, we obtain a quasi static complex system

$$(\underline{\underline{K}}_1 + i\omega\underline{\underline{K}}_2 - \omega^2 \underline{\underline{K}}_3)\,\underline{\phi} = y_m i\omega \underline{P}_1$$

with $\underline{\underline{K}}_1 = \int_S \left(\frac{\partial \underline{N}^T}{\partial x}, \frac{\partial \underline{N}^T}{\partial y}\right)\begin{pmatrix}\frac{\partial \underline{N}}{\partial x} \\ \frac{\partial \underline{N}}{\partial y}\end{pmatrix} dS$ coming from the continuity condition on the field.

$\underline{\underline{K}}_2 = \frac{1}{c}\int_{\Gamma_4} \underline{N}^T\, \underline{N}\, d\Gamma_4$ representing the radiation condition $\underline{\underline{K}}_3 = \frac{1}{g}\int_{\Gamma_1} \underline{N}^T\, \underline{N}\, d\Gamma_1$

explaining the free surface condition body motion. $\underline{P}_1 = \int_{\Gamma_3} \underline{N}^T \sin\theta\, d\Gamma_3$ introducing the sollicitation due to

In the equation (2) of the discretized problem, we uncouple the nodal values vector in their real and imaginery parts ($\phi = \phi_1 + i\phi_2$) we obtain the expression : $\left[(\underline{\underline{K}}_1 - \omega^2\underline{\underline{K}}_3) + i\omega\underline{\underline{K}}_2\right](\underline{\phi}_1 + i\underline{\phi}_2) = iy_m\omega \underline{P}_1$

If we separate real and imaginary parts, we obtain :

$$\begin{bmatrix}(\underline{\underline{K}}_1 - \omega^2 \underline{\underline{K}}_3) & -\omega\underline{\underline{K}}_2 \\ -\omega\underline{\underline{K}}_2 & -(\underline{\underline{K}}_1 - \omega^2 \underline{\underline{K}}_3)\end{bmatrix}\begin{bmatrix}\underline{\phi}_1 \\ \underline{\phi}_2\end{bmatrix} = \begin{bmatrix}\underline{0} \\ -y_m \omega \underline{P}_1\end{bmatrix}$$

It is this system that must be resolved by classical algorithms of the finite elements method. With the ϕ_1 and ϕ_2 well known values, we can compute the forces and pressures in all the space and then also heaving hydrodynamic coefficients.

b. Rolling

The studied motion takes the following form $\alpha = \alpha_m e^{i\omega t}$ that gives for the body surface condition the following contribution $\frac{\partial \phi}{\partial n} = \vec{n}\cdot\vec{V} = \alpha_m \cdot i\omega(x \sin\theta - y \cos\theta)e^{i\omega t}$ where x and y are the coordinates of the body considered point in rotation.

The other limiting conditions are the same that in the case of heaving and we obtain the same system with for second member $\begin{bmatrix}0 \\ -\alpha_m\cdot\omega\cdot\underline{P}_2\end{bmatrix}$ with $\underline{P}_2 = \int_{\Gamma_3} \underline{N}^T(x \sin\theta - y \cos\theta) d\Gamma_3$

c. Wave sollicitations on a fixed body

The computation of this problem takes not the same way in the bi or three dimensional case. Of course, the radiation condition is given by $\frac{\partial \phi}{\partial n} + \frac{1}{c}\frac{\partial \phi}{\partial t} = \frac{\partial \phi_I}{\partial n} + \frac{1}{c}\frac{\partial \phi_I}{\partial t}$

In the bi-dimensional case, the wave (ϕ_I) follows of course the condition

$$\frac{\partial \phi_I}{\partial n} + \frac{1}{c}\frac{\partial \phi_I}{\partial t} = 0$$

and we obtain the system

$$\begin{bmatrix} (\underline{\underline{K}}_1 - \omega^2 \underline{\underline{K}}_3) & -\omega \underline{\underline{K}}_2 \\ -\omega \underline{\underline{K}}_2 & -(\underline{\underline{K}}_1 - \omega^2 \underline{\underline{K}}_3) \end{bmatrix} \begin{bmatrix} \phi_1 \\ \phi_2 \end{bmatrix} = \begin{bmatrix} 0 \\ 0 \end{bmatrix}$$

The system solution is meanwhile not $\phi_1 = \phi_2 = 0$ because we must also artificially impose the $\phi_1 + i\phi_2 = \phi_I$ values at the boundaries corresponding at the incident wave.

In the three dimensional case, the complete discretized system will be $(\underline{\underline{K}}_1 - \omega^2 \underline{\underline{K}}_3) + i\omega \underline{\underline{K}}_2 \, \underline{\phi} = \underline{P}_4$

or

$$\begin{bmatrix} (\underline{\underline{K}}_1 - \omega^2 \underline{\underline{K}}_3) & -\omega \underline{\underline{K}}_2 \\ -\omega \underline{\underline{K}}_2 & -(\underline{\underline{K}}_1 - \omega^2 \underline{\underline{K}}_3) \end{bmatrix} \begin{bmatrix} \underline{\phi}_1 \\ \underline{\phi}_2 \end{bmatrix} = \begin{bmatrix} \text{Re } \underline{P}_4 \\ \text{Im } \underline{P}_4 \end{bmatrix} \quad \text{with} \quad \begin{aligned} \text{Re } \underline{P}_4 &= \int_{S_4} \underline{N}^T f_1(x,y,z) dS_4 \\ \text{Im } \underline{P}_4 &= \int_{S_4} \underline{N}^T f_2(x,y,z) dS_4 \end{aligned}$$

$f = f_1 + i f_2$ where the f function represents the contribution $\frac{\partial \Phi_I}{\partial n} + \frac{1}{c} \frac{\partial \Phi_I}{\partial t}$ due to the wave sollicitation

7. COMPUTATION EXAMPLES

First example

To improve the programme performances, we have realized a complete computation of a rectangular caisson and we have compared these results with scale model results in a wave tank. The studied caisson is 120 cm length, 20 cm breadth 12,5 cm depth and has a draft of 4 cm.

The caisson is placed transversaly to the canal axe during the trials, its breadth being parallel at this axe. We only study rolling and heaving motions, sheering being stopped and the other motions inexisting. Of course, the wave front is parallel at the longitudinal caisson inertia axe and the sollicitation is zero for pitching, Yawing and surging motions. We will study a very long caisson to be able to only use a bi-dimensional finite elements study, the strip theory being applicable. The heaving and rolling hydrodynamic coefficients are in this case the same in bi-or three dimensional study. The used discretization is represented in figures 8 and 9 reproducing in detail the black part of the figure 8.

It has 238 elements with 6 or 8 knots, so in total 705 knots. As the problem is symmetrical we only use half the fluid space during the study of the forced rolling and heaving motions. In the case of diffraction, the used discretization is the double that this one represented. The mesch system exists up to the canal bottom, at 2,18 m from the free surface. The obtained results are reproduced in figures 10 and 11. For heaving the Y_A motion amplitudes are divided by the sollicing wave amplitude H while the rolling angles α_A by the maximum angle of the tangent at the free surface wave,

$$\alpha_H = \frac{\pi H}{L}$$

The periods variation is represented in the diagrams by the reduced frequency (which is a non-dimensional coefficient). Its value is

$$\bar{\omega} = \frac{\omega^2 b}{g}$$

where ω is the motion frequency
       b is half the breadth of the caisson

The experimental study has been realized in the wave tank of the Hydrodynamic laboratorium of Liege University. It is possible to generate waves amplitudes of around 2 to 15 cm, with periods intervals between 0,5 and 3 seconds.

The heaving an rolling motions measures are obtained by an optical system.

Picture 1 gives a vue of the realized experiences and the optical system. We can also see supple cables necessary to prevent the model drift.

The concordance between numerical and experimental results is excellent, except at the rolling natural frequency where the computed amplitudes are too big and when the viscosity has an influence.

Nevertheless the computed rolling natural period is the same that its experimental value : it is very important for the design of structures for which we always must avoid the resonance.

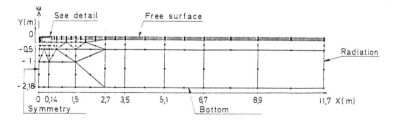

Figure 8

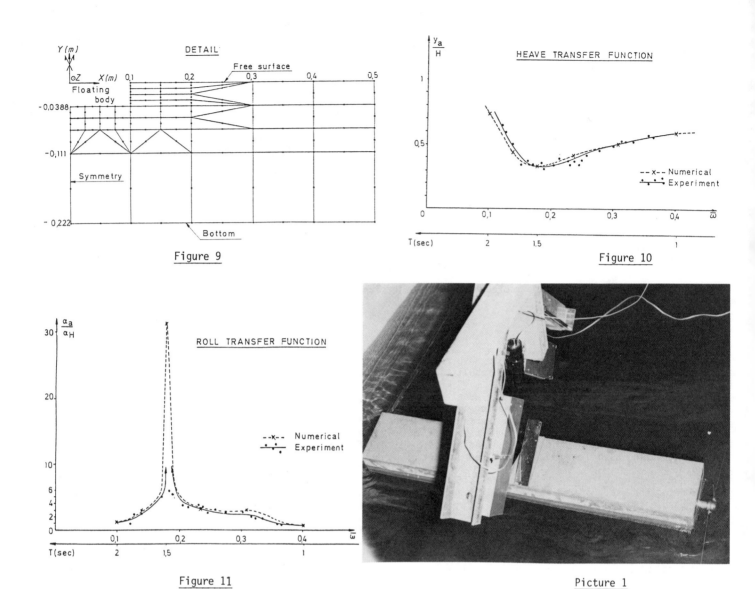

Figure 9

Figure 10

Figure 11

Picture 1

Second example
Another way to improve the programme performances is to measure heaving hydrodynamic coefficients (C_m and C_A) on scale model. The results are compared with computed results.

We have realized such a study with a cylindrical caisson. This caisson is 26,4 cm high, has a diameter of 50 cm and a draft of 7,2 cm. A vertical periodic movement called "forced heaving motion" is imposed in still water. Both amplitude and frequency of the motion are known. Force $F(t)$ we need to impose "forced heaving motion" is measured by a quartz force link (see picture 2). $F(t)$ takes the place of F_4 (wave force) in the dynamic equation so that C_m and C_A are known.
The used discretization has 142 elements with 8 knots, so in total 249 knots. It is a tridimensional discretization representing a quarter of the fluid space.
The obtained results are reproduced in figures 12 and 13.
The concordance between numerical and experimental results is good for C_m curve, not for C_A curve. That's because we have neglected viscosity in the computed results and because the damping force is small in this case so that measures accuracy is not sufficient.

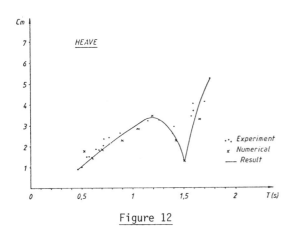

Figure 12

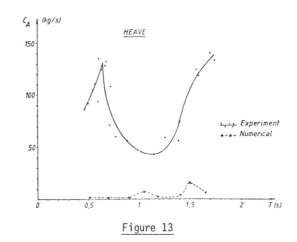

Figure 13

Picture 2

DRIFT FORCES AND DAMPING IN NATURAL SEA STATES

A CRITICAL REVIEW OF

THE HYDRODYNAMICS OF FLOATING STRUCTURES

H. LUNDGREN
Technical University of Denmark
& Danish Hydraulic Institute

STIG E. SAND
Technical University of Denmark

J. KIRKEGAARD
Danish Hydraulic Institute
Denmark

SUMMARY

1. Drift and damping forces are of paramount importance to the design of large floating structures (TLPs, semisubmersibles, etc.). A scrutiny of recent literature shows that the determination of these forces still may need *clarification*.

2. Since the drift forces depend so much on wave directionality and grouping, there is a great need for *three-dimensional recordings of natural sea states* in various locations. These recordings should include the determination of the phases. It is not safe to assume that the phases are random.

3. The first order description of recorded *sea states in the North Sea* is mentioned in Sec. 2. The recordings cover frequencies, directions and phases.

4. The *second order description*, as derived from the first order description, is discussed in Sec. 3, including frequencies, directions and phases. These low frequency waves are of significance for the drift forces.

5. For the determination of drift and damping forces one should always observe the *basic principle* of f i r s t combining the various flow components, i.e. waves, currents and body motions.

6. A survey of the physics of *drift forces*, involving 8 different types and their relative importance, is given in Sec. 4.

7. A survey of *damping forces*, involving 7 different types, is given in Sec. 5.

8. A discussion in Sec. 6 of the *Mathieu instability* problem results in the conclusion that wave directionality and grouping are of great significance.

9. *Model testing techniques* are mentioned in Sec. 7. Tests with offshore structures should be made in 3-D reproduced natural sea states. With an oscillating carriage, forces from irregular waves and currents can be studied at greater scale in 2-D flow. In this connection it is possible to compensate for dominant inertial pressures.

1. INTRODUCTION

It is well known that drift and damping forces are decisive for the design of slowly oscillating systems with low damping, such as a vessel connected to a single point mooring (SPM), a moored semisubmersible or a tension leg platform (TLP).

While the developments of SPMs and semisubmersibles have been gradual, so that it has been possible to utilize previous experience for each step into deeper water or worse environmental conditions, the application of TLPs will represent a rather *discontinuous development* into a technologically virginal area. As a consequence of this, and in order to reduce the risk of failures, extensive efforts have been spent, and are being spent, by the offshore industry and by hydrodynamic institutions, on investigations of many kinds, including numerical and physical models, hydrodynamic theories, etc.

A perusal of recent literature reveals, however, that there still exist considerable *difficulties* in:

(1) The description of the environmental factors.

(2) The relative role, as well as the magnitude, of the various hydrodynamic forces.

These difficulties are discussed below.

The *conclusion* is that a large amount of investigation is still needed in order to eliminate the uncertainties inherent in the design of large floating structures, which could otherwise be in a risky situation or require excessive costs.

1.1 Environmental Factors

The *wind velocity profile and turbulence* during extreme gales are reasonably well known. The local distributions above the troughs and the crests of a natural sea state, however, require further studies, in as much as the gusts contribute to the slow drift forces. These local distributions may be accounted for in a physical model, if a natural, gusty wind is superimposed upon the waves in the area around the structure.

The *tidal currents* at a given location are known with good accuracy from numerical models of the oceans. The turbulence characteristics, which influence the drift forces, may require local measurements because of stratification.

The *wind-driven currents* in fairly shallow areas like the North Sea may be predicted with good accuracy from simple numerical models, cf. for example, the design forces for the Danish North Sea gas pipeline (Brink-Kjær, et al, 1982). For wind-driven oceanic currents the development of a satisfactory numerical model is needed.

In a given situation the drift forces from winds and currents are stochastically independent and may be added accordingly (Wichers, 1979).

The most important, and least known, drift forces originate from the *waves*. The difficulties encountered are due to deficient descriptions of the natural sea states.

Historically, sea states have been described on the following levels:

1. Periodic, 2-D design waves.
2. Irregular, 2-D waves with a certain spectrum (infinitely wide fronts).
3. Irregular, 3-D waves with a certain directional spectrum (short-crested waves).

It is not until recently that physical and numerical models have moved onto Level 3.

A number of numerical models have been developed for the forecast or hindcast of the wave motion on Level 3 (see, for example, Brink-Kjær, et al, 1982).

It is evident that the *wave directionality* has great influence on the low frequency motions of moored structures (surge, sway, yaw). This has been demonstrated by model tests with a bulk carrier exposed to a deterministic reproduction of storm waves recorded in nature (Kirkegaard, et al, 1980).

It has often been assumed that natural waves have random phases. This is most unlikely because of the high order, nonlinear interactions. In 2-D tests it has been demonstrated, as could be expected, that the *wave grouping* has great influence on the surge motions (Spangenberg and Kofoed Jacobsen, 1981).

The *grouping of waves over varying directions* has a particular influence on the yaw motions, which could be critical for the stability of a TLP, cf. Sec. 6 below. Such grouping is available from a storm in the Danish sector of the North Sea, based on a 3-D recording system (see Secs. 2 and 3 below). There is a great need, however, for many more 3-D records from various areas, before it may become possible to extract some 'typical' natural sea states.

The influence of the 3-D structure of waves on the *Mathieu instability* problem is described in Sec. 6 below.

The directional spread of wave energy observed at one point has its counterpart in a *directional concentration*, which is known to generate 3-D plungers in the middle of the ocean. Though such plungers have limited influence on the low frequency motions, they may be important for the survival.

1.2 Hydrodynamic Forces

The difficulties inherent in numerical and physical models may be illustrated as follows:

(a) Some numerical models consider only *potential flow* for the determination of drift and damping forces, neglecting the viscous effects.

(b) Other numerical models consider only *viscous forces*, neglecting the potential flow forces.

(c) The value of the Morison *drag coefficient*, C_d, is often taken from hydrodynamic conditions that differ widely from the situation considered.

(d) The drift and damping forces should always de determined a f t e r proper *superposition* of the various flow components, i.e. wave motion (including wave flow separation), current, and body motion. If one does not adhere to this principle, the result may be off by orders of magnitude.

(e) The *scale effects* in physical models are mainly due to the low values of Reynolds number when drag or friction is involved. This applies to both waves and currents.

The difficulties concerning the drift and damping forces will be discussed in Secs. 4 and 5 below. In some cases the scale effects, which exaggerate drag and friction in physical models, increase the damping forces more than the drift forces, which is likely to lead to results on the unsafe side.

2. NATURAL SEAS - FIRST ORDER DESCRIPTION

It is generally recognized that the directionality of natural waves is important for the design of offshore structures. The three-dimensional character of ocean waves can be described by two fundamentally different methods. One is *stochastic* and based on the application of a spectrum with random phases and a parameterized spreading function. The other is *deterministic* and utilizes all information - also the phases - from site wave records. Here emphasis is placed on analyses attached to the latter method.

The directional measurements have been obtained by means of an echo sounder giving the surface elevations, η, and an electromagnetic current meter giving the two horizontal components of orbital velocities, u and v. The necessary information could as well have been obtained by means of a pitch and roll buoy measuring the elevations, η, and the slopes η_x and η_y. Application of the deterministic theory is, however, demonstrated for the η, u and v records taken in one vertical.

It is assumed that the wave motion in the natural sea is linear and composed of a large number of sinusoidal waves having infinitely wide fronts. When the three simultaneous time series η, u and v are decomposed into Fourier series, it is possible to utilize the fact that u and v form a velocity vector while η reveals whether the component considered has a crest or a trough attached to this vector. Thus, in narrow frequency bands it is possible to determine the directional distributions. The mathematical principles applied to obtain such results are described in Sand, 1979. When the large number of directional components are determined, and the phases have been stored, the wave field can be calculated in an area around the point of recording. In this respect the theory has been checked by measuring also the velocities u_1, v_1 at another point $1.3\,\bar{\lambda}$ away, $\bar{\lambda}$ being the mean wavelength. Satisfactory agreement was obtained between measured and calculated quantities.

In order to compare directional distributions obtained under different conditions, the spread is often fitted to a functional form. This has mostly in the literature been $\cos^{2s}((\theta-\theta_m)/2)$, with $s(f)$, or a normal distribution, $N(\mu,\sigma^2)$. It has been shown how well the empirical variations of the spreading parameter, $s(f)$, agree for measurements in the Baltic Sea, the Atlantic Ocean, and in the Gulf of Mexico (Sand, 1982b). However, the mean direction as a function of frequency, $\theta_m(f)$, (around which the waves are spread with parameter $s(f)$) shows very different variations, depending on the type of storm, fetch, etc.

It is easier to fit and to visualize the directional distributions when they are represented by *normal distributions*, $N(\theta_m(f), \sigma^2(f))$. For illustration, two normal distributions with $\theta_m = 0$ and σ^2 equal to 0.1 rad^2 and 1 rad^2, respectively, are shown in Fig. 1. The corresponding standard deviations in degrees are also given.

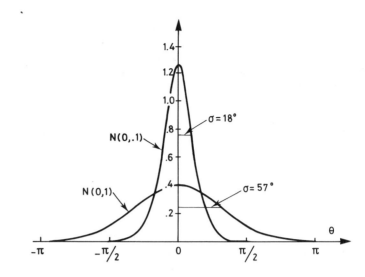

Fig. 1
Normal distributions with standard deviations given in degrees.

Normal distributions have been applied to *directional measurements* taken during storms in the North Sea and in the Baltic Sea. For narrow frequency intervals the distribution is fitted to the actual spreading. This results in series of frequency dependent mean directions, $\theta_m(f)$, and angular variances, $\sigma^2(f)$. Fig. 2 shows, like the comparisons referred to above, a striking agreement between the variances obtained in different locations and under quite different conditions. The frequencies are normalized by the frequency $f_{1,-1} = \sqrt{m_1/m_{-1}}$, where m_n is the n'th moment of the one-dimensional spectrum, $S_{\eta\eta}(f)$. The well-known tendency is seen, viz. the variance increases with frequency, except at the lowest wave frequencies.

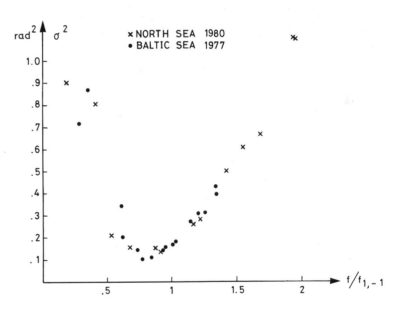

Fig. 2
Angular variance σ^2 as a function of dimensionless frequency $f/f_{1,-1}$ for two storms.

The records were taken during a storm from SE in the Baltic Sea and one from NNW in the North Sea. The significant wave heights were 2.4 m and 6.2 m, respectively. The *mean direction* as a function of frequency, $\theta_m(f)$, is shown in Fig. 3, which also reflects the development of the history of the storms.

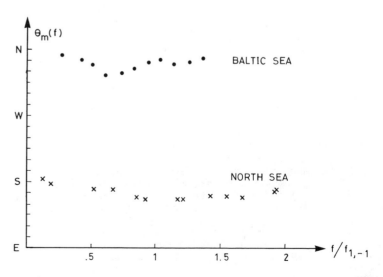

Fig. 3
Mean direction θ_m as a function of dimensionless frequency $f/f_{1,-1}$ for storms from SE (Baltic Sea) and NNW (North Sea), respectively.

The normal distribution used for the directional spread is on the usual form

$$N(\theta_m, \sigma^2) = \frac{1}{\sigma\sqrt{2\pi}} \exp\{-(\theta-\theta_m)^2/2\sigma^2\} \qquad (2.1)$$

Furthermore, in the range of frequencies where essential energy is present, cf. Fig. 2, the following linear relationship is found

$$\sigma^2(f) = 0.72\, f/f_{1,-1} - 0.54 \qquad (2.2)$$

Hence, so far as a linear description of natural seas is concerned the directional spectrum is given by

$$S_{\eta\eta}(f,\theta) = S_{\eta\eta}(f)\, N\bigl(\theta_m(f), \sigma^2(f)\bigr) \qquad (2.3)$$

Of course, for a deterministic description the phase information obtained from the site wave records has to be applied. Apart from this, however, a *natural sea state* is defined by

(i) The one-dimensional spectrum, $S_{\eta\eta}(f)$.

(ii) The mean direction as a function of frequency, $\theta_m(f)$.

(iii) The variance, $\sigma^2(f)$, of the energy around θ_m as a function of frequency, cf. for instance (2.2).

An attempt to illustrate the directional spectrum is made in Fig. 4, in which some of the angular standard deviations from Fig. 2 are indicated along the one-dimensional spectrum from the North Sea measurements. The lowest frequency part of this spectrum is discussed in the following section.

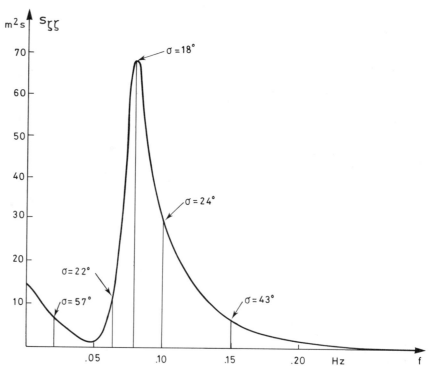

Fig. 4
Spectrum measured in the North Sea with indication of angular standard deviation σ at various frequencies. The second order low frequency part has been exaggerated by a factor 4 for clarity.

3. NATURAL SEAS - SECOND ORDER DESCRIPTION

In a second order description of a natural sea state two types of additional terms, both solutions to the Laplace equation, have to be introduced in order to satisfy the nonlinear surface conditions. So, in addition to the well-known linear solution expressed in regular sine components of wavenumber vectors $\vec{k}$, terms with wavenumbers $\vec{k}_n + \vec{k}_m$ and $\vec{k}_n - \vec{k}_m$ occur, cf. Fig. 5. Obviously, the $\vec{k}_n + \vec{k}_m$ components have very short wavelengths (corresponding to second harmonic waves). The practical importance of these components is in the wave shapes. With a view to the drift forces, the discussion may be confined to the long wave components $\vec{k}_n - \vec{k}_m$.

The *low frequency* part, $S_{\zeta\zeta}$, of the total directional spectrum, $S_{\zeta\zeta}(f,\theta)$, is shown schematically in Fig. 6. A measured second order low frequency spectrum from the North Sea is seen in Fig. 4. From radiation stress considerations it can be deduced that the second order components

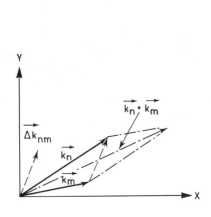

Fig. 5
Second order solution to the Laplace equation: a short wave component $\vec{k}_n + \vec{k}_m$, and a long one $\Delta\vec{k}_{nm} = \vec{k}_n - \vec{k}_m$.

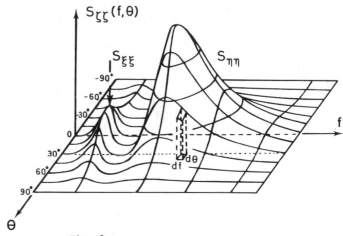

Fig. 6
Directional spectrum containing both the wave frequencies, $S_{\eta\eta}$, and the low frequencies, $S_{\xi\xi}$, with $S_{\zeta\zeta} = S_{\eta\eta} + S_{\xi\xi}$.

follow the groups, with troughs beneath the higher waves and crests in between. Thus, the low frequency waves are a group-induced phenomenon, and, for a group composed of two waves, f_n and f_m, the low frequency component is determined by the difference $f_n - f_m$, while direction and wavenumber are found from $\vec{k}_n - \vec{k}_m$.

There are some interesting characteristics of the second order low frequency waves. Firstly, the fact that the direction of travel is determined by the difference of two first order wavenumber vectors implies that the *angular spread* is much larger than that of the first order waves, as demonstrated by Fig. 5. For the North Sea measurements, Fig. 4, a typical value of the standard deviation was $\sigma = 57°$.

Secondly, it can be seen that the 3-D character of the first order waves results in a considerable *reduction of the amplitudes* of the second order low frequency waves, as compared with a 2-D situation. Transfer functions from the first to the second order waves have been derived (Sand, 1982a). These show, for example, that the low frequency waves are smaller in the (natural) 3-D case than in a (theoretical) 2-D case. Consider, for example, two components each of amplitude 5 m, travelling in the directions $+15°$ and $-15°$ with the frequencies $f_n = 1/14$ Hz and $f_m = 1/19$ Hz, respectively, and a water depth of, say, 450 m. Such a simple wave field induces a second order wave of period 53 seconds, travelling in the direction $42°$ with a wave height of 0.11 m. For comparison, it can be mentioned that in the corresponding 2-D situation the height would be larger by a factor of 3.

Finally, due to the vectorial wavenumber difference the 3-D low frequency waves have larger wavenumbers and, hence, *smaller wavelengths* than corresponding 2-D waves.

Since the second order waves can be calculated, by means of transfer functions, from the first order waves it is also possible to determine the *low frequency spectrum*, $S_{\xi\xi}$. However, as previously described (Sand, 1982b) it is important to note that the *phases are needed* for this computation. As mentioned earlier, the phases (attached to $S_{\eta\eta}(f,\theta)$) can either be obtained from site wave records or they can be randomly selected. Naturally, this ambiguity leads directly into the present intense debate about stochastic and deterministic approaches.

In principle, it is possible to attach an infinity of phase combinations to a given first order spectrum, $S_{\eta\eta}(f,\theta)$. Each combination will give its own wave grouping and, hence, drift forces. If, therefore, a time series is produced on the basis of the spectrum and a random process for the phases, the result is just one of many possibilities. In addition, it has never been proved that the natural sea state that we want to simulate is characterized by a random phase distribution. Thus, it may be feared that some of the important, but as yet unrevealed *features of nature* could be *eliminated* by the random phase concept.

Out of such fear the Danish Hydraulic Institute has for ten years worked on a deterministic basis from *site wave records*. Thereby it is assumed that important properties of natural waves are retained. Hopefully, essential properties are also preserved when site records are scaled up in time and amplitudes in order to represent, for example, a 100 year situation. Since the discussion of the phase distribution in nature has not come to a final conclusion, the present philosophy is that a reproduction based on information from nature by no means could be inferior to just o n e arbitrary realization of a stochastic process. Incidentally, the deterministic concept has been supported by vertical breakwater tests, where the application of time series with randomly selected phases implied no shock forces, whereas natural wave records (with an identical spectrum) gave reasonable shock force distributions, independent of the length of the wave flume.

It should be noted that the emphasis in this paper on the deterministic approach is related to the present *state of the art* with respect to wave grouping, etc. It would be highly desirable in the years to come if the probability distribution of phases could be determined, perhaps as a function of water depth, fetch, etc. When such profound knowledge become available, time series might be produced from the directional spectrum combined with the phase distribution. In the necessary future research two *recently developed concepts* may be involved. One is the group spectrum derived from η^2, called SIWEH (Funke and Mansard, 1979 and 1981). The other is the low frequency, second order wave spectrum, $S_{\xi\xi}(f)$, described above (Sand, 1982a).

4. DRIFT FORCES

4.1 Wave Related Drift Forces

In contrast to the first order wave forces, which oscillate between large positive and negative values at the wave frequencies, the wave related drift forces are *characterized by*:

- (a) having small values,
- (b) having mostly a mean value different from zero,
- (c) varying with much lower frequencies than the wave frequencies, and
- (d) giving rise to horizontal motions (of moored structures) much larger than the first order motions.

The most *important drift forces* are:

I. Viscous flow forces:
  A. Wave drag drift.
  B. Current-wave drag.
  C. Current-wave friction.

II. Potential flow forces:
  D. Wave elevation drift.
  E. Velocity head drift.
  F. Body translation drift.
  G. Body rotation drift.
  H. Second-order-wave drift.

The physics of these 8 drift forces will be explained below. A *common feature* of these forces is that they are of second or higher order. A second order term is defined as a term that contains the product of two first order terms. As first order terms are understood:

(1) Wave elevations, velocities, and pressures.
(2) Current velocity.
(3) Body displacements due to first order waves.

4.2 Wave Drag Drift

Though the Morison formula gives an imperfect picture of the hydrodynamic situation in connection with wave drag, it will be used for illustration, applied to a vertical cylinder of diameter d and draught D placed in a small wave with horizontal velocities u:

$$F_d = \frac{1}{2} \rho C_d d u |u| \cdot D \tag{4.1}$$

For a small sine wave the mean value of $u|u|$ over the period T vanishes. Hence, there is no essential drift.

For a *large wave*, however, it is necessary to take the variation, $\eta(t)$ of the wave elevation into consideration. Consider, for example, one of the columns in a TLP, where the heave is negligible compared with the wave heights. Then, the factor D in (4.1) must be replaced by $D + \eta$, giving

$$F_d = -\rho C_d d u |u| \cdot (D + \eta) \tag{4.2}$$

Now, the mean value of F_d over the period is positive because positive values of u are accompanied by positive values of η, and negative values of u by negative η.

As will be seen below, the mean value of (4.2) over the period is the *dominant drift force* on a TLP (and a semisubmersible). This has been stated by Ferretti and Berta, 1981 (without indicating the C_d value applied).

The wave drag drift has been applied in several *numerical models*:

(i) Pijfers and Brink, 1977, where the C_d-value has been chosen as a weighted average between the Reynolds dependent steady-flow value, $C_d(Re)$, involving the sum of the current velocity and the wave (Stokes) drift velocity, and the Keulegan-Carpenter dependent, but Reynolds independent, wave flow value, $C_d(KC)$, involving the horizontal velocity.

(ii) Denise and Heaf, 1979 (without indicating the C_d-value used).

(iii) Sebastiani, et al, 1981, where reference is made to Ferretti and Berta, 1981, mentioned above.

Other numerical models exclude the wave drag drift on the grounds that it is a cubic term (u^2, η). Indeed, it seems to be a *quartic term*, because C_d is approximately proportional to the wave height (see below).

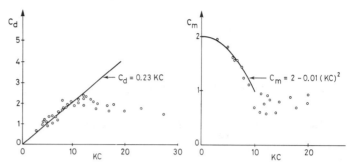

Fig. 7
Subcritical C_d- and C_m-values found in an oscillating water tunnel (from Bearman and Graham, 1979).

As an illustrative *example*, one of the columns in the TLP discussed by Gie and de Boom, 1981, has been considered (with the significant wave as a regular wave):

$$d = 16.9 \text{ m}, \quad D = 35 \text{ m}, \quad H = H_s = 18.1 \text{ m}, \quad \overline{T} = 16.7 \text{ s} \tag{4.3}$$

For this wave, $KC = 3.36$ and, according to Fig. 7, the subcritical C_d-value is about 0.67. A supercritical value of 0.35 has been chosen as a rough estimate. Fig. 8 shows the variation of the

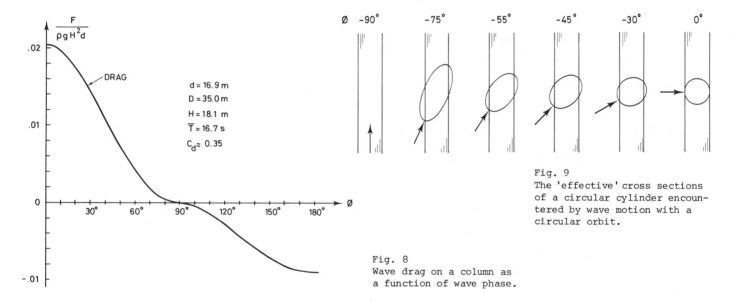

Fig. 9
The 'effective' cross sections of a circular cylinder encountered by wave motion with a circular orbit.

Fig. 8
Wave drag on a column as a function of wave phase.

drag (integrated from $z = -35$ m to $z = \eta$) for a sine wave. After integration over the period the mean value is found to be

$$F_{dr,A} = 0.0024 \, \rho \, g \, H^2 \, d \tag{4.4}$$

(If the second order wave profile is used, the increase is only 1%.) Originally, the Morison formula was tentatively proposed for application in shallow water, where the wave orbits are very long ellipses. To use it in deep water where the orbits are circles, is most questionable, cf. Fig. 9, which demonstrate that the 'effective' cross sections encountered by a wave with crest at $\phi = 0°$ are elliptic most of the time. Hence, it is possible that the actual C_d-values are less than indicated in Fig. 7. Anyway, it is likely that $C_d \approx 0$ at the beginning of the wave crest ($\phi = -90°$), where separation starts along the rear generatrix of the cylinder.

At the *stern of a ship* exposed to head waves there is also a wave drag drift because of the wave flow separation in connection with the (elevated) crest, as compared with the nearly potential flow in the (depressed) trough.

Further studies are required in order to determine the C_d-values for a column in the prototype, where the flow is *supercritical*. In addition, there are great difficulties in determining the *low frequency wave drag drift* in a natural sea state with its changing groups and directions, particularly because the development of the 'drag coefficient' depends upon the wave height, the wave period, etc.

4.3 Current-Wave Drifts

If a small current velocity (or Stokes drift), V_c, is added to u in (4.1), the *current-wave drag* is found as the mean value over the period:

$$F_{dr,B} = \frac{1}{2} \rho C_d d \frac{4}{\pi} u_m V_c D \tag{4.5}$$

where, for simplicity, the maximum horizontal orbital velocity, u_m, has been assumed to be constant over the draught D, as in (4.1). If the variation of the wetted surface is considered, there will be a minor change of the drift force.

The *current-wave friction* has some influence on large ships in head waves because the wave friction factor, f_w, is much larger than the current friction factor. For small currents the resulting friction drift is proportional to f_w, u_m and V_c, and it is active along the sides and the bottom of the ship. The force is analogous to the wave friction damping; reference is made to Fig. 11 and pertaining text.

4.4 Potential Flow Drift Forces

The most *thorough discussion* of the 5 types of potential flow drift forces on floating bodies has been given by Pinkster, 1980, to which reference is made for all details.

The forces may be divided into 3 groups:

(i) 'D. Wave elevation drift' and 'E. Velocity head drift' may, roughly, be said to originate from the reflection of the first order waves from the body. This is illustrated by Fig. 10, where the drift force on a restrained ship in beam sea is seen to diminish as the wavelength λ increases relatively to the ship length L_{pp}.

(ii) 'F. Body translation drift' and G. Body rotation drift' originate from the displacements of the body in combination with the pressures on the body from the first order waves.

(iii) 'H. Second-order-wave drift' is the force exerted on a (fixed) body by the low frequency, second order waves discussed in Sec. 3 above.

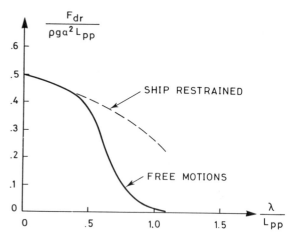

Fig. 10
Mean wave drift forces on loaded 130,000 dwt tanker in beam sea (from Løken and Olsen, 1979).

For uni-directional irregular waves a considerable amount of work has been done in order to determine the *quadratic transfer function* from first order waves to drift forces (Newman, 1975; Faltinsen and Løken, 1979). Seemingly, only random phases have been considered. Hence, the influence of natural phases remains to be seen. This also applies to the influence of directional sea states.

The *wave elevation drift* on a body is the integral around its waterline of the force $\frac{1}{2} \rho g \eta^2$, where η is the first order wave elevation. This expression represents for a first order pressure distribution the excess triangular force at still water level that does not vanish when averaged over time. The force is taken as a horizontal pressure force perpendicular to the waterline. (For a sloping side of the body there is also a vertical force; this is without interest for the drift forces.) The diffracted waves are included in η.

The *velocity head drift* on a body is the integral over its wetted surface of the pressure $p_v = -\frac{1}{2} \rho V^2$, where V is the velocity vector at the body surface due to the first order wave motion, including the waves diffracted around the body. p_v is the pressure drop corresponding to the velocity term in the Bernouilli equation.

As an *example* will be taken the column given by Eq. (4.3). The exact determination of the drift forces requires the application of the 3-dimensional, potential flow source method. Instead, an approximate approach is used because it gives a simpler picture of the problem. To begin with, the draught is assumed to be infinite. Then there is an explicit solution along the following lines (cf., for example, Isaacson, 1977 or Skovgaard and Jonsson, 1981): The incoming wave, propagating in the direction $\theta = 0$, is developed in a series of Bessel functions. Because the interest is in the drift force in the wave direction, only the terms involving $J_0(kr)$ and $J_1(kr) \cdot \cos\theta$ contribute, where k is the wavenumber and $r = a = d/2$ is the surface of the cylinder. In the reflected wave the corresponding Hankel functions are involved.

Since the diameter of the cylinder is small compared with the wavelength, it suffices to take the first terms in the power series of the Bessel and Hankel functions. The result is a wave elevation drift

$$F_{dr,D} \approx \frac{\pi^2}{16} (ka)^3 \rho g H^2 d \qquad (4.6)$$

The velocity head drift is $F_{dr,E} = -\frac{1}{2} F_{dr,D}$, thus reducing the total drift on a fixed cylinder to 50% of (4.6). If, as a rough estimate, the limited draught, $D = 35$ m, is taken into consideration by a reduction factor, $[1 - \exp(-2kD)] = 0.64$, a potential flow drift force of

$$F_{dr,D+E} = 0.00036 \rho g H^2 d \qquad (4.7)$$

results. It will be noted that this is only 15% of the wave drag drift in (4.4), the latter being based on an estimate of the supercritical C_d-value.

The *body translation drift* is associated with the 3 translational displacements x_i ($i = 1,2,3$) that will shift the body into a slightly different pressure field because of the pressure gradient $\partial p/\partial X_i$, due to the first order wave motion, including the diffracted waves. Hence, the product $x_i \cdot \partial p/\partial X_i$ has to be integrated over the wetted surface as an additional pressure.

The *body rotation drift* is associated with the 3 angular displacements x_j ($j = 4,5,6$) that will shift the directions of the pressures on the sides of the body due to the first order wave motion, including diffraction. For example, a roll angle, x_4, will tilt the bottom of a ship so that the total wave pressure, P_3, on the bottom will give a sway force component, $F_2 = -x_4 \cdot P_3$, in the X_2-direction. Since the wave pressures integrated over the body surface can be expressed by the body accelerations, the body rotation drift is found to be $x_j \times (M \ddot{x}_i)$, where M is the mass of the body, and the summation extends over all i and j.

Usually, the body displacement drifts contribute to a reduction of the total drift force, cf. the two curves in Fig. 10.

The *second-order-wave drift* is due to the pressure gradient in the second order waves. Since these waves have low frequencies and, hence, very long wavelengths, the pressure gradient will normally correspond to the surface slope of the second order wave, unless the body dimensions are very large. The pressure gradient has to be applied to the body plus added mass.

As an *example*, the second-order-wave drift has been calculated for the column in Eq. (4.3) for a directional sea state with a significant wave height, $H_s = 18.1$ m, and a mean period, $\overline{T} = 16.7$ s, corresponding to a peak period, $T_p = 18.0$ s. The sea state used has been scaled up from a natural sea state recorded three-dimensionally in the North Sea. The result is a significant drift force (corresponding roughly to second order waves with $H_s \approx 0.15$ m and $\lambda \approx 1500$ m)

$$F_{dr,H} \approx 0.0008 \rho g H_s^2 d \qquad (4.8)$$

determined from the transfer functions (Sand, 1982a). This force is about twice the mean drift (4.7).

5. DAMPING FORCES

5.1 Hydrodynamic Damping Forces

The most *important damping forces* are:

I. Wave-related, viscous flow forces:
   L. Wave drag damping.
   M. Wave friction damping.

II. Current-related, viscous flow forces:
   N. Current drag damping.
   O. Current friction damping.

III. Other viscous flow forces:
   P. Pure drag damping.
   Q. Pure friction damping.

IV. Potential flow forces:
   R. Radiation damping.

The physics of these damping forces will be explained below. Most of the damping forces are, at least approximately, proportional to the drift velocity, V_{dr}. In Groups I and II they increase, in addition, with some other velocity (wave orbital velocity or current velocity). Only in Group III the damping force is quadratic in the sense that it is proportional to V_{dr}^2. These forces occur in still water.

Theoretical considerations, as well as experiments (Wichers and van Sluijs, 1979), show that the wave-related, viscous damping dominates. Next comes the current-related, viscous damping. The potential flow damping is often negligible.

5.2 Wave-Related Damping Forces

For a regular wave the *wave drag damping*, F_L, may be found from a formula similar to (4.5), replacing V_c by V_{dr}. Thus the damping coefficient becomes

$$b_L = \frac{2}{\pi} \rho C_d d \cdot (u_m D) \tag{5.1}$$

assuming u_m constant over the draught D. In this connection it should be remembered that C_d increases with KC (Fig. 7).

Wave friction damping, F_M, is of importance along the sides and bottom of a ship moored to an SPM, as illustrated by Fig. 11. Assuming a circular orbital motion, the water particle motion relative to the ship will exhibit an orbit opened by the amount $\Delta X = V_{dr} T$ after the wave period T.

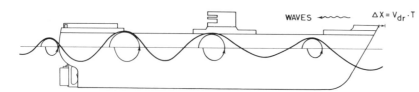

Fig. 11
Wave friction damping along the sides of a ship.

If V_{dr} is small compared with the orbital velocity, the friction will be dominated by the wave friction coefficient, f_w. For a smooth wall and fully turbulent flow, the friction factor to be applied to the orbital velocity is (Fredsøe, 1981b, Fig. 4):

$$f_w = 0.032 \, RE^{-0.15} \quad \text{with} \quad RE = u_m a/\nu \tag{5.2}$$

where a = orbital radius (orbital amplitude under ship's bottom). Thus, along the ship's side, the damping coefficient becomes

$$b_M = \pi \rho f_w \cdot (a \, u_m D) \tag{5.3}$$

assuming a and u_m constant over the draught D. A similar formula may be derived for the ship's bottom.

For the combination of waves and currents reference is made to Fredsøe, 1981a, and for rough walls to Fredsøe, 1981b, and Jonsson, 1980.

5.3 Other Damping Forces

If the current drag is $\frac{1}{2} \rho C_d A_d V_c^2$, where A_d is the cross-sectional area, and V_c is much larger than V_{dr}, the *current drag damping* coefficient becomes

$$b_N = \rho C_d A_d V_c \tag{5.4}$$

Similarly, for the *current friction damping* with a friction coefficient, f_c, and a surface area, A_s,

$$b_O = \rho f_c A_s V_c \tag{5.5}$$

The *pure drag damping* force $F_P = \frac{1}{2} \rho C_d A_d |V_{dr}| V_{dr}$ and the *pure friction damping* force $F_Q = \frac{1}{2} \rho f_c A_s |V_{dr}| V_{dr}$ are, in contrast to the other damping forces, quadratic in the drift velocity. Some numerical models have taken only the quadratic damping forces into consideration. However, since these forces correspond to still water, they are without significance for structures in waves or currents.

For a given frequency of an oscillating structure the *radiation damping* force is linear in V_{dr}. The energy loss corresponds to the energy flux in the waves generated by the oscillating structure and radiated towards infinity. If the drift forces originate from waves or currents, the radiation damping coefficient may often be neglected. An exception is a ship that is exposed to waves other than head waves.

6. MATHIEU INSTABILITIES

In the equation of motion of compliant offshore structures, e.g. a TLP, a *time varying stiffness* term can occur due to the wave forces. For simple harmonic stiffness variations the system is characterized by Mathieu's equation, which under certain conditions produces unstable solutions. Reference is made to Rainey, 1978 and 1981, to Richardson, 1979, and to Jefferys and Patel, 1981. The analysis for natural sea states requires either a time-domain numerical or a physical model.

An instability is a constantly growing oscillation as a result of energy steadily being put into the system faster than it can be dissipated. In the Mathieu case the time varying stiffness term may do net work on the system faster than it can be dissipated by a linear damping. As mentioned below, three sources may lead to varying stiffness of the system and thereby to the possibility of instabilities.

Consider, for example, the constant part of the stiffness for the surge motion, although the phenomenon discussed could as well appear in the sway or yaw modes. The stiffness is determined as the sum of the tether tensions, T, divided by the tether length, ℓ. The *first*, obvious, time-varying *contribution* is due to the heave force introducing the term $(F_3/\ell) \cos \omega t$. With a surge force F_1 the equation of motion for the TLP is

$$(M + M_a^1) \ddot{x}_1 + b \dot{x}_1 + \frac{1}{\ell}(T + F_3 \cos \omega t) x_1 = F_1 \sin \omega t \qquad (6.1)$$

where M_a^1 is the added mass in the surge mode. The interesting solutions, as far as instabilities are concerned, are the transients, i.e. terms that decay under the influence of a relatively small damping. These are found for zero horizontal load, $F_1 = 0$, in (6.1). The substitution $r = x\,e^{at}$ with $a = \frac{1}{2} b/(M + M_a^1)$ leads to the equation

$$\ddot{r} + \omega_1^2 (1 + Q \cos \omega t) r = 0 \qquad (6.2)$$

where ω_1 is the natural surge frequency with

$$\omega_1^2 = \frac{T}{\ell(M + M_a^1)} - a^2 \quad \text{and} \quad Q = \frac{F_3}{\ell(M + M_a^1)\omega_1^2} \qquad (6.3)$$

This is Mathieu's equation and, by means of tabulated solutions, Fig. 12 can be drawn. It gives Q as a function of the ratio, ω/ω_1, between the wave frequency and the natural frequency. It appears that twice the natural surge frequency is a very critical wave frequency.

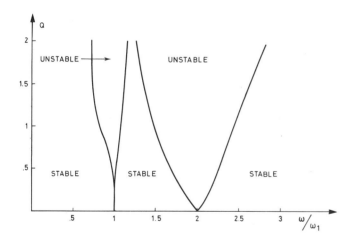

Fig. 12
Regions of instability of Mathieu type
(from Richardson's discussion, p. 74 in Rainey, 1978).

A *second contribution* to the time varying stiffness is found from the change of the surge force due to the displacement x_1 of the structure. Since this force varies in the x_1-direction, the platform is exposed to the force $F_1^1 = F_1 \sin(\omega t - k x_1)$. Taking into account that $k x_1 \ll 1$, allows the linearized version $F_1^1 = F_1 \sin \omega t - k F_1 x_1 \cos \omega t$ to be introduced. The cos-term may be included in the stiffness term in (6.1) because the two terms are in phase with each other. This will again lead to the Mathieu equation (6.2) but for the value of Q to be used in Fig. 12:

$$Q = \frac{F_3 + k \ell F_1}{\ell(M + M_a^1)\omega_1^2} \qquad (6.4)$$

The unstable solutions grow with an exponential time constant, which decreases with increasing Q. Thus, the stability of the structure depends on the damping coefficient b in such a fashion that the oscillations will continue to grow until sufficient damping is mobilized by the motion of the platform. The wave frequencies in the unstable regions of Fig. 12 are very low, e.g. around twice the natural frequency ω_1, and the instability depends, of course, on the extent to which such low frequency waves occur in natural seas.

Using the example of the TLP in Gie and de Boom, 1981, the surge period is 106 s, corresponding to low frequency wave periods around 53 s for the most critical instability region. This is just a typical period for the group-induced second order waves. With 3 - 5 waves in a group the wave period should lie in the interval 10 - 18 s, which is reasonable for a natural sea state. Hence, it seems worthwhile to consider the second order waves. It was earlier found that a typical amplitude is of the order of 0.1 m. In principle, this should be sufficient to trigger the instability since the instability region in Fig. 12 extends down to Q = 0 for ω/ω_1 = 2. However, a definite answer will require further investigations.

The *third source* of instability appears when two wave trains, with a frequency difference equal to the natural frequency of the structure, are considered. In this case the equations will produce what Rainey, 1978, has called subharmonic resonance. It is, indeed, a resonance situation although the oscillations are still governed by damping as in the former instability cases. From the equations it is possible to conclude that the worst resonance appears in a cross-sea, i.e. the situation where the two wave trains travel at right angles to each other. In uni-directional waves the phenomenon is found to vanish. Thus, with regard to this third kind of instability, obviously, the directionality of the natural sea becomes important.

Because of the stiffness variations occurring for a TLP, some potential sources of instability have been discussed above. Evidently, the next problem is to evaluate the damping in order to determine the behaviour of the structure, and to assess to which extent the transients of the possible instabilities will grow. It is well-known that physical model tests with a TLP require special precautions since the hydrodynamic damping increases with decreasing model scale, and the instability phenomena are highly dependent on the damping coefficients. However, in order to obtain realistic results it can be *concluded* that the model tests should reproduce the natural, directional sea, as well as the correct second order, group-induced waves (cf. Ottesen Hansen, et al, 1981).

7. MODEL TESTING

In offshore research *numerical and physical models* are expected to supplement each other for the next decade or two. It has previously been demonstrated that a satisfactory deterministic reproduction of a natural sea state may be obtained by means of only seven, vertically hinged wave generators (Sand, 1979, Lundgren, et al, 1979 or Sand and Lundgren, 1981).

A new *wave basin* (Fig. 13) with 60 wave generators for the 3-D *reproduction of natural sea states* (including approximate reproduction of second order waves) is being planned at the Danish Hydraulic Institute for the study of offshore structures. It is designed for scales 1:80 to 1:100 with a maximum wave height in the model of 0.4 m. With a width of 0.5 m of each flap the system is capable of generating a wide range of sea states. The limiting direction of essential energy at wave frequencies has been chosen as 60° with a minimum period of 0.76 s (λ = 0.89 m). The design and the software of the control system are based on the experience with the pilot system described in the above references.

For thorough studies of 2-D wave forces on individual cylinders, riser bundles, hydroelasticity, etc. the technique of the *oscillating carriage* has been gradually improved over the last seven years in close cooperation between the Institute of Hydrodynamics and Hydraulic Engineering (Technical University of Denmark) and the Danish Hydraulic Institute. The carriage shown in Fig. 14 is installed at the former institute and has been used in a number of basic and applied projects.

As seen in Sec. 4, the wave drag is a dominant drift force on the columns of a TLP, a semisubmersible, etc. At the same time the KC-values are so low that the drag may be 5% or less of the inertia force.

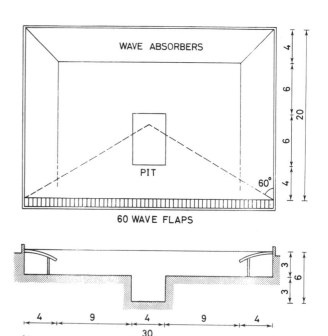

Fig. 13
Wave basin for 3-D reproduction of natural sea states.

With the *arrangement of pressure transducers*, T, shown in Fig. 15, there is complete compensation for the inertia pressure corresponding to C_m = 2. Hence, the range of the transducers can be fully utilized for recording the drag only. In this connection it is natural to consider any difference between the total pressure and the inertial pressure for C_m = 2 as being due to separation and vortex shedding in accordance with the flow history (Maull and Milliner, 1978, and Lundgren, et al, 1979).

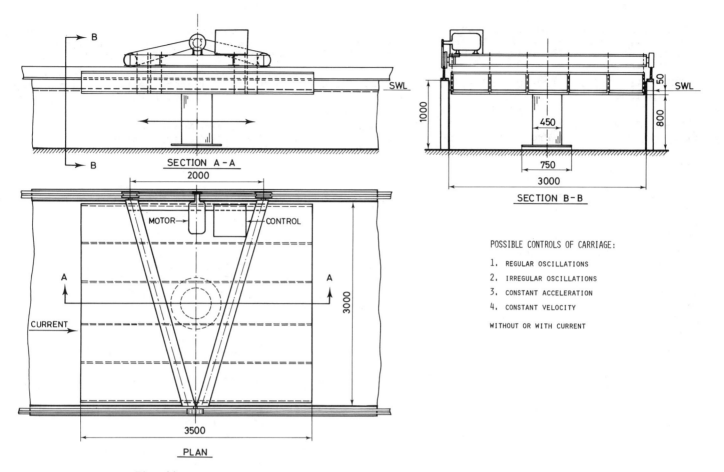

Fig. 14
Oscillating carriage for the study of wave and current forces in 2-D flow.

The oscillating carriage will allow *Reynolds numbers* of 200,000 or more. Compared with the oscillating water tunnel it has the advantage of being able to simulate irregular wave motion, also in combination with a current.

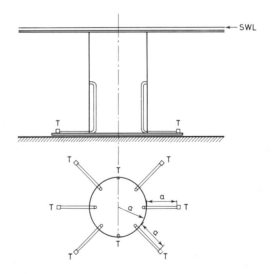

Fig. 15
Arrangement of pressure transducers, T, with compensation for the inertial pressure.

8. REFERENCES

BEARMAN, P. W. and J. M. R. GRAHAM, 1979, "Hydrodynamic Forces on Cylindrical Bodies in Oscillatory Flow," *Proceedings, 2nd International Conference on the Behaviour of Off-Shore Structures*, London, England, 1979, Volume 1, Paper 24, pp. 309-322. Cranfield, Bedford, England: BHRA Fluid Engineering.

BRINK-KJÆR, O., J. DIETRICH and K. MANGOR, 1982, "Wave and Current Study for the Danish North Sea Gas Pipeline," *Proceedings, ASCE & ECOR International Symposium Directional Wave Spectra Applications '81*, Berkeley, California, 1981.

DENISE, J.-P. F. and N. J. HEAF, 1979, "A Comparison Between Linear and Non-Linear Response of a Proposed Tension Leg Production Platform," *Proceedings, 11th Annual Offshore Technology Conference*, Houston, Texas, 1979, Volume III, Paper OTC 3555, pp. 1743-1754.

FALTINSEN, O. M. and A. E. LØKEN, 1979, "Slow Drift Oscillations of a Ship in Irregular Waves," *Applied Ocean Research*, Volume 1, Number 1, pp. 21-31.

FALTINSEN, O. M. and F. C. MICHELSEN, 1975, "Motions of Large Structures in Waves at Zero Froude Number," *International Symposium on The Dynamics of Marine Vehicles and Structures in Waves*, London, England, 1974, Paper 11, pp. 91-106. London, England: The Institution of Mechanical Engineers.

FERRETTI, C. and M. BERTA, 1981, "Viscous Effect Contribution to the Drift Forces on Floating Structures," *Proceedings, International Symposium on Ocean Engineering-Ship Handling*, Gothenburg, Sweden, 1980, Paper 9, pp. 1-10. Gothenburg, Sweden: Swedish Maritime Research Centre, SSPA.

FREDSØE, J., 1981a, "Mean Current Velocity Distribution in Combined Waves and Current," *Institute of Hydrodynamics and Hydraulic Engineering, Technical University of Denmark*, Lyngby, Denmark, Progress Report 53, pp. 21-26.

FREDSØE, J., 1981b, "A Simple Model for the Wave Boundary Layer," *Institute of Hydrodynamics and Hydraulic Engineering, Technical University of Denmark*, Lyngby, Denmark, Progress Report 54, pp. 21-28.

FUNKE, E. R. and E. P. D. MANSARD, 1979, *On the Synthesis of Realistic Sea States in a Laboratory Flume*, National Research Council Canada, Report LTR HY 66.

FUNKE, E. R. and E. P. D. MANSARD, 1981, "On the Meaning of Phase Spectra," *Proceedings, International Symposium on Hydrodynamics in Ocean Engineering*, Trondheim, Norway, 1981, Volume 1, pp. 49-69. Trondheim, Norway: The Norwegian Institute of Technology.

GIE, T. S. and W. C. DE BOOM, 1981, "The Wave Induced Motions of a Tension Leg Platform in Deep Water," *Proceedings, 13th Annual Offshore Technology Conference*, Houston, Texas, 1981, Volume III, Paper OTC 4074, pp. 89-98.

ISAACSON, M. DE ST. Q., 1977, "Shallow Wave Diffraction Around Large Cylinder," *Journal of the Waterway, Port, Coastal and Ocean Division*, Volume 103, Number WW1, pp. 69-82. New York: American Society of Civil Engineers.

JEFFERYS, E. R. and M. H. PATEL, 1981, "Dynamic Analysis Models of the Tension Leg Platform," *Proceedings, 13th Annual Offshore Technology Conference*, Houston, Texas, 1981, Volume III, Paper OTC 4075, pp. 99-107.

JONSSON, I.G., 1980, "A New Approach to Oscillatory Rough Turbulent Boundary Layers," *Ocean Engineering*, Volume 7, Number 1, pp. 109-152 plus Tables in Number 4, pp. 567-570.

KIRKEGAARD, J., S. E. SAND, N.-E. OTTESEN HANSEN and M. HVIDBERG-KNUDSEN, 1980, "Effects of Directional Sea in Model Testing," *Proceedings, Ports '80*, Norfolk, Virginia, pp. 597-613. New York: American Society of Civil Engineers.

LUNDGREN, H., O. BRINK-KJÆR, S. E. Sand and V. JACOBSEN, 1979, "Improved Physical Basis of Wave Forces," *Proceedings of the Specialty Conference Civil Engineering in the Oceans IV*, San Francisco, California, 1979, Volume 1, pp. 1-16. New York: American Society of Civil Engineers.

LØKEN, A. E. and O. A. OLSEN, 1979, "The Influence of Slowly Varying Wave Forces on Mooring Systems," *Proceedings, 11th Annual Offshore Technology Conference*, Houston, Texas, 1979, Volume IV, Paper OTC 3626, pp. 2325-2335.

MAULL, D. J. and M. G. MILLINER, 1978, "Sinusoidal Flow Past a Circular Cylinder," *Coastal Engineering*, Volume 2, Number 2, pp. 149-168.

NEWMAN, J. N., 1975, "Second-order, Slowly-varying Forces on Vessels in Irregular Waves," *International Symposium on The Dynamics of Marine Vehicles and Structures in Waves*, London, England, 1974, Paper 19, pp. 182-186. London, England: The Institution of Mechanical Engineers.

OTTESEN HANSEN, N.-E., S. E. SAND, H. LUNDGREN, T. SORENSEN and H. GRAVESEN, 1981, "Correct Reproduction of Group-Induced Long Waves," *Proceedings, 17th Coastal Engineering Conference*, Sydney, Australia, 1980, Volume I, Chapter 48, pp. 784-800. New York: American Society of Civil Engineers.

PIJFERS, J. G. L. and A. W. BRINK, 1977, "Calculated Drift Forces of Two Semisubmersible Platform Types in Regular and Irregular Waves," *Proceedings, 9th Annual Offshore Technology Conference*, Houston, Texas, 1977, Volume IV, Paper OTC 2977, pp. 155-164.

PINKSTER, J. A., 1980, *Low Frequency Second Order Wave Exciting Forces on Floating Structures*, Netherlands Ship Model Basin, Wageningen, Netherlands, Publication Number 650.

RAINEY, R. C. T., 1978, "The Dynamics of Tethered Platforms," *Transactions of The Royal Institution of Naval Architects*, Meeting, Royal Institution of Naval Architects, 1977, Volume 120, pp. 59-80.

RAINEY, R. C. T., 1981, "Parasitic Motions of Offshore Structures," *The Royal Institution of Naval Architects*, Volume 123, pp. 177-194.

RICHARDSON, J. R., 1979, *Mathieu Instabilities and Response of Compliant Offshore Structures*, National Maritime Institute, Feltham, England, Report NMI R49.

SAND, S. E., 1979, *Three-Dimensional Deterministic Structure of Ocean Waves*, PhD Thesis, Institute of Hydrodynamics and Hydraulic Engineering, Technical University of Denmark, Series Paper 24.

SAND, S. E., 1982a, "Long Waves in Directional Seas," *Coastal Engineering*, in press.

SAND, S. E., 1982b, "Short and Long Wave Directional Spectra," *Proceedings, ASCE & ECOR International Symposium Directional Wave Spectra Applications '81*, Berkeley, California, 12 pp.

SAND, S. E., 1982c, "Long Wave Problems in Laboratory Models," Submitted to *Journal of the Waterway, Port, Coastal and Ocean Division*. New York: American Society of Civil Engineers.

SAND, S. E. and H. LUNDGREN, 1981, "Selection and Three-Dimensional Reproduction of Wave Records that give Maximum Ship Motions," *Proceedings, International Symposium on Hydrodynamics in Ocean Engineering*, Trondheim, Norway, 1981, Volume 1, pp. 101-120. Trondheim, Norway: The Norwegian Institute of Technology.

SEBASTIANI, G., A. D. GRECA and G. BUCANEVE, 1981, "Characteristics and Dynamic Behaviour of Technomare's Tension Leg Platform," *Proceedings, International Symposium on Hydrodynamics in Ocean Engineering*, Trondheim, Norway, 1981, Volume 2, pp. 947-961. Trondheim, Norway: The Norwegian Institute of Technology.

SKOVGAARD, O. and I. G. JONSSON, 1981, "Computation of Wave Fields in the Ocean Around an Island," *International Journal for Numerical Methods in Fluids*, Volume 1, pp. 237-272.

SPANGENBERG, S. and B. KOFOED JACOBSEN, 1981, "The Effect of Wave Grouping on Slow Drift Oscillations of an Offshore Structure," *Proceedings, International Symposium on Ocean Engineering-Ship Handling*, Gothenburg, Sweden, 1980, Paper 8, pp. 1-14. Gothenburg, Sweden: Swedish Maritime Research Centre, SSPA.

WICHERS, J. E. W., 1979, "Slowly Oscillating Mooring Forces in Single Point Mooring Systems," *Proceedings, 2nd International Conference on Behaviour of Off-Shore Structures*, London, England, 1979, Volume 3, Paper 27, pp. 661-692. Cranfield, Bedford, England: BHRA Fluid Engineering.

WICHERS, J. E. W. and M. F. VAN SLUIJS, 1979, "The Influence of Waves on the Low-Frequency Hydrodynamic Coefficients of Moored Vessels," *Proceedings, 11th Annual Offshore Technology Conference*, Houston, Texas, 1979, Volume IV, Paper OTC 3625, pp. 2313-2324.

ON THE LARGE AMPLITUDE AND EXTREME MOTIONS OF
FLOATING VESSELS

S.N. Smith
Seaforth Maritime Ltd., U.K.

M. Atlar
Glasgow University, U.K.

SUMMARY

This paper considers the large amplitude and extreme motions of floating vessels in particular by solving the frequency domain equation of motion in the time domain. Experimental and theoretical heave response-amplitude-operators for a semi-submersible are given. The use of a group of design-waves is proposed and from the application of this, it is shown that the conventional use of such steady-state response-amplitude-operators is not completely justified. The simulated response is shown to be dependent on the preceeding conditions and is also influenced by factors such as the proximity to resonance. The technique can be used to estimate some of the uncertainty in the motions which will be required knowledge for the application of semi-probabilistic design methods. Other uncertainties are identified including the variation of added-mass and damping coefficients with depth/draft ratio. Results which show the same general trends are given for a SWATH-type vessel.

1. INTRODUCTION

The estimation of large amplitude and extreme motions is of considerable practical interest in ocean engineering. Such motions contribute to the structural loads in the vessel itself and are also of great importance in the design of cranes, diving-system recovery equipment, flare-booms, and so on. Much work has been reported on the response of floating vessels with variations in the formulation and solution of the problem and the inclusion of various non-linearities in the prediction methods. Similarly the variation of added-mass and damping coefficients with the various non-dimensional numbers has been widely investigated both theoretically and experimentally. The effect of prediction method including the variation in coefficients has been investigated, (Paulling, 1981) and the mathematics of extreme-value statistics is well developed.

Despite the above there is still a lack of data on the degree of uncertainty in the response which is one of the requirements for the application of semi-probabilistic design methods. Such methods acknowledge the degree of uncertainty in the characteristic requirements and capability of the system (e.g. load and strength), and have been used for ships' structures, (Faulkner and Sadden, 1979) and more recently have been introduced for other marine structures, (Faulkner, 1981). Semi-probabilistic methods have also been suggested for capsize safety, (Kure, 1979).

In this paper the heave response is calculated for wave-groups using a step-by-step integration of the equations of motion. The results are compared with the usual steady-state response which shows that the use of steady-state transfer functions or response-amplitude-operators is not necessarily justified. However becayse of the conceptual nature of the wave-groups, the use of linear theory, and other uncertainties, it is suggested that the results be used as representing a degree of uncertainty rather than a deterministic "correct" solution.

2. REGULAR WAVE RESPONSE: THEORY AND EXPERIMENT

Herein the frequency domain motion response to regular waves is calculated by the approach used by Oo and Miller, (1977) based on the work of Hooft, (1971). Thus the vessel heave is represented by the well-known equation

$$(m+am)\ddot{z} + C\dot{z} + Kz = F \cos \omega t \tag{1}$$

where m = vessel mass
am = added mass
C = damping coefficient
K = restoring force coefficient
F = exciting force
ω = wave frequency (rads/sec)

The complete solution consists of a particular integral plus a complimentary function and takes the form

$$z = \left\{ Ae^{\lambda_1 t} + Be^{\lambda_2 t} \right\} + \frac{F/K \sin(\omega t - \phi)}{\sqrt{\left(1 - \left(\frac{\omega}{\omega_n}\right)^2\right)^2 + 4\gamma^2 \left(\frac{\omega}{\omega_n}\right)^2}} \tag{2}$$

where $\phi = \tan^{-1}\left\{ \dfrac{2\gamma\left(\frac{\omega}{\omega_n}\right)}{1 - \left(\frac{\omega}{\omega_n}\right)^2} \right\}$ and $\omega_n = \sqrt{\dfrac{K}{(m+am)}}$ \qquad (3)

A and B are constants depending on the initial conditions while λ_1 and λ_2 are the roots of the characteristic equation

$$\lambda^2 + 2\gamma\omega_n\lambda + \omega_n^2 = 0 \tag{4}$$

The first term in Eq. 2 (the transient part) is the term for damped free vibration while the second term is for the forced vibration and gives the steady-state solution. (The steady-state solution is the theoretical equivalent of the regular wave response.)

Figure 1, from a vibration text, (Van Santen, 1958) shows the build-up of vibration, given by Eq. 2, for a mass-springer-damper system starting from rest subject to a sinusoidal excitation. The interference between the transient term and the steady-state part of the solution depends on the relationship between the exciting frequency and the natural frequency but the importance of including it is apparent. Similarly for other changes in motion there is a transient term which should be considered. However, for applying the linear theory of superposition only the steady-state part of the solution is considered, (St. Denis and Pierson, 1953).

The steady-state theoretical response was calculated for a rectangular hull semi-submersible model, Fig. 2, (scale approximately 1:113) and compared with the regular wave experimental response, Fig. 3, thus verifying that reasonable agreement is obtained.

Figure 4 shows the full-scale regular wave response-amplitude-operators taken from the literature for different semi-submersibles operating around the world from which it can be seen that during storm conditions, with wave frequencies in the range 0.3 - 0.4 rads/sec, such vessels will be operating in the vicinity of their resonant heave frequencies with resultant high motion amplitudes.

3. REPRESENTATION OF THE SEAWAY

Since the influential paper by St. Denis and Pierson, (1953), it has become accepted practice to consider the seaway as consisting of a sum of Fourier components. The seaway is thus represented by a spectrum allowing the calculation of statistical averages of responses for linear or linearised systems through the use of response-amplitude-operators. An alternative view of the seaway is obtained by dividing the wave record into the zero-crossing period with each wave having an associated wave height. This provides the basis for the design-wave approach, although the parameters of the design-wave can be determined from consideration of the sea-spectra. That is, a wave of a certain height and with a unique frequency is obtained. The design-wave approach is justifiably popular with designers and is widely used because it is simple to apply and gives easily recognisable parameters against which to work.

Recent work on wave-groups has started to define the environment in which large waves occur (see Appendix), but here the concept of a group of design-waves is proposed. The groups that will be used contain the single-frequency design-wave and also other sinusoidal waves of the same frequency, but lower amplitude, juxtaposed to give the general shape of an amplitude modulated wave system. This is not mathematically rigorous because a discontinuity at the start of each cycle is introduced. However since the design-wave is taken as being a valid representation, this group of design-waves must also have some validity. (In a recent paper, (St. Denis, 1980), a wave model was set up by juxtaposition of third order Stokes waves to produce a statistical description of moderately severe seas. The resulting discontinuity of orbital velocities was treated by fairing process).

In addition it should also perhaps be noted that the use of the Fourier representation with frequency dependent coefficients in the equations of motion is slightly anomalous since it involves the necessity of the different coefficients co-existing at any instant in time.

4. RESPONSE IN WAVE GROUPS

If the heave response is calculated for any particular wave on the basis of regular wave response-amplitude-operators then, as will be shown below, there may be considerable error since in reality the response will depend on the preceeding conditions. Recent work on wave-groups, as discussed in the Appendix, suggests that the ratio of the largest wave amplitude may be around 1:0.6 and, furthermore, that the zero-crossing periods may be approximately the same. The details of the groups used here are given in Fig. 5.

The response in the wave-groups was calculated by solving the usual equation of motion in the time domain using the Runge-Kutta-Nystrom method which is applicable for such an equation. Since a time step of $t=T/40$ gave no appreciable difference to $t=T/20$ the latter was used.

Figure 6 shows the frequency dependent added-mass and damping coefficients for a section of the hull obtained using a close-fit method program developed in the department, (Atlar, 1981). From this it can be seen that the calculated added mass coefficients are resonably constant but that the calculated damping coefficients are not. However it has been shown, (Paulling, 1981) that the use of potential-flow damping alone, i.e. ignoring viscous effects, will lead to over prediction near the resonant frequency so in this paper the experimental values, obtained from heave decay tests, have been used for all frequencies since they do include these viscous effects and also 3D effects. Similarly model scale wave amplitudes were used.

The theoretical coefficients are strictly for small amplitude oscillations about the mean position and were calculated for another two depth/draft ratios which represent positiions through which the section could oscillate during large amplitude motions. This shows the possibility of uncertainty from using a constant coefficients througout the cycle but is not further considered.

The motions resulting from the wave-group are shown for the rectangular hull semi-submersible at two different frequencies in Figs 7 and 8. From these it can be seen that in the groups used, for a frequency very close to resonance the steady-state solution tends to overestimate the response. In this case the simulated heave is about 20% less than the steady-state response. Further from resonance the simulated response can overshoot the steady-state response, by about 10% in Fig. 8. Both results will have significance for the required underdeck clearance for instance but the second case may be more important because it could give rise to higher accelerations than anticipated which would have serious implications for certain items of structure. (Such behaviour is not a function of the numerical method but is a characteristic of the equation of motion. Similar results were obtained on an analogue computer).

The response was also calculated for a three-hulled SWATH-type ship, (Smith, 1982a, 1982b). Such vessels will typically be smaller than conventional semi-submersibles and will therefore encounter resonant conditions especially when operating in following seas. As for the semi-submersible the simulation results suggest that at the resonant frequency the steady-state solution overestimates the response. The maximum heave amplitude in Fig. 9 for an eleven cycle group is some 15% less than the steady-state response to the largest wave. Since the extreme motions will be associated with resonant frequencies they will tend to be overestimated by conventional techniques, however as before the maximum acceleration may be under-estimated. The heave exciting force is included in Figs. 9 and 10. For $\omega/\omega_n = 1$ the motions lags the force by $90°$ but it is interesting to note that the maximum heave occurrs not $90°$ but $450°$ after the maximum force. The response in the second half of the group is greater than that in the first half because of the intial conditions. In fact if the response to a_3 and a_q is examined then for a_3 the steady-state solution overestimates the response by a certain quantity, denoted by ε_3, but in the second half of the group the steady-state solution underestimates the response by ε_q. (The subscript denotes the position in the group.) Interestingly the average simulated response to a_3 and a_q is 0.101 m whereas the steady-state response is 0.112 m. If two groups of design-waves are considered, as in Fig. 10, then it can be seen that although the average response may be higher the maximum response in the second group is no higher than in the first. Thus for the given parameters there is a steady-state response to the entire group. This may be of relevance if an appropriate representation of the seaway could be devised.

5. DISCUSSION

Although based on a recent design the semi-submersible is not a scale replica of any particular vessel and similar results would be obtained for other goemetries, as is illustrated by the SWATH case. No particular claim is made as to the absolute accuracy of the results of the simulation and as mentioned in the Introduction they are best considered as representing a degree of uncertainty rather than a deterministic correct solution. The groups of design-waves perhaps represent an extreme case, but any other wave-train or initial conditions could be used as desired, (and the infinity of possibilities is the drawback to time-domain simulations). Two such possibilities are shown in Figs 11 and 12 for the semi-submersible. These consist of a regular wave train which is sufficiently long for the vessel to reach the steady-state solution before encountering the high waves. Changing the number of high waves from 1 to 3 (as in Figs 11 and 12 respectively) produces a broad band of possible solutions ranging from about 10% less than, to 15% greater than the steady-state solution. However, if extreme cases such as the proposed groups are used then some bounds can be estimated for the uncertainties.

Other uncertainties not considered here include lack of agreement between theory and experiment, non-linearities due to viscous effects, higher order wave theories, effect of moorings, uncertainies in extreme-value statistics, etc.

The paper maybe suggests that use should be made of the impulse response function as was considered twenty years ago, (Cummins, 1962). The adopted procedure does however have the advantage of simplicity and indeed for the semi-submersible both the response-amplitude-operators and the simulated motions were calculated on a micro-computer. Thus even the smallest design-office would be able to estimate these uncertainties for any particular case.

6. CONCLUSIONS

The simulated response in the proposed design-wave-groups has been shown to differ from the steady-state response. The solution arrived at depends on the preceeding conditions and the proximity to resonance but from the examples considered for a semi-submersible and a SWATH ship the simulated response may range from 20% less, to 15% greater than the steady-state response. The simulated response at resonant frequencies is less than the steady-state response which suggests that the conventional design-wave technique may overestimate the extreme motions.

However the overshoot at frequencies greater than resonance has more serious implications because it could lead to higher acceleration induced loads than expected. The method can be used to estimate some of the uncertainty in the response for use in semi-probabilistic design methods.

7. ACKNOWLEDGEMENTS

The authors would like to thank Professor D. Faulkner and Mr. N. S. Miller, Department of Naval Architecture and Ocean Engineering, Glasgow University, for making this work possible. However, the opinions expressed are those of the authors alone. The work was funded by the Science and Engineering Research Council, Marine Technology Programme, with financial assistance for the second author from the Turkish Ministry of Education.

8. REFERENCES

ATLAR, M., 1981, Report in Preparation, Glasgow University, Department of Naval Architecture and Ocean Engineering Report NAOE-HL-81

CUMMINS, W.E., 1962, "The Impulse Response Function and Ship Motions", Schiffstechnik, Heft 47, June, 1962

DAVIDAN, I.N., Y.M. KUBLANOV, L.I. LOPATUKHIN, and V.A. ROZHKOV, 1974, "The Results of Experimental Studies on the Probabilistic Characteristics of Wind Waves", International Symposium on the Dynamics of Marine Vehicles and Structures in Waves, 1974

FAULKNER, D. and J. A. SADDEN, 1979, "Toward a Unified Approach to Ship Structural Safety", Transactions, RINA, 1979 Volume 121

FAULKNER, D., 1981, "Semi-Probabilistic Approach to Design of Marine Structures", Proceedings, Extreme Loads Response Symposium, Arlington, Virginia, 1981: SSC-SNAME

GODA. Y., 1976, "On Wave Groups", Proceedings Conference on Behaviour of Offshore Structures (BOSS), Trondheim, 1976

HOOFT, J.P., 1971, "A Mathematical Method of Determining Hydrodynamically Induced Forces on a Semi-Submersible," Tranasactions SNAME, 1971, Volume 79

KURE, K., 1979, "Capsize Safety", Proceedings, 4th Ship Technology and Research (STAR) Symposium, 1979: SNAME

MOLLO-CHRISTENSEN, E., and A. RAMAMONJIARSO, 1978, "Modelling the Presence of Wave Groups in a Random Wave Field", Journal of Geophysical Research, Volume 83, No. C8, August 20, 1978

OO, K.M., and N.S. MILLER, 1977, "Semi-Submersible Design: The Effect of Differing Geometries on Heaving Response and Stability", Transactions, RINA, 1977, Volume 119

PAULLING, J.R., 1981, "The Sensitivity of Predicted Loads and Responses of Floating Platforms to Computational Methods", Proceedings, 2nd International Conference on the Integrity of Offshore Structure, Glasgow, 1981

St. DENIS, M. and W.J. PIERSON, 1953, "On the Motions of Ships in Confused Seas", Transactions SNAME, 1953

St. DENIS, M., 1980, "On the Statistical Description of Seaways of Moderate Severity", Proceedings, 5th Ship Technology and Research (STAR) Symposium, 1980: SNAME

SMITH, S.N., 1982a, "Design and Hydrodynamic Performance of a Small Semi-Submersible (SWATH) Ship", Spring Meetings RINA, 1982

SMITH, S.N., 1982b, Design and Hydrodynamic Assessment of Small Semi-Submersible (SWATH) Research Vessel, Ph.D. Thesis, University of Glasgow, 1982

Van SANTEN, G.W., 1958, Introduction to the Study of Mechanical Vibration, Philips Technical Library, 1958

APPPENDIX: Waves and Wave-Groups

The wave amplitude spectrum alone does not give a unique description of the wave record associated with it since a combination of the same amplitude spectrum with different phase spectra can result in entirely different wave trains with varying amount of grouping. The wavw-group literature, reviewed elsewhere, (e.g. Goda, 1976) shows that the actual length of wave groups is greater than would be expected from random theory. Various papers exists in, for example, the J. of Geophysical Research and it has been suggested, (Mollo-Christensen and Ramamonjiarson, 1978) that the wave field does not consist of independently propogating Fourier components but wholly or in part of wave-groups of permanent type, (Stokes wave packets). Under certain circumstances such "envelope solitons" will be left unchanged by collisions with other groups. Other measured data, (Davidan et al, 1974) indicate that the number of waves in a group amount, on the average, to 5 - 6 and, in addition, the values of a_{max-1}/a_{max} and $\omega_{max-1}/\omega_{max}$ have mean values of 0.6 and 1 respectively. The amplitude values can be compared with Fig. 5 and the frequency value suggests that using a single frequency group is not completely unrealistic.

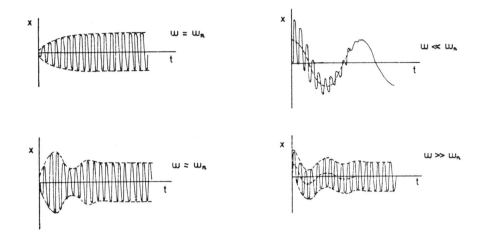

Figure 1 Build Up Of A Forced Vibration

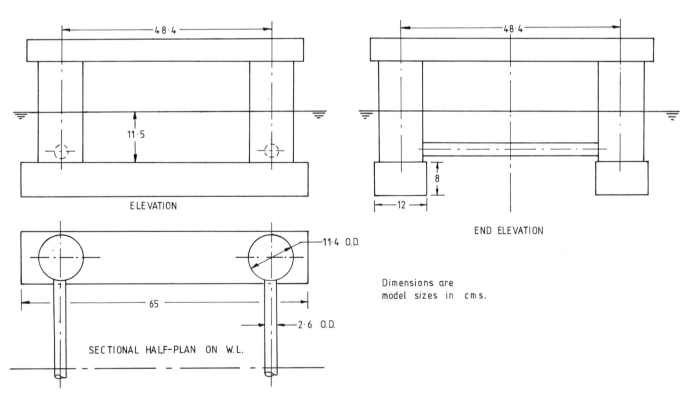

Figure 2 Twin Rectangular Hull Semi-Submersible Model

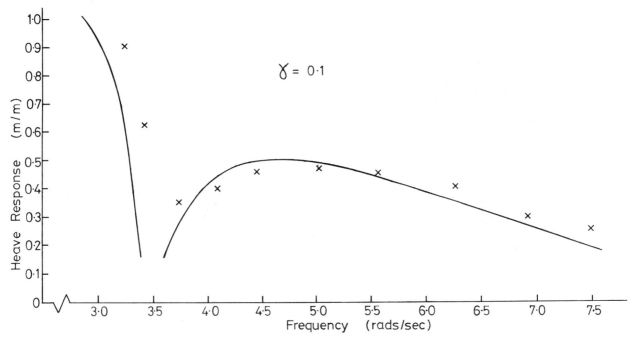

Figure 3 Semi-Submersible Model Heave Response : Theory and Experiment

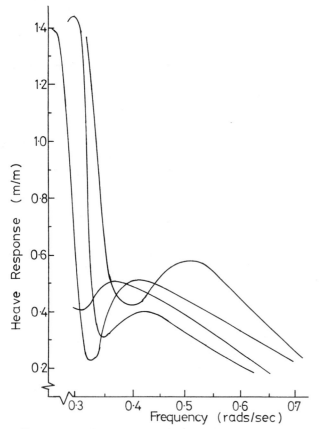

Figure 4 Semi-Submersible Heave Responses

WAVE AMPLITUDES (cms)					
Fig. No.	7,8	9,10		11	12
No. of Wave			No. of Wave		
a_1	3	2.5			
a_2	6	5.0	a_{n-m}	7	7
a_3	9	7.5	⋮		
a_4	12	10.0	a_{n-2}	7	7
a_5	9	12.5	a_{n-1}	7	7
a_6	6	15.0	a_n	12	12
a_7	3	12.5	a_{n+1}	7	12
a_8		10.0	a_{n+2}	7	12
a_9		7.5	a_{n+3}	7	7
a_{10}		5.0	⋮	⋮	⋮
a_{11}		2.5			
$\dfrac{a_{max-1}}{a_{max}}$	0.75	0.8		0.58	0.58

Figure 5   Group   Wave-Amplitudes

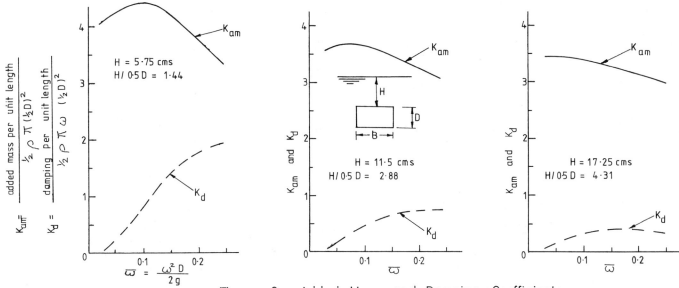

Figure 6 Added Mass and Damping Coefficients

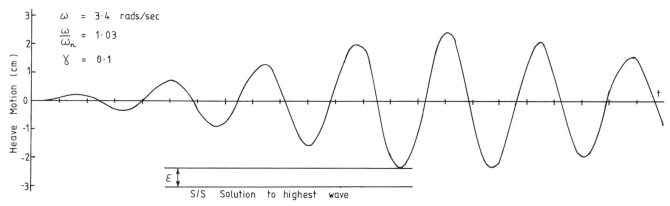

Figure 7 Simulated Response of Semi-Submersible Model

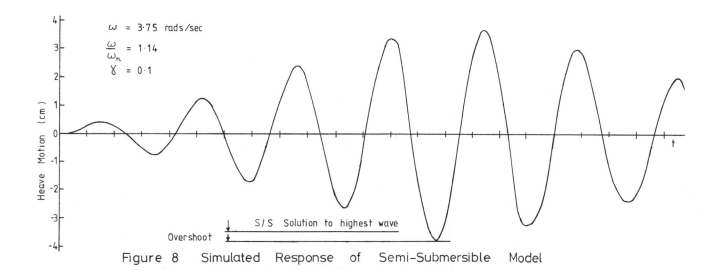

Figure 8 Simulated Response of Semi-Submersible Model

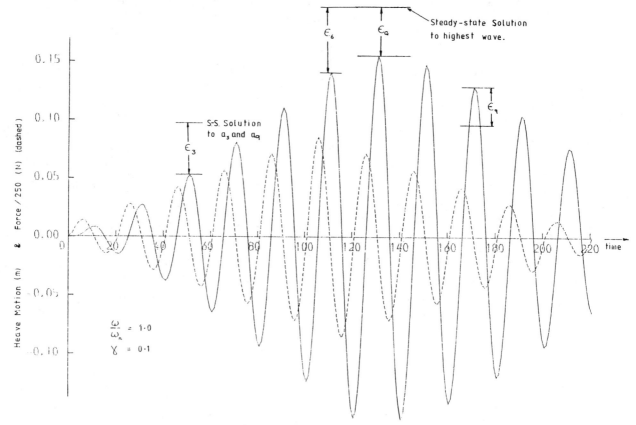

Figure 9   Simulated Response of SWATH Model

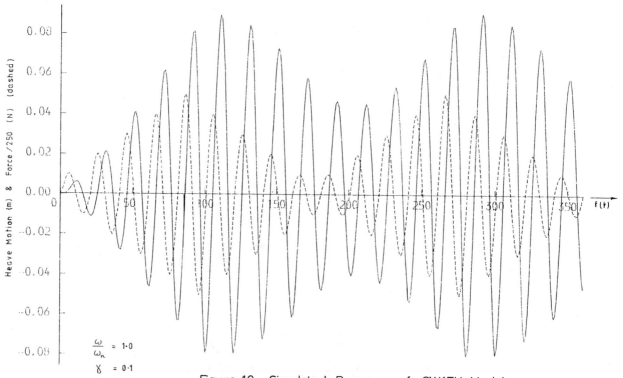

Figure 10   Simulated Response of SWATH Model

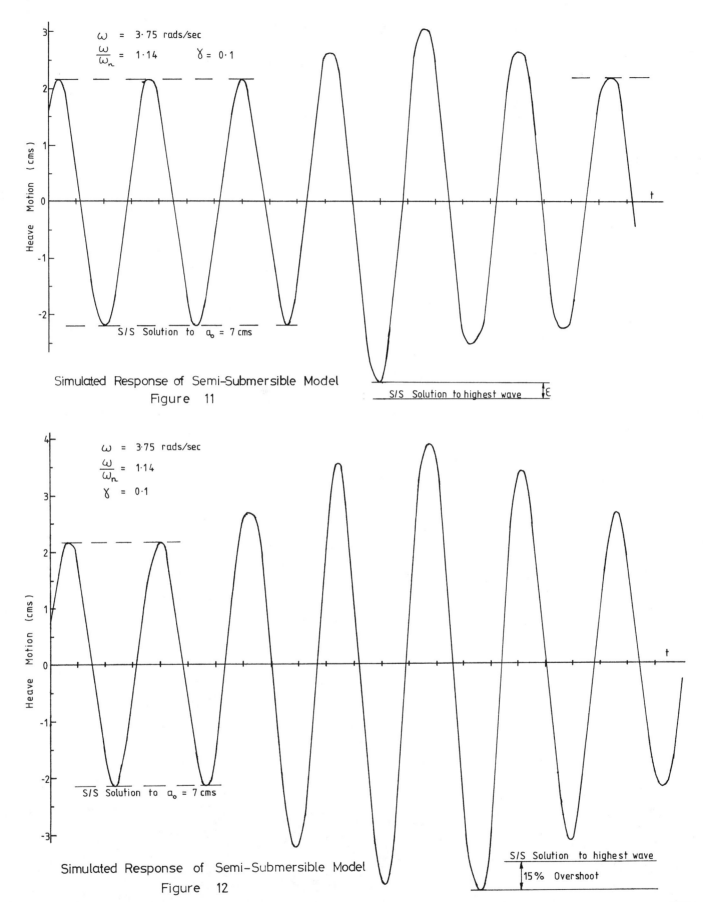

Simulated Response of Semi-Submersible Model
Figure 11

Simulated Response of Semi-Submersible Model
Figure 12

Technical Poster Sessions
HYDRODYNAMICS

Risers and Mooring Systems

NON-LINEAR ANALYSIS OF MOORING SYSTEMS

C. Ganapathy Chettiar and Satish C. Nair
Ocean Engineering Centre
Indian Institute of Technology
Madras, India

SUMMARY

The analysis and design of moorings is carried out using different techniques such as the method of imaginary reactions, the methods formulated by Leonard Pode, Wilson and so on. In this present contribution, the cables are analysed by applying the geometrically non-linear finite element procedure for both static and dynamic conditions. Some of the important requirements for these mooring systems are to limit excursion on station or in depth; to be statically and dynamically stable and to be rugged and long lived. In compliance with the severe limitations of motion, it is necessary that the mooring systems be of several branches or legs. The configuration of multi-point mooring systems subjected to varying current field is determined by this analysis. To ensure a safe and economical engineering design, one needs to be able to predict the mooring line motion and tension, as a result of oscillating wave forces. The analysis therefore, also provides the forces coming on the cables and the force components existing at the anchor and at the bottom of a surface buoy or a ship.

The cable array is discretised into a number of finite elements. An iterative technique is adopted, by assuming an initial configuration of the cable system and correcting the configuration, by successive solutions applying the non-linear finite element analysis (NONSAP) with the position dependent hydrodynamic loading. The analysis can also take into account the external parameters such as force from spaced buoyant floats, different types of loadings, different elastic support conditions of anchors etc. The entire analysis is carried out in the system IBM 370/155.

By using the above procedure, the analysis of a single and triple point mooring array is carried out; several parametric effects of the subsystems involved are studied. The analysis also takes into account, the dynamic characteristics of moorings, such as mode shapes, frequencies and vibration periods of such cable systems.

INTRODUCTION

Moored cable systems become important because of their use in areas such as environmental data gathering through moored or drifting buoys, towing of underwater bodies, sensing acoustical parameters for defence purposes and laying or recovery of submarine transmission cables. In order to design a moored cable system, the designer must exactly know the external and internal forces that come upon the superstructure, the cables and anchors. The external forces that come upon the system are dependent on the wave, current and wind characteristics. Various formulations incorporating the normal, tangential and transverse components of the hydrodynamic forces and other force modifying features such as variation of current profile, bare or faired cables, cable strumming force from spaced buoyant floats etc are used. The internal forces developed in the moored cable system are dependent on the elastic or inextensible analysis, linear or nonlinear behaviour of cable material, dynamic nature of the loading functions and the deterministic or stochastic modelling of the sea surface.

Various mathematical procedures have been proposed for the static/dynamic, linear/nonlinear response analysis of single point or multipoint moored cable systems. Besides the tables (Pode, 1951) and (Wilson, 1967) for cables with non-stretch behaviour, many other methods have been proposed. Some of them are method of imaginary reactions (Skop and O'Hara, 1968, 1970), (Skop, 1970), Finite Element Method (Eiriz, 1973), (Chhabra, 1974), Newton Raphson procedures (Gormally, 1966), Adam-Moulton predictor-corrector method with Runge Kutta starter (Blendendender, 1970), Runge Kutta procedure (Schreiber, 1973) and the Kutta Merson variant (Elizabeth Cutthil, 1968), CSMP (Seidl, 1971), Lumped mass approach (Paquethe and Henderson, 1965).

FORCES ON CABLES

The forces to consider for the study of mooring lines immersed in the ocean are gravity forces, wave forces, current forces etc.

Gravity Forces

These may act either upwards or downwards depending on the difference between the weight and buoyancy. If only small currents are present, the mooring line of a surface buoy hangs almost vertically from the buoy, the tension at any point in the line being approximately equal to the immersed weight of the line and equipment below that point. In oceanic applications, however, strong currents are often present. Considerable drag forces are then applied to the mooring line, which assumes a new equilibrium configuration. Tension increases and strong anchoring becomes necessary.

The gravity force per unit length of line is given by

$$P = W - B \text{ Kg/m,}$$

where W is the weight per metre of the line and B the weight of water displaced by one metre of the line.

Wave Forces

In computation of wave forces, Airy's wave theory (a first order theory) is used to define the kinematics of the wave. The approach representing the wave forces on the cylinder as the sum of the drag and inertia components, is widely accepted in engineering practice. Each of these forces is defined in terms of the kinematics of the water particles beneath a wave. In this approach, (Morrison, 1950) separated the wave force prediction into two phases.

(a) Calculation of kinematic flow field from the wave force characteristics.

(b) Relating the forces to kinematic flow field by means of two hydrodynamic force coefficients C_D and C_M.

$$df_T = df_D + df_I \qquad (1)$$

$$df_T = \tfrac{1}{2} C_D \rho D(u-u_o)|u-u_o|ds + C_M \rho \left(\frac{\pi D^2}{4}\right)(\bar{u} - \bar{u}_o)ds \qquad (2)$$

where u_o and $\bar{u}_o$ are the velocity and acceleration of the cable.

The wave velocity of the water particle u and the acceleration $\bar{u}$ are calculated using the formulae given below:

$$u = \frac{a\,\sigma\cosh(k'z)\,\cos(\sigma t)}{\sinh(k'H)} \qquad (3)$$

and $\bar{u} = \dfrac{-a\sigma^2 \cosh(k'z) \sin(\sigma t)}{\sinh(k'H)}$ (4)

where

- a = amplitude wave measured from M.S.L.
- k' = wave number
- σ = wave angular frequency
- σt = phase angle
- z = height in water above mudline
- H = depth of water
- ds = element length.

Current Forces

The forces on cables due to current can be obtained from the equations (3) and (4), using the value of u pertaining to the current velocity.

Forces on a Cable Element

With reference to the Figure 1, $\bar{u}$, $\bar{v}$, $\bar{w}$ be a set of cartesian unit vectors at any point P along the cable and $\bar{u}$ be in the direction of the cable at that point. Let $\bar{i}$, $\bar{j}$, $\bar{k}$ be the cartesian unit vectors in the Global coordinate system. If ψ is the angle between the horizontal plane and the vector $\bar{u}$ and φ the angle between the projection of $\bar{u}$ in the horizontal plane and the vector i, then the relation between the cable vector set and the earth vector set is given by

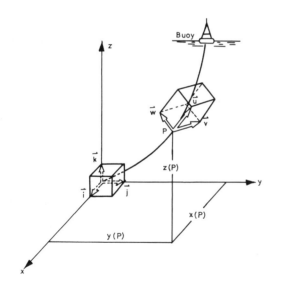
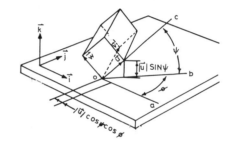

FIG.1.

$\bar{u} = \cos\psi\cos\varphi\,\vec{i} + \cos\psi\sin\varphi\,\vec{j} + \sin\psi\,\vec{k}$ (5)

$\bar{v} = -\sin\varphi\,\vec{i} + \cos\varphi\,\vec{j}$ (6)

$\bar{w} = -\sin\psi\cos\varphi\,\vec{i} - \sin\psi\sin\varphi\,\vec{j} + \cos\psi\,\vec{k}$ (7)

If ds $\vec{u}$ be the cable element length of infinetesimal length located at P, the components of the cable element length in the x, y and z directions are then given by

$dx = ds\,\vec{u}\cdot\vec{i} = ds\,\cos\psi\cos\varphi$ (8)

$dy = ds\,\vec{u}\cdot\vec{j} = ds\,\cos\psi\sin\varphi$ (9)

$dz = ds\,\vec{u}\cdot\vec{k} = ds\,\sin\psi$ (10)

Gravity Forces

If ps is the difference between weight and buoyancy per unit of cable length, then the resultant gravity force on the cable element is given by $-p\,ds\,\vec{k}$.

Cable Forces

The forces on the cable element are the cable element immersed weight, the hydrodynamic resistance on the element, and the tension forces at both ends of the element. The expression of these forces is hereafter derived.

Let 3-dimensional velocity field V be represented by

$$\vec{V} = V_x \vec{i} + V_y \vec{j} + V_z \vec{k} \tag{11}$$

where $V_x = V_x(x,y,z)$, $V_y = V_y(x,y,z)$ and $V_z = V_z(x,y,z)$.
The tangential component of the resistance $\vec{F}$ is given by

$$\vec{F} = F\,ds\,\vec{u} = \frac{1}{2}\,\varsigma\,C_{DT}\,\pi d\,(\vec{V}\cdot\vec{u})^2\,ds\,\vec{u} \tag{12}$$

with $\vec{V}\cdot\vec{u} = V_x \cos\psi \cos\varphi + V_y \cos\psi \sin\varphi + V_z \sin\psi$ and C_{DT} the tangential drag coefficient of the cable. The normal components $\vec{G}$ and $\vec{H}$ of the resistance are respectively given by

$$\vec{G} = G\,ds\,\vec{v} = \frac{1}{2}\,\varsigma\,C_{DN}\,d(\vec{V}\cdot\vec{v})^2\,ds\,\vec{v} \tag{13}$$

$$\vec{H} = H\,ds\,\vec{w} = \frac{1}{2}\,\varsigma\,C_{DN}\,d(\vec{V}\cdot\vec{w})^2\,ds\,\vec{w} \tag{14}$$

with $\vec{V}\cdot\vec{v} = -V_x \sin\varphi + V_y \cos\varphi$

$\vec{V}\cdot\vec{w} = -V_x \sin\psi \cos\varphi - V_y \sin\psi \sin\varphi + V_z \cos\varphi$

and C_{DN} the normal drag coefficient of the cable. The quantities F, G and H are space dependent and must be evaluated at each point P along the cable.

Since the direction of the current velocity for this study assumed is in the horizontal direction along y-axis, we may write,

$$\vec{F} = \frac{1}{2}\,\varsigma\,C_{DT}\,\pi d\,(V_y^2 \cos^2\psi \sin^2\varphi)\,ds\cdot\vec{u} \tag{15}$$

$$\vec{G} = \frac{1}{2}\,\varsigma\,C_{DN}\,d(V_y^2 \cos^2\varphi)\,ds\cdot\vec{v} \tag{16}$$

and $$\vec{H} = \frac{1}{2}\,\varsigma\,C_{DN}\,d(V_y^2 \sin^2\psi \sin^2\varphi)\,ds\cdot\vec{w} \tag{17}$$

Now, the forces in the x-direction are given by,

$$F_x = \frac{1}{2}\,\varsigma\,C_{DT}\,\pi d(V_y^2 \cos^2\psi \sin^2\varphi)\,(\cos\psi \cos\varphi)\,ds \tag{18}$$

$$G_x = \frac{1}{2}\,\varsigma\,C_{DN}\,d(V_y^2 \cos^2\varphi)\,(-\sin\varphi)\,ds \tag{19}$$

$$H_x = \frac{1}{2}\,\varsigma\,C_{DN}\,d\,(V_y^2 \sin^2\psi \sin^2\varphi)\,(-\sin\psi \cos\varphi)\,ds \tag{20}$$

Forces in the y-direction are,

$$F_y = \frac{1}{2}\,\varsigma\,C_{DT}\,\pi d(V_y^2 \cos^2\psi \sin^2\varphi)\,\cos\psi \sin\varphi\,ds \tag{21}$$

$$G_y = \frac{1}{2}\,\varsigma\,C_{DN}\,d(V_y^2 \cos^2\varphi)\,\cos\varphi\,ds \tag{22}$$

$$H_y = \tfrac{1}{2} \rho\, C_{DN}\, d(V_y^2 \sin^2\psi \sin^2\varphi)\, (-\sin\psi \sin\varphi)\, ds \qquad (23)$$

Forces in the z direction are

$$F_z = \tfrac{1}{2} \rho\, C_{DT}\, \pi d(V_y^2 \cos^2\psi \sin^2\varphi)\, \sin\psi\, ds \qquad (24)$$

$$H_z = \tfrac{1}{2} \rho\, C_{DN}\, d(V_y^2 \sin^2\psi \sin^2\varphi)\, \cos\psi\, ds \qquad (25)$$

ANALYSIS

Cables are usually treated as elements that resist only tension in the longitudinal direction. The complexity of the analysis is due to the fact that it involves the solution of an extensively nonlinear structural response problem.

The nonlinearities are caused due to the following factors:

(1) The large changes in configuration which moored-cable structures undergo, in response to applied loads (geometric nonlinearity).

(2) The nonlinear load-deformation behaviour of cables, which also includes the lack of any significant-stiffness in compression (material nonlinearity).

(3) The dependency of the load on the position, shape and orientation of the structure.

(4) The drag loading which is a function of the velocity squared

and (5) The limiting surface and seafloor constraints which are dependent on the cable response.

In this present contribution, the cables are analysed by applying the geometrically nonlinear finite element procedure for both static and dynamic conditions, Nonlinear Structural Analysis Program (NONSAP).

Apreprocessor has been developed for determining the loads due to current and waves. The cable array is discretised into a number of finite elements. An iterative technique is adopted, by assuming an initial configuration of the cable system and correcting the configuration, by successive solutions applying the nonlinear finite element analysis with the position dependent hydrodynamic loading. The analysis has the advantage that it can exactly take into account the external parameters such as force from spaced buoyant floats, different types of loadings, different elastic support conditions of anchors etc. The entire analysis is carried out in the system IBM 370/155.

FINITE ELEMENT PROCEDURE

The Finite Element method has proven to be a very effective discretization procedure for structural analysis. The non-linear behaviour due to large displacements, strains and/or material behaviour is studied using the program NONSAP.

MOORING SYSTEM ANALYSED

A few of the buoy mooring systems used are shown in Fig.2. The multipoint mooring systems have the following advantages: (a) reduced horizontal motion of the floating structure, (b) enough space below the floating structure to deploy instrumentation, (c) use of the moorings for close spacing of oceanic sensors, (d) reduction of the chances of kinking and fouling to occur, (e) increased life expectancy and (f) increased probability of recovering the systems. However, the multileg systems are expensive and difficult to deploy.

A single point mooring shown in Fig.3 is analysed using the programme. The cable is discretised into fifty elements. The element chosen is a truss element and the type of non-linear analysis is that of updated lagrangian formulation. A nonlinear elastic material model is used. The Fig.4 gives the forces and displacements in the mooring system. A buoyant force of 300 kg was applied at the middle of the cable and the configuration of the cable is also shown in Fig.4. The mode shapes and natural frequency are given in Fig.5.

A three point mooring system shown in Fig.6 is analysed. The results are shown in Fig.7.

CONCLUSIONS

The preprocessor developed along with NONSAP has been used for the analysis of single point and multipoint mooring systems. The static and dynamic characteristics of the cable system can be determined using the programmes. It can also incorporate any of the realistic conditions at anchor and at the surface, subsurface buoyant forces etc.

ACKNOWLEDGEMENTS

The authors wish to express their thanks to the authorities of Indian Institute of Technology for the facilities offered for carrying out the study.

REFERENCES

1. Bathe, K.T. et al, 1974, NONSAP A Structural Analysis Program for Static and Dynamic Response of Non-linear Systems, University of California, Division of Structural Engineering and Structural Mechanics, Report No.UCSESM 74-3.

2. Berteaux, H.O., 1975, Buoy Engineering, Woods Hole Oceanographic Institution, Woods Hole, Massachusetts.

3. Blendendender, J.W., 1970, "Three Dimensional Boundary Valve Problems for Flexible Cables", Offshore Technology Conference, Houston, Texas, Paper 1281.

4. Chhabra, N.K., 1974, Mooring Mechanics - A Comprehensive Computer Study, C.S. Draper Laboratory, Cambridge, Mass.

5. Eiriz, H.H., 1973, Ocean Systems Structural Analysis Computer Capabilities, Lockheed Missile and Space Co., Sunnyvale, California.

6. Elizabeth Cuthill, 1968, A Fortran IV Program for the calculation of a Flexible Cable in a Uniform Stream, Naval Ship Research and Development Centre, Report No.2531, Bethesda, Md.

7. Gormally, J.M., 1966, The Analysis of Mooring Systems and Rigid Body Dynamics for Suspended Structures, Technical Report No.14, Whippany, N.J.

8. Morison, J.R. et al, 1950, "The Force exerted by Surface Waves on Piles", Petroleum Transactions, American Institute of Mechanical Engineers, Vol. 189.

9. Paquethe, R.G. and Henderson, B.E., 1965, The Dynamics of Simple Deep Sea Buoy Moorings, Technical Report No.TR 65-79, G.M. Research Laboratory, Santa Barbara, California.

10. Pode Leonard, 1951, Tables for computing the equilibrium configuration of a flexible cable in a uniform stream D.T.M.B. Report No.687.

11. Schreiber, M., 1973, Computer Program for the analysis of moored cable systems, Lockheed Missile and Space Co., Sunnyvale, California.

12. Seidl, L.H., 1971, "On the dynamics of a mooring line", Look Lab, Hawaii, Vol.2, No.1, pp 11-22.

13. Skop, R.A. and O'Hara, G.J., 1968, "The Static equilibrium configuration of cable arrays by use of method of imaginary reactions", Naval Research Laboratory, Washington D.C., Report No.6819.

14. Skop, R.A. and O'Hara, G.J., 1970, "The Method of imaginary reactions - a new technique for analysing structural cable systems", Marine Technology Society Journal, Vol.4, No.1, pp 21-30.

15. Skop, R.A., 1970, "The method of imaginary reactions: applications to N-Point Moors', Marine Technology Society, Washington, DC, Vol.1, pp 1-22.

16. Wilson, B.W., 1967, "Elastic Characteristics of Moorings", ASCE Journal of Waterways and Harbours Division, Proceedings, pp 27-56.

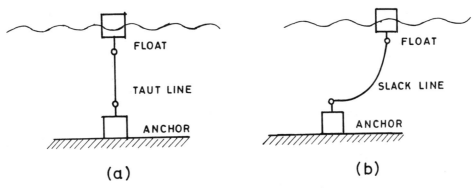

(a)   (b)

Single point systems

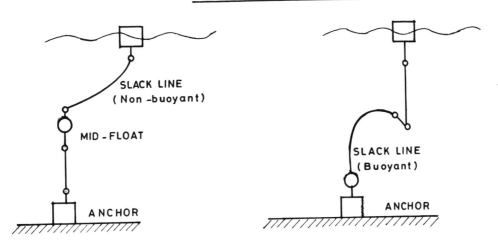

(c) Two buoy systems   (d) Two line systems

Two point systems

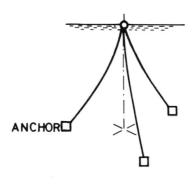

FIG. 2.   (e)   TRI-MOOR SYSTEM

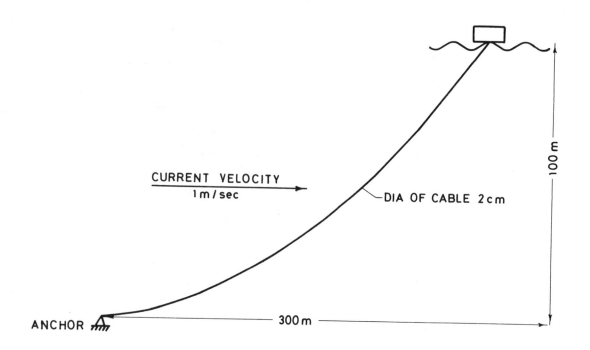

FIG.3. SINGLE POINT MOORING

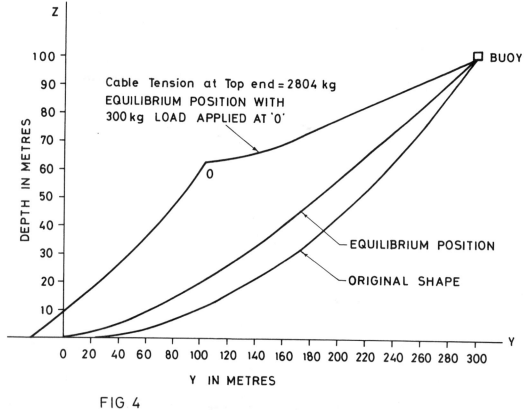

FIG 4

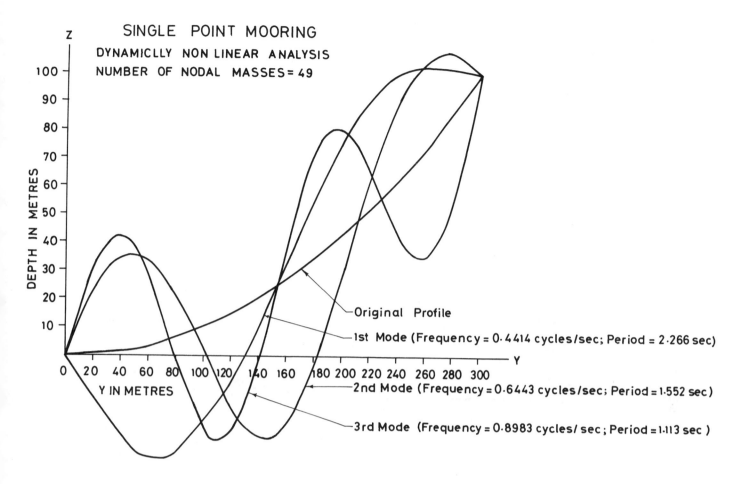

FIG. 5. SINGLE POINT MOORING

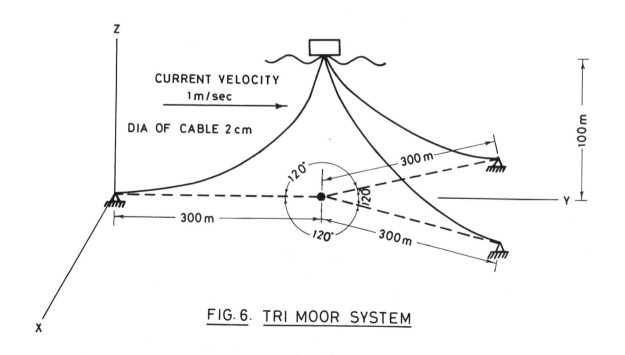

FIG. 6. TRI MOOR SYSTEM

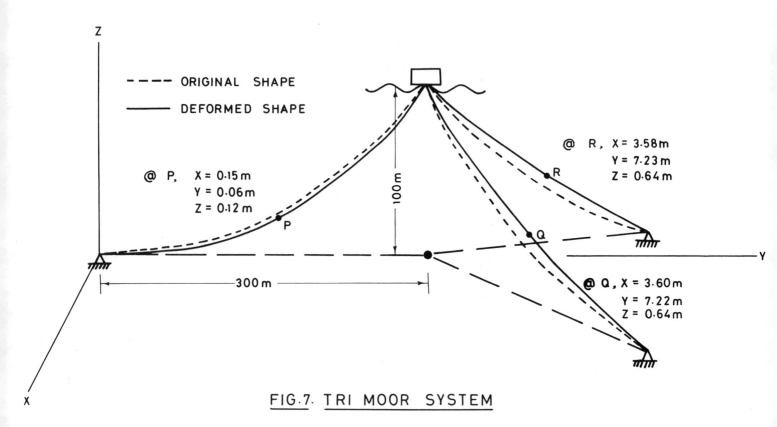

FIG. 7. TRI MOOR SYSTEM

THE DYNAMIC BEHAVIOUR OF
TAUT-MOORED
SINGLE-POINT MOORING BUOYS

A. Roy Halliwell
Offshore Engineering Dept.
Heriot-Watt University
Edinburgh, Scotland

Iain J.A. Jardine
Shell International
The Hague, Netherlands
(formerly Research Student,
Heriot-Watt University)

SUMMARY

The paper describes the results of a series of hydraulic scaled-model tests on some taut-moored single-point mooring buoys. The tests were carried out as part of an extensive study to identify the reasons for the failures that occurred with two such buoys installed in the Mediterranean Sea, off the coast of Libya.

The model tests showed that the surface-riding tension-leg buoy system in comparatively shallow water (<30m) suffers from two major disadvantages which prevent the realisation of the potential advantages of this type of single-point mooring. Firstly the motions which the buoy experiences during wave conditions are at times large and involve complex combinations of surge, sway and yaw. Secondly, snatching of the tensioned anchor chains occurs frequently, even during moderately severe sea conditions. The combination of these two disadvantages produces further difficulties: the resulting wear causes unequal pretensions in the chains and lower mean values, the effect of both of these being to make either the motion or the snatching worse.

The motions of the vertical anchor leg mooring system are greatly influenced by its natural periods of surge, sway and yaw. To give the best performance these natural periods should be at least three times greater than the expected wave periods. However, except for large water depths, this specification cannot be met and therefore, unfortunately, the system is likely to be unsatisfactory for most practical sites.

1. INTRODUCTION

The last two decades have seen the introduction and development of various single-point mooring (SPM) systems, primarily for use by large oil tankers when loading or discharging their cargoes. The rapid developments of offshore oil and gas resources during the last decade has accelerated the use and development of SPM's and the success of them, in all types of environmental conditions, has been a notable engineering achievement.

Flory and Poranski (1977), in a review-type paper, stated that "...SPM design...is still an emerging science" and went on to encourage the continued interchange of ideas and engineering experience by adding "Communications of ideas among SPM designers has helped to bring the science of SPM design to the stage it is today, and hopefully will continue to advance the science.....". The communication of ideas has been primarily through the considerable number of technical papers and notes which have been published, and it is natural that these papers have concentrated on the success stories, describing new proposals and concepts, recent designs and analysis techniques, which it is claimed can improve the performance and advance the understanding of the behaviour of SPM systems. However, it is profitable for engineers and engineering in general to know about failures and to have some indication as to why certain proposals have not come to full, operating fruition. It is therefore unfortunate, at least in the opinion of the authors, that very few of the technical papers have concentrated on the difficulties or failures encountered by designers and research workers. This paper attempts to highlight the difficulties associated with a particular SPM concept - the vertical anchor leg mooring (VALM) system.

The VALM concept was first proposed in the mid-sixties and actually developed to the stage where two such buoys were constructed in the early seventies but as far as the authors are aware, the system has not been used since. In its basic form the system consists of a surface-riding buoy which is moored to the seabed by a number of vertical chains or wires. The length of the vertical anchor legs or lines is such that the buoy is drawn down into the water below its free-floating position and therefore each chain is pre-tensioned and the system behaves as an inverted pendulum. The concept of vertical moorings provides a number of potential advantages over the commonly used catenary anchor leg mooring (CALM) arrangement. The most obvious is the considerable savings arising from the greatly reduced length of anchor chain required: the total length of chain required for each catenary of the CALM system is approximately ten times the depth of water at the installation. If the tanker moves too close to the buoy and over-rides the buoy, with the CALM system the tanker is likely to interfere with the catenary anchor moorings and cause considerable damage, whereas with the VALM system little or no damage is likely to occur. There are also potential advantages associated with the underbuoy hoses. Firstly, because the buoy has much less overall movement than its counterpart in the CALM system, it follows that the overall length of the underbuoy hose is reduced. Secondly, the pitching and rolling movement of the buoy in the CALM system produces considerable wear and tear on both the underbuoy and surface hoses and (at least until the introduction of a special universal-type joint around 1975) is the reason for many underbuoy hose failures. The removal of this undesirable movement of the buoy should therefore give considerable savings, not only because of the direct savings associated with the cost of hoses but also the more important savings arising from the reduction of down-time. The latter is especially important because the replacement of underbuoy hoses can be delayed for weeks by prolonged moderate seas in which tankers could operate at the SPM but which are severe enough to prevent the divers working to replace the hoses. Of course, the VALM system incorporating a surface-riding buoy is sensitive to changes in water depth and it follows that its use is limited to locations experiencing comparatively small tidal ranges. However, the potential advantages are considerable and such that it might be expected that the concept would be used at, at least, some of the large number of locations where the tidal range is not a restriction.

This paper presents a brief description of the results obtained during an investigation of some VALM systems, using hydraulic scaled-model tests. These tests have shown that the potential advantages of the VALM system cannot be realised at most practical locations because of a number of disadvantages associated with the dynamic behaviour of this type of surface-riding taut-moored buoy.

2. DESCRIPTION OF THE BUOYS AND SOME ELEMENTARY CONSIDERATIONS OF THEIR BEHAVIOUR

The behaviour of SPM systems has been an active research interest at Heriot-Watt University since 1973. Primary interest has centered upon the low-frequency oscillations of the tanker when moored at an SPM buoy and subjected to wave loading (Owen and Linfoot, 1976 and Halliwell, et al., 1982) but there has also been a general interest in the design and performance of SPM systems (Halliwell, et al., 1975). This general interest was maintained in 1976 by the start of an investigation into the dynamic behaviour of VALM systems. This new investigation was prompted by the failure of two such SPM systems, off the coast of Libya in the Mediterranean Sea, in late 1974.

The two SPM systems were very similar, the larger one had been designed to handle tankers up to 100,000DWT in 26m water depth and the smaller one to handle tankers up to 30,000 DWT in 20m water depth. Figure 1 gives the general arrangement and pertinent dimensions for the larger buoy:

the smaller buoy had the same depth of hull (3.7m) but its diameter was 9.4m, the pitch circle diameter (P.C.D.) for the four vertical chains was 7.3m and there were some other detail design differences. Each SPM system consisted of essentially a large cylindrical steel buoy which was anchored to the seabed by means of four heavy steel chains. For each chain the attachment on the buoy was made through a rubber shock absorber with each chain passing through holes in the buoy chamber and through the centre of the shock absorbers to be held by a fid pin passing through the centre of a chain link. The chain load was transferred via the fid pin and a steel plate through the shock absorbers to the body of the buoy.

The failure of the two systems occurred during storm conditions. The magnitude of the storm was such that some damage to the superstructure of the buoys and to the underbuoy and floating hoses was to be expected, but additionally one of the anchor chains, on each buoy, became disconnected and serious damage occurred to some of the shock absorbers. The immediate interest in the failure of the two SPM systems led the general investigation of VALM systems (Jardine, 1979) to concentrate attention on the survival condition of these types of buoy, although a small number of tests were also made for the operational mode.

Before describing the dynamic behaviour of the buoys, as measured in the hydraulic scaled-model tests, and highlighting the difficulties surrounding the VALM design it is useful to outline some features of the motion response which, based on elementary considerations alone, might be expected from a simple VALM system during the passage of regular waves. The response of any dynamic system to a harmonically varying forcing function is greatly influenced by its own natural or free oscillations. For a simple VALM buoy similar to that illustrated in Figure 1 the periodic time of undamped free oscillations in surge or sway (T_s) is given by the equation,

$$T_s = 2\pi \sqrt{\frac{(M+KM')H}{nP_o}} \qquad (1)$$

where M is the mass of the buoy (including the chains),
M' is the displaced mass of water,
K is added mass coefficient for the buoy,
nP_o is the total pretension in the chains, i.e., P_o in each of n chains,
H is the effective length of the chains or legs, i.e., the distance between the free-hinge connection at the seabed and at the buoy,
and g is the gravity constant (9.81 m.sec^{-2})

Remembering that,

$$nP_o = (M'-M)g = A\rho g b \qquad (2)$$

equation (1) can be rewritten as

$$T_s = 2\pi \sqrt{\frac{H}{g} \left\{ (1+K) \frac{M}{A\rho b} + K \right\}} \qquad (3)$$

where A is the area of the water-line plane of the buoy,
ρ is the density of the water,
and b is the drawdown produced by the pretensioning, i.e., the distance between the actual water level and the water line corresponding to zero tension in the chains.

Although the swinging pendulum-type mode of oscillation is probably the most important, the VALM system has an obvious torsional mode of oscillation which is likely to influence the response of the system to wave loading. The equation for the periodic time (T_t) of free torsional oscillations can be written in a form similar to either equation (1) or equation (3), i.e.

$$T_t = 2\pi \sqrt{\frac{(M+K_tM')H}{nP_o} \cdot \frac{r_o^2}{R^2}} \qquad (4)$$

or,

$$T_t = 2\pi \sqrt{\frac{H}{g} \cdot \frac{r_o^2}{R^2} \left\{ (1+K_t) \frac{M}{A\rho b} + K_t \right\}} \qquad (5)$$

where K_t is the added mass coefficient for torsional motion of the system,
R is the radius of the pitch circle of the chains,
and r_o is essentially the radius of gyration of the system about its vertical axis but with some allowance made for the effects of added mass on the torsional inertia.

While the assessment of the performance of the design will be based on a large number of factors, two particularly important aspects are the motion response of the system and the magnitude of the loads in the anchor chains. For satisfactory motion response it seems obvious that any resonance should be avoided and therefore that the natural periods for surge, sway and torsion (yaw) must be large enough to avoid the possibility of coinciding with the periods of the wave environment. It follows that for most practical sites the periodic times of the free oscillations must be greater than say 13 sec. For the larger prototype buoy with the design pretension of 20 tonnes per chain, the values of T_s and T_t were estimated to be about 22 and 14 sec. respectively. The corresponding values for the smaller buoy, in shallower water and with lower pretension, were both a little smaller. It is obviously desirable that these natural periods be larger but it is difficult to achieve this when the water depth (and therefore the value of H) is less than about 30m. From equations (1) - (5) the obvious way to increase T_s and T_t is to decrease the pretension and thereby reduce the value of b: the alternative is to increase the mass and therefore the size and cost of the buoy. However, any reduction in drawdown increases the possibilities of the anchor chains becoming slack during the passage of waves, and subsequently snatch loading occurring as the anchor lines snap tight.

If the buoy remained stationary it would be a simple matter to determine the size of the wave which would produce slack chains, i.e., any wave in which the trough is larger than the drawdown b. However, the response of the buoy to the waves can improve the situation because the upthrust of the surface-riding buoy is increased as the buoy swings. If the swinging of the buoy is phased relative to the wave in such a way that when the trough passes, the buoy is more-or-less at the limit of its pendulum-type oscillation, then there is a greater chance that the chains will remain taut. Such an oscillation is presented pictorially in Figure 2. The force causing the swinging action of the buoy arises from the orbital velocities and accelerations of the water particles due to the wave action, and the amplitude and phase of the pendulum motion are determined by the period and magnitude of the water particle orbits along with the value of T_s.

If the anchor chains do not go slack their tension during the passage of a wave can be evaluated approximately, using elementary considerations, as

$$\frac{P}{P_o} = \left(1 + \frac{\eta}{b}\right)\left(1 - \frac{\theta^2}{2}\right) + \frac{H}{b}\frac{\theta^2}{2} \tag{6}$$

where η is the height of the water surface above the MWL,
 θ is the angular displacement from the vertical of the chains,
and both will vary harmonically during the passage of regular waves. Consideration of this equation shows that if the anchor chain is not to become slack during the passage of a wave, either $\eta/b \not< -1$, i.e., the magnitude of the trough must never exceed the drawdown b, or when $\eta/b < -1$ the value of $(H/2b)\theta^2$ must be sufficiently large to compensate, and maintain the upthrust on the buoy by it being displaced sufficiently from the vertical during the passage of the wave trough. In practice it is impossible to produce a reasonable design for most practical sites in water depths <30m for which $\eta/b \not< -1$, and therefore the designer is relying on the swinging motion of the system to prevent snatching of the anchor chains. Because the passage of waves will inevitably, on occasions, produce troughs when the buoy is more-or-less vertically above its seabed anchor points, some snatching has to be anticipated and shock absorbers are therefore incorporated into the design.

Returning to the question of increasing the values of T_s and T_t to reduce the chances of resonance: decreasing the value of P_o and thereby b clearly increases the possibilities of snatching. While therefore in principle the pretension allows the system to be "tuned" to provide the best response to a particular mean period of waves, in practice any possible variations are small and are likely to be accompanied by increased snatching of the anchor chains. Another possibility for increasing the values of T_s and T_t is to design the buoy so that the value of A in equations (3) and (5) is small. This leads to a design more like a TLP, with a fully submerged hull supporting a working deck above the MWL by columns having comparatively small cross-sectional area. In somewhat deeper water, say > 35m, this arrangement has some possibilities but unfortunately in relatively shallow water little can be gained by this technique because the reduction in the value of H for such a design is usually considerable and can even result in a reduced value of T_s and T_t for the new design.

3. THE MODEL TEST PROGRAMME

The test programme was arranged to examine the reasons for the failure of the prototype buoys and to consider the effects of various design parameters on the dynamic behaviour of VALM systems which incorporate a surface-riding tension-leg buoy. A comprehensive series of tests was therefore carried out using scaled models of each of the prototype buoys and the test conditions were selected to enable the following aspects to be investigated,
  (i)   the anchor chain loads and buoy motions in regular waves of various heights and periods;
  (ii) the chain loads and buoy motions in irregular waves;
  (iii) the effect of reducing the pretension loads below the design values;
  (iv) the effect of uneven pretension loads;
and, (v) the influence of water depth.
The tests were also extended to allow a parametric study of the VALM system with attention focusing on the effects of the stiffness and position of the shock absorbers, the shape of the hull of the

buoy and the number and arrangement of the anchor chains. These tests also led to some new design proposals, based on the basic VALM concept, and preliminary tests were made of a number of these.

4. THE MODELS AND TEST FACILITIES

4.1 Wave basin facility. All the model tests were carried out at Heriot-Watt University in a 9m x 9m wave basin which has a maximum working depth of about 800mm. Beaches, constructed from fibreglass panels, extend over three sides of the basin. The fourth side contains the wave generation system, which consists of seventeen independent wavemaker modules. Each module contains an oscillating paddle driven by an electric servo-motor through a worm and pinion drive mechanism. The servo-motor moves in response to the power signal from a servo amplifier which itself is fed with appropriate analogue input signals. Since each wavemaker module has its own set of electronic equipment associated with it, it is possible to route different signal inputs to each servo-amplifier, thus making it possible to generate a wide variety of sea conditions, including short-crested seas and seas with angled wave fronts. For the purposes of all the model tests described in this paper the wave basin facility was used to generate long-crested regular and irregular waves, travelling parallel to the sides of the wave basin and therefore the features of independent wavemaker control were not required.

4.2 The models. The models of all the buoys were constructed in PVC: the larger prototype buoy to a linear scale ratio of 1:48 and the smaller buoy to a scale of 1:46. These reasonably large models had the advantage that the mooring loads in the models were large enough to be measured with accuracy while at these scales the required wave conditions were still within the capabilities of the generation system. The anchor chains were simple brass linked chain of the correct submerged weight per unit length. The shock absorbers were represented by compression springs on the body of the buoy having the correct spring stiffness at the model scale. No attempt was made to include either the underbuoy or the floating hoses in the model. When making this decision it was argued that the hoses would increase the system's resistance to motion and therefore probably improve its overall behaviour somewhat, however the system might well need to survive storm conditions either before the hoses have been connected or at a time when they have been temporarily disconnected.

A number of other models were used for the parametric studies. These models were usually simplified, so no details of the superstructure were included and some details of the hull shape were ignored. The simplified form of the buoys allowed changes to shock absorbers, mass distribution of the hull and the arrangement of anchor chains to be easily introduced.

4.3 Simulation of waves. The regular waves in the basin are produced using an accurate sine-wave signal generator to provide the inputs to the servo-amplifier driving the wave generators.

For irregular waves a pseudo-random analogue signal of suitable spectral and statistical characteristics is used. The analogue signal for a particular sea-state is synthesized using a Nova computer and a DAC (digital to analogue convertor) sub-system and stored on an instrumentation tape recorder. The irregular waves simulated in the tests were based on the ITTC form of the Pierson-Moskowitz wave spectrum. The ITTC spectrum is defined only in terms of the significant wave height ($H_{1/3}$) and the average wave period ($\bar{T}$) is given by

$$\bar{T} = 3.86 \sqrt{H_{1/3}} \qquad (7)$$

The spectra of the measured (simulated) and theoretical seas are given in Figure 3 corresponding to a significant wave height of 5.1m at a model scale of 1:48. Axes are marked in prototype units.

4.4 Measurement techniques. The transducers and associated equipment for measurements and the system for data collection and analysis has been developed over a number of years, since the basin was constructed in 1976. In these tests measurements were taken of the following quantities,
  (i) the water-surface elevation (wave heights);
  (ii) the load in each of the four anchor chains;
and (iii) the surge, sway and yaw motion of the buoy.

Wave height measurement is carried out using electrical resistance probes. Each gauge is calibrated prior to and following a sequence of tests by noting the change in bridge output associated with a prescribed displacement of the probes. Chain loads were measured using a simple tension ring cell which acted as one of the links in an anchor chain. The ring was made from phosphor bronze and had a number of miniature strain gauges attached. The load cells were placed near the seabed end of the anchor chains: this enabled the wires from the strain gauges to be taken along the seabed and out of the water to the balancing bridge without interfering hydrodynamically with the motion of the buoy. The load cells were calibrated statically using a simple scale arrangement for applying loads: the calibrations were checked before and after each set of tests. Figure 4 indicates the numbering scheme employed for the load cells.

Motions of the buoy were sensed by means of three accelerometers, mounted in the positions indicated in Figure 4. Accelerometer 2 sensed, primarily, surge motions; Accelerometer 1 sensed, primarily, sway motions and Accelerometer 3 sensed both surge and yaw motions since it was located at a distance from the axis of the buoy. The displacements of the buoy were obtained by integration of the digitized information associated with the analogue outputs of the three accelerometers. Each accelerometer was calibrated prior to a sequence of test runs. The interpretation of the accelerometer information provided some difficulties when the motion of the buoy was large and violent, mainly arising from the accelerometers responding to the pitch, roll and heave motions of the buoy. Consideration was given to using the SELSPOT system instead: this system is based on opto-electronic principles using infra-red light emitting diodes (LED's) placed on the object whose motion is to be studied. This system has been very successfully used for much of the SPM research carried out in the basin, especially to study the low frequency motions of the tanker Halliwell et al., 1982). However, the major difficulty for the present study was that from time to time the VALM buoys were completely submerged and it was feared that the LED's might be damaged.

5. THE MAIN FEATURES OF THE DYNAMIC BEHAVIOUR

The scaling of the models and the interpretation of the results were all made on the basis of the Froude law of similitude. In this short paper it is only possible to present a few of the results and these have been chosen to illustrate the main features of the behaviour observed in the model tests. The test results quoted are all presented in equivalent prototype values.

5.1 **Tests with regular waves.** Figure 5 presents some examples of the variation of wave elevation and chain load during the basic tests with the larger (11.7m dia.) buoy. At low values of wave height little or no snatching took place and the motions of the buoy were small. Figure 5a shows that the peak values of a chain load occur with the same periodicity as that of the waves and the variations of load are as would be expected from elementary considerations of the surge response of the system (equation (6)). However, during tests with wave heights greater than 3m, significant snatch loads occur on all chains during the passage of each wave. The value of the snatch load varies considerably from one cycle to the next, see for example Figure 5c, and the motion although still broadly regular exhibits irregular features.

Table 1. Summary of test results for the larger buoy with regular waves.
(water depth 26.16 m, model scale 1:48)

Wave Height m	Wave Period secs	Anchor chain loads in tonnes				Chain pretensions in tonnes				Total excursion of buoy		
		Chain 1 F_{max}	Chain 2 F_{max}	Chain 3 F_{max}	Chain 4 F_{max}	Chain 1	Chain 2	Chain 3	Chain 4	Surge m	Sway m	Yaw deg
3	7.5	30	46	30	40	19	20	20	15	-	-	-
3	9.0	56	97	53	72	20	19	23	10	-	-	-
3	10.5	56	106	53	63	18	20	20	20	-	-	-
3	12.0	52	64(96)	52	82	23	18	22	19	1.48	0.67	3.7
5.1	7.5	154	121	102	116	29	11	20	13	5.14	1.63	12.9
5.1	9.0	207	259	158	272	19	23	19	23	1.73	2.74	24.0
5.1	10.5	229	197	158	143	21	21	24	21	0.96	1.39	6.6
5.1	12.0	49	179	80	145	22	22	22	19	1.78	0.43	2.0
6.5	7.5	84	124(187)	91	83(135)	19	20	20	20	6.3	2.93	50.6
8.4	9.0	169(373)	131(225)	114(144)	174	24	21	23	20	1.54	5.38	23.8
8.1	10.5	203	210(251)	158	195	19	21	20	23	5.04	5.76	37.3
8.4	12.0	148(404)	197	140	200	20	21	19	23	0.91	3.31	11.7
5.1	9.0	184	207	175	249	8	33	11	33	1.63	3.26	24.7
							Mean = 21					

Table 1 presents a summary of the basic regular-wave tests carried out for the larger buoy and shows the maximum loads in the anchor chains during each test and the total amplitude of the surge, sway and yaw motions of the buoy. In certain cases exceptionally high snatch loads occurred during the first few wave cycles, before the system had settled into its roughly regular response, and these values are given alongside the highest snatch loads encountered during the remainder of the test. The results show that in some instances the peak values of the snatch loads are quite sensitive to the wave period. For example, with a wave height of 5.1m and period 7.5 sec., the maximum value of chain load that occurred during the test was 154 tonnes. This can be compared with the value of 272 tonnes during the test with the same wave height of 5.1m but with a wave period of 9 seconds. The records for the first part of these tests are presented in Figures 5b and 5c and it is clear that the general level of snatching is much less for the 7.5 sec. period waves than for the 9.0 sec. period waves. When the wave period is 12 sec. the chain loads are smaller again and comparable with those occurring at 7.5 sec. period. It would seem, therefore, that there is a small range of wave periods for which the snatching characteristic is particularly serious.

The motion response is very interesting and shows that the sway and yaw motions of the buoy are sometimes large. Broadly speaking, the large loads occur when the sway motion is large, or larger than the surge motions, but it is difficult to be absolute about the relationship because the buoy motions are complex and generally involve a combination of surge, sway and yaw. The table also shows that in some wave conditions large surge motion of the buoy is associated with comparatively lower chain loads; in the absence of sway and yaw this feature is consistent with the elementary theory. Large yawing motions do not of themselves produce large chain loads, although of course this type of motion may well result in damage to components attached to the buoy.

An interesting and somewhat unexpected result which emerged from the regular tests is that the buoy motion and chain loads are not increased in any consistent manner by an increase in waveheight above about 5.0m. It would seem therefore that the waveheight is important up to the region where snatching occurs, but once this situation has been established the magnitude of the waveheight is of no great significance.

5.2 <u>Tests with irregular waves.</u> For sea conditions below $H_{1/3}$ equal to 3m there is little evidence of snatching (as can be seen from Figure 6a) and for much of the time the chain loads remain at or near their pretension values. Comparison of Figures 5 and 6 shows that the load histories are noticeably different in character from those found in the regular-wave tests. The snatch loads are still present with the more severe wave conditions, however, they no longer occur on a regular basis but in a random fashion, associated with the lower wave troughs.

Table 2. Frequency of load exceedance during irregular tests with the larger buoy in 26.16m water depth.

Sig. Wave Height $H_{1/3}$ m	Average Period T secs	Test Duration (equive. Prototype mins)	Chain Pretension (tonnes)	Number of occasions of load exceedance during test duration															
				Chain 1				Chain 2				Chain 3				Chain 4			
				>250	>200	>150	>100 tonnes	>250	>200	>150	>100 tonnes	>250	>200	>150	>100 tonnes	>250	>200	>150	>100 tonnes
3	6.7	15	21	0	0	0	0	0	0	0	0	0	0	0	0	0	0	0	1
4	7.7	23	21	0	0	0	4	0	0	6	11	0	0	0	2	0	0	0	10
5.1	8.7	25	21	0	0	1	16	0	1	16	26	0	0	3	19	0	0	1	21
5.1	8.7	24	10	0	3	15	24	1	4	16	37	0	0	3	22	0	1	3	15

The frequency of the occurrence of large snatch loads is best seen from Table 2. When interpreted in terms of a prototype time of one hour, the results of the test with a significant wave height of 5.1m and the design pretension values of 20 tonnes in each chain imply,
   at least 2 occasions > 200t.;
   at least 50 occasions > 150t., but excluding those greater than 200t.;
   at least 196 occasions > 100t., but excluding those greater than 150t.
The maximum snatch load encountered in this test was 203 tonnes and this value would, of course, be exceeded if the duration of the test was increased.

The results for the larger buoy showed that if the pretensions are reduced from their design value of 20 tonnes the snatch loads increase. The test with just 10 tonnes pretension in each chain gave maximum snatch loads in each of the four chains of 221, 311, 174 and 205 tonnes compared with 151, 203, 163 and 156 tonnes for the equivalent test with 21 tonnes pretension. However, the tests with the smaller buoy showed that decreasing the pretension value below its design value (10 tonnes per chain) did not noticeably increase the snatching although the motion response of the buoy was worse. Presumably in this case the advantages arising from the resulting increase in natural period for the system outweighed the disadvantages associated with the reduced drawdown. Overall the results showed that the behaviour of a VALM system is sensitive to the pretension and while an optimum value might be obtained for any buoy in a particular depth of water, any changes of pretension due to variations of mean water depth or wear of the anchor chains are almost certain to cause the overall performance of the system to deteriorate.

6. GENERAL AND PARAMETRIC STUDIES

The basic tests showed that the performance of the VALM system is markedly affected by the relative magnitude of the input wave period to its own natural periods of surge, sway and yaw. It was also clear that for the buoys tested there was complex coupling between the surge, sway and yaw motions at certain frequencies. To examine some of the effects in detail a careful study was undertaken of the motion response of a VALM system which was very similar to the prototype buoys but which allowed the response to be measured for wave periods up to and a little above its own surge natural period. To do this using the same test facilities required the size of the buoy to be changed a little to give a lower absolute value of T_s: the model tested had values of T_s and T_t of 15 and 11.2 sec. respectively, when interpreted with a linear scale of 1:40. The motion response of the system (as measured by the accelerometers) is shown in Figure 7: the experimental points were obtained at wave periods corresponding to 5-18 sec. at 1 sec. intervals. The figure shows that the surge motion of the system does indeed reach resonance when the wave period coincides with the surge natural period and that there is a wide range of wave periods for which the response is appreciable. More surprising and crucial for the design is the second region of large surge motion that occurs around $T=0.5T_s$. The sway response is also interesting and serious, showing that there is strong coupling between the surge and sway over the range $0.35<T/T_s<0.7$. It is interesting to note that there is almost no sway response in the region of surge resonance at $T=T_s$. The torsional response also shows two resonance peaks; the motions at the 2nd harmonic are particularly violent, resulting in high angular accelerations of the buoy. Consideration of these response curves shows that for this type of surface-riding taut-moored buoy undesirable motions are likely during the passage of waves for which the periods are greater than $0.33T_s$ (about). Therefore, for a satisfactory design T_s should be greater than $3\overline{T}$, but this is impossible to achieve for water depths less than about 30m and moderately severe seas.

Various changes to the design have been examined during the parametric studies but while some improvements in the performance have been obtained, in general the motions and snatch loading of the system continue to be unsatisfactory for some parts of the range of wave periods 6-12 sec., with moderate size waves. The conflict or frustration of the design is well illustrated by the effect of the mean pretension as discussed in Section 2 and illustrated in Section 5. Similar conflicts were found to be associated with the shock absorbers. Tests without shock absorbers showed high snatch loading and gave convincing proof of the need for them. However, when they were introduced the motion of the buoy became worse and although the magnitude of the snatch loading was reduced the incidence of snatching was not changed. Attempts to improve the motions of the buoy often resulted in increased snatching and vice versa. The authors are therefore forced to the conclusion that this type of surface-riding taut-moored buoy cannot be designed for satisfactory operation at most practical sites in water depths less than 30m.

REFERENCES

FLORY, J. F. and P. F. PORANSKI, 1977, "The Design of Single Point Moorings," Proceedings, Ninth Annual Offshore Technology Conference, Houston, Texas, USA 1977, Vol. II OTC 2827, pp. 169-176.

HALLIWELL, A. R., B. T. LINFOOT, P. C. MACHEN and D. G. OWEN, 1975, <u>Single Point Moorings: An Appraisal of the Present Position and An Assessment of the Feasibility of a Computer-Based Design-Analysis</u>. Institute of Offshore Engineering, Heriot-Watt University, Project Report No. 74/4.

HALLIWELL, A. R., R. HARRIS and D. G. OWEN, 1982, <u>A Parametric Physical Model Study of the Low-Frequency Motions of Single-Point Mooring Systems</u>. Department of Offshore Engineering, Heriot-Watt University, Project Report No. 82/1.

JARDINE, I.J.A., 1979, <u>A Study of the Tension-Leg Concept for Single-Point Mooring Applications</u>, Ph.D. Thesis, Heriot-Watt University, Department of Offshore Engineering.

OWEN, D. G. and B. T. LINFOOT, 1976, "The Development of Mathematical Models of Single-Point Mooring Installations," Proceedings, Eighth Annual Offshore Technology Conference, Houston, Texas, U.S.A. 1977, Vol. I OTC 2490, pp. 689-700.

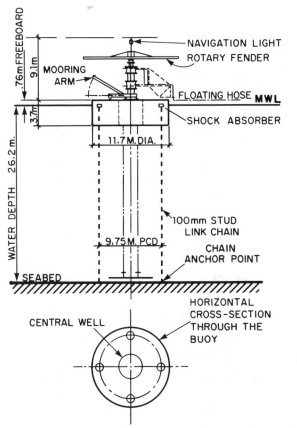

Figure 1. General arrangement and dimensions for the larger buoy.

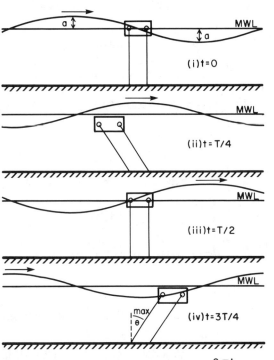

HEIGHT OF WATER ABOVE MWL = $a \sin \frac{2\pi t}{T}$
HORIZONTAL VELOCITY OF WATER PARTICLES = $u_o \sin \frac{2\pi t}{T}$
SWING OF BUOY = $\theta_{max} \sin \left(\frac{2\pi t}{T} - \pi\right)$

Figure 2. Illustration of the possible response of the taut-moored buoy to regular waves.

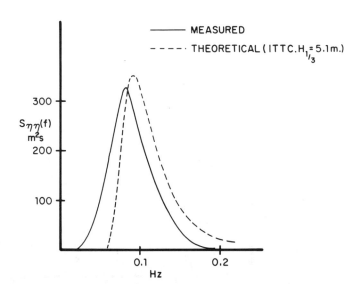

Figure 3. The measured and theoretical wave spectra for the irregular waves.

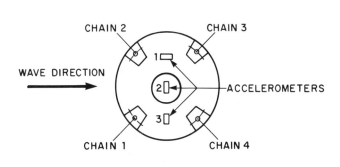

Figure 4. Location of accelerometers and position of anchor chains.

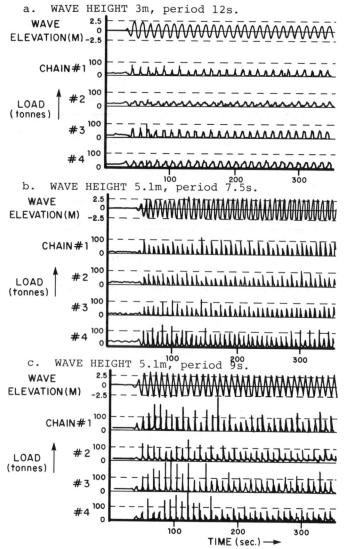

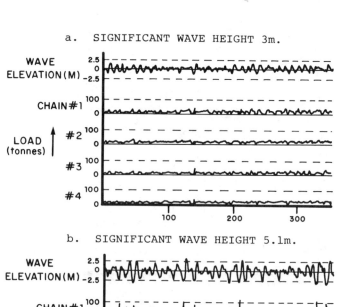

Figure 5. Regular waves: variation of wave elevation and chain loads with time for the larger prototype buoy. (pretension 20t. per chain)

Figure 6. Irregular waves: variation of wave elevation and chain loads with time for the larger prototype buoy. (pretension 20t per chain)

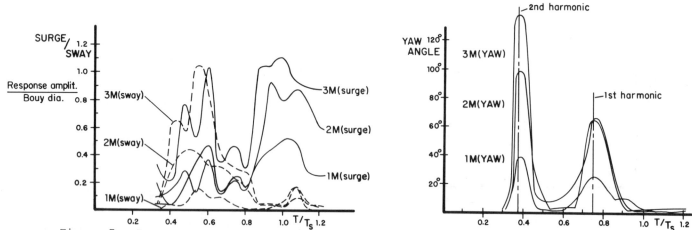

Figure 7. Response curves for a taut-moored surface riding buoy. buoy diameter 9.8m and depth 4.9m; water depth 25m; pretension 20t. per chain (4); shock absorbers 13.65 KN/mm natural period of surge 15s., of sway 15s., of torsion 11.2s.

VIBRATIONS OF PIPE ARRAYS IN WAVES

N.-E. Ottesen Hansen
LICconsult, Consulting Engineers Ltd.,
Denmark. Institute of Hydrodynamics
and Hydraulic Engineering, Technical
University of Denmark, Denmark.

SUMMARY

Hydroelastic vibrations of pipe arrays exposed to waves are considered. A pipe group is exposed to both inertial forces and drag forces from the surrounding wave motion. Superimposed on these forces are the hydroelastic effects such as vortex shedding, selfexcited vibrations (wake galloping) and turbulence. The paper deals with the vortex shedding aspect.

To present the method of analysis a simple pipe array of two pipes, one behind the other, is considered. The pipes are more and less aligned with the flow with the rear pipe in lee of the front pipe. The response to vortex shedding is analysed by 1) determining the force per unit length experimentally and 2) applying these forces in the equations of motion for the individual pipes in the array. Taking into account the correlation of forces along the length of the pipes the response of the full scale pipe array is then found.

For Keulegan-Carpenter numbers above 30-40 it is found that locking-on only appears when the velocity exceeds one half the amplitude of the relative velocity during the passage of waves. Further, provided that this criterion is satisfied, locking-on for cross vibrations only appears if the reduced velocity of the surrounding flow is between 4.8 and 7. The correlation length of vortex shedding along the pipes is found to be of the order of three diameters at small vibrations, but it is rapidly growing with increasing amplitudes. For amplitudes above half a diameter the vortex shedding is fully correlated over the length of the front pipe. On the rear pipe the correlation length is of the order of three diameters if placed 7-10 pipe diameters from the front pipe. Increasing the spacing the correlation length will gradually approach the values for free standing pipes.

A simple linear analytical model is suggested for calculating the response of fullscale pipe groups exposed to vortex shedding. It is shown, that the amplitude of the transverse force, generated by a vortex cell, is determined by a constant cross force coefficient $C_L \simeq 0.9$. The hydrodynamic damping is decreased in the locking-on region due to the interaction between vortex shedding and pipe vibrations. Applying the force and the damping with the correlation length of the vortex shedding the response of the individual pipes is found. It is shown, that in practice locking-on to vortex shedding is much less severe in waves than in steady current. The results indicate, that it is possible to design pipe arrays to sustain the locking-on in waves.

INTRODUCTION

Pipe arrays exposed to waves will under certain circumstances be vulnerable to hydroelastic vibrations. An example of a pipe array is a multibore riser connecting a production platform with the manifold at the sea bottom, Fig. 1. The pipe array will be exposed to two different types of vibrations:

1. The response directly to the drag and inertia forces arising from the wave motion. This response is shown by the fully drawn line of Fig. 1 and follows either the wave frequency or the natural frequency of the main riser string.

2. Superimposed on the direct response there can be hydroelastic response of the individual flowlines in some of the subspans. This response is indicated by the dotted lines of Fig. 1.

The hydroelastic response is caused by either vortex shedding locking-on, selfexcited vibrations (wake galloping), turbulence or by combinations of all these mechanisms. In either case they are able to produce stresses which are large enough to be considered in design.

The paper presents an analysis of the effects of vortex shedding on the individual pipes in a pipe group exposed to waves.

METHODOLOGY

For the moment the analysis of the interaction between pipes in a pipegroup when exposed to wave loads has to be based on empirical data to a certain degree. In this paper the following methodology is used:

1. The hydroelastic forces and responses per unit length of the pipe array are determined for the appropriate range of Reynolds numbers, Re, and Keulegan-Carpenter numbers, KC. This is done experimentally with short sections of the pipe array to a suitable scale.

2. The relative velocity between the pipes and the surrounding flow in the full scale pipe group system is determined. This is done for instance by standard computer programmes for analysis of risers exposed to waves. To do this analysis it is necessary to know the drag coefficient C_D and mass coefficient C_m for the pipe group.

3. With the relative velocity known the local Reynolds numbers, $Re(z)$ and Keulegan-Carpenter numbers, $KC(z)$, along the pipe group is found for each wave period.

4. With local $Re(z)$ and $KC(z)$ known the forces for the hydroelastic effects found in 1 are used to determine the forces over the depth of the full scale pipe groups. To do this correctly it is essential to know the correlation between the loads along the pipes.

5. Expressing the equations of motion for the individual pipes in the pipe group and applying the force distribution determined in point 4 the hydroelastic responses of the different pipes in the pipegroup are found.

In concept the method is rather straight forward but it involves a heavy experimental and computational work. To present the procedure of analysis in so simple a way as possible a more simple pipe group than that in Fig. 1 is considered in the following. The pipe group consists of two pipes placed, one behind the other, at a small angle to an oscillatory flow, Fig. 2. Of course some of the hydroelastic effects occurring in more complex pipe configurations will be lost, but the more basic ones are maintained, such as shielding of a rear pipe and vortex shedding interaction between pipes. The relative magnitude of these effects determined in the simple system will indicate what to expect in more complex systems.

EXPERIMENTS

The forces per unit length of the simple two pipe system of Fig. 2 are determined in an experimental set-up with flexibly mounted rather short, smooth pipes in an oscillatory flow. The test set-up is shown on Figs. 3 and 4 and is described in detail in Appendix 1. In this set-up forces, deflections, accelerations, correlation lengths of vortex shedding and pressures along the pipes are determined for different spacings of the cylinders, c, for different elasticities of the mounting of the cylinders, for different structural damping ratios, β, for different Reynolds numbers and for different KC numbers.

It should be noted that the test set-up only produces one-dimensional oscillatory flow. The effects due to the orbital motions in wave flow are not included. It is believed, however, that the test set-up produces sufficiently reliable results for the hydroelastic effects, at least for the higher KC numbers, because these effects only seem to depend on the velocity component normal to the pipe axis (at least so long as the velocity vector has an angle larger than 45^0 with the pipe, (King, 1977). Further the Reynolds number range is limited, but at least for locking-on effects this is no problem. It is demonstrated previously, (King, 1977), that vortex shedding effects at extreme high Reynolds numbers in nature can be accurately reproduced in laboratory tests even when the Reynolds number there is only a few thousands.

Typical results of the tests are shown in Figs. 5 and 6 for the two pipe groups exposed to steady current flow and oscillatory flow. It is seen that the pipes are vibrating violently due to vortex shedding locking-on. They vibrate with frequencies exactly equal to the eigenfrequency of the pipes. The responses of the front pipe are in all cases the more severe, probably because the vortices shed from the front pipe hit the rear pipe rather irregularly thus destroying the regular pattern of vortex shedding on this pipe. It is further seen from the figures that the locking-on starts when the velocity exceeds half the maximum velocity and that it stops when the velocity drops below the same value. In general the results which are valid for KC > 30-40 can be summarized as follows:

1. In oscillatory flow the front pipe is the more heavily exposed to vortex shedding locking-on, whereas the response of the rear pipe is weaker. The response of the front pipe corresponds closely to a free standing pipe without interaction so long as the spacing is above 7. D. The same general results are reached with steady current flow.

2. Vortex shedding locking-on only occurs in the range of relative velocity, V_r, of $4.7 < V_r < 7$. The frequency of the vibrations corresponds closely to the eigenfrequency of the pipes. Further the local velocity has to be above ½ of the peak velocity in a half period before locking-on can occur. This result was found in previous tests with multibore risers (Ottesen Hansen, et al, 1979).

3. Generally, the vortex shedding response in oscillatory flow is lower than in the similar steady state flow situation, because it takes time to develop the vibrations, Figs. 5 and 6. This is due to the fact that the system only gradually accumulates energy from the surrounding flow when building up the resonance response. Further the lack of correlation of the forces along the pipe will delay the growth of the vibration. In the present tests the latter effect is minor since the pipe sections used in the tests were short.

4. In-line locking-on did occur in the tests at lower reduced velocities, but the response was small and could be considered negligible for practical purposes.

A comprehensive description of the test results are presented in a detailed report,(Ottesen Hansen and Hvidbjerg Knudsen, 1982).

To apply the above results in the extrapolation to full scale pipe arrays exposed to waves the results summarized in points 1-4 have to be presented in a form where they can be applied in practical engineering vibration analysis.

The most simple approach is to assume that the problem is described by the normal harmonic equation, i.e. the displacements _transversely_ to the flow of the i'th pipe is given by:

$$m_i \ddot{x}_i + 2\beta_i(t) m_i \omega_{0i} \dot{x}_i + m_i \omega_{0i}^2 x_i = F_i(t, x_i, x_j) \tag{1}$$

in which m_i is the virtual mass, ω_{0i} the natural cyclic frequency of the i'th pipe for vibrations transverse to the flow. $\beta_i(t)$ is the damping ratio of the i'th pipe and $F_i(t,x_i,x_j)$ is the force acting on the pipe. t is the time. This is a more simple model than usually used as for instance the wake oscillator model, (Blevins, 1977). The justification for applying it is naturally that it fits the data. Using it in air has produced excellent results, (Hansen, 1980).

The damping ratio $\beta_i(t)$ consists of three contributions.

$$\beta_i(t) = \beta_{is} + \beta_{ih} - \beta_{iv} \tag{2}$$

in which β_{is} is the structural damping, β_{ih} is the hydrodynamic damping and β_{iv} is a negative damping arising from the interaction between the vortex shedding and the vibrating pipe at locking-on.

The individual terms in eq. 2 are given by the following expressions:

$$\beta_{is} \simeq \text{Constant, which is known} \tag{3}$$

$$\beta_{ih} \simeq \frac{\rho DL\, C_D}{4\pi\, m_i} \int_0^{2\pi/\omega_{0i}} \sqrt{\left(u(t) + \dot{u}(t)(\tau-t)\right)^2 + \dot{Y}_i^2(t)\sin^2(\omega_{0i}\tau)}\, \sin^2(\omega_{0i}\tau)\, d\tau \qquad (4)$$

$$\beta_{iv} \simeq \frac{\rho DL\, C_v}{\omega_{0i}\, m_i}\, u(t) \qquad (5)$$

in which ρ is the density of water, D the pipe diameter, L the pipe length, $C_D = C_D(c/D)$ the drag coefficient for the pipe, $C_v = C_v(c/D)$ a coefficient describing the negative damping between the vortex shedding and the vibration of the pipes. $u(t)$ is the relative velocity between the pipes and the surrounding flow and Y_i is the amplitude of the vibration. The eq. 4 may seem rather complicately but it only expresses that the damping of the vibration is caused by fluid drag, which acts in the direction of the instantaneous velocity and is proportional to the instantaneous relative velocity squared. For small vibrations and gradually varying flow $\dot{Y}$ and $\dot{u}(t)$ are small and eq. 4 takes the normal form of:

$$\beta_{ih} = \frac{\rho DL\, C_D}{\omega_{0i}\, m_i}\, u(t) \qquad (6)$$

Eqs. 4 and 6 are moving averages of the hydrodynamic damping (averaged over $2\pi/\omega_{0i}$).

Applying eqs. 1-6 it is found that the experimental results are described satisfactory by a value of $C_L \simeq 0.9$ and a value of $C_v \simeq 0.7$ for the front pipe. For the rear pipe the same values can be used although the response is smaller. This is contradictory, of course, but the reason is that the vortices are shed in cells which break up when convected downstream with the main flow, see Fig. 7. This implies that the rear cylinder is hit by vortices with different time lags over the length. The force coefficient for the transverse force amplitude is defined in eq. 16.

Inside each vortex cell it is reasonable to assume that C_L and C_v still maintain their values of 0.9 and 0.7 respectively. The time lag is taking into consideration by expressing the statistical average of the damping over the length and eq. 5 is therefore replaced by:

$$\beta_{iv} \simeq \frac{\rho D\, C_v}{\omega_{0i}\, m_i}\, u(t) \sqrt{\int_0^L \int_0^L \exp\left(-2\frac{|z_1-z_2|}{\ell_{ci}}\right) dz_1\, dz_2} \qquad (7)$$

in which $\ell_{ci} = \ell_{ci}(Y_i, \dot{Y}_i, KC/(c/d), Re)$. Y_i is the amplitude of the upstream pipe. The expression underneath the square root is simply the autocorrelation function, which is assumed to be given by an exponential function. The variation of ℓ_{ci} with the different parameters are shown in Fig. 8. There were found no systematic variation with the vibrational amplitude of the upstream pipe, nor with the Reynolds number, Re. The ratio $KC/(c/D)$, however, is important. For small values there is no effect, for large values the length scale is decreased. The expression in Fig. 8 is rather simplistic but is at least applicable in the presented c/D range ($c/D < 20$).

PIPE ARRAYS IN THE SEA

The experimental results for the forcing functions F_i are used to find the response in the full scale pipe arrays. The equation of motion of the i'th pipe transverse to the flow is given by

$$\frac{\partial^2}{\partial z_i^2}\left(EI\frac{\partial^2 x_i}{\partial z_i^2}\right) - \frac{\partial}{\partial z_i}\left(N\frac{\partial x_i}{\partial z_i}\right) + m_i\frac{\partial^2 x_i}{\partial t^2} + c_k\frac{\partial x_i}{\partial t} = F_i\left((t,z_i),Y_j\right) \qquad (8)$$

in which EI is the stiffness of the i'th pipe, N the normal force and c_k a coefficient describing the structural damping, which is assumed to be of the viscous type. Y_j is the amplitude of the other pipe. Assuming that the deformation can be expressed in terms of orthogonal eigenfunctions $\psi_k(z_i)$ the solution to eq. 8 has the form:

$$x_i = \sum_{k=1}^{\infty} Y_{ik}(t)\, \psi_k(z_i) \qquad (9)$$

Inserting eq. 9 into eq. 10, multiplying with $\psi_k(z_i)$ and integrating over the length of the pipes gives:

$$\ddot{y}_{ik} + 2\beta_{ik}(t)\omega_{oik}\dot{y}_{ik} + \omega_{oik}^2 y_{ik} = F_{ik}/m_{ik} \tag{10}$$

in which

$$m_{ik} = \int_0^{L_k} m_{ik}\psi_k^2 \cdot dz_i \tag{11}$$

$$\omega_{oik}^2 = \frac{1}{m_{ik}} \int_0^{L_k} \left(\frac{\partial^2}{\partial z_i^2}\left(EI \frac{\partial^2 \psi_k}{\partial z_i^2}\right) - \frac{\partial}{\partial z_i}\left(N \frac{\partial \psi_k}{\partial z_i}\right)\right)\psi_k dz_i \tag{12}$$

$$\beta_{ik} = \beta_{iks} + \beta_{ikh} - \beta_{ikv} \tag{13}$$

$$\beta_{ikh} \simeq \frac{\rho D}{4\pi m_{ik}} \int_0^L c_D \psi_k^2 \overline{\int_0^{\frac{2\pi}{\omega_{oik}}} \sqrt{\left(u(t,z) + \dot{u}(t,z)(\tau-t)\right)^2 + \dot{Y}_i^2(t)\psi_k^2 \sin^2(\omega_{oik}\tau)} \sin^2(\omega_{oik}\tau) \, d\tau} \, dz_i \tag{14}$$

$$\beta_{ikv} = \frac{\rho D C_v}{\omega_{oik} m_{ik}} \sqrt{\overline{\int_{L_{1k}}^{L_{2k}} \int_{L_{1k}}^{L_{2k}} u(t,z_1)\psi_k(z_1) u(t,z_2)\psi_k(z_2) \exp\left(-\frac{2|z_1-z_2|}{\ell_{ci}}\right) dz_1 dz_2}} \tag{15}$$

in which eq. 14 and eq. 15 are further extensions of eq. 4 and eq. 7 to modal analysis. Eq. 15 expresses that β_{ikv} is the statistical expectation of the different vortex action down along the pipe. $L_{1k}(t)$ is the lower boundary of the vortex shedding region for the k'th mode and $L_{2k}(t)$ is the upper boundary of the locking-on region for the pipe as defined in point 2 of the summary of the experimental results in the preceeding chapter. ℓ_{ck} varies as depicted in Fig. 8. β_{iks} is the structural damping ratio, which is assumed known. F_{ik} is just like the negative damping ratio $-\beta_{ikv}$ dependent on the correlation between the vortex shedding over the length of the pipe considered. It is therefore natural to express the force by a similar type of expression as β_{ikv}:

$$F_{ik} = \sqrt{\overline{\int_{L_{1k}}^{L_{2k}} \int_{L_{1k}}^{L_{2k}} \tfrac{1}{2}\rho C_{Li}(z_1) u^2(t,z_1) D \tfrac{1}{2}\rho C_{Li}(z_2) u^2(t,z_2) \cdot D \exp(-2|z_1-z_2|/\ell_{ci}) dz_1 dz_2}} \sin(\omega_{oik}t) = F'_{ik}(t) \sin(\omega_{oik}t) \tag{16}$$

The $\sin(\omega_{oik}t)$ stems from the fact that the pipe vibrates with its natural frequency at locking-on. C_{Li} is the force coefficient for the amplitude of the transverse force.

Solving the system of eqs. 8-16 it has to be done for each mode. In practice a long series of different modes has to be considered as shown on Fig. 9 and each of them have to be analysed by means of the above system of equations. Not all modes will consist of orthogonal eigenfunctions but still the results of the method are sufficiently good approximations.

$\beta_{ik}(t)$ in eq. 10 is in vortex shedding a rather small quantity usually never above a value of 0.2. Assuming $\beta_{ik}(t)$ small the following approximative solution to eq. 10 can be defined:

$$Y_{ik} = \exp\left[-\int_{t_b}^{t} \beta_{ik}(\tau)\omega_{oik}\,d\tau\right] \left\{\int_{t_b}^{t} \frac{F'_{ik}(t')}{2m_{ik}\omega_{oik}}\right.$$

$$\left.\exp\left[\int_{t_b}^{t'} \beta_{ik}(\tau)\omega_{oik}\,d\tau\right] dt' + Y_{ik}(t_b)\right\} \sin(\omega_{oik}t) \tag{17}$$

in which t_b is the time at which the locking-on starts and Y_{ik} is the amplitude of the vibrations. Eq. 17 is easily calculated numerically.

Eq. 17 is only valid for the locking-on regime. Outside the locking-on regime it is not valid. On the other hand the right hand side of eq. 10 is zero in this case. Assuming again that $\beta_{ik}(t)$ is a reasonable small quantity the decay of the vibrations outside the locking-on region can be found to:

$$Y_{ik} = Y_{ik}(t_s) \sin(\omega_{oik}t) \exp\left\{\int_{t_s}^{t} 2\omega_{oik} \int_{t_s}^{t'} \beta'_{ik}(\tau) \sin\left(2\omega_{oik}(t'-\tau)\right)\right.$$

$$d\tau\,dt' + \frac{1}{2\omega_{oik}^2} \left.\frac{d\beta'_{ik}}{dt}\right|_{t=t_s} \left(1-\cos(2\omega_{oik}(t-t_s))\right)$$

$$\left. + \frac{1}{2\omega_{oik}^2} \beta'_{ik}(t_s) \sin\left(2\omega_{oik}(t-t_s)\right)\right\} \tag{18}$$

t_s is the time at which the locking-on stops. $\beta'_{ik}(\tau)$ is defined by:

$$\beta'_{ik} = \beta_{iks} + \beta_{ikh} \tag{19}$$

Eqs. 17 and 18 determine the development of vibrational amplitudes at the passage of waves.

The proposed calculation model is rather simple and in fact it fits the experimental values remarkable well.

In Fig. 10 a comparison is made between current induced vibrations of pipes with the present theory. Applying the expressions for the correlation length shown in Fig. 8 and assuming that the slenderness ratio of piles used in the tests was $L/D \simeq 20-30$ the fit is rather perfect. The important parameter here is the correlation length. It should further be noted that an increase in C_D takes place at locking-on which of course is important for the damping.

To determine the relative importance of the different effects two calculation examples are made.

Calculation Example 1

A pipe group consisting of two pipes with diameter $\emptyset$ 100 mm, pipe wall thickness $t = 12$ mm and spacing $c = 1.0$ m is suspended vertically from a platform at a site with 100 m water depth. It is exposed to waves with height $H = 28.0$ m and period $T = 17.0$ s. The tension in each of the pipes is 600 kN. The ratio of structural damping β_{iks} is assumed to be 0.003.

It is assumed that the pipe group vibrates in a mode of 6 sine functions down the depth as shown on Fig. 11. It is assumed that the relative velocity between the pipe and the flow is approximately 0.75 times the absolute velocity in the waves when calculated by the first order Airy theory.

The analysis of the vibration is made for three different expressions for the correlation length ℓ_c in order to determine the effect.

$$\ell_c = \infty \cdot D \tag{20}$$

ℓ_c as shown in Fig. 8 (expression 1) $\tag{21}$

ℓ_c varying 10 times slower with Y_i than shown in Fig. 8 (Expression 2) (22).

The results are shown in Fig. 11 and they show the major importance of an accurate description of the correlation length. Applying the expressions for the correlation length of the front pipe and the rear pipe as presented in Fig. 8 the amplitude of vibration developes as shown in Fig. 12. It is seen that the amplitudes obtained are far below those experienced in a similar constant current, where the amplitude $Y \simeq 2D$. The main reasons are that locking-on only takes place over part of the depth due to the hyperbolic cosine velocity profile and that the development of the vibrations is slowed down by the lack of correlation.

Calculation Example 2

The environmental conditions are the same as in Example 1, but now the two pipes are connected with spacers as shown on Figs. 3 and 13. The distance between the spacers is 16.67 m. The results are shown in Figs. 13 and 14. Here the response is somewhat higher than in Example 1 because the locking-on takes place over the full length of the mode.

CONCLUSION

The vortex shedding locking-on in pipe arrays can be described by a simple linear calculation model. The decisive parameter for practical use is the correlation length of the vortex shedding.

APPENDIX 1

The experiments with oscillatory flow and steady state flow were made with a motor driven, 10 HP, carriage running on rails along a 3 m wide flume, Figs. 3 and 4. The pipes were flexibly suspended in the carriage by means of an arm system which held the pipes vertically but allowed free movements in the two horizontal directions. The pipes were held by a strong spring system, which could work in the two horizontal directions. The structural damping of the spring system could be adjusted. The tests were made by oscillating the carriage back and forth.

REFERENCES

BLEVINS, R.D., 1977, Flow-induced Vibration. New York: Van Nostrand Company.

HANSEN, S.O., 1980, Wind Loads on Chimneys, Part 1 and 2 (in Danish), The Technical University of Denmark, Structural Research Laboratory, Report No. R 124.

KING, R., 1977, "A Review of Vortex Shedding Research and its Application", Ocean Engineering, Volume 4, pp. 141-171.

OTTESEN HANSEN, N.-E., V. JACOBSEN, and H. LUNDGREN, 1979, "Hydrodynamic Forces on Composite Risers and Individual Cylinders", Proceedings, 1979 Offshore Technology Conference, Houston, Texas, Volume 3, Paper OTC 3541, pp. 1607-1621.

OTTESEN HANSEN, N.-E., V. JACOBSEN and G. LARNÆS, 1981: "Hydro-elastic Instability of Pipe Arrays in Waves", Proceedings, HOE'81, Trondheim, Norway, 1981, Volume 1, pp. 289-306; Trondheim, Norway: The Norwegian Institute of Technology.

OTTESEN HANSEN, N.-E., 1982, Hydro-elastic Vibration of Flexible Pipe Arrays in Oscillatory Flow. The Technical University of Denmark, Institute of Hydrodynamics and Hydraulic Engineering, Series Paper No. 31.

SARPKAYA, T. and F. RAJABI, 1979, "Dynamic Response of Piles to Vortex Shedding in Oscillating Flows", Proceedings, 1979 Offshore Technology Conference, Houston, Texas, Volume 4, Paper OTC 3647, pp. 2523-2528.

SAPRKAYA, T., F. RAJABI, M. ZEDAN and F.J. FISCHER, 1981, "Hydroelastic Response of Cylinders in Harmonic and Wave Flow, Proceedings, 1981 Offshore Technology Conference, Houston, Texas, Volume 1, Paper OTC 3992, pp. 383-390.

ACKNOWLEDGEMENTS

The tests were funded by The Danish Council for Scientific and Industrial Research.

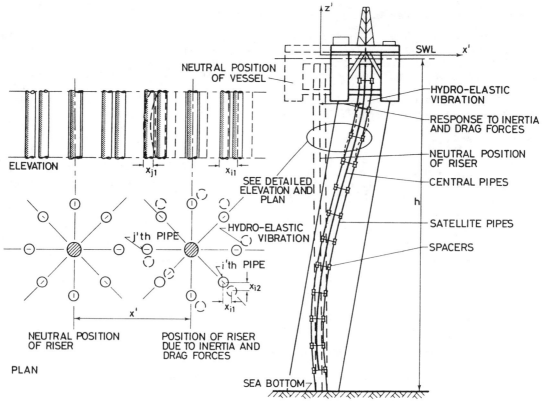

Fig. 1. Multibore production riser exposed to wave loads.

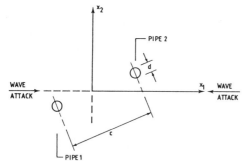

Fig. 2. Simple pipe group used in the calculations and presentation of results.

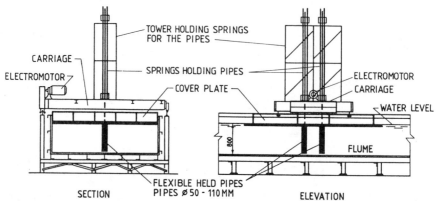

Fig. 3. Test set-up. The pipes are elasticly supported by large springs and held vertically by an arrangement of arms placed in the cover plate (the arms are not shown).

Fig. 4. Test set-up seen from flume. The illuminated pipes beneath the cover plate is the test pipes.

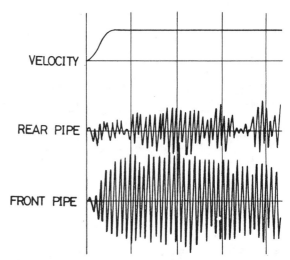

Fig. 5. Response crosswise to flow of a two pipe system exposed to current. Angle $10°$, Re $=5\ 10^4$, c/D = 7.2, L/D = 15.

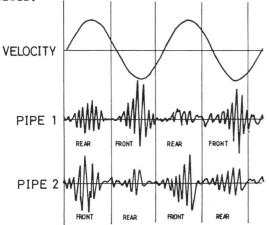

Fig. 6. Response of a two-pipe system exposed to oscillatory flow. Angle $10°$, Re = $5\ 10^4$, KC = 88, c/D = 7.2, L/D = 15.

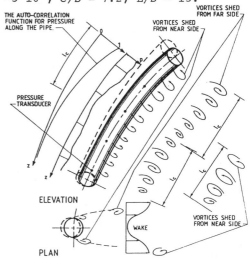

Fig. 7. The correlation length of vortex shedding.

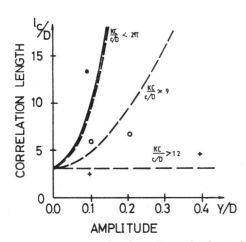

Fig. 8. The correlation length for the two-pipe system of Fig. 2.   Front pipe, Rear pipe. Limits KC/(c/D) applies to rear pipe.

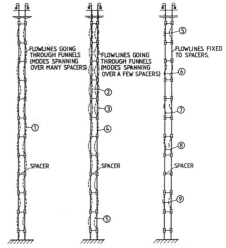

Fig. 9. Typical natural modes in a two-pipe system.

649

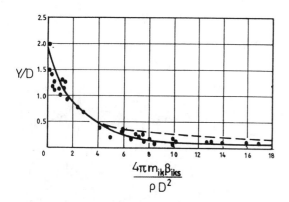

Fig. 10. Comparison between measured vortex shedding response of piles in current (King,1977) and theory presented in this paper. —— l_c shown in Fig. 8. — — $l_c = \infty$.

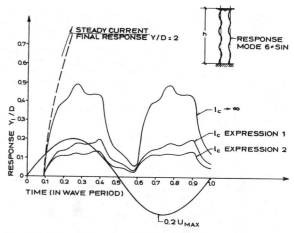

Fig. 11. Calculation example 1. The dependance of l_c on the response. Expressions 1 and 2 are defined in eqs. 20 and 21.

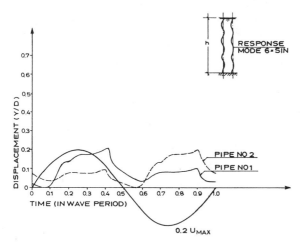

Fig. 12. Calculation example 1. Response of pipes for l_c shown in Fig. 8.

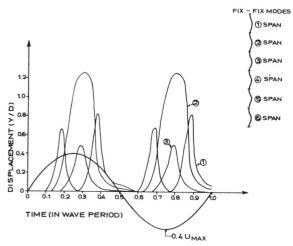

Fig. 13a. Calculation example 2. The response in the different spans. Span 1 is the uppermost span.

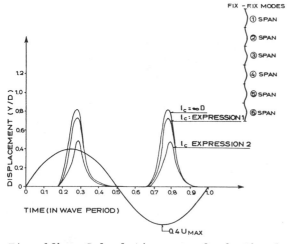

Fig. 13b. Calculation example 2. The dependance of l_c on the response. Expressions 1 and 2 are defined in eqs. 20 and 21.

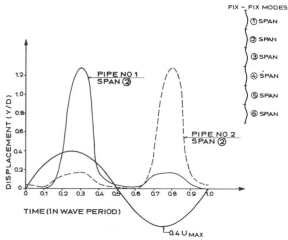

Fig. 14. Calculation example 2. Response of pipes for l_c shown in Fig. 8.

DYNAMIC BEHAVIOUR OF ANCHOR LINES

Carl M. Larsen, dr.ing.
The Norwegian Institute of Technology

Ivar J. Fylling, dr.ing.
The Ship Research Institute of Norway

SUMMARY

This paper deals with dynamic analysis of anchor lines. A general discussion of the problem is given, and computer programs developed for dynamic analyses are briefly described.

The paper is based on results obtained during the last five years through research and commission work at the Norwegian Institute of Technology and The Ship Research Institute of Norway.

Examples of comparison between analytical and experimental results are given and a simple way of treating geometrical nonlinearities is demonstrated. A series of results describing the dynamic behaviour of wire lines demonstrates that design stage estimates of dynamic loads in such lines can be obtained in a simple manner.

A semi-linear method for analysing the dynamic nonlinear behaviour of a clump weight system is described. The method is based upon a time domain integration and correction of final results by use of the known static nonlinear behaviour.

Results from a fatigue study of the wire in a clump weight system is reported, and the dynamic behaviour of this sytem is discussed in detail.

1. INTRODUCTION

1.1 Background and previous work

The dynamic behaviour of anchor lines has traditionally not been regarded as an important aspect of the practical design and operation of mooring systems. In most analyses of such systems the anchor line is modelled as a static element, i.e. the cable force is determined by the terminal point position only. This socalled quasi-static model is in most cases sufficient in order to describe the vessel motions. It is, however, in current practice also applied to determine mooring line design loads. This load is not to exceed a specified portion of the line breaking strength. The margin between allowable, quasi-static load and breaking strength shall cover 1) dynamic loads in addition to the quasi-static load, and 2) uncertainty of the actual line strength. The first of these items is the subject of this paper.

The dynamic behaviour of anchor lines has gained an increasing interest in the last few years. This is due to several reasons.

- It is desired to reduce the frequency of anchor line breakages. During the years 1979-1980 the breakage frequency in the North Sea was approximately one breakage pr. platform pr. year /1/.

- New types of anchor systems are being proposed incorporating new components such as clump weights, buoys, or new materials. These systems are expected to show a dynamic behaviour different from that of traditional systems.

- Ancor systems are proposed for permanent production facilities. The relative importance of dynamic loads will - particularly due to fatigue - increase with increasing service life.

- With increasing awareness of risks involved in the offshore petroleum industry the demand for reliability assessment is increasing.

- Development of computation tools has brought the cost of dynamic analysis of non-linear, multi-degree of freedom systems down to a moderate level.

A large number of papers have been written about methods for analyzing dynamic cable response. Surveys of available literature can be found in /2,3,4/. The non-linear nature of the problem has, however, made it difficult to give comprehensive and generally valid results. For the same reason general verification of theoretical results and computer programs are difficult to obtain. Examples of experimental results, partly compared to analytical results are given in /5,6,7/. All of these works are concerned with mooring chain for which the geometric deformation is much more important than elastic deformation. All available results indicate that typical mooring lines have a large damping due to lateral drag resistance. Thus, the hydrodynamic drag forces are generally more important than the inertia forces. A typical picture is that the dynamic force for a given amplitude increases uniformly with increasing frequency within the frequency band of wave excitations.

1.2 Scope and content of the paper

The present paper is based on results obtained during the last five years through research and commission work at the Norwegian Institute of Technology and the Ship Research Institute of Norway.

First a brief review of the nature of the dynamic cable problem is given as a background for the presentation of the computer programs that have been developed.

A few examples of comparison between analytical and experimental results are given and a simple way of treating geometrical non-linearities is demonstrated.

A series of results describing the dynamic behaviour of wire lines demonstrates that design stage estimates of dynamic loads in such lines can be obtained in a simple manner.

A fatigue analysis of the wire in a clump weight system for a deepwater floating production platform is presented, and the last section is devoted to a more detailed discussion of the dynamic behaviour of this system.

2. CHARACTERISTICS OF THE ANCHOR LINE PROBLEM

Figure 1 gives a principal view of important effects for a typical anchor line as applied for mooring of offshore structures. At a given static (mean) position, the line will have a static configuration defined by the anchor and vessel position, line properties and forces from current. The line will, however, also experience tension variations caused by vessel motions and dynamic effects from current and waves. The static configuration is the basis for the dynamic analysis discussed in the following.

2.1 External dynamic effects

The vessel motions will cause tension variations in the mooring line. Normally first order vessel motions are calculated by methods assuming the mooring lines to be linear springs on the vessel. This can be done with acceptable accuracy as dynamic and non-linear effects in the lines will have negligible effects on the motions. Consequently, these motions are known previous to the dynamic anchor line analysis and can be considered as given displacements at upper end.

Concerning slowly varying vessel motion due to non-linear wave forces and wind, the non-linear restoring force/displacement relation can be significant. Such motions will therefore often be calculated by assuming a non-linear quasi-static mooring line characteristic. The periods of slow-drift oscillations are normally much longer than the natural periods of a mooring line (minutes compared to Õ 10 seconds). A consequence of this is that slowly varying vessel motions need not be considered in a dynamic mooring line analysis, but should be regarded as a change of static position.

Current is normally regarded as a static phenomenon. For two reasons, however, current should be considered in a dynamic analysis. Firstly, the quadratic drag term in the wellknown Morison's formula requires all velocity components to be known in order to calculate dynamic drag forces. Secondly, current will give vortex shedding and hence lateral vibrations in the cable. Such vibrations can further cause a dramatic increase of the drag coefficient. An example of this effect is that in dynamic analysis of anchor lines, in-line drag coefficients are typically in the range 1.5-2.5 rather than 1.0-1.2 as often used for force on rigid cylinders.

Velocities and accelerations in water caused by waves will give lateral forces and hence dynamic cable response. Because of small cable diameters the accelerations related forces can be neglected.

2.2 Non-linear effects

The aim of a dynamic analysis of an anchor line is normally to calculate the line tension variations. This response is non-linear due to several effects. Most important is the non-linear function relating tension to displacement at upper end. This non-linearity is easily demonstrated by discussing the static problem. Figure 2 shows a line at a given position with end tension T_0, and the non-linear function relating tension to change of horizontal position x.

The stiffness K at x = 0 has the following contributions

- elasticity of the cable
- geometric change of the free hanging cable
- increase of free cable length

The contribution from elasticity of cable resting on bottom is dependant on the friction coefficient. By neglecting friction, the total bottom line will be tensioned and hence give an unrealistic large flexibility. Most synthetic fibre ropes will have non-linear elasticity, and the two last contributions will always give non-linear effects.

It the cable was fixed at the contact point for x = 0, the stiffness would increase as indicated by the dotted line on Figure 2b. This increase is caused by the loss of flexibility associated with the line resting on bottom and change of position of the bottom contact point.

Another important non-linear effect is the water/structure interaction. For the cable element shown in Figure 3, forces are found from Morison's equation and given by:

$$F_n = F_d + F_i$$
$$F_d = \frac{1}{2} \rho C_D D \ | v_{n,rel} | \ v_{n,rel} \tag{1}$$
$$F_i = C_M \frac{\pi D^2}{4} a_{n,f} - \rho (C_M - 1) \frac{\pi D^2}{4} a_{n,s}$$

where:

F_n : Total force normal to element axis

F_d : Drag force

F_i : Inertia force
ρ : Density of water
D : Diameter of cable
C_D : Drag coefficient
$v_{n,rel}$: Relative velocity normal to element
C_M : Inertia coefficient
$a_{n,f}$: Acceleration of fluid normal to element
$a_{n,s}$: Acceleration of structure normal to element

Friction effects for motions along the element axis is neglected. As previously mentioned, the first term of F_i is normally neglected, while the second term is known as "added mass".

The drag force includes the square term which means that hydrodynamic damping cannot be described by a conventional damping coefficient or matrix in a differential equation to describe dynamic equilibrium. It is also important to note that both F_d and F_i will show a non-linearity caused by the fact that the forces are dependant on velocities and accelerations <u>normal</u> to the element. Consequently, dynamic changes in the tangent angle α (see Figure 3) will cause variations in drag forces from a constant current and also effect added mass related to the global axes x and z.

2.3 Finite element model

The mathematical model used for dynamic mooring line analysis in the present work is based upon the finite element method, and a time domain step-by-step integration procedure is applied.

The differential equation describing dynamic equilibrium can then be written:

$$M \ddot{x} + C \dot{x} + K x = F(t, \dot{x}) \tag{2}$$

where:

- M : Mass matrix of system, including structural mass and hydrodynamic mass. A lumped mass matrix is here used, which means that M is a diagonal matrix.

- C : Damping matrix of system. Hydrodynamic damping is the dominant damping contribution and is here treated as forces. Consequently, C will include internal damping in the material and geodynamic damping in the sea floor. For most problems, C can be omitted.

- K : Stiffness matrix. The element used is the conventional bar element included geometric stiffness proportional to the tension in the element.

- $\ddot{x}, \dot{x}, x$: Acceleration, velocity and displacement vector respectively.

In the present work all matrices on the left hand side of Eq. (2) are constant. The only non-linear term considered is the response dependant drag forces. This gives certainly a favourable situation with regard to computing time, specially for stochastic time simulations, but will introduce errors. Within a linear solution procedure some effort can however be done in order to reduce this error.

An important feature of a finite element model is stiffness K relating tension variation to displacement at upper end, see Figure 2. If the element model is terminated at the bottom contact point and fixed there, this stiffness will be identical to the tangent of the dotted line on the figure, K'.

By introducing elements on the bottom and linear springs as indicated on Figure 4, the flexibility of the model will be increased. By a proper choice of the magnitude of the bottom spring stiffness, a correct initial stiffness for the finite element model can be obtained.

The non-linear T/x-relation shown on Figure 2 cannot be directly accounted for. The static tension is, however, easily obtained for any position of upper end, and can be used in order to correct the results. This correction is illustrated on Figure 5.

The static line tension is given by T_s, and dynamic variations are calculated for given displacements at upper end, x, and other external effects. The result from a dynamic analysis is given by ΔT_{DT} and have two contributions

ΔT_K : a quasistatic contribution defined by the linear stiffness

ΔT_D : a dynamic contribution caused by drag forces and inertia effects.

ΔT_K is obviously incorrect and can easily be substituted by the correct quasistatic value ΔT_Q. The total force will normally be

$$T_T = T_S + \Delta T_{DT} \tag{3}$$

but by introducing this correction, we have

$$T_{TC} = T_S + \Delta T_Q + \Delta T_D \tag{4}$$

The solution will still have errors from the non-linearity of the cable subjected to lateral inertia and drag forces and the effect of tension variations on the stiffness, but the proposed correction is certainly an important improvement.

3. BRIEF DESCRIPTION OF COMPUTER PROGRAMS

Three computer programs have been developed. All are based upon the finite element method using a conventional truss/bar element with elastic and geometric stiffness /8/. A time domain step-by-step integration procedure known as Newmark's β-family /9/ is applied. The static configuration for an unloaded line is found by a rapid transfer matrix iteration technique /10/ and is the basis for the dynamic analysis. Non-linear drag forces are accounted for by a simple double step procedure described in /11/. This procedure has shown to give results of acceptable accuracy compared to iterative techniques.

LINDYN /12/ performs a harmonic, two-dimensional analysis of an anchor line. Linear wave theory is applied to analyse effects from waves propagating in the plane defined by the line. In addition to a dynamic analysis, eigenfrequencies and mode shapes can be calculated. The program makes no use of external storage.

STOCCA /13/ is a further development of LINDYN and performs a stochastic, two-dimensional analysis. Stochastic motions of upper end can be taken from files generated by other programs or generated by the program by transfer functions and wave speatra. Series for tension variation are further used in a fatigue evaluation using the Miner-Palmgren approach for cumulative fatigue damage. Fatigue curves provided by British Ropes /14/ are built into the program. These curves give fatigue life (number of cycles) for varying amplitudes, mean tension and line breaking load. Slowdrift oscillations can be accounted for as a change in mean tension, and the quasi-static correction described in Sec. 2.3 is performed taking the total motions into consideration.

HATCAN /15/ is also developed from LINDYN but in another direction than STOCCA. HATCAN can perform a harmonic, three-dimensional analysis of a line and include current and waves from arbitrary directions. Buoys and concentrated weights can be considered at any position along the line. Eigenfrequencies and mode shapes can be calculated, and friction between the line and the sea floor is accounted for in the static as well as in the dynamic analysis.

4. EXAMPLES OF RESULTS OBTAINED FROM EXPERIMENTS AND ANALYSIS USING LINDYN

4.1 Horizontally suspended submerged cable - comparison with model tests

In order to give a base for verification of dynamic analysis program, a series of model tests was carried out in the small basin at the Ship Model Tank in Trondheim /16/. The model setup is shown in Figure 6.

A 10 m span of 10 mm diameter steel wire was suspended between a fixed point and an excitation mechanism. The cable was subjected to harmonic end-point oscillations covering a range of frequencies from 0 to 2 H_z and amplitudes of 50, 75 and 100 mm. Referring to Figure 7 it can be seen that while 50 mm amplitude gives only a moderate geometrical non-linearity, the 100 mm amplitude gives a strong non-linearity.

Analytical results were obtained by means of the finite element program LINDYN described in Chapter 3.

Comparison of experimentally and theoretically determined cable force /17/ indicated that the theoretical model gives satisfactory results for a wide range of frequencies and amplitudes.

Figure 8 shows results for the largest amplitude. Referring to Figure 7 the cable behaviour is strongly non-linear at this amplitude. The theoretical model seems, however, to yield a fairly good description of the frequency dependence of the force even at this amplitude. At zero frequency the experiment reflects the static behaviour of the cable. With increasing frequency, the maxima increase up to the range of the applied force transducers. The minima decrease towards zero, indicating that the cable becomes slack. Negative tension at high frequencies are due to finite bending stiffness of the cable giving a small compression instead of the ideal slack condition.

For moderate amplitudes and frequencies, the correspondence is good for both minima and maxima. As the dynamic stress range with increasing frequency approaches twice the static load, both minima and maxima appear to be under-estimated. The total dynamic stress range appears, however, to be fairly well reflected by the analytic model.

It could be expected that the validity of the linear stiffness model of LINDYN would be limited to cases with nearly symmetric responses, i.e. maxima and minima symmetric about the average force. It appears, however, that the theoretically predicted maxima coincide with the experimental results far beyond this range, although the theoretical model gives compression in the cable.

The analytical model does not include the non-linearity of the stiffness characteristic. This is evident from the large-amplitude case, Figure 8. A simple correction for this phenomenon has been included in the latest version of the program (see Section 2.3). This is simply a compensation for the static non-linearity as expressed in the static force-displacement characteristic. Figure 8 indicates that this correction will relax the amplitude-restriction and allow larger non-linearities.

4.2 Dynamic behaviour of full-scale anchor lines

In order to increase the knowledge about the dynamic behaviour of anchor lines, a series of analyses of wire- and chain lines at various water depth and tension levels have been carried out. The results presented here were obtained as part of a research project sponsored by Det norske Veritas and the Norwegian Rig Owners Association /18/.

Wire lines

The analysis is performed for ordinary 76 mm steel wire. Breaking strength for this dimension is typically 3700 kN. Currently recommended maximum load for anchoring systems, is 123 tons in a normal working conditions, and 185 tons in survival conditions. These levels are based on traditional quasi-static analysis, i.e. static response of the line subjected to forced end-point motions.

In the present analyses water depths from 70 to 300 m and tension levels from 70 to 220 tons were included.

Figure 9 shows a typical result. The response appeared to be nearly proportional to the amplitude, so the results are given as response to excitation amplitude ratio. Starting at a level corresponding to a static response, the dynamic force increases uniformly up to a level corresponding to a purely elastic deformation. This is due to the hydrodynamic resistance against lateral motions. Another interesting feature of the results is the difference between top- and bottom forces. The increasing difference with increasing frequency indicates the effect of longitudinal inertia forces.

The apparent elastic behaviour over the interesting frequency range has been utilized in preparing Figure 10 as a summary of the behaviour in 150 m water depth for low to high tension levels. It is interesting to note that the elastic behaviour seems to cover the most important frequency range of wave-induced motions for all tension levels. Thus, the dynamic force is only weakly influenced by the static tension level in the wire line.

The "irregularity" of the curves at high frequencies is due to axial dynamic effects. A drop in the upper end tension indicates a tendency to resonance: a smaller force is required to drive the oscillation with a given amplitude. A corresponding drop may not occur at the bottom end. In a resonant situation with forced motion the tension may be significantly larger at the fixed end than at the moving end. In this frequency range the dynamic behaviour can not be properly observed by measurements on the platform. This fact may cause problems for the design and monitoring of deep water (more than 300 m) anchoring systems.

A preliminary conclusion is that in moderate water depths (up to 300 m) a conservative estimate of dynamic response is obtained by simply neglecting the geometric deformation. This approach is even more simple than the traditional quasi-static approach since the dynamic part of the response is linear, and vessel motion spectra may be converted directly to cable force spectra.

The difference between a quasi-static analysis and an elastic dynamic analysis is illustrated in Figure 11. As the total stiffness, K, asymptotically approaches the elastic stiffness, K_E, with increasing load, the difference between the two estimates of maximum force, F_{max}, decreases with increasing value of the average force.

Chain lines

A series of analyses were also carried out for 76 mm chain lines. These lines are used at a much lower stress level than wire lines, and the elastic deformation gives only a small contribution to the static force-displacement characteristic.

Examples of results are given in Figure 12. The results demonstrate that the elastic stiffness plays an important role in determining the range of dynamic loads. The response is, however, considerably more non-linear than for wire lines, and no "flat" range corresponding to purely elastic deformation is observed.

The two line lengths used in Figure 12 reflects the consequence of including or excluding the elastic deformation of 400 m of chain assumed to be resting on the bottom.

The study of the dynamic behaviour of chain lines is being continued in order to study computation methods and evaluate a better base for design criteria.

5. FATIGUE ANALYSIS OF CLUMP WEIGHT SYSTEM USING STOCCA

5.1 Description of the system

The work reported in this section was carried out as a commission work for "Norwegian Contractors" and published with their allowance. The work is entirely reported in /19/.

The purpose of the work was to perform a fatigue analysis of the anchor lines used for a concrete production platform, CONPROD, on 350 meters water depth.

The motions of the platform were found in a separate analysis /20/ and was supplied as data files with time series for low- and high-frequency motions generated from identical waves. 9 seastates were analysed with significant wave heights increasing from 2 to 15 meters. Effects from wind and current were included as a change of mean platform position.

The anchor system of the platform consists of 16 lines. Each line has 3 segments with main properties displayed in Table 1. Figure 13 gives a view of the mooring line and the static relation between line tension and offset of the platform. The second segment acts as a clump weight and is built up from 8 heavy chains in parallell. The intention of the design is that the clump weights on all lines should be partly in contact with the sea floor when no horizontal forces are acting on the platform.

5.2 Fatigue analysis

Dynamic cable analyses were carried out for 5 directions of wind, waves and current relative to the mooring line of concern, see Figure 14. The motions for center of gravity of the platform was transformed to vertical and horizontal motions at upper cable end (fairlead) in the plane defined by the cable. The static position of upper end was found from the static platform offset and wave/wind/current direction. Hence, direction 1 on Figure 14 always gave maximum line tension for a given sea state, while minimum was found for direction 5. For direction 3, vertical motions only were present as motions normal to the line plane were neglected. (Two-dim. analysis).

In the dynamic analyses high-frequency platform motions were used as upper end displacements. The results were modified in order to take into account the non-linear tension/displacement characteristic as previously described. In this modification the actual low-frequency motions were added to the static position and hence a tension including all effects was found. Simulation period was 30 minutes.

Data for fatigue capacity of the wire gave fatigue life in terms of number of cycles before failure for varying amplitudes and mean tension levels. In the fatigue evaluation the amplitudes were defined as the contribution from high-frequency motions only, using the low-frequency motions to define mean tension at every cycle. In addition tension cycles from low-frequency motions were counted separately, using the mean platform position to define mean tension.

Fatigue damage for the line in a sea state S and wave direction α was found using the well-known Miner-Palmgren's hypothesis:

$$D(S,\alpha) = \sum_{i,j} \frac{n_i | \sigma_{m_j}}{N_i | \sigma_{m_j}} \qquad (5)$$

where:

$D(S, \alpha)$: Fatigue damage for simulated period, sea state S and direction α.

$n_i | \sigma_{m_j}$: Number of cycles in tension range interval i and mean tension j.

$N_i | \sigma_{m_j}$: Fatigue capacity (cycles before failure) for the same tension range and mean tension.

Accumulated damage for the platform lifetime T (20 years) was then found by summing up the damage from all sea states and wave directions, taking the probability of occurrence for the specified sea state into account:

$$D_{TOT} = \frac{T}{\Delta t} \sum_{i,j} D(S_i, \alpha_j) \cdot P(S_i, \alpha_j) \qquad (6)$$

where:

Δt : Length of simulation period.

$D(S_i, \alpha_j)$: Damage for sea state S_i and direction α_j.

$P(S_i, \alpha_j)$: Probability of occurrence for the same sea state and direction.

The result indicates a fatigue damage for the proposed system of the order of magnitude 10^{-3}. From this very low figure one should conclude that fatigue of the wire line itself is not of vital interest when designing mooring systems of the actual type. However, important uncertainties exist, which will be discussed in the following.

5.3 Discussion of results

The method for dynamic cable analysis applied in the present study was two-dimensional, semi-non-linear, and forces from waves and current on the line were neglected. These effects will be discussed more in detail in the last section of this paper, where it is concluded that the applied method as such should give results of acceptable accuracy for a fatigue analysis.

Uncertainties are mainly related to the applied data for fatigue capacity and use of Miner-Palmgren's hypothesis. The fatigue curves are given with the breaking load of the wire as a linear scaling factor and established on the basis of experiments with far smaller wire cross sections than the actual one. The uncertainty related to this scaling is unknown. No bending effects are considered in the analysis. If bending variations will occur at fairlead, this will certainly reduce the fatigue life. This effect calls for improved fatigue data and a detailed study of the wire behaviour at fairlead. The use of Miner-Palmgren's hypothesis is normally accepted for steel structures, but has so far not been verified for wire lines. Such verification can only be obtained from experimental work.

Effect from corrosion is not included in the fatigue data applied. Normally, corrosion will decrease the fatigue capacity significantly, but the effect can to some extent be reduced by adequate protection. This point is, however, important and need further evaluation.

Fatigue data are given for the wire itself and not for the applied termination details. This point will always be the weakest part of the wire. Data applied in an analysis should therefore be valid for termination details in sea water. The fatigue analysis can, however, be carried out in the same way as was presented in the actual study.

6. DYNAMIC BEHAVIOUR OF CLUMP WEIGHT SYSTEMS

All results presented in this section are obtained by use of the computer program HATCAN, and the analyses are carried out for the same 3-segment anchor line as described in Section 5.

6.1 Finite element model

As described in Sect. 2.3 the finite element model must use linear springs to simulate bottom contact. The stiffness of such springs will affect the tension/displacement relation (stiffness) at upper end, which calls for a carefully selection of spring values. Figure 15 demonstrates the influence from bottom springs on the stiffness of the finite element model. Ideally, the model stiffness should be tangents to the nonlinear line characteristic. From the figure it can be seen that correct stiffness can be obtained, but not by use of the same spring values at any tension level. At position D on the fiure, the softest springs have to be used, while stiffer springs give the best results for position A and B.

It should be noted that errors in the model stiffness will not give errors of the same order in the results as the correction described in Sect. 2.3 will substitute the linear stiffness contribution to end tension with values taken from the nonlinear curve. The model stiffness will, however, also afffect the motions of the cable and hence drag and inertia effects. The total solution is therefore dependant on a proper choise of bottom spring values.

Tension versus horizontal displacement was here used as criterion for correct values of bottom springs. This is, however, not the only reasonable criterion. Tension versus vertical displacement is an alternative if vertical motions of upper end are dominating. Vertical displacement of bottom nodes is another alternative. This criterion can be relevant for clump weight systems as vertical motions of the clump weight is a key parameter to describe dynamic effect of the system. Evaluation of bottom springs on the basis of alternative criteria has not been attempted.

6.2 Presentation of results

Natural periods are calculated for the four first modes and presented as functions of upper end position and tension on Figure 16. The first and third modes were always found to be oscillations out of plane, while second and fourth modes were in-plane oscillations. The dramatic increase of the first natural period at x = 10.5 m is caused by the fact that the clump weight is without bottom contact from this point. Increase of tension will therefore cause a significant increase of line length lifted from bottom and hence allow for larger motions of the heavy clump weight. Variation of bottom spring stiffness within the same range as used for the model discussion gave neglectable effect on the presented natural periods.

Dynamic response versus exitation period is presented on Figure 17. Upper end is given a circular in-plane motion with radius 1.0 meters, and mean position is at x = 0.0 (See Fig. 14). Relevant natural periods are indicated, and total dynamic response (ΔT_{tot}) as well as response from drag and inertia (ΔT_{dyn}) are given. The curves show a somewhat confusing picture for periods between 4 and 8 seconds with no dominant peak at the second natural period. The total dynamic response are smaller than the quasistatic response for periods exceeding 16 seconds. This is caused by a phase angle between dynamic and quasistatic response. The maximum dynamic amplification for this load condition is approximately 2.0 for periods of interest.

The importance of the nonlinear stiffness correction described in Sect. 2.3 is dependant on the amplitude and the pattern of the line characteristic at the actual mean position. At point C on Fig. 15 the nonlinear stiffness is significant. The results presented on Fig. 18 are obtained for an analysis of a line at this particular position with exitation at upper end in horizontal direction only. The deviation from linear to nonlinear stiffness is significant and clairly illustrates the need for the correction. Comparison between the corrected curve and results from a complete nonlinear analysis could have verified weather this semi-nonlinear method gives results of acceptable accuracy even for strongly nonlinear systems.

The nonlinear behaviour of the system is further demonstrated on Fig. 19. Dynamic response is presented as function of amplitude for horizontal exitation of upper end. Total dynamic response ΔT_{tot}, quasistatic response ΔT_q and response from drag and inertia ΔT_{dyn} are given. The effect from change of bottom spring stiffness is also indicated, and is seen to be moderate. For small amplitudes the curves for total and quasistatic response are fairly close, but for increasing amplitudes the gap is increasing and clairly demonstrates the need for dynamic analysis. The dynamic response curves are nonlinear, but the strongest nonlinearity is found close to origin. The dotted lines on the figure indicates a possible linearization with moderate errors within an amplitude interval of 1-6 meters.

A simple demonstration of nonlinear dynamic behaviour is shown on Figure 20. Two analyses are performed with 1.0 m amplitude exitation in horizontal and vertical direction respectively. The response obtained by linear superposition of these results is compared to the response from simultaneous exitation. The response amplitudes are seen to be slightly different. More dominant is, however, the deviation of curve patterns. The curve for simultaneous exitation has a more non-harmonic pattern that the superposition curve. This is caused by the nonlinear drag term, which will give an increasing contribution to a response at 3 x exitation frequency for increasing amplitudes. This type of response is clairly seen from the curves.

Effect from current is illustrated on Figure 21. Current direction is from the anchor against the platform, with a velocity at surface of 1.5 knots (0.75 m s^{-1}). Response caused by inertia and drag effects are given, and from the figure it can be seen that stress amplitudes are unchanged. Current is seen to give an unsymmetric response, and hence an increase of maximum response. From this one can conclude that current is of minor importance for a fatigue analysis, but should be considered when extreme values for line tension are calculated.

Figure 22 illustrated the effect from waves on the tension. Wave height is 30 meters and the period is 15 seconds. The line response is given for a line fixed at upper end at surface and at 30 meters water depth as is the case for the actual platform. The response is seen to be small and significantly reduced by the lowering of the temination point. The response is also given for a 3.0 meter amplitude horizontal exitation alone and in combination with the same wave. The effect from the wave is dependant on the phase angle ε between wave and motion, and is seen to increase the response for ε = 180 degrees and decrease for ε = 0.0. In both cases the effect is small and can certainly be neglected in fatigue analyses.

Three-dimensional effects can be analysed to a limited degree by the actual computer program. This is caused by the fact that the line has no elastic stiffness normal to its plane, but geometric stiffness only. Forces normal to the plane will therefore cause geometric changes and no elastic deformation. Consequently, motions out of plane can be calculated, but not the associated tension variations. An upper limit for dynamic tension can be estimated by calculating the arc length of the line for the deformed geometry. This analysis is not theoretically correct as it is an uncorrect combination of linear and nonlinear theory, but if a small response is found this indicates that the true response will be small. The most unfavourable situation for out-of-plane response is exitation at the natural frequency for a line with the clump weight free from bottom. This case was analysed for the actual line, and the results indicated a motion amplitude at the clump weight of 90% the exitation amplitude. Calculation of the arc length at the extreme response indated a line response less than 2% of the response from the same amplitude in horizontal direction. From these calculations one can conclude that exitation normal to the line plane is of minor interest for clump weight systems.

7. CONCLUSIONS

Some aspects of dynamic behaviour and analysis of deepwater mooring lines have been discussed in the preceding chapters. The main conclusions are:

- A constant-stiffness dynamic analysis appears to give satisfactory results for deepwater mooring lines without clump weights.

- Analysis of wire line systems indicates that the dynamic behaviour in the typical wave-frequency range, 4-15 seconds period, is governed by the purely elastic deformation.

- Within this frequency range the dynamic response is nearly linear.

- Chain systems do not show a corresponding linear elastic response within the wave frequency range, but the elastic stiffness determines the range of dynamic loads.

- For many wire line systems the dynamic response can be taken care of by a simple modification of the commonly used quasistatic response analysis.

- For chain systems or wire line systems with strongly non-linear responses more elaborate analyses are required, for instance time-domain analysis with irregular forced excitation of correct amplitude.

- Fatigue analysis based upon semi-linear time domain analysis can be performed with reasonable computer costs. The main uncertainties for fatigue evaluation is today the fatigue capacity in sea water for the wire and termination details.

- The dynamic response in a clump weight system is significant. A quasistatic analysis will not give results of acceptable accuracy neither for fatigue nor for extreme tension analysis.

- Even if the anchor line system is strongly nonlinear, the presented semilinear method gives promising results. A verification is, however, needed and can be obtained by comparison of results from a complete nonlinear analysis.

- Effects from current on line tension is significant, but the dynamics tension range is almost unaffected. Current need therefore not to be considered in a fatigue analysis, but should be present in an extreme tension calculation.

- Wave forces on the cable have a moderate effect compared to exitation at upper end for the present system. The relative importance of waves is, however, dependent on many factors such as platform motions and water depth at upper line end. A general conclusion can therefore not be drawn.

- Oscillations normal to the line plane seems to have only small effects on line tension variations. This phenomena should, however, be studied more in detail by use of a complete nonlinear method for analysis.

8. REFERENCES

1. Furuholt, E.: "Brudd i ankerliner, data innsamling", OR 21281.12. The Ship Research Institute of Norway, Trondheim 1981 (Commission report, not published).

2. Casarella, M.J., Parsons, M.: "Cable Systems under Hydrodynamic Loading", Marine Technology Society Journal, August 1970.

3. Chro, Y., Casarella, M.J.: "A Survey of Analytical Methods for Dynamic Simulation of Cable-Body Systems", Journal of Hydronautics, Vol. 7, Oct. 1973.

4. Migliore, H.J., Webster, R.L.: "Current Methods for Analyzing Dynamic Cable Response", Shock and Vibration Digest, June 1979.

5. Slujes, M.F. van, Blok, J.J.: "The Dynamic Behaviour of Mooring Lines", OTC Paper no. 2281, Offshore Technology Conference, Houston 1977.

6. Shimada, K., Kobayashi, M., Hineno, M.: "Dynamic Effect on Tension of Mooring Lines", Transactions of the West Japan Society of Naval Architects, No. 60, Aug. 1980.

7. Suhara, T., Koterayama, W., Tasai, F., Hiyama, H.: "Dynamic Behaviour and Tension of Oscillating Mooring Chain", OTC Paper no. 4053, Offshore Technology Conference, Houston 1981.

8. Felippa, C.A.; "Refined Finite Element Analysis of Linear and Non-linear two-dimensional Systems", University of California, Berkeley, U.S.A. 1966.

9. Clough, R.W. and Penzien, J.: "Dynamics of Structures", McGraw Hill Book Company, 1975.

10. Furuholt, E.: "ANKAN", Users manual. The Ship Research Institute of Norway, Trondheim 1981.

11. Larsen, C.M.: "Static and Dynamic Analysis of Offshore Pipelines during Installation", Division of Marine Structures, NTH, Trondheim 1976.

12. Larsen, C.M. and Furuholt, E.: "LINDYN", A Program for Dynamic Analysis of Mooring Lines. User's Manual. The Ship Research Institute of Norway, 1975.

13. Larsen, C.M. and Mørch, M.: "STOCCA", A Program for Stochastic Cable Analysis. To appear.

14. "Ropes for Deep Water Moorings", British Ropes 1979.

15. Larsen, C.M. and Holtan, P.E.: "HATCAN", A Program for Harmonic Three-dimensional Analysis of Cables and Anchor Lines. To appear.

16. Eggen, S.: "Kabeldynamikk modellforsøk", P.R. 4340003, Norges Skipsforskningsinstitutt, Trondheim 1978. (In Norwegian, unpublished).

17. Fylling, I.J., Wold, P.T.: "Cable Dynamics - Comparison of Experimental and Analytical Results", Report R-89.79 Norges Skipsforskningsinstitutt, Trondheim 1979.

18. Fylling, I.J., Ormberg, H.: "Dynamic Analysis of Mooring Lines", Teknologidagene i Rogaland, Stavanger 1981.

19. Larsen, C.M.: "Fatigue and extreme tension calculation of mooring lines for the CONPROD platform". SINTEF Report STF71 F80003. Division of Structural Engineering. Trondheim 1980.

20. Sand, Ø.; "Calculation of motions for the CONPROD platform. Commission report. Chr. Michelsens Institute, Bergen 1980.

Segment no.	Material	Length m	Submerged weight N/m	Stiffness EA N
1	wire	540	820	$8.2 \cdot 10^8$
2	8 chains	40	40000	$100 \cdot 10^8$
3	chain	500	3480	$13.2 \cdot 10^8$

Table 1. Mooring line data.

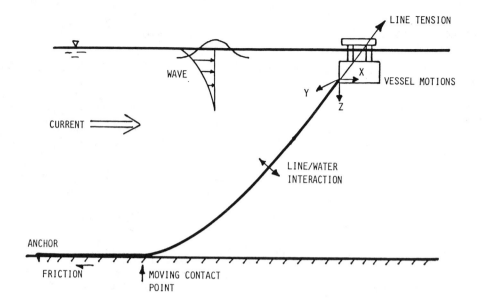

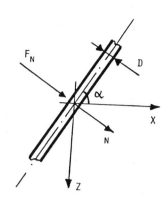

Figure 3 Inclined cable element.

Figure 1 Important effects for a typical anchor line.

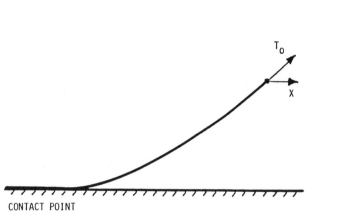

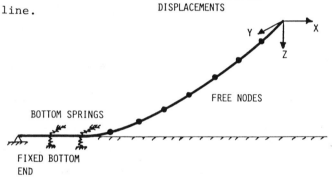

Figure 4 Model for FEM analysis.

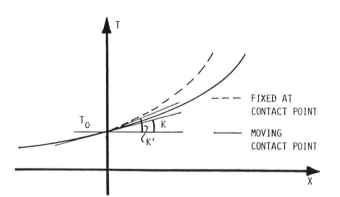

Figure 2 Line tension versus horizontal position.

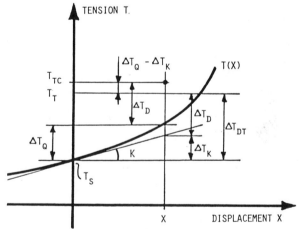

Figure 5 Correction of results from a linear dynamic analysis by use of the nonlinear cable stiffness.

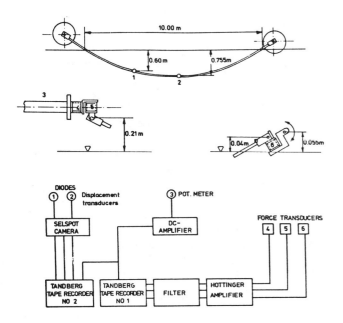

Figure 6. Model test set-up for horizontally suspended, submerged line.

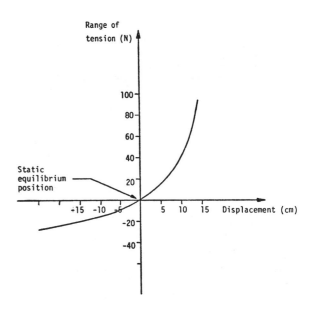

Figure 7. Static force - displacement characteristic of horizontally suspended model line

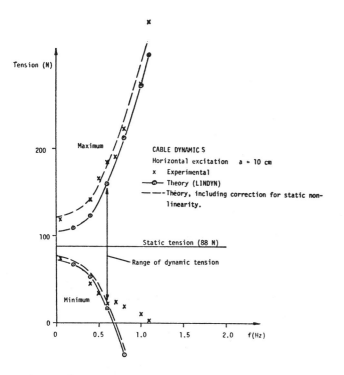

Figure 8. Comparison of analytical and experimental dynamic response in horizontally syspended model line

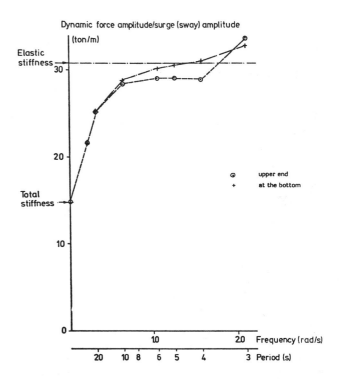

Figure 9. Dynamic response in wire line at 70 m water depth. Static tension 70 tonnes.

663

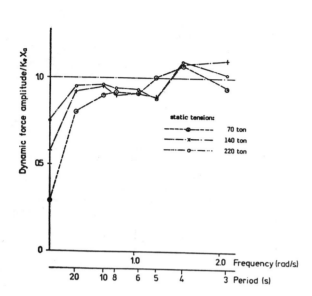

Figure 10. Dynamic response in wire line at 150 m water depth. Dynamic tension in upper end relative to tension under purely elastic static deformation.

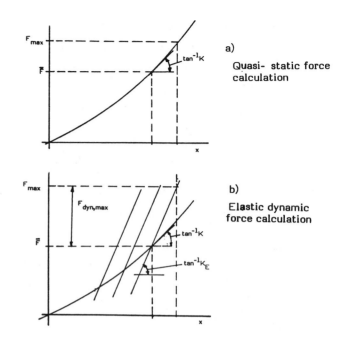

Figure 11. Simplified dynamic force estimates for deep-water wire line systems.

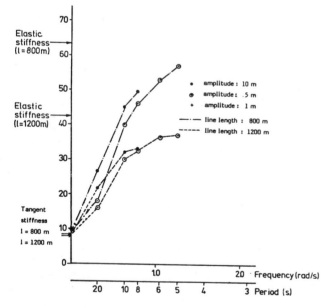

Figure 12. Dynamic response in chain line at 150 m water depth. Static tension 140 tonnes.

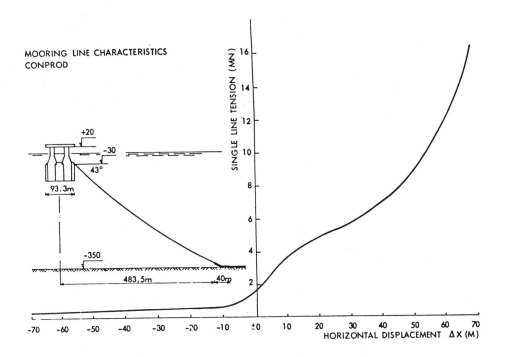

Figure 13 Mooring line characteristics.

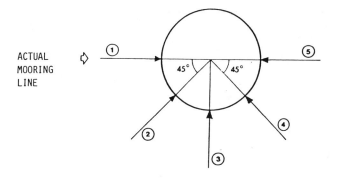

Figure 14 Directions of waves, wind and current relative to analysed mooring line.

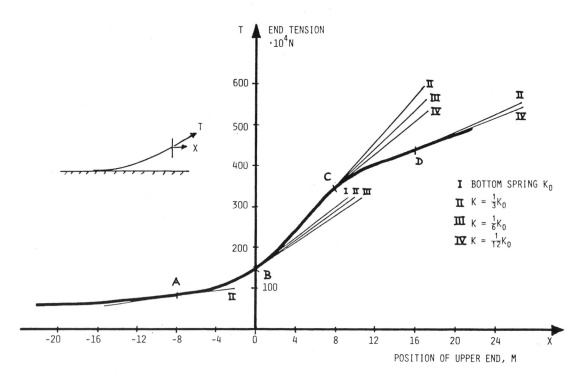

Figure 15 Stiffness of finite element model compared to mooring line characteristics.

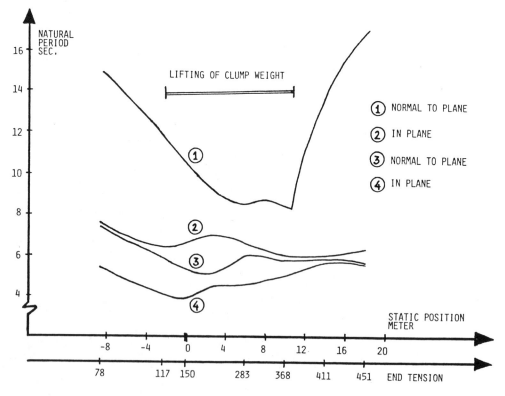

Figure 16 Natural periods as function of upper end position and line tension.

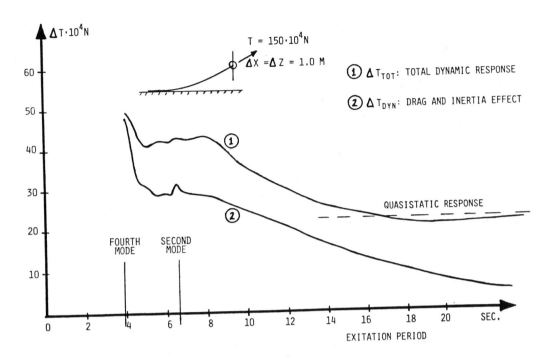

Figure 17 Dynamic response versus exitation period.

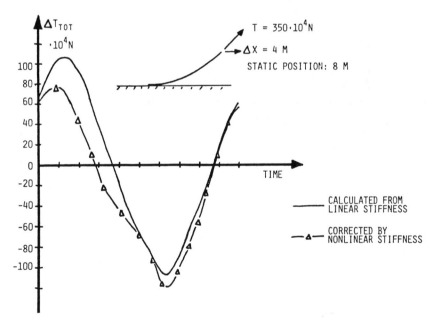

Figure 18 Effect from nonlinear correction on linear results.

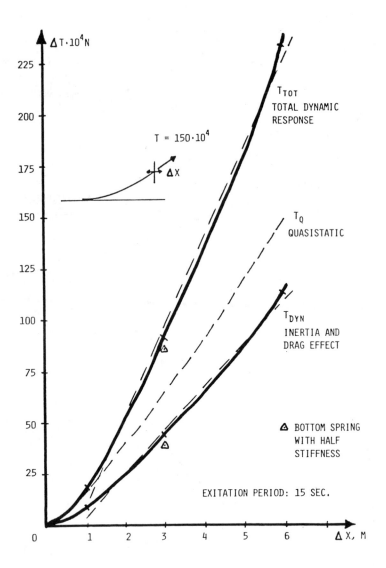

Figure 19 Dynamic response versus amplitude of upper end motion.

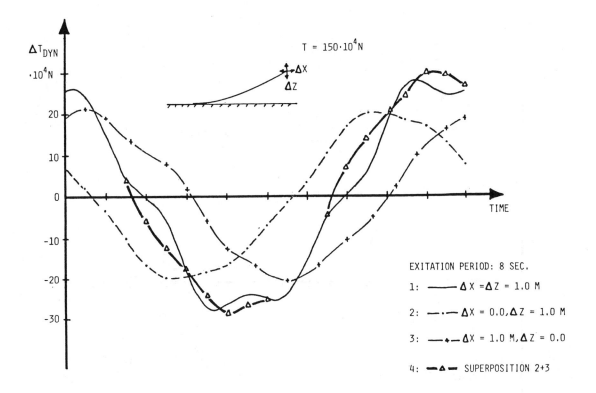

Figure 20 Demonstration of non-linear behaviour.

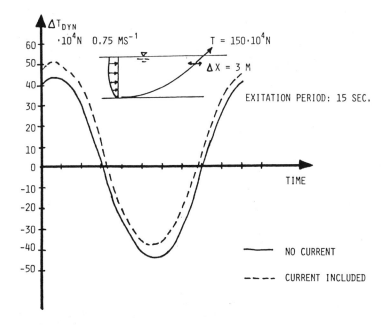

Figure 21 Effect from current.

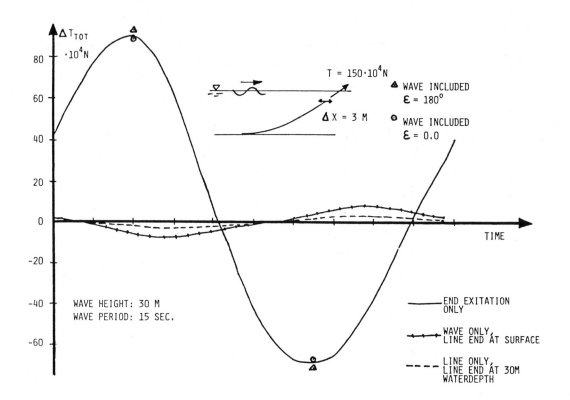

Figure 22   Effect from waves.

THE RESPONSE AND THE LIFT-FORCE ANALYSIS OF AN ELASTICALLY-MOUNTED CYLINDER OSCILLATING IN STILL WATER

K. G. McConnell
Department of Engineering Science and Mechanics
Iowa State University

Y. S. Park
The Korea Advanced Institute of Science and Technology

SUMMARY

An elastically-mounted cylinder is driven with sinusoidal motion in a fluid at rest and is free to respond in a direction perpendicular to the input motion. The data are presented in a somewhat radical manner that leads to an integrated picture of this type of fluid structure interaction problems. Large amplitudes of motion occur when the ratio of structural natural frequency to excitation frequency is an integer, giving closed Lissajous paths of structural motion. A lock-on phenomenon was clearly present for lower A/D ratios. In addition, larger amplitudes of response were obtained than predicted by previous theories. The results from the elastically-mounted cylinder tests clearly indicate the importance of allowing structural response to occur if meaningful results are to be obtained.

INTRODUCTION

The urgent demands for optimum design of off-shore structures have raised interest in fluid-structure interaction problems, especially on structures exposed to periodic fluid flow. It is extremely difficult to obtain theoretical solutions for problems of this type. The difficulty is partly due to our incomplete knowledge of the flow field around the structure, as well as to the problems associated with the coupling of structural oscillations and fluid flow. Consequently, most studies on these fluid-structure interaction topics have been done experimentally. The ultimate question is what happens to the structure when fluid flows around it.

In the last three decades, many research studies have been concerned with a fixed or an elastically-mounted cylinder in a steady fluid flow field. Most of these studies have been based on the Morison equation (Morison et al., 1950) to explain in-line fluid forces, while some cases have examined the so-called lift force. It is known that the vortex sheddings behind the cylinder give a periodic lift force. Several research studies have shown that the lift force plays an important role in determining the behavior of a cylinder due to fluid flow not only because of its magnitude, but also because of its frequency.

Relatively few studies have considered periodic fluid flow around a fixed or an elastically-supported cylinder. In this case, the flow characteristics are much more complicated than in the steady flow case since the fluid particles, the wake, and the cylinder continuously interact, creating a fluid flow pattern of great complexity.

Hamann and Dalton (1971), Garrison et al. (1977), Sarpkaya (1975, 1976), Keulegan and Carpenter (1958), and others have measured fluid forces on a fixed cylinder exposed to periodic fluid flow. Of these investigators, Sarpkaya (1976) has done the most systematic study where fluid in-line and lift forces were measured over the wide range of Reynolds numbers. He concluded for periodic fluid flow around a fixed cylinder that the fluid forces are strongly related to the Keulegan-Carpenter number $U_m T/D = 2\pi A/D$, as well as the Reynolds number $U_m D/\nu$ where U_m is the maximum fluid oscillating velocity, T is the fluid oscillating period, A is the fluid oscillating amplitude, ν is the fluid kinematic viscosity, and D is the cylinder diameter.

It is believed, for an elastically-mounted cylinder, that the cylinder response and the fluid forces interact and affect each other. Consequently, the fluid force estimation based on stationary cylinder data must be in error compared to the actual situation.

The present study assumes that the relative motion between the fluid and cylinder can form a closed Lissajous figure under certain conditions (Fig. 1). A locked-on phenomenon occurs in this case where the vortex-shedding frequency (f_v) and the structural-natural frequency (f_n) are related to the oscillating-fluid frequency (f_d) in an integer manner. This locked-on phenomenon causes an increase in the lift force and the structure response due to a synchronized motion and vortex shedding from the cylinder. It is helpful to visualize the vortices as "trapped" within the Lissajous path (Fig. 1). Note how the cylinder moves naturally around the "trapped vortices," reinforcing the vortices on each pass. As the cylinder driving amplitude (A) increases, there is more time for the vortices to dissipate and move out of the pattern before the cylinder returns. Consequently, the cylinder response amplitude decreases since this simple pattern of behavior begins to break down, and there is less chance for lock-on to occur.

While the above description is helpful in visualizing a typical cylinder moving relative to the fluid, it is also believed that the case of periodic flow about an elastically-mounted cylinder involves a complex interaction. The relationship between the cylinder natural frequency (f_n), the cylinder response frequency (f_r), the vortex shedding frequency (f_v), and the fluid oscillating frequency (f_d) is more important in determining fluid forces and structural responses than the conventional non-dimensional parameters described by the Reynolds and Keulegan-Carpenter numbers.

McConnell and Park (1980) measured fluid lift forces and the cylinder responses for A/D ratios of 6, 7, and 8 when an elastically-mounted, 2-inch diameter cylinder was driven sinusoidally in a fluid at rest. They found that there are three dominant frequency components in the lift forces called lower sideband, vortex shedding frequency, and upper sideband. They also found that the three components are determined by the frequencies of the vortex shedding and the periodic fluid flow.

In this study the fluid lift forces and the cylinder responses due to oscillating fluid flow around an elastically-mounted, 1-inch diameter cylinder were measured experimentally for extended A/D ratios. Then the frequency components of the lift forces and the cylinder responses were closely examined to obtain a relationship between the frequency components.

It will be seen that this apparently radical departure from conventional practice for presenting experimental results leads to an integrated picture of fluid-structure interaction problems for an elastically-mounted cylinder in oscillating fluid flow.

EXPERIMENT

In order to maintain strict control of the fluid oscillating amplitudes and frequencies, it was decided to move the cylinder sinusoidally in a fluid at rest, rather than oscillating the fluid around a circular cylinder in order to obtain the desired relative motion between fluid and cylinder. As shown in Fig. 2, a 1-inch diameter cylinder was mounted below plate C. This model was driven in a still fluid with a sinusoidal motion in the input direction of

$x = A\sin(\omega_d t)$ where x is the displacement, A is the amplitude of cylinder driving motion, ω_d is the angular frequency of driving motion, and t is time. Both plate C and the cylinder model can either respond or be constrained by constraint cables when in place in the output direction (y-direction), as shown in Fig. 2. Plate C was mounted on plate B by aluminum flexures in order to provide the y-direction flexibility. The fluid lift forces acting in the output direction (y-direction) and the cylinder acceleration were measured using two force transducers. The amplitude of the cylinder response motion was measured from strain gages mounted on the supporting aluminum flexures.

In order to systematically obtain data, the fluid lift forces (F_L) and the y-direction cylinder response amplitude (y) were monitored using a Norland Model 3001 data analysis unit. The cylinder driving frequency (f_d) was varied from 0.067 Hz to 0.67 Hz while the cylinder driving amplitude (A) varied from 3 inches (76mm) to 19 inches (0.48 m). These measured signals were analyzed, using the Norland, to obtain the frequency components of interest (vortex shedding frequency [f_v], the cylinder response frequency [f_r], the cylinder driving frequency [f_d]) and the root mean square (RMS) values of the fluid lift force and the cylinder response amplitude. The detailed method to obtain all the frequency components of interest is given below.

1) f_{ns} (standard natural frequency of the cylinder in y-direction)

The natural frequency of the cylinder in y-direction was obtained from observations of the y-direction free-damped oscillation signal of the cylinder in still fluid. It was found that the natural frequency is dependent on the amplitude of the cylinder response, i.e., the natural frequency decreases as the cylinder response amplitude increases. For convenience, the standard natural frequency (f_{ns}) was defined as the natural frequency when the cylinder RMS response amplitude is around 0.1 inch (2.5mm). The measured standard natural frequency of the 1-inch diameter cylinder was 1.585 Hz when the cylinder length submerged in water (ℓ) was 22 inches (0.56 m).

2) f_v (vortex shedding frequency)

The vortex shedding frequency (f_v) was obtained by taking the power spectral density of the measured fluid lift force signal. Among the several peaks in the power spectral density, the highest peak was considered to correspond to the vortex shedding frequency (f_v).

3) f_d (cylinder driving frequency)

The cylinder driving frequency (f_d) was constantly monitored during the experiment using the on-off signal from a microswitch placed at one end of the input motion. A Hewlett-Packard Model HP 5280 frequency counter measured the driving frequency which was maintained within 0.5 percent during the experiment.

4) f_r (cylinder response frequency)

The cylinder response frequency (f_r) was obtained by taking the power spectral density of the measured cylinder response signal. The highest peak of the power spectral density was chosen as the cylinder response frequency (f_r).

The details about the experimental apparatus and the data analyzing method can be found in Park (1979, 1981).

RESULTS

The present study emphasizes frequency analysis of the fluid lift forces and response amplitudes when an elastically-mounted cylinder is exposed to a periodic flow in order to understand the subtle interactions that take place. The detailed discussion about each frequency component and its relationships are as follows:

1) The vortex shedding frequency (f_v)

The vortex shedding frequency (f_v) plays a very important role in determining the cylinder response and the fluid lift forces acting on the cylinder. The Strouhal number (S_t) is widely used to describe both steady and unsteady vortex shedding phenomenon. For a circular cylinder in a periodic fluid flow, the Strouhal number (S_t) is usually defined as

$$S_t = \frac{f_v D}{U_m} = \frac{f_v D}{2\pi f_d A} = \frac{f_v}{f_d K} \tag{1}$$

where D is the cylinder diameter, U_m is the peak fluid velocity ($=A\omega_d$), and K is the Keulegan-Carpenter number ($=2\pi A/D$). The obtained f_v/f_d ratio was plotted as a function of K (Fig. 3) for several different cylinder driving frequencies. The cylinder was both fixed and released for these measurements. A released cylinder is free to move in the y-direction, while a fixed cylinder is constrained from moving in the y-direction. From Eq. (1), it is seen that a constant Strouhal number plots as a straight line. For reference, two constant Strouhal number lines of 0.15 and 0.20 were drawn in Fig. 3, where several important facts can be noted.

First, in the case of a fixed cylinder the Strouhal number is bounded in the range from 0.15 to 0.20 for all cylinder driving frequencies and different Keulegan-Carpenter numbers. This fact matches quite well with Sarpkaya's result (1976). In the released cylinder case, it was found that the Strouhal number is also in the 0.15 to 0.20 range with one exception, the so-called locked-on region.

Second, the vortex shedding frequency (f_v) increases with increasing values of the Keulegan-Carpenter number for both the fixed and released cylinder.

Third, for the low Keulegan-Carpenter number region (low A/D ratio < 12), an integer f_v/f_d ratio was always found. On the other hand, for the high Keulegan-Carpenter number (high A/D ratio > 12), a fractional f_v/f_d ratio appeared. In the low A/D ratio case, a shed vortex has neither significantly dissipated nor moved relative to its spatial relationship to the other vortices in Fig. 1. In this case, the vortex is in proper position to interact with the cylinder when it returns to that location. Then, as shown in Fig. 1, a strong and consistent vortex shedding spatial pattern is formed. This spatial pattern corresponds to an integer f_v/f_d ratio. In comparison, a large A/D ratio produces a relatively inconsistent vortex shedding pattern. In this case, the vortex wakes tend to either dissipate or move away by the time the cylinder returns, giving fractional f_v/f_d ratios.

Fourth, the most important factor for lock-on to occur is for $f_n/f_d = f_v/f_d =$ integer. More details about this lock-on phenomenon will be discussed in a later section.

2) The cylinder response frequency (f_r)

In Fig. 4 a non-dimensional number which includes the cylinder response frequency (f_r) $Df_r/2Af_d = [f_r/f_d]/K$ is plotted against the f_{ns}/f_d ratio. In the figure the straight lines for different A/D ratios show the case where the cylinder responds in the y-direction with the cylinder standard natural frequency (f_{ns}), i.e., $f_r = f_{ns}$. It can also be seen that most of the points lie on the straight line with one exception, the locked-on region. This locked-on behavior corresponds to the strong tendency for the f_r/f_d ratio to be an integer, especially when the A/D ratio is less than 10 and f_{ns}/f_d ratio is near an integer value. In other words, the cylinder response frequency (f_r) changes in such a way that the f_r/f_d ratio remains an integer even when the cylinder response frequency (f_r) is different from the cylinder natural frequency (f_n). This strong tendency to form closed motion paths around integer f_n/f_d ratios (Fig. 1) can be viewed as a type of locked-on behavior. This type of locked-on behavior was only observed for low A/D ratios (A/D < 10). For high A/D ratios, it was difficult to obtain a consistent vortex shedding situation.

3) The relationship between the vortex shedding frequency (f_v), the cylinder natural frequency (f_n), and the cylinder response frequency (f_r)

If the cylinder driving frequency (f_d) and the A/D ratio are known, then the vortex shedding frequency (f_v) determines the corresponding Strouhal number ($S_t = [Df_v]/[2\pi Af_d]$), the cylinder response frequency (f_r) determines V_R ($= [Af_d]/[Df_r]$), and the standard natural frequency (f_{ns}) determines V_R^* ($= [Af_d]/[Df_{ns}]$). In order to directly compare all three frequency components (f_v, f_n, and f_r), three quantities (S_t, $1/[2\pi V_R]$, and $1/[2\pi V_R^*]$) were plotted together against the f_{ns}/f_d ratio in the upper part of Fig. 5 for a low A/D ratio (A/D = 4), and Fig. 6 for a high A/D ratio (A/D = 15). A careful examination of the figures reveals that the graphs can be divided into three distinct regions: lower, middle, and upper parts of the f_{ns}/f_d ratio.

For the lower f_{ns}/f_d ratio region, the vortex shedding frequency is always higher than the cylinder response frequency and the cylinder natural frequency. In this region, the relationship between f_d, f_v, and f_r was observed to be $f_r = f_v - 2f_d$.

For the middle range of the f_{ns}/f_d ratio, the vortex shedding frequency always matches with the cylinder response frequency and the cylinder natural frequency (f_n). It was also observed that the vortex shedding frequency changes to make the Strouhal number stay in the range of 0.15 to 0.20. In this region, the vortex shedding frequency excites the cylinder at its natural frequency. The resulting cylinder motion and the vortex shedding strongly interact and support each other to give large amplitudes of cylinder responses. The locked-on behavior occurs in this mid-range region.

For the upper range of the f_{ns}/f_d ratio, the vortex shedding frequency is always lower than the cylinder response frequency except for several data points which correspond to a locked-on condition. In most of these cases, the vortex shedding frequency adjusts to make the Strouhal number between 0.15 to 0.20. This result is the same as that obtained for a stationary cylinder exposed to an oscillating fluid flow. The Strouhal number behavior in the upper region of the f_{ns}/f_d ratio appears similar to that for the stationary cylinder case since little cylinder response occurs, giving a nearly constrained condition. The strength of vortex shedding seems to be weaker than in the middle and lower f_{ns}/f_d ratio regions. In turn, this causes weaker cylinder responses so that the cylinder response frequency becomes less correlated with the vortex shedding frequency.

4) The cylinder response amplitude in y-direction (y)

The RMS value of the cylinder responses (y) was obtained from the cylinder response signal. The y_{RMS}/D ratio was plotted against the f_{ns}/f_d ratio in the lower part of Fig. 5 for a low A/D ratio (A/D = 4) and Fig. 6 for a high A/D ratio (A/D = 15). As explained previously, it is easier to form a closed cylinder moving path for the low A/D ratio case than for the high A/D ratio case. In the low A/D ratio case, however, the change in the cylinder response frequency from the cylinder natural frequency corresponds to a sharp decrease in both the amplitude of cylinder response and the fluid lift force. Note that in Fig. 5 (low A/D ratio), steep peak and valley variations can be observed. On the other hand, that kind of sharp decrease and increase cannot be found in Fig. 6 (high A/D ratio) because it is difficult for the locked-on condition to occur for high A/D ratios.

It was also found that the f_n/f_d ratio and the f_r/f_d ratio play an extremely important role in determining the cylinder responses. When $f_n/f_d = f_r/f_d = f_v/f_d$ are equal and integer values, the highest cylinder response amplitude occurs. The cylinder response amplitudes are self-limiting as in the steady flow case due to the size of vortices and relative spatial locations. The maximum cylinder response amplitude is about 1.0 D to 1.15 D peak to peak and tends to decrease as the Keulegan-Carpenter number increases.

5) The fluid lift force coefficient ($C_{L[pk]}$ and $C_{L[RMS]}$)

The peak and RMS fluid lift force coefficients were defined as below

$$C_{L(pk)} = \frac{\text{Max Lift Force}}{0.5\rho D \ell (A\omega_d)^2} \qquad (2)$$

and

$$C_{L(RMS)} = \frac{\text{RMS Lift Force Value}}{0.5\rho D \ell (A\omega_d)^2} \qquad (3)$$

where ρ is the fluid density. The calculated lift force coefficients were plotted against the f_{ns}/f_d ratios in Fig. 7 for an A/D ratio of 4 and in Fig. 8 for an A/D ratio of 15. The upper part of the figures show the peak lift force coefficients ($C_{L[pk]}$) while the lower part shows the RMS lift force coefficients ($C_{L[RMS]}$). As seen in the figures, the RMS force coefficients exhibit the same functional dependence as the peak values.

Sarpkaya (1976) studied identically defined fluid lift force coefficients for a fixed cylinder in an oscillating fluid flow. He concluded that the fluid lift forces depend on Keulegan-Carpenter and Reynolds numbers. On the other hand, this study deals with elastically-mounted cylinders, and it was found that the f_{ns}/f_d ratio plays an important role in determining the force coefficient, while both the Reynolds and Keulegan-Carpenter numbers appear to play a secondary role.

Even though the experimental conditions are different, it is worthwhile to compare these results with the Sarpkaya (1976) data. Figure 7 shows the case where the Keulegan-Carpenter number is constant (A/D = 4; K = 25.1), and the Reynolds number was varied by changing the cylinder driving frequency. The peak lift force coefficients which Sarpkaya obtained for each corresponding Reynolds and Keulegan-Carpenter number were plotted as a dotted line in the upper part of Fig. 7. The dotted line matches quite well with the value of the $C_{L(pk)}$ in this work, when the f_{ns}/f_d ratio is around 3.3, 4.6, 5.6, and 7.4. It was found in each of these regions where the Sarpkaya data match with this experiment that very low cylinder response amplitudes occurred as shown in the lower part of Fig. 5. In other words, the lift force coefficients for an elastically-mounted cylinder with very low response amplitudes are in close agreement with those of a fixed cylinder. It was also found that the cylinder vibration in the y-direction generally increases the lift force coefficients. As shown in Figs. 7 and 8, the trend is that integer f_{ns}/f_d ratios give high cylinder response amplitudes and also a sharp increase of lift force coefficient, especially for the low A/D ratio region. It was observed that the lift force coefficient increases up to twice of that for a fixed cylinder in certain cases.

CONCLUSION

This study investigated what happens when an elastically-mounted circular cylinder is exposed to periodic fluid flow. The experiment showed that the cylinder responses in the y-direction affect vortex formation behind the cylinder and, in turn, influence the fluid lift forces acting on the cylinder. Previously, it was known that the fluid lift forces acting on a fixed cylinder due to periodic fluid flow are dependent on Reynolds and Keulegan-Carpenter numbers. On the other hand, new facts found in this experiment for an elastically-mounted cylinder in an oscillating fluid flow field suggest that the f_{ns}/f_d ratio is an extremely important dimensionless ratio for determining the corresponding locked-on phenomenon for Reynolds numbers under 40,000 and K < 120. A more unified picture of the subtle interaction among frequency components was examined. Further studies on larger structures and much higher Reynolds numbers need to be undertaken so that Reynolds numbers effects and f_{ns}/f_d ratio effects can be clearly delineated.

ACKNOWLEDGMENT

The support of the Iowa State Engineering Research Institute for this project is gratefully acknowledged.

REFERENCES

1. GARRISON, C. J., J. B. FIELD and M. D. MAY, 1977, "Drag and Inertia Forces on a Cylinder in Periodic Flow," Proceedings of ASCE Journal of the Waterway, Port, Coastal and Ocean Division, May, Volume 103, Number WW2, pp. 193-204.

2. HAMANN, F. H. and C. DALTON, 1971, "The Forces on a Cylinder Oscillating Sinusoidally in Water," Transcripts of ASME Journal of Engineering Industry, Volume 93, November, pp. 1197-1202.

3. KEULEGAN, G. H. and L. H. CARPENTER, 1958, "Forces on Cylinders and Plates in an Oscillating Fluid," Journal of Research of the National Bureau of Standards, Volume 60, Number 5, May, pp. 423-440.

4. McCONNELL, K. G. and Y. S. PARK, 1980, "The Frequency Components of Fluid Lift Forces Acting on a Cylinder Oscillating in Still Water," Presented at the 4th SESA International Conference, Boston, Massachusetts, May.

5. MORISON, J. R., M. R. O'BRIEN, J. W. JOHNSON and S. A. SCHAAF, 1950, "The Forces Exerted by Surface Waves on Piles," Petroleum Transactions, Vol. 189, pp. 148-154.

6. PARK, Y. S., 1979, Experimental Apparatus for Measuring Fluid Forces Acting on a Cylinder Driven Sinusoidally in Still Water, M.S. Thesis, Iowa State University, Department of Engineering Science and Mechanics.

7. PARK, Y. S., 1981, The Response and the Lift Force Analysis of a Cylinder Oscillating in Still Water, Ph.D. Thesis, Iowa State University, Department of Engineering Science and Mechanics.

8. SARPKAYA, T., 1975, "Forces on Cylinders and Spheres in a Sinusoidally Oscillating Fluid," Transactions of ASME Journal of Applied Mechanics, Volume 42, Number 1, March, pp. 32-37.

9. SARPKAYA, T., 1976, "Vortex Shedding and Resistance in Harmonic Flow About Smooth and Rough Circular Cylinders at High Reynolds Numbers," Naval Postgraduate School, Paper No. NPS-59SL76021, February.

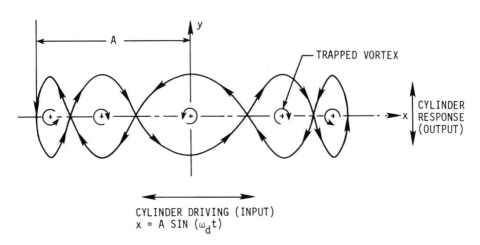

Fig. 1. An example of Lissajous loop of cylinder motion when $f_r/f_d = 5$.

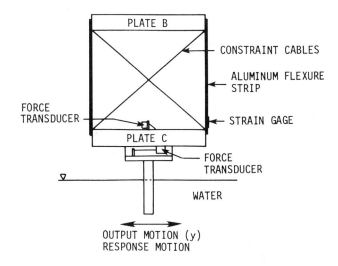

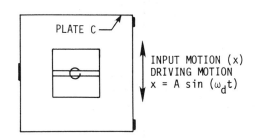

Fig. 2. Schematic of cylinder structure showing input (driving) and output (response) directions.

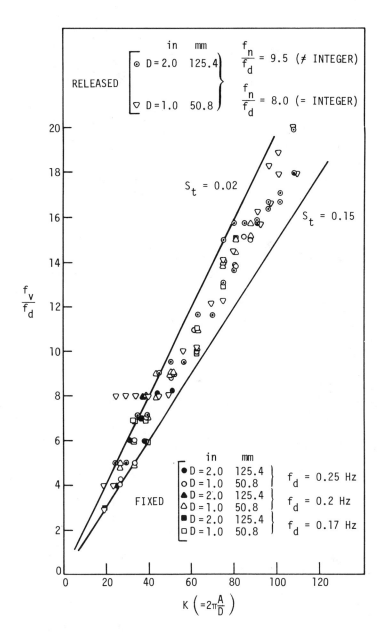

Fig. 3. f_v/f_d vs k (= $2\pi A/D$)

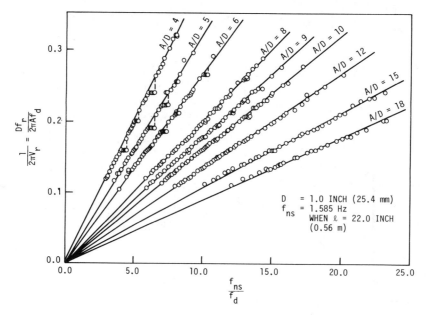

Fig. 4. $\frac{1}{(2\pi V_r)} = \frac{Df_r}{2\pi A f_d}$ vs f_{ns}/f_d for $D = 1$ inch and varying A/D ratio.

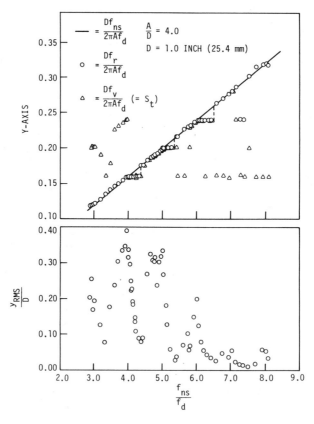

Fig. 5. Frequency relationships between f_r, f_n, and f_y and cylinder response amplitudes (y_{RMS}/D) for $D = 1$ inch and A/D = 4.

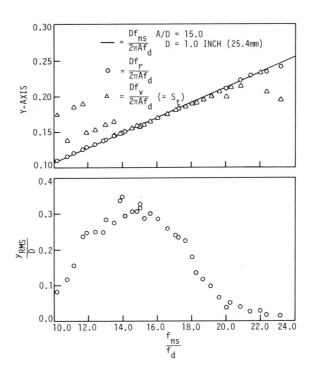

Fig. 6. Frequency relationship between f_r, f_n, and f_v and cylinder response amplitudes (y_{RMS}/D) for $D = 1$ inch and $A/D = 15$.

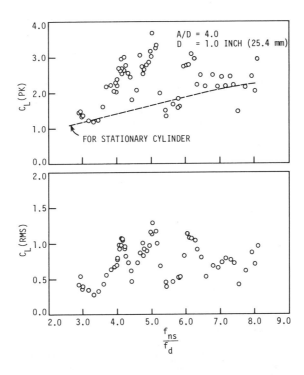

Fig. 7. Lift force coefficient $C_{L[pk]}$, and $C_{L[RMS]}$) vs f_{ns}/f_d when $D = 1$ inch and $A/D = 4$.

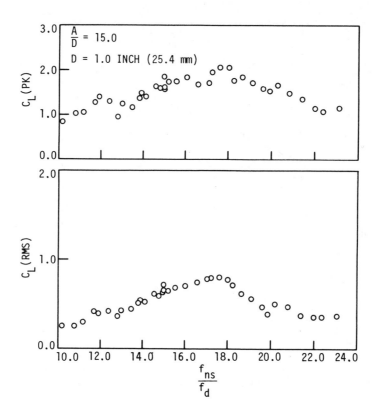

Fig. 8. Lift force coefficient ($C_{L[pk]}$, and $C_{L[RMS]}$) vs f_{ns}/f_d when D = 1 inch and A/D = 15.

MEASUREMENT OF FLCTUATING FORCES ON AN OSCILLATING CYLINDER IN A CROSS FLOW

Mark J. Moeller
Room 5-021
Mass. Inst. of Tech.
77 Mass. Ave.
Cambridge, Mass.

Patrick Leehey
Room 1-207
Mass. Inst. of Tech.
77 Mass. Ave.
Cambridge, Mass.

SUMMARY

A force transducer to measure the unsteady forces associated with vortex shedding from a cylinder in a cross flow has been developed. The transducer measured the force on a small isolated section of the test cylinder. The relative displacement of the active section of the transducer to the test cylinder was sensed by a capacitive pickup. The transducer was calibrated by a magnet and coil calibrator. The motion of the transducer was sensed by an accelerometer attached to the test cylinder. The accelerometer signal was used to compensate for the effective mass of the transducer. The force transducer allowed a simultaneous measurement of the local force and the cylinder motion.

The force transducers were used in the MIT Marine Hydrodynamics Laboratory's closed circuit water tunnel. The test cylinders were suspended on a yoke that was driven by a Bruel and Kjaer shaker. The test cylinders were 1.6 cm in diameter and 41 cm long. The tests were conducted at a Reynolds number of 19,300. The transducer was used to investigated the effect of cylinder motion on the local force. The test cylinders were driven sinusoidally at amplitude to diameter ratios from 0.05 to 0.5 in the reduced frequency range $0.1 < fd/U_\infty < 0.3$.

I. Introduction:

A right circular cylinder in a cross flow undergoes vortex shedding at moderate Reynold's numbers. The vortex shedding induces unsteady forces on the cylinder transverse to the direction of the flow as well as steady and unsteady forces in the flow direction. These unsteady forces can cause destructive levels of vibration in many common cases such as power lines (Blevins 1977), towing cables (Blevins 1977), and trashracks (Crandall et al 1975). The cylinder motion plays an important role in the shedding process.

The shedding of vortices by a cylinder in a crossflow is a narrowband random process. The vortices are shed in a band of frequencies around the Strouhal frequency. The shedding process is also a spatially random process. Although the geometry is two-dimensional, the shedding of vortices at two points on a cylinder is uncorrelated if the spatial separation of the points is great.

The vortex shedding process is sensitive to several different disturbances such as the freestream turbulence level (Gerrard 1965), surface roughness of the cylinders (Szechenyi 1975), cylinder end conditions (Cowdery 1962), (Standsby 1974), and motion of the supporting cylinder (Bishop and Hassan 1964), (Sarpkaya 1977), (Schargel 1980), (Tanida et al 1973), (Crandall et al 1975). The motion of the cylinder is of particular interest. Cylinder motion can cause the frequency of the vortex shedding to change, lock-on, to the frequency of vibration of the cylinder (Bishop and Hassan 1964). The magnitude of the fluctuating forces is influenced by the cylinder motion. The power spectra of the fluctuating forces is also influenced by the cylinder motion.

The effect of cylinder motion was investigated experimentally using a force measuring technique described in Moeller and Leehey (1982). The experiments were conducted in the recirculating water tunnel in the MIT Marine Hydrodynamics Laboratory using the equipment of the MIT Acoustics and Vibration Laboratory. The force transducer and its calibration are discussed in Section Two. The experiments on a cylinder oscillated harmonically in a steady flow are discussed in Section Three. The conclusions are presented in Section Four.

II. The Force Transducer

2.1 Transducer Design

The method adopted in this investigation was to develop a transducer to measure the local force at different stations on the same cylinder. The transducer was originally developed to be used in a wind tunnel (Moeller and Leehey 1982). The adaptation of the force transducer to a marine environment is described in this paper.

The elements of the force transducer are shown in Figure 1. The transducer consists of a 1.27 cm diameter aluminum ring, .051 cm thick, supported on a 1.91 cm long cantilever beam element. The beam element is aluminum, 0.08 cm thick. The beam element is rigidly fixed in a linen laminated phenolic base. Copper plating was epoxied to the base such that when the transducer was assembled there was a 0.0127 cm gap between the spring element and the copper plate. The ring element was epoxied to the beam element. The transducer electronics were fixed in place and the probe was wired up. The assembled probe was epoxied in the supporting cylinder maintaining a 0.025 cm gap between the ring element and the supporting cylinder. The transducer was sensitive to the displacement of the beam element relative to the copper plates. It was primarily sensitive to the force in the normal direction because it was much stiffer in the cross direction compared to the normal direction.

The probes were waterproofed using a commercially available RTV silicone elastomer that did not produce a corrosive by-product in curing. This provided a waterproof bond to the aluminum ring and to the steel cylinder. The test cylinder was filled with a two component RTV.

2.2 Electronic Design

The transducer senses the displacement of the spring element relative to the copper plates by means of a capacitive voltage divider. The copper plates and the beam element act as two variable capacitors with the capacitance depending on the position of the beam element. The circuit for sensing the change in capacitance is shown in Figure 2. The probe was driven with a 700 kHz sine wave from a Colpitts oscillator. The probe output was buffered by a built in preamp. The probe output was buffered to reduce its output impedance and improve its signal to noise. The signal from the probe was demodulated by a diode ring demodulator (Buck 1974). The resulting output was a voltage linearly related to the relative displacement of the beam element to the copper plates.

2.3 Calibration of the Transducers

A magnet and coil calibrator was developed to calibrate the force transducer. A 0.32 cm diameter permanent magnet, 1.27 cm long was rigidly attached to the ring element. The magnet was immersed in a coil a known, controlled distance. A current was put through the coil resulting in an electromotive force on the magnet proportional to the current. The magnet and coil calibrator was calibrated using an impedance head.

The magnet and coil calibrator was used to do a broad band calibration of the force transducer. A white noise generator was used to drive the magnet and coil calibrator. The data were analyzed using a Hewlett-Packard Structural Dynamics Analyzer 5423A. The analyzer has two analog to digital converters. The digitized data were fast fourier transformed. Spectral estimates were averaged to obtain stable estimates of the desired frequency domain quantities, the cross spectrum and the power spectra for both channels of data. These spectra were used to calculate the transfer function and the coherence.

The frequency range of interest in water was from 3 Hz to 50 Hz. The probes were calibrated in air using the magnet and coil calibrator. The probes had a resonance in air at 1.5 kHz. In water the resonance was lower than this but still well above the frequency range of interest. The probes were calibrated in the frequency range of interest in the calibration stand. The transfer function between the force input and the transducer response was flat and in phase.

III. Fluctuating Forces on an Oscillating Cylinder

3.1 The Water Tunnel

The water tunnel in the MIT Marine Hydrodynamics Laboratory is a closed circuit water tunnel with a 50.8 cm by 50.8 cm square test section, 137 cm long. The contraction ratio of the water tunnel is 4.5 to 1. There is a honeycomb of one inch diameter acrylic tubes upstream of the contraction to reduce the freestream turbulence levels. The water velocity could be varied from 0.12 m/sec to 10 m/sec.

The freestream flow conditions were measured using a laser doppler velocimeter and the freestream turbulence was measured using a hot film anemometer. The mean velocity profile across the test section was measured with laser doppler velocimeter. The mean velocity profile was uniform across the test section. The freestream turbulence level was measured using a cylindrical hot film probe, TSI model 1212. The probe was connected to a TSI model 1750 constant temperature anemometer. The anemometer output was linearized using an Analog Devices 433B Multifunction Module. The mean velocity was measured on a digital multimeter and the root mean square velocity was determined on a B+K true rms meter. The freestream turbulence level was 0.9%.

3.2 Yoke and Shaker

The apparatus used to oscillate the cylinder in the flow was designed by Schargel (1980). His apparatus was modified by the author to include endplates on the support struts and to accomodate a 1.6 cm cylinder.

The test cylinder was supported in a yoke, shown in Figure 3. The yoke consisted of a pair of aluminum struts of rectangular cross-section 1/4"x 1 1/2" and 25" long. The struts were supported in a steel frame. The frame consisted of two 1/4" aluminum triangular pieces (one per side). Two threaded rods held the bottom of the frame and two aluminum angle irons were used as the upper braces to provide lateral stiffness to the frame. Mounted on the inside of the frame were four 1/2" ball bushings (two per side) aligned so that the yoke could oscillate only in the verticle direction. The shafts were 1/2" diameter stainless steel rods. Two 1/2" diameter steel tubes connected the top and bottom of the stainless steel rods together. The aluminum struts were connected to the steel tubes. This arrangement allowed verticle motion while eliminating horizontal motion.

The struts extended vertically downward through the plexiglass window of the test section. The window was sealed using 3/8" thick neoprene compression seals. The struts were machined to an air foil shape to minimize their drag and were bent so that a higher aspect ratio cylinder could be used.

The yoke was attached to a B+K model 4801 shaker using a B+K model 4814 mode study shaker head. The power amplifier was a B+K model 2707. The shaker has a 2.54 cm peak to peak displacement limit and a force limit of 381 Newtons. The shaker was used to oscillate the cylinder. The total mass of the yoke plus the test cylinder was 2.4 Kg. At low frequencies the displacement limit of the shaker determined the maximum amplitude that the test cylinder could be oscillated at. At high frequencies the available force determined the maximum amplitude.

3.3 Stationary Cylinder Results

The test cylinder spanned the water tunnel and was supported by the struts. The endplates attached to the struts were 10 diameters in diameter and were designed as specified by Cowdery (1962). The endplates were used to provide repeatable boundary conditions for the experiments.

The struts were blocked so that the cylinder was stationary and the flow was turned on and allowed to stabilize. The flow velocity was set and the transducer output was tape recorded on a FM tape recorder. The data were also analyzed in real time using the Structural Dynamics Analyzer. The rms sectional lift coefficient and the shedding frequency were determined from the spectrum levels of the lift signal. The spectrum level of the lift force is shown in Figure 4 for a Reynold's number of 19300. The peak is associated with the vortex shedding. The Strouhal number, $f_s = fd/U_\infty$, is 0.19 and the rms sectional lift coefficient is 0.43.

3.4 Oscillating Cylinder Experiments

The test cylinder spanned the water tunnel with the transducer on the center line of the water tunnel. It was supported in a yoke that allowed only verticle motion. The yoke was oscillated harmonically by the B+K shaker. An accelerometer on the yoke sensed the motion of the yoke and test cylinder. The result was a simultaneous measurement of the local force and acceleration.

The experiments were performed by setting the flow velocity and then setting the desired driving frequency on the oscillator. The cylinder motion was changed from stationary to the desired amplitude of motion by turning up the gain potentiometer on the shaker. The flow was allowed to stabilize and then the data were taken. The data were tape recorded on an FM tape recorder and analyzed using the Structural Dynamics Analyzer.

The force transducer was acceleration sensitive because it was a physical device and had mass. When the cylinder was oscillated in air the force transducer output was a signal equal to the effective mass times the cylinder acceleration.

The transducer responds to its own inertia as well as to the fluid loading on it. When the cylinder was stationary, the transducer responded to the lift forces on it due to the vortex shedding. These forces occurred in a narrow band around the Strouhal frequency, f_s. The cylinder was oscillated in the mean flow at the forcing frequency, f_f, (different from f_s), at a fixed amplitude. The response was observed at the forcing frequency, f_f, and at the Strouhal frequency, f_s. The excitation was a pure tone at the forcing frequency and the acceleration was nearly pure tone. The acceleration and force were analyzed in the frequency domain. The coherence was high only at the forcing frequency. The transfer function between the acceleration and force was meaningful only at the forcing frequency. It was used to determine the relative phase of the acceleration and the force at the forcing frequency. The magnitude of the force at the Strouhal frequency was also monitored. The force at the driving frequency can be broken down into two components, the force in phase with the velocity and the force in phase with the acceleration. The force in phase with the acceleration included the transducer response due to its effective mass. The force in phase with the velocity was the work producing force.

A phenomenon that has been observed before is that the shedding frequency would change from the Strouhal frequency to the driving frequency under certain conditions. This phenomenon was termed lock-in (Bishop and Hassan 1964). The lock-in phenomenon has been observed for both oscillations in line with the flow as well as transverse to the flow (Crandall et al 1975).

3.5 Oscillating Cylinder Results

The lock-in effect was investigated for discrete points in the nondimensional amplitude, nondimensional frequency space. The amplitude displacement, δ, was nondimensionalized on cylinder diameter, $\delta^* = \delta/d$. The frequency was nondimensionalized on the Strouhal frequency, $f^* = f/f_s$. The lock-in effect was investigated in the region $0.0 < \delta^* < 0.5$ and $0.6 < f^* < 1.4$. This was limited by the shaker displacement and peak power. The data shows which frequencies and amplitudes were required for lock-in to occur. The trend observed was that the closer to the shedding frequency the smaller the amplitude for lock-in to occur. And at a fixed frequency the larger the amplitude of motion the more likely lock-in is to occur. Both the locked-in behavior and non locked-in behavior were investigated. The region of $\delta^* - f^*$ space where lock-in was observed is shown in Figure 5. The lock-in region is the region inside the lines in Figure 5. Outside this region lock-in behavior was not observed.

Non Locked-In Behavior

In the non locked-in region the response of the force transducer showed two peaks. Figure 8 shows the non locked-in response for $f^* = 0.72$ and $\delta^* = 0.41$. The response of the force transducer contains two peaks, one at the forcing frequency and one at the Strouhal frequency. The peak at the forcing frequency includes the response due to the effective mass of the transducer. At the forcing frequency at amplitudes, δ^* up to 0.4, the fluid was predominately mass like. The motion of the cylinder at these frequencies did not completely disrupt the shedding of vortices at the shedding frequency. There was still the same general character to the force spectrum as when the cylinder was stationary at the same Reynold's number.

Locked-In Behavior

In the lock-in region the shedding frequency changed from the Strouhal frequency to the forcing frequency and only one peak occurred. An example of lock-in behavior is shown in Figure 7, for an $f^* = 1.0$ and $\delta^* = 0.38$. The spectrum levels of the stationary cylinder and the locked in behavior can be compared in Figure 7. The cylinder motion has organized the flow as can be seen from the narrowness of the peak for the locked in case. The vortex shedding has become almost pure tone for the locked-in case. An example of non lock-in and lock-in behaviour at the same forcing frequency is shown in Figure 8, $f^* = 1.12$ and $\delta^* = 0.05$ and $\delta^* = 0.34$. This shows that at a fixed frequency the flow can be locked-on or not locked-on depending on amplitude of motion.

IV. Results And Conclusions

1. The force transducer is an effective method of measuring the local force on an oscillating cylinder undergoing vortex shedding.

2. Boundaries of where lock-in occurred were mapped out in a frequency-displacement space for a Reynolds number of 19,300.

3. The non lock-in force on a cylinder oscillated transverse to the flow consists of an inertia-like force at the driving frequency and a narrow band force near the Strouhal frequency.

Acknowlegements

The funding for this work was provided by the Office of Naval Research, Contract No. N00014-76-6-0696.

References

Blevins, R.D., 1977, Flow Induced Vibration. New York: Von Reinhold Co.

Crandall, S.H., Vigander, S., and March, P.A., 1975, "destructive Vibration of Trashracks due to Fluid Structure Interaction," Journal of Engineering for Industry, pp. 1359 - 1365.

Cowdery, C.F., 1962, A Note on the Use of Endplates to Prevent Three Dimensional Flow at the Ends of Bluff Cylinders, NPL AERO Report 1025.

Gerrard, J.H., 1965, "A Disturbance-sensitive Renolds Number Range of the Flow Past a Circular Cylinder," Journal of Fluid Mechanics, Volume 22, part 1, pp. 187 - 196.

Moeller, M.J., and Leehey, P., 1982, "Measurement of Fluctuating Forces on a Cylinder in a Cross Flow," Proceedings, 28th International Instrumentation Society, Las Vegas, Nevada, may 1982.

Sarpkaya, T., 1977, Transverse Oscillations of a Circular Cylinder in a Uniform Flow, Part 1, Naval Postgraduate School, NPS-69SL77071.

Shargel, R.S., 1980, The Drag Coefficient for a Randomly Oscillating Cylinder in a Uniform Flow, Masters Thesis, Massachussetts Institute of Technology, Department of Ocean Engineering.

Szechenyi, E., 1975, "Supercritical Reynolds Number Simulation for Two-dimensional Flow Over Circular Cylinders, Jornal of Fluid Mechanics, Volume 70, part 3, pp 529-542.

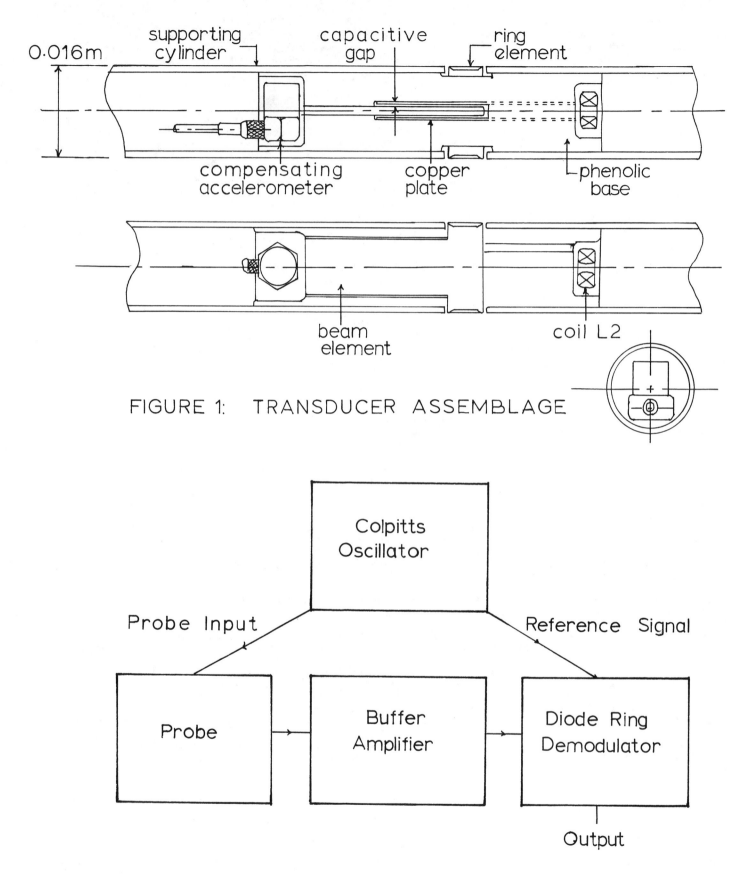

FIGURE 1: TRANSDUCER ASSEMBLAGE

FIGURE 2: Transducer Electronics

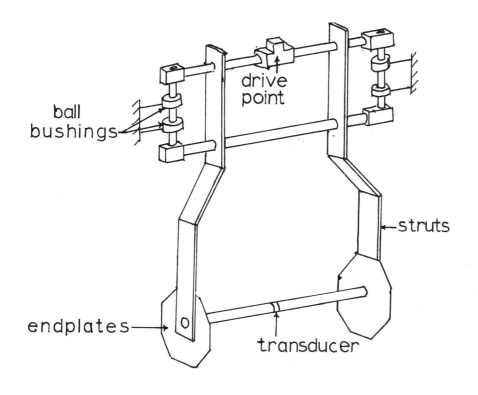

FIGURE 3: TEST APPARATUS

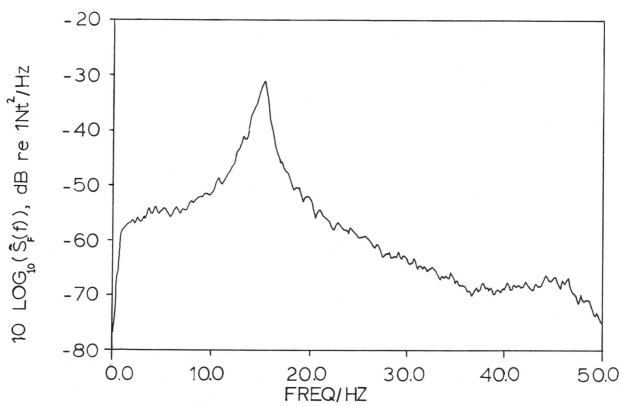

FIGURE 4: SPECTRUM LEVEL LIFT FORCE, STATIONARY CYLINDER
Re = 19,300 , C_L = 0.43

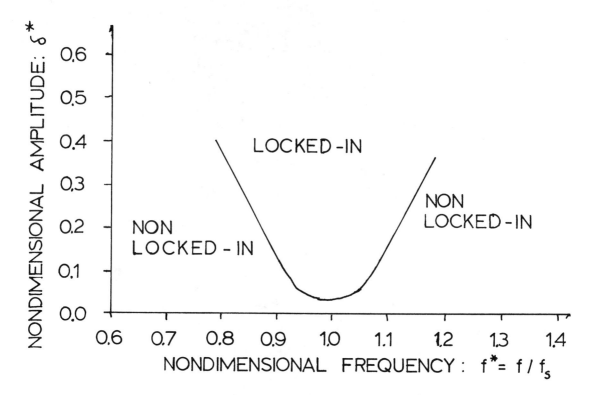

FIGURE 5: LOCK-IN BOUNDARIES IN δ^*-f^* SPACE RE = 19,300

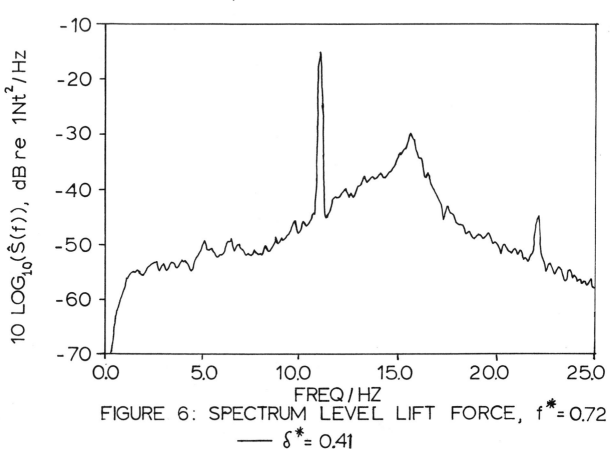

FIGURE 6: SPECTRUM LEVEL LIFT FORCE, f^*= 0.72
—— δ^*= 0.41

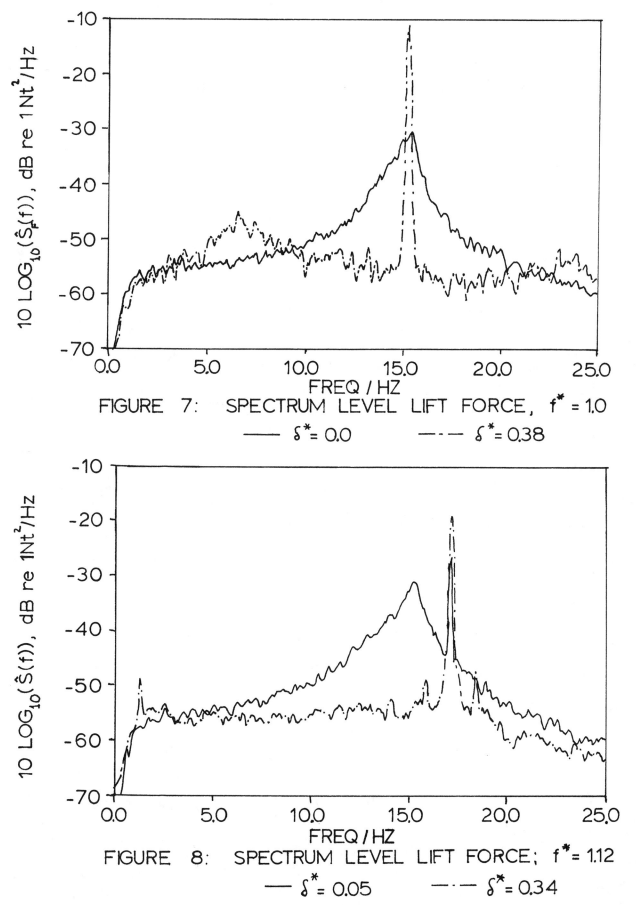

FIGURE 7: SPECTRUM LEVEL LIFT FORCE, $f^* = 1.0$
—— $\delta^* = 0.0$ —·— $\delta^* = 0.38$

FIGURE 8: SPECTRUM LEVEL LIFT FORCE; $f^* = 1.12$
—— $\delta^* = 0.05$ —·— $\delta^* = 0.34$

OSCILLATIONS OF CYLINDERS IN WAVES AND CURRENTS

R. L. P. Verley
River and Harbour Laboratory
Trondheim,
Norway.

D. J. Johns
University of Technology,
Loughborough,
U.K.

SUMMARY

Two research investigations have been conducted, into the vibrations of vertical flexible cylinders in waves and into the forces on cylinders oscillated in still water or steady currents.

In the first set of experiments flexible and similar rigid cylinders were mounted vertically side by side in a wave flume. Reactions on the clyinders were measured and compared. Under some circumstances vibrations of the flexible cylinder occurred and criteria have been established for the onset and amplitude of the vibrations.

The vibrations were caused by response to vortex shedding and were correctly predicted by a mathematical dynamic model of the flexible cylinder, where the input force data was obtained from the rigid cylinder and damping was taken as the experimental still water value. When hydrodynamic damping, obtained by taking into account relative velocity effects, was included, the vibrations were underestimated. The use of the modified Morison's formula grossly underestimated the vibrations, partly due to the non-inclusion of vortex induced forces and partly due to the over-prediction of damping.

In the second set of experiments a horizontal cylinder was attached to a massive pendulum and forced to oscillate in still water or in-line with various currents. Forces on the cylinder were measured and from these steady drag coefficients and oscillatory drag and inertia coefficients were calculated. The vibration free nature of the forcing enabled extremely accurate results to be obtained over at least two orders of magnitude of the important parameters.

Large variations in the coefficients were observed and in particular the steady component of drag could, under different circumstances, be negative, or up to twice its normal value. The apparent non-applicability of the relative velocity drag force in the first experiments was explained.

Flow visualization experiments were conducted and the variations in the various coefficients have been explained in terms of the vortex shedding and wake characteristics.

The sometimes dramatic changes observed in the vortex shedding patterns and in the wake have been fully explained by considering the inherent instability of wake geometries other than the von Karman type alternating wake, the symmetric vortices which develop after flow reversal occurs, and the distance travelled by the cylinder relative to the fluid in the time between one flow reversal and the next.

These second experiments gave an explanation of the results obtained in waves in the first experiments.

The results obtained in this work provide a step forward in the understanding of two problems in the offshore industry.

(i) The fluid-induced damping of cylindrical structures or members vibrating in a flowing fluid.

(ii) The forces on a rigid cylindrical member due to a combination of waves and currents.

In addition the results provide an advancement in the understanding of the fluid dynamics of time dependent flows.

INTRODUCTION

The past thirty years has seen an immense development in the number and type of platforms for the offshore oil industry and the task of designing for severe environmental loadings is one of the most exacting. Of particular importance are vibration and fatigue which apply both in terms of the movement of a platform as a whole at its lower natural frequencies, and in terms of vibrations of individual slender members at their natural frequencies.

Direct excitation of the platform at its fundamental frequency by high frequency waves or by higher harmonics of the wave forces introduced by non-linear effects may occur.

Excitation of slender circular members by vortex shedding in currents or waves may occur Collapse due to vortex excited vibrations in steady currents is known to have occurred (Wootton et al (1972)) and correspondingly fatigue failure in waves has been attributed to periodic vortex shedding forces (Wiegel et al (1957)).

In all such vibration problems the prediction and determination of the damping is most important yet this remains one of the least known factors of offshore platform design.

Another important unknown in platform design is the effect of simultaneously applied waves and currents. Whilst the analyses required for this problem are well understood, there has been virtually no research into the magnitude of the forces acting on structures due to this complex flow.

The first study to be reported is concerned directly with the problem of vibrations of slender cylindrical members in waves. The second involves a more fundamental investigation into the forces on cylinders oscillated in line with a current, the results of which may be applied to problems of the damping of vibrations in waves or currents, or of forces due to combined waves and currents.

The first investigation was conducted at the British Hydromechanics Research Association (B.H.R.A.), Cranfield, U.K., and the second at the River and Harbour Laboratory (V.H.L.), Trondheim, Norway. This paper is based on a thesis successfully submitted by the first author for the degree of Ph.D. to the Loughborough University of Technology (U.K.). (Verley (1980)).

FIRST INVESTIGATION
VIBRATIONS OF ISOLATED CYLINDERS IN REGULAR WAVES

Description of Apparatus

Due to the variation in values of coefficients obtained by various researchers on rigid cylinders in waves and the small amount of data available, it was considered advisable to compare directly the forces on a flexible cylinder with those on a rigid cylinder mounted beside it, but with a sufficient distance between the cylinders that no interactions should occur. It was anticipated that the rigid cylinder would provide reference data against which to compare the flexible cylinder, thus isolating the effects due to flexibility and avoiding uncertainties in the estimated forces for rigid cylinders based on uncertain coefficient values and inaccuracies in estimated flow kinematics.

Because of the difficulty in varying the structural properties of an elastic cylinder, it was decided to use a rigid cylinder flexibly mounted as an inverted pendulum, and such is meant throughout this investigation when "flexible cylinder" is referred to.

The flexible cylinder was pin-jointed to a dummy cylinder at its bottom end and helical springs on long wires were attached to the upper end to give the natural frequencies required. The springs were anchored above the water level to stiff cantilevers and forces were measured by strain-gauging one of the cantilevers as a half-bridge. Natural frequencies were easily altered by changing the helical springs. Initially motion was permitted only in the direction parallel to the wave propagation, which is referred to as the "in-line" direction. Long inextendible wires were employed to restrain motion in the direction perpendicular to the wave propagation, the "cross-flow" direction. For later tests, in the cross-flow direction, the sets of restraining wires and helical springs were changed around to permit cross-flow motion and force measurement and to prevent in-line motion.

The decision to restrain one motion whilst looking at the other motion was taken because it simplified the test rig considerably and eased the analysis. King (1974), looking at steady flow induced hydroelastic vibrations, found that restraining motion in one direction had little or no effect on the response in the other direction.

It is recognised that the restriction of vibration to one direction of motion at a time is a simplification which may alter the flow field.

The rigid cylinder was also pin-jointed at its lower end to a dummy cylinder, and forces were measured on it by a half bridge strain-gauged stiff P.V.C. cantilever mounted inside the dummy cylinder at the upper end. This method was employed for the rigid cylinder throughout all tests. The natural frequency of the rigid cylinder was 16.5 Hz in water.

The wave flume used for the tests was 20 m long, 1.5 m deep and 1 m wide. A hinged flap type hydraulically driven regular wave maker was installed at one end, the amplitude and frequency of movement being controlled by a simple oscillator. The test cylinders were placed about half-way down the length of the flume. The beach had a slope of 20° and was covered with a 5 cm thick porous rubber matting. The reflection coefficient was estimated to be between 2 and 5% over the range of the wave frequencies used.

The tests were conducted with the top of the test cylinders 0.08 below the water surface in 1.2 m of water. The length of the test sections of the cylinders was 0.76 m. Above and below the test cylinders dummy cylinders were mounted to minimise end effects, there being a gap of approximately 0.5 mm between the test sections and the dummy sections.

All wave measurements were taken using a twin wire resistance type wave gauge. The wave gauge was self-compensating for changes in temperature and salinity, and allowed specific calibration factors to be selected at will.

Outputs recorded

Initially outputs were taken from the strain gauges at the top of the two cylinders and fed through amplifiers into a U.-V. recorder. Preliminary tests in the in-line direction indicated that the output from the flexible cylinder was the same as from the rigid cylinder, but (sometimes) with a higher frequency component super-imposed. To look at the higher frequency component the two signals were subtracted electrically and the resultant "difference reaction" trace displayed on the U.-V. recorder. This difference trace represented the effects due to flexibility, and this was the basic variable quantified. This difference trace was not used when testing in the cross-flow direction. The strain gauge outputs were calibrated in terms of force applied at the top of each cylinder, and there was found to be less than 1% deviation in the calibrations from the best linear fits. The measurements in the tests thus represent the reactions measured at the top of the cylinders.

Various methods of quantifying the vibration reaction have been used. In initial tests in the in-line direction, the highest peak-to-peak value of the difference reaction in a 15 second recording period was taken by eye from the U.-V. traces. In later tests, r.m.s. values were taken using a true reading r.m.s. meter set with a time constant of 15 seconds. This r.m.s. value is denoted R_x.

In the cross-flow direction it was not possible to subtract the rigid cylinder reaction from the flexible cylinder reaction. This was because the variation due to lift forces, on top of which was superimposed vibration, was sometimes similar to and sometimes the mirror image of that on the rigid cylinder, depending, presumably, on whether vortex shedding occurred from the same or opposite sides of the two cylinders. (This made no difference in the in-line tests). Therefore, in order to compare the results with those of the in-line direction, the average amplitude of the vibration component of the flexible cylinder reaction was estimated by eye and converted to r.m.s. using a form factor of 1.11 (as for a sinusoid). The r.m.s. value calculated is denoted R_y. The results are thus inherently less accurate than the in-line results.

Tests conducted

Several test series have been conducted for both the in-line and cross-flow directions. The cylinder diameter, D, and still water values of natural frequency, n, and logarithmic damping, δ_{sw}, have been varied over the following ranges.

In-line tests		Cross-flow tests	
D	.0190, .0254, .0508 m	D	.0254 m
n	from 1.7 to 6.7 Hz	n	from 2.2 to 5.4 Hz
δ_{sw}	from .05 to .3	δ_{sw}	from .08 to .7

The wave frequency, n_w, was varied in the range .5 to 1.0 Hz, and the wave height between .05 and .25 m.

Various damping factors were obtained by having one of the springs controlling the flexible cylinder natural frequency passing through a bath of silicone fluid. By using various silicone fluids the damping could be varied over the range δ_{sw} = .05 - .7, and was perfectly linear over the range of amplitudes of vibration encountered.

Experimental Procedure

For each wave frequency, at which each flexible cylinder natural frequency and damping combination was to be tested, the procedure was as follows.

After the previous test, the water in the wave flume was allowed to settle until there was no visible movement of the surface (about 10 minutes). The oscillator driving the wave generator was set for the appropriate wave frequency and the lowest wave height and the wavemaker was started. Recording of the traces was started after about 10 waves had passed the cylinders, and continued for a period of about 20 seconds. The amplitude of the wavemaker was then increased to the next wave height (without stopping the wavemaker), and results again recorded after about 10 waves at the new height had passed. The four lowest wave heights were tested in this manner, i.e. without stopping between wave heights. For the higher wave heights, however, the water in the flume was allowed to settle between tests at each wave height. Testing was conducted at 7 wave heights at each of 5 wave frequencies.

Results

Initial testing in the in-line direction showed that for low wave heights there was no visible difference between the traces from the rigid and flexible cylinders. For larger wave heights the flexible cylinder trace resembled that from the rigid cylinder, but with a higher frequency component superimposed which was at, or very near, the flexible cylinder natural frequency. It was noted that the average number of vibration cycles in the difference trace per wave cycle is equal to the integral multiple of the wave frequency nearest to the natural frequency.

The initial results have been quantified in Fig. 1 against U_m/nD, where U_m is the maximum horizontal fluid velocity calculated at the top of the cylinder using linear wave thoery. Experimental points are plotted for n = 6 and 1.65 Hz.

It is of interest to note that vibrations occurred even for the highest ratio of natural to wave frequency, 6/.6 = 10, the criterion $U_m/nD \geq 1.0$ apparently being the only criterion to be satisfied for vibration to occur.

For a given value of U_m/nD more response occurred for higher values of n. For the higher natural frequencies there is little scatter from the curves shown in Fig. 1; however for lower natural frequencies there is, for any particular natural frequency, greater response for some wave frequencies than for others. The greater response appeared to occur when the natural frequency was nearer to an integral multiple of the wave frequency.

In later results the natural to wave frequency ratio is expressed as a parameter $|(1 - rn_w/n)|$, where r expresses the number of oscillations of the difference trace per wave cycle, i.e. r is an integer such that rn_w/n is as near to 1.0 as possible. Thus the nearer the parameter $(1 - rn_w/n)$ is to 0.0, the nearer the natural frequency is to an integral multiple of the wave frequency and the greater response expected.

The results are plotted as R_x/nD^3 for the in-line direction and R_y/nD^3 for the cross-flow direction in Fig. 2. Average curves are plotted for the points in various ranges of δ_{sw} and $|(1 - rn_w/n)|$.

The parameters R_x/nD^3 and R_y/nD^3 are not non-dimensional, but are used for convenience as they were found to collapse the results for all cylinder diameters, natural frequencies and wave frequencies.

From the results it is seen that vibration first starts at $U_m/nD \simeq 1.0$ for both the in-line and cross-flow directions. Greater response occurs for lower values of $|(1 - rn_w/n)|$, but the parameter has less effect for higher values of δ_{sw}. Increased damping reduces the response, but not by very much. The cross-flow results are similar to the in-line except that the response is about twice as great.

Conclusions

Reactions were measured at the top of the two cylinders and compared. Under some circumstances vibration of the flexibly mounted cylinder occurred. The major results are summarised as follows: (i) Vibrations were due to the response to vortex shedding and could occur even if the natural frequency of the flexible cylinder was many times higher than the wave frequency. (ii) Parameters controlling the onset of vibration and quantifying the magnitude of vibration have been isolated. (iii) The overall amplitude of the reaction of the rigid cylinder was predicted well by applying Morison's formula and by using coefficient values obtained in oscillatory flow distributed over the cylinder according to the local values of Keulegan-Carpenter number. The smaller variations in the reaction were not predicted by this model. (iv) The following mathematical models were used to try to predict the experimentally obtained flexible cylinder vibration data for vibrations parallel to the direction of wave advance. (a) The experimental reaction measured on the rigid cylinder was used as input to a dynamic mathematical model of the flexible cylinder, which used still water values of added mass and damping. (b) The experimental rigid cylinder reaction was used as input to a model which included additional "hydrodynamic" damping due to relative velocity effects. (c) Morison's formula was used to predict the rigid cylinder input reaction to a model using still water added mass and damping. (d) The modified Morison's formula was used to predict both the input reaction to a model and to give additional hydrodynamic damping. (v) Models (c) and (d) do not include any vortex shedding forces. Model (a) adequately predicted the experimental vibrations parallel to the direction of wave advance and also vibrations in the direction perpendicular to wave advance. Models (b), (c) and particularly (d) underpredicted the vibrations. The results of the models imply that fluid forces were unaltered by the cylinder vibrations, that hydro-elastic vibration did not occur, and that a full knowledge of the forcing on a rigid cylinder would be sufficient to predict the response of flexible cylinders. Furthermore, the results suggested that the use of the modified Morison's formula in the dynamic analysis of offshore platforms may not correctly predict vibrations. (vi) A quasi-steady model of vortex shedding in oscillatory and wave flows has been studied where the instantaneous shedding frequency is considered to be proportional to the instantaneous velocity, and the instantaneous force amplitude proportional to the square of the instantaneous velocity. The model gave a good qualitative prediction of lift traces found in wave and oscillatory flow experiments, and of the variation in the spectral content of the lift forces with varying Keulegan-Carpenter number.

SECOND INVESTIGATION

OSCILLATIONS OF CYLINDERS IN STILL WATER AND IN-LINE WITH CURRENTS

Description of the apparatus

The basic experimental apparatus consisted of a massive pendulum which spanned a current flume and to which a horizontal cylinder was attached and submerged in either still or flowing water. The pendulum was sufficiently massive that the damping was small, there being at most a 2% decrease in amplitude per cycle. The position of the cylinder was monitored by a linear spring and force transducer, and from this signal the damping was calculated and this could be related to the oscillatory drag coefficient. In addition, the forces on the cylinder in the horizontal direction were measured, and from these force time histories could be calculated the oscillating drag and inertia coefficients and steady component of drag coefficients. The end plates on the horizontal cylinder were intended to ensure 2-dimensionality of the flow.

The basic pendulum rig had a natural frequency of 0.4764 Hz in air with no cylinder attached. A number of tests was also conducted with a corresponding frequency of 1.409 Hz. With various cylinders attached the lowest frequencies were 0.4717 Hz and 1.397 Hz respectively.

The flume used was specially built for the experiements. The main part of the channel was 10 m long, 0.5 m wide, and 0.7 m deep, with a water depth of .55 m used in the experiments. The water level was held constant for the various flow rates. The cylinder and mounting arm combination was hinged to the main horizontal beam at the upper end and held to a bar attached to a beam by a ring-type force transducer. The force transducer was mounted such that it measured forces on the cylinder perpendicular to the line between the pivot points of the pendulum and the centre of the cylinder. Thus when the pendulum was deflected the force measured was not quite horizontal. The maximum deflection was $12°$ giving a maximum error in horizontal force measurement of about 2%.

The FM tape recorder used had a built in amplifier which could give amplification of 1, 2, 5 or 10 times.

Tests Conducted

Seven horizontal cylinders were tested with diameters from .01275 m to .0701 m and flow velocities up to 0.445 m/s.

The maximum amplitude of motion of the cylinders in the experiments was about 0.15 m. Data was obtained for amplitudes down to about 0.001 m. The ranges of various major parameters were

V/nD	.6 to 47
$V/\dot{x}_o$	.05 to 50
$\beta = nD^2/\nu$	80 to 3500

Experimental procedure

In order to obtain the force coefficients for the various cylinders it is necessary to conduct the experiments with the cylinders and also with no cylinder attached to the pendulum. The data obtained from the rig with no cylinder attached is subtracted from that from tests with the various cylinders to yield effects due to the cylinders. The variation in the amplitude of motion for any one test condition is about two orders of magnitude, and forces vary by even more, so it is evident that it is necessary to vary the sensitivity of the amplifiers in the course of the experiments. Flow visualisation experiments have also been conducted.

Analysis of Results

The results have been analysed using two equations. In the first the force per unit length of the cylinder is considered to be given by the relative velocities and accelerations (fluid acceleration is zero however), and written

$$f(t) = .5 \rho D\, C_{D3}(t)\, (V - \dot{x})\, |(V - \dot{x})| - .25 \rho \pi D^2\, C_a(t)\, \ddot{x} \qquad (1)$$

where C_{D3} and C_a are time dependent drag and added mass coefficients. There is in this situation no Froude-Krylov force, and the inertia coefficient which would apply with the fluid oscillating rather than the cylinder would be given by $C_M = C_a + 1$. N.B. $\dot{x}$ and $\ddot{x}$ are the structural velocity and acceleration, respectively.

The second equation used assumed that the forces due to the steady current and the cylinder oscillations could be written independently, i.e.

$$f(t) = .5 \rho D\, C_{D1}(t) V^2 - .5 \rho D\, C_{D2}(t)\, \dot{x}\, |\dot{x}| - .25 \rho \pi D^2\, C_a(t)\, \ddot{x} \qquad (2)$$

Here C_{D1} and C_{D2} are drag coefficients associated with the steady and oscillatory drag forces respectively.

C_{D3}, C_{D1}, C_{D2} and C_a are in general functions of time. In this paper time averaged coefficients will be used.

Two methods are commonly used to determine time averaged coefficients in wave or oscillatory flows. In the first the oscillatory drag and inertia coefficients are related to the first two Fourier coefficients of the time series, that is to say the sine and cosine coefficients at the fundamental, oscillatory frequency. This method was used by Keulegan and Carpenter (1958). The other commonly used method is a least squares fitting of the two coefficients to the force measured in a cycle, as used by Sarpkaya (1976, etc.).

The following methods are used in this paper. The oscillatory drag force is determined from the damping measured, and used to calculate time averaged oscillatory drag coefficients from (1) and (2). The steady component of drag force measured is equated to the steady components of equations (1) or (2), to give time averaged steady drag coefficients. Time averaged oscillatory drag and inertia coefficients are determined from the time histories of force on the cylinder using the Fourier analysis method used by Sarpkaya (1976). A detailed derivation of the coefficients is given in Verley (1980) where it is also shown that all the coefficients are equivalent to those determined by equating to Fourier coefficients, as used, for example, by Matten (1976) in his re-analysis of the data of Mercier (1973).

In equation (1) the steady and oscillatory drag forces are mutually dependent and the time-averaged oscillatory drag and steady drag coefficients determined are denoted CDDEP and STCDDEP, respectively.

In equation (2) the steady and oscillatory drag forces are independent, and the time-averaged coefficients are denoted CDIND and STCDIND, respectively.

The time-averaged added mass coefficient is the same for both equation (1) and (2), and is called CA. The coefficient which will be plotted, however, is the inertia coefficient, CM = CA + 1.

CDDEP and STCDDEP are thus the time-averaged oscillatory and steady components of $C_{D3}(t)$ in equation (1) and CDIND and STCDIND are the time-averaged values of $C_{D1}(t)$ and $C_{D2}(t)$ in equation (2), respectively. CDDEP and CDIND represent contributions over and above the viscous contribution calculated from Stokes' equations (and nearly always very much smaller).

Thus, using time averaged coefficients, equation (1) may be rewritten as

$$f(t) = .5 \rho D (\text{STCDDEP}) \cdot (\text{steady component of } (V - \dot{x})\, |(V - \dot{x})|)$$
$$+ .5 \rho D (\text{CDDEP}) \cdot (\text{oscillatory component of } (V - \dot{x})\, |(V - \dot{x})|)$$
$$- .25 \rho \pi D^2 (\text{CA})\, \ddot{x} \qquad (3)$$

Similarly equation (2) may be rewritten as

$$f(t) = .5 \rho D(STCDIND) V^2 \\ + .5 \rho D(CDIND) \dot{x} |\dot{x}| \\ - .25 \rho \pi D^2 (CA) \ddot{x} \qquad (4)$$

The above equations apply to the experimental condition of a cylinder oscillating in a steady current. For a condition of a stationary cylinder in a combined steady current, of velocity V, and oscillatory current, of instantaneous velocity U, equations (3) and (4) would still apply, but with CM (= 1 + CA) replacing CA, -U replacing $\dot{x}$ and -U replacing $\ddot{x}$.

A time-averaged oscillatory drag coefficient from equation (2) was also determined by the method of Fourier averages applied to the force time histories. This coefficient is called CDIND'. Values of CDIND and CDIND' were found to be virtually identical when compared from tests in still water, but there was rather more scatter in the results of CDIND'. Therefore only results of CDIND, as obtained from the damping, are presented by Verley (1980).

Example Results

In Fig. 3 are shown examples of the damping, C_{cyl}, for various cylinders attached in various currents. It is seen that there is virtually no scatter except for results at the very highest flow velocities, when the scatter is still only about $\pm$ 5%.

The steady force, $\overline{F}_t$, for the rig with various cylinders attached in various currents also shows very little scatter in the results.

The scatter in the oscillatory drag coefficients, e.g. for CDIND· x_o/D versus $V/\dot{x}_o$ is very small other than for the highest flow velocities at the smallest amplitudes of oscillation, when the scatter is about $\pm$ 10%. It is nevertheless in this area where the method of equating the oscillatory drag force to the measured damping is most advantageous. The oscillatory drag force is extremely small compared to the steady drag force and small compared to the inertia force, and the results for CDIND', calculated from the force time histories, exhibit very much more scatter because of the smallness of this component. The smallness of the oscillatory drag force compared to other forces does not, however, imply that it is unimportant, precisely because of its relationship to damping.

For the steady drag coefficient, STCDDEP, there is very little scatter for high values of V/nD, and rather more, though still relatively little at lower V/nD. For V/nD $\leq$ 1.5 the scatter was about $\pm$ .3 (in STCDDEP), between various tests at similar values of V/nD.

For the inertia coefficient, CM, the scatter is small for low flow velocities, particularly at low x_o/D where the inertia forces dominate. For higher velocities of flow, there is more scatter, and for very high velocities, Fig. 4, there is extremely large scatter, particularly for low x_o/D. The reason for this scatter at high V/nD is that the inertia force is then extremely small compared to the steady force, particularly at low x_o/D and is comparable in magnitude with the variations in force due to turbulence, variations in flow velocity and the alternating forces from vortex shedding. It is not very important to quantify the inertia coefficient in this range precisely because of its small relative magnitude, and the inertia coefficient has not been plotted in the results where it exhibits large scatter.

The time averaged oscillatory drag coefficient based on equation (1), CDDEP, is plotted in Fig. 5 and based on equation (2), CDIND, in Figs. 6 and 7, for β = 200 - 500. In this range of β the results for the still water oscillatory drag coefficient versus x_o/D are virtually identical for all β, and the values of CDDEP and CDIND in currents are also the same for all β in this range.

The experimental apparatus consisted of a horizontal cylinder which was attached horizontally below a massive pendulum. Various cylinders were thus caused to oscillate in still water and in-line with various currents and the forces on the cylinders, as well as the logarithmic decrement of the pendulum motion, were recorded.

Conclusions

Due to the nature of the rig there was no extraneous vibration and only a very small decrease in amplitude per cycle. Extremely good accuracy in the measurement of forces and the subsequent calculation of coefficients was obtained. Large ranges of all the parameters of importance were covered, except for Reynolds number which was always subcritical. The major results in still water may be summarized as follows:

(i) For oscillations of a cylinder in still water, the drag and inertia coefficients have been accurately determined as functions of x_o/D $\beta = nD^2/\nu$, where x_o = amplitude of oscillation, D = cylinder diameter, n = frequency of oscillation and ν = kinematic viscosity. It has been found advisable, particularly at low amplitudes, to separate the viscous drag contribution, given approximately by Stokes' unseparated viscous flow equations, from the total drag coefficient. (ii) For very small amplitudes of vibration, the forces on the cylinders were accurately given by the application of Stokes' equations, and an expression has been found for the amplitude up to which the forces may be reasonably approximated by these equations.

The major results in currents may be summarized as follows: (i) It has been found that the force on a cylinder oscillating in line with a steady current is best described by a relative velocity type equation

$$f_{(t)} = .5 \rho D C_D(t) (V - \dot{x}) |(V - \dot{x})| - .25 \rho \pi D^2 C_a(t) \ddot{x} \qquad (1)$$

where V = steady current velocity, $\dot{x}$, $\ddot{x}$ = instantaneous cylinder velocity and acceleration, $C_D(t)$ and $C_a(t)$ = time dependent drag and added mass coefficients and ρ = fluid density. By taking averages of this equation over a cycle, three time-averaged coefficients are obtained for the steady component of drag, the oscillatory component of drag, and the added mass. The variations of these coefficients with respect to various parameters are presented. (ii) For values of a parameter $V/nD > 17$, the flow becomes quasi-steady, and the steady and oscillatory drag coefficients are equal for all amplitudes of oscillation. For $V/nD > 30$ the value is equal to the steady drag coefficient on a stationary cylinder. At lower values of V/nD values of the time averaged steady and oscillatory drag coefficients differ. (iii) The time-averaged oscillatory drag coefficient varies smoothly, other than for $x_o/D \leq .2$ when $V/nD \simeq 2.0$ or 3.0, from its value in still water at low V/nD, where it is dependent upon x_o/D and β, to the stationary cylinder steady drag coefficient at high V/nD. (iv) The variations in the time-averaged steady drag coefficient are complex, however, the three most important observations are: (a) A minimum occurs at $x_o/D \simeq 1.0$, which, for $V/nD < 1.6$, is negative. (b) A maximum, with value $\simeq 1.0$, occurs at $x_o/D \simeq 2.3$. (c) A maximum occurs for $V/nD > 3$ for $V/\dot{x}_o$ somewhere between 1 and 2, where $\dot{x}_o$ is the velocity amplitude of oscillations. This maximum may take a value up to twice the stationary cylinder steady drag coefficient value. (v) The time-averaged added mass coefficient varies considerably and is closely, but in an inverse way, related to the time-averaged steady drag coefficient.

Flow visualisation experiments have been conducted and the variations in the coefficients have been explained in terms of vortex shedding and wake characteristics. (Verley (1980)).

The sometimes dramatic changes observed in the vortex shedding patterns and in the wake have been fully explained by considering the inherent instability of wake geometries other than the von Karman type alternating wake, the symmetric vortices which develop after flow reversal occurs, and the distance travelled by the cylinder relative to the fluid in the time between one flow reversal and the next.

These second experiments gave an explanation of the results obtained in waves in the first experiments.

OVERALL CONCLUSIONS

The results obtained in this work provide a step forward in the understanding of two problems in the offshore industry. (i) The fluid-induced damping of cylindrical structures or members vibrating in a flowing fluid. (ii) The forces on a rigid cylindrical member due to a combination of waves and currents. In addition the results provide an advancement in the understanding of the fluid dynamics of time dependent flows.

REFERENCES

1. KEULEGAN, G.H. & CARPENTER, L.H., 1958, "Forces on cylinders and plates in an oscillating fluid". J. Res., Nat. Bureau Standards, 60, No. 5, pp. 423 - 440.

2. KING, R., 1974, Vortex excited structural oscillations of a circular cylinder in flowing water. Ph.D. Thesis, Loughborough University of Technology, Department of Transport Technology.

3. MATTEN, R. B., 1976, "Calculation of drag forces on a circular cylinder in a combined plane oscillatory and uniform flow field using Morrison's equation". N.M.I., Draft Report, April, 1976.

4. SARPKAYA, T., 1976, "Vortex shedding and resistance in harmonic flow about smooth and rough circular cylinders at high Reynolds numbers". Naval Postgrad. School, Monterey, Report NPS-59 SL 76021.

5. VERLEY, R. L. P., 1980, <u>Oscillations of Cylinders in Waves and Currents</u>. Ph.D. Thesis, Loughborough University of Technology, Department of Transport Technology.

6. **WIEGEL, R. L.**, et al, 1957, "Ocean wave forces on circular cylindrical piles". J. Hydr. Div., ASCE, <u>83</u>, HY2, pp. 1 - 34.

7. WOOTTON, L. R. et al, 1972, "Oscillations of piles in marine structures". CIRIA, Report 40, England.

ACKNOWLEDGEMENTS

The organisations whose financial, technical and supervisory support made the work described possible are: British Hydromechanics Research Association (BHRA), Cranfield, U.K.; Department of Energy (Offshore Fluid Loading Advisory Group), U.K.; University of Technology, Loughborough, U.K.; River and Harbour Laboratory (VHL), Trondheim, Norway; Royal Norwegian Council for Scientific and Industrial Research; Norwegian Institute of Technology.

Many individuals also contributed in various ways including: Mr. Michael Prosser, Mr. John Sheehan, Mr. George Young of BHRA; Professor Geir Moe, Dr. Alf Torum, Mr. Sveun Refseth, Mr. Johan Rechsteiner and Mr. Ivar Dybvik of VHL.

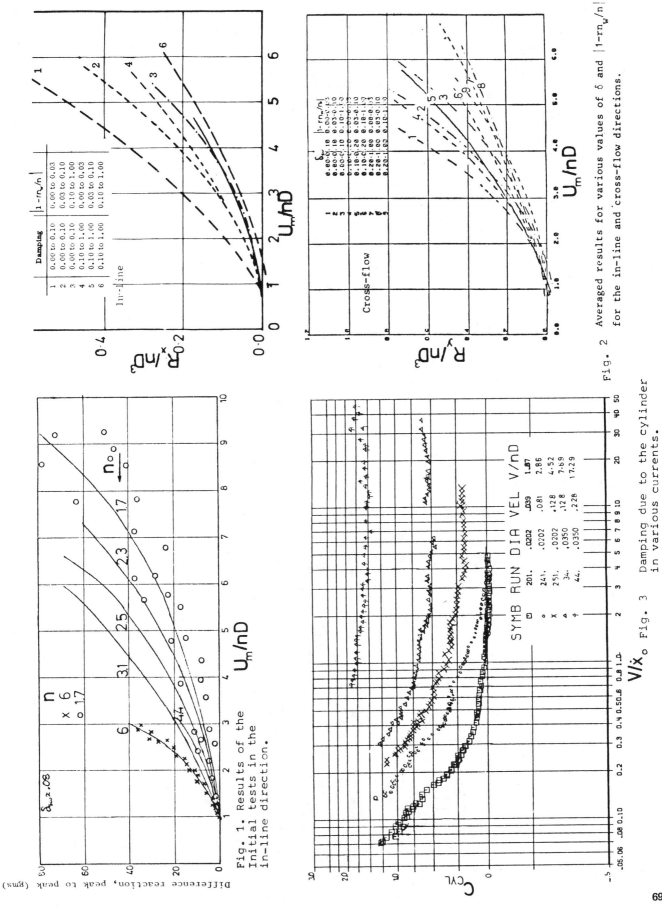

Fig. 1. Results of the initial tests in the in-line direction.

Fig. 2 Averaged results for various values of δ and $|1-rn_w/n|$ for the in-line and cross-flow directions.

Fig. 3 Damping due to the cylinder in various currents.

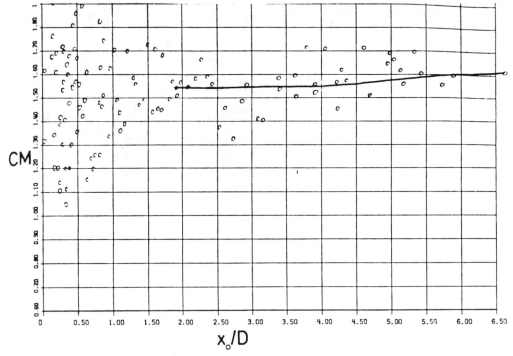

Fig. 4 Example results for CM versus x_o/D, very high V/nD.

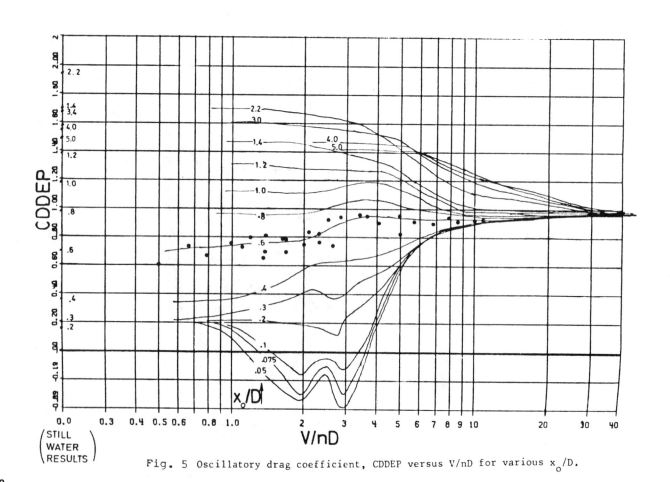

Fig. 5 Oscillatory drag coefficient, CDDEP versus V/nD for various x_o/D.

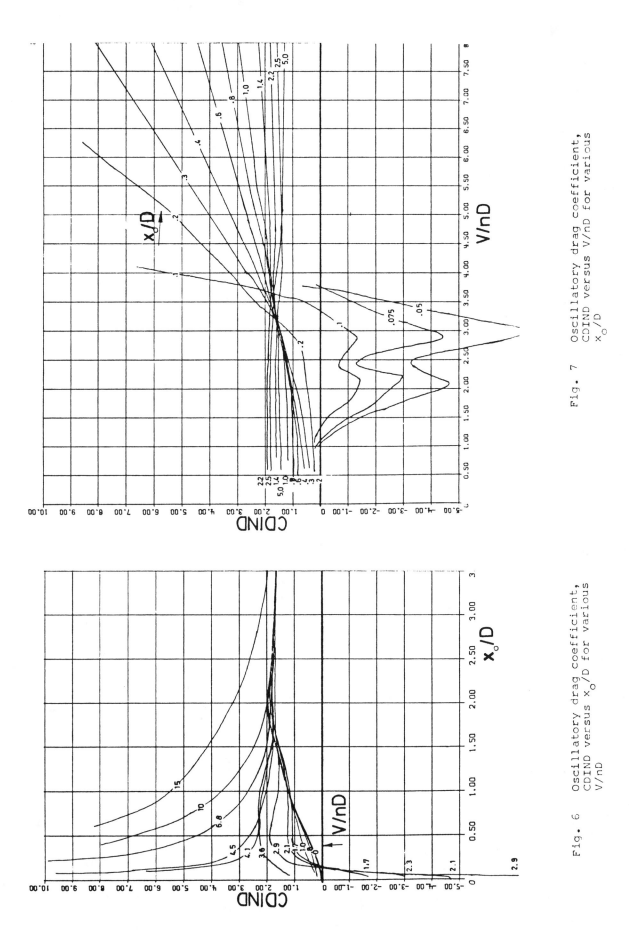

Fig. 6 Oscillatory drag coefficient, CDIND versus x_o/D for various V/nD

Fig. 7 Oscillatory drag coefficient, CDIND versus V/nD for various x_o/D

Technical Poster Sessions

STRUCTURES

Behaviour of Tubular Members

GLOBAL BUCKLING DESIGN CRITERIA FOR RISERS

Michael M. Bernitsas
Assistant Professor
Department of Naval Architecture
and Marine Engineering
The University of Michigan
Ann Arbor, Michigan, U.S.A.

Theodore Kokkinis
Graduate Student, Research Assistant
Department of Naval Architecture
and Marine Engineering
The University of Michigan
Ann Arbor, Michigan, U.S.A.

SUMMARY

The purpose of this work is (1) to clarify the issue of global Euler buckling of risers which are in tension over their entire length and (2) to establish design rules for proper distribution of the riser supporting forces between the top tension and the buoyancy modules in order to prevent global buckling without overdesigning the structure.

Customary Euler analysis is used to formulate the eigenvalue problem which consists of a fourth order linear differential equation with variable coefficients and four boundary conditions. Two sets of boundary conditions are considered which model a riser hinged at its lower end and either hinged or movably hinged at its top. Two solutions have been developed. One is expressed in terms of Airy functions and the other in terms of power series. The combined numerical results of the two methods cover the entire domain of practical interest of the two design variables β, the dimensionless effective weight per unit length, and τ, the dimensionless effective overpull at the riser's lower end.

Two design rules that are currently used to estimate the riser top tension are studied and compared to the derived critical buckling loads. The first rule which states that: "The tension to be applied at the top of the riser should be a multiple of the weight of the riser and contents in water" is a conservative one. The second rule which states that: "The tension to be applied at the top of the riser should be a multiple of the weight of the riser in vacuum minus the net buoyancy of the buoyancy modules" may result in global buckling of risers even if they are in tension over their entire length.

The actual τ_{crit}, the dimensionless critical top tension and the tensions which the two design rules yield are plotted versus β, and compared. These graphs clearly show the advantages and disadvantages of the currently used design rules as well as the region of buckling of risers in tension. An alternative design rule is proposed.

Finally, numerical examples are worked out in order to demonstrate the applicability of the rules and to show that existing risers may actually buckle even while the actual tension is positive over their entire length.

INTRODUCTION

Marine drilling, production and mining risers are open-ended long slender tubular columns which are subject to distributed load. The load consists of the riser weight, tension exerted at the top, internal fluid static pressure due to the drilling mud in the riser and external hydrostatic pressure. This load may cause global buckling of a riser. The purpose of this study is to predict the structural instability of these columns for two different sets of boundary conditions of practical importance which model a hinged-hinged and a hinged-movably hinged riser. Particular emphasis is given to the analysis of the effect of internal pressure on the stability of risers. It has been proved by Bernitsas and Taylor (1982) that internal static pressure due to a fluid in a gravity field may destabilize tubular columns which are in tension along their entire length.

The riser buckling problem is formulated and solved and the results are used in order to derive proper global buckling design criteria and evaluate the two widely used practical rules for computation of the tension required at the top of a riser. The results of this work are illustrated by two numerical examples which show the effects of the design rules which are currently used as well as the proper distribution of supporting riser forces between the tension required at the top and the buoyancy of the buoyancy modules. The general dimensionless results and the two numerical examples clearly show that risers may buckle in tension due to internal pressure.

PROBLEM FORMULATION

Customary Euler analysis is used to compute the buckling loads of a riser which is subject to the distributed load described above. These are the eigenvalues of the problem which is formulated in this section. The problem consists of a fourth order linear homogeneous differential equation and four homogeneous boundary conditions.

Differential Equation

A linearized, small slope, small deflection, small shear, Bernoulli-Euler beam model can be used for the computation of the buckling loads. This model is a special case of the general three dimensional nonlinear dynamic model developed by Bernitsas (1982b). Attention should be paid to the integration of the internal static pressure forces due to the circulating drilling mud in the riser and the external hydrostatic pressure forces (Bernitsas 1980). The differential equation for the Euler global buckling analysis is:

$$\frac{d^2}{dz^2}\left[EI\frac{d^2U}{dz^2}\right] - \frac{d}{dz}\left[(W_e z + P_e(0))\frac{dU(z)}{dz}\right] = 0 \qquad (1)$$

where

$$W_e = W_{st} + W_m - B_w - B_m \qquad (2)$$

is the effective weight per unit length,

$$W_{st} = \rho_{st} g \frac{\pi}{4}(D_o^2 - D_i^2) \qquad (3)$$

is the weight of the riser tube per unit length, ρ_{st} is the mass density of steel, D_o is the outer riser tube diameter, D_i is the inner riser tube diameter,

---

NOMENCLATURE

A_i	Airy function of the first kind	$T(z)$	Actual tension at z
B_i	Airy function of the second kind	TTR	Tension at the top of the riser
B_m	Net buoyancy of buoyancy modules per unit length	U	Lateral displacement of riser
B_w	Displaced weight of water by a unit length of riser	W_e	Riser effective weight per unit length
		W_m	Weight of drilling mud per unit length
D_i, D_o	Inner and outer riser diameters	W_{st}	Weight of riser tube per unit length
E	Young's modulus	z	Vertical coordinate along the riser
g	9.81 m/sec²		Greek Letters
I	Riser cross-sectional area moment of inertia	β	Dimensionless W_e
		δ	Dimensionless TTR
L	Length of riser	ρ_m	Drilling mud mass density
p	Dimensionless vertical coordinate along the riser	ρ_{st}	Mass density of steel
		ρ_w	Mass density of water
$P_e(z)$	Effective tension at z	τ	Dimensionless effective overpull at the lower end of the riser
Q	Shear force	τ_{crit}	Dimensionless eigenvalues

$$W_m = \rho_m g \frac{\pi}{4} D_i^2 \qquad (4)$$

is the weight of the drilling mud in the riser per unit length, ρ_m is the drilling mud mass density,

$$B_w = \rho_w g \frac{\pi}{4} D_o^2 \qquad (5)$$

is the displaced weight of water by a unit length of riser, B_m is the net buoyancy of the buoyancy modules per unit length, z is the vertical coordinate along the riser, U is the lateral displacement of the riser, E is Young's modulus,

$$I = \frac{\pi}{64} (D_o^4 - D_i^4) \qquad (6)$$

is the moment of inertia of the cross sectional area,

$$P_e(z) \equiv T(z) + \rho_w g \frac{\pi D_o^2}{4}(h_w - z) - \rho_m g \frac{\pi D_i^2}{4}(h_m - z) \qquad (7)$$

is the effective tension which is defined by equation (7) and satisfies equation (8)

$$\frac{dP_e(z)}{dz} = W_e, \qquad (8)$$

L is the length of the riser, h_w is the depth of the water which is approximately equal to L, h_m is the depth of the drilling mud, which is approximately equal to L, from its free surface to $z=0$ which corresponds to the lower end of the riser, and $T(z)$ is the the actual tension in the riser.

Assuming that the properties of the riser are constant along the riser, that is independent of z, we can rewrite equation (1) in the following dimensionless form

$$\frac{d^4 U}{dp^4} - (\beta p + \tau)\frac{d^2 U}{dp^2} - \beta \frac{dU}{dp} = 0 \qquad (9)$$

where

$$\beta = \frac{W_e L^3}{EI} \qquad (10)$$

is the dimensionless effective weight,

$$\tau = \frac{P_e(0) L^2}{EI} \qquad (11)$$

is the dimensionless overpull at the lower end of the riser and,

$$p = \frac{z}{L} \qquad (12)$$

is the dimensionless coordinate along the riser.

The advantage of the dimensionless form of equation (9) is that all the design, material and loading variables are lumped in two dimensionless variables, i.e. β and τ (Bernitsas 1981).

Boundary Conditions

For a general column analysis eight sets of boundary conditions which are considered of practical importance have been studied by the authors (Bernitsas et al 1983, Bernitsas and Kokkinis 1983a and 1983b). For the case of risers and the analysis of the problem of buckling of risers in tension two cases are considered of practical importance and are studied in this work. These correspond to a riser with a hinged lower end and a hinged or movably hinged top.

a. **Hinged-hinged riser**

For the lower end of the riser, i.e. for $p=0$ the boundary conditions are

$$U(0) = 0 \qquad (13) \qquad \frac{d^2 U(0)}{dp^2} = 0 \qquad (14)$$

For the upper end of the riser, i.e. for $p=1$, we have

$$U(1) = 0 \qquad (15) \qquad \frac{d^2 U(1)}{dp^2} = 0 \qquad (16)$$

b. **Hinged-movably hinged riser**

For the lower end of the riser, which is hinged,

$$U(0) = 0 \qquad (17) \qquad \frac{d^2 U(0)}{dp^2} = 0 \qquad (18)$$

For the upper end of the riser, which is hinged but is allowed to move horizontally the boundary conditions are (see Bernitsas et al 1983)

$$\frac{d^3 U(1)}{dp^3} - (\beta+\tau)\frac{dU(1)}{dp} = 0 \qquad (19) \qquad \frac{d^2 U(1)}{dp^2} = 0 \qquad (20)$$

Differential equation (9) along with either one of the two sets of boundary conditions, constitute an eigenvalue problem. This problem can be solved and yield the critical curves in the $\beta-\tau$ plane. Either β or τ can be considered as the independent variable leaving the other one as the dependent or critical buckling load variable. In this work β is considered independent. Thus τ_{crit} is defined by the eigenvalues of the problem and gives the buckling loads.

SOLUTION AND RESULTS

The eigenvalue problem defined in the previous section has been solved by Bernitsas (1981) in two ways. The need to develop more than one solution is dictated by the fact that the numerical implementation of either methods does not yield answers over the entire domain of practical interest. Both algorithms fail for high values of β, τ or certain combinations of β and τ. In the $\beta-\tau$ plane the two solutions yield results over different but highly overlapping regions. The combined result of the two methods cover the entire domain of practical interest (Bernitsas et al 1983, Bernitsas and Kokkinis 1983a).

a. **Closed form solution**

The solution to equation (9) is

$$U(p) = \beta^{-1/3}\{c_1 U_1(x) + c_2 U_2(x) + c_3 U_3(x) + c_4\} \qquad (21)$$

where c_1, c_2, c_3 and c_4 are constants of integration

$$U_1(x) = \int^x Ai(\zeta)\,d\zeta \qquad (22) \qquad U_2(x) = \int^x Bi(\zeta)\,d\zeta \qquad (23)$$

$$U_3(x) = -\int^x Ai(\zeta)\int^\zeta Bi(\zeta)\,d\eta\,d\zeta + \int^x Bi(\zeta)\int^\zeta Ai(\eta)\,d\eta\,d\zeta \qquad (24)$$

$$x = \beta^{1/3} p + \beta^{-2/3}\tau \qquad (25)$$

and $Ai(x)$ and $Bi(x)$ are the Airy functions of the first and second kind.

Using equation (21) and any of the two sets of boundary conditions which were presented in the previous section we can compute the critical buckling loads τ_{crit} versus β. These computations require numerical evaluation of the Airy functions at $p=0$ and $p=1$, that is for $x_0 = \beta^{-2/3}\tau$ and $x_1 = \beta^{1/3} + \beta^{-2/3}\tau$. However, for high values of x_0 and x_1 the Airy function of the first kind becomes very small and the Airy function of the second kind very large. Actually (see Bender and Orszag 1978)

$$Ai(x) \underset{x\to\infty}{\sim} \frac{1}{2} \pi^{-1/2} x^{-1/4} \exp[-2x^{3/2}/3] \tag{26}$$

and
$$Bi(x) \underset{x\to\infty}{\sim} \pi^{-1/2} x^{-1/4} \exp[2x^{3/2}/3] \tag{27}$$

Operations between Ai and Bi for high values of x_0 and x_1 make the algorithm unstable even in double precision arithmetic. However, this algorithm gives solutions for high values of β and high buckling mode order where the second method fails.

b. <u>Series solution</u>

The solution to equation (9) can be expressed in the following form

$$U(p) = \sum_{n=1}^{\infty} a_n p^{n-1} \tag{28}$$

Substitution of equation (28) into equation (9) yields a recursive relation between the coefficients a_n. Equivalently a_n can be expressed in terms of the first four coefficients in the series since (9) is a fourth order linear differential equation. Thus four boundary conditions (13) to (16) or (17) to (20) and equation (28) can be used to calculate τ_{crit} as function of β. This method yields numerical answers for high β values and low buckling mode order where the first method fails.

c. <u>Results</u>

The results of the previous analysis, that is the critical loads τ_{crit} for the first buckling mode are presented versus β in figures 1 and 2. Figure 1 depicts the τ_{crit}-β curve for a hinged-hinged riser and figure 2 for a hinged-movably hinged riser. From τ_{crit} we can compute the effective bottom overpull $P_e(0)$ for a given riser. However, it is more convenient to evaluate the critical loads in terms of the tension which is required at the top of the riser to prevent buckling. Let TTR be the tension at the top of the riser. That is

$$TTR = T(L) \tag{29}$$

From equations (2), (7) and (8) we get:

$$T(z) = T(0) + (W_{st} - B_m)z \tag{30}$$

Equation (30) is correct only if the riser tubes have constant internal and external diameters along the entire length of the structure. The general expression for $T(z)$ has been derived in reference (Bernitsas 1982a).

Equation (7) for $z=0$ yields

$$P_e(0) = T(0) + (B_w - W_m)L \tag{31}$$

Using equations (29), (30) and (31) we get

$$TTR = P_e(0) + W_e L \tag{32}$$

In dimensionless form (32) becomes

$$\delta = \tau + \beta \tag{33} \qquad \text{where} \qquad \delta = \frac{TTR\, L^2}{EI} \tag{34}$$

The critical values of δ for which buckling may occur are also plotted in figures 1 and 2. In the same figures the values of δ which are computed according to the available design rules-discussed later in this paper-are also plotted for comparison.

BUCKLING IN TENSION DUE TO INTERNAL PRESSURE

It has been proved theoretically and explained with the aid of discrete analog structural models that slender tubular columns may buckle even if the actual tension is positive along their entire length due to the action of internal fluid static pressure (Bernitsas and Taylor 1982). In the case of risers where the internal mud static pressure may become significantly greater than the external hydrostatic pressure this phenomenon needs further investigation.

Buckling may occur if

$$\tau < \tau_{crit} \qquad (35) \qquad \text{or} \qquad P_e(0) < P_{e_{crit}}(0) \qquad (36)$$

Due to (31) inequality (36) becomes

$$T(0) + (B_w - W_m)L < P_{e_{crit}}(0) \qquad (37)$$

For a partially buoyed riser, which is usually the case in practice, if $T(0) > 0$ then $T(z)$ is greater than zero for all values of z between 0 and L. Consequently if the following double inequality holds the riser may buckle even if the actual tension in the riser is positive along the entire riser length

$$(B_w - W_m)L < T(0) + (B_w - W_m)L < P_{e_{crit}}(0) \qquad (38)$$

This inequality can be satisfied even for moderately high values of W_m, the drilling mud weight per unit length. This observation is very important for the evaluation of the tension required at the top of the riser in order to prevent global buckling.

A phenomenon similar to the above has been reported by Goodman and Breslin (1976). They have proved that vertical cables heavier than water may not collapse in water because of the action of the external hydrostatic pressure. They have explained this phenomenon using the Poisson effect and the cable material compressibility property. It can be seen from equation (31) and inequality (37) that the hydrostatic pressure sustains the riser and may actually prevent buckling.

GLOBAL BUCKLING DESIGN CRITERIA

For a riser system of given material properties, dimensions and drilling mud density the tension to be exerted at the top of the riser to prevent global buckling must be defined. Two different rules are often used to calculate TTR. In this section these rules are discussed and compared to the actual buckling constraint depicted in figures 1 and 2 for a hinged-hinged and a hinged-movably hinged riser respectively.

a. First Rule

If we do not take into consideration safety factors this practical rule states that "The tension at the top of the riser is equal to the weight of the riser and contents in water." That is TTR is equal to the weight of the riser, plus the weight of the drilling mud in the riser minus the weight of water displaced by the riser tubes, minus the net buoyancy of the buoyancy modules. Consideration of safety factors will change the numerical values but not the substance of this analysis. In mathematical form this rule yields

$$TTR > W_e L \qquad (39)$$

In dimensionless form inequality (39) becomes

$$\delta > \beta \qquad (40)$$

This rule is plotted in figures 1 and 2. It gives vales of TTR in the stable region of the $\delta-\beta$ plane. However, adoption of this rule, especially with a safety factor, results in unnecessary high values for TTR.

b. Second Rule

The purpose of this rule is to make the riser be under tension at all points along its length. To this effect, the rule stipulates that the tension at the top be equal to the weight of the tube in vacuum minus the net buoyancy of the buoyancy modules. This rule is based on the fact that when the riser is in a vertical position and has uniform properties, that is independent of z, no tension is required to support the drilling mud weight and no buoyancy force is exerted on the riser due to the water displaced by the riser tubes.

In mathematical terms this rule becomes

$$TTR > (W_{st} - B_m)L \qquad (41)$$

or in dimensionless form

$$\delta > \eta\beta \qquad (42)$$

where

$$\eta = \frac{W_{st} - B_m}{W_e} \qquad (43)$$

Equation (42) is also plotted in figures 1 and 2 for

$$\eta = 0.8 \qquad (44)$$

This value is used as an example and does not affect the generality of this discussion. It is obvious from figures 1 and 2 that this second rule can give values both in the stable and unstable region of the δ-β plane. This means that the rule is unsafe unless an appropriate safety factor is used. Actually the regions between lines [D] and [2] in figures 1 and 2 which are marked with the letter R are the regions in the δ-β plane where risers may buckle due to internal mud static pressure even if they are in tension along their entire length.

c. Proper Design Rule

This is set by the solution to the eigenvalue problem presented in this paper. Actually line [D] in the δ-β plane provides the stability boundary. In mathematical terms the requirement that TTR be in the stable region for the riser is expressed as

$$\delta > \delta_{crit} \qquad (45)$$

NUMERICAL EXAMPLES

Two numerical examples are worked out in this section in order to (1) demonstrate that risers may globally buckle due to internal pressure even if the actual tension is positive along their entire length and (2) show the proper distribution of supporting riser forces between TTR and the buoyancy of the buoyancy modules. All the information required for this analysis is included in figures 1 and 2 in dimensionless form.

a. Variation of TTR with the Riser Length

Consider a riser composed of circular tubes with the following properties

$$D_o = 20" \ (50.8 \text{ cm}) \qquad (46)$$

$$D_i = 18.75" \ (47.6 \text{ cm}) \qquad (47)$$

$$B_m = 0.3 \ B_w \qquad (48)$$

$$\rho_{st} = 7{,}850 \text{ kg/m}^3 \qquad (49)$$

$$E = 30 \cdot 10^6 \text{ psi} \ (2.07 \cdot 10^{11} \text{ Nt/m}^2) \qquad (50)$$

In addition we assume that

$$\rho_m = 1.5 \ \rho_w \qquad (51)$$

Using the above data and equation (43) we get

$$\eta = 0.663 \qquad (52)$$

The results are shown in figures 3 and 4 for a hinged-hinged and a hinged-movably hinged riser respectively. These figures show the variation of the required TTR with the riser length. All three design rules are used for computation of TTR. It is clear from figures 3 and 4 that the first design rule results in unnecessarily high TTR values and the second one will cause buckling of risers in tension. Actually hinged-hinged and hinged-movably hinged risers with the above properties will buckle in tension for the following values respectively

$$L \gtrsim 260 \text{ m} \qquad (53)$$

$$L \gtrsim 160 \text{ m} \qquad (54)$$

b. Distribution of Supporting Forces between TTR and B_m

Consider a riser of length

$$L = 500m \ (1640 \text{ ft}) \qquad (55)$$

which has the properties of the riser of the previous example except for relation (48). Using the three rules discussed in the previous section we can compute the distribution of supporting forces between TTR and the buoyancy of the buoyancy modules. The results of the computations are shown in figures 5 and 6 for a hinged-hinged and a hinged-movably riser respectively. It becomes clear from these figures that the first rule yields unnecessarily high supporting forces and that the second rule will result in buckling of a hinged-hinged riser in tension for

$$B_m/B_w \lesssim 1.23 \qquad (56)$$

and in buckling of a hinged-movably hinged riser in tension for

$$B_m/B_w \leq 1.25 \tag{57}$$

CONCLUSIONS

The contributions of this work can be summarized in the following

1. Dimensionless buckling loads have been computed for a hinged-hinged riser. The critical loads are plotted in the form of dimensionless effective overpull at the lower end of the riser or dimensionless tension at the top of the riser versus the dimensionless effective weight β.

2. Dimensionless buckling loads have been computed for a hinged-movably hinged riser as well in the form described in the first conclusion.

3. The two practical rules used for calculation of the tension required at the top of the riser are evaluated by comparison with the exact values derived by solving the linearized Euler buckling problem.

4. It has been proved theoretically and demonstrated with a numerical example that internal pressure may actually destabilize risers which are in tension over their entire length.

5. Finally it has been illustrated with a numerical example how the dimensionless buckling loads can be used to calculate the proper distribution of the riser supporting forces between the tension at the top of the riser and the buoyancy provided by the buoyancy modules.

REFERENCES

BENDER, C. M. and S.A. ORSZAG, 1978, *Advanced Mathematical Methods for Scientists and Engineers*. New York, McGraw Hill Book Co.

BERNITSAS, M. M., 1980, "Riser Top Tension and Riser Buckling Loads," *American Society of Mechanical Engineers, Applied Mechanics Division*, Computational Methods for Offshore Structures, Volume 37, pp. 101-109.

BERNITSAS, M. M., 1981, *Static Analysis of Marine Risers*, The University of Michigan, Department of Naval Architecture and Marine Engineering, Report No. 234.

BERNITSAS, M. M., T. KOKKINIS, and W. FALLER, 1981, *Buckling of Slender Columns under Distributed Load*, The University of Michigan, Department of Naval Architecture and Marine Engineering, Report No. 240.

BERNITSAS, M. M., 1982a, "Problems in Marine Riser Design," *Marine Technology*, Volume 19, Number 1, pp. 73-82.

BERNITSAS, M. M., 1982b, "A Three Dimensional, Nonlinear, Large Deflection Model of Dynamic Response of Risers, Pipelines and Cables," *Journal of Ship Research*, Volume 26, Number 1.

BERNITSAS, M. M. and J. E. TAYLOR, 1982, "Stability of Slender Tubes under Static Pressure and Tension," *Journal of Structural Mechanics*, (in press).

BERNITSAS, M. M., T. KOKKINIS, and W. FALLER, 1983, "Buckling of Columns with Nonmovable Boundaries," Transactions, American Society of Civil Engineers, *Journal of Engineering Mechanics Division* (to appear).

BERNITSAS, M. M. and T. KOKKINIS, 1983a, "Buckling of Columns with Movable Boundaries," Transactions, American Society of Civil Engineers, *Journal of Engineering Mechanics Division*, (to appear).

BERNITSAS, M. M. and T. KOKKINIS, 1983b, "Comparisons of Buckling Loads of Columns," Transactions, American Society of Civil Engineers, *Journal of Engineering Mechanics Division*, (to appear).

GOODMAN, T. R. and J. R. BRESLIN, 1976, "Statics and Dynamics of Anchoring Cables in Waves," *Journal of Hydronautics*, Volume 10, Number 4, pp. 113-120.

SUGIYAMA, Y. and K. ASHIDA, 1978, "Buckling of Long Columns under their Own Weight," Transactions, *Japanese Society of Mechanical Engineers*, Volume 21, Number 158, pp. 1228-1235.

ACKNOWLEDGEMENTS

This work has been supported by the Horace H. Rackham School of Graduate Studies of the University of Michigan under Faculty Research Grant #387565.

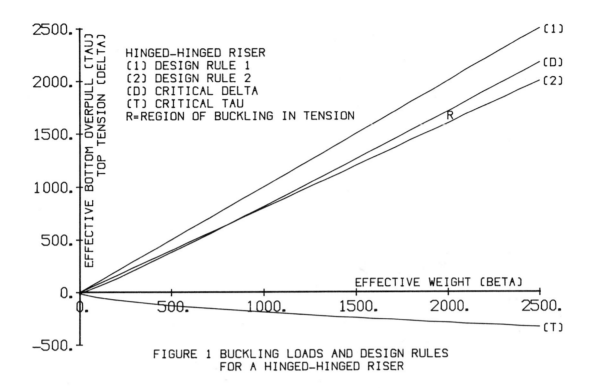

FIGURE 1 BUCKLING LOADS AND DESIGN RULES
FOR A HINGED-HINGED RISER

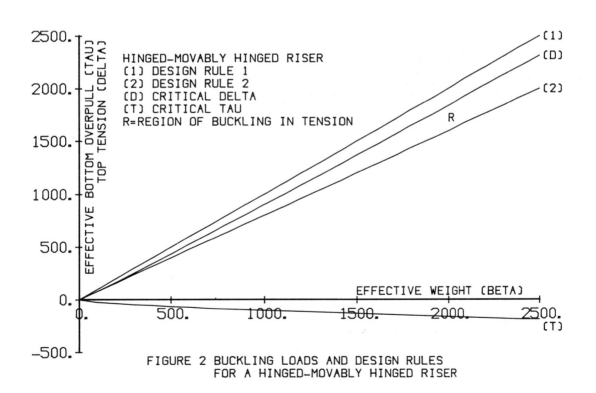

FIGURE 2 BUCKLING LOADS AND DESIGN RULES
FOR A HINGED-MOVABLY HINGED RISER

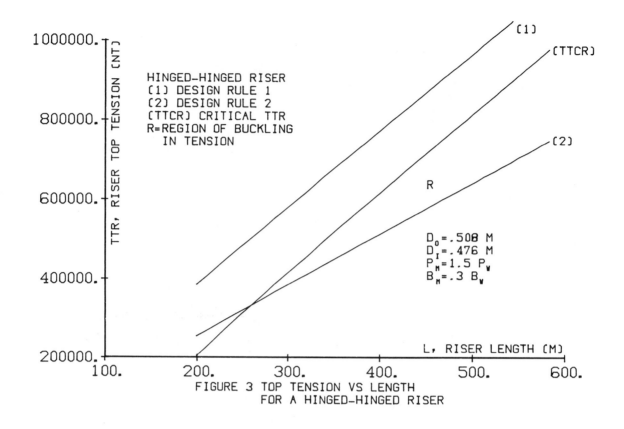

FIGURE 3 TOP TENSION VS LENGTH
FOR A HINGED-HINGED RISER

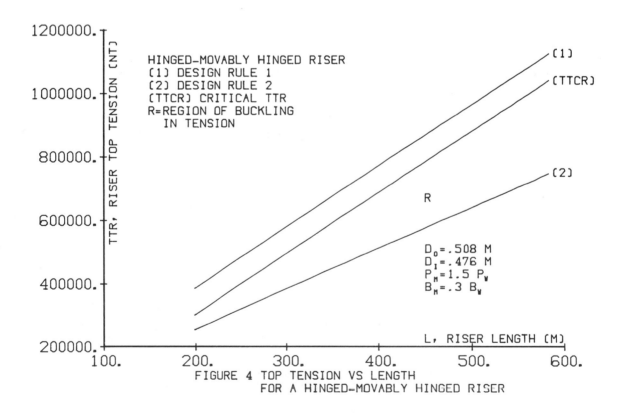

FIGURE 4 TOP TENSION VS LENGTH
FOR A HINGED-MOVABLY HINGED RISER

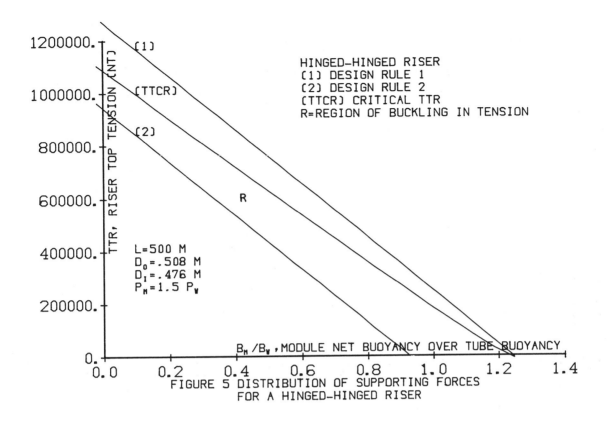

FIGURE 5 DISTRIBUTION OF SUPPORTING FORCES FOR A HINGED-HINGED RISER

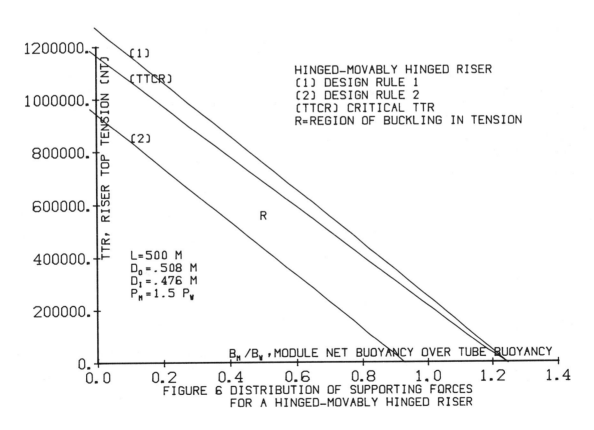

FIGURE 6 DISTRIBUTION OF SUPPORTING FORCES FOR A HINGED-MOVABLY HINGED RISER

NON LINEAR BEHAVIOUR OF STEEL STRUCTURAL CONNECTIONS

A. COLSON
Laboratoire de Mécanique et Technologie - CACHAN - FRANCE

SUMMARY

The non linear behaviour of steel structural connections has a lot of origins (residual stresses of welding, stress concentrations, local secondary effects ...) which induce local yieldings. These local yieldings contribute to the stiffness decrement of the connection all along a monotonic loading and induce the non linearity.

Thermodynamics of irreversible processes is used to define variables and to precise the general form of constitutive equations.

The three parameters used for describing a monotonic loading or a cyclic loading are :
- the initial stiffness of the connection computed from the elastic behaviour of the material by usual methods in simple cases or finite element method in more complex ones,
- the ultimate solicitation of the connection computed from a perfect plastic behaviour of the material
- an internal variable taking the manufacturing conditions into account and inducing the non linearity.

The general equations are written in a three-dimensional form, while the applications and the experimentations are shown in a uni-dimensional form. The interaction circumstances are specified.

Such a modelization is interesting since :
- it enables to introduce the manufacturing conditions into the structural design
- for the cyclic loading, it describes the softening feature of the behaviour and it allows to estimate energy dissipation
- it can lead to the optimization of structures by the utilization of connections situated between fixed-connections and hinged connections.

The studies concerning the structural connections carried out up to now have two principal objectives.
- The first objective is to set up calculation methods to design the connection components so as to reach the strength level required (6,1,2). The solicitations are computed from a structure calculation in which the girders are supposed to have perfect elastic behaviour or a perfect elastic plastic behaviour. In the same way, the connections are supposed to have a perfect behaviour ($\Omega = 0 \, \forall M$ for a fixed-connection, $M = 0 \, \forall \Omega$ for a hinged connection and $M = k \, \Omega$ for an elastic connection)
- The second objective is to make sure that the connections designed that way, give a sufficient capacity of plastic deflexion in order to allow for the formation of plastic hinges (2).

Tests carried out on fixed connections which meet the two previous requirements show us the non linear load-deflexion behaviour all along the loading (2,10). Although the maximal load capacity and the maximal deflexion capacity are reached, there appears a deformability greater than estimated from the beginning of the loading (see Fig. 1).

Such a remark leads us to think that perfect connections do not exist and that the calculations made with this modelization are wrong.

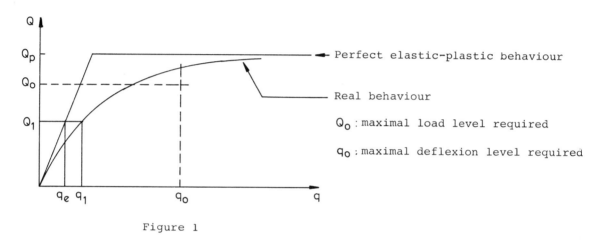

Figure 1

On the other hand tests carried out on hinged connections show us a real strength capacity which would be used for structure optimization.

For all the connections we see that the knowledge of the real behaviour of the connections would allow to describe the total behaviour of the structure properly. The setting up of such a study makes it necessary for the non-linear behaviour to be modelized. This non linear behaviour is imputable to a lot of imperfections in manufacturing such as : residual stresses, more especially of welding, stress concentrations, local secondary effects, ... etc which induce local yielding. These local yieldings contribute to the stiffness decrement of the structure along the loading. The model proposed hereafter takes into account these local yieldings by means of one parameter, an internal variable, which added to the two well known parameters - initial stiffness and ultimate solicitation - allows for the complete modelization of the behaviour in case of monotonic loading or cyclic loading (4).

I. NOTATIONS AND DEFINITIONS

In this paper the connection is considered an unknown box for which only input and onput solicitation and displacement are used. For a three-dimensional description of the behaviour of the connections the following notations are used according to Figure 2.

The solicitation vector is the same as the generalized stresses vector in beam theory and the components of displacement vector are the translations and the rotations according to the x_1, x_2, x_3 axis.

$$\text{Solicitation vector } (F) = \begin{bmatrix} S_1 \\ S_2 \\ S_3 \\ M_1 \\ M_2 \\ M_3 \end{bmatrix} \qquad \text{Ultimate solicitation vector } (F_p) = \begin{bmatrix} S_{p1} \\ S_{p2} \\ S_{p3} \\ M_{p1} \\ M_{p2} \\ M_{p3} \end{bmatrix}$$

$$\text{Displacement vector } (D) = \begin{bmatrix} U_1 \\ U_2 \\ U_3 \\ \Omega_1 \\ \Omega_2 \\ \Omega_3 \end{bmatrix} \qquad \text{Initial stiffness matrix } [R] = \begin{bmatrix} Q_1 & & & & & \\ & Q_2 & & & 0 & \\ & & Q_3 & & & \\ & & & K_1 & & \\ & 0 & & & K_2 & \\ & & & & & K_3 \end{bmatrix}$$

$$\text{Reduced solicitation matrix } [f] = \begin{bmatrix} s_1 & & & & & \\ & s_2 & & & 0 & \\ & & s_3 & & & \\ & & & m_1 & & \\ & 0 & & & m_2 & \\ & & & & & m_3 \end{bmatrix} \quad \text{such as} \quad (F_p)^T [f] = (F)$$

$$\text{with} \quad s_i = \frac{S_i}{S_{pi}} \quad \text{and} \quad m_i = \frac{M_i}{M_{pi}}$$

The initial stiffness matrix is computed from the perfect connection with a linear behaviour of the material. Perfect connection means real connection without imperfections such as residual stresses, stress concentration .... The ultimate solicitation vector is computed from a perfect plastic behaviour of the material. It can be obtained by a limit analysis theory for example.

In case of very complex connections these two parameters can be obtained by identification from test results.

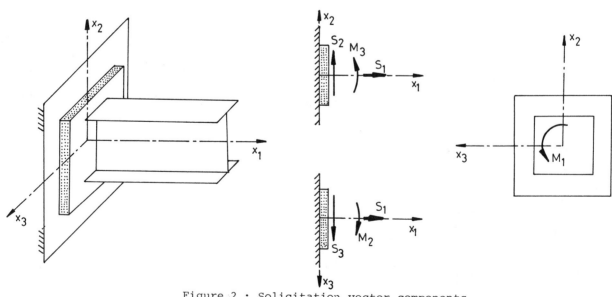

Figure 2 : Solicitation vector components

II. THERMODYNAMICS SCHEME-POTENTIAL EXPRESSIONS-CONSTITUTIVE EQUATIONS

Thermodynamics of irreversible processes is used to define variables and to precise the general form of constitutive equations (5). The general scheme used hereafter has been presented by Jean Lemaitre concerning damage theory (8).

The local elastic-plastic hypothesis for the material enables us to obtain the total elastic strain (D_e) associated with a loading value. The existence of local yieldings gives a greater total strain by addition of an unelastic strain (D_p) such as $(F)^T(\dot{D}_p)$ is the total power dissipation. We traditionaly write $(D) = (D_e) + (D_p)$. The imperfections which produce the local yieldings are particularized by the scalar internal variable "a" and its associated variable "α" such as the product $a\dot{\alpha}$ is the mobilized power for the modification of the residual stresses. The intrinsic dissipation is $W = a\dot{\alpha} + (F)^T(\dot{D}_p)$.

The complete set of variables for three-dimensional problems within the hypothesis of simple loading (or radial loading), uniform temperature, without effect of strain rate is the following :

	Observable variables	Internal variables	Associated variables
elastic strain	(D_e)		(F)
unelastic strain		(D_p)	(F)
imperfections		a	α

Let us take free energy as thermodynamical potential function of all observable and internal variables

$$\Psi = \Psi((D_e),(D_p),a) \quad \quad (1)$$

Let us call Ψ^* the LEGENDRE FENSCHEL transformation of Ψ

$$\Psi^* = \Psi^*((F),a) \quad \quad (2)$$

There the state equations are obtained by

$$(D_p) = \frac{\partial \Psi^*}{\partial (F)} \quad \text{and} \quad \alpha = \frac{\partial \Psi^*}{\partial a} \quad \quad (3)$$

The thermodynamical approach (Germain 1973) justifies the existence of a pseudo-potential of dissipation Φ from which complementary constitutive equations for internal variables are obtained. Φ is a convex function of all observable and internal variables and their first time derivative.

$$\Phi = \Phi((D_e),(D_p),a,(\dot{D}_p),\dot{a})$$

and Φ^* is the LEGENDRE - FENSCHEL transformation of Φ

$$\Phi^* = \Phi^*((F),a,(\dot{F}),\dot{a})$$

Then the complementary constitutive equations are given by

$$(\dot{D}_p) = \frac{\partial \Phi^*}{\partial (F)} \quad \ldots \ldots \ldots (4)$$

$$\dot{\alpha} = \frac{\partial \Phi^*}{\partial a} \quad \ldots \ldots \ldots (5)$$

$$(D_p) = \frac{\partial \Phi^*}{\partial (\dot{F})} \quad \ldots \ldots \ldots (6)$$

$$\alpha = \frac{\partial \Phi^*}{\partial \dot{a}} \quad \ldots \ldots \ldots (7)$$

POTENTIAL EXPRESSIONS

The general forms suggested for the free energy and the pseudo-potential of dissipation are :

$$\Psi^* = (F_p)^T \cdot \left(((F_p)^T \cdot [R]^{-1})^T \cdot [\mathbb{F}] \right) \quad \ldots \ldots \ldots (8)$$

and $$\Phi^* = (F_p)^T \cdot \left(((F_p)^T \cdot [R]^{-1})^T \cdot [\mathbb{G}] \right) \quad \ldots \ldots \ldots (9)$$

with $$[\mathbb{F}] = \sum_{n=1}^{\infty} \frac{[f]^{na+2}}{na+2} \quad \ldots \ldots \ldots (10)$$

and $$[\mathbb{G}] \sum_{n=1}^{\infty} \frac{[f]^{na+1}}{na+1} \cdot [\dot{f}] \quad \ldots \ldots \ldots (11)$$

$[\mathbb{F}]$ and $[\mathbb{G}]$ are matrix defined by convergent serial matrix. Effectively all the components of the reduced solicitation matrix are strictly less than 1.

INTERACTION CONDITIONS FOR SOLICITATIONS

These conditions can be introduced in free energy potential and pseudo potential of dissipation at different levels.

1) By the modidification of the initial stiffness matrix $[R]$. After some test results, it seems that it is the situation for a Shear Force-Bending moment interaction.
2) By the modification of the ultimate solicitation vector (F_p). It is probably the case of a Normal force - Bending moment interaction.
3) By the modification of the matrix $[f]$ which could be transformed into a non diagonal matrix. This last hypothesis would be the most difficult to make use of because the calculation of the serial matrix would be very difficult to carry out.

III. UNIDIMENSIONNAL CASE : BENDING MOMENT - ROTATION ABOUT x_3 AXIS

All applications presented in this section correspond to this position and test results have shown that interaction effects can be neglected.

Let $M_{p_3}, M_3, K_3, m_3, \Omega_3$ be respectively noted M_p, M, K, m and Ω.

In this case, the free-energy and pseudo-potential of dissipation expressions are :

$$\Psi^* = \frac{M_p^2}{K} \sum_{1}^{\infty} \frac{m^{na+2}}{na+2} \quad \ldots \ldots \ldots (12)$$

$$\Phi^* = \frac{M_p^2}{K} \sum_{1}^{\infty} \frac{m^{na+1}}{na+1} \dot{m} \quad \ldots \ldots \ldots (13)$$

After differenciating we obtain $\Omega p = \frac{Mp}{K} \sum m^{na+1}$ and with definition of the initial stiffness $\Omega_e = \frac{M}{K} = \frac{M_p}{K} m$.

Then $\Omega = \Omega_e + \Omega_p = \frac{Mp}{K} m (1 + .. + m^{na})$

$$\Omega = \frac{M}{K} \frac{1}{1-m^a} \quad \quad (14)$$

Otherwise $\quad \alpha = \frac{\partial \Psi^*}{\partial a} = \frac{Mp^2}{K} \sum_1^\infty \frac{nm^{na+2}}{na+2} (\frac{1}{na+2} - \ln m)$ .... (15)

and $\quad \dot{\alpha} = \frac{\partial \Psi^*}{\partial a} = \frac{Mp^2}{K} \sum_1^\infty nm^{na+1} \ln m . \dot{m}$ .... (16)

MONOTONIC LOADING

The only interesting constitutive equation is (14) which gives the following behaviour (Figure 3).

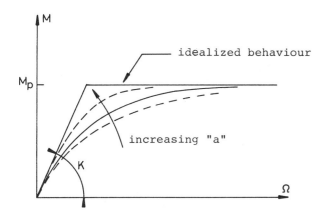

The initial stiffness is depicted by the graph tangent at the origin. The ultimate solicitation is depicted by the asymptot (Mp). For a great value of "a" we obtain a quasi perfect behaviour, more especially near the origin with a quasi elastic behaviour. This is the case of a connection without initial imperfections.
For a smaller value of "a", particularly near 1, unelastic strains are important and there is a lot of initial imperfections due to manufacturing conditions. At this point up to now known test results give "a" value between 1 and 5 approximatively.

CYCLIC LOADING

The cyclic behaviour is depicted by the same law as in monotic loading with the following modification of parameters at each loading inversion.
- modification of ultimate solicitation
- modification of the value of "a" taking into account the modification of the residual stresses.

On the other hand the initial stiffness stays at the same value. The representation in the $A_K \Omega$, $A_K M$ axis (Figure 4) is :

$$\Omega = \frac{M}{K} \frac{1}{1-m_k^{a_k}} \quad \quad (17)$$

with $\quad m_k = \frac{M}{M_{pk}}$; $\quad |M_{pk}| = |M_{k-1}| + |M_p| \quad$ and $\quad a_k = f(|\Omega_p|)$

The evolution law $a_k = f(|\Omega_p|)$ is obtained by the study of the dissipation $\int \alpha \dot{a} dt$ all along the loading. $|\Omega_p|$ denotes the cumulative strain unelastic deformation since the beginning of the loading. This study gives us the new value of the internal variable "a" at each loading inversion.

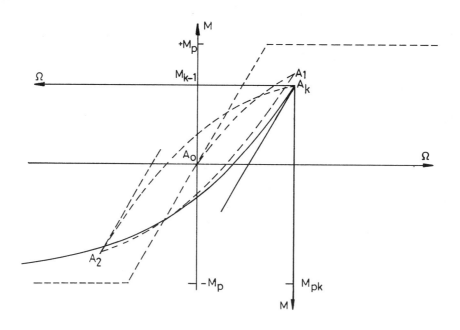

Figure 4 : Cyclic loading description

APPLICATION TO A BEAM-TO-COLUMN CONNECTION

The connection studied is described in Figure 5. The initial stiffness is computed, using the hypothesis that only the upper angle is subjected to strain (4). Effectively, the lower angle is blocked by the columnflange. The strain of the upper angle is shown in Figure 6 and the deflexion $\delta = \frac{Fl_1^3}{12EI_1}$ gives the initial stiffness:

$$K = \frac{12\ EI_1 h^2}{l_1^3} = 4.65 \times 10^6 \text{ m.N/rd} \quad \dots\dots\dots\dots\dots\dots\dots\dots\dots\dots\dots\dots\dots\dots\dots (18)$$

The ultimate solicitation is computed using the classic scheme of limit analysis with three plastic hinges as shown in Figure 7. Hence $M_p = 31.68$ mKN.

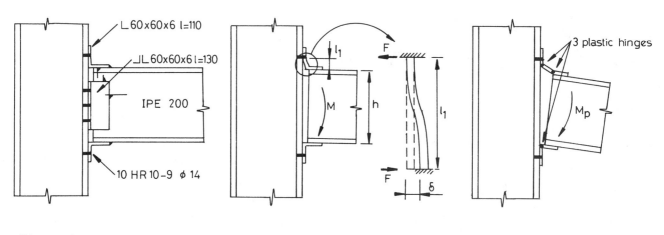

Figure 5

Figure 6

Figure 7

Test results on cyclic loading are given in Figure 8. An automatic identification founded on a last square method gives us the parameter values for the first loading. The results are :

$K = 4.7 \times 10^6$ mN/rd ; $M_p = 32$ mKN   and $a = 1.56$

The application of the method proposed in the earlier section leads to results given in Figure 9.

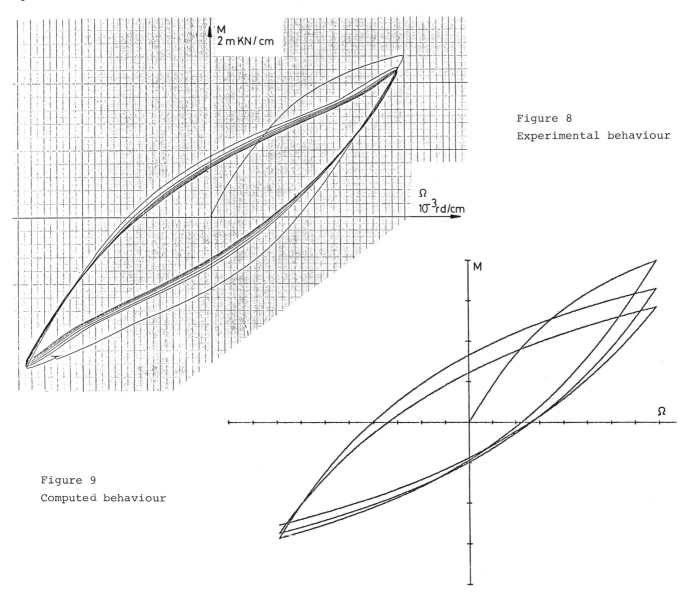

Figure 8
Experimental behaviour

Figure 9
Computed behaviour

In case of more complex connections (bolted beam to column connection for example) the calculations of initial stiffness and ultimate solicitation are more difficult. Initial stiffness can be computed using a finite element method (Breysse, Béré, 1981). For fixed connections the ultimate solicitation can be taken to be equal to the ultimate solicitation of the beam ($M_p = Z\ \sigma_e$).

If necessary the three parameters M_p, K and a can always be found by an identification after test results.

Other test results and computed results are given in Reference (4).

IV. APPLICATION TO A TWO BRACE TUBULAR K JOINTS

Test carried out at Delft University (12) on such connections give us a lot of results for monotonic loading. From these results, we have kept those of the test described in Figure 10. Test results and automatic identification results are given in Figure 11.

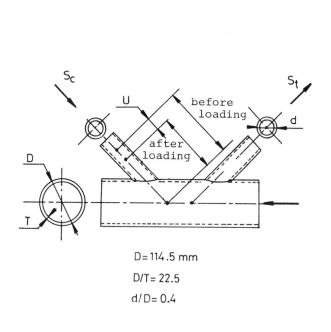

Figure 10

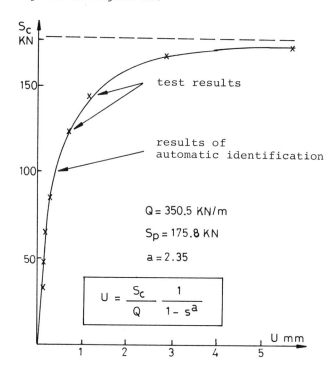

Figure 11

V. CONCLUSION

The suggested model seems to be very interesting to us because it allows the real behaviour of the connections to be described as accurately as possible. Its analytical form allows it to be used in automatic calculation.

It seems that it could also be used to describe the behaviour of the following connections
- Steel concrete (lower part of the steel columns)
- Connections between reinforced concrete elements (7,11)

REFERENCES

1. AGERSKOV H., 1977, "Analysis of bolted connections subject to prying" Journal of the structural Division, ASCE Vol. 103, ST 11 November 1977, pp. 2145-2163

2. BOUAZIZ J.P., 1977, "Assemblages poutre-poteau soudés. Dimensionnement et comportement experimental dans le domaine élasto-plastique" Construction Métallique n° 4, pp. 37-50

3. BREYSSE D.,BERE S., 1981, "Détermination des caractéristiques élastiques d'un assemblage métallique à l'aide d'un calcul par éléments finis", Internal report, Laboratoire de Mécanique et Technologie, Cachan, France

4. COLSON A., 1981, "Modélisation du comportement non linéaire des assemblages", Construction Métallique, N° 2, pp. 41-49

5. GERMAIN P., 1973,"Cours de Mécanique des milieux continus", Paris, MASSON

6. JOHNSON, CANNON, SPOONER, 1959, "High tensile preloaded bolted joints", British welding Association, Report D 1/6/1959, pp. 560-568

7. KUCZYNSKI W., GOSZCZYNSKI S., 1980, "Behaviour of hyperstatic reinforced concrete beams subject to increasing loads", Archiwum Inzynierii Ladowej, Tome XXVI, N° 1, 1980, pp. 79-94

8. LEMAITRE J, CHABOCHE J.L., 1978, "Aspect phénoménologique de la rupture par endommagement" Journal de Mécanique appliquée, Vol. 2, n° 3, 1978, pp. 315-365

9. PARFIT J., CHEN W.F., 1976, "Tests of welded steel beam-to-column moment connections" Journal of the structural division ASCE Vol. 102, ST1, pp. 189-202

10. RENTSCHLER GP., CHEN W.F., DRISCOLL G.C., "Tests of beam-to-column web connections" Journal of the structured division, ASCE, Vol. 106 ST5, pp. 1005-1022

11. SOLEIMANI D., POPOV E.P., BERTERO V., "Hysteritic behaviour of reinforced concrete beam-column subassemblages" Journal of the american concrete institute, N° 11 November 1979, pp. 1179-1195

12. WARDENIER J, 1977, "Investigation into the static strength of welded warren type joints made of circular hollow sections", Report n° 6-77-5, Stevin Laboratory, Delft University

THE EFFECT OF VARIABLE CHORD WALL THICKNESS ON THE
STRESSES IN A LIGHT, CAST 90°-45° K TUBULAR JOINT

C. D. Edwards
Department of Mechanical Engineering,
University of Nottingham, U.K.

H. Fessler
Department of Mechanical Engineering,
University of Nottingham, U.K.

SUMMARY

Four three-dimensional, frozen-stress photoelastic models have been tested. Axial tension has been applied separately to the 90° and 45° braces of models which are identical, except for the position of the chord bore relative to the outer surfaces. External, tapered collars on the chord around each brace and internally tapered brace walls blending with large fillet radii produce very strong shapes which are easy to cast.

Comprehensive stress distributions show that, for single brace loading, making the chord core eccentric reduces the peak stresses. There are no significant stress concentrations in balanced loading. Comparison with similar work with models of welded joints shows greatly increased fatigue life prediction.

INTRODUCTION

Cast steel nodes are stronger than welded ones because of their much better design in the critical fillet areas and because they have no fillet welds, being connected to the tubular members of the structure by butt welds in low-stress regions.

The design shown in Fig. 1 has external taper collars on the chord surrounding each brace and internal tapers of the brace stubs. These tapers make it easy to cast and increase the strength at the highly stressed intersections. Large fillet radii (external R, and internal r) further improve the castability and reduce the stress concentrations. The brace spacing, $g = -0.6T$ gives the largest overlap of braces which does not produce excessive chord thickness where the braces are closest together. The brace diameter/chord diameter ratio, d/D, = 0.64 has been chosen to make it easy to cast strong corner joints. The wall thickness parameters, $D/T = 24$ and $t/T = \frac{1}{2}$ are typical. The brace stubs are as short as possible for good welding access. Tapers, fillets and core-support holes, defined in Fig. 1, are our Mark 2 design, which has been tested in out-of-plane bending (Edwards and Fessler, 1981).

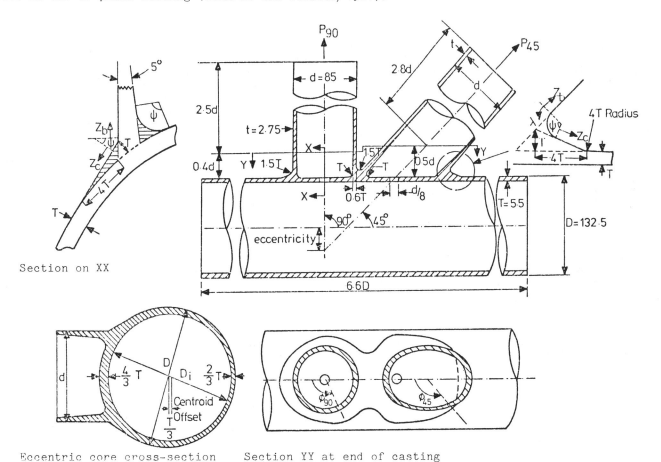

Fig. 1 Shapes, dimensions in mm, and nomenclature of models

The most highly stressed part of the chord can be strengthened by thickening without increasing the weight of the joint by making the chord core eccentric to the outside, as shown in Fig. 1. The maximum sensible eccentricity makes the minimum node chord wall thickness equal to the thickest tubular member. A survey of platforms for water depths up to 300 m shows $36 < D/T < 120$. Selecting $D/T = 36$ results in an eccentricity of $T/3$ which makes $T_{max}/T_{min} = 2$ and thereby also improves the castability without changes to the pattern and with only a small change to the core box.

This paper presents complete results for the Mark 2 shape of K joint under axial loading of each brace with concentric and eccentric chords. The frozen-stress photoelastic technique (Fessler and Little, 1978 and Heywood, 1952) has been used.

EXPERIMENTAL WORK

The models were very accurate (Edwards and Fessler, 1981) precision castings (Fessler et al, 1979b) in Araldite CT200 with Hardener 901. Each model consisted of a node, cemented to chord and brace tubes where the steel casting is butt-welded to the steel tubes.

Each model was simply supported by solid, circular plugs in the ends of the chord and loaded by one axial force P applied by a plug cemented into the brace. One of the chord supports was restrained to exert a thrust in the RH part of the chord (see Fig. 1) to react the component of P_{45} along the chord.

The loading system ensured that P acted along the appropriate brace axis, that no stresses were set up due to the weight of the model and that the model could expand and contract freely during the stress-freezing cycle. The loads were P_{90} = 41N and P_{45} = 60N, chosen to produce adequate fringe orders in 1.5 and 2.5 mm thick slices. Established experimental techniques (Fessler and Little, 1978) were used for stress-freezing, slicing, calibration and photoelastic analysis of surface stresses, using oblique incidence observations (Heywood, 1952). The mean stress in each brace was obtained as the mean of the inner and outer surface stresses in the $\varphi = 0°, 90°, 180°$ and $270°$ planes in the tubular member 1.3d for the 90° and 1.1d for the 45° brace, from the ends of the brace taper. In these regions the variations of fringe orders through the brace walls were small. These values differed from $P/\pi(d - t)t$ by 4% to 7%.

RESULTS

Only one brace of each model was loaded, as defined in Table 1. Variable chord thickness values from Models C01 and C02 are always shown with the corresponding constant wall thickness results from Models B02 and B05. For each model, all surface stresses (σ) are presented as multiples of the mean axial stress (σ_{nom}) in its loaded brace. At any one position, only the major principal stress is presented. Except for Fig. 5, the directions of the major principal stresses were as expected, in the plane of the graphs in Figs. 2 to 6 and in the plane of maximum curvature of the fillets.

For the 90° brace loadings, results are shown for the plane of symmetry ($\varphi = 0° = 180°$, see Fig. 1) in Fig. 2, for the $\varphi_{90} = 90°$ plane in Figs. 4 and 6, and around the brace and chord fillets in Fig. 7.

For the 45° brace loadings, Figs. 3a and 3b present results for the plane of symmetry and Fig. 5 the stresses in the $\varphi_{45} = 90°$ plane which intersects the brace axis at the outside of the chord i.e. parallel to the $\varphi_{90} = 90°$ plane, but inclined to the 45° brace axis. These fillet stresses are shown in Fig. 8.

For 'balanced' loading ($P_{90} \sin 45°$), i.e. no resultant force perpendicular to the chord axis, the stress distributions around both fillets are shown in Fig. 9 as multiples of the nominal stress in the 45° braces (σ/σ_{nom45})

90° Brace Loading

In the plane of symmetry (see Fig. 2) in the unreinforced chord, the mean of the outer and inner surface stresses approximates to the nominal chord bending stress, shown chain-dotted. The latter depends on the length of the model and is usually unimportant because most joints are subjected to balanced loading. The inner surface stresses are tensile like the outer surface ones, showing that, for this length of chord, the bending of the whole chord as a beam is more important than local bending of the chord wall. The greatest stress occurs at the chord fillet of the (loaded) 90° brace of model B02; it is 2.1 times the local, nominal, chord bending stress at this position. The variable-chord thickness stresses are similar to, but generally lower than, the corresponding constant-thickness values.

The stresses in the $\varphi_{90} = 90°$ plane in the chord (Fig. 4) and in the brace (Fig. 6) are generally similar for the two different chord thicknesses. Peak stresses occur in the outside and inside brace fillets; both are reduced by making the chord thickness variable. These high stresses occur in the transfer of load from the brace stub, which is stiff in axial tension, to the chord, which is flexible in bending. This causes significant ovalisation of the chord which produces high stresses in the fillets to maintain compatibility of displacement at the joint. Fig. 4 shows bending of the chord walls to be more extensive in the variable-thickness chord than in the constant-thickness model. There is little bending at the end of the brace taper, making this a good position for the butt weld with the brace.

Fig. 7 shows the fillet stress distributions around both braces. The junction of the chord fillets is called J. Complete chord fillets are shown in the diagrams, although the tapers are tangential to the chord wall between $60° < \varphi < 120°$. Making the chord thickness variable generally reduces the fillet stresses slightly.

45° Brace Loading

These normalised stresses are generally much lower than those due to P_{90} because the length

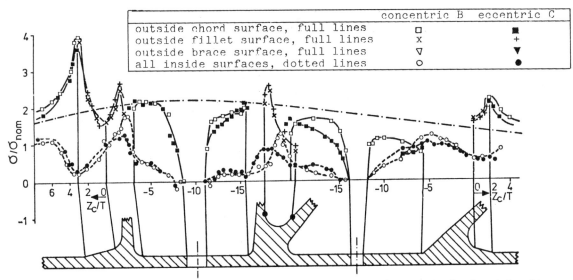

Fig. 2 Stresses in the plane of symmetry due to P_{90}. The nominal chord bending stress is shown chain-dotted.

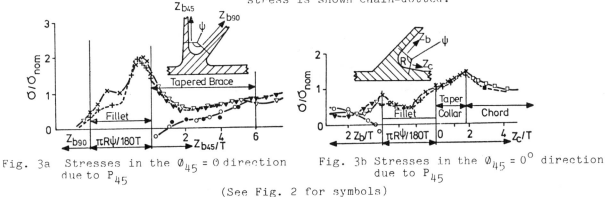

Fig. 3a Stresses in the $\phi_{45} = 0$ direction due to P_{45}

Fig. 3b Stresses in the $\phi_{45} = 0°$ direction due to P_{45}

(See Fig. 2 for symbols)

of the brace-chord junction is $\sqrt{2}$ times greater in the 45° brace and the force perpendicular to the chord axis is $\sqrt{2}$ times smaller than for the 90° brace loading. Furthermore, compression is introduced into the chord by this loading which partially counteracts the tensile stresses set up by chord bending. The chord stresses in Fig. 5 are about half of the corresponding values in Fig. 4 as could have been predicted from the above. However the peak stresses in the brace fillet in the $\phi_{45} = 90°$ and 180° directions are 0.35 to 0.4 times the corresponding values in the 90° brace loading. In the $\phi_{45} = 0°$ direction the ratio is 0.75.

Figs. 3a and 3b do not show the complete distribution in the plane of symmetry (unlike Fig. 2) because the positions not shown had similar, but 50% lower values of stress than the peak shown. Exceptional to this was the collar-chord junction of the 90° brace at $\phi_{90} = 0°$ where peak stresses of 1.4 σ_{nom} for the eccentric and 1.6 σ_{nom} for the concentric chord model were found. The peak stress is at the brace-brace junction ($\phi_{45} = 0$) not the chord-collar junction as in 90° brace loading, and is 2.00σ_{nom} for B05, 1.95σ_{nom} for C02.

Fig. 8 shows the fillet distribution around both braces. The peak stresses occur at $0 \leq \phi_{45} \leq 30°$. This is due to the locally increased wall thickness which stiffens the chord. Significantly, the brace stress remote from the fillet at $\phi_{45} = 0$ is higher than at $\phi_{45} = 180°$. The distributions for the constant and variable chord walls are very similar to each other, with the latter showing slightly lower peak stresses except in the critical region ($\phi_{45} < 30°$).

Balanced Loading

Table 2 and Fig. 9 show that all the high fillet stresses caused by single brace loading are substantially reduced in the usual situation of opposed brace loads with no resultant force perpendicular to the chord. The greatest combined stresses occur at the 90° brace fillet and are only 1.5 times the nominal 45° brace stress. In this loading both chord cross-sections have the

Table 1 Fillet Stress Concentrations due to single Loads and their Positions

Tensile load of Chord Shape Model Ref.	90° Brace				45° Brace			
	Concentric		Eccentric		Concentric		Eccentric	
	B02		C01		B05		C02	
	value	position	value	position	value	position	value	position
Outside surface		λ/ψ		λ/ψ		λ/ψ		λ/ψ
Brace $\varphi_{90} = 90°$	3.74	0.3	3.40	0.3	1.4	0.3	1.10	0.2
Brace $\varphi_{90} = 0$(1)	2.65	0.1	2.65	0.1	2.00	0.2	1.95	0.2
Brace $\varphi_{45} = 90°$	1.25	0.5	1.0	0.6	1.45	0.5	1.20	0.4
Chord $\varphi_{90} = 0$	3.90	≈0.5	3.60	≈0.5	1.6	≈0.5	1.4	≈0.5
Chord J (2)	2.0	≈0.5	2.0	≈0.5	1.2	≈0.5	-	-
Chord $\varphi_{45} = 180°$	2.20	≈0.5	2.15	≈0.5	1.5	≈0.5	1.45	≈0.5
Inside surface								
Brace $\varphi_{90} = 90°$	-2.05	0.2	-1.85	0.2	-1.1	0.2	-1.1	0.2
Brace $\varphi_{45} = 90°$	-0.82	0.3	-0.87	0.3	-1.2	0.2	-1.0	0.2
Core support hole								
$\varphi_{90} = 0$	3.0	-	-	-	1.4	-	1.2	-
$\varphi_{90} = 90°$	2.85	-	2.3	-	1.4	-	1.4	-
$\varphi_{90} = 180°$	2.0	-	2.4	-	-	-	-	-

Notes: 1. The brace fillet position $\varphi_{45} = 0$ coincides with $\varphi_{90} = 180°$ - see Fig. 1
2. J is the junction of the external reinforcements - see Fig. 8.

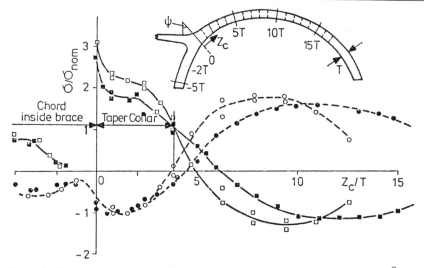

Fig. 4 Stresses due to P_{90} in the chord in the $\varphi_{90} = 90°$ plane

Chord shape	concentric B	eccentric C
outside chord surface, full lines	□	■
outside fillet surface, full lines	x	+
outside brace surface, full lines	∇	▼
all inside surfaces, dotted lines	○	●

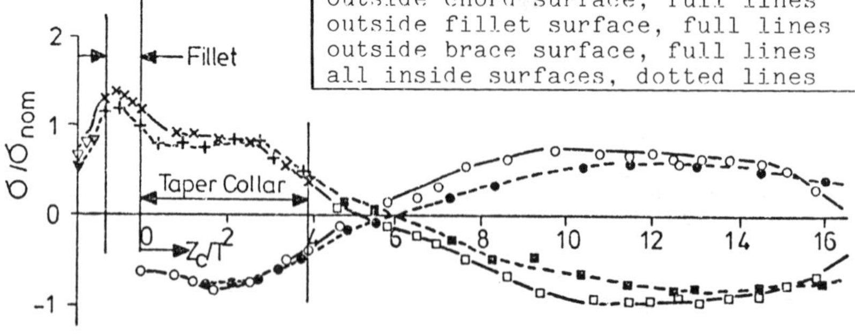

Fig. 5 Stresses due to P_{45} in the $\varphi_{45} = 90°$ plane

same peak values, but in the variable-thickness chord it occurs at $\varphi_{90} = 60°$ instead of $\varphi_{90} = 90°$.

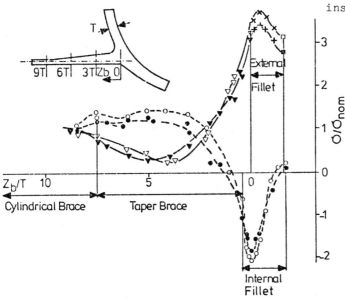

Fig. 6 Stress due to P_{90} in the brace and fillets in the $\varphi_{90} = 90°$ plane (Fig. 5 for symbols).

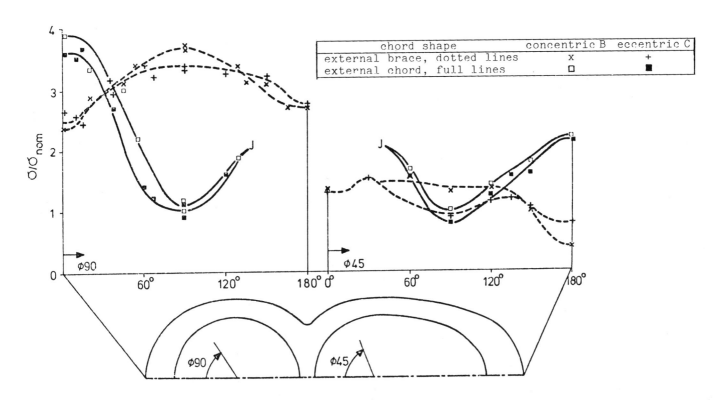

Fig. 7 Fillet stresses due to P_{90}.

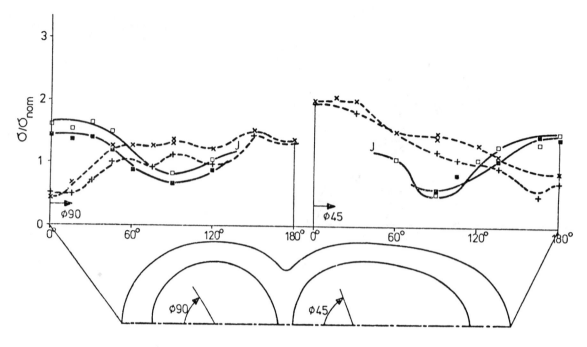

Fig. 8 Fillet stresses due to P_{45}. (Fig. 7 for symbols)

Table 2 Stress Concentrations due to balanced Loading as Multiples of the 45° Brace Nominal Stress

Chord Shape	Concentric, B		Eccentric, C	
	value	position	value	position
Outside Surface		λ/ϕ		λ/ϕ
Brace $\phi_{90} = 0$	– 1.5	0.2	– 1.4	0.1
$\phi_{90} = 90°$	– 1.5	0.3	– 1.35	0.3
$\phi_{45} =$	1.3	0.2	1.2	0.2
Chord $\phi_{90} = 0$	– 1.1	≈ 0.5	– 1.0	≈ 0.5
Inside Surface				
Brace $\phi_{90} = 90°$	0.5	0.2	0.3	0.2
$\phi_{45} = 90°$	– 0.7	0.2	– 0.5	0.2
Core Support Hole				
$\phi_{90} = 0$	– 0.7	–	– 0.5	–
$\phi_{90} = 90°$	– 0.6	–	– 0.3	–

DISCUSSION

The peak stresses in these cast nodes are very much lower than in welded construction (Fessler and Little, 1978 and Fessler et al, 1979a). Fig. 10 shows, for balanced loading, a comparison of fillet stress distributions of non-overlapped and overlapped welded joints and the concentric cast joint. There is some similarity of the stress distributions for the non-overlapped joint and the casting in Fig. 10. Although much lower than the non-overlapped, the overlapped joint still has a peak stress twice that of the casting.

The non-dimensional parameters for these joints, together with predictions of Stress Concentration Factors (SCF) for the welded ones are presented in Table 3. The non-overlapped K-joint SCFs were calculated using Kuang's formulae (Kuang et al, 1977).

$$SCF_c = 1.37 \, (T/D)^{-0.5} \, (d/D)^{-0.235} \, (t/T)(g/D)^{0.094} \sin^{1.28}\theta$$

$$SCF_b = 3.736 \, (T/D)^{-0.065} \, (d/D)^{-0.642} \, (t/T)^{.433} \, (g/D)^{0.048} \sin^{0.82}\theta$$

where suffices b and c refer to the brace and chord ends of the fillet weld. The overlapped KO joint SCFs are from Marshall, (1978).

$$SCF_c = 1.27 \, (t/T)(D-T/T)^{\frac{1}{2}}\sin\theta$$

$$SCF_b = 1 + 0.6 \, (1 + (t/T)^{\frac{1}{2}}(d/D)^{-\frac{1}{2}})SCF_c) \exp(-(0.5T + t)/(0.5dt)^{\frac{1}{2}})$$

Table 3 Comparison of Fillet Stress Concentrations due to balanced Loading in welded and cast Joints

Joint Type Geometry Ref.	Welded Non-Overlapped KN	Welded Overlapped KO	Cast K9045 B
D/T	25.6	25.6	24.1
d/D	0.53	0.53	0.64
t/T	0.50	0.50	0.50
g/D	0.11	-0.39	-0.025
R/T	0.25	0.25	1.50
Measured SCF_b	-5.15	3.10	-1.5
Calculated SCF_b	-4.61	-3.20	-
" SCF_c	-3.30	-2.43	-
For cast shape parameters:-			
Calculated SCF_b	-3.87	-3.10	-
" SCF_c	-2.75	-2.31	-

Table 3 also shows SCF values calculated for KN and KO joints with the same non-dimensional parameters as the casting. It can be seen that the experimental values agree well with prediction (especially for the overlapped joint) and that the predictions vary by less than 20% for the different d/D values. This suggests that it is valid to compare the experimental results for cast and welded shapes.

With a typical load history (Dover, 1981) and using Miner's Rule and the DOE-Q S-N relationship (UK Dept. of Energy, 1978), $\log_{10}N = 14.62 - 4.13 \log_{10}S$, the fatigue life of the casting is found to be 10 times better than the welded joint even though the fatigue relationship is probably conservative for the former, and in the absence of any data for the casting, the in-plane bending SCFs are assumed the same for both.

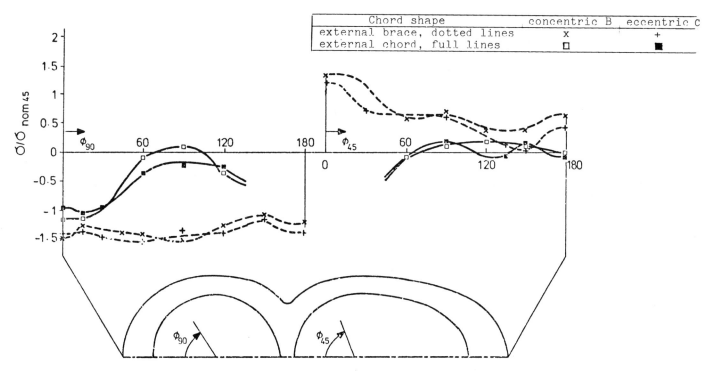

Fig. 9 Fillet stresses due to balanced loading with nominal stress taken as that in 45° brace.

There is little data published enabling comparison to be made between different castings. Analysis of a waisted K60-60 jiont under balanced loading (Gibstein, 1981), but with quite different design features, shows maximum SCFs of 2.36 on the inside usrface and 1.86 on the outside. These figures are much lower than those predicted for welded joints of comparable geometry.

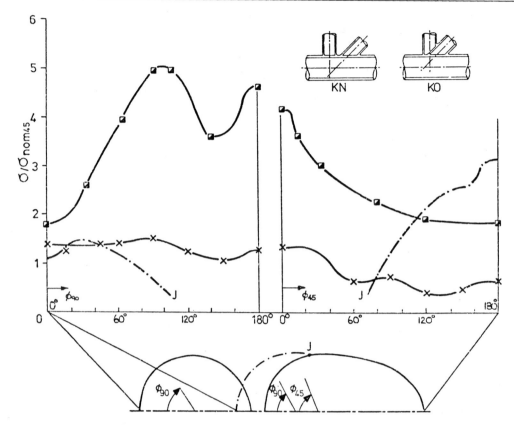

Fig. 10 Fillet stresses due to balanced loading

Tables 1 and 2 show the maximum inside surface stresses obtained with two configurations. As would be expected, their signs are opposite to those of the outside surface stresses. Under single-brace loading, the inside surface peaks are all about 0.5 to 0.6 of the maximum in that model, whilst, with balanced loading, they are less (35% to 45%). This finding, which is similar to that for out-of-plane bending (Edwards and Fessler, 1981), suggests that the outside surface is the critical one for this design.

Fig. 2 shows chord surface stresses in the plane of the paper dropping to zero at the core-support holes. Stresses tangential to the hole perimeters have been measured and the highest of these are presented in Tables 1 and 2. Although these stresses are significant in single brace loading (usually about 70% of the fillet maxima), they are much reduced in balanced loading and are always less than half the fillet maxima.

It is clear from Figs. 7, 8 and 9 that making the chord wall thickness variable produces little change in the fillet stresses. Thus it appears, especially in balanced axial loading, that the stress distribution depends more on the relative dimensions of the whole chord and brace reinforcements and the fillets. Although varying chord wall thickness does not reduce the stresses much, it may improve castability. It may also be advantageous for T and Corner T joints as lower stresses were found in each of the single brace loadings.

CONCLUSIONS

The cast K joints analysed in this paper contain very low stresses under balanced, axial loading. The peak stresses are in the fillets surrounding the $90°$ brace and are 1.5 times the nominal stress in the $45°$ brace. There is little change in stress around the $90°$ brace fillet, showing this to be an efficient design. Stresses on inside surfaces and around core-support holes are much lower than those on outside surfaces.

Making chord wall thickness variable has little effect under balanced loading but under single brace loading it produces lower peak stresses.

The maximum SCF of 1.5 is half that in a comparable, overlapped, welded joint and 0.3 of that in a non-overlapped joint. The halving of SCF produces an order of magnitude improvement in fatigue life even when conservative assumptions are made about the SCFs in bending and the fatigue strength of the cast joint.

ACKNOWLEDGEMENTS

This work is supported by an SERC grant. The authors wish to thank the Department's technicians for their skilled assistance and Messrs. G. Marston, E. F. Walker and A. M. Wood of British Steel Corporation for their advice on casting procedures.

REFERENCES

1. HEYWOOD, R. B., 1952, Designing By Photoelasticity, Chapman & Hall.
2. KUANG, J. et al, 1977, "Stress Concentrations in Tubular Joints," SPE Journal, August, pp 287-300.
3. DOVER, W. D., 1981 "Novel Methods and Techniques," Paper 13, A Short Course on Fatigue of Offshore Steel Structures, NEL, East Kilbride, Scotland.
4. FESSLER, H. and LITTLE, W. J. G., 1978, "Fillet Stresses in Tubular Joints Obtained by Photoelastic Techniques." International Conference on Integrity of Steel Offshore Structures, Institute of Engineers and Shipbuilders, Glasgow, Scotland.
5. FESSLER, et al, 1979a "Elastic Stresses Due to Axial Loading of Tubular Joints With Overlap," 2nd International Conference on Behaviour of Offshore Structures, London, England.
6. FESSLER, et al, 1979b, "Precision Casting of Epoxy Resin Models Using Expandable or Re-usable Moulds," 8th All-Union Conference on Photoelasticity, Tallin, USSR.
7. FESSLER, H. and EDWARDS, C. D., 1981, "Design and Stress Analysis of a Light, Cast $90°-45°$ K-Joint," International Conference on Steel in Marine Structures, Volume 1, Paper ST4.4, Paris, France.
8. U.K. Department of Energy, 1978, (2nd Impression) Offshore Installations : Guidance on Design and Construction, London, England, HMSO.
9. MARSHALL, P. W., 1978, A Review of Stress Concentration Factors in Tubular Connections, Shell Oil Company, Houston, Report CE-32.
10. GIBSTEIN, M., 1981, Photoelastic Stress Analysis of a Cast Steel K-Node, Det Norske Veritas, Report No. 81-0023.

AN EXPERIMENTAL AND ANALYTICAL INVESTIGATION OF
THE FATIGUE BEHAVIOUR OF MONOPOD TUBULAR JOINTS

S. S. Gowda
Engineering and
 Applied Science
Memorial Univ.
 of Newfoundland

D. V. Reddy*
Coastal &
 Oceanographic Eng.
Univ. of Florida
Gainesville, FL

M. Arockiasamy
Engineering and
 Applied Science
Memorial Univ. of
 Newfoundland

D. B. Muggeridge
Engineering and
 Applied Science
Memorial Univ. of
 Newfoundland

P. S. Cheema
College of
 Trades and
 Technology
St. John's, Nfld

SUMMARY

This paper describes an experimental and analytical investigation of the static and fatigue behaviour of monopod tubular joints. Crack initiation and propagation, leading to final failure of the joints subjected to (i) constant amplitude sinusoidal, and (ii) pseudo-random loadings have been studied. Strains were measured under both static and fatigue loadings; for the case of pseudo-random loading hot-spot strains were recorded continuously, and the r.m.s. value of the stress range obtained. Life estimates for the pseudo-random loading case, using S-N Curve, Fracture Mechanics, and Weibull approaches with assumed values of material constants, correlate reasonably well with the experimental findings.

---

*On leave from the Faculty of Engineering and Applied Science, Memorial University of Newfoundland, St. John's, Nfld.

1. INTRODUCTION

Most of the offshore drilling and production platforms are of the jacket-type, Fig. 1 (Masubuchi, 1970). For shallow ice-infested waters, such as the Cook Inlet and the Beaufort Sea, gravity platforms are more appropriate, Figs. 2 and 3. Recently, a monopod structure supported by three branch members has been proposed for more than 350 m (1,000 ft.) of water by Heerema Engineering Company, Fig. 4. Analytical and experimental investigations of stresses for monopod tubular joints subjected to both static and fatigue loads are scarce in the literature, compared to typical T, K and Y joints of framed platforms on which extensive research has been carried out by several investigators (Dover, et al, 1978, Bouwkamp, 1966, and Beale and Toprac, 1967). This paper describes studies of monopod tubular joints under constant amplitude and random loadings.

The measured stress concentration factors are compared with finite element analysis. Life estimates of the joint for random load fatigue are made using the S-N curve, Fracture Mechanics, and the Weibull Distribution approaches and compared with the experimental results.

2. TEST PLAN AND METHODOLOGY

2.1 Design and Fabrication of Models

The material used for the specimens was of ASTM A53 Grade B specification with a yield strength of 53 ksi. The dimensions were 1.828 m high x 323.80 mm O.D. x 9.52 mm (72 in. x 12 3/4 in. tk. X 0.375 in.) for the main member, and 141.20 mm x 6.98 mm tk. (5 9/16 in. O.D. x 0.275 in.) for the branch members. The branch members, spaced at 120° to each other, intersect the main chord at one third of its height. To avoid stress concentrations due to proximity, the axial and lateral loadings were located at distances of more than three times the main chord diameter from the weld intersection. The ends of all the branch tubes and the main chord were welded to a base plate bolted to the heavy duty 762 mm (30 in.) thick test floor. All welds were of the fillet type with full penetration; the electrodes used were of AWS-ASTM E7018 specification. The welds were checked by magnetic particle inspection as well as by radiographic examination using Gamma rays to ensure that no cracks or major weld defects exist. The welds for the first two joints were ground to enable comparison with other joints which will be tested without grinding of the weld toes.

2.2 Test Set-up and Instrumentation (Fig. 5)

For loading purposes, a reaction frame was suitably adapted for both vertical and horizontal loading conditions with hydraulic actuators of capacities 222.5 kN (55,000 lb.) and 667.5 kN (150,000 lb.) respectively. Special connections were provided for static vertical and cyclic horizontal loading. The horizontal loading to the main chord was applied with a 'wrap-around' semi-circular plate assembly to circumferentially distribute the load on the stiffening sleeve in both compression and tension. The semi-circular plates were connected to the horizontal ram through end plates. Great care was taken to achieve the verticality of the 'dead weight simulating' hydraulic ram to reduce the load eccentricity to the irreducible minimum.

For proper instrumentation of strain gauges to measure strains at critical 'hot-spot' locations under static and fatigue loadings, stress-coat using brittle lacquer coating was applied to the joint portion covering a length of about 1 1/2 times the main tube diameter and a distance equal to the branch tube diameter. The choice of the coating was based on the temperature and humidity in the laboratory. A Tens-Lac brittle lacquer undercoat, U-10, consisting of a mixture of aluminum powder and a carrier solvent, was first sprayed on to the surface of the joint to obtain uniform reflectivity. Brittle lacquer was then applied over the undercoat at intervals of 2 min. between each coating, except the last one, which was applied after 20 min. The coating was cured at about 75°F for 24 hours, and brought down to the testing temperature of 65°F. A vertical step-wise static loading was first applied and no cracks were found. After releasing the vertical load, a step-wise horizontal load was applied and cracks in the coating were marked on the surface at the incremental loading levels. The spread of the lacquer cracks around the joint gave a picture of the overall strain distribution and the severity of the strains at critical locations. These observations were used to establish the gauge locations. The coating crack distribution is shown in Fig. 6.

Each specimen was instrumented with a total of 70 electrical resistance strain gauges and the strains were recorded with a strip-chart recorder at specified cycle numbers. The signals from the load cell and the rosette gauges at the critical location, were recorded with a HP 3968 tape recorder. The data was later transferred to a digital tape for analysis.

2.3 Finite Element Analysis

Hot-spot stresses depend on the joint configuration and type of loading. Values for typical T, K, Y and X joints for different loading conditions and geometries are available in the literature (Kuang, et al, 1975; Marshall, 1974; Wordsworth, 1975). In the present investigation, finite element analysis was used to determine the hot-spot stress concentration factors (hot-spot stress/nominal stress in the brace). The joints were modelled using eight-noded superparametric general shell elements. Each node of the shell element had six degrees of freedom - three translational and three rotational. The joint mesh (nodal coordinates and element nodal connectivities) was determined by automatic mesh generation, Fig. 7. Appropriate boundary conditions were imposed to simulate the restraint provided by the welds around the tubes connected to the base plate. Displacements, maximum principal strains and stresses were obtained from the analysis.

2.4 Load Input

The load magnitudes were scaled down values of dead load and ice forces measured on gravity monopod structures in Cook Inlet, Alaska (Peyton, 1968 and Blumberg and Strader, 1969). The modelling criterion was based on equal axial and critical bending stresses.

The first stage in the testing procedure was the application of incremental static loading up to the peak amplitudes of the prescribed fatigue loading in the vertical and horizontal directions. For the constant amplitude fatigue test, the vertical and horizontal loads were 22.24 kN (5 kips) and 66.72 kN (15 kips) respectively, and for random loading 177.92 kN (40 kips) and ±62.27 kN (±14 kips). All the measured strains were converted to geometric strain concentration factors (ratio of measured strain to nominal strain in the brace).

The load combination for the first model was a harmonically varying load of amplitude 66.72 kN with a frequency of 0.25 Hz, and a static vertical load of 22.24 kN. A positive sine wave, symmetric about the zero reference, was generated by appropriate dynamic loading selection in the MTS machine.

The load-time history of the random load signal was recorded on magnetic tape and played from a LYREC tape recorder. For continuous operation, a special purpose loop attachment device was used to provide for a 750 cm long, 1.25 cm wide (300 in. long, 1/2 in. 1 Mil) instrumentation tape. The tape, adjustable to any desired length with the loop idler, was activated at the desired speed of 15 cm/sec (6 in./sec.). The signal was filtered with a cut-off frequency of 20 Hz. and fed to the controller unit of the MTS machine. The controller unit was an electro-hydraulic testing system, with servocontrol, fail safe, and readout functions. The hydraulic actuator applies the mechanical input to the specimen and to a transducer. The transducer conditioner supplies AC excitation to its associated transducer and provides a DC output proportional to the mechanical input. A digital indicator continuously monitored and displayed the voltage input representing the load levels. The load level for the horizontal hydraulic actuator was selected from the Digital Indicator corresponding to the maximum peak load of ±62.27 kN (±14 kips). The load-time and strain-time histories are shown in Figs. 8 and 9 respectively. The input signal has a central frequency of about 1 Hz. The power spectral density of the load signal is shown in Fig. 10. The load cell and strain gauge signals at the hot-spot were recorded continuously to monitor the joint behaviour during crack propagation. Also, strains at important locations were recorded intermittently on a strip-chart recorder.

3. FATIGUE ANALYSIS

Continuous recording of the hot-spot strain indicated the strain distribution before crack initiation and during its propagation. The loading used was a broad band stationary random signal with an approximate Gaussian probability density function. The central frequency of the spectrum was approximately 1 Hz. The power spectral density of the load signal is shown in Fig. 10.

3.1 Fatigue Crack Growth Development

The first specimen was tested under constant amplitude sinusoidal loading with a frequency of 0.25 Hz. and a maximum load of ±66.72 kN (±15 kips). The first crack was observed at approximately 40,000 cycles, at the critical hot-spot near the weld toe in the main chord above the branch tube on the horizontal hydraulic actuator side, Fig. 5. No attempt was made to measure the crack depth during testing of this first joint. The crack spread very rapidly around the weld for a distance of 138 mm (5.52 in.). The number of cycles at this stage was about 50,000. The crack propagation slowed down after this and almost ceased at about 65,000 cycles. However, side cracks started on both sides of the main tube almost symmetrically, Fig. 11. The test was terminated when the final crack length was about one half of the circumference of the main chord.

The second specimen was tested under random loading with a maximum peak load of ±62.27 kN (±14 kips). The load signal with 48 cycles was repeated at approximately one minute intervals. The load and strain signals are shown in Figs. 8 and 9. The fatigue crack growth behaviour in this case was different from the first joint. The first crack appeared on the right side of the main tube at approximately 238,000 cycles. It was shallow approximately 1 cm (0.394 in.) long and formed on the weld toe of the main member. A second crack was noticed on the left side of the main tube weld toe at about 302,000 cycles. These first and the second cracks were 20 mm (0.787

in.) and 40 mm (1.57 in.) away from the critical hot-spot location respectively. Gamma ray examination of the weld indicated slight slag at these locations of crack initiation. The critical hot-spot location showed no defects in the weld toe. This indicates that minor weld defects which act as stress raisers are the main cause for crack initiations rather than the hot-spot stress. As the number of cycles increased, these cracks propagated towards the hot-spot and joined to form a long crack. The crack around the weld on the right side started spreading through the main member wall at about 290,000 cycles, and that on the left side at 707,650 cycles. Cracking around the weld slowed down very much but became faster on either side of the branch tube 1, in the main chord wall, almost exactly at the same height. The crack initiations at increasing numbers of cycles and the spread of the main crack into the chord member can be seen in Fig. 12. A very important observation in the crack growth behaviour under random loading was the very rapid rate of crack propagation on the main chord compared to that around the weld.

4. FATIGUE LIFE ESTIMATION

Analytical values for the fatigue life of the specimens tested are obtained based on three formulations: 1) the well-known S-N curve procedure, 2) Fracture Mechanics approach, and 3) the Weibull method. The appropriate constants for the first and third formulations were determined from test values from a group of tubular specimens with varying geometric parameters reported by the U. K. Dept. of Energy: (Snedden, 1981), and for the second formulation from values reported by Gurney, 1979. This procedure had to be used as the current test series of the authors is still in progress.

4.1 S-N Approach

The relationship between stress range, S, and number of cycles to failure, N, is of the form:

$$\ln N = \ln k + m \ln S \tag{1}$$

The parameters, k and m, determined using linear regression analysis of the data, shown in Table 1, are 10.27 and 0.68 respectively.

For the stress range of 136.44 N/mm² of the present investigation, the 90% confidence interval of the life of the joint is:

$$4.17 \times 10^5 < N < 1.66 \times 10^6 \text{ cycles}$$

The experimental value of 9.5×10^5 cycles is well within this range.

4.2 Fracture Mechanics Approach

The stress intensity factor K representing the state of stress at the crack tip is given by:

$$K = Y \sigma \sqrt{\pi c} \tag{2}$$

where

σ = applied stress,

Y = correction factor,

and

c = half-crack length.

The crack growth rate in a material is a function of the stress intensity factor which varies with the applied load. This is given by the Paris Law:

$$\frac{dc}{dN} = C(\Delta K)^m \tag{3}$$

where

C and m are material constants,

ΔK is the stress intensity factor range,

and

N is the number of cycles to failure

The correction factor, Y, is a function of the geometry of the crack; for a surface crack it is equal to 1.12 (Dover, et al, 1978, Becker, et al, 1972, Pan and Plummer, 1976, Hartt, et al, 1980).

Substituting Eq. 2 in Eq 3, gives,

$$\frac{dc}{dN} = C(1.99\Delta\sigma)^m c^{m/2} \tag{4}$$

On integration, Eq. 4 becomes,

$$N = \frac{1}{C(1.99\Delta\sigma)^m} \left(\frac{1}{1-m/2}\right) \left[c_f^{1-m/2} - c_o^{1-m/2}\right] \tag{5}$$

where

$\Delta\sigma$ = the hot-spot stress range,

c_o = initial flaw length,

and

c_f = final crack length.

According to Maddox, 1974, and Tomkins and Scott, 1978, the typical crack-like surface defects of the order of $2c_o$ = 0.20 - 2.0 mm are always present in fillet welds. This is also stated in the codes (DnV Report, 1977), for welds improved by grinding. In the present analysis, the values were: $2c_o$ = 0.2 mm, measured final crack length, $2c_f$ = 440 mm, and the r.m.s. stress range, $\Delta\sigma$ = 136N/mm² based on a correction factor of $\sqrt{3}$. Since the stresses induced in the specimen are of a pseudo-random type, the effective stress range is assumed to be $\sqrt{3}$ times the r.m.s. value (Nibbering and Faulkner, 1978). The material constants, C, and m are evaluated on the basis of the linear relationship of log C to m (Gurney, 1979), established for steels under plane strain conditions. The material constant, m, can vary from 2 to 6 (Teramoto, et al, 1975, Thebault, et al, 1980). In this study a value of m = 2.52, can be reasonably assumed to fit the experimental data, which gives a value of C = 4.79 x 10^{-12}.

Substituting the above values in Eq. 5, the number of cycles to failure is 9.32 x 10^5 cycles, which compares well with the recorded total of 9.5 x 10^5 cycles.

4.3 Weibull Approach

The numbers of cycles to failure for the group in the U.K. D.O.E., 1981 Report (Table 1), were used in the Weibull distribution to obtain a straight line plot of the scattered data.

According to the Weibull distribution, the fatigue lives of group of specimens can be represented by a family of frequency distribution functions such as:

$$f(N) = \frac{b}{N_a - N_o} \left[\frac{(N - N_o)}{(N_a - N_o)}\right]^{b-1} \exp\left\{-\left[\frac{N - N_o}{N_a - N_o}\right]^b\right\} \tag{6}$$

where

N = specimen life,

N_o = minimum life parameter,

N_a = characteristic life parameter occuring at 63.2% failure,

and

b = Weibull shape or slope parameter.

In dealing with life phenomena, it is reasonable to assume N_o = 0, that is the expected minimum of the population life to be zero (Lipson and Sheth, 1973, ASTM, 1963). Then, the cumulative distribution function for the fraction of population failing prior to life N is

$$F(N) = 1 - \exp\left\{-\left[\frac{N}{N_a}\right]^b\right\} \tag{7}$$

Eq. 7 can be written in the form

$$y = bx + c \qquad (8)$$

where

$$y = \ln \ln \frac{1}{1-F(N)}$$

$$c = b \ln N_a,$$

$$x = \ln N,$$

$$b = \frac{n\Sigma xy - (\Sigma x)(\Sigma y)}{n\Sigma x^2 - (\Sigma x)^2},$$

and

$$c = \frac{\Sigma y - b\Sigma x}{n}.$$

The main basis of the analysis is the consideration of the hot-spot stress range and the number of cycles to failure.

For the data in Table 1, the following least square lines have been obtained using Eq. 8:

Median Ranks: $y = 3.45\chi - 48.31$ (9)

50% Ranks: $y = 5.16\chi - 73.25$ (10)

95% Ranks: $y = 2.73\chi - 37.46$ (11)

The graph of Eq. 9 and the data points are shown in Fig. 14. The 90% confidence values for the Weibull parameters and the mean numbers of cycles are:

Slope: $1.93 < \hat{b} < 4.97$

Characteristic life: $9.1 \times 10^5 < \hat{N}_a < 1.5 \times 10^6$

Mean: $8 \times 10^5 < \mu < 1.4 \times 10^6$

The experimental fatigue life of the joint in the present investigation is 9.5×10^5 cycles which checks well with the above values, for a 90% confidence interval of the mean.

5. CONCLUSIONS

(i) The initiation of fatigue cracks under random loading, occurs at the defect location in the hot-spot region, but not necessarily at the true hot-spot point.

(ii) The initiation and propagation of the crack consumes the major life of the joint, i.e. until the crack fronts on either side of the hot-spot region reach the main chord wall. After this only a small amount of life is left for the crack length to become equal to half the circumference of the main tube. At this stage, for all practical purposes the joint is considered to have completely failed. The initiation of the visible crack occurred at about 25% of the life of the second specimen i.e. 2.5×10^5 cycles. The test was terminated at 9.5×10^5 cycles, when the crack length was more than half the main chord circumference.

(iii) Estimated fatigue life values by the S-N curve, Fracture Mechanics, and Weibull methods seem to correlate reasonably well with the experimental findings in the current investigation, in spite of the approximate values used in the constants.

REFERENCES

ASTM, 1963, A Guide for Fatigue Testing and the Statistical Analysis of Fatigue Data, ASTM, Philadelphia, Committee E-9.

Beale, L. A. and Toprac, A. A., 1967, Analysis of In-Plane T, Y and K Welded Tubular Connections Univ. of Texas, Department of Civil Engineering, S.F.R.L. Report No. P550-9.

Becker, J. M., Gerberich, W. W., and Bouwkamp, J. G., 1972, "Fatigue Failure of Welded Tubular Joints," American Society of Civil Engineers, Structural Division, Vol. 98, No. ST1, pp. 37-59.

Bouwkamp, J. G., 1966, Tubular Joints Under Static and Alternating Loads, University of California, Berkeley, Department of Civil Engineering, Report No. 66-15.

Dnv, 1977, Rules for the Design Construction and Inspection of Offshore Structures, Appendix C.

Dover, W. D., Hibbered. R. D., and Holdbrook, S. J., 1978, "A Fracture Mechanics Analysis of the Fatigue Failure of T-Joints Subject to Random Loading," Proceedings, International Symposium on Integrity of Offshore Structures, Institution of Engineers and Shipbuilders in Scotland, Glasgow, pp 3/1-3/31.

Gurney, T. R., 1979, Fatigue of Welded Structures, Cambridge, U.K., Cambridge University Press.

Hartt, W. H., Henke, T. E., and Martin, P. E., 1980, Influence of Sea Water and Cathodic Protection upon Fatigue of Welded Steel Plates, as Applicable to Offshore Structures, Florida Atlantic University, Department of Ocean Engineering, Boca Raton, Florida, Rept. No. TR-OE-80-1.

Jazrawi, W., and Davies, J. F., 1975, "A Monopod Drilling System for the Canadian Beaufort Sea," The Society of Naval Architects and Marine Engineers, New York, pp. E-1-E-27.

Kuang, J. G. Potvin, A. B., and Heick, R. D., 1975, "Stress Concentrations in Tubular Joints," Proceedings 7th Annual Offshore Technology Conference, Houston, Texas, OTC 2205, pp. 593-615.

Lipson, C. and Sheth, N. J., 1973, Statistical Design and Analysis of Engineering Experiments, McGraw-Hill, Inc.

Maddox, S. J., 1974, "A Fracture Mechancis Approach to Service Load Fatigue in Welded Structures," Welding Research International, Vol. 4, No. 2, pp. 2-29.

Marshall, P. W., 1974, "Basic Considerations for Tubular Joint Design in Offshore Constructions," Proceedings, International Conference on Welding in Offshore Construction, Welding Institute, Newcastle, 1974.

Masubuchi, K., 1970, Materials for Ocean Engineering, Cambridge, Mass., MIT Press.

Nibbering, J. J. W., and Faulkner, D., 1978, "Some Considerations for Fracture-Safe Design of Maritime Structures," Proceedings, International Symposium on Integrity of Offshore Structures, Institution of Engineers and Shipbuilders in Scotland, Glasgow, pp. 1/1-1/22.

Pan, R. B. and Plummer, F. B., 1976, "A Fracture Mechanics Approach to Nonoverlapping Tubular K-Joint Fatigue Life Prediction," Proceedings, 8th Offshore Technology Conference Houston, Texas, 1976, OTC 2645, pp. 317-331.

Snedden, N. W., 1981, Offshore Installations: Guidance on Design and Construction - Proposed New Fatigue Design Rules for Steel Welded Joints in Offshore Structures, (Compiled), U. K. Dept. of Energy.

Teramoto, S., Kawasaki, T., Kaminokado, S., and Matoba, M., 1975, Fatigue Strength of Welded Tubular Joints in Offshore Structures Mitsubishi Technical Bulletin No. 100.

Thebault, J., Olangon, M., and Barnouin, B., 1980, "Fatigue Analysis of a Tubular Joint Broken Under Waves and Current Loadings," Proceedings, 12th Offshore Technology Conference Houston, Texas, OTC 3698, pp. 197-206.

Tomkinds, B. and Scott, P., 1978, "An Analysis of the Fatigue Endurance of Tubular T-Joints by Linear Elastic Fracture Mechanics," Proceedings, European Offshore Steels Research Seminar, U.K. Dept. of Energy - Commission of European Communities, 1978, Paper No. 20, PP. 492-508.

Wordsworth, A. C., 1975, "The Experimental Determination of Stresses at Tubular Joints," Proceedings, BSSM/RINA Joint Conference on Measurement in the Offshore Industry, Edinburgh, 1975.

ACKNOWLEDGEMENTS

The authors would like to thank Prof. C. D. diCenzo and Dr. G. R. Peters, Dean and Associate Dean of the Faculty of Engineering and Applied Science, and Dr. I. Rusted, Vice-President, Memorial University of Newfoundland for their continued interest and encouragement. Appreciation is expressed to Dr. H. Wang, Chairman of the Department of Coastal and Oceanographic Engineering, University of Florida for encouraging the second author (DVR) to complete the work on this project. The assistance given by Mr. Cran, Sales Engineering Manager, Stelco Canada in the form of material and fabrication costs, and the support by the Natural Sciences and Engineering Research Council of Canada Grant A-8119 (DVR) are gratefully acknowledged.

Table 1: EXPERIMENTAL RESULTS FOR TUBULAR JOINTS (Snedden, 1981)

Hot-Spot Stress Range N/m^2	Cycles to Failure
176	7.1×10^5
198	7.8×10^5
242	9.4×10^5
198	1.0×10^6
179	1.1×10^6
179	1.3×10^6
225	1.7×10^6

Fig. 1 Jacket-Type Structure.
(Masubuchi, 1970)

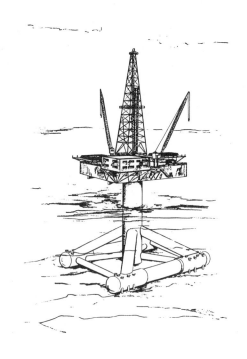

Fig. 2 Monopod Platform.
(Masubuchi, 1970)

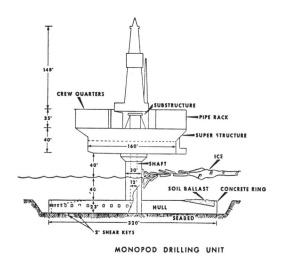

Fig. 3 Monopod Drilling Unit.
(Jazrawi and Davies, 1975)

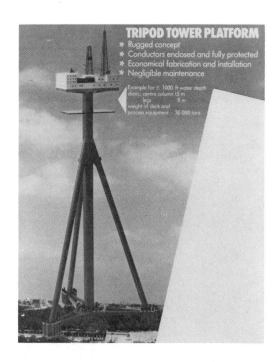

Fig. 4 Typical Tripod Tower Platform.
(Offshore Engineering, 1981)

Fig. 5 Setup for 'Dry' Testing.

Fig. 6 Brittle Lacquer Coating Crack Pattern around Joint Intersection.

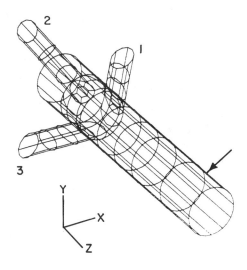

Fig. 7 Finite Element Mesh Generation for the Joint.

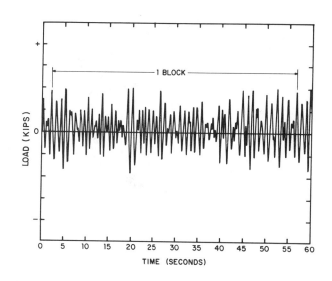

Fig. 8 Pseudo-random Input Load Time-history.

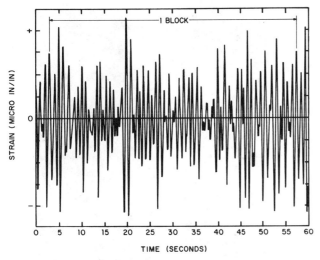

Fig. 9 Hot Spot Strain Time-history.

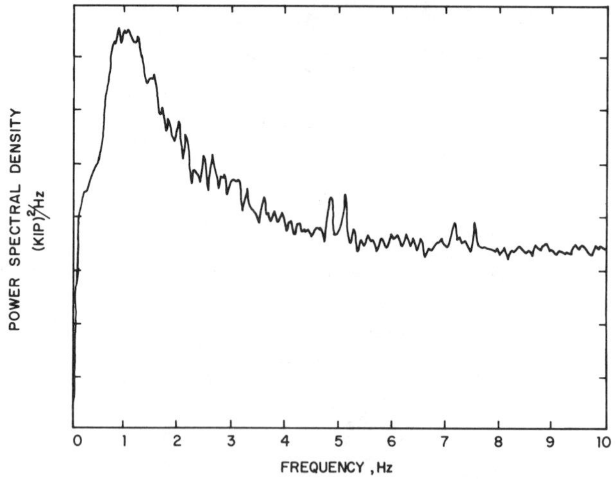

Fig. 10 Power Spectral Density of Loading.

Fig. 11 Crack Growth and Failure Pattern for the Harmonic Load Test.

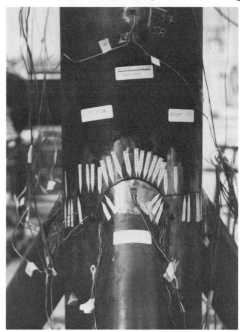

Fig. 12 Crack Growth and Failure Pattern for the Pseudo-random Load Test.

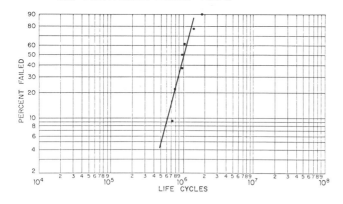

Fig. 13 Weibull Fatigue Life Plot.

ON THE SIGNIFICANCE OF NON-LINEAR RANDOM WAVE LOADING IN THE FATIGUE ANALYSIS OF JACKET TYPE OF STRUCTURES

H.B. Kanegaonakar
Indian Institute of Technology,
Bombay, India.

R. M. Belkune
Indian Institute of Technology,
Bombay, India.

C. K. Ramesh
Indian Institute of Technology,
Bombay, India.

SUMMARY

For a reliable behavioural analysis, an accurate representation of loading on offshore structures, exposed to random wave climate, is the most important aspect. Due to the randomness and quadratic nature of drag forces, the governing stochastic differential equations for dynamic response calculations are non-linear in character. In this paper, the solution of these equations is achieved by transforming them into equations with relative displacements between surrounding water and structure as unknowns and expanding the corariance functions of the response into a power series. Only the first two terms are included for the response calculations. Results from above analysis are compared with the response values from equivalent linearized solutions for a jacket type of structure. Fatigue damage is estimated using both the above mentioned response calculation techniques.

It is concluded that non linearity in drag force term significantly affects the predicted estimates of the fatigue life of the tower. Equivalent linearization of the drag underestimates the extreme values of responses and overestimates the fatigue life of the tower, and hence for reliable estimation of **fatigue** damage, it is essential to retain nonlinearity of the drag during solution process.

NOMENCLATURE

A	−	Lumped projected area at the node. in m^2
A. D.	−	Accumulated Fatigue Damage
b	−	Slope of S-N Curve
$[c]$	−	Structural Damping matrix
C_d, C_m	−	Drag and Inertia Coeff. in Morisson Equation
$G(.)$	−	Power series defined in equation I - 17
Hs	−	Significant wave height in. m.
$H(w)$	−	Impulse response function in frequency domain
$[k]$	−	Reduced stiffness Matrix for the test structure
L	−	Length of the chord
$[\hat{m}]$	−	Mass matrix for the structure
N	−	Number of modes considered
$N(s)$	−	Number of stress cycle for failure at stress level s
p	−	Natural frequency of the tower
$p_1(s)$	−	Probability density function
R	−	Mean radius of chord
r_b	−	Mean radius of brace
$r, \dot{r}, \ddot{r}$	−	Relative displacement, velocity and acceleration resp.
$R_{r_1 r_2}(.)$	−	Covariance function between r_1 and r_2
s	−	Stress range
$S_{r_1 r_2}(.)$	−	Power spectral density function between r_1 and r_2
t	−	Wall thickness of Brace
T	−	Duration of a sea state
T_c	−	Wall thickness of chord
T_s	−	Ambient period of tower zero up crossing
$u, \dot{u}, \ddot{u}$	−	Water particle displacement, velocity, acceleration rep.
V	−	Lumped projected volume at the node in m^3
β^1	−	Hydrodynamic damping
$\rho_{r_1 x_1}$	−	Correlation coefficient between r_1 and x_1
ν	−	Mean zero up crossing frequency
$[\phi]$	−	Eigenvector matrix
$\langle . \rangle$	−	Expected value

INTRODUCTION

Offshore structures subjected to random sea waves experience fatigue problems which often govern the design of many structural elements. Reliable prediction of fatigue damage calls for realistic representation of sea wave forces and structural responses. A nondeterministic analysis is now well established for a linearized damping system. For the jacket type of platforms, the drag force which is nonlinear in character, has significant contribution to the total wave loading. For a linear spectral analysis, therefore this nonlinear drag term is linearized by equivalent linearization technique. Borgman[1] applied a method of non-linear transformations to obtain a series expansion of covariance functions to include non-linear drag effect in the calculation of hydrodynamic forces on objects. This method was utilized by Tung[3] to develop a response calculation technique. Pierson-Holmes distribution was used by Burrows[2] to investigate into the effects of non-linearity of drag on extreme responses.

This paper concerns with the spectral methods to demonstrate that the effect of retention of nonlinearity in drag term on the fatigue damages, is significant. The governing nonlinear stochastic equations are solved by using first two terms of the series expansion of the covariance functions and iterative procedure. The longterm wave climate is assumed to be modelled by several sea states, each of which is considered to be stationary, ergodic, narrow banded and Gaussian. Each sea state is modelled through the Pierson-Moskowitz spectrum. Linear wave theory is used to derive statistical expressions for water particle kinematics and wave forces determined using Morrison equation. Stress histories and fatigue life are obtained using both equivalent linearization and series expansion with cubic approximation for drag term.

MATHEMATICAL MODEL

An idealized tower structure in two-dimensions as shown in Fig. 1 is used for investigations (structure used in Ref. 2). The natural period of this structure in water is 0.62 sec. The structure is idealized by lumping masses at 13 nodes. The reduced stiffness matrix is determined by using static condensation technique. Structural damping is assumed to be 5 % of the critical damping. For the purpose forcing function calculations, the projected areas and volumes are lumped at 13 nodes and are shown in Table 1 with no forces at deck level. The structure is analysed for only horizontal transverse loading due to random waves in the direction of wave propagation. A sea state characterized by Pierson Moskowitz spectrum defined by significant wave height (H_s) and mean zero up crossing period (T_z) given by,

$$S_{hh}(w) = (H_s/2)^2 \frac{692}{w(1.09\, wT_z)^4} \exp\left(\frac{-692}{(1.09\, wT_z)^4}\right) \qquad \ldots (1)$$

is used. Linear wave theory is used to obtain wave kinematics. The standard equation of motion for the tower is

$$[m]\{\ddot{x}\} + [c]\{\dot{x}\} + [k]\{x\} = \rho C_m [V]\{\ddot{u}\} + \rho C_d [A]\{(\dot{u}-\dot{x})|(\dot{u}-\dot{x})|\} \qquad \ldots (2)$$

where $m = \hat{m} + \rho(C_m - 1)[V]$

This equation is then transformed into equation with relative response as an unknown. The mathematical details to obtain structural response using series expansion of the drag term is given in APPENDIX I

Dynamic response of the tower is determined for each sea state block. Standard deviation of the displacement at each node is used to determine the standard deviation of the forces at the nodes using condensed stiffness matrix. Standard deviation of the forces and moments for each member are computed using the above nodal forces and performing static analysis. Stress concentration factors are calculated by using empirical formula by Kuang, Potvin and Leick[6]

For chord $\quad SCF = 1.981 \cdot \alpha^{0.057}\, e^{-1.2\beta^3}\, \gamma^{0.808}\, \tau^{1.333} \qquad \ldots (3)$

For brace $\quad SCF = 3.751 \cdot \alpha^{0.12}\, e^{-1.35\beta^3} \cdot \gamma^{0.55} \cdot \tau \qquad \ldots (4)$

where $\alpha = L/R, \quad \beta = r_b/R, \quad \gamma = R/T_c, \quad \tau = t/T_c$

The response is assumed to be narrow banded both in linear and nonlinear analysis. Stress-cycle counting is done by using expected rate of platform zero up crossing. Thus

$$T_s = 2\pi \left[\frac{\int_{-\infty}^{\infty} S_{xx}(w)\, dw}{\int_{-\infty}^{\infty} w^2 S_{xx}(w)\, dw}\right]^{1/2} \qquad \ldots (5)$$

Within each sea state block the hot-spot stresses are assumed as Rayleigh distributed. Fatigue damage is calculated for all members using AWS - x curve shown in Fig.2. Rayleigh curve piecewise integration is truncated at eight standard deriation where the effect of stress-range cycles on fatigue damage is found to be negligible. Average fatigue damage of the tower is calculated as the average of the damages of the members meeting at the node number seven. Using Miner rule for each sea state, accumulated damage is estimated as

$$A.D. = \frac{1}{T_s} \cdot T \int_0^\infty \frac{p(s)}{N(s)} ds. \qquad \ldots (6)$$

$$p(s) = \frac{s}{m_o} \exp(-s^2/2m_o) \qquad \ldots (7)$$

$$m_o = \int_0^\infty S_{ss}(w) \, dw \; ; \; N(s) = N_1 \left(\frac{S_1}{s}\right)^b \qquad \ldots (8)$$

S_1, N_1 - a point on S - N curve

b - slope of the S - N curve

The damage in each sea state is summed over the long term wave climate.

NUMERICAL RESULTS AND DISCUSSION

The results of the response analysis are shown in Fig.3. Table 2. defines the wave climate showing the percentage occurrence of the sea states considered over the fatigue life of the tower, with the significant wave height and mean zeroupcrossing period for each sea state.

Sea state 5 for twelve hour storm durations with significant wave height 6.9 m and mean zero upcrossing period of 9.3 secs is considered by Burrows[2] for the dynamic analysing of the same tower as in Fig.1, using probabilistic method and retaining nonlinearity in the drag term using Pierson-Holmes distribution. The mean extreme displacement of the deck (Node 6) obtained using spectral method and equivalent linearization technique is found to be about 32°/. less than the value quoted by Burrows[2], using nonlinear probabilistic analysis upto the fourth moment. For the same sea state, the mean extreme deck deflection obtained using nonlinear analysis with series expansion and cubic approximation is 8°/. higher than the corresponding equivalent linearized analysis. Hence, as far as the first excursion probability estimates are concerned, analysis using non-Gaussian distribution for wave loading seems to be a better solution.

The results presented in Fig. 3. show that the difference between extreme response values increases as mean zero crossing period decreases approaching the fundamental period of the tower. As the wave frequency increases, the contribution of the non-linear cubic term in the drag-force spectrum increases. At the highest sea state, the difference in the linear and nonlinear responses is 8°/. which increases gradually to 13.4°/. difference at the lowest sea state. This is quite in agreement with the observation by Moe[5].

Since the fatigue damage is less dependent on the extreme responses, the effect of nonlinearity in drag on fatigue life is amply demonstrated using present method of analysis. Fig. 4 shows the difference between fatigue damages by linear and nonlinear analysis for each state, for the member between node 1 and 2. For this member linear analysis overestimates fatigue life by 42°/.. The average cumulative fatigue damage at node number seven is understimated by 43°/. using linear analysis.

Present analysis is only for two dimensional structures and one dimensional sea states. Based upon the results presented herein, and from the literature, it may be presumed that the percentage increase in fatigue damage would increase if directionality and local wave actions are included in the analysis.

CONCLUSIONS

Results of this study show the importance of non-linearity of the drag in the prediction of fatigue damages for jacket type of structures, wherein the drag forces predominate. The linearized analysis significantly underestimates the fatigue damage and for reliable prediction of fatigue life of the jacket type of platforms, the effects of nonlinearity should be included by performing nonlinear analysis.

REFERENCES

1. Borgman L.E., 1967, Random Hydrodynamic Forces on Objects, *The Annals of Mathematical Statistics*, Vol.38 No. 1 pp 37-51.

2. Burrows R., 1979 "Probabilistic Description of the Response of Offshore Structures to Random Wave Loading", <u>Mechanics of Wave Induced Forces on Cylinders ed. T.L. Shaw</u>, Pitman Advanced Publishing program, U.K.

3. Tung C.C., 1979. "On Response of Structures to Random Waves", <u>Applied Ocean Research,</u> Vol. 1 No. 4 pp 209 - 212

4. Wade B.G., Langey Earnest, Wu S.C. 1978, "The Effects of Current on Probabilistic Fatigue Analysis of Fixed Offshore platforms", <u>Tenth Annual Offshore Technology Conference, Houstan Taxas.</u>, Volume II paper 3164 pp 1011 - 1014

5. Moe G., 1977, "Stochestic Dynamic Analysis of Jacket Type Platforms." Nonlinear drag Effects and « Effective » Nodal Areas , <u>Safety of Structures under Dynamic Loading,</u>Norwegion Institute of Technology, Trondhenim.

6. Netherlands Steering Committee on Offshore Structures Problems, 1980, "<u>Fatigue Behaviour of Joints in Jacket Constructions</u>", StuPOC -IV-4 Netherlands Industrial Council for Oceanology, Netherlands.

APPENDIX I

The standard equation of motion of a tower before linearization is

$$[m]\{\ddot{x}\} + [c]\{\dot{x}\} + [k]\{x\} = \rho C_m[V]\{\ddot{u}\} + \rho C_d[A]\{(\dot{u} - \dot{x})|(\dot{u} - \dot{x})|\} \quad \ldots \text{I-1}$$

where $\quad [m] = [\hat{m}] + \rho(C_m - 1)[V] \quad \ldots \text{I-2}$

Substituting $u-x = r$, $\dot{u}-\dot{x} = \dot{r}$, $\ddot{u}-\ddot{x} = \ddot{r}$ in Eq. I-1, we get

$$[m]\{\ddot{r}\} + [c]\{\dot{r}\} + [k]\{r\} = [m^1]\{\ddot{u}\} + [c]\{\dot{u}\} + [k]\{u\} - \rho C_d[A]\{\dot{r}|\dot{r}|\} \quad \ldots \text{I-3}$$

Transforming above equation into normal co-ordinates using eigenvector matrix $[\phi]$, eq. I-3 can be rewritten as,

$$[M]\{\ddot{r}\} + [C]\{\dot{r}\} + [K]\{r\} = [\phi]^T[[m^1]\{\ddot{u}\} + [c]\{\dot{u}\} + [k]\{u\}] - [\phi]^T \rho C_d[A]\{\dot{r}|\dot{r}|\} \quad \ldots \text{I-4}$$

in which

$\quad [M] = [\phi]^T[m][\phi]$ - Generalized mass matrix $\quad \ldots \text{I-5}$

$\quad [C] = [\phi]^T[c][\phi]$ - Generalized damping matrix $\quad \ldots \text{I-6}$

$\quad [K] = [\phi]^T[k][\phi]$ - Generalized stiffness matrix $\quad \ldots \text{I-7}$

Let $\quad [\phi]^T[[m^1]\{\ddot{u}\} + [c]\{\dot{u}\} + [k]\{u\}] = \{f(t)\} \quad \ldots \text{I-8}$

and $\quad \rho C_d[\phi]^T[A]\{\dot{r}|\dot{r}|\} = C_D\{\dot{r}|\dot{r}|\} \quad \ldots \text{I-9}$

Assuming ergodic Gaussian process for water particle motion and structural responses, the correlation matrix of the response is

$$[R_{rr}(\theta)] = \langle \{r(t)\}\{r(t+\theta)\}^T \rangle$$

$$= \sum_{p=1}^{N} \sum_{q=1}^{N} \int_{-\infty}^{\infty} \int_{-\infty}^{\infty} h_p(\alpha_1) h_q(\alpha_2) \langle f_p(t-\alpha_1) f_q(t+\theta-\alpha_2) \rangle d\alpha_1 d\alpha_2$$

$$- \int_{-\infty}^{\infty} \int_{-\infty}^{\infty} h_p(\alpha_1) h_q(\alpha_2) C_{D_p} \langle f_q(t+\alpha_2+\theta) \dot{r}_p(t-\alpha_1)|\dot{r}_p(t-\alpha_1)| \rangle d\alpha_1 d\alpha_2$$

$$- \int_{-\infty}^{\infty} \int_{-\infty}^{\infty} h_p(\alpha_1) h_q(\alpha_2) C_{D_q} \langle f_p(t-\alpha_1) \dot{r}_q(t+\theta-\alpha_2)|\dot{r}_q(t+\theta-\alpha_2)| \rangle d\alpha_1 d\alpha_2$$

$$+ \int_{-\infty}^{\infty} \int_{-\infty}^{\infty} h_p(\alpha_1) h_q(\alpha_2) C_{D_p} C_{D_q} \langle \dot{r}_p(t+\theta-\alpha_2)|\dot{r}_p(t+\theta-\alpha_2|\dot{r}_q(t-\alpha_1)|\dot{r}_q(t-\alpha_1)| \rangle d\alpha_1 d\alpha_2$$

$$\ldots \text{I-10}$$

where $h(\alpha)$ is the unit impulse response function.

From Eq. I-8

$$S_{f_p f_q}(w) = \sum_{i=1}^{N_c} \sum_{j=1}^{N_c} \phi_{ip}\phi_{jq}[m_i^1 m_j^1 S_{\ddot{u}_i \ddot{u}_j}(w) + c_i c_j S_{\dot{u}_i \dot{u}_j}(w) + k_i k_j S_{u_i u_j}(w) + m_i^1 c_j S_{\ddot{u}_i \dot{u}_j}(w) +$$
$$+ m_i^1 k_j S_{\ddot{u}_i u_j}(w) + c_i m_j^1 S_{\dot{u}_i \ddot{u}_j}(w) + c_i k_j S_{\dot{u}_i u_j}(w) + k_i m_j^1 S_{u_i \ddot{u}_j}(w) + k_i c_j S_{u_i \dot{u}_j}(w)]$$
... I-11

where i and j denote the node numbers and N_c is total number of nodes. Using Gaussion nature of the forcing function and the response

$$< \dot{r}^2(t-\alpha_1)|\dot{r}(t-\alpha_1)| > = R_{\dot{r}f}(\theta - \alpha_2 + \alpha_1)/\sigma_{\dot{r}}^2$$
... I-12

and $$< f(t + \theta - \alpha_2) \dot{r}(t-\alpha_1)|\dot{r}(t-\alpha_1)| > = \sqrt{8/\pi} \, \sigma_{\dot{r}} R_{\dot{r}f}(\theta - \alpha_2 + \alpha_1)$$
... I-13

Taking fourier transform of the correlation function between $\dot{r}$ and f

$$S_{\dot{r} f} = \sum_{p=1}^{N} \sum_{q=1}^{N} \frac{\overline{H_p}(w) S_{f_p f_q}}{1 + \overline{H_q}(w)\sqrt{8/\pi} \, C_{D_q}/\sigma_{\dot{r}}}$$
... I-14

Also we have the relation $R_{f\dot{r}}(\theta) = R_{\dot{r}f}(-\theta)$
... I-15

The expected value of the quadratic form is

$$< \dot{r}(t + \theta - \alpha_2)|\dot{r}(t + \theta - \alpha_2)|\dot{r}(t-\alpha_1)|\dot{r}(t-\alpha_1)| > = \sigma_{\dot{r}}^4 \, G(R_{\dot{r}\dot{r}}(\theta - \alpha_2 + \alpha_1)/\sigma_{\dot{r}}^2)$$
... I-16

where $G(x) = 1/\pi[(4x^2 + 2)\sin^{-1}x + 6x\sqrt{1-x^2}]$
... I-17

Therefore

$$R_{\dot{r}\dot{r}}(\theta) = \sum_{p=1}^{N} \sum_{q=1}^{N} \int_{-\infty}^{\infty} \int_{-\infty}^{\infty} \dot{h}_p(\alpha_1) \dot{h}_q(\alpha_2) R_{f_p f_q}(\theta - \alpha_2 + \alpha_1) \, d\alpha_1 d\alpha_2$$

$$- \frac{8}{\pi} C_D \sigma_{\dot{r}} \int_{-\infty}^{\infty} \int_{-\infty}^{\infty} \dot{h}_p(\alpha_1) \dot{h}_q(\alpha_1) [R_{\dot{r}_p f_q}(\theta - \alpha_2 + \alpha_1) + R_{f_q \dot{r}_p}(\theta \pm \alpha_2 + \alpha_1)] \, d\alpha_1 d\alpha_2$$

$$+ C_{D_p} C_{D_q} \sigma_{\dot{r}}^4 \int_{-\infty}^{\infty} \int_{-\infty}^{\infty} \dot{h}_p(\alpha_1) \dot{h}_q(\alpha_2) G(R_{\dot{r}\dot{r}}(\theta - \alpha_2 + \alpha_1)/\sigma_{\dot{r}}^2) \, d\alpha_1 d\alpha_2$$
... I-18

Expanding $G(x)$ into a power series as

$$G(x) = \frac{1}{\pi}[8x + 4/3 x^3 + \ldots]$$
... I-19

and using perturbation technique

$$R_{\dot{r}\dot{r}}(\theta) = R_1(\theta) + \varepsilon R_2(\theta)$$
... I-20

wherein $\varepsilon = 1/6$, Comparing Eq. I-20 with equation I-18 and equating first power of ε to nonlinear term and rest terms with zeroth power of ε we get $R_1(\theta)$ and $R_2(\theta)$. Taking Fourier transform of these we get

$$S_1(\dot{r} \, \dot{r}) = \sum_{p=1}^{N} \sum_{q=1}^{N} \left\{ \frac{H_p(w) \overline{H_q}(w) S_{f_p f_q} [1 - 2\sqrt{8/\pi} \, C_{D_p} \sigma_1 \, \text{Re}(\frac{\overline{H_p}(w)}{1 + \overline{H_q}(w)\sqrt{8/\pi} \, C_{D_q} \sigma_{\dot{r}}})]}{1 - 8/\pi \, C_{D_p} C_{D_q} \sigma_1^2 \, H_p(w) \overline{H_q}(w)} \right\}$$
... I-21

Here, $\overline{H}_p(w) = \dfrac{iw/M_p}{p^2 - w^2 + 2i(\beta_p + \beta_p^1)wp}$; $\beta_p^1 = \dfrac{\sqrt{8/\pi}\, C_{D_p}\sigma_1}{2p \cdot M_p}$ and $\beta_p = \dfrac{C_p}{2p \cdot M_p}$ ... I-22

'p' is the natural frequency of the tower in p^{th} mode and $\overline{H}_p(w)$ is complex conjgate of $H_p(w)$

$$S_2(\dot{r}\,\dot{r}) = \sum_{p=1}^{N}\sum_{q=1}^{N} H_p(w)\overline{H}_q(w)\left[-8/\pi\, C_{D_p}\dfrac{\sigma_2^2}{\sigma_1} S_{f_p f_q} R_e\left(\dfrac{\overline{H}_p(w)}{\overline{H}_q(w)\sqrt{8/\pi}\,C_{D_q}\sigma_1 + 1}\right)\right.$$
$$\left. + \dfrac{8}{\pi} C_{D_p} C_{D_q} \sigma_2^2 S_{1\dot{r}_p\dot{r}_q}(w) + \sigma_1^{1/2} S_{1\dot{r}_p\dot{r}_q}(w) * 3 \right]/(1 - 8/\pi\, C_{D_p} C_{D_q} \sigma_1^2 H_p(w)\overline{H}_q(w)) \quad \text{... I-23}$$

$$\sigma_1^2 = \int_{-\infty}^{\infty} S_1(w)\, dw, \qquad \sigma_2^2 = \int_{-\infty}^{\infty} S_2(w)\, dw \qquad \text{... I-24}$$

$S_1(w) * 3$ is the three fold convolution of $S_1(w)$

To solve equations I-21 and I-23, iterative process is necessary.

Knowing $S_{rf} = \dfrac{1}{iw} S_{r\dot{f}}$, The coefficient of correlation between r and f can be calculated using

$$\dfrac{\int |S_{rf}|}{\sqrt{\int S_{ff}} \cdot \sqrt{\int S_{rr}}} = \rho_{rf} \qquad \text{... I-25}$$

Coefficient of correlation between water particle velocity and force is

$$\dfrac{\int |S_{uf}|}{\sqrt{\int S_{uu}}\sqrt{\int S_{ff}}} = \dfrac{\int [Im^1 S_{u\ddot{u}} + c S_{u\dot{u}} + kS_{uu}]}{\sqrt{\int S_{uu}} \cdot \sqrt{\int S_{ff}}} = \rho_{fu} \qquad \text{... I-26}$$

From these

$$\rho_{ru} = \rho_{rf} \cdot \rho_{fu}, \quad \text{and} \quad \rho_{xu} = \rho_{ru}$$

Substituting Eq. I-12, I-14, I-20, I-21, I-23 into Eq. I-10 after taking inverse fourier transforms and then taking Fourier transform of Eq. I-10 we get,

$$S_{rr}(w) = \sum_{p=1}^{N}\sum_{q=1}^{N} H_p(w)\overline{H}_q(w) S_{f_p f_q}\left[1 - 2\sqrt{8/\pi}\, C_{D_p}\sigma_{\dot{r}} R_e\left(\dfrac{\overline{H}_p(w)}{\overline{H}_q(w)\sqrt{8/\pi}\,C_{D_q}\sigma_{\dot{r}} + 1}\right)\right.$$
$$\left. + \sum_{p=1}^{N}\sum_{q=1}^{N} C_{D_p} C_{D_q} \sigma_{\dot{r}}^2 [8/\pi\, S_{\dot{r}_p \dot{r}_q}(w) + 4/3\pi\, \sigma_1^{1/4} S_{\dot{r}_p \dot{r}_q}(w) * 3] H_p(w)\overline{H}_q(w) \right. \qquad \text{... I-27}$$

Using the relation, $\sigma_r^2 = \sigma_u^2 + \sigma_x^2 - \sigma_{xu}^2 - \sigma_{ux}^2$ ... I-28

mean square response for the structure can be found out. Mean extreme response is calculated using

$$E[x(t)_{max}] = \sigma_x \left(\sqrt{2 \log \nu T} + \dfrac{0.5772}{\sqrt{2 \log \nu T}}\right) \qquad \text{... I-29}$$

in which $\nu = \dfrac{1}{2\pi}\left[\dfrac{\int_{-\infty}^{\infty} w^2 S_{xx}(w)\, dw}{\int_{-\infty}^{\infty} S_{xx}(w)\, dw}\right]^{1/2}$ ... I-30

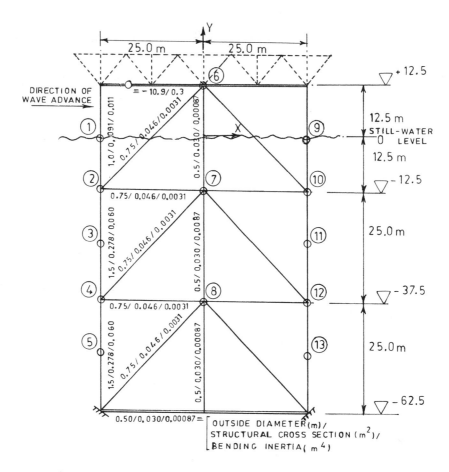

FIGURE-1. IDEALIZED TOWER IN TWO DIMENSIONS.

TABLE-1. LUMPED VOLUMES AND AREAS

NODE	1	2	3	4	5	6	7	8	9	10	11	12	13
V(m³)	4.9	50.4	22.1	56.6	22.1	0.0	52.7	52.7	4.9	50.4	22.1	56.6	22.1
A(m²)	6.3	74.2	18.8	77.3	18.8	0.0	93.7	93.7	6.3	74.2	18.8	77.3	18.8

TABLE-2. SEA STATES OCCURRING OVER THE FATIGUE LIFE OF STRUCTURE.

SEA STATE	1	2	3	4	5
SIGNIFICANT WAVE HEIGHT H_s (m)	0.65	1.25	2.5	3.7	6.9
MEAN ZERO UPCROSSING PERIOD T_z (sec)	2.8	4.0	5.6	6.8	9.3
PERCENT OCCURRENCE	58.2	30	11	0.79	0.01

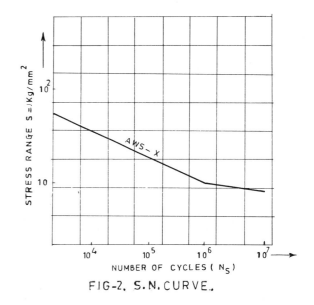

FIG-2. S.N. CURVE.

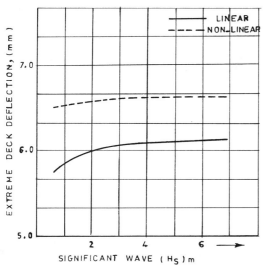

FIG.3. MEAN EXTREME DECK (NODE-6) DEFLECTIONS.

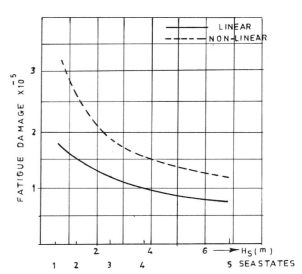

FIG.4. FATIGUE DAMAGE IN EACH SEA-STATE FOR A MEMBER BETWEEN NODE 1,2.

AUTOMATION IN UNDERWATER WELDING

K. Masubuchi
Massachusetts Institute of Technology
U.S.A.

V. J. Papazoglou
Massachusetts Institute of Technology
U.S.A.

SUMMARY

In recent years a rapid growth in the offshore industry with a concurrent increase in operational depths has been witnessed. As a consequence, underwater operations, including inspection, maintenance, repair and construction, will have to be performed at ever increasing depths. To successfully perform these operations automatic devices that will aid the skilled workers/divers in their jobs or, optimally, diver-alternative work wystems will have to be developed. This paper addresses the subject of automation in one of the aforementioned operations, namely underwater welding.

First a brief summary of the present state-of-the-art of underwater welding is given. Both dry and wet processes are discussed together with their applications and the major problems associated with them. This is followed by a description of recent research efforts at the Massachusetts Institute of Technology aimed at developing simple fully automated and integrated ("instamatic") welding systems for marine applications. Two of these systems, one for underwater stud welding, and the other based on the concept of cartridge-based packaging of welding, are described in some detail. Finally, future possibilities in more advanced underwater welding automation are described, including the use of supervisory controlled undersea teleoperators.

INTRODUCTION AND BACKGROUND

Offshore oil drilling and other ocean-related industries have grown tremendously in recent years and they are expected to grow in the future. As the number of ocean engineering structures has increased, there has been an increased demand for underwater construction, inspection, and repair of these structures.

An important characteristic of the recent developments is an increase in operational depths. Offshore production facilities are now operating to depths in excess of 1,000 feet; pipelines are being laid to depths of 2,000 feet; and exploratory drilling is reaching 5,000 feet and beyond. At the same time the U.S. Navy has shown an increased interest in deep water operations, as can be seen by the recent establishment of a National Research Council Committee to study and identify potential applications in this area. As a consequence underwater operations, including inspection, maintenance, and construction, will have to be performed at ever increasing depths.

To meet the demand for the construction and repair of offshore oil drilling rigs and other structures, the technology of underwater welding has advanced significantly during the last decade. Despite these advances, however, the overwhelming majority of underwater welding operations are currently performed manually by skilled welders/divers.

Presently, divers can work competitively at depths up to 350 ft of seawater depending on the task (Eppig, 1981). Beyond these depths bottom time is limited largely due to the lengthy compression/decompression time required. Furthermore, factors having to do with depth per se (such as life support equipment) become increasingly costly, whereas at the same time personal safety is more and more difficult to maintain (as evidenced by the rather alarming mortality figures for commercial divers in the North Sea). Finally, as operational depths increase, divers need to be assisted by machines in order to perform welding jobs requiring certain levels of skill; this is due to, among other factors, the reduction of the diver's dexterity and ease of vision.

It thus becomes obvious that the need exists for alternatives to divers for the successful performance of underwater welding in the offshore environment. Optimally these diver-alternative welding systems should have the dexterity and decision-making capabilities of the skilled welder/diver if deep sea operations are to be carried out in a satisfactory manner. In other words, there is a need for underwater welding systems having fully adaptive automatic control capabilities. To arrive at such a system a lot more research has to be done. Nevertheless, the first steps towards the realization of it have already been started and are described in this paper.

After a brief summary of the present state-of-the-art of underwater welding technology is offered, two relatively simple underwater automatic welding systems that have recently been developed by researchers at the Massachusetts Institute of Technology (M.I.T.) are described. Finally, a discussion of future possibilities in the automation of underwater welding operations by fully utilizing the robotic and electronic technologies will be given.

PRESENT STATE-OF-THE ART OF UNDERWATER WELDING TECHNOLOGY

Underwater welding processes which are currently in use may be classified into five groups, as shown in Table 1, depending upon the environment in which the welding operation takes place. In all five groups the arc welding processes, including shielded metal arc (SMA), gas metal arc (GMA), and gas tungsten arc (GTA), are the predominant ones. Today actual underwater welding jobs are alsmot exclusively done manually using skilled diver/welders, as previously mentioned. M.I.T. researchers, however, believe that it is quite possible to introduce automation in the welding processes aimed at improving the weld quality and reducing the fabrication cost.

The dry chamber welding processes are generally capable of producing high quality welds; they are, however, very expensive due primarily to the high cost of deploying specially designed dry chambers and supporting facilities, including the life support equipment. It is believed that these costs can be significantly reduced if welding operations can be performed by using robots and other mechanical means.

Compared to the dry chamber processes, the wet welding processes, especially the wet manual SMA processes, are simpler, less expensive, and more versatile; weld quality, however, is rather poor so that their applications are limited mainly to repairs. Robots can again be used to assist the welder/diver in making better quality welds under less stressed working conditions.

Applications. Joining of important structural members, such as strength members of oil drilling rigs and underwater pipelines, is normally done using the dry processes. A recent paper discusses the state-of-the-art of the hyperbaric underwater welding process and applications of the process primarily performed by the Taylor Diving and Salvage Company (Delaune, 1979).

Although most underwater welding jobs are done in shallow waters, it is possible to conduct welding in deep waters too. A major limitation in performing wet manual SMA welding in deep waters comes from the diving systems. When conventional diving equipments are used, it is difficult to successfully perform welding operations beyond a depth of 200 ft. Using a saturation diving system, however, it is possible to perform wet SMA welding up to a depth of 1000 ft or, perhaps, even deeper. It has been reported that under simulated test conditions oxygen cutting and wet welding have been successfully performed at depths up to 1000 ft and welds have been made at depths in excess of 1200 ft (Delaune, 1979).

The dry chamber processes can also be used for deep sea applications. For example, the one-atmosphere welding can be performed at any depth as long as a necessary diving system is developed. The hyperbaric chamber processes can also be used in deep sea. Weld quality is believed to be generally good, because welds are made in the dry environment. The cost, however, becomes extremely high, as previously mentioned. Much of the information regarding actual applications of the dry chamber arc welding processes are unfortunately kept as a commercial secret. Two recently published examples of such applications are:

(1) Dry habitat welding in the open sea at a depth in excess of 1000 ft performed by the Taylor diving and Salvage Company (Delaune, 1979).

(2) A three-year experimental, deep-water welding study conducted in the North Sea near Stavanger, Norway in which welds were made in 300 m (1000 ft) water depths. This study, completed in 1978, aimed at demonstrating the feasibility of welding at great depths under atmospheric pressure (Anon, 1978).

Problems Associated With Underwater Welding. There are a number of technical and operational problems associated with underwater welding that have to be addressed and solved before any meaningful automation of the processes can be made. It is not possible to discuss these problems in detail in this paper because they are quite diversed depending not only upon the environment (dry versus wet, with or without additional external pressure) but also on the particular welding processes (SMA, GMA, etc.). In this section a short discussion of the problems encountered in the wet arc welding processes will be given.

The unique feature of wet welding is that the metals to be joined are completely immersed in water, a fact causing much more severe problems than in the case of dry processes. The major effects of water on arc welding are:

(1) Due to the quenching effect of water, a weldment cools rapidly, resulting in a hard and brittle weld.

(2) Bubbles are formed due to the intense heat of the welding arc, increasing the probability of a porous weld metal.

(3) Hydrogen in the bubbles and in the surrounding water may cause hydrogen-induced cracking.

The above effects can be enhanced by the increasing pressure as one goes deeper and by the salinity of the water. Unfortunately, only a limited number of documents have been published describing the above phenomena, with the majority of underwater welding related publications describing primarily practical and operational aspects of the processes.

Three M.I.T. Sea Grant reports (Brown, et al., 1974; Tsai, et al., 1977; Masubuchi, 1981a) and two recent papers (Masubuchi, 1980; Masubuchi, 1981b) present detailed discussions on various subjects related to the above phenomena, including: (1) formation of bubbles surrounding the welding arc; (2) underwater welding arc physics and metal transfer; (3) heat transfer in underwater welds; (4) effects of pressure on underwater welds; (5) polarity effects in underwater welding; (6) waterproofing of electrode coating; (7) microstructures and hardness of underwater welds; (8) crack sensitivity of underwater welds; and (9) porosity and other types of defects in underwater welds.

It is hoped that further research efforts will enable us to better understand the fundamentals of the underwater welding processes and hence to improve their performance and utilization through automation.

AUTOMATIC UNDERWATER WELDING SYSTEMS DEVELOPED AT M.I.T.

Systematic research on underwater welding has been conducted during the last 12 years in the Department of Ocean Engineering, M.I.T., resulting in the generation of much valuable information on the subject (Brown, et al., 1974; Tsai, et al., 1977; Masubuchi, 1981a). During the last few years this research effort has convinced M.I.T. investigators that fully automated and integrated ("instamatic") underwater welding systems which can be operated by people with no welding skill can be developed. Such a development would be desirable for two reasons. First, in some underwater welding jobs, especially those performed in deep sea, it becomes very difficult to secure workers who are qualified in both diving and welding. And second, in deep-sea operations even a qualified diver/welder may not be able to perform high-quality welding jobs, because welding must be performed under adverse conditions.

The latest research effort at M.I.T. towards the realization of these "instamatic" welding systems for marine applications has led to the development of an interesting concept, that of an enclosed welding unit containing a piece to be welded in a "cassette". This idea is a rather radical departure from the current practice, where the plates to be welded are first assembled together and then a welder or a welding machine comes to the weld location to execute welding. Compared to this, the new concept is based on a cartridge welding device which contains the piece to be welded and which automatically performs the desired weld in one operation (for details see below).

Some of the relatively simple examples that have been studied for application using the instamatic welding systems are as follows (see Figure 1):

(1) Stud welding of a bar to a flat plate, as shown in Figure 1a. The feasibility of instamatic stud welding systems has already been proven (Masubuchi, et al., 1978). Recently, a floodable stud welding system has been developed (Schloerb, 1982) that has a flexible magnetic base and which can also be used for welding a stud to a curved plate or a pipe (see Figure 1b). Details of this system are described in a later subsection.

(2) Joining of a flat plate to another flat plate or of a pipe to a flat plate by fillet welding, as shown in Figures 1c and 1d. It has already been demonstrated that this can be done using the flux shielded process (Masubuchi, 1981a). An attempt is currently being made to develop a special flux that will enable welding in all positions. The currently available process can be used in the downhand position only.

(3) Efforts have been successful in developing devices of integrated and automated systems for lap welding a cover plate to a flat plate (patch welding), as shown in Figure 1e (Gustin, 1982). Details of the system are given in a later subsection.

(4) Various ways for replacing a pipe section, as shown in Figure 1f, have been studied. The major difficulty in this case is that welding must be performed in all positions unless the pipe can be rotated during welding. Then, welding conditions must be changed as welding progresses and the welding position changes. In order to successfully weld a pipe in all positions a machine probably needs in-process sensing and control capabilities. An automatic welding machine which has such an advanced adaptive control system may be called a "smart" welding machine.

Potential Uses of Instamatic Welding Systems. Efforts have been made to identify potential uses of instamatic welding systems. Some of these uses are as follows:

(1) Certain repair jobs on board a ship and an offshore oil drilling rig which must be performed when no skilled welder is available. It is important to have the capability of performing some repair works without having a skilled welder.

(2) Rehabilitation of steel structures, especially offshore oil drilling rigs and other ocean engineering structures. Many ocean structures were built during the last 10 years, especially since the oil embargo by the OPEC nations. Some of them need various types and degrees of repairs. Unlike ships which can be brought to docks for repair, many repair jobs of ocean engineering structures must be performed under water. Instamatic welding systems may be used for some repair jobs, primarily minor repairs.

(3) There are a number of jobs requiring attachment of various pieces to ocean structures. For example, the attachment of anodes for corrosion protection is a major maintenance job for an ocean structure. There are many other minor jobs for attaching metal pieces with various shapes and sizes for a variety of purposes to ocean structures. Many of these welding jobs may be done by instamatic welding systems.

(4) Instamatic welding systems can be extremely useful for many welding jobs required in salvage operations. An integrated device may be developed to lift a sunken object. A system also may be developed to close an opening by welding so that air can be pumped into the sunken structure to obtain enough buoyancy.

(5) Certain welding jobs which must be performed in a compartment where sparks from the welding arc may cause fires or explosion. In fact, many fires and explosions in ships and on oil drilling rigs occurred during welding. Many of these fires and explosions can be prevented when welding is performed in an enclosed box.

(6) Certain welding jobs which must be performed in hazardous environment, such as leaking chemical products and radioactive materials. In some cases it is difficult or impossible to perform welding by human welders.

Underwater Stud Welding. Stud welding in the dry environment has been used for many applications. However, the use of stud welding underwater had not been studied seriously until M.I.T. researchers discovered that stud welding could be successfully used underwater (Masubuchi, et al., 1978).

Two basic reasons make underwater stud welding a very good candidate for automation, especially for deep-sea applications:

(1) A key factor in underwater welding is how to remove water from the weld zone. In stud welding the stud to be welded can have direct contact with the plate, thus only a small amount of water can exist near the surfaces to be joined. The effect of water pressure is considerably less in stud welding than in an ordinary arc welding in which metal particles are transferred through the arc plasma.

(2) Virtually no skill is required. The stud can be placed inside the welding gun so that the whole operation can be performed in one single step.

A recent research study at M.I.T. has resulted in the successful development of an automatic floodable arc stud welding gun (Schloerb, 1982). Figure 2 presents a schematic diagram of the system which illustrates the key features of the stud gun.

The system incorporates a conventional diesel stud welding generator and a conventional electronic controller which are located out of the water. The trigger switch, which initiates the weld cycle, is also located on the surface for safety reasons. This switch is incorporated in a knife switch which assures that electrical power to the stud gun is only on when a weld is actually being made. The knife switch is operated by the diver's tender on command from the diver. An electrical umbilical connects the surface equipment to the submersible arc stud welding gun.

The heart of the stud gun is the stud lifting mechanism. This mechanism lifts the stud a predetermined distance in order to produce a precise arc gap between the end of the stud and the work surface. The arc gap is adjusted by moving the lift mechanism along the support framework.

The force to lift the stud is generated by a solenoid which is activated by the control unit. The control unit initiates the welding arc at the same time as the stud is lifted. The arc causes the end of the stud and a portion of the work surface to melt. After a predetermined time, the control unit stops the arc and de-energizes the solenoid. At this point, a spring forces the stud back against the work surface. A disposable ceramic collar, called a ferrule, is held in place around the base of the stud to prevent weld spatter. The molten metal solidifies quickly leaving the stud welded to the work surface.

The system employs conventional ferrules and a conventional ferrule clamp. The ferrules are press-fit on the ends of the studs, using steel wool prior to the dive in order to make it easier for the diver to handle them. The steel wool also serves as an arc initiator.

Considerable work has been done to make the entire system as simple as possible. As a result, the lifting mechanism has only one moving part--the lifting rod which transmits force to the stud. Another key feature is the fact that the lift mechanism is "water-floodable" rather than waterproof. A plastic case prevents electrolysis of the lift mechanism components and helps to reduce the extent of the electric field which surrounds the gun.

A major part of the work done in developing the stud gun has been an investigation of the electric field in the water around the gun. Preliminary calculations anticipated that this field would present a hazard for the diver who is operating the gun. This was confirmed by measurements which found the magnitude of the electric field near the gun to be more than 100 times the recommended safe limit. A conventional underwater welding stinger, on the other hand, was found to produce a field whose magnitude was just equal to the safe limit.

It has been proposed that a grounded aluminum shield, placed around the stud, could be used to reduce the field to a safe level. Development of the shield is not complete. A preliminary shield design was found to reduce the magnitude of the electric field by a factor of 20.

In addition to the components described above, the stud gun consists of the following basic parts: a chuck holder, which adapts to conventional stud welding chucks, connects the lifting rod to the stud and serves as the "hot" electrical connection; and, a permanent magnetic base employed to hold the gun in place during the welding operation. The base will attach to both flat and cylindrical surfaces.

The developed system has been successfully tested underwater at pressures up to 300 psi.

Cartridge-Based Packaging of Welding Systems. A "packaged" welding system is one in which the various subtasks involved in the welding operation (e.g., joint positioning, arc direction, weld travel speed) are predetermined by the hardware design of the welding equipment. In other words, the welded joint that can be produced by a particular packaged system is defined by the specific geometrical configuration of that system. For devices of this type, the operator is only required to place the equipment on the worksite and then operate on/off switches.

Packaged welding systems, although necessarily of a limited applicability range, appear to be a desirable alternative to current welding practices in certain occasions. In particular, in cases where the geometry of the weld is regular and can thus be prespecified, such equipment could be used very efficiently. This is especially true when ordinary manual welding has to be performed in dangerous environments, as for example in nuclear reactors where high radiation levels are present. Several examples of welding geometries for which packaged systems are envisioned are shown in Figure 1 and were discussed in a previous subsection.

Automatic packaged cartridge welding devices have three salient characteristics in addition to the packaging of the weld process:

(1) Because of the limitations designed into the devices, they are low in cost. Based upon literature received from manufacturers of welding robots and general automatic welding equipment, it is estimated that packaged equipment could cost as little as 10% of the cost of more sophisticated equipment capable of making the same welds.

(2) A particular type of weld can be reliably performed. This is true whether the machine is used repeatedly in the same situation (such as would be the case in a mass production environment), or occasionally, in diverse situations (for instance, emergency repairs).

(3) Devices considered herein require minimal operator skill. Ideally, the operator need only (a) insert the cartridge, (b) position the welder, and, (c) turn power on and off.

To explore the feasibility of such systems, a representative one was designed, constructed, and tested (Gustin, 1982). Figure 3 shows the schematic diagram of the developed device which is capable of welding a circular "cartridge" of mild carbon steel to a base plate of similar material (circular patch weld).

With reference to Figure 3, the elements of the design can be summarized as follows:

(1) A GMAW (MIG) gun is mounted on a shaft so that rotation of the shaft causes the gun to traverse the desired circular path.

(2) A cartridge holder is mounted concentrically on the shaft. The cartridge holder accepts the patch to be welded and mechanically positions the patch against the base plate. A bearing component of the cartridge holder allows the shaft to rotate without requiring holder rotation.

(3) The enclosure shown provides structural support for the other components. In addition, the enclosure provides the potential for using the device as an isolated welding system.

Commercial welding equipment (e.g., power supply, shielding gas, cooling water systems, wire feed equipment, etc.) is used in conjunction with the illustrated equipment to perform the desired circular patch weld.

The designed system was tested successfully in a laboratory environment. Good quality welds were obtained judging from their bead and heat effect uniformity, good fusion characteristics, absence of any burn-through, and adequate weld penetration.

In general we believe that one of the applications where packaged systems may possibly be especially effective is their use in conjunction with robots. The combination of instamatic and robotic capabilities offers a functional flexibility which neither one has by itself.

FUTURE POSSIBILITIES IN UNDERWATER WELDING AUTOMATION

In the previous section the development of simple fully automated and integrated ("instamatic") welding systems for marine applications was described. These systems are believed to be only the first step towards more sophisticated automatic underwater welding systems. The recent advances in microcomputer and microprocessor technologies have convinced M.I.T. investigators that as a next step systems capable of performing simple underwater welding operations by remote manipulation techniques should be and can be developed (Masubuchi and Papazoglou, 1981).

Initials efforts towards this goal are currently under way. They involve the integration of the instamatic underwater welding technology with the technology of underwater telemanipulators under supervisory control developed at the M.I.T. Man-Machine Systems Laboratory (Sheridan and Verplank, 1978; Yoerger and Sheridan, 1981). Teleoperators are defined to be general purpose submersible work vehicles controlled remotely by human operators and with video and/or other sensors, power and propulsive actuators for mobility, with mechanical hands and arms for manipulation and possibly a computer for a limited degree of control autonomy. Based on this definition, then, a manned submersible is not a teleoperator vehicle, but the attached manipulators are certainly teleoperators, requiring control through a viewing port or through closed-circuit video. Supervisory control is a hierarchical control scheme whereby a system (which could be a teleoperator, but could also be an aircraft, power plant, etc.) having sensors, actuators and a computer, and capable of autonomous decision-making and control over short periods and in restricted conditions, is remotely monitored and intermittently operated directly or reprogrammed by a person (Sheridan and Verplank, 1978).

The distinction between direct human control of a teleoperator and supervisory control of a teleoperator is made graphically in Figure 4. Under direct control the operator's control signals are sent directly to the remote manipulator, and sensor information is fed directly back to the operator. The "hand control" can be a master-slave positioning replica or a control joystick. Under supervisory control the operator's control signals are relayed through a local computer to the remote computer, which then processes the signals and acts on the information. The relayed signals are not necessarily the raw signals generated by the operator. In fact, the signal is usually a coded instruction of high information density which must be interpreted to be utilized. The operator's input could range from a purely manual analogic command to a highly abstract symbolic command. The remote computer not only interprets the local computer's messages but also acts on the sensor information available to it about its environment. The remote computer only relays information to the operator which is deemed important and necessary for effective supervision--the responsibility for the specific details of control is usually left to the subordinate computer.

Generally speaking, the supervisor control system enables the human operator to:

(1) plan the actions of the remote system in a way which greatly reduces the need for continuous manual control,

(2) monitor the system as it carried out the planned actions,

(3) intervene if problems arise with which the remote system cannot deal.

It is expected that the initial efforts of this research will focus on relatively simple welding tasks, primarily using dry underwater welding processes. The reason for such a direction is that any sophisticated welding automation scheme or the utilization of wet processes will make the interfacing and integration of the two aforementioned technologies much more difficult. At later stages, and after the initial efforts prove to be successful, one can put emphasis on upgrading the automation of the welding operation itself, possibly by using in-process sensing and control.

Anticipated Benefits. We believe that the offshore industry and other ocean related industries will receive direct benefits from the automation of underwater welding operations. The most far-reaching benefit will involve the enhancement of the capabilities of underwater work systems. Currently there exist several manned undersea work vehicles equipped with manipulators. The capabilities of these systems, however, are rather limited and involve relatively simple tasks. Their potential will be realized even further if they can also be fitted by devices capable of performing underwater welding.

Furthermore, given the fact that underwater operations are being performed at ever increasing depths and that as depths increase the economics of the currently available systems make them prohibitively expensive, it becomes evident that untethered, unmanned underwater work systems have to be developed. Although such systems do not exist at the present time, we believe that the developmental efforts currently underway in various parts of the world will soon make such systems possible. At that time we believe that we should be ready to equip them with several operational capabilities. Automatic underwater welding devices must be one of them.

REFERENCES

ANON, 1978, "Atmospheric Welding at 300 M Water Depth," Ocean Industry, August 1978, pp. 61-63.

BROWN, A. J., BROWN, R. T., TSAI, C. L., and MASUBUCHI, K., 1974, Fundamental Report on Underwater Welding, Massachusetts Institute of Technology, Sea Grant Program, Report No. MITSG 74-29.

DELAUNE, P. T., as told to WEBER, J. D., 1979, "Hyperbaris Underwater Welding - The State of the Art. Parts I and II," Welding Journal, Volume 59, Numbers 8 and 9, pp. 17-25 and 28-35.

EPPIG, S. H., 1981, "Vehicle Maneuverability Augments Remote Controlled Manipulator Task Capability," Proceedings, Oceans '81 Conference, Boston, Massachusetts, U.S.A., 1981, Volume 2, pp. 1143-1149. Piscataway, New Jersey, U.S.A.: Institute of Electrical and Electronic Engineers.

GUSTIN, H., 1982, Cartridge-Based Packaging of Automatic Welding Systems, S.M. Thesis, Massachusetts Institute of Technology, Department of Mechanical Engineering.

MASUBUCHI, K., OZAKI, H., and CHIBA, J., 1978, "Underwater Stud Welding," Proceedings, Oceans '78 Conference, Washington, D.C., U.S.A.

MASUBUCHI, K., 1980, "Underwater Factors Affecting Welding Metallurgy," Proceedings, Seminar on Underwater Welding of Offshore Platforms and Pipelines, New Orleans, Louisiana, 1980, Miami, Florida, U.S.A.: American Welding Society (to be published).

MASUBUCHI, K., 1981a, Development of Joining and Cutting Techniques for Deep-Sea Applications, Massachusetts Institute of Technology, Sea Grant Program, Report No. MITSG 81-2.

MASUBUCHI, K., 1981b, "Review of Underwater Welding Technology," Proceedings, Oceans '81 Conference, Boston, Massachusetts, U.S.A., 1981, Volume 2, pp. 649-651. Piscataway, New Jersey, U.S.A.: Institute of Electrical and Electronic Engineers.

MASUBUCHI, K., and PAPAZOGLOU, V. J., 1981, Underwater Welding and Cutting by Remote Manipulation Techniques, Massachusetts Institute of Technology, Department of Ocean Engineering, proposal submitted to Sea Grant.

SCHLOERB, D., 1982, Development of a Diver-Operated Underwater Arc Stud Welding System, S.M. Thesis, Massachusetts Institute of Technology, Department of Mechanical Engineering.

SHERIDAN, T. B., and VERPLANK, W. L., 1978, Human and Computer Control of Undersea Teleoperators, Massachusetts Institute of Technology, Man-Machine Systems Laboratory, Report to Office of Naval Research under Contract N00014-77-C-9256.

TSAI, C. L., OZAKI, H. MOORE, A. P., ZANCA, L. M., PRASAD, S., and MASUBUCHI, K., 1977, <u>Development of New, Improved Techniques for Underwater Welding</u>, Massachusetts Institute of Technology, Sea Grant Program, Report No. MITSG 77-9.

YOERGER, D. R., and SHERIDAN, T. B., "Development of a Supervisory Manipulation System for a Free-Swimming Submersible," <u>Proceedings, Oceans '81 Conference</u>, Boston, Massachusetts, U.S.A., 1981, Volume 2, pp. 1170-1174. Piscataway, New Jersey, U.S.A.: Institute of Electrical and Electronic Engineers.

ACKNOWLEDGEMENTS

The authors would like to thank the Office of Sea Grant, National Ocean and Atmospheric Administration, United States Department of Commerce for its generous financial support. We would also like to thank the following companies for providing matching funds: Hitachi Shipbuilding and Engineering Co., Ishikawajima-Harima Heavy Industries, Kawasaki Heavy Industries, Mitsui Engineering and Shipbuilding Co., Nippon Kokan Kaisha, and Sumitomo Heavy Industries.

Table 1 Classification of underwater welding processes currently in use

A. <u>Dry Chamber Processes</u>. Welding takes place in a dry environment.

   1. <u>One-Atmosphere Welding</u>. Welding is performed in a pressure vessel in which the pressure is reduced to approximately one atmosphere independent of depth.

   2. <u>Hyperbaric Dry Habitat Welding</u>. Welding is performed at ambient pressure in a large chamber from which water has been displaced. The welder/diver does not work in diving equipment.

   3. <u>Hyperbaric Dry Mini-Habitat Welding</u>. Welding is performed in a simple open-bottom dry chamber which accomodates the head and shoulders of the welder/diver in full diving equipment.

B. <u>Portable Dry Spot Process</u>. Only a small area is evacuated and welding takes place in the dry spot.

C. <u>Wet Process</u>. Welding is performed in water with no special device creating a dry spot for welding. In manual wet welding the welder/diver is normally in water.

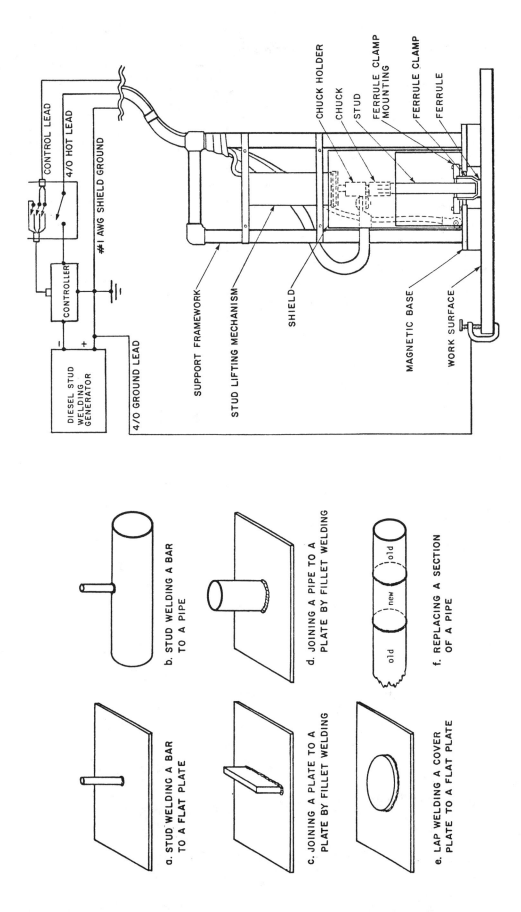

Figure 1 Several basic types of joints investigated for welding by instamatic systems

Figure 2 Diver operated underwater arc stud welding system

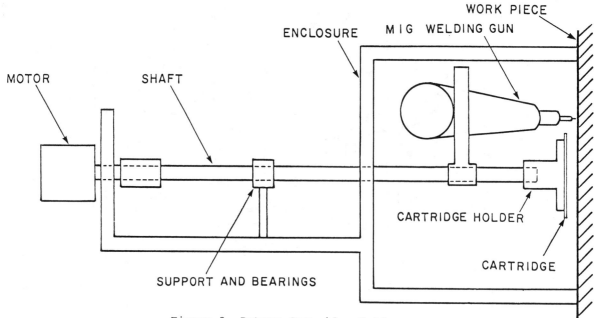

Figure 3 Rotary Cartridge Welder

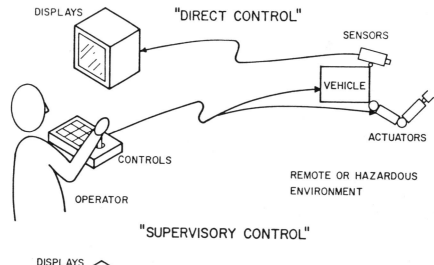

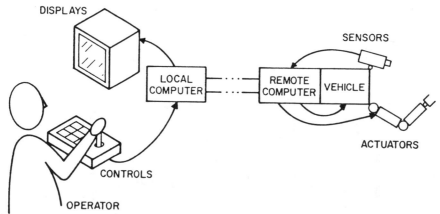

Figure 4 Direct manual control and supervisory control of a teleoperator

Technical Poster Sessions
STRUCTURES

Structural Response under Extreme Loads

RELIABILITY OF STIFFENED STEEL CYLINDERS
TO RESIST EXTREME LOADS

| P.K. Das | P.A. Frieze | D. Faulkner |
| University of Glasgow | University of Glasgow | University of Glasgow |

SUMMARY

This paper describes a procedure which can be used to rationalise the evaluation of partial factors appropriate to limit state structural design. The procedure is illustrated through examination of stringer-stiffened cylinders subjected to axial compression.

The first part of the paper reviews the clauses relating to cylinders in the DnV Rules for Offshore Structures and appraises the limit state format and partial factors adopted therein. The advanced Level II reliability analysis method used in the study and the strength formulation adopted are described. Some uncertainties associated with extreme loads are defined. Modelling uncertainty is quantified by calibrating the strength model with available experimental data.

The remainder of the paper is concerned with the stiffened cylinder reliability analysis. The sensitivity of the strength model to the design parameters is evaluated, as are the appropriate partial coefficients. Variability of the safety index with load is described and, finally, the reliability of stiffened cylinders under extreme axial loads determined.

INTRODUCTION

With the current emphasis in offshore oil activity towards the emergence of novel semi-buoyant platforms for marginal field developments, the need to save structural weight is becoming very critical. The use of particular structural configurations such as orthogonally stiffened cylinders goes some of the way to satisfying this requirement. Probably of equal importance, however, is the particular design rule adopted and the associated partial safety factors which must be adequate but not excessive. The rule having the greatest direct influence on weight is that concerned with ultimate collapse under extreme loads.

This interest in semi-buoyant platforms is reflected in the current efforts to devise rational rules for TLP's and similar structures, a process hampered by the dearth of experimental data concerning stiffened cylinders having geometries appropriate to offshore structures, and of reliable but straight-forward corresponding design procedures. The work reported in this paper is a first step towards establishing a reliable design approach. A recently proposed design method will be calibrated against experimental data via a modelling parameter, and then an advanced first-order second-moment reliability analysis technique will be used to quantify reliability under different extreme loads.

REVIEW OF DnV RULES

The DnV Rules for the design, construction and inspection of offshore structures, 1977 (DnV, 1978), provide the most comprehensive set of limit state recommendations for offshore structure design. Through a non-mandatory Appendix, characteristic strengths and interaction formulae are presented to provide guidance for proportioning member sizes where instability is the governing structural action. The present work is concerned with the buckling of stiffened cylinders, although unstiffened cylinders, beam-columns, frames, and plane unstiffened and stiffened plates are also covered by the Rules.

Buckling of Cylindrical Shells

The Rules' recommendations on shells are contained within Section C3 of Appendix C. Stiffened cylinders are divided into three groups, ring-stiffened, ring- and stringer-stiffened (orthogonally stiffened) and stringer-stiffened cylinders. All possible buckling modes are considered, and two methods are advanced to deal with them. One consists of ensuring the components of shells are of adequate stiffness so that the mode in question will not occur. The other approach is to provide formulae for estimating component elastic critical buckling stresses and to modify these as appropriate to account for imperfection effects, geometry influences and plasticity.

The former method is applied predominantly to the design of stiffening elements so that neither they nor their components such as flanges and webs will suffer local buckling of any sort. This is mainly torsional buckling although in the case of webs of flanged sections, it will also include compression buckling. Limitations are given for both rings and stringers although it does appear to be a little inconsistent that flat-bars can be used as stringers but not effectively as rings. (Section C3.7 does not specifically exclude rings but the restriction on second-moment of area about a radial axis through a ring leads to impractical proportions for flat bars, typically depth to thickness ratios of less than three.) It also applies to the buckling of ring-frames in their own plane. The Rules recommend that general (orthotropic) buckling should not precede either local instability of the shell between rings in the case of ring-stiffened cylinders, or stringer-stiffener buckling in the case of orthogonally stiffened cylinders. When calculating the in-plane stiffness of ring-frames, an effective width of shell is specified to act with the frame.

Buckling of curved panels, of the shell of ring-stiffened (and unstiffened) cylinders, and of the overall cylindrical beam-column are the only instability modes dealt with directly in the Rules. No guidance is provided on the buckling stresses of stringer-stiffened cylinders other than to treat them conservatively as plane stiffened panels, or on the general buckling of orthogonally stiffened cylinders should a designer wish this to be a possible buckling mode. To design for either of these modes, it is indicated that experimental or computer solutions are required.

Buckling formulae are presented generally using the classical form of the equation. Values of the buckling coefficients, which account for the effect of geometrical parameters on buckling stress, can be found either from charts or from explicit expressions presented in the Rules. For particularly imperfection-sensitive structural components, such as axially and/or pressure loaded ring-stiffened cylinders, 'knockdown' factors are provided to reduce the elastic critical buckling stresses to those more appropriate to practical structures.

Overall buckling is treated using the beam-column design recommendations of Appendix C1. Because of possible interaction between overall and local buckling when the column reduced slenderness exceeds 0.2, additional local bending stresses have to be considered. These are calculated using parameters determined in the beam-column analysis.

The buckling formulae presented cover single loading actions only. For curved panels, these are axial compression, circumferential compression and in-plane shear. For combined loading a sum-of-squares interaction formula is used, the stresses due to bending and torsion being added to those of compression and shear respectively. For ring-stiffened (or unstiffened) cylinders, the load actions considered are axial compression, bending, torsion, external radial pressure and external hydrostatic pressure. Interaction is handled in the same manner as for curved panels.

Plasticity is accounted for using a Merchant-Rankine-type interaction formula:

$$\frac{1}{R_k^2} = \frac{1}{f_y^2} + \frac{1}{f_e^2} \tag{1}$$

where R_k is the elasto-plastic buckling stress (the Rules' characteristic resistance), f_y is the characteristic material resistance, and f_e is the elastic critical buckling stress. Introducing a reduced slenderness parameter

$$\lambda = (f_y/f_e)^{\frac{1}{2}} \tag{2}$$

equation (1) can be rearranged as

$$R_k = \phi f_y \tag{3} \qquad \text{where } \phi = \frac{1}{(1 + \lambda^4)^{\frac{1}{2}}} \tag{4}$$

Partial Factors

The partial factor format for control of instability in the Rules is

$$S_d \leq R_d = \frac{R_k}{\gamma_m} \frac{\psi}{\kappa} \tag{5}$$

where S_d is the design loading, R_d and R_k are the design and characteristic resistances, γ_m is the material partial coefficient and has a value of 1.15, ψ is a factor to reflect post-buckling behaviour and takes values of 1.0 or 0.9 depending on whether redistribution is possible or not, and κ is a factor which reflects the uncertainties associated with slenderness in structures prone to instability and assumes values of 1.0 for $\lambda < 0.5$, 1.3 for $\lambda > 1.0$, and $0.7 + 0.6\lambda$ for $0.5 \leq \lambda \leq 1.0$.

For structural actions involving instability, each possible mode is considered separately, hence the above format. For other actions, the above expression is expressed in a more general form

$$S(\Sigma F_i \gamma_{fi}) \leq \frac{R_k}{\gamma_m} \cdot \frac{\psi}{\kappa} \tag{6}$$

where $S(\Sigma F_i \gamma_{fi})$ is now the design loading effect, γ_{fi} being the partial load coefficients associated with the characteristic loads F_i.

The relationship between R_k and the partial coefficients implied by equation (5) is linear, i.e. the characteristic resistance is proportional to yield stress and slenderness. Expanding equation (3)

$$R_k = \frac{f_y}{(1 + f_y^2/f_e^2)^{\frac{1}{2}}} \tag{7}$$

shows these relationships are not necessarily linear. Consider a case where $\lambda = 1.0$ so that $f_e = f_y$. In this case $\kappa = 1.3$ and $R_k = f_y/\sqrt{2}$ so that the design resistance, equation (5) becomes, for $\psi = 1.0$

$$R_d = f_y/\sqrt{2} \times 1.15 \times 1.3 = f_y/2.11 \tag{8}$$

However, γ_m is a material coefficient which should strictly only apply to f_y: likewise, κ to f_e. Introducing γ_m and κ to (7) accordingly, the design resistance now is

$$R_d = (f_y/1.15)/\{1 + (f_y/1.15)^2/(f_e/1.3)^2\}^{\frac{1}{2}} = f_y/1.74 \tag{9}$$

which is 21% higher than the result in equation (8).

Thus a strict application of partial coefficients shows that the Rules can be 21% conservative for steel structures having elastic critical buckling stresses equal to the characteristic yield stress. For structures of other slendernesses, the difference will be smaller and will be insignificant for structures where instability effects are negligible. A similar strict interpretation was made (Flint, et al, 1981) in the derivation of partial coefficients for the draft new British Steel Bridge Code of Practice, Part 3 (BSI, 1979).

Material Coefficient γ_m

The actual value for the material coefficient γ_m adopted in the Rules of 1.15 is also worth examining. The Rules indicate that the characteristic strength of steel may be assumed to be the guaranteed minimum yield strength. For mild steel this can be taken as 245 N/mm^2 so that the design value of yield strength is 245/1.15 = 213 N/mm^2 assuming γ_m is a partial coefficient on the characteristic value as implied by the Rules.

Surveys on yield stress used by Flint, et al, 1981, indicate a mean value of 285 N/mm^2 and a standard deviation of 20 N/mm^2 (coefficient of variation = 7.0%) are appropriate for mild steel. This is a slightly greater variability than the 5% experienced by DnV through their own material testing services (Fjeld, 1978). The mean value was reduced by 15 N/mm^2 for the calibration process reported by Flint, et al, 1981 to allow for the strain rate effect. The outcome of this calibration process was a value for the material coefficient of 1.08 providing a design value of yield stress of 227 N/mm^2 some 6½% greater than the DnV Rules design yield stress. The central material partial factors (mean yield strength ÷ design strength) are 1.19 and 1.27 in Flint, et al, 1981 and the Rules respectively. The proximity of 1.15 to 1.19 suggests that the γ_m as adopted in the Rules may be a central safety factor rather than the partial coefficient on characteristic yield strength as described in the Rules.

Alternatively, γ_m could be considered as a combination of γ_{m1}, a real material coefficient, and γ_{f3}, the I.S.O. Standard 2394 factor to account for inaccurate assessment of the effects of loading. The fact that the Rules' load coefficients are noticeably less than those adopted in other limit state codes, e.g. the ECCS recommendations on steel construction (ECCS, 1978), but that designs based on the Rules still have reliability levels similar to those founded on other codes accepted for offshore structure design (Fjeld, 1978) lends support to this suggestion.

Possible further evidence lies in the fact that the same γ_m is used in the Rules' design for buckling of bars and frames. For the design of these elements, strength curves are proposed which appear to have been directly based on the ECCS Column Curves (ECCS, 1978). The latter were deduced after an extensive experimental and numerical investigation which was statistically evaluated and led to a material-modelling factor (γ_{m1}, γ_{m2}) of 1.0, i.e. the statistical parameters defining the experimental results, and the numerical modelling when the measured uncertainties of yield stress were taken into account, were practically identical. The strength curves were based on the 2.3% fractile of results in contrast with the 5% recommended for use in the Rules. Strictly, therefore, there appears to be no justification for the use of γ_m = 1.15 in the Rules for the design of bars and frames unless γ_m includes some effect of loading.

The column design curves incorporated in the draft new British Steel Bridge Code (BSI, 1979) appear to be 'Perry Equation' equivalents of the ECCS Column Curves. The correlation exercise reported by Flint, et al, 1981 found that the partial coefficients for material (γ_{m1}) and modelling (γ_{m2}) appropriate to the column design curves were 1.08 and 0.98 respectively. These represent a joint factor of 1.06 which was reduced to 1.03 when γ_{m1} was fixed at a value of unity. In either case the accuracy of the design curves was confirmed indicating little need for a γ_m factor greater than about 1.05 for these particular structural elements.

Modelling Coefficient κ

κ as adopted in the Rules seems to play the same role as γ_{m2} of the I.S.O. Standard in that it is intended to cover possible weaknesses in a structure arising from causes other than those due to material. This is often interpreted as a factor to quantify inaccuracies in design models which could thus vary from one structural element to another.

κ was introduced into the Rules' partial factor format seemingly to reflect the imperfection sensitivity of slender structures susceptable to instability. This included bars and frames which, as discussed above, are designed using column curves based on statistically reliable data. The data was derived for reduced slendernesses in the range 0.6 to 1.4. Consequently, since the uncertainties associated with these structures have been quantified across a fairly wide range of slendernesses, there seems little need here for a modelling parameter which varies with slenderness.

Design clauses for doubly-curved shells subjected to external pressure are not included specifically in the Rules. They have, however, an imperfection sensitivity not dissimilar to that of axially compressed cylinders so can be treated in design in the manner of the Rules. Faulkner has examined the variability of torisphere strength in the reduced slenderness range 0.40 to 0.88 and found no evidence that this increases with slenderness (Faulkner, 1979). Numerical studies using a sophisticated elasto-plastic buckling analysis technique have indicated that hemispheres having imperfections equal to the tolerances (approximately 0.6 x shell thickness) specified in two codes concerned with externally pressurised structures show a greater sensitivity to imperfection when they are stocky ($\lambda = 0.51$) than when they are slender ($\lambda = 1.39$), i.e. the reverse of the effect of κ (Papadimitriou and Frieze, 1981).

Externally pressurised ring-stiffened cylinders have also been examined using the Rules and the effect of κ was found to 'skew' the design curve the wrong way (Odland and Faulkner, 1981).

Buckling Modes not considered by the Rules

The Rules indicate that buckling modes not considered in Appendix C can be designed only by validated and documented methods. However, where such omissions exist within the Rules, DnV attempt to provide appropriate procedures which normally go through several phases of development and verification. A design model for closely-spaced stringer-stiffened cylinders subjected to axial load and external pressure is currently passing through these phases. The model was first presented by Valsgård and Steen, 1980, and further developed by Valsgård and Foss, 1981. It is understood to now be a permitted design method in that structures designed in accordance with the procedure satisfy the requirements of the Rules.

The model, however, has only received limited validation, of the deterministic kind, because of the dearth of experimental data on stiffened cylinders subjected to axial loading: validation appears not to have been attempted for the combined loading case because of the complete absence of data. No attempt has yet been made to quantify errors in the model itself, errors which are inevitable in any sort of numerical or analytical modelling but which are probably at their greatest in design models.

This paper is concerned with a preliminary reliability assessment of this design model. It represents a first step in the development of a procedure for the rational ultimate limit state analysis of major sections of offshore platforms. Although codified strength models are not necessarily the best available, they represent accepted and understood methods. Such models can be compared with reliable test data and modelling errors quantified. An advanced first-order second-moment (Level II) reliability method can then be used to economically quantify the variation in reliability index with load. This Level II method can also provide central coefficients for the different design variables, as well as partial coefficients for these same variables for use with their characteristic rather than mean values. Characteristic values represent those values necessary to achieve a particular (chosen) level of success, e.g. 95%, and tend to be used for those variables to which the strength model is most sensitive, of which yield stress is a good example.

RELIABILITY ANALYSIS METHOD

The procedure used for this work is an advanced first-order second-moment method. It has been discussed in detail (CIRIA, 1977, and Baker and Flint, 1978) and was used in the derivation of partial coefficients for the new British Steel Bridge Code BS5400 (Flint, et al, 1981). A brief outline of the method is given below.

If $x_1, x_2, \ldots x_n$ are the n independent variables involved in a structural design problem, a general form of any limit state equation for the structure is

$$Z = g(x_1, x_2 \ldots x_n) > 0 \qquad (10)$$

where the nature of g depends on the structural type and limit state under consideration. The failure surface is given by $Z = 0$ and a linear approximation to this can be found by using the Taylor series expansion

$$Z \simeq g(x_1^*, x_2^*, \ldots x_n^*) + \sum_1^n (x_i - x_i^*) g_i'(\underline{x}^*) \qquad (11)$$

where $g_i'(\underline{x}^*) = \partial g/\partial x_i$ evaluated at the unknown design point $\underline{x}^* = (x_1^*, x_2^*, \ldots x_n^*)$. If m_i and σ_i represent the means and standard deviations of the basic variables x_i, the mean value of Z is

$$m_Z \simeq \sum_1^n (m_i - x_i^*) g_i'(\underline{x}^*) \qquad (12) \quad \text{and the standard deviation} \quad \sigma_Z \simeq \left[\sum_1^n \{g_i'(\underline{x}^*) \sigma_i\}^2 \right]^{\frac{1}{2}} \qquad (13)$$

σ_Z may be expressed as a linear combination of σ_i's as follows

$$\sigma_Z = \sum_1^n \alpha_i g_i'(\underline{x}^*) \sigma_i \qquad (14)$$

where $\alpha_i = \dfrac{g_i'(\underline{x}^*)\sigma_i}{\left[\sum\limits_{j=1}^{n}\{g_j'(\underline{x}^*)\sigma_j\}^2\right]^{\frac{1}{2}}}$ (15)

If the reliability index β of the design is defined as m_z/σ_z, then from equations (12) and (14)

$$\beta = \frac{\sum\limits_{1}^{n}(m_i - x_i^*)g_i'(\underline{x}^*)}{\sum\limits_{1}^{n}\alpha_i g_i'(\underline{x}^*)\sigma_i}$$ (16)

from which it follows

$$\sum\limits_{1}^{n} g_i'(\underline{x}^*)(m_i - x_i^* - \alpha_i\beta\sigma_i) = 0$$ (17)

The solution of this equation is

$$x_i^* = m_i - \alpha_i\beta\sigma_i \quad \text{for all } i$$ (18)

and x_i^* is referred to as the 'design point'. It corresponds to the point of maximum probability of failure density when all the variables are normally distributed. For given values of m_i, σ_i and β, equation (18) can be solved in conjunction with equation (15).

Finally, the probability of failure for the structure is

$P_f = \Phi(-\beta)$ where Φ is the standardised normal distribution function.

If any of the design variables have non-normal distributions, the following transformation is adopted

$$m_i^N = x_i^* - \Phi^{-1}\{F(x_i^*)\}\sigma_i^N$$
$$\sigma_i^N = \frac{f^N\left[\Phi^{-1}\{F(x_i^*)\}\right]}{f(x_i^*)}$$ (19)

where m_i^N, σ_i^N are the mean and standard deviation of the equivalent normal distribution, F is the cumulative distribution function of x_i, f is the probability density function of x_i, and f^N is the normal probability density function. This action has the effect of equating the cumulative probabilities and the probability densities of the actual and approximating normal distribution at the design point x_i^*.

By replacing σ_i by $m_i v_i$ where v_i is the coefficient of variation, equation (18) can be written as

$$x_i^* = m_i(1 - \alpha_i\beta v_i)$$ (20)

the term in parenthesis is the central partial coefficient.

The above procedure has been computerised using the algorithm described in CIRIA, 1977. A further development of this purports to have the advantage of finding stationary values of β (Baker and Flint, 1978), but the results are numerically the same. However, it does not allow for design points to be determined for different safety indices and so the former procedure was preferred.

STRENGTH MODELLING

Except in cases where extensive experimental and numerical studies have been conducted, errors of modelling will exist in design strength formulations. CIRIA, 1977, suggests that to account for such errors a modelling parameter should be introduced relating actual to predicted strengths and the mean and variance of this established by evaluation against sufficient experimental data. Because of limitations of present theoretical techniques and on testing capabilities, such calibration of the modelling parameter will be limited to components only of offshore platforms. It might be expected, however, that, with time, configurations of several components will be within modelling capability thus allowing redistribution and redundancy to be investigated rationally.

The use of a single modelling parameter would have to be considered carefully in any study particularly where instability may be involved. It is not difficult to imagine that an instability design model could show coincidence with test data at low slenderness where ultimate limit state and yield coincide but show increasing differences with slenderness. The modelling parameter in this case is likely to have both a significant mean and variance. Even evaluating the parameter over limited sections of the slenderness range would lead to structures whose reliability varied with slenderness because of the large variances obtained for structures at the slender end of the range. Alternatively, two modelling parameters could be introduced, one which varied with slenderness and therefore effectively rotated the predicted values relative to the experimental data, and the other which would quantify the now much reduced scatter between the predicted and experimental results. Evaluation of the slenderness dependent parameter might be performed deterministically or possibly statistically prior to the reliability analysis. In either case, adequate experimental data is required.

In the calibration process, the modelling parameter was treated deterministically to provide a unique value of actual to predicted loads for each model under consideration. For the reliability analysis, the modelling parameter was first treated as a statistical variable so that the sensitivity (α_i) could be determined. Subsequent calculations treated it deterministically. The g function (equation 10) is therefore

$$Z = X_m R_k - Q_o \quad \text{when } X_m \text{ is the modelling parameter, and } Q_o \text{ the characteristic extreme load.} \tag{21}$$

Stiffened Cylinder Design Model

Briefly, the DnV design model for stiffened cylinders subjected to axial load and external pressure assumes that collapse will occur when the stress at mid-span, arising from the axial load plus bending, reaches yield. Bending arises from amplification of the initial column-like distortion and external pressure deflection in the effective plate-stiffener combination under the action of the axial load, the amplification being derived from a first-order buckling solution to the problem of a beam-column on an elastic foundation: the elastic foundation arises from curvature of the shell plating. The stringer-stiffened cylinder buckling stress was derived using potential energy assuming one half wave would always form between rings.

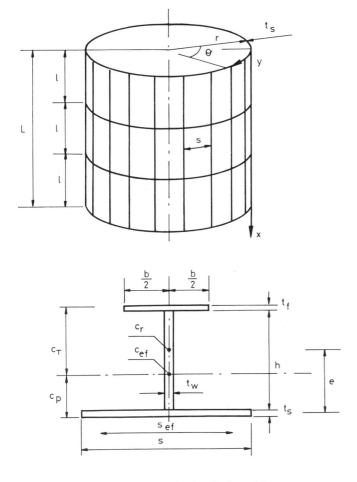

FIGURE 1 Geometrical details

The most recent version of the model is described in Valsgård and Foss, 1981. In the present work, only axial load will be considered. The model is outlined in Appendix A, geometrical details being presented in Fig. 1.

In the model, several of the variables have been subjected to different interpretations during DnV's correlation studies with test data. This applies in particular to residual stress, f_R and ψ_1 described as a correlation factor. Strictly ψ_1 should be used to differentiate between internal and external stiffening. In Vasgård and Foss, 1981, the best correlation was obtained by assuming $\psi_1 = 1.0$, that f_R was zero, and that the shell plating was fully effective when acting in combination with stiffener, i.e. α(equation A1) is unity. These same assumptions were made in the present study. The proposed method predicts experimental data with greater accuracy then the Rules (Valsgård and Steen, 1980).

Therefore, R_k in equation (21) is given by equation (3) in which Φ is as defined by equation A1 or, since from the above assumptions $\alpha = \eta = 1$, more simply by equation A2.

SELECTION OF CHARACTERISTIC EXTREME LOADS

An experiment represents the situation where equation (21) is identically zero, i.e. the characteristic values of resistance presuming $X_m = 1$ and load coincide with their mean values. When $X_m \neq 1$, $X_m R_k$ becomes the mean value.

For assessing the reliability of the stiffened cylinders for loads less than that necessary to cause collapse, the selection of possible characteristic loads is required. The DnV Rules provide one such method using the format of equation (6) to give an estimate of the characteristic load effect as follows

$$Q_k = \frac{R_k \psi}{\gamma_f \gamma_m \kappa} \tag{22}$$

This could represent an upper bound on characteristic loads since R_k (strictly $X_m R_k$) is not a characteristic value in that it represents some fractile of expected resistances, but is to be considered as the mean of a set of data having very little variability. The use of the measured (mean) value of yield stress rather than some fractile of statistically assessed yield stress measurements also implies that R_k as determined by equations (3) and (A1) will represent an upper limit on resistance in this case. This overestimate on load will possibly be counterbalanced by the presence of κ which has values greater than 1.0 for the geometries to be considered.

Most fixed structures are relatively insensitive to the frequency content (wave periods) of the excitation. Design is then based on the single regular design wave approach with a sufficiently high recurrence interval, e.g. 100-year. The main and perhaps only virtue in such an approach to its simplicity. Such assumptions, however, are quite dubious for compliant structures which are much more sensitive to wave periods and wind gustiness. Thus, modified regular design wave approaches have been used and are being more formally established, for example, for tension leg platforms. These attempt to identify wave height and period combinations over a credible range of probabilities for a given recurrence interval which will yield the most extreme forces to be used to establish characteristic design values.

TABLE 1
Extreme Tether Forces (expressed in RMS values) for differing Storm Durations and Probabilities of Exceedence

Exceedence Probability ε	Storm Duration (hours)		
	1	2	3
0.632+	3.31	3.63	3.99
0.500*	3.42	3.75	4.12
0.0100	4.49	4.73	5.01
0.0010	4.98	5.19	5.45
0.0001	5.42	5.62	5.86

Wave spectral peak period = 16.5 s;
Mean period of response = 15.06 s;
Bandwidth parameter = 0.508
+ mode   * mean

However, because of weight criticality, more rational approaches based on irregular seas and extremal statistics are being sought. These should take account of spreading and other factors which would represent a large penalty if ignored in compliant structures. The results for the dynamic component of wave-induced mooring tensions (and hence column forces) for a Tension Leg Platform are illustrated in Table 1. The mean square forces were computed in linear frequency domain using transfer functions derived from model tests in the extreme 100-year storm irregular sea environment. Ochi's 1973 method was then used to establish relationships between the RMS mooring tensions ($\sqrt{r_o}$) and the expected extreme tensions (t_e) for various storm durations (T hours) and exceedence probabilities (ε)

$$t_e = \sqrt{2\ln\frac{(60)^2 T}{2\pi}\sqrt{\frac{r_2}{r_o}}} \sqrt{r_o} \qquad (23)$$

where r_o is the area under the response spectrum and r_2 is its second-moment. The method is useful in catering for arbitrary spectral bandwidths <0.9 but the relations are only valid for small ε. For ships Ochi, 1973, suggests $\varepsilon = 0.01$ which would then be regarded as a characteristic load.

The mode and mean values are also shown in Table 1 for three storm durations. The average coefficients of variations derived for the equivalent normal distributions (for the same t_e and ε) for the 1,3 and 12 hour storms are 0.15, 0.13, and 0.10 respectively with an overall average of 0.126. These are objective uncertainties inherent in the process and, as expected, they diminish (but will never vanish) as the storm becomes fully developed.

To complete the distribution of maximum load effects for reliability analysis it is then necessary to "add" in the other objective uncertainties, such as those arising from drag and inertial effects and also the subjective uncertainties arising from the imperfect modelling of extreme seas. These latter are difficult to assess, but Faulkner, 1981, has suggested 0.15 for small ships of 90-120m length. Ignoring drag and other objective uncertainties (see later) would then suggest a total wave-induced uncertainty of about $(0.13^2 + 0.15^2)^{\frac{1}{2}} = 0.20$. However, if directionality is ignored, and allowing for other attenuation effects in offshore structures an extreme wave-induced uncertainty of about 0.15 is probably more appropriate. Such considerations affect the choice and definition of characteristic loads and will be reported more fully later.

CALIBRATION OF STRENGTH MODEL

Test Data
   Data on axial compression tests on small-scale stiffened cylinder models reported by Walker, 1979, have been used by DnV in the process of substantiation of their design model. This data will also form the basis of the present calibration.

   Table 2 lists the geometrical parameters and the material properties of these cylinders.

Modelling Coefficient
   As indicated, for calibration X_m were derived deterministically and they are presented in column 2 of Table 3 for the models considered. The coefficient is seen to vary significantly, ranging from 1.121 in the case of model UC5 to 0.857 for model UC9.

TABLE 2
Model Data for Small-Scaled Narrow Panelled Cylinders

Model	r/t_s*	ℓ/r	h/t_w	t_w/t_s	e/t_s	a	Imperfection w_1/t_s	w_1/ℓ	Yield stress f_y (N/mm^2)	E (kN/mm^2)
UC2	200	1.11	8	1.0	-4.5	0.25	0.5	1/444	320	210
UC5	200	1.11	16	1.0	-8.5	0.51	0.6	1/370	338	203
UC7	280	1.11	16	1.6	-8.5	0.35	1.2	1/259	311	211
UC9	360	1.11	16	1.0	-8.5	0.36	2.3	1/174	340	211
B1	280	1.55	16	1.0	-8.5	0.36	1.2	1/364	313	210
B5	280	1.56	10.7	1.5	-8.5	0.55	1.2	1/364	318	210

* $t_s = 0.81$ mm   The number of stringers in each case is 40

TABLE 3
Variation of Safety Index (β) with Means and Coefficients of Variations of Extreme Characteristic Loads

Model	$\gamma_f = 1.0$						$\gamma_f = 1.3$					
	X_m	V_p 10%	V_p 20%	X_m	V_p 10%	V_p 20%	X_m	V_p 10%	V_p 20%	X_m	V_p 10%	V_p 20%
UC2	1.107	2.81	1.87	1	2.01	1.31	1.107	4.92	3.43	1	4.10	2.81
UC5	1.121	1.79	1.16	1	0.91	0.58	1.121	3.89	2.64	1	2.96	1.97
UC7	0.986	2.48	1.62	1	2.59	1.70	0.986	4.62	3.17	1	4.72	3.26
UC9	0.857	3.23	2.13	1	4.52	3.06	0.857	5.44	3.75	1	6.76	4.80
B1	0.928	3.04	2.01	1	3.65	2.45	0.928	5.21	3.61	1	5.84	4.10
B5	0.953	3.16	2.08	1	3.57	2.36	0.953	5.38	3.71	1	5.79	4.03

Since the basic buckling mode for these models is that of a stringer-stiffened cylinder, it might be expected that uncertainties in the prediction of the buckling stress could exhibit some correlation with the modelling coefficient. A non-dimensional form for buckling stress which also accounts for variation in yield stress is $\sqrt{f_y/f}$ where f is the elastic critical stress for the buckling mode under consideration.

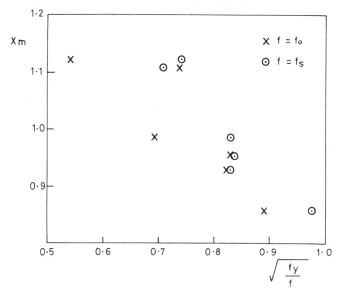

FIGURE 2 Variation of modelling parameter with unstiffened and stiffened cylinder buckling stress

Valsgård and Steen, 1980, list values of stringer-stiffened cylinder buckling stresses for the models under consideration. They are based on the 'smeared' stiffener analysis reported by Hutchinson and Amaziego, 1967. Denoting these stresses by f_o, Fig. 2 shows the variation of X_m with the factor $\sqrt{f_y/f_o}$. The scatter is seen to be significant particularly for those cylinders with X_m values around 1.1: this corresponds to test stresses slightly greater than yield. Use of the 'neutral equilibrium stress' derived by Walker and Sridharan, 1980, in the process of devising a prediction method for stringer-stiffened cylinder strengths in place of f_o had little effect on the scatter.

Examining X_m in relation to the basic cylinder geometrical parameters (Table 2) shows that the modelling parameter varies inversely with r/t_s and thus, possibly, the elastic critical buckling stress of the shell only under axial compression. Denoting this value by f_s (= $E\, t_s/r\sqrt{3(1 - \nu^2)}$) Fig. 2 shows the variation of X_m with $\sqrt{f_y/f_s}$. Considerably less scatter results from the use of this relationship compared with that arising from the first attempt.

Useful as this may be for cylinders of similar geometry, through extrapolation X_m would appear to require very large values for stiffened cylinders of lesser R/t_s's: these would appear to be unrealistic. Obviously insufficient data is available to derive even a deterministic relationship between X_m and some parameter whereas, of course, a statistical assessment is naturally required. It is to be hoped that current research programs on stiffened cylinders will go some of the way to providing this additional data.

RELIABILITY ANALYSIS

Probability Distributions

It seems generally agreed (CIRIA, 1977 and Fjeld, 1978) that resistance variables should be represented by log-normal distributions. Elastic modulus and yield stress were thus treated in this way, (Table 4). Most geometric parameters can be represented by normal distributions. Dead

TABLE 4 Basic Variables, Statistical Data, and Results of the Analysis of Cylinder B1

Basic variable	Distribution type	Mean	C.O.V.	Standard deviation	Design point	Sensitivity factor	Partial safety factor
t_w	normal	0.81 mm	4%	0.0324 mm	0.803	+0.0586	
h	normal	12.96 mm	4%	0.5184 mm	12.82	+0.0740	
t_s	normal	0.81 mm	4%	0.0324 mm	0.783	+0.231	0.967(1.034)
r	normal	226.8 mm	5%	11.34 mm	217.63	+0.224	0.960(1.042)
ℓ	normal	353.808 mm	4%	14.152 mm	353.06	+0.0146	
w_1	normal	0.972 mm	5%	0.0486 mm	0.977	-0.0320	
P	normal	210148 N	20%	42029 N	333470	-0.812	1.586
E	log-normal	210,000 N/mm^2	4%	8400 N/mm^2	209210	+0.0205	
f_y	log-normal	313 N/mm^2	8%	25.04 N/mm^2	272.11	+0.474	0.869(1.150)
X_m	deterministic	0.928	-	-	-	-	

$\beta = 3.61$ $p_f = 1.53 \times 10^{-4}$

loading is well represented by a normal distribution while live loading is probably more appropriately modelled by an extremal-type distribution (Fjeld, 1978, and Flint, et al, 1981). Environmental loads, waves, currents, wind, etc. are usually specified by some estimate of their extreme value. Various approaches are available for defining the distributions of the different extreme load types and Gumbel, Weibull and extremal have been suggested (CIRIA, 1977, Fjeld, 1978, and Bouma et al, 1979) as being appropriate.

However, it should be remembered that the distribution of the extreme value expected to occur in a certain number of wave encounters or period of time is the one required for ultimate load reliability estimates (Faulkner, 1981). This will have a Poisson type distribution about the most likely or modal value (with about 63.2% chance of being exceeded) which in turn can be derived from the "initial" or "long-term" probability density function. Moreover, most offshore structural components have loading effects from many sources which tend to normalize the distribution. The closeness of the mean and modal values in Table 1 supports the use of a normal distribution for the present purpose and this has been assumed.

The coefficient of variation for yield stress reportedly ranges between 5% and 10%. An extensive study by Baker, 1972, found it to vary between 6% and 7%. 8% was used in the present investigation since no indication of the statistical variability of the material used in the construction of the models was reported (Walker and Sridharan, 1980). In the case of elastic modulus, half this value was adopted (Table 4).

Variation in geometrical parameters tends to be small particularly where inspection can readily check items against the specification as in the case of dimensions, e.g., frame spacing. Thicknesses of elements at critical sections reportedly have standard deviations of the order of 1% of the nominal value (Flint, et al, 1981) which can probably be ignored. Bouma, et al, 1979, suggests dimensions can be treated deterministically. However, until at least some reported data becomes available it seems prudent to assume coefficients of variations of some 4% to 5% for the geometrical variables. Imperfections will normally show more variability than this but, because in design one would usually be dealing with an imperfection approximately equal to the tolerance, i.e. around the 90% to 95% fractile of distortion magnitudes, the variation in this 'upper limit' will be smaller. It was assumed here to be 5%.

Dead load variability is usually less than 5%. Live load representing moveable equipment, stored materials etc. will show a similar dispersion although, for offshore structures, a 'growth factor' for this type of load needs serious consideration. Wave and current forces as represented by drag and inertia coefficients shows variabilities between 14% and 28% (Kim and Hibbard, 1975, Fjeld, 1978, Baker and Wyatt, 1979, and Bouma, et al, 1979), although values about twice these levels have been recorded on single components in oscillating flow. Uncertainty in wind loading appears to vary between 10% and 35% (Fjeld, 1978, and Baker and Wyatt, 1979). However, the basis for the basic design loads in each of these cases is different and is probably the cause of this significant variation.

For combinations of all these loads, it might be expected that dead and live load dominated situations would have a coefficient of variation up to 10%, whereas 20% to 30% would appear to be more appropriate where extreme environmental forces play a significant role after allowing for some attenuation as mentioned earlier.

Sensitivity to Design Variables

Table 4 contains details of the basic variables and their distributions and the results of the reliability analysis of cylinder B1 when subjected to an axial load calculated from equation (22) by taking $\psi = 0.9$, $\gamma_f = 1.3$, $\gamma_m = 1.15$ and κ as determined via equation A3: the load coefficient of variation is 20%. The sensitivity factors α_i (equation 15) are listed in the penultimate column. The sensitivity of the model to each design variable is directly proportional to the absolute value of α_i, while a positive value indicates the variable is a resisting type, negative implies a loading or disturbing type.

For this case, load is seen to have the greatest effect on the model followed by yield stress. Of lower and almost equal influence are the geometry variables of radius and thickness. Initial distortion is noteable for its lack of influence.

This ranking of sensitivities can be qualitatively supported by re-analysing the structure while incorporating small changes in the means and in the standard deviations of each of the variables in turn and examining how the failure probability of the structure varies with these changes. The results of such an investigation are presented in Figs. 3 and 4 in respect of mean and standard deviation respectively, the abscissa representing ratios of the modified to the nominal (Table 4) values of the variables. Fig. 3 clearly indicates which are the resisting and which are the disturbing variables, while Fig. 4 reflects the ranking of influence most effectively.

Not unexpectedly, Fig. 3 shows that increases in the mean value of resisting variables improves the reliability significantly. Likewise Fig. 4 indicates that a reduction in the dispersion of all variables leads to an enhancement of safety. With particular reference to yield stress where supplies from different steel mills can show significant differences with respect to both mean and variability of the material, the figures show it is clearly preferable from a safety viewpoint to select a supplier with steel having a mean on the high side which also shows little variability.

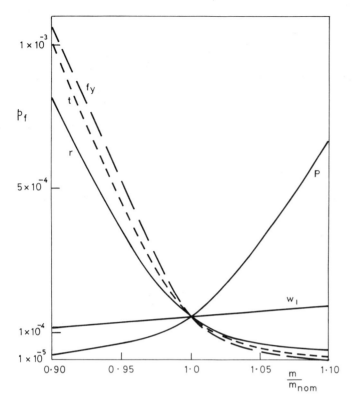

FIGURE 3   Sensitivity to changes in mean (Model B1)

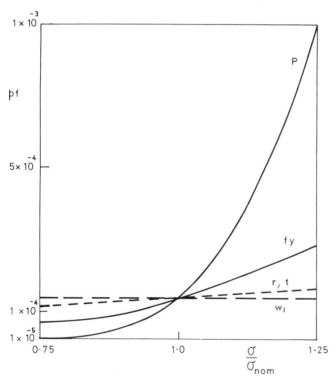

FIGURE 4   Sensitivity to changes in standard deviation (Model B1)

Repeating the analysis with the load coefficient of variation at 10% has the effect of reducing the model sensitivity to load and increasing that due to yield stress to -0.632 and 0.630 respectively. Clearly the large coefficient of variation on load in the first analysis contributed significantly to the sensitivity shown to load by the model. For the second case the sensitivity factors on thickness and radius increased to 0.302 and 0.305 respectively.

Also listed in Table 4 are the values of the variables at the design or, as in this case, the failure point. These represent the point on the failure surface having the largest failure probability which is given at the bottom of the table together with the corresponding value of the safety index.

As indicated earlier, the sensitivity of the design model to X_m was examined. It was found to have an influence on the model very similar to that of yield stress, highlighting the importance of firstly, the need to introduce such parameters into limit state models and, secondly, that they need to be reasonably accurately defined.

Partial Safety Factors

As illustrated by equation (20), the central partial safety factor for each variable can be determined readily from the reliability analysis: it represents the ratio of the design point value of a variable to its mean value. The central factors of Model B1 are listed in Table 4 for the four variables to which the strength model shows the greatest sensitivity. Strictly the load has been derived as a characteristic one. Although this will not effect the reliability analysis, it will effect the interpretation of the meaning of this partial factor in relation to load.

For design variables to which a strength model shows particular sensitivity, it is suggested that characteristic values rather than mean ones should be used particularly in the design context. The partial factors appropriate for this thus relate the design and the characteristic values, and can be derived easily from the central partial factor since

$$x_{ik} = m_i (1 \mp k_i v_i) \qquad (24)$$

where x_{ik} is the variable characteristic value, k_i is a coefficient depending on the fractile represented by x_{ik}, and the minus or plus sign is used depending on whether x_i is a resisting or loading variable respectively. Recalling equation (20), the ratio of design to characteristic values is

$$\frac{x_i^*}{x_{ik}} = \frac{1 - \alpha_i \beta v_i}{1 \mp k_i v_i} \qquad (25)$$

This is the required partial safety factor in which the numerator represents the previously defined central partial safety factor. Therefore, as derived in this present example, the partial factor on load is that defined by equation (25) whereas those appropriate for the remaining variables are the central factors as given by the numerator. In the table, the numbers in parenthesis are just the reciprocals of the central factors. These have been included since they represent the form in which resisting partial factors are usually quoted.

Equation (25) shows that for a loading variable, the central partial factor will be numerically greater than the partial factor on the characteristic value. Consequently, a value higher than the 1.586 would have been found as the partial factor on load if this had been a central safety factor. The partial factor on load is thus considerably greater than on any of the other variables.

The results presented in Table 4 are typical of those derived for all the models although not insignificant differences were found between the sets of partial safety factors.

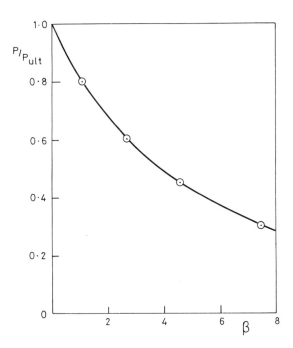

FIGURE 5 Variation of design load with safety index (Model B1)

Variation of Safety Index with Load

By changing the load considered in the reliability analysis, different reliabilities will be obtained. This was done in the case of model B1 (Table 4). The results are presented in Fig. 5. The ratio of applied (characteristic) load to mean predicted (and experimental) collapse load (P/P_{ult}) has been plotted against the reliability index. The curve initiates at the collapse load where the probability of failure equals unity ($\beta=0$) and reduces with increasing chance of success. Thus for any load ratio, the corresponding reliability can be established and vice versa. This curve is useful in that simply for any desired level of safety, the overall central safety factor can be readily established so that once any resistance partial factors such as those on material and for modelling have been taken into account, the load partial factor can be deduced. Alternatively, if different applied loads are derived through the use of varying factors on the load components, i.e., dead, live, environmental or accidental, the resulting changes in reliability can be quickly assessed.

Reliability under Extreme Loads

Using the characteristic extreme loads derived from equation (22) when $\psi = 0.9$, $\gamma_m = 1.15$, κ as appropriate to the geometry and $\gamma_f = 1.3$, the reliability of the models under this loading can be assessed. The results are presented in Table 3 for X_m equal to 1.0 (corresponding to the Rules predicted strength) and to the value derived from the calibration process, and for two values of coefficient of variation on loading, viz. 10% and 20%. Safety indices corresponding to a dead or live load dominated situation, $\gamma_f = 1.0$, are included for comparison. The apparent lower values of reliability found under this loading scheme are not realistic, the results just reflect the fact that equation (22) leads to a higher applied load for small values of γ_f.

Table 3 shows that under all conditions, the reliability of the models varies significantly. This is most pronounced when the modelling parameter is ignored, as would be the case in practice, and the variability on load is high. For example, the probabilities of failure corresponding to the extremes of the safety indices listed in column 13 are 2.44×10^{-2} and 7.9×10^{-7}. It might have been expected that through the inclusion of the modelling parameter, the present analysis would have led to consistent levels of reliability for all models. The only reason seen for the disparity is that applied loads have been derived by a process involving a parameter that is a function of geometry, viz, κ (equation 22). The lack of consistency arising from the use of this parameter goes some way to supporting the contentions raised earlier concerning its inappropriateness.

The effect of the modelling parameter can be highlighted by comparing the relative reliabilities of models UC2 and UC9 under extreme loading when v_p is 20%. Introducing X_m increases the reliability of the first model, (β from 2.81 to 3.43, p_f from 2.47×10^{-3} to 3.01×10^{-4}) but decreases it in the second case (β from 4.80 to 3.75, p_f from 7.9×10^{-7} to 8.84×10^{-5}). Clearly modelling parameters need evaluating if reliability is to be improved.

CONCLUSIONS

A critical review of the safety format adopted in the DnV Rules for Offshore Structures for limit state design has shown them to lead to inconsistent levels of reliability through the use of their partial coefficients on material and modelling (on instability-prone structures). It would appear to be more rational to apply the material factor to yield stress only and not to the characteristic strength as at present, and that the modelling coefficient be deleted.

A recently developed DnV model for the design of closely-spaced stringer-stiffened cylinders subjected to axial compression and external pressure has been calibrated against test data on axially loaded models. The resulting model parameter was shown to vary with cylinder geometry and demonstrated less scatter when plotted against the elastic critical buckling stress of the shell only rather than the stringer-stiffened geometry, both normalised with respect to yield stress.

Sensitivity studies conducted on the design model using an advanced first-order second-moment reliability analysis technique showed it to be most sensitive to load, yield stress and modelling coefficient. Radius and thickness had not insignificant influences but it was noteable that initial distortion, of the stringers between ring-frames, had virtually no effect.

Reliability indices derived for the models for characteristic extreme loads calculated in accordance with the DnV Rules safety factor format demonstrated a considerable spread: nearly all of this was due to the Rules instability modelling parameter. Nevertheless, these results highlighted the need to introduce rationally derived modelling parameters.

The approach to reliability assessment adopted in the reported work involving calibration of design models against test data allows partial factors between load and resistance to be derived for any desired level of safety or, alternatively, the reliability of a design to be established once the applied load has been determined. It also allows partial factors on load, material and modelling to be rationally derived whether based on mean or characteristic values.

Reliability is very dependent upon the loading assumptions and a more rational approach for this is being pursued. Better modelling in both extreme load effects and in ultimate strength are necessary for the efficient design of weight critical structures such as tension leg platforms. Adequate but not excessive safety is required with notional safety indices perhaps in the range 3.5 to 4.0. Studies are in hand in an attempt to achieve all these objectives.

APPENDIX A

Outline of Stiffened Cylinder Design Model

The characteristic average buckling resistance is given by $R_{Nk} = \phi f_y$ where ϕ is defined by

$$\phi = \alpha \beta_m \eta \tag{A1}$$

where $\alpha = \dfrac{A_{ef}}{A_t}$ is the effective area coefficient (see fig. 1),

$\beta_m = \dfrac{f_N}{f_k}$ is the characteristic mean axial strength ratio, and

$\eta = \dfrac{f_k}{f_y}$ is the characteristic strength coefficient.

Where

$$\frac{f_N}{f_k} = \frac{(1 + \mu_1 + \lambda_1^2) - \{(1 + \mu_1 + \lambda_1^2)^2 - 4\lambda_1^2\}^{\frac{1}{2}}}{2\lambda_1^2} \tag{A2}$$

$f_k = f_y - f_R/2$ is the characteristic reduced yield stress as effected by welding residual stresses f_R,

$\lambda_1^2 = \alpha \dfrac{f_k}{f_e}$ is the effective reduced slenderness,

$\mu_1 = \dfrac{f_E \, c\omega}{f_e \, i_{ef}^2}$ is the imperfection parameter allowing for the shell curvature effect,

$c = c_p$ or c_T is the neutral fibre distance of the effective stringer-shell combination for plate induced or stringer induced failure,

$\omega = \omega_1 + c_p \dfrac{1 - \alpha}{\alpha}$ or ω_1 is the effective imperfection for plate or stringer induced failure

ω_1 is the initial stringer deflection in its own plane between ring-frames.

$$f_e = \frac{\alpha \pi^2 E \, i_{ef}^2}{\ell^2} + \frac{\psi_1}{1 + a} \frac{E t_s}{\{3(1 - \nu^2)\}^{\frac{1}{2}} r} \tag{A3}$$

is the elastic critical buckling stress of the stringer-stiffened cylinder between ring-frames, f_e is the Euler buckling stress of the effective stringer-shell combination, i_{ef} is the radius of gyration of the effective stringer-shell combination, E is Young's modulus, ψ_1 is a correlation factor to be experimentally and/or numerically derived, a is the ratio of stringer area to shell area, and ν is Poisson's ratio.

The buckling mode assumed in the model was one half-wave in the axial direction and n full-waves in the circumferential direction where

$$n^2 = \pi\{12(1-\nu^2)\}^{\frac{1}{4}} \frac{r}{\ell}(\frac{r}{t_s})^{\frac{1}{2}} - \pi^2(\frac{r}{\ell})^2$$

The geometric variables are defined in Fig. 1.

REFERENCES

BAKER, M.J., 1972, Variability in the Strength of Structural Steels - a Study in Structural Safety, Part 1: Material Variability, CIRIA, Tech. Note 44.

BAKER, M.J. and FLINT, A.R., 1978, "Safety Approaches for Structures Subjected to Stochastic Loads", Proceedings, International Symposium on Integrity of Offshore Structures, Glasgow, 1978, Paper 10. Glasgow: Instn. of Engrs and Shipbuilders in Scotland.

BAKER, M.J. and WYATT, T.A., 1979, "Methods of Reliability Analysis for Jacket Platforms", Proceedings, 2nd International Conference on Behaviour of Offshore Structures, London, 1979, Vol. 2, Paper 84, pp.499-520. Cranfield, Bedford: BHRA Fluid Engineering.

BOUMA, A.L., MONNIER, Th and VROUWENVELDER, A., 1979, "Probabilistic Reliability Analysis", Proceedings, 2nd International Conference on Behaviour of Offshore Structures, London, 1979, Vol.2, Paper 85, pp.521-542. Cranfield, Bedford: BHRA Fluid Engineering.

British Standards Institution, 1979, BS5400: Steel, Concrete and Composite Bridges, Draft Part 3: Code of Practice for Design of Steel Bridges. London: British Standards Institution.

Construction Industry Research and Information Association, 1977, The Rationalisation of Safety and Serviceability Factors in Structural Codes, CIRIA, Report 63.

Det norske Veritas, 1978, Rules for the Design, Construction and Inspection of Offshore Structures, 1977. Oslo: Det norske Veritas.

European Convention of Constructional Steelwork, 1978, European Recommendations for Steel Construction. ECCS-EG 77-2E.

FAULKNER, D., 1979, "The Safe Design and Construction of Steel Spheres and End Closures of Submersibles, Habitats and Other Pressured Vessels", Proceedings, 2nd International Conference on Behaviour of Offshore Structures, London, 1979, Vol. 2, Paper 86, pp.543-556. Cranfield, Bedford: BHRA Fluid Engineering.

FAULKNER, D., 1981, "Semi-probabilistic Approach to the Design of Marine Structures", Proceedings, Extreme Loads Response Symposium, Arlington, 1981. Society of Naval Architects and Marine Engineers.

FJELD, S. 1978, "Reliability of Offshore Structures", Petroleum Technology, pp.1486-1496.

FLINT, A.R., SMITH, B.W., BAKER, M.J. and MANNERS, W., 1981, "The Derivation of Safety Factors for Design of Highway Bridges", in The Design of Steel Bridges, Ed. Rockey and Evans. London: Granada.

HUTCHINSON, J.W. and AMAZIEGO, J.E., 1967, "Imperfection Sensitivity of Eccentrically Stiffened Cylindrical Shells", AIAA, Vol. 5, pp.392-401.

KIM, Y.Y. and HIBBARD, H.C., 1975, "Analysis of Simultaneous Wave Force and Water Particle Velocity Measurements", Proceedings of Offshore Technology Conference, Houston, 1975, Paper 2192.

OCHI, M.K., 1973, "On Prediction of Extreme Values", Ship Research, Vol. 17, No. 1, pp.29-37.

ODLAND, J. and FAULKNER, D., 1981, "Buckling of Curved Steel Structures - Design Formulations", in Integrity of Offshore Structures, Edited. Faulkner. Cowling and Frieze. London:Applied Science.

PAPADIMITRIOU, A. and FRIEZE, P.A., 1981, "Numerical Prediction of Hemisphere Strength", Proceedings, International Conference on Buckling of Shells in Offshore Structures, London, 1981, Paper 9. London: Imperial College of Science and Technology.

VALSGÅRD, S. and STEEN, E., 1980, "Simplified Strength Analysis of Narrow Panelled Stringer Stiffened Cylinders under Axial Compression and Lateral Load", Det norske Veritas, Tech. Report No.80-0590.

VALSGÅRD, S. and FOSS, G., 1981, "Buckling Research in Det norske Veritas", Proceedings, International Conference on Buckling of Shells in Offshore Structures, London, 1981, Paper 19. London: Imperial College.

WALKER, A.C. and SRIDHARAN, S., 1978, "Buckling of Compresses Longitudinally Stiffened Cylindrical Shells", Proceedings, 2nd International Conference on Behaviour of Offshore Structures, London, 1979, Vol. 2, Paper 72, pp.341-356. Cranfield, Bedford: BHRA Fluid Engineering.

WALKER, A.C. and SRIDHARAN, S. 1980, "Analysis of the Behaviour of Axially Compressed Stringer Stiffened Cylindrical Shells, Proceedings, Institution of Civil Engineers, Part 2, Vol. 69, June, pp.447-472.

ACKNOWLEDGEMENT

The authors wish to acknowledge the Science and Engineering Research Council, Marine Technology Directorate, for their financial support of the first author and of the research program on reliability underway in the Department.

EXTREME WAVE DYNAMICS OF DEEPWATER PLATFORMS

R. D. LARRABEE
Shell Development Company, U.S.A.

SUMMARY

A simulation-based methodology is presented for selecting design values for the dynamic, response of fixed, template-type platforms in extreme, random seas. An efficient, time-domain computer algorithm, which includes a pure modal solution, simplified wave force areas and volumes, and 3-D random wave kinematics, is an important feature. The simulated probability distributions of dynamic and static base shear and overturning moment provide a rational basis for selecting dynamic amplification factors (DAF) for design. A larger estimate of DAF is obtained when the extreme platform response is assumed to be Gaussian. The feasibility of the methodology for design is illustrated for an example platform and directional wave spectrum. Safety factors for dynamic response are discussed.

INTRODUCTION

The first requirement for the in-place performance of a fixed platform is to resist the extreme lifetime sea state. For platforms that respond statically, the extreme design can be carried out using a design wave and a quasi-static structural and foundation analysis. For fixed platforms in deeper water--those with periods exceeding 3 to 4 seconds--dynamic inertial forces associated with the mass of the structure must be considered as well. The degree of dynamic response is closely related to the frequency content of the exciting force. Random wave theories should be used.

For wave force problems that can be successfully linearized, spectral methods present a convenient method of using random waves. For fixed, template platforms, the resulting spectra of response provide sufficient data for fatigue damage estimates. However, extreme wave forces on fixed, template-type platforms are underpredicted by a linearized, frequency domain solution (see, for example, Hackley, 1979). The nonlinearities are in the wave force and are due to both the velocity squared drag term in Morison's equation and the inundation effect as the wave surface moves from trough to crest. As shown below, even the seemingly simpler task of finding the ratio of extreme dynamic to extreme static response cannot be done completely with just spectra. The analyst cannot avoid time domain simulation of the extremes of the dynamic response.

SCOPE

This paper will describe a time domain procedure for extreme wave dynamics. This procedure is a method that adds to the regular wave static solution extra forces to account for the dynamic inertial loads. These forces are based on dynamic amplification factors computed by random wave analyses. The overall procedure is viable for the design of a major structure. This paper will focus on how to select the dynamic amplification factor (DAF).

When the loading is random waves, there is wave-to-wave variability in the structure's response. Obtaining statistically significant extreme values of the response of all members to random wave forces is computationally expensive. It is feasible to simulate only selected global responses (such as base shear, overturning moment, deck displacement) with random waves in the time domain. Designing members for forces at the instant in time when a global response is maximum is a well accepted engineering approach. Thus, the time domain simulations may be limited to global response.

Two other features are desirable for a design procedure. The first is that the new procedure must be related to existing, extreme static wave procedures. Due to linear wave theory, directional spreading and techniques for kinematics in wave crests, the wave force coefficients that were derived for regular wave kinematics cannot be used directly. Even with re-derived coefficients, the applied force from random waves may not be consistent with the applied force from regular waves. This is a problem if the random wave results are to be used directly, but is avoided in an amplification factor method.

The second consideration is practical: a truly random time domain analysis will give slightly different answers each time it is done, complicating design checking or verification. Limiting the random simulation to the determination of amplification factors reduces this difficulty.

THE DYNAMIC AMPLIFICATION FACTOR

There are several ways to arrive at dynamic amplification factors (DAF) from time domain results. Certain potential methods should be approached with caution. One of these methods is simulating the response to large regular waves with various periods. The result would be a plot of DAF versus wave period. These DAF are not usually realistic due to the false frequency components of applied force that are created by a particular regular wave theory. Even if these effects were not present, the analyst must still choose a design value. The DAF at the most likely wave period or mean wave period are candidates, as are the average of DAF over some range of periods. No choice is completely rational.

With the results from random wave, time domain analyses, one has many samples of large static and dynamic responses to use to estimate a DAF. Following a deterministic viewpoint, one could take the ratio of the maximum dynamic response in a time segment to the maximum static response in the same segment, and average these DAF over all segments simulated.

An alternative DAF that is both rational and applies to either simulated or theoretical responses can be defined. For convenience, this definition will be labeled the "probability of exceedance DAF" or $DAF(p_e)$. Like any DAF, it is the ratio of a dynamic response to the quasi-static value of the same response. Each response is selected from its own cumulative probability distribution such that the responses have the same probability of being exceeded. One way to choose that probability is to identify a duration, T, of the sea state and then specify the median of the maximum during this duration as the design value. The probability of exceedance would then be 0.5 in a CDF of maximums during T and

$$\text{DAF}(p_e) = \frac{\text{max. in T dynamic response at } p_e = 0.5 \text{ probability}}{\text{max. in T static response at } p_e = 0.5 \text{ probability}} \tag{1}$$

In this probability of exceedance definition, a maximum dynamic value is not coupled with the maximum static event in the same wave or simulation segment. Rather, it is paired with a static event with the same probability of being exceeded. This probability can be identical to the probability that the design wave height is based on--for example, the median maximum wave in a 3-hour storm. Thus, for any static response level, the dynamic response obtained by multiplying the static response by $\text{DAF}(p_e)$ will have the same probability of being exceeded.

For the Gaussian process that results from linear frequency domain analysis, the $\text{DAF}(p_e)$ can be derived analytically. Letting the probability of exceeding some level r at least once in time T be approximated by the upcrossing rate of level r, $\nu^+(r)$, times T; that is,

$$p_e = \Pr\begin{bmatrix}\text{exceeding r at}\\ \text{least once in time T}\end{bmatrix} \approx \nu^+(r)T = \nu_o^+ T \exp(-(r-m)^2/2\sigma^2) \tag{2}$$

Applying equation (2) to the static (subscript s) and dynamic (subscript d) processes and solving for r for a given p_e, equation (1) becomes

$$\text{DAF}(p_e) = \frac{m_d + \sqrt{-\ln\left(\frac{p_e}{\nu_{od}^+ T}\right) 2} \, \sigma_d}{m_s + \sqrt{-\ln\left(\frac{p_e}{\nu_{os}^+ T}\right) 2} \, \sigma_s} \tag{3}$$

where ν_o^+ is the zero upcrossing rate, m is the mean, σ is the standard deviation, and p_e is the probability of exceedance. If the means are zero, then for a range of extreme values of interest, $\text{DAF}(p_e)$ is well approximated by the ratio of the standard deviations of the dynamic to the static response. The use of equation (3) for non-Gaussian responses will be tested with the results of the following simulation example.

EXAMPLE

The following example illustrates the steps in the proposed extreme wave dynamics methodology. The example problem is a 700-foot, template-type, drilling and production platform with a broadside period of 5 seconds and an end-on period of 4 seconds. The damping in the fundamental modes is set to 4.5%. The static design criteria for the site are augmented by a design storm that contains additional detail needed for a dynamic solution. The design storm is specified by a single directional spectrum, a storm duration, and a parallel current velocity profile. The wave spectrum is shown in Figure 1. Only one storm approach angle is considered in this example. The average approach angle was about 26 degrees from true broadside; the directional spreading is frequency dependent and based on hindcast models; the current profile tapers linearly from 4 feet per second at the surface to 0 at -150 feet; and the energy in the wave spectrum including the wind driven, high frequency tail is equivalent to a 23-foot significant wave height.

Time Domain Simulation

The initial step is the organization of an efficient procedure to simulate response to random waves. Some essential features for efficiency are, first, the generation of horizontal wave kinematics time series by the fast fourier transform. Vertical water acceleration and velocity are neglected as being of secondary importance to lateral response.

Second, the areas and volumes of the structural members are lumped together into a few hundred points. Some further approximations to the conventional projected areas and volumes are required for simulated kinematics from a directional wave spectrum. These wave force points are positioned in vertical columns to preserve the 3-D character of the complete model and provide a fine discretization only in the upper part of the structure. Figure 2 illustrates the position of the wave force columns at the waterline in the simplified model of the example platform (Row B is broadside).

Third, a pure modal dynamic solution is used for the dynamic response and sums of the applied forces are the only static responses available. The applied forces computed at each wave force point are distributed to neighboring structural joints. Using the mode shapes obtained from the complete model eigenvalue solution, the modal force time series is constructed. Finally, by keeping all of the data for each 128-second simulation segment in core, the cost of reading and writing data to temporary storage is avoided.

For the example structure, an hour of real time can be simulated in about 20 minutes of computer time. This is over 100 times faster than the random wave, pure modal solution for the same responses using the complete 700-joint model. Although not done here, this time could be cut to 10 minutes by judiciously excluding kinematic segments based on the maximum wave amplitude in the segment. A linear transfer function of the dynamic response may yield a more discriminating tool to reject those segments where the maximum response is below the level of interest. Further work is warranted.

Even with these simplifications for efficiency, accurate values of static and dynamic base shear and overturning moment are obtainable. The simplified model solution was compared to a state-of-the-art, linear elastic, time domain program that uses a static-plus-modal-dynamic technique for stress solution. Because both programs use the same mode shapes, it is most instructive to compare the transfer function between wave amplitude and total applied force. At the maximums of random waves, the total applied force agrees to within 0.5 percent. But as the transfer functions in Figure 3 show, the applied force is overpredicted at some frequencies near the platform period. However, the maximum dynamic base shear in two random wave segments showed the simplified solution to be biased low by 4 and 5 percent.

Results

The simulation was done in 128-second segments and the maximum values were saved in histograms and in ordered lists. For ease in interpretation, these histograms were converted to plots of the median maximum value in some interval T (by simply assuming the 128-second segments are mutually independent). Figure 5 summarizes 21 hours of simulation in which no wave segments were excluded for insufficient height. Figure 5 shows both static and dynamic base shear for the broadside direction. If we select 2 hours for T (the duration of the sea state), one can read from the graph that the median maximum dynamic base shear is 10,500 kips. Taking the ratio with the median maximum static value, the DAF(p_e) is 1.45. From this same simulation, the DAF for broadside overturning moment is 1.33, DAF for end-on base shear is 1.3, and DAF for end-on overturning moment is 1.2.

With these simulation results the balance of the design can be done with a quasi-static, regular wave analysis. Four extra joint load sets can be computed, two broadside and two end-on. The vertical distribution of the joint loads can be proportional to the first and second bending modes in each direction multiplied by the mass at the level of the joint. The joint loads can be scaled so that when applied at the time when the applied wave force reaches its maximum value, the base shear and overturning moment DAF's are obtained.

DISCUSSION

The numerical results from the example highlight some of the reasons for the proposed method. A linearized simulation was also conducted to illustrate the results one would obtain from a frequency domain analysis. Unidirectional waves were simulated from the spectrum in Figure 1 and applied directly broadside to the simplified model. Table 1 summarizes the second moment statistics from the linearized and fully nonlinear wave force simulations. For the linearized case, wave amplitudes were suppressed and the value of a constant C_D was selected iteratively so that the standard deviation of static base shear would equal the nonlinear case. Figure 6 compares the power spectra of the linear and nonlinear models. Even for this simple linearization, the linear and nonlinear power spectrum are about the same.

For the linear model, the presence of current has no effect on the spectrum since the current must be constant. For the nonlinear wave force model, current added to the wave velocity before the absolute value product increases the oscillating force and so increases the energy in the applied force spectrum. The spectrum in Figure 7 compared to Figure 6 shows the increase to be spread over all frequencies. While not done here, there is likely some linearization of the drag force that could reproduce the nonlinear spectrum with current.

The spectra of response can be obtained, either by simulation or in the frequency domain, much more economically than the simulated extremes of response. By assuming the response is Gaussian, extremes can be predicted from the spectra and mean values. The solid lines in Figures 4 and 5 are the simulations of broadside base shear (static and dynamic) for without and with current cases, respectively, of the example. Using the mean, standard deviation, and zero-upcrossing rate obtained during these (nonlinear) simulations (Table 1), the Gaussian solution (equation 2) is also plotted. As expected, the Gaussian solution falls substantially below the simulated solution for the durations shown. (The 4-hour maximums are about a mean plus 4 standard deviation level for the Gaussian processes.)

While the absolute value of a Gaussian solution is greatly in error, the relative values are, at least for this case, more reasonable. On the bottom of Figures 4 and 5 are the ratios of the dynamic to the static response as a function of duration. For the Gaussian solutions the DAF is practically constant with duration T (from equation 3). Note that the Gaussian DAF exceeds the simulated DAF--1.7 to 1.60 for no current and 1.62 to 1.45 for current for the range of durations shown. However, in a variation of the example not reported here (in which a greater proportion of the energy in the spectrum of dynamic response was at the platform period), the Gaussian solution gave DAF = 2.4 while the simulated extremes gave DAF = 1.7. Before a general conclusion of the conservative nature of equation (3) can be drawn, additional parameter variations would need to be done. Parameters of interest would be the significant wave height, the wave energy at the platform period, and the model for wave kinematics in the crest of the waves.

SAFETY FACTORS FOR DYNAMIC RESPONSE

The safety factors in design guidelines and rules account for many items that the designer does not consider explicitly during each design. These items include uncertainties in applied loads (and strength), bias in nominal load values (and strengths), and bias and error in engineering models and analysis methods. Safety factors are intended for use with a specific set of load and strength definitions and design procedures which all together define a safety recipe. If there is dynamic response, the safety recipe appropriate for quasi-static response needs to be re-evaluated. One solution is to use worst-case estimates for dynamic parameters. Another would be to use average (and consistently defined) parameters, but change safety factors for dynamic response. The differences between dynamic and quasi-static response that can lead to changes in the safety recipe fall into two categories.

First, there are new parameters not present in quasi-static response. These include (1) platform period, (2) modal damping, (3) the phasing in time of several dynamic modes with the quasi-static response, and (4) the high frequency tail of the wave spectrum near the platform period. The impact of the uncertainty and bias associated with these parameters will differ depending on whether the dynamic analysis is being done for fatigue or extreme waves.

The second category is differences in the analysis methods. Designers and writers of design guidelines must compare the relative quality of regular wave, quasi-static analysis with random wave, dynamic analysis that will be used in design. This is complicated by the fact that random wave analysis (necessary for dynamic analysis) includes explicitly two uncertainties that are only included implicitly in regular wave analysis. These are the uncertainty of total applied force given a wave height and the uncertainty of maximum member stress given a maximum (dynamic or static) base shear. Such explicit modeling could justify a partial reduction of safety factors (or other component of the safety recipe) for random wave dynamic analysis.

Investigations of probability based design guidelines for offshore structures (e.g., Moses, 1981) provide an impetus to rationalize the safety recipe for fixed platforms that respond dynamically. To develop a dynamic guideline that produces safety consistent with the static guideline, the following format is suggested: (1) eliminate excessively conservative bias in period and damping estimates, (2) use the same probability of exceedance for dynamic as for static response, and (3) augment the load factor on internal structural forces to account for the differences discussed above.

CONCLUSIONS

The use of the probability of exceedance definition for the dynamic amplification factor (DAF) allows several approaches for finding the design DAF for the extreme dynamic response of fixed platforms. The most rigorous approach available is to simulate in the time domain the extreme static and dynamic response. This is a viable approach for design when the simulation is limited to the global responses. Member stresses can be estimated with a quasi-static wave analysis and extra static loads that reproduce the selected DAF for base shear and overturning moment.

REFERENCES

Hackley, M. B., 1979, "Wave Force Simulation Studies in Random, Directional Seas," _Proceedings, 2nd International Conference on the Behavior of Off-Shore Structures_, London, England, 1979, Paper 30, pp. 371-382.

Moses, F., 1981, _Guidelines for Calibrating API RP2A for Reliability-Based Design_. Dallas, Texas: American Petroleum Institute, Production Research Report PRAC-80-22.

Table 1

SELECTED PARAMETERS OF SIMULATED BROADSIDE BASE SHEAR IN RANDOM WAVES

	Static Base Shear			Dynamic Base Shear		
	Mean kips	S.D. kips	Zero Cross sec^{-1}	Mean kips	S.D. kips	Zero Cross sec^{-1}
	Waves Unidirectional, Wave Force Linear					
Current = 0	0	841		0	1427	
Current = 4 fps	514	841		520	1427	
	Waves Unidirectional, Wave Force Nonlinear					
Current = 0	72	844		72	1491	
Current = 4 fps	1021	1209		1032	2253	
	Waves Multidirectional, Wave Force Nonlinear					
Current = 0	50	711	.0905	50	1170	.147
Current = 4 fps	871	998	.0905	881	1696	.149

Note: S.D. = standard deviation.

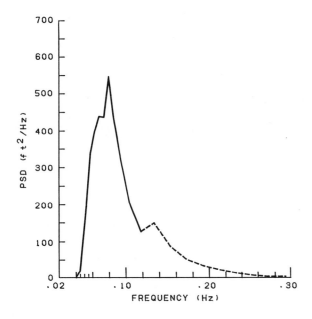

Fig. 1 - Wave Spectrum for Example Sea State

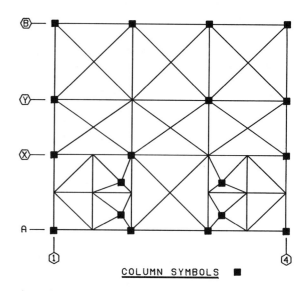

Fig. 2 - Locations of Columns of Lumped Area and Volume for Members at the Waterline of the Example Platform

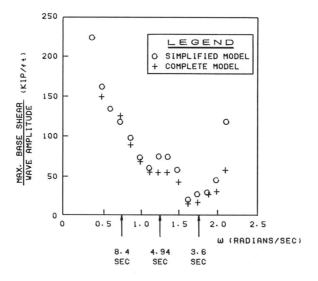

Fig. 3 - Broadside Wave to Static Base Shear Transfer Function by Individual Regular Waves

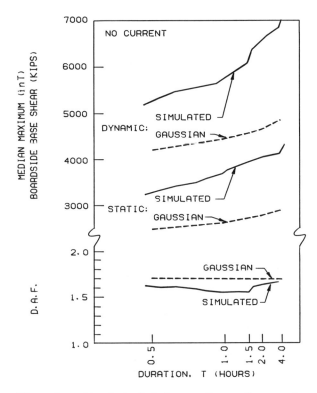

Fig. 4 - Simulated Extreme Base Shears for Example Sea State without Current

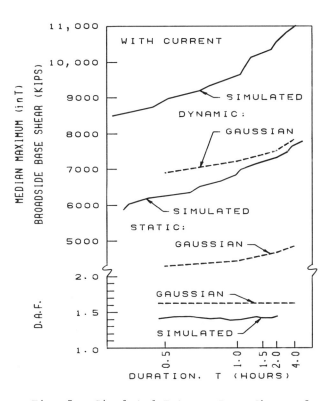

Fig. 5 - Simulated Extreme Base Shears for Example Sea State with Current

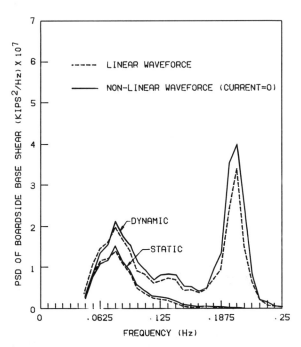

Fig. 6 - PSD of Base Shear for Unidirectional Broadside Waves

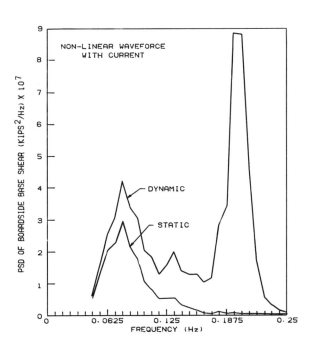

Fig. 7 - PSD of Base Shear for Unidirectional Broadside Waves with Current

THE GENERALIZED MODULAR ANALYSIS IN OFFSHORE ENGINEERING STRUCTURES

Lin Shao-pei   Qiu Chun-hang   Ji Zen
Research Institute of Engineering Mechanics
Dalian Institute of Technology, China

SUMMARY

The analysis of offshore structure is performed by the so-called generalized modular analysis, in which "module" concept is expanded to deal with the structural-composition, loading-composition and the loading combination corresponding to the environmental condition by means of mathematical programming model, besides that, substructuring technique is combined with the group theoretic method to treat various symmetric offshore structures.

An example of reinforced concrete offshore fixed platform is examined to show the availability of this technique to the complicated multi-cylindrical submerged structures.

1. INTRODUCTION

The analysis of offshore engineering structures are characterized not only by the complexity of structural composition but also by the extremely cruelty of environmental condition which causes tremendous combination of loading cases as to make a simplified reasonable safety design is impossible.

Toward this end, the generalized modular analysis of structure is developed on the bases of "structural module" and "loading module" concepts, then the multi-level substructuring technique and multi-level loading assemblagement are used to represent the whole complexity of structure and environmemtal loadings. Furthermore, by introducing the group theory and combining it with the substructuring technique, simplifications have also been made, by which the symmetric offshore structures under arbitrary loads can be treated efficiently.

2. SUBSTRUCTURAL MODULE AND GENERALIZED DISPLACEMENTS

In applying matrix method of analysis to large offshore structures, the number of structural elements very often exceeds the capability of available computer programs and consumes a large amounts of computer times, consequently, some measures of structural partitioning must be employed. Structural partitioning means the division of the complete structure into a number of substructures, the boundaries of which may be specified arbitrarily. However, it is preferable for convenience to make structural partitioning correspond to constructive composition of the structure. It is well known that if the displacements or forces on the boundaries of structures have been found, each substructure can then be analyzed separately under known substructure boundary displacements or forces.

Deffering from that of the above mentioned conventional philosophy of substructuring technique, the substructural module concept is developed for describing all the repeated parts of structure in one pass, it states: The substructural module is a sample of structural model with definite structural configuration, definite geometrical topology and definite boundary nodes (although, the constraints of these nodes can be changed arbitrarily) and its structural stiffness can be varied according to an arbitrary proportional factor even it is a negative value. It is important to point out that the substructural module is just a description of a real structure, it can be realized only if a calling sequence which is executed to this substructuring module by a higher order substructural module located on the "constructive tree" of the structure, has been performed[1]. Therefore, any kinds of structure can be composed by various types of basic elements as well as different kinds of substructural modules hierarchically. As a matter of fact, the whole structure can be regarded as the root of the tree and each branch represents the substructural module being called/or different types of basic elements being assembled, this rule is valid for every level and finally a so-called "constructive tree" of structure is formed (Fig.3). The analysis of structure thus can be divided into a hierarchy of substructural module analysis in different levels. The number of substructural modules and the levels of "constructive tree" are actually unlimited. The referring of substructural modules to form the whole structure according the "constructive tree" is more flexible and less restrained.

Any kinds of element can be installed into a substructural module with weak interface if its displacements can be interpolated by the boundary node displacements. Toward this end, a generalized displacement idea[1] is developed such as to deal with the structural complexity by means of "master-slave relationship" of the node displacements.

Let's denote $\{u_m\}$, the displacement vector of master node M, and $\{u_s\}$, the displacement vector of the slave node S, it is clear

$$\{u_s\} = [T_s][I_d]\{u_m\} + [I_r]\{u_s\} \tag{2.1}$$

where $[T_s]$; rigid-arm transformation matrix

$$[T_s] = \begin{bmatrix} 1 & 0 & 0 & 0 & dz & -dy \\ 0 & 1 & 0 & -dz & 0 & dx \\ 0 & 0 & 1 & dy & -dx & 0 \\ 0 & 0 & 0 & 1 & 0 & 0 \\ 0 & 0 & 0 & 0 & 1 & 0 \\ 0 & 0 & 0 & 0 & 0 & 1 \end{bmatrix} \tag{2.2}$$

in which, $dx = x_s - x_m$; difference between abscissas of nodes M and S along x direction, etc. d and r; integers for the subscript, we have $d+r=6$. At the same time

$$[I_d] + [I_r] = [I] \tag{2.3}$$

where [I]; identity matrix of order 6.

Then, the displacements **of a slave node** depend on those of a master node/or only a subset of it depends upon those of a master node. The subordinate displacements will be excluded from the global displacement vector. Hence we must classify the displacements into five kinds, i.e. the independent one; the dependent one; defined to be zero; defined to be non-zero and that of the free displacement. The six displacements of a node $u, v, w, \theta_x, \theta_y$ and θ_z may be qualified individually according to the needs for the idealisation of structures, and every node has its own qualification number.

As a result of introducing the "master-slave relationship", the number of unknown independent displacements will be decreased considerably. In addition, the ill-condition of the global stiffness matrix is avoided and the accuracy of the analysis can be assured.

With respect to the moving parameters of the substructural modules, three translation parameters u_o, v_o, w_o and three rotational Euler's angles ψ, θ, φ are introduced. By using moving parameters a substructural module can be moved to an arbitrary position in space and installed onto the exact designed position.

3. LOADING MODULE COMBINATION AS A RESULT OF SUBSTRUCTURING

As we know that the loads are acting throughout the whole structure, most of them, in certain cases, are similarily distributed over the similar structures, then a question has been naturally arised: Can the load data necessary for describing the load conditions of a structure be simplified? The answer is positive. Since a substructural module is defined by a series of data representing the structural information physically, at the same time, every load is acted on the structural portion, thus, the load conditions can be naturally described by means of the combination of loading modules.

The loading module concept is developed to describe all the repeated patterns of loading in one pass, it states: The loading module is a sample of load model with different combination of loads acting on a definite substructural module. A loading module is composed by two parts of loadings: one is contributed by the loads directly acting on the nodes of the substructural module itself:

$$M_1, M_2, M_3 \cdots \cdots M_{LDMDL(ISB)}$$

where M; loading module acting on the nodes of substructural module being examined.
LDMDL(ISB); subscript, representing the number of the loading module involved in the investigated substructural module ISB.

the other part of the loads is contributed by the loading modules of lower level substructural modules by a load calling sequence hierarchically, i.e.

$$P_1, P_2, P_3 \cdots \cdots P_{ICASE(ISU)} \cdots \cdots$$

where P; loading module referred from those of the lower level substructural modules.
ICASE(ISU); subscript, representing the number of loading cases of the super-element ISU just called.

$$P(ISB) = \sum_i^{LDMDL(ISB)} \lambda_i M_i + \sum_{j=1}^{ISON(ISB)} \sum_{k=1}^{ICASE(ISU)} \lambda_{j,k} P_k(ISU(j)) \tag{3.1}$$

where ISON(ISB); number of super-element called by the investigated substructural module.
$\lambda_i, \lambda_{j,k}$; factors of the loads for the loading module itself and for those loading modules referred from lower level substructural modules.

Thus, we may represent the loads as the summation of two parts as shown in eq.(3.1).

4. GENERALIZED SYMMETRIC MODULAR ANALYSIS BY GROUP THEORY

In offshore engineering practice, it is frequently encountered that the artificial structures contain a variety symmetric properties. The symmetry of a structure can be described by symmetry group G of the structure. Any moving element $g \in G$ is called symmetric rigid motion if the final position of this structure completely coincide its initial position under an action of such a motion $g \in G$. As an example of offshore structure, Fig.1 shows the symmetry of a multi-cylindrical submerged structure, its symmetry consists of twelve symmetric motions, i.e.

$$G = \{e, c_6^1, c_6^2, c_6^3, c_6^4, c_6^5, \sigma_{aa}, \sigma_{cc}, \sigma_{bb}, \sigma'_{aa}, \sigma'_{cc}, \sigma'_{bb}\} \tag{4.1}$$

where e denotes a zero motion (fixed), $c_6^1, c_6^2, c_6^3, c_6^4$ and c_6^5 are anti-clockwise rotations through angles $\pi/3, 2\pi/3,\ldots$, and $5\pi/3$ about axis OZ, respectively, σ_{aa} is the reflection in the plane a-a etc. Thus, a structure is naturally divided into n symmetric regions (or super elements) in which n denotes the number of moving elements (that is rigid symmetric motions) in symmetry group G of the structure. If we choose any symmetric region as a substructural module, then the whole structure can automatically be assembled by referring this substructuring module through rigid symmetric motions $g \in G$.

Because all the motions which are executed during assembly are only the moving elements g belonging to the symmetry group G, it is possible to take full advantage of the symmetry of the structure. Of course, we should take new generalized displacements $\tilde{\delta}_{\ell \mu m}$ as independent unknown variables instead of original displacements δ_k (k=1,2,...K). The relationship between them is given by

$$\{\tilde{\delta}\} = [U]^{t*} \{\delta\} \tag{4.2}$$

$$\{\delta\} = [U] \{\tilde{\delta}\} \tag{4.3}$$

where $\{\tilde{\delta}\}$ and $\{\delta\}$ denote new and original generalized displacement sub-vector consisting of $\tilde{\delta}_{\ell \mu m}$ ($\ell=1,2,\ldots c$, $\mu=1,2,\ldots,\lambda_\ell, m=1,2,\ldots,d_\ell$) and δ_k (k=1,2,...K), respectively. [U] is an unitary matrix, $[U]^{t*}$ is conjugate transpose of matrix [U] The elements of [U] can be determined from

$$\sum_{d=1}^{d_\ell} [U]_{k,\ell u d} [\tau_\ell(g)] = \sum_{k'} [T(g)]_{k,k'} [U]_{k',\ell \mu m} \quad \text{(for all } g \in G\text{)} \tag{4.4}$$

in which [T] is a linear representation of symmetry group G, τ_ℓ is the ℓ-th irreducible representation of group G, [T(g)] and [$\tau_\ell(g)$] are the representation matrices of T and τ_ℓ for the $g \in G$, respectively.

As an example of offshore structure shown in Fig.1, the symmetry group of which is C_{6v}, so that there are six irreducible representations $\tau_1, \tau_2, \ldots, \tau_6$. The first four of them are one dimensional and the two others are two dimensional. Corresponding to the six displacements $u_1, u_2, \ldots, u_6$ (or six θ_x's etc.) of symmetric nodes on $\theta = 0, \pi/3, \ldots, 5\pi/3$, the representation matrices of T can be presented as follows

$$[T(e)] = \begin{bmatrix} 1 & 0 & 0 & 0 & 0 & 0 \\ 0 & 1 & 0 & 0 & 0 & 0 \\ 0 & 0 & 1 & 0 & 0 & 0 \\ 0 & 0 & 0 & 1 & 0 & 0 \\ 0 & 0 & 0 & 0 & 1 & 0 \\ 0 & 0 & 0 & 0 & 0 & 1 \end{bmatrix}, \quad [T(C_6^1)] = \begin{bmatrix} 0 & 1 & 0 & 0 & 0 & 0 \\ 0 & 0 & 1 & 0 & 0 & 0 \\ 0 & 0 & 0 & 1 & 0 & 0 \\ 0 & 0 & 0 & 0 & 1 & 0 \\ 0 & 0 & 0 & 0 & 0 & 1 \\ 1 & 0 & 0 & 0 & 0 & 0 \end{bmatrix}, \ldots, [T(\sigma'_{bb})] = \begin{bmatrix} 0 & 0 & 0 & 0 & 0 & 1 \\ 0 & 0 & 0 & 0 & 1 & 0 \\ 0 & 0 & 0 & 1 & 0 & 0 \\ 0 & 0 & 1 & 0 & 0 & 0 \\ 0 & 1 & 0 & 0 & 0 & 0 \\ 1 & 0 & 0 & 0 & 0 & 0 \end{bmatrix} \tag{4.5}$$

From eq.(4.4) we have

$$[U] = \frac{1}{\sqrt{3}} \begin{bmatrix} 1/\sqrt{2} & 1/\sqrt{2} & 1 & 0 & 1 & 0 \\ 1/\sqrt{2} & -1/\sqrt{2} & \cos 2\pi/3 & \sin 2\pi/3 & \cos \pi/3 & \sin \pi/3 \\ 1/\sqrt{2} & 1/\sqrt{2} & \cos 4\pi/3 & \sin 4\pi/3 & \cos 2\pi/3 & \sin 2\pi/3 \\ 1/\sqrt{2} & -1/\sqrt{2} & \cos 6\pi/3 & \sin 6\pi/3 & \cos \pi & \sin \pi \\ 1/\sqrt{2} & 1/\sqrt{2} & \cos 8\pi/3 & \sin 8\pi/3 & \cos 4\pi/3 & \sin 4\pi/3 \\ 1/\sqrt{2} & -1/\sqrt{2} & \cos 10\pi/3 & \sin 10\pi/3 & \cos 5\pi/3 & \sin 5\pi/3 \end{bmatrix} \tag{4.6}$$

Substituting this relation into eq.(4.2) we obtain the expression of new six generalized displacements:

$$\begin{aligned}
\tilde{u}_{111} &= \frac{1}{\sqrt{6}} \sum_{k=1}^{6} u_k \\
\tilde{u}_{211} &= \frac{1}{\sqrt{6}} (u_1 - u_2 + u_3 - u_4 + u_5 - u_6) \\
\tilde{u}_{511} &= \frac{1}{\sqrt{3}} (u_1 + u_2 \cos\frac{2\pi}{3} + u_3 \cos\frac{4\pi}{3} + u_4 + u_5 \cos\frac{8\pi}{3} + u_6 \cos\frac{10\pi}{3}) \\
\tilde{u}_{512} &= \frac{1}{\sqrt{3}} (u_2 \sin\frac{2\pi}{3} + u_3 \sin\frac{4\pi}{3} + u_5 \sin\frac{8\pi}{3} + u_6 \sin\frac{10\pi}{3}) \\
\tilde{u}_{611} &= \frac{1}{\sqrt{3}} (u_1 + u_2 \cos\frac{\pi}{3} + u_3 \cos\frac{2\pi}{3} - u_4 + u_5 \cos\frac{4\pi}{3} + u_6 \cos\frac{5\pi}{3}) \\
\tilde{u}_{612} &= \frac{1}{\sqrt{3}} (u_2 \sin\frac{\pi}{3} + u_3 \sin\frac{2\pi}{3} + u_5 \sin\frac{4\pi}{3} + u_6 \sin\frac{5\pi}{3})
\end{aligned} \tag{4.7}$$

Let $[\tilde{R}], [\tilde{r}]$ be the global stiffness matrix and sub-matrix corresponding to new generalized vector $\{\tilde{\Delta}\}$ and sub-vector $\{\tilde{\delta}\}$, respectively, then the total displacement deformation energy is given by

$$\frac{1}{2} [\tilde{\Delta}]^{t*} [\tilde{R}][\tilde{\Delta}] = \frac{1}{2} \sum \{\tilde{\delta}\}^{t*} [\tilde{r}] \{\tilde{\delta}\} = \frac{1}{2} \sum \{\delta\}^{t*} [U][\tilde{r}][U]^{t*} \{\delta\} \tag{4.8}$$

It is clear that under the action of any linear symmetry transformation $[T(g)]$, there will be an invariant for energy expression and then from Schur's lemma of group theory, we have

$$[\tilde{R}]_{\ell\mu m,\ell'\mu'm'} = 0 \text{ for all } \ell \neq \ell' \text{ or } m \neq m' \tag{4.9}$$

in which sets of values of ℓ, μ, m and ℓ', μ', m' denote a row and a column number in matrix $[\tilde{R}]$, respectively. From eq.(4.9) $[\tilde{R}]$ is expressed in the block diagonalized form, i.e., the problem has been divided into a series of sub-problems.

It is worth to notice that the whole assemblage processes can be completed by the application of group theory and this algorithm only deals with boundary independent displacements of the substructural module (the interior displacements have been eliminated by substructuring treatment).

5. LOADING PROGRAMMING TECHNIQUE

As the loading module concept is a result consequent to the substructuring technique and the combination of loading modules as well as the substructural modules are hierachically independent with each other, then a load linear programming model of structural analysis can be introduced to seek the maximum internal forces or responses of offshore structures under environmental conditions.

Instead of the traditional idea of specified load cases combination, we suggest a linear programming model of loads which is characterized by introducing the so called "loading module space" L_m and "internal force space" F_m.

A loading module space L_m represents the whole possible collections of actual load combinations and is constructed by the existing loading modules of the structure (or substructural module) as its basic vectors $\{M_i\}$; we have

$$\{L_c\} = \alpha_1\{M_1\} + \alpha_2\{M_2\} + \cdots + \alpha_n\{M_n\} = \sum_i^n \alpha_i \{M_i\} \qquad (n=1,2,\ldots \text{LDMDL}) \tag{5.1}$$

the actual load case $\{L_c\}$ is just a vector in the loading module space L_m, and from (5.1) we recognize $\{L_c\}$ is a linear combination of LDMDL loading modules, $\{L_c\} \in L_m$. Similarily, an internal force module space can be understood as a projective space of L_m, we have

$$\{F_c\} = \beta_1\{N_1\} + \beta_2\{N_2\} + \cdots + \beta_n\{N_m\} = \sum_j^m \beta_j \{N_j\} \qquad (m=1,2,\ldots \text{LDMDL}) \tag{5.2}$$

the actual internal force response of the structure is just a vector in the internal force space F_m, and from eq.(5.2) it is shown that $\{F_c\}$ is a linear combination of LLMDL independent internal force modules, $\{F_c\} \in F_m$.

The load linear programming model can be illustrated by denoting the maximum internal forces as the objective function which is varied by design variables x——factors of loading module combination under constraint equations in load formation. Mathematically, the objective function F_{max} and the constraint equations are linear to the design variables x, i.e.

$$F_{max}(x) = (N,X) = N^T X \tag{5.3}$$

where N and X are the internal force response vector and design variable (factors of the load combination) vector respectively. $N, X \in E^n$. The constraint equations are

$$A X \leq B \qquad X \geq 0 \tag{5.4}$$

where $B \in E^m$ and A is a m×n constraint matrix. The set of feasible solution is defined by

$$M = \{x \mid x \in E^n, Ax \leq B, x \geq 0\} \tag{5.5}$$

and the load linear programming problem consists of finding an extreme value of $F_{max}(x)$ on M, herein, E^n is just the load module space L_m in LDMDL dimensions. The programming procedure proceeds for each "concerned internal force", thus, a series of maximum internal forces can be found from (5.5) alternatively.

6. PRACTICAL EXAMPLE

Consider an offshore platform with multi-cylindrical submerged tanks shown in Fig.1, the structural discretization by finite element approach gives the total number of nodes and displacements are equal to 1140 and 4704, respectively (in which the number of independent displacements is equal to 3984). As the **investigated structure** posses C_{6v} group symmetry, then advantages can be taken by using group theoretic method combined with substructuring technique.

Fig.2 shows the different substructural modules for partitioning and representing the original structure, evidently, there are six substructural modules used in the analysis. It is mentioned that in order to reserve group symmetry of the structure, the upper platform of investigated structure is modelized as a symmetric hexagonal plate in the analysis, for one will emphasize and focus their attentions on structural complexity at the submerged portions.

Fig.3 illustrates the "constructive tree" of this structure, which shows the compositions of the structure by multi-level assemblage of substructural modules. Detail information of generalized modular analysis of this offshore platform can be found in Table 1.

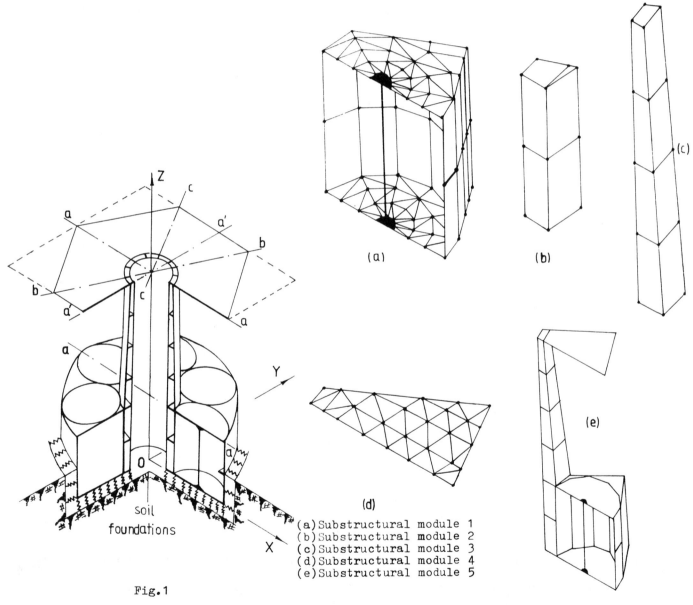

(a) Substructural module 1
(b) Substructural module 2
(c) Substructural module 3
(d) Substructural module 4
(e) Substructural module 5

Fig.1

Fig.2

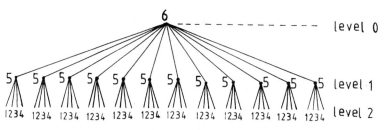

Fig.3

Table 1

NO. of substru. module	Number of basic elements	Number of super elements	Number of loading modules	Number of load cases	Number of boundary indep. unknown displacements	Number of interior unknown displacements	The method of assemblage
1	138 5 types	-	30	2	113	154	direct input
2	18 3 types	-	-	-	45	0	direct input
3	21	-	10	4	60	0	direct input
4	35	-	2	3	60	60	direct input
5	-	4	-	5	236	0	Assemblage by moving parameters $u_o, v_o, w_o, \psi, \theta$ and φ.
6	-	12	-	2	0	1416 each sub-problem: 147; 89; 118; 118; 236; 236; 236; 236.	Assemblage by group theory automatically The whole pro--blem is reduced in eight sub-problems

REFERENCES

1 ZHONG Wan-xie, 1977, "A General Purpose Program for Structural Analysis--JIGFEX"," Journal of Dalian Institute of Technology, NO.3, pp.19-42, NO.4, pp.14-35.

2 ZHONG Wan-xie and LIN shao-pei, 1978, " Functions and Application of JIGFEX Program,"Dalian Institute of Technology, Research Institute of Engineering Mechanics, Report NO.78-116.

3 LIN Shao-pei, 1980, "Load Linear Programming Model for Structural Analysis,"Journal of Dalian Institute of Technology, NO.4,pp.19-25.

4 QIU Chun-hang, 1980,"The Analysis of Generalized Symmetric Structures Consisting of Multi-Type Finite Elements", Journal of Dalian Institute of Technology, NO.4,pp.27-44.

NUMERICAL SIMULATION FOR THE COLLAPSE ANALYSIS
OF LARGE STIFFENED STEEL ASSEMBLAGES

Dr. R.S. Puthli   Ir. F.S.K. Bijlaard   Ir. H.G.A. Stol

Institute for Building Materials and Building Structures (IBBC)
of the Netherlands Organization for Applied Scientific Research (TNO), The Netherlands

SUMMARY

The paper describes some approximations and assumptions used to simulate the combined geometric and material non-linear behaviour of large stiffened steel plated structures up to collapse. These methods are used in a finite element computer program CASPA developed for the collapse analysis of three dimensional structures assembled with eccentrically stiffened plates. To elucidate the capabilities of this approach, an experimental study on a complex, large scale box girder with an intermediate diaphragm is simulated with the finite element method. These results were made available by British researchers (Einarsson and Dowling, 1979; Crisfield, 1980). Comparisons between the results of tests and the computer analysis are illustrated.

INTRODUCTION

Experimental work for investigating the non-linear behaviour of realistically sized stiffened plates up to collapse is prohibitively expensive, so that studies are limited in number, if at all practicable. Large, general purpose computer programs are cheaper than experimental work, but tend to be very expensive for parametric studies. The authors present a special purpose numerical tool that is less expensive for parametric studies than general purpose programs, and show that it is general enough to give acceptable accuracy in representing the behaviour of complex stiffened plates.

Thin, stiffened plate elements collapse with a complex interaction between buckling and plastic flow. Consequently, an accurate numerical simulation requires a complex description of the stiffness matrix that includes all the geometrical non-linear effects (Puthli, 1981) as well as division of the plates into several layers through the plate depth when considering plasticity. For modelling plasticity up to eleven integration stations have been used through the depth (Marcal, 1970). A complete stress history has to be stored at all these numerous integration stations throughout each plate layer. Studies carried out using these techniques give highly accurate results. However, the heavy penalty on computer storage and time restricts the analysis to small areas of a structure, especially when parametric studies are considered (von Sättele, et al, 1981). For these reasons, the approach to collapse analysis of structures so far has been largely on isolated individual panels or plates with idealized boundary conditions, each component being analyzed separately with various levels of sophistication.

The approach described in this paper is to include large sections of the structure, or even the complete structure. The use of artificial boundaries in the regions close to the area of interest and simplified boundary conditions do not therefore require to be given. Some of the ideas used in this method of approach have been discussed earlier (Crisfield and Puthli, 1978). In order to consider such comparatively large models, only the important contributions have been selected in the non-linear finite element formulation used. The theory is fully described elsewhere (Puthli, 1980) and therefore only the relevant details are given here. The only drawback in this approach is a relatively small but acceptable loss of accuracy in the behaviour of the real structure.

The present theoretical model is applicable to isotropic materials such as structures made of hot rolled structural steel plates and structural sections, possessing a pronounced yield point, followed by a significant yield plateau. It may be pointed out that the assumptions and approximations in the model are by no means the optimum relationship between accuracy of representation, computer storage and time. The authors merely demonstrate the use of such ideas that can nonetheless represent the collapse behaviour of real stiffened steel structural assemblages to an acceptable accuracy.

BASIC ASSUMPTIONS

- Thin plate theory is used, where normals to the mid-surface of the plate and stiffener remain normal and do not change in length after deformation. Therefore transverse shear deformation and all strain components normal to the mid-surface are ignored.

- A Total (or Initial) Lagrangian formulation is assumed, where the strains are referenced to the original undeformed coordinates.

- The large deflection effects are assumed to apply only to the strains and not to the curvatures, restricting the geometrical non-linear effects to moderately large deflections.

- The material non-linear behaviour is restricted to a linear-elastic, perfectly plastic stress-strain relationship. Strain hardening is ignored and yield is defined by an approximate von Mises based criterion discussed in the following section.

- Geometric and material non-linearities are ignored for torsional effects in the stiffener, so that torsional instability (tripping) of the stiffener is not modelled.

APPROXIMATIONS

1) For the membrane behaviour of the plate, the simplest first order rectangular element is chosen, based only on displacements at the corner nodes (Crisfield, 1973, Zienkiewicz, 1971). Bilinear in-plane functions are used. For the stiffener, an in-plane displacement function coinciding with that of the plate elements at the sides is chosen (Puthli, 1980). The flexural behaviour of the plate is represented by a displacement function (Zienkiewicz and Cheung, 1968) which is the well known non-conforming, restricted quartic polynominal for a quadrilateral. The stiffener behaviour in bending is given by a one-dimensional version of the displacement function for the plate (Puthli, 1980).

2) Reduced integration techniques are used selectively on sections of the stiffness matrix, (Pawsey and Clough, 1971; Crisfield and Puthli, 1978), which not only reduce time in forming the matrices, but also improve the accuracy when self straining is caused by spurious shear energy terms, producing an over stiff element (or locking).

3) The approximate yield criterion for the plate is based upon a single layer assumption, so that the variables are six stress resultants (N_x, N_y, N_{xy}, M_x, M_y, M_{xy}) in the plate depth rather than three stresses (σ_x, σ_y, σ_{xy}) in various layers of the plate depth and instantaneous plastification of the plate section is assumed. The approximate yield function is defined as a function of the form:

$$F_i(Q_t, Q_m, Q_{tm}) = 0$$

where: $Q_t = \dfrac{1}{t^2 \sigma_o^2}(N_x^2 + N_y^2 - N_x N_y + 3N_{xy}^2)$, $\qquad Q_m = \dfrac{1}{t^2 \sigma_o^2}(M_x^2 + M_y^2 - M_x M_y + 3M_{xy}^2)$

$$Q_{tm} = \dfrac{4}{t^3 \sigma_o^2}(M_x N_x + M_y N_y - \tfrac{1}{2} M_x N_y - \tfrac{1}{2} M_y N_x + 3M_{xy} N_{xy}),$$

σ_o = uniaxial yield stress, $\quad t$ = plate thickness

Robinson (Robinson, 1971) reviewed the various linear approximations to Ilyushin's exact full section yield criterion (Ilyushin, 1956).

$F_1 : Q_t + Q_m = 1,$ $\qquad\qquad F_2 : Q_t + \sqrt{Q_m} = 1,$

$F_3 : Q_t + Q_m + \dfrac{1}{\sqrt{3}}|Q_{tm}| = 1,$ $\qquad\qquad F_4 : Q_t + Q_m + 2|Q_{tm}| = 1,$

If P_i denotes the limit load of yield surface F_i and P_o the limit load for the exact yield surface, Robinson (Robinson, 1971) gave the following bounds:

$$0.995\ P_o \leq P_1 \leq 1.155\ P_o, \qquad\qquad 0.833\ P_o \leq P_2 \leq P_o$$
$$0.939\ P_o \leq P_3 \leq 1.034\ P_o, \qquad\qquad 0.8\ P_o \leq P_4 \leq P_o$$

Robinson also reviewed two higher order approximations due to Ivanov (Ivanov, 1967), where:

$$F_5 : Q_t + \tfrac{1}{2} Q_m + \sqrt{(\tfrac{1}{4} Q_m^2 + Q_{tm}^2)} = 1,$$

$$F_6 : Q_t + \tfrac{1}{2} Q_m - \tfrac{1}{4} \frac{(Q_t Q_m - Q_{tm}^2)}{Q_t + 0.48\ Q_m} + \sqrt{(\tfrac{1}{4} Q_m^2 + Q_{tm}^2)} = 1$$

The bounds for these are:

$$0.955\ P_o \leq P_5 \leq P_o, \qquad\qquad 0.999\ P_o \leq P_6 \leq 1.005\ P_o$$

A modification of Ivanov's higher order yield function F_6 is used by Crisfield (Crisfield, 1979). For the present work, the linear approximation F_3 is used since it is the closest linear optimum. It may be pointed out, however, that whichever of the approximate yield functions is used in the formulation, the accuracy of the results is not greatly affected. The drawback is that the loss in plate stiffness between first fibre yield and full section yield is not represented, giving a slightly stiffer structure in the plastic regions, until full section (approximate) yield is achieved. The collapse loads are within acceptable accuracy even when using surface F_1 (Martins and Owen, 1981). Whichever full section yield surface is chosen, large savings are made in both computer storage and time.

4) The approximate yield criterion is not used for the stiffener elements, since the depth of the stiffener is much larger than the plate thickness. The loss of accuracy is therefore much larger and sometimes unacceptable (Little, 1976). A layered approach has to be used so that the stress history in each layer is stored. This would be the correct approach so that a flow rule could be used with an incremental stress strain relationship. Unloading of the stiffener from the yield surface is then accounted for. When the analysis described in this paper was carried out (Puthli, et.al., 1981), this was not possible, but has now been incorporated into the formulation. The disadvantage of the layered solution for the stiffener is not as severe as for the plate, because only one dimension and only one stress, σ_x, is involved.

For the stiffened box girder analysis, deformation theory associated with total strains was used (Puthli, 1980). Some savings are made in storage, because fewer variables require to be stored than with the layered approach. However, any elastic unloading from the yield surface is not represented and could lead to inaccurate results. This, however, did not affect the present analysis significantly.

5) The structure is subdivided into various regions. Normally, the region close to the area of interest or high stresses is idealised by many more finite elements than in regions removed from this area. Such regions (region A in Figs. 1, 2 and 3) exhibit combined material and geometric non-linearity, requiring a fine mesh of finite elements to model the complex behaviour accurately. Regions further removed (region B in figs. 1, 2 and 3) have a coarser finite element mesh, dependent upon the shape of the geometric non-linearity expected. Even further away (region C in figs. 1, 2 and 3) the behaviour of the structure is predominantly

linear-elastic. Classical deep beam theory may therefore be used to derive a "special beam element" (Puthli, 1980) which is compatible with the finite element in-plane displacements at the common boundary between regions B and C and has only two degrees of freedom representing downward deflection and a rotation at the other extremity of region C. The whole region C is thus modelled with one such "special beam element".

NUMERICAL COMPARISON WITH EXPERIMENT (EINARSSON AND DOWLING, 1979) USING COMPUTER PROGRAM CASPA ON A LARGE SCALE BOX GIRDER

In order to test the accuracy and capabilities of the formulation in computer program CASPA, apart from simple tests reported elsewhere (Puthli, 1980), a complex structure was sought that could test all the assumptions and approximatons used and also provide a calibration on the accuracy of the results. The above mentioned test model has components in compression, in-plane bending, shear and localized eccentric loading, Furthermore, it represents an assemblage typical to offshore deck structures.

A few idealizations were made in the analysis that had a slight bearing on the results. – First, residual stresses were ignored, although they were high between and close to the stub stiffeners (Fig. 1).
- Secondly, at the eccentric bearing supporting the diaphragm, no transverse restraint was assumed.
- Thirdly, for convenience in modelling, positions of the eccentric flange stiffeners were adjusted transversely so that they coincided with the the stub stiffeners without leaving the designed gap of 40 mm (see Fig. 1).

Initial imperfections (measured out-of- plane displacements) were accounted for (Puthli, et al, 1981). The growth of imperfection at and between the eccentrically and concentrically welded stiffeners that are depicted with thicker lines, can be observed in the panels of the box girder at collapse (see Fig. 8).

Fig. 4 compares the relationship between total applied load on the diaphragm bearings and the indentation deflection of the diaphragm. Indentation deflection may be defined as the difference in in-plane displacement between the bearing-diaphragm interface and other identification points within the diaphragm. The stiffness is accurately simulated up to about 1100 kN, after which the numerical model exhibits a stiffer behaviour, largely because of ignoring residual stresses, but partly also because of the approximate yield criterion. For these two reasons, the collapse load is also therefore somewhat higher (2850 kN) than the experimental observation (2620 kN), a difference of less than 9 %. The experimental values for stresses are based upon 'nominal' stresses (strain ϵ multiplied by modulus of elasticity E), as if the material remains elastic, whereas analytical stresses are based upon Ilyushin yield. Also, the high initial residual stresses in the diaphragm were ignored in the analysis. Therefore, some discrepancies are observed in the stress comparisons at collapse load (see Fig. 5a and 5b). Because of frictional resistance between the bearing and box girder, the transverse flange stresses near the bearing are quite different in the analysis, because no lateral restraint was assumed (Fig. 5c). Away from the bearing where the three idealizations mentioned in the analysis (see above) have little influence, the results were good (Fig. 5d and 5e). Fig. 6 shows good agreement in the axial stresses in the load bearing stiffener (see Fig. 1). The analytical bending stresses are a little smaller than those obtained experimentally, because of the third mentioned idealization (see previous paragraph) where the flange stiffener provides a higher restraint to the diaphragm stiffeners. Probably because of the influence of residual stresses, the comparison of yield patterns predicted from total strains at collapse of both models (Fig. 7) are not conclusive, although the tendency towards a yield mechanism is the same. Fig. 8 gives an exaggerated computer plot of computed deformations at collapse.

CONCLUDING REMARKS

It has been shown that even for complex steel plated structures, the use of assumptions and approximations gives answers that are acceptable to engineering accuracy. Such a tool can therefore be used for two purposes. First for conducting parametric studies on stiffened panels with only a few confirmatory experimental verifications. Secondly, for analysing the collapse behaviour of individual and specific structures, such as discussed in the present paper.

Apart from the assumptions and approximations in the structural behaviour, improvements to the solution procedure also reduce analysis time. Recently, since the first successful attempt (Riks, 1979) at faster convergence methods in iterative processes, some work has been reported on fast iterative methods especially describing the post-limit point response of structures (Crisfield, 1981; Ramm, 1981) where much computer time may be saved with a better "arc length control" convergence criterion. Such methods, together with automatic incrementations, would save computer time and speed up analysis precedures. Also, post limit point response is necessary in deciding the stability of post collapse behaviour. Such a method is therefore currently being tested with computer program CASPA, and first results are very promising. As a postcript to improvements in analysis procedures, approximate analysis techniques (Noor and Peters, 1981) are also being investigated to reduce computational effort, where the finite element equations are reduced to a small system of algebraic equations. This artifice is only possible with linear stress-strain relationships and therefore not general enough to merit consideration for the analysis of offshore steel structures.

Since carrying out the above analysis, CASPA has been extended to allow unloading of the stiffeners from the yield surface and including residual stresses in the plate and stiffeners. When the arc length control and automatic choice of incremental steps is fully operational, work will commence on a series of parameter studies on eccentrically stiffened simply supported panels under uniaxial compression. They will be used to extend the Dutch steel design rules and include offshore structures.

ACKNOWLEDGEMENTS

The authors wish to express their thanks to Mr. B. Einarssen and Professor P.J. Dowling of Imperial College, London, and Dr. M.A. Crisfield of T.R.R.L., U.K., for providing the reports on their box girder tests and for useful discussions. Special thanks are due to Mr. B. Einarsson for kindly supplying further data for comparison that was not available in their report. The work presented here forms part of the Marine Technology Reseach Program sponsored by the Dutch government and steel industry.

REFERENCES

CRISFIELD, M.A., 1973, Large deflection elasto-plastic buckling analysis of plates using finite elements, Transport and Road Research Laboratory, Crowthorne, England, Report No. LR 593

CRISFIELD, M.A. and R.S. PUTHLI, 1978, "Approximations in the non-linear analysis of thin plated structures, "Proceedings, International Conference on Finite Elements in Non linear Solid and Structural Mechanics, Geilo, Norway, 1977, Volume 1, pp 373-392, Trondheim, Norway: Tapir Publishers.

CRISFIELD, M.A., 1979, Ivanov's yield criterion for thin plates and shells using finite elements, Transport and Road Research Laboratory, Crowthorne, England, Report No. LR 919.

CRISFIELD, M.A., 1980, Theoretical and experimental behaviour of lightly stiffened box-girder diaphragms, Transport and Road Research Laboratory, Crowthorne, England, Report No. LR 961.

CRISFIELD, M.A., 1981, "A fast incremental/iterative solution procedure that handles 'snap-through'", Computers and Structures, Volume 13, pp 55-62.

EINARSSON, B. and P.J. DOWLING, 1979, Steel Box Girders - Tests on simply stiffened rectangular diaphragms - model 1, Engineering Structures Laboratories, Imperial College, London, U.K., CESLIC Report BG 54.

IL'YUSHIN, A.A., 1956, Plasticity (in Russian), Moscow, 1948: Gostekhizdat and Plasticité (in French), Paris: Editions Eyrolles.

IVANOV, G.V., 1967, "Approximating the final relationship between the forces and moment of shells under the Mises plasticity condition". Inzhenernyi Zhurnal Mekhanika Tverdogo Tela, Volume 2, Number 6, pp 74-75.

LITTLE, G.H., 1976, "Discussion on full-range analysis of steel plates and stiffened plating under uniaxial compression by M.A. Crisfield "Proceedings of the Institution of Civil Engineers, Volume 61, Part 2, June, pp 445-452.

MARCAL, P.V., 1970, "Large defelction analysis of elastic-plastic shells of revolution" Journal of the American Institution of Aeronautics and Astronautics, Volume 8, pp 1627-1634.

MARTINS, R.A.F. and D.R.J. OWEN, "Elastioplastic and geometrically nonlinear thin shell analysis by the semiloof element", Computers and Structures, Volume 13, pp 505-513.

NOOR, A.K. and J.M. PETERS, 1981, "Tracing post-limit-point paths with reduced basis technique", Computer Methods in Applied Mechanics and Engineering, Volume 28, pp 217-240.

PAWSEY, S.F. and R.W. CLOUGH, 1971, "Improved numerical integration of thick shell finite elements" International Journal for Numerical Methods in Engineering, Volume 3, pp 575-586.

PUTHLI, R.S., 1980, "Collapse analysis of three dimensional assemblages of eccentrically stiffened hot rolled steel plates and shallow shells", HERON volume 25, Number 2.

PUTHLI, R.S., 1981, "Geometrical non-linearity in collapse analysis of thick shells, with application to tubular steel joints", HERON, Volume 26, Number 2.

PUTHLI, R.S., F.S.K. BIJLAARD and H.G.A. STOL, 1981, Comparison of computer program CASPA with large scale test results on stiffened steel plated assemblages - experimental results from Imperial College, London, on rectangular box-girders with an intermediate diaphragm, Institute TNO for Building Materials and Building Structures, Rijswijk, The Netherlands, Report No. B-81-88/63.6.0678.

RAMM. E., 1981, "Strategies for tracing non-linear response near limit-points", pp 63-89, Non linear Finite Element Analysis in Structural Mechanics, Editors - Wunderlich, W., Stein, E. and Bathe, K.J., New York: Springer

RIKS, E., 1979, "An incremental approach to the solution of snapping and buckling problems", International Journal for Solids and Structures, Volume 6, pp 529-551.

ROBINSON, M., 1971, "A comparison of yield surfaces for thin shells" International Journal of Mechanical Sciences," Volume 13, Number 4, pp 345-354.

VON SÄTTELE, J.H., E. RAMM and M. FISCHER, 1981, "Traglastkurven einachsig gedrückter Rechteckplatten mit Seitenverhältnissen $\alpha \leq 1$ bei vorgegebenen geometrischen Imperfectionen." Der Stahlbau, Volume 7, pp 205-210.

ZIENKIEWICZ, O.C. and Y.K. CHEUNG, 1968, "The finite element method for analysis of elastic isotropic and orthotropic slabs". Proceedings of the Institution of Civil Engineers, Volume 28, Paper No. 6726, August, pp 471-488.

ZIENKIEWICZ, O.C., 1971, The Finite Element Method in Engineering Science, London: Mc. Graw-Hill.

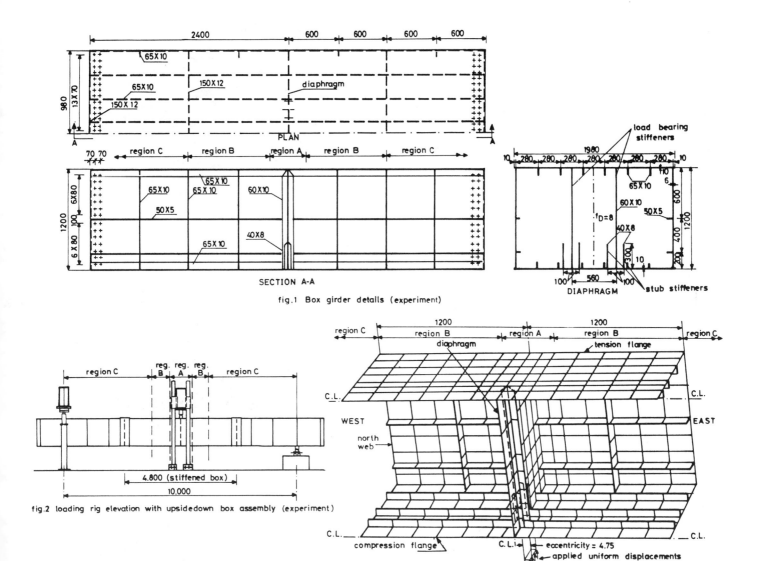

fig.1 Box girder details (experiment)

fig.2 loading rig elevation with upsidedown box assembly (experiment)

fig.3 finite element mesh idealization for the regions A and B (half box analyzed using symmetry)

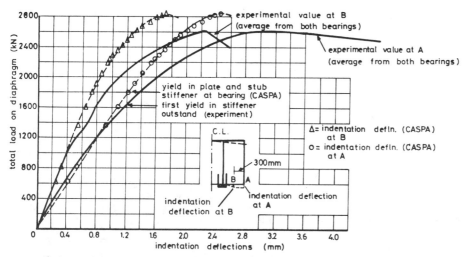

fig.4 relationship between load and indentation deflection

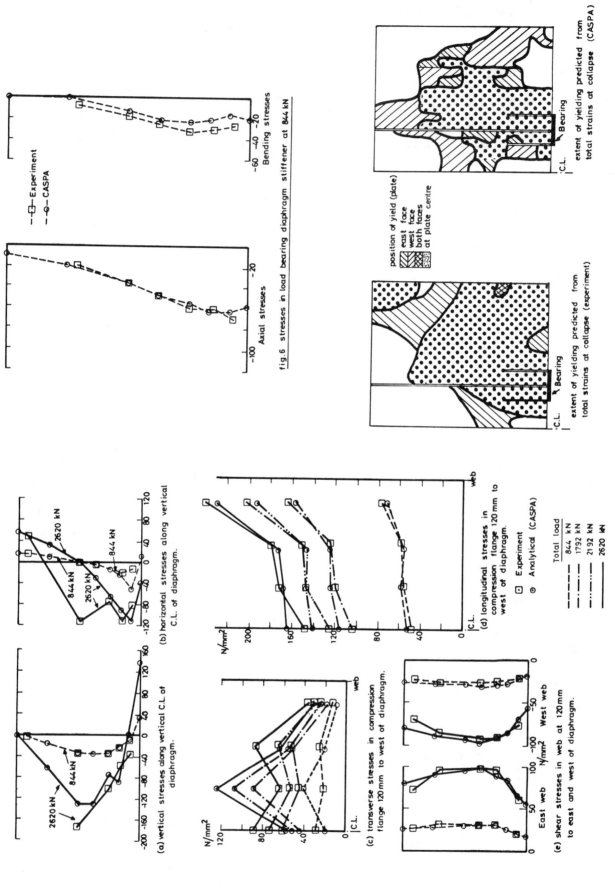

fig.6 stresses in load bearing diaphragm stiffener at 844 kN

fig.7 comparison of yield at collapse (experiment and analysis)

fig.5 some stress comparisons in the diaphragm, flange and web of the box girder.

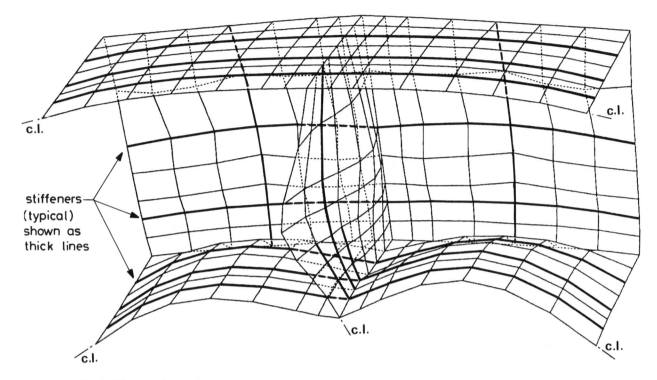

fig. 8. Three dimensional computer plot of total deformations at collapse.
(all displacements are shown to a scale of 30 times the box dimensions)

THE HYDROELASTIC RESPONSE OF THE SUBMERGED STRUCTURES
WITH ARBITRARY LOCATED MULTI-CYLINDRICAL PILES OF DIF-
FERENT DIAMETERS, SUBJECTED TO EARTHQUAKE AND VIBRATION

Qu Nai-si
Research Institute of Engineering Mechanics
Dalian Institute of Technology

Abstract

Owing to the urgent needs of offshore development in China, the earthquake and vibration analysis of submerged structures with arbitrary located multi-cylindrical piles of different diameters are studied on the base of reference 1 . The main points of this paper in comparison with that of the ref. 1 are as follow:
(1) Revision has been made in the determination of the initial frequencies and modes and considerations in three dimensional effects have been adopted in calculation;
(2) The iteration approach for determining hydrodynamic pressure on multi-cylindrical pile structures is replaced by an effective method in which the actual vibrating modes with water are reduced by the combination of free vibrating modes without water. Then, the frequencies and the modes can be obtained in one pass and the hydrodynamic pressure of multi-cylindrical piles can easily be determined.

The (1) and (2) are the key points of abovementioned statements. It is well known that sufficient number of frequencies and modes is necessary for seismo-resistance analysis of structures. The application of the equivalent modes instead of the actual one with the coupling effects of the added mass of water make the economic and effective use of computer time possible. The effectiveness of the method presented in this paper has been proved in the calculation provided in reference 2 .

1. INTRODUCTION

It is well known that the response of an offshore structures must condider the action of ocean surface waves, currents, winds gravity loading and other loads. In the complex systems, for example, the multi-cylindrical structure with different diameters as shown in Fig.1, the fluid-structure-soil interaction must be simulated to obtain a realistic structural model.

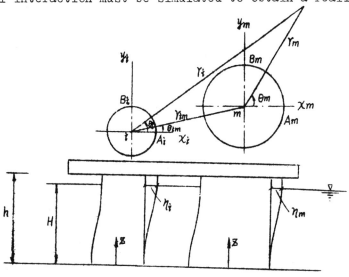

Fig.1

Based on ref. 1, the hydridynamic pressure on arbitrary i-th pile per unit length along x-direction is

$$P_i^x = 2\pi a_i^2 \frac{\rho_0}{g} \omega^2 \sum_{k=1}^{\infty} G'_{ik} D^x_{ik} \cdot \frac{1}{H} \int_0^H Y_i^x(\xi)\cos\lambda_k\xi d\xi \cdot K_1(\eta'_k a_1)\cos\lambda_k Z \sin\omega t$$
$$- \frac{2\pi\rho_0}{g}\omega^2 \sum_{m\neq i} a_m a_{1K} \sum_{K=1}^{\infty} G'_{mk} I_1(\eta'_k a_1)[D^x_{mk}\{k_0(\eta'_k \gamma_{1m})+K_2(\eta'_k \gamma_{1m})\cos 2\theta_{1m}\}\cdot \frac{1}{H}\int_0^H Y_m^x(\xi) \qquad (1)$$
$$\cos\lambda_k\xi d\xi + D^y_{mk} K_2(\eta'_k \gamma_{1m})\sin 2\theta_{1m}\cdot \frac{1}{H}\int_0^H Y_m^y(\xi)\cos\lambda_k\xi d\xi]\cdot \cos\lambda_k Z \sin\omega t$$

where
$$G'_{mk} = \frac{4\lambda_k H}{\eta'_k a_m(\sin 2\lambda_k H + 2\lambda_k H)(K_0(\eta'_k a_m)+K_2(\eta'_k a_m))} \qquad (2)$$

$Y_i^x(Z)$ and $Y_i^y(Z)$ denote the elastic amplitudes of i-th vibration mode in x and y directions. D^x_{mk} and D^y_{mk} denote the coefficients of hydrodynamic pressure in x and y directions. H represents the depth of water; ρ_0/g is the density of water; g is the gravity acceleration and E_ν is in units of 23200 kg/cm².
$K_n(\ell'_k \gamma_m)$ and $I_n(\ell'_k \gamma_m)$ denote the first and second kind Bessel function with image arguments e.g. the revised Bessel function.

where λ_k value can be ontained from following equation
$$\lambda_k \cdot ctg\lambda_k H = -\delta\lambda_k H \qquad (3)$$

where $\eta'_k = \sqrt{\lambda_k^2 - C_0^2}$, $C = \sqrt{gE\nu/\rho_0}$; $C_H = C_0 H = \omega H/C$, $C_0 = \omega/C$, $\delta = gH/(C^2 \cdot C_H^2)$ (4)

Assuming that h=H then we have
$$Y_m^x(z) = \sum_{k=1}^{\infty} B^x_{mk}\cos\lambda_k z$$
$$Y_m^y(z) = \sum_{k=1}^{\infty} B^y_{mk}\cos\lambda_k z \qquad (5)$$

We will neglect P_i^y for vibration along y direction, because it is similar to x direction. P_i^x and P_i^y can be obtained once D^x_{ik} and D^y_{ik} are determined.

$$D^y_{1k} + \sum_{m\neq i}(\frac{B^x_{mk}}{B^y_{1k}}\frac{K_2(\eta'_k\gamma_{1m})\sin 2\theta_{1m}}{K_0(\eta'_k a_1)+K_2(\eta'_k a_1)}\cdot D^x_{mk} + \frac{B^y_{mk}}{B^y_{1k}}\frac{K_0(\eta'_k\gamma_{1m})-K_2(\eta'_k\gamma_{1m})\cos 2\theta_{1m}}{K_0(\eta'_k a_1)+K_2(\eta'_k a_1)}D^y_{mk}) = 0 \qquad (6)$$
$$D^x_{1k} + \sum_{m\neq i}(\frac{B^x_{mk}}{B^x_{1k}}\frac{K_0(\eta'_k\gamma_{1m})+K_2(\eta'_k\gamma_{1m})\cos 2\theta_{1m}}{K_0(\eta'_k a_1)+K_2(\eta'_k a_1)}D^x_{mk} + \frac{B^y_{mk}}{B^x_{1k}}\frac{K_2(\eta'_k\gamma_{1m})\sin 2\theta_{1m}}{K_0(\eta'_k a_1)+K_2(\eta'_k a_1)}D^y_{Hk}) = 1$$

Since the natural frequencies and mode shape $Y_i^x(Z)$, $Y_i^y(Z)$ and ω are involved in P_i^x and P_i^y, an iterative approach must be used for calculating the hydrodynamic pressure per unit length on i-th pile as presented in ref. 1.

2. A refined method for determining the hydrodynamic pressure

The vibrating frequencies and modes of submerged multi-cylindrical pile structures is presented in this paper. It is recognized that it will be impossible to evaluate the hydrodynamic pressure without knowing the vibrating frequencies and modes of the submerged multi-cylindrical pile structures, it is even very tedious and cumberous to obtain each frequency and mode through iteration approach for the dynamic analysis and earthquake response study. Therefore it is beneficial to develop a simplified approach presented in this paper, the stratagy of which is to determine the actual frequencies and modes of submerged structure by the conbination of those frequencies and modes obtained from that of the case without the presence of added mass of water. Then the hydrodynamic pressure as well as the frequencies and modes of the structure can be determined in one pass.

The theoretical derivations developed in this paper are limited to an elastic cylinders mainly subjected to flexual vibration, all the principles presented are suitable for the case of shearing vibration. If m-th cylinder is vibrating flexually along x direction with the vibrating mode of $\bar{Z}_m^x$ then the following expression between $\bar{Z}_m^x$ and circular frequency ω must be satisfied

$$EJ_m \frac{d^4 \bar{Z}_m^x}{dZ^4} - \frac{\rho_m F_m}{g} \omega^2 \bar{Z}_m^x + \sigma_{rm}^x = 0 \qquad (7)$$

where σ_{rm}^x is the hydrodynamic pressure per unit length along x-direction. From equ.(1) we have

$$\sigma_{rm}^x = P_m^x = \frac{2\pi\rho_0 \omega^2}{g} f_m^x = \frac{2\pi\rho_0 \omega^2}{g} \{a_m^2 \sum_{k=1}^{\infty} G'_{mk} D_{mk}^x \frac{1}{H}\int_0^H \bar{Z}_m^x(\zeta)\cos\lambda_k\zeta d\zeta \cdot K_1(\eta'_k a_m)\cos\lambda_k Z - \sum_{i \neq m}^{l} a_m a_i$$

$$\sum_{k=1}^{\infty} G'_{ik} I_1(\eta'_k a_m)(D_{ik}^x \cdot (K_0(\eta'_k \gamma_{im}) + K_2(\eta'_k \gamma_{im})\cos 2\theta_{im}) \frac{1}{H}\int_0^H \bar{Z}_i^x(\zeta)\cos\lambda_k \zeta d\zeta + D_{ik}^y K_2(\eta'_k \gamma_{im})\sin 2\theta_{im}$$

$$\frac{1}{H}\int_0^H \bar{Z}_i^y(\zeta)\cos\lambda_k\zeta d\zeta)\cos\lambda_k Z\} \qquad (8)$$

F_m------Cross section of m-th cylinder ; EJ_m ----- *flexual rigidity of m-th cylinder*

ρ_m------relative weight of material of m-th cylinder

ρ_0------relative weight of water.

the submerged vibrating mode of m-th cylinder $\bar{Z}_m$ can be expressed by the following expansion of modes Z without the presence of added mass of water.

$$\bar{Z}_m^x = \sum_{n=1}^{\infty} A_n Z_{mn}^x \qquad (9)$$

where A_n---coefficients to be determined.

As it is well known that the vibrating mode Z_{mn}^x without the presence of added mass of water and corresponding circular frequencies ω_{on} must satisty the following governing equation:

$$EJ_m \frac{d^4 Z_{mn}^x}{dZ^4} - \frac{\rho_m F_m}{g} \omega_{on}^2 Z_{mn}^x = 0 \qquad (10)$$

since (9) and (10), we have:

$$EJ_m \frac{d^4 \bar{Z}_m^x}{dZ^4} - \frac{\rho_m F_m}{g} \sum_{n=1}^{\infty} \omega_{on}^2 A_n Z_{mn}^x = 0 \qquad (11)$$

substituting (8),(10) and (11) into (7), then

$$\frac{\rho_m F_m}{g} \sum_{n=1}^{\infty} \omega_{on}^2 A_n Z_{mn}^x - \frac{\rho_m F_m}{g} \omega^2 \sum_{n=1}^{\infty} A_n Z_{mn}^x + \frac{2\pi\rho_0}{g} \omega^2 f_m^x = 0 \qquad (12)$$

Multiplying $Z_{mn}^x dx$ to equation (12) and integrating from 0 to H along each cylinder, due the orthogonality of Z with mass, then we obtain:

$$\sum_{m=1}^{l} \frac{1}{H} N_{mn} \omega_{on}^2 A_n - \sum \frac{1}{H} \omega^2 N_{mn} A_n + \frac{2\pi\rho_0}{g}\omega^2 \sum_{m=1}^{l} F_m^x = 0 \qquad (13)$$

here

$$F_m^x = a_m^2 \sum_{K=1}^{\infty} G'_{mK} D_{mk}^x \sum_{h=1}^{\infty} \frac{A_h}{H}\int_0^H Z_{mh}^x(\xi)\cos\lambda_k\xi d\xi \cdot K_1(\eta'_k a_m) \frac{1}{H}\int_0^H Z_{mn}^x(Z)\cos\lambda_k Z dZ - \sum_{i \neq m}^{l} a_m a_i \sum_{k=1}^{\infty} G'_{ik} I_1(\eta'_k a_m)$$

$$D_{ik}^x (K_0(\eta'_k \gamma_{im}) + K_2(\eta'_k \gamma_{im})\cos 2\theta_{im}) \sum_{h=1}^{\infty} \frac{A_h}{H}\int_0^H Z_{ih}^x(\xi)\cos(\lambda_k\xi)d\xi + D_{ik}^y K_2(\eta'_k r_{im})\sin 2\theta_{im}$$

$$\sum_{h=1}^{\infty} \frac{A_h}{H} \cdot \int_0^H Z_{ih}^y(\xi)\cos\lambda_k\xi d\xi] \cdot \frac{1}{H}\int_0^H Z_{mn}^x(Z)\cos\lambda_k Z dZ \qquad (14)$$

Denote

$$N_{mn} = \frac{1}{H}\int_0^H \frac{\rho_m F_m}{g}[Z_{mn}^x(Z)]^2 dZ \qquad (15)$$

$$M_{mnk} = \frac{1}{H}\int_0^H Z_{mn}^x(Z)\cos\lambda_k Z dZ \qquad (16)$$

---

*) Symbol denotes that i=i,2,... until i=l except i=m (there are totally l cylinders)

substituting (16) into (14)

$$F_m^x = \sum_{h=1}^{\infty}\sum_{k=1}^{\infty} a_m^2 G'_{mk} D^x_{mk} A_h M_{mhk} K_1(\gamma'_k a_m) M_{mnk} - \frac{1}{2}\sum_{i \neq m} a_m a_i \sum_{k=1}^{\infty} G'_{ik} I_1(\gamma'_k a_m)[D^x_{ik} \cdot (K_0(\gamma'_k \gamma_{im})$$
$$+ K_2(\gamma'_k \gamma_{im})\cos 2\theta_{im})\sum_{h=1}^{\infty} A_h M_{mhk} + D^y_{ik} K_2(\gamma'_k \gamma_{im})\sin 2\theta_{im} \sum_{h=1}^{\infty} A_h M_{ihk}] M_{mhk}$$ (17)

The governning equation of vibration in y direction can be derived in similar way.

Each of the above mentioned equations (14),(15),(16) and (17) has its own physical meaning. It is not difficult to understand that in equation (7) the coefficients A_n are symmetric and to express these coefficients in matrix form are necessary for computer calculation. We first write eq. (13) as

$$\sum_{m=1}^{\infty} \frac{\omega_{on}^2}{H} N_{mn} A_n - \omega^2 [\sum_{m=1}^{\infty} A_n - \frac{2\pi \rho_0}{g}\sum_{m=1}^{\infty} F_m^x(A_h, M_{mnk}, M_{mhk})] = 0$$ (18)

Assume that $$F_m^x(A_h, M_{mnk}, M_{mhk}) = \sum_{h=1}^{n} H^x_{mh} A_h$$ (19)

then $$H^x_{mh} = a_m^2 \sum_{h=1}^{\infty}\sum_{k=1}^{\infty} G'_{mk} D^x_{mk} M_{mhk} K_1(\gamma'_k a_m) M_{mnk} - \frac{1}{2}\sum_{i \neq m} a_m a_i \sum_{K=1}^{\infty} G'_{ik} I_1(\gamma'_k a_m)[D^x_{ik} \cdot$$ (20)

$$(K_0(\gamma'_k \gamma_{im}) + K_2(\gamma'_k \gamma_{im})\cos 2\theta_{im}) M_{mhk} + D^y_{ik} K_2(\gamma'_k \gamma_{im})\sin 2\theta_{im} M_{ihk}] M_{mnk}$$

Substituting (20) and (19) into eq. (18)

$$\sum_{m=1}^{\infty} \frac{\omega_{on}^2}{H} N_{mn} A_n - \omega^2 [\sum_{m=1}^{\infty} \frac{N_{mn}}{H} A_n - \sum_{m=1}^{\infty}\sum_{h=1}^{n} \frac{2\pi \rho_0}{g} H^x_{mh} A_h] = 0$$ (21)

Denoting that

$$K_{nn} = \sum_{m=1}^{\infty} \frac{\omega_{on}^2}{H} N_{mn} ; \qquad \delta_{nn} = \sum_{m=1}^{\infty} \frac{2\pi \rho_0}{g} H^x_{mh}$$ (22)

Then (21) can be written

$$[K]\{A\} = \omega^2 [M]\{A\}$$ (23)

here

$$[K] = \begin{bmatrix} K_{11} & & & & \\ 0 & K_{22} & & \text{SYM} & \\ 0 & & \ddots & & \\ \vdots & & & \ddots & \\ 0 & 0 & 0 & & K_{nn} \end{bmatrix}$$ (24)

$$[M] = \begin{bmatrix} (\frac{K_{11}}{\omega_{o1}^2} - \delta_{11}) & & & \text{SYM} \\ \delta_{21} & (\frac{K_{22}}{\omega_{o2}^2} - \delta_{22}) & & \\ \vdots & & \ddots & \\ \delta_{n1} & \cdots & & (\frac{K_{nn}}{\omega_{on}^2} - \delta_{nn}) \end{bmatrix}$$ (25)

$$\{A\} = \{A_1 \; A_2 \; A_3 \; \ldots \; A_n\}^T$$ (26)

The Solution of eq. (17) gives the n-th circular frequencies and n-th vibrating modes $A^{(n)}$, then the n-th vibrating modes of the multi-cylinder structure with added mass of water Y can be obtained through (7) and (9). Eq. (9) can also be expressed in matrix form as

$$[Y^x_{m(n)}] = [Z^x_{m(n)}][A^{(n)}]$$ (27)

where

$$[Y^x_{m(n)}] = \begin{bmatrix} Y^x_{m1(1)} & Y^x_{m1(2)} & & Y^x_{m1(n)} \\ Y^x_{m2(1)} & Y^x_{m2(2)} & \cdots & Y^x_{m2(n)} \\ \vdots & \vdots & & \vdots \\ Y^x_{mn(1)} & Y^x_{mn(2)} & & Y^x_{mn(n)} \end{bmatrix}$$ (28)

$$[Z^x_{m(n)}] = \begin{bmatrix} Z^x_{m1(1)} & Z^x_{m1(2)} & & Z^x_{m1(n)} \\ Z^x_{m2(1)} & Z^x_{m2(2)} & \cdots & Z^x_{m2(n)} \\ \vdots & \vdots & & \vdots \\ Z^x_{mn(1)} & Z^x_{mn(2)} & & Z^x_{mn(n)} \end{bmatrix}$$ (29)

$$[A_n] = \begin{bmatrix} A_1^{(1)} & A_1^{(2)} & \cdots & A_1^{(n)} \\ A_2^{(1)} & A_2^{(2)} & \cdots & A_2^{(n)} \\ \vdots & \vdots & & \vdots \\ A_n^{(1)} & A_n^{(2)} & \cdots & A_n^{(n)} \end{bmatrix}$$ (30)

It is to be noted that eq. (17) is a diagonal matrix and N_{mn} involoed in the diagonal element can been determined easily by integrating the vibrating modes $Z_m^x(n)$ in absence of water. Although matrix M is symmetric in nature, each H_{ms}^x involved in the element has to be solved through eq. (25), that is the essential difference between multi-cylindrical piles and single pile.

$$\begin{rcases} [K][\bar{A}_K] = [M][A_{K-1}] \\ [K]^* = [\bar{A}_K]^T [K][\bar{A}_K] \\ [M]^* = [\bar{A}_K]^T [M][\bar{A}_K] \\ [K]^*[A_K'] = [M]^*[A_K'][\Omega^2] \\ [A_K] = [\bar{A}_K][A_K'] \end{rcases} \quad (31)$$

The seismic and dynamic analysis of multi-cylindrical submerged structures can be performed by using conventional approach in space framework dynamic analysis, once the frequencies, vibrating modes and the coupling mass of water reduced by hydrodynamic pressure are known. In order to assure the accuracy of calculation in computer the Bessel function with image arguments are used.

3. Numerical examples

Two examples are presented for illustration purpose(Fig. 2,3), it includes:
(1) The calculation of natural frequencies, modes and hydrodynamic pressure of two cylindrical pield structures. Denote that the related weight of material ; the depth of water H; the moment of inertia of cross section area I; the polar moment of intertia J; the area of cross section A; modulus of elasicity E and shear modulus G, thus, we give

$H=120m$, $A=84.32m^2$, $I=118.2m^4$, $J=3236.4m^4$, $E=3*10^6 T/M^2$, $G=1.08 t/m^2$, $\rho=2.5t/m^3$,

The natural frequencies, vibrating modes, hydrodynamic pressure of structure and structural response due to earthquake and vibration can be calculated:

NUMERICAL RESULTS

First frequency
$\qquad f_1=0.50$ Hz (by computer)
$\qquad f_1=0.48$ Hz (by experiment)

The comparison of theoretical results with the experimental results is favorable. Internal forces and deflections of structure, subjected to earthquake in two directions are shown in Tab.1.

(2) A four cylindrical piled structure is studied by using original data as following:

$H=2.0m$, $A_1=0.005M^2$, $A_2=0.01M^2$, $I_1=0.0001M^4$, $I_2=0.000804M^4$,

$J_1=0.000201M^4$, $J_2=0.001608M^4$, $E=2.1*10^7 t/m^2$, $G=0.808 \times 10^7 T/M^2$, $\rho=7.8T/M^2$

NUMLRICAL RESULTS

Natural feaquencies
$\qquad f_1=8.11$ Hz (by computer)
$\qquad f_2=9.52$ Hz (by computer)

Internal forces and deflections of structure, subjected to earthquake in two directions are shown in Tab.2.

It is assumed in the analysis that the platform of structure is regarded as a rigid body.

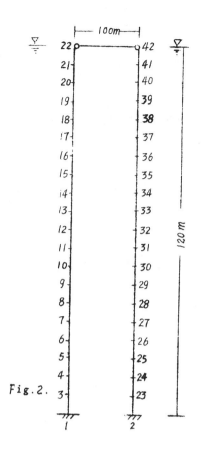

Fig. 2.

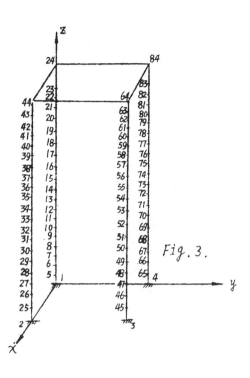

Fig. 3.

Unit: T;M;S.   Table 1

EL.	NO.	$u \times 10^6$	$v \times 10^6$	$w \times 10^6$	$\sigma_x \times 10^6$	$\sigma_y \times 10^6$	$\sigma_z \times 10^6$	N_x	N_y	N_z	M_x	M_y	M_z
I	1	0	0	0	0	0	0	153	1294	770	41	59544	74983
	3	1037	843	0	166	137	0	153	1293	770	41	51295	59456
II	3	1037	843	0	166	137	0	158	1272	767	40	51300	59427
	4	2246	1850	11	235	198	0	158	1273	767	41	47282	51792
III	4	2246	1850	11	235	198	0	165	1243	762	41	47284	51784
	5	3839	3206	11	294	254	0	165	1243	762	41	43341	44325

Unit: T;M;S. Table 2

EL.	NO.	$u \times 10^6$	$v \times 10^6$	$w \times 10^6$	$\theta_x \times 10^6$	$\theta_y \times 10^6$	$\theta_z \times 10^6$	N_x	N_y	N_z	M_x	M_y	M_z
I	1	0	0	0	0	0	0	0.402	0.110	0.095	0.013	0.100	0.112
	5	1	1	0.5	5	1	1	0.402	0.110	0.095	0.013	0.091	0.101
II	5	1	1	0.5	5	1	1	0.402	0.110	0.095	0.013	0.091	0.101
	6	1	1	1	7	7	1	0.402	0.110	0.095	0.013	0.086	0.096
III	4	0	0	0	0	0	0	0.407	0.126	0.516	0.016	0.557	0.140
	65	0	1	0.3	1	1	1	0.407	0.126	0.516	0.016	0.506	0.128
IV	65	0	1	0.3	1	1	1	0.407	0.126	0.516	0.016	0.506	0.128
	66	0	2	0.5	1	1	1	0.407	0.126	0.516	0.016	0.480	0.121

4. CONCLUSION

The revision of the iterative method presented in paper 1 make the computational effort for calculation the earthquake response of structure considerably reduced.

It is recognized by the example that the revised method above is very effective. Besides that it is efficient even in statistical and hydro-elastic based analysis of structures. Nevertheless simplified method used today for random or coupling problem analysis seems improper, however, this paper presents an approach to take these two aspects into consideration simultaneously.

References

1. Seimo KOTSUBO and Teruhiko TAKANISHI, 1978, " A Method of analysis of Multi-piles Foundation with Different Diameters and Arbitrary Position of Piles," Proceedings of the Japanese Society of Civil Engineers, NO. 276, pp. 1-12.

2. Qu Nai-si, 1979,"The Earthquake Hydro-elastic Response Analysis of Liquid Storage Tanks," Journal of Dalian Institute of Technology, NO.3, pp.86-100.

3. Qu Nai-si, 1980, "Stochastic Response of Offshore Platform," Journal of Dalian Institute of Technology, NO.4, pp.167-168.

4. Gao,M.,1981, Probabilistic Dynamic Analysis of Pile-Supported Rigid Platform, Nanking Water Conservancy Scientific Research Institute, Technical Report NO. 08.

LONG-TERM RANDOM SEA-STATE MODELLING

S. Shyam Sunder
Massachusetts Institute of Technology
U.S.A.

Jerome J. Connor
Massachusetts Institute of Technology
U.S.A.

SUMMARY

This paper presents a new long-term random sea-state model that traces the changes in energy and frequency content of the sea surface elevation in time. The model assumes that it suffices to account for variations in characteristic wave height, either significant or visual, usually one of the short-term sea-state descriptors defining the wave spectrum. The parametric form of the wave spectrum is considered to be time-invariant.

The modelling strategy involves transforming the Weibull or lognormally distributed characteristic wave height trajectory into a zero-mean, unit-variance, Gaussian random process. A spectral density function is then estimated for the derived random process using a classical Fourier transform based spectral estimator. A separate power spectrum is obtained for each of the four seasons representing a year. Synthetic characteristic wave height traces are simulated by generating pseudo-random time-histories of the zero-mean, unit-variance Gaussian process from the derived spectra and then applying an inverse transformation to obtain the characteristic wave heights. The synthetic time-histories appear to be realistic and in reasonable agreement with the observed data. Finally, a sample analysis of long-term foundation stiffness degradation for a steel jacket platform is presented.

The new modelling strategy presented in this paper is significantly superior to the two earlier models presented in Shyam Sunder, et al, 1979, and Angelides, et al, 1981, for the following several reasons: (i) the need for a much smaller database for deriving the model, (ii) the ability to allow for seasonal sea-state variability with greater confidence, (iii) the independence of the model from a threshold defining "storms", (iv) elimination of the "threshold downcrossing" problem, and (v) elimination of the limitation associated with the choice of the number of harmonics in the Fourier series approximation to the nonstationary storm trace in the 1979 model.

Seven years of hourly visual data collected by Ocean Weather Ship "M" in the North Atlantic is used to develop the model.

INTRODUCTION

The sea surface elevation is often modelled as a zero-mean random function, locally stationary in time and locally homogeneous in space. Important information about this function is contained in the mean power spectral distribution of the random process of time only, obtained by fixing the spatial coordinates. The wave spectrum which is usually parametrized in terms of significant wave height, H_s, average zero-crossing period, T_z, etc., is nearly time-invariant over periods of time ranging from one to several hours (Cartwright, 1974, Kinsman, 1965, and Nordenstrøm, 1971).

For most design purposes, linear elastic steady-state analysis with spectral parameters fixed to unfavorable values is considered sufficient. The main output of the analysis is a conservative estimate of the mean power spectrum of various response quantities (displacements, forces, etc.). This stationary, semi-probabilistic model (with the spectral parameters being treated as deterministic constants) is inadequate for the calculation of long-term reliability.

Accurate calculation of reliability during long intervals of time is possible from knowledge of only the joint distribution of the spectral parameters at a generic point in time, provided that the final state of damage is assumed not to depend on the order in which different seastates follow one another. Consequently, serial representation of loads is not essential. This approach is commonly adopted in the fatigue analysis of offshore platforms (Kinra and Marshall, 1979, and Moan, et al, 1976), and involves the use of the Palmgren-Miner linear cumulative damage hypothesis. However, in other situations such as in the assessment of reliability against foundation fatigue, the order in which loads are applied cannot be neglected since, in this case, the applicability of the Miner's rule of damage accumulation is questionable (Van Eekelen, 1977).

The long-term, nonstationary sea-state model developed herein accounts for the probabilistic time dependence of the characteristic wave height trajectory. The remaining spectral parameters are assumed to be deterministically related to the characteristic wave height value, while the parametric form of the wave spectrum is considered to be time-invariant. The modelling approach is purely empirical in that the form and parameters of the model are determined from historical wave data without consideration of the underlying physical mechanisms (formation and evolution of wind storms, transfer of energy from the boundary layer to the sea, etc.). Directional distribution of wave energy is ignored.

Hourly information from Ocean Weather Ship "M" in the North Atlantic (N66° 02°E) is used to develop the model and to obtain numerical estimates for the parameters. The data covers a seven year period from 1961 to 1968, but includes no record of significant wave height. The model is therefore constructed in terms of visual wave height, H_v. One may consider H_s and H_v to be equal, or one may convert H_v to H_s by using more accurate relationships, e.g., those based on equiprobability functions (Nordenstrøm, 1971).

The model is best used in a simulation mode as a means of generating pseudo-random time-histories of characteristic wave height (as opposed to time-histories of sea surface elevation) at any given site. By using these time-histories and the several available deterministic relations between the characteristic wave height and the remaining spectral parameters, one can simulate the random evolution in time of the wave spectrum. This series of wave spectra becomes input information for reliability analysis. An application of the model to the assessment of off-shore foundation degradation is given at the end of the paper. Two earlier long-term random sea-state models are reviewed before the new model is presented.

REVIEW OF EARLIER MODELS

Two long-term random sea-state models have been developed to account for the probabilistic nature and random time variation of the spectral parameters, assuming that the parametric form of the wave spectrum remains time-invariant (Shyam Sunder, et al, 1979, and Angelides, et al, 1981). The models also assume that it suffices to account for variations in characteristic wave height.

The general modelling strategy involves characterizing the sea by a serious of storms considered to occur when the characteristic wave height exceeds a predefined threshold. The threshold is selected such that: (1) platform response is represented accurately (errors may come from neglecting the action of waves during interstorm calm periods), thus the threshold should be set below some critical level of platform response sensitivity; and (2) storms that are different on physical grounds are identified as separate storms. This would imply the choice of relatively large values of threshold. On account of these conflicting requirements, threshold values between 1m and 5m are considered to be appropriate for reliability analysis against different types of failure events, e.g., smaller values for fatigue failure and higher values for extreme elastic response. The storms themselves are described by a duration, an average intensity, and a nonstationary random process which traces the evolution of the characteristic wave height from initial build-up to the final decay. The 1979 model represents the nonstationary storm trace as the sum of a deterministic component associated with the memory capacity of the waves and a nondeterministic component, assumed random and uncorrelated, associated with highly localized environmental changes and noise in the data attributable either to the recording instrument or to the human observer. A limited number of terms in the Fourier series are used to approximate the deterministic component. The parameters of the Fourier series are probabilistically defined for storms of a given duration and average intensity. The 1981 model represents the nonstationary storm trace in terms of a time-varying mean value function for the characteristic wave height and a mean power spectral density function for the fluctuation of the characteristic wave height about its mean value function. The fluctuation is taken to be a portion of a stationary, zero-mean Gaussian process. The mean value function of the characteristic wave height and the spectral density function of the fluctuations are functions of the storm duration and average intensity.

These two models are nonstationary at the microscale of each storm, but are stationary on a large scale (storm arrivals and gross storm characteristics such as duration and average intensity). Macroscale stationarity and an unusually long historical record, such as that from Ocean Weather Ship "M", allow one to use large sample sizes for the inference of parameters and to reduce statistical estimation uncertainty.

In what follows, the major limitations of the two existing models are briefly summarized to justify the need for an improved modelling strategy.

Size of Data Base: Both the 1979 and 1981 models were derived with fourteen years of hourly data. Since fourteen years of wave data is a luxury seldom available in practice, the possible effects of a reduced data base need to be understood. Although a reduction in the size of the data is unlikely to result in significant changes in estimates of parameters for the joint distribution of storm duration and average intensity, the consequent reduction in the number of storms in a given duration-average intensity class is likely to affect estimates of parameters describing the nonstationary time variation of characteristic wave height. This will be more true for those classes which have a small sample size even with the large data set used.

Seasonal Sea-State Variations: Both the models developed do not allow for seasonal variations in sea-state characteristics. Although they can easily be generalized to allow for seasonal variations in the parameters, the consequent reduction in sample size for any given season due to a fixed record length would introduce significant uncertainties in the estimation of the model parameters.

Selection of Threshold: The 1979 and 1981 models were derived for a specific value of threshold of 1.5m in order to study foundation degradation under cyclic loading. For the selected threshold, 2933 storms were identified covering on the order of half the total data length of 126,117 samples for the fourteen years. This sensitivity of the model parameters to the particular choice of a threshold level is undesirable, since for applications requiring a different threshold level, a completely new model has to be derived.

Threshold Downcrossing Problem: The characteristic wave heights were observed to have values less than the threshold in about five to ten percent of the simulated pseudo-random storm time-histories using both models. This phenomenon was more prnounced at the beginning and at the end of storms, and thus indicated the inability of the models to approximate the transient behaviour adequately.

In addition to these limitations, the 1979 model has an additional limitation associated with the choice of the number of harmonics in the Fourier series approximation to the nonstationary storm trace.

OVERVIEW OF PROPOSED MODELLING STRATEGY

Given the one-sided power spectral density function of a zero-mean, stationary, random process $x(n)$, $G_x(\omega)$, it is possible to generate a pseudo-random time-history for $x(n)$ as follows:

$$x(n) = \sum_{k=0}^{N/2} a(k) \cos[\omega_k Tn + \phi(k)] \qquad n = 0,1,\ldots,N-1 \qquad (1)$$

with

$$a(k) = \sqrt{2 \, G_x(\omega_k) \Delta\omega} \qquad (2)$$

$$\omega_k = \frac{2\pi}{NT} k \quad \text{(radians/sec)} \qquad (3)$$

$\phi(k)$ is a random phase angle uniformly distributed over $(0,2\pi)$, T is the data sampling period and is equal to one hour for the Ocean Weather Ship "M" data, and $\Delta\omega = 2\pi/NT$ is the incremental frequency in radians/sec. According to the Central Limit Theorem, the $x(n)$ generated according to

Eq. (1) represents a zero-mean, stationary, and Gaussian random process.

The modelling process would be straightforward if the characteristic wave heights were normally distributed. In this case, the data analysis procedure would be as follows: (a) obtain the mean value of the characteristic wave heights, (b) generate a new time series with zero mean by subtracting the mean value from the original data, and (c) estimate the one-sided power spectral desnity function for the zero-mean, stationary, Gaussian random process. Synthesis of the characteristic wave height time-histories from the estimated spectra would involve two steps: (a) simulation of a zero-mean, stationary, Gaussian process from the estimated power spectral density function using Eqs. (1)-(3), and (b) addition of the mean value of characteristic wave height to the simulated zero-mean random process.

However, this procedure cannot be applied directly because the characteristic wave heights are not normally distributed. First, the distribution is not symmetrical about its mean, but tends to be positively skewed, and second, the distribution has a definite lower bound. That is, the characteristic wave height is always greater than or equal to zero.

The general modelling strategy developed here consists of four basic steps: (a) apply a transformation, T, to the characteristic wave height data such that the resulting sequence, $x_1(n)$, is normally distributed with zero-mean and unit variance, (b) estimate the one-sided power spectral density function for $x_1(n)$, (c) synthesize a zero-mean, unit variance, stationary, Gaussian random process, $x_2(n)$, with Eq. (1) using the estimated spectrum, and (d) apply an inverse transformation, T^{-1}, to $x_2(n)$ to obtain the simulated characteristic wave height time-history. This procedure is schematically illustrated in Fig. 1.

MODEL DESCRIPTION AND PARAMETER ESTIMATION

Data Transformation

Although considerable attention has been given to statistical information of long-term wave height, the probability distribution applicable to long-term characteristic wave height is the focus of much criticism (Ochi, 1978). Some researchers believe that the data can be fitted by the Weibull distribution, while others claim that the lognormal distribution is more appropriate. The estimation of parameters describing the Weibull distribution using the method of moments often leads to a poor fit of the distribution to the data. On the other hand, if the parameters are estimated empirically using a least-squares fit, the fit tends to be very good (Shyam Sunder, 1979). However, even with a good Weibull fit, the transformation of the data from a Weibull distribution to a normal distribution is quite complex.

If the lognormal distribution is adopted, the transformation, T, is very simple and may be expressed as:

$$x_1(n) = T[H_v(n)] = \frac{h(n) - m_h}{\sigma_h} \qquad (4)$$

where $h = \ln H_v$, m_h and σ_h are the mean and standard deviation of h. The inverse transformation, T^{-1}, follows from Eq. (4):

$$H_v(n) = \exp[m_h + x_2(n)\sigma_h] \qquad (5)$$

The lognormal distribution is a reasonably good approximation to the data and is adopted in this work.

The estimation of m_h and σ_h is now considered. The data from Ocean Ship "M" is first divided according to season: Winter (December - February), Spring (March - May), Summer (June - August) and Fall (September - November). Since seven years of data is used, seven time-histories of wave data are available per season. The mean and standard deviation of H_v, i.e., m_{H_v} and σ_{H_v}, are then estimated for each season by combining all seven years of data. Finally, m_h and σ_h are obtained with:

$$\sigma_h = \sqrt{\ln(V_{H_v}^2 + 1)} \qquad (6)$$

$$m_h = \ln m_{H_v} - \sigma_h^2/2 \qquad (7)$$

where $V_{H_v} = \sigma_{H_v}/m_{H_v}$. Table 1 presents the values of the estimated parameters by season. These results indicate that there is significant seasonal dependence of the wave data, and that the winter season is most severe, followed by fall, spring and summer in decreasing order of severity.

Based on the method of moments parameter estimators, the lognormal distribution has been shown to be much better than the Weibull distribution for the Ocean Weather Ship "M" data (DeLeeuw, 1981). Unfortunately, even the lognormal distribution does not provide a good fit to the data at the extremes. In order to improve the curve fit at the upper extremes of characteristic wave height, it is found that an empirical change of σ_h is necessary. Such a change causes a rotation of the linear plot on lognormal paper about the fifty percent cumulative probability level. The empirically obtained values of the standard deviation, σ_h^1, are also listed in Table 1. Figures 2a-d present the lognormal approximations to the observed cumulative probability distributions for the four seasons of the year. As a consequence of empirically changing σ_h, the probabilities associated with characteristic wave heights less than about one meter are underestimated. However, this issue is relatively unimportant since the fatigue damage contribution from these heights is usually low enough to be neglected in a reliability analysis.

The empirical change in σ_h alters only the total energy in the spectrum but not the spectral shape. Thus, while Eq. (4) remains unchanged, Eq. (5) is modified to:

$$H_v(n) = \exp[m_h + x_2(n) \sigma_h^1] \qquad (8)$$

and is based on σ_h^1. Equations (4) and (8) indicate that at the same probability level an alternative height is defined.

$$x_1(n) = \frac{h(n) - m_h}{\sigma_h} = \frac{h^1(n) - m_h}{\sigma_h^1} = x_2(n) \qquad (9)$$

where $h^1(n)$ is the simulated logarithmic value of the characteristic wave height and is normally distributed since it is derived from the zero-mean, unit-variance, Gaussian random process $x_2(n)$. However, $x_1(n)$ is derived from observed data and is not strictly a Gaussian process. But if the assumption is made that the power spectral density function of $x_1(n)$ and $x_2(n)$ are the same, then $x_2(n)$ may be generated from $G_{x_1}(\omega)$ with reasonably good results.

In summary, therefore, the transformation, T, is applied according to Eq. (4), the power spectral density function $G_{x_1}(\omega)$ is estimated for $x_1(n)$, the random process $x_2(n)$ is simulated according to Eqs. (1)-(3), and finally, the inverse transformation, T^{-1}, is applied using Eq. (8). It now remains to derive the spectrum.

Spectral Estimation

The procedure for estimating the power spectral density function of $x_1(n)$ is based on averaging periodograms derived for every three month period representing a season. Although a typical three month period yields a data length on the order of 2100-2200 points, for convenience in applying the Fast Fourier Transform (FFT) algorithm attention is restricted to an N of 2048 per season. Such a data sequence is taken to represent one sample of a stationary random process. Consequently, consideration of sequences spanning the same three month period over seven years yields an ensemble of seven time-histories. The periodogram for each sample is generated using the FFT with

$$\tilde{G}_{x_1}(\omega_k) = \frac{T}{2\pi} \cdot \frac{1}{N} \left| \sum_{n=0}^{N-1} x(n) e^{-j(2\pi kn/N)} \right|^2 \qquad k = 0,1,\ldots,N-1 \qquad (10)$$

Then the spectral estimate $G_{x_1}(\omega_k)$ is the average of the seven periodograms for any given season and has units of $(\ln \text{meter})^2 - \sec$. Due to the discrete-time implementation of the Fourier transform, the spectrum is uniquely defined only up to the Nyquist frequency which corresponds to $k = N/2$. In order to derive the one-sided spectral density function from the two-sided spectrum estimated with Eq. (10), it is necessary to multiply all spectral ordinates up to the Nyquist frequency by a factor of two. Furthermore, the estimated spectrum is checked to ensure that the spectral ordinate at zero frequency is zero and that the area under the spectral density function is unity, since the spectrum represents a zero-mean, unit-variance random process. Note that the area under the one-sided spectral estimate of a zero-mean random process should be equal to the variance of the random process provided that aliasing errors are negligible. The spectral estimates for each of the four seasons are presented in Figs. 3a-d. If necessary, a spectrum may be estimated for the entire year by averaging the four seasonal spectra. Such a spectrum is shown in Fig. 4. Note that these figures only show the spectra up to a frequency of $\omega_k = 0.7 \times 10^{-5}$ rad/sec, i.e., $k = 82$. As can easily be seen, the spectra tend to decay rapidly and approach zero at higher frequencies. This confirms that aliasing errors are negligible.

A comparison of the four seasonal spectra shows that they are remarkably similar. In fact, if the four spectra are overlapped, the rapid fluctuations in the individual spectra may readily be associated with statistical variability. Thus for all practical purposes it may be sufficient to consider just the averaged yearly spectrum of Fig. 4. An analytical approximation to this spectrum is obtained with an equation describing a Type II extreme-value distribution:

$$G_{x_1}(\omega) = \frac{k}{u}\left[\frac{u}{\omega}\right]^{k+1} \exp\left\{-\left[\frac{u}{\omega}\right]^k\right\} \qquad \omega \geq 0 \tag{11}$$

where $k = 0.654$ and $u = 8.81 \times 10^{-6}$.

While the development presented here has made use of an ensemble of samples for estimating the seasonal spectrum, it should be pointed out that a long-term random sea-state model could theoretically be derived with just one season's length of data. Such a model has, in fact, been implemented with success by making the ergodicity assumption and applying a windowing method for smoothing the raw spectral estimate for the data. However, details of this modelling strategy are not discussed here.

STORM SIMULATION AND ASSESSMENT OF FOUNDATION DEGRADATION

Pseudo-random time-histories of visual wave height have been generated using Eqs. (1)-(3) and Eq. (8) with the actual seasonal spectra (not the analytical yearly spectrum of Eq. (11)), appropriate values of m_h and σ_h^1 from Table 1, and $N = 2048$. Since more than one season's length of data was simulated, different sets of random phase angles were generated for each record representing a season. In this way, three years of records were obtained by appropriately combining the records generated for a particular season.

Typical simulations of visual wave height for the Ocean Weather Ship "M" site are shown in Figs. 5a-d. (Note that the figures are plotted to different scales). A comparison of the simulated time-histories with the historically observed data shows that the synthetic time-histories are reasonable. However, the simulated data has a higher degree of fluctuations which may be associated with the fact that the historical data is recorded in discrete intervals of 0.5m while the simulated data is not similarly restricted. Further, as expected, the lower wave heights ($\leq$ 1m) are suppressed in the simulated data. An analysis of the probability distributions of the simulated visual wave heights showed that the simulations accurately replicate the lognormal distributions based on m_h and σ_h^1. Application of the long-term random sea-state model to the assessment of foundation stiffness degradation is discussed in what follows.

Clays subjected to repetitive loads start degrading in their mechanical properties when cyclic stress exceeds a threshold value. The reason for this degradation is the development of excess pore pressure with consequent reduction of effective stress and soil stiffness. In the case of off-shore foundations, one may expect degradation to occur during (major) storms, possibly followed by partial recovery in the intervals between storms.

For the present application, a steel-jacket platform is considered and soil degradation is measured in terms of loss of horizontal foundation stiffness (details of the structural configuration, dimensions, structural model, and foundation model can be found in Angelides, 1978). Spatial phasing of the different tubular elements that comprise the platform is taken into account in the evaluation of wave forces and in the estimation of structural and foundation response. The horizontal stiffness is assumed to evolve in time according to a Markov process, discrete in state (each state corresponds to a small range of stiffness values), and continuous in time. It is assumed that during a storm no recovery of the soil occurs. Between successive storms, two extreme cases of recovery are considered: (1) Return to the initial state ("total recovery model"); and (2) Continuous degradation without recovery ("no recovery model"). The lowest stiffness state is modelled as a "trapping state", so that at any given time foundation failure is associated with the probability accumulated in that state.

Numerical results are obtained for the case of 15 states and 577 storms based on a threshold of 1.6m. The initial state probability vector based on arbitrarily selected probability values, and the terminal state probabilities after 170 days and 365 days are listed in Table 2 for the case of no recovery between storms. With increasing time, a shifting of the peak of the state probability vector towards the higher states is observed indicating a degradation in foundation stiffness. When the results are compared with those obtained using the historical storms it appears that the results are conservative, but less so than for the earlier two models.

CONCLUSIONS

A new long-term stochastic model for sea-states has been developed to simulate the random evolution in time of the wave spectrum. The series of wave spectra thus generated becomes input information for reliability analysis in which sequencing of loads has to be taken into account. As an example, the model has been applied in assessing the reliability of platform foundations due to stiffness degradation. Results indicate that the simulated sea-states are realistic and in reasonable agreement with the observed data. It has been shown that the new model is significantly superior to the two earlier models developed in 1979 and 1981, and, in particular, overcomes the problems and shortcomings associated with those models.

ACKNOWLEDGEMENTS

This work was funded in part by INTEVEP, S.A. We are most appreciative of their support. The contribution of David DeLeeuw and Kong-Ann Soon in the numerical implementation of the long-term sea-state model is gratefully acknowledged.

REFERENCES

ANGELIDES, D.C., 1978, Stochastic Response of Fixed Offshore Structures in Random Sea, Ph.D. Thesis/Research Report R78-37, Massachusetts Institute of Technology, Department of Civil Engineering.

ANGELIDES, D.C., et al, 1981, "Random Sea and Reliability of Offshore Foundations," Journal of the Engineering Mechanics Division, Proceedings of the American Society of Civil Engineers, Volume 107, Number EM1, pp. 131-148.

CARTWRIGHT, D.E., 1974, "Theoretical and Technical Knowledge: Paper 1 - The Science of Sea Waves after 25 Years," The Dynamics of Marine Vehicles and Structures in Waves, London, England.

DELEEUW, D.N., 1981, A Long-Term Stochastic Model of Ocean Storms, S.M. Thesis, Massachusetts Institute of Technology, Department of Civil Engineering.

KINRA, R.K. and P.W. MARSHALL, 1979, "Fatigue Analysis of the Cognac Platform," Proceedings, 11th Offshore Technology Conference, Houston, Texas, 1979, Paper OTC 3378, pp. 169-185.

KINSMAN, B., 1965, Wind Waves, Their Generation and Propagation on the Ocean Surface, Englewood Cliffs, New Jersey: Prentice-Hall, Inc.

MOAN, T., et al, 1976, Stochastic Dynamic Response Analysis of Gravity Platforms, The Norwegian Institute of Technology, Division of Ship Structures, Report SK/M33.

NORDENSTRØM, N., 1971, Methods for Predicting Long Term Distribution of Wave Loads and Probability of Failure for Ships, Det norske Veritas, Report No. 71-2-S.

OCHI, M.K., 1978, "Wave Statistics for the Design of Ships and Ocean Structures," Transactions, The Society of Naval Architects and Marine Engineers, Volume 86, pp. 47-76.

SHYAM SUNDER, S., 1979, Stochastic Modelling of Ocean Storms, S.M. Thesis/Research Report R79-7, Massachusetts Institute of Technology, Department of Civil Engineering.

SHYAM SUNDER, S., et al, 1979, "A Stochastic Model for the Simulation of a Non-Stationary Sea," Proceedings, 2nd International Conference on the Behavior of Off-Shore Structures, London, England, 1979, Volume 1, Paper 9, pp. 95-106. Cranfield, Bedford, England: BHRA Fluid Engineering.

VAN EEKELEN, H.A.M., 1977, "Single-Parameter Models for Progressive Weakening of Soils by Cyclic Loading," Geotechnique, Volume 27, Number 3, pp. 357-368.

TABLE 1: STATISTICAL PARAMETER ESTIMATES FOR OCEAN WEATHER SHIP "M" DATA, 1961-68.

Season	m_{H_v}	σ_{H_v}	m_h	σ_h	σ_h^1
Winter	3.044	1.727	0.974	0.528	0.462
Spring	2.447	1.402	0.753	0.533	0.456
Summer	1.786	0.911	0.464	0.481	0.413
Fall	2.548	1.430	0.798	0.523	0.436

TABLE 2: STATE PROBABILITIES AT END OF 170 DAYS AND 365 DAYS OF SEA EXPOSURE FOR CASE OF NO RECOVERY BETWEEN STORMS.

State	Horizontal Stiffness ($k \times 10^{-7}$ lbf/ft)	State Probabilities		
		Initial	After 170 days	After 365 days
1	1.2	0.7	$\cong 0$	$\cong 0$
2	1.1	0.2	$\cong 0$	$\cong 0$
3	1.0	0.50×10^{-1}	$\cong 0$	$\cong 0$
4	0.9	0.30×10^{-1}	0.22×10^{-5}	$\cong 0$
5	0.8	0.10×10^{-1}	0.51×10^{-2}	0.12×10^{-4}
6	0.7	0.40×10^{-2}	0.64×10^{-1}	0.26×10^{-2}
7	0.65	0.10×10^{-2}	0.23	0.46×10^{-1}
8	0.6	0.10×10^{-2}	0.21	0.19
9	0.55	0.10×10^{-2}	0.91×10^{-1}	0.59×10^{-1}
10	0.525	0.10×10^{-2}	0.29	0.37
11	0.475	0.10×10^{-2}	0.89×10^{-1}	0.23
12	0.45	0.50×10^{-3}	0.16×10^{-1}	0.99×10^{-1}
13	0.425	0.30×10^{-3}	0.96×10^{-3}	0.93×10^{-2}
14	0.415	0.10×10^{-3}	0.24×10^{-3}	0.18×10^{-2}
15	0.4	0.10×10^{-3}	0.10×10^{-3}	0.10×10^{-3}

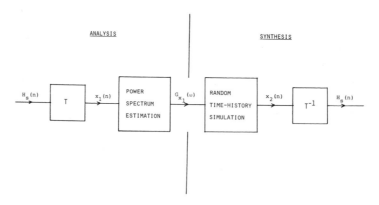

FIG. 1 SCHEMATIC OF LONG-TERM RANDOM SEA-STATE MODEL

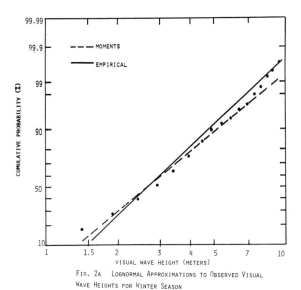

Fig. 2a Lognormal Approximations to Observed Visual Wave Heights for Winter Season

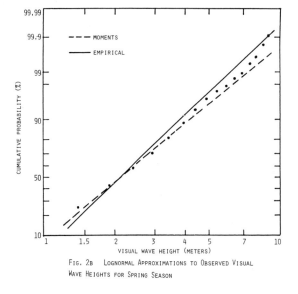

Fig. 2b Lognormal Approximations to Observed Visual Wave Heights for Spring Season

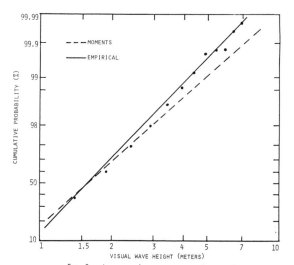

Fig. 2c Lognormal Approximations to Observed Visual Wave Heights for Summer Season

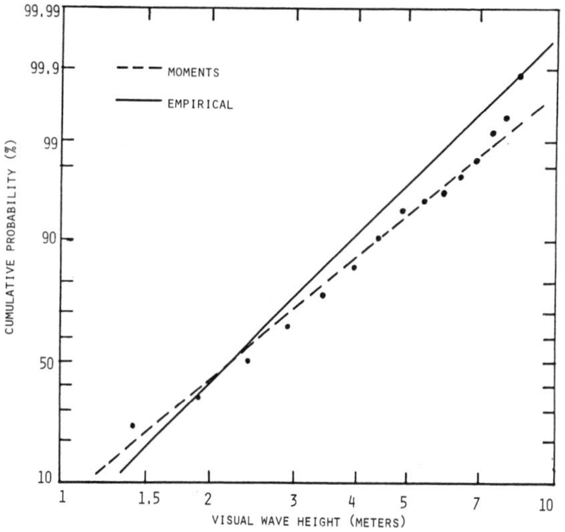

Fig. 2d Lognormal Approximations to Observed Visual Wave Heights for Fall Season

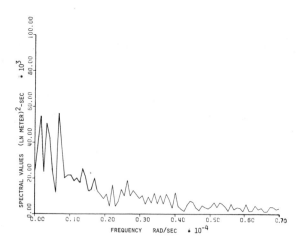

Fig. 3A Estimated Power Spectral Density Function for Winter

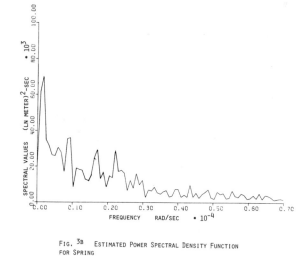

Fig. 3B Estimated Power Spectral Density Function for Spring

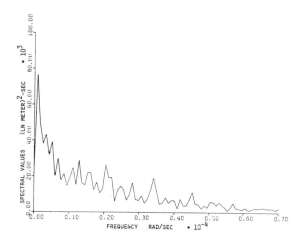

Fig. 3C Estimated Power Spectral Density Function for Summer

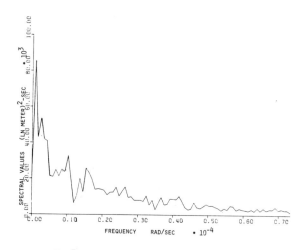

Fig. 3D Estimated Power Spectral Density Function for Fall

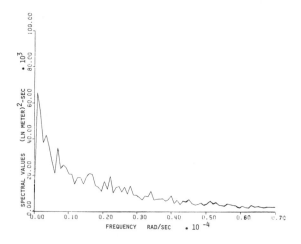

Fig. 4 Averaged Estimate of the Power Spectral Density Function for the Year

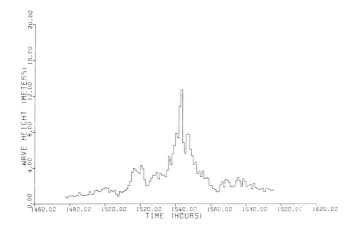

Fig. 5A Example of Simulated Visual Wave Height Time-History for the Winter Season

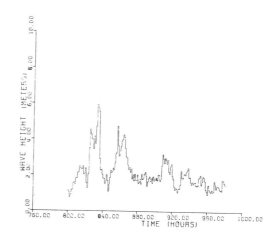

Fig. 5B Example of Simulated Visual Wave Height Time-History for the Spring Season

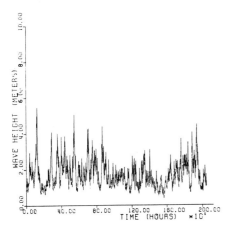

Fig. 5C Example of Simulated Visual Wave Height Time-History for the Summer Season

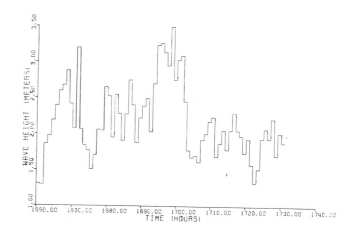

Fig. 5D Example of Simulated Visual Wave Height Time-History for the Fall Season

Technical Poster Sessions
—
STRUCTURES

Design and Reliability Issues for Alternate Structural Concepts

NONDESTRUCTIVE TESTING OF OFF-SHORE STRUCTURES BY SHEAROGRAPHY

Dr. Y. Y. Hung *
Industrial Holographics, Incorporated
Auburn Heights, Michigan, U.S.A.

Dr. Ralph M. Grant
Industrial Holographics, Incorporated
Auburn Heights, Michigan, U.S.A.

ABSTRACT

This paper describes a new optical method of nondestructive testing referred to as "Shearographic Nondestructive Testing" (SNDT). SNDT seems to be well suited for nondestructive inspection of off-shore structures. Principle of the method as well as it applications to detection of debonds in composite structure, cracks in pressure vessels and pipes, and evaluation of weld joints are presented.

* Dr. Y. Y. Hung is also an associate professor at Oakland University, presently on leave.

1. INTRODUCTION

Failures of off-shore structures are costly and could be disastrous. One avenue to minimize such occurences is to take preventive measures by inspecting the integrity of critical members. The existing techniques of nondestructive testing which are commonly used include visual inspection, radiography, magnetic particle, dye penetrant, eddy current, ultrasonics, infra-red, and holography. This paper presents a new optical method which will be referred to as "Shearographic Nondestructive Testing" (SNDT). It has been demonstrated that it is feasible to employ SNDT in a production environment.

2. HOW DOES SNDT DETECT FLAWS

SNDT is an interferometric method which allows surface strains in objects to be measured. It involves a double exposure photographic recording with the objects under inspection being deformed between the exposures. The processed photograph yields a interference fringe pattern which is a contour map of the surface strain due to the deformation. Since flaws in objects usually induce strain concentrations, SNDT reveals flaws by identifying the strain-induced concentrations. Strain concentrations produce anomolies in the fringe pattern which can easily be identified. Both surface and internal flaws are detectable with SNDT, as internal flaws also influence the surface deformation.

SNDT is a full-field method which does not require scanning or contacting. Thus, it is more suitable for production inspections that normally require a high inspection rate. SNDT also allows the criticality of flaws to be directly assessed; hence, it facilitates the accept-reject decision-making.

3. SNDT VS OTHER NDT TECHNIQUES

One major difference between shearography and other NDT techniques is the method by which the flaw is revealed. Other NDT techniques detect surface flaws by visual means. Dye penetrant and magnetic particles techniques are really enhanced visual techniques. Radiography and ultrasonics techniques detect internal flaws by identifying the inhomogeneities due to discontinuities in materials. Not all flaws are critical. The criticality depends on flaw size and flaw shape as well as the flaw location. These techniques lack the ability of differentiating between critical flaws and cosmetic flaws. Thus, they may lead to false rejection.

Premature failures of components are usually caused by excessive stress level (which are related to strain) induced by flaws. SNDT reveals flaws by identifying the regions experiencing excessive strains induced by the flaws. If the stress mode in the testing is similar to that in an actual object, SNDT can be used to detect the critical flaws only (i.e. flaws which create unsafe level of stress); cosmetic flaws can be ignored and false rejections can be avoided. Since a flaw in a low stress region does not weaken the component, it will not produce an alarming stress level to be detected by SNDT.

Holography detects flaws in a manner similar to SNDT except that holography measures displacements whereas shearography measures strains. It is easier to correlate imperfections with strain anomalies rather than displacement anomalies. Besides, SNDT does not require the special vibration isolation needed by holography. Thus, SNDT is more suited for inspections in production plant environments.

4. PRINCIPLE OF SNDT

SNDT utilizes an image-shearing technique recently developed for strain measurements.[1] **Fig. 1** shows a schematic diagram of a typical set up for performing shearographic nondestructive testing. The object under test is illuminated by an expanded laser beam and it is imaged by an image-shearing camera. A photographic film in the image plane of the camera is first exposed with the object in its undeformed state and it is subsequently exposed after the object is deformed. The processed photographic film yields a fringe pattern which depicts the gradients of the surface displacements (i.e. strains) due to the deformation. Defects including the internal flaws, which induce stress concentrations produce anomalies in the fringe pattern. The fringe anomalies usually appear as localized areas of high fringe density.

As shown in **Fig. 1**, the image-shearing camera is a camera equipped with a shearing device. The shearing device is a glass wedge located in the iris plane of the lens and it covers half of the lens aperture. Without the wedge, rays scattered from an object point, say $P_0(x,y)$ and received by the two halves of the lens will converge to a point in the image plane, i.e., a point on the object is mapped into a point on the image plane. The glass wedge is an angle prism which deviates rays passing through it. In the presence of the wedge, the rays from $P_0(x,y)$ focused by the half of the lens covered by the wedge are deviated; a point $P_0(x,y)$ in the object is mapped into two points P_1' and P_2' in the image plane. Thus, a pair of laterally-**sheared** images are observed in the image plane and hence the name **Shearography.** In other words the image-shearing camera brings the rays scattered from one point on the object surface to meet with those scattered from a neighboring point in the image plane. If the wedge is so oriented that the rays are deviated in the x-z plane, rays from a point $P(x,y)$ in the object will be brought to meet the rays from a neighboring point in the x-direction ($P(x+ x,y)$). This is illustrated in **Fig. 2**. Since the object is illuminated by coherent light (light that has the ability to interfere), the rays from the two neighboring points interfere with each other producing a random interference pattern (commonly known as a speckle pattern). When the object is deformed, a relative displacement between the two point occurs. This relative displacement produces a relative optical path change and hence a phase change which slightly modifies the interference speckle pattern. Superposition of the two interference speckle patterns (deformed and undeformed) by double exposure produces a fringe pattern which is similar to the beat produced by interference of two periodic signals of slightly different frequency. This fringe pattern measures the relative displacement between two neighboring points. If the separation between the two neighboring points is small, the relative displacements approximately represent the displacement gradient which is directly related to the strain.

5. EXPERIMENTAL DEMONSTRATION

5.1 Detection of Debonds in Composite Structures

With the increasing emphasis on the weight reduction, the use of composite material has become more popular because of its high strength to weight ratio. One major flaw in composite structures is debonds. **Fig. 3** indicates the presence of a debond in a fibre-reinforced composite cylinder detected by SNDT.

Debonds are characterized by a double bullseye fringe pattern. The area covered by the double bullseye approximately represents the size of the separations and the fringe density is related to the depth (distance from the surface) of the separations. The separation closer to the surface has higher fringe density and vice versa.

5.2 Detection of Cracks in Pressure Vessels and Pipes

SNDT is very well suited to the detection of flaws such as cracks in pressure vessel. Pressurization, which is the actual in service loading, is a convenient means of stressing the pressure vessel during the testing. **Fig. 4** shows a fringe pattern indicating the presence of an internal crack in a pressure vessel. **Fig. 5** reveals a crack in a pipe.

5.3 Evaluation of Welds

SNDT is an effective means for the evluation of weld joints. **Fig. 6** shows an imperfection of a butt weld joining two steel plates. The imperfection is a skip in the weld.

6. CONCLUSION

The work presented in this paper only represents a preliminary exploration of SNDT. SNDT can be used to evaluate the integrity of virtually any structural components built of any materials (plastics, rubber, metals, ceramics, composite materials, etc.). Further explorations of the SNDT for productions inspection are in progress. Full capability of SNDT still awaits to be explored. It appears that SNDT can be developed into a practical tool for on-site field tests of off-shore structures.

7. REFERENCE

(1) Y. Y. Hung and R. M. Grant, U.S. Patent Number 4139302 - Method and Apparatus for Interferometric Deformation Analysis.

8. ACKNOWLEDGEMENT

The authors gratefully acknowledge the support of Industrial Holographics, Inc., Auburn Heights, Michigan, 48057, U.S.A. which manufactures the equipment used to carry out this study.

9. FIGURES

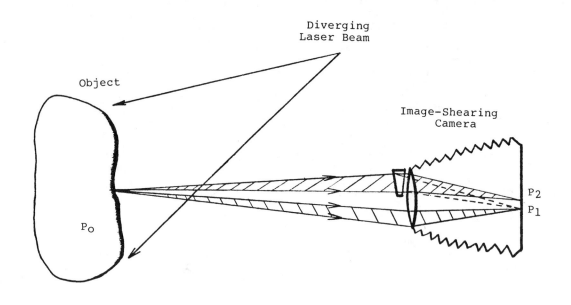

FIG. 1: SCHEMATIC DIAGRAM FOR SHEAROGRAPHIC NONDESTRUCTIVE TESTING

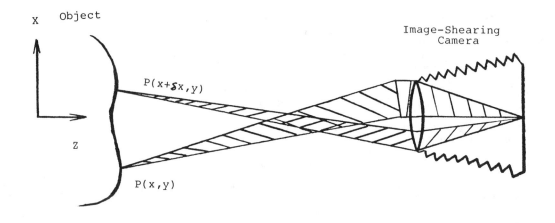

FIG. 2: RAYS FROM TWO NEIGHBORING POINTS ARE BROUGHT TO INTERFERE AT THE IMAGE PLANE

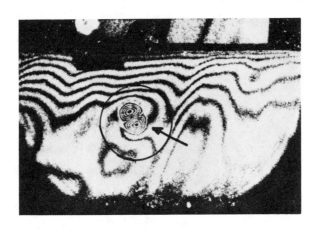

FIG. 3: A DEBOND IN A FIBRE-REINFORCED COMPOSITE CYLINDER IS REVEALED BY THE PRESENCE OF A DOUBLE BULLSEYE FRINGE PATTERN

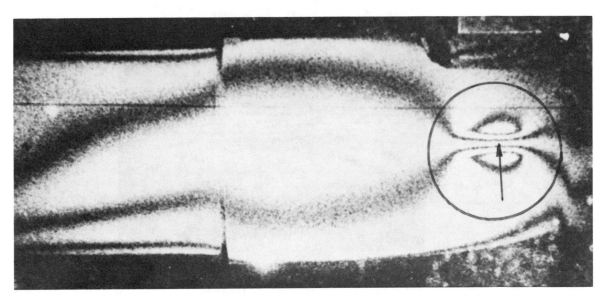

FIG. 4: THE FRINGE ANOMALY REVEALING AN INTERNAL CRACK IN A SHELL

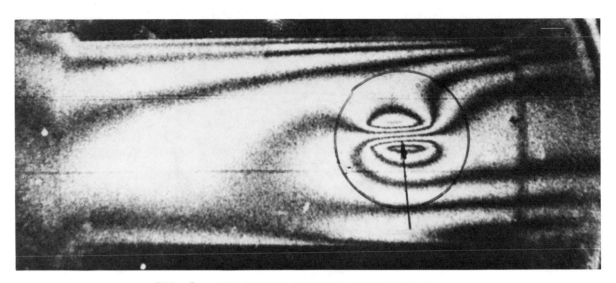

FIG. 5: THE FRINGE ANOMALY REVEALING AN INTERNAL CRACK IN A PIPE

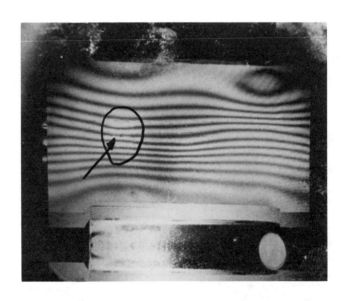

FIG. 6: THE PERTURBED FRINGE AREA INDICATES A SKIP IN A BUTT WELD

PROBABILISTIC STRENGTH OF STEEL FIBER REINFORCED CONCRETE

M. Matsuishi
Hitachi Shipbuilding and Engineering Co.

S. Iwata
Hitachi Shipbuilding and Engineering Co.

SUMMARY

The paper is concerned with the prediction of the probabilistic strength of concrete reinforced with short lengths of steel fibers. The first part deals with the probabilistic analysis of the number of fibers intersecting a cross section of the concrete with one-dimensional, two-dimensional or three-dimensional fiber distribution. In the analysis, the restraint of fiber distribution by surfaces of the concrete are fully considered. The expected number of fibers intersecting the cross section can be increased by aligning the fibers with the loading direction.

The second part deals with pull-out tests of steel fiber reinforced concrete subjected to tensile loading. The effects of fiber orientation, embedment length and the number of fibers upon the pull-out loads and the peak loads are clarified.

The third part deals with the probabilistic strength of the tensile strength of concrete reinforced with short and discontinuous steel fibers. The failure of steel fiber reinforced concrete is assumed to result from an accumulation of pull-out and/or tensile fracture of the individual fibers. A theoretical model for the load-carrying-capacity of each fiber is based on a series of pull-out tests on steel fiber reinforced concrete subjected to tensile loading. The expected strength of steel fiber reinforced concrete can be increased by aligning the fibers with the loading direction. The strength near the surfaces decreases due to the restraint of the fiber distribution by the formwork placed at the surface of the concrete.

1 INTRODUCTION

Offshore structures should have a high degree of safety and durability. A huge offshore structure, if made of steel, entails enormous cost to prevent corrosion of steel. The utilization of concrete for such an offshore structure gives an advantage of low maintenance cost. However, it is difficult to secure water tightness, once concrete cracking develops.

Recent studies have shown that concrete reinforced with short lengths of steel fibers possesses some improved characteristics when compared to ordinary concrete. Ref. (1),(2),(3) The safety and durability of offshore structures is ensured, if steel fiber reinforced concrete (SFRC) is placed over the surface of steel structures, or if SFRC is used at the region of concrete structures where large tensile stress is induced.

The authors carried out both experimental and theoretical researches on the probabilistic strength of SFRC: Pull-out test were carried out to study the effects of fiber orientation, embedment length and the number of steel fibers. A probabilistic theory is presented to predict the tensile strength of concrete reinforced with short, discontinuous steel fibers randomly oriented and uniformly dispersed in a cement-based matrix. The theoretical analysis shows how the expressions obtained from the pull-out behaviour of a single fiber may be adapted to describe the probabilistic ultimate-load-carrying-capacity of SFRC.

In the present paper, fiber orientation and dispersion are defined as follows.
A one-dimensional random distribution (1D distribution): Fibers are uniaxially aligned parallel to the tensile stress and dispersed uniformly. (Fiber inclination θ and ϕ in Fig. 1 are zero.)
A two-dimensional random distribution (2D distribution): Fibers are aligned randomly in planes parallel to the tensile stress. (Fiber inclination ϕ is zero.)
A three-dimensional random distribution (3D distribution): The orientation of fibers is random in three dimensions.

2 PROBABILITY OF FIBERS TO INTERSECT A CROSS SECTIONS

If one assumes a completely uniform distribution of fibers in a space shown in Fig.2, the probabilities of the fiber centroid x_g, y_g and z_g to be between $(x, x+dx)$, $(y, y+dy)$ and $(x, z+dz)$ are

$$\left. \begin{array}{l} P_x(x)\ dx = \dfrac{dx}{(x_1-X)} \\[4pt] P_x(y)\ dy = \dfrac{dy}{(y_1-Y)} \\[4pt] P_x(z)\ dz = \dfrac{dz}{(z_1-Z)} \end{array} \right\} \quad (1)$$

where $X = l_f \sin\theta \cos\phi$, $Y = l_f \sin\theta \sin\phi$, $Z = l_f \cos\theta$

The fiber inclination θ and ϕ shown in Fig. 1 will take on all values between 0 and pi. In the case of 2D distribution, the probability of fiber inclination θ to be between $(\theta, \theta+d\theta)$ is

$$P_\theta(\theta)\ d\theta = d\theta/\pi \quad (2)$$

In the case of 3D distribution, the probability of fiber inclination θ and ϕ to be between $(\theta, \theta+d\theta)$ and $(\phi, \phi+d\phi)$, respectively, are

$$P_{\theta\phi}(\theta, \phi)\ d\theta\ d\phi = \sin\theta\ d\theta\ d\phi/2\pi \quad (3)$$

2.1 The number of fibers intersecting a cross section

As the probabilities, Eqs.(1) to (3), are statistically independent, the probabilities of a fiber to intersect a cross section are given for 1D, 2D and 3D distribution, respectively

$$\left. \begin{array}{l} P_1 = \iiint P_x P_y P_z\ dx\ dy\ dz = \int 1/(z_1-Z)\cdot dz \\[4pt] P_2 = \iiiint P_x P_y P_z P_\theta\ dx\ dy\ dz\ d\theta = \iint 1/\pi(z_1-Z)\ dz\ d\theta \\[4pt] P_3 = \iiiint\!\!\int P_x P_y P_z P_{\theta\phi}\ dx\ dy\ dz\ d\theta\ d\phi = \iint \sin\theta/2\cdot(z_1-z)\ dz\ d\theta \end{array} \right\} \quad (4)$$

where the variables θ and ϕ take all values between 0 and pi, and z between z_s and z_e, respectively. Estimated means of the number of fibers intersecting the cross section at $z=z_\beta$ are given

$$E_i = N_f P_i \quad (5)$$

where N_f is the total number of fibers and subscript i denotes 1D, 2D and 3D distribution, respectively.

The limits Z_s and z_e of the integral with respect to z in Eq.(4) are given as follows.
For $0 \leq z_\beta \leq Z$

$$\left. \begin{array}{l} Z_s = l_f \cos \theta/2 \\ z_e = z + l_f \cos \theta/2 \end{array} \right\} \quad (6)$$

$$\left. \begin{array}{l} E_1 = N_f z_\beta/(z_1-l_f) \\ E_2 = 2N_f/\pi [2z_1/\sqrt{z_1^2-l_f^2} \tan^{-1}\sqrt{(z_1+l_f)/(z_1-l_f)} + \cos^{-1}(z_\beta/l_f) - 2(z_1-z_\beta)/\sqrt{z_1^2-l_f^2} \\ \tan^{-1}\sqrt{(z_1+l_f)(l_f-z_\beta)/(z_1-l_f)(l_f+z_\beta)} - \frac{\pi}{2}] \\ E_3 = N_f[z_\beta/l_f \cdot \log(z_1-z_\beta)/(z_1-l_f) + z_1/l_f \cdot \log z_1/(z_1-z_\beta) - z_\beta/l_f] \end{array} \right\} \quad (7)$$

For $Z \leq z_\beta \leq z_1 - Z$

$$\left. \begin{array}{l} z_s = z_\beta - l_f \cos \theta/2 \\ z_e = z_\beta + l_f \cos \theta/2 \end{array} \right\} \quad (8)$$

E_i are obtained by substituting $z_\beta = l_f$ in Eq.(7).

For $z_1 - Z \leq z_\beta \leq z_1$

$$\left. \begin{array}{l} z_s = z_\beta - l_f \cos \theta/2 \\ z_e = z_1 - l_f \cos \theta/2 \end{array} \right\} \quad (9)$$

E_i are obtained by replacing z_β in Eq.(7) by $z_1 - z_\beta$.

Expected values of the average number of fibers intersecting the cross section, n_i, are defined as follows.

$$n_i = E_i / A \quad (10)$$

where A is the sectional area of SFRC $(=x_1 y_1)$.

2.2 The number of fibers per unit area

The expected number of fibers per unit area intersecting a cross section are given.

1D distribution $\quad e_1 = \int N_f P_x P_y P_z \, dx \, dy \, dz/dx \, dy = \int N_f/x_1 y_1 (z_1-l_f) \, dz \quad (11)$

2D distribution $\quad e_2 = \iint N_f P_x P_y P_z P_\theta \, dx \, dy \, dz \, d\theta/dx \, dy = \iint N_f/\pi (x_1-X)(y_1-Y)(z_1-Z) \, dz \, d\theta \quad (12)$

3D distribution $\quad e_3 = \iiint N_f P_x P_y P_z P_{\theta\phi} \, dx \, dy \, dz \, d\theta \, d\phi/dx \, dy = \iiint N_f \sin \theta/2\pi (x_1-x)$
$(y_1-Y)(z_1-Z) \, dz \, d\theta \, d\phi \quad (13)$

By integrating from 0 to pi with respect to θ and ϕ, and from z_s to z_e with respect to z, the expected number of fibers per unit area are obtained.

(1) 1D distribution

The integral limits with respect to z is the same as Eqs.(6), (8) and (9). Thus

$$e_i = n_i \quad (14)$$

(2) 2D distribution

The limits z_s and z_e in Eq.(12) are dependent on both the fiber inclination θ and the cross section z_β, and can be defined in nine regions shown in Fig. 3. As a typical example, the expected values in a space V_2 shown in Fig. 4 are given as follows.

For $0 \leq z\beta \leq l_f$

$$\left.\begin{array}{l} z_e = z_\beta + l_f \cos\theta/2 \\ z_s = z_\beta - x/\tan\theta + l_f \cos\theta/2 \end{array}\right\} \quad \text{for } 0 \le z_\beta \le l_f \cos\theta \tag{15}$$

$$\left.\begin{array}{l} z_e = z_\beta + l_f \cos\theta/2 \\ z_s = z_\beta - l_f \cos\theta/2 \end{array}\right\} \quad \text{for } l_f \cos\theta \le z_\beta \le z_1 - l_f \cos\theta \tag{16}$$

$$e_2 = 2N_f/\pi \, [I_1 \, z_\beta/y_1 + I_2 \, l_f/y_1] \tag{17}$$

where I_1 and I_2 are given in Ref.(3).

For $l_f \le z_\beta \le z_1 - l_f$, e_2 is obtained by substituting $z_\beta = l_f$ in Eq.(17).

For $z_1 - l_f \le z_\beta \le z_1$, e_2 is obtained by replacing z_β in Eq.(17) by $z_1 - z_\beta$.

(3) 3D distribution

The limits z_s and z_e in Eq.(13) are dependent on both the fiber inclinations θ and ϕ, and the cross section z_β, and are defined in each region shown in Fig.5. As a typical example, the expected values in V_3 shown in Fig.5 are given as follows.

For $0 \le z_\beta \le l_f$

$$\left.\begin{array}{l} z_s = l_f \cos\theta/2 \\ z_e = z_\beta + l_f \cos\theta/2 \end{array}\right\} \quad \text{for } 0 \le z_\beta \le l_f \cos\theta \tag{18}$$

$$\left.\begin{array}{l} z_s = z_\beta - l_f \cos\theta/2 \\ z_e = z_\beta + l_f \cos\theta/2 \end{array}\right\} \quad \text{for } l_f \cos\theta \le z_\beta \le z_1 - l_f \cos\theta \tag{19}$$

$$e_3 = 4\left[\int_0^{\theta'}\int_0^{\pi/2} f_e \, l_f \cos\theta \, d\theta \, d\phi + \int_{\theta'}^{\pi/2}\int_0^{\pi/2} f_e \cdot z_\beta \, d\theta \, d\phi \right] \tag{20}$$

where $f_e = N_f \sin\theta/2\pi(x_1-X)(y_1-Y)(z_1-Z)$, $\theta' = \cos^{-1}(z_\beta/l_f)$

For $l_f \le z_\beta \le z_1 - l_f$, e_3 is obtained by substituting $z_\beta = l_f$ in Eq.(20).

For $z_1 - l_f \le z_\beta \le z_1$, e_3 is obtained by replacing z_β in Eq.(20) by $z_1 - z_\beta$.

Eq.(20) is integrated numerically.

2.3 Numerical calculation and discussion

When the fiber length l_f is infinitesimally small as compared with the length z_1, we obtain the expected number of fibers per unit area.

$$e_1/B = n_1/B = 1, \quad e_2/B = n_2/B = 2/\pi, \quad e_3/B = n_3/B = 1/2 \tag{21}$$

where $B = p/a_f$. a_f and p are the sectional area and the volume fraction of fibers, respectively.

The calculated results of n_i and e_i are shown in Fig.6. As seen in the figure, the expected values of the number of fibers near the surfaces of SFRC are smaller than those apart from the surfaces. This is because the integral limits, z_s and z_e, and the fiber orientation, θ and ϕ, are restrained by formwork placed on the surface of SFRC. The expected values of the number of fibers can be increased by aligning the fibers with the loading direction.

3. EXPERIMENT OF FIBER PULL-OUT

A series of pull-out tests on SFRC subjected to tensile loading was carried out to study the variation of test results from fiber orientation, embedment length and the number of fibers. Both deformed and round fibers were employed in the tests.

From the large amount of data obtained from the tests, some of the results of the deformed fibers produced by shearing sheets (25x0.5x0.25 mm) are summarized in Figs. 7 - 10.

Fig.7 shows load vs pull-out distance curves for various embedment lengths of single fibers parallel to the tension. Fig.8 represents variation of observed peak load with the embedment length. It can be seen that for the emebedment length up to 5.8 mm, the peak load increases proportionally to the embedment length and attains a constant value corresponding to the tensile breaking strength of the fiber for the embedment length greater than 5.8 mm.

Fig.9 represents the peak loads of the pull-out tests on two fibers which were symmetrically oriented with respect to the loading direction. It can be seen that the measured peak loads coincide well with the calculated ones given by the following equation.

$$F_\theta = F_0 \cos \theta \qquad (22)$$

where F_0 is a peak load on a fiber aligned parallel to the applied load.

Fig.10 shows the peak load of pull-out tests on four fibers; two of them were symmetrically oriented with respect to the loading direction and the other two were aligned parallel to the applied load. It is found that the measured peak loads coincide well with the calculated ones given by the following equation.

$$F_c = (N_0 F_0 + N_\theta F_\theta)/(N_0 + N_\theta) \qquad (23)$$

where F_0 and N_0: the peak load and the number of fibers aligned to the tension

F_θ and N_θ: the peak load and the number of fibers lying at an oblique angle, θ, to the tension

4. ANALYSIS OF PROBABILISTIC TENSILE STRENGTH

Because of the relatively small cracking strain of Portland cement concrete or mortal, initial cracking occurs at a relatively early stage. The matrix crack is bridged by discontinuous fibers. The reinforcing action by the fibers occurs through the fiber-matrix interfacial shear stress. As the result, a significant contribution of the fibers occurs after the matrix starts cracking.

In the present paper, SFRC's failure is assumed to result from an accumulation of pull-out and/or tensile fracture of the individual fibers. The load-carrying-capacity of each fiber is dependent on the fiber orientation θ relative to the direction of the applied tension, and the length of the fiber withdrawn in pull-out. Fiber/fiber interaction is not considered.

4.1 Case-1: neglecting the effect of pull-out length

The load carried by a fiber can be represented in the form

$$F = F_0 f(l_e) \cos \theta \qquad (24)$$

where F_0: min. (pull-out load, tensile breaking strength) of fiber aligned parallel to the direction of the tensile stress
$f(l_e)$: the function representing the effect of fiber pull-out length, l_e

If we neglect the effect of pull-out length of individual fibers intersecting a crack surface, Eq.(24) becomes

$$F = F_0 \cos \theta \qquad (25)$$

The expected value of the tensile strength of SFRC are obtained, from Eqs.(5), (11), (12) (13) and (25).

(1) Expected values of mean strength

The expected values of mean strength per unit area at $z=z_\beta$ under tension are obtained by multiplying the probability of fibers intersecting the cross section by the load carried by fibers.

$$s_1 = \frac{F_0 N_f}{A} \int \frac{dz}{z_1 - l_f} \qquad (26)$$

$$s_2 = \frac{F_0 N_f}{\pi A} \iint \frac{\cos \theta}{z_1 - z} dz\, d\theta \qquad (27)$$

$$s_3 = \frac{F_0 N_f}{2A} \iint \frac{\sin \theta \cos \theta}{z_1 - z} dz\, d\theta \qquad (28)$$

where the variables, θ and z, take values between 0 and pi, and between z_s and z_e, respectively.

For $0 \leq z_\beta \leq l_f$, Eq. (26)-(28) becomes for the specific fiber distribution,

1D distribution $\quad s_1 = \dfrac{F_0 N_f z_\beta}{A(z_1-l_f)}$ (29)

2D distribution $\quad s_2 = \dfrac{F_0 N_f}{2A} f_2$ (30)

3D distribution $\quad s_3 = \dfrac{F_0 N_f}{3A} f_3$ (31)

where
$$f_2 = \frac{4}{\pi} [\frac{2 z_1^2}{l_f \sqrt{z_1^2-l_f^2}} \tan^{-1} \sqrt{\frac{z_1+l_f}{z_1-l_f}} + 2 \frac{z_1}{l_f} \tan^{-1} \sqrt{\frac{l_f-z_\beta}{l_f+z_\beta}} + \frac{\sqrt{l_f^2-z_\beta^2}}{l_f} - \frac{2 z_1(z_1-z_\beta)}{l_f \sqrt{z_1^2-l_f^2}}$$
$$\tan^{-1} \sqrt{\frac{(l_f-z_\beta)(z_1+l_f)}{(l_f+z_\beta)(z_1-l_f)}} + \frac{z_\beta}{l_f} \cos^{-1}(\frac{z_\beta}{l_f}) - \frac{\pi}{2} \frac{z_1}{l_f} - 1]$$

$$f_3 = 3 [\frac{z_1 z_\beta}{l_f^2} \log(\frac{z_1-z_\beta}{z_1-l_f}) + \frac{z_\beta(z_\beta-l_f)}{l_f^2} + \frac{z_1^2}{l_f^2} \log(\frac{z_1}{z_1-z_\beta}) - \frac{z_1 z_\beta}{l_f^2} - \frac{z_\beta^2}{2l_f^2}]$$

For $l_f \leq z_\beta \leq z_1-l_f$, s_i are obtained by substituting $z_\beta = l_f$ in Eqs. (29),(30) and (31).

For $z_1-l_f \leq z_\beta \leq z_1$, s_i are obtained by replacing z_β by z_1-z_β in Eqs. (29), (30) and (31).

(2) Expected values of strength per unit area

The expected values of strength per unit area of SFRC under tension are obtained by multiplying the probability of fibers intersecting the cross section by the load carried by the fibers.

$$\left. \begin{array}{l} r_1 = F_0 N_f \int \dfrac{dz}{x_1 y_1 (z_1-l_f)} \\[6pt] r_2 = \dfrac{F_0 N_f}{\pi} \iint \dfrac{\cos\theta\ dz\ d\theta}{(x_1-X)(y_1-Y)(z_1-Z)} \\[6pt] r_3 = \dfrac{F_0 N_f}{2\pi} \iiint \dfrac{\sin\theta\ \cos\theta\ dz\ d\theta\ d\phi}{(x_1-X)(y_1-Y)(z_1-Z)} \end{array} \right\} \quad (32)$$

By integrating Eq.(32) from 0 to pi with respect to θ and ϕ, and from z_s to z_e with respect to z, we have the expected values of strength for the specific fiber distribution.

1D distribution: r_1 is equal to s_i of Eq.(29).

2D distribution: The limit z_s and z_e in Eq.(32) are dependent on both the fiber inclination θ and the cross section z_β, and can be defined in the nine regions shown in Fig.3. As a typical example the expected values of strength in the space V_2 shown in Fig.4 are obtained by using z_e and z_s of Eqs.(15) and (16).

For $0 \leq z_\beta \leq l_f$, $\quad r_2 = \dfrac{F_0 N_f}{2A} [I_3 z_\beta + I_4 l_f]$ (33)

where
$$I_3 = \frac{4}{\pi l_f D} [z_1 \log \frac{(z_1+l_f)t_0^2+(z_1-l_f)}{z_1-l_f} + z_1 \log \frac{x_1}{x_1 t_0^2 - 2 l_f t_0 + x_1} + \frac{2 x_1 z_1}{\sqrt{z_1^2-l_f^2}}$$
$$\tan^{-1} \frac{(z_1+l_f)t_0}{\sqrt{z_1^2-l_f^2}} - 2\sqrt{x_1^2-l_f^2} (\tan^{-1} \frac{x_1 t_0-l_f}{\sqrt{x_1^2-l_f^2}} + \tan^{-1} \frac{l_f}{\sqrt{x_1^2-l_f^2}})]$$

$$I_4 = \frac{4}{\pi l_f^2 D} [D \log(1+t_0^2) - D \log 2 + (x_1^2-l_f^2) \log \frac{2(x_1-l_f)}{x_1 t_0^2 - 2 l_f t_0 + x_1} + z_1^2 \log$$
$$\frac{2z_1}{(z_1+l_f)t_0^2+(z_1-l_f)} - 2z_1 \sqrt{x_1^2-l_f^2} (\tan^{-1} \sqrt{\frac{x_1-l_f}{x_1+l_f}} - \tan^{-1} \frac{x_1 t_0 - l_f}{\sqrt{x_1^2-l_f^2}}) +$$
$$\frac{2 x_1 z_1^2}{\sqrt{z_1^2-l_f^2}} (\tan^{-1} \sqrt{\frac{z_1+l_f}{z_1-l_f}} - \tan^{-1} \frac{(z_1+l_f)t_0}{\sqrt{z_1^2-l_f^2}})]$$

3D distribution: The limits z_s and z_e in Eq.(32) are dependent on both the fiber inclination θ and ϕ, and the cross section z_β, and are defined in each region shown in Fig.5. As a typical example, the expected values in the space V_3 shown in Fig.5 are obtained by using Eqs.(18) and (19).

For $0 \leq z_\beta \leq l_f$,
$$r_3 = \frac{F_0 N_f}{3A} \left[\int_0^{\theta_0} \int_0^{\pi/2} f\, l_f \cos\theta\, d\theta\, d\phi + \int_{\theta_0}^{\pi/2} \int_0^{\pi/2} f\, z_\beta\, d\theta\, d\phi \right] \tag{34}$$

where
$$\theta_0 = \cos^{-1}(z_\beta/l_f)$$
$$f = \frac{6A \sin\theta \cos\theta}{\pi(x_1-X)(y_1-Y)(z_1-Z)}$$

For $l_f < z_\beta \leq z_1 - l_f$, r_2 and r_3 are obtained by substituting $z_\beta = l_f$ in Eqs.(33) and (34), respectively.

For $z_1 - l_f \leq z_\beta \leq z_1$, r_2 and r_3 are obtained by replacing z_β in Eqs.(33) and (34) by $z_1 - z_\beta$, respectively.

(3) Numerical calculation and discussion

When the fiber length is infinitesimally small as compared with the length of SFRC, we obtain the expected values of the strength under tension.

$$r_1/C = s_1/C = 1, \quad r_2/C = s_2/C = 1/2, \quad r_2/C = s_3/C = 1/3 \tag{35}$$

where $C = F_0\, p/a_f$

The calculated results of the expected values of the strengths are shown in Fig.11. As shown in the figure, the expected strengths of SFRC under tension can be increased by aligning the fibers with the loading direction. The strength near the surfaces decreases due to the restraint of the fiber distribution by the formwork placed at the surface of SFRC.

4.2 Case-2: considering the effect of pull-out length

The pull-out behaviour of the individual fiber is assumed to be elastic-perfectly-plastic. The load carried by the fiber can be represented in the form

$$f(\delta) = \begin{cases} (F_y/\delta_y)\delta & 0 \leq \delta \leq \delta_y \\ F_y & \delta_y \leq \delta \leq l_e \\ 0 & \delta > l_e \end{cases} \tag{36}$$

where δ : the pull-out length of fiber corresponding to a crack width of SFRC
l_e : maximum pull-out length of the fiber represented in the form

$$l_e = \min.(l_1, l_f - l_1)$$
$$l_1 = l_f/2 - z_G + z_\beta, \quad 0 \leq z_\beta \leq l_f, \quad l_f/2 \leq z_G \leq z_\beta + l_f/2$$

F_y: peak pull-out load of fiber represented in the form

$$F_y = \begin{cases} a_1\, l_e & 0 \leq l_e \leq l_0 \\ F_0 & l_e > l_0 \end{cases}$$

δ_y: pull-out length under the pull-out load F_y

l_0: pull-out length when pull-out load reaches F_0

The strength of SFRC under tension is dependent on the pull-out length of fibers. The pull-out length is so determined that the expected strength takes a maximum value.

As a typical example, the expected strength of SFRC with one dimensional fiber distribution is calculated.

$$r_1 = \max.\left(\int \frac{f(\delta)\, N_f\, dz}{(x_1-X)(y_1-Y)(z_1-Z)} \right) \tag{37}$$

The expected value of the strength is obtained by solving Eq.(37) under the condition in which the load supported by each fiber takes the maximum value.

For $0 \leq z_\beta \leq l_0$,
$$r_1 = \frac{F_0 N_f z_\beta^2}{2 l_0 x_1 y_1 (z_1-l_f)} \tag{38}$$

For $l_0 \leq z_\beta \leq l_f-l_0$
$$r_1 = \frac{F_0 N_f (2 z_\beta-l_0)}{2 x_1 y_1 (z_1-l_f)} \tag{39}$$

For $l_f-l_0 \leq z_\beta \leq l_f$,
$$r_1 = \frac{F_0 N_f (z_\beta+l_f-2 l_0)}{2 x_1 y_1 (z_1-l_f)} \tag{40}$$

For $l_f \leq z_\beta \leq z_f-l_f$, r_1 is obtained by substituting $z_\beta=l_f$ in Eq.(40).

The calculated results of the expected strength is shown in Fig.11. It can be seen that the expected strength of SFRC under tension is reduced by considering the effect of pull-out lengths of fibers upon their load-carrying capacities.

5. CONCLUDING REMARKS

The authors carried out both theoretical analyses of the probabilistic strength of concrete reinforced with short lengths of steel fibers (SFRC). Pull-out tests were also carried out on SFRC. The important information obtained in this study is as follows.

(1) The expected number of fibers intersecting a cross section is increased by aligning the fibers with the loading direction.
(2) When the length of the fiber is infinitesimally short as compared with the length of fibers the number of fibers of aligned SFRC is reduced to $2/\pi$ of its value by randomizing in two dimensions and to one half of its value by radomizing in three dimensions.
(3) The measured pull-out load of steel fiber increases proportionally to the embedment length and attains a constant value corresponding to the tensile breaking strength of the fiber for the embedment length greater than a critical value determined for each type of steel fiber.
(4) The measured pull-out load of inclined fiber decreases proportionally to $\cos \theta$, where θ is an angle between fiber direction and the direction of the applied load.
(5) The calculated expected-values of the strength of SFRC near the surface decreases due to the restraint of the fiber distribution by the formwork placed at the surface of SFRC.
(6) When the length of the fiber is infinitesimally short as compared with the length of the concrete, the expected value of the calculated strength of aligned SFRC is reduced to one half of its value by randomizing in two dimensions and to one third of its value by randomizing in three dimensions.

REFERENCES

1) Batson, G., E. Jenkins and R. Spatney, 1972, "Steel Fibers as Shear Reinforcement in Beams," Journal of American Concrete Institute, October, pp640-644.

2) Henager, C. H. and T. J. Doherty, 1976, "Analysis of Reinforced Fibrous Concrete Beams," Proceedings of the American Society of Civil Engineering, Volume 102, Number ST1, pp177-188.

3) Matsuishi, M. and S. Iwata, 1982, "Strength of Concrete Reinforced with Discontinuous Steel Fibers," Journal of the Kansai Society of Naval Architects, Number 184, September. (to be published)

4) Swmy, R. N. and P. S. Mangat, 1974, "A Theory For The Flexural Strength Of Steel Fiber Reinforced Concrete, " Cement and Concrete Research, Volume 4, pp313-325.

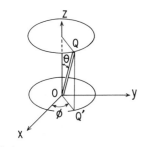
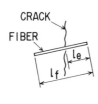

(a) FIBER ORIENTATION   (b) PULL-OUT LENGTH

FIG. 1 STEEL FIBER

	x-COORD	y-COORD	z-COORD
G_1	X_0	Y_0	Z_0
G_2	X_0	Y_0	$z_1 - Z_0$
G_3	$x_1 - X_0$	Y_0	$z_1 - Z_0$
G_4	$x_1 - X_0$	Y_0	Z_0
G_5	X_0	$y_1 - Y_0$	Z_0
G_6	X_0	$y_1 - Y_0$	$z_1 - Z_0$
G_7	$x_1 - X_0$	$y_1 - Y_0$	$z_1 - Z_0$
G_8	$x_1 - X_0$	$y_1 - Y_0$	Z_0

$X_0 = l_f \cdot \sin\theta \cdot \cos\phi / 2$
$Y_0 = l_f \cdot \sin\theta \cdot \sin\phi / 2$
$Z_0 = l_f \cdot \cos\theta / 2$

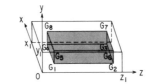

FIG. 2 FEASIBLE SPACE OF FIBER CENTROID

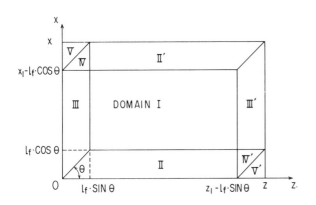

FIG. 3 DOMAINS OF INTEGRAL LIMITS FOR 2D DISTRIBUTION
($0 \leq \theta \leq \pi/2$)

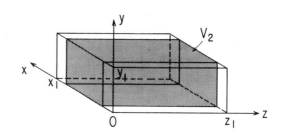

$V_2 \begin{cases} l_f \leq x \leq x_1 - l_f \\ 0 \leq y \leq y_1 \\ 0 \leq z \leq z_1 \end{cases}$

FIG. 4 SPACE V_2 FOR 2D DISTRIBUTION

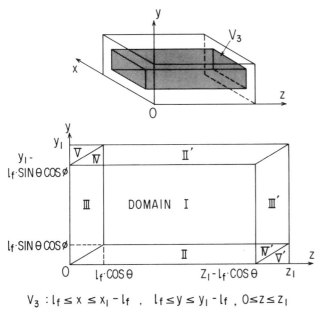

$V_3 : l_f \leq x \leq x_1 - l_f , \quad l_f \leq y \leq y_1 - l_f , \quad 0 \leq z \leq z_1$

FIG. 5 SPACE V_3 FOR 3D DISTRIBUTION

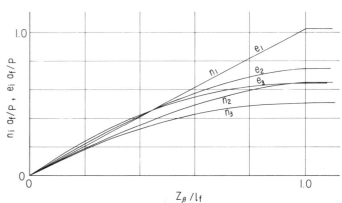

FIG. 6 EXPECTED VALUES OF THE NUMBER OF FIBERS
($l_f = 25$, $x_1 = y_1 = 100$, $z_1 = 1000$ mm)

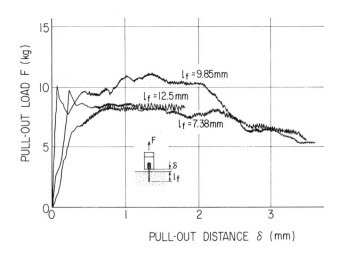

FIG. 7 PULL-OUT LOAD VS PULL-OUT LENGTH

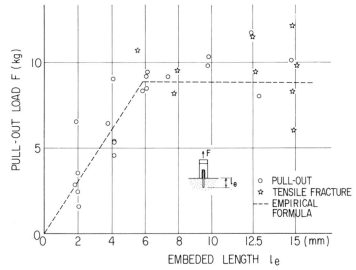

FIG. 8 PEAK LOAD VS EMBEDED LENGTH

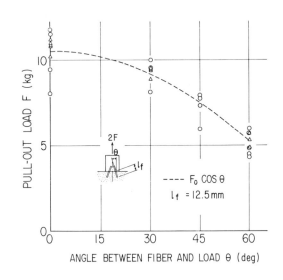

FIG. 9 PEAK PULL-OUT LOAD OF INCLINED FIBERS

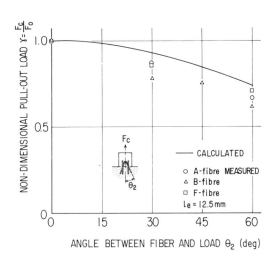

FIG. 10 PEAK PULL-OUT LOAD OF INCLINED FIBERS WITH $\theta_1 = 0$ AND θ_2

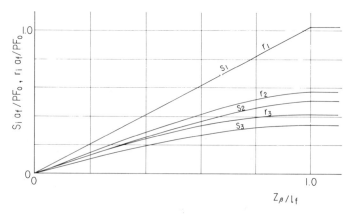

FIG. 11 EXPECTED VALUES OF STRENGTH
($l_f = 25$, $x_1 = y_1 = 100$, $z_1 = 1000$ mm)

DEVELOPMENT OF THE TRIPOD TOWER PLATFORM DESIGN

F.C. Michelsen
Heerema Engineering Service
The Netherlands

J. Meek
Heerema Engineering Service
The Netherlands

SUMMARY

Described in this paper are details of the structural design of the Tripod Tower Platform (T.T.P.) which is a fixed steel structure intended for supporting oil and gas exploitation activities in the regions of the North Sea where water depth is between 180-375 m approximately, as well as in other areas of similar depth and severe environmental conditions.

The structural concept of the T.T.P., made possible by recent advances in the technologies of steel making, fabrication and construction, is shown to deviate substantially from current platform design practices. Principally it comprises a vertical centre column supported by three inclined legs, all manufactured as large diameter tubulars of heavy wall thickness. Background of design and description of general features are presented, as are results of the analysis of wave loading and the associated static and dynamic stress levels. Special attention has been paid to the dynamic behaviour of the T.T.P. since this has a direct bearing on its fatigue life, which for fixed deep water structures plays a governing role. The last part of the paper deals with the analysis of the pre-installed piled foundation and the installation of the complete structure excluding deck and modules.

INTRODUCTION

In man's search for hydrocarbon reservoirs worldwide he has discovered that a good number of these are located below the sea bottom. In regions where the depth of water is relatively shallow and weather conditions not too severe, such as close to the coast in the Gulf of Mexico, the establishment of production from offshore oil and gas fields did not present problems which could not be overcome by means of well known technology. Thus offshore exploration of oil and gas got underway in the Gulf of Mexico as early as the 1940's. For that purpose simple structures were in the beginning built on piles driven into the sea bottom. As depth of water increased one saw the development of the submersible and the jack-up, although for the production facilities a space frame structure made up of tubulars and called a jacket was generally adopted. The technology of jacket platform construction was thus gradually developed to a fairly high degree of sophistication.

When commercial reservoirs of oil and gas were located in the North Sea in 1969 a new frontier of the offshore exploration and exploitation of hydrocarbons came into existance. Much of the technologies suitable for the Gulf of Mexico climatology were soon found to fall short in the hostile environments of the North Sea and new developments were called for. The economic exigencies brought about by the oil crisis of 1973, and those to follow, provided an added impetus in that direction.

Although the concrete gravity platform gained an acceptance, most platforms installed in the North Sea have so far been of the steel jacket type. On account of the extremely hostile climatology of the North Sea and the large topside facilities required, these structures have reached a size and complexity which only some years ago would hardly have been considered feasible. To realize these jackets it has been necessary to establish special fabrication facilities and construct purpose built large transportation and launch barges that can handle jacket weights of up to 30 000 tonne. Hand in hand with this the development of super pile driving hammers evolved which are capable of driving the large piles called for to full penetration in all soil conditions generally found in the North Sea. The development of crane lifting capacity was not any less spectacular. Within a span of nine years this capacity went from 500 s.t. to 5000 s.t., which is now available in a tandem lift with the two cranes of the semi-submersibles the "Balder" and the "Hermod". These vessels have substantially reduced the time required to install fixed steel platforms in the North Sea and can furthermore do so on a year around basis.

So far the maximum depth of water conquered in the North Sea is approximately 160 m. A reason for this is that enough fields were found within this depth range to provide an economical basis upon which an orderly build-up of the offshore exploration of oil and gas could be established. Taxation policies adopted by the countries having jurisdiction over the oil and gas fields have undoubtedly also had an influence on the developments in this regard. Having said this, it must be emphasized, however, that the state-of-the-art has yet to reach the point where one can decide within an acceptable level of risk that a field located in 300 m water depth, for example, is to be developed; that is to say, design and specifications are ready to be drawn up and contracts for construction and installation can be entered into.

Surveys indicate that a significant number of promising geological formations, with a high probability of containing commercial quantities of recoverable oil and gas, are located in water depths of up to 400 m in the North Sea. If the regions north of the 62° parallel are included then indications are that recoverable reserves within these depths are greater than what can be recovered in the shallower parts of the North Sea. Nothing more needs to be said to emphasize that the incentive to develop the design of a fixed steel platform suitable for installation in the deeper parts of the North Sea is prevalent.

Concepts for such platforms have been presented in the literature. Two jackets have already been installed in approximately 300 m water depth in the Gulf of Mexico. These particular structures would, in their present configuration, not stand up to the North Sea environment. It has become known that deep water jackets for this region are now being studied, but based upon preliminary investigations it is the opinion of the authors that the weight and complexity of a jacket designed for 350 m and North Sea conditions will become prohibitive on account of the governing fatigue criteria. Inspection and maintenance is in this respect also important. Construction and especially installation in areas of soft soil conditions may in addition present installation operations involving an unacceptably high degree of risk. For these reasons it was decided at Heerema to approach the problem of the deep water fixed steel platform design afresh, without being constrained by our experience with existing design concepts and taking full advantage of recent developments which had taken place in steel making, welding and construction equipment both onshore and offshore. The result of this endeavour is the Tripod Tower Platform (T.T.P.), and the subject of this paper is the presentation of its essential design features as a structure.

Mention should be made of other types of platforms designed for operation in deep water, such as the Tension Leg, Articulated Tower, Semi-Spar and others. Although the conquer of water depth is a common objective of these concepts they frequently differ significantly in regard to other main characteristics. But so do the characteristics of the oil and gas fields so that what is a viable platform for the exploration of one field may not be so for another.

Fundamentally the deep water platform concepts can be divided into two main categories, depending upon which of the following two principles the design has been based upon, i.e. the platform is either designed to be stiffly connected to the sea floor or to be more flexibly linked to the same and thus move more freely with the waves.

The two principles differ fundamentally in regard to dynamic response characteristics of the platforms as they are being subjected to the action of the waves. To avoid high stress levels due to the peak value of the dynamic amplification factor it is necessary that natural periods of vibrations of the structure fall outside the range of wave periods of the wave spectra where the waves have a significant energy level. In the case of the stiff structure this implies that the first natural period must be relatively low. For the North Sea it should preferably be somewhat less than 5 sec., depending upon the damping factor.

The flexibly linked platform is designed to have a natural period which is considerably longer than that of the waves encountered. This will, however, not necessarily prevent the structure from being excited in its first natural mode by the waves, since the slowly varying second order wave forces have a spectrum which cover a wide range of long periods. Wind forces may also come into play. Additionally it is important for these platform structures to pay special attention to the higher modes of vibration.

At the outset of the T.T.P. design project it was realized that the success of obtaining a first natural period of oscillation of less than 5 sec. for a "stiff" platform designed for a water depth of more than 1000 ft. was far from guaranteed. Nevertheless, considering the overall aspects of its structural design, as well as problems associated with its construction and installation, it was decided to go for that type of platform structure; a contributing factor being a design topside weight of facilities in the order of 25 - 30 000 tonne.

BACKGROUND OF DESIGN

In 1978, when the T.T.P. design project was initiated, the large platforms now in operation in the North Sea were under construction. The special launch barges required for transportation and launching were soon to be delivered and the first heavy lift semi-submersible crane vessels had just entered service. The offshore industry had indeed reached a high level of technology in a short period of time. A great deal of experience had been gained and one had learned how to generally master the harsh climatology encountered in the North Sea.

Indications that the industry was getting ready for a push into new and untried regions were becoming evident, with special emphasis being placed on an increased water depth capability. The literature reported on research on tension leg design concepts whereas in the Gulf of Mexico the Cognac platform was being installed. The latter demonstrated the successful use of underwater equipment, such as pile driving hammers and positioning equipment in a depth of water of 300 m. It failed to do so convincingly in regard to the jacket type of platform itself, however. It appeared, in fact, that the industry's attention was being turned towards deep water platforms other than the fixed steel jacket type. This tendency must be seen in the light of reports on occurrences of local fatigue failures of jackets already in use in the North Sea resulting in high costs of surveys and maintenance over the life time of these structures. In general terms, the deep water jacket for the North Sea will be extremely heavy and very costly to build and maintain. But why should a fixed steel structure necessarily be a jacket? Was it possible that the industry had become tradition bound and that other structural configurations would prove more suitable for the purposes intended? When a field has been found to be commercial, it is of course tempting and expedient to decide upon reasonable extrapolations of proven platform design concepts. To develop anything new from scratch might take years of lead time with no guarantee of success, and therefore not to be expected from a business point of view unless it became absolutely necessary.

Asking the questions raised above, and considering its future engagements in the offshore construction industry, Heerema decided in 1978 that the time was right for the initiation of an in-house project directed towards the development of a piled steel platform designed for water depths in the excess of 300 m.

At the outset of the project the following design objectives were adopted :

- The basic design concept should be as simple as possible.

- The structure was to have a piled foundation to allow for differences in bottom topology and soil conditions.

- Steel weight should be less than the estimated steel weight of a hypothetical jacket designed for the same water depth.

- Construction costs per unit steel weight should be lower than those of other steel structures.

- Construction should be based on existing facilities with only relatively small modifications being required.

- Installation should be based on existing techniques, or small extensions thereof, and on existing equipment and vessels.

- Wave loading should be as low as possible to reduce dynamic stress levels and thus ensure an adequate fatigue life.

- The design should be such that costs of monitoring, surveys and maintenance of the platform structure over its life time are minimized.

- Cognizant of the problems associated with the risers and conductors installed on deep water platforms the contemplated platform design should provide maximum protection possible for these members against fatigue and corrosion damage.

DESCRIPTION OF GENERAL FEATURES

The general configuration of the T.T.P. is shown in fig. 1. Without bracings, as depicted in fig. 2, the structure can be installed in water depths of up to approximately 300 m in the North Sea. The most characteristic feature of the T.T.P. is its simplicity. Its main structure comprises only four members; one vertical centre column, or tower, and three inclined legs arranged in the pattern of the tripod, hence the name Tripod Tower Platform.
All main members are manufactured as heavy walled tubulars without internal stiffeners. The choice of tubulars is obvious from considerations of hydrodynamic forces. The elimination of internal stiffeners is desirable for two reasons; firstly, it will increase the fatigue life of the structure and secondly, it reduces its costs. The latter is contingent upon the availability of fabrication facilities capable of rolling and welding steel plates of 125 mm thickness to a diameter of 15 m for the column and 160 mm thickness to a diameter of 8 m for the tripod legs. Fortuitously this capability has become available during the last few years, and structural components similar to those proposed for the T.T.P., and which meet the approval of the regulatory bodies, have already been delivered to be used in offshore structures. These developments, together with that of the steel mills capacity to deliver steel plates of the required thickness and of high through-thickness quality, were key factors in making the T.T.P. a feasible design concept.

The dimensions of the T.T.P. are chosen on the basis of a balancing of several design parameters which frequently have opposite effects. The 160 mm wall thickness of the legs is, for instance, necessary to provide sufficient cross-sectional area at a leg diameter of 8 m for their axial stiffness to reach a value required for the natural period criterion for the total structure. If one was to increase the diameter and decrease the wall thickness then the added mass and hydrodynamic loads would increase. Furthermore the hydrostatic collapse pressure of the tubulars would decrease such that one would either have to incorporate internals in the design or fill the legs with water to a higher level than currently proposed. The total result would be an undesirable increase of the natural periods.
The centre column serves as support for the topside facilities and transmits that load to the legs at the node. It also provides most of the torsional stiffness of the structure and, in addition, gives protection to conductors and risers against the action of hydrodynamic forces and corrosion. The choice of wall thickness and diameter has been arrived at through an optimization process similar to that used in the case of the legs where the height of water level inside the column has also been accounted for.

The main platform structure is being supported by a pre-installed piled foundation. This choice of foundation was made for several reasons. The weights and sizes of the T.T.P. and its separate members are such that it is not practical, or even possible, to construct and assemble the complete structure onshore and launch it for towing to location. Assembly must therefore take place with the structure afloat so that buoyancy forces can be taken advantage of. It was found that to fit the foundation structures to the main structure before installation at the field is difficult and has, furthermore, several disadvantages. In many areas of the North Sea it is found that the sea bed is extremely soft so that the stability of a platform structure before being piled and grouted is dangerously low. This problem can be circumvented if a pre-installed foundation is used and that option was consequently adopted for the T.T.P.

A most important part of the T.T.P. structure is the node where the legs are welded to the column. Details of this node are shown in fig. 3. Note that the transition section is designed such that the loads in the column can be smoothly transferred to the legs which are designed to support the total weight of the structure including topside facilities. Details of the analysis of the node are given in the section on structural analysis where it is shown that also in this important region the stresses are sufficiently low to satisfy the fatigue life criterion. An advantageous feature of the node structure is that its principal structural members are located in the interior of the column in a region where, during the lifetime of the structure, access under dry and atmospheric conditions for the purpose of N.D.T. surveys is always possible.

The location of the node is dictated by dynamic response considerations, as is shown in the section on structural analysis. Not considering the torsion, the two first modes of the structure can be described as a bending mode and a sway mode. Through a detailed analysis of damping it is found that the bending mode is far less damped than the sway mode. This implies that the dynamic amplification factor of the bending mode is higher, and it is, therefore, imperative that the natural period of that mode should be as low as possible so that maximum fatigue life is obtained. The relatively low position of the node also serves to reduce the magnitude of the wave excitation forces. The length of the waves exciting the natural modes are in fact so small that the wave pressures acting on the node are essentially equal to zero. Furthermore, one finds that the resultant forces of these waves acting on the column are small since the diameter of the column is not small compared to the wave lengths so that wave forces are those predicted by the wave diffraction theory.

The node is built as an integral part of the column. The joining of the legs to the node is made by field welds, i.e. these welds are not made in the construction yards. Where and when they are completed will depend upon the adopted method of assembly of the structure, as described in more detail in the section on installation. Even for the case where the T.T.P. is being assembled at its final location the welds, joining the legs to the column will be made in the dry and under atmospheric conditions.

During the installation at location the legs of the T.T.P. are set into receptacles built into the pre-installed and piled foundation and then grouted. The T.T.P. has purposely been designed with no mechanical joints of any kind. The use of divers has been reduced to a minimum and no heavy deep water work to be performed by divers should be necessary.

No discussion on or details of the deck structure are included in this paper except to say here that its design would be more or less conventional. A single column support is not unusual and has been incorporated in other designs as well. In regard to safety a single column structure may in certain respects be better than a two or even three column structure in that the probability of accidental impact with ships or other floating objects is reduced. The T.T.P. has, futhermore, been found to have a relatively high resistance to possible damage from such impacts.

The ease of inspection of the T.T.P. during its life time is greatly enhanced by the simplicity of its design. The legs can, for example, be inspected for structural failures by means of remotely controlled ultrasonic scanners which are lowered down the inside of the legs. Structural parts, except for foundations, are easily accessible for inspection and maintenance.

STRUCTURAL ANALYSIS

The environmental conditions used as basis for the various analyses of the T.T.P. which have been performed can be considered as typical for the northern North Sea. Characteristic data are shown in table 1.
Further design criteria are a topside weight of 30 000 tonne, water depths of 315 m and 340 m and an air gap of 25 m. The soil conditions and their influence on the stiffness of the foundation are dealt with later. The two types of T.T.P.'s presented in this paper will be referred to as TTP 1033 for a water depth of 315 m (1033') and TTP 1115 for a water depth of 340 m (1115').

Dynamic Analysis and Fatigue

The dynamic analyses, as they have been carried out for various water depths and geometries have primarily been focussing on obtaining natural periods sufficiently low to meet fatigue life requirements. These specific requirements will be dealt with later. The modal shapes of the lowest frequencies, being determining for the fatigue life of the structure, are a "bending mode" and a "sway mode". The "bending mode" is defined as the modal shape resulting in bending of the structural components of the T.T.P. and virtually zero foundation displacement. The "sway mode" can be defined as a lateral displacement of the tower, including significant foundation displacement, with relatively little bending of the structure taking place. The modal shapes for two geometries and associated water depths are shown in figs. 4 and 5.

The amount of damping of the oscillating structure has in the initial calculations been assumed to be 2% of critical, which at that time was considered a conservative assumption. However, as this assumption was frequently queried during further studies, an assessment of the real values of the damping was considered necessary.

An elegant method of approach which was used to check whether 2% of critical damping was in fact conservative is based on a paper published by Prof. J. Kim Vandiver (Vandiver, 1980). His proposed method is very useful when the dynamically amplified response at a natural frequency is of concern, as in this case. It is shown that the response is not inversely proportional to the total damping, but is proportional to the ratio of the radiation damping to the total damping, this being due to the fact that radiation damping and the wave exciting forces are not independant quantities. The paper shows the existance of an upper bound to the response and it, furthermore, provides a method by which one can make a quick assessment of the dynamic response of a structure like the T.T.P.
The Vandiver approach led to dynamic stress levels approximately twice as high as derived from the "conventional" calculations assuming 2% of damping for all modes. This result precipitated the evaluation of the following more accurate calculations of the various damping parameters for the bending and sway modes.

The magnitudes of the damping experienced by the T.T.P. when oscillating in a bending mode or in a sway mode have been found to differ significantly. The main reason for this will be shown to be the contribution of the soil damping to the total damping. The total damping is composed of:

- wave radiation damping
- viscous damping
- structural damping
- soil damping

The resulting wave radiation damping force in deep water is considered to act at a water depth of $L/2\pi$ where L is the wave length. In terms of the period (T) of the modal shape of interest $L = 1.56T^2$. Given the natural period and the modal shape, the equivalent mass of the T.T.P. placed at the level of the resultant damping force can be calculated (Hooft, 1970). By using this simple equivalent single degree of freedom system an estimate can be made of the magnitude of the critical damping. For the TTP 1033 (315 m) the above approach results in a damping ratio for the bending mode of 0.5% and for the sway mode of 0.25% of critical.

For both bending and sway mode the viscous damping per unit length of oscillating member has an average value of 0.02% (TTP 1033). It should be noted that the influence of an ambient current on the viscous damping has been disregarded. This influence can be significant.

The structural damping which the structure is experiencing is set at 0.06%. This low value is a consequence of the geometrical simplicity of the fully welded structure and the generally low stress levels reached.

For the establishment of the magnitude of the foundation damping R.V. Whitman's publication "Soil Platform Interaction" has been used (Whitman, 1976). In our case the spring stiffness of the foundation is composed of contributions from piles and "mudmats", the latter being the flat bottom of the foundation, see figs. 6 and 7. Only the mudmats are considered to dissipate damping energy in the soil half space. Energy absorption occurs due to vertical, horizontal and rotational displacement of foundations. As will be shown later the vertical and rotational stiffness are relatively high compared to the lateral stiffness. The horizontal motion of the foundation is in most cases a principal contributor to the total soil damping imposed on the structure.
For the TTP 1033 this approach results in the following estimated foundation damping ratio's: 1.75% for the sway mode and 0.20% for the bending mode. The major contribution to the soil damping of the sway mode (1.50%) originates from the horizontal foundation displacements. For the bending mode, with its virtually zero horizontal foundation displacement, the major contributing damping (0.10%) is derived from the rotational movements of the foundation.

The above calculations, as carried out for the TTP 1033, result in a total estimated damping of 0.7% of critical for the bending mode and of 2.0% of critical for the sway mode. As the initial dynamic and fatigue calculations had been carried out with a total damping of 2% of critical, the above findings resulted in a fatigue life of approximately one tenth of that originally estimated. This reduction can be fully contributed to the very small damping of the bending mode.

Subsequent sensitivity studies carried out to investigate the influence of the natural periods on the fatigue life showed that, for the bending mode, a period of close to 3 seconds would be required to meet the fatigue life requirements. At this stage of the T.T.P. project the minimum required fatigue life was set at 100 years. The natural period of the sway mode, having a less significant influence on the fatigue life than the period of the bending mode, should in that case preferably not be higher than 4.0 to 4.5 seconds. These criteria, imposed on the simple (= unbraced) geometry will probably reduce its depth capability to approximately 300 m (1000'). The actual depth capability will have to be established for each case by detailed calculations, however. Further analysis was in view of this depth limitation concentrated on a T.T.P. designed for a water depth of 340 m, the decision to do so being prompted by recent promising oil and gas findings in the northern North Sea.

It is evident from what has been stated already that the necessity of satisfying the natural period requirements associated with an adequate fatigue life asks for additional measures to be applied to the T.T.P. structure designed for the greater water depths. Simultaneously an inshore assembly of the T.T.P. structure was being considered as an alternative installation procedure further to the initially proposed offshore installation. The assembled tower would subsequently be floated out and, once at final location, be upended and lowered into a pre-installed piled foundation. Such an assembly and tow require a bracing system to support legs and column.

Further studies thus called for an additional bracing system serving the purposes of :

- reducing the bending mode natural period
- increasing the water depth capability
- providing structural strength and stability during assembly and tow

These analyses led to the concept of the TTP 1115 (340 m water depth), with natural periods of 3.11 seconds for bending and 4.37 seconds for sway, as compared to 4.9 seconds and 4.1 seconds for the TTP 1033, respectively. See figs. 4 and 5.

Although the basic geometry of the TTP is extremely simple, a number of parameters can, nevertheless, be varied and used to influence its dynamic characteristics, i.e. the modal shapes and related natural periods.

Such parameters are :

- elevation of tetrahedron node
- inclination of legs
- diameter/wall thickness ratio of tubulars
- amount of water in the tubulars
- stiffness of foundation
- number and type of braces (if incorporated in the design)

Furthermore the design topside weight will highly influence the final geometry and its dynamic characteristics. Other important criteria and restrictions are the bending mode natural periods of the separate structural members, maximum wall thickness of tubulars (fabrication), hydrostatic collapse and local buckling, structural steel weight optimization, etc. Optimizing the many design parameters involved, some of which are having opposite effects, led in fact to the described T.T.P. design concepts.

Of prime importance for dynamic response and fatigue life of the structure are the wave exciting forces. As stated earlier, one of the design objectives of the concept was the minimizing of wave loads acting on the structure. The waves with periods close to the T.T.P. natural periods have wave lengths which are of the order of the diameter of the centre column, e.g. T = 3s, L_3 = 1.56 x 9 = 14m (centre column O.D. = 15 m) and if T = 5s, L_5 = 1.56 x 25 = 39 m. Consequently the three dimensional diffraction theory according to MacCamy and Fuchs should be applied to calculate the wave excitation forces on the structure (MacCamy and Fuchs, 1954). For the waves of interest as regards fatigue, this approach leads to a reduction of wave loads to about one half of the predictions by Morison (see fig.8) and to a corresponding increase in fatigue life by a factor of approximately 7.5. It should be mentioned that the position of the node is such that the influence on the loads due to waves of periods coinciding with the natural periods of the structure is for all practical purposes equal to zero. The dynamic stress levels in various sea states for one of the highest stressed nodes (middle of the leg span of the TTP 1033) are shown in table 2. The maximum deck acceleration is estimated to have a r.m.s. value of 0.1 m/s^2.

In the above sensitivity of the fatigue life to natural periods, modal shapes and the associated damping ratio's has already been stressed. Stress concentrations at critical joints will, however, also have a significant bearing on the fatigue life. In this respect the tetrahedron node design is of special interest. To a certain extent, such platform characteristics can be controlled by proper design. This is equally true, although to a lesser extent, for the wave excitation forces. Beyond the control of the designer are the wave statistics and the applicable SN-curves, however.
The original fatigue calculations for the TTP 1033 have been carried out on the basis of the probability distribution of sea states as shown in table 1. Re-analysis of the measurements performed in the northern North Sea resulted in a wave scatter diagram which shows, as compared to the previously used sea state distribution, a shift of the periods of the scatter diagram of approximately one second towards the natural periods of the T.T.P. This revised scatter diagram now forms the basis for the fatigue analysis of the T.T.P. design. The result of this adaptation was a reduction of the fatigue life of the TTP 1033 by about 50%.

Although much is presently known about the fatigue of steel joints and plates, including the influence of sea water and cathodic protection, information about the fatigue life of thick plates (centre column wall 125 mm, wall thickness of legs 160 mm) is rather scarce. Originally, the SN-curve D according to the DnV Rules for the Design, Construction and Inspection of Offshore Structures 1977 has been used for establishing the fatigue life. Subsequently the SN-curves as proposed by the Department of Energy in its "Guidance Notes" have been applied (fig.9). These are more stringent and have been extrapolated to apply to 125 mm and 160 mm thick plates in a manner which may be conservative. Extensive thick plate testing is still in progress and more accurate data are expected to become available shortly.

Design of the Tetrahedron Node

Having shown that the T.T.P. design is fatigue dominated, it will be clear that the design of the tetrahedron node imposes severe requirements on the magnitude of the "hot spot" stresses. At the outset of the design of the node a maximum stress concentration factor of 2 was aimed for. In this respect the stress concentration factor (s.c.f) is defined as the maximum hot spot stress divided by the nominal stress in a leg.

A first estimate of the stress concentration for a tubular connection has been carried out using conventional formulas (Kuang, et al) resulting in a s.c.f. of about 10. Although the validity of such formulas for the T.T.P. dimensions and design could be queried, it was clear that special attention should be paid to the node design to ensure that the s.c.f. would not exceed 2 significantly. As local stiffening of the centre column would inevitably result in hard spots with high stresses, the application of plates and stiffeners was not given any further consideration.

A very elegant and possibly adequate solution would be to insert a (stiffened) sleeve into the node where it should be grouted to the centre column wall. The state-of-the-art of grouted joint design does not as yet, however, give all the answers to the effects of high tensile forces and to the fatigue behaviour in particular.

A third and tentatively preferred solution for the node design is presented in fig.3. This configuration was the result of discussions with staff of Shell Internationale Petroleum Maatschappij B.V., and was subsequently investigated by a finite element analysis under a joint contract with SIPM. The analyses were carried out by Det norske Veritas. The configuration is primarily the result of an attempt to avoid hard spots and obtaining "hard lines" instead, which results in a considerable increase in steel area for the actual load transfer and thus a reduction in stress levels. Local bending of the centre column wall is minimized by this solution in which the legs are fitted with a transition piece to accommodate the joining up with the column along the straight boundary lines of a rectangle. The centre column itself is equipped with vertical and horizontal plates to enable a direct and smooth stress transfer from leg to column and leg to leg. The studies led relatively quickly to the s.c.f. of approximately 2 as aimed for. It is felt, that even a further reduction of the s.c.f. can be obtained through optimization of structural detailing.

Static Analysis

Unlike the dynamics and fatigue investigations the static analysis of the T.T.P. is rather a check calculation than a governing design case. To illustrate this it suffices to say that the 100-year design wave, plus maximum wind and current, load the structure to approximately 50% of the allowable stresses (assuming grade 50 steel). The total horizontal load on the tower is in this case approximately 10 000 t resulting in a deck displacement of 0.31 m. As far as hydrostatic collapse, local buckling, etc., are concerned the relevant API and DnV recommendations have been adhered to.

Redundancy and Accidental Loads

Of special interest is the redundancy of the structure and its resistance against accidental loads, such as ship collisions, falling objects, fire and earthquake. The design criteria to be used in the cases of ship collision and earthquake are presently under review by the regulatory authorities. Preliminary investigations have been carried out by Det norske Veritas. DnV has analysed the column and found the platform favourable as compared to other platforms when considering ship collision.

It is expected that in certain areas of the North Sea the structure must be capable of withstanding shock loads caused by earthquakes. In view of the load carrying capacity of the T.T.P., and its structural configuration, it is the opinion of the authors that it can sustain the effects of such loads due to earthquakes of a severity to be expected in these waters.

Foundation Analysis

The T.T.P. has been designed to be supported by a piled foundation. A principle reason for this is that soil conditions differ greatly from place to place, varying from dense sands and hard clays to under-consolidated soils. The latter can be expected to present some difficult problems in the case of a gravity structure, which also places stringent requirements on the level quality of the sea bottom, presence and size of boulders, etc.
The driving of heavy piles at a water depth of 350 m or more can now be considered a fairly routine task following the recent introduction of large capacity hydraulic underwater hammers which are being handled by the cranes of semi-submersible crane vessels. These operations require, furthermore, a minimum of diver support.
The soil characteristics assumed for a study of the load carrying capacity and stiffness of the foundation of the T.T.P. are shown in fig. 11. As can be seen the soil is very soft to North Sea standards, however, considered typical for large areas of the Norwegian Trench.

For the purposes of the axial bearing capacity calculations of the piles, the average cohesive shear strength profile of clay layers down to 65 m has been used. Below that level the results of the soil investigation were so arbitrary that two possible cases were studied; a) alternating sand and clay layers of thicknesses less than two times the pile diameter, and b) a sand clay mixture. The axial bearing capacity has been calculated for the two cases in accordance with procedures described in API RP 2A paragraphs 2.6.4 and 2.6.5. The results obtained for a pile of 3.0 m diameter and 63 mm wall thickness are shown in fig. 11.
The total number of piles in each of the leg foundations is ten 3 m diameter piles plus two 2 m diameter piles which are used for leveling purposes. Similarly the centre column foundation has nine 3 m diameter piles plus three 2 m diameter piles. Preliminary designs of the foundations are shown in figs. 6 and 7.

The horizontal load deflection behaviour of a single pile was calculated in accordance with the API RP 2A. Considering the pile to be fixed to the foundation at the mudline, the horizontal deflection of a single pile at that level due to a horizontal load of 300 tonne was found to be 0.12 m, i.e. the stiffness of the single pile is 2500 tonne/metre.

The mutual interaction between the laterally loaded piles in a pile group results in overall reduction of the total lateral stiffness of the group. Poulos has formulated an elastic theory which describes this phenomenon (Poulos, 1971a, 1971b). On the basis of this theory it was found that the reduction in horizontal stiffnesses of the pile group forming the T.T.P. leg foundation as compared to the summed single pile stiffnesses amounts to approximately 2.25. This value has been used for the actual non-elastic case so that the total stiffnesses of the leg foundation becomes (10 x 2500 + 2 x 1250)/2.25 = 12 200 tonne/metre. On account of the soft soil condition this stiffness is lower than the value required to obtain an acceptably low value of the natural periods of the T.T.P. The 3 m diameter piles have, therefore, been given a batter of 1:5. The stiffness of the foundation due to this feature has been calculated considering the foundation together with the piles as an elastic structure. On that basis the radial stiffness of a leg foundation is found to be equal to 39 000 tonne/metre, whereas for the column foundation the lateral stiffness becomes 22 000 tonne/metre. Under the assumption that the total stiffness of a foundation is the sum of the two effects described above, the total numerical values of the foundation elements of the TTP 1033 can be calculated and are presented in table 3.

The driveability analysis has been based upon the performance characteristics of the Menck MHU 1700 hydraulic underwater hammer. The pile considered has a diameter of 3 m and a wall thickness of 63 mm. The method of analysis used has been developed inhouse (Heerema 1979a, 1979b, 1981). This method has been used extensively for several years to predict driveability of a large number of platform installations and has been thoroughly checked by post-analyses of pile driving records.
Considering 150 blows per foot to represent refusal, probably a pile penetration of more than 90 m can be obtained, whereas the minimum predicted attainable penetration is 77 m. The ultimate bearing capacity required amounts to 4500 t, which needs a penetration of 75 m (fig. 11).

INSTALLATION

As the offshore structures have increased in size and weight so have the problems which the offshore construction industry have had to tackle in installing them. The tools which this industry has developed to handle the tasks presented to it have come to play a central role in the realization of the complex production platforms installed in the North Sea during the last few years.

According as the industry moves into the deep parts of the North Sea it will discover that otherwise attractive platform concepts cannot be realized on account of problems associated with offshore installation which cannot be solved within the present level of available technology. One is therefore well advised to place installation on the critical path of the network of design activities of any deepwater platform. The reason why this is apparently not so in many cases is that much of the knowhow sought by designers is primarily available only to a few installation contractors and for obvious reasons not well published. The authors have, however, been in the fortunate position to have had access to much of the existing expertise in the field of offshore platform installation technology. It may in fact be said that the initiation of the T.T.P. design project is a direct result of the technological jump that took place in this field in 1978 with the introduction of large semi-submersible crane vessels.

The space available in this paper does not permit a detailed treatment of the installation schemes developed as an integral part of the design process of the T.T.P. Only the broad outlines of these schemes will therefore be described here which, it is hoped, will nevertheless give the reader sufficient information for an overall evaluation of the feasibility of the platform concept.

Broadly two distinguishable installation methods have been devised. An offshore installation procedure is proposed for the TTP 1033 wheras an inshore assembly of main structural members to be followed by a tow-out for the final installation offshore, has been worked out for the TTP 1115. Both TTP's are proposed to be erected on a pre-installed piled foundation which consists of four separate sections. A centre foundation segment is first installed over a pre-drilling template, being located by docking pins, and subsequently piled. Each of the three separate foundations for the tripod legs is fitted with a location frame which is stabbed over pins located on the centre foundation (figs. 1 and 6). After pile driving of these, the total assembly forms an integral foundation. Piles will be driven with an underwater hammer and may be given a batter. The foundation can be adapted to a wide range of soil conditions. Furthermore, it allows for unevenness of the seabed since differences in the levels of the individual foundations can be compensated for by adjusting the lengths of the legs in the fabrication yard.

In the TTP 1033 offshore installation procedure the centre column and the three legs are towed to the offshore site in a horizontal floating position and will be upended by selective flooding and subsequently lowered into a receptacle built into their respective foundations. The centre column is still maintained floating (in a vertical position) while it is being horizontally restrained at the seabed level by the sleeve of the foundation. The legs will be pulled towards the centre column and made to rest there in saddles located below the node. The lower ends of the legs have a spherically shaped surface which is in contact with a mating surface of the corresponding foundation. Locking lugs are provided at the lower ends of the legs to produce an immediate capability of the legs to transfer tension forces to the foundations. The centre column is then lowered slightly by flooding to allow the upper ends of the legs to reach into the tetrahedron node structure where they meet up with the sleeves, to which they are welded under dry atmospheric conditions (fig. 10). At this stage the centre column is not yet supported by its foundation but is resting on the three legs, resulting in a geometrically unconstrained configuration.
After welding of the legs to the column the receptacle space of the foundations around the leg footings will be grouted to provide a fixed joint. The centre column will be grouted to its foundation after installation of deck and modules. The centre column will be filled with water up to the -80 m level whereas the legs will be filled up to the -190 m level.

The inshore assembly as worked out for the TTP 1115 deals with a more complicated geometry since , in addition to the legs and centre column, there is now also a bracing system. This bracing structure is to be fabricated as a separate sub-assembly in a construction basin which can be flooded. The floating bracing structure will be towed to the inshore site of assembly, as will the legs and centre column (Fabrication of these components could have taken place at various yards simultaneously). The maximum required draft for the above transports is approximately 11 m to permit the passing of the node of the centre column. Upon arrival at the installation site the bracing structure will be upended and partially supported by two cranes. In this position it can be mated with the horizontally floating centre column and joined either by welding or by means of grouting. Subsequently the legs can be connected to the centre column and bracing structure. The partly assembled structure requires to be rotated twice to facilitate the making of the welded/grouted connections in the dry at water level.
Finally, the completed structure is towed to its offshore location where it is upended by flooding and lowered into the pre-installed foundation.
A deep water fjord and towing route with a minimum water depth of approximately 165 m should preferably be available for the proposed installation procedure.

CONCLUDING REMARKS

The T.T.P. design concept is another example of the rapid development of the technologies of the offshore industry that has taken place during the last decade. Engaging in the mental exercise of translating the T.T.P. design into its physical dimensions, the picture that emerges is that of a structure which only a few years ago could be considered fiction. The reason that this is not the case today is a direct result of man's search for increasingly scarce energy sources. The costs of this search are also increasing as are the complexities of the technological problems encountered. A driving force behind the endeavours to overcome these obstacles is the confirmed location of large reservoirs of hydrocarbons in deeper waters than those being exploited today. In the North Sea in particular a degree of urgency exists in developing these fields at a time when an overall infrastructure of the offshore industry is still present in this area. Exactly how this will be done is continuously under study. The fixed steel platform will, however, in the authors' opinion come to play a significant role in the future extraction of oil and gas from reservoirs located in deep water.

ACKNOWLEDGEMENTS

The prominent role that the late Mr. P.S. Heerema came to play in the offshore construction industry is well characterized by the words vision and daring, which are synonymous with the manner by which he shaped many of the important technological developments of that industry. The T.T.P. design exemplifies his approach to one of the many challenges before us and as he conceived it, and it was the authors' privilege to have the opportunity to work on that design under his inspiring guidance and support.

Members of the Heerema staff contributing measurably to the T.T.P. design are Mr. R. Uittenbogaard and Mr. H. Heikens Pleizier who were much engaged in the project during its early stages when the T.T.P. concept evolved, and Mr. Franz Groen and Mr. Kees Willemse, who have carried out many of the calculations required and have participated in several of the special studies which have been conducted.
The authors also wish to acknowledge the input provided by Det norske Veritas with the execution of special finite element, dynamic and fatigue computer analyses. Consultations with the staff of the British Welding Institute have been instrumental in establishing viable welding procedures whereas the information required in regard to manufacturing of the tubulars for legs and columns have been generously supplied by Mr. Th. Schmeitz of S.I.F., Helden, Holland.
Recognition is, furthermore, given to the invaluable benefits and inspiration derived from the constructive criticism and technical suggestions provided by the staff of Shell Internatinale Petroleum Maatschappij B.V., The Hague. Examples are a more detailed consideration of damping components and of the fatigue behaviour. Moreover the idea of investigating possible in-shore assembly methods was first suggested by them.

REFERENCES

HEEREMA, E.P., 1979, "Relationships between Wall Friction, Displacement, Velocity and Horizontal Stress in Clay and in Sand for Pile Driveability Analysis"
Ground Engineering, January 1979

HEEREMA, E.P., 1979, "Predicting pile driveability : Heather as an Illustration of the "Friction Fatigue" theory"
European Offshore Petroleum Conference, London, Paper No. 50, October 1979.
Reprinted in "Ground Engineering", May 1980.

HEEREMA, E.P., 1981, "Dynamic Point Resistance in Sand and Clay for Pile Driveability Analysis"
Ground Engineering, September 1981.

HOOFT, J.P., 1970, "Oscillatory Wave Forces on Small Bodies"
Int. Shipbuilding Progress, April 1970.

MAC-CAMY, R.S. & FUCHS, R.A., 1954, "Wave Forces on Piles : A Diffraction Theory",
U.S. Army Corps of Engineers, Beach Erosion Board, Techn. Memo No. 69, Washington 1954.

POULOS, Harry G., 1971, "Behavior of laterally loaded piles : I - single piles",
Journal of the Soil Mechanics and Foundations Division, May 1971.

POULOS, Harry G., 1971, "Behavior of laterally loaded piles : II - pile groups",
Journal of the Soil Mechanics and Foundations Division, May 1971.

VANDIVER, J.Kim, 1980, "Prediction of the Damping-Controlled Response of Offshore Structures to Random Wave Excitation"
Society of Petroleum Engineers Journal, February 1980.

WHITMAN, Robert V., 1976, "Soil-Platform Interaction",
Boss '76, Behaviour of offshore structures, August 1976.

API RP 2A, January 1980
API Recommended Practice for Planning, Designing and Constructing Fixed Offshore Platforms. American Petroleum Institute.

DnV, 1977
Rules for the design, construction and inspection of offshore structures, Det norske Veritas, Oslo 1977.

	wave		current m/s			wind m/s
	H(m)	T(s)	1	2	3	1 hr. sust.
max. storm	31	17.7	1.2	1.2	0.6	40
summer storm	17	12.5	1.2	1.2	0.6	24

Note: 1 = at surface
2 = 100m below surface
3 = at sea bed

Sea state	H_s	T_z	PROB.
1	1.10	5.50	.1700
2	2.20	5.50	.2010
3	2.20	6.50	.1920
4	3.30	6.50	.1620
5	3.30	7.50	.0950
6	4.30	7.50	.0830
7	4.30	8.50	.0170
8	5.40	8.50	.0570
9	6.80	10.00	.0180
10	8.50	11.50	.0043
11	10.30	12.50	.0004
12	12.50	13.50	.0002
13	15.00	14.50	.0001

Sea state table for fatigue analysis

H_s = significant wave height
T_z = zero up crossing period

TABLE 1: DESIGN DATA, ENVIRONMENTAL CONDITIONS

Most probable largest stress range in node 33 in a sea state duration of 3 hours

Condition A σ = 364 kgf/cm^2
          B σ = 217 kgf/cm^2
          C σ = 299 kgf/cm^2

A: max. storm cond. H_s = 15.5m T_z = 10 sec.
B: summer storm cond. H_s = 8.5m T_z = 8   sec.
C: max. operating cond. H_s = 7.4m T_z = 6.5 sec.

TABLE 2: DYNAMIC STRESS LEVELS IN NODE 33 (TTP 1033, middle of leg span)

Foundation	Direction	Stiffness
Leg	Radial	51200 t/m
	Tangential	20000 t/m
	Vertical	1955200 t/m
	Rotational	1.15×10^8 tm/rad
Column	Horizontal	35000 t/m
	Vertical	1879300 t/m
	Rotational	2.00×10^8 tm/rad

TABLE 3: STIFFNESS OF FOUNDATION ELEMENTS (TTP 1033)

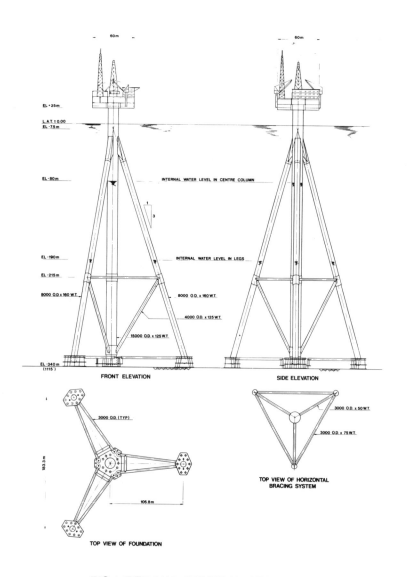

FIG 1: TTP 1115 GENERAL ARRANGEMENT

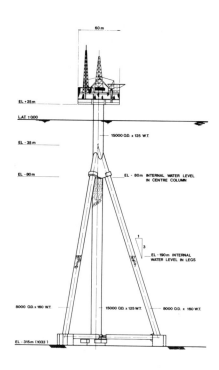

FIG. 2: TTP 1033 GENERAL ARRANGEMENT

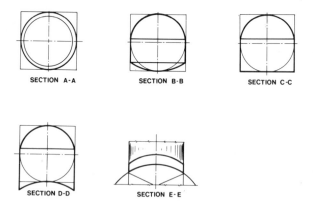

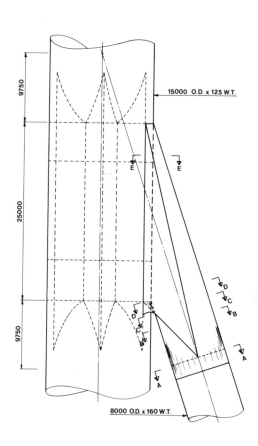

FIG. 3: TETRAHEDRON NODE TTP 1115

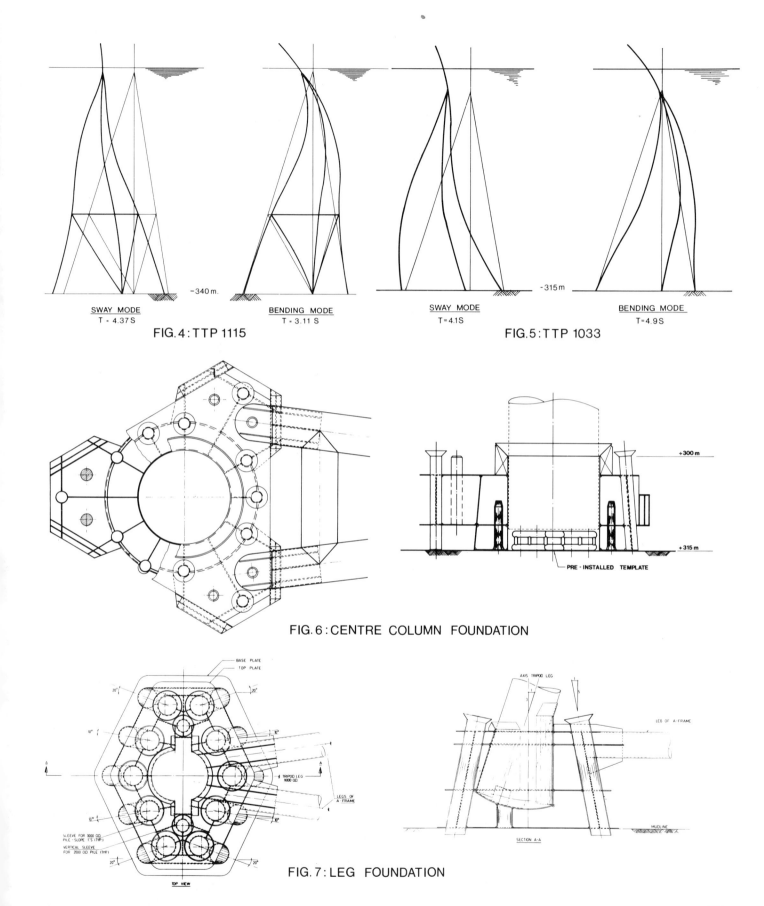

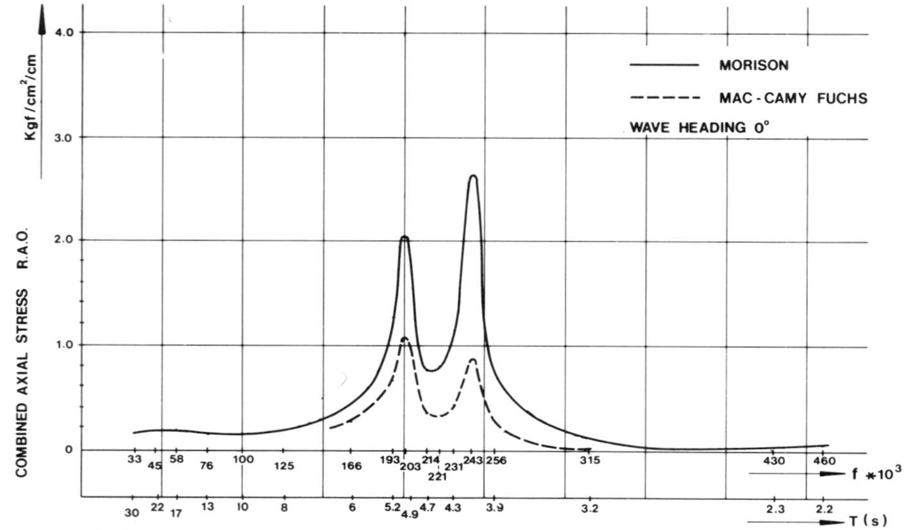

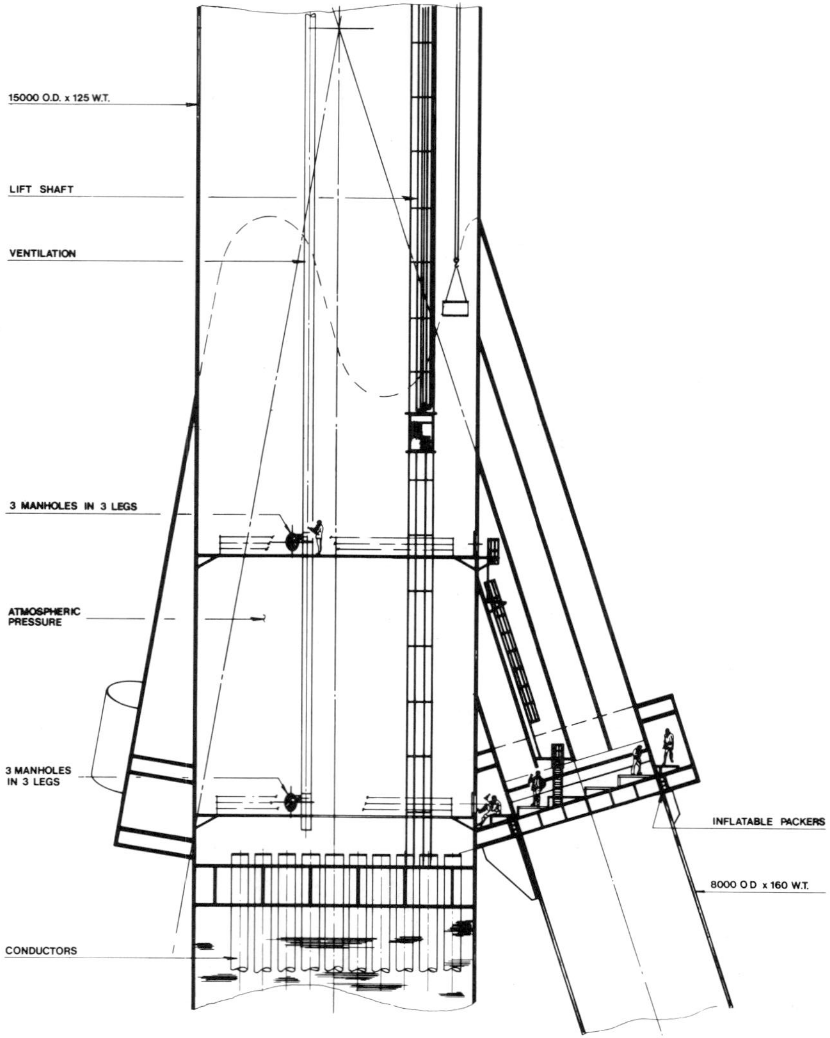

FIG.10: FIELD WELDING OF LEG TO COLUMN

FIG. 8: STRESS AMPLITUDE TRANSFER FUNCTIONS T.T.P. 1033 NODE 33

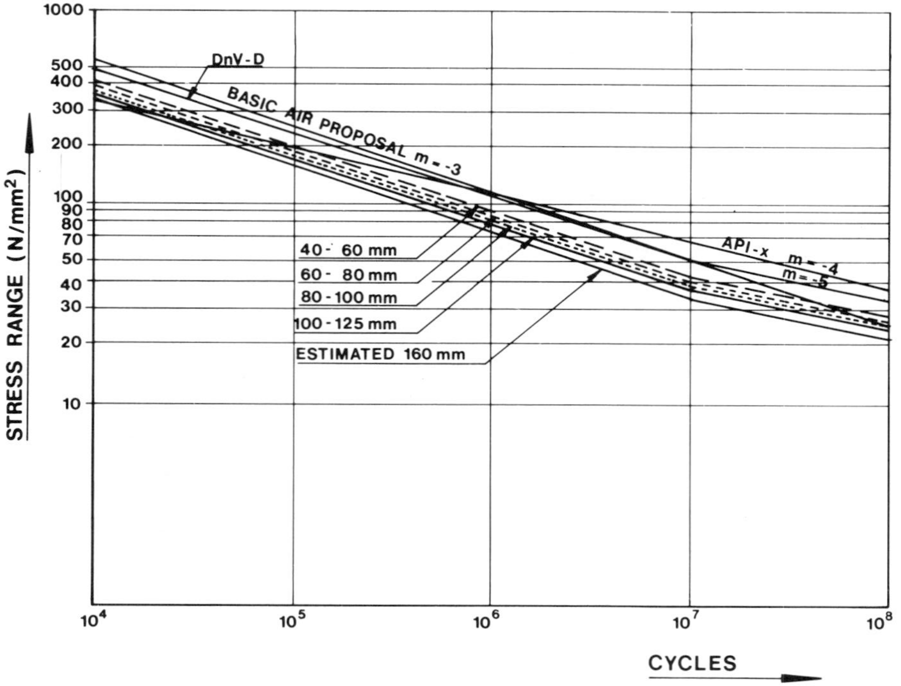

FIG. 9: DRAFT SN-CURVES

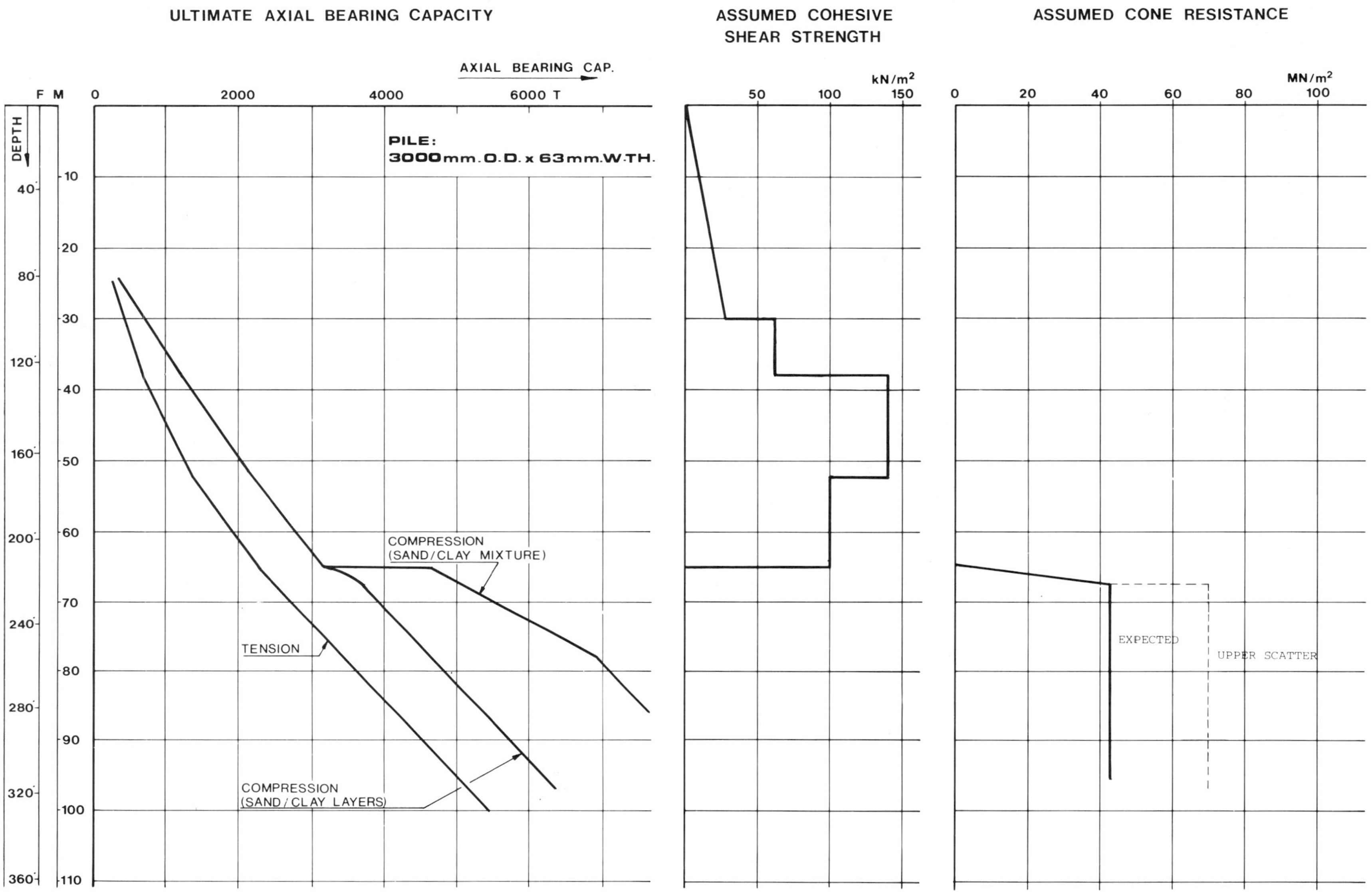

FIG.11: SOIL AND FOUNDATION CHARACTERISTICS

NONLINEAR ANALYSES AND EXPERIMENTS OF REINFORCED CONCRETE
PLATES SUBJECTED TO CONCENTRATED MOMENTS

D. G. Morrison

McDermott Inc., New Orleans, Louisiana, U.S.A.

SUMMARY

The primary objective of this paper is to present a grid model for the nonlinear analysis of plates. The model, composed of a grid of beam elements responding in flexure and torsion, is desscribed in the following sections.

A further purpose is to provide data on the behaviour of reinforced concrete plates to static and dynamic (rate of strain significant) loading.

To verify the modelling approach calculated and measured response to two plate-column connections are compared. This comparison follows a description of the grid model, and an explanation of the tested connections. The connections are of reinforced concrete, differing in slab reinforcement ratio, and the modelling accounts for a number of nonlinearities.

By modelling the plate as a grid of beam elements, relatively simple stress-analysis programs may be used to approach the nonlinear behaviour in a series of linear steps.

Reinforced concrete slabs have been proposed in various offshore structures such as concrete gravity platforms and LNG offshore structures, and the response into the nonlinear range is of importance.

INTRODUCTION

As observed by Graff and Chen (1981), many simplifications must be made in the analysis of nonlinear response of reinforced concrete elements in offshore structures. Comprehension and analysis of the material non-linearities involved -- such as creep, shrinkage, cracking, and bar-slip -- are still developing. Information on the flexural behaviour of reinforced concrete slabs subjected to concentrated moment without shear failure is sparse (Park and Gamble, 1980), and the presented study should provide a contribution.

Although a large number of "static" tests are documented (Hawkins, 1975), relatively little is reported on the response of slabs to dynamically applied loading.

Reinforced concrete structures have been used and proposed (Dybwad, et al, 1980) in the North Sea, and it is expected (Graff and Chen, 1981) that like structures will be used in waters off the North American shoreline. Basic research and development in reinforced concrete is of importance to engineers in both regions.

The bottom supported reinforced concrete platforms used in the North Sea (Graff and Chen, 1981), (FIP Report, 1979), (Kjekstad and Stub, 1978), have made extensive use of slab systems, most notably in the base of the structure. The base is very costly and rigorous analysis of it is necessary. Most base structures (FIP Report, 1979), (Kjekstad and Stub, 1978) are provided with vertical skirts at the edge of cantilevered base slabs. Condeep Stratford A and Frigg TCP2 (Kjekstad and Stub, 1978) have good examples of this detail. These skirts may rest on local hard points on the seabed causing confined loads on the slab. The design of the slab is usually governed (FIP Report, 1979) by these loads. An objective of this paper is to assist the designer and researcher in understanding the complex behaviour of these slab systems.

Reinforced concrete structures designed to store and transport liquid fuels (Mahin and Matsunaga, 1980), (Marshall, 1978), also use slab assemblies. These slabs are often proposed for secondary protection of the containment structure. The presented work could be useful for column-slab connections in these structures.

As experience with reinforced concrete offshore structure increases more demands will be made on understanding the complex response of this material in challenging applications.

THE GRID MODEL

The model and the assumed member properties are described, indicating the development of the nonlinear model.

Description of the Grid

In matching the behaviour of a portion of slab, the grid method replaces this portion by beams. Equivalent stiffness properties (flexural and torsional) of the beams are derived by matching the curvature of the plate portion for a given force with that of the beams under a statically equivalent force; this paring considers the curvature due to pure flexure in two directions and the curvature due to pure torsion. A concise description of the procedure applied to the analysis of elastic plates is available in Yettram, et al (1965).

In the nonlinear model a square grid pattern was used, with a square portion of plate represented by a beam along each of its four boundaries. For a linear elastic analysis the effect of assuming a Poisson's ratio of zero and a square grid pattern is equal flexural and torsional stiffness for the beam members (obtained from the stiffness of the uncracked section). A representation of a quarter of the specimen (idealized in Fig. 2) by the grid elements is shown in Fig. 1.

The model of the slab-column connection resists lateral load by three force quantities: flexure of the beams framing into the front face of the column, torsion of the beams representing the plate to the side of the column, and eccentricity of shear reactions at the end of beams framing into the column from both directions (Fig. 3).

Modelling the reinforced concrete plate specimens as a grid of beams has a number of advantages: beam behaviour may be easier to explain than plate action, bar-slip phenomenon may be more readily incorporated, application to material nonlinearity is possible with simple crack and yield criteria, and the change in the relationship between flexural and torsional stiffness after cracking can be handled conveniently.

Assumed Member Properties

Load-deformation properties of members were calculated from material properties. The relationship between the flexural and torsional stiffnesses in a beam before and after cracking was different. Certain assumptions regarding bar-slip were made for ease of calculation.

Steel and concrete properties were based on measurements (Table 1). To account approximately for the uncertain affects of shrinkage and construction stresses on cracking moments, moduli of rupture were reduced to one half of the mean measured moduli. A simple trilinear moment-curvature relationship was calculated for each beam (needed to determine flexural and torsional grid stiffnesses) using the simplified moment-curvature expressions given by Cardenas and Sozen (1968) (cracking and yield moment from elastic theory). These simplified curves are shown in comparison with curves calculated continuously in Fig. 4.

The torsional moment was treated as two principal moments of opposite sign working at right angles to each other, at an angle of $45°$ to the torsional moment. The isotropic reinforcement layout used resulted in equal flexural strength in any direction -- hence pure torsion and pure moment had like strength per unit width (Fig. 5a). The flexibility of the beams representing a portion of plate in torsion was altered after cracking bearing in mind that the relative orientation of cracks to reinforcement orientation influences the stiffness in torsion (expressions in Cardenas and Sozen, 1968). Torsional flexibility is a function of the angle between the vector representing torsion and the direction of one of the reinforcement axes (the reinforcement axes were mutually orthogonal) (Fig. 5b). Even though this angle varied over the slab, it was assumed to be $45°$ in determining torsional flexibility. The beam torsional stiffness was thus about a half of the beam flexural stiffness after cracking.

Reinforcement pullout at the column face was based on the model in Fig. 6 and an assumed average bond stress of 4.0 MPa. The relative rotation due to bar-slip may be thought of as a spring connection for the "flexural" beam at the support boundary.

TEST PROGRAM

Experimental data from the testing of plate-column connections are presented to provide a basis for evaluating the grid model, and to develop fundamental information on the behaviour of flat-plates.

Outline of Experimental Work

In all, tests of 9 slab-column specimens are documented (Morrison and Sozen, 1981). Of the nine specimens, five (S1 to S5) were subjected to static horizontal load reversals and four were tested dynamically by subjecting the base of the column to a strong motion simulating one horizontal component of a (measured earthquake) acceleration record. The dynamic tests were initiated by a pilot specimen (DP) and followed by a series D1 to D3 with properties comparable to those of S1 to S3.

Overall dimensions for the monolithic plate-column assemblies are summarized in Fig. 7. Nominally the plate measured 1.8 m (6 ft.) and was 76 mm (3 in.) thick. The centrally located column was 0.3 m (1.0 ft.) square in cross section. The distance between the supporting hinge and the position on which the added mass (dynamic tests) rested or the jack (static tests) worked was 1.12 m (3 ft. 8 in.).

Measured mean material properties are listed in Table 1. Concrete was mixed using Type III Portland Cement, river sand, and a gravel of 10 mm (3/8 in.) maximum size. The entire slab was cast, with the column positioned vertically, from a single batch of concrete. Each specimen and associated test cylinders were cured moist for seven days. Strengths in Table 1 represent means of values obtained from three cylinder tests.

Column reinforcement provided by four No. 8 bars was the same in all specimens. Slab reinforcement, cut from deformed No. 2 bars, varied in amount as indicated in Table 1. Spacing of slab reinforcement was uniform in both directions throughout the width of the slab. Each yield stress listed in Table 1 represents mean of three tensile coupon tests.

Description of Experiments

In the statically tested specimens the load was applied cyclically. The amplitude of the cycles was increased in 4 steps. At each of these levels a number of cycles of constant amplitude took place. In the final cycle the specimen was pushed to its limit, or the limit of the stroke of the jack (75mm). The overall responses of specimens S1 to S3, and the position of the displacement measurement, are indicated in Fig. 8.

Three types of dynamic tests were performed in a recurring sequence for each specimen (Specimens D1 to D3). These were: (1) a free-vibration test to determine dynamic characteristics at very low amplitudes, (2) a "random-motion" test obtained by having the platform of the simulator move to develop an acceleration history simulating one component of a measured record (N component of El Centro 1940 with the time scale compressed by 2.5) and (3) steady-state tests obtained by having the platform move sinusoidally in a series of frequencies to measure change in response.

In Fig. 9, a through c, the measured moment-displacement responses in the dynamic random-motion runs are compared directly with the envelopes, in one quadrant, to the moment-displacement curves obtained in the static tests. It is observed that in every case the dynamically measured strength was higher than that measured in the statically loaded specimen. Some of the difference is due to differences in material properties. To eliminate that effect in the comparison, the measured maximum moments were divided by the calculated flexural capacity of the entire width of the slab section. The results are plotted in Fig. 9d from which two conclusions may be drawn.

Dynamic testing resulted in an increase in strength which ranged from one-fifth (for $\rho = 0.006$) to one-third (for $\rho = 0.013$) of the comparable statically loaded specimen. The observed increase in strength is not inconsistent with reported increases of upper-yield stress in reinforcement (Bertero, et al, 1974), (Criswell, 1970) at comparable strain rates. Negative implications of this observation, such as increased shear stress, should be taken into account in design. However, any positive implication must be ignored for general application not only because of the sparsity of observations but also because a full-scale structure is likely to have an effective period of oscillation longer than that in the tests.

Figure 9d also indicates a reduction in measured maximum moment as a ratio of the potenial yield moment with increase in reinforcement, for both the statically and dynamically tested specimens. It is important to note (a) that the observed reduction is a function of the potential strength of the plate and would not be sensed by a model based on linear stiffness properties alone (or a model based on linear response of gross or cracked slab section) and (b) that the reduction in strength was not associated with a drastic change in the mechanism of failure. As will be demonstrated numerically, premature "yielding" in the specimens with larger amounts of uniformly distributed reinforcement could be attributed to the flexibility of the connection.

In general, crack patterns in specimens S1 to S3 were similar. However, there was a difference in the trajectories of the major cracks (shaded areas in Fig. 10). The inclination, in the horizontal plane with respect to the lateral axis, increased with the amount of reinforcement (Fig. 10a to c). A similar trend was observed in the dynamically tested specimens.

BEHAVIOUR OF THE MODEL

In this section calculated results from the grid model are compared with like observed results from the tested connections. Further comparisons between the calculated responses of two specimens analyze the behaviour of the model with different parameters.

Comparison of Calculated and Observed Response

The nonlinear responses of the slabs were calculated by a routine "step-by-step" linear procedure. The "Polo-Finite" (Lopez, et al) system at the University of Illinois (Urbana-Champaign) was used for each step. The beam properties were changed when the moment exceeded the values shown for cracking and yielding moments.

For each of the two specimens the calculated response of the slab is compared with the following observed quantities: (1) Moment-rotation curves (Fig. 11); (2) the twist of the slab about the lateral centerline to the side of the column (Fig. 2 and 12); (3) the vertical slab displacement along the longitudinal centerline (Fig. 2 and 13); (4) the measured strain in bars to the side of the column monitoring the progress of yield across the plate.

The comparison of measured and calculated moment-rotation curves for specimen S1 are given in Fig. 11a. A change of stiffness caused by initial cracking can be deduced from the calculated points; the progressive reduction in stiffness seen in the calculated curve in Fig. 11a is due to spreading of cracking in the slab shortly after load cycle 21 was reached, yielding of the slab at the front column-face was calculated and is indicated in Fig. 11; in Fig. 11 the calculated yielding of elements across the width continued -- with the points shown at which torsional yielding (to the side of the column) and yielding across the full width of the plate were calculated.

The calculated and measured moment-rotation curves for the heavily reinforced specimen, S3, are presented in Fig. 11b. As was done with the calculations of specimen S1, the positions at which the various major changes in stiffness occur are noted (from cracking to yielding).

The next quantity compared was the measured twist of the portion of slab to the side of the column (Fig. 12a). This twist was obtained from the difference in the vertical displacement of the slab measured along two parallel lines divided by the distance between the two lines (Fig. 2). Matching of these calculated and measured twisting angles would enhance the credibility of the model used. Figure 12a shows the comparison of the observed and calculated values at various loads for specimen S1. (The positions of these loading stages are indicated on the moment-rotation curves of Fig. 11a, and 9). A favorable comparison was obtained over a wide range of loads.

The calculated twisting angles, as presented in Fig. 12b, bear a good resemblance to the measured values for specimen S3.

Figure 13a compares measured and calculated vertical displacements along the longitudinal axis for various loading stages. The measured slab deflection along the longitudinal centerline to the back of column (W) and those measurements to the front of the column (E) were the same for loading steps 1.0, 11.0 and 21.0. However, for loading step 31.0 the measurements to the back of the column (31W) were significantly larger than those measurements to the front of the column (31E). The calculated values compared well with the measurements to loading stages 21. At loading stage 31, calculation results corresponded closely to the average values (31 avg.).

A favorable correspondence is also present in the measured and calculated vertical deflections along the longitudinal axis of specimen S3 (Fig. 13b).

As pointed out in Fig. 11a, the lightly reinforced specimen S1 yielded across the full width of the plate. This statement is supported by strain gauge readings that implied yielding of the bars across the full slab width. At other loading points the calculated spread of yielding was verified by the strain gauge readings; by loading 31 yielding had been calculated for elements to the front of the column and this was also evident in the gauge readings.

The strain gauge readings of specimen S3 support the extent of yielding calculated in the grid model.

Comparison of Calculated Responses of S1 and S3

In the calculated response, the lightly reinforced specimen S1 utilized the full slab width, but the more heavily reinforced specimen S3 did not (Fig. 9d). In each case, a good correlation between calculated and measured values was obtained. The calculated response of each specimen is compared to investigate the factors causing the differences in the response of the two specimens. The contributions of the "front face" and "side face" to the overall stiffness of each specimen can be gathered from Fig. 14a (specimen S1) and Fig. 14b (specimen S3).

The changes in incremental stiffness also influenced the increment in moment carried by the grid beams across the width of the slab. The moments in defined beams across the width of specimen S1 and S3 are indicated for increasing slab-column connection rotation in Fig. 15a (specimen S1) and Fig. 15b (specimen S3). The key in each figure gives the position of the beams. For each beam the moment was normalized by its yield moment.

At the time of yielding in beam "30" the moment in beam "34" was about 40% of its yield moment in specimen S1, but only 20% of its yield moment in specimen S3. The overall stiffness of specimen S1 was sufficient to enable beam "34" to mature at about the same time that the plate to the side face of the column yielded in torsion. This happened at a connection rotation of about 2.5% (position "D", Fig. 15a). When the portion of plate to the side face of the column yielded in torsion, the moment in beam "34" of specimen S3 had only reached 30% of the yield moment. The subsequent low stiffness of specimen S3 (Fig. 14b) restricted the beam "34" from developing its potential strength within a reasonable connection rotation.

CONCLUSIONS

Results from a series of three reinforced concrete plate-column assemblies tested dynamically were compared with those from three similar specimens subjected to static load reversals in order to develop fundamental information pertaining to the behaviour of reinforced concrete flat-plate structures subjected to lateral loads. On the basis of the results and the analyses reported, the following conclusions can be made about the reinforced concrete plate-column test specimens and the grid model:

(1) The higher the reinforcement ratio, the lower was the ratio of the moment developed by the slab to the full yield-moment capacity of the slab. The observed trend was similar for statically and dynamically loaded specimens (Fig. 9d). The higher strength of the dynamically loaded specimens was attributed primarily to the strain rate and may not be projected to behaviour of such connections in structures of which effective fundamental frequency would be lower than 3 Hz. With respect to strength, the effective slab width ranged from 100 percent (at $\rho = 0.0065$) to 60 percent (at $\rho = 0.013$) of the full width of the slab.

(2) Nonlinear response of the plate-column specimens was well modelled by a relatively simple grid of beams responding in flexure and torsion. The "monotonic" moment-rotation curve calculated using this model matched closely the envelopes of the curves obtained under cyclic loading (static) in post-yield as well as initial ranges of loading.

(3) Difference in response of the specimens (inability of the relatively heavily

reinforced specimens to achieve yielding across the full slab width) could be explained by flexural action, and without invoking the phenomena associated with shear failures.

ACKNOWLEDGEMENT

This research was sponsored by the National Science Foundation under Grant PFR-78-16318 at the Civil Engineering Department of the University of Illinois at Urbana-Champaign. The paper is based on the writer's doctoral dissertation submitted to the Graduate College of the University of Illinois under the guidance of Prof. Mete A. Sozen. The panel of consultants for the project included M. H. Eligator, A. E. Fiorato, W. D. Holmes, R. G. Johnston, J. Lefter, W. P. Moore, Jr., and Adolf Walser.

REFERENCES

1. Bertero, V. V., Rea, D., Mahin, S., and Atalay, M. B., "Rate of Loading Effects on Uncracked and Repaired Reinforced Concrete Members", *Proc. 5th World Conference on Earthquake Engineering*, Vol. 2, Rome, 1974, pp. 1461-1470.

2. Cardenas, A., and Sozen, M. A., "Strength and Behaviour of Isotropically and Nonisotropically Reinforced Concrete Slabs Subjected to Combinations of Flexural and Torsional Moments", Civil Engineering Studies, Structural Research Series No. 336, University of Illinois, Urbana, May, 1968.

3. Criswell, M. E., "Static and Dynamic Response of Reinforced Concrete Slab-Column Connections", Shear in Reinforced Concrete, ACI Special Publication SP-42, SP42-31, p. 721.

4. Criswell, M. E., "Strength and Behaviour of Reinforced Concrete Slab-Column Connections Subjected to Static and Dynamic Loadings", Dec., 1970, U.S. Army Engineer Waterways Experiment Station, Vicksburg, Mississippi, Technical Report N-70-1 (Final Report).

5. Dybwad, K., Nes, K., and Kure, G., "Floating Concrete Platform for Deep Water Oil Production and Storage", *Proceedings of the European Offshore Petroleum Conference and Exhibition*, EUR 268, London, 1980.

6. FIP State of Art Report: "Foundations of Concrete Gravity Structures in the North Sea", FIP/6/2, Cement and Concrete Association, Wexham Springs, Slough SL3 6PL, August, 1979.

7. Graff, W. J., and Chen, W. F., "Bottom-Supported Concrete Platforms: Overview", Proc., *ASCE, Journal of the Structural Division*, June, 1981, p. 1060.

8. Hawkins, N. M., "Shear Considerations for Concrete Ships and Offshore Structures", *Proceedings of the Conference on Concrete Ships and Floating Structures*, Berkeley Campus, University of California, Sept. 1975, p. 119, Published by University Extension.

9. Kjekstad, O., Stub, F., "Installation of the ELF TCP2 Condeep Platform at the Frigg Field", EUR 14, *European Offshore Petroleum Conference and Exhibition*, London, October, 1978, p. 121.

10. Lopez, L. A., Dodds, R. H., Rehak, D. R., and Urzua, J., "POLO-FINITE, A Structural Mechanics System for Linear and Nonlinear Analysis", Civil Engineering Systems Laboratory, University of Illinois at Urbana-Champaign.

11. Mahin, S. A., and Matsunaga, E., "Analysis of Reinforced and Prestressed Concrete LNG Offshore Structures", OTC 3843, *Offshore Technology Conference*, Houston, 1980.

12. Marshall, A. L., "Concrete for Marine Transport and Storage of Cryogenic Liquids", EUR 29 *European Offshore Petroleum Conference and Exhibition*, London, October, 1978, p. 225.

13. Morrison, D. G., and Sozen, M. A., "Response of Reinforced Concrete Plate-Column Connections to Dynamic and Static Horizontal Loads", Civil Engineering Studies, Structural Research Series No. 490, University of Illinois, Urbana, April, 1981.

14. Park, R., and Gamble, W. L., "Reinforced Concrete Slabs", John Wiley and Sons, New York, 1980.

15. Yettram, A., and Husain, H. M., "Grid Framework Method for Plates in Flexure", Proc. *ASCE, Journal of the Engineering Mechanics Division*, June, 1965, p. 53.

SPECIMEN	REINFORCEMENT RATIO ρ	TENSILE STRENGTH OF CONCRETE f_t [MPa]	COMPRESSIVE STRENGTH OF CONCRETE f'_c [MPa]	YIELD STRESS OF STEEL f_y [MPa]	EFFECTIVE DEPTH OF STEEL* d
S1	0.0065	2.9	45.8	323	64
S2	0.0098	2.3	35.1	330	64
S3	0.0131	2.2	33.9	335	64
S4	0.0098	2.2	34.9	320	64
S5	0.0098	2.4	35.2	340	64
D1	0.0065	3.3	36.3	290	64
D2	0.0098	2.3	33.9	327	64
D3	0.0131	3.0	36.5	355	64
DP	0.0079	3.0	48.0	340	64

*Measured mean values for both top and bottom reinforcement.

TABLE 1. MATERIAL PROPERTIES

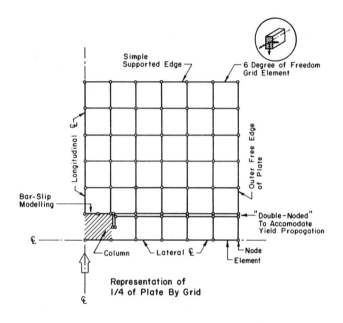

FIG. 1. REPRESENTATION OF ¼ OF PLATE BY GRID OF BEAMS

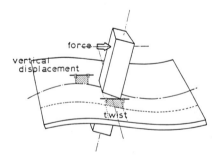

FIG. 2. IDEALIZED REPRESENTATION OF SPECIMENS

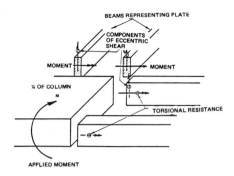

FIG. 3. FORCE COMPONENTS RESISTING APPLIED MOMENT AT THE COLUMN-SLAB CONNECTION

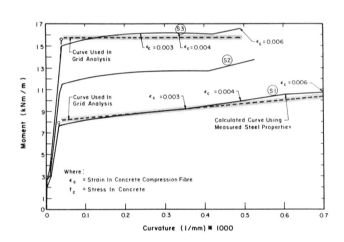

FIG. 4. MOMENT-CURVATURE RELATIONSHIPS

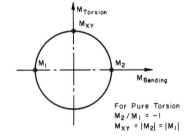

FIG. 5. (a) COMPONENTS OF TORSION

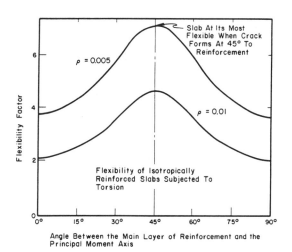

(b) FLEXIBILITY IN TORSION

FIG. 5. TORSION RELATIONSHIPS

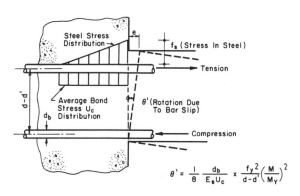

$$\theta' = \frac{1}{8} \frac{d_b}{E_s U_c} \times \frac{f_y^2}{d-d'} \left(\frac{M}{M_Y}\right)^2$$

FIG. 6 BAR-SLIP MODELLING

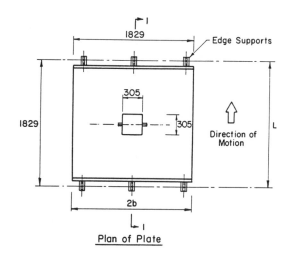

Plan of Plate

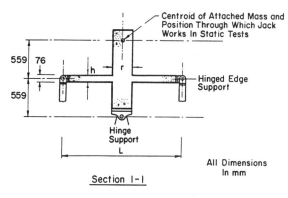

Section I-I

FIG. 7. SECTION AND PLAN OF ASSEMBLY

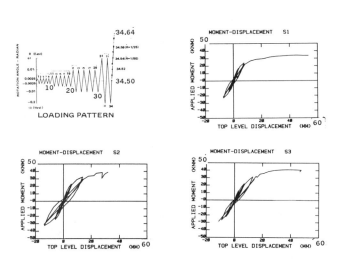

FIG. 8. MOMENT-DISPLACEMENT RELATIONSHIPS (S1 TO S3), AND LOADING PATTERN

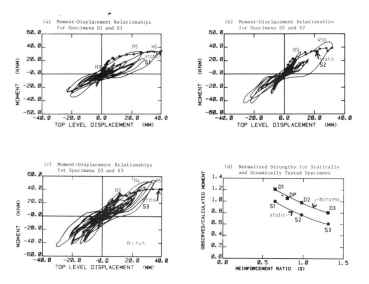

FIG. 9. COMPARISONS OF DYNAMICALLY AND STATICALLY TESTED SPECIMENS

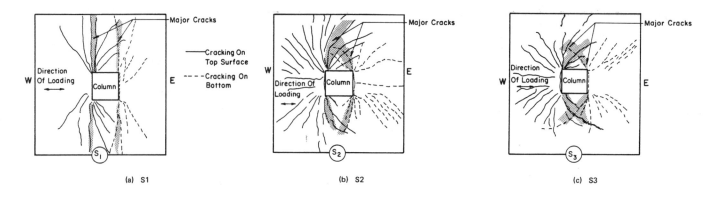

FIG. 10. IDEALIZED CRACK PATTERNS

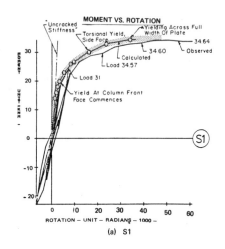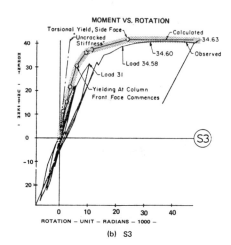

FIG. 11. MEASURED AND CALCULATED MOMENT-ROTATION RELATIONSHIPS (S1 AND S3)

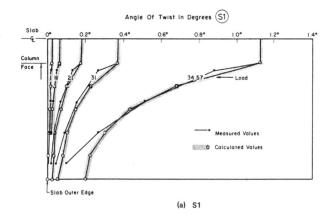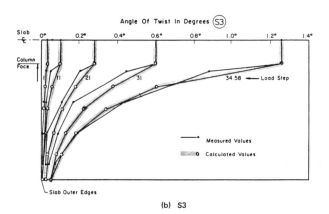

FIG. 12. MEASURED AND CALCULATED ANGLES OF TWIST

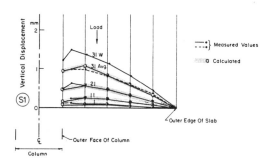
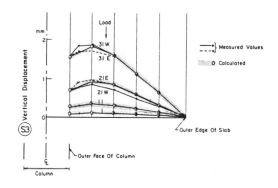

(a) S1

(b) S3

FIG. 13. MEASURED AND CALCULATED VERTICAL DISPLACEMENTS

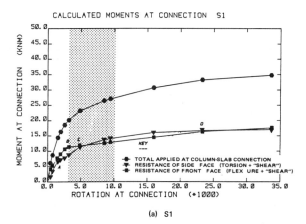
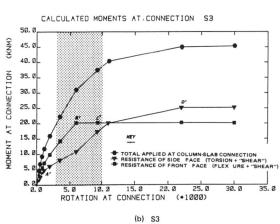

(a) S1

(b) S3

FIG. 14. COMPONENTS OF CALCULATED MOMENT-ROTATION RELATIONSHIPS

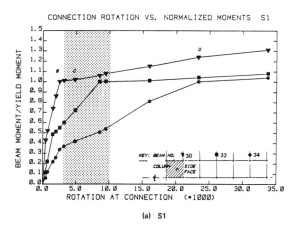

(a) S1

(b) S3

FIG. 15. RESPONSE OF SPECIFIC GRID-BEAMS NORMALIZED BY THEIR RESPECTIVE YIELD MOMENTS.

873

RESISTANCE OF PRESTRESSED CONCRETE SLABS TO EXTREME LOADS

S.H. Perry   I.C. Brown
Department of Civil Engineering, Imperial
College, London, U.K.

SUMMARY

Types of concentrated loading are classified and discussed in relation to the extreme loading of slabs. Static tests, and tests simulating hard impact loading, have been carried out on bonded prestressed concrete slabs. Slabs were 1.5m square and post-tensioned to give either a uniform or approximately triangular distribution of prestress. Static loading was applied centrally by means of a hydraulic activator, and impact loading by means of a dropped mass. Contact loads have been recorded during impact.

1. INTRODUCTION

Previously, work on prestressed concrete slabs has been confined mostly to unbonded slabs subjected either to a uniformly distributed load or to concentrated load simulating column punching behaviour. In the USA unbonded slabs are more commonly used for buildings than bonded slabs, being both cheaper and quicker to erect.

However, recent developments, particularly in offshore engineering, have led to an increasing interest in bonded slabs, where the bonding both provides corrosion protection to the steel and also improves post-yield behaviour. Also, attention is being paid to concentrated loads, such as impact loading, as well as to uniformly distributed loading which is often assumed for initial design. Concentrated loads will be more likely to result in local damage and penetration through a structure than a uniformly distributed load, but it is necessary to identify the different types of such loading before any description of structural response can be attempted.

The classes of concentrated loading may be conveniently grouped according to cause, as follows.

Class 1. <u>Static loads and quasi-static impact loads.</u> This class covers all conditions where the structure is able to mobilise all reserves of strength including the stiffness of associated members such as edge beams and adjacent slabs and columns, the flexibility of the whole structure and, for a slab, membrane action.

Included in this category would be "soft" impacts caused by vehicle or ship collisions. Such collisions would often be accompanied by substantial damage to the impacting body, and the loading duration is long compared with the fundamental period of the structural element that has been hit.

Class 2. <u>Accidental impact loads caused by dropped objects.</u> Until the advent of 200-300m high marine structures the only objects likely to be dropped accidentally onto onshore structures, except during construction, would have low mass or density. In either case damage is likely to be minor, even at quite high approach velocities.

However the advent of massive marine structures with large oil storage tanks and of manned capsules on the sea bed, and the relatively high incidence in offshore operations of dense dropped objects, increases the risk of accidental impact. In deep water it is possible that the object that has been dropped could reach terminal velocity before reaching the sea bed. In these cases the resulting structural damage, and possible heavy pollution, loss of life and high cost of repairs or recovery may be disproportionately high compared with the nature of the accident.

Class 3. <u>Aircraft impact.</u> Aircraft impact comes into a class of its own, partly because no other class of concentrated loading involves such a high mass at such high speed, and partly owing to the extreme nature of the possible damage should such an impact occur on a nuclear installation.

Class	Description of loading	Hard or soft	Examples	Typical Velocities m/s	Relative Incidence of extreme loading	Principal cause of concern
1	a) static	h	machinery vehicles, ships	0	rare	buildings
	b) quasi-static	h/s		0 - 10	frequent at < 2 m/s	bridge parapets and piers, marine structures
2	Dropped objects	h/s	containers, steelwork, pipes, tornado borne debris	5 - 40	frequent	buildings during construction, offshore marine structures, sea bed capsules.
3	Aircraft	a) s b) h	overall parts	200 - 300	rare	nuclear installations
4	Ballistic	h	bullets missiles shells	1000	frequent in certain areas	military and certain civilian structures
5	Nuclear		beyond the scope of this study			

Table 1 Classification of concentrated loading according to cause

Class 4. <u>Ballistic impacts.</u> Ballistic impact differs from accidental impact both in the magnitude of approach velocity and also because, with ballistics, the impact may be accompanied by an explosion.

Class 5. <u>Nuclear blast.</u> Nuclear blast either caused deliberately, or as the result of an accident in a power station, is beyond the scope of the present study. In general the blast can be represented by a uniformly distributed load of high intensity. However secondary damage could be caused by flying objects dislodged by the blast, and result in concentrated impact in one of the classes described above.

This paper describes the response of model prestressed concrete bonded square slabs to concentrated loading of the types described in groups 1A (static) and 2 (dropped objects).

2. DESCRIPTION OF TEST SLABS

The general arrangement of slabs is shown in Figure 1. Well graded flint-type river sand and gravel, of 10mm maximum size, were used for the concrete, designed for a nominal 28 day strength of 50 N/mm^2 (7500 lb/in^2). Prestressing strands were either 5mm or 7mm diameter normal relaxation prestressing wires, with strengths of 33.5 kN and 62.0 kN respectively. Horizontal ducts for prestressing wires were created by means of 10mm diameter, PVC sleeved, mild steel bars. The bars and sleeving were rotated at 2 hours and 3 hours after casting, and removed at 4 hours, thus leaving straight horizontal ducts. Slabs were moist cured for 7 days, usually stressed at 7 to 14 days and tested at about 21 days.

Both concentrically and eccentrically prestressed slabs have been tested. Ignoring self-weight effects the former do not deflect when stressed, but when an eccentric prestressing force is applied to a prismatic slab, the slab deflects upwards (unless its weight is sufficient to counteract this effect) and a trapezoidal or triangular compressive stress distribution is imposed throughout its section.

Each wire was stressed twice, on each occasion given an initial force equivalent to 70% of its characteristic strength. The second stressing was necessary to counteract, by shimming at the anchorage of each wire, the decrease in steel strain due to draw-in and elastic shortening. Ducts were grouted immediately after stressing. An average prestress loss of 20% was recorded before each slab was tested (Okafor and Perry, 1982).

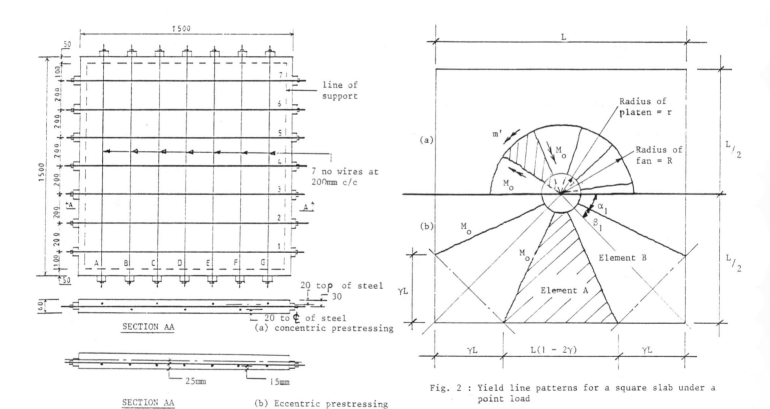

Fig. 1 : Layout of prestressed slabs

Fig. 2 : Yield line patterns for a square slab under a point load

(a) Fan mechanism

(b) Overall flexural mechanism

3. STATIC CONCENTRATED LOAD TESTS (Class 1a)

3.1 Theoretical Work

For a square slab subjected to a concentrated load failure can occur either through a fan mechanism, or an overall flexural mechanism arising from yield lines radiating from the load and extending to the edges of the slab (Figure 2). The collapse load of an under-reinforced slab is normally higher than the classic Johansen's yield load. This is because the yield-line theory is based on the pure moment capacity of a slab cross-section. It does not, therefore, take into account any strain hardening of the steel or in-plane forces in the slab resulting from changes in geometry. Many authors have analysed the case of simply supported slabs with uniformly distributed loading, taking into account the membrane effect after the yield-line mechanism. Such slabs under concentrated loading, with large shear forces and bending moments at the point of application of the load, have not been considered.

It has been shown (Okafor and Perry, 1982) that Kemp's (1965) approach to the analysis of post yield behaviour under a UDL can be modified for the condition $a \leq a'$, where a = central deflection of a slab, a' = limiting deflection beyond which the neutral axis lies outside the slab section in the central region of the slab, to show that the collapse load, P, is given by

$$P = 4M_o \left\{ \frac{\tan\alpha_1}{\left\{\frac{1}{2} - \frac{r}{L}\right\}} + \frac{\tan\beta_1}{\left\{\frac{1}{2} - \frac{r}{L}\frac{\cos\alpha_1}{\cos\beta_1}\right\}} \right\} \left\{ 1 + \beta \left(\frac{a}{a'}\right)^2 \right\} \tag{1}$$

where M_o = positive yield moment per unit width of slab, and α_1 and β_1 = half angles subtended by the yield lines at the centre of the slab.

Furthermore, it can be shown that for all practical values of $\frac{r}{L}$, equation (1) reduces to

$$\frac{P}{P_y} = 1 + \beta\left(\frac{a}{a'}\right)^2 \tag{2}$$

where P_y = theoretical yield load of a slab and β = a constant dependent upon the geometry and steel and concrete properties. Similarly, for $a \geq a'$, it can be shown that

$$\frac{P}{P_y} = 1 + \beta\left[2\sqrt{\frac{a}{a'}} - 1\right]^2 \tag{3}$$

Since
$$a' = 8\beta \frac{M_o}{T_o}$$

where T_o = yield force in tensile reinforcement per unit width of slab, equations (2) and (3) may be written

$$\frac{P}{P_y} = 1 + \frac{a^2}{64\beta} \frac{T_o}{M_o} \tag{2a}$$

$$\frac{P}{P_y} = 1 + \beta\left\{\sqrt{\frac{a}{2}\frac{T_o}{M_o}} - 1\right\}^2 \tag{3a}$$

Although these equations are, in fact, identical to those obtained by Kemp (1965) for a UDL, it should be noted that the evaluation of the axial forces and yield moments used to derive the above expressions neglected shear displacement between elements A and B (Fig. 2 - only true when the elements are of equal area), in-plane shear along yield lines, the effect of the platen on the yield line crack patterns and the effect of the inclined shear cracks on the rigid elements.

3.2 Test Arrangements

The loading system consisted of a steel frame carrying a 200 kN hydraulic actuator connected to a servo-control system. This allowed either load control or displacement control during testing. Slabs were simply-supported on a sturdy steel rig high enough (900mm) to allow examination of the bottom face of a slab under test.

During test, the deflection of the slab was recorded at several points using both dial gauges and linear transducers, and both prestressing steel and concrete strains were recorded by means of electric resistance gauges fixed at various critical points (Okafor and Perry, 1982).

A series of tests was carried out with various platen sizes (102mm, 152mm and 203mm) and prestressing steel ratios (0.245% to 0.641%).

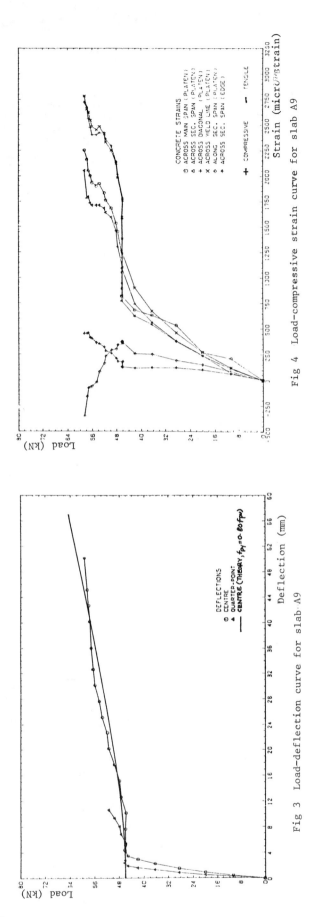

Fig 3 Load-deflection curve for slab A9

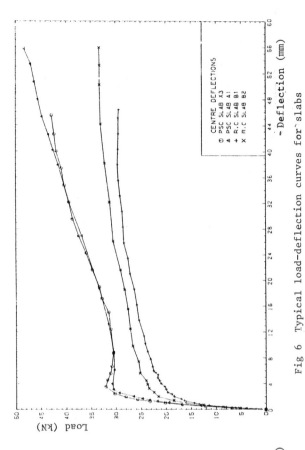

Fig 4 Load-compressive strain curve for slab A9

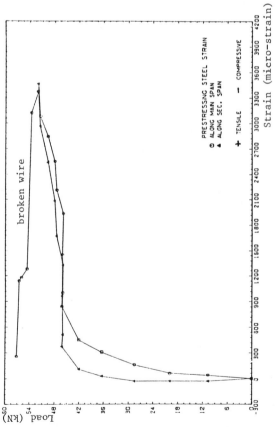

Fig 5 Load-strain curve for central wires of slab A9

Fig 6 Typical load-deflection curves for slabs

3.3 Experimental Results

Load-deflection and load-strain curves for a typical slab are shown in Figures 3, 4 and 5. It was noted that

3.3.1 Deflections

1. load-deflection curves for all slabs, especially those with a low percentage of steel, displayed distinct yield plateaux at loads coincident with the corresponding yield mechanism (Figure 3),

2. every part of a prestressed slab was yielding simulataneously, confirmed by the profiles of the corner and quarter-point deflections being similar to that of the centre-point,

3. both smaller platen size and larger steel ratio gave rise to smaller ultimate deflections,

4. slabs loaded up to as much as 60% of their eventual ultimate load, recovered almost completely (typically 0.5mm residual deflection) and showed only a small decrease in stiffness on reloading,

5. maximum deflections, just before maximum load, were 40-60mm, with residual values of 2-8mm after rebound at failure.

3.3.2 Strains

The load-strain curves for the prestressing steel (measured on a wire below the edge of the platen) and, for the concrete, the load-tangental compressive strain curves (that is, strains measured across yield lines of the slab) and the load-radial strain curves (strains measured along assumed yield lines) all display clear yield plateaux. (Figures 4 and 5). These are followed by rising portions to failure except for the radial strain curves which, after the yield mechanism, always show a progressive reduction in value.

3.3.3 Post-yield behaviour

1. All slabs failed violently, after development of yield-line mechanisms, by punching out of a plug of concrete in the form of a frustrum of a cone. The level of violence was higher for high steel ratios.

2. Failure was immediately followed by rebound, this tended momentarily to throw the compression face of the slab into tension, with the result that the radial yield lines penetrated to the top face.

3. At about 95% of ultimate load, cracks, approximately circular and concentric with the platen, appeared on the compression face of the slabs, at spacings of 2h to 3h. These closed without trace after rebound at failure.

4. For slabs loaded with 102mm platens (c/h = 1.7) the critical zone of punching was always around the periphery of the platen, for larger platens (152 or 203mm, c/h = 2.5 or 3.4) the critical zone of punching was outside the platen periphery.

3.4 Conclusions

The results, which have been explained in detail elsewhere (Okafor and Perry, 1982), show that yield-line theory gives better results for under-reinforced prestressed slabs than for ordinarily reinforced slabs. After the yield mechanism, tensile membrane action commences as a means of supporting further load until terminated through collapse through a local fan mechanism. On the basis of the results of these tests and those reported elsewhere, it seems reasonable to conclude that for simply supported under-reinforced concrete slabs, both ordinarily reinforced and prestressed, subjected to concentrated loading, <u>the Johansens yield load constitutes a 'lower' upper bound solution for the collapse load while the local fan mechanism constitutes an 'upper' upper bound solution.</u>

Furthermore, for design purposes the results suggest that,

i. If the prestressing steel of a bonded under-reinforced prestressed slab is given an initial prestress of $0.7f_{pu}$ and has an effective prestress, after losses, of at least $0.5f_{pu}$, it is likely to attain a stress of $0.8f_{pu}$ at the yield-line mechanism of the slab. This stress is equivalent to the yield stress, f_y, of steel in an ordinary reinforced concrete slab. Although this steel stress has been obtained for a nominal concrete strength of $50N/mm^2$ (7500 lb/in^2) only, it is likely to apply to all concrete strengths (30 to $60N/mm^2$, 4500 to 9000 lb/in^2) used in prestressed slab construction, as the yield moment is relatively insensitive to concrete strength.

ii. The steel stress, f_{pb}, at the failure of a bonded prestressed slab is closely predicted by the modified ACI (1977) equation 18.3;

$$f_{pb} = f_{pu}\left[1 - 0.6\rho_{ps}\frac{f_{pu}}{f_{cu}}\right]$$

iii. The present tests suggest that a spacing of prestressing steel of up to six times the effective depth of a slab will give good service load behaviour.

4. HARD IMPACT LOAD TESTS (Class 2)

4.1 Test Arrangements

Impact tests, to investigate the response of prestressed slabs to dropped objects, have been carried out on model slabs of the same form as those used for static tests.

A general view of the test rig is shown in Fig.7. The impacting body is dropped freely down the inside of a tube manufactured from two battened steel channels. The maximum height of drop available depends on the length of the impacting body, but is about 5m. This limits the approach velocity to 10m/s. If faster approach velocities are required than an alternative method of propelling the impacting body would have to be incorporated.

At present, the impacting body is a steel bar, 127mm diameter and 1.2m long. However, there is no difficulty in changing this to increase or reduce the mass, provided that the body can still fit inside the guide tube. Also, a "soft" impact, rather than a "hard" impact, can be simulated by including in the dropped mass a yielding or buckling thin-walled cylinder, or a suitable dash pot.

Fig 7 General view of test rig.

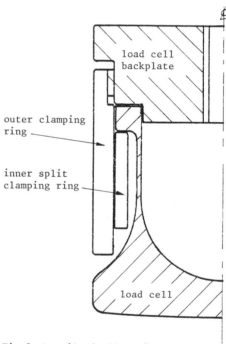

Fig 8 Detail of 150mm diameter load cell assembly

4.2 Instrumentation

A major aim of the project has been to try to measure the contact force during impact. For this purpose a load cell is fitted to the front end of the dropped mass. The arrangement of the load cell is shown in Fig. 8. A D2 tool steel was used for the construction. It was specified that after final heat treatment the steel should have a hardness of 56 on the Rockwell 'C' scale. This results in a steel of very high strength which, although less hard than the steel for a machine tool (where a hardness of Rockwell 'C' 62 might be specified), is much tougher. The load cell should therefore be reasonably abrasion resistant but also not too prone to brittle failure during repeated impacts. Semiconductor gauges have been used in the load cell. These gauges have a very fast response and also a much higher output than foil gauges, thus reducing the amount of signal amplification that is needed. It has been assumed that a static calibration for the load cell will be adequate because there is evidence that the elastic modulus of steel does not change as the rate of loading is increased (Mainstone, 1975). The natural frequency of the load cell assembly is such that it might prove a limitation to its use for some impacts. However, for these particular tests the response is sufficiently good. This point is discussed more fully later in the paper.

Fig 9 View of underside of slab showing complete penetration of impacting body and fracture of prestressing wires.

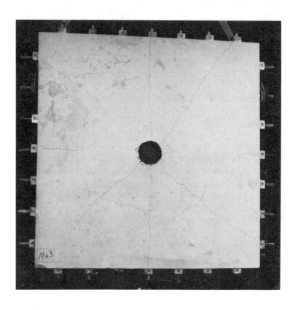

Fig 10 Top of slab after impact.

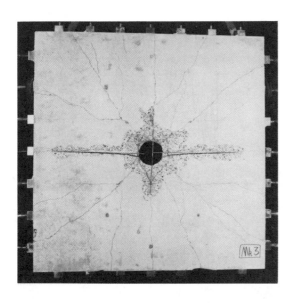

Fig 11 Underside of slab after impact.

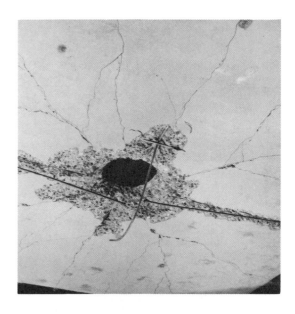

Fig 12 Detail of underside of slab after impact.

Measurements are also made of strains in the prestressing wires and on the concrete surface, and of displacements of the slab. Displacements are obtained by numerical double-integration of the output from accelerometers on the slab, because it was thought that the response of LVDT transducers would be too slow. Difficulty was experienced in devising a suitable method for collecting the experimental data. Initially it was intended to use a recording oscilloscope. However there are problems with accurate triggering of the oscilloscope, especially as the record length (one timebase sweep) is so short. Also, the resulting traces need digitising if further processing is required. The scheme that has been adopted is to pass each individual channel of information through an analogue-to-digital converter and to store the results in static memory. Subsequently the information from these static memories is read into a micro-computer for processing.

4.3 Results

Preliminary tests have been carried out. Figs. 9-12 show details of one particular test. It is hoped that more results will be available at the conference, but, although only tentative conclusions are possible at this stage, some discussion follows.

Fig. 13 shows the contact load that was recorded in three successive impacts carried out at progressively faster approach speeds on one prestressed slab. Fig. 14 shows a detailed trace, with magnified time axis, of the initial part of the contact load during impact on another slab. On these traces periodic fluctuations of the load can be identified at three distinct frequencies. The faster fluctuation, clearly visible on Fig. 14, with a period of about 200 microseconds, is caused by stress wave reflections within the load cell. This variation is unwanted, in that it is a function of the measuring system and occurs only in the laboratory test, not in a prototype impact. However, because the impact duration is relatively long, these fast vibrations of the load cell are not too troublesome and are not even discernible on the traces of Fig. 13. The other vibration seen in Fig. 14, also just visible in Fig. 13, with a period of about 2 milliseconds, is caused by stress wave reflections in the steel bar which forms the dropped mass. This form of load variation will occur in both test and prototype. Finally, the first trace of Fig. 13 shows the slowest fluctuation, with a period of about 6 milliseconds, which is caused by transverse vibration of the slab.

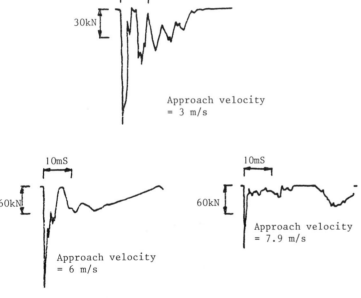

Fig 13 Contact load history for successive impacts on a prestressed concrete slab

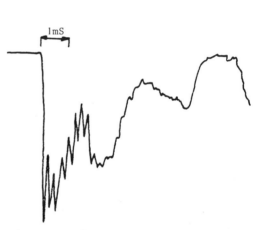

Fig 14 Detail of initial contact load history for an impact on a reinforced concrete slab

Looking at Fig. 13, a suggested explanation for the sequence of events during an impact can be given. The initial sharp, peaked loading is the response of the concrete in the vicinity of the contact area. The form of this part of the loading is affected by local conditions in the slab including damage caused by previous impacts, but is insensitive to support conditions. It is during this period that cracks in the contact zone are formed. In the first two traces on Fig. 13 contact is lost briefly after the initial impact and re-established as the slab undergoes transverse vibrations. The form of the following part of the loading depends on how damaged the slab is. In the first case, with low approach velocity and little damage, the slab remains a complete structural unit and its transverse vibrations are indicated throughout the load trace. In the second case, at higher approach velocity, very considerable damage occurs during the initial impact and it is postulated that, subsequently, the major connection between the centre of the slab, supporting the dropped mass, and the outside of the slab is by means of the prestressing wires acting as a net. Finally, in the third trace, at even higher approach velocity, complete penetration of the slab occurs with fracture of the prestressing wires and tearing of the wires through the concrete cover on the underside of the slab.

4.4 Conclusion

Until more test results are available there is not much point in attempting any analysis of the problem. At that stage, the most promising analytical methods can be investigated, but it may be that, because impact loading and response are so complicated, the only sensible method for predicting response will be by full-scale or model testing. The authors have presented, at the previous BOSS conference, a discussion of various methods of analysis suitable for describing impact (Brown and Perry, 1979).

5. FURTHER WORK

Experimental work has already been completed on ordinary reinforced slabs (Figure 6) and prestressed slabs with corner restraint, both subjected to static concentrated load (Okafor, 1981); and the present impact work is being extended to cover the behaviour of both reinforced and prestressed shells and domes. In addition, the enhancement to both strength and ductility resulting from the triaxial compression induced by both discrete multidirectional prestressing and spiral mesh reinforcement is under investigation (Perry et al, 1982). Beyond this, it is hoped that funding will become available to allow investigation of

5.1 the response of all types of slabs to Class 3 loading (for example, simulating an aircraft engine), with measurement of all transient effects including contact load

5.2 the behaviour of other structural forms, including beams, one-way slabs and structures comprising clusters of elements (such as beam/slab systems), where part of the system is subjected to a concentrated impact load.

6. REFERENCES

American Concrete Institute, 1977, Building Code Requirements for Reinforced Concrete, ACI 318-77, Michigan.

BROWN, I.C. and S.H. PERRY, 1979, "Transverse Impact on Beams and Slabs", Proceedings, 2nd International Conf. on the Behaviour of Off-Shore Structures, London, England, 1979, Vol. 2, Paper 73, pp.357-368, Cranfield, Bedford, England BHRA Fluid Engineering.

KEMP, K.O. 1967, "Yield of a Square Reinforced Concrete Slab on Simple Supports, Allowing for Membrane Forces", The Structural Engineer, Volume 45, Number 7.

MAINSTONE, R.J. 1975, "Properties of Materials at High Rates of Straining or Loading", Materiaux et Constructions, Volume 8, Number 44, pp. 102-116.

OKAFOR, H.O. 1980, "Bonded prestressed Concrete Square Slabs subjected to Concentrated Loading", Ph.D. thesis, Imperial College of Science and Technology, University of London.

OKAFOR, H.O. and S.H. PERRY, 1982, "Post-Yield Behaviour of Bonded Prestressed Concrete Square Slabs subjected to Concentrated Loading", The Structural Engineer, Volume 60B, Number 2.

PERRY, S.H. W.E. ARMSTRONG and A.J. HARRIS, 1982, "Axially loaded concrete prisms with non-uniform lateral confinement", to be published.

7. ACKNOWLEDGEMENTS

The experimental work described in this paper was funded by the British Science and Engineering Research Council (SERC), Marine Technology Directorate, and would not have been possible without the skills and enthusiasm of the technical staff of the Concrete Laboratories, Imperial College, London.

DESIGN ASPECTS OF ARTIFICIAL SAND-FILL ISLANDS

W.M.K. TILMANS and K. den BOER
Delft Hydraulics Laboratory
The Netherlands

J. LINDENBERG
Delft Soil Mechanics Laboratory
The Netherlands

SUMMARY

The paper deals with the present investigation techniques regarding the design of artificial sand-fill islands, applied by the Delft Hydraulics Laboratory and the Delft Soil Mechanics Laboratory. The two laboratories, together with Rijkswaterstaat, have joint their efforts in view of the necessity of an integrated hydraulic and soil mechanical approach in designing such islands. A description is given of specific investigation techniques concerning the geotechnical and morphological stability of the slopes of the sand-fill and the structural stability of sea defence constructions. Special attention is given to wave reducing structures.

1. INTRODUCTION

Artificial islands should be economically feasible and technically safe. To meet these requirements, a large number of hydraulic, structural and soil mechanical aspects have to be considered [1]. As the objective of artificial islands, and also their environment, shows a large range of varieties, it is impossible to develop general design criteria.

In this paper, a description is given of the present investigation techniques that are applied by the Delft Hydraulics Laboratory and the Delft Soil Mechanics Laboratory, to establish design criteria for the geotechnical, morphological and structural stability of the shore front of artificial sand-fill islands. These islands would consist of solid fill, extending a certain distance above sea level, and protected by natural sand slopes or constructed embankments of rock or artificial armour. Attention is focussed on open sea conditions. Ice forces, which form an important part of the external load in arctic regions, and the permafrost behaviour of the soil are therefore not considered.

The principal hydraulic conditions for the design of sand-fill islands are described in detail in Chapter 2. Geotechnical investigation techniques which can be used to analyse the underwater slope stability are discussed in Chapter 3. Depending on the design conditions of the island, various sea defence systems can be distinguished which differentiate in slope, height and composition of the island shore front. In this paper three main types of sea defence have been considered:
• sand beaches
• slope defence constructions
• vertical face structures.
Design aspects of sand beaches are discussed in Chapter 4. In reducing wave run-up and resisting displacement both wave reducing structures and slope defence constructions can be considered. Chapter 5 deals with available techniques to reduce wave attack on the island shore, with special emphasis on the use of submerged breakwaters. In Chapter 6 the state of the art of slope protection has been given, while in Chapter 7 attention has been focussed on vertical face structures.

2. HYDRAULIC DESIGN CONDITIONS

In the design of the island shore front, the hydraulic conditions, which are anticipated in the lifetime of the island, have to be taken into account. In this respect data on waves, currents and water levels should be known and analysed.

2.1 Waves

Wave data include long term wave statistics and short term descriptions of wave records corresponding to selected sea states, e.g. operational conditions and extreme conditions, for the design of structural elements.

If wave data are inadequate for a determination of wave statistics, synoptic weather charts can be used in combination with a wave forecasting technique. Together with Rijkswaterstaat and the Royal Netherlands Meteorological Institute, the Delft Hydraulics Laboratory has further improved the wave-forecasting model GONO, developed by the Royal Netherlands Meteorological Institute [2]. So far, the model shows a good agreement between measured and computed data. At present, the model is applied for the design of North Sea structures.

Refraction and shoaling. When waves are affected by the bottom, wave refraction and shoaling analyses may be required. Sophisticated methods are nowadays available to perform these wave transformation computations [3].

Breaking. When depths are less than 2 to 3 times the significant wave height, highest waves will break. Seelig [4] has determined graphs to compute the maximum significant wave heights in shallow water in case of no reflection. Results will also be applicable for sloping face structures with a reflection coefficient up to 40%.
For vertical face structures, reflection coefficients are such that the maximum incident significant wave height is reduced to about 75% of the value found for the non-reflecting condition.

2.2 Water level

The water level is determined by [5]:
• astronomic tides (tidal information is given in [6,7])
• barometric pressures (of importance in cyclone areas)
• wind set-up (of importance in shallow water)
• wave set-up (of importance in shallow water, shoreward of the breaker line).

2.3 Currents

Tidal currents are of importance in enclosed seas and bays, where tidal gradients between neighbouring shore stations may cause high currents. Along open oceans currents are normally weak. Tidal observations and current measurements are required to determine the existing flow pattern. Model tests or computations can be applied to obtain values for the future situation after completion of island structures (see Section 4.3).

Wind induced currents are caused by the wind shear force and are highly stochastic therefore. Long term measurements are required to obtain statistically reliable data. It can be demonstrated that for steady state conditions the wind induced current will only be 2 to 5% of the wind speed. This figure is in general too low to cause problems in coastal engineering.

Wave induced currents are caused by waves breaking on the shore. The velocities within these currents vary strongly with the angle of wave incidence.

3. GEOTECHNICAL SLOPE STABILITY ANALYSIS

3.1 Introduction

In principle the geometry of an artificial sand-fill island is quite simple. The alignment of the horizontal upper surface follows directly from the objective and environmental conditions of the island. To establish the slopes towards the sea bottom, geotechnical aspects have to be considered in addition to practical ones.

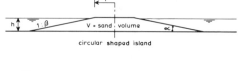

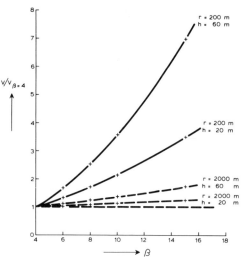

In many cases for economic reasons it will be tried to design the slopes as steep as possible in order to reduce the quantities of sand-fill involved, see Figure 1. This is particularly the case when proper fill material is only available at a large distance.

The designers preference for steep slopes is limited by practical and geotechnical conditions: "which steepness is still feasible" and "which slope is sufficiently safe". The steepest slopes can be realized by hydraulic filling inside bunds made beforehand and by using controlled placement techniques with the pipe reaching the seabed. In general the steepness will be limited to 1:4 ($\alpha \sim 14°$) depending among other things on the kind of material used.

From the geotechnical point of view relatively coarse sand will be preferred as sand-fill material. Furthermore, the percentage of fines must be limited. Other important data needed for the geotechnical stability analysis are the in situ density of the sand and information about those boundary conditions which may change in future. In fact, even a 1:4 slope consisting of loose sand is sufficiently safe if no external changes act upon it, because a 1:4 slope demands an angle of internal friction of the soil of only 14°.

Fig. 1 Volume of sand-fill as a function of underwater slope, for different island geometries

3.2 Sand-fill densities and field measurements

Without densification hydraulic filling will result in relatively loose packings. The relative densities generally are in the range of 0.2 - 0.5, depending on the kind of sand used, the applied dumping technique, the production during dredging, the hydraulic regime etc. However, only some tendencies of the governing relations are known and there still remain a lot of uncertainties. Therefore in many cases density data are required and field measurements have to be carried out. A number of frequently used measuring techniques can be distinguished, among these Standard Penetration Tests (SPT) and Cone Penetration Tests (CPT). At the Delft Soil Mechanics Laboratory a lot of experience has been built up with the CPT measuring technique in a great variety of circumstances.

In the North Sea coastal areas a large number of measurements in loose to dense sands have been executed, expecially for the foundation analysis of the storm surge barrier in the Eastern Scheldt in The Netherlands. For these Delta Works in particular, the density measuring technique was developed to a routine instrument gving reliable results. Coupled to CPT measurements it leads to very helpful additional information.

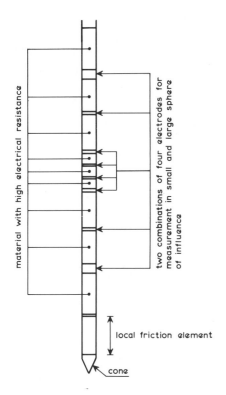

Fig. 2 Combined CPT-density probe

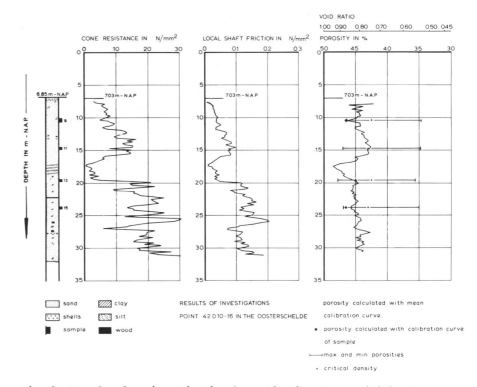

Fig. 3 Example of an investigation for estimating the sensibility for flow slides

In Figure 2 the combined CPT-density probe is shown, which is based on the difference in electrical resistivity between soil and pore water. In Figure 3 an example is given of a measurement. The expected relationship between porosities and CPT-values is clearly seen. For nearshore and offshore purposes the measurement can be carried out from barges or with a submersible working chamber. Such a diving bell, being available at the Delft Soil Mechanics Laboratory, is shown in Figure 4. This diving bell can operate in water depths up to 200 m.

a. Exterior view

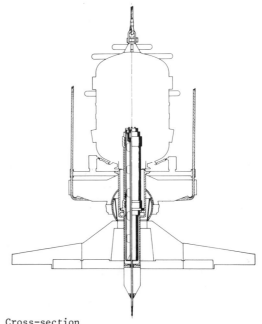

b. Cross-section

Fig. 4 Diving bell

3.3 Liquefaction analysis and slope stability

A liquefaction analysis normally involves evaluation of field measurements and laboratory tests on rebuilt sand samples. In many cases the connection between field and laboratory tests is found by using empirical graphs or formulae. In this way, a relative density can be derived from CPT- or SPT-data. Often, such an estimation will be too rough and not acceptable. In those cases density measurements deliver additional information with which the reliability of the analysis will increase. As shown in Figure 3, the critical porosity (from laboratory tests with quasi-static or cyclic loading), plotted in a graph of field porosity versus depth, results in a clear picture of liquefaction susceptibility of the bottom profile.

As mentioned before, the relatively loose sand of a hydraulic-fill island with slopes 1:4 or flatter is theoretically safe in case of static conditions. But, at sea, the conditions are always changeable. Currents and waves will cause erosion which often results in locally steepened slope parts. The expected influence of erosion must be taken into account in the slope stability calculations. During storms the varying wave pressure at the slope produces cyclic stresses in the soil. In loose deposits this may introduce liquefaction and sand slides, involving large masses of sand material. In The Netherlands in the past years much attention has been paid to the wave induced pressures and effective stress changes in the subsoil. Calculation methods have been developed [8,9] and calculated results have been compared with measured data in large-scale tests [10].

Flow slides in loose deposits

In the south-west part of The Netherlands extensive flow slides of sand occur regularly. Sometimes millions of cubic metres of sand are involved. In many cases these slides take place during extreme tides but otherwise calm sea conditions. It is believed that these slides are caused by temporarily more intensive erosion. In fact, these quasi-static flow slides occur all over the world where loose sands are deposited relatively fast. Well-known cases are the slides along the Mississippi river and in the Norwegian fjords.

Some years ago an experimental flow slide study was carried out in a flume of the Delft Hydraulics Laboratory. With these tests the understanding of the mechanisms involved could be extended. Some of the results have been presented in [11]. With the increased insight in the flow slide phenomenon, together with field and laboratory investigations, a practical strategy was developed which could be used in many cases in nature.

Seismic conditions

In seismic areas an earthquake analysis is required. For this reason often a horizontal acceleration is added to the soil mass, involved in a series of stability calculations. However, for relatively loose sand this is often insufficient because of the build up of pore water pressure due to shear stress cycles. For this reason the design earthquake must be defined in more detail. The number of cycles and the cycle amplitude have to be estimated realistically before a decisive pore pressure build up can be determined in the laboratory by means of triaxial and simple shear tests. Subsequently, the Delft Soil Mechanics Laboratory introduces this pore pressure, or a part of it, in the slope stability calculations, together with the earthquake acceleration factor.

Nowadays, in many soil mechanics institutes more fundamental seismic analyses are developed in which the earthquake shaking is introduced at the base boundary only. In these calculations a more fundamental soil behaviour is assumed. Pore pressures in soil layers above the base and liquefaction of the soil in the ultimate case originate in a more realistic way. However, although these new developments are very promising, for practical slope stability analyses they are not yet suitable.

In some cases the geotechnical analysis leads to additional measures, for instance slope protections or densification of the sand-fill. In case the design is focussed on relatively steep sand slopes, the designer always must consider the necessity of densification in that stage, because normally the sand-fill density and liquefaction potential cannot be predicted sufficiently before the hydraulic-fill works are completed.

4. SAND BEACHES

4.1 Introduction

An artificial sand-fill island will be subject to wave and current attack. Hence, it sould be constructed in such a way that the stability of the island in this erosive environment is guaranteed. Sand beaches represent one of the most economical types of sea defence, especially with respect to capital cost. However, the maintenance of such unprotected slopes under wave and current attack requires periodic beach nourishment and can therefore cost more over a period of time than other types of sea defence. Consequently, in the design of sand beaches, possible erosion due to storm surges and tidal and wave induced longshore currents, has to be taken into account.

Depending on the island geometry and its location with respect to the mainland, the island will influence the nearshore current and wave conditions of the mainland. A change in these conditions in its turn will have an impact on the coastal morphology of the mainland.

4.2 Profile changes due to storm surges

The beach profile of an unprotected sand-fill island is mainly determined by the onshore-offshore movement of sediment under wave motion. The effective direction and intensity of this transverse transport is dependent on the wave conditions, the geometry of the beach profile, and the characteristics of the bed material.

Variations in wave characteristics lead to changes of the beach profile. Along normal coasts, with a relatively flat foreshore, the beach will generally accrete at moderate wave conditions and recede under severe wave attack. For a small island, rebuilding of the coastal profile will not be a matter of course, due to the relatively steep foreshore and the small scale of the sand transport system. Consequently, the design of the island beach and foreshore should be based on the possible profile changes during storm surges, which may be expected in the lifetime of the island.

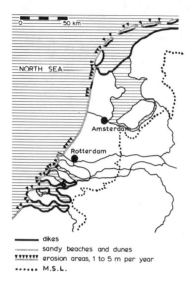

Fig. 5 Dutch coast, erosional areas

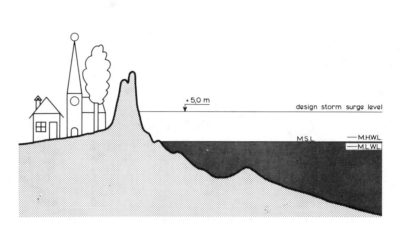

Fig. 6 Typical Dutch coastal profile

For The Netherlands the sand beaches and dunes are of vital importance as a primary sea defence system, as more than half of the country is lying below mean sea level (Figures 5 and 6). Since the severe storm surge of 1953, as a result of which large areas in the western part of The Netherlands were inundated, it is the aim of the Dutch Government that all sea defence works are designed to withstand a storm surge with a frequency of occurrence of once in 10,000 years. A main problem to be solved is the assessment of the rate of dune erosion, subject to this design storm surge, in order that justified reinforcement measures can be taken against a possible breakthrough of the dunes.

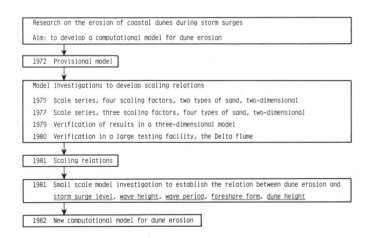

Fig. 7 Research programme on dune erosion

The Delft Hydraulics Laboratory has put great effort in the research on beach and dune erosion during storm surges. Since 1972 an extensive research programme has been carried out to quantify the beach erosion and the extent of profile changes due to storm surges (see Figure 7). A large number of two-dimensional and three-dimensional model tests has been performed to investigate the process of dune erosion and to develop the relevant scaling relations.

In view of the everlasting doubts about the reliability of the small-scale models, it was decided to check the results in the large-scale model facility, the Delta flume, of the Delft Hydraulics Laboratory. Nearly full-scale tests were performed with random waves up to 2.1 m significant height (Figure 8). These tests confirmed the validity of the developed scaling relations. Moreover, the model results agreed well with known field data [12].

Fig. 8   Large-scale test on dune erosion in the Delta flume

The test results indicated that the rate of dune erosion is controlled by the difference between the initial profile and the (quasi) equilibrium profile. Once the storm profile has developed, the process of dune erosion will come to an end. The dune erosion quantity as a function of the steepness of the initial profile is shown in Figure 9.

Moreover, the test results showed that during storm surge conditions a uniform profile will develop, which configuration is dependent on the wave characteristics and the grain size of the bed material, but which is independent of the pre-storm profile. The development towards an equilibrium profile during a full-scale storm surge is shown in Figure 10. This figure illustrates that the sand, which is eroded from the dunes, is distributed over a relatively short distance. The post-storm waterdepth at the distribution limit is about 0.9 H_s below storm surge level. The relation between the profile steepness and the fall velocity of the bed material can be found from [12]:

$$\text{tg } \alpha_2 = \text{tg } \alpha_1 \ (w_2/w_1)^{0.56}$$

in which $\text{tg}\alpha_1$ and $\text{tg}\alpha_2$ are the slopes of beaches with sand having fall velocities w_1 and w_2 respectively, under identical wave conditions.

As a result of the extensive research activities on dune erosion, the Delft Hydraulics Laboratory is very well equipped to give advice on the design of soft sea defence systems and their behaviour under storm conditions.

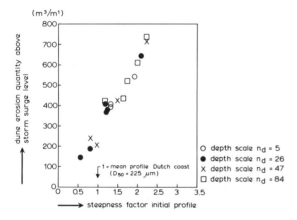 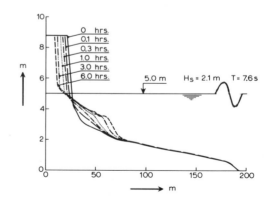

Fig. 9   Dune erosion as a function of the steepness of the initial profile

Fig. 10   Development of the initial profile towards an equilibrium profile during a full-scale storm surge

4.3 Erosion due to tidal, wind- and wave-driven currents

As already stated, the morphological development of sand-fill islands will be determined by the wind, wave and current conditions in their vicinity. A detailed analysis of these nearshore phenomena demands for a three-dimensional model in which all physical processes have been modelled explicitly. However, in homogeneous situations, where the vertical structure of the flow is unimportant, depth-averaged models can be used.

The Delft Hydraulics Laboratory has available a two-dimensional numerical model FRIMO which reflects a balance between the gradient of the radiation stresses of the waves, the bottom and surface stresses and the gradient of the mean water level. Wave transformation from deep water towards the shore and dissipation of energy due to wave breaking are incorporated in the model. The model operates within a scheme of finite elements.

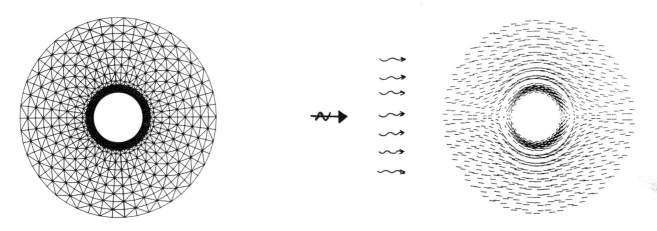

a. finite element grid   b. computed flow regime

Fig. 11 Two-dimensional model FRIMO

In Figure 11 the computation grid and the computed flow regime around an island have been given for both waves and current. As no turbulent exchange and convection are incorporated in the FRIMO-model, the flow regime in the lee zone of the island cannot be modelled correctly. The reproduction of the flow in this area demands for a much more complicated model which accordingly is more expensive in operation.

As a result of a close scientific co-operation between the National Hydraulics Laboratory of Chatou in France and the Delft Hydraulics Laboratory, such a sophisticated model, called BICOC, is by now available and fully operational [13]. The model is based on the complete Reynolds equations and operates within a finite difference scheme. An example of a computation with this model, for current only, is given in Figure 12.

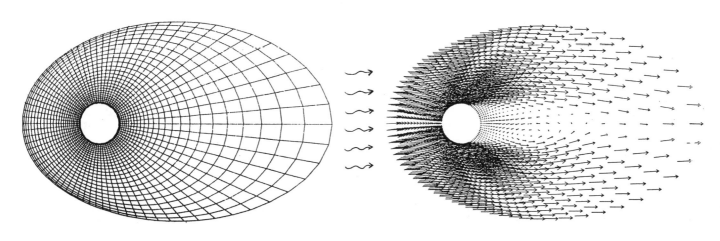

a. finite difference grid   b. computed flow regime

Fig. 12 Two-dimensional model BICOC

On the basis of the resulting flow regime, together with the computed nearshore wave climate, the longshore transport of sediment, both inside and outside the surfzone, can be determined via a sediment transport formula which involves both waves and current influences. The resulting net sediment drift can then be computed by taking into account the relevant percentages of occurrence of the difference wave and current conditions on a yearly basis. A change of the net drift along the island boundary will induce local erosion or accretion.

4.4 Impact on the mainland

Depending on the dimensions of the island, the wave energy coming from deep water will be partly dissipated and reflected. Diffraction effects may also be encountered. As a result of the latter, in the lee zone of the island transfer of energy along the wave crests can occur, which generates waves in that zone. Hence, the wave energy approaching the coast of the mainland from deep water will on the one hand be decreased through reflection and dissipation, and on the other hand be spreaded through diffraction.

Independent of the presence of the island the refraction phenomenon plays an important role in wave transformation towards the coast of the mainland. Due to this phenomenon the wave rays of the incoming waves will be bent towards a direction perpendicular to the depth contours due to a decreasing depth towards the shore.
Apart from the waves approaching the coast of the mainland from deep water, waves may also be locally generated by wind in the area between the island and the mainland. The combined effect of the diffraction and refraction phenomena on these wave fields will ultimately determine the nearshore wave conditions of the mainland. A change in these conditions will in its turn have an impact on the coastal morphology of the mainland. The extent of this influence depends on the dimensions of the offshore island and the distance between the island and the mainland, together with the angle of wave incidence relative to the shoreline.

To illustrate the above described phenomena, in Figures 13 and 14 the computed influence of an industrial island in the North Sea, at a distance of 30 km to the Dutch coast, has been shown with respect to the nearshore wave conditions and coastal morphology. In this case the island, with a length of 12 km, was schematized as a system of two semi-infinite breakwaters.

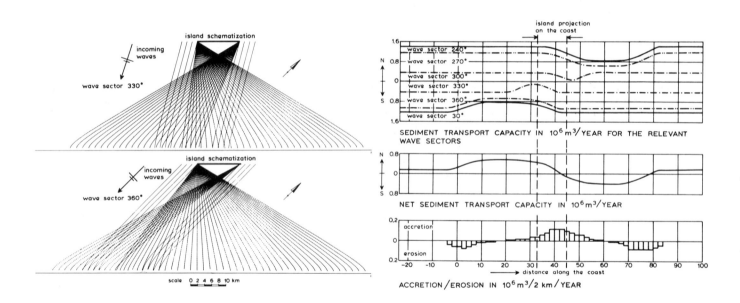

Fig. 13 Computation of nearshore wave conditions of the mainland. Wave refraction/diffraction diagram

Fig. 14 Sediment transport capacity and pattern of accretion and erosion for the coast of the mainland

5. STRUCTURES TO REDUCE WAVE ATTACK

5.1 General

There are a number of ways in which wave energy may be dissipated. The principal ways are described below.

Forced instability. Due to the configuration of a structure, changes in water depth or a counter current, instability and wave breaking may be initiated. Violent interaction with a structure results in shock forces of considerable magnitude. If waves are induced to break over a reasonable distance on a gradually sloping face, the forces applied to the structure will be spreaded over a greater period of time. The required power for the generation of a counter current of sufficient magnitude will be considerable.

Surface damping. This kind of damping is produced by skin friction resistance, e.g. by a thin flexible membrane on the surface, artifical seaweed fronds or a number of closely spaced vertical plates in the longitudinal direction. In all cases a very large surface area is required.

Out of phase damping. A change of phase is introduced by a structure so that the transmitted wave has a phase different from that of the incident wave. This may be achieved by a floating structure that has a natural period of oscillation which is long compared to the wave periods.

Interference with orbital motion. This sort of device serves to impede the orbital motion of the water particles, thereby absorbing wave energy. In this case a large volume of surface area is required.

Wave reducing structures, based on one or more of the foregoing principles can be roughly divided as follows:

fixed structures : (1) pile structures; (2) vertical barriers; (3) horizontal platforms; (4) sloping barriers; (5) pneumatic and hydraulic breakwaters.

floating structures : (1) floating structures with a big draught; (2) solid floating structures with a small draught; (3) flexible floating structures with a small draught.

As an example, wave reduction by submerged breakwaters is described in the next section.

5.2 Wave reduction by submerged breakwaters

In Figure 15 the plan and cross-section of a typical submerged breakwater in front of an artificial island is shown. Tests have been carried out to measure wave reduction by this type of submerged breakwater. Results are shown in Figure 16. The area where breaking waves were present is shown in Figure 17.

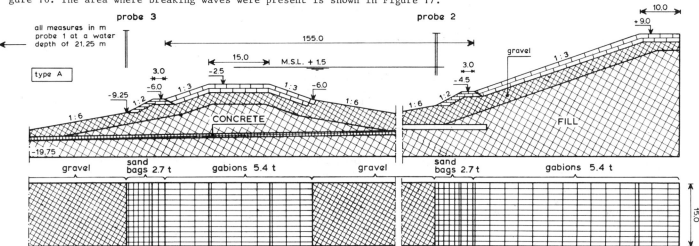

Fig. 15 Plan and cross-section of a typical submerged breakwater in front of an artificial island

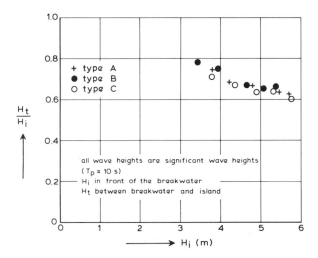

Fig. 16 Wave reduction by a submerged breakwater

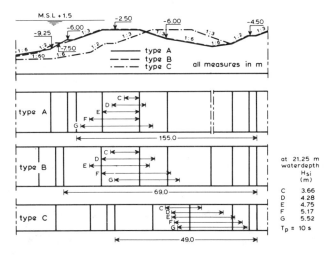

Fig. 17 Area of breaking waves

6. SLOPE PROTECTION

Many types of slope protection can be applied to prevent erosion of beach material. A rough indication of the type of slope protection, depending on the water depth and wave height, is shown in Figure 18. The different alternatives will be discussed briefly below.

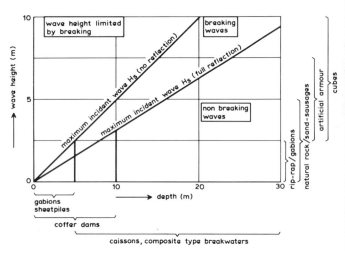

Fig. 18 Application area for various types of slope protection

Fig. 19 Model test on the stability and wave run-up of gabions

Sand-sausages. As sand in general is one of the cheapest and usually most available materials, sand-sausages may be an attractive solution for the protection of sand-fill islands. The Delft Hydraulics Laboratory has studied the stability of mattresses composed of sand-sausages with a diameter of 0.9 m on a circular island with slope 1:3. With peak periods, T_p, of the wave spectrum of 8 and 10 s the lifting of mattresses started at H_s = 2 m. With H_s = 5 m and T_p = 10 s the mattresses were lifted up about 0.8 m and with H_s = 5 m and T_p = 8 s the lift was about 0.5 m. Maximum wave run-up amounted to about 6.5 m with H_s = 2.3 m and T_p = 10 s, and about 7.5 m with H_s = 3.0 m and T_p = 10 s.

Gabions. Gabions consist of wire-mesh containers, filled with sand bags or coarse material. Gabions have been tested on stability and wave run-up, see Figure 19. Results of wave run-up on a 1:3 slope with T_p = 10 s are given in Table 1. It can be concluded that maximum wave run-up on a slope protected with gabions is 60 to 65% of maximum wave run-up on a slope protected with sand-sausages. Some results on the stability of gabions with dimensions 0.9 x 0.9 x 3.6 m^3, a mass of 5,400 kg and a mass density of 1,860 kg/m^3, are given in Table 2.

H_s (m)	2.4	3.1	3.7	4.3
maximum wave run-up (m)	3.8	5.0	6.3	7.5

Table 1 Maximum wave run-up with gabions

Table 2 Significant wave heights (m) at which damage is initiated

T_p	number of layers	slope 1:2	slope 1:2.5	slope 1:3
two-dimensional tests				
10	2	2.2	-	2.9
8	2	2.3	-	3.0
8	1*	-	2.9	-
three-dimensional tests				
10	2	-	2.9	-
8	2	-	3.5	-

*) After initiation, damage increased very fast

Gravel. During more than ten years the Laboratory has carried out a research programme on equilibrium profiles and longshore transport of gravel under regular and irregular wave attack. A description of this research has been presented in [14,15].

In describing profile formation and longshore transport, two groups of parameters may be composed:
a. The external parameters, characterizing the wave attack and the initial beach geometry, viz. significant wave height and wave length on deep water (H_{sd}, L_{sd}), significant wave period (T_s), depth of foreshore (h_f), height of beach top (k), initial slope (tgα), grain diameter (D) and angle of wave approach on the foreshore (φ_f). Because

the larger grains of a grain distribution seem to be determinant for the bed load transport, the D_{90} was chosen to characterize this grain size distribution. D_{90} means that 90% of the weight of a sample has a diameter smaller than D_{90}.

b. <u>The internal parameters</u>, characterizing the resulting equilibrium profile and the sediment transport during the formation of the profile. These parameters are shown in Figure 20.

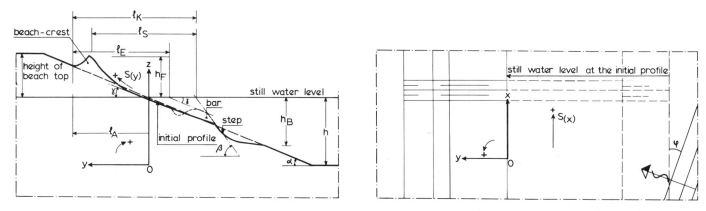

Fig. 20 Internal parameters of the equilibrium profile of gravel beaches

Summarizing all tests, the following conclusions can be drawn:
- It appears to be possible to determine an optimum initial beach profile in y-direction which shows minimum erosion and accretion for a given wave condition.
- It is possible to give the value of all the equilibrium profile parameters shown in Figure 20 as a function of the external parameters.
- The total longshore transport of material in x-direction can be described in terms of grain diameter and design wave conditions by the relationships between internal and external parameters, as shown in Figure 21.

For a project in Italy (see Figure 22) good results were obtained in predicting the longshore transport rate along a natural gravel beach.

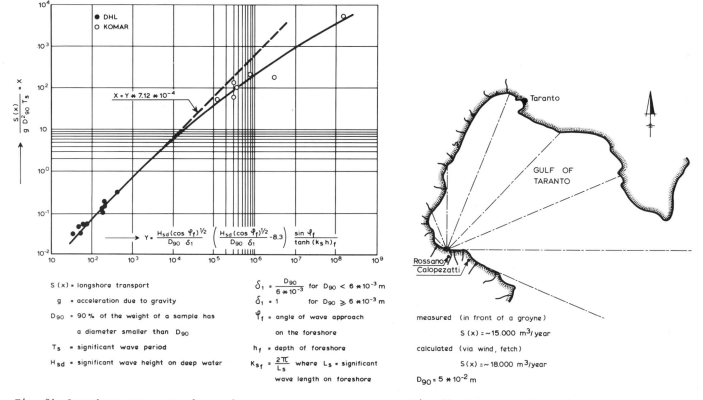

Fig. 21 Longshore transport of gravel

Fig. 22 Rossano beach, Italy

Rip-rap. Rip-rap is graded quarrystone. When natural rock is available rip-rap slope protections may be applied. Tests at the Delft Hydraulics Laboratory indicate that the behaviour of rip-rap under wave attack is very similar to that of gravel.

Uniform rock. If only limited deformations of the slope protection are allowed, use of uniform rock can be considered. Formulae describing the stability limit of uniform rock are summarized in [16].

Artificial armour units. When natural rock is not available, artificial armour units can be used as slope protection. Numerous types of armour units have been developed. Some of these are non-interlocking and derive their stability mainly from their mass. A second group of artificial armour units are interlocking units, like Dolosse and Tetrapods.

For these types of units traditional modes of failure - e.g. displacement of units - are no longer decisive, but in many cases internal stresses will be the limiting factor.

The Delft Hydraulics Laboratory has in several cases applied a method to record the position of the units by means of single-frame film exposures, taken prior to and after wave uprush. The initiation of motion and the degree of rocking is easily established in this way. Results of model tests have been correlated to breakage of 15t Dolosse observed in nature after a storm with a maximum significant wave height H_s = 7.1 m.

Recently the Delft Hydraulics Laboratory has studied the stability of a breakwater-section in 40 m water depth in the Delta flume, see Figure 23. In that study accelerations of two units have been measured, as well as the initiation of motion and the degree of rocking.

Fig. 23 Large-scale model test on the stability of artificial armour units.

7. VERTICAL FACE STRUCTURES

As follows from morphological considerations, the horizontal upper surface of the island must be protected against storm wave conditions. For this reason a vertical face structure will be chosen in many cases. The following types of vertical face structures can be distinguished:

• gabions   • sheet piles, coffer dams   • concrete blocks   • caissons.

The geotechnical stability of caissons, often placed on filter material, must be judged based on a realistic prediction of the decisive load boundaries. In The Netherlands in this respect a lot of experience has been built up during the last 40 years. Many caisson alternatives have been considered in the Delta Works period, The Eastern Scheldt closure works particularly.
Geotechnical calculation methods were developed and regularly the applicability was tested with measurements of large-scale experiments. A detailed description of one of these test studies is presented in a Boss '82 paper [10]. In these tests, carried out in the Delta flume, a 1700 kN concrete caisson was loaded by real water waves up to a wave height of 2.5 m. The behaviour of the two layer foundation bed with a total height of 2.75 m was intensively measured. Figure 24 shows a longitudinal section of the test set-up.

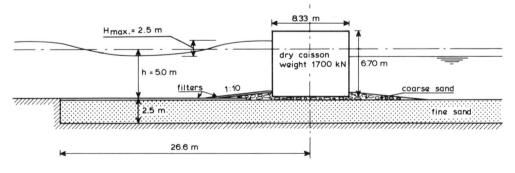

Fig. 24 Longitudinal section of set-up of large-scale stability test on a caisson

In order to reduce the hydraulic loadings, it is important to avoid wave breaking against the structure, which causes wave impacts on the vertical face. Structures in front of the vertical face to reduce wave attack may be of importance in this respect. In order to determine the hydraulic loadings two different approaches can be followed:
• determination of local pressures
• determination of total forces.

Recently the Delft Hydraulics Laboratory has developed a computer programme for calculating pressures due to non-breaking waves against a vertical wall. Figure 25 shows a comparison of pressures calculated with this programme and pressures measured in the Delta flume. Total forces can be measured in model tests as shown in Figure 26.

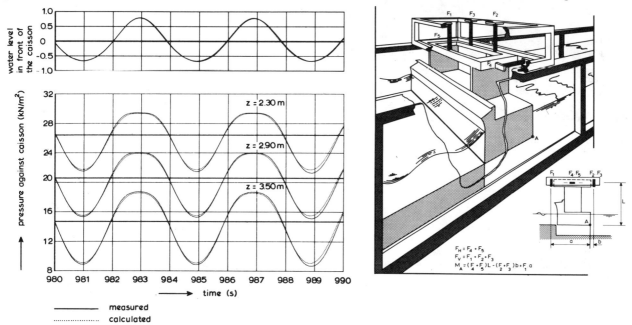

Fig. 25 Comparison of measured and calculated pressures against a caisson

Fig. 26 Force metering frame

Numerous types of parapets can be made in order to prevent overtopping of a vertical face structure, see Figure 27. For several types the amount of overtopping is shown in Figure 28.

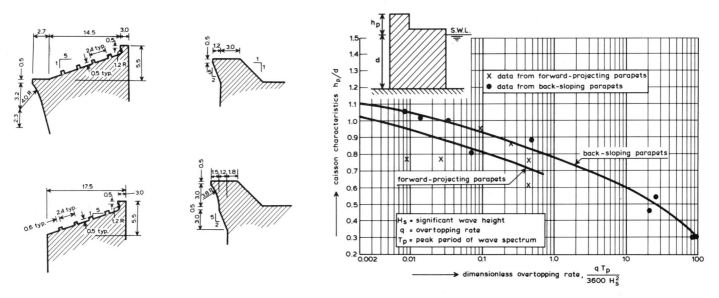

Fig. 27 Types of parapets

Fig. 28 Typical overtopping rate diagram

Fig. 29 Model test on the stability of the apron in front of a caisson

Attention should be paid to the stability of the apron in front of a vertical face structure, see Figure 29. This apron is required to prevent erosion of the foundation. Reference [17] gives calculation methods for estimating the required weight of the apron. This estimation can be verified by means of model tests.

REFERENCES

1   PIANC, 1977, "Technical and economic aspects of artificial offshore islands for the reception of large ships", Annex to PIANC Bulletin no. 28, vol. III

2   DE VOOGT, W.J.P., KOMEN, G.J., BRUINSMA, J., 1981, "The KNMI operational wave prediction model GONO", Proceedings, IUCRM Symposium on wave dynamics and radio probing of the ocean surface, May 1981

3   BERKHOFF, J.C.W., 1976, Mathematical models for simple harmonic linear water waves, Delft Hydraulics Laboratory, Publication no. 163

4   SEELIG, W.A. et al, 1980, Estimating nearshore conditions for irregular waves, U.S. Army, Corps of Engineers, CERC, Fort Belvoir

5   EUGENIE LISITZIN, 1974, Sea level changes, Elsevier Oceanographic Series 8, Amsterdam

6   Admiralty Tide Tables, Hydrographer of the Navy, London

7   International Hydrographic Bureau, Tides. List of Harmonic Constants, Monaco, since 1930

8   YAMAMOTO, T., KONING, H.L., SELLMEYER, J.B., HIJUM, E. van, 1978, "On the response of a poro-elastic bed to water waves", Journal of Fluid Mechanics, vol. 87, part. 1, pp. 193-206

9   DE GROOT, M.T., SELLMEYER, J.B., 1979, "Wave induced pore water pressures in a two-layer system", LGM-mededelingen, part XX, no. 2, 3 and 4, pp. 67-79

10 LINDENBERG, J., SWART, J.H., KENTER, C.J., DEN BOER, K., 1982, "Wave induced pressures underneath a caisson. A comparison between theory and large-scale tests", Proceedings, 3rd International Conference on the Behaviour of Off-Shore Structures, August 1982

11 HEIJNEN, W.J., 1980, "Soil Mechanics Research in Coastal and Offshore Engineering", Proceedings, Symposium "Future needs for hydraulic and soil mechanic research in coastal and offshore engineering", The Netherlands, pp. 95-112, August 1980

12 VELLINGA, P., 1982, "Beach and dune erosion during storm surges", to be published in Coastal Engineering, Elsevier

13 BISCH, A.M., 1981, Modèle bi-dimensionnel d'écoulement autour d'obstacles, Electricité de France, Direction des études et recherches, Laboratoire National d'Hydraulique, Chatou, rapport HE 041/81/14

14 HIJUM, E. van, 1974, "Equilibrium profiles of coarse material under wave attack", Proceedings, 14th Conference on Coastal Engineering, Copenhagen, 1974

15 HIJUM E. van, 1976, "Equilibrium profiles and longshore transport of coarse material under wave attack", Proceedings, 15th Conference on Coastal Engineering, Honolulu, 1976

16 PIANC, 1976, "Final Report of the International Commission for the Study of Waves", Annex to PIANC bulletin no. 25, vol. III

17 CERC, 1975, Shore protection manual

Subject Index

Adfreeze phenomenon, 1:82
AFDA (Approximate Full Distribution Approach), 2:152
AIDJEX (Arctic Ice Dynamics Joint Experiment), 1:76
Airy functions, 2:708
Airy's wave theory, 1:361-362; 2:15, 622-625
AISC buckling formula, 2:41
Alaska:
    North Slope of, 1:75, 76
    west coast of, use of first buoyant caisson on, 1:7
Alaskan Oil and Gas Association (AOGA):
    project monitoring by, 1:73
    Project No. 96, 1:82
    Project No. 152, 1:83
ALS (antiliquefaction system), 1:269
Alternate structural concepts, design and reliability issues for, 2: 829-898
Alternate systems, design issues for:
    deep water concepts for northern North Sea, 1:29-39
    gravity and jacket-type structures, 2:147-165
    guyed tower platforms, 2:123-146
Aluminum model piles, experiments with, 2:387
AMDS (Arctic Mobile Drilling Structure), 1:86
American Petroleum Institute (API):
    Bulletin 2N (1981), 1:73, 78, 81
    code and recommendations of: comparison with Norwegian Institute of Technology, 1:215-218
        for earthquake excitations, 1:22-23; 2:40
        inelastic behaviour of tubular braces, 2:42
        in North Sea pile foundation analysis, 1:97
        on soil stratigraphy, 2:309
    development of advisory standards by, 1:5
    PRAC-15 project, 2:11-15, 22
    PRAC-22 committee, 2:15
American Welding Society, guidelines developed by, 1:5
Anchor lines, dynamic behaviour of, 2:651-670
Anchors embedded in sand, uplift capacity study and design procedure for, 2:451-463
Anemometry, laser doppler, 2:544
Anisotropic Elasto-Plasticity Theory, 2:338-341
ANSR-1 program, 2:47
Antiliquefaction system (ALS), 1:269

AOGA (see Alaskan Oil and Gas Association)
API (see American Petroleum Institute)
APLA (see Arctic Production and Loading Atoll)
APOA (see Arctic Petroleum Operators Association
Approximate Full Distribution Approach (AFDA), 2:152
Arctic, hydrocarbon extraction in, 1:71-89
Arctic Ice Dynamics Joint Experiment (AIDJEX), 1:76
Arctic Mobile Drilling Structure (AMDS), 1:86
Arctic Petroleum Operators Association (APOA):
    project monitoring by, 1:73
    Report Project 99, 1:75
Arctic Production and Loading Atoll (APLA), 1:85
Ardersier (U.K.), 1:29
ARMSM group on Plancoet pile, 1:142-157
Articulated column, fatigue analysis of, 1:25
Articulated concrete tower, Woodrow's, 1:32
Articulated platform, northern North Sea feasibility studies on, 1:32-33
Articulated riser, feasibility and design of, 1:51-57
Artificial islands:
    in Arctic, 1:84-85
    sand-fill: design of, 2:884-898
        impact on mainland of, 2:892
Axially loaded piles, laboratory study of, 1:105-121
Axisymmetric flow pattern, 1:668-672

Barge-launched jacket, 1:6
Bauschinger effects, 2:41, 42, 43
Beaufort Sea, 1:73, 75-77, 87
Bering Sea, 1:74-77
Berkeley shaking table, 2:41
Bernoulli-Euler beam model, 1:567; 2:706
Bernoulli's equation, 1:669; 2:181, 585, 600
Bessel functions, 1:473; 2:601
BICOC model, 2:891
Biot's equations of poro-elasticity, 1:358, 363
Biot's Theory of Consolidation, 2:342
Blowout prevention stack, 1:87
Blowout plumes, hydrodynamic structure of, 1:660-684
Bo-hai Gulf (see Gulf of Bo-hai)

Boltzmann superposition principle in foundation displacement, 1:301
Borehole geophysics logging, 1:133-134
    procedures in, 1:269-271; 2:421
Borgman's formulation, 2:15
Borneo, 2:22
BOSOR-5 program, 2:62
Boston blue clay, 2:338, 342, 344, 346-347, 352-354
Braced structures, 2:46-47
"Breakout" in lifting large objects from seabed, 1:364-366
British Research Station, 1:195-198
Buckling:
    of cylindrical shells, 2:770-771
    propagating, 1:187-199
    of risers, 2:705-715
    in ship/platform impacts, 2:262-264
Buoyant towers:
    in deep water, 1:127, 134
    wave-frequency response measurement of, 1:425-443
Bureau of Land Management, 1:6

CA (conventional anchored) drillship, 1:23
Caissons:
    mobile Arctic, 1:86
    stability test on, 2:896-898
    wave-induced pressure under, 1:337-357
    wave-induced stresses near, 1:363-364
California coast, jacket-type platform on, 1:4
CALM (catenary anchor leg mooring), 2:632
Candelabra model, 2:15-16
CANMAR drill ships in Arctic zone, 1:87
Cartesian coordinate axes, 1:428
CASPA program, 2:803-804, 807
Catamaran, semisubmersible, 1:435
Catenary anchor leg mooring (CALM), 2:632
    in northern North Sea, 1:37-38
Cavity expansion method, 2:341, 345, 346
Cavity expansion theory in pile displacement, 1:287-288
Ceradine's dynamic shakedown theorem, 1:320-325
"Cerveza Liguera" platform, 1:9
"Cerveza" platform, 1:8-9
Characteristic wave force (see One-hundred-year storm)
Charpy-V testing, 1:25
China:
    collapse of flare jacket off, 2:401-414
    earthquake and vibration analysis in, 2:810-816
Chukchi Sea, 1:75
Clay:
    flare jacket collapse and, 2:404, 412, 413
    glacial, testing of in North Sea, 1:195-198
    in shear testing, 2:364-371
    strength tests of low plasticity types of, 2:438-450
    (See also Boston blue clay; Sabine clay)

Clumpweight, 2:130-134, 144-146
    in anchoring of catenary moored platform, 1:38
    dynamic behaviour of, 2:658-660, 662-664, 667-668
    fatigue analysis of, 2:657-658
    in guyed tower design, 1:530, 552, 555
Coefficient of variation (COV) of parameters, 1:131-132
"Cognac" platform, 1:7-8, 29; 2:11, 15, 25
Collision of ships against platform, 2:257-278
    impact mechanics, 2:258-259
Colpitts oscillator, 2:682
Conat group articulated concrete tower, 1:32
Concrete:
    prestressed, subject to extreme loads, 2:874-883
    reinforced with steel fibers, strength of, 2:837-848
        subject to concentrated moments, non-linear analysis of, 2:864-873
    technology of, 1:26
    (See also Concrete gravity platform; Concrete off-shore structures)
Concrete gravity platform (Condeep T300), 1:30-31
    corrosion protection of, 1:26
    drainage system of, 2:373-382
    reinforcement of, 1:26
Concrete off-shore structures:
    in biaxial cyclic compression, 2:235-253
    implications of fatigue in, 2:203-221
    uniaxial tensile strength of, 2:223-234
Condeep platforms, 2:376
    Beryl A, 2:374
    Brent B, 1:249; 2:374
    Stratford A, 2:865
    (See also Concrete gravity platform)
Cone penetration test (CPT), 1:23, 175-177; 2:418-421, 886-888
    on Plancoet pile, 1:142
    (See also Penetrometer)
Cone structures in Arctic, 1:86-87
Conoco:
    theoretical model of TLP of, 1:411-423
    world's first TLP designed by, 1:34, 530
CONSOL operation, 1:349
Conventional anchored (CA) drillship, 1:23
Cook Inlet (Alaska), 1:86
    ice in, 1:10, 73
Copper River Prodelta, 2:364
Coriolis force, 1:76, 567
Coulomb friction, 2:90-91
Coulomb's equation, 1:145
Coulomb's failure equation, 1:223
COV (see Coefficient of variation)
Cowden Till (U.K.), 1:194
CPT (see Cone penetration test)
Cranes (see Off-shore platform cranes)
Critical stress level, 1:230-232

Cross Island, 1:74
Crout's factorisation, 1:431
Cyclic displacement parameters, 1:251-255
Cyclic loading tests, 1:145-149, 245-260, 297-312; 2:721-723
"Cyclic pseudo creep" method, 1:146, 149; 2:392
Cyclic shear strain amplitude, 1:245
Cylinder strength meter test, 2:467-471

DAF (see Dynamic Amplification Factor)
D'Alembert's principle, 2:524-527
Damping (see Drift forces)
Danish Hydraulic Institute, 2:597
Darcy's law, 1:358, 359, 365
Davis Strait, 1:75
DCDT (direct current displacement transducer), 1:106
Deep water platforms:
   jacket: design and construction of, 1:3-17
      extreme wave dynamics of, 2:784-791
      piled steel, in northern North Sea, 1:20-30
   in northern North Sea extreme conditions, 1:18-49
   pile foundation design for, 1:125
   techniques for, 1:7-10
Deep water structures:
   hydrodynamics of, 2:477-548
   pollution caused by, 1:10
Delft Hydraulics Laboratory, 1:337, 343, 2:884-898
Delft Soil Mechanics Laboratory, 1:337, 343; 2:884-898
Delta flume, 1:339-356; 2:889-890, 896-897
DEMTRA model, 2:351-355
DENTA program, 2:268
Design storm assumptions for off-shore gravity platforms, 1:255
   (See also One-hundred-year storm)
Deterministic method of wave characterization, 2:594
DIFRAC-R program, 1:474
Direct current displacement transducer (DCDT), 1:106
Direct simple shear soil stress measurement, 2:363-371
Dirichlet condition, 1:429
DNP (Den Norske Pelecomité; NPD, Norwegian Petroleum
      Directorate), 1:215-218, 244
DnV Rules, 2:770-777
Dome Petroleum:
   Arctic production loading atoll of, 1:85
   first caisson island of, 1:85
Dowel theory, 2:405-414
Drammen clay, 1:244, 253, 255-256
Drift forces:
   on body structures, 1:467-489
   and damping in natural sea states, 2:592-607
   on multi-component bodies, 1:432-435
   on semisubmersibles in waves, 1:447-466
DRIVE-10 program, 1:128-130
"Druck" transducer, 1:343
DYNA3S, 2:15, 18

Dynamic Amplification Factor (DAF), 2:785-791
Dynamic positioned drillship, 1:23

Earthquakes:
   API code on, 1:22-23, 2:40
   behaviour of guyed tower in, 2:127
   on deep water structures, 1:131
   (See also Seismic data; Seismic waves)
East China Sea, 2:507, 515, 521
Eigenvalues, 2:706-709, 711
Elastic pressure ovalization curve in off-shore pipelines, 2:172-173
Elastically mounted cylinder, 2:671-680
Elastomeric materials in articulated platforms, 1:37
Envelope process, 1:516-525
Equilibrium equations for helical rod, 2:77-99
Equivalent caisson method, 1:111
Equivalent single tube analysis, 1:601-602
Euler analysis, 2:706
Euler's momentum equation for steady inviscid flow, 1:667-669
Exxon Corporation:
   guyed towers of, 1:33-34, 530
   proposed floating caisson of, 1:87
   sea ice studies of, 1:78, 82

Fast Fourier Transform (FFT), 2:491, 541, 821-822
   in drag analysis of cylinders, 1:392
FASTRUDL program system in North Sea pile foundations, 1:97
Fatigue analysis:
   effect of damping on, 2:286
   effect of wave spreading on, 2:286-290
   of gravity-type structures, 2:147-165
   of guyed towers, 1:25-26
   model damping ratio in, 2:282-286
   of monopod tubular joints, 2:738-739
   of off-shore platforms, 1:24-28
      concrete, 2:204-211
      steel jacket, 2:3-25, 147-165, 747-755
      tripod tower, 2:851-855
   of planar joints, 2:29
   in stranded cables, 2:77-99
   of tensile strength of concrete, 2:223-234
   of tubular joints, 2:29-30, 31
Fatigue damage during towing of deep water structure, 1:29
Fatigue life:
   of marine risers, 1:622-629
   of monopod tubular joints, 2:739-741
   sensitivity estimates of, 2:279-291
   (See also Wave spectral shape and directional variabilities)
Fatigue strength of welded connections in North Sea, 2:26-36
FEAP finite element program, 1:303
FFT (see Fast Fourier Transform)

"Field of hardening moduli," 2:392-393
Finite element analysis, 2:342, 625, 654-655
    for delta flume, 1:349-355
    around driven piles, 1:286
    in shakedown analysis of off-shore structures, 1:322-324
Finite element formulation for fluid-saturated porous medium, 1:315-325
First-order, second-moment theory of structural reliability, 2:4
First-order uncertainty analysis, 2:479-480
Flare jacket, collapse of, 2:401-414
Flexural mode, 2:125-126
Floating box and cylinder, semisubmersible, 1:434
Floating production facilities, environmental design criteria for, 1:643-658
Floating systems in Arctic, 1:87
Floating vessels:
    hydrodynamics of, 2:551-617
    large amplitude and extreme motion of, 2:608-617
Force coefficient evaluation for inclined members, 1:373-386
Foundations, 1:93-372; 2:128-129, 295-473
    (*See also* Pile foundations)
Fourier series, 2:594
Fourier's theorem, 1:453, 473; 2:610
Fracture mechanics, 1:25; 2:739-740
Frigg field, geotechnical instrumentation of, 1:262-280
FRIMO model, 2:891
Froude-Kriloff force, 1:415; 2:501, 583, 586
Froude number, 1:570, 571, 604, 681, 684
Fuller finite element analysis, 1:602-603

Gabions, 2:894, 896
Galerkin's method of integration, 2:580, 587
Galloping of multiple risers, 1:620, 621
Garrison theory, 1:22
Gaussian parent distribution, 1:648, 662
Gaussian process, 1:516-525, 662; 2:147-148, 817, 820-821
Generalized modular analysis, 2:792-798
Glacial clay testing in North Sea, 1:195-198
"Glory hole," installing mudline blowout preventive stack in, 1:87
Goodman diagram, 2:232
Gravity platforms:
    concrete (*see* Concrete gravity platform)
    design principles for, 1:243-260
    fatigue analysis of, 2:147-165
    steel, in northern North Sea, 1:31-32
Gravity structures (Frigg field), geotechnical instrumentation of, 1:262-280
Greenland, 1:75
Green's function for single fluid region, 1:427-428, 431
GTP (*see* Guyed tower platforms)
Gulf Canada, mobile Arctic caisson of, 1:85-86
Gulf of Alaska, clay of, 2:364-371

Gulf of Mexico, 1:8-10, 29, 132, 162; 2:13, 15, 24, 25
    clay of, 2:364-371
    eight-pile structure in, 1:9
    extreme criteria for, compared to North Sea, 1:41
    installation of "Cognac" in, 1:7-8
    Ocean Test Structure (OTS) program in, 1:373-386
    soil testing procedures in, 1:162
Gulf of Paria, 1:162-163
Gulf of Taranto, gravel beach of, 2:895
Guyed tower platforms (GTP), 1:4, 10
    alternate design concepts and strategies for, 2:123-146
    in deep water, 1:127, 134
    design considerations for, 1:529-545
    earthquake behaviour of, 2:127
    fatigue analysis of, 1:25-26
    mooring lines of, 1:546-561
    in northern North Sea, 1:33-34

Hammer, hydraulic, design and driving capacity of, 2:315-324
Hankel function, 1:473; 2:601
Hasselmann theory, 2:543
HATCAN program, 2:655, 658-660
Heerema proposed steel tripod tower, 1:32
Hilbert transform technique, 1:517
"Hockey puck" failure, 1:84
Holocene sand in Netherlands, 1:299, 306
Holography in detecting flaws in off-shore structures, 2:832
"Hondo" platform, 1:4, 7
Hooke's law, 1:359
Horizontal cylinders:
    heavily roughened in waves, 1:387-406
    semisubmersible, 1:433-434
Hostun sand, 2:394-395, 398-399
Hummock field, 1:74-75
"Hurricane-proof" concept in design of "Cognac," 1:8
Hutton Field (U.K. waters), TLP design and construction in, 1:20, 34
Hybrid islands, Arctic, 1:85-86
Hydraulic actuator, 2:316, 738
Hydraulic hammer, design and driving capacity of, 2:315-324
Hydroelastic vibrations in pipe arrays, 2:641-650
Hysteresis:
    in strand displacement behaviour, 2:89-91
    in TLP structure, 2:77-99

IBM 370/155 system, 2:621, 625
Ice, sea, 1:73-77
Ice loads:
    on fixed structures, 1:78-83
    and flare jacket collapse, 2:401-414
IMPACT program, 1:83-84
Incremental permanent strain method, 1:301-303

Initiation pressure, 2:188
Installation engineering, procedure for, 1:4
Islands (*see* Artificial islands; Hybrid islands)
ITTC (International Towing Tank Conference) report on hydrodynamic loading, 1:425
Ivanov's higher yield function, 2:802-803

Jacket-type structures:
  in deep water, 1:127, 134
    (*See also* Deep water platforms, jacket)
  fatigue characteristics of, 2:478-485
Jack-up drilling unit (JDU), model for hydrodynamic evaluation of, 2:572-579
Johansen's yield load, 2:877, 879
JONSWAP wave spectrum, 1:21, 605
  in laboratory wind-wave tests, 2:542-547
J-tube, 1:27

Keller differential transducer, 1:343
Kelvin's method of stationary phase, 1:473
Keulegan-Carpenter number, 1:388, 395, 571, 603, 605, 606, 616-617; 2:103, 527, 599, 642-644
  elastically mounted cylinders and, 2:672-675
  and wave forces on inclined members, 1:377-378
Kinematics:
  envelope method used to predict extreme responses in, 1:516-525
  of helix, 2:79-80
  of multi-component bodies, 1:426-436
  random wave, non-linear properties of, 1:493-515
  of submarine pipeline, 2:171-179
  of wire layer continuum, 2:85-86
Kistler transducer, 1:343
Kvaerner proposal for articulated platform, 1:32-33

La Roche Chalais clay, 2:397
Laboratory waves, wind-wave spectra modeling in, 2:541-548
  (*See also* Wave forces)
Labrador, 1:75
Lagrangian formulation, 2:260
Lake Maracaibo, drilling in, 1:4
Laplace distribution, 2:27, 32
Laplace equation, 2:596
Laser doppler anemometry, 2:544
Lena field, guyed tower platform construction in, 1:19, 22, 33
Limit State Design (LSD), 2:4
LINDYN program, 2:655-657
Linear radial consolidation analysis, 2:342
Linear random wave theory (LRWT), 1:499-505
Linear voltage displacement transducer (LVDT), 2:189, 191-192, 230, 553, 557, 882

Lissajous figure, 2:672, 676
Longitudinally stiffened cylinders, 2:59-73
Longuet-Higgins distribution, 1:651, 652; 2:214
Louisiana coast, 1:4
LRWT (linear random wave theory), 1:499-505
LSD (Limit State Design), 2:4
LVDT (linear voltage displacement transducer), 2:189, 191-192, 230, 553, 557, 882
"LW"-theory, 1:414

McCormack Predictor-Corrector method, 2:543
Macro fabric features in soil fabric studies, 1:182-190
Magnus Field, construction of deep-water jacket in, 1:19
MAHIAK transducers, 1:269
  (*See also* Piezometer)
Marine Hydrodynamics Laboratory:
  recirculating water tunnel of, 2:682
  underwater welding systems developed at, 2:759-762
Marine risers, 1:563-639
  (*See also* Multiple risers; Production risers)
Markov process, 2:822
Marsh-McBurney current meter, 1:394
Maruo's formula, 1:414, 415
Mathieu instabilities, 1:413, 548
  definition of, 2:603-604
Maureen Field steel gravity platform, 1:31-32
MAX curve, 2:106
MCC (*see* Modified Cam-Clay)
Measuring transducer, LVDT (*see* Linear voltage displacement transducer)
Mechanical damping system for Tension Leg Platform, 2:497-522
Mediterranean Sea, mooring buoy study in, 2:631
Melan's static shakedown theorem, 1:314, 319-320
Microcracking, 2:237
Micro fabric features in soil fabric studies, 1:183
MIN curve, 2:106
Mindlin's equations, 2:349, 384
  in North Sea pile foundation analysis, 1:98
Miner-Palmgren hypothesis, 2:658
Miner's rule, 2:151, 155, 733
Miner's sum, 2:211, 216
Mode Acceleration Method, 1:529
Model scaling parameters, 1:625
Modified Cam-Clay (MCC), 2:339-342, 346, 351
  in skin friction tests, 1:285-287
Modified Walkley-Black Method in Orinoco clay tests, 1:167
Mohr-Coulomb yielding condition, 2:392-393, 395, 397
Mohr-Coulomb yield surface, 1:329
Mohr's circle state of stress, 1:286; 2:365-370
Monotonic loading, 2:721, 724
Mooring systems:
  of guyed towers, 2:129-134, 141-143
  non-linear analysis of, 2:621-630

Mooring systems (*Cont.*):
    single point, 2:631-640
    vertical anchor leg, 2:632-640
Morison equation:
    drag force linearization in, 1:21-22
    elastically mounted cylinders and, 2:672
    in model for drilling risers, 1:603
    in particle kinematics, 1:394
    in prediction of cylinder input reaction, 2:694
    in TLP behaviour, 1:412, 531; 2:102, 103, 104, 105
    in unsteady wave flow, 1:389
    velocity squared drag term in, 2:785
    in wave force analysis, 1:374-379; 2:147, 149, 532-539, 580-581, 749-750
Mudmats and stability during towing, 1:30
Multi-component bodies, wave-induced responses in, 1:424-443
Multiple risers, current-induced motion in, 1:618-639

National Research Council, geotechnical Arctic report of, 1:76
NC (*see* Normal consolidated soils)
NESSI program, 1:254
Netherlands:
    Oosterschelde storm-surge barrier in, 1:299-304, 306
    sea defense system of, 2:888-898
Netherlands Ship Model Basin, 1:447
Newman's relations, 1:414, 471
Newmark's alpha = 1/4 method, 1:529
Newton-Raphson method, 1:532; 2:47
Newton's Second Law, 1:426
NGI (*see* Norwegian Geotechnical Institute)
Nigg Bay (U.K.), 1:29
Nonlinear Structural Analysis Program (NONSAP), 2:625
Non-zero coefficient of skewness, 1:494, 497
Normal consolidated (NC) soils, 1:223-224, 232, 233
North Sea:
    analysis of pile foundation system for, 1:95-103
    base drainage systems of concrete gravity platforms in, 2:373-382
    drift forces and damping in, 2:592-607
    drilling platform in, 1:95-103
    fatigue strength of welded connections in, 2:26-36
    Frigg field gravity structures in, 1:262-280
    industrial island in, 2:892-893
    maximum wave calculations for (*see* One-hundred-year storm)
    northern part of: clay samples of, 1:184-186
        current-induced riser motion in, 1:621
        deep water concepts for extreme conditions in, 1:18-49
        guyed tower feasibility in, 1:33-34
        low-plasticity soils in, 2:439-446
        non-recommendation of "Cognac" platform for, 1:29
        one-hundred-year storm-surge event in, 1:650
        TLP design project for, 1:644-658
    tripod tower platforms (TTP) in, 2:847-863
    wind speed/wave height relationship in, 1:646-658
    off-shore platform cranes in, 2:551-560
    soil fabric studies in, 1:181-198
    time and cost planning for soil investigation in, 2:417-424
    (*See also* Concrete gravity platforms; Norwegian Sector)
North Slope (Alaska), 1:75, 76
Norton Sound (Alaska), 1:73, 76
Norwegian Continental Shelf, mapping of slide areas of, 1:24
Norwegian Geotechnical Institute (NGI), 1:95, 103, 243, 411; 2:363, 364
Norwegian Hydrodynamics Laboratories, envelope method of, 1:516
Norwegian Institute of Technology:
    on current-induced motions in multiple risers, 1:618-639
    off-shore pile behaviour prediction by, 1:203
Norwegian "Safety Off-shore" program, 1:22
Norwegian Sector (North Sea):
    drag coefficient method in, 1:21, 46
    earthquake risks in, 1:22-23
    installation of eight-legged steel jacket in, 1:96
Norwegian Trench, 1:19, 21-23
NO-TIDE loading pattern, 1:303
"Notional" probability of failure, 2:4
Novo-Technik transducer, 1:343
NPD (Norwegian Petroleum Directorate; DNP, Den Norske Pelecomité), 1:215-218, 244
NTH (Norges Tekniske Høgskole), 1:215, 218
NTNF (*see* Royal Norwegian Council for Scientific and Industrial Research)
Nutcracker tests, 1:83
Nyquist frequency, 2:821

OC (*see* Overconsolidated soils; Overconsolidation ratio)
Ocean Test Structure (OTS), 1:373-386
Oedometer, 1:167-171, 206-207
Oedopath (*see* Oedo-triaxial test)
Oedo-triaxial test, 1:206-208
Off-shore operations, U.S. regulation of, 1:6
Off-shore pipelines:
    hydroelastic vibrations of, in waves, 2:641-650
    initiation of propagating buckle in, 2:187-199
    submarine, analysis of collapse of, 2:169-186
Off-shore platform cranes, dynamic response in models of, 2:551-560
Off-shore platforms:
    articulated, 1:32-33
    design considerations for, 1:5-6
    generalized modular analysis of, 2:792-798
    shearographic testing in, 2:831-836
    ship collisions against, 2:257-278
    (*See also* Platforms)

Off-shore structures:
    force coefficient evaluation for inclined members of, 1:373–386
    poro-elastic seabed interaction with, 1:358–370
Ohkusu's damping coefficient, 1:432
One-hundred-year storm (100-year wave force), 1:20–21, 244–245
    in design of guyed tower, 2:128
    in northern North Sea, 1:650
    North Sea compared to Gulf of Mexico, 1:41
Oosterschelde storm-surge barrier (Netherlands), 1:299–304, 306
Orinoco clay, cone penetration and engineering properties of, 1:161–179
Orinoco Delta, 1:162–163
Orthogonally stiffened cylinders, 2:59–73
Orthotropic sheet, 2:82
Oscillating cylinder:
    in cross flow, 2:681–689
    in still water, 2:671–680
    in waves and currents, 2:690–701
OTS (Ocean Test Structure), 1:373–386
Outer Continental Shelf Lands Act of 1953, 1:6
Overconsolidated (OC) soils, 1:223, 233–234
Overconsolidation ratio (OCR), 1:223, 227; 2:308, 328–331, 342, 370
    in gravity platforms, 1:244–251
Over-water drilling for oil, historical development of, 1:4–5

Pacific Illite, 2:364–371
Palmgren-Miner Linear Damage Summation Rule, 2:30
Paris Law, 2:739
Penetrometer, 2:441–444, 450, 465–466
Permafrost, 1:76–77
Permanent displacement in cyclic foundations, 1:297–312
Piers in Eastern Scheldt estuary, 1:338
Pierson-Moskowitz spectrum, 2:154, 214, 749–750
Piezometer, 1:164, 269, 271
Piled steel jacket in northern North Sea, 1:20, 29–30
Pile foundations:
    for deep water fixed structures, design of, 1:125–140
    in North Sea, analysis of, 1:95–103
    in silty soils, 1:141–157
    under static loads, 2:295–359
Pile groups:
    subject to earthquake and vibration, hydrostatic response of, 2:810–816
    subject to lateral loads, analysis of, 2:384–390
Piles:
    aluminum model, 2:387
    calculating axial capacity of, 2:325–337
    installation of, 2:341
    laterally loaded, design of, 2:348–359
    off-shore behaviour of, 1:203–219
    rod shear interface of, 2:307–314
    skin friction analysis of, 1:221–240
    soil, interaction analysis of, 1:209–215
    (See also Pile foundations; Pile groups; Pile shafts)
Pile shafts, consolidation analysis of, 2:338–347
Pipeline systems (see Off-shore pipelines)
Planar oscillatory flow tests, 1:604–605
Plancoet pile, 1:129
    ARMSM findings on, 1:142–157
Plasticity theory, applied to submarine mudslides, 2:465
Plastic yield line theory, 2:257
Platforms:
    buoyant, 1:424–443
    inspection methods for, 1:27–28
    shakedown analysis of, 1:313–333
    (See also Deep water platforms; Deep water structures; Gravity platforms; Guyed tower platforms; Off-shore platforms; Tower structures; Tripod Tower Platform)
Poisseuille flow, 1:365
Poisson's ratio, 1:98, 549; 2:394
Pore pressure:
    around driven piles, 1:288
    in Frigg field gravity structure, 1:265–280
    laboratory study of, 1:105–121
    pile-driving resistance of, 2:295–304
    reduction of, beneath concrete gravity platforms, 2:373–382
    in seabed, 1:346–356
        poro-elastic, 1:358–370
Pressuremeter, 1:23
    self-boring (PAF) type, 1:142
Production risers, analysis and design of, 1:599–617
Proofloading, 1:28
Propagating buckle, 2:187–199
Propagation pressure, 2:188
Prudhoe Bay (Alaska), 1:75
P-Y curve approach, 2:353–355, 384–389

Quasi-static wave balancing, 2:210

RAO (Response Amplitude Operator), 2:480–481
Rayleigh-Normal distribution, 2:478–485
Rayleigh statistics, 1:651
    distribution of, 2:13, 27, 28, 214, 478–485
    method of, 2:153
Redcar (U.K.), 1:192–193
"Requirements for Verifying the Integrity of OCS Platforms," 1:6
Response Amplitude Operator (RAO), 2:480–481
Reynolds number, 1:625, 672, 681, 683; 2:103, 577–579
    in analysis of wave force on inclined members, 1:377–378
    as drag force coefficient, 1:388, 397
    elastically mounted cylinders and, 2:672–675

Reynolds number (*Cont.*):
    hydroelastic forces and, 2:642–644
    in riser-type systems, 1:570–572
Rigid cylinders, experiments on, 1:566
Ring stiffened cylinders, 2:59–73
Risers:
    for catenary moored platform, 1:38
    global buckling design criteria for, 2:705–715
    (*See also* Articulated riser; Production risers)
Riser-type systems, dynamic behaviour of, 1:565–598
Rod shear interface, 2:305–314
Rod shear test, 2:306–308
Royal Norwegian Council for Scientific and Industrial Research (NTNF), 1:219, 244
Rubble field, 1:74–75
Runge-Kutta-Nystrom method, 2:610
Runge-Kutta time integration scheme, 2:104
Rykswaterstaat Deltadienst, 1:299, 337

Sabine clay, 2:353–355, 357
Safety index β_s, 2:4
Sand sausages, 2:894
Santa Barbara Channel, deep water jacket platform in, 1:4, 7
Scatter diagram, 1:20–22, 26
SCF (Stress Concentration Factor), 2:6
Schmidt number, 1:681, 683
Schur's lemma of group theory, 2:796
Seabed, geotechnical issues in, 2:417–473
Sea ice in Arctic, 1:73–77
Sea of Genkai, 2:500
Sea-state modeling, long-term random, 2:817–827
Sea surface elevation, model of, 2:817–827
Seaquakes (*see* Earthquakes; Seismic data; Seismic waves)
Second order intermittent theory (SOIRWT), 1:503–504
Second order random wave theory (SORWT), 1:494–502, 505
SEEP program, 1:339
Seismic data:
    in northern North Sea, 1:22–23
    in resistance analysis of structures, 2:810–816
Seismic waves, case histories and analyses of, 2:561–571
Self-floating jacket, 1:6–7
Semi-infinite fluid domain, 1:427
Semi-submersible drilling unit (SSDU), model for hydrodynamic evaluation of, 2:572–579
Semisubmersibles:
    low-frequency motions in waves of, 1:447–466
    model responses of, 2:613–617
Shakedown analysis of off-shore platform, 1:313–333
"Shakedown" settlement, definition of, 1:262
Shallow water:
    jacket installation in, 1:6
    shallow analysis methods for platforms in, 2:124
SHANSHEP test program and results, 1:171–172

Shearographic Nondestructive Testing (SNDT), 2:831–836
Shear strength:
    MIT tests and experiments in, 1:106–120
    of mud, 2:464–473
    (*See also* Vane test)
Shear stress ratio, 1:249
Shear testing, 2:363–372
Shear transfer function, 2:15
Shell Oil, COSSAC concept of, 1:86
Ship/platform collisions, 2:257–278
    impact mechanics of, 2:258–259
SING-A computer program, 1:474
Single degree-of-freedom (SDOF) system, constant-parameter, 1:532–533
Single-point mooring (SPM) buoys, dynamic behaviour of, 2:631–640
"Sink-source" techniques, 1:414, 475
SIWEH function, 2:598
Skin friction studies, 1:210–211, 219; 2:327–328
    around driven piles, 1:283–293
    on piles, 1:221–240
Skirt pile, 1:4
"Slip velocity" in blowout study, 1:664
Slow drift oscillation, 1:644
S-N curve, 1:25, 2:4–11, 13, 22, 154, 219, 739–741
    in wire tether deterioration, 1:35
    (*See also* Fatigue analysis)
SNDT (Shearographic Nondestructive Testing), 2:831–836
Soil fabric studies, 1:181–198
Soil flows, 2:425–437
Soil investigations, 1:164–179, 2:417–424
    factors influencing efficiency of, 2:419–421
    weather downtime and, 2:421–422
SOIRWT (second order intermittent theory), 1:503–504
Sonar scanning systems:
    in analysis of ice flows, 1:75, 77
    wave-profiler, 1:391
SORWT (second order random wave theory), 1:494–502, 505
SOSEI II, 2:500
SOWM (Spectral Ocean Wave Model), 2:478
Spectra family concept, 2:478
Spectral fatigue analysis, mathematical model for, 2:148–152
Spectral Ocean Wave Model (SOWM), 2:478
SPLICE in North Sea pile foundation analysis, 1:97
SPM (*see* Single-point mooring buoys)
SPONS program:
    prototype finite element consolidation and, 1:339
    soil properties in, 1:349–352, 354–356
SPT (*see* Standard Penetration Test)
Spudcan foundation in guyed towers, 1:53; 2:125
SSDU (*see* Semi-submersible drilling unit)
STAGS-C1 program, 2:62

Standard Penetration Test (SPT), 2:886-888
Statfjord Field, production platform installation in, 1:19
Statoil, 1:18-20, 27
    criteria for fatigue analysis of, 1:25
    recommendations for jacket design by, 1:29
    suggested drag force coefficients of, 1:21
Steady flow theory for blowout plumes, 1:663
Steel gravity platform in northern North Sea, 1:31-32
Steel jacket structures, fatigue design and strength of (*see* Fatigue analysis; Structural stability)
Steel structural connections, non-linear behaviour of, 2:716-724
Steel tripod tower (Heerema), proposed design for, 1:32
Stiffened cylinders, extreme load resistance and, 2:769-783
Stiffened steel plates, collapse analysis of, 2:799-809
STOCCA program, 2:655, 657-658
Stochastic method of wave characterization, 2:594
Stokes theory, 2:580-582, 586, 599, 600
Storm duration, design assumptions in, 1:255-256
    (*See also* One-hundred-year storm)
Strain hardening, 2:260
Strain path method, 2:341, 345, 346
Strand, axial stiffness of, 2:88-89
Strand theory in Tension Leg Platform, 2:78-99
Stress Concentration Factor (SCF), 2:6
Stress-strain variables:
    allowable, for fatigue design, 2:3-25
    on circular tubes, 2:61
    in off-shore gravity platforms, 1:244-260
    of piles, 1:225-232
    shear-testing measurements of, 2:363-372
    stress-parameter control and, 2:28
    Weibull cumulative distribution function for, 2:13
    (*See also* Weibull distribution)
Strouhal frequency:
    in oscillating cylinder experiments, 2:684
    in steady state towing of cylinder, 1:390, 397
Strouhal number, 1:571; 2:672-675, 682, 684
    drag-coefficient use of, 1:398-399
Structural concepts, alternate, design and reliability issues for, 2:829-898
Structural stability:
    of circular tubes, 2:59-73
    of steel off-shore structures, 2:39-58
Structures, 2:3-292, 705-898
    (*See also* specific types of structures)
Su (undrained shear strength), 1:23, 272
Submarine pipelines, analysis of collapse of, 2:169-186
Submarine slides:
    plasticity theory and, 2:465
    soil flows generated by, 2:425-436
Sverdrup Basin (high Arctic), 1:75
SWATH-type ship, 2:611
Sway mode, 2:125-126

TAPS (Trans-Alaska Pipeline System), 1:73, 88
Target lifetime safety index, 2:4
Tarsiut prospect, 1:85
TBP (tethered buoyant platform), 2:101, 109
Tecnomare Steel Gravity (TSG) platform, 1:31-32
"Tensioned beam" computational method, 1:601-602
Tension Leg Buoy, 2:500
Tension Leg Platform (TLP), 1:4
    in deep water, 1:127, 134
    laboratory studies of behaviour of, 1:411-423
    North Sea construction of, 1:530
        northern North Sea, 1:34-37
    Response Amplitude Operator (RAO) of, 2:480-481
    response analysis of, with mechanical damping system in waves, 2:497-522
    structure of, 1:127-130, 133-134, 644-658; 2:101-102
        design strategy, 2:593-594
        dynamic response, 2:100-120
        interwire slippage and fatigue prediction in tethers, 2:77-99
        theoretical model, 1:411-423
Tension piles in silty soils, 1:141-157
Terzaghi's consolidation equation, 1:358, 360
Tethered buoyant platform (TBP), 2:101, 109
Tethered production platform (TPP), 2:101, 109
Tether forces, 1:104-106
Tetrahedron node in TTP design, 2:854
Thermodynamics of irreversible processes, 2:719-720
Three-dimensional diffraction theory, 1:454
TLP (*see* Tension Leg Platform)
TNO Institute for Building Materials and Building Structures, 2:147-169
Torvane testing, 1:166, 172-174
Total probability theorem, 2:152
Tower structures:
    in the Arctic, 1:86
    Woodrow's, 1:32
    (*See also* Buoyant towers; Guyed tower platforms)
TPP (tethered production platform), 2:101, 109
Trans-Alaska Pipeline System (TAPS), 1:73, 88
Transverse instability, 1:619-620
Tripod Tower Platform (TTP):
    design development of, 2:847-863
    tetrahedron node and, 2:854
Tromsøplateau, types of soil in, 1:23-24
TRS-80 microcomputer, 2:15, 18
TSG (*see* Tecnomare Steel Gravity platform)
TTP (*see* Tripod Tower Platform)
Tubular braces, strength of, 2:39-58
Tubular K joint:
    effect of monotonic loading on, 2:724
    effect of variable chord wall on, 2:726-735
Tubular members and joints:
    fatigue behaviour of, 2:736-746
    in ship/platform impacts, 2:261

Tubular members and joints (*Cont.*):
   structural welding of, 1:5
"Turbulent entrainment process," 1:664
Two-layer Dowel theory, 2:410-414

Unconsolidated undrained triaxial compression (UUC), 1:172
Underwater welding:
   automation in, 2:756-766
   design guidelines for, 1:5
Undrained shear strength (su), 1:23, 272
United Kingdom Department of Energy, 1:25
United States Coast Guard (U.S.C.G.), 1:6
United States Geological Survey, 1:6, 2:364
United States inland lakes, oil drilling in, 1:4
UNIVAC computer, 1:208; 2:268
Unstiffened tubulars, 2:59-73
UUC (unconsolidated undrained triaxial compression), 1:172

Vane test, 1:23, 107-121; 2:444-446, 450
   definition of, 2:465
Vertical anchor leg mooring system (VALM), dynamic behaviour of, 2:632-640
Vertical cylinders, semisubmersible, 1:432-433
Vertically moored platform (VMP), 2:101, 102, 109
Vertical mode, 2:126
Vertical-sided indentures, ice loads on, 1:78-79
Vertical "stacking" procedure, 1:10
Very deep water, oil and gas production facilities for, 1:50-70
Viscoelastic formulation in displacement of foundations, 1:300-301, 303, 306
Von Mises criterion, 2:801-803
Vortex shedding effect:
   in catenary moored platform, 1:38
   in flexibly mounted cylinders, 1:566
   in marine risers, 1:605-606
   in multiple risers, 1:619, 620, 628-629
   in oscillating cylinders, 2:671-701
   in pipe arrays, 2:641-650
   in Tension Leg Platform, 1:35

Wake coupling, 1:620
Wake interference, 1:620
Ward-Hunt ice shelf, 1:75

Water, use of, as drive cushion, 2:319
Wave Energy Converter (WEC), 2:212-221
Wave energy devices, 1:432-443
Wave forces:
   on buoyant platforms, 1:432-434
   cross-spectral densities of on inclined cylinders, 2:486-496
   displacement for, 1:253-255
   effect of marine growth on, 1:387-407
   effect of on design and analysis of marine structures, 2:477-485
   near fixed caisson, 1:363-364, 370
   on floating structures using finite element method, 2:580-591
   irregular, wind-generated pressures on circular cylinder, 2:532-540
   on Ocean Test Structure (OTS), 1:373-386
   one-hundred-year characteristic (*see* One-hundred-year storm)
   on semisubmersibles, 1:435, 447-489
   stochastic method of characterization of, 2:594
   wind-generated, 2:532-540
   (*See also* Laboratory waves)
Wave profiler, 1:391
Wave spectral shape and directional variabilities, 2:477-485
WEC (Wave Energy Converter), 2:212-221
Weibull distribution:
   in northern North Sea extreme conditions, 1:21, 648
   for reliability analysis, 2:153-154
   with scatter diagram, 2:214
   in sea-state modeling, 2:817, 820-821
   in spectral fatigue analysis, 2:13, 149, 739-741
Welding (*see* Underwater welding)
Wind-generated waves, 2:532-540
Wind speed/wave height relationships in North Sea, 1:646-658
Wind-wave spectra, 2:541-548
WIP test, 1:164, 174
Wirsching correction factor, 2:284
Woodrow's articulated concrete tower, 1:32

X-braced systems, 2:39-58

Young's modulus, 1:567

Zone of Established Flow (ZOEF), 1:662-666, 679-681, 683
Zone of Surface Flow (ZOSF), 1:666, 681

Author Index

Aas, P. M., 1:95, 243
Alm-Paulsen, A., 1:18
Almeland, I. B., 1:95
Amdahl, J., 2:257
Amundsen, T., 2:417
Anastasiou, K., 1:493
Andenaes, E., 1:243; 2:373
Andersen, K. H., 1:243
Andresen, A., 2:373
Angelides, D. C., 2:100
Arockiasamy, M., 1:313; 2:736
Atlar, M., 2:608
Aubrey, D., 2:391
Audibert, J. M. E., 1:125
Azzouz, A. A., 1:161

Babcock, C. D., 2:187
Balfour, J. A. D., 2:551
Baligh, M. M., 1:161; 2:338
Bamford, S. R., 2:305
Bardis, L., 1:467
Baron, G., 2:541
Basu, A. K., 1:529
Bea, R. G., 1:125
Belkune, R. M., 2:747
Bernitsas, M. M., 2:705
Bijlaard, F. S. K., 2:799
Bliault, A., 1:643
Bliek, A., 1:546
Bogard, D., 1:105
Bouckovalas, G., 1:297
Bowcock, A. O., 2:551
Bradley, M. S., 1:643
Brown, I. C., 2:874
Bungum, H., 2:561
Buyukozturk, O., 2:235

Chaplin, J. R., 1:493
Chateau, G. M., 1:50

Cheang, L., 1:105
Cheema, P. S., 2:736
Chen, C.-Y., 2:100
Chen, H., 2:477
Chryssostomidis, C., 1:565
Clausen, C. J. F., 1:95
Colson, A., 2:716
Connor, J. J., 2:348, 817

Das, P. K., 2:769
Datta, M., 2:295
de Campos, T. M. P., 2:438
de Winter, P. E., 2:169
Dello Stritto, F. J., 1:373
den Boer, K., 1:337; 2:884
Deo, M. C., 2:486
Dover, A. R., 1:125; 2:305
Dowling, P. J., 2:59
Dutta, A., 1:529
Duval, P. S., 2:123
Dyvik, R., 2:363

Eatock Taylor, R., 1:424
Edgers, L., 2:425
Edwards, C. D., 2:726
Eide, O., 2:373
Ellery, G. D., 2:315
Elliott, R. M., 2:315

Fadl, M. O., 2:451
Faltinsen, O. I., 1:411
Faulkner, D., 2:769
Fessler, H., 2:726
Finlay, T. W., 2:451
Fournier, J., 2:325
Frieze, P. A., 2:769
Fylling, I. J., 1:411; 2:651

Ganapathy Chettiar, C., 2:621
Gens, A., 2:438
Gowda, S. S., 2:736
Grande, L., 1:203; 2:401
Grant, I., 2:541
Grant, R. M., 2:831
Grilli, S., 2:580
Gudmestad, O. T., 1:18

Haldar, A. K., 1:313
Halliwell, A. R., 2:631
Hambly, E. C., 2:203
Harding, J. E., 2:59
Hariharan, M., 2:383
Hicher, P. Y., 1:262; 2:325
Hight, D. W., 2:438
Hobbs, R. E., 2:77
Hoffait, Th., 2:580
Holmes, R., 2:26
Hove, K., 2:561
Huijsmans, R. H. M., 1:447
Hujeux, J. C., 2:391
Hung, Y. Y., 2:831
Huslid, J. M., 1:18

Ivanov, A. V., 2:572
Iwata, S., 2:837

James, E. L., 2:315
Janbu, N., 1:203; 2:401
Jardine, I. J. A., 2:631
Ji, Z., 2:792
Jiang, C.-W., 1:373
Jizu, X., 2:401
Johns, D. J., 2:690
Jonsrud, R., 2:373

Kanegaonkar, H. B., 2:747
Kaplan, P., 1:373
Karadeniz, H., 2:147
Kardomateas, G., 1:546
Karlsrud, K., 2:425
Katayama, M., 2:497
Kavvadas, M. J., 2:338, 348
Kenter, C. J., 1:337
Kerr, J., 2:26
Kirkegaard, J., 2:592
Kokkinis, T., 2:705
Kokkinowrachos, K., 1:467
Kulikova, A. N., 2:572
Kumarasamy, K., 2:383
Kyriakides, S., 2:187

Lacasse, S., 1:243
Ladd, C. C., 1:161
Lagoni, P., 1:221
Larsen, C. M., 2:651
Larrabee, R. D., 2:784
Lassoudiere, F., 2:391
Lauritzsen, R., 2:417
Lee, G. C., 1:3
Leehey, P., 2:681
Lejeune, A., 2:580
Lejeune, P., 2:580
Leverette, S. J., 1:643
Lin, S., 2:792
Lindenberg, J., 1:337; 2:884
Love, M. A., 1:181
Lundgren, H., 2:592
Luyties, W. H., 2:3

Mahin, S. A., 2:39
Marchal, J., 2:580
Marr, W. A., 1:297
Marshall, P. W., 2:3
Marsland, A., 1:181
Martins, J. P., 1:283
Masubuchi, K., 2:756
Matlock, H., 1:105
Matsuishi, M., 2:837
Mavrakos, S., 1:467
McConnell, K. G., 2:671
Meek, J., 2:847
Mei, C. C., 1:358
Meimon, Y., 2:391

Michelsen, F. C., 2:847
Milgram, J. H., 1:659
Miwa, E., 2:497
Mizikos, J. P., 1:262; 2:325
Moan, T., 2:257
Moe, G., 1:618
Moeller, M. J., 2:681
Morrison, D. G., 2:864
Muggeridge, D. B., 2:736

Naess, A., 1:516
Nair, D., 2:123
Nair, S. C., 2:621
Narasimhan, S., 2:486
Nath, J. H., 1:387
Nordal, S., 1:203

Ottesen Hansen, N.-E., 2:641
Overvik, T., 1:618

Papazoglou, V. J., 2:756
Park, Y. S., 2:671
Patel, M. H., 1:599
Patrikalakis, N. M., 1:565
Perry, S. H., 2:874
Pinkster, J. A., 1:447
Pollalis, S. N., 2:348
Popov, E. P., 2:39
Potts, D. M., 1:283
Price, W. I. J., 2:203
Prince, A., 1:181
Puech, A. A., 1:141
Puthi, R. S., 2:799

Qu, N., 2:810
Qui, C., 2:792
Qui, D., 2:523

Raman, H., 2:532
Ramesh, C. K., 2:747
Rao, P. S. V., 2:532
Raoof, M., 2:77
Reddy, D. V., 1:313; 2:736
Reinhardt, H. W., 2:223
Ropers, F., 2:325

Sand, S. E., 2:592
Sarohia, S., 1:599
Selnes, P. B., 2:561
Shyam Sunder, S., 2:817
Smith, G., 2:541
Smith, S. N., 2:608
Søreide, T. H., 2:257
Stol, H. G. A., 2:799
Suarez, L. F., 2:305
Sutherland, H. B., 2:451
Swart, J. H., 1:337

Taby, J., 2:257
Takahashi, M., 2:438
Tarvin, P. A., 2:464
Teigen, P. S., 1:411
Tickell, R. G., 1:493
Tilmans, W. M. K., 2:884
Triantafyllou, M. S., 1:546
Tricklebank, A. H., 2:203

Unoki, K., 2:497
Urzua, A., 1:297

Vallejo, L. E., 2:464
van Hooff, R. W., 1:411
Van Houten, R. J., 1:659
van Manen, S., 2:147
Vandiver, J. K., 2:279
Verley, R. L. P., 2:690
Vrouwenvelder, A., 2:147

Watt, B. J., 1:71
Will, S. A., 2:100

Zayas, V. A., 2:39
Zietsman, J., 1:424
Zimmie, T. F., 2:363
Zisman, J. G., 2:235
Zuo, Q., 2:523